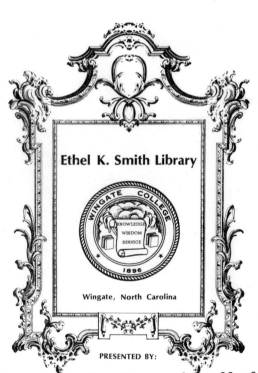

Ethel K. Smith Library

Wingate, North Carolina

PRESENTED BY:
Administration, Faculty, and Staff of
Wingate College
IN HONOR OF: Robert D. Billinger, Sr.,
Father of Robert D. Billinger, Associate
Professor of History, Wingate College

PHYSICAL CHEMISTRY

This book is part of the

Allyn and Bacon Chemistry Series

Consulting Editors:　　Daryle H. Busch
　　　　　　　　　　　　Harrison Shull

PHYSICAL CHEMISTRY

J. Philip Bromberg

ALLYN and BACON, INC.

Boston
London
Sydney
Toronto

To Nancy, Ronna, and Mark

Copyright © 1980 by Allyn and Bacon, Inc., 470 Atlantic Avenue, Boston, Massachusetts 02210. All rights reserved. No part of the material protected by this copyright notice may be reproduced or utilized in any form or by any means, electronic or mechanical, including photocopying, recording, or by any information storage and retrieval system, without written permission from the copyright owner.

Library of Congress Cataloging in Publication Data

Bromberg, J Philip
   Physical chemistry.

   (Allyn and Bacon chemistry series)
   Bibliography: p.
   Includes index.
   1. Chemistry, Physical and theoretical. I. Title.
II. Series.
QD453.2.B76   541'.3   79-10480
ISBN 0-205-06572-4

10 Mar 81

Printed in the United States of America.

Manufacturing Buyer: Karen Mason
Production Editor: Judith Fiske

080958

CONTENTS

Preface ix
To the Student xiii

1 Introduction 1

PART I THE GAS LAWS

2 The Gas Laws 15
3 The Ideal Gas in the Kinetic-Molecular Model 44

PART II PHENOMENOLOGICAL THERMODYNAMICS

4 Conservation of Energy and the First Law of Thermodynamics 57
5 Applying the First Law of Thermodynamics. Energetics of Ideal Gas Processes 75
6 The Practical Enthalpy Scale. Thermochemistry 93
7 The Second Law of Thermodynamics: Heat Engines, Entropy, and the Direction of Chemical Change 104
8 Entropy Calculations and Absolute Entropies. The Third Law of Thermodynamics 128
9 Work, Free Energy, and Equilibrium 141

PART III APPLICATIONS OF THERMODYNAMICS

10 The Equilibrium Constant for Gas Reactions 167
11 Phase Equilibria 189
12 Colligative Properties of Ideal Solutions 202
13 Phase Diagrams 230

14 Activities of Non-Electrolyte Solutions 257
15 Ions in Solutions 275
16 Activities of Ions. The Debye-Hückel Limiting Theory of Ionic Activities 296
17 Electrochemical Cells 311
18 Gravitational, Electrical, Magnetic, and Surface Work 339
19 Macromolecules 366

PART IV THE STATISTICAL APPROACH

20 Introduction to Statistical Methods 391
21 The Maxwell-Boltzmann Distribution of Molecular Velocities 404
22 Collisional and Transport Properties of Gases 423
23 Statistical Mechanics 445
24 Applications of Statistical Mechanics 464

PART V THE QUANTUM MECHANICS OF ATOMS AND MOLECULES

25 Corpuscles, Waves, and the Nuclear Atom 477
26 Preliminaries to Quantum Mechanics 499
27 The Postulates of Quantum Mechanics. Applications to Simple Systems 523
28 Rotations and Vibrations of Molecules 542
29 Statistical Mechanics of Diatomic and Polyatomic Molecules 565
30 The Hydrogen Atom 577
31 Approximate Methods, The Helium Atom, and Selection Rules 599
32 Electron Spin and More Complicated Atoms 618
33 Molecules and Chemical Bonding 643

PART VI ATOMS AND MOLECULES IN CONDENSED STATES

34 The Crystalline State 673
35 X-Ray Diffraction Studies of Crystals 690
36 Thermal Properties of Crystals 726
37 Metals, Semiconductors, and Insulators 739
38 Electric and Magnetic Properties of Atoms and Molecules 762

PART VII THE RATES OF CHEMICAL REACTIONS

39 Phenomenological Rates of Chemical Reactions   791
40 Reaction Mechanisms   810
41 Some Theoretical Approaches to Chemical Kinetics   824
42 Photochemistry   840

Bibliography   867

Index   873

PREFACE

My intent in writing this book was to teach physical chemistry to students at the mid-college level, not to supply them with a reference book on physical chemistry. Therefore I have tried to explain concepts as simply and informally as possible, with as much of a conversational style as the editors would allow to pass their critical scrutiny. The result, hopefully, is a book that the students would want to use even if their instructor had selected a different text to use for the course.

In writing a book one must make many decisions regarding order and topics. In all cases I have made these decisions from the point of view of what I consider pedagogical soundness. Thus I have chosen classical or phenomenological thermodynamics to precede statistical mechanics and quantum mechanics. I have further chosen *not* to commingle macroscopic and microscopic thermodynamics in the early parts of the book dealing with macroscopic thermodynamics; I think it is easier to develop an understanding of entropy based on heat and work considerations alone. In doing so, I have taken the liberty of inventing new terms such as "work capacity" and discussing entropy in terms of the destruction of "work capacity."

One question we as instructors must ask ourselves is what we wish the students to accomplish in the physical chemistry course and what level of achievement we can expect them to attain. This is a very difficult question, and one which unfortunately is all too rarely asked. I think we can expect the students to be able to undertake real thermodynamic calculations at the conclusion of the course; I do not think we can expect them to undertake real quantum mechanical calculations. This means that the discussions dealing with thermodynamics must be more complete than those dealing with quantum mechanics. The students should be able to compute the equilibrium constant from thermochemical data; I would not expect them to be able to compute the energy levels of a new molecule.

Students should also acquire a basic understanding of how a spectrum is related to the energy levels of atoms and molecules. They need not have a thorough understanding of selection rules based upon group theory or of time dependent perturbation theory, but they should know enough to understand how the Boltzmann distribution affects intensities in rotational spectra.

The range of topics covered is relatively standard, though some topics included here are not generally found in most physical chemistry texts. Some of these are the material on Legendre transformations and the discussions of electrical, magnetic, and gravitational effects in thermodynamics. Students should recognize that thermodynamics is also applicable to other forms of work besides pressure-volume work, and that such things as the piezoelectric effect are of interest to chemists and not just physicists and electrical engineers.

Those of you who have written books are aware of space constraints imposed by publishers. This means that many topics must be eliminated and choices must be made. An example of this is Raman spectroscopy, which I feel is a very important field today, especially in view of the experimental advances associated with the advent of lasers. On the other hand, vibrational and rotational spectroscopy are also important, and even more important from the pedagogical point of view since they are easier to understand. We must make a choice. I think the students are better served if they are provided with a good understanding of vibrational and rotational spectroscopy, and no knowledge of Raman spectroscopy, rather than a poor understanding of all spectroscopy. With this philosophy in mind, our discussion of Raman spectroscopy is limited to a short note and one of the problems, and this statement also applies to a number of other topics, such as the magnetic analog of the Clausius Clapeyron equation, magnetorestriction, recursion relations for Hermite polynomials, and ladder operators.

Some 800 problems are included in this book. In view of the present concern with energy and environmental problems, many of these, particularly in the thermodynamics sections, deal with electric power generation and environmental problems. I hope that a student who has completed this course would understand the meaning of the statement "energy can be conserved, but it cannot be recycled." Many of the problems dealing with energy production are stated in terms of English units such as tons, Btu/lb, etc. This is done out of recognition of the paradoxical situation in the United States, where these obsolete units are used. Students should know how to deal with them. It is hoped that by the time the next edition of this book is being prepared, these can be converted to SI and SI based units to conform with the rest of the problems in the book.

Some of the problems are stated in terms of dollars; it is perhaps appropriate for students to recognize that in the real world dollars are important, and it costs more to provide heat energy through an intermediary electric power generating plant than to provide the heat energy directly. Many of the problems contain the results of experiments in tabulated form, and the students are asked to treat the experimental data properly to get the desired result—an absolute entropy, a standard electrode potential, a polymer average molar mass, Avogadro's number, or other constant. One consistent weakness students have is their lack of intuitive and quantitative feeling for experimental errors. I have tried to compensate for this somewhat in Chapter 20, which contains an introduction to statistical methods, and also in several problems. While a complete discussion of least squares fits and chi square tests is inappropriate to a book of this type, we have made several references to these procedures.

The references cited in the footnotes and bibliography are for the most part the references I used in putting this book together. Many of these are books I have collected over the years and are sitting on my shelf. Thus many of the references are to older books such as Pauling and Wilson and other early quantum mechanical books. I saw no reason to update these; most of the basic references cited here will be available in almost all science collections, whereas some of the newer references will be harder to find.

Biographical sketches of many scientific notables mentioned in the book have been included; I believe that this makes for a more interesting book. In addition, the students (and the instructors) should know something about the people they see mentioned. I am quite certain that if I were to start again, my choices would be different, but I do feel that I have selected a representative group.

As an author, I carefully prepared a typed manuscript and a complete set of figures, many rough drawn, for this book. These were shipped off to the publisher and somehow were reviewed and came back first as a reviewed manuscript, then as an edited manuscript, then as galley proofs and page proofs and finally the finished book, by some process which remains a total mystery to me. For their efforts in seeing this book to its final form I must express my sincere thanks to William Roberts, John Shine, and Judy Fiske of Allyn and Bacon.

Several individuals have been of assistance in preparing this book. Sanford Levy, M.D., Edward Curtiss, M.D., and Joseph Maroon, M.D. have provided valuable advice and suggestions on those aspects of the book dealing with medical problems and procedures, and their assistance is gratefully acknowledged. Others have graciously provided figures as acknowledged in the text. While writing this book I was for part of the time a grantee of the National Science Foundation under Grant No. GP-23517X2 and also of the Pennsylvania Science and Engineering Foundation of the Pennsylvania Department of Commerce under Grant No. PSEF-298. While neither of these grants was connected with writing this book, some of the research done in electron scattering and environmental problems under these grants found its way into the book, and the assistance of the sponsoring organizations and of Robert Laughlin of PSEF is gratefully acknowledged. All of the line drawings representing plots of data and many of the drawings of mathematical functions such as the hydrogen wave functions were prepared on the computer of the Mellon Institute NMR Facility for Biomedical Research sponsored by the National Institute of Health under Grant RR-00292.

I was especially fortunate in having been associated with E.N. Lassettre and the late Thomas G. Fox while at Carnegie-Mellon. Both have had substantial effects upon the direction of my work. I feel privileged to consider them among my friends, and thank them for their efforts in my behalf.

I should also like to express my heartfelt thanks to the rest of the senior faculty of the chemistry department who were my associates at Carnegie-Mellon. Their active support and encouragement were a source of inspiration to myself and my contemporaries on the faculty. Without their exemplary efforts in our behalf, our work would surely have foundered.

Finally, and most importantly, I must thank my wife and children for putting up with me during the trying time of putting this book together. They have been a constant source of support these past few years. I can now look forward to spending more time with them and less on book writing.

TO THE STUDENT

At many colleges and universities the physical chemistry course has acquired a reputation for being a "hard" course, and you have probably heard stories to this effect. Forget the stories for the moment. Enjoy yourself, and come to your own conclusion.

The prerequisites for an adequate comprehension of the material are rather minimal. I have assumed that (1) you have already studied elementary chemistry and know how to calculate the quantity of sulfer in a ton of sulfuric acid; (2) you have studied elementary calculus and know how to compute the derivative of x^n; and (3) you have studied elementary physics and know that $F = ma$, that velocity is the time derivative of position, that acceleration is the time derivative of velocity, and that Ohm's law holds for the passage of electricity through a resistor.

You should be aware that unless you are a very rare genius you will not master the material by simply reading it and coming to class. It is of vital importance to work out a substantial number of the problems placed at the ends of chapters. Your ability to do well on exams will be directly related to the amount of problem-solving practice you have.

I have tried to include some additional information in many of the problems, and have tried to orient some of them toward environmental, energy, biological, and everyday situations. Many of them are two hour problems—unless you stop to think about them for five minutes, in which case you may be able to work them out in two minutes. A small number of the problems are necessarily quite lengthy and involve substantial arithmetical computation, but can be worked quite quickly with the use of a calculator.

You probably will not work out all of the problems, but you should at least glance through them. Many of them contain additional information related to the text material. You may even find some problems you will want to solve out of curiosity. Do not be afraid to use some common sense. Unless it is specifically stated that a gas is not ideal, you should assume it is ideal. If a temperature is required and none has been specified select a reasonable one, such as room temperature, or 37 °C if the problem deals with a biological application. Learn to recognize answers that are obviously incorrect, and try to approximate answers before going ahead with the actual computation. If you come up with an answer of 300 °C as the final temperature of a sample of ice, you should suspect that you have made an error, and you should recheck your work.

After studying the text and working the problems you may find that there are subjects you wish to pursue further. Footnotes in each chapter and a bibliography (arranged by topic) at the end of the book provide a convenient source of references. You may find it helpful to peruse these references for additional information or for a different perspective on the material.

Your journey through physical chemistry begins with a review of the properties of gases. From there you will move into the mainstream of the material, where you will develop an understanding of the fundamental principles of thermodynamics on a macroscopic level. After you are acquainted with the basic material for equilibrium in chemical systems you will learn to apply the thermodynamic principles to a host of problems involving solutions, gases, and phase equilibria. Next you will develop an appreciation for the fine structure of matter, and form a model of the universe based on atomic and molecular theory. You will find that the final result is the same regardless of whether you adopt a microscopic or macroscopic point of view. At the final stop you will examine chemical kinetics and the rate at which equilibrium is established.

Throughout the journey you will explore how basic principles are exemplified in the real world and investigate the effects of technology on our environment. You will study the production of work from the combustion of fossil fuels and examine the direct production of electrical work from chemical reactions in batteries.

In most cases I have developed the various concepts in as rigorous a manner as possible. You will be alerted, however, when concepts are presented in a nonrigorous manner. When I use analogies you should bear in mind that an analogy is not reality; it is an artificial concept designed to aid comprehension.

The main objective of this book is to help you understand the methodology of science and technology, and not to overwhelm you with a large body of diverse facts. It is my hope that you will be able to apply these general methods to many types of problems, regardless of subject area.

CHAPTER ONE

INTRODUCTION

1.1 SCIENCE

Science is the branch of knowledge that concerns itself with the study and elucidation of natural or observable phenomena. Therefore science is observational. This term observational holds the key to why science is different from other fields, such as philosophy and theology. A much debated problem during the Middle Ages was the number of angels that could dance on the head of a pin. Our definition excludes such speculation from the purview of science, since the dancing of angels on the head of a pin is not a physical observable.

Things that cannot be observed at first often become observable as time passes and phases of intellectual activity are transferred from the realm of philosophy to the realm of science. The atomic theory of Democritus was philosophy rather than science, since the atomic concept was a physical nonobservable at the time of ancient Greece. Atomism did not attain the status of science until some 200 years ago, when Daniel Bernoulli and John Dalton refined the concept of the atom by defining experimentally measurable quantities. With Bernoulli, the quantity was the pressure of a gas; with Dalton, the quantity was the relative masses of atoms.[1] Discrete atoms are not observable, but their effects are. Just because we cannot see atoms, we do not discredit their existence. We believe atoms exist, because only if we assume that atoms exist can a host of *observed* phenomena be satisfactorily explained.

Science consists of two concomitant aspects, observation and elucidation. Without observation there is no need for elucidation and no science. Without elucidation science cannot advance, for it is doomed to be little more than an expanding collection of uncorrelated facts. The observational aspect of science is usually called experimental science; the elucidation is usually called theoretical science.

Science usually progresses in a direction that leads from observation to elucidation. We emphasize at the outset that observations (or experiments) are fixed and theories are variable. The experiment, assuming that it is properly done, cannot be

> DEMOCRITUS (born about 465 B.C.), Greek philosopher, is considered by many to be the Aristotle of the fifth century. According to Democritus, atoms are eternal and invisible. They are so small that their size cannot be decreased, hence the name *atomos*, which means "indivisible." They are incompressible and completely fill the space they occupy. Atoms are eternal and uncaused, like motion, but the compounds of atoms may increase or decrease, corresponding to birth and death. The modern concept of the conservation of matter is reflected in the ancient Greek doctrine that "nothing can arise from nothing."

[1] See Chapter 3 for Bernoulli, Chapter 25 for Dalton.

> DANIEL BERNOULLI (1700–1782) was a member of a distinguished Swiss family of scientists and mathematicians. He is best known as one of the founders of the field of probability. He applied his results on probability to economics and other practical matters, like gambling. In his important work *Hydrodynamica,* he helped to establish hydrodynamics. It was in this work that he proposed his derivation of Boyle's law. He was one of the earliest workers in jet propulsion, proposing an ingenious method for propelling ships by ejecting water from the stern.

altered. The experiment may be refined as advancing technology enables us to make more precise measurements. Theories, on the other hand, are considered valid only so far as they properly fit the observed data, and they must often be changed either drastically or subtly to meet the demands of new or finer measurements. The preferred theory is the simplest one that can explain all the observations. Theory and experiment go hand in hand. Experiments often contain inherent uncertainties, which are often removed only after theory suggests that they are wrong.

1.2 THE VALUE OF OBSERVATION; STEREOISOMERISM AND THE TETRAHEDRAL BOND

Suppose we examine, as a specific example of the development of a theory, the interrelated concepts of optical isomerism and the tetrahedral bond. By 1848, two different isomers had been isolated from tartaric acid residues produced by the fermentation of grapes. One of the isomers was optically active; the other was optically inactive and was called racemic acid. In that year, Louis Pasteur was engaged in his extensive researches on grape fermentation; he noted that crystals of the optically active sodium ammonium tartrate were asymmetric to the extent that they could not be superimposed with their mirror images. Some years earlier, crystallographers had attributed optical activity to this kind of asymmetry. Thus, for example, quartz was known to exist in two different crystalline forms, one the mirror image of the other. The two different types of quartz rotated plane-polarized light by the same amount but in opposite directions.

Pasteur, recalling the earlier work on quartz, prepared crystals of the optically inactive tartrate, fully expecting them to be completely symmetrical in the sense that they would be identical with their mirror images. When he examined his collection of crystals under the microscope, he was surprised to discover two different kinds of crystals. One kind was identical to the optically active tartrate. The other was a hitherto unknown crystal, which was the mirror image of the first type. Pasteur sep-

> JOHN DALTON (1766–1844), English chemist, physicist, and meteorologist, is considered by many to be the "father" of the atomic theory of matter, although *grandfather* is perhaps a more appropriate term. He did some work on color blindness, an abnormality from which he suffered. In 1803, he published his paper "Absorption of Gases by Water and Other Liquids," in which he presented what is now known as Dalton's law of partial pressures. He was led to his theory of atomism by his studies of gases. In one of his papers published in 1805, he said, "Why does not water admit its bulk of every kind of gas alike? . . . The circumstance depends on the weight and number of the ultimate particles of the several gases." John Dalton was led to the atom by reflecting that different gases had different values of the constant of Henry's law.

> LOUIS PASTEUR (1822–1895), French chemist and microbiologist, made numerous fundamental contributions to both chemistry and biochemistry. His work was not restricted to basic research, but also covered practical matters such as the manufacture of wine, beer, and vinegar; silkworm disease; preservation of foods; and the prophylaxis of diseases in general. His interest in practical problems evolved naturally from his basic research interests, particularly those in fermentation. He demonstrated the biological rather than chemical origin of the fermentation process, and this logically implied a biological or "germ" theory of disease. Pasteur's work with molecular asymmetry led him to suggest in 1860 that the atoms of a right-handed compound might be arranged in the form of a spiral or a tetrahedron. He never developed this suggestion, and it was left to Le Bel and van't Hoff (1874) to link Pasteur's work with the tetrahedral carbon atom of Kekulé, from which emerged the asymmetrical carbon atom model. Pasteur's name rapidly became a household word both in connection with his treatment of rabies and in connection with the process of pasteurization. He discovered this process while working to disprove the theory of spontaneous generation; its first practical application was in the preservation of wine. In his later years, Pasteur often expressed regret at having abandoned his early researches in the relationship between asymmetry and life. He felt that had he continued this work, and devoted less time to practical industrial chemistry, he would have become the Newton or Galileo of biology.

arated the two types by hand and found that when solutions were prepared, each rotated plane-polarized light by the same amount but in opposite directions. Further, when a solution was prepared with equal quantities of each type, the resulting solution was optically inactive.

Since quartz and other active crystals such as sodium chlorate lose their activity in solution, the optical activity of these compounds could not be associated with an atomic or a molecular property. Their activity was attributed to the arrangement of the atoms in the crystal. Tartaric acid, on the other hand, retained its optical activity in solution, and Pasteur correctly concluded that in these compounds the arrangement of the atoms in the molecules caused the activity. Since the fundamental theory of the fourfold valency of carbon was still ten years in the future, it was not possible then to relate optical activity to molecular structure. Recall that 1848 was the year in which Gmelin first noted that the essential difference between organic and inorganic compounds was that organic compounds always contain carbon. Not until 1858 did Kekulé and Couper independently propose that carbon had a valence of four, setting the stage for further developments.

By 1874, additional optically active pairs had been discovered. In that year two independent papers appeared, one by van't Hoff, and the other by Le Bel, which were to affect the development of chemistry significantly. Each paper pointed out that in every known case of optical isomerism there existed at least one carbon atom that was bonded to four different groups. The scientists had carried this observation one step further, noting that if the four different groups about the carbon atom were placed at the corners of a tetrahedron, with the carbon atom at the body center of the tetrahedron, then two different arrangements were possible. The two resulting molecular arrangements were mirror images of each other. This was the same type of asymmetry that was found necessary for the optical activity of crystals. The tetrahedral nature of the bonds about the carbon atom had been inferred decades before the advent of newer techniques such as X-ray crystallography, which made it possible to determine these bonds directly.

So far our discussion has been directed towards the question, What is the geomet-

> FRIEDRICH AUGUST KEKULÉ VON STRADONITZ (1829–1896), German chemist, was the scion of a noble Czechoslovakian family that traced its ancestry back to the fourteenth century. He began his studies with the intention of becoming an architect but chose chemistry after taking Liebig's chemistry course. In 1857, Kekulé proposed his tetravalent theory for carbon, not only for simple molecules such as CH_4, CH_3Cl, and CCl_4 but also for molecules containing more than one carbon; thus he laid the foundation for structural organic chemistry. He took the initiative in organizing the first International Congress of Chemists, which met in Karlsruhe in 1860. (It is difficult to overestimate the importance of this congress for the study of chemistry thereafter.) Its purpose was to reduce the confusion then existing in chemical terminology and in atomic weights. At this conference the Avogadro molecular hypothesis was firmly established by Cannizzaro. In addition, many old atomic weights such as 20 for Ca, 43.8 for Sr, and 63.5 for Ba were corrected to their modern values. Without these corrections, Mendeleev could not have arrived at his periodic table of the elements.
>
> Kekulé's insight into structural organic chemistry and his ability to infer correctly the ring structure of benzene may well be due to his early training in architecture. The ring structure came to Kekulé in a vision in which he saw a chain of carbon atoms closing itself like a snake biting its tail. He then quickly worked out the closed skeleton of six carbon atoms joined by alternating single and double bonds, and he spent the next several years at the arduous task of experimentally confirming his theory.

rical configuration of the carbon bond? When it has been shown to be a tetrahedron, an additional question remains to be answered. That question may be stated, Why is the bond tetrahedral? The why question was not answered until the 1930s, following the development of quantum mechanics and the theory of molecular orbitals; a theoretical basis for the existence of the tetrahedral bond could then be formulated (see Chapter 33). The sequence of events in this discussion is important in showing how a scientific theory is developed, so let us carefully review the steps.

1. Observations were made on numerous optically active pairs of isomers.
2. Van't Hoff and Le Bel correlated these observations by noting that each carbon atom was surrounded by four different groups, and they proposed the tetrahedral carbon bond.
3. A rationale for the tetrahedral bond was proposed, based on a set of universal principles, namely quantum mechanics.

Step 1 constitutes the observation or experiment. It resulted in a mass of data about the existence of optically active isomers. In step 2, the data are correlated in a law. (In other instances, the correlation will be expressed by a mathematical relation; Kepler's laws of motion in astronomy furnish an example). For the carbon atom

> JACOBUS HENRICUS VAN'T HOFF (1852–1911), Dutch physical chemist, received the first Nobel Prize in chemistry in 1901 for "the discovery of the laws of chemical dynamics and of osmotic pressure." Van't Hoff was one of the early developers of the laws of chemical kinetics, developing methods for determining the order of a reaction; he deduced the relation between temperature and the equilibrium constant of a chemical reaction. In 1874, van't Hoff (and also J. A. Le Bel, independently) proposed what must be considered one of the most important ideas in the history of chemistry, namely the tetrahedral carbon bond. Pasteur's work had demonstrated that optical isomerism existed because of isomers that were mirror images of each other. Van't Hoff carried this idea to asymmetry on the molecular level, and asymmetry required bonds tetrahedrally distributed about a central carbon atom. Structural organic chemistry was born.

> JOHANN KEPLER (1571–1630), German astronomer, is considered the founder of physical astronomy. He was assistant to Tycho Brahe at Prague, and succeeded to Tycho's position when Tycho died in 1601. Kepler's three laws mark one of the most significant contributions in the history of science. They are:
>
> 1. The orbits of the planets are ellipses with the sun at one focus.
> 2. Lines joining the planets to the sun sweep out equal areas in equal times.
> 3. The square of a planet's period is proportional to the cube of its distance from the sun.
>
> Circular motion now gave way to elliptical motion. Kepler, in his early years, published several religious works in which he tried to show that the birth of Jesus took place some five years before the commonly accepted date.

the law expressed a geometrical arrangement of atoms. The relations, or laws, are based on a particular physical picture of the universe—in this case, the fourfold valency of the carbon atom. This picture we call a *model*. Models are extremely important in the development of scientific theories and we shall meet them repeatedly in our studies. A model is the simplest physical picture of a system that will adequately explain the observations.[2] In step 3, we have the final elucidation of the observed phenomena in the form of a theory that is generally valid, applying to every similar system in the universe. The words *general validity* are very important. The theory of molecular bonding goes beyond simply explaining the existence of the tetrahedral bond in carbon. The theory applies to all molecules regardless of the geometrical configuration.

Science is not static. As better measuring instruments were developed, it was discovered that in many instances the geometrical configuration of carbon compounds deviated from a perfect tetrahedron. This necessitated certain adjustments in the theory to account for these deviations. In a certain sense, observation and theory form a closed-loop, feedback system. Simple theories are developed to explain simple observations. As observational techniques are refined, the laws and theories must be adjusted to take these improved experiments into account. Scientific "truth" does not exist in a vacuum. It exists in a framework of certain basic assumptions. Since these assumptions may change from one generation to another, scientific truth itself can change. In the early 1800s, it was assumed that the formula for water was HO. This led to the conclusion that an oxygen atom weighed eight times as much as a hydrogen atom. Quantitative chemical analysis developed quite rapidly, even with this erroneous conclusion. Not until Avogadro's law was generally accepted were this error and many others corrected, causing many phases of chemistry to be simplified.

In all observation, it is important to recognize how cause and effect are connected. Consider the following: It has been conclusively demonstrated that the beating of drums will bring back the sun after a solar eclipse.

1.3 PHENOMENOLOGICAL AND POSTULATIONAL APPROACHES

Science usually progresses by observations of events of nature. We may call these observations experiments. Scientists draw conclusions from these observations and

---

[2] The intimate relation between the law and the model is very important. It would have been possible to explain optical isomerism by a molecular configuration in which the carbon atom was situated outside the tetrahedron, since the same type of asymmetry could be present in this situation. This configuration could not be reconciled, however, with a fourfold valency of carbon and the absence of disubstituted isomers.

> AMEDEO AVOGADRO, Conte Di Quaregna (1776–1856), Italian physicist, is remembered exclusively for his enunciation in 1811 of what we now call Avogadro's law. This law states simply that at the same temperature and pressure equal volumes of all gases contain the same number of molecules, regardless of whether the molecules consist of one atom or more than one. Avogadro held knowledge ahead of his time. His bold hypothesis held the key to many perplexing problems, but his idea lay in limbo, totally neglected, until it was resurrected some fifty years later by his countryman Cannizzaro at the Karlsruhe Conference in 1860.

produce "laws" of nature. We accept such laws as valid predictors of events so long as they accurately predict events. This we call the *phenomenological* approach.

The basic laws of optics were laid down by a series of experiments that established the law of reflection and the law of refraction. By these laws we can predict the passage of light beams through various media; also by these laws, microscopes, cameras, and other optical instruments could be designed and built. Since this system of optics is based on experiments, it can be called a phenomenological approach.

A different approach to the subject exists, however. One can simply postulate some basic principle. In optics this would be Fermat's principle of least time. Fermat's principle states that when a light beam travels between two points, it will travel along a path in such a way that the time it takes to get from one point to the other will be the minimum. This postulate leads to the laws of reflection and of refraction; thus this single postulate is responsible for a consistent science of optics. This approach is the *postulational* approach.

Historically, the phenomenological approach usually is first. Subsequently, postulational methods are developed in which an entire discipline of science can be concisely based on a small set of fundamental postulates. During our study, we shall witness both approaches. For thermodynamics, we shall use the phenomenological approach; for quantum mechanics, we shall use the postulational approach, tempered by phenomenological observations.

1.4 UNITS

Science is ultimately based on measurements, and a unified system of units is far preferable to a system in which each country uses a different set. The three basic units are time, mass, and length. Time, we know, we measure with a clock; length is what we measure with a ruler; and mass we measure with a balance. Time and length are simple concepts that were well known to ancient civilizations; the accomplishments of such civilizations in astronomy and geometry attest to their knowledge. The distinction between mass and weight (as a force) is the product of the late Middle Ages. Mass can be introduced by Newton's law of gravitation,

$$F = \frac{Gm_1m_2}{d^2}$$

which measures the attractive force between two masses separated by a distance d. The proportionality constant is given by $G = 6.67 \times 10^{-11}$ m^3 sec^{-2} kg^{-1}. Mass, length, and time are primary units, and others are defined by them; thus force is defined by Newton's equation $F = ma$. Charge is another primary unit.

The unit of time is the second (s), now defined as 9.192631770×10^9 periods of the

SEC. 1.4 UNITS

> ISAAC NEWTON (1642–1727), English scientist, was perhaps the greatest genius in the annals of science. In the third decade of his life he discovered what we now call the binomial theorem, and he conceived the differential calculus, which he called fluxions. He soon discovered what he called the inverse method of fluxions, or the integral calculus, and he postulated the gravitational force. His laws of motion are familiar to anyone who has taken a course in physics. He served as a member of Parliament and as master of the mint during his last 27 years. In that office, Newton invented the outer ridge on coins to prevent the clipping off of bits of the metal. There was some controversy about whether Newton or Leibniz first invented the differential calculus. It is now accepted that Leibniz invented the calculus later than Newton and independently of him.

radiation of a particular spectral line of the cesium 133 atom. The unit of length is the meter (m), now defined as 1.65076373×10^6 wavelengths of the radiation of a particular spectral line in the krypton 86 spectrum. The unit of mass is the kilogram (kg), which is defined as the mass of a platinum block stored at the International Bureau of Weights and Measures in France.[3] These three units form the first three of the so-called *base units* of the International System of Units (the *Système International*, abbreviated SI).[4] The SI units are further subdivided by prefixes such as *deci-* (d), *centi-* (c), and *milli-* (m) for the values 0.1, 0.01, and 0.001. These units and abbreviations are summarized in Appendix I. We shall mostly use the SI units. Sometimes, however, we shall use non-SI units, when the traditional unit prevails. Thus, for example, we shall often use the liter as a volume unit. The liter is defined as the volume that a mass of 1 kg of pure water occupies at its maximum density and at standard atmospheric pressure. The liter differs from the cubic decimeter by about 28 parts in 10^6.

Quantities will be either dimensionless or dimensioned. Consider the following example of operating with dimensioned numbers. We shall calculate the difference in potential energy associated with raising a 1-kg weight a distance of 1 m in the earth's gravitational field. The relevant equation is $V = mgh$, where m is the mass, $g = 9.8$ m s^{-2} is the gravitational acceleration, and h is the height. We get

$$V = (1 \text{ kg})(9.8 \text{ m s}^{-2})(1 \text{ m}) = 9.8 \text{ kg m s}^{-2} \text{ m}$$
$$= 9.8 \text{ kg m}^2 \text{ s}^{-2} = 9.8 \text{ joules}$$

Any valid equation must have identical units on both sides of the equality sign.

Historically, the unit of heat has been the calorie; this was first defined as the amount of heat necessary to raise the temperature of 1 g of water 1 °C. When it became apparent that the heat capacity of water was a function of temperature, it was necessary to redefine the calorie and to specify a particular temperature. The specified temperature was 15 °C. Since electrical measurements are capable of much greater precision than other types of measurements, the Ninth International Conference on Weights and Measures recommended that the joule (J) should be used as the unit of heat. The joule is 1 volt-coulomb, or the energy a coulomb of electricity gains

---

[3] The National Bureau of Standards in Washington, D.C., maintains three duplicates of the prototype standard, one of the three being designated the primary standard for this country. The care exercised in working with these is exemplified by the fact that in the early 1970s, when the Washington standards were washed (they apparently get dirty), the procedure took eight months.

[4] Other base units include the ampere for electric current, the Kelvin for thermodynamic temperature, the mole, and the candela for luminous intensity. A brief but complete description of the SI system is available in NBS Special Publication 300, *The International System of Units (SI)* (Washington: U.S. Government Printing Office, 1974).

as it passes through a potential difference of 1 volt. The National Bureau of Standards has adopted a *defined calorie,* which is equal to exactly 4.184 J. The calorie as a unit of heat is still popular with many chemists because of their previous training, but it is rapidly being phased out in favor of the joule. The National Bureau of Standards currently reports and tabulates all calorimetric data in joules.

The SI unit of power is the *watt.* Power is defined as the rate at which work is done, and the watt is 1 J s^{-1}. The watt-hour or kilowatt-hour is a unit of energy. The horsepower, a common unit in many countries, equals 746 watts. The unit of heat called the Btu is still common in the United States in engineering circles (particularly in reference to heating, air conditioning, and electric power generation) and is equivalent to about 1055 J (or 252 calories).

1.5 MATHEMATICAL REVIEW

Following is a brief review of some elementary mathematical formulas and definitions. It is assumed that you have already studied all the material presented here. These pages are designed only to refresh your memory and will mainly present only final results. If you find that any points are unclear, please refer to a calculus textbook.

Operations with exponents should be almost effortless.

$$x^{-a} = \frac{1}{x^a} \qquad (x^a)(x^b) = x^{(a+b)} \qquad x^a y^a = (xy)^a \qquad \frac{x^a}{x^b} = x^{(a-b)}$$

We shall often deal with logarithms. Suppose that M, N, and b are positive numbers, $b \neq 1$. If $a = b^d$, then the logarithm of a to the base b is given by $\log_b(a) = d$.

$$\log_b(MN) = \log_b(M) + \log_b(N) \qquad \log_b\left(\frac{M}{N}\right) = \log_b(M) - \log_b(N)$$

$$\log_b(M^p) = p \cdot \log_b(M) \qquad \log_b\left(\frac{1}{M}\right) = -\log_b(M)$$

$$\log_b(b) = 1 \qquad \log_b(1) = 0$$

To change the base of logarithms, we have for $c \neq 1$

$$\log_b(M) = \log_c(M) \cdot \log_b(c) = \frac{\log_c(M)}{\log_c(b)}$$

When $b = 10$, we have ordinary logarithms to the base ten. Natural logarithms are to the base $b = e = 2.7183 \ldots$ and will be abbreviated by ln. We shall usually write the expression e^x in the form $\exp(x)$. It is also worth recalling that

$$\log_e(x) = \ln(x) = 2.303 \cdot \log_{10}(x) \qquad \text{or} \qquad \log_{10}(x) = 0.434 \cdot \ln(x)$$

We shall often have to solve equations. An equation in one variable can always be reduced to the form $f(x) = 0$. The quadratic equation $ax^2 + bx + c = 0$ has for its solution

$$x = \frac{-b \pm \sqrt{b^2 - 4ac}}{2a}$$

The two roots are either both real or both imaginary. If in working out any physical problem in this course you come up with an imaginary solution, either we or you have made a mistake, but it is probably you. The proper solution of a problem will often provide one physically meaningful and one meaningless root. It should be apparent that a root that yields a negative mass is meaningless. Complicated equations are best solved by successive approximations. Values of x are chosen and used to calculate $f(x)$. The values of x are increased until the sign of $f(x)$ changes. This means that the root has been bracketed by two successive increments. The magnitude of the increment is then reduced, and the process repeated until the desired degree of accuracy is achieved. The availability of minicomputers and programable calculators can greatly lessen the time needed for the calculation.

The derivative of $f(x)$ is given by the limit of $[f(x + \Delta x) - f(x)]/\Delta x$ as Δx approaches zero. It has a geometrical meaning in that it is equal to the slope of the curve. Derivatives of some simple functions are

$$\frac{d}{dx}(uv) = u\frac{dv}{dx} + v\frac{du}{dx} \qquad \frac{d}{dx}(x^a) = ax^{(a-1)}$$

$$\frac{d}{dx}(\ln u) = \frac{1}{u}\frac{du}{dx} \qquad \frac{d}{dx}(e^u) = e^u \frac{du}{dx}$$

Integration is the opposite of differentiation. Its geometrical meaning is that of the area under a curve.

$$\int du = u + \text{constant} \qquad \int_{u_1}^{u_2} du = u_2 - u_1$$

$$\int x^a\, dx = \frac{1}{a+1} x^{(a+1)} \qquad \int \frac{dx}{x} = \ln x \qquad \int e^x\, dx = e^x$$

In the last three equations the constants of integration have been omitted.

If you have read thus far, you should be saying to yourself, "Why bother to put such trivial stuff in here?" If you do, indeed, say that to yourself, then go ahead. If you have not said that, and are having trouble with some of the equations thus far presented, then you will soon run into great difficulty with the material in this book. Stop, do not pass GO, do not collect $200. Go back to your elementary math books and review this material.

If u is a function of more than one variable,

$$u = u(x, y, \ldots, z) \tag{1.1}$$

it is often important to know how u will change relative to changes in any one of the variables, say x, while the remaining variables stay constant. The function that specifies the change is the *partial derivative*. In (1.1), the partial derivative of u with respect to x and with y and z constant is written

$$\left(\frac{\partial u}{\partial x}\right)_{y,z}$$

Second partial derivatives are defined likewise, with the differentiation carried out twice. There are two possibilities. For the "pure" derivatives such as $(\partial^2 u/\partial x^2)$, $(\partial^2 u/\partial y^2)$, we just differentiate twice with respect to the indicated variable, holding all other variables constant. The other possibility is the "mixed" partial derivative, in which we differentiate first with respect to one variable, then with respect to the other; thus,

$$\frac{\partial^2 u}{\partial x\, \partial y} = \left[\frac{\partial}{\partial x}\left(\frac{\partial u}{\partial y}\right)_x\right]_y \tag{1.2}$$

and the other equivalent mixed derivative

$$\frac{\partial^2 u}{\partial y\, \partial x} = \left[\frac{\partial}{\partial y}\left(\frac{\partial u}{\partial x}\right)_y\right]_x \tag{1.3}$$

The two mixed derivatives of Equations (1.2) and (1.3) are equal. This will prove to be a very useful piece of information.

Suppose that in our original function, $u = u(x, y, \ldots, z)$, we introduce infinitesimal changes in the variables of magnitude $dx, dy, \ldots, dz$. Then du, the total change in u arising from these changes in the variables, is given by

$$du = \left(\frac{\partial u}{\partial x}\right)_{y,z,\ldots} dx + \left(\frac{\partial u}{\partial y}\right)_{x,z,\ldots} dy + \cdots + \left(\frac{\partial u}{\partial z}\right)_{x,y,\ldots} dz \tag{1.4}$$

The value du represents the *total differential* of u.

Consider a function of two variables, $u = u(x, y)$. The total differential can be written in the form

Or if we write
$$du = \left(\frac{\partial u}{\partial x}\right) dx + \left(\frac{\partial u}{\partial y}\right) dy \tag{1.5}$$

$$M(x, y) = \frac{\partial u}{\partial x} \quad \text{and} \quad N(x, y) = \frac{\partial u}{\partial y} \tag{1.6}$$

Equation (1.5) can be rewritten as

$$du = M\, dx + N\, dy \tag{1.7}$$

where M and N are functions of x and y.

Suppose we start from an initial position (x_1, y_1) and go to the final position (x_2, y_2). For a certain class of functions, the change in u in going from the initial to the final position will be independent of the path taken, and Δu will be given by

$$\Delta u = u(x_2, y_2) - u(x_1, y_1) = u_2 - u_1 \tag{1.8}$$

Such a differential is called an *exact*, or *perfect*, differential. It is an implication of exactness that u is a function only of the state of the system, and the term *state function* is used to describe u. The condition that du as written in (1.7) should be exact is simply that

$$\left(\frac{\partial M}{\partial y}\right)_x = \left(\frac{\partial N}{\partial x}\right)_y \tag{1.9}$$

Equation (1.9) expresses the necessary and sufficient condition that du should be exact; this criterion is often called the Euler condition for exactness.

An equivalent necessary and sufficient condition that du be exact is given by the statement that the integral around any closed path (meaning that the initial and final points of the integrating path are the same) should be zero, or

$$\oint du = 0 \tag{1.10}$$

Equation (1.9), of course, implies Equation (1.10), and conversely, Equation (1.10) implies (1.9). The thermodynamic functions we shall deal with will usually be state functions.

Some very useful relations between partial derivatives are given in the next three equations:

$$\left(\frac{\partial x}{\partial y}\right)_z = \frac{1}{(\partial y/\partial x)_z} \tag{1.11}$$

$$\left(\frac{\partial x}{\partial y}\right)_z = \frac{(\partial x/\partial w)_z}{(\partial y/\partial w)_z} \tag{1.12}$$

$$\left(\frac{\partial x}{\partial y}\right)_z = -\frac{(\partial z/\partial y)_x}{(\partial z/\partial x)_y} \tag{1.13}$$

Equation (1.13) can be derived quite simply. From the functional relation $z = f(x, y)$, we can write the total differential of z as

$$dz = \left(\frac{\partial z}{\partial x}\right)_y dx + \left(\frac{\partial z}{\partial y}\right)_x dy \tag{1.14}$$

The condition of constant z is obtained by setting $dz = 0$; and by rearranging the terms in (1.14), we get (1.13) directly. Let's now consider a simple example.

Science strives for efficiency. Measurements are kept to a minimum, and the simplest measurement is the best. For a quantity of liquid, three variables of interest are the mass m, the volume v, and the density ρ. There is no point in measuring all three. If we measure any two, we can calculate the third from the relation $\rho = m/v$. We shall often be interested in variables that are difficult or even impossible to measure. But we can often relate the difficult quantity to easily measurable quantities. Consider the following case.

The *coefficient of thermal expansion* α is defined as

$$\alpha = \frac{1}{v}\left(\frac{\partial v}{\partial T}\right)_p \tag{1.15}$$

The relation gives the fractional increase in volume per unit increase in temperature at constant pressure. The value α is easily measurable. Just measure the volume change on heating at constant pressure. The *compressibility* β is defined by

$$\beta = -\frac{1}{v}\left(\frac{\partial v}{\partial p}\right)_T \tag{1.16}$$

The compressibility gives the fractional increase in volume for pressure changes at constant temperature. The negative sign is placed in the definition as a matter of convenience, to make β a positive number. The compressibility is also easily measurable. Just measure the effect of pressure on volume in a thermostat.

Now suppose we want to measure the derivative $(\partial p/\partial T)_v$, in other words, how the pressure changes with temperature at constant volume. For a gas, the quantity is easily measurable. Just enclose the gas in a rigid vessel and measure the change in pressure with temperature. The measurement is not so simple with liquids and solids. It is very difficult to maintain a constant volume as the temperature changes, because the concomitant volume changes of the container will grossly affect the measurements. We can get around the difficulty by finding a relation between this derivative and the other easily measurable quantities α and β. Since $v = f(p, T)$, we can use (1.13) to write

$$\left(\frac{\partial p}{\partial T}\right)_v = -\frac{(\partial v/\partial T)_p}{(\partial v/\partial p)_T} = \frac{\alpha}{\beta} \tag{1.17}$$

Note that we could have gotten the same result by writing the total differential of v, setting it equal to zero, and following the procedure used in (1.14). To complete the example, let us look up the values for α and β for mercury. They are $\alpha = 2 \times 10^{-4}$ K^{-1} and $\beta = 4 \times 10^{-6}$ atm^{-1}. Thus for a mercury thermometer whose entire volume is filled at 50 °C, the internal pressure at 51 °C will be 50 atm. The glass cannot withstand this pressure and will break. Hence the warning against cleaning fever thermometers in hot water. (Note that this computation assumes no volume change of the glass.)

PROBLEMS

1. How much sulfur is in a ton of sulfuric acid?
2. What is the derivative of x^2? of x^5? of $x^{1.4}$?
3. What is the voltage drop across a 9-ohm (Ω) resistor when a current of 12 amperes (A) passes through the resistor? How much energy is dissipated by the resistor per minute? What wattage rating should the resistor have to ensure that it does not burn out?
4. In engineering circles in the United States, the heating value of coal is often given in units of Btu/lb. A Northern Appalachian coal has a heating value of 9769 Btu/lb. What is the heating value in joules per kilogram and calories per gram?
5. When a bank of capacitors is discharged to set off a photoflash lamp, 1 kJ of energy is dissipated in 10 μs. What power level does this correspond to?
6. (a) The mass of the earth is 6.0×10^{24} kg. The mass of the sun is 2.0×10^{30} kg. The distance between the earth and the sun is 1.5×10^{11} m. Calculate the gravitational force between the earth and the sun.
   (b) A proton has a mass of 1.67×10^{-24} g. An electron has a mass of 9.1×10^{-28} g. In a hydrogen atom, the proton and the electron are separated by a distance of 5.3×10^{-9} cm. Calculate the gravitational force between the proton and the electron.
7. What is the total differential dw of the function $w = x^2 + y^2 + z^2$?
8. For the function of Problem 7, evaluate the change in w in going from the point (1, 1, 1) to the point (1.01, 1.01, 1.01) (a) by actually evaluating w at the two points and taking the difference; and (b) by using the expression for the total differential. (c) Are they the same? Why or why not?
9. Now consider the function of Problem 7 in cylindric coordinates,

   $$x = r \cos \theta \quad y = r \sin \theta \quad z = r$$

   Show that $dw = 4r\, dr$.
10. Assume that the earth is a perfect sphere with a circumference of 25,000 miles. A string circling the equator in contact with the earth is thus 25,000 miles long. A second string circles the earth so that it is at all times 1 in. above the surface of the earth. How much longer is the second string than the first?
11. At a maximum or a minimum of a function, $f(x, y)$, we must simultaneously satisfy the conditions $\partial f/\partial x = 0$ and $\partial f/\partial y = 0$. Find the value of z at the minimum for the function

    $$z = f(x, y) = x^2 - xy + y^2 + 2x + 2y - 3$$

12. (a) Show that the differential $dz = (x^2 + y^2)dx + 2xy\, dy$ is exact.
    (b) Show that the differential $dz = xy\, dx + xy\, dy$ is inexact.
13. Evaluate the change in z, or Δz, in going from the point (0, 0) to the point (1, 1) for each of the differentials of Problem 12 for each of the following paths:
    (a) Go from (0, 0) to (0, 1) keeping x constant and equal to 0; then to (1, 1) keeping y constant and equal to 1.
    (b) Go from (0, 0) to (1, 0) keeping y constant and equal to 0; then to (1, 1) keeping x constant and equal to 1.
    (c) Go along the path $y = x$.
    (d) Go along the path $y = x^2$.
    The four paths for the exact differential should give the same answers. (*Note:* By Δz we mean $\int_{0,0}^{1,1} dz$.)
14. Derive Equation (1.17) starting with the total differential dV.
15. Pressure has the units of force per unit area. Show that the product of pressure and volume has the same dimensions as energy or work.
16. A large 1000-megawatt electric power generating station consumes 3 million tons of coal a year. The coal it burns has the calorific value 12,438 Btu/lb and contains 2.73% sulfur and 14.9% ash.
    (a) How many joules of heat are produced in burning the coal?
    (b) Assuming that all the sulfur is converted to SO_2, how much SO_2 gas is produced in a year?
    (c) The U.S. Environmental Protection Agency (EPA) has established new source performance standards (NSPS), which restrict the emissions of new coal-fired power plants to 1.2 lb SO_2 per million Btu. How much sulfur would have to be removed from the coal for it to meet this standard?

PART ONE

THE GAS LAWS

CHAPTER TWO

THE GAS LAWS

2.1 WHAT IS AN IDEAL GAS?

Modern experimental science began with studies of how gases behaved; these studies comprised numerous "pneumatic" experiments in the latter half of the seventeenth century in Europe and in England. The behavior of gases was fundamentally important in establishing the combining ratios of atoms and hence the table of atomic masses. Let's start at the end of the story, and examine the functional relation between the pressure and the volume of gases. We shall take one mole of various gases and measure the volume as a function of pressure at a constant temperature of 300 K. We can display the data in several ways. One way is to plot the volume against the pressure on what is known as a P-V diagram. Another way is suggested by the results of Problem 15 in Chapter 1, wherein the product PV has the dimensions of energy, or work. We shall plot this product PV as a function of the pressure, measuring P in atmospheres and V in liters. We shall, in addition, divide the product by RT, where R is a universal constant known as the *gas constant* (0.08205 liter atm mol^{-1} K^{-1}) and T is the temperature. The ordinate $Z = PV/RT$ is known as the *compressibility factor;* it is a dimensionless quantity.

In Figure 2.1, we have plotted Z vs P for several gases out to quite large pressures. The behavior of the curves seems complex and differs substantially from gas to gas. On the other hand, all the curves appear to be converging at small pressures. This suggests that we repeat the experiment, restricting our attention to the small-pressure region. The results of this experiment are shown in Figure 2.2, and the whole verifies our conclusion that the various curves are converging. All the curves converge to $Z = 1.000$ in the limit as P approaches zero. In this limit, $Z = PV/RT = 1.000$, or $PV = RT$ for one mole of gas. All gases have the same pressure-volume dependence in this limit. We call this limiting case an *ideal gas*.

Notice that in Figure 2.2, each gas approaches the limit as a straight line with a finite slope. At these small pressures, we can write

$$Z = \frac{PV}{RT} = 1.000 + b'P \tag{2.1}$$

where the slope b' is different for each gas (and will further depend on the temperature). This means that a gas will deviate from ideal gas behavior at any finite pressure, however small. On the other hand, b' is usually small; therefore these deviations will be small at small pressures. For N_2 the deviation is only 0.02% at 1 atm, whereas for CO_2, the worst case on the diagram, it is 0.5%. This means that for moderate pressures of the order of an atmosphere or so, we can use the ideal gas equation without introducing a very large error. It is instructive, and may also prove entertaining, to examine the development of the ideal gas equation.

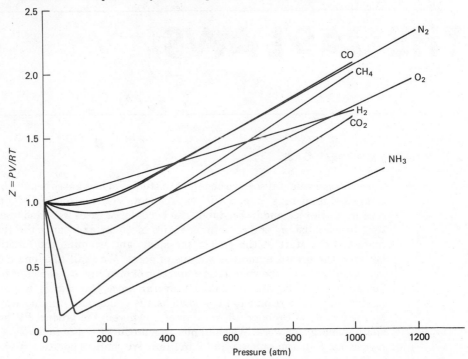

FIGURE 2.1 Compressibility factor vs pressure for several gases.

2.2 PRESSURE

The middle of the seventeenth century marked a milestone in scientific thought. Informed people began to understand the concept of pressure in essentially the same terms as we understand it today. During the first half of that notable century, people spoke loosely about "nature abhorring a vacuum" in much the same way that they were later to speak of phlogiston, caloric, spontaneous generation, and many other concepts that they did not completely understand. Every water-well engineer knew that a simple, single-stage suction lift pump would not work at depths exceeding 34 ft. In 1643, Torricelli built the world's first barometer and answered the "why" of the water-well enigma.

Consider the situation shown in Figure 2.3. In part (a) we have a simple U tube open at both ends and half-filled with mercury. At equilibrium, the height in each arm of the tube is the same. We shall use a tube with a uniform circular cross-sectional area of 1 cm²; also note that mercury has a density of 13.6 g cm^{-3}. In part

---

EVANGELISTA TORRICELLI (1608–1647), Italian physicist and mathematician, was inspired by the work of Galileo in mechanics. In 1641 he went to Florence, where he met Galileo and acted as his amanuensis during the last three months of Galileo's life. He is best known for his invention of the barometer in 1643. He also worked on fluid motion, the theory of projectiles, pulleys, and the properties of the cycloid.

SEC. 2.2 PRESSURE

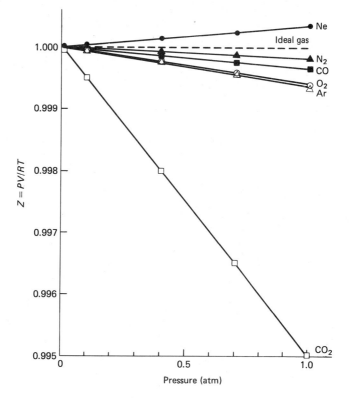

FIGURE 2.2 Compressibility factors as a function of pressure at small pressures at 300 K. (Values from *Tables of Thermal Properties of Gases,* National Bureau of Standards Circular 564, 1955).

FIGURE 2.3 A U-tube manometer under different conditions. (a) Both sides open to the atmosphere. (b) One side evacuated. (c) Extra weight added to one side.

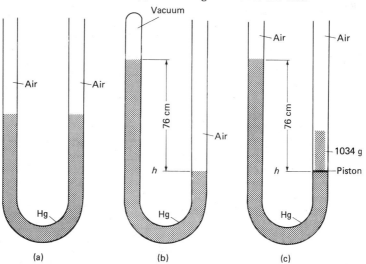

(b), the left limb of the tube has been evacuated and sealed. The mercury rises in that limb until the height is 76 cm above the height in the right limb, varying slightly with atmospheric conditions and the height above sea level.

In part (c), we place a massless piston in the right arm in contact with the mercury surface, and place a 1034-g weight on the piston. We observe that the mercury in the left arm rises to 76 cm above the level of the mercury in the right arm, the same distance as in (b). The column of mercury has a mass of $76 \times 13.6 = 1034$ g. The pressure due to atmospheric air thus equals the pressure a column of mercury 76 cm high exerts. Air can support a column of mercury of that height, and since mercury is 13.5 times as dense as water, the atmosphere will support a column of water 1026 cm (or 34 ft) high. That is why no one had been able to build a single-stage simple lift pump to raise water from a well by more than 34 ft, since the lift pump operates by creating a vacuum above the surface of the water.

A device such as that in Figure 2.3b can serve as an admirable *barometer* or gas-pressure-measuring device. The usual laboratory barometer more closely resembles Torricelli's original device, shown in Figure 2.4. A sealed tube is filled with mercury and inverted in a pool of mercury. The pressure is then determined by measuring the height h of the mercury in the sealed tube above the level in the pool. The main difference between Torricelli's device and the modern instrument lies in the ease and accuracy with which h can be measured.

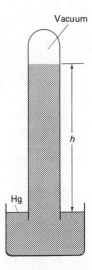

FIGURE 2.4 Torricelli barometer. The pressure is given by the height h.

The French philosopher, Pascal, reasoned that by analogy with hydrostatic pressure, the atmosphere could be considered a *sea of air,* and pressure should vary with height—high at the bottom of a mountain and lower at the top. Like the modern professor who sends his graduate students to do the work, Pascal sent his brother-in-law, Florin Perier, to climb the mountain. Perier carried out the experiment under Pascal's instructions on a grand scale in 1648. He measured the barometric pressure at the foot and peak of Puy de Dome, a mountain in Auvergne in central France (elevation 1465 m). The success of the experiment contributed to the rapid acceptance of Pascal's ideas.

> BLAISE PASCAL (1623–1662), French religious philosopher, mathematician, and scientist, is perhaps best known for his *Pensées,* a theological philosophical work. Pascal was one of the founders of the science of hydrodynamics, and he showed that the height of the mercury column in a barometer decreases as the barometer is carried up a mountain. He was a mathematical prodigy, completing a work on conic sections before he was sixteen. He was one of the early workers in probability, along with Fermat; the correspondence between Pascal and Fermat about the best procedure for betting in games of chance is said to have created the science of probability. Pascal carried Torricelli's work on the cycloid a bit further, solving the difficult problem of the quadrature of the cycloid. He solved various problems dealing with the centers of gravity of solids, and this work was intimately associated with the development of the differential calculus.

2.3 DIGRESSION ON NONSENSE

In many fields of endeavor, whether politics, economics, sociology, psychology, education, or science, new facts appear, crying out to be explained by some theory. Many of the postulated theories will be preposterous. Consider the experiment Torricelli performed, and the vacuum space above the mercury column known as the Torricellian vacuum. The current universally accepted explanation has already been given. Consider the following explanation Linus proposed in the middle of the seventeenth century. Linus claimed that the space above the mercury column in the tube contained an invisible cord, which he called a funiculus (derived from the same root as *funicular,* or cable railway; it was also called the umbilicus). Linus claimed that this cord had properties that allowed it to extend from the top of the closed tube to the surface of the mercury column. It was held sufficiently strong to support a column of mercury 29 in. high. If you don't believe the funicular theory, consider the following argument. Instead of sealing the end of the glass tube in a Torricellian barometer, close the end with your finger, and then invert the tube in a pool of mercury. The column in the tube will drop to a height of 29 in., leaving a vacuum between the top of the mercury column and your finger. You will then feel the flesh of your finger being pulled into the vacuum space. "Aha!" cries Linus, "That is the funiculus pulling your finger into the tube."

Robert Boyle and others replied that it was the pressure of the atmosphere outside the tube that was pushing the finger in. Boyle was an excellent experimentalist, and by numerous experiments he was able to demolish Linus's views. He built vacuum pumps, and pressure pumps, and was able to adjust the pressure outside the barometer, thus varying the height of the column of mercury that could be supported. He dropped feathers in vacuum to study the effect of air on falling bodies, and rang bells in vacuum systems to study the effect of air on sound waves. Boyle even tried to cut the hypothesized umbilicus by manipulating a knife within the Torricellian vacuum. It was during these studies that he discovered what we now call *Boyle's law.*[1]

---

[1] James Bryant Conant, ed., *Harvard Case Histories in Experimental Science.* Case 1. *Robert Boyle's Experiments in Pneumatics,* ed. James Bryant Conant (Cambridge: Harvard University Press, 1950). This little book contains a fascinating description of the early experiments dealing with gases.

ROBERT BOYLE (1627–1691), an English scientist, was born in Ireland, the fourteenth child of Richard Boyle, the great Earl of Cork. He spent much of his youth studying in Europe, and he returned to England in 1644 to devote his life to study and experimental research. He was one of the founding members of the Royal Society of London. When he was elected its president in 1680, he declined the honor because of his scruples against taking oaths. Boyle was introduced to the study of gases by Otto von Guericke's air pump, and with the help of Robert Hooke (of Hooke's law fame) he improved the machine to the point of having at his disposal the finest vacuum pumps of the day. An account of his research with the *machina Boyleana*, or pneumatic engine, was published in 1660 under the title *New Experiments Physico-Mechanical Touching the Spring of Air and Its Effects*.

Besides working on gases, Boyle studied the propagation of sound waves in air and the expansive force of freezing water, and about crystals, refraction, and electricity. His favorite subject was chemistry, and he did some work on the transmutation of the elements. His book *The Sceptical Chemyste* was published in 1661. He was much interested in philosophy and theology, and among his lesser works are such gems as *A Free Discourse against Customary Swearing* (1695), *High Veneration Man Owes to God* (1685), *Excellence of Theology Compared with Natural Philosophy* (1664), and *Occasional Reflections upon Several Subjects* (1665). Jonathan Swift satirized this last work in *A Pious Meditation upon a Broomstick*, and Samuel Butler did the same in *An Occasional Reflection on Dr. Carlton's Feeling a Dog's Pulse at Gresham College*. In his will, Boyle provided for the Boyle Lectures for proving the Christian religion against "notorious infidels, viz. atheists, theists, pagans, Jews, and Mahommedans." The grant stipulated that controversies between Christians were not to be mentioned.

2.4 THE SPRING OF AIR

During the decade that started in 1660 Robert Boyle carried out a series of experimental investigations that were particularly important from two aspects. Firstly, his investigations led to an important empirical equation relating the pressure and the volume of a gas. Secondly, his experiments were perhaps the *first* scientific research carried out in the modern tradition. He executed a carefully controlled experiment in which he set up a system with two variables, scrupulously measuring one of these variables, the volume, as he varied the other, the pressure. He was finally able to express all his results in a simple equation, which is now known as Boyle's law.

His apparatus is shown in Figure 2.5. It consists of a glass tube of uniform bore in the form of the letter J, sealed at the short arm, and open in the long arm. Some gas is trapped in the closed, short arm by partially filling the tube with mercury. The pressure of the gas in the tube can then be varied by adding or removing mercury from the long arm. The gas volume at any pressure can be measured by taking the distance from h, the top of the mercury column in the closed arm, to the sealed top of that arm, and multiplying by the cross-sectional area of the tube. The pressure can be measured by taking the distance $(h' - h)$ between the tops of the two mercury columns. To this distance must be added the pressure due to the atmosphere.

Table 2.1 reproduces some of the data from Boyle's original work. Boyle did not actually measure the volume. What he did was to divide the short arm of the J tube into equal lengths, and then to adjust the pressure so that the top of the mercury column coincided with one of these equal division lines. Since the bore was uniform, the length of the air column was proportional to the volume. The entries in column A are thus in arbitrary volume units. The pressure measurements were made in inches.

SEC. 2.4 THE SPRING OF AIR

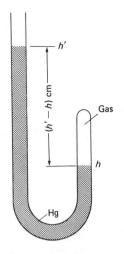

FIGURE 2.5 Boyle's experiment. The pressure of the gas is given by $(h' - h)$ plus the atmospheric pressure. The pressure is varied by adding or removing mercury from the open side.

TABLE 2.1 Boyle's original data.

(A)	(B)	(C)	(D)	(E)	
No. of equal spaces in shorter leg that contained the same parcel of air diversely extended	Height of mercurial cylinder in longer leg that compressed the air into those dimensions	Height of mercurial cylinder that counter-balanced the pressure of the atmosphere	Aggregate of the two last columns, (B) and (C), exhibiting the pressure sustained by included air	What that pressure should be according to the hypothesis, which supposes pressures and expansions to be in reciprocal proportion	Product of column (A) and column (D)
48	00		$29\frac{2}{16}$	$29\frac{2}{16}$	1398
46	$01\frac{7}{16}$		$30\frac{9}{16}$	$30\frac{6}{16}$	1406
44	$02\frac{13}{16}$		$31\frac{15}{16}$	$31\frac{12}{16}$	1405
42	$04\frac{6}{16}$		$33\frac{8}{16}$	$33\frac{1}{7}$	1407
40	$06\frac{3}{16}$		$35\frac{5}{16}$	35	1413
38	$07\frac{14}{16}$		37	$36\frac{15}{19}$	1406
36	$10\frac{2}{16}$		$39\frac{5}{16}$	$38\frac{7}{8}$	1415
34	$12\frac{8}{16}$		$41\frac{10}{16}$	$41\frac{2}{17}$	1415
32	$15\frac{1}{16}$		$44\frac{3}{16}$	$43\frac{11}{16}$	1414
30	$17\frac{15}{16}$		$47\frac{1}{16}$	$46\frac{3}{5}$	1412
28	$21\frac{3}{16}$		$50\frac{5}{16}$	50	1409
26	$25\frac{3}{16}$		$54\frac{5}{16}$	$53\frac{10}{13}$	1412
24	$29\frac{11}{16}$	$29\frac{2}{16}$	$58\frac{13}{16}$	$58\frac{2}{8}$	1412
23	$32\frac{3}{16}$		$61\frac{5}{16}$	$60\frac{18}{23}$	1410
22	$34\frac{15}{16}$		$64\frac{1}{16}$	$63\frac{6}{11}$	1409
21	$37\frac{15}{16}$		$67\frac{1}{16}$	$66\frac{4}{7}$	1408
20	$41\frac{9}{16}$		$70\frac{11}{16}$	70	1414
19	45		$74\frac{2}{16}$	$73\frac{11}{19}$	1408
18	$48\frac{12}{16}$		$77\frac{14}{16}$	$77\frac{2}{3}$	1402
17	$53\frac{11}{16}$		$82\frac{12}{16}$	$82\frac{4}{17}$	1407
16	$58\frac{2}{16}$		$87\frac{14}{16}$	$87\frac{3}{8}$	1406
15	$63\frac{15}{16}$		$93\frac{1}{16}$	$93\frac{1}{5}$	1396
14	$71\frac{5}{16}$		$100\frac{7}{16}$	$99\frac{6}{7}$	1406
13	$78\frac{11}{16}$		$107\frac{13}{16}$	$107\frac{7}{13}$	1402
12	$88\frac{7}{16}$		$117\frac{9}{16}$	$116\frac{5}{8}$	1411

SOURCE: Data in the first five columns were taken from J. B. Conant, *On Understanding Science* (New Haven: Yale University Press, 1947), p. 54.

The last column of the table lists the product of the first and fourth columns, and you can see that these entries are remarkably constant. The constancy of these entries led Boyle to formulate his law that the product of the volume and the pressure is a constant for any fixed quantity of gas, or

$$PV = \text{constant} \tag{2.2}$$

Scientific data are usually displayed graphically, and we have done this in Figures 2.6 and 2.7 for Boyle's original data in two different ways. In Figure 2.6 we plot pressure against volume, and in Figure 2.7 against reciprocal volume. In both figures the squares indicate the experimental points and the solid lines are smooth curves drawn through the experimental points. The straightness of the curve in Figure 2.7 shows that this is a more satisfactory way to plot the data to indicate clearly the simplicity of Boyle's law.

Note that the entries in the last column of Table 2.1 are not exactly constant, whereas they should be so according to Boyle's law. In fact, they vary randomly about some average value. Every experiment contains some inherent errors, which can be minimized only to a certain extent, depending on available technology and measuring instruments. It is informative to mention some of the errors involved in Boyle's measurements.

Initially we have the fact that distances were reported only to the nearest sixteenth inch, and seventeenth-century instruments were less accurate than their modern counterparts. Then there is the fact that the cross-sectional area of the glass tube was not perfectly uniform, nor the end sealed in a perfectly flat manner; thus some error is introduced by the assumption that the column of gas was a perfect cylinder. In addition, the temperature as well as the atmospheric pressure may have varied during the experiment.

There are statistical criteria that can be applied to Boyle's measurements to tell us what the expected uncertainty of the final product is, although we shall not go into them. If one computes the average and the standard deviation of the 25 PV en-

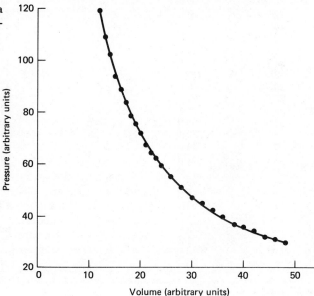

FIGURE 2.6 Boyle's data on a plot of pressure vs volume.

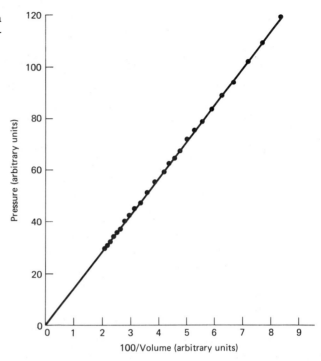

FIGURE 2.7 Boyle's data on a plot of pressure vs inverse volume.

tries in Table 2.1, the result is 1408 ± 5; the standard deviation is less than 0.5%, which is respectable precision.[2]

There are also statistical criteria to determine the straight line that "best" represents a set of experimental points. That line is the one for which the sum of the squares of the distances from the points to the line is a minimum. The procedure for finding the line is known as the method of *least squares*. This method was used to generate the straight line in Figure 2.7. The equation of this straight line is[3]

$$P = 1403(V^{-1}) + 0.2 \qquad (2.3)$$

2.5 PRESSURE UNITS

Pressure is defined as force per unit area. In the *Système International* (SI) units, force is measured in *newtons* (N). The newton is a so-called *derived* unit. In SI base units, the newton is 1 m kg s^{-2}. It is the force required to impart an acceleration of 1 m s^{-2} to a mass of 1 kg. The SI unit of pressure is the pascal (Pa), which equals 1 newton per square meter (N m^{-2} or kg m^{-1} s^{-2}). The unit one atmosphere (atm) is 1.01325×10^5 newtons per square meter, or 1.01325 dynes per square centimeter. This equals the force exerted by a column of mercury 760 mm high. The unit millimeter (mm) is often used in the laboratory. In work with vacuum systems, the pressures usually encountered are less than 10^{-3} mm and the unit micron (1 micron = 10^{-6} m) is often used. *Micron* is abbreviated by the Greek letter μ.

---

[2] Hugh D. Young, *Statistical Treatment of Experimental Data* (New York: McGraw-Hill Book Co., 1962). This short book contains an excellent and simple introduction to the treatment of experimental data.
[3] *Ibid.*, p. 115, for a discussion of linear least-squares fits.

Strictly speaking, using the millimeter or any other unit of linear dimension for pressure is not correct, since such units are not dimensionally valid. To correct this situation, a new unit for pressure has been adopted, the *torr,* in honor of Torricelli. The torr is defined as 101325/760 Pa. In other words, it is defined such that 1 atm is equal to 760 torr. The torr is thus the old unit millimeter (which is dimensionally incorrect). The old unit micron is now the millitorr (abbreviated mtorr). This new unit, although not an SI unit, is extensively used in high-vacuum technology.

2.6 DIGRESSION ON BAROMETRIC INSTRUMENTATION

Aside from instruments used to measure the fundamental quantities of time, length, and mass (clock, ruler, and balance), the barometer was the first instrument to find its way into the scientific laboratory. The standard mercury barometer used to measure atmospheric pressure in the laboratory is the simple Torricelli type shown in Figure 2.4. Several kinds of pressure-measuring devices are common in modern laboratories, some specifically for measuring low pressures.

The aneroid barometer is the most common barometric instrument in homes. Some air is sealed into a bellows, which can then expand or contract lengthwise, depending on whether the pressure outside the bellows is lower or higher than the pressure inside. As shown in Figure 2.8, the bellows can be connected so as to drive a pointer that indicates the pressure. This type of instrument can be constructed to cover a wide range of pressure and can be bought for about $5, although precision aneroid barometers may cost hundreds of dollars. A more sophisticated precision "manometer" is shown in Figure 2.9. The simple U-tube manometer has been depicted in Figure 2.3. If the sealed end is opened to a gas reservoir at constant pressure, then the instrument operates as a differential manometer; it measures the pressure *difference* between the unknown pressure in the system and the known pressure in the reference side. This kind of manometer can be made more sensitive by using oil instead of mercury. Because oil has a much lower density than mercury, the height difference between the two columns will be larger, with oil rising 15 times higher than mercury. Capillarity and meniscus effects, density, and volatility can affect the choice of manometric fluid.

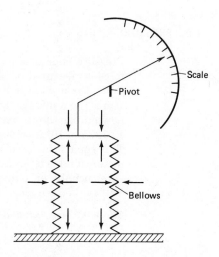

FIGURE 2.8 Bellows type of barometer. Some gas is sealed in the thin-walled bellows at a fixed pressure. The bellows expands or contracts with changing atmospheric pressure, moving the pointer about the pivot point.

FIGURE 2.9 Schematic of a precision-type bellows manometer. Here the "bellows" is a spiral of thin-walled quartz tubing called a *Bourdon tube*. (Fused quartz as a Bourdon-tube material has the attributes of chemical inertness, stability, almost perfect elasticity, negligible hysteresis, and very low thermal expansion.) As the differential pressure across the walls of the Bourdon tube changes, the mirror at the bottom of the tube rotates. The angle of rotation is detected optically by reflecting a beam of light from the mirror. Readout is automated by using a servomechanism. The instrument can be purchased to operate over a range of pressure, and it has a resolution of better than 0.0001% of the maximum pressure in the range. (Courtesy of Texas Instruments, Inc., Houston, Texas.)

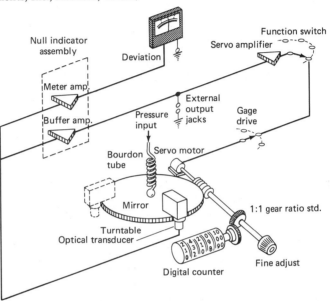

In the laboratory, pressures of the order of 0.01 millimeter of mercury (mmHg) and much lower are often measured, and many schemes have been designed for this. The McLeod gauge, based on Boyle's law, was the earliest very low pressure manometer and is a pressure amplification device; it is shown in Figures 2.10 and 2.11. The gas in the bulb is at the same pressure as the gas in the system. The mercury in the bottom flask will rise in the McLeod gauge when the pressure above it is raised by admitting gas into the flask through the stopcock at the side. When the mercury gets to the point indicated by the cutoff level, all the gas at the originally low pressure in the bulb is trapped. As the mercury rises still higher in the gauge, the large volume of gas in the bulb is compressed into the small-bore capillary tube at a much higher pressure. The final pressure is given by the height difference $(h' - h)$. We shall call the final pressure P_{fin}. (The extra side reference tube is made of the same small-bore capillary tubing as the capillary tube at the top of the bulb to eliminate capillary depression errors.) The initial volume of the gas is V_{init}, or the volume of the bulb above the cutoff level as measured before the gauge is assembled. The final volume of the gas V_{fin} is found by measuring the distance from h to the top of the sealed capillary tube and multiplying this distance by the known cross-sectional area of the capillary tube. By Boyle's law, we have

$$P_{init}V_{init} = \text{constant} = P_{fin}V_{fin} \tag{2.4}$$

FIGURE 2.10 Diagram of McLeod gauge. Mercury is indicated by the shaded region. The mercury is raised by admitting air through the stopcock and lowered by pumping through the stopcock. A scale is attached to the gauge to facilitate measuring the distance $h' - h$, which is the final pressure.

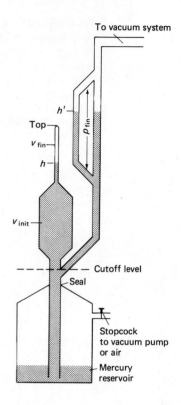

The only unknown is P_{init}, the quantity to be measured, and we know that

$$P_{\text{init}} = \frac{P_{\text{fin}} V_{\text{fin}}}{V_{\text{init}}} \qquad (2.5)$$

The sensitivity of the gauge can be increased by increasing the volume of the bulb or by decreasing the diameter of the capillary tube. Pressures as low as 10^{-6} torr can be measured with this kind of gauge.

The McLeod gauge suffers from several inherent problems. It is inconvenient and does not allow continuous pressure measurement. The mercury in the reservoir must be raised and lowered each time a measurement is made. This is time-consuming. The gauge is subject to large errors when it is used with gases that can condense in the bulb as the pressure is raised. The device also suffers from an inherent error that was just recently discovered. Mercury has a measurable vapor pressure at room temperature, and it is necessary to interpose a liquid nitrogen trap between the gauge and the system to prevent the mercury from contaminating the system. This sets up a stream of mercury that evaporates from the reservoir at room temperature and condenses in the trap. This stream acts as a diffusion pump, pumping the gauge to a pressure below that of the system. This error, which can be 10% or more, is known as the *mercury streaming error*.[4] In addition, the liquid nitrogen trap renders the gauge useless for measuring the pressure of condensable gases.

---

[4] H. Ishii and K. Nakayama, *Transactions of the Eighth Vacuum Symposium and Second International Congress*, vol. 1 (New York: Pergamon Press, 1961), p. 519.

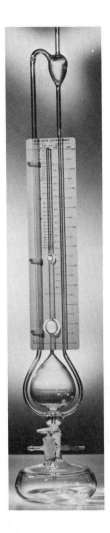

FIGURE 2.11 A commercial McLeod gauge. (Photo courtesy of CVC Products, Inc., Rochester, N.Y.)

Figures 2.12 and 2.13 depict a diaphragm manometer in which the right side is connected to the pressure to be measured, and the left side to a reference pressure. When the instrument is used as an absolute manometer, the reference pressure is supplied by a high-vacuum system. The manometer measures the pressure difference between P_{meas} and P_{ref}. The unit is calibrated by measuring the displacement of the diaphragm from the equilibrium position at various known pressures. In the most sensitive instruments of this type, pressures as low as 10^{-5} torr can be measured. The displacement is not measured directly; rather, a property of the system that depends on distance is measured. In the *capacitance manometer,* the diaphragm forms one plate of a capacitor; the other plate is fixed. The relevant measurement is capacitance, which is a function of the distance between the plates. The pressure can be continuously displayed on a strip chart recorder.

The thermocouple gauge depends for its operation on the proclivity of a gas to conduct heat at a rate proportional to the pressure. If a fixed current is passed through a resistor in a vacuum, the temperature of the resistor will be inversely proportional to

FIGURE 2.12 Diagram of a diaphragm manometer.

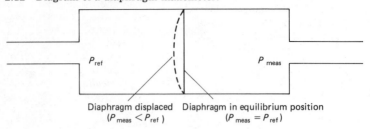

FIGURE 2.13 A commercial capacitance manometer. On the right is the "head," which contains the diaphragm and is connected to the vacuum system. On the left is the electronic circuitry that detects the changes in the capacitance. (MKS Baratron® Capacitance Manometers. Courtesy of MKS Instruments, Inc., Burlington, Mass.)

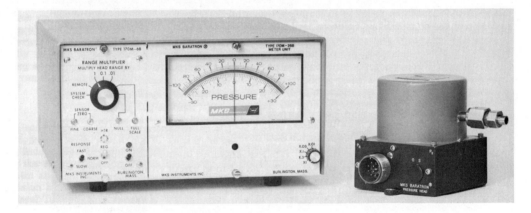

the pressure. Thermocouple gauges are useful in the range 10^{-3} to 10^{-1} torr, but their readings depend on the gas used.

The ionization gauge is based on the principle that if a gas is bombarded by electrons, ions will be produced, with the number of ions a function of the pressure. The relevant measurement here is the magnitude of the positive ion current. Ionization gauges are useful in the range 10^{-11} to about 10^{-4} torr, and again, the readings depend on the nature of the gas.[5]

2.7 TEMPERATURE

New discoveries in a science usually spur a rapid increase in research activity in that field. Boyle's discovery was no exception, giving rise to a rapidly expanding activity in the area of pneumatic research. Many experimenters carefully studied

---

[5] Saul Dushman, *Scientific Foundations of Vacuum Technique,* 2d ed., rev. members of the research staff, General Electric Research Laboratory, ed. J. M. Lafferty (New York: John Wiley & Sons, 1962).

> HUMPHRY DAVY (1778–1829), English chemist, started his illustrious career at the Pneumatic Institution in Bristol, where he discovered the intoxicating effect of nitrous oxide, or "laughing gas." He recommended that the gas could be used as an anesthetic for surgery, but this recommendation was never taken seriously, and instead, breathing laughing gas for intoxication became the rage in England. This work brought Davy to the attention of Count Rumford, who invited him to the Royal Institution, where his career began in earnest. Davy's book *Elements of Agricultural Chemistry* (1813) was the outgrowth of his lectures on the subject given over ten years; and it remained the standard book on the subject for fifty years. Davy, in his first work in electrochemistry in 1800, proved that water when electrolyzed does not yield acid and base but only hydrogen and oxygen, and he proposed that "chemical" attraction and electrical attraction arise from the same causes. The ionic bond was born. Napoleon had developed a great interest in galvanic experiments, and this work won for Davy the medal Napoleon offered for the best work in this field. In 1807, Davy demonstrated that sodium and potassium hydroxides were not elements by electrolyzing these compounds and isolating sodium and potassium. In biology he demonstrated that photosynthesis was carried on by marine vegetation and that venous blood contained carbon dioxide. He was much interested in the practical applications of science as evidenced by his book on agricultural chemistry and his invention of the Davy lamp (1816), a device that proved invaluable in coal mining. Davy wrote poetry throughout his life, and the quality of this work is evidenced by the statement of the poet Coleridge that if Davy "had not been the first chemist, he would have been the first poet of his age."

the effect of pressure on volume and verified that the PV product was constant. It soon became evident that this constant PV value changed from day to day and from season to season. It was realized before long that the qualifying phrase "at constant temperature" had to be added to Boyle's law to make it valid. In contradistinction to the barometer, which Torricelli essentially invented in one step, the thermometer took many years to evolve into the modern instrument in form and scale.[6] Sir Humphrey Davy once said, "Nothing tends so much to the advancement of science as the application of a new instrument." This was particularly true of the thermometer.

The ancient Greeks developed numerous toys and mechanical devices that depended on the expansion of gases for their operation. Extending these toys to the measurement of "degree of hotness," or temperature, apparently first occurred to Galileo about 1600. His instrument was a glass bulb filled with air, with a long stem extending downward into some water. As the temperature changed, the air in the bulb contracted or expanded, and the change was indicated by the height of the water in the tube. The device was actually a thermoscope rather than a thermometer; it was a temperature-indicating device and not a temperature-measuring device. Its most severe drawback was that the change in liquid level was affected by changes in barometric pressure as well as in temperature. Sanctorius, a colleague of Galileo, calibrated such a device using snow and a burning candle to provide two fixed points. Such were the modest beginnings of thermometry.

The first liquid thermometer was built in 1631 by Jean Rey, a French physician, who used the device for taking the temperature of patients. Since the stem of the device was open, it suffered from the serious defect of having the water evaporate. The first thermometer with a sealed stem was built in 1641 by the Grand Duke Ferdinand II of Tuscany, who used alcohol instead of water. In principle, this is identical

---

[6] James Bryant Conant, ed., *Harvard Case Histories in Experimental Science*. Case 3. *The Early Development of the Concepts of Temperature and Heat; The Rise and Decline of the Caloric Theory,* by Duane Emerson Roller (Cambridge: Harvard University Press, 1950). This little book contains a fascinating history of the development of the thermometer.

> GALILEO GALILEI (1564–1642) began by studying medicine at Pisa, but he turned to mechanics, mathematics, and astronomy. He discovered the periodicity of the pendulum while watching a lamp swinging in the Cathedral of Pisa in 1581. He wrote a treatise on the center of gravity of solids (an important problem in those days), and during 1589 to 1591 carried out a series of experiments by which he established many principles of dynamics. During this period he did his famous experiment of dropping various weights from the Leaning Tower of Pisa, showing that the velocity of fall is independent of weight. He also demonstrated that the path of a projectile is a parabola. His greatest contribution was to astronomy, for which he laid the groundwork for the Copernican view of the solar system. He improved the telescope to the point at which it became a highly useful instrument. His first attempt had a magnifying power of 3, but he soon improved this to 32. Galileo manufactured hundreds of telescopes with his own hands, and these were in great demand throughout Europe. He observed the lunar mountains and demonstrated that the Milky Way was a collection of countless stars. He discovered the satellites of Jupiter, greatly strengthening the Copernican view. The publication of his famous *Dialogo dei Due Massimi Sistemi del Mondo* in 1632 was acclaimed in every part of Europe but one. The views were in direct conflict with those of the Church at the time, and Galileo was handed over to the Inquisition in 1633. He "recanted" his views, spending his remaining eight years in seclusion, a requirement the Inquisition imposed. He was engaged in work with his disciple Torricelli when he died.

to the modern thermometer; the chief difference is that the modern thermometer has greater uniformity of the bore and hence greater accuracy and reproducibility.

The calibration of thermometers involved determining a fixed upper point and a fixed lower point, then dividing the scale into fixed intervals. The earliest thermometers used the lowest winter cold and the highest summer heat for the two fixed points. It soon became apparent that this was unsatisfactory. It was impossible to compare the temperatures of Siberia with temperatures of southern Italy. In 1669, Honoré Fabri adopted the temperature of melting snow for the lower fixed point.

In 1688, Dalence suggested that the upper fixed point should be changed to the melting point of butter. This is a relatively fixed temperature, but it depends on the quality of the milk from which the butter is made. Not until 1694 did Carlo Renaldini of Padua propose the freezing and boiling points of water as the two fixed points. When it was realized that atmospheric pressure had an effect on these points, it was further stipulated that the calibration pressure should be one atmosphere.

The size of the "degree" in any temperature scale depends on the temperature values assigned to the calibration points. In the Fahrenheit system, the value 32 is assigned to the freezing point of water and 212 to the boiling point. In the Celsius system, the values 0 and 100 are assigned to these same two points. There are thus 180 Fahrenheit degrees and 100 Celsius degrees between the freezing and boiling points of water. The Celsius degree therefore is $\frac{180}{100} = 1.8$ times the Fahrenheit. To convert from one system to the other, one must take this factor into account and likewise remember that the zero points of the two systems do not coincide. The relation is given by

$$F = \tfrac{9}{5}C + 32 \tag{2.6}$$

where F refers to degrees Fahrenheit and C to degrees Celsius.

When we study the second law of thermodynamics, we shall see that only one fixed temperature is needed to establish a temperature scale. The modern temperature scale is defined by setting the temperature of a fixed point, the triple point of water, at exactly 273.1600. . . . This choice establishes the unit of temperature so that there are 100 units between the freezing point and the boiling point of water. The symbol °C (read "degree Celsius") is used for the Celsius scale. No degree symbol is used with the thermodynamic temperature scale, and temperatures are simply written as K (read "Kelvin").

2.8 CHARLES'S LAW (SOMETIMES CALLED GAY-LUSSAC'S LAW)

Suppose we take a fixed quantity of gas and measure its pressure and volume at a series of different temperatures. We find that the product PV will be a constant at each temperature and that the value of this constant varies in a monotonic[7] manner with temperature. If we plot the points as a function of temperature, as shown in Figure 2.14, we get a straight line that goes through the origin at -273.15 °C. At that temperature, the value of the product PV is zero. Note that the low-temperature part of the curve is an extrapolation from higher temperatures, since all gases condense to liquids before reaching -273.15 °C. The equation of the straight line in Figure 2.14 is

$$PV = B(t + 273.15) \tag{2.7}$$

where P is the pressure, V is the volume, t is the temperature in Celsius degrees, and B is a constant, which is given by the slope of the straight line. In fact, B is a composite constant, having two parts; one part tells how much gas is present, and the second part is a universal constant of nature, the *gas constant*.

Equation (2.7) seems much simpler if we define a new temperature scale, which has its origin, or zero point, at -273.15 °C, and which uses a unit of the same magnitude as the Celsius unit. On this new temperature scale, water has a freezing point of

FIGURE 2.14 Plot of the product PV vs temperature for various quantities of gas. The intercepts are always the same, namely -273.15 °C.

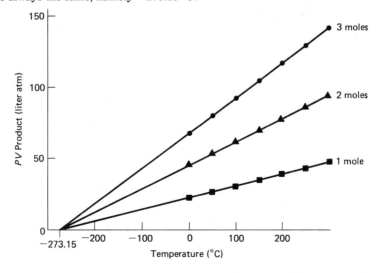

[7] If y is a function of x, and if y always increases as x increases, then y is said to be a *monotonically* increasing function of x; if y always decreases as x increases, y is said to be a *monotonically* decreasing function of x.

> JOSEPH LOUIS GAY-LUSSAC (1778–1850), French chemist, was a pioneer in balloon ascensions. In 1804, Gay-Lussac alone and sometimes with Jean Biot made several balloon ascensions to altitudes as high as 7000 m, where he made observations on magnetism, temperature, humidity, and the composition of air. He could not find any variation of composition with height. In 1809, he published his memoir on the combination of gases. In it he pointed out that gases combine in simple proportions by volume; and this is still called Gay-Lussac's law. He then turned his attention to what students call "wet chemistry," isolating and studying potassium and using potassium to isolate boron from boric acid. Gay-Lussac's work on chlorine brought the scientist into controversy with Sir Humphry Davy. Gay-Lussac assumed chlorine to be an oxygen-containing compound, while Davy correctly considered it an element, a view that Gay-Lussac eventually accepted. He showed that prussic acid contained hydrogen but no oxygen. Lavoisier had insisted that oxygen was the critical constituent of acids, and Gay-Lussac's work on prussic acid was an important cornerstone for the hydrogen acid theory. The prefix *hydro-* for oxygen-free acids is attributable to Gay-Lussac.

+273.15 and a boiling point of +373.15. To convert from Celsius temperature to the new scale, we just add 273.15 to the temperature on the Celsius scale. We shall call this new temperature scale the *ideal gas temperature scale,* and rewrite Equation (2.7):

$$PV = BT \qquad (2.8)$$

The symbol T is the temperature on this new scale. The zero point of this ideal gas temperature scale is called *absolute zero*. It is seen to have the following meaning in the light of Equation (2.8): if the pressure of a gas is kept constant, then the absolute zero is that temperature at which the volume of an ideal gas would equal zero. It is evident that this is a purely mathematical definition, since all real gases condense at temperatures well above the absolute zero. We shall subsequently show that this ideal gas temperature scale is identical to the thermodynamic temperature scale (Chapter 7). Thus, since we may be expecting this result, we have used an uppercase T in Equation (2.8) to denote temperature.

2.9 THE IDEAL GAS EQUATION

The value of the constant B in Equation (2.8) must be proportional to the amount of gas present so we can replace it by nR, where n is the number of moles of gas in the sample and R is a universal constant known as the *gas constant*. Boyle's law and Charles's law can now be combined into one grand equation known as the *ideal gas equation:*

$$PV = nRT \qquad (2.9)$$

The ideal gas equation is the monumental result of the efforts of many people working in many parts of the world over two hundred years or so, starting with Robert Boyle. An understanding of the behavior of gases was necessary for the development of the table of atomic masses. That table is another monumental result of the efforts of many people, and its tale is intimately associated with the ideal gas equation. We shall only say about it here that determining the quantity of any material that constitutes a *mole* of that material and realizing that equal moles of different materials contain the same number of molecules (or particles) constituted perhaps the greatest single achievement of chemistry during the nineteenth century, providing the base for further advances.[8] Among the significant results of that work was the numerical evaluation of the gas constant R.

$$R = 0.0820575 \text{ liter atm K}^{-1} \text{ mol}^{-1}$$
$$= 82.0575 \text{ cm}^3 \text{ atm K}^{-1} \text{ mol}^{-1} \qquad (2.10)$$

Since the liter atmosphere has the dimensions of energy, we can also write

$$R = 8.31431 \text{ J K}^{-1} \text{ mol}^{-1} = 1.98717 \text{ cal K}^{-1} \text{ mol}^{-1} \qquad (2.11)$$

The application of Equation (2.9) should be familiar from previous studies. Essentially there are four variables—P, V, n, and T. With any three of these given, the fourth is uniquely determined. Some applications of this equation are given in the problems. We might consider as a specific example how the ideal gas equation is useful to cardiologists.

2.10 EXAMPLE: OXYGEN CONSUMPTION AND CARDIAC OUTPUT

An important measurement in treating certain types of heart disease is the *cardiac output*, or the rate at which the heart pumps blood. The value can be measured directly by the insertion of flow meters into the proper blood vessels. This requires a surgical procedure. The rate can be measured safely, quickly, and indirectly by analyzing the oxygen content of the inhaled air and the exhaled air and also the oxygen content of both the arterial blood and the venous blood.

Suppose that the inhaled air contains 20.94% O_2 and the exhaled air contains 16.89% O_2. The patient breathes at the rate of 7.77 liters min^{-1} at 21.5 °C and an ambient pressure of 731 torr. In practice, all the data are reduced to values at standard temperature and pressure (STP = 1 atm, 0 °C). The patient's respiration rate is therefore

$$\frac{731}{760} \times \frac{273}{294.5} \times 7.77 = 6.73 \text{ liters min}^{-1} \text{ at STP}$$

The patient's O_2 consumption is now computed to be

$$(0.2094 - 0.1689) \times 6.73 = 0.273 \text{ liter min}^{-1} O_2 \text{ at STP}$$

Analysis of the oxygen content of the patient's blood indicates that the arterial blood contains 17.63 cm^3 of O_2 per 100 cm^3 of blood and the venous blood contains 12.61 cm^3 of O_2 per 100 cm^3 of blood, with the O_2 volumes measured at standard temperature and pressure. We assume that the entire O_2 consumption shows up in the difference between arterial and venous O_2 content. Since the blood transports $10(17.63 - 12.61)/1{,}000 = 0.0502$ liter O_2 per liter of blood,

$$\text{cardiac output} = \frac{0.273}{0.0502} = 5.44 \text{ liters of blood/min}$$

This diagnostic test is much safer and more convenient than surgical procedures. (*Note:* In practice, the final measure used is the *cardiac index*, which is the cardiac output per square meter of body surface area.)

---

[8] Part of the story is to be found in Chapter 25. The complete story can be found in the reference given in footnote 1 of Chapter 25.

2.11 GAS MIXTURES

In the cardiac output example, we used the fact that the ideal gas equation applies to mixtures of gases as well as to pure gases. This fact, originally based on experimental investigation, is known as *Dalton's law of partial pressures*. The law states that the total pressure of any gas mixture equals the sum of the pressures that each constituent gas would exert if placed in the vessel alone. Thus we can write

$$P_{tot} = P_1 + P_2 + P_3 + \cdots + P_n$$
$$= \sum_i P_i \qquad (2.12a)$$

where P_i are the partial pressures of the individual gases. The ideal gas law contains no quantities characteristic of the type of gas involved. In a sense, any molecule in the gas is unaware of the existence of other molecules. The ideal gas equation can be separately applied to the individual partial pressures:

$$P_{tot} = n_1 \frac{RT}{V} + n_2 \frac{RT}{V} + n_3 \frac{RT}{V} + \cdots + n_n \frac{RT}{V}$$
$$= \frac{RT}{V} \sum_i n_i$$
$$= \frac{nRT}{V} \qquad (2.12b)$$

where n is the total number of moles of gas in the vessel.

Concentrations of gases in mixtures can be expressed in several ways. The pressure fraction will be the ratio of the partial pressure to the total pressure, P_i/P_{tot}, while the mole fraction is the ratio of the number of moles of the constituent species to the total number of moles, $X_i = n_i/n$. Since the pressure of a gas is directly proportional to the number of moles, the pressure fraction must equal the mole fraction,

$$\frac{P_i}{P_{tot}} = \frac{n_i(RT/V)}{n(RT/V)} = \frac{n_i}{n} = X_i \qquad (2.13)$$

Using this information, we can calculate the average molecular mass of air, assuming air to be composed of N_2, O_2, and Ar, with individual partial pressures of 0.78, 0.21, and 0.01, and molecular masses of 28, 32, and 40 g. The simplest way to proceed is to calculate the mass of 1 mol air; it is

N_2: (0.78)(28) = 21.8 g
O_2: (0.21)(32) = 6.7 g
Ar: (0.01)(40) = 0.4 g
Total mass = 28.9 g = average mol. mass of air

2.12 REAL GASES AND THE VAN DER WAALS EQUATION OF STATE

As measuring techniques improved over the years, it was not long before it was discovered that the ideal gas law was not as universally applicable as it had originally been thought. The product PV at constant temperature was not quite constant; indeed, it deviated substantially from constancy for many gases, with different devia-

SEC. 2.12 REAL GASES AND THE VAN DER WAALS EQUATION OF STATE

> JOHANNES DIDERIK VAN DER WAALS (1837–1923), Dutch physicist, followed an unusual academic path, viewed in the light of modern practice. He taught physics at a *Hochschule* until 1877, when he was appointed professor of physics at the University of Amsterdam. His work on the kinetic theory of gases led to further work on the continuity of liquid and gaseous states; this in turn led to the derivation of his equation of state in an effort to describe critical phenomena. His statement of the law of corresponding states provided much of the information necessary for Dewar and Ohnes to liquefy many of the permanent gases. Van der Waals received the Nobel prize for physics in 1910.

tions for different gases. A convenient value for measuring the deviation from ideality is the compressibility factor $Z = P\bar{V}/RT$, which we noted in the first section of this chapter. The bar over V indicates molar volume. The amount by which Z deviates from unity is a measure of the lack of ideality in a *real*, or *imperfect* or *nonideal*, gas. One point was, however, quite clear. In the limit as the pressure approached zero, *all* gases obeyed the ideal gas law, as shown in Figure 2.1.

Two properties are inherent in an ideal gas:

1. The gas molecules occupy no volume.
2. There are no interaction forces between the gas molecules.

For any real gas, neither of these characteristics is valid at finite pressure. For any gas, both of these become more nearly valid in the limit of zero pressure. This can be seen from the following simplified argument: If the gas molecules do occupy a volume because of their finite size, then this volume will be approximately independent of pressure. This effective volume is known as the *excluded volume,* and we can calculate an order of magnitude value for it. Assuming a spherical molecule with a diameter d, the molecular volume is given by $\frac{4}{3}\pi(d/2)^3$. The actual excluded volume is four times this value, or $\frac{2}{3}\pi d^3$, as you can see by referring to Figure 2.15. Molecules have diameters of the order of 3×10^{-8} cm; monatomic molecules such as He are somewhat smaller, and large organic molecules somewhat larger. Thus the total excluded volume per mole is just this number times Avogadro's number, or $(6 \times 10^{23})(\frac{2}{3})(\pi)(3 \times 10^{-8})^3 = 34$ cm^3 mol^{-1}. At room temperature the molar volume of an ideal gas is about 24,000 cm^3 at 1 atm, and the excluded volume is about 0.1% of the total volume. At 10 atm, at which the total molar volume is 2400 cm^3, $V_{excluded}$ amounts to 1% of the total volume, whereas at 100 atm it is 10% of the total volume. The relative error in neglecting $V_{excluded}$ increases substantially as the pressure in-

FIGURE 2.15 Excluded volume for a real-gas molecule.

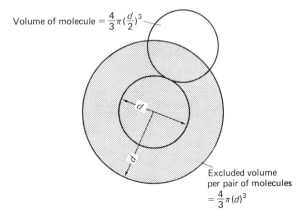

creases. On the other hand, as the pressure decreases, this relative error decreases, and it disappears in the limit as $P \to 0$. Van der Waals first suggested this analysis and proposed that the volume term in the ideal gas equation $PV = nRT$ must be corrected by this term. If we denote the molar excluded volume by b, the equation takes the form

$$P(V - nb) = nRT \qquad (2.14)$$

Note that this equation has the same form as Equation (2.1) for one mole of gas with $b' = b/RT$.

The effect of intermolecular attraction in a real gas can be dealt with in the following way. A gas molecule in the interior of the vessel will be attracted to other adjacent gas molecules by very weak forces called *van der Waals forces*. Since the molecule will be completely surrounded by neighbor molecules, there will be no net force arising from the van der Waals attraction. At the instant a molecule strikes the wall it can only be attracted to molecules in the interior of the vessel. These interior molecules will exert a net force that tends to reduce the impact the molecules make in colliding with the wall. The average reduction in impact for one molecule will depend on the average number of molecules in its vicinity, which in turn is proportional to n/V. In addition, the total pressure reduction will be proportional to the number of collisions per unit area of wall, and this too will be proportional to n/V. Thus the factor n/V enters twice, and we can write the reduction in total pressure caused by the molecular attractions as $P_{red} = a(n^2/V^2)$, where a is the proportionality constant. The corrected pressure can now be written as

$$P = \frac{nRT}{(V - nb)} - a\frac{n^2}{V^2} \qquad (2.15)$$

where we have used Equation (2.14) for P. This equation can be rearranged as

$$\left(P + a\frac{n^2}{V^2}\right)(V - nb) = nRT \qquad (2.16)$$

Equation (2.16) was first obtained by van der Waals in 1873, and it is known as the *van der Waals equation*. Table 2.2 lists the van der Waals constants a and b for a number of gases.

TABLE 2.2 Van der Waals constants.

Gas	a (lit^2 atm mol^{-1})	b (lit mol^{-1})
NH_3	4.17	0.0371
Ar	1.35	0.0322
CO_2	3.59	0.0427
CO	1.49	0.0399
Cl_2	6.49	0.0562
C_2H_6	5.49	0.0638
He	0.0341	0.0237
H_2	0.244	0.0266
Kr	2.32	0.0398
CH_4	2.25	0.0428
Ne	0.211	0.0171
N_2	1.39	0.0391
O_2	1.36	0.0318
C_3H_8	8.66	0.0845
SO_2	6.71	0.0564
Xe	4.19	0.0511
H_2O	5.46	0.0305

The procedure we have just outlined for justifying the van der Waals equation suffers from several deficiencies. In the excluded volume argument, we assumed that the molecules are "hard" spheres, when in reality they can be distorted, particularly at high pressures and also high temperatures; under such conditions, the molecules undergo severe collisions. Also, the intermolecular attraction is related to the intermolecular distance, and this too will be affected by the density of the gas. Despite these deficiencies, the equation does agree reasonably with experiment over a wide range of pressures (see Table 2.3). Further details on additional correction terms to equations of state can be found in the references listed in the Bibliography at the end of the book.

TABLE 2.3 Compressibility factors for several gases calculated by van der Waals equation compared with experimental values.

Pressure (atm)	CO_2		Ar		CO	
	calc	exp[a]	calc	exp	calc	exp
1	0.9980	0.9982	0.9997	1.0000	0.9998	1.0002
4	.9918	.9927	.9989	0.9999	.9994	1.0010
7	.9857	.9871	.9981	0.9999	.9989	1.0017
10	.9794	.9815	.9974	0.9998	.9985	1.0025
40	.9145	.9252	.9906	1.0002	.9957	1.0113
70	.8446	.8697	.9859	1.0022	.9962	1.0225
100	0.7707	0.8155	0.9829	1.0057	1.0001	1.0359

Pressure (atm)	H_2		N_2		O_2	
	calc	exp	calc	exp	calc	exp
1	1.0006	1.0005	0.9999	1.0003	0.9997	1.0000
4	1.0023	1.0020	0.9996	1.0011	.9988	1.0001
7	1.0041	1.0034	0.9994	1.0020	.9980	1.0001
10	1.0059	1.0048	0.9992	1.0029	.9972	1.0002
40	1.0237	1.0193	0.9984	1.0129	.9897	1.0016
70	1.0420	1.0339	1.0006	1.0248	.9842	1.0042
100	1.0602	1.0486	1.0061	1.0383	0.9803	1.0079

NOTE: Temperature is 400 K.

[a] Experimental values from *Tables of Thermal Properties of Gases*, National Bureau of Standards Circular 564 (1955).

2.13 CRITICAL CONSTANTS AND THE VAN DER WAALS EQUATION

The values of the constants in the van der Waals equation can be determined by experimentally measuring the PV product for any gas and finding the values of a and b that best fit the experimental data. The constants are also related to what are called the *critical constants*. We can write the van der Waals equation in the form

$$\overline{V}^3 - \left(b + \frac{RT}{P}\right)\overline{V}^2 + \frac{a}{P}\overline{V} - \frac{ab}{P} = 0 \qquad (2.17)$$

which is a cubic equation in the volume.

Now we examine Figure 2.16, which shows a series of isotherms for CO_2 near the *critical region*. We start at the point labeled (1) on the 29.929 °C isotherm in the gas-phase region. As the pressure increases, the volume decreases until we reach the point (2), where the gas starts to liquefy; the curve is then horizontal until point (3). In the horizontal region, the gas liquefies at the constant vapor pressure, and only

FIGURE 2.16 Isotherms of CO_2 near the critical point. The volume is relative to $V = 1.000$ at 0 °C and 1.000 atm. The dashed van der Waals curves are "idealized" curves. Actual van der Waals curves are shown in Figure 2.17. (Michels, Blaisse, and Michels, *Proc. Roy. Soc. A* **160** (1937):367.

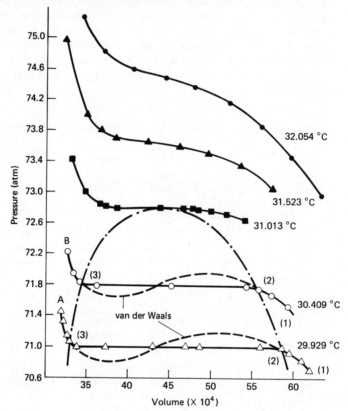

the volume changes. At point (3), the last of the gas liquefies and the isotherm curve rises sharply, since liquids are relatively incompressible. The situation for the 30.409 °C isotherm is similar; the only difference is that points (2) and (3) have moved closer to each other. As the temperature of the isotherm is increased, these two points come closer and closer, until we reach the isotherm at 31.013 °C. On this isotherm, the two points have merged into one point. There is no longer any discontinuity in the curve. This point, which is now a point of inflection, is called the *critical point*. The temperature is called the critical temperature T_c, the pressure the critical pressure P_c, and the volume the critical volume V_c.

If one mole of a gas is placed in a sealed glass tube whose volume equals V_c, and if the temperature is below T_c, then a sharp boundary is observed separating the liquid from the vapor. As the tube is warmed, this boundary becomes less distinct as the densities and hence the refractive indices of the vapor and liquid become more nearly equal. At the critical point, the density of the gas becomes equal to the density of the liquid, and the boundary between the two disappears completely. Above T_c there is no clear distinction between liquid and vapor.

A cubic equation such as (2.17) must have at least one real root and may have three. The dashed curves in Figure 2.16 indicate plots of the van der Waals equation

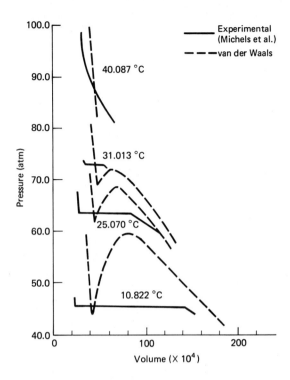

FIGURE 2.17 Isotherms of CO_2 showing real van der Waals curves. (Michels, Blaisse, and Michels, *Proc. Roy. Soc. A* 160 (1937):367.)

below the critical curve; it is a cubic, and has one maximum and one minimum. (The dashed curves in Figure 2.16 are approximations to the van der Waals curves. Actual van der Waals curves are shown in Figure 2.17.) At the critical temperature, the maximum and the minimum coalesce into an inflection point. The condition for an extremum, either minimum or maximum, is that the first derivative should vanish, $(\partial P/\partial V)_T = 0$. If the second derivative is less than zero, then the extremum is a maximum; if it is greater than zero then the extremum is a minimum. The condition for a point of inflection is that the second derivative should vanish, or $(\partial^2 P/\partial V^2)_T = 0$. We can thus write the following three equations that are valid at the critical point:

$$P_c = \frac{RT_c}{V_c - b} - \frac{a}{V_c^2} \tag{2.18}$$

$$\left(\frac{\partial P}{\partial V}\right)_T = 0 = \frac{-RT_c}{(V_c - b)^2} + \frac{2a}{V_c^3} \tag{2.19}$$

$$\left(\frac{\partial^2 P}{\partial V^2}\right)_T = 0 = \frac{2RT_c}{(V_c - b)^3} - \frac{6a}{V_c^4} \tag{2.20}$$

These can be solved as a set of three simultaneous equations, to give

$$T_c = \frac{8a}{27bR} \qquad V_c = 3b \qquad P_c = \frac{a}{27b^2} \tag{2.21}$$

or written in another way

$$b = \frac{V_c}{3} \qquad a = 3P_c V_c^2 \qquad R = \frac{8P_c V_c}{3T_c} \tag{2.22}$$

Using these values for a, b, and R, we can write the van der Waals equation in the form

$$\left[\frac{P}{P_c} + 3\left(\frac{V}{V_c}\right)^{-2}\right]\left[\frac{V}{V_c} - \frac{1}{3}\right] = \frac{8}{3}\frac{T}{T_c} \tag{2.23}$$

The ratios of the actual values of the temperature, pressure, and volume to their critical values are known as *reduced variables,* and are written as

$$T_R = \frac{T}{T_c} \qquad P_R = \frac{P}{P_c} \qquad V_R = \frac{V}{V_c} \tag{2.24}$$

Rewriting Equation (2.23) in terms of the reduced variables gives the simpler form

$$\left(P_R + \frac{3}{V_R^2}\right)\left(V_R - \frac{1}{3}\right) = \frac{8}{3}T_R \tag{2.25}$$

Notice that Equation (2.25) is an equation of state that does not contain any empirical constants explicitly; hence all gases will approximately follow the same equation of state when expressed in terms of the reduced variables. That is to say, if any two gases have the same values for any two of the reduced variables, they will have the same values for the third. They are then said to be in *corresponding states*. This rule is known as the *law of corresponding states*. Note that although we have discussed the law in terms of the van der Waals equation, the law of corresponding states has a

FIGURE 2.18 Compressibility factors as a function of reduced state variables. (Reprinted with permission from Gouq-Jen Su, *Industrial and Engineering Chemistry* 38 (1946):803. Copyright by the American Chemical Society.)

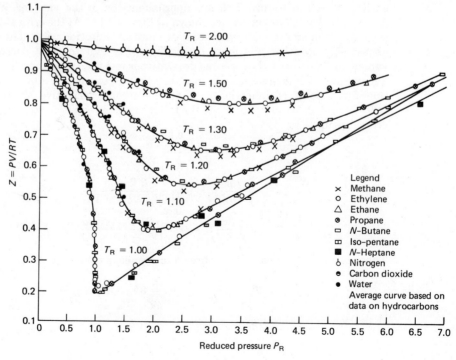

SEC. 2.14 OTHER EQUATIONS OF STATE

TABLE 2.4 Critical constants.

Gas	T_c (K)	P_c (atm)	V_c (cm^3 mol^{-1})
NH_3	405.6	111.8	72.5
Ar	151	48.1	75.2
CO_2	304.3	73.1	95.7
CO	134.2	35.0	90.1
Cl_2	417.2	76.2	123.8
C_2H_6	305.3	47.2	143
He	5.3	2.26	57.8
H_2	33.3	12.8	65.0
Kr	210	54	107
CH_4	190.7	45.9	98.8
Ne	44.5	25.9	41.7
N_2	126.1	33.6	90.1
O_2	154.4	49.7	74.4
C_3H_8	368.8	41.6	....
SO_2	430.4	77.5	123
Xe	289.8	58.3	113.6
H_2O	647.2	217.7	45

general basis that transcends that equation (see Problem 32). The law is illustrated in Figure 2.18, where the compressibility factor Z is plotted against the reduced pressure P_R for several gases at a series of reduced temperatures T_R. Critical constants for a number of gases are listed in Table 2.4.

2.14 OTHER EQUATIONS OF STATE

Some other equations have been proposed to provide a better fit to the experimental data. Among these is the Dieterici equation,

$$P = \frac{nRT \exp(-na/VRT)}{V - nb} \tag{2.26}$$

and the Berthelot equation

$$\left(P + \frac{n^2a}{TV^2}\right)(V - nb) = nRT \tag{2.27}$$

The most general equation of state is the *virial equation,* one form of which is

$$\frac{PV}{nRT} = 1 + B\frac{n}{V} + C\left(\frac{n}{V}\right)^2 + D\left(\frac{n}{V}\right)^3 + \cdots \tag{2.28}$$

The coefficients $B, C, D, \ldots$ are functions of the temperature, with B known as the second virial coefficient, C the third virial coefficient, and so on. One of the results of statistical mechanics is that the second virial coefficient can be related to the intermolecular potential $V(r)$ by the equation

$$B(T) = 2\pi N_0 \int_0^\infty \left[1 - \exp\left(\frac{-V(r)}{kT}\right)\right] r^2 \, dr \tag{2.29}$$

where N_0 is Avogadro's number, k is the Boltzmann constant, and r is the distance between molecules. Equation (2.29) is useful for getting information about intermolecular forces. The derivation of Equation (2.29) and further applications of its use can be found in the references listed in the Bibliography.

PROBLEMS

1. An ideal gas occupies 1 liter at a pressure of 7 atm. What volume will the gas occupy at a pressure of 0.8 atm at the same temperature?
2. A pressure is measured with a McLeod gauge. The gauge has a bulb volume of 150 cm^3, and the diameter of the capillary tube is 1 mm. When the mercury comes to equilibrium, it reaches a level 6.7 cm below the sealed end of the bulb capillary and rises to 13.6 cm above the sealed end in the side capillary. Calculate the pressure.
3. Calculate the standard molar volume. That is, calculate the volume of 1 mol gas at a pressure of 1 atm and a temperature of 0 °C.
4. If a person breathes at the rate of 14 respirations a minute, and each respiration involves 0.5 liter of air, calculate the total weight of air breathed in a day. (Take the average molecular weight of air to be 29 g.)
5. An amount 2.1 g Dry Ice (which is solid CO_2) is placed in a sealed 1-liter vessel. Using the ideal gas equation, calculate the final pressure for room temperature.
6. Water boils at 100 °C. The density of liquid water is 1 g cm^{-3}. Calculate the percentage increase in volume as water passes from a liquid to a gas at 100 °C.
7. A 1.000-liter flask weighs 42.3486 g when evacuated. When it is filled with a gas at 24 °C at a pressure of 0.83 atm, the flask weighs 43.9154 g. Calculate the molecular weight of the gas. (Ans. = 46 g.)
8. Dinitrogen tetraoxide (N_2O_4) dissociates according to the equation $N_2O_4 = 2NO_2$. A 1-liter bulb is filled with 2.05 g N_2O_4 gas at 298 K. At equilibrium, the final pressure is 1.00 atm. Calculate the degree of dissociation.
9. Gaseous PCl_5 decomposes at high temperatures according to

    $$PCl_5(g) = PCl_3(g) + Cl_2(g)$$

    At equilibrium at 1 atm and 250 °C, the density of a sample of PCl_5 was 4.0 g per liter. Calculate the degree of dissociation.
10. A tank of compressed O_2 gas is 6 ft high and has a diameter of 8 in. The tank is filled with O_2 to a pressure of 3000 psi. How many moles of O_2 does the tank hold, and what volume would the gas occupy at 1 atm (15 psi) and 300 K? (Assume that the ideal gas law holds, although it really does not at these high pressures.)
11. A new kilomegawatt power-generating plant is designed for a large city. The plant will consume about 6000 tons of coal a day. Assume that the coal is pure carbon. How many tanks a day of O_2 are needed to burn the coal (get data from Problem 10)?
12. Assume that the coal of Problem 11 contains 2% sulfur, which produces SO_2 on burning. Estimate the area of the city you live in. If all the SO_2 produced in a day were evenly distributed over the surface of your city to a height of 200 meters (m), calculate the concentration of SO_2 in the air you breathe in parts per million (ppm). Calculate the partial pressure of SO_2. From the data of Problem 4, how much SO_2 would you breathe in a day? The Environmental Protection Agency (EPA) has established a primary standard concentration limit of 80 μg m^{-3} for SO_2. Would the air meet this standard? (*Note:* Since the SO_2 is blown by the wind, the actual concentration would be less than you calculated.)
13. Under normal quiet breathing, a man will breathe at the rate of 14 respirations a minute, and take in 0.5 liter air with each breath. The air he inhales is 20.9% O_2 and 0.04% CO_2 by volume. The air he exhales is 16.3% O_2 and 4.38% CO_2. The remainder of the gases are water vapor, N_2, and other gases.
    (a) What is the weight of O_2 a man consumes in a day for his physiological processes?
    (b) What is the weight of CO_2 he adds to the atmosphere in a day?
    (c) If one exhaled breath is collected, how many milliliters of 0.01 molar (M) Ca(OH)$_2$ is required to titrate the CO_2?
    (d) How many people does it take to equal the CO_2 production of the power plant of Problem 11?
14. Human blood contains O_2 and CO_2 besides many other constituents. The O_2 is combined with the hemoglobin, and the CO_2 occurs in the form of HCO_3^- ions. It is found that every 100 cm^3 arterial blood contains the equivalent of 20 cm^3 O_2 and 38 cm^3 CO_2, whereas venous blood contains the equivalent of 12 cm^3 O_2 and 45 cm^3 CO_2. Assuming that a human has a total of 10 liters of blood divided equally between venous and arterial kinds, calculate the number of moles of O_2 and CO_2 present in the human bloodstream.
15. Consider natural gas to be pure methane, CH_4. A homeowner uses 10^5 ft^3 of natural gas in a year to heat his home. How much does the gas weigh?
16. A goat weighing 20 kg produces about 375 cm^3 gas per hour in his intestines. Assume that the gas is methane. How long does it take the goat to produce an amount of intestinal gas equal to his own weight?
17. A spherical weather balloon has a diameter of 10 m when it is filled with helium gas at a pressure of 1.05 atm. What is the total weight the balloon can lift (including the weight of the balloon material)?
18. At 500 °C and a pressure of 699 torr (or mmHg), sulfur vapor has the density 3.71 g liter^{-1}. What is the molecular formula of sulfur?
19. Suppose scientists had decided to call the temperature at which water freezes 100 on an absolute scale. What would be the boiling point of water on this scale?

PROBLEMS

20. The values of atomic masses are based on the definition of a mole as that amount of ideal gas contained in 22.414 liters at 1.00 atm pressure and 273.15 K. Calculate the molecular weight of O_2 based on defining the mole as 30.000 liters of gas at 1 atm and 273.15 K.

21. Redo Problem 10 for a real gas using the van der Waals equation. (A tank of compressed O_2 gas is 6 ft high and has a diameter of 8 in. The tank is filled with O_2 to a pressure of 3000 psi. How many moles of O_2 does the tank hold, and what volume would the gas occupy at 1 atm (15 psi)?) Compare with the results for the ideal gas.

22. Find the minimum in the P-V curve for 1 mol CO_2 at 0 °C.

23. The density of water vapor at 100 °C and 1 atm pressure is 0.0005970 g cm^{-3}. Calculate the molar volume. How does this compare with the ideal gas molar volume? Calculate the compressibility factor.

24. At lower pressures, the ideal gas value for the molar volume, $V = RT/P$ can be used in the correction term to the pressure without introducing any great error. In addition, when the terms are multiplied, cross terms containing the product ab can be neglected. Within this approximation, show that the van der Waals equation can be written in the form
$$PV = RT\left\{1 + \left[\frac{b}{RT} - \frac{a}{(RT)^2}\right]P\right\}$$

25. Using the expression of Problem 24, derive an expression for the compressibility within that approximation. The slope of the compressibility curve is given by $(\partial Z/\partial P)$ and indicates how fast the gas deviates from ideality at low pressures. Calculate the slope and evaluate it for O_2, N_2, CO_2, NH_3, and He at room temperature.

26. The temperature at which the curve of Z vs P starts tangent to the ideal gas line is called the *Boyle* temperature. Derive an expression for the Boyle temperature using the van der Waals equation, and calculate this temperature for the gases of Problem 25.

27. Using the van der Waals constants of Table 2.2, calculate the critical constants for CO_2, Ar, He, and O_2 and compare with the values in Table 2.4.

28. Show that the van der Waals equation can be rearranged to get
$$\frac{PV}{nRT} = \frac{1 - (a/RT)(n/V) + (ab/RT)(n/V)^2}{1 - b(n/V)}$$
Now expand the denominator of this expression using the binomial theorem
$$(1-x)^{-1} = 1 - x + x^2 - \cdots$$
to show that the second and third virial coefficients can be written in terms of the van der Waals constants as
$$B = b - \frac{a}{RT} \qquad C = b^2$$

29. The potential for hard sphere molecules of diameter σ can be written as
$$V(r) = \begin{cases} 0 & \text{for } r > \sigma \\ \infty & \text{for } r < \sigma \end{cases}$$
Use Equation (2.29) to show that for this type of potential the second virial coefficient is given by $B(T) = \frac{2}{3}\pi N_0 \sigma^3$. Show that this result is the same as the result obtained from the van der Waals equation (see Problem 28) if the intermolecular attraction term a/V^2 is neglected.

30. Show that according to the van der Waals equation the compressibility factor at the critical point is given by
$$\frac{P_c V_c}{RT_c} = \frac{3}{8} = 0.375$$

31. Show that Berthelot's equation of state leads to the critical constants
$$V_c = 3b \qquad T_c = \left(\frac{8a}{27bR}\right)^{1/2} \qquad P_c = \left(\frac{aR}{216b^3}\right)^{1/2}$$
and that $a = 3P_c V_c^2 T_c$ and $b = V_c/3$. Thus at the critical point
$$\frac{P_c V_c}{RT_c} = \frac{3}{8} = 0.375$$
which is the same result obtained with the van der Waals equation.

32. Show that the law of corresponding states is also a consequence of the Berthelot equation.

CHAPTER THREE

THE IDEAL GAS IN THE KINETIC-MOLECULAR MODEL

3.1 A SIMPLIFIED TWO-DIMENSIONAL GAS MODEL

The laws describing the behavior of ideal gases were based on experimental observations. We shall derive these laws from a simple model that treats a gas as a collection of moving molecules. Daniel Bernoulli first made the connection between an atomistic theory and observation in the early 1700s, although the significance of what he had done was not fully recognized until more than a hundred years later. Before we go into the details of Bernoulli's treatment, we shall examine a two-dimensional representation of molecular motion.

We assume that the gas molecules act as billiard balls moving randomly on the surface of a billiard table; the cushions of the table correspond to the walls of a gas container. We further assume that the balls and the table are perfect, that all frictional resistance is absent, and that collisions between balls and between balls and cushions are perfectly *elastic;* that is to say, that no kinetic energy is lost during a collision. We place our collection of balls on the table, as shown in Figure 3.1, and impart a fixed quantity of kinetic energy to the collection. As the balls collide with each other, individual balls may be speeded up or slowed down. There is an average speed of the balls; it is determined by the total energy introduced into the collection and the size of the collection. Some balls will move with greater speeds than this average and some with lesser.

Pressure is a three-dimensional quantity that is defined as force per unit area. Our two-dimensional analog of pressure will be force per unit length. As the balls collide with the walls of the table, they exert a force on the walls (Figure 3.2). The average force on a given wall divided by the length of the wall will give the force per unit length, the two-dimensional analog of pressure. If our billiard table is such that one of the walls is movable, we have an analog of a piston in a cylinder. In three dimensions we adjust the volume; in two dimensions, the area. In both instances we keep the total number of molecules (balls) fixed.

Suppose we move the wall and reduce the area to one-half the original area. Note that we have to move the wall so *slowly* that no momentum is transferred to the balls. (Alternatively, we can move the wall quickly if we make sure that no ball approaches while it moves.) The average speed of the balls remains the same. Since the number of balls per unit area is increased by a factor of 2, there will be twice as many collisions with the wall in unit time. The force, and hence the "pressure," will also increase by a factor of 2. On the other hand, if we double the area, the pressure is re-

FIGURE 3.1 The two-dimensional kinetic-molecular model. Billiard balls on a billiard table. The arrows on the balls indicate the random directions of motion. The wall on the right side is movable, so that the total available area is adjustable. The situation at two different times is shown; they are separated by a small interval Δt. (a) Billiard table at time $t = 0$. (b) Same table at time $t = \Delta t$. Some of the balls in (a) have undergone collisions.

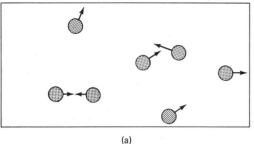
(a)

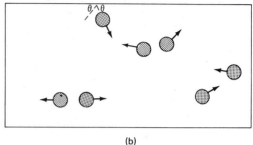
(b)

duced to one-half its original value. The pressure is inversely proportional to the area of the table for a fixed number of balls with a fixed energy. Recall that Boyle's law states that the pressure of a gas is inversely proportional to the volume at constant temperature.

Suppose that we keep the area constant and adjust the speed of the balls. If we double the average speed, the balls will strike the wall twice as often per unit time. In addition, each impact will have twice the original force, and hence the so-called pressure will increase by a factor of 4. If we triple the speed, the number of times of impact will be tripled and so will the force of each impact, increasing the pressure by a factor of 9. The pressure is hence proportional to the square of the speed. Since the kinetic energy is $\frac{1}{2}mv^2$, the pressure is proportional to the kinetic energy. Recall that for an ideal gas, the pressure is proportional to the temperature.

Our simple model suggests that the pressure of a gas is caused by collisions of the molecules with the walls of the vessel, and that the kinetic energy of the molecules is associated with the temperature. Both of these suggestions are correct. There are other things we can do with this simple billiard table analog. Suppose we place a barrier in the center of the table. If we put red balls on one side and blue balls on the other, we can then remove the barrier and study diffusion in two dimensions as the red and blue balls are intermingled.

We can apply the analog to a simplified view of evaporation by sloping the walls of the table, as shown in Figure 3.3. There is a particular average velocity, but not all

FIGURE 3.2 A collision of a ball with the wall. Note that only the component of motion perpendicular to the wall is changed; it is, in fact, reversed. The other component is unchanged.

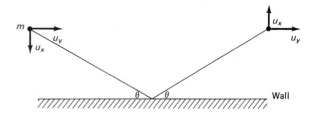

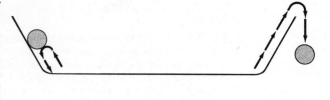

FIGURE 3.3 An analog of evaporation; a billard table with sloped walls. The ball on the right has enough energy to overcome the potential energy barrier of the wall; it rolls over the top and leaves the table. The ball on the left is traveling more slowly and does not have enough energy to overcome the barrier; it rolls part way up the sloped wall, and then back down onto the table.

balls have that velocity. Most have velocities near the average, but some will have greater velocities. Most of the balls will roll part way up the sloped wall and roll right back down. A few, however, will have enough energy to get over the top and leave the table. This is a particularly significant consequence, for every time a ball with greater-than-average energy leaves the table, the average energy of the remaining balls is lowered. The sloped wall constitutes a potential energy barrier that can be considered the analog of the energy barrier at the surface of a liquid. Only a few molecules have enough energy to overcome this barrier (at least at temperatures much below the boiling point). Further, the lowering of the average velocity of the remaining molecules is indicated by a cooling of the liquid that remains, since liquids do cool as they evaporate. Remember, however, that a gas is not a collection of billiard balls on a billiard table. An analog is not reality; it simply is an aid to understanding the real world. Let us now expand our model to three dimensions and derive Boyle's law from a purely mechanical standpoint, in the manner of Daniel Bernoulli in 1738.

3.2 BOYLE'S LAW DERIVED FROM THE KINETIC-MOLECULAR THEORY

We first consider a cubic box of side a that contains one gas molecule of mass m, as shown in Figure 3.4. The x component of the velocity is u_x. The molecule approaches the wall, which is perpendicular to the x axis, with a momentum component $p_x = mu_x$. After the molecule collides with the wall, its motion is reversed, and it leaves the wall with an x component of velocity, given by $-u_x$; its momentum is changed to $p_x = -mu_x$. Each collision with the wall introduces a momentum change of $\Delta p_x = -2mu_x$. The number of collisions the molecule makes with the wall is $u_x/2a$; that the particle must traverse the box twice to get back to the original wall is responsible for the factor 2. According to Newton's laws of motion, force equals the time rate of change of momentum. The total force this one molecule exerts on the wall is the momentum change for each collision multiplied by the number of collisions per unit time, or

$$f_x = (2mu_x)\left(\frac{u_x}{2a}\right) = \frac{+mu_x^2}{a} \tag{3.1}$$

FIGURE 3.4 Molecule in a cubic box. Here we see the coordinate system and the velocity components of a molecule.

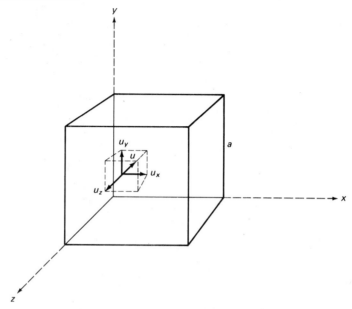

The value mu_x has a positive sign, since the force on the wall is opposite in direction to the force on the molecule. If we have N identical molecules in the box, then the total force exerted on the wall is

$$F_x = f_{x1} + f_{x2} + \cdots + f_{xN}$$
$$= \sum_{i=1}^{N} f_{xi} = \sum \frac{mu_{xi}^2}{a} = \frac{m}{a} \sum u_{xi}^2 \quad (3.2)$$

We now define a quantity called the *mean square* velocity component

$$\overline{u_x^2} = \frac{\Sigma u_x^2}{N} \quad (3.3)$$

It is the average value of the *square* of the velocity u_x^2. Note very carefully that it is not the square of the average velocity. The average velocity is $\overline{u_x} = \Sigma x_i/N$, and $\overline{u_x} = 0$, since at any instant half the molecules travel in the $+x$ direction and the other half travel in the $-x$ direction. The square of the average velocity (sometimes called the squared mean velocity) is denoted by $(\overline{u_x})^2$; in this case $(\overline{u_x})^2 = 0 \neq \overline{u_x^2}$. We can write F_x in terms of the mean square velocity:

$$F_x = \left(\frac{Nm}{a}\right) \overline{u_x^2} \quad (3.4)$$

Since pressure is defined as force per unit area, we have $P = F_x/a^2$, or

$$P = \left(\frac{Nm}{V}\right) \overline{u_x^2} \quad (3.5)$$

where V is the volume of the box. Now for each molecule, it is true that

$$u_i^2 = u_{xi}^2 + u_{yi}^2 + u_{zi}^2 \tag{3.6}$$

There are N equations of the form of (3.6). If we take the sum of all the equations and divide by N, we have

$$\overline{u^2} = \frac{\Sigma \overline{u_i^2}}{N} = \frac{\Sigma \overline{u_{xi}^2}}{N} + \frac{\Sigma \overline{u_{yi}^2}}{N} + \frac{\Sigma \overline{u_{zi}^2}}{N}$$

$$= \overline{u_x^2} + \overline{u_y^2} + \overline{u_z^2} \tag{3.7}$$

We next recognize that x, y, and z directions are equivalent, since molecular motion is random, and

$$\overline{u_x^2} = \overline{u_y^2} = \overline{u_z^2} \tag{3.8}$$

in which case we have

$$\overline{u^2} = 3\overline{u_x^2} \tag{3.9}$$

and Equation (3.5) becomes

$$P = \frac{Nm\overline{u^2}}{3V} \tag{3.10}$$

or

$$PV = \tfrac{1}{3}Nm\overline{u^2} \tag{3.11}$$

This is Boyle's law; the product PV is constant. Dalton's law of partial pressures also follows from this analysis, as you will show in Problem 16.

Before going ahead, we especially note two assumptions we have made that distinguish the ideal gas used in this molecular model from a real gas:

1. There is no long-range interaction energy between molecules, since no intermolecular forces exist. In a sense, each molecule is completely "unaware" of the existence of any other molecule except at the instant of collision.
2. Gas molecules occupy no volume.

These two assumptions are simplifications, since real molecules interact and also occupy volume. In the limit of zero pressure, the errors introduced by these assumptions become vanishingly small, and all gases behave as ideal gases.

3.3 VELOCITIES AND TRANSLATIONAL KINETIC ENERGIES OF GAS MOLECULES

If we compare Equation (2.11) with the ideal gas law $PV = nRT$, we are led to the important result that

$$\tfrac{1}{3}Nm\overline{u^2} = nRT \tag{3.12}$$

which relates the temperature, a macroscopically observable quantity, to the molecular velocity, a microscopic quantity. For one mole of gas, $n = 1$, $N = N_0 = 6.02 \times 10^{23}$, and $Nm = N_0 m = M$, where M is the molar mass. Thus

$$\tfrac{1}{3}M\overline{u^2} = RT$$

or

$$\overline{u^2} = \frac{3RT}{M} \tag{3.13}$$

SEC. 3.3 VELOCITIES AND TRANSLATIONAL KINETIC ENERGIES OF GAS MOLECULES

The square root of the mean square velocity is called the *root-mean-square velocity* (or in shortened form, rms velocity)[1]

$$u_{rms} = (\overline{u^2})^{1/2} = \sqrt{\frac{3RT}{M}} \tag{3.14}$$

In calculating molecular velocities, it is important to keep the units straight. If SI base units are used for the molar mass and the velocity (kilograms and meters per second), then the gas constant must be used in joules. If grams and centimeters per second are used, then the gas constant must be used in ergs. For an example, let us calculate the rms velocity of a nitrogen molecule at 25 °C. The molar mass is 0.028 kg, the temperature is 298 K, and the gas constant is 8.314 J K^{-1} mol^{-1}, and we get

$$u_{rms} = \sqrt{\frac{(3)(8.314)(298)}{0.028}} = 515 \text{ m s}^{-1} \quad (=1152 \text{ mph})$$

Root-mean-square velocities of several gases are shown in Table 3.1.

Gas molecules travel at high velocity. Indeed, this conclusion of the kinetic theory of gases contributed to the lack of its acceptance at first proposal. By the theory, if a bottle containing a noxious substance like H_2S were opened at one end of a room, the odor should be immediately obvious to a person at the other end of the room, since molecules travel at such high velocity. Yet it was known that in such circumstances, it took some time for the odor to arrive. The solution to the paradox has the explanation that at normal atmospheric pressure, the density of molecules is so great that they make billions of collisions with other molecules before getting to the other side. With each collision, the direction of travel of the molecule is changed and often even reversed. The speed of the molecules is very great, but the directions of the molecules keep changing and they make very slow progress in any given direction. (We shall discuss this point further in Chapter 22.)

On the other hand, Equation (3.14) did explain the phenomenon of *effusion*. In 1846, Graham observed that for two different gases enclosed in a vessel, if a vacuum pump were connected to the vessel through a very small orifice, the ratio of the rates

TABLE 3.1 Root-mean-square (rms) velocities of various gas molecules.

Gas	Molar mass (g)	100 K	298 K[a]	1000 K
H_2	2	1117	1928	3532
He	4	790	1363	2498
CH_4	16	395	682	1249
H_2O	18	372	643	1177
N_2	28	299	515	944
O_2	32	279	482	883
CO_2	44	238	411	753
Kr	84	172	298	545
Xe	131	138	238	436
Hg	201	111	192	352
UF_6	352	84	145	266

NOTE: Velocities are expressed in meters per second (m s^{-1}).

[a] 298 K = room temperature = 25 °C.

---

[1] When we consider the actual shape of the velocity distribution curve (Chapter 21), we shall see that the rms velocity is larger than the average velocity (or more correctly, the average "speed") by a factor of 1.09.

at which the gases left the vessel was inversely proportional to the ratio of the square roots of the molar masses, or

$$\frac{r_1}{r_2} = \left(\frac{M_2}{M_1}\right)^{1/2} \tag{3.15}$$

This is known as Graham's law of effusion. Since the rate at which a gas effuses through an orifice is expected to be proportional to the velocity of the gas molecules, Graham's law follows from Equation (2.14).

The total kinetic energy of a gas is given by

$$E_t = \tfrac{1}{2} N m \overline{u^2} \tag{3.16}$$

where we have used a subscript t to denote translational energy. If we compare this with Equation (3.12), we find that for an ideal gas

$$E_t = \tfrac{3}{2} n R T \tag{3.17}$$

or $E_t = \tfrac{3}{2} RT$ per mole of gas. It is sometimes convenient to work with molecular energies instead of molar energies. For a gas sample containing N molecules, the number of moles of gas is

$$n = \frac{N}{N_0}$$

where N_0 is Avogadro's number. Substituting this in Equation (3.17), we get

$$E_t = \frac{3}{2} \frac{N}{N_0} RT \tag{3.18}$$

It is convenient to define a new constant called the *Boltzmann constant*:

$$k = \frac{R}{N_0} \tag{3.19}$$

In terms of this constant, Equation (3.18) becomes

$$E_t = \tfrac{3}{2} N k T \tag{3.20}$$

The Boltzmann constant is the gas constant per molecule. The kinetic energy of one molecule can now be written as

$$\epsilon_t = \tfrac{3}{2} k T \tag{3.21}$$

The Boltzmann constant is equal to 1.38062×10^{-23} J K^{-1}. This constant appears frequently in equations that arise from molecular models.

3.4 MECHANICAL DEGREES OF FREEDOM AND THE EQUIPARTITION OF ENERGY

The mechanical concept of a *degree of freedom* refers to an independent mode of motion (a translation, a rotation, or a vibration) in one of three mutually independent directions in space. The total number of degrees of freedom of a mechanical system equals the number of variables required to specify the motion of the system. Thus for example, a mass point has three degrees of freedom. Its position at any instant is completely specified if values of the x, y, and z coordinates of the mass point are given at that instant. Its motion is completely specified by three equations of motion, one equation for each of the three cartesian coordinates. For a mechanical system with more than one mass point, the total number of degrees of freedom is $3A$, where A is the number of mass points. We need a set of three equations to define the motion of each of the mass points.

If we consider a molecule as a mechanical system of A atoms, then each of the atoms will contribute three degrees of freedom, for a total of $3A$ degrees. These $3A$ degrees of freedom may be distributed among the three types of motion—translation, rotation, and vibration.

1. For any molecule, there will *always* be three degrees of freedom associated with the translational motion of the center of mass of the molecule. One degree of freedom is associated with each of the three directions in space.
2. For a molecule with more than one atom, there will be rotational degrees of freedom associated with rotational motion about the three mutually perpendicular axes, as shown in Figure 3.5. If the molecule is linear (as all diatomic mole-

FIGURE 3.5 Translational, rotational, and vibrational modes of motion of some simple molecules.

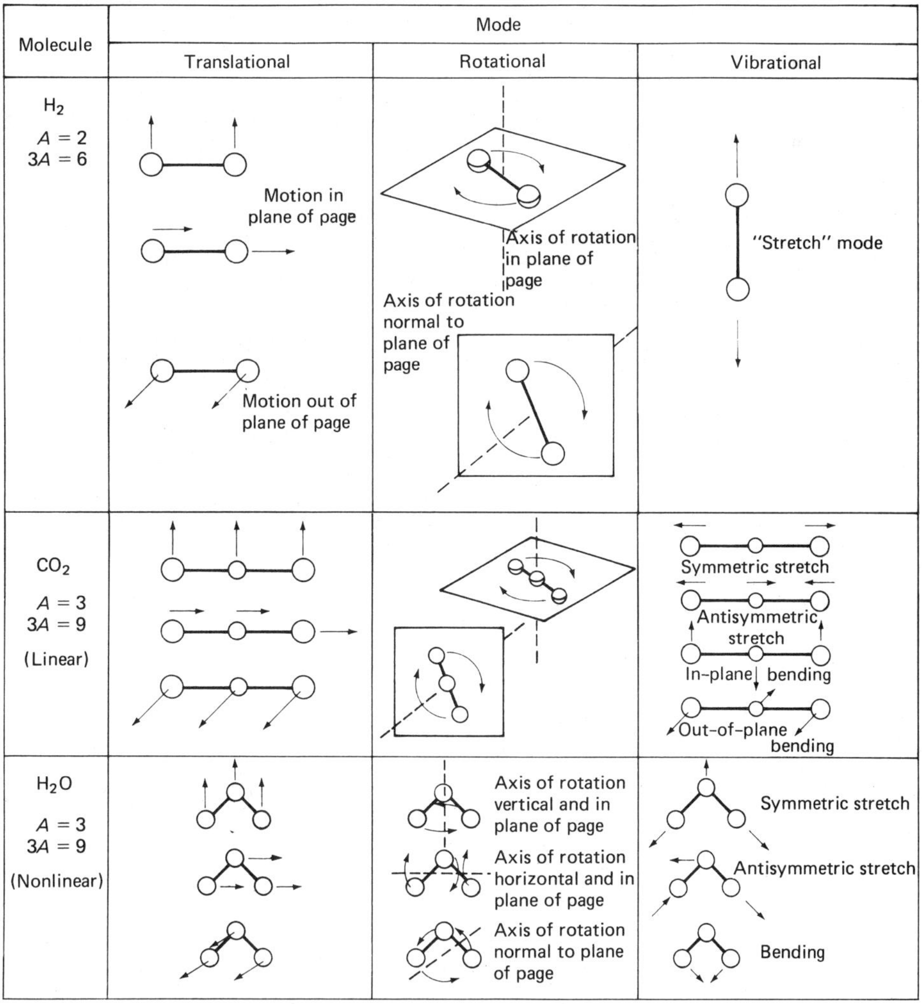

cules and polyatomics such as CO_2 and acetylene are), then there will be two rotational degrees of freedom. These are associated with rotations about each of the two mutually perpendicular axes normal to the molecular axis. Since the atoms are considered point masses, we do not consider rotation about the molecular axis itself.[2] For polyatomic molecules that are not linear, we consider rotational motion about each of the three mutually perpendicular directions in space; there are three rotational degrees of freedom associated with a nonlinear polyatomic molecule.

3. *All* the remaining degrees of freedom are assigned to vibrational modes of motion. For a linear molecule, the number of vibrational degrees of freedom is $3A - 5$, the 5 arising from two rotational plus three translational degrees of freedom. For a nonlinear molecule, the number of vibrational degrees of freedom is $3A - 6$, the 6 arising from the three rotational plus the three translational degrees of freedom. Representations of vibrational motion are shown in Figure 3.5.[3]

The distribution of vibrational degrees of freedom is summarized in Table 3.2.

Using Equations (3.7), (3.16), and (3.17), we can write

$$E_t = \tfrac{3}{2}RT = \tfrac{1}{2}M\overline{u^2} = \tfrac{1}{2}M(\overline{u_x^2} + \overline{u_y^2} + \overline{u_z^2}) \tag{3.22}$$

for one mole of gas. This suggests that for one mole of a monatomic gas, we can associate $\tfrac{1}{2}RT$ thermal energy with each degree of freedom. This is a special case of the *equipartition of energy principle*. This principle states that for a sample containing n moles of gas:

1. For each translational degree of freedom of a molecule, there is a contribution of $\tfrac{1}{2}nRT$ to the thermal energy of the gas.
2. For each rotational degree of freedom of a molecule, there is a contribution of $\tfrac{1}{2}nRT$ to the thermal energy of the gas.
3. For each vibrational degree of freedom of a molecule, there is a contribution of nRT to the thermal energy of the gas.

TABLE 3.2 Degrees of freedom, total thermal energy, and ideal heat capacity of various molecules.

Number atoms per molecule	Total number degrees of freedom	Translational degrees of freedom	Rotational degrees of freedom	Vibrational degrees of freedom[a]	Total thermal energy per mole[b]	Ideal heat capacity per mole
1	3	3	0	0	$\tfrac{3}{2}RT$	$\tfrac{3}{2}R$
2	6	3	2	1	$\tfrac{7}{2}RT$	$\tfrac{7}{2}R$
3 (linear)	9	3	2	4	$\tfrac{13}{2}RT$	$\tfrac{13}{2}R$
3 (nonlinear)	9	3	3	3	$\tfrac{12}{2}RT$	$\tfrac{12}{2}R$
A (linear)	$3A$	3	2	$3A - 5$	$(3A - \tfrac{5}{2})RT$	$(3A - \tfrac{5}{2})R$
A (nonlinear)	$3A$	3	3	$3A - 6$	$(3A - 3)RT$	$(3A - 3)R$

[a] The number of vibrational degrees is given by tot − (trans + rot).
[b] For total energy, we take $\tfrac{1}{2}RT$ for each translational and rotational degree of freedom, and $\tfrac{2}{2}RT$ for each vibrational degree of freedom.

[2] The reason is that linear molecules have zero moments of inertia about the line joining the atoms if we assume that the atoms are point masses. This approach also neglects the possibility of electronic angular momentum.

[3] Each of the vibrational modes is known as a *normal mode*. Normal modes are considered further in Part V in connection with the vibration of molecules and infrared spectroscopy.

SEC. 3.5 THE HEAT CAPACITY ANOMALY

Note that each vibrational degree of freedom contributes twice as much thermal energy as a translational or a rotational degree of freedom does. The total thermal energies of gas molecules with various configurations are given in Table 3.2.

3.5 THE HEAT CAPACITY ANOMALY

The heat capacity at constant volume is

$$C_v = \left(\frac{\partial E}{\partial T}\right)_v \quad (3.23)$$

It can be determined by measuring the quantity of heat required to increase the temperature of the gas by a unit amount at constant volume. Some experimental heat capacities are shown in Figure 3.6. Except for the monatomic gas He, the results are inexplicable in the light of the discussion of the equipartition of energy.

By the equipartition of energy principle, the molar heat capacities of gases with A atoms per molecule are as follows:

Monatomic gases ($A = 1$): $C_v = 3(\tfrac{1}{2}R) = \tfrac{3}{2}R$

Linear molecules: $C_v = 3(\tfrac{1}{2}R) + 2(\tfrac{1}{2}R) + (3A - 5)R = (3A - \tfrac{5}{2})R$

Nonlinear molecules: $C_v = 3(\tfrac{1}{2}R) + 3(\tfrac{1}{2}R) + (3A - 3)R = (3A - 3)R$

These results are listed in Table 3.2. Clearly the equipartition of energy predicts a heat capacity that is independent of temperature. Just as clearly, all diatomic and polyatomic molecules shown in Figure 3.6 have temperature-dependent heat capacities; and further, at lower temperatures all the heat capacities are much lower than predicted. Helium, on the other hand, has the predicted heat capacity. This was

FIGURE 3.6   Plots of C_v (in units of R, the gas constant) as a function of temperature for several gases.

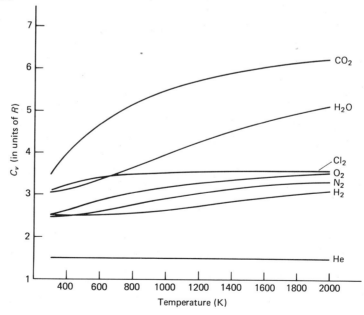

a particularly perplexing point around 1900, and it could not be satisfactorily explained until the development of quantum mechanics. There are good reasons for this anomalous behavior, to be explained subsequently (Chapter 29). We are deferring a substantial body of information to a later portion of the book. Meanwhile, we shall ignore the existence of atoms and molecules as we turn our attention to thermodynamics.

PROBLEMS

NOTE: In this problem set, when we say velocity, we mean rms velocity.

1. Calculate the velocity of a He molecule at 4.2 K (the boiling point of He), 298 K, and 1000 K. For each temperature, calculate the total translational energy of a single molecule and of a mole of gas. Do the same calculation for Xe.

2. Some N_2 gas is in a cubic container 1 m on an edge at 300 K at a pressure of 1 atm. Estimate the force from the single impact of a single molecule that strikes the wall perpendicularly.

3. A particle of water mist weighs 10^{-10} g. What is its velocity at 300 K based on the kinetic-molecular theory?

4. A number of great convenience is n^*, the number of molecules per cm^3. Calculate n^* for pressures of 1 atm, 1 torr (or $\frac{1}{760}$ atm), 1 millitorr, and 10^{-10} torr. At what pressure is n^* equal to one molecule per cm^3? (Take $T = 300$ K.)

5. A 1-liter box contains 2.4×10^{22} molecules of H_2 at a pressure of 1 atm. How fast are the molecules traveling? At what temperature would CH_4 molecules be traveling at the same speeds? What is the total translational energy of the molecules in the box?

6. Both H_2O and CO_2 have three atoms per molecule, yet the two do not have the same heat capacities. Which heat capacity is larger, and why?

7. A gas particle weighs 10^{-22} g; it is in a 1-cm^3 container and traveling with a velocity of 10^6 cm s^{-1}. What is its kinetic energy? What temperature does this correspond to? If there were 10^{22} such particles in the box, what would the pressure be?

8. A 1-liter cubical box contains He at a pressure of 10^{-10} atm at 298 K. Calculate the total number of collisions per second that the molecules make with the walls of the box (disregarding that the molecules may collide with each other). (*Hint:* Assume that the molecules travel only along the axes of the box; that is, one-third of the molecules travel along the x axis, one-third along the y axis, and one-third along the z axis. This assumption is not strictly according to fact, but it gives an answer sufficiently close to the true answer.)

9. The box of Problem 8 is placed in a total vacuum, and a hole with a diameter of 0.01 mm is pierced in the wall of the box. How long will it take the pressure to fall to one-half its original value? (Use the same assumptions as in Problem 8. This is a process called molecular effusion. This problem is discussed in Chapter 22, which deals with further aspects of the kinetic theory of gases, and you can find the solution there. Many students should be able to solve it at this point without referring ahead.)

10. The sloped wall of the billiard table of Figure 3.3 is 1 m long and is sloped at an angle of 30°. How fast must a point mass travel to be able to leave the billiard table?

11. For an object to escape from the earth's gravitational field, it must have the escape velocity $v_e = \sqrt{2gR}$, where g is the gravitational acceleration constant, 980 cm s^{-2}, and R is the radius of the earth, 6.4×10^8 cm. At what temperature does a molecule of H_2 attain the escape velocity? At what temperature does a molecule of N_2 attain the escape velocity? Work out the same problem for the moon, for which $g = 170$ cm s^{-2} and $R = 1.7 \times 10^8$ cm.

(*Note:* When we consider the distribution of velocities, we shall be able to compute the fraction of molecules that have velocities exceeding these velocities. You should now have some hint of why the earth's atmosphere contains N_2 and not H_2 and why the moon has no atmosphere. It is worth mentioning here that the fraction of molecules traveling with velocity greater than 5 times the rms velocity is about 10^{-10}; the fraction at $10 u_{rms}$ is 10^{-43}, and at $20 u_{rms}$ is about 10^{-173}. We shall actually compute these numbers in Chapter 21.)

12. A spherical rocket ship has a diameter of 30 ft and is traveling through space at a velocity of 25,000 miles per hour. For a pressure in outer space of 10^{-14} torr and a temperature 100 K, how many molecules does the rocket ship collide with in a second?

13. A mole of He is in a flask at 0 °C. How much heat must be added to the flask to raise the temperature of the gas to 100 °C?

14. Derive an expression that will convey the fractional increase in thermal energy of a monatomic gas if the temperature is raised by one degree.

15. Derive an expression that will convey the increase in velocity of a monatomic gas if the temperature is increased by a small amount ΔT.

16. Take a gas mixture containing N_1 molecules of mass m_1, and N_2 molecules of mass m_2, and so on. By summing all the pressures from each type of molecule, show that Dalton's law of partial pressures is a result of the kinetic-molecular theory of gases.

PART TWO

PHENOMENOLOGICAL THERMODYNAMICS

CHAPTER FOUR
CONSERVATION OF ENERGY AND THE FIRST LAW OF THERMODYNAMICS

The first law of thermodynamics is nothing more than a restatement of the principle of the conservation of energy developed in classical physics. This classical principle is based on the realization that while energy may manifest itself in a variety of forms (for example, kinetic energy, gravitational potential energy, and electrostatic energy), the sum of all the different forms of energy in any particular system is fixed. Energy can be transformed from one type to another, always subject to this condition that the total energy must be constant. Thermodynamics extends this conservation principle to include one additional type of energy, heat. The first law of thermodynamics arose from a series of experiments, carried out during the nineteenth century, that demonstrated that work could be quantitatively converted to heat. In the elementary applications of thermodynamics, we shall be primarily interested in these two manifestations of energy—heat and work—and in particular, the work done in expanding a gas against a pressure. We shall concern ourselves with the interconversion of work and heat under a wide variety of circumstances. We shall see that whereas the first law of thermodynamics is based on the observation that work can be quantitatively converted to heat, the second law predicts that the converse is not necessarily so. There are processes in which heat cannot be quantitatively converted to work, and therein lies a fundamental source of inefficiency in electric-power-generation facilities.

4.1 CLASSICAL MECHANICS AND THE CONSERVATION OF ENERGY

The principle of conservation of energy is a cornerstone in the development of the laws of physics. It did not arise in one instant, the result of one brilliant idea, but slowly evolved over many centuries. The basic principles of classical mechanics (the motion of objects, for instance) were established before 1700 by a series of observers, starting with the ancient Greeks, continuing to Galileo and Kepler, and culminating in the brilliant work of Newton. The first person to recognize a conservation principle was Leibnitz (1693), who noted that the sum of kinetic energy ($\frac{1}{2}mv^2$) and potential energy (mgh) of masses in gravitational fields was a constant. In any system, there was no restriction about how the total energy was divided between kinetic and potential energy; the restriction applied to the sum of the two. It is thus a simple matter to calculate the velocity of a ball dropped from a rooftop. At the instant the ball is released, it has no kinetic energy, but does have potential energy, given by mgh, where h is the height of the building. When the ball hits the ground at $h = 0$, this potential energy must have been converted to kinetic energy $\frac{1}{2}mv^2$, allowing a solution for the velocity by solving the equation

$$\tfrac{1}{2}v^2 = gh$$

The process can be reversed. If we start the ball upward with that velocity, it will rise to a height h; at that point it will have zero velocity, all the kinetic energy having been converted to potential energy. The total energy remains constant.

There is no natural zero point of potential energy. We can take it to be any convenient place. We are concerned only with differences in potential energy and not with the absolute value. We shall find that in thermodynamics many of the functions we deal with, such as internal energy, enthalpy, free energy, and the work function, are similar to the potential energy in that they have no natural zero points, only artificial zero points that are fixed as a matter of convenience. These functions are often referred to as "potentials", and under certain circumstances they measure the useful work one can get from a system. We shall also see that one function we deal with is unique in that it has a natural zero; that function is the entropy, and its natural zero is the subject of the third law of thermodynamics.

Initially, conservation of energy referred to the sum of kinetic energy and potential energy in a gravitational field. As additional forms of energy were discovered, these new forms were likewise included in the conservation of energy principle. These new energies included the energy of compression or expansion of a spring, $V_{\text{spring}} = \frac{1}{2}kx^2$, where k is the spring constant and x is the displacement, and the potential energy associated with an electric charge in an electric field, $V_{\text{el}} = qE$, where q is the charge and E the electric potential. Applying the conservation of energy principle to the motion of bodies in these potentials should be familiar to you from your elementary physics course. The discovery that heat was another form of energy and therefore subject to conservation of energy was another link in the chain of discoveries leading to the first law of thermodynamics.

4.2 MACROSCOPIC AND MICROSCOPIC APPROACHES

There are two distinct ways in which we can view 18 g of water, whether it is in liquid, solid, or gaseous form. We can view it simply as 18 g of stuff, regardless of its fine structure, or we can view it as a collection of 6×10^{23} molecules of H_2O, each molecule containing two atoms of hydrogen and one atom of oxygen. The first view, which we call the *macroscopic* view, considers only the bulk properties of various substances. This is the approach we shall use at first. The second view, which leads to the *microscopic* approach, is conceptually more difficult. One must start with the basic equations of motion for each of the 6×10^{23} molecules, and thence by examination and calculation finally arrive at the bulk properties of various substances. It makes no difference which approach one uses, since all valid scientific pictures of the real world must be coherent regardless of the approach to each. Scientists believe that the microscopic approach is valid because it gives the same answers as the macroscopic approach. The microscopic approach is developed in statistical mechanics.

4.3 HEAT AND DEGREE OF HOTNESS

A proper understanding of the distinction between the sensation of heat on the one hand and degree of hotness (or temperature) on the other is only about 150 years old.[1] If you were to insert your hand into a beaker of water at 98.6 °F, you would feel no particular sensation except a slight wetness on the surface of your skin. If you were to insert your hand into a beaker of water at 200 °F, however, the severe pain would cause you to withdraw your hand immediately. Heat is a form of energy *in transit*. It appears only at the boundaries of systems. A system can have energy, but it cannot have heat. Heat can appear only when energy moves from one part of the universe to another. In the above experiment, heat moves from the hot water to the cold hand. Both beakers of water have heat capacities; they have the ability to "heat" colder objects; but they **do not have heat.**

On the molecular level, temperature can be associated with the motion of molecules, and it is related to the thermal energy. Temperature is an **intensive** property. It is independent of the amount of material present. A drop of water from the hot beaker will have the same temperature as the rest of the beaker. Thermal energy is an **extensive** property. It depends on the amount of material present. If the amount is doubled, the total thermal energy is doubled. Since heat is energy in transit, the amount of heat that can be transferred from a body also depends on the amount of material present. Little pain is associated with a small drop of hot water splashing on a hand.

The first person to distinguish clearly between heat and temperature was Joseph Black, a Scottish physician and chemist, who carried out a series of calorimetric experiments during the years 1759–1762. He discovered the concepts we now call heat capacity and latent heat of fusion.

Black showed that if equal volumes of water at 100 °F and 150 °F were mixed, the resulting temperature was midway between the two extremes, or 125 °F. If warm mercury was used in place of the warm water, however, the resulting final tempera-

---

JOSEPH BLACK (1728–1799) was a Scottish chemist and physician. With his earliest work (1754), he antedated Lavoisier in discovering the existence of a distinct gas separate and apart from common air. He studied the reaction limestone = lime + CO_2 quantitatively, using a balance; and he was the founder of quantitative "pneumatic" chemistry. He discovered that his "fixed air" was produced by fermentation, respiration, and the burning of charcoal. He soon, however, turned his attention to studying heat, leaving to Lavoisier the work on the discovery of the oxidation process. Black was the first person to appreciate the difference between heat and temperature, and he effectively discovered the concepts we now call heat capacity, latent heat of fusion, and latent heat of vaporization. James Watt, the inventor of the practical steam engine, was a friend and student of Black. Although Black successfully engaged in a variety of original researches of the most fundamental nature, he published only three papers in his lifetime. Apparently the aphorism "Publish or perish" was inapplicable in those days. The scientific community was much smaller then, and Black's work was successfully communicated to the world at large.

---

[1] An excellent elementary summary of the events leading to the first law of thermodynamics is given in James Bryant Conant, ed., *Harvard Case Histories in Experimental Science.* Case 3. *The Early Development of the Concepts of Temperature and Heat; The Rise and Decline of the Caloric Theory,* by Duane Emerson Roller (Cambridge: Harvard University Press, 1950).

> COUNT RUMFORD (1753–1814) was born Benjamin Thompson in the American colonies. He led one of the most hectic and fascinating lives in the annals of science. A poor storekeeper's apprentice, he made his fortune at the age of 19 by marrying the richest widow in the colonies; she was 32 at the time. He chose the wrong side during the Revolutionary War and left his home just ahead of the tar and feather gang. From then on, his life was like an adventure movie. In London, he was made an undersecretary in one of the ministries, and his share of the graft is estimated at 7000 pounds a year. Under suspicion of betraying the British, he fled London. He served in the Austrian army in a Turkish campaign and then entered the service of the elector of Bavaria. In Bavaria, he served as minister of war, minister of the interior, and chief scientist to the elector; he also was a secret agent for the British, but the Bavarians did not know this. He was eventually forced out of Bavaria. In 1804, he married Lavoisier's widow.
>
> Count Rumford founded the Royal Institution of London with his own funds; and he designed the central heating system for the building, a unique amenity in those days. His scientific achievements covered an amazing range. When he was 14, he could calculate solar eclipses to within four seconds of accuracy. He developed techniques for measuring the explosive force of gunpowder; and he invented the ballistic pendulum for measuring the velocity of rifle bullets. He perceived certain aspects of the kinetic theory of gases. His main contribution concerned heat, and in particular the heat produced by friction, which was as he termed it inexhaustible. He demolished the caloric theory of heat as a material substance. As he put it, "It is hardly necessary to add that anything which any insulated body or system of bodies can continue to furnish without limitation cannot possibly be a material substance." He called heat a "vibratory motion taking place among the particles of the body." He even foresaw the energy crisis, with his paper "Description of a New Boiler with a View to the Saving of Fuel."

ture was not midway between the two extremes, but rather 115 °F, indicating that water has a greater "capacity" for heating than mercury does. It takes more heat to raise the temperature of a given volume of water by one unit than it does to raise the temperature of the same volume of mercury.

In further experiments on heats of fusion, Black determined the heat of fusion of water to be 139 Btu lb^{-1}. Considering the crude calorimetric procedures available to Black, his result agrees remarkably well with the modern value 144 Btu lb^{-1} (or 4.18 J g^{-1}).

It still remained to be discovered that heat was a form of energy. The caloric theory of heat, in which heat was thought to be a material substance that passed from one object to another, gained wide acceptance in scientific circles. Count Rumford, who was born Benjamin Thompson in the American colonies, tried to "weigh" heat, and he determined that if heat had any weight at all it was less than 1 ppm.[2] He concluded his paper describing these experiments with the statement, "I think we may safely conclude that all attempts to discover any effect of heat upon the apparent weights of bodies will be fruitless."

4.4 THERMODYNAMIC PROCESSES

In the experiment in which hot water and cold water were mixed to produce water with a final temperature midway between the two extremes, a change took place. We shall use the generic term *thermodynamic process* to describe situations in which

---

[2] Rumford's work on heat is given in Roller, *Early Development*. An excellent short review of his fascinating life and work is given by M. Wilson, Count Rumford, *Scientific American* 203, no. 4 (October 1960): 158.

changes occur. Further, we shall differentiate between natural and unnatural processes. A *natural process* is one that will naturally occur in the absence of constraints. When hydrogen is mixed with oxygen at ordinary temperatures, these elements react to form liquid water. When a ball rolls off a table, it falls to the floor. These are natural processes. An *unnatural process* is the reverse of a natural process.

This does not mean that unnatural processes do not take place. They simply do not take place by themselves. If one applies the proper constraints to the system, meaning the proper "driving forces," then unnatural processes become natural processes. Water can be decomposed into its elements by electrolysis with a battery. When a ball rolls off a table, it can rise to the ceiling if it is hit with a bat.

The first law of thermodynamics concerns itself only with the thermodynamic process itself, without regard to whether it is natural or unnatural. The second law of thermodynamics concerns itself with the distinction between natural and unnatural processes, and provides proper criteria to differentiate between them. Thermodynamics will not, however, concern itself with the time it takes to complete the process. Thermodynamics will not differentiate between the case of hydrogen and oxygen in a sealed container achieving final equilibrium only after many years, and the case of the same two gases ignited by a spark and achieving final equilibrium in a short time. A natural process is a *spontaneous* process; it occurs spontaneously. An unnatural process is a *nonspontaneous* process; it does not occur spontaneously.

A very important process in our development of thermodynamics is the expansion of a gas. This is conveniently carried out by enclosing the gas in a sealed cylinder fitted with a movable piston. The process

$$\text{ideal gas (2 atm)} \longrightarrow \text{ideal gas (1 atm)}$$

is shown in Figure 4.1. Two different methods for carrying out the process are indicated. In Figure 4.1a, the external pressure is 1 atm, and a weight has been placed on the piston to produce a total pressure in the cylinder of 2 atm. If we remove an infinitesimal amount of mass from the weight, the piston will move up by an infini-

FIGURE 4.1 Reversible and irreversible expansions of a gas. (a) Reversible expansion. The gas is expanded by removing infinitesimal amounts from the weight; the system is in equilibrium at the beginning, the end, and all intermediate points. (b) Irreversible expansion. The rod that holds the piston in place is removed, and the piston rapidly moves up; the system is in equilibrium only at the beginning and the end, not at the intermediate points.

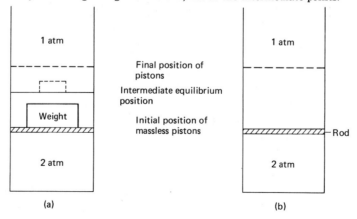

tesimal amount, the volume will increase by an infinitesimal amount, and the pressure will decrease by an infinitesimal amount. As we remove more and more weight, the gas slowly expands, until the last bit of weight is removed and the final pressure is 1 atm. Figure 4.1b depicts a similar situation, except that now the piston is held in place by a rod that protrudes into the cylinder. When we pull out the rod, the piston rapidly moves up until the final pressure is 1 atm. Note that these methods are only two of infinitely many different ones for carrying out the expansion.

The process shown in Figure 4.1a is particularly important. Each individual step of that process is called an *infinitesimal process*. An infinitesimal process is a process that takes place to the extent that there is at most an infinitesimally small change in *every* macroscopic property of the system.

4.5 INSULATING AND CONDUCTING WALLS

The processes described in Figure 4.1 can be carried out under two extreme conditions of heat flow. If the walls of the system are made of a good heat conductor such as copper and the system is placed in contact with an infinitely large source of heat (a thermostat), then the entire system will always be at this constant temperature. Heat can enter or leave the gas through the walls of the system to maintain a constant temperature. This process is an **isothermal** process.

At the other extreme, the walls of the cylinder can be made of a heat-insulating substance such as asbestos. The flow of heat will be totally restricted, and no heat will flow either from the gas or to it. A process carried out with no heat flow is called an **adiabatic** process. During an adiabatic process the temperature of the system may change.

The two possibilities are distinct, and the law of conservation of energy (or the first law of thermodynamics) often predicts different results depending on whether the process is carried out isothermally or adiabatically.

4.6 SYSTEM AND SURROUNDINGS

Again consider the example of Figure 4.1. Our attention is focused on the gas in the cylinder. The gas in the cylinder is the *system,* but the gas is not all that concerns us. In an isothermal process, heat may flow into or out of the gas. If it flows out, it must go somewhere; if it flows in, it must come from somewhere. This somewhere is the *surroundings*. For an isothermal process, the surroundings will be the thermostat. For an adiabatic process, there is no surroundings for heat. The system is isolated from heat flow. In the illustration Figure 4.1, as the gas expands (or contracts), the piston moves against an external force, and work is done either by the system or on the system. The work source is the surroundings, this time relative to work.

It is important to differentiate clearly between system and surroundings, since the proper delineation of system and surroundings may provide most of the solution to a given problem. The type of boundary of the system determines how the system may interact with the surroundings. A heat-conducting boundary allows the flow of heat into or out of the system, whereas an insulating boundary completely restricts the flow of heat. A boundary with a movable wall (such as a piston) enables the system to do work on the surroundings; rigid walls make expansion work impossible.

It is important to remember that the conservation of energy principle applies to the entire universe. The energy of the universe is constant. Thus, when the conservation of energy principle is applied to any given situation, it must apply to the system plus its surroundings. The sum total of energy in the system *plus* its surroundings must remain constant, but there does not need to be any restriction on flow of energy between system and surroundings so long as the total is constant. Read that last sentence again, and once more. It is the essence of half of the first law of thermodynamics.

4.7 EQUILIBRIUM

Now consider the initial and final states of the two processes depicted in Figure 4.1. The internal forces are exactly counterbalanced by the external forces. Neither system has any tendency to undergo any change. The state of each system can be precisely specified by the temperature and pressure of the gas. The temperature, pressure, density, and other properties are uniform throughout the gas. We say that the system is in *mechanical equilibrium* with its surroundings. There is no tendency for the system to interact with the surroundings. On the other hand, if we consider the process depicted in Figure 4.1b, there is a tendency for the gas to expand after we remove the rod. While the piston is moving rapidly from initial to final position, the system is undergoing change. During the instant following the removal of the rod, the gas near the piston undergoes a slight expansion and cools slightly, while the gas at the far end remains stationary. During the rapid expansion, the system interacts with the surroundings. The temperature, pressure, and other properties are not uniform during the expansion. The system is not in equilibrium with its surroundings.

Suppose that the cylinder is at temperature T and we place it in a thermostat at this same temperature T. There is no tendency for heat to flow between the cylinder and the thermostat. We say that the system is in *thermal equilibrium* with its surroundings. If the cylinder were at a temperature T' different from T, then heat would tend to flow between the system and its surroundings. The system would not be in thermal equilibrium with its surroundings (until the temperatures were equalized).

In thermodynamics we deal exclusively with what we call *equilibrium states*. In an equilibrium state, all the variables of the system are constant and uniform throughout. There is no tendency for change. All thermal, mechanical, chemical, and any other properties of the system must be fixed and unchanging.

4.8 REVERSIBLE PROCESSES

A *reversible* process is a finite process that can be considered the limit of a sum of infinitesimal processes. A reversible process is one that is in equilibrium at every stage of the process. A reversible process goes from the initial to the final state through an infinite series of equilibrium states. A process that is not reversible is *irreversible*. The distinction is important.

Figure 4.1 shows the difference between the reversible expansion of a gas and the irreversible expansion. In 4.1a we proceeded by way of an infinite number of infinitesimal processes. The system is in equilibrium at every stage. Each time we remove an infinitesimally small bit of mass from the weight, the system moves from one equilibrium state to another equilibrium state that is slightly different. This expansion is a reversible expansion. The reversible process is idealized. It assumes the absence of friction between the piston and the walls of the cylinder. It is impossible to realize idealized conditions in practice but they can be acknowledged in principle. The idea is akin to the frictionless pulleys, massless rods, and point masses in classical mechanics. It is impossible to realize them in practice either, but they are legitimate in principle.

On the other hand, the expansion depicted in Figure 4.1b is irreversible. To be sure, the initial and the final states of the system are equilibrium states. But once the rod is pulled, the system is no longer in equilibrium for it is undergoing the expansion. It is not in equilibrium at any step in the process.

A basic difference between reversible expansion and irreversible expansion of a gas is the following. During the reversible process, the pressure inside the cylinder is at all times equal to the external pressure. (That is the total pressure. The pressure exerted by the weight must be added to the atmospheric pressure.) During the irreversible process, the internal pressure is not always equal to the external pressure. In this case they are equal only at the start and at the finish (again remembering that the initial pressure exerted by the rod must be added to the external atmospheric pressure). We shall see that the work obtained in going from some initial state to a final state is intimately connected with the way in which the process is carried out.

4.9 WORK

Mechanical work is done whenever an applied force moves over a distance. The work done by a force F moving a distance dr in the direction of the applied force is given by

$$dw = F\, dr \qquad (4.1)$$

If the component of F along the direction of motion is $F \cos \theta$, then the work is given by

$$dw = F \cos \theta\, dr \qquad (4.2)$$

If the force moves a finite distance, the work is given by the integral

$$w = \int_{r_1}^{r_2} F\, dr \qquad (4.3)$$

If the force is constant in both direction and magnitude, then Equation (4.3) can be directly integrated to give

$$w = F \int_{r_1}^{r_2} dr = F(r_2 - r_1) \qquad (4.4)$$

where r_2 is the final position and r_1 the initial position.

Consider the work done in raising a 1-kg weight a distance 1 m in the earth's gravitational field. The force is mg, where m is the mass and g is the gravitational acceleration constant; and

$$\begin{aligned} w &= mg \int_{r_1}^{r_2} dr = mg(r_2 - r_1) \\ &= (1 \text{ kg})(9.8 \text{ m s}^{-2})(1 \text{ m}) \\ &= 9.8 \text{ kg m}^2 \text{ s}^{-2} = 9.8 \text{ J} \end{aligned} \qquad (4.5)$$

We can take as an example of a nonconstant force the stretching of a spring. The force is directly proportional to the amount of stretching as given by Hooke's law,

$$F = -kr \qquad (4.6)$$

where k is the spring constant and r is the displacement from the equilibrium position of the spring. The sign is negative because the force is a restoring force and acts in a direction opposite to the displacement. The work dw done in stretching the spring a distance dr is

$$dw = -kr\, dr \qquad (4.7)$$

The work done in stretching the spring from its equilibrium position ($r = 0$) to some position r is

$$w = -\int_{r=0}^{r} kr\, dr = -\tfrac{1}{2}kr^2 \qquad (4.8)$$

From Equation (4.5), we see that work has the same units (joules) as energy. Work is a form of energy. The work required to lift a 1-kg mass a distance 1 m exactly equals the difference in potential energy of the mass as it is raised that distance. This is a general theorem of mechanics. To increase the potential energy of a system, work must be done on that system and the work must equal the increase in potential energy. Similarly, the mass can do useful work as it falls but only at the expense of its potential energy. Work, like heat, is a manifestation of energy.

4.10 THE SIGN CONVENTION FOR WORK

Like heat, work appears only when the system interacts with the surroundings. Either the system does work on the surroundings, or the surroundings does work on the system. Which of these two situations shall we call *positive* work?

Consider the spring in the situation in which the spring is the system and the person stretching it is the surroundings. The energy of the spring is in the form of potential energy, given by $\tfrac{1}{2}kr^2$.[3] When the spring is stretched, work is done **on** the spring, and the energy of the spring is increased by an amount equal to the work w. When the spring is eased back to its equilibrium position, work is done **by** the spring on the surroundings, and the potential energy of the spring is decreased by w. The surroundings referred to is the hand doing the pulling. The basic reference point is the system. When work is done by the system, work is done on the surroundings; when work is done on the system, work is done by the surroundings.

There is universal agreement about the sign convention regarding heat flow. Heat that flows from the system to the surroundings is negative heat, and conversely, heat that flows from the surroundings to the system is positive heat. There is at present no universal agreement on the sign convention about work. It makes no difference which convention one adopts so long as one is consistent. Here we shall adopt the convention:

Work done on *the system is positive.*

Work done by *the system is negative.*

---

[3] The force of the spring is a conservative force. A conservative force is one that is derivable from a potential through the equation $F = -\partial V/\partial r$. Differentiating the expression for the potential energy in this **manner** yields Equation (4.6).

Before about 1960, the opposite convention was the most common in chemistry textbooks. That convention arose naturally from the early work in thermodynamics. Much of that work was carried out on the practical applications of steam engines. In that view, an important variable was the work output of the engine, that is the work done *by* the system; at that time it was only natural to call it positive work. There has been a distinct change over the past two decades, and while the older convention is still in use, there has been a gradual shift to the convention we have adopted.[4] This convention views all signed quantities from the point of view of the system. Anything going into the system is termed positive. Many physical chemistry textbooks use the older convention. Therefore, when you are perusing any book on thermodynamics, you must first ascertain which convention is being used; it makes no difference which, so long as it is used consistently.

4.11 EXPANSION WORK

An important kind of work in thermodynamics is the work systems do when they expand (or contract) against an opposing pressure, as in the examples of Figure 4.1. We shall be concerned with this kind of work almost exclusively during our initial discussions. Consider the situation in Figure 4.2. The gas (the system) is contained in a cylinder at a pressure P_{int} by a movable piston, which we take to be weightless. The external pressure is P_{ext}, and the opposing pressure is $P_{op} = P_{ext}$. Since pressure is force per unit area, the opposing force is $F_{op} = P_{op}A$, where A is the cross-sectional area of the piston. We now allow the piston to move a small distance dx.

Since there is motion against an applied force, work is done; the differential amount of *expansion work* is

$$dw_{exp} = -F_{op}dx = -P_{op}A\,dx \tag{4.9}$$

The negative sign arises by virtue of our convention. Since $A\,dx$ is the differential volume change dV of the system, the differential work is[5]

$$dw_{exp} = -P_{op}\,dV \tag{4.10}$$

For a finite expansion, the work is the integral of (4.10), or

$$w_{exp} = -\int_{V_{init}}^{V_{fin}} P_{op}\,dV \tag{4.11}$$

Consider as an example the work done when a mole of liquid water freezes at 0 °C and 1 atm. Since P_{op} is constant, we have from Equation (4.11)

$$-w_{exp} = \int_{V_{init}}^{V_{fin}} P_{op}\,dV = P_{op}(V_{fin} - V_{init})$$

where V_{init} is the molar volume of liquid water, 18 cm³, and V_{fin} is the molar volume of ice, 19.6 cm³. The work is

$$-w = 1.0(19.6 - 18) = 1.6 \text{ cm}^3 \text{ atm} = 0.162 \text{ J} = 0.039 \text{ cal}$$

Now let us examine the difference between reversible and irreversible expansion work of a gas.

---

[4] Our convention is the convention the IUPAC now recommends. Physicists have apparently been using this convention uniformly for many decades. Chemistry textbooks are still split, particularly at the elementary chemistry level, where many books still use the older convention.

[5] When the opposite sign convention is used for work, Equation (4.10) is written $dw_{exp} = P_{op}\,dV$.

FIGURE 4.2 Expansion work of a gas.

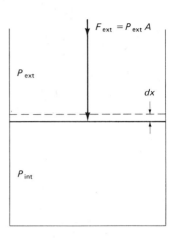

4.12 ISOTHERMAL REVERSIBLE EXPANSION WORK OF IDEAL GASES

Expansion work is *always* given by Equation (4.11). Specifying the process only establishes how that integral is evaluated. The final result usually depends on how the process is stipulated. As noted previously, a basic characteristic of a reversible process is that the internal pressure always equals the external opposing pressure. Further, for an ideal gas, $PV = nRT$, and we can write an analytical expression for the pressure, $P_{op} = P_{ext} = P_{int} = nRT/V$. Under the *isothermal* condition that T is constant, Equation (4.11) becomes

$$-w_{exp} = \int_{V_{init}}^{V_{fin}} P_{op}\, dV = \int nRT \frac{dV}{V}$$

$$= nRT \int \frac{dV}{V}$$

$$= nRT \ln \frac{V_{fin}}{V_{init}} \quad (4.12)$$

For 1 mol of gas, $w = -RT \ln(V_{fin}/V_{init})$. This is true only for an isothermal reversible expansion.

Consider as an example the isothermal reversible expansion of 1 mol of an ideal gas at 300 K from an initial volume of 20 liters to a final volume of 40 liters. The work is

$$-w = RT \ln \tfrac{40}{20} = RT \ln 2$$
$$= (8.314)(300)(0.693) = 1728 \text{ J}$$
$$= (1.987)(300)(0.693) = 413 \text{ cal}$$
$$= (0.082)(300)(0.693) = 17 \text{ liter atm}$$

where the different units of work correspond to the different units used for the gas constant. According to Boyle's law, $P_1V_1 = P_2V_2$ at constant temperature; thus for an ideal gas we can write

$$w = -nRT \ln \frac{V_2}{V_1} = -nRT \ln \frac{P_1}{P_2}$$

Some reflection on the principle of the conservation of energy indicates that there is more to this than we have put down. For an isothermal process, the temperature is constant. The system does work on the surroundings during an expansion. The energy for this work must come from somewhere. We shall see shortly that there is also a heat flow involved when work is done in an isothermal expansion, whether it is reversible or irreversible.

4.13 ISOTHERMAL IRREVERSIBLE EXPANSION WORK OF IDEAL GASES

We can take as an example of an irreversible expansion of an ideal gas the expansion shown in Figure 4.1b. Here the gas expands from V_{init} at pressure P_{init} to V_{fin}, but the opposing pressure is the constant P_{fin}. The starting point is the same as for a reversible process, namely, $-w = \int P_{op} dV$. The difference lies in the substitution we make for P_{op}, which in this case is constant. The work is

$$-w_{exp} = \int_{V_{init}}^{V_{fin}} P_{op} dV = P_{fin} \int_{V_{init}}^{V_{fin}} dV$$
$$= P_{fin}(V_{fin} - V_{init}) \tag{4.13}$$

As the expression now stands, one has to calculate initial and final volumes, and the units of the answer will be liter atmospheres, assuming that the pressure is given in atmospheres and the volume in liters. It is more convenient to delay numerical substitution, and instead substitute for V_{fin} and V_{init} their equivalents, namely nRT/P_{fin} and nRT/P_{init}, to get

$$-w = P_{fin}(V_{fin} - V_{init}) = P_{fin}\left(\frac{nRT}{P_{fin}} - \frac{nRT}{P_{init}}\right)$$
$$= nRT\left(1 - \frac{P_{fin}}{P_{init}}\right) \tag{4.14}$$

Consider as an example the irreversible expansion at 300 K of 1 mol of an ideal gas from an initial pressure of 2 atm to a final pressure of 1 atm against a constant opposing (external) pressure of 1 atm. The work is

$$-w_{exp} = RT(1 - 0.5) = 0.5RT$$
$$= (0.5)(8.314)(300)$$
$$= 1247 \text{ J} = 298 \text{ cal}$$

If the same expansion had been carried out reversibly so that the internal pressure was at all times equal to the external pressure, the work would be

$$-w_{exp} = RT \ln \frac{P_1}{P_2} = RT \ln 2$$
$$= 0.693RT$$
$$= (0.693)(8.314)(300)$$
$$= 1729 \text{ J} = 413 \text{ cal}$$

A substantial difference exists between the reversible and the irreversible expansion. When we study the second law of thermodynamics, we shall see that the reversible process always produces more work than the irreversible process when a system is taken from the same initial to the same final state.

4.14 WORK AND P-V DIAGRAMS: CYCLIC PROCESSES

Figure 4.3 shows a plot of pressure vs volume for an ideal gas at constant temperature. The work done in going from A to B along the curve is $-\int_A^B P\,dV$, which is just the area under the curve. This will be true regardless of the shape of the curve. The work done in going from B to A along the *same* curve is $w = -\int_B^A P\,dV = +\int_A^B P\,dV$, which is the same area but with the reverse sign. The process in going from A to B and then back to A is cyclic. A *cyclic process* is any process in which the system ends at the same point as that from which it starts. In the special case in which we go from A to B and then back to A along the *same* path, the total work done is the sum of the two, which is zero.

Figure 4.4 indicates a more general path for a cyclic process. Here we go from A to B through point C, then back to the starting point A through point D. The work in going from A to B is given by the area indicated by the 45° hatch lines. The work in going from B back to A is indicated by the −45° hatch lines. The total work is the

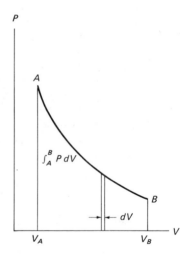

FIGURE 4.3 Expansion work on a *P-V* diagram. Single-stage process.

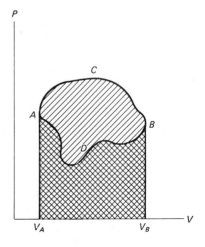

FIGURE 4.4 Expansion work on a *P-V* diagram. Cyclic process.

area enclosed by the curve $ABCD$. It is apparent that the work in any cyclic process is not necessarily zero. This is because the work done by expansion of the gas between the same initial and final states is not the same for all processes.

If we consider, instead of PV work, the work done in lifting masses in the earth's gravitational field, we find that the opposite situation holds. There the work done in any cyclic process (that is, the initial and final states are the same) equals zero. That is because the work done in going from A to B is independent of the path. In mechanics, this is characteristic of the work associated with conservative fields. This is a property of the mathematical entity we call the perfect, or exact, differential. Equivalently, it is the property of what is called a state function. A state function has a numerical value dependent only on the state of the system and not on the path taken to achieve that state. Many of the thermodynamic functions we shall deal with are state functions. State functions (or exact differentials) have been discussed in the mathematical review of Chapter 1.

4.15 WORK IS HEAT AND HEAT IS WORK

The acceptance of heat as another form of energy that can be included in the conservation of energy principle along with kinetic energy, potential energy, electrostatic energy, magnetic energy, and any other type of energy came from a series of experiments. These were carried out during the first half of the nineteenth century and showed that work was converted to heat. These experiments were concerned with determining the "mechanical equivalent of heat." Heat was produced from a wide variety of work sources, including electrical work, various types of mechanical work, and even horses. It was found that the amount of heat produced from a given amount of work was always the same, regardless of the type of work used in the conversion.

The first person to demonstrate clearly that work was converted to heat was Benjamin Thompson of Massachusetts, later Count Rumford of the Holy Roman Empire. In 1798, Thompson was working for the Duke of Bavaria supervising the boring of cannons at the Munich Arsenal. He noted the large amount of heat generated and set about measuring the heat. This he did by arranging the boring lathe in such a way that the brass cannon blank and the borer were immersed in a water bath during the operation. From the heat capacity of water, the amount of water used in the bath, and the temperature rise of the water, he was able to compute the heat evolved in the process. Since a horse furnished the motive power to drive the machinery, Thompson estimated the heat produced by a horse working for an hour. The value he got for the mechanical equivalent of heat was, in modern units, 0.183 calorie per joule. The modern accepted value is 0.239 calorie per joule. Despite the large frictional losses in the gear train and bearings and the unknown physical condition of the horse, his value agrees adequately with the modern value.

The effect of the experiment at the time was extraordinary. In Thompson's own words:

It would be difficult to describe the surprise and astonishment expressed in the countenances of the bystanders on seeing so large a quantity of cold water heated, and actually made to boil, without any fire. Though there was, in fact, nothing that could justly be considered as surprising in this event, yet I acknowledged fairly that it afforded me a degree of childish pleasure, which were I ambitious of the reputation of a grave philosopher, I ought most certainly rather to hide than to discover. [6]

---

[6] "An Inquiry Concerning the Source of Heat Which Is Excited by Friction," by Benjamin Thompson (Count Rumford), read to a meeting of the Royal Society in January, 1798, and later published in *Philosophical Transactions*, Vol. 88 (1798) p. 80.

> JAMES PRESCOTT JOULE (1818–1889), English physicist, had the strength of mind to put science ahead of beer. He owned a large brewery but neglected its management to devote himself to scientific research. His name is associated with Joule's law, which states that the rate at which heat is dissipated by a resistor is given by I^2R. He was the first to carry out precise measurements of the mechanical equivalent of heat; and he firmly established that work can be quantitatively converted into heat.

The proponents of the caloric theory of heat claimed that the heat was liberated because the metal in the form of fine shavings had a lower heat capacity than the bulk metal did. Thompson did the experiment again using a blunt borer, and he got the same result with fewer but coarser shavings. The result was independent of the size of the shavings. The opposition then claimed that the heat effect came from the action of air on the surface of the brass, each time the old surface was removed and a new surface exposed. In 1799, Sir Humphry Davy rubbed two pieces of ice against each other in a vacuum and found that the frictional dissipation of work was able to provide heat to melt the ice in the absence of air. The conversion of mechanical work into heat had been demonstrated.

The final experiments in this series were carried out by James Joule during the 1840s. Joule converted work into heat in a variety of ways and with an accuracy unapproached by previous measurements. He converted work into heat by electrical heating, and by compressing gases, forcing liquids through fine capillary tubes, and rotating paddle wheels in liquids. The last method converted gravitational work into heat, since the paddle wheel was turned by a falling weight, and the heat was produced by frictional dissipation in the liquid. He found that the heat produced by a given amount of work was independent of the kind of work. Joule, in his paper "On the Mechanical Equivalent of Heat," which described these results, concluded that 772 ft-lb of work produced an amount of heat necessary to heat 1 lb of water 1 °F. This corresponds to 0.241 calorie per joule in modern units, which is within 1% of the modern value of 0.239 calorie per joule.

If a given amount of work could be converted into a given amount of heat, then both phenomena must be manifestations of the same thing, namely energy. It was then a logical extension of the principle of conservation of energy to include heat along with the other forms of energy in the sum total of energy that was conserved.

4.16 THE FIRST LAW OF THERMODYNAMICS

We postulate a function that we call the internal energy and give it the symbol E (in many books it is given the symbol U). We focus our attention on a portion of the universe that we divide into two constituent parts, system and surroundings. Energy can flow between the system and the surroundings, but according to the conservation of energy law, energy cannot leave that portion of the universe we have established for ourselves. The total energy of the system *plus* the surroundings must be constant. Further, there is no restriction on the size of the surroundings.

Whenever a process takes place, energy can flow between system and surroundings subject to the condition that the total energy change in the system plus the surroundings is zero:

$$0 = \Delta E_{tot} = \Delta E_{syst+surr} = \Delta E_{syst} + \Delta E_{surr} \qquad (4.15)$$

Thus the conservation of energy principle requires that

$$\Delta E_{syst} = -\Delta E_{surr} \quad (4.16)$$

In applying these equations, we must consider the two forms of energy, heat and work. The energy change of the system equals the heat that enters (or leaves) the system q, plus the work done on the system w, and we can write[7]

$$\Delta E_{syst} = q + w \quad (4.17)$$

From Equation (4.16), we have

$$\Delta E_{surr} = -\Delta E_{syst} = -q - w \quad (4.18)$$

Now let's see what happens in a *cyclic* process. We take a system from state A to state B, then back to state A. There are two possibilities. Either ΔE_{syst} equals zero, or it does not equal zero. We first consider the possibility that it does not equal zero in a cyclic process and examine the consequences of that assumption. We break the cyclic process into two steps, $A \to B$ followed by $B \to A$. We calculate the energy change for the process as follows:

Step 1 $\quad \Delta E_{syst(A\to B)} = q_1 + w_1 \quad \Delta E_{surr(A\to B)} = -q_1 - w_1 \quad$ (4.19a)
$(A \to B)$

Step 2 $\quad \Delta E_{syst(B\to A)} = -q_2 - w_2 \quad \Delta E_{surr(B\to A)} = +q_2 + w_2 \quad$ (4.19b)
$(B \to A)$

Now let us adjust Step 2 so that $q_2 = q_1$ and calculate the sum of the two steps, that is, ΔE for the cyclic process $A \to B \to A$. We get

$$\Delta E_{syst(A\to B\to A)} = w_1 - w_2 \neq 0 \quad \text{(according to initial assumption)} \quad (4.20a)$$

$$\Delta E_{surr(A\to B\to A)} = -w_1 + w_2 \neq 0 \quad \text{(according to initial assumption)} \quad (4.20b)$$

Our assumption that ΔE_{syst} is not equal to zero has not led us into any violation of the principle of conservation of energy, since ΔE_{tot}, which is the sum of ΔE_{syst} and ΔE_{surr}, is equal to zero. A closer examination of Equation (4.20), however, points to the fundamental absurdity of our original assumption. Assume $w_1 > w_2$. (If this turns out not to be true, simply reverse the process.) What we should have is a cyclic process, in which work is done by the surroundings on the system with *no other changes in the universe*. The cyclic process could then be carried out over and over again, continually extracting work from the surroundings with no other attendant costs. This is *perpetual motion*. (In fact, it is called perpetual motion of the first type.) The work so obtained could be used to drive automobiles, propel ships and airplanes across the seas, and generate electricity. It is the solution to the energy crisis. Unfortunately, it is impossible.

It is the fundamental postulate of the first law of thermodynamics that such a situation is impossible. Our original assumption is incorrect, because it leads to results that are contrary to human experience. The correct assumption must be that for *any cyclic process*

$$\Delta E_{syst(cyclic)} = 0 \quad (4.21)$$

---

[7] When the opposite sign convention is used for work, the first law of thermodynamics is written $\Delta E = q - w$.

It is a consequence of Equation (4.21) that E is a function only of the state of the system. That is to say, in going from A to B

$$\Delta E_{\text{syst}(A \to B)} = E_B - E_A \qquad (4.22)$$

The energy change for any process equals the difference in energies between the final and the initial state. In mathematical terms, E is a state function; the differential of E, or dE, is an exact differential. (We have examined some properties of these types of functions in Chapter 1.)

To summarize

$$\boxed{\Delta E = E_B - E_A = q + w} \qquad (4.23)$$

The energy change for any process that takes a system from state A to state B is equal to the energy difference between the two states; it also equals the energy that flows into (or out of) the system. The energy change can consist of two component parts, heat and work. The items q and w are not functions of the state of the system, but the quantity $q + w$ is a state function. Neither dq nor dw are exact differentials, but $d(q + w)$ is an exact differential.

One question remains unanswered. The question is, How do we know that Equation (4.23) is correct? That question does not have a really satisfactory answer in the usual sense. It can be applied to any fundamental basic postulate and the answer is the same. We know it is correct because it works. When the results of Equation (4.23) are applied to real systems undergoing real processes, the results we predict are the ones that we see.

PROBLEMS

The following conversions will be helpful in some of the following problems:

The basic unit of energy is the joule, which is 1 kg m^2 s^{-2}.
The erg is 1 g cm^2 s^{-2}, or 10^{-7} J.
In electrical units, the joule is 1 volt-coulomb.
The defined calorie is exactly 4.184 J.
The Btu is about 252 calories.
The unit of power is the watt (W), which is defined as 1 joule per second (J s^{-1}).
The horsepower is 746 W.
The ampere is a flow of current equal to 1 coulomb per second (C s^{-1}).

1. You have a hi-fi set that draws 250 W of power at 110 V. Assume that the amount of work produced is negligible. How much heat is associated with the operation of the hi-fi set? (The work is associated with moving the loudspeaker. You now know why large hi-fi sets must be provided with cooling vents in the cabinet.)
2. How much work is required to lift a 1-kg weight a distance of 100 m? If the weight is then dropped, how fast is it traveling when it hits the ground (if air resistance is neglected)?
3. The weight of Problem 2 falls into a tub of water containing 1 liter of water, and all the energy is dissipated into heat from frictional effects. The heat capacity of water is 1 cal K^{-1} g^{-1}. What is the temperature increase of the water? (Assume that no water splashes out.)
4. Since voltages, resistances, and currents can be measured very precisely, the measurements of electrical work can be conveniently made with a greater accuracy than what is possible with any other type of work. Therefore electrical work is often used to calibrate calorimetric instruments. A 42.6-ohm (Ω) resistor is placed in a calorimeter, and a current of 1.386 amperes (A) is passed through the resistor for 400 s. The temperature of the calorimeter increases by 2.479 °C. Calculate the heat capacity of the calorimeter.
5. A spring has a spring constant of 100 dyn cm^{-1}. It is compressed a distance of 10 cm. Calculate the work required to compress the spring.
6. After the spring of Problem 5 is compressed, it is tied with a piece of catgut and placed in a beaker of HCl, which dissolves the spring. Has the energy of the spring in the form of potential energy of compression been destroyed? (Think about this for a bit before you look for the answer, which is given in Problem 10.)

7. Just as a compressed gas in a piston can expand in a reversible or an irreversible manner, a compressed spring can be decompressed in a reversible or an irreversible manner. Consider the following two possibilities. Two identical springs are compressed by placing 1-kg weights on them; they are compressed a distance of 10 cm. These weights will be the analog of a constant atmospheric pressure and will be considered permanently attached to the springs. An additional 1-kg weight is then placed at the top of each of the springs, and each spring is compressed an additional 10 cm.

One spring is reversibly decompressed by slowly removing infinitesimally small amounts of weight from the extra 1-kg weight until the entire 1 kg is removed, and the spring slowly stretches the distance 10 cm back to the starting position. The second spring is irreversibly decompressed by suddenly removing the entire 1-kg weight, and the spring expands against the constant weight of the original 1-kg mass. Calculate the work done by each of the two springs during the decompression.

In the compression, the energy originally put into the spring to compress it was the same in both cases, and to satisfy the conservation of energy, we must get back the same amount of energy from both springs. What happens in the irreversible decompression? (See next problem.)

8. In actual fact, the extra energy appears as kinetic energy, since the 1-kg weight will be moving with a finite speed in the irreversible decompression. Calculate the speed.

9. A real spring has internal friction, and the motion of the spring in the irreversible decompression will eventually cease because of internal friction. The energy will be converted into heat. Calculate the amount of heat.

10. The energy of compression in Problem 6 must appear as extra heat. If we compare the heat evolved by two identical springs, one compressed and the other uncompressed, the compressed spring must evolve more heat in the reaction. Calculate the heat difference. If the beaker contained 100 ml HCl, what is the difference in final temperature of the HCl for the compressed and uncompressed springs? (Assume that the HCl has the same heat capacity as pure water.)

11. From the data given in the text on Black's experiments, calculate a value for the heat capacity of mercury, and compare this value with the currently accepted value.

12. The amount 0.3 mol of an ideal gas is reversibly compressed from an initial pressure of 1 atm to a final pressure of 5 atm at a constant temperature of 298 K. Calculate the work.

13. After the compression of Problem 12, the gas is allowed to expand irreversibly back to the starting point against a constant external pressure of 1 atm from an initial pressure of 5 atm at 298 K. Calculate the work.

14. You have a 10,000-Btu hr^{-1} air conditioner in your bedroom. You have a light fixture in your bedroom with three 100-W light bulbs. The lights are on while the air conditioner is running. What fraction of the capacity of the air conditioner is being used to remove the heat generated by the lights? Assuming 50% efficiency, what amount of horsepower does the electric motor need to achieve 10,000 Btu hr^{-1}?

15. Use the same data for bedroom and air conditioner as in Problem 14. Assume that you eat 2500 calories of food per day, and remember that the food calorie of the diet books is really the kilocalorie (kcal) that scientists use. Thus a reasonable estimate for the heat generated by a human is about 2500 kcal a day. If you are in the bedroom while the air conditioner is running, what fraction of the capacity of the air conditioner is used to remove the heat you generate?

16. You exhale at the rate of 14 breaths a minute, each breath containing 0.5 liter air. You exhale against a constant external pressure of 1 atm. Calculate the total amount of expansion work you do against the earth's atmosphere in a day. What fraction of a day's energy intake (2500 kcal) is this? (*Note:* It's not as bad as it looks. Remember that each time you exhale your body volume contracts, so the atmosphere is simultaneously doing work on you.)

17. Your heart beats at the rate of 70 times a minute, each beat pumping about 60 cm^3 of fluid against a pressure of 120 mmHg. How much expansion work does the heart do in a day? (*Note:* The heart actually does much more work than this. The heart muscles are contracting, the blood is accelerated, and there are frictional losses and other factors involved.)

18. The heat capacity of water is 75.3 J K^{-1} mol^{-1}. The heat of fusion of ice is 6025 J mol^{-1}. How much ice would you have to eat to burn up the equivalent of 500 food calories (that is, 500 kcal) of ice cream? (*Note:* The ice diet is not recommended. The large quantities of water consumed can cause severe problems.)

19. A mole of ideal gas is isothermally and reversibly expanded from an initial pressure of 2 atm to a final pressure of 1 atm at 298 K. Calculate the work (a) by the usual formula; (b) by plotting the PV isotherm and graphically integrating the area under the curve.

20. Using the approximation of Problem 24 in Chapter 2, calculate an expression for the isothermal expansion work of a van der Waals gas. The amount 0.4 mol CO_2 is contained in a piston fitted with a cylinder, and the gas is kept at a constant temperature of 273 K. The gas is expanded from an initial volume of 11.2 liter to a final volume of 22.4 liter at constant temperature. Use your expression to compute the work. What would be the work if the gas behaved ideally?

CHAPTER FIVE

APPLYING THE FIRST LAW OF THERMODYNAMICS. ENERGETICS OF IDEAL GAS PROCESSES

5.1 EXPERIMENTAL MEASUREMENT OF ΔE AS CONSTANT-VOLUME HEAT

Consider the usual expression for the first law of thermodynamics,

$$\Delta E = q + w \tag{5.1}$$

We shall restrict our discussion to include only one particular type of work that we call pressure-volume work, or expansion work, $dw = -P_{op}\, dV$. Equation (5.1) can be written in differential form:

$$dE = dq + dw = dq - P_{op}\, dV \tag{5.2}$$

For a constant-volume process, $dV = 0$ and $dw = 0$. Thus

$$dE = dq_v \tag{5.3}$$

where q_v is the heat absorbed at constant volume. Equation (5.3) can be integrated to yield

$$\Delta E = q_v \tag{5.4}$$

Equation (5.4) provides a convenient basis for the experimental measurement of energy changes involved in chemical reactions. We carry out the reaction at constant volume and measure the heat involved. The heat directly yields a value of ΔE. Experiments of this kind are carried out in an instrument called a *bomb calorimeter*, a schematic of which is shown in Figure 5.1. Let us consider the reaction

$$C(\text{graphite}) + O_2(g) = CO_2(g)$$

A weighed piece of graphite is placed in the bomb so that it can be ignited by the electrical wires. The bomb is sealed and immersed in a temperature bath containing a known volume of water. After the air is removed, oxygen is admitted to the bomb under excess pressure. The initial temperature is noted and the carbon is ignited. Heat is produced by the reaction at constant volume, and the temperature of the bath increases. The system is the graphite plus the oxygen in the bomb; the sur-

FIGURE 5.1 Bomb calorimeter.

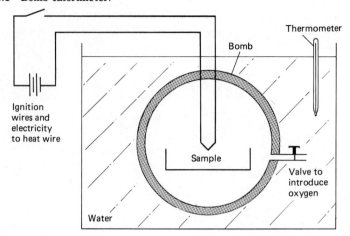

roundings include the bath and the walls of the bomb. From the temperature rise, and the heat capacities of the water plus the material of the bomb, the heat of the reaction q_v is determined. This heat directly yields the energy change for the process $\Delta E = q_v$. In this case, the surroundings get hotter. Heat leaves the system; therefore q is negative. If heat enters the system, then the surroundings get colder, and q is positive.

In a bomb calorimeter, we measure ΔE for one particular path, namely the path for which the volume is constant. We do this because it is experimentally convenient. Energy is a state function. It makes no difference how we measure ΔE for the reaction. So long as the initial and final states are the same, we always get the same results.

5.2 ENTHALPY AS CONSTANT-PRESSURE HEAT

Given an initial state and a final state, the energy difference between them for *any* process that takes the system from the initial to the final state can be determined by measuring the heat evolved (or absorbed) in one particular process, that is, a constant-volume process that takes the system from the initial to the final state. Thus, the heat evolved at constant volume gives us an experimental measure of ΔE for *any* process involving the same initial and final conditions. What about the heat evolved in a constant-pressure process? Does that give us an experimental measure of a state property? Reactions are normally carried out in open beakers in the laboratory at essentially a constant pressure. Large distillation towers used in manufacturing gasoline operate at constant pressure, and so do the giant blast furnaces of steel companies. Let us now calculate the changes accompanying a process at constant pressure.

We call the constant-pressure heat q_p and write

$$\Delta E = E_2 - E_1 = q_p + w = q_p - \int P\, dV = q_p - P(V_2 - V_1) \tag{5.5}$$

when no work besides $P\, dV$ work is done. We can rearrange Equation (5.5) to get

$$(E_2 + PV_2) - (E_1 + PV_1) = q_p \tag{5.6}$$

Now we define a new function called the *enthalpy*,
$$H = E + PV \tag{5.7}$$
so that
$$\Delta H = H_2 - H_1 = (E_2 + P_2V_2) - (E_1 + P_1V_1) = q_p \tag{5.8}$$
when $P_1 = P_2$, as in the present case. Since $E, P,$ and V are functions only of the state of the system, it is apparent from Equation (5.7) that the enthalpy is also a function only of the state of the system. Therefore ΔH is independent of the path of the process; it depends only on the initial and final states.

The differential form of Equation (5.7) is
$$dH = dE + d(PV) = dE + P\,dV + V\,dP \tag{5.9}$$
For any process at constant pressure $dP = 0$, and
$$dH = dE + P\,dV \tag{5.10}$$
The difference between dE and dH is just the work term $P\,dV$. Equation (5.10) can be written in the integrated form
$$\Delta H = \Delta E + P\,\Delta V \quad \text{(constant } P) \tag{5.11}$$

5.3 HEAT CAPACITIES

Heat capacities measure the heat required to change the temperature of a substance. They are commonly measured under two special conditions, constant volume and constant pressure:
$$C_v \equiv \frac{dq_v}{dT} \qquad C_p \equiv \frac{dq_p}{dT} \tag{5.12}$$

These heat capacities can be related to E and H in the following manner; the energy can be written as a function of the volume and temperature, $E = E(V, T)$.[1] As a mathematical consequence of this, we can write
$$dE = \left(\frac{\partial E}{\partial T}\right)_V dT + \left(\frac{\partial E}{\partial V}\right)_T dV \tag{5.13}$$

At constant volume ($dV = 0$), we have seen that $\Delta E = q_v$, or $dE = dq_v$ (note that dq_v is an inexact differential). Since $dV = 0$, we can write $dE = (\partial E/\partial T)_V\, dT$, which leads to
$$C_v \equiv \frac{dq_v}{dT} = \left(\frac{\partial E}{\partial T}\right)_v \tag{5.14a}$$

We could likewise have started with $H = H(P, T)$ to find that
$$C_p \equiv \frac{dq_p}{dT} = \left(\frac{\partial H}{\partial T}\right)_p \tag{5.14b}$$

---

[1] There are really three variables—V, T, and P. An equation of state provides an equation relating these three, however, enabling us to eliminate one of the three variables. We could equally well write $E = E(P, T)$ or $E = E(P, V)$. Our choice of $E = E(V, T)$ is one of convenience.

At constant volume, no work is done. All the heat is used to warm the substance. For constant pressure, there is a work term; and the inflowing heat divides in two parts. One part is used to warm the sample, and the other part is used for expansion work. Since some of the heat is "wasted" on work, it takes more heat per unit temperature rise at constant pressure than at constant volume. Hence, $C_p > C_v$. The difference between C_p and C_v is very small for liquids and solids, since the ΔV terms are small; but the difference is substantial for gases.

We can calculate the difference $(C_p - C_v)$ in the following way. By Equation (5.13),

$$dq + dw = dE = \left(\frac{\partial E}{\partial T}\right)_V dT + \left(\frac{\partial E}{\partial V}\right)_T dV$$

If the work is restricted to pressure-volume work,

$$dq - P\,dV = C_v\,dT + \left(\frac{\partial E}{\partial V}\right)_T dV$$

and if it is specified that P = constant,

$$dq_p = C_p\,dT = C_v\,dT + \left(\frac{\partial E}{\partial V}\right)_T dV + P\,dV$$

$$= C_v\,dT + \left[\left(\frac{\partial E}{\partial V}\right)_T + P\right] dV$$

Now we can divide by dT to get

$$C_p - C_v = \left[P + \left(\frac{\partial E}{\partial V}\right)_T\right] \left(\frac{\partial V}{\partial T}\right)_P \tag{5.15}$$

The derivative $(\partial E/\partial V)_T$ has the dimensions of pressure and is called the *internal pressure*. As early as 1843, Joule tried to measure the internal pressure of a gas by allowing a gas under pressure to undergo a *free expansion* into a vacuum.

5.4 THE JOULE EXPANSION
(THERMODYNAMIC DEFINITION OF AN IDEAL GAS)

The principle of Joule's famous expansion experiment is indicated in Figure 5.2. Two bulbs are connected by a stopcock and immersed in water. A thermometer measures the temperature of the water. Bulb A is filled with gas under pressure, and bulb B is evacuated. The experiment consists in opening the stopcock, allowing the gas from bulb A to expand into the vacuum, and measuring the temperature change. Joule found no temperature change. He concluded from this experiment that the energy of the gas depended only on the temperature and not on the volume.

If there really is no temperature change on expansion into the vacuum, then it must be true that $\Delta E = q + w = 0$. This is so because both q and w are zero. That $q = 0$ is obvious, since there is no heat flow to the surroundings—that $\Delta T = 0$ makes this evident. That w is also zero becomes clearer if we examine the equivalent experiment shown in Figure 5.3. A cylinder fitted with a massless piston has been substituted for the two bulbs. One half of the cylinder is filled with gas, while the other half is evacuated; the piston is held rigidly in place by a rod. The act of opening the stopcock is replaced by the act of removing the rod. In this experiment the gas ex-

SEC. 5.4 THE JOULE EXPANSION (THERMODYNAMIC DEFINITION OF AN IDEAL GAS)

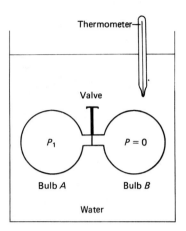

FIGURE 5.2 Experimental setup for Joule expansion.

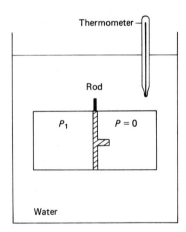

FIGURE 5.3 Equivalent arrangement for Joule expansion.

pands against the piston; but there is no external force, since $P_{op} = 0$. Thus $w = -P_{op}\,\Delta V = 0$. Since E is constant,

$$dE = \left(\frac{\partial E}{\partial V}\right)_T dV + \left(\frac{\partial E}{\partial T}\right)_V dT = 0 \tag{5.16}$$

Since no temperature change was found, $dT = 0$, and it therefore follows that $(\partial E/\partial V)_T\,dV = 0$. Since we know that $dV \neq 0$, we can conclude that

$$\left(\frac{\partial E}{\partial V}\right)_T = 0 \tag{5.17}$$

Note very carefully that the preceding paragraph opened with the words "If there really is no temperature change. . . ." Equation (5.17) is valid *only* if $\Delta T = 0$. It does happen that there is a small temperature change when a real gas undergoes a free expansion. Joule could not detect this change in his experiment, since the heat capacity of the water was so much larger than the heat capacity of the air.[2] On the

---

[2] In connection with the discussion of the opening section of Chapter 1, this is an example of an experiment being "wrong."

other hand, Equation (5.17) is valid for an ideal gas. When we study the second law of thermodynamics, we shall see that Equation (5.17) follows from the fact that the equation of state of an ideal gas is $PV = nRT$. We can take Equation (5.17) as the thermodynamic definition of an ideal gas. An ideal gas is a gas whose energy depends only on the temperature.

Since $(\partial E/\partial V)_T = 0$, Equation (5.15) can be greatly simplified for an ideal gas. Since $V = RT/P$, therefore $(\partial V/\partial T)_p = R/P$, and Equation (5.15) reduces to

$$C_p - C_v = R \qquad \text{(ideal gas)} \qquad (5.18)$$

5.5 JOULE-THOMSON EXPANSION

Several years after Joule's original expansion experiment, Joule and Thomson jointly developed a better method for measuring the temperature change on expansion. Their apparatus is shown schematically in Figure 5.4. The apparatus is insulated, so that the process is adiabatic; $q = 0$. A mole of gas on the high-pressure side with molar volume $\overline{V}_1$ at pressure P_1 is forced through the porous plug to the low-pressure side at molar volume $\overline{V}_2$ and pressure P_2.[3] On the high-pressure side, the work done *on* the gas is $P\,\Delta V = P_1\overline{V}_1$. The work done *by* the gas on the low pressure side is $P_2\overline{V}_2$. The net work is the difference $w_{\text{net}} = P_1\overline{V}_1 - P_2\overline{V}_2$, and

$$\Delta E = E_2 - E_1 = q + w_{\text{net}} = 0 + w_{\text{net}} = w_{\text{net}}$$
$$= P_1\overline{V}_1 - P_2\overline{V}_2$$

Rearranging, we get

$$E_2 + P_2\overline{V}_2 = E_1 + P_1\overline{V}_1$$

or $H_2 = H_1$. The Joule-Thomson expansion is a constant-enthalpy process.

The quantity measured is the temperature change resulting from a pressure drop.

WILLIAM THOMSON, Lord Kelvin (1824–1907), Irish-born British physicist, occupied the chair of natural philosophy at the University of Glasgow with distinction for 53 years, starting in 1846. In 1848, Thomson proposed his absolute scale of temperature, which is independent of the thermometric substance. While Thomson is best known in chemical circles for his contributions to thermodynamics, his work encompassed many important areas. In one of his earliest papers dealing with heat conduction of the earth, Thomson showed that about 100 million years ago, the physical condition of the earth must have been quite different from that of today. He did fundamental work in telegraphy and aided in the development of undersea telegraph cables. This led to work in navigation, including an improved marine compass, sounding apparatus, tide gauges, tide predictors, and simplified navigation tables. For his services in trans-Atlantic telegraphy, Thomson was raised to the peerage, with the title Baron Kelvin of Larg. There was no heir to the title, and it is now extinct.

---

[3] In the experiment, the plug was made of meerschaum, a porous mineral, hydrous magnesium silicate ($H_4Mg_2Si_3O_{10}$), which is used in making bowls for fine pipes.

FIGURE 5.4 Joule-Thomson experiment.

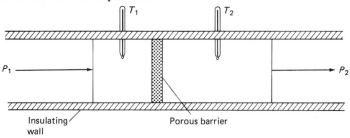

The *Joule-Thomson coefficient* is defined as

$$\mu \equiv \left(\frac{\partial T}{\partial P}\right)_H \tag{5.19}$$

If μ is positive, then the gas cools on expansion; if μ is negative, the gas warms on expansion. The temperature at which the sign changes is called the *inversion temperature*. The Joule-Thomson expansion is important in liquefying gases. Most gases have positive Joule-Thomson coefficients, and hence they cool on expansion at room temperature. A demonstration of the cooling effect can be seen by carefully opening the valve of a tank of compressed CO_2 gas, allowing the gas to expand from the high tank pressure to 1 atm. The cooling effect is often enough to condense water, or even ice, on the valve. This experiment is shown in Figure 5.5. Liquefaction of gases is accomplished by a succession of Joule-Thomson expansions.

The inversion temperature for H_2 is -80 °C. Above the inversion temperature, μ is negative. This means that at room temperature hydrogen warms on expansion. Hydrogen must first be cooled below the inversion temperature (usually with liquid nitrogen) so that it can be condensed to the liquid state by further Joule-Thomson expansion cooling. Joule-Thomson coefficients are functions of both temperature and pressure. Values of μ for several gases are indicated in Table 5.1. For an ideal gas, $\mu = 0$.

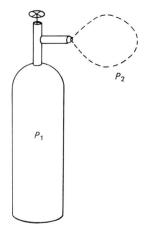

FIGURE 5.5 Expansion of a gas from a high-pressure tank to atmospheric pressure through a valve. The dotted line separates the system from the surroundings. In practice, the dotted line may be a balloon.

TABLE 5.1 Joule-Thomson coefficients of several gases at different temperatures.

T (°C)	He	Ar	N_2	CO_2
300	−0.0060	0.064	0.014	0.265
250	−.0063	0.098	0.033	0.308
200	−.0064	0.138	0.056	0.377
150	−.0065	0.185	0.087	0.489
100	−.0064	0.241	0.129	0.649
50	−.0063	0.322	0.186	0.895
0	−.0062	0.431	0.266	1.290
−50	−.0062	0.596	0.397	2.413
−100	−.0058	0.861	0.649	
−150	−0.0052	1.812	1.266	

NOTE: Entries are μ_{JT} at 1 atm in units of Kelvins per atmosphere. Selected values from Smithsonian physical tables.

5.6 ISOTHERMAL EXPANSION OF AN IDEAL GAS

For an ideal gas, $(\partial E/\partial V)_T = 0$; the energy depends only on the temperature. This means that *for any isothermal process involving an ideal gas*, $\Delta E = 0$. When an ideal gas expands (or contracts) from an initial volume V_1 to a final volume V_2 at constant temperature T, then

$$\Delta E = q + w = 0 \tag{5.20}$$
or
$$q = -w \tag{5.21}$$

This is the starting point. Now let's examine the application of this result to reversible and irreversible processes.

5.7 ISOTHERMAL *REVERSIBLE* EXPANSION OF AN IDEAL GAS

By Equation (5.21), the heat is equal to the work. We have already derived an expression for reversible expansion work in Equation (4.12), and

$$q = -w = \int_{V_1}^{V_2} P \, dV = \int_{V_1}^{V_2} nRT \, \frac{dV}{V}$$
$$= nRT \ln \frac{V_2}{V_1} = nRT \ln \frac{P_1}{P_2} \tag{5.22}$$

Now consider the process from the point of view of the conservation of energy. The energy of the gas is unchanged. The system does work, and the energy for this work must come from somewhere. Where it comes from is the surroundings, and energy enters the system as heat. This is a process in which heat energy is converted into work energy. A numerical example applying Equation (5.22) has been given in Chapter 4. Figure 5.6 indicates a number of *isotherms* on a *P-V* diagram. The work, and also the heat, is equal to the areas under each of the curves.

Now we consider the possibility of using such a conversion for the continual generation of useful work, say to provide the motive power for an automobile. If the system were made infinitely large, then we could in principle construct an infinitely large piston containing an infinitely large amount of gas, whereby we could in one expansion step continually generate work. Any real process must, however, be finite, and the system must be small for a practical automobile engine. Real engines are

FIGURE 5.6 Isotherms of an ideal gas at several temperatures.

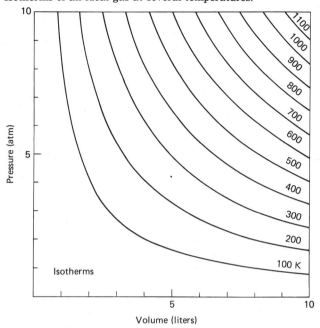

based on cyclic processes, whereby the engine goes through an expansion cycle, then a compression cycle, back to the starting point. The process is repeated over and over, each step producing a certain amount of useful work. We cannot generate useful work from a cyclic process involving isothermal expansion and compression of an ideal gas, because the net useful work is zero. That this is so can be seen by examining Equation (5.22). In the reverse process, the gas is brought from V_2 back to its starting point V_1. The reverse work is the negative of the original forward work, $w_{\text{rev}} = -w_{\text{for}}$, and the reverse heat is likewise equal and opposite to the forward heat, $q_{\text{rev}} = -q_{\text{for}}$. The net work is the sum of w_{for} and w_{rev}, and the net heat is the sum of q_{for} and q_{rev}. The values w_{net} and q_{net} are both zero; we can produce no net useful work in this manner.

5.8 ISOTHERMAL *IRREVERSIBLE* EXPANSION OF AN IDEAL GAS

The starting point for the isothermal irreversible expansion of an ideal gas is the same as for the reversible isothermal process, namely $\Delta E = 0$. Again the heat equals the work, and the arguments relating to the conversion of heat into work are the same. The gas is expanded from V_1 to V_2 at constant temperature T. If the expansion is against the constant pressure P_2 (the final pressure), we have

$$q = -w = \int_{V_1}^{V_2} P_{\text{op}}\, dV = P_2 \int_{V_1}^{V_2} dV = P_2(V_2 - V_1) = P_2\left[\frac{nRT}{P_2} - \frac{nRT}{P_1}\right]$$
$$= nRT\left(1 - \frac{P_2}{P_1}\right) = nRT\left(1 - \frac{V_1}{V_2}\right) \tag{5.23}$$

A numerical example applying Equation (5.23) has been given in Chapter 4.

Two additional points should be made about the irreversible expansion. Firstly, it should be apparent that an irreversible isothermal compression at *constant pressure* is impossible. The force opposing the motion of the piston is the pressure exerted by the gas contained in the cylinder, and this pressure cannot remain constant during the compression. It is the *opposing* pressure that is used in calculating the work.

Secondly, it can be shown that in any process that is going from any given initial state to a given final state, the maximum work is produced in the reversible path. We shall demonstrate this for the general process when we study the second law of thermodynamics. This fact can, however, be demonstrated quite simply for the two specific processes just described by comparing Equation (5.22) to Equation (5.23) and computing the ratio w_{rev}/w_{irr} for the isothermal expansions. We have, letting X be the ratio V_2/V_1,

$$\frac{w_{rev}}{w_{irr}} = \frac{\ln(V_2/V_1)}{1 - V_1/V_2} = \frac{\ln X}{1 - 1/X} = \frac{\ln X}{(X-1)/X} = \frac{X \ln X}{X - 1} \tag{5.24}$$

This last quantity is always greater than unity, as we can see by examining the power series expansion for $\ln X$ (note that for an expansion, $V_2 > V_1$ and $X > 1$):

$$\ln x = \frac{x-1}{x} + \frac{1}{2}\left(\frac{x-1}{x}\right)^2 + \frac{1}{3}\left(\frac{x-1}{x}\right)^3 + \cdots + \frac{1}{n}\left(\frac{x-1}{x}\right)^n + \cdots \tag{5.25}$$

$$\left(x > \frac{1}{2}\right)$$

Hence $w_{rev} > w_{irr}$ for these particular expansions.

5.9 ADIABATIC EXPANSION OF A GAS

For an adiabatic expansion, $q = 0$. No heat flows across the boundary between the system and the surroundings. Interaction between the system and the surroundings can take place only through pressure-volume work. The starting point for *any* adiabatic process is

$$\Delta E = q + w = 0 + w = w \tag{5.26}$$

We can further write

$$dE = nC_v \, dT = dw = -P \, dV \tag{5.27}$$

where n is the number of moles of gas.

5.10 REVERSIBLE ADIABATIC EXPANSION OF AN IDEAL GAS

Equation (5.27) applies to any adiabatic expansion of any gas. If the expansion is reversible, then the internal pressure equals the external pressure, and for an ideal gas we can substitute nRT/V for P. For one mole of ideal gas, Equation (5.27) becomes

$$C_v \frac{dT}{T} = -R \frac{dV}{V} \tag{5.28}$$

SEC. 5.10 REVERSIBLE ADIABATIC EXPANSION OF AN IDEAL GAS

If we take C_v to be constant (a reasonably accurate assumption for real gases, and absolutely accurate for ideal gases) and integrate between the initial and final states, we get

$$C_v \ln \frac{T_2}{T_1} + R \ln \frac{V_2}{V_1} = 0 \tag{5.29}$$

Remembering that for an ideal gas $C_p - C_v = R$, and defining the ratio C_p/C_v to be γ, we can write

$$(\gamma - 1) \ln \frac{V_2}{V_1} + \ln \frac{T_2}{T_1} = 0 \tag{5.30}$$

or

$$\frac{T_1}{T_2} = \left(\frac{V_2}{V_1}\right)^{(\gamma-1)} \tag{5.31}$$

Using Equation (5.31), we can calculate the quantities needed to compute the work from Equation (5.27). For any ideal gas, $T_1/T_2 = (P_1 V_1)/(P_2 V_2)$, and we can put Equation (5.31) into the convenient form

$$P_1 V_1^\gamma = P_2 V_2^\gamma \tag{5.32}$$

For an adiabatic reversible expansion of an ideal gas

$$PV^\gamma = \text{constant} \tag{5.33}$$

In an adiabatic expansion, the gas is cooled. In the reverse process of an adiabatic compression, the signs on all the quantities are reversed; work is done *on* the gas, and the temperature of the gas increases.[4] The expansion process is used in refrigeration systems, in which a cooling effect is desired. The compression process is used in the diesel engine, in which a high compression ratio raises the temperature of the gasoline-air mixture above the ignition temperature.

Consider as an example the reversible adiabatic expansion of one mole of ideal gas from $V_1 = 20$ liters to $V_2 = 40$ liters, starting from an initial temperature $T_1 = 300$ K. We shall take the gas to be monatomic, with $C_v = \frac{3}{2}R$. Using Equation (5.29), we get

$$\frac{3}{2} R \ln \frac{T_2}{300} + R \ln \frac{40}{20} = 0$$

$$\ln \frac{T_2}{300} = -\frac{2}{3} \ln 2 = -0.4621$$

$$\frac{T_2}{300} = \exp(-0.4621) = 0.63$$

$$T_2 = 189 \text{ K}$$

You should verify that Equation (5.31) gives the same result. The values P_1 and P_2 can be found from the ideal gas law, since we now know the initial and final volumes

---

[4] The temperature changes attending adiabatic expansions and compressions were well known in the eighteenth century. According to the old caloric theory, every atom was surrounded by an atmosphere of caloric, the amount of caloric being proportional to the temperature. The atoms attracted each other, whereas the caloric forces were repulsive. The squeezing of the caloric was associated with the temperature increase during an adiabatic compression.

FIGURE 5.7 Adiabats of an ideal gas. The numbers at the ends of the curves are absolute temperatures. The curves are for *reversible* adiabatic processes.

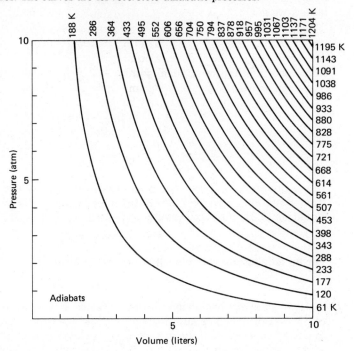

and temperatures. The internal energy change is calculated from Equation (5.27) to be

$$\Delta E = \int_{T_1}^{T_2} dE = nC_v \int_{T_1}^{T_2} dT = C_v(T_2 - T_1)$$
$$= \tfrac{3}{2}(8.314)(189 - 300)$$
$$= -1384 \text{ J} = -331 \text{ cal}$$

Since the work is ΔE, we get $w = -1384$ J. A series of *adiabats* for the reversible adiabatic expansion of an ideal gas is shown in Figure 5.7.

5.11 IRREVERSIBLE ADIABATIC EXPANSION OF AN IDEAL GAS

The ideal gas, initially at P_1, V_1, and T_1, expands irreversibly and adiabatically against a constant external pressure P_2 to the final state P_2, V_2, T_2. The starting point of the solution to the irreversible expansion is the same as for the reversible expansion, namely Equation (5.27), $nC_v \, dT = -P \, dV$. Since P and C_v are constant, we have for one mole of ideal gas

$$C_v(T_2 - T_1) = -P_2(V_2 - V_1) = -P_2\left(\frac{RT_2}{P_2} - \frac{RT_1}{P_1}\right)$$
$$= -R\left(T_2 - \frac{T_1 P_2}{P_1}\right) \tag{5.34}$$

Since we know P_1, P_2, T_1, and C_v, the equation is easily solved for T_2.

For many irreversible processes, it is difficult to calculate ΔE. It is usually easier to perform calculations for reversible processes, and we know that for state functions it makes no difference how we get from any initial state to a final state. The process can be reversible or irreversible, or any combination of the two. The value ΔE will be the same for all processes. If we can dream up a reversible process to accomplish the same change, we can do our calculations on that "new" process, knowing full well that the computed value of ΔE will be equally valid for the original process. Sometimes it is convenient to use the sum of a reversible process and a trivial irreversible process as a two-step path for the "new" process.

Consider as an example the following process equivalent to the irreversible adiabatic expansion noted above. Instead of going to the final state directly, first go to an intermediate state by a free expansion against a vacuum, $P_{ext} = 0$. In this step, $w = 0$. For an ideal gas, $q = 0$, $\Delta E = 0$, and therefore $\Delta T = 0$. The intermediate state will be at V'_1, P'_1, T_1. Since the first step is isothermal $P_1 V_1 = P'_1 V'_1$. Then carry out a reversible adiabatic expansion from the intermediate state to the final state, P_2, V_2, T_2. The value of ΔE calculated for this second part of the two-step "new" process will equal ΔE for the original one-step direct process. In the example of this section, it is easier to calculate the irreversible work directly. In many instances, however, it is impossible to calculate the values directly from the irreversible process, and it is necessary to convert the irreversible process into an equivalent reversible process.

5.12 EXAMPLE: HEATING AT CONSTANT VOLUME AND AT CONSTANT PRESSURE

Consider a mole of ideal gas initially at 1 atm pressure and 300 K. The constant-volume heat capacity is $\tfrac{3}{2}R$; the constant-pressure heat capacity is $C_p = C_v + R$, or $\tfrac{5}{2}R$. We shall approximate R in this problem as 2.0 cal K^{-1} mol^{-1}. Thus $C_v = 3$ cal K^{-1} mol^{-1} and $C_p = 5$ cal K^{-1} mol^{-1}.[5]

Figure 5.8 shows a scheme for adding heat to the mole of gas at constant volume

FIGURE 5.8 (a) Constant-volume heating. $P_{init} = 1$ atm; $T_{init} = 300$ K; $V_{init} = V_{fin}$ (you can calculate it); $T_{fin} = 400$ K; $P_{fin} =$ (you can calculate it; see Problem 5). (b) Energy flow for (a).

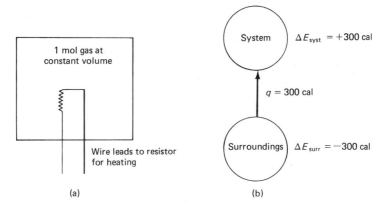

[5] For arithmetical simplicity, we use the gas constant in calorie units rather than joule units in this section. In calories, R can be rounded to 2 cal K^{-1} mol^{-1}, and all the resulting numbers are such that the arithmetic can be done in one's head.

using a resistor. Current is passed through the resistor until 300 cal of heat are dissipated. (Note: Although the resistor is pictured inside the vessel, it is part of the surroundings.) The energy change ΔE is given by $\Delta E = q + w$. Since $w = 0$, $\Delta E = q = 300$ cal $= C_v \, dT$. The final temperature is 400 K.

Now consider the process in which we achieve the same temperature change but at constant pressure. This is shown in Figure 5.9. At constant pressure we need 500 cal of heat to produce the same temperature change, $q = 500$ cal. Now we compute ΔE for the constant-pressure process. As always, $\Delta E = q + w$, so $\Delta E = 500 + w$, and we must compute w. (Note that even though this expansion is against a constant 1 atm external pressure, it is still a reversible process, since it is in equilibrium during every stage; the internal pressure is always equal to the external opposing pressure. The heat can in principle be added very slowly by using small currents.) Since the pressure is constant,

$$dw = -P \, dV = -d(PV)$$

Since $PV = nRT = RT$ for 1 mol gas, we have

$$dw = -d(RT) = -R \, dT$$
and
$$w = -R(T_2 - T_1) = -200 \text{ cal}$$

Thus $\Delta E = 500 - 200 = 300$ cal, exactly the same as the constant-volume case, as it must be, since E is a function only of temperature for an ideal gas. For the enthalpy change, we have

$$\Delta H = \Delta E + \Delta(PV) = \Delta E + \Delta(RT)$$
$$= 300 + 200 = 500 \text{ cal}$$

which is the same as the constant-pressure heat. It is also the same as $C_p \, \Delta T$.

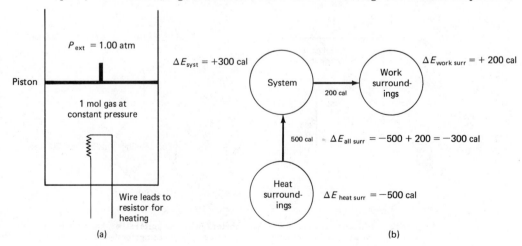

FIGURE 5.9 (a) Constant-pressure heating of an ideal gas. $P_{init} = P_{fin} = 1$ atm; $T_{init} = 300$ K; $T_{fin} = 400$ K; V_{init} = value of V_{init} calculated for Figure 5.8; V_{fin} = (you can calculate it; see Problem 5). (b) Energy-flow diagram for (a). The surroundings have been broken up into two parts, a heat surroundings (the resistor) and a work surroundings (the external air pressure).

Let's consider the energy flow in the example. At constant volume there is no work. We are concerned only with the internal energy of the gas, and the heat flow. Energy in the form of heat flows from the surroundings to the system. The result of this is that E_{surr} decreases by 300 cal and E_{syst} increases by 300 cal; $\Delta E_{net} = 0$. The energy flow is shown in Figure 5.8b.

At constant pressure there is an additional work term of 200 cal. This energy flow is shown in Figure 5.9b. Here 500 cal of heat flow from the surroundings to the system. Initially the surroundings lose 500 cal, and the system gains 500 cal energy. But the system then does 200 cal of work on the surroundings. The net energy gain of the system is still 300 cal; the net loss of the surroundings is still 300 cal.

A change that is the same as that of the constant-pressure process can be accomplished by the following two-stage process. First carry out the constant-volume heating from 300 to 400 K as in the first part of this example. The pressure rises to P'. Then carry out an isothermal reversible expansion from P' to a final pressure of 1 atm. The arithmetic of this two-step process is left for one of the problems.

We could also have carried out the equivalent of the constant-pressure change by a two-step partially irreversible process. First do the constant-volume heating as above for which

$$\Delta E = C_v(\Delta T) = 300 \text{ cal}$$

Then carry out an irreversible free expansion ($P_{ext} = 0$) until the internal gas pressure is 1 atm. By Joule's law, there is no work and no heat involved in the free expansion. The energy change ΔE for this part is zero, and the total ΔE is the same 300 cal. The enthalpy change ΔH is still 500 cal. It makes no difference how the process is carried out. The heat q and the work w may be different for different processes, but ΔE and ΔH must be the same no matter how the process is carried out so long as the initial states are the same for each process and the final states are also the same. Further, dE is always given by $C_v\, dT$, and dH by $C_p\, dT$ regardless of whether the process is at constant volume or constant pressure. (This last statement is correct *only* for ideal gases. There is an additional term arising from $(\partial E/\partial V)_T$ in Equation (5.16) for solids, liquids, and real gases. The derivative is zero only for ideal gases.)

5.13 HEATING REAL SUBSTANCES

Heat capacities of real substances are not independent of temperature, and the heat capacities of thousands of substances have been measured over wide ranges of temperature. The masses of heat capacity data are conveniently handled by equations of the form[6]

$$C_p = A + BT + CT^2 + DT^3 + ET^4 \tag{5.35}$$

where the values of the constants are chosen to provide the best agreement with the experimental values over the relevant temperature range. The constants for numerous substances are listed in Table 5.2. From this kind of tabulation it is possible

---

[6] There is no implication of a special theoretical basis for Equation (5.35). The equation is simply a convenient way of reproducing the smooth curve of experimental heat capacities vs temperature in a brief tabular form.

TABLE 5.2 Heat capacities of substances in the form $C_p = A + BT + CT^2$ (J K^{-1} mol^{-1}) at 1 atm.

Substance	A	$B \times 10^3$	$C \times 10^7$
GASES[a]			
CH$_3$COCH$_3$ (acetone)	22.472	205.97	-635.20
CH$_3$CH$_2$CH$_2$CH$_3$ (n-butane)	18.230	303.56	-926.54
CO$_2$	25.999	43.50	-148.31
CO	26.861	6.97	-8.19
C$_2$H$_6$	9.184	160.17	-460.27
H$_2$S	28.719	16.12	32.84
CH$_4$	14.146	75.50	-179.90
N$_2$	27.372	5.23	-0.03
O$_2$	25.723	12.98	-38.61
C$_3$H$_8$ (propane)	9.447	241.15	-736.12
H$_2$O	30.359	9.61	11.84
H$_2$	29.066	-0.83	20.12
Cl$_2$	31.696	10.14	-40.37
Br$_2$	35.241	4.07	-14.86
HCl	28.166	1.81	15.47
HBr	27.521	4.00	6.61
SO$_2$	32.217	22.2	-34.7
NH$_3$	25.895	33.0	-30.46
SOLIDS[b]			
C (graphite)	$C_p = 11.18 + 0.0109T - 489109/T^2$		
C (diamond)	$C_p = 9.05 + 0.0128T - 545175/T^2$		
Cu	22.76	6.1	
CuS	44.35	11.1	
Pb	24.14	48.8	
PbBr$_2$	75.86	13.	
PbCl$_2$	66.44	35.	
HgCl	46.23	15.5	
HgCl$_2$	64.02	43.1	
Ag	23.43	62.8	
AgBr	35.90	59.0	
AgCl	40.17	38.9	
Na	20.96	22.4	
NaBr	49.12	9.8	
NaCl	45.15	17.6	
S (rhombic)	15.19	26.8	
S (monoclinic)	18.33	18.4	
I$_2$	40.12	49.79	
LIQUIDS[c]			
Hg	27.66		
Br$_2$	113.68		
I$_2$	80.33		
H$_2$O	75.48		

SOURCE: Selected values calculated from tabulations of H. M. Spencer, *J. Amer. Chem. Soc.* 67 (1945):1859; and H. M. Spencer and J. L. Justice, *J. Amer. Chem. Soc.* 56 (1934):2311.

[a] The values for gases are valid for the temperatures 273 K to 1500 K.
[b] The values for solids are valid from 273 K to the melting point.
[c] The values for liquids are valid from melting point to boiling point.

to compute the heat needed to raise the temperature of a substance from T_1 to T_2 at constant pressure. For a mole of substance, the heat is

$$q_p = \Delta H = \int_{T_1}^{T_2} C_p \, dT = \int_{T_1}^{T_2} (A + BT + CT^2 + DT^3 + ET^4) \, dT$$

$$= A(T_2 - T_1) + \frac{B}{2}(T_2^2 - T_1^2) + \frac{C}{3}(T_2^3 - T_1^3)$$

$$+ \frac{D}{4}(T_2^4 - T_1^4) + \frac{E}{5}(T_2^5 - T_1^5) \tag{5.36}$$

PROBLEMS

1. Three moles of ideal gas are isothermally and reversibly expanded from an initial pressure of 10 atm to a final pressure of 1 atm at 300 K. Calculate the heat, work, ΔE, and ΔH.

2. Three moles of ideal gas are isothermally and irreversibly expanded from an initial pressure of 10 atm to a final pressure of 1 atm against a constant external pressure of 1 atm. The temperature is 300 K. Calculate q, w, ΔE, and ΔH.

3. For the reaction

$$H_2(g) + \tfrac{1}{2}O_2(g) = H_2O(l) \quad (1 \text{ atm}, 298 \text{ K})$$

285.84 kJ of heat are evolved per mole of water produced if the reaction is carried out at constant pressure. How much heat would be evolved if the reaction were carried out at constant volume?

4. The heat of vaporization of benzene is 30 kJ mol^{-1}. For what length of time must a current of 1.12 A pass through a resistor of 50 Ω to vaporize 47 g benzene that is kept at the boiling point?

5. Do the arithmetic for the reversible two-step equivalent process at the end of the example in the next-to-last section of the text.

6. (a) At 0 °C, the density of ice is 0.9168 g cm^{-3}, and the density of liquid water is 0.9998 g cm^{-3}. For the process in which 1 mol ice is melted, which is bigger, ΔE or ΔH? By how much?
(b) At 100 °C the density of liquid water is 0.9584 g cm^{-3}. One mole of water is converted to vapor at 100 °C. Which is larger, ΔE or ΔH? By how much? (Assume that the vapor is an ideal gas.)

7. How much work is done on the surroundings if 10 g water is vaporized at the boiling point in a reversible manner at constant pressure (a) at sea level, where the pressure is 760 torr and water boils at 100 °C? (b) in Los Alamos (elevation 2.3 km), where the pressure is 580 torr and water boils at 93 °C?

8. An amount 37 g of N_2 is heated at a constant pressure of 1 atm from 300 K to 500 K. Using the expression for the heat capacity $C_p = A + BT + CT^2$, with the values of the constants taken from Table 5.2, calculate the heat q, the work w, ΔH, and ΔE for the process.

9. An ideal gas has a constant-volume heat capacity $C_v = 1.5R$. From an initial temperature 300 K it undergoes an adiabatic expansion during which its temperature falls to 250 K. How much work did the gas do during the expansion? Calculate ΔE, ΔH, and q for the process.

10. Consumption of oxygen gas has skyrocketed of late for use in the space program as a fuel and in many new steel plants that use oxygen-fired furnaces. Oxygen is manufactured most economically by distilling liquid air. Assume that air is composed of 80 mol % N_2 and 20 mol % O_2. The production of liquid N_2 as a by-product of the process explains the large drop in prices of liquid N_2 over the past 25 years. Assume the value of $C_p = \tfrac{7}{2}R$ is independent of temperature for N_2 and O_2 gas. Go to the library and find (1) the boiling points of N_2 and O_2 and (2) the heats of vaporization of N_2 and O_2.
At a constant pressure of 1 atm, from an initial 273 K in the gaseous state:
(a) How much heat must be removed from 1 mol O_2 to liquefy it?
(b) How much heat must be removed from 1 mol N_2 to liquefy it?
(c) How much heat is involved in liquefying an amount of air that can produce enough oxygen to burn a ton of coal (pure carbon)?
(d) What fraction of this last heat is wasted on the N_2 by-product? (*Note:* it is not completely wasted, since the cold N_2 can be used to precool additional air.)

11. The adiabatic flame temperature is defined as the temperature that would result if all the heat of the reaction went into heating the product gases. Calculate what this temperature is when $CH_4(g)$ is burned; the reaction evolves 890 kJ per mole of CH_4 burned. The products are CO_2 and H_2O, both assumed to be gases. (Remember that air is 80% N_2.) Use the heat capacity data of Table 5.2, omitting the terms in T^2.

12. One mole of ideal gas is initially at 300 K. It is expanded reversibly and adiabatically from an initial pressure of 10 atm to a final pressure of 1 atm. Calculate the final temperature (a) if $C_v = \tfrac{3}{2}R$; (b) if $C_v = \tfrac{5}{2}R$.

13. One mole of ideal gas is initially at 300 K. It is expanded irreversibly and adiabatically from an initial pressure of 10 atm to a final pressure of 1 atm against a constant opposing pressure of 1 atm. Calculate the final temperature if (a) $C_v = \frac{3}{2}R$; (b) $C_v = \frac{5}{2}R$.

14. The compression ratio in a diesel engine is 16:1; that is, the gas in the piston is compressed to $\frac{1}{16}$ its original volume. Take C_p of air to be 27 J K^{-1} mol^{-1} independent of temperature. Assuming the compression stroke of a diesel engine to be reversible and adiabatic, calculate the temperature increase of air in the cylinder during a compression stroke.

15. Assume that the Joule-Thomson coefficients of Table 5.1 are independent of pressure. You have a cartridge containing CO_2 at 70 atm and 30 °C. The cartridge is opened, and the CO_2 expands adiabatically against the constant 1 atm pressure. What is the final temperature? Explain any approximations you made.

16. For an ideal gas, ΔH (and also ΔE) is zero for any isothermal process. The enthalpy change ΔH will usually change for an isothermal process involving a real gas. The Joule-Thomson coefficient for a van der Waals gas is given by

$$\mu_{JT} = \frac{(2a/RT) - b}{C_p}$$

Calculate ΔH for an isothermal compression of 1 mol N_2 from 1 to 250 atm at 300 K. Take C_p to be $4.5R$, independent of temperature.

17. After the compression of Problem 16, the N_2 at 250 atm and 300 K undergoes a Joule-Thomson expansion back to 1 atm. What is the final temperature? Take N_2 to be a van der Waals gas.

18. The boiling point of N_2 is 77 K. What is the required starting pressure to reduce the temperature of N_2 from 300 K to the boiling point by a one-step Joule-Thomson expansion? Use the expression for μ_{JT} of Problem 16.

19. An electric-power generating plant has a capacity of 1000 megawatts (MW). How much work is generated if the plant operates continuously for one year? Suppose you want to produce the same amount of work by a single-stage isothermal expansion of an ideal gas from $P_1 = 1000$ atm to $P_2 = 1$ atm at 300 K. How many moles of gas would you need? If you used a cylinder with diameter 100 m, how long would the cylinder have to be? If the cylinder walls were made of 0.5-cm-thick steel, how much would the cylinder weigh if the density of the steel were 8 g cm^{-3}?

20. The barometric formula gives the pressure as a function of height above a reference point usually taken at sea level. (We shall derive the formula in our discussion of the kinetic theory of gases in Chapter 21.) The formula is

$$P = P_0 \exp\left(\frac{-Mgh}{RT}\right)$$

where P_0 is the pressure at the reference level (1 atm if the reference is at sea level), M is the molecular weight of the gas, h is the height above the reference level, g is the gravitational acceleration constant, R is the gas constant, and T is the temperature. As a mass of warm air rises, it expands. We may consider the expansion to be adiabatic. Further assuming that the expansion is reversible, calculate the change in temperature of an air mass as it rises 1 km, starting from sea level. (This effect is associated with the clouds surrounding mountain peaks. Warm air at lower altitudes rises as it moves over a mountain, and it adiabatically expands as it rises. The cooling effect of the adiabatic expansion is often enough to cool the air to below the dew point, at which time fog will appear. On the other side of the mountain the air mass falls, and the process is reversed. The air is adiabatically compressed, the air warms, and the fog clears. See Philip E. Stevenson, *J. Chem. Ed.* 47 (1970):272.)

21. One mole of ideal gas is contained in a cylinder with a diameter of 20 cm. The gas is initially at 1 atm and 300 K. The gas is contained in the cylinder with a massless piston, and the cylinder is kept with its axis vertical. The cylinder is placed in a large thermostat at 300 K. A 3000-kg weight is gently placed on the piston, and the gas is isothermally and irreversibly compressed. Heat in the amount of 3000 J is absorbed by the thermostat during the compression.
(a) Calculate the initial and final volumes of the gas.
(b) How far does the piston move?
(c) How much work was done on the gas?
(d) As the problem is stated, why is the final volume irrelevant to determining the work in part (c)?
(e) Pressure-volume work is normally given by $dw = -P_{op} dV$. For the work done on the gas, what would P_{op} be? (*Note:* The problem indicates no method of evaluating P_{op}.)
(f) How much gravitational work is done on the 3000-kg weight?
(g) Suppose the 3000 kg had been added to the piston in infinitesimally small increments. How much work would be done in this case?
(h) Carry out an energy-flow balance sheet for the irreversible process and show that there is no net energy change for the system plus the surroundings. (*Note:* Make sure you include gravitational energy for the weight.)
(i) What changes would be introduced into your answers if the weight had been dropped onto the piston from a height of several meters?

CHAPTER SIX

THE PRACTICAL ENTHALPY SCALE. THERMOCHEMISTRY

6.1 HEATS OF REACTION

The heat evolved (or absorbed) in any chemical reaction is an important experimental quantity. The heat evolved by the reaction $C + O_2 = CO_2$ is important to electric-power companies. They are engaged in the business of burning coal and converting the resulting heat into work. The heat of this reaction lets them calculate the amount of coal they must purchase to produce a given amount of electrical work.[1]

In the wine industry, the important reaction is fermentation, in which sugar is converted into alcohol and CO_2 by wine yeasts. The reaction evolves heat. In the large vats used in modern wineries, this heat cannot be dissipated to the atmosphere fast enough to prevent a substantial temperature rise in the vat. This temperature rise has a deleterious effect on the yeast; and if it gets high enough, it will halt the fermentation entirely. Yeast operates best at 60–70 °F. Cooling must therefore be provided by the manufacturer. The heat of the reaction must be known so that the optimum cooling can be provided. Too small a unit will yield poor results, and too large a unit will waste resources.

We have seen that the energy change ΔE for a chemical reaction equals the heat of the reaction at constant volume q_v. This heat can be conveniently measured in a constant-volume bomb calorimeter. We also know that we could invent a new state function, the enthalpy $H = E + PV$; ΔH provides a direct measure of the heat evolved at constant pressure. It is the enthalpy change that is of greater *practical* importance, since most reactions are carried out at constant pressure. We now must face the following problem. For the advancement of science and technology, it is necessary to measure and tabulate the heats evolved by many different types of reactions. Should these tables be published as tables of ΔE or as tables of ΔH? It should be clear (or will eventually become clear) that tables of ΔH will be more useful than tables of ΔE.

We could measure ΔH directly in a constant-pressure calorimeter, as depicted in Figure 6.1. The temperature rise of the water enables us to calculate the heat of the reaction at constant pressure q_p. This procedure would give a direct measurement of

---

[1] For simplicity we shall often assume that coal is pure graphite. Coal is really a complex substance composed of free carbon, organic compounds, and minerals. The heat of combustion of coal (or its *calorific value*) is usually stated in units of Btu lb^{-1}. The calorific value of coals used in the United States varies over the range 7000–14,000 Btu lb^{-1}.

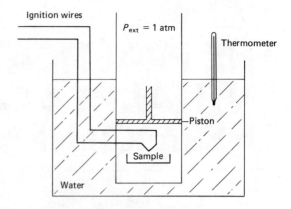

FIGURE 6.1 Constant-pressure calorimeter.

ΔH for the reaction. This experimental procedure is inherently much less accurate than the bomb calorimeter experiment, however. The piston would have to be large, and the frictional resistance between the piston and the cylinder walls would be large. In principle the scheme is fine, but in practice it is much too inconvenient. Our problem then is essentially that we want to publish tables of ΔH but can conveniently measure only ΔE. The solution is simple. We measure ΔE and use the defining relation $H = E + PV$ to convert the value of ΔE to a value of ΔH.

Consider this reaction that is important to diet-conscious people, the combustion of alcohol,

$$C_2H_5OH(l) + 3O_2(g) = 2CO_2(g) + 3H_2O(l)$$

at 298 K. If this reaction is carried out in a bomb calorimeter, the heat evolved is 1364.4 kJ mol^{-1}. Thus $\Delta E = q_v = -1364.4$ kJ mol^{-1}. To convert this number to a value of ΔH, we start with the defining relation $H = E + PV$, and proceed to

$$dH = dE + d(PV) \tag{6.1}$$

The difference is in the PV term. For the moment we shall neglect the effect of solids and liquids on this term and consider only the gaseous products and reactants. If we further assume that all the gases can be treated as ideal gases, then Equation (6.1) becomes

$$dH = dE + d(nRT) \tag{6.2}$$

Since R is a constant, and the temperature can be kept constant, we can write

$$dH = dE + RT\,dn$$

which can be integrated, to yield

$$\Delta H = \Delta E + RT\,\Delta n \tag{6.3}$$

Δn is the change in the number of moles of gases in the reaction. It is equal to the number of moles of *gaseous* products less the number of moles of *gaseous* reactants. In the combustion of alcohol, there are 2 moles of gaseous products and 3 moles of gaseous reactants; thus $\Delta n = 2 - 3 = -1$, and we get

$$\begin{aligned}\Delta H &= -1364.4 + (0.008314)(298)(-1.0) \\ &= -1366.9 \text{ kJ mol}^{-1}\end{aligned} \tag{6.4}$$

This is the heat we should get if alcohol were burned in a constant-pressure calorimeter. Incidentally, this is the reaction that effectively takes place in physiological systems. Tables of calories for dieters contain the results of these kinds of measurements and calculations. The unit that dieticians call the calorie is what we call the kilocalorie:

$$1 \text{ food calorie} = 1 \text{ thermodynamic kilocalorie} = 4.184 \text{ kJ}$$

The correction term we calculated amounts to only about 0.2% of the total number. Any approximations we made apply only to this 0.2%. If the approximation error is as large as 1%, it will introduce an error of only 0.002% to our final number. This is well below the limit of accuracy of the usual experiment. If greater accuracy is required, the correction term can be calculated without these approximations. One of the approximations we made was to neglect the effect of solids and liquids on the P-V term. We can justify that on the basis of the extremely small size of that term relative to gases.[2]

The negative sign of ΔH indicates that heat flows from the system to the surroundings. The surroundings get hotter. We call this an *exothermic* reaction in contradistinction to an *endothermic* reaction, in which ΔH is positive, heat flows from the surroundings to the system, and the surroundings get colder. Finally we may reverse the reaction and write

$$2CO_2(g) + 3H_2O(l) = C_2H_5OH(l) + 3O_2(g)$$

For this reaction, $\Delta H = +1366.9$ kJ mol^{-1}. The signs are simply changed.

At constant volume there is no work, and we measure q_v. At constant pressure there is work in addition to the direct heat of the reaction. The system shrinks in volume, thus the surroundings do work on the system. The work term is counterbalanced by an additional heat flow to the surroundings in the amount $RT\,\Delta n$. For a system that expands, the work and extra heat terms are in the opposite directions. For reactions in which the number of moles of reactant gases equals the number of moles of product gases, ΔH and ΔE will be essentially equal, since the volume does not change very much (see Problem 6 in Chapter 5).

During the reaction the temperature inside the calorimeter may reach thousands of degrees. The final results are reported at 298 K. Is any error introduced by this fact? The answer is a definite unequivocal *No*. The experiment measures ΔE, which is independent of the path linking the initial and the final states. In the alcohol experiment, the initial state is 1 mol of C_2H_5OH (l) and 3 mol of O_2(g) at 298 K. The final state is 2 mol of CO_2(g) and 3 mol of H_2O(l) at 298 K. The path linking the two states is immaterial; intermediate states are immaterial. We are concerned only with the initial and final states.[3] Henceforth when we say "heat" of a reaction, we mean the enthalpy change for the reaction ΔH.

---

[2] A mole of gas occupies about 20,000 cm^3 at normal temperatures and pressures, while a mole of liquid or solid occupies about 20 cm^3. In a reaction in which there is a change from a solid or a liquid to a gas, ΔV is about (20,000 − 20) cm^3. The volume of the solid or liquid can thus be neglected without introducing much of an error.

[3] In actual practice there is a slight warming effect, about 2–5 °C in a bomb calorimeter experiment. This can be corrected for.

6.2 IMPOSSIBLE EXPERIMENTS; THE ADDITIVITY OF ENTHALPY CHANGES

It is impossible to measure many heats of reaction for one reason or another, but the values of the heats are important. For example, there are two forms of carbon, graphite and diamond. Of the two, graphite is the more stable. Graphite is the equilibrium state of carbon at 1 atm and room temperature. Diamond slowly converts to graphite, but we don't worry about this when we purchase diamonds. The rate of the reaction is measured in millions of years. How then can we measure the heat of the reaction C(diamond) = C(graphite)? It is really quite simple. First we burn diamond in a calorimeter and measure its heat of combustion. Then we do the same with graphite, and measure its heat of combustion. The two heats are not the same.

$$C(\text{diamond}) + O_2(g) = CO_2(g) \qquad \Delta H = -395.409 \text{ kJ mol}^{-1} \qquad (6.5)$$
$$C(\text{graphite}) + O_2(g) = CO_2(g) \qquad \Delta H = -393.513 \text{ kJ mol}^{-1} \qquad (6.6)$$

These two can be combined by first reversing (6.6),

$$CO_2(g) = O_2(g) + C(\text{graphite}) \qquad \Delta H = +393.513 \text{ kJ mol}^{-1} \qquad (6.7)$$

and then "adding" (6.5) and (6.7), to get

$$C(\text{diamond}) + O_2(g) + CO_2(g) = C(\text{graphite}) + O_2(g) + CO_2(g)$$
$$\Delta H = (-395.409 + 393.513) \text{ kJ mol}^{-1}$$
$$C(\text{diamond}) = C(\text{graphite}) \qquad \Delta H = -1.896 \text{ kJ mol}^{-1} \qquad (6.8)$$

Consider the reaction

$$C(\text{graphite}) + \tfrac{1}{2}O_2(g) = CO(g)$$

as a second example. It is also impossible to measure this heat of combustion, since we cannot burn carbon to yield only carbon monoxide. Most of the product will be CO_2. We can, however, take pure $CO(g)$ and burn that to measure the heat of combustion of CO:

$$CO(g) + \tfrac{1}{2}O_2(g) = CO_2(g) \qquad \Delta H = -282.990 \text{ kJ mol}^{-1} \qquad (6.9)$$

or when products and reactants are interchanged,

$$CO_2(g) = CO(g) + \tfrac{1}{2}O_2(g) \qquad \Delta H = +282.990 \text{ kJ mol}^{-1} \qquad (6.10)$$

Now we can "add" Equations (6.10) and (6.6) to get

$$C(\text{graphite}) + \tfrac{1}{2}O_2(g) = CO(g) \qquad \Delta H = -110.523 \text{ kJ mol}^{-1} \qquad (6.11)$$

If we could burn 1 mole of graphite to yield CO at constant pressure, 110.523 kJ of heat would be evolved.

Consider what we have done in this CO example. We started from an initial state of 1 mole of graphite and 1 mole of oxygen. The final state was 1 mole of CO_2 gas, but we did not get from the initial state to the final state in one step. We first went to an intermediate state of 1 mole of CO gas and $\tfrac{1}{2}$ mole of oxygen gas. The fact that we went to the final state in two stages was irrelevant as far as the overall value of ΔH was concerned. If we call ΔH_1 the enthalpy change in going from the initial state to the intermediate state, and ΔH_2 the enthalpy change in going from the intermediate state to the final state, then the total enthalpy change is $\Delta H_{\text{tot}} = \Delta H_1 + \Delta H_2$. The value ΔH_{tot} is the heat of combustion of graphite and is given in (6.6). The value ΔH_2

is the heat of combustion of CO gas and is given by (6.9). The value ΔH_1 is what we want to find. Thus

$$\Delta H_{\text{tot}} = \Delta H_1 + \Delta H_2$$
$$-393.513 = -282.990 + \Delta H_1$$

and $\Delta H_1 = -393.513 + 282.990 = -110.523$ kJ mol^{-1}, exactly the number we got in (6.11). This procedure, in which we add chemical reactions algebraically, is known as *Hess's law*. It enables us to compute heats of reactions for many different reactions from a few measured heats of reactions.

Equation (6.11) is a reaction in which a compound is formed from its elements. The heat of such a reaction is called a *heat of formation* and has the symbol ΔH_f. If *all* products and *all* reactants are in the so-called *standard state* of 1 atm pressure, a superscript ° is appended, to give ΔH_f°. Further, the temperature may be specified by an additional subscript, to wit, $\Delta H_{f,298.15}^\circ$. When there is no specifying temperature, it is understood that the temperature is the standard temperature of 298.15 K. These quantities are of vital practical importance.

6.3 "NONREACTIVE" REACTIONS

The word *reaction* is usually associated with a chemical equation in which one group of chemicals (the reactants) is converted into a different group of chemicals (the products). The heat of a reaction is a much more general concept than this. Thermodynamics is concerned with processes, regardless of type, and ΔH as a thermodynamic quantity applies to any process. Equation (6.8), in which diamond is converted to graphite, is a reaction in which the formulas of the chemical species are identical on the product and the reactant sides. Yet we consider this reaction to have properties entirely analogous to the normal kind of reaction. The diamond undergoes a transition to a different crystalline form of carbon, and ΔH is a *heat of transition*. Similarly, the process in which ice is converted to liquid water is a reaction in which water undergoes a change in physical state from solid to liquid; the heat of transition in this "reaction" is a *heat of fusion*. The heat of the reaction in which water is taken from the liquid to the gaseous state is a *heat of vaporization*. An equation for a general type of thermodynamic process could be written as

$$A(P_1, V_1, T_1) = B(P_2, V_2, T_2) \tag{6.12}$$

where the left-hand side of the equation represents the initial state and the right-hand side the final state.

6.4 IMAGINARY STATES

Now we want to construct a table of enthalpy changes that will provide a large amount of information in a small space. It is not necessary to list ΔH for every conceivable reaction, since Hess's law enables us to calculate many reaction heats from a few tabulated heats. We still need a convention for the tabulating. Consider the two reactions

$$H_2O(s, P_1, T_1) = H_2O(l, P_2, T_2) \tag{6.13}$$
$$H_2O(l, P_1, T_1) = H_2O(g, P_2, T_2) \tag{6.14}$$

The number of possible choices of P_1, P_2, T_1, and T_2 for these two reactions is infinite. We want to reduce this infinite number to two. Obviously it helps to make $T_1 = T_2$ and $P_1 = P_2$ so that the initial and final temperatures and pressures are always the same in our tabulating system. A convenient pressure to choose is 1 atm. A convenient temperature is room temperature, or 25 °C = 298.15 K. These are the conditions for our standard tabulating procedure, and they are the temperature and pressure for any substance in the so-called **standard state.** Then we shall develop a series of equations that enable us to calculate values of ΔH at any other temperature or pressure besides these standard state values.

This poses certain conceptual problems. Consider the reaction for the vaporization of water, in which all species are in the standard state:

$$H_2O(l, 1 \text{ atm}, 298 \text{ K}) = H_2O(g, 1 \text{ atm}, 298 \text{ K}) \qquad \Delta H° = 44.014 \text{ kJ mol}^{-1} \qquad (6.15)$$

This is an impossible reaction. Liquid water can exist at 1 atm and 298 K, but water vapor cannot exist at 1 atm and 298 K. This state of gaseous water, which we take to be the standard state, exists purely in our imagination. But the value of ΔH for real states can be calculated from this. A real reaction for the vaporization of water can be written

$$H_2O(l, 1 \text{ atm}, 298 \text{ K}) = H_2O(g, \text{vp}, 298 \text{ K}) \qquad (6.16)$$

where vp is the vapor pressure of water at 298 K. We can calculate ΔH for this reaction from (6.15) in the following two-step process:

$$H_2O(l, 1 \text{ atm}, 298 \text{ K}) = H_2O(g, 1 \text{ atm}, 298 \text{ K}) \qquad \Delta H = 44.014 \text{ kJ mol}^{-1} \qquad (6.15)$$
$$H_2O(g, 1 \text{ atm}, 298 \text{ K}) = H_2O(g, \text{vp}, 298 \text{ K}) \qquad \Delta H = 0 \qquad (6.16a)$$

In Equation (6.16a) we took $H_2O(g)$ to be an ideal gas. This means that $\Delta H = 0$ since the process is isothermal. In this trivial example, the real Equation (6.16) is just the sum of the imaginary Equations (6.15) and (6.16a), yielding the value $\Delta H = 44.014$ kJ mol^{-1} for the real vaporization of water at 298 K. Most cases will not be this simple.

6.5 THE ENTHALPY SCALE

In what follows we take all products and reactants to be in the standard state of 1 atm and 25 °C. Once again we examine Equation (6.6),

$$C(\text{graphite}) + O_2(g) = CO_2(g) \qquad \Delta H_f° = -393.5 \text{ kJ mol}^{-1} \qquad (6.6)$$

where we have written $\Delta H_f°$ because it is a heat of formation, and all products and reactants are in the standard state. If we arbitrarily assign the value $H = 0$ to each of the reactants (elemental) in (6.6), we can write

$$\Delta H_f°(CO_2) = H(CO_2) - H(C) - H(O_2) = H(CO_2) - 0 - 0$$

The enthalpy of $CO_2(g)$ on this artificial scale is just the heat of formation of $CO_2(g)$, which is given in Equation (6.6).

As the enthalpy scale is set up, every element is assigned the value $H = 0$ in its standard state of 1 atm and 25 °C. If there is more than one possible physical state of the element at those conditions, the standard state is taken as the most stable state. The most stable state of carbon is graphite, thus the value $H = 0$ is assigned to C(graphite). For bromine, the standard state is $Br_2(l)$, and for iodine it is $I_2(s)$. Every

SEC. 6.5 THE ENTHALPY SCALE

compound is then assigned an enthalpy value equal to the enthalpy change for the reaction in which the compound is formed from its elements in the standard state. Since these enthalpies are *heats of formation,* we call them by that name. The procedure can be shown by constructing a very brief table of enthalpies (or heats of formation) for the substances C(diamond), $CO_2(g)$, $CO(g)$, and $H_2O(l)$.

The combustion reaction of Equation (6.6) directly provides

$$\Delta H_f^\circ(CO_2) = -393.5 \text{ kJ mol}^{-1}$$

From (6.8), we get $\Delta H_f^\circ(\text{diamond}) = +1.9 \text{ kJ mol}^{-1}$. Finally, (6.11) provides $\Delta H_f^\circ(CO) = -110.5 \text{ kJ mol}^{-1}$. We can now construct the starting table as shown in Table 6.1. The first three entries are elements in their standard states, and by definition $\Delta H_f^\circ = 0$ for these three. In fact, these entries will usually be omitted in the tables. The last entry in the table is for the reaction $H_2(g) + \frac{1}{2}O_2(g) = H_2O(l)$. Our standard heats of formation can be treated as though they were "absolute" enthalpy values. Consider the reaction

$$CO(g) + \tfrac{1}{2}O_2(g) = CO_2(g)$$

at standard temperature and pressure.

$$\Delta H = H(CO_2) - H(CO) - H(O_2) = \Delta H_f^\circ(CO_2) - \Delta H_f^\circ(CO) - \tfrac{1}{2}\Delta H_f^\circ(O_2)$$
$$= -393.5 - (-110.5) - 0.000$$
$$= -283.0 \text{ kJ mol}^{-1}$$

which is the same as the value given in Equation (6.9). The table can now be expanded by adding new reactions and by applying Hess's law. Suppose we want to add alcohol, $C_2H_5OH(l)$, to the table. What we want is ΔH for the reaction

$$2C(\text{graphite}) + 3H_2(g) + \tfrac{1}{2}O_2(g) = C_2H_5OH(l)$$

This reaction is the sum of the following three,

$2C + 2O_2 = 2CO_2$	$\Delta H = 2(-393.5) \text{ kJ}$	(6.17a)
$3H_2 + \tfrac{3}{2}O_2 = 3H_2O(l)$	$\Delta H = 3(-285.9) \text{ kJ}$	(6.17b)
$2CO_2 + 3H_2O(l) = C_2H_5OH(l) + 3O_2$	$\Delta H_{comb} = +1366.9 \text{ kJ}$	(6.17c)

Here the first two reactions are the formation reactions for $CO_2(g)$ and $H_2O(l)$; the third is the combustion reaction for alcohol written in reverse. The sum of the three reactions yields

$$\Delta H_f^\circ(C_2H_5OH) = -\Delta H_{comb}^\circ(C_2H_5OH) + 2\Delta H_f^\circ(CO_2) + 3\Delta H_f^\circ(H_2O) - 3\Delta H_f^\circ(O_2)$$
$$= +1366.9 + 2(-393.5) + 3(-285.9) - 3(0.0)$$
$$= -277.8 \text{ kJ mol}^{-1} \tag{6.18}$$

Table 6.2 lists standard enthalpies of formation obtained in this way.

TABLE 6.1 Starting table for enthalpies of formation.

Compound	$\Delta H_{f,298}^\circ$ (kJ mol^{-1})
$O_2(g)$	0.000
$H_2(g)$	0.000
C(graphite)	0.000
C(diamond)	+ 1.9
CO(g)	−110.5
$CO_2(g)$	−393.5
$H_2O(l)$	−285.8

CHAP. 6 THE PRACTICAL ENTHALPY SCALE. THERMOCHEMISTRY

TABLE 6.2 Standard enthalpies of formation for various substances.

Substance	ΔH_f° (kJ mol^{-1})	Substance	ΔH_f° (kJ mol^{-1})
$O_2(g)$		$P_2(g)$	141.50
$H_2(g)$		$P_4(g)$	54.89
$H_2O(g)$	-241.8264	C(graphite)	
$H_2O(l)$	-285.8400	C(diamond)	1.8962
$F_2(g)$		CO(g)	-110.5233
$Cl_2(g)$		$CO_2(g)$	-393.5127
HCl(g)	-92.312	$CH_4(g)$	-74.848
$Br_2(l)$		$C_2H_2(g)$	226.748
$Br_2(g)$	30.71	$C_2H_4(g)$	52.283
HBr(g)	-36.23	$C_2H_6(g)$	-84.667
$I_2(c)$		$C_2H_5OH(g)$	-235.31
$I_2(g)$	62.241	$C_2H_5OH(l)$	-277.634
HI(g)	25.94	Pb(c)	
$H_2S(g)$	-20.146	$PbCl_2(c)$	-359.20
$SO_2(g)$	-296.90	$PbBr_2(c)$	-277.02
NO(g)	90.374	Hg(l)	
$NO_2(g)$	33.853	$Hg_2Cl_2(c)$	-264.93
$N_2O(g)$	81.55	Ag(c)	
$N_2O_4(g)$	9.661	AgCl(c)	-127.035
$NH_3(g)$	-46.19	AgBr(c)	-99.50
P(g)	314.55	S(rhombic)	
P(c, white)		S(monoclinic)	0.297

SOURCE: *National Bureau of Standards Circular 500.*
NOTE: Standard conditions 1.000 atm and 25 °C obtain.

6.6 EFFECT OF TEMPERATURE ON ΔH

Using Table 6.2, we can calculate enthalpy changes for numerous reactions provided that we are content to restrict ourselves to 298 K. Suppose we wish ΔH for a different temperature. What we need is an expression for the derivative $(\partial \Delta H/\partial T)_p$.

The general case can be indicated as follows:

$$\text{At } T_2: \quad \text{Reactants} \xrightarrow{\Delta H_2} \text{Products}$$
$$\Delta H^r \uparrow \quad \text{Path 2} \quad \downarrow \Delta H^p \quad (6.19)$$
$$\text{At } T_1: \quad \text{Reactants} \xrightarrow{\Delta H_1} \text{Products}$$
$$\text{Path 1}$$

We are given ΔH_1 for the reaction at temperature T_1, and we want to find ΔH_2 for the reaction at T_2. In addition we know that the heat capacities of the reactants and products are given by C_p^r and C_p^p. If we carry out the process by the two different paths indicated in (6.19), then the enthalpy change for the two paths must be the same, and

$$\Delta H_1 = \Delta H_2 + \Delta H^r + \Delta H^p \qquad (6.20)$$

ΔH^r is the heat required to take all the reactants from T_1 to T_2, and ΔH^p is the heat required to take all the products from T_2 to T_1. Remembering that $C_p = (\partial H/\partial T)_p$, we have

$$\Delta H^r = \int_{T_1}^{T_2} C_p^r \, dT; \qquad \Delta H^p = \int_{T_2}^{T_1} C_p^p \, dT \qquad (6.21)$$

In these two steps, the products and reactants are just being heated or cooled. Combining (6.20) and (6.21), we get

$$\Delta H_2 = \Delta H_1 + \int_{T_1}^{T_2} (C_p{}^p - C_p{}^r)\, dT = \Delta H_1 + \int_{T_1}^{T_2} \Delta C_p\, dT \tag{6.22}$$

where ΔC_p is the change in heat capacity. It equals the sum of the heat capacities of all the products less the sum of the heat capacities of all the reactants. Equation (6.22) can be used to calculate ΔH_2 if we know the heat capacities of the products and reactants.

Without going through the three-step process, we could have derived Equation (6.22) directly from the definition of enthalpy changes and heat capacities, as follows:

$$\Delta H = H_\text{products} - H_\text{reactants} \tag{6.23}$$

If both sides of this equation are differentiated with respect to T at constant pressure, we get

$$\left(\frac{\partial\, \Delta H}{\partial T}\right)_p = \left(\frac{\partial H_\text{prod}}{\partial T}\right)_p - \left(\frac{\partial H_\text{react}}{\partial T}\right)_p$$
$$= C_p{}^p - C_p{}^r = \Delta C_p \tag{6.24}$$

which is an equivalent form of (6.22).

For an example, let's calculate the heat of fusion of water at 263 K, given that the heat of fusion at 273 K is 6.010 kJ mol^{-1}. If we take C_p for water and ice to be independent of temperature and equal to 75 and 33 J K^{-1} mol^{-1} respectively, then $\Delta C_p = 75 - 33 = 42$ J K^{-1} mol^{-1}, and Equation (6.22) becomes

$$\Delta H_{263} = \Delta H_{273} + \Delta C_p \int_{273}^{263} dT = 6010 + 42(263 - 273)$$
$$= 6010 - 420$$
$$= 5590 \text{ J mol}^{-1}$$

In most instances there will be more than one product and one reactant, and we need the sums of the heat capacities. In addition, the heat capacities generally depend on temperature, and we must carry out the indicated integration using expressions of the form $C_p = A + BT + BT^2$. An important and practical illustration of the procedures for determining enthalpy changes is the synthesis of ammonia.

6.7 EXAMPLE: MANUFACTURE OF AMMONIA

In the industrial process for making ammonia, nitrogen is caused to react with hydrogen at elevated temperatures to produce the ammonia gas. For both practical and theoretical applications it is important to know the heat of the reaction

$$\tfrac{1}{2}N_2(g) + \tfrac{3}{2}H_2(g) = NH_3(g)$$

at elevated temperatures. We shall calculate the heat of the reaction for 500 °C (or 773 K) using the enthalpies of formation of Table 6.2 and the heat capacities of Table 5.2. At 298 K, the enthalpy change for the reaction is

$$\Delta H^\circ_{298} = \Delta H^\circ_f(NH_3) - \tfrac{1}{2}\Delta H^\circ_f(N_2) - \tfrac{3}{2}\Delta H^\circ_f(H_2)$$
$$= -46.19 - 0 - 0$$
$$= -46.19 \text{ kJ mol}^{-1}$$

From the heat capacity data,

$$\Delta C_p = C_p(NH_3) - \tfrac{1}{2}C_p(N_2) - \tfrac{3}{2}C_p(H_2)$$
$$= 25.895 + 33(10^{-3})T - 30.46(10^{-7})T^2$$
$$\quad - \tfrac{1}{2}[27.372 + 5.23(10^{-3})T - 0.03(10^{-7})T^2]$$
$$\quad - \tfrac{3}{2}[29.066 - 0.83(10^{-3})T + 20.12(10^{-7})T^2]$$
$$= -31.4 + 32(10^{-3})T - 61(10^{-7})T^2$$

$$\int_{298}^{773} d(\Delta H^\circ) = \int_{298}^{773} \Delta C_p \, dT$$

$$\Delta H^\circ_{773} - \Delta H^\circ_{298} = -31.4 \int_{298}^{773} dT + 32(10^{-3}) \int_{298}^{773} T \, dT$$
$$\quad - 61(10^{-7}) \int_{298}^{773} T^2 \, dT$$
$$= -31.4(773 - 298) + (\tfrac{1}{2})32(773^2 - 298^2) \times 10^{-3}$$
$$\quad - (\tfrac{1}{3})61(773^3 - 298^3) \times 10^{-7}$$
$$= (-14{,}900 + 8000 - 900) \times 10^{-3} \text{ kJ mol}^{-1}$$
$$\Delta H^\circ_{773} - (-46.2) = -7.8$$
$$\Delta H^\circ_{773} = -54.0 \text{ kJ mol}^{-1}$$

In problems like this, it is important to keep track of the decimal places. Heat of formation is usually tabulated in kilojoules (or kilocalories) and heat capacity is usually tabulated in joules (or calories). Thus there is a difference of a factor of 1000 between the enthalpies calculated in the first part and the correction to the enthalpy calculated in the second part. In addition, many of the old-style tabulations are oriented towards the older engineering practices, and Btu lb^{-1} or other odd units may be used.

6.8 ENTHALPY AND THE SPONTANEITY OF REACTIONS

Thomsen and Berthelot believed that chemical reactions would proceed spontaneously if ΔH for the reaction were negative. This is true for almost all known reactions If the criterion is to be valid, however, it must be valid in all cases. The reaction

$$Ag(s) + \tfrac{1}{2}Hg_2Cl_2(s) = AgCl(s) + Hg(l)$$

is known to proceed as written, yet $\Delta H = +5.36$ kJ mol^{-1}. This one exception suffices to exclude the sign of ΔH as a criterion for the spontaneity of a chemical reaction. The first law does not contain within it any criterion for the directionality of natural processes. The first law requires that the total energy remain constant, and it does not in any way specify how the energy flows. It is the second law of thermodynamics that specifies how energy flows between system and surroundings and also the state of the system at equilibrium.

PROBLEMS

1. All the following gases can be used as fuels: H_2, CH_4, C_2H_2, C_2H_4, and C_2H_6. When they are burned, the products are $CO_2(g)$ and $H_2O(l)$. Calculate the heat of combustion of each. Which fuel produces the most heat per mole of fuel? per kilogram of fuel?

PROBLEMS

2. Calculate the energy changes in the combustion reactions of Problem 1.
3. From the data in Table 6.2, calculate the enthalpy changes $\Delta H°_{298}$ for the following reactions:

   $S(\text{rhombic}) + O_2(g) = SO_2(g)$
   $2NO_2(g) = N_2O_4(g)$
   $C_2H_4(g) + H_2O(g) = C_2H_5OH(g)$
   $2CH_4(g) = C_2H_6(g) + H_2(g)$
   $H_2S(g) + \tfrac{3}{2}O_2(g) = H_2O(l) + SO_2(g)$
   $H_2(g) + PbCl_2(c) = 2HCl(g) + Pb(c)$

4. Calculate the enthalpy changes $\Delta H°_{298}$ for the following reactions:

   $C_2H_2(g) + H_2(g) = C_2H_4(g)$
   $C_2H_2(g) + 2H_2(g) = C_2H_6(g)$
   $C_2H_4(g) + H_2(g) = C_2H_6(g)$

5. Assume that coal is pure graphite. Coal is burned in the furnace of a power station. The flue gases consist of 99% CO_2 and 1% CO (excluding other constituents such as N_2). What would be the percentage increase in available heat if the air flow were improved so that all the coal was burned to give CO_2. If the coal sells for $20 a ton, and the power station uses 6000 tons a day, how much would it save in a year by improving the air flow? (Use room temperature values of enthalpies.)

6. The amount 0.5000 g benzene, C_6H_6, is burned in a constant-volume bomb calorimeter at 25 °C. The products are $CO_2(g)$ and $H_2O(l)$. Heat is evolved in the amount 20.892 kJ. Calculate the heat of combustion ΔH per mole of benzene burned. Calculate $\Delta H°_{f,298}$ for benzene.

7. At 25 °C, acetylene is converted into benzene in a catalytic flow:

   $3C_2H_2(g) = C_6H_6(l)$

   The process is operated at a rate that produces 100 kg benzene per hour. How much heat must be added or removed per hour to maintain the temperature 25 °C. (*Note:* The standard heat of formation of benzene(l) is 49.028 kJ mol^{-1}.)

8. The formula of sucrose is $C_{12}H_{22}O_{11}$. Sucrose weighing 1.0825 g was burned in a bomb calorimeter, and the temperature of the calorimeter rose by 3.4662 °C. When a current of 1.000 amperes (A) was passed through a resistor of 10 ohms (Ω) in the calorimeter for 1000 seconds (s), the temperature rise of the calorimeter was observed to be 1.9410 °C. Calculate the heat capacity of the calorimeter, the heat of combustion of sucrose, and the standard enthalpy of formation of sucrose. (Ans. $\Delta H°_f = -2206$ kJ mol^{-1})

9. Every afternoon a man eats a candy bar containing 3 oz sucrose. At night he has a drink of whiskey containing 2 oz alcohol. He wants to lose some weight. Is he better off giving up the candy bar or the whiskey? (*Note:* 1 ounce = 28.3 g.) See Problem 8 for sucrose data.

10. Wine is produced when sugars in fruits ferment by the action of various yeasts. Consider the overall reaction to be

    $C_{12}H_{22}O_{11}(\text{aq}) + H_2O(l)$
    $\qquad = 4C_2H_5OH(l) + 4CO_2(g)$

    Calculate the heat of reaction. The yeast operates most efficiently at about 20 °C. A winery plans to build a large fermentation vat in which the production of alcohol will reach 15 kg per hour. How large a refrigeration unit will be needed to remove the heat generated by the process? (Assume that the standard heat of formation of sucrose in solution is the same as for the pure crystal. See Problem 8.)

11. Calculate the heat of the reaction $H_2(g) + \tfrac{1}{2}O_2(g) = H_2O(g)$ at 1500 K.

12. From Table 6.2, calculate the heat of vaporization of water at 25 °C. Using the heat capacity data, calculate the heat of vaporization of water at 100 °C.

13. Calculate $\Delta H°_{1000}$ for the reaction

    $\tfrac{1}{2}H_2(g) + \tfrac{1}{2}Br_2(g) = HBr(g)$

14. In normal, quiet respiration, a person will breathe at the rate of 14 breaths a minute, and take in 0.5 liter air with each breath. The air inhaled is 20.94% O_2 by volume. The air exhaled is 16.30% O_2. The rest of the gases are water vapor, CO_2, N_2, and inert gases. Assume that a person derives all the energy required daily by burning sucrose (this is a gross simplification; other carbohydrates, fats, and other processes are involved).
    (a) How much sucrose does the individual burn in a day?
    (b) How much heat is produced in a day from the process?
    The relevant data are given in Problem 8. (The correct answer is within a factor of about 1.5 of the actual value of 2000–2500 kcal per day.)

CHAPTER SEVEN

THE SECOND LAW OF THERMODYNAMICS: HEAT ENGINES, ENTROPY, AND THE DIRECTION OF CHEMICAL CHANGE

7.1 HEAT ENGINES

The first law of thermodynamics was developed by extending the principle of the conservation of energy to include heat as a form of energy. The relevant observations dealt with the quantitative conversion of work into heat. The second law of thermodynamics arose from a study of the reverse process, the conversion of heat into work.

The practical conversion of heat into useful work requires a cyclic process if work is to be produced continuously. The first person to convert heat into work successfully was James Watt; he built the first practical steam engine in 1769 and laid the foundation for the rapid industrialization of Europe and America during the nineteenth century.[1] The steam engine is conceptually a simple machine. Heat is generated by burning coal or any other convenient fuel. The heat is used to convert water into steam, and the steam is expanded against a piston, doing work in the process. We need not concern ourselves with the actual mechanical components of the engine such as valves, linkages, and other matters important to engine designers. We shall, however, be concerned with the following aspects of the engine:

Boiler, or hot reservoir: the high-temperature region through which heat is introduced into the system.

---

[1] In Watt's engine, steam was introduced into the cylinder and condensed, creating a partial vacuum. The power stroke was provided as the atmosphere pushed against the lower pressure in the cylinder. Useful work is obtained by connecting the piston to another moving object, such as the drivewheel of a locomotive, through a system of cogs, gears, and other mechanical linkages. One big technological problem Watt solved was the practical conversion of linear motion into circular motion. The piston moves up and down, and this linear motion had to be converted into the circular motion of the wheel. Solving this problem was an important contribution by Watt, and it enabled him to construct the engine.

> JAMES WATT (1736–1819), Scottish engineer, improved the steam engine to the point of making it a practical device. He was an intimate friend of Joseph Black, the discoverer of latent heat. In Watt's engine, steam was introduced into the cylinder and condensed, creating a partial vacuum. The power stroke was provided by the pushing of the atmosphere against the lower pressure in the cylinder. In the modern steam engine, the power stroke is provided by the high-pressure steam.

Condenser, or cold reservoir: the low-temperature region in which the steam is condensed.

Working fluid: steam (in the case of the steam engine) or any fluid capable of expansion.

Engine: for our purposes this will be a "black box," which takes heat from the hot reservoir, converts some of it into work, and discards the remainder to the cold reservoir.

The final aspect, of critical importance, is the *efficiency e* of the engine. This is defined as

$$e = \frac{w_{\text{output}}}{q_{\text{input}}} \tag{7.1}$$

Efficiency is the ratio of the work output to the heat put into the system.

The operation of a general heat engine is shown in Figure 7.1. Heat q_2 is taken from the hot reservoir at temperature t_2 and introduced into the engine. The engine does work w and discards heat q_1 to the cold reservoir at temperature t_1. Conservation of energy requires that

$$-w = q_2 - q_1 \tag{7.2}$$

and the efficiency of the engine is then

$$e = \frac{-w}{q_2} = 1 - \frac{q_1}{q_2} \tag{7.3}$$

FIGURE 7.1 Heat engine. Heat q_2 is passed from the hot reservoir to the engine. The engine delivers work $-w$ and discards heat q_1 to the cold reservoir.

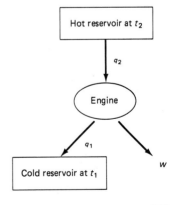

Conservation of energy then requires that $e \leq 1$. The maximum possible efficiency is unity; this requires that $q_1 = 0$ (or $q_2 = \infty$). We shall soon see that this is impossible and that the efficiency of a *cyclic* heat engine must always be less than unity.

Here, to indicate temperature, we have used lowercase t and not the uppercase T as in previous chapters. This is because we have not yet defined a thermodynamic temperature scale.[2] The second law of thermodynamics enables us to define a natural temperature scale based on the efficiencies of steam engines. When we have shown that the natural temperature scale is identical to the ideal gas temperature scale used previously, we shall revert to uppercase T for temperature.

An engine is a device that converts energy from one form to another. A heat engine converts heat to work. Not all engines are heat engines. The electric motor is a cyclic device that converts electrical work to mechanical work. The human body converts chemical work to mechanical work. The battery converts chemical work to electrical work. The water wheel and the weight-driven grandfather clock convert potential energy to mechanical work. The electric generator converts mechanical work to electrical work. In all cases, the efficiency is defined as the ratio of useful work output to energy input. The expressions for the theoretical efficiencies may be different for different types of energy-conversion devices. In some cases, the theoretical efficiency may be unity. For the moment, we shall restrict our discussion to heat engines and the consequences of the efficiency of this particular type of engine.

A refrigerator is a heat engine in reverse. Suppose we reverse all the arrows in Figure 7.1. We get the result shown in Figure 7.2. Now we put work *into* the engine. This takes heat q_1 from the cold reservoir and puts out heat q_2 into the hot reservoir. Again $q_2 = -w + q_1$, but here the heat flows from the cold to the hot reservoir. This is what occurs in the usual household refrigerator. The cold reservoir is the inside of the refrigerator. The engine is the compressor, and the source of work input to the compressor is the electric motor. The hot reservoir is the kitchen. The engine causes heat to flow from the cold reservoir to the hot reservoir, and the inside of the refrigerator gets colder at the expense of heating the kitchen. Now let's examine the flow of heat between hot and cold objects.

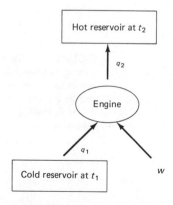

FIGURE 7.2 The heat engine in reverse, acting as a refrigerator or a heat pump. Work w enters the engine and causes heat to flow from the cold reservoir to the hot.

---

[2] The ideal gas temperature scale is a practical scale that is based on a particular equation of state. The natural temperature scale that results from the second law of thermodynamics is independent of the thermometric material.

7.2 IMPOSSIBLE HEAT FLOW

Consider two blocks of copper, each of equal mass. Let one of them be at 100 °C and the other at 0 °C, and let us assume that the heat capacity is independent of temperature. The heat capacities of the two blocks are thus the same. What happens when the two blocks are brought into thermal contact? Relative to the first law of thermodynamics, there are many possibilities for the final result, as for example,

$$Cu(0°) + Cu(100°) = Cu(-10°) + Cu(110°) \tag{7.4}$$
$$Cu(0°) + Cu(100°) = Cu(10°) + Cu(90°) \tag{7.5}$$
$$Cu(0°) + Cu(100°) = 2Cu(50°) \tag{7.6}$$

The store of human experience indicates that the first possibility given by Equation (7.4) is impossible. The cold block does not of itself get colder at the expense of the hot block's getting hotter. The second possibility, indicated by (7.5), is possible provided that the blocks are separated before they achieve equilibrium. The third possibility, expressed by (7.6), is also possible, and it is the final result if the blocks are left in contact for a long time. All three possibilities are clearly valid within the first law of thermodynamics, but the first is invalid within human experience.

Suppose we reverse the three processes. It should be apparent that the reverse of the impossible process is a possible process, whereas reversing the two possible processes results in impossible processes. What we are looking for is some valid criterion lying outside the first law of thermodynamics that will tell us the order in which to write the relevant processes so that they describe possible spontaneous processes.

A process in which we start with two blocks of Cu, each at 50 °C, and end with one block at 40° and the other at 60° does not occur by itself. It is not a spontaneous, but an unnatural process. We can, however, achieve this result by introducing outside work. Consider two equal ideal gas samples in thermally isolated cylinders, both initially at 50 °C and at different pressures. Suppose that the total heat capacity of the gas samples equals the heat capacity of the copper blocks. Then

1. Carry out a reversible adiabatic expansion on one of the gas samples so that its temperature falls to 30 °C. It does work $w = C_v \, \Delta T$.
2. Use the work to compress the gas in the second cylinder reversibly and adiabatically. Its temperature will rise to 70 °C.
3. Bring each of the gas samples into thermal contact with one of the blocks.

The final result is that the first gas sample and one of the blocks is at 40 °C, while the second gas sample and the other block are at 60 °C. The process is outlined in Figure 7.3 and discussed further in Problem 1.

We have accomplished what we set out to do. We caused one block to fall in temperature to 40 °C while the other rose to 60 °C. But that is the end. To complete the cycle, we must introduce work to bring the gases back to their original state; this would be a *practical* refrigerator. The aim is to reduce this work to as low a level as possible, in fact to zero. This leads to the principle of Clausius, which states the impossibility of reducing the work to zero:

> *It is impossible to construct a* cyclic *engine that will produce no other effect than to transfer heat from a cold object to a hot object.*

The forced flow of heat from a cold body to a hot body by a *cyclic* process must involve converting some work into heat. The key word is *cyclic;* we have just seen that the change is feasible with a one-stage process.

FIGURE 7.3 Method for cooling one block and heating a second. (IG = ideal gas.) The two copper blocks have equal masses and equal heat capacities. The cylinders contain equal numbers of moles of ideal gas, and the heat capacity of the ideal gas is the same as the heat capacity of the copper blocks. (a) Initial state. (b) The gas of higher pressure is allowed to expand adiabatically and reversibly while compressing the gas of lower pressure. The pressure of IG^1 decreases from P_a^1 to P_b^1 and its temperature falls from 50 to 30 °C. At the same time, the pressure of IG^2 increases from P_a^2 to P_b^2 while its temperature increases from 50 to 70 °C. The copper blocks are unchanged. (c) Each of the cylinders is brought into thermal contact with one of the copper blocks, resulting in a final temperature of 40 °C for IG^1 and one of the copper blocks and of 60 °C for IG^2 and the other copper block. In the final step, the volumes of the cylinders are kept constant, and the final pressures will be P_c^1 and P_c^2. The two cylinders are thermally insulated in (a) and (b). The insulation is then removed for the final step in (c). (See Problem 1.)

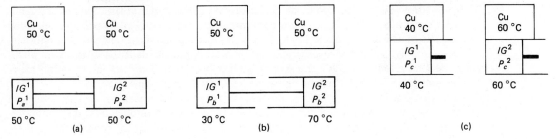

7.3 IMPOSSIBLE HEAT ENGINES

Consider a steam engine with the boiler and the condenser at the same temperature. Is such a device possible? What are the ramifications? A steamship, call it the S.S. *Thermo*, could sail across the ocean by extracting heat from the ocean water at temperature t and converting the heat to work in the ship's cyclic engine. The process does not violate the first law of thermodynamics. It would be the solution to the energy crisis, since the ocean is an infinite source of heat (being replenished by solar radiation). Unfortunately it will not work. Many have tried to find such a process; all have failed. We shall take it as a fundamental principle that such an engine is impossible. This impossibility is expressed by the principle of Thomson:

> *It is impossible to construct a* cyclic *engine that will produce no other effect than to extract heat from a reservoir and convert it into an equal amount of work.*

Such an impossible engine is known as a *perpetual motion machine of the second kind*. The United States Patent Office has received patent applications for many examples of unsuccessful attempts to construct engines of this type.[3]

7.4 THE SECOND LAW OF THERMODYNAMICS

Either the principle of Clausius or the principle of Thomson can be used as an expression of the second law of thermodynamics. Either one will enable us to derive quantitative statements about the theoretical efficiencies of heat engines and to ob-

---

[3] The United States Patent Office no longer accepts patent applications for perpetual-motion machines.

> RUDOLF JULIUS EMMANUEL CLAUSIUS (1822–1888), German mathematical physicist, is perhaps best known for his statement of the second law of thermodynamics in the form "Heat cannot of itself pass from a colder to a hotter body," which he presented to the Berlin Academy in 1850. He also made fundamental contributions to the field of the kinetic theory of gases and anticipated Arrhenius by suggesting that molecules in electrolytes continually exchange atoms.

tain that most elusive of thermodynamics functions, the *entropy*. The entropy is what provides a measure of the spontaneity of processes such as those expressed in Equations (7.4) to (7.6).

We have written the principles of Clausius and of Thomson (and hence the basic statement of the second law of thermodynamics) as though they should be obvious to the meanest intellect. This assumption is not necessarily true, and if they do not appear obvious to you at this time don't worry—at least not yet. We shall adopt a plan. Let us for the moment assume that the two principles are indeed correct and examine the consequences of this assumption. If the consequences we arrive at agree with what we observe in the real world, then we can proceed under the assumption that the two principles are correct. If at any stage we find a real contradiction, then we shall have to go back and adjust our starting hypothesis. This is the way any scientific theory proceeds.

We really are not interested in steam engines. We are interested in processes such as

$$\tfrac{1}{2}N_2(g) + \tfrac{3}{2}H_2(g) = NH_3(g) \tag{7.7}$$

Which way does this reaction proceed spontaneously, and to what extent? How does one adjust the reactants to obtain the highest yields of NH_3? These are vital chemical concerns, and we shall reach a point at which we can answer these questions. While the path may be devious, let us not lose sight of our final goal. We must first consider the consequences of the second law of thermodynamics as applied to the efficiencies of steam engines.

7.5 THE CARNOT CYCLE

The first attempt to describe quantitatively the process in which heat is converted into useful work was in 1824 by the French engineer Sadi Carnot. In the *Carnot engine,* heat is transferred from a hot reservoir at temperature t_2, partly converted to work, and partly discarded into a cold reservoir at t_1. It is a cyclic engine; the engine

> NICOLAS LÉONHARD SADI CARNOT (1796–1832) was a French military engineer. His only published work was *Réflexions Sur la Puissance Motrice du Feu et sur les Machines Propres à Développer cette Puissance* (1824), in which he discussed the conversion of heat into work and laid the foundation for the second law of thermodynamics. He was the scion of a distinguished French family that was very active in political and military affairs. His nephew, Marie François Sadi Carnot (1837–1894), was the fourth president of the Third French Republic.

FIGURE 7.4 Four steps of a Carnot cycle. (a) *Step 1*, Isothermal expansion at t_2. (b) *Step 2*, Adiabatic expansion to t_1. (c) *Step 3*, Isothermal compression at t_1. (d) *Step 4*, Adiabatic compression to initial t_2 and V_1.

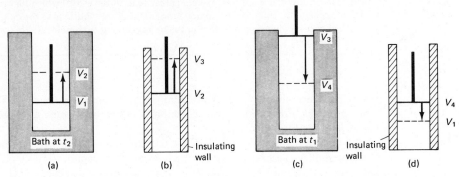

is returned to its initial state after one cycle. We use a gas as the working substance of our engine; thus the work is ordinary expansion work. We assume the process to be reversible and consider the cycle to be composed of four reversible steps:[4]

Step 1. Isothermal expansion at temperature t_2.
Step 2. Adiabatic expansion to temperature t_1.
Step 3. Isothermal compression at temperature t_1.
Step 4. Adiabatic compression back to the original t_2.

The four steps are shown in Figure 7.4. In Step 1, the cylinder, in thermal contact with the hot reservoir at t_2, is reversibly and isothermally expanded from its initial volume V_1 to V_2; the gas absorbs heat q_2 from the reservoir and does work w_1 in this first step. In Step 2, the gas is reversibly and adiabatically expanded from V_2 to V_3, $q = 0$, and the temperature falls from t_2 to t_1; the work done by the gas in the second step is w_2. In Step 3, the gas is reversibly and isothermally compressed from V_3 to V_4. Heat q_1 is absorbed by the cold reservoir, and work w_3 is done *on* the gas.[5] In Step 4, the final one, the gas is reversibly and adiabatically compressed, the temperature rising to the original t_2, and the volume going from V_4 to the original V_1; $q = 0$, and the work done *on* the gas is w_4. Since the engine and the working fluid are returned to the original state, the process is cyclic and can be repeated. There is no restriction on the initial volume. Likewise, V_2 and V_3 are not subject to any other constraints except practicality. The volume V_4, however, must be chosen so that the system will return to its initial state in Step 4. The complete cycle is indicated on a P-V diagram in Figure 7.5 for the particular case of the ideal gas Carnot cycle. The total work is the area enclosed by the isotherms and adiabats. Now let's see how the first law of thermodynamics relates to the energetics of the Carnot cycle.

---

[4] The condition reversibility is required in this discussion because the reversible process always produces more work than the irreversible process; hence the efficiency will be greater for the reversible process. We have not yet demonstrated this fact in general, but we have demonstrated it for expansion work involving an ideal gas (see Equation (5.24)). We shall show the general case in Chapter 9.

[5] We have written q_1 as a positive quantity. Remember that if we refer to a heat flow with respect to the system, the heat is a negative quantity.

SEC. 7.6 ENERGETICS OF THE CARNOT CYCLE

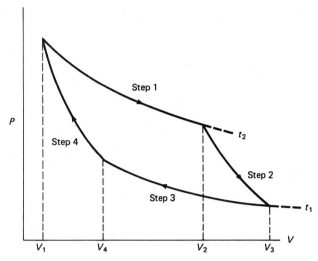

FIGURE 7.5 Carnot cycle on a *P-V* diagram.

7.6 ENERGETICS OF THE CARNOT CYCLE

Throughout this discussion remember that the system is the working fluid. The surroundings comprise the two temperature reservoirs, hot and cold, and the mechanical system to which the piston is connected to produce work. This mechanical system we can call the *work source;* for reversible work, we call it the *reversible work source.* The cylinder and the piston compose the boundary between system and surroundings. Since the process is cyclic, the initial and final states of the system are the same, and for one complete cycle

$$\Delta E_{\text{syst}} = 0 \tag{7.8}$$

Remembering that q_2 was the heat flow to the system from the hot reservoir, and q_1 was the heat flow from the system to the cold reservoir, we find that the total heat flow to the system is

$$q = q_2 - q_1 \tag{7.9}$$

The total work done by the system in one cycle is

$$w = -w_1 - w_2 + w_3 + w_4 \tag{7.10}$$

Thus for one cycle

$$\Delta E = 0 = q + w = q_2 - q_1 + w \tag{7.11}$$

or
$$-w = q_2 - q_1 \tag{7.12}$$

The net work done by the engine is just the heat taken from the hot reservoir less the heat returned to the cold reservoir. The efficiency is

$$e = \frac{-w}{q_2} = \frac{q_2 - q_1}{q_2} = 1 - \frac{q_1}{q_2} \tag{7.13}$$

Now let's try to build another engine that operates between the same two temperatures but has a higher efficiency. The implications of our failure will be very significant.

7.7 IMPOSSIBLE CARNOT ENGINES

We call the engine discussed in the previous section the *original engine*. Its efficiency is given in Equation (7.13). Its operation is indicated schematically in Figure 7.1; and Figure 7.2 shows the engine operating in reverse as a refrigerator.

IMPOSSIBLE ENGINE NO. 1; HEAT FLOW

Now we construct a second engine that operates between the same two temperatures. We assume that this second engine has a greater efficiency than the original engine, perhaps because of better mechanical design or a better working fluid. Quantities associated with this second engine are denoted by primes. This second engine has efficiency $e' > e$. We operate both engines in sequence for one cycle—the new engine in the forward direction, and the original engine in the reverse direction. Further, we arrange the steps of the cycle so that the work of each engine is the same, $w = w'$. Consider the result.

Step 1. New engine in the forward direction:

$$-w' = -w = q_2' - q_1' \qquad (7.14a)$$

Step 2. Old engine in the reverse direction:

$$+w = -q_2 + q_1 \qquad (7.14b)$$

The net result is the sum of both steps, or

$$-w' + w = 0 = -(q_2 - q_2') + (q_1 - q_1') \qquad (7.15)$$

Our original assumption that $e' > e$ requires that

$$\frac{-w'}{q_2'} = \frac{-w}{q_2'} > \frac{-w}{q_2}$$

This is possible *only* if $q_2' < q_2$; thus $(q_2 - q_2')$ in (7.15) must be a positive quantity, and the negative sign preceding it indicates that it is a quantity of heat that flows from the engine to the hot reservoir. On the other hand, the quantity $(q_1 - q_1')$ is a positive quantity, and it indicates heat being transferred from the cold reservoir to the engine.

The net result of coupling the two engines is to transfer a quantity of heat $q = q_1 - q_1' = -(q_2 - q_2')$ from a cold reservoir to a hot reservoir by a *cyclic* process with *no other change in the universe*. No work has been expended. We could repeat the process over and over, continually causing heat to flow "uphill." Think of the consequences. We could use the refrigerator as the cold reservoir and the stove as the hot reservoir. Without expending any work we could operate the refrigerator and the stove! We can tell the electric company to disconnect our power. Unfortunately, the scheme is impossible. It violates human experience. It violates the second law of thermodynamics as embodied in the principle of Clausius. Our original assumption was not valid. **We cannot build a more efficient engine** in this way.

IMPOSSIBLE ENGINE NO. 2; PERPETUAL MOTION

Let us make one more attempt to build a more efficient engine. This time we assume that the heat input into each engine is the same, $q_2' = q_2$. For our hypothetical new engine, $e' > e$; therefore $w' > w$. Again we operate the engines in sequence. First the new, more efficient engine in the forward direction, then the original engine in reverse. Again we consider the result.

Step 1. New engine in the forward direction:
$$-w' = q_2' - q_1' = q_2 - q_1' \tag{7.16a}$$

Step 2. Original engine in reverse:
$$+w = -q_2 + q_1 \tag{7.16b}$$

Adding the two steps, we get
$$-w' + w = (q_1 - q_1') > 0 \tag{7.17}$$

The net result of coupling the two engines is to extract a quantity of heat $q = q_1 - q_1'$ from a reservoir at constant temperature and convert it into useful work $-(w' - w)$ by a *cyclic* process with *no other change in the universe*. Such an engine is impossible. It violates human experience. It contradicts the second law of thermodynamics as embodied in the principle of Thomson. Again we reach an absurd result and must conclude that our original assumption that $e' > e$ must be incorrect. **We cannot build a more efficient engine** in this way.

7.8 THE EFFICIENCY OF A CARNOT ENGINE AND THE THERMODYNAMIC TEMPERATURE SCALE

The results of the previous section lead inexorably to the conclusion that

all reversible Carnot cycles operating between the same two temperatures must have the same efficiency.

For reversible cycles this efficiency will be the maximum possible. It is independent of the working fluid; it is independent of the individual steps in the cycle; it depends only on the temperatures of the hot and cold reservoirs. We can therefore write

$$e = f(t_2, t_1) = \frac{-w}{q_2} = \frac{q_2 - q_1}{q_2} \tag{7.18}$$

where $f(t_1, t_2)$ is some function of the two temperatures. Since $e = 1 - q_1/q_2$, the ratio q_1/q_2 is also a function only of the temperature:

$$\frac{q_1}{q_2} = 1 - e = g(t_2, t_1) \tag{7.19}$$

Now consider the three isotherms in Figure 7.6 at temperatures t_3, t_2, and t_1. We can operate three different Carnot cycles using these isotherms: one between t_3 and t_2, a second between t_2 and t_1, and a third between t_3 and t_1. Each of the three cycles is indicated by a loop on the figure. For the cycles, we can write

Cycle 1. $\quad \dfrac{q_3}{q_2} = g(t_3, t_2)$

Cycle 2. $\quad \dfrac{q_2}{q_1} = g(t_2, t_1)$

Cycle 3. $\quad \dfrac{q_3}{q_1} = g(t_3, t_1)$

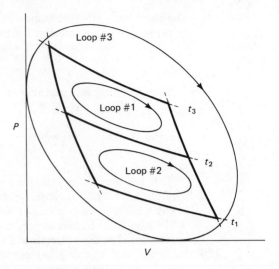

FIGURE 7.6 Two Carnot cycles sharing the isotherm t_2.

Thus
$$g(t_3, t_1) = g(t_3, t_2) \cdot g(t_2, t_1) \qquad (7.20)$$

Equation (7.20) must be valid for any arbitrary value of t_2. This is possible only if $g(t_i, t_j)$ has the form

$$g(t_i, t_j) = \frac{G(t_i)}{G(t_j)} \qquad (7.21)$$

where G is a function of the single variable t. Thus we can write

$$\frac{q_2}{q_1} = \frac{G(t_2)}{G(t_1)} \qquad (7.22)$$

Equation (7.22) can be used as the basis for a *thermodynamic temperature scale*.[6] Lord Kelvin was the first to formulate this basis; he simply took the function $G(t)$ to be the "temperature" and called it T. Thus we have $q_2/q_1 = T_2/T_1$, and (7.18) can be rewritten as

$$e = \frac{q_2 - q_1}{q_2} = \frac{T_2 - T_1}{T_2} \qquad (7.23)$$

This temperature scale has a natural zero-point origin. It is the temperature at which a cold reservoir would have to be maintained so that an efficiency of unity for a Carnot engine could be achieved. When we consider the third law of thermodynamics, we shall see that this zero temperature is unattainable.

Only one point must be defined for this thermodynamic temperature scale. Once that point is fixed, the size of the unit degree is determined. The present scale is defined so that the triple point of water is at exactly 273.16000. . . K. The triple point of water is that point at which liquid, solid, and gaseous water are in equilibrium.

---

[6] This temperature scale, although it is fundamentally important, is not a practical measuring scheme. One could, in principle, compare temperatures by constructing a Carnot engine and measuring the efficiency of the engine when operating between an unknown temperature and a particular standard temperature. In practice, the problems are insurmountable. We could not build the perfect engine. Even if we could, the effect of the engine on any finite temperature reservoir would be such as to alter the temperature.

7.9 THE IDEAL GAS CARNOT CYCLE

When we consider the phase rule, we shall see that the temperature of this point is indeed fixed. The normal freezing point of water is about 0.01 K lower than the triple point. By introducing T as the symbol for temperature on the thermodynamic temperature scale, we have been looking ahead. We have also used T as the symbol for temperature on the ideal gas scale. We are now in a position to demonstrate that the two scales are identical.

7.9 THE IDEAL GAS CARNOT CYCLE

Since the efficiencies of all Carnot engines operating between the same two temperatures are the same regardless of the working fluid, we can simplify matters by taking the working substance to be an ideal gas. The relevant formulas for reversible isothermal and adiabatic expansions of ideal gases have been derived in Chapter 5. Applying these formulas to the four-step Carnot cycle, we get:

Step 1. Isothermal expansion: $-w_1 = q_2 = RT_2 \ln \dfrac{V_2}{V_1}$

Step 2. Adiabatic expansion: $-w_2 = -\int_{T_2}^{T_1} C_v \, dT; \, q = 0$

Step 3. Isothermal compression: $w_3 = q_1 = RT_1 \ln \dfrac{V_4}{V_3}$

Step 4. Adiabatic compression: $w_4 = \int_{T_1}^{T_2} C_v \, dT; \, q = 0$

The total work done by the gas is then

$$-w = -w_1 - w_2 + w_3 + w_4 = -w_1 + w_3 = RT_2 \ln \frac{V_2}{V_1} + RT_1 \ln \frac{V_4}{V_3} \quad (7.24)$$

Remembering that for a reversible adiabatic expansion of an ideal gas

$$C_v \ln \frac{T_{\text{fin}}}{T_{\text{init}}} + R \ln \frac{V_{\text{fin}}}{V_{\text{init}}} = 0$$

we have

$$C_v \ln \frac{T_1}{T_2} + R \ln \frac{V_3}{V_2} = 0 = C_v \ln \frac{T_2}{T_1} + R \ln \frac{V_1}{V_4} \quad (7.25)$$

Hence $V_4/V_1 = V_3/V_2$, and $V_2/V_1 = V_3/V_4$. The total work is now

$$-w = R(T_2 - T_1) \ln \frac{V_2}{V_1}$$

The efficiency of an ideal gas Carnot engine is then given by

$$e = \frac{-w}{q_2} = \frac{T_2 - T_1}{T_2} \quad (7.26)$$

which is identical to (7.23). We can conclude that therefore the ideal gas temperature scale is identical to the thermodynamic temperature scale.

7.10 THE ECONOMICS OF POWER GENERATION

What are some aspects of the economics of converting heat into work? In electric-power-generating stations, the cold reservoir is usually a large body of water such as a river or a lake at a temperature of about 300 K. If the hot reservoir (the steam) is at 400 K, the theoretical efficiency is (400 − 300)/400, or 25%. That means that under ideal conditions we can at best get 25 J of electrical work for every 100 J of heat input. The efficiency would actually be less than 25%, since we cannot build an ideal reversible Carnot engine. If the temperature of the hot reservoir were raised to 600 K, the theoretical efficiency would be raised to (600 − 300)/600, or 50%. With a boiler temperature of 900 K, the efficiency becomes (900 − 300)/900, or 66.7%. It is apparent why an important objective in designing power generation equipment is to attain higher and higher operating temperatures. An increase of 1% in efficiency will save millions of dollars in fuel over the life of the equipment for a modern large kilo-megawatt power station. The limitations on temperature are metallurgical. It is difficult to design metal alloys able to withstand these extreme operating temperatures.

In a modern steam turbine, T_2 can be as high as 811 K, and T_1 is about 310 K; thus the maximum theoretical efficiency is about 62%. But the inherent properties of the steam cycle do not allow the heat to be introduced at a constant T_2. Other factors such as irreversibility reduce this value to a maximum theoretical efficiency of about 53%. Modern steam turbines achieve about 89% of that value, or about 47% net efficiency, in converting thermal energy to mechanical energy. To get the overall efficiency of the production of electricity by the steam turbine, this value must be multiplied by the efficiencies of other energy converters in the chain from fuel to electricity. Modern large boilers convert about 88% of the chemical energy in the fuel into usable heat. (The remaining 12% is "lost" to the surroundings—atmosphere, furnace walls, and the like.) The generators are remarkably efficient and convert about 99% of the mechanical energy from the turbine into electrical energy. The overall efficiency in the chain is thus 0.88 (boiler) × 0.99 (generator) × 0.47 (turbine), or about 41%. For every 100 J of fuel burned at the power station, only 41 J of electrical energy are delivered to the user, not counting transmission losses.[7]

The steam locomotive has an efficiency of only about 10%, whereas a diesel engine has an efficiency of some 37%. This explains the conversion of the railroad system from steam to diesel power. The automobile engine has about 25% efficiency, which explains why large trucks use diesel engines.

Now let's consider the problem of converting our private transportation system to electric battery–powered cars. Electricity must be generated to charge the batteries. The efficiency of delivering electricity to the user is about 41%, as we have just seen. The efficiency in charging the batteries (conversion of electrical energy to chemical) is about 70%. The efficiency with which the battery can deliver electricity to the motor (conversion of chemical energy to electrical) is also about 70%. Finally, the efficiency of the electric motor (conversion of electrical energy to mechanical) is about 85%. The overall efficiency is about 0.41 × 0.7 × 0.7 × 0.85, or 17%, which is less than the 25% efficiency of the internal combustion engine. At current rates of use, it is estimated that if all cars were electrically powered, we should have to increase the power-generating capacity of the United States by some 75%. Such is the importance

---

[7] The September 1971 issue of *Scientific American* was devoted to energy and power, and it is well worth reading. This issue contains an article by C. M. Summers, starting on p. 148, "The Conversion of Energy," which is of particular interest in connection with this discussion.

of the automobile. There are some benefits, however. If the electricity were to be generated in coal or nuclear stations, we could save massive quantities of petroleum. We could eliminate a large portion of our refining capacity, and also eliminate the carbon monoxide, nitrogen oxides, and other pollutants produced by automobiles in our cities. This last advantage would be partially offset by the pollutants produced in the countryside at the site of the generating plant.

The inefficiencies of the various processes produce what we may term waste heat. This is heat that is unavailable to do useful work.[8] It is this wasted heat that is a prime source of *thermal pollution*. This waste heat from power stations is delivered to the cold reservoir, usually a river or a stream. The heat is enough to raise the temperature of the stream, often with disastrous results to wildlife, especially fish. (Attempts to use this waste heat for "farming" lobsters have been made in the New England states. Thus far the procedure has not been particularly successful.)

7.11 ENTROPY: A NEW STATE FUNCTION

Equation (7.23) can be rewritten to yield

$$\frac{q_2}{T_2} - \frac{q_1}{T_1} = 0 \tag{7.27}$$

In other words, for a *reversible* Carnot engine operating over one complete cycle,

$$\sum \frac{q_{\text{rev}}}{T} = 0 \tag{7.28}$$

This is the criterion for a state function; thus q_{rev}/T is a state function for a reversible Carnot cycle. If we can generalize this result to any process, then we shall have uncovered a new general state function. We can do this by demonstrating that any cyclic process can be decomposed into innumerable infinitesimally small Carnot cycles, which we call "cyclets."

Figure 7.7a shows a general cyclic process QRS on a P-V diagram. The area enclosed by QRS is the work. A series of adiabats have been superimposed upon the cycle, and a series of isotherms connecting the adiabats have been drawn to approximate the circumference. These constitute a set of small reversible Carnot cyclets. For each cyclet,

$$\frac{dq_2}{T_2} + \frac{dq_1}{T_1} = 0$$

Now we refer to Figure 7.7b, which shows one of the Carnot cyclets on a larger scale. The path followed by the cyclet is $abcda$. The actual path follows the line ef at the upper temperature, and gh at the lower temperature. In the limit as the adiabats are drawn more and more closely together, the area enclosed by the curve $abcda$ approaches the area of the closed curve $efghe$, and this is true for every cyclet. Since the total work done in the real path is just the area enclosed by the curve QRS, the work

---

[8] Note that this is "waste heat" only relative to the original process of converting heat to mechanical energy. It is, of course, available for other useful purposes such as heating buildings. The concept of *cogeneration* is much used in some countries, notably Sweden, to improve the overall efficiency of energy consumption. In cogeneration, electric-power-generating stations are operated in conjunction with systems requiring direct heat, such as buildings and factories. While the efficiency of the electric-power-generating station is reduced, the overall efficiency of producing work *plus* heat is increased.

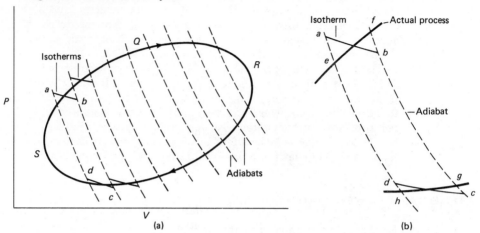

FIGURE 7.7 (a) General reversible cyclic process divided into a series of Carnot cyclets. (b) Large-scale view of Carnot cyclet.

will also equal the sum of the closed curves *efghe*. The work done by the cyclets thus equals the work done by QRS in the limit as the adiabats approach each other. Now what about the heats?

If dw' is the work done along ab, then in the limit as the adiabats approach each other $dw = dw'$, where dw is the actual work along the path ef. Since ab is an isothermal process, $-dw' = dq'_2$, and hence

$$-dw = -dw' = dq'_2 = dq_2$$

where dq'_2 is the heat along the Carnot cyclet isotherm ab, and dq_2 is the heat along the actual path ef. By applying the same arguments to the lower part of the cyclet, we see that $dq_1 = dq'_1$.

In the limit as the adiabats are drawn more and more closely together, the reversible cycle QRS can be replaced by the sum of the individual cyclets, and $\Sigma\, dq/T$ can be replaced by an integral. Equation (7.28) then becomes

$$\oint \frac{dq_{\text{rev}}}{T} = 0 \qquad (7.29)$$

This is true for any reversible process, since we have just shown that any reversible cyclic process can be decomposed into an infinite number of Carnot cyclets. Equation (7.29) expresses the condition for a state function. We denote this new function by S and give it the name *entropy*.

$$dS = \frac{dq_{\text{rev}}}{T} \qquad (7.30)$$

For any change,

$$\Delta S = S_{\text{fin}} - S_{\text{init}} = \int_{\text{init}}^{\text{fin}} \frac{dq_{\text{rev}}}{T} \qquad (7.31)$$

$1/T$ is an integrating factor for the differential dq_{rev}. That is to say, it converts dq_{rev}

into an exact differential. The integral $\int_a^b dq_{\text{rev}}$ is a function of the path, but $\int_a^b dq_{\text{rev}}/T$ is independent of the path.

$$\oint dS = \int_a^b dS + \int_b^a dS = S_b - S_a - S_b + S_a = 0$$

Note that when we define $dS = dq_{\text{rev}}/T$, we are very careful to write the subscript "rev." If dq is not reversible, then dq/T is not entropy. It is something else, and we do not even honor it with a special name. Thus we must be very careful in computing entropy changes for irreversible processes. We must devise a reversible process to take us from the same initial to the same final state. Only then can we compute the entropy change based on dq_{rev} for the equivalent reversible process.

7.12 CLAUSIUS'S INEQUALITY

Since the efficiency of any *irreversible* Carnot cycle is less than the efficiency of a reversible Carnot cycle operating between the same two temperatures, we have

$$\frac{q_2 - q_1}{q_2} < \frac{T_2 - T_1}{T_2} \quad \text{or} \quad \left(1 - \frac{q_1}{q_2}\right) < \left(1 - \frac{T_1}{T_2}\right) \quad \begin{array}{l}\text{(irreversible} \\ \text{Carnot cycle)}\end{array}$$

Thus
$$\frac{q_2}{T_2} - \frac{q_1}{T_1} < 0 \quad \text{(irreversible Carnot cycle)} \quad (7.32)$$

Now consider a general cyclic process into which irreversibility enters to the slightest extent. When we carry out the analyses of Figure 7.7, Equation (7.32) will hold for any cyclet that is irreversible. The sum of all the cyclets is then

$$\oint \frac{dq}{T} < 0 \quad \text{(irreversible)} \quad (7.33)$$

This is true for any cyclic process that is irreversible to the slightest extent. A more complete equation can be written as

$$\oint \frac{dq}{T} \leq 0 \quad (7.34)$$

where the equality holds for a completely reversible process, and the "less than" for the irreversible process.

7.13 ENTROPY AND THE DIRECTION OF CHEMICAL CHANGE

We are now in a position to provide a partial answer to the main question, namely, in which direction does a given process proceed spontaneously? The answer is to be found in the statement that **the entropy of an isolated system always increases during an irreversible process.** First we shall prove this with the aid of Clausius's inequality, and then we shall examine the consequences.

Consider the cyclic process shown in Figure 7.8. We *isolate* the system and take it from A to B *irreversibly*. Then we return the system to the initial state A by a reversible process. Since the complete process is cyclic and at least partially irreversible,

FIGURE 7.8 Cyclic process in which the system is isolated and taken from $A \to B$ irreversibly, then reversibly returned to the initial state A.

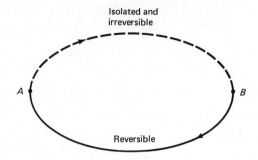

Clausius's inequality applies and $\oint dq/T < 0$, or

$$\int_A^B \frac{dq_{\text{irrev}}}{T} + \int_B^A \frac{dq_{\text{rev}}}{T} < 0 \qquad (7.35)$$

The first integral in (7.35) vanishes, since $dq = 0$ for any isolated process. (That is what we mean by *isolated*. There is no heat flow or any other kind of interaction across the boundary.) The second integral is $\Delta S_{\text{rev}} = S_A - S_B$; thus $(S_A - S_B) < 0$, or

$$\Delta S_{\text{irrev}} = S_B - S_A > 0 \qquad (7.36)$$

and the proof is completed.

All naturally occurring (that is, spontaneous) processes are irreversible; thus the criterion for directionality can be stated as follows: If we remove all constraints from any system and isolate it, the system will change so that $\Delta S > 0$. In another form, if the energy of a system is kept fixed, the system will tend towards the highest possible entropy. The condition of constant energy is implied by the condition of isolation, since the energy in any isolated system is constant.

7.14 EXAMPLE: HEAT TRANSFER

Suppose we have two temperature baths that are infinitely large, a hot bath at 400 K and a cold bath at 300 K, as shown in Figure 7.9. We thermally connect the two baths with a copper rod until 100 J of heat flows across the rod; then we disconnect the two baths. Does heat spontaneously flow from the hot to the cold bath, or from the cold to the hot bath? Let us assume that it flows from the hot to the cold bath. If this assumption is correct, then as shown in the previous section, we should get $\Delta S_{\text{tot}} > 0$ for the process.[9] Remember, though, that this criterion applies to the entropy change for the system *plus* surroundings. Further, since $dS = dq_{\text{rev}}/T$, it is necessary to compute the heat for a reversible process that accomplishes the same result.

To carry out the process in an equivalent reversible manner, we introduce the proper surroundings. We bring an ideal gas sample in contact with the hot bath and reversibly and isothermally expand the gas until 100 J of heat flow from the bath to the gas. Similarly, we bring another ideal gas sample in contact with the cold bath and reversibly and isothermally compress the gas until 100 J of heat flow from the

---

[9] Note that the principle of Clausius forbids heat to flow spontaneously from a cold bath to a hot bath.

SEC. 7.14 EXAMPLE: HEAT TRANSFER

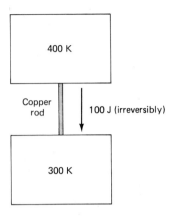

FIGURE 7.9 Irreversible transfer of 100 J of heat from a hot bath to a cold bath by connecting the baths with a copper rod.

gas to the cold bath. The net result is to remove 100 J reversibly from the hot bath and insert 100 J reversibly into the cold bath. The surroundings are the two cylinders of gas at the two different temperatures. This process, which accomplishes the same result in a reversible manner, is shown in Figure 7.10. The net result for the reversible process can be summarized as follows:

Hot bath
(400 K)

$$\Delta S_{\text{syst}} = \frac{q_{\text{rev}}}{T} = \frac{-100}{400} = -\frac{1}{4} \text{ J K}^{-1}$$

$$\Delta S_{\text{surr}} = \frac{q_{\text{rev}}}{T} = \frac{+100}{400} = +\frac{1}{4} \text{ J K}^{-1}$$

Cold bath
(300 K)

$$\Delta S_{\text{syst}} = \frac{q_{\text{rev}}}{T} = \frac{+100}{300} = +\frac{1}{3} \text{ J K}^{-1}$$

$$\Delta S_{\text{surr}} = \frac{q_{\text{rev}}}{T} = \frac{-100}{300} = -\frac{1}{3} \text{ J K}^{-1}$$

For the complete process

$$\Delta S_{\text{syst}} = -\tfrac{1}{4} + \tfrac{1}{3} = +\tfrac{1}{12} \text{ J K}^{-1} \qquad \Delta S_{\text{surr}} = +\tfrac{1}{4} - \tfrac{1}{3} = -\tfrac{1}{12} \text{ J K}^{-1}$$

$$\Delta S_{\text{tot}} = \Delta S_{\text{syst}} + \Delta S_{\text{surr}} = +\tfrac{1}{12} - \tfrac{1}{12} = 0 \quad \text{(reversible heat transfer)}$$

FIGURE 7.10 The change of Figure 7.9 accomplished by a reversible process. An amount 100 J of heat is reversibly removed from the hot bath by an isothermal reversible expansion, while 100 J of heat is added to the cold bath by an isothermal reversible compression.

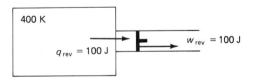

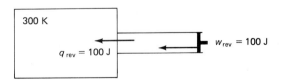

The total entropy change for the reversible process is zero. To be sure, the entropy of the system has increased, but there has been a corresponding decrease in the entropy of the surroundings that exactly balances the increase for the system.

What about the irreversible heat transfer? The system is again the two baths. Since entropy is a state function, ΔS_{syst} for the irreversible heat transfer must be the same as ΔS_{syst} for the reversible heat transfer, or $+\frac{1}{12}$ J K^{-1}. But what about the surroundings? In the irreversible transfer, the two baths are simply connected. There is no surroundings. Neither heat nor work crosses the boundary of the system. For the irreversible heat transfer, $\Delta S_{\text{surr}} = 0$, and

$$\Delta S_{\text{tot}} = \Delta S_{\text{syst}} + \Delta S_{\text{surr}} = +\tfrac{1}{12} + 0 = +\tfrac{1}{12} \text{ J K}^{-1}$$

The entropy change ΔS_{tot} is a positive quantity, and this is the criterion for a spontaneous process. Our original assumption that heat flows from the hot bath to the cold bath was correct. If we had originally assumed that the heat flowed from the cold bath to the hot bath, we should have obtained $\Delta S_{\text{tot}} = -\frac{1}{12}$, a negative quantity. That would have indicated an unnatural process, implying that we had made the incorrect assumption.

If the two baths are at the same temperature, then bringing the two in thermal contact would allow heat to pass reversibly from one bath to the other. This would result in $\Delta S_{\text{tot}} = 0$. When the constraints are removed from a system and the energy is held constant, the system will change spontaneously so as to increase its entropy to the maximum possible value. When two objects at different temperatures are brought into thermal contact, heat flows from the hot object to the cold object, causing the entropy of the system (the two objects) to increase. There is no surroundings, thus $\Delta S_{\text{surr}} = 0$, and $\Delta S_{\text{tot}} = \Delta S_{\text{syst}} > 0$. The entropy increases until it has reached its maximum possible value, at which time the system is at equilibrium. In working problems of this type, it is very important to separate carefully the system from the surroundings. Likewise, it is important to realize that we are free to bring in any valid surroundings to achieve reversibly the equivalent of a given process that takes place irreversibly.

To summarize, for any process in a system at *constant energy:*

If for that process $\Delta S_{\text{tot}} > 0$, then the process will spontaneously proceed as written. The process is a natural process.

If for that process $\Delta S_{\text{tot}} < 0$, then the process is an unnatural one and will not take place as written. We have written it the wrong way and must reverse everything.

If for that process $\Delta S_{\text{tot}} = 0$, then the initial state is in equilibrium with the final state.

7.15 THE MINIMUM ENERGY PRINCIPLE

Another problem of directionality concerns a falling weight. When a ball rolls off a table, does it fall to the floor or up to the ceiling? We know that the ball falls spontaneously to the floor, so let us try to apply the entropy condition to the problem. The first step is to achieve the same change by a reversible process so that we can calculate the entropy change.

An equivalent reversible process is shown in Figure 7.11. A massless string is attached to the weight, and the string is wrapped tightly about the shaft of a perfect electric generator. As the weight slowly falls, it turns the generator, thereby gen-

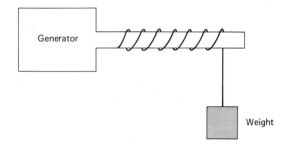

FIGURE 7.11 Reversible conversion of gravitational potential energy into electrical energy. As the weight slowly falls, it turns the generator.

erating electrical work. (This process is one in which gravitational work is converted into electrical work; the conversion may theoretically have an efficiency of unity.) There is no heat flow, $q_{rev} = 0$, and therefore $\Delta S_{syst} = 0$, $\Delta S_{surr} = 0$, and $\Delta S_{tot} = 0$.

The result seems paradoxical. We know that the process is spontaneous; thus the entropy change should be greater than zero. Yet we have calculated it to be zero. But our entropy change criterion applies to systems at constant energy, and this system is not at constant energy. As the weight falls, its energy in the earth's gravitational field decreases. It is a principle of classical mechanics that systems always tend towards minimum energy. We can state the energy minimum principle as "A system at constant entropy will tend to the minimum energy."

7.16 ENTROPY AS UNUSED "WORK CAPACITY"

Let's take another look at the two previous processes dealing with heat transfer and falling weights from the point of view of useful work obtainable. In the first example dealing with the two temperature baths, we can get useful work out of the system when the process is carried out reversibly as follows. The 100 J of heat from the hot bath is passed to the reversible work source (the cylinder with the piston containing an ideal gas) and the reversible work source does 100 J of work on its surroundings. The reversible work source is then adiabatically and reversibly expanded, doing more work on its surroundings until its temperature falls to the temperature of the cold bath. The reversible work source is then isothermally and reversibly compressed in thermal contact with the cold bath at 300 K until it passes the 100 J to the cold bath. In this step, the surroundings of the work source does 100 J of work on the work source. The total work is the sum of the three steps. The first and third steps cancel, and the total useful work obtainable from the process is the work of the second step, the adiabatic expansion. For this reversible process, $\Delta S_{tot} = 0$. Note that this process is not a cyclic process. The process is composed of the first three steps of a four-step Carnot cycle.

When the process is carried out irreversibly by simply bringing the two baths into thermal contact, the heat passes from the hot to the cold bath, and no useful work is obtained. For this process, we have calculated that $\Delta S_{tot} > 0$. The work that was originally obtainable is no longer obtainable. We might call this obtainable work "work capacity." In the irreversible process, this work capacity is destroyed. The entropy change is a measure of the destruction of work capacity. Every natural (that is, irreversible) process is capable of doing useful work as the system approaches equilibrium. If the process is carried out irreversibly, $\Delta S_{tot} > 0$, and we do not get the full measure of the work available. In fact, we may get no work for the totally irreversible process. If the process is carried out reversibly, then we get the full measure of the available work. The entropy change ΔS_{syst} is the same for the reversible and irreversible processes. The difference lies in ΔS_{surr}. For a reversible process, the surroundings are brought into the scheme of events; and ΔS_{surr} is equal and opposite to ΔS_{syst}, so that $\Delta S_{\text{tot}} = 0$. For the irreversible process, there is no compensating entropy change in the surroundings, and $\Delta S_{\text{tot}} > 0$.

In the second example, an analogous situation holds. When the weight falls reversibly, its potential energy is converted into electrical energy, and no work capacity has been destroyed. It has simply been converted to a different form. On the other hand, if the process is allowed to take place irreversibly, the weight falls freely to the floor, and this work capacity is destroyed. We have already calculated the entropy change for the weight in this process and found it to be zero, $\Delta S_{\text{syst}} = 0$. If work capacity has indeed been destroyed, then there should be an increase in entropy. Where is it? It is in the surroundings! When the weight hits the floor, its kinetic energy is dissipated in the form of heat. This heat enters the surroundings, and we can compute the entropy change from a reversible process that passes that amount of heat to the surroundings. It will be positive, thus $\Delta S_{\text{surr}} > 0$, and therefore $\Delta S_{\text{tot}} > 0$ since $\Delta S_{\text{syst}} = 0$.

Once this work capacity is lost, it is lost forever. To get it back from the bath arrangement, we must do work to take the 100 J from the cold bath and put it back in the hot bath. In the weight experiment, we must do work on the weight to put it back on the table and get back to our starting point.

7.17 BUT WHAT IS ENTROPY?

Of all the thermodynamic concepts, entropy is the most difficult for students. There is nothing to actually grasp in one's mind that is related to previous experience. *Heat* and *work* are words that we have been using since childhood. Although the true meaning of these words did not become part of our general knowledge until much later, the words themselves and some concepts about their meaning have always been with us. The same is true of *energy*. *Enthalpy* is something new, but conceptually it is sufficiently close to *energy* that it poses no great difficulty. *Entropy* is a different matter. There seems to be something artificial about taking the reversible heat, dividing by the temperature, and saying, "Aha! I have a new state function." But we have shown that it is indeed a state function. Entropy change is a measure of the destruction of work capacity. Later we shall see that the entropy change (or more correctly $T \Delta S$) of a process measures that portion of the energy change of the process that is unavailable for producing useful work; it measures the energy that is irrevocably lost when processes are harnessed to produce work. This ascribes rather negative qualities to entropy.

For the moment we take the entropy change to be just the reversible heat divided by the temperature. No more, no less. If you are unhappy with that at this point, bear up; you will probably become happier as time goes on.[10] If you are occasionally tempted to wail "But what is entropy!" as countless students have wailed be-

---

[10] In many treatments of thermodynamics, a microscopic view of entropy based on probability and statistics is presented at this point in the development. We are restricting our present discussion to the macroscopic view. The microscopic view of entropy is presented in Chapters 23 and 24. The interested student is referred to those chapters for a microscopic discussion.

fore, let the still small voice be heard calling from the wilderness, "$dS = dq_{rev}/T$," and for any change

$$\Delta S = S_{fin} - S_{init} = \int_{init}^{fin} \frac{dq_{rev}}{T}$$

7.18 THE ENERGY-ENTROPY EXTREMALS

We now find ourselves at the crossroads of two extremum principles:

At constant energy, entropy tends to a maximum.
At constant entropy, energy tends to a minimum.

Where does that leave us? Do the two principles conflict or are they complementary? In fact, they are complementary. A similar complementary situation between two extremal conditions exists in the familiar isoperimetric problem of plane geometry. A circle can be defined by either of two extremal conditions. A circle is a two-dimensional plane figure so constructed that:

At constant perimeter, the area is a maximum.
At constant area, the perimeter is a minimum.

These two extremal conditions are equivalent and apply to every circle, yet they provide two separate ways of generating a circle. Regardless of which condition is used to generate the circle, the final circle must satisfy *both* conditions for the final values of area *and* perimeter.

In the real world, we deal not with systems that have constant energy or constant entropy but with systems that have constant variables such as temperature and pressure. The next question becomes: If the temperature and pressure are kept constant, and the energy and entropy allowed to vary, what happens? We shall come back to this point in a little while. For the moment, let us simply say that the solution to this question is to be found in a new thermodynamic state function that we shall call the free energy. First, we must develop some schemes for calculating entropy changes and for establishing the absolute entropy scale.

PROBLEMS

1. With reference to Figure 7.3, suppose that each copper block contains 1 mol copper. Take the heat capacity of copper as 23 J K^{-1} mol^{-1}.
   (a) Using an ideal gas with a heat capacity given by $C_v = \frac{3}{2}R$, determine how much gas is needed in the cylinders so that the heat capacity of the gas will be the same as the heat capacity of the copper block.
   (b) Suppose we start with initial pressures $P_a^1 = 50$ atm and $P_a^2 = 1$ atm. Calculate P_b^1 and P_b^2.
   (c) Calculate the ratios of the diameters of the cylinders so that the two pistons move the same amount in the process leading from (a) to (b).
   (d) Calculate the final pressures P_c^1 and P_c^2.
   (e) In going from (a) to (b), how much work does IG^1 do on IG^2?
   (f) For step (c), calculate the heat passing between the cylinders and the copper blocks.

2. The vapor pressure of water at 150 °C is about 5 atm. The vapor pressure of water at 310 °C is about 100 atm. Compare in efficiency a steam engine operating at 150 °C with a steam engine operating at 310 °C. In both instances, the condenser is at 30 °C. (*Note:* Be sure that you understand that the pressures are irrelevant in this problem. The higher pressure is only of concern when you are designing a steam engine and must make the components strong enough to withstand

the pressure. It is only the temperatures that enter into a calculation of efficiency.)

3. A modern steam turbine operates at a temperature of 811 K. The cold reservoir is at 310 K. Calculate the theoretical efficiency of the turbine. For every 100 J heat input to the turbine at the hot reservoir, how many joules of work are produced, and how many joules of heat are ejected to the cold reservoir? To increase the theoretical efficiency by 1%, to what temperature would the hot reservoir have to be raised if the temperature of the cold reservoir were constant?

4. Heat in the amount 100 J flows from a hot reservoir at 811 K to a cold reservoir at 310 K. The reservoirs are large enough that their temperatures are unaffected by the heat transfer. Calculate the entropy change.

5. Show that two adiabats on a P-V diagram cannot cross. (*Hint:* Show that if they did cross, an inventive person could construct a cyclic engine that would violate the second law.)

6. One mole of an ideal gas with $C_v = 30$ J K^{-1} mol^{-1} is used as the working substance in a Carnot engine. The engine starts from a high pressure of 10 atm and a high temperature of 600 K. The gas is isothermally expanded to a pressure of 1 atm and then adiabatically expanded to reach a low temperature of 300 K. It is then isothermally compressed followed by an adiabatic compression back to the initial state.
(a) Calculate the heat and work of each stroke of the cycle.
(b) Calculate the efficiency and the total work output per cycle.
(c) Repeat the calculation for an initial pressure of 100 atm, all other factors remaining the same.

7. Plot the Carnot cycle of Problem 6 on a graph of P vs V. The net work delivered per cycle equals the area enclosed by the curves. Graphically evaluate this area, and compare your answer with that of Problem 6b.

8. Sketch a Carnot cycle on a graph of T vs S.

9. Suppose the triple point of water were defined as exactly 300.000 K instead of 273.1600 K. Calculate the boiling point of water on this scale.

10. A refrigerator is powered by a ½-hp motor. The interior of the refrigerator is to be maintained at -30 °C. The maximum kitchen temperature is 35 °C. What is the maximum heat leak into the refrigerator that can be tolerated?

NOTE: The answers you will get for Problems 11 through 14 will be only roughly correct but should indicate the large quantities involved. We do not take into account such things as boiler inefficiencies, transmission losses, and irreversibility or any other such variables.

11. A new electric-power-generating station is planned; it will have a capacity of 10^6 kilowatts (1 kilo-megawatt) and use a steam turbine plant. The turbine will operate at a temperature of 811 K, and a stream at 310 K will provide the low-temperature reservoir. Assume that the steam turbine will operate at its theoretical efficiency and that the efficiencies of the boiler and generator are 100%. When the plant is operating at capacity, how many tons of coal will it use in a year? Figure for both kinds of coal: (a) Assume that the coal is pure graphite. (b) Take the coal to be Northern Appalachian coal with a calorific value of 12,549 Btu lb^{-1}.

12. To keep the fish in the stream of Problem 11 in good health, it is imperative that the temperature increase of the water in the stream be kept to less than 5 °C. How large a flow must the stream have to keep within this limit?

13. The power company plans to burn high-sulfur coal in the plant of Problem 11. Conservationists raised a hue and cry on hearing about the plans, and the board of directors of the power company are looking into the possibility of using natural gas as a fuel. Assuming that natural gas is pure methane, calculate the amount of natural gas required for a year's operation of the power plant. (Calculate it in tons, liters, and cubic feet). Call your local coal company and your local gas company to find the cost of gas and coal. Compare what the yearly fuel bill for the power plant would be if it operates with coal with the cost for operating with natural gas.

14. Call your local utility company to find out how much energy is needed to heat a home electrically. (If you live in the tropics skip this problem.) How many homes could be heated electrically by the output of the power station? How many homes could be heated with gas heaters by delivering the gas directly to the homes?

15. The performance of an electric-power-generating station is normally denoted the *heat rate* and is quoted in units of Btu kWh^{-1} by engineers. A reasonably efficient plant has a heat rate of some 10,000 Btu kWh^{-1}. What is its actual efficiency?

16. Figure 7.1 seems to imply that the Carnot cycle process is carried out with four different cylinders. Describe how the four-step process can be achieved with only one cylinder.

17. In the accompanying figure the dashed curves are adiabats and the solid curves are isotherms at three different temperatures. Calculate the en-

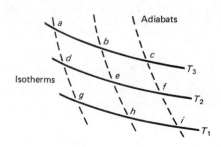

tropy change in the direction from a to i along the following paths:

(a) Path *abcfi*. (b) Path *adghi*.
(c) Path *adehi*. (d) Path *abcfedghi*

Show that the four answers are the same.

18. You have 1 mol liquid water at 90 °C. The outside air temperature is 5 °C. How much work can you get from the water by using the water as a hot reservoir and the air as a cold reservoir? (Remember that the hot reservoir is continually cooling during the heat extraction, since it is a finite reservoir. The temperature of the hot reservoir will approach the temperature of the cold reservoir as the system comes to equilibrium.)

19. In one of the most famous quotations of thermodynamics, Clausius wrote (*Pogg. Ann.* 125 (1865):400):

   Die Energie der Welt ist constant. Die Entropie der Welt strebt einem Maximum zu. ("The energy of the world is constant. The entropy of the world tends towards a maximum.")

   Explain Clausius's statement.

20. In his famous paper "On the Equilibrium of Heterogeneous Substances," *Trans. Conn. Acad. Sci.* 3 (1875):108, J. Willard Gibbs noted the following:

   The criterion of equilibrium for a material system which is isolated from all external influences may be expressed in either of the following entirely equivalent forms:

   I. *For the equilibrium of any isolated system it is necessary and sufficient that in all possible variations of the state of the system which do not alter its energy, the variation of its entropy shall either vanish or be negative.* If ϵ denotes the energy, and η the entropy of the system, and we use a subscript letter after a variation to indicate a quantity of which the value is not to be varied, the condition of equilibrium may be written

   $$(\delta\eta)_\epsilon \leq 0$$

   II. *For the equilibrium of any isolated system it is necessary and sufficient that in all possible variations in the state of the system which do not alter its entropy, the variation of its energy shall either vanish or be positive.* This condition may be written

   $$(\delta\epsilon)_\eta \geq 0$$

   Explain these statements.

21. Explain the meaning of the statement "Energy can be conserved, but it cannot be recycled."

22. Suppose that your next-door neighbor has never studied science and knows nothing about thermodynamics. How would you explain the statement of Problem 21 to him?

CHAPTER EIGHT

ENTROPY CALCULATIONS AND ABSOLUTE ENTROPIES. THE THIRD LAW OF THERMODYNAMICS

8.1 ENTROPY CHANGES FOR SOME SIMPLE PROCESSES

The relation $dS = dq_{rev}/T$ implies that we cannot compute the entropy change directly for an irreversible process, since we need q_{rev}. If we can dream up a reversible process to achieve the same change, however, then we can calculate q_{rev} for that process to get ΔS_{syst} for the change. As a first example, consider the *isothermal* expansion of an ideal gas. The terms w and q depend on whether the expansion is carried out reversibly or irreversibly, but ΔS will be the same for both. We can always get from the initial state to the final state by a reversible expansion, for which

$$q_{rev} = nRT \ln \frac{V_2}{V_1} \tag{8.1}$$

Since ΔS is just q_{rev}/T,

$$\Delta S = nR \ln \frac{V_2}{V_1} \tag{8.2}$$

for *any* isothermal expansion of an ideal gas, reversible or irreversible. For an expansion, $V_2 > V_1$, and $\Delta S > 0$; for a compression, $V_2 < V_1$, and $\Delta S < 0$. This is as it should be, for if we isolate the system and remove all constraints (the constraint is the external opposing pressure), the system should spontaneously move in the direction of increasing entropy, that is, the direction of increasing volume.

For a second example, consider the process in which a substance is heated from T_1 to T_2 at constant pressure. The reversible heat is

$$dq_{rev} = C_p \, dT \tag{8.3}$$

where C_p is the constant-pressure heat capacity. The entropy change ΔS is then

$$\Delta S = \int_{T_1}^{T_2} C_p \frac{dT}{T} \tag{8.4}$$

If the heat capacity is constant over this temperature range, we can write

$$\Delta S = C_p \int \frac{dT}{T} = C_p \ln \frac{T_2}{T_1} \tag{8.5}$$

An analogous expression holds for constant-volume heating, with C_v substituted for C_p.

For the third simple example, consider the entropy change associated with a change of state such as melting or vaporization. Here the reversible heat is $q_{rev} = \Delta H_{trans}$, where ΔH_{trans} is the heat of transition; thus

$$\Delta S_{trans} = \frac{\Delta H_{trans}}{T} \tag{8.6}$$

A slightly more difficult problem is posed by the isothermal mixing of two ideal gases.

8.2 ENTROPY OF MIXING

An apparatus for mixing two ideal gases is shown in Figure 8.1. A barrier is placed in a cylinder between the two ends. The volume on the left side of the barrier is V_1 and on the right side V_2. Initially, n_1 moles of gas A are in the left side and n_2 moles of gas B are in the right side. The pressure and the temperature are the same on each side of the barrier. The barrier is removed and the two gases intermix until each is uniformly distributed throughout the cylinder. What is the entropy change associated with the mixing of the two gases?

According to the law of partial pressures, each gas can be treated separately as though the other were not present. Gas A expands from an initial volume V_1 to a final volume $(V_1 + V_2)$. The entropy change is

$$\Delta S_A = R n_1 \ln \frac{V_1 + V_2}{V_1} \tag{8.7a}$$

and similarly

$$\Delta S_B = R n_2 \ln \frac{V_1 + V_2}{V_2} \tag{8.7b}$$

Noting that $V_1/(V_1 + V_2) = n_1/(n_1 + n_2)$ is X_1, the mole fraction of component 1, we get the total entropy change as

$$\Delta S = \Delta S_A + \Delta S_B = -R n_1 \ln X_1 - R n_2 \ln X_2 \tag{8.8}$$

For a final solution of one mole of the mixture, we divide by $n_1 + n_2$, the total number of moles and get

$$\Delta S_{mix} = -R(X_1 \ln X_1 + X_2 \ln X_2) \tag{8.9}$$

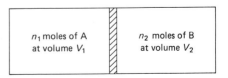

FIGURE 8.1 Mixing gases A and B by removing the partition between them.

per mole. By using a process of successive mixings, one can generalize Equation (8.9) to any number of components as

$$\Delta S_{mix} = -R \sum_i X_i \ln X_i \qquad (8.10)$$

The volume composition of air is about 79% N_2, 20% O_2, and 1% Ar. The entropy of mixing these constituents to get the composition of air is

$$\Delta S = -R[0.79 \ln 0.79 + 0.20 \ln 0.20 + 0.01 \ln 0.01]$$
$$= +4.6 \text{ J K}^{-1} \text{ mol}$$

Note that since the mole fraction X must be less than unity in any mixture, the entropy of mixing is always a positive quantity. This is as it should be, since the process takes place spontaneously.

8.3 MATHEMATICAL GAMESMANSHIP

By using what we have learned thus far about the first and second laws of thermodynamics, and some elementary properties of partial derivatives, we can derive some important results that apply to entropy and energy.

For any reversible process with expansion work only, we can write the first law as

$$dE = dq_{rev} - P\,dV \qquad (8.11)$$

Since $dq_{rev} = T\,dS$, this becomes

$$dE = T\,dS - P\,dV \qquad (8.12)$$

This is a very important result that is valid for any system at constant composition in which the work is restricted to expansion work. If we consider E a function of S and V, then $E = E(S, V)$, and we can also write

$$dE = \left(\frac{\partial E}{\partial S}\right)_V dS + \left(\frac{\partial E}{\partial V}\right)_S dV \qquad (8.13)$$

Comparing (8.12) with (8.13) yields

$$\left(\frac{\partial E}{\partial S}\right)_V = T \qquad (8.14)$$

and

$$\left(\frac{\partial E}{\partial V}\right)_S = -P \qquad (8.15)$$

In these two equations, the intensive variables T and P are given in terms of the extensive variables S, E, and V.

In Equation (8.13), we wrote dE in terms of dS and dV. What if we write it in terms of the variables T and V? From $E = E(T, V)$, we get

$$dE = \left(\frac{\partial E}{\partial T}\right)_V dT + \left(\frac{\partial E}{\partial V}\right)_T dV$$
$$= C_v\,dT + \left(\frac{\partial E}{\partial V}\right)_T dV \qquad (8.16)$$

SEC. 8.3 MATHEMATICAL GAMESMANSHIP

We can also write

$$dq_{rev} = dE + P\,dV = T\,dS \tag{8.17}$$

Equations (8.16) and (8.17) can be combined to yield

$$dS = \left(\frac{C_v}{T}\right)dT + \frac{1}{T}\left[P + \left(\frac{\partial E}{\partial V}\right)_T\right]dV \tag{8.18}$$

Another expression for dS is

$$dS = \left(\frac{\partial S}{\partial T}\right)_V dT + \left(\frac{\partial S}{\partial V}\right)_T dV \tag{8.19}$$

If we now compare (8.18) with (8.19), we get

$$\left(\frac{\partial S}{\partial T}\right)_V = \frac{C_v}{T} \tag{8.20}$$

and

$$\left(\frac{\partial S}{\partial V}\right)_T = \frac{1}{T}\left[P + \left(\frac{\partial E}{\partial V}\right)_T\right] \tag{8.21}$$

Suppose we now take the partial derivative of (8.20) with respect to V, and the partial derivative of (8.21) with respect to T. From Equation (8.20), we get

$$\left[\frac{\partial}{\partial V}\left(\frac{\partial S}{\partial T}\right)_V\right]_T = \frac{\partial^2 S}{\partial V\,\partial T} = \left[\frac{\partial}{\partial V}\left(\frac{C_v}{T}\right)\right]_T = \frac{1}{T}\left(\frac{\partial C_v}{\partial V}\right)_T \tag{8.22}$$

and from Equation (8.21), we get

$$\left[\frac{\partial}{\partial T}\left(\frac{\partial S}{\partial V}\right)_T\right]_V = \frac{\partial^2 S}{\partial T\,\partial V} = \frac{\partial}{\partial T}\left\{\frac{1}{T}\left[P + \left(\frac{\partial E}{\partial V}\right)_T\right]\right\}_V$$

$$= -\frac{P}{T^2} - \frac{1}{T^2}\left(\frac{\partial E}{\partial V}\right)_T + \frac{1}{T}\left(\frac{\partial P}{\partial T}\right)_V + \frac{1}{T}\left(\frac{\partial^2 E}{\partial T\,\partial V}\right) \tag{8.23}$$

Since the mixed derivatives are equal, the right-hand sides of (8.22) and (8.23) are equal. It is also true that

$$\frac{\partial^2 E}{\partial T\,\partial V} = \frac{\partial^2 E}{\partial V\,\partial T} = \left[\frac{\partial}{\partial V}\left(\frac{\partial E}{\partial T}\right)_V\right]_T = \left(\frac{\partial C_v}{\partial V}\right)_T \tag{8.24}$$

Now we can combine (8.22), (8.23), and (8.24), to arrive at

$$\left(\frac{\partial E}{\partial V}\right)_T = T\left(\frac{\partial P}{\partial T}\right)_V - P \tag{8.25}$$

This is an important result; it allows us to predict the change in internal energy associated with an isothermal change for any substance if we know the equation of state for the substance. For an ideal gas, the equation of state is $P = nRT/V$, and the application of Equation (8.25) yields

$$\left(\frac{\partial E}{\partial V}\right)_T = T\left[\frac{\partial(nRT/V)}{\partial T}\right]_V - P = \frac{nRT}{V} - P = 0 \tag{8.26}$$

This is the same result that Joule found for the free expansion of an ideal gas. We have previously stated that the validity of Equation (8.26) was the experimental result of Joule's expansion experiment. We can now also state that this result is required by the second law of thermodynamics by virtue of the equation of state of an ideal gas.

By following analogous procedures, we can also show that the following similar results hold,

$$dH = T\,dS + V\,dP \tag{8.27}$$

$$\left(\frac{\partial S}{\partial T}\right)_P = \frac{C_p}{T} \tag{8.28}$$

$$\left(\frac{\partial H}{\partial P}\right)_T = V - T\left(\frac{\partial V}{\partial T}\right)_P \tag{8.29}$$

and that for an ideal gas $(\partial H/\partial P)_T = 0$.

8.4 ENTROPY CHANGES IN IDEAL GAS PROCESSES

Recall that for an ideal gas $dE = C_v\,dT$; and if we start with Equation (8.12), $dE = T\,dS - P\,dV$, we can write

$$dS = \frac{C_v\,dT}{T} + \frac{P\,dV}{T}$$

For an ideal gas, $P = nRT/V$, and this becomes

$$dS = C_v\frac{dT}{T} + nR\frac{dV}{V} \tag{8.30}$$

The entropy change for a finite process is

$$\Delta S = S_2 - S_1 = \int_{T_1}^{T_2} C_v\,d(\ln T) + \int_{V_1}^{V_2} nR\,d(\ln V)$$

$$= C_v \ln\frac{T_2}{T_1} + nR \ln\frac{V_2}{V_1} \tag{8.31}$$

provided that C_v is constant over this temperature range. At constant volume, the second term in (8.31) is zero, and we are left with the constant-volume version of Equation (8.5). If the temperature is constant, the first term is zero; we are then left with the same result as for the isothermal expansion of Equation (8.2).

For an adiabatic expansion, both the temperature and the volume may change, and both terms of (8.31) are needed. If the adiabatic expansion is reversible, then $dq_{\text{rev}} = 0$, and $\Delta S = 0$. The left-hand side of (8.31) is zero, which is identical to (5.29). The entropy change ΔS must be zero for *any* reversible adiabatic process.

8.5 EXAMPLE

Consider the isothermal reversible expansion of 1 mole of ideal gas from 5 atm to 1 atm at 300 K. The process is

$$IG(5\text{ atm}, 300\text{ K}) \longrightarrow IG(1\text{ atm}, 300\text{ K})$$

For this process, $\Delta E = 0 = q + w$, and

$$q = -w = RT \ln 5 = (8.314)(300)(1.609) = 4013\text{ J}$$

SEC. 8.5 EXAMPLE

Since the process is reversible, the heat is reversible heat, and $q_{rev} = 4013$ J. The entropy change is

$$\Delta S = \frac{q_{rev}}{T} = \frac{4013}{300} = 13.38 \text{ J K}^{-1}$$

We could have obtained this same result by directly applying

$$\Delta S = R \ln \frac{V_2}{V_1} = R \ln \frac{P_1}{P_2}$$

which is valid for any *isothermal* process involving an ideal gas.[1]

Now what about the surroundings? The heat q_{rev} flowed from the surroundings to the system. Since

$$q_{rev,syst} = -q_{rev,surr}$$

we have

$$\Delta S_{surr} = -\Delta S_{syst} = -13.38 \text{ J K}^{-1}$$

The *total* entropy change is

$$\Delta S_{tot} = \Delta S_{syst} + \Delta S_{surr} = 0$$

Now suppose we carry out the process irreversibly against a constant pressure of 1 atm. Entropy is a state function; and since the initial and final states of the system are the same as in the reversible expansion, ΔS_{syst} must be the same, $\Delta S_{syst} = +13.38$ J K^{-1}. Now what about the surroundings? The entropy change for the surroundings in the irreversible process will be different from that in the reversible process because the initial and final states of the *surroundings* will be different for the irreversible process. We must devise an equivalent method to accomplish the same process. For the irreversible expansion against a constant pressure,

$$q_{irr} = w_{irr} = \int P \, dV = P \int dV = P_2(V_2 - V_1)$$

$$= P_2 \left(\frac{RT}{P_2} - \frac{RT}{P_1} \right) = RT \left(1 - \frac{P_2}{P_1} \right)$$

$$= 0.8 RT = (0.8)(8.314)(300)$$

$$= 1995 \text{ J}$$

To calculate ΔS, we can use the following two-step equivalent process:

Step 1. $IG(5 \text{ atm}, 300 \text{ K}) \xrightarrow{\text{reversible}} IG(P^*, 300 \text{ K})$ ($q_{rev} = 1995$ J)

Step 2. $IG(P^*, 300 \text{ K}) \xrightarrow[\substack{\text{expansion against} \\ P_{op} = 0}]{\text{free}} IG(1 \text{ atm}, 300 \text{ K})$ ($q_{irr} = 0$)

The pressure P^* is some intermediate pressure between 1 and 5 atm such that the heat is 1995 J, and this will also be the work done on the surroundings. We need not concern ourselves with the actual numerical value of P^*. In Step 1,

$$\Delta S_{syst} = +\frac{1995}{300} = 6.65 \text{ J K}^{-1}$$

and

$$\Delta S_{surr} = -\Delta S_{syst} = -6.65 \text{ J K}^{-1}$$

---

[1] For beginners working out problems, it is usually sound technique to start with first principles, instead of trying to remember a specific formula or looking it up in the book.

To calculate ΔS_{syst} for Step 2 directly, we should have to calculate P^*. But we can calculate it indirectly, since we know that the entropy change for the sum of the two steps must be 13.38 J K^{-1}. The value ΔS_{syst} for Step 2 is thus $13.38 - 6.65 = 6.73$ J K^{-1}.

Now what about the surroundings for Step 2? There is no interaction with the surroundings, hence there is no change in the surroundings; $\Delta S_{surr} = 0$ for Step 2. The entropy change ΔS_{surr} for the entire irreversible process is just ΔS_{surr} for Step 1, which is -6.65 J K^{-1}. The total entropy change for the irreversible process is now

$$\Delta S_{tot} = \Delta S_{syst} + \Delta S_{surr} = +13.38 - 6.65 = +6.73 \text{ J K}^{-1}$$

Let's take note of what has occurred. In the reversible process, we got the full measure of available work, and $\Delta S_{tot} = 0$. In the irreversible process, we did not get the full measure of the available work, getting only about half of what we could have got if we had carried out the process reversibly. The difference in work between the reversible and the irreversible process is $4013 - 1995 = 2018$ J, and this much "work capacity" has been destroyed forever.

8.6 ENTROPY CHANGES ON HEATING A PURE SUBSTANCE; EXPERIMENTAL METHOD

Suppose we want to determine the entropy change associated with heating a substance from T_1 to T_2 at constant pressure. Since $dS = dq_{rev}/T$, and $dq_{rev} = C_p \, dT$,

$$dS = C_p \frac{dT}{T} \tag{8.32}$$

For a finite change, the entropy difference is

$$\Delta S = S_2 - S_1 = \int_{T_1}^{T_2} C_p \frac{dT}{T} \tag{8.33}$$

A possible experimental procedure is to measure the heat capacity at various temperatures and determine ΔS by graphical integration. Figure 8.2 shows plots of C_p vs T for three different substances. For AgCl(s) and Ag(s), we have plotted C_p for 1 mole of material. For Cl$_2$, we have plotted C_p for $\frac{1}{2}$ mole of material, for reasons which will shortly become obvious. In Figure 8.3, the same data have been plotted somewhat

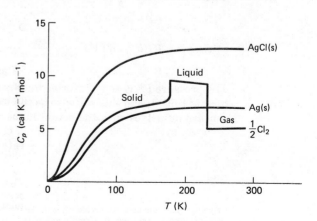

FIGURE 8.2 Heat capacities C_p vs temperature T for Ag, AgCl, and $\frac{1}{2}$Cl$_2$.

SEC. 8.7 THE THIRD LAW OF THERMODYNAMICS AND THE FIDUCIAL POINT OF ENTROPY

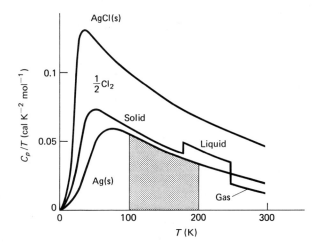

FIGURE 8.3 Data of Figure 8.2 plotted as C_p/T vs T. Graphical integration to get absolute entropies. The value ΔS for heating Ag(s) from 100 to 200 K is given by the shaded region.

differently, namely C_p/T vs T. For any of our three substances, ΔS for heating the substance from T_1 to T_2 is given by the area under the respective curve between the two temperatures, as shown in Figure 8.3. This area for Ag(s) is indicated by the shaded area. For Cl_2, the situation is complicated by the discontinuities in the curve at the boiling and freezing points. This is taken into account by adding to the area under the curve the values of ΔS_{trans} at each transition point where a change in state occurs.

If we had some scheme by which we could establish a reference point of zero entropy for various substances, the procedure we have just described would enable us to determine *absolute* entropies at any temperature by calculating ΔS from the fiducial point. We could then calculate ΔS for reactions such as

$$\text{Ag(s)} + \tfrac{1}{2}\text{Cl}_2(g) = \text{AgCl(s)}$$

by just taking the differences in absolute entropies

$$\Delta S = S_{\text{AgCl}} - S_{\text{Ag}} - \tfrac{1}{2}S_{\text{Cl}_2}$$

The scheme for establishing the fiducial point is provided by the third law of thermodynamics.

8.7 THE THIRD LAW OF THERMODYNAMICS AND THE FIDUCIAL POINT OF ENTROPY

Equation (8.28) gives the variation of S as a function of T, $(\partial S/\partial T)_p = C_p/T$. Since both temperature and heat capacity are positive quantities, the derivative is also positive.[2] Therefore S must be a monotonically increasing function of the temperature. Now, what happens in the limit as T approaches zero? If C_p remains finite or goes to zero much more slowly than T, then the derivative becomes infinitely large as $T \to 0$. If C_p approaches zero at a rate much higher than T, then the derivative goes to zero as $T \to 0$.

---

[2] This is always true for macroscopic systems. For microscopic collections of molecules in which temperature can be defined by a Boltzmann distribution, it is possible to get "negative temperatures" by exciting the molecules to produce population inversions. Such situations are excluded from the present discussion.

The approach to zero began in the 1860s, when the design of heat exchangers became advanced enough to allow the liquefaction of gases by Joule-Thomson expansions. About then, it became possible to liquefy air and to separate N_2 and O_2 by distillation. Liquid N_2 boils at 77 K and liquid O_2 boils at 90 K. Modern steelmaking facilities use huge quantities of O_2 gas for their oxygen-fired furnaces. The O_2 is produced by distillation of liquid air. Since air is only about 20% O_2 and 80% N_2, large quantities of liquid N_2 are produced as a by-product. This accounts for the large price reduction in liquid N_2 over the past 20 years.

To liquefy hydrogen gas, it must first be cooled to below the inversion temperature. This is usually done with liquid N_2. Liquid H_2, which boils at 20 K, was first liquefied by James Dewar in 1898. Liquid He boils at 4.2 K and was first produced by Kammerlingh-Onnes in Leiden in 1908; he used liquid H_2 to precool the He gas. Temperatures as low as 0.8 K can be attained by pumping on liquid He, but huge vacuum pumps are required to carry off the large volumes of gas.

Refrigeration by adiabatic demagnetization was first accomplished by Giauque in 1933. This is a process in which the magnetic moments of electrons in paramagnetic salts are aligned with a magnetic field, and then the field is removed.[3] By 1950, temperatures as low as 0.001 K had been achieved by this technique. Attempts to get still lower temperatures have involved using the magnets associated with the magnetic moments of nuclei rather than of electrons. The present lowest limit is about 2×10^{-5} K.

We have reached within 2×10^{-5} of $T = 0$.[4] Can we get the rest of the way? All indications point to No. The answer is the third law of thermodynamics.

> *The absolute zero of temperature is unattainable by any process, no matter how idealized, in a finite number of steps.*

What does this tell us about the fiducial point of entropy?

Consider the process $a \to b$. At any temperature T,

$$S_a = {}^\circ S_a + \int_0^T C_a \, d(\ln T) \tag{8.34a}$$

$$S_b = {}^\circ S_b + \int_0^T C_b \, d(\ln T) \tag{8.34b}$$

Where ${}^\circ S_a$ and ${}^\circ S_b$ are the entropies of a and b at 0 K; C_a and C_b are the respective heat capacities. Now let's consider the process $a(T') \to b(T'')$, which we carry out *reversibly* and *adiabatically*. Since $q_{\text{rev}} = 0$, therefore $\Delta S = 0$, and $S_a = S_b$. From Equation (8.34), we have

$${}^\circ S_a + \int_0^{T'} C_a \, d(\ln T) = {}^\circ S_b + \int_0^{T''} C_b \, d(\ln T)$$

or

$${}^\circ S_b - {}^\circ S_a = \int_0^{T'} C_a \, d(\ln T) - \int_0^{T''} C_b \, d(\ln T) \tag{8.35}$$

---

[3] Adiabatic demagnetization is discussed in Chapter 18.

[4] The difference between 10^{-3} and 10^{-5} on a scale of T does not seem very substantial. On the other hand, it is a factor of 100, and if we consider instead a scale of T^{-1}, we are looking at values of 10^3 and 10^5 on an inverse temperature scale. In a certain sense, a scale of T^{-1} is more fundamental for entropy than a scale of T, since $dS = q_{\text{rev}}(T^{-1})$. Also, since $dS = C_p \, d(\ln T)$, we might consider $\ln T$ the fundamental temperature unit for third law considerations.

Now, if $°S_b > °S_a$, we could set $T'' = 0$ and find some T' that would then satisfy (8.35). This process would then enable us to reach $T = 0$. But this is impossible! Therefore we must conclude that $°S_b = °S_a$, and that

Every substance must have the same molar entropy at the absolute zero of temperature.

(This statement should be qualified by the condition that the substance must be in its lowest energy state. The significance of this qualification will become apparent shortly.) Since we are interested in entropy differences, it does not matter what we set this value to be. The simplest choice is zero, and we shall set the entropy of every pure substance in its lowest energy state to $S = 0$ at $T = 0$.[5] Now we can use the third law of thermodynamics to construct a table of absolute entropies.

8.8 THIRD LAW ENTROPIES

The procedure for establishing a table of absolute entropies is now straightforward. The term C_p is measured as a function of temperature, and the data are plotted as in Figure 8.3. The absolute entropy is then given by the total area under the curve up to the desired temperature. The symbol $S°_{298.15}$ is assigned to an absolute entropy determined in this manner at the standard conditions of 1 atm and 298.15 K. The only problem is measuring C_p at very low temperatures. From Figure 8.3 we see that the low-temperature contribution is quite small. In Chapter 36 we shall show that at low temperature the heat capacity of a crystal is given by

$$C_p = AT^3 \qquad (8.36)$$

where A is a constant that is characteristic of the crystal. In practice, C_p is measured over a range of temperatures down to some convenient lowest temperature, perhaps 20 or 4 K. The value of A is then determined by using the lowest experimental value of C_p, and Equation (8.36) is then used to generate that portion of the heat capacity curve that was not actually measured. Table 8.1 lists the third law entropies for numerous substances determined in this way.

Again we may raise the question, How do we know that the third law is correct? The answer is again that we know it is correct because it works. We can calculate ΔS for any reaction from the absolute entropies of the products and reactants. For the reaction $Ag(s) + \frac{1}{2}Cl_2(g) = AgCl$, we get $\Delta S° = -57.95 \pm 1.05$ J K^{-1} mol^{-1} by this procedure. This value is based only on the third law and experimental heat capacities. When we study free energy, we shall see that there are other methods, independent of the third law, whereby we can measure $\Delta S°$ for reactions. One such method is based on constructing a battery in which the relevant reaction produces electrical work, and measurement of the voltages yields a value for ΔS. By this method, we find $\Delta S° = -57.45 \pm 0.42$ J K^{-1} mol^{-1} for the same reaction, agreeing with the value obtained from the third law. The smaller uncertainty in the second method reflects the greater accuracy with which ΔS can be measured by that technique. We could fill the next 50 pages with examples like this. In every case, ΔS is the same within the experimental uncertainty for the two methods. That is what we mean by "correct."

---

[5] In the view of macroscopic thermodynamics, the choice of $S = 0$ at $T = 0$ is arbitrary. Later, when we study microscopic thermodynamics, we shall see that this choice can be justified.

CHAP. 8 ENTROPY CALCULATIONS AND ABSOLUTE ENTROPIES. THE THIRD LAW

TABLE 8.1 Standard entropies of various substances.

Substance	$S°$ (J K^{-1} mol^{-1})	Substance	$S°$ (J K^{-1} mol^{-1})
$O_2(g)$	205.029	$P_2(g)$	218.11
$H_2(g)$	130.587	$P_4(g)$	279.91
$H_2O(g)$	188.724	C(graphite)	5.6940
$H_2O(l)$	69.940	C(diamond)	2.4389
$F_2(g)$	203.3	CO(g)	197.9074
$Cl_2(g)$	222.949	$CO_2(g)$	213.639
HCl(g)	186.678	$CH_4(g)$	186.19
$Br_2(l)$	152.3	$C_2H_2(g)$	200.819
$Br_2(g)$	245.346	$C_2H_4(g)$	219.45
HBr(g)	198.476	$C_2H_6(g)$	229.49
$I_2(c)$	116.7	$C_2H_5OH(g)$	282.0
$I_2(g)$	260.580	$C_2H_5OH(l)$	160.7
HI(g)	206.330	Pb(c)	64.89
$H_2S(g)$	205.64	$PbCl_2(c)$	136.4
$SO_2(g)$	248.53	$PbBr_2(c)$	161.5
NO(g)	210.618	Hg(l)	77.4
$NO_2(g)$	240.45	$Hg_2Cl_2(c)$	195.811
$N_2O(g)$	219.99	Ag(c)	42.702
$N_2O_4(g)$	304.30	AgCl(c)	96.11
$NH_3(g)$	192.51	AgBr(c)	107.11
P(g)	163.09	S(rhombic)	31.88
P(c, white)	44.4	S(monoclinic)	32.55
$N_2(g)$	191.489		

SOURCE: National Bureau of Standards Circular 500.

NOTE: Standard conditions 1.000 atm and 25 °C obtain.

There are, however, a few paradoxes in which the different methods give different results. In each case, it has been possible to find the source of the discrepancy. The sources usually fall into one of two categories. The first category includes compounds for which the true plot of C_p/T appears as shown in Figure 8.4. If C_p is measured down to only 20 K and extrapolated to zero by Equation (8.36), then an error will be

FIGURE 8.4 Anomalous low-temperature heat-capacity behavior. The solid curve indicates the true curve. The dashed curve indicates the T^3 approximation. If heat capacity measurements are extended down to only 20 K, an error equal to the shaded area will be introduced into the absolute entropy.

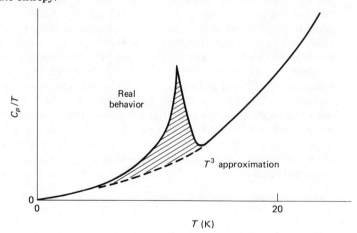

> JOSIAH WILLARD GIBBS (1839–1903), American scientist, was professor of mathematical physics at Yale University from 1871 until his death. His series of papers "On the Equilibrium of Heterogenous Substances," published in the *Transactions of the Connecticut Academy of Sciences* (1876–1878) was one of the most important series of papers in the history of thermodynamics. He is also considered one of the founders of statistical mechanics. The Copley Medal of the Royal Society of London was presented to him as "the first to apply the second law of thermodynamics to the exhaustive discussion of the relation between chemical, electrical, and thermal energy and capacity for external work."

introduced equal to the area shown in the shaded region. Such spikes are due to phenomena such as antiferromagnetic transitions. This type of discrepancy is resolved when the heat-capacity measurements are extended to lower temperatures, and the region of the spike is encompassed by the experimental data.

In the second category, the source of the discrepancy lies in the entropy of mixing. The third-law fiducial point applies to perfectly ordered pure crystals. Glasses and disordered crystals (not to mention solutions) are excluded from the general rule that $S° = 0$ at $T = 0$. Consider, for example, the N_2O molecule, which may be written NNO. If we examine two adjacent molecules in a crystal, they may be oriented parallel (NNO NNO) or antiparallel (NNO ONN). The two orientations have only a slight energy difference, and at low temperatures the random orientations are effectively frozen in. Thus there appears an extra contribution due to the entropy of mixing, which in this case is

$$S° = -R \sum_i X_i \ln X_i = -R(\tfrac{1}{2} \ln \tfrac{1}{2} + \tfrac{1}{2} \ln \tfrac{1}{2})$$

$$= R \ln 2 = 5.76 \text{ J K}^{-1} \text{ mol}$$

This is the same as the discrepancy between the two methods to within the experimental uncertainty.

So much for entropy by itself. We now go on to that magnificent invention of J. Willard Gibbs, the *free energy*, and therein lies the solution to the problem of equilibrium at constant temperature and pressure.

PROBLEMS

1. Three moles of an ideal gas are isothermally expanded from an initial pressure of 2 atm to a final pressure of 1 atm at 25 °C.
   (a) Devise a reversible process to carry out the change and calculate ΔS_{syst}, ΔS_{surr}, and ΔS_{tot}.
   (b) Devise an irreversible procedure to carry out the change so that no work is done. Calculate ΔS_{syst}, ΔS_{surr}, and ΔS_{tot}.
   (c) Now carry out the expansion irreversibly against a constant external pressure of 1 atm. Calculate ΔS_{syst}, ΔS_{surr}, and ΔS_{tot}.

2. Glitch has a heat capacity $C_p = 17$ J K^{-1} mol^{-1}. The heat capacity is independent of temperature. Calculate the entropy change on heating glitch from 300 to 400 K.

3. Glotch has a heat capacity given by $C_p = 17 + 10^{-3} T - 10^{-6} T^2$ J K^{-1} mol^{-1}. Calculate the entropy change on heating glotch from 300 K to 400 K.

4. One kg ice at -10 °C is converted to steam at 100 °C. Calculate the entropy change. (The heat of vaporization of water is 40.67 kJ mol^{-1}; the heat of fusion of water is 6.01 kJ mol^{-1}; the heat capacities of ice and liquid water are 33 and 75 J K^{-1} mol^{-1}. Assume that the heat capacities are independent of temperature.)

5. Ice in the amount 10 g at 0 °C is added to 20 g liquid water at 80 °C in a Dewar flask. The relevant thermodynamic data on water is given in Problem 4. Ignoring the heat capacity of the Dewar flask and assuming that ΔH_{fus} is independent of temperature:
   (a) Determine the final temperature.
   (b) Devise a reversible process for carrying out

the change. Compute ΔS_{syst}, ΔS_{surr}, and ΔS_{tot} for the reversible process.

(c) Compute ΔS_{syst}, ΔS_{surr}, and ΔS_{tot} for the direct irreversible addition of the ice to the water.

6. For a reversible adiabatic expansion, since $q_{rev} = 0$, there must be no entropy change. There are two effects in the adiabatic process, one from the cooling of the gas, and a second from the expansion of the gas. Show that for the reversible adiabatic expansion, these two contributions to ΔS are equal and opposite, yielding $\Delta S = 0$. Show that if the expansion is adiabatic and irreversible, the two contributions are not equal and opposite, and in fact ΔS must be greater than zero.

7. An ideal gas has $C_p = \frac{5}{2}R$. Three moles are taken from an initial state of 10 °C and 2 atm to a final state of -30 °C and 0.2 atm. Calculate ΔS for the change.

8. Two flasks are connected with a stopcock. The first flask contains 1 mol He at 4 atm. The second flask contains 1 mol He at 0.25 atm. The temperature is held constant at 300 K, and the stopcock is opened. Calculate ΔS. (*Hint:* You had better devise a reversible process to carry out this change if you want the correct answer.)

9. Suppose that in Problem 8 the second flask contained Ar at 0.25 atm instead of He at 0.25 atm. How would that affect your answer?

10. Using the data of Problem 4, calculate the absolute third-law entropy of water vapor at 100 °C and 1 atm. (The room-temperature value for liquid water is listed in Table 8.1.)

11. The accompanying table lists values of the heat capacity C_p of silver as a function of temperature. (P. F. Meads, W. R. Forsythe, and W. F. Giauque, *J. Amer. Chem. Soc.* 63 (1941):1902.) Assume that $C_p = AT^3$ between 0 K and 15 K and that $C_p = 0$ at 0 K. Evaluate A; and then by plotting the data, calculate the absolute entropy of Ag at 100, 200, and 298.1 K. (*Ans.*: 17.17, 32.76, and 42.70 at 100 200, and 298.1 K. The units are J K^{-1} mol^{-1}.) Suppose that there is a 10% error in extrapolating from 15 K to $T = 0$. How large an error would this introduce in the room-temperature value?

T (K)	C_p (J K^{-1} mol^{-1})	T (K)	C_p (J K^{-1} mol^{-1})
15	0.67	120	21.60
20	1.72	140	22.59
30	4.77	160	23.30
40	8.39	180	23.90
50	11.65	200	24.27
60	14.31	220	24.57
70	16.33	240	24.89
80	17.89	260	25.21
90	19.13	280	25.39
100	20.96	298.1	25.49
		300	25.50

12. An isolated system consists of a block of A and a block of B, with heat capacities C_A and C_B that are independent of temperature. The initial temperatures of the blocks are $T_{i,A}$ and $T_{i,B}$. The two blocks are brought into thermal contact, and the final temperatures are given by $T_{f,A}$ and $T_{f,B}$. Show how you could deduce that $T_{f,A} = T_{f,B}$ from the fact that the entropy of the system is the highest when the final temperatures are equal.

13. Fill in all the steps leading to the derivations of Equations (8.27), (8.28), and (8.29).

14. Show that for an ideal gas $(\partial H/\partial P)_T = 0$.

15. Show that for a van der Waals gas $(\partial E/\partial V)_T = a/V^2$.

16. By using procedures analogous to those in the text, show that

$$\left(\frac{\partial S}{\partial V}\right)_T = \left(\frac{\partial P}{\partial T}\right)_V$$

(*Hint:* Start with $S = S(T, P)$; $dS = (\partial S/\partial T)_V dT + (\partial S/\partial V)_T dV$. Note: In Equation (1.17) we showed that $(\partial P/\partial T)_V = \alpha/\beta$, where α is the coefficient of thermal expansion and β is the compressibility.)

17. Show that $dS = (C_p/T) dT - V\alpha\, dP$. (*Hint:* Start with $S = S(T, P)$.)

18. When gases are dealt with, the entropy effects due to compression and expansion can be quite large. For liquids and solids, they are quite small. For liquid water at room temperature, $\alpha = 2 \times 10^{-4}$ K^{-1} and $\beta = 5 \times 10^{-5}$ atm^{-1}. The density is 1 g cm^{-3}. Calculate the entropy change involved in compressing 1 mol water isothermally from 1 atm to 1000 atm. (*Hint:* Use the result of Problem 17.)

19. Carry out the compression of Problem 18, now doing it reversibly and adiabatically. Calculate the temperature increase of the water.

20. Another equation of general validity is

$$dS = \frac{C_v}{T} dT + \frac{\alpha}{\beta} dV$$

Derive this equation. (*Hint:* This was started in Problem 16.)

21. Show that for the special case of the ideal gas, the equation of Problem 20 is equivalent to (8.30).

22. Compute the entropy of mixing in producing a mole of methane–helium mixture that is 75% helium.

23. In forming 1 mol of a binary mixture consisting of component A and component B, there are an infinite number of possible concentrations ranging from pure A to pure B. Show that of all the possible concentrations, the one consisting of $\frac{1}{2}$ mol of each component is the one with the largest entropy.

24. At the absolute zero of temperature, orthohydrogen can exist in any one of nine quantum states. Calculate the residual entropy of orthohydrogen at the absolute zero if the concentrations of all the different states are equal.

25. Evaluate P^* in the example in the text. Show that ΔS_{syst} for Step 2 is 6.73 J K^{-1} by direct calculation.

CHAPTER NINE

WORK, FREE ENERGY, AND EQUILIBRIUM

9.1 REVERSIBLE WORK IS MAXIMUM WORK

On several occasions we have noted that the work obtained from a reversible process is always greater than the work obtained from an irreversible process. We are now in a position to demonstrate this for the general case. First we must take the previous discussion of Clausius's inequality one step further.

Consider a cyclic process in which a system is taken irreversibly from state a to state b and then returned reversibly to state a. Using Equation (7.33), we have

$$\oint \frac{dq}{T} = \int_a^b \frac{dq_{\text{irr}}}{T} + \int_b^a \frac{dq_{\text{rev}}}{T} < 0$$

or

$$\int_a^b \frac{dq_{\text{irr}}}{T} + \int_b^a dS < 0$$

If we interchange the limits on the second integral,

$$\int_a^b \frac{dq_{\text{irr}}}{T} - \int_a^b dS < 0$$

and

$$\int_a^b dS > \int_a^b \frac{dq_{\text{irr}}}{T}$$

If the change from a to b is infinitesimal, then $dS > dq_{\text{irr}}/T$, or

$$T\, dS > dq_{\text{irr}} \tag{9.1}$$

Now we consider a process for which the energy change is $dE = dq + dw$. The value dE will be the same regardless of whether the process is carried out reversibly or irreversibly. The value dq for the reversible and irreversible processes is

$$dq_{\text{rev}} = dE - dw_{\text{rev}} = T\, dS \tag{9.2a}$$
$$dq_{\text{irr}} = dE - dw_{\text{irr}} < T\, dS \tag{9.2b}$$

If Equation (9.2a) is subtracted from (9.2b), we are left with

$$dw_{\text{rev}} - dw_{\text{irr}} < 0 \tag{9.3}$$

Equation (9.3) states that dw_{rev} must have a greater negative value than dw_{irr}. Recalling that work done *by* the system is a negative quantity, we can conclude that the work obtained from a reversible process is always greater than the work obtained from an irreversible process.

9.2 MAXIMUM USEFUL WORK

Suppose we consider a general process A → B. If it is carried out at constant pressure, there is a necessary work term, $P\,dV$, and this work is wasted. Thus if fuel is burned at constant pressure to provide energy for a Carnot engine, expansion work will be done against the atmosphere if the system experiences an increase in volume. This work is lost. The heat of the process can be passed to a Carnot engine, where it is converted into *useful* work. We can generally write

$$w = w_{\text{useful}} + w_{\text{exp}} \tag{9.4}$$

where w_{exp} is the expansion work done against the surroundings. The term w_{useful} gives the net useful work, over and above the expansion work produced by an energy converter. There is no restriction on w_{useful}, and it can be electrical or any other type of work, including expansion work, which can be harnessed to useful purposes. Any process carried out at constant volume implies that $w_{\text{exp}} = 0$, and all the work is useful work. Now suppose we examine the production of useful work under various boundary conditions.

MAXIMUM USEFUL WORK AT CONSTANT S AND V; ENERGY

The first law of thermodynamics can be written

$$dE = dq + dw = dq + dw_{\text{useful}} + dw_{\text{exp}}$$
$$= dq + dw_{\text{useful}} - P\,dV$$

For maximum work we must have a reversible process, in which case $dq = T\,dS$, and

$$dE = dw_{\text{useful}} + T\,dS - P\,dV \tag{9.5}$$

The boundary conditions of constant S and V imply that $dV = 0$ and $dS = 0$, or

$$dw_{\text{useful}} = dE \tag{9.6}$$

The maximum useful work obtainable from a process at constant S and V is given by the energy change of the system. Since work done *by* the system is negative by our convention, this means that when the system does work on the surroundings at constant T and V, the energy of the system *decreases*. This decrease in energy is what measures the maximum useful work obtainable from the process at constant S and V, or

$$dw_{\text{useful}}(\text{by the system}) = -dE \tag{9.6a}$$

Now let's try constant S and P.

MAXIMUM USEFUL WORK AT CONSTANT S AND P: ENTHALPY

We can write the differential of the enthalpy as

$$dH = dE + d(PV) = dE + P\,dV + V\,dP$$
$$= dq + dw_{\text{useful}} + dw_{\text{exp}} + P\,dV + V\,dP \tag{9.7}$$

If we wish the maximum useful work, the process must be reversible, in which case $dq = T\,dS$. Also, since $dw_{\exp} = -P\,dV$, those two terms will cancel, and

$$dH = dw_{\text{useful}} + T\,dS + V\,dP$$

If we now impose the boundary condition of constant S and P, then

$$dw_{\text{useful}} = dH$$

and things are beginning to look a bit interesting. The maximum useful work obtainable from a process at constant pressure and entropy is given by the enthalpy decrease for the process.

We could have defined enthalpy in this way by starting with Equation (9.5). If reversibility and constant entropy are imposed, (9.5) becomes

$$dE = dw_{\text{useful}} - P\,dV \qquad (9.8)$$

At constant P, we can rearrange Equation (9.8):

$$dw_{\text{useful}} = (dE + P\,dV) = d(E + PV) = dH \qquad (9.9)$$

If we had not previously defined enthalpy, (9.9) would be a perfectly good definition. Let's continue this procedure.

MAXIMUM USEFUL WORK AT CONSTANT T AND V: HELMHOLTZ ENERGY

By requiring reversibility and constant V, Equation (9.5) takes the form

$$dE = T\,dS + dw_{\text{useful}}$$

If constant temperature is now imposed, we can write

$$dw_{\text{useful}} = (dE - T\,dS) = d(E - TS) = dA \qquad (9.10)$$

The function $E - TS$ is called the *work function*, or the *Helmholtz free energy*, or simply the *Helmholtz energy*. In this book, we shall use *Helmholtz energy*. It is symbolized by the letter A. The decrease in A measures the maximum useful work obtainable from a process at constant T and V.

HERMANN LUDWIG FERDINAND von HELMHOLTZ (1821–1894), German scientist, worked in areas spanning the range from physics to physiology. His paper *Uber die Erhaltung der Kraft* ("On the Conservation of Force," 1847) was one of the epochal papers of the century. Along with Mayer, Joule, and Kelvin, he is regarded as one of the founders of the conservation of energy principle. His *Physiological Optics* was in its time the most important publication ever to have appeared on the physiology of vision. In connection with these studies he invented the ophthalmoscope in 1851, still a fundamental tool of every physician. His *Sensations of Tone* (1862) established many of the basic principles of physiological acoustics. Helmholtz, in addition to his work on thermodynamics and physiology, made important contributions to electrodynamics, meteorology, and hydrodynamics.

MAXIMUM USEFUL WORK AT CONSTANT T AND P: FREE ENERGY

At constant T and P, Equation (9.5) becomes

$$dE = T\,dS + dw_{useful} - P\,dV$$

which can be rearranged to

$$\begin{aligned}dw_{useful} &= (dE - T\,dS + P\,dV)\\ &= d(E - TS + PV) = dG\end{aligned} \qquad (9.11)$$

The function $(E - TS + PV)$ is variously called the *Gibbs function*, the *Gibbs potential*, the *Gibbs free energy*, *free energy*, or the *Gibbs energy*.[1] In this book we shall refer to it as the *free energy*. Its symbol is G.

$$G = E - TS + PV \qquad (9.12a)$$
$$G = H - TS = A + PV \qquad (9.12b)$$

The decrease in free energy of a process is a measure of the maximum useful work obtainable from that process at constant temperature and constant pressure. Since both free energy G and the Helmholtz energy A are functions of state functions, they are state functions themselves.

9.3 FUNDAMENTAL RELATIONS

We now have five thermodynamic state functions: $E, H, A, G,$ and S.[2] In this section we shall develop several extremely useful expressions, which we shall use over and over again as we progress through the material. No new concepts are being introduced; we are merely rewriting old ideas in new forms.

We can write the equations for H, G, and A in differential form as

$$dH = dE + P\,dV + V\,dP$$
$$dA = dE - T\,dS - S\,dT$$
$$dG = dE - T\,dS - S\,dT + P\,dV + V\,dP$$

In what follows we shall restrict the discussion to expansion work. Using Equation (8.12), we can write

$$dE = T\,dS - P\,dV \qquad (9.13)$$

Substituting this value for dE in the three previous equations, we get

$$dH = T\,dS + V\,dP \qquad (9.14)$$

---

[1] The Commission on Symbols, Terminology, and Units of the Division of Physical Chemistry of the International Union of Pure and Applied Chemistry (IUPAC) has recommended that the term *Helmholtz energy* be used for A, and that the term *Gibbs energy* be used for G. (M. L. McGlashan, *Pure and Applied Chemistry* 21 (1970):1.) We have adopted their recommendation for A. The older term *free energy* is still entrenched in chemical usage, particularly elementary chemistry textbooks, hence we use the older term for G.

[2] This list of five does not exhaust the total number of possibilities. For ordinary processes we can ignore the effects of gravitational fields, electric and magnetic fields, and others. We are now concerned only with expansion work. There are processes in which nonexpansive work is important. These introduce additional work terms that must be added to the energy, and new functions can be invented that measure the total useful work available from say a process at constant temperature and magnetic field. We shall examine these in Chapter 18. See also Problem 4.

SEC. 9.3 FUNDAMENTAL RELATIONS

$$dA = -S\,dT - P\,dV \qquad (9.15)$$
$$dG = -S\,dT + V\,dP \qquad (9.16)$$

If we set the left-hand sides of Equations (9.13)–(9.16) equal to zero, we get

$$\left(\frac{\partial S}{\partial V}\right)_E = \frac{P}{T} \qquad (9.17)$$

$$\left(\frac{\partial S}{\partial P}\right)_H = \frac{-V}{T} \qquad (9.18)$$

$$\left(\frac{\partial V}{\partial T}\right)_A = \frac{-S}{P} \qquad (9.19)$$

$$\left(\frac{\partial P}{\partial T}\right)_G = \frac{S}{V} \qquad (9.20)$$

For an ideal gas, since both energy and enthalpy depend only on the temperature, Equations (9.17) and (9.18) give the isothermal dependence of entropy with volume and pressure respectively. For one mole of gas at constant temperature, Equation (9.17) leads to

$$dS = \frac{P\,dV}{T} = \frac{R\,dV}{V}$$

$$\Delta S = R \int_{V_1}^{V_2} \frac{dV}{V} = R \ln \frac{V_2}{V_1} = R \ln \frac{P_1}{P_2} \qquad \text{(ideal gas)}$$

which is identical to the previously determined expression for the isothermal expansion of an ideal gas given in (8.2). We could have arrived at the same result from (9.18).

We can write expressions for the four functions in functional form as

$$E = E(S, V) \qquad (9.21)$$
$$H = H(S, P) \qquad (9.22)$$
$$A = A(T, V) \qquad (9.23)$$
$$G = G(T, P) \qquad (9.24)$$

In differential form, the functions can also be written as

$$dE = \left(\frac{\partial E}{\partial V}\right)_S dV + \left(\frac{\partial E}{\partial S}\right)_V dS \qquad (9.25)$$

$$dH = \left(\frac{\partial H}{\partial P}\right)_S dP + \left(\frac{\partial H}{\partial S}\right)_P dS \qquad (9.26)$$

$$dA = \left(\frac{\partial A}{\partial V}\right)_T dV + \left(\frac{\partial A}{\partial T}\right)_V dT \qquad (9.27)$$

$$dG = \left(\frac{\partial G}{\partial P}\right)_T dP + \left(\frac{\partial G}{\partial T}\right)_P dT \qquad (9.28)$$

If we now compare Equations (9.25)–(9.28) with (9.13)–(9.16) and equate coefficients, we get

> JAMES CLERK MAXWELL (1831–1879), British physicist, presented his first scientific paper to the Royal Society of Edinburgh at the age of 15. In chemistry he is best known for his Maxwell distribution and his contributions to the kinetic theory of gases. In physics his name is most often associated with his Maxwell equations for electromagnetic fields. He was the first holder of a newly established professorship of experimental physics at Cambridge University. At Cambridge, Maxwell directed the preliminary plans for the Cavendish Laboratory.

$$\left(\frac{\partial E}{\partial V}\right)_S = -P \qquad \left(\frac{\partial E}{\partial S}\right)_V = T$$

$$\left(\frac{\partial H}{\partial P}\right)_S = V \qquad \left(\frac{\partial H}{\partial S}\right)_P = T$$

$$\left(\frac{\partial A}{\partial V}\right)_T = -P \qquad \left(\frac{\partial A}{\partial T}\right)_V = -S \qquad (9.29)$$

$$\left(\frac{\partial G}{\partial P}\right)_T = V \qquad \left(\frac{\partial G}{\partial T}\right)_P = -S$$

Applying the Euler conditions for an exact differential to Equations (9.13)–(9.16), we obtain a set of useful relations called *Maxwell's equations*.

$$\left(\frac{\partial T}{\partial V}\right)_S = -\left(\frac{\partial P}{\partial S}\right)_V \qquad (9.30)$$

$$\left(\frac{\partial T}{\partial P}\right)_S = \left(\frac{\partial V}{\partial S}\right)_P \qquad (9.31)$$

$$\left(\frac{\partial P}{\partial T}\right)_V = \left(\frac{\partial S}{\partial V}\right)_T \qquad (9.32)$$

$$\left(\frac{\partial V}{\partial T}\right)_P = -\left(\frac{\partial S}{\partial P}\right)_T \qquad (9.33)$$

9.4 EQUILIBRIUM UNDER VARIOUS BOUNDARY CONDITIONS

We are now ready to answer the question we set out to solve, namely, What are the equilibrium conditions for a system, and in particular what are the equilibrium conditions for the boundary conditions of constant temperature and constant pressure? A brief glance at Equations (9.13)–(9.16) indicates that:

At constant S and V, $dE = 0$
At constant S and P, $dH = 0$
At constant T and V, $dA = 0$
At constant T and P, $dG = 0$.

Thus, at constant T and P, the equilibrium state will be specified for that state for which $dG = 0$.

These four equilibrium conditions state simply that the respective functions are extremals, either minima or maxima. They are in fact minima, as has been noted for energy. To elegantly demonstrate that they are minima, we should have to show that the second differentials d^2E, d^2H, d^2A, and d^2G are greater than zero.[3] We shall approach the problem in a less elegant fashion.

From a simple point of view, one can say that if a system is displaced from equilibrium, it can do work as it returns to equilibrium. We have previously shown that the maximum useful work obtainable from a system equals the decrease in energy, enthalpy, Helmholtz energy, or free energy, depending on the boundary conditions. If these functions are to decrease as the system does work while returning to equilibrium, the relevant functions must have increased when the system was displaced from equilibrium. This can occur only if the equilibrium was a minimal extremal position. We can also show that they are minima by using Clausius's inequality. We shall do this only for the free energy; the others are very similar. We start with the definition $G = E + PV - TS$, and write it in differential form as $dG = dE + P\,dV + V\,dP - T\,dS - S\,dT$. Applying the boundary conditions of constant P and T, and the first law of thermodynamics, $dE = dq + dw$, we get

$$dG = dq + dw + P\,dV - T\,dS$$

Restricting ourselves to expansion work, $dw = -P\,dV$, this becomes

$$dG = dq - T\,dS \tag{9.34}$$

Clausius's inequality tells us that $dq \leq T\,dS$, where the equality applies to a reversible process, and the "less than" applies to an irreversible process. We can thus write

$$dG \leq 0 \tag{9.35}$$

Here the equality sign applies to a system in equilibrium, and the "less than" applies to nonequilibrium conditions. Thus as a system tends towards equilibrium from a nonequilibrium starting position, the free energy decreases to a minimum.

9.5 EXPERIMENTAL DETERMINATION OF ΔG FOR CHEMICAL REACTIONS

Equation (9.12b) immediately suggests a convenient calorimetric method for determining free energy changes for chemical reactions. Since $G = H - TS$,

$$dG = dH - T\,dS - S\,dT \tag{9.36}$$

At constant temperature this becomes

$$dG = dH - T\,dS$$

which can be integrated to

$$\Delta G = \int dG = \int dH - \int T\,dS = \int dH - T\int dS$$
$$\Delta G = \Delta H - T\,\Delta S \tag{9.37}$$

If we know ΔH and ΔS for the reaction at the specified temperature, we can get ΔG from Equation (9.37). For the reaction

$$H_2(g) + \tfrac{1}{2}O_2(g) \rightleftharpoons H_2O(l)$$

The standard enthalpy of formation $\Delta H_f^\circ = -285.840$ kJ mol^{-1} and $\Delta S^\circ = -163.16$ J K^{-1} mol^{-1} at standard temperature and pressure. The free energy of formation ΔG_f°

---

[3] See for example H. B. Callen, *Thermodynamics* (New York: John Wiley & Sons, 1966), chap. 6.

TABLE 9.1 Free energies of formation of various substances.

Substance	ΔG_f° (kJ mol^{-1})	Substance	ΔG_f° (kJ mol^{-1})
$O_2(g)$		$P_2(g)$	102.9
$H_2(g)$		$P_4(g)$	24.35
$H_2O(g)$	−228.5958	C(graphite)	
$H_2O(l)$	−237.1918	C(diamond)	2.8660
$F_2(g)$		CO(g)	−137.2683
$Cl_2(g)$		$CO_2(g)$	−394.3830
HCl(g)	−95.265	$CH_4(g)$	−50.794
$Br_2(l)$		$C_2H_2(g)$	209.200
$Br_2(g)$	3.142	$C_2H_4(g)$	68.124
HBr(g)	−53.22	$C_2H_6(g)$	−32.886
$I_2(c)$		$C_2H_5OH(g)$	−168.62
$I_2(g)$	19.37	$C_2H_5OH(l)$	−174.77
HI(g)	1.30	Pb(c)	
$H_2S(g)$	−33.020	$PbCl_2(c)$	−313.97
$SO_2(g)$	−300.37	$PbBr_2(c)$	−260.41
NO(g)	86.688	Hg(l)	
$NO_2(g)$	51.84	$Hg_2Cl_2(c)$	−210.66
$N_2O(g)$	103.60	Ag(c)	
$N_2O_4(g)$	98.286	AgCl(c)	−109.721
$NH_3(g)$	−16.636	AgBr(c)	−95.939
P(g)	279.11	S(rhombic)	
P(c, white)		S(monoclinic)	0.096

SOURCE: National Bureau of Standards Circular 500.

NOTE: Standard conditions 1.000 atm and 25 °C obtain.

is thus −237.19 kJ mol^{-1}. The superscript ° indicates that all products and reactants are in their standard states, and the subscript f indicates that it is a free energy of formation. Table 9.1 lists standard free energies of formation for numerous compounds. You will use this table like the table of enthalpies of formation.

For a system at constant pressure and temperature in which the initial and final states are in equilibrium, $dG = 0$. For our water example, we know that the products and reactants are not in equilibrium at 1 atm and 25 °C, and of course ΔG is not equal to zero. The problem of finding the equilibrium conditions thus reduces to finding the conditions under which ΔG will be zero. In other words, under what pressure conditions will the products and reactants be in equilibrium at 25 °C? We shall return to this particular point in the next chapter.

9.6 SOME PRACTICAL WORK CONSIDERATIONS

Let's examine two methods by which useful work can be obtained from the reaction $H_2 + \tfrac{1}{2}O_2 = H_2O(l)$. In the first method, we burn the H_2 in the furnace of a Carnot engine at constant pressure. The heat is the enthalpy change, $\Delta H = -285.8$ kJ mol^{-1}. The heat of this reaction can be passed to the Carnot engine, where part of it is converted into work and the rest discarded as "waste heat" to the cold reservoir. We have seen that in a modern large steam turbine installation with the hot reservoir at 811 K and the cold reservoir at 310 K, the theoretical maximum efficiency is 62%. Under these conditions, the maximum theoretical useful work obtainable from the reaction is $0.62 \times 285.8 = 177$ kJ mol^{-1} H_2 burned.

In the second method, we get work from the reaction by directly converting the chemical energy into electrical work at constant pressure and temperature in a *fuel cell*. With this method, the maximum useful work obtainable is given by the decrease in free energy of the reaction, or 237.2 kJ of work per mole of H_2 consumed. Thus, in theory one can get about 34% more work in the fuel cell than in the steam-power plant, with a reduction of many of the attendant pollution problems of the steam plant. This explains the present interest in fuel cells in an energy-short technological society.

When H_2 reacts with O_2 at constant T and P to give $H_2O(l)$, the maximum useful work is $\Delta G = \Delta H - T\,\Delta S$. The term ΔH measures the heat energy involved, but not all the heat is available for useful work. The $T\,\Delta S$ term is a measure of the energy that is unavailable for useful work (in a cyclic process) under the most favorable circumstances.

9.7 DIGRESSION: HOW DIFFERENT ARE THE THERMODYNAMIC FUNCTIONS?

Before continuing with the mainstream of our discussion, it will be instructive to consider the interrelations between the thermodynamic functions from a different perspective. We have constructed five functions. Entropy (or more correctly $T\,dS$) is in a class by itself in that it measures the *heat* of a reversible process. The remaining four are similar to each other in that they provide a measure of the available *work* for reversible processes under various boundary conditions.

In mechanics a conservative field is one for which a force is derivable from a potential. An analogous situation exists in thermodynamics in which an intensive property is obtained from the various thermodynamic functions, thus $P = -(\partial E/\partial V)_S$. Work terms arise from the product of an intensive property with its associated extensive property, for example, $P\,dV$. When we compare these work terms with the definition of work, $F\,dx$, the intensive properties such as P take the form of generalized forces. For this reason, the functions $E, H, A,$ and G are often referred to as *potentials*. Heat is also measured by the product of an intensive and an extensive property; in the expression $T\,dS$, the term T is the intensive and S the extensive property.

Now recall Equations (9.21)–(9.24), in which the four potentials are written in terms of their natural variables. The energy E is a function of the *extensive* properties of the system, S and V. What we have accomplished in constructing $H, A,$ and G from E is to substitute an intensive property for its associated extensive property. Enthalpy is generated from energy by replacing the extensive property V by its associated intensive property P. The Helmholtz energy A is generated by replacing S by T; and G is generated by simultaneously replacing S and V by T and P. These can be regarded as analogous to coordinate transformations such as the transformation from cartesian to polar coordinates. Here the transformation involves replacing an extensive property by its associated intensive property. When viewed in this light, the functions $H, A,$ and G are simply the energy transformed into a different set of variables. In mathematics, such a transformation is known as a *Legendre transformation*.

9.8 DIGRESSION ON LEGENDRE TRANSFORMATIONS[4]

The procedure by which the energy function is transformed into enthalpy, Helmholtz energy, and free energy is a special case of a general kind of transformation that the French mathematician, Legendre (1752–1833), discovered in his studies on solving differential equations. The transformation starts with a given function F of n variables $u_1, u_2, \ldots, u_n$:

$$F = F(u_1, u_2, \ldots, u_n) \tag{9.38}$$

A new variable v is introduced such that

$$v_i = \frac{\partial F}{\partial u_i} \tag{9.39}$$

and a new function G is defined such that

$$G = \sum_i^n u_i v_i - F \tag{9.40}$$

Since Equation (9.39) allows us to solve for the u_i as functions of the v_i, we can write G as

$$G = G(v_1, v_2, \ldots, v_n) \tag{9.41}$$

We now consider the infinitesimal variation of G produced by arbitrary infinitesimal variations in the v_i,

$$\delta G = \sum_i \left(\frac{\partial G}{\partial v_i}\right) \delta v_i \tag{9.42}$$

Using the form for G expressed in (9.40), we can write the variation as

$$\delta G = \sum_i (u_i\, \delta v_i + v_i\, \delta u_i) - \delta F \tag{9.43}$$

Since

$$\delta F = \sum_i \left(\frac{\partial F}{\partial u_i}\right) \delta u_i \tag{9.44}$$

(9.43) can be written as

$$\delta G = \sum_i \left[u_i\, \delta v_i + \left(v_i - \frac{\partial F}{\partial u_i}\right) \delta u_i\right] \tag{9.45}$$

From Equation (9.39) we note that the quantity in parentheses, $[v_i - (\partial F/\partial u_i)] = 0$; thus

$$\delta G = \Sigma u_i\, \delta v_i \tag{9.46}$$

---

[4] This section, while relevant, is not necessary for our development of thermodynamics. It is placed here for the interested student. If you are confused by this section, just forget it. If you do read this, and your interest in the subject is aroused, a more complete presentation is in the reference of footnote 3, and also in C. Lanczos, *The Variational Principles of Mechanics* (Toronto: University of Toronto Press, 1957), p. 161 et. seq.

SEC. 9.8 DIGRESSION ON LEGENDRE TRANSFORMATIONS

> ADRIEN MARIE LEGENDRE (1752–1833), French mathematician, is intimately associated with the development of elliptic functions, a subject to which he devoted some forty years of his life. His name is most familiar to chemists for solving differential equations by Legendre polynomials; these polynomials form the basis for the solution of the wave equation in quantum mechanics. He was the first to publish the method of least squares (though Gauss may have anticipated him). The method of least squares is invaluable in analyzing experimental data, and it provided the basis for the rapid advances in astronomy in the early nineteenth century. Legendre's textbook on geometry, *Eléments de Géométrie*, which first appeared in 1794, was one of the most successful mathematics textbooks ever published.

Comparing (9.42) and (9.46) gives the remarkable result that

$$u_i = \frac{\partial G}{\partial v_i} \tag{9.47}$$

Comparing this with Equation (9.39) demonstrates the beautiful symmetry in the transformation. In tabular form, the scheme resembles the accompanying chart.

Old		New
$u_1, u_2, \ldots, u_n$	Variable	$v_1, v_2, \ldots, v_n$
$F = F(u_1, u_2, \ldots, u_n)$	Function	$G = G(v_1, v_2, \ldots, v_n)$
$v_i = \frac{\partial F}{\partial u_i}$	New variable	$u_i = \frac{\partial G}{\partial v_i}$
$G = \Sigma u_i v_i - F$	New defined function	$F = \Sigma u_i v_i - G$
$G = G(v_1, v_2, \ldots, v_n)$	New function	$F = F(u_1, u_2, \ldots, u_n)$

Old system and *new system* lose their meaning in this symmetrical transformation. *Old* and *new* are equivalent. The *new* variables are the partial derivatives of the old function with respect to the old variables, but the *old* variables are the partial derivatives of the new functions with respect to the new variables. It makes no difference which way we go. The new function is the Legendre transformation of the old function, but the old function is the Legendre transformation of the new function. We can go in either direction (compare Equation (9.29)).

We have showed the transformation for the case in which all the variables are changed. It is not necessary to change all the variables. In the thermodynamic transformation, we may only change one or two of the variables.

If we assume that F is a function of two sets of variables u_i and w_i, we can write

$$F = F(u_1, u_2, \ldots, u_n; w_1, w_2, \ldots, w_m) \tag{9.48}$$

Here the w_i are independent of the u_i. They do not participate in the transformation; and they occur in F simply as parameters. The new function G will likewise contain them. The w_i are the *passive* and the u_i the *active* variables of the transformation. When one carries through the analysis, the previous results (that is, Equation (9.47)) will remain unchanged. There will be one additional equation (again compare Equation (9.29)),

$$\frac{\partial F}{\partial w_i} = \frac{\partial G}{\partial w_i} \tag{9.49}$$

In thermodynamics, the enthalpy is the partial Legendre transform of the energy that replaces the volume by the pressure. The Helmholtz energy is the partial Legendre transform that replaces the entropy by the temperature as the independent variable. The Gibbs free energy is the transform that simultaneously replaces the entropy by the temperature and the volume by the pressure as the independent variables. To see how the scheme works, we shall tabulate the transformation for the Helmholtz energy in both directions (see accompanying table). The variable V in this transformation is the passive variable. The application of (9.49) tells us that

$$\left(\frac{\partial E}{\partial V}\right)_S = \left(\frac{\partial A}{\partial V}\right)_S$$

	$E \rightarrow A$	$A \rightarrow E$
Start	$E = E(S, V)$	$A = A(T, V)$
New variable	$T = \left(\dfrac{\partial E}{\partial S}\right)_V$	$-S = \left(\dfrac{\partial A}{\partial T}\right)_V$
New function defined	$A = E - TS$	$E = A + TS$
New function	$A = A(T, V)$	$E = E(S, V)$

In a previous section we showed that the minimum Helmholtz energy principle was nothing more than the minimum energy principle in a different guise. This follows immediately from the properties of the Legendre transform. If the Helmholtz energy is the transform of the energy, then the energy is the transform of the Helmholtz energy. Energy, enthalpy, Helmholtz energy, and free energy are essentially different ways of looking at the same thing.

Before leaving this digression, it is worth mentioning that the Legendre transformation is useful in classical mechanics. The Lagrangian is defined as a function of the $2n$ variables of coordinates q and velocities v:

$$L = L(v_1, v_2, \ldots, v_n; q_1, q_2, \ldots, q_n)$$

The generalized momenta are defined as the derivatives

$$p_i = \frac{\partial L}{\partial v_i}$$

A new function called the Hamiltonian is defined by

$$-H = L - \sum_i p_i v_i$$

where the velocities are now given by the derivative of H with respect to the p_i. This transformation in mechanics is useful because a set of n simultaneous second-order differential equations is converted to a set of $2n$ first-order differential equations that is easier to handle. Additional information on these equations is available in textbooks on mechanics.

9.9 SOME SIMPLE FREE-ENERGY CALCULATIONS

ISOTHERMAL EXPANSION OF AN IDEAL GAS

For the process $IG(P_1, V_1, T) \rightarrow IG(P_2, V_2, T)$, we find that $\Delta E = 0 = q + w$, and $\Delta H = 0$.

SEC. 9.9 SOME SIMPLE FREE-ENERGY CALCULATIONS

$$q_{\text{rev}} = -w_{\text{rev}} = \int P\, dV = nRT \ln \frac{V_2}{V_1} = nRT \ln \frac{P_1}{P_2}$$

$$\Delta S = \frac{q_{\text{rev}}}{T} = nR \ln \frac{V_2}{V_1}$$

$$\Delta G = \Delta H - T\,\Delta S = 0 - nRT \ln \frac{V_2}{V_1}$$

$$\Delta G = -nRT \ln \frac{V_2}{V_1} = -nRT \ln \frac{P_1}{P_2} \tag{9.50}$$

The free energy change ΔG for the isothermal expansion of an ideal gas is the reversible work. We have written w_{rev} and q_{rev}, but this does not imply that (9.50) is valid only for a reversible process. Since G is a state function, ΔG must be the same for any type of process. We perform the calculation for the path that takes us from the initial state to the final state in the way that can be most easily calculated; this path is the reversible path. For an isothermal process, $\Delta A = \Delta E - T\,\Delta S$. Since $\Delta E = 0$, it is always true that $\Delta A = \Delta G$ for the isothermal expansion of an ideal gas.

NORMAL BOILING

In this example we consider a liquid boiling at its normal boiling point (the temperature at which the vapor pressure is 1 atm). For water the process is

$$\text{H}_2\text{O}(l,\ 100\ °\text{C},\ 1\ \text{atm}) \longrightarrow \text{H}_2\text{O}(g,\ 100\ °\text{C},\ 1\ \text{atm})$$

Again, $\Delta G = \Delta H - T\,\Delta S$. Here ΔH is the heat of vaporization, since the pressure is constant; ΔS is $\Delta H_{\text{vap}}/T$. Therefore

$$\Delta G = \Delta H_{\text{vap}} - \frac{T\,\Delta H_{\text{vap}}}{T} = 0$$

as we knew it should be before we even started the calculation. After all, liquid water is in equilibrium with gaseous water at 100 °C and 1 atm. The condition for equilibrium at constant temperature and pressure is that $\Delta G = 0$. This is a general result. For any phase transition, $\Delta G_{\text{trans}} = 0$ at the transition temperature and pressure, where the two phases are in equilibrium.

CALCULATION OF VAPOR PRESSURE

Consider the vaporization of water at 25 °C:

$$\text{H}_2\text{O}(l,\ 25\ °\text{C},\ 1\ \text{atm}) \longrightarrow \text{H}_2\text{O}(g,\ 25\ °\text{C},\ P)$$

P is the vapor pressure of liquid water at 25 °C, and the problem is to find P. Since the two phases are in equilibrium, ΔG must be zero. Call it $\Delta G_{\text{I}} = 0$. We can solve the problem by the following two-step equivalent procedure.

$$\text{H}_2\text{O}(l,\ 25\ °\text{C},\ 1\ \text{atm}) \xrightarrow{\Delta G_{\text{I}} = 0} \text{H}_2\text{O}(g,\ 25\ °\text{C},\ P)$$
$$\searrow_{\Delta G_{\text{II}}} \qquad \nearrow_{\Delta G_{\text{III}}}$$
$$\text{H}_2\text{O}(g,\ 25\ °\text{C},\ 1\ \text{atm})$$

We can calculate ΔG_{III} if we hold that H_2O vapor behaves as an ideal gas; it is just ΔG for an isothermal expansion:

$$\Delta G_{iii} = -RT \ln \frac{P_1}{P_2} = +RT \ln \frac{P_2}{P_1} = RT \ln P$$

We can get ΔG_{ii} from a table of standard free energies. It is $\Delta G_{f,g}^\circ - \Delta G_{f,l}^\circ$ or $+8596$ J mol^{-1}. Since $\Delta G_i = \Delta G_{ii} + \Delta G_{iii} = 0$, we have[5]

$$RT \ln P = -8596$$

$$\ln P = -\frac{-8596}{8.314 \times 298.16} = -3.4675$$

$$P = 0.0312 \text{ atm} = 23.7 \text{ torr}$$

This is a special case of an equilibrium constant. (*Note:* The value for ΔG_f° of $H_2O(g)$ got into the table from someone's measurement of vapor pressure. We have worked the procedure backwards.)

9.10 MATHEMATICAL MANIPULATIONS OF THERMODYNAMIC FUNCTIONS

PRESSURE DEPENDENCE OF THE FREE ENERGY

Using one of the results of Equation (9.29), we can write

$$\left(\frac{\partial G}{\partial P}\right)_T = V \qquad (9.51)$$

For an isothermal change, $dG = V\, dP$, and for an ideal gas

$$\Delta G = \int dG = \int_{P_1}^{P_2} V\, dP = -RT \ln \frac{P_1}{P_2} \qquad \text{(ideal gas only)}$$

This is identical to Equation (9.50).

TEMPERATURE DEPENDENCE OF THE FREE ENERGY

Another equation in (9.29) is $(\partial G/\partial T)_p = -S$. Since $G = H - TS$, we have $S = (H - G)/T$, and this derivative can be written[6]

$$\left(\frac{\partial G}{\partial T}\right)_P = \frac{G - H}{T} = -S \qquad (9.52)$$

Since

$$\frac{d}{dT}\left(\frac{G}{T}\right) = \frac{1}{T}\frac{dG}{dT} - \frac{G}{T^2}$$

---

[5] There is a subtle point here. We seem to be taking the logarithm of a dimensioned quantity, since we write $\ln P$. Actually we are taking the logarithm of the dimensionless quantity $P = P_2/P_1$. Since P_1 is unity, we have simply omitted it from the expression. On several occasions, we shall write expressions such as $\ln P$; in all such instances, P will represent the dimensionless ratio of two pressures.

[6] Since the entropy equals zero at the absolute zero of temperature, (9.52) inplies that G and H must be equal at $T = 0$.

Equation (9.52) can be written

$$\left[\frac{\partial(G/T)}{\partial T}\right]_P = -\frac{H}{T^2} \qquad (9.53)$$

or in another equivalent form—

$$\left[\frac{\partial(G/T)}{\partial(1/T)}\right]_P = H \qquad (9.54)$$

Equations (9.52)–(9.54) are known as the *Gibbs-Helmholtz equations*. By recognizing that in going from state A to state B,

$$\Delta G = G_B - G_A \qquad \Delta H = H_B - H_A \qquad \Delta S = S_B - S_A$$

and applying the Gibbs-Helmholtz equations to G_A and G_B, we get

$$\left(\frac{\partial \Delta G}{\partial T}\right)_P = -\Delta S = \frac{\Delta G - \Delta H}{T} \qquad (9.55)$$

from Equation (9.52). If we start with (9.54) instead, the result is

$$\left[\frac{\partial(\Delta G/T)}{\partial(1/T)}\right]_P = \Delta H \qquad (9.56)$$

Equation (9.56) provides a convenient operational procedure for determining ΔH. Many reactions are not amenable to experimental measurement of ΔH but are amenable to measurement of ΔG. Equation (9.56) stipulates that for ΔG measured as a function of temperature, if $\Delta G/T$ is plotted against T^{-1}, the slope of the curve is given by ΔH.[7]

ENTROPY VARIATION

Two additional relations given in Equation (9.29) are

$$\left(\frac{\partial G}{\partial P}\right)_T = V \qquad \text{and} \qquad \left(\frac{\partial G}{\partial T}\right)_P = -S$$

Since the mixed derivatives are equal, we can take

$$\left[\frac{\partial}{\partial T}\left(\frac{\partial G}{\partial P}\right)_T\right]_P = \left[\frac{\partial}{\partial P}\left(\frac{\partial G}{\partial T}\right)_P\right]_T$$

and get

$$\left(\frac{\partial S}{\partial P}\right)_T = -\left(\frac{\partial V}{\partial T}\right)_P$$

which is identical to (9.33). Recognizing that $(\partial V/\partial T)_P = \alpha V$, where α is the coefficient of thermal expansion, we can write

$$\left(\frac{\partial S}{\partial P}\right)_T = -\alpha V \qquad (9.57)$$

---

[7] In the most general case, the slope of the curve is not constant, and ΔH is determined by constructing a tangent to the curve. If ΔH is independent of T, then the curve is a straight line whose slope is ΔH.

For the temperature variation of entropy, we shall work with two columns, one for constant volume and one for constant pressure.

Constant Volume *Constant Pressure*

$$\left(\frac{\partial S}{\partial T}\right)_V = \left(\frac{\partial S}{\partial E}\right)_V \left(\frac{\partial E}{\partial T}\right)_V = \frac{C_v}{T} \qquad \left(\frac{\partial S}{\partial T}\right)_P = \left(\frac{\partial S}{\partial H}\right)_P \left(\frac{\partial H}{\partial T}\right)_P = \frac{C_p}{T}$$

$$dS = \frac{C_v\, dT}{T} \qquad\qquad dS = \frac{C_p\, dT}{T}$$

$$\Delta S = \int_{T_1}^{T_2} C_v\, d(\ln T) \qquad\qquad \Delta S = \int_{T_1}^{T_2} C_p\, d(\ln T)$$

We have seen these results previously, and have used them in our discussion of the third law of thermodynamics.

9.11 REDUCTION OF DERIVATIVES

Certain kinds of data are easily measurable, whereas others may involve a great investment of time and effort. Thus, for example, we may be engaged in high-pressure experiments in which we need to know how the enthalpy changes with pressure at constant temperature. We need, in other words, the derivative $(\partial H/\partial P)_T$. We could determine the derivative by undertaking a series of calorimetric experiments over a range of pressure values, but calorimetric experiments are difficult and time-consuming. We want to determine the information in the easiest possible way. We have meters to measure temperature, pressure, and volume. We do not have entropy meters, enthalpy meters, free-energy meters, or Helmholtz-energy meters. Variables that are easily measurable include compressibilities, coefficients of thermal expansion, and heat capacities.

The rules of the game are quite simple. We start with the thermodynamic functions S, E, H, A, and G, and the variables P, V, and T, and create a derivative of the form $(\partial X/\partial Y)_Z$, where X is any one of the eight listed functions, Y is any of the remaining seven, and Z is any of the remaining six. The end result is to be a simple expression involving only either those quantities such as T, P, and V that can be easily measured, or quantities such as C_p, α, and β that can be found in standard tables. The starting point will be one of the various expressions in Equations (9.13)–(9.33). It will be useful to recall that

$$C_p = \left(\frac{\partial H}{\partial T}\right)_p = T\left(\frac{\partial S}{\partial T}\right)_p \qquad \alpha = \frac{(\partial V/\partial T)_p}{V}$$

$$C_v = \left(\frac{\partial E}{\partial T}\right)_v = T\left(\frac{\partial S}{\partial T}\right)_v \qquad \beta = \frac{-(\partial V/\partial P)_T}{V}$$

$(\partial E/\partial V)_T$

We derived an expression for this derivative in Equation (8.25). Now let us get the same expression in a different manner and take the result a few steps further.

SEC. 9.11 REDUCTION OF DERIVATIVES

$$\left(\frac{\partial E}{\partial V}\right)_T = \left(\frac{\partial (A + TS)}{\partial V}\right)_T$$

$$= \left(\frac{\partial A}{\partial V}\right)_T + T\left(\frac{\partial S}{\partial V}\right)_T$$

$$= -P + T\left(\frac{\partial P}{\partial T}\right)_V \tag{9.58}$$

In Equation (1.17) we showed that $(\partial P/\partial T)_v = \alpha/\beta$. Using this result, we get

$$\left(\frac{\partial E}{\partial V}\right)_T = -P + T\frac{\alpha}{\beta} \tag{9.59}$$

This derivative is zero for the ideal gas; it is generally not zero for real gases, liquids, or solids.

The variation of H with pressure is obtained in a similar manner:

$$\left(\frac{\partial H}{\partial P}\right)_T = \left(\frac{\partial (G + TS)}{\partial P}\right)_T = \left(\frac{\partial G}{\partial P}\right)_T + T\left(\frac{\partial S}{\partial P}\right)_T$$

$$= V - T\left(\frac{\partial V}{\partial T}\right)_p \tag{9.60}$$

$$\left(\frac{\partial H}{\partial P}\right)_T = V(1 - T\alpha) \tag{9.61}$$

For an ideal gas $(\partial H/\partial P)_T$ is also zero, since $\alpha = T^{-1}$.

$C_P - C_V$

From the definitions of C_p and C_v, the difference is given by

$$C_p - C_v = \left(\frac{\partial H}{\partial T}\right)_p - \left(\frac{\partial E}{\partial T}\right)_v$$

and if $H = E + PV$ is substituted for H, we get

$$C_p - C_v = \left(\frac{\partial E}{\partial T}\right)_p + P\left(\frac{\partial V}{\partial T}\right)_p - \left(\frac{\partial E}{\partial T}\right)_v$$

Since $dE = (\partial E/\partial V)_T \, dV + (\partial E/\partial T)_v \, dT$, we can divide the equation by dT at constant pressure, to get

$$\left(\frac{\partial E}{\partial T}\right)_p = \left(\frac{\partial E}{\partial V}\right)_T \left(\frac{\partial V}{\partial T}\right)_p + \left(\frac{\partial E}{\partial T}\right)_v$$

Thus $$C_p - C_v = \left[P + \left(\frac{\partial E}{\partial V}\right)_T\right]\left(\frac{\partial V}{\partial T}\right)_p \quad (=R \text{ for an ideal gas})$$

Using (9.58) for $(\partial E/\partial V)_T$ yields

$$C_p - C_v = T\left(\frac{\partial P}{\partial T}\right)_v \left(\frac{\partial V}{\partial T}\right)_p$$

and finally

$$C_p - C_v = \frac{\alpha^2 VT}{\beta} \tag{9.62}$$

This is a very useful result since it is always possible to measure C_p experimentally, but except for gases it is almost impossible to measure C_v. Equation (9.62) provides numerical values for C_v from easily measurable quantities.

JOULE-THOMSON COEFFICIENT

$$\mu = \left(\frac{\partial T}{\partial P}\right)_H$$

$$dH = \left(\frac{\partial H}{\partial P}\right)_T dP + \left(\frac{\partial H}{\partial T}\right)_P dT = 0$$

at constant H; thus

$$\left(\frac{\partial T}{\partial P}\right)_H = -\frac{(\partial H/\partial P)_T}{(\partial H/\partial T)_P} \quad \text{(Could have been written directly from (1.13))}$$

$$= -\frac{1}{C_p}\left(\frac{\partial H}{\partial P}\right)_T$$

Thus we can write

$$\mu = \left(\frac{\partial T}{\partial P}\right)_H = \frac{V(T\alpha - 1)}{C_p} \quad (= 0 \text{ for an ideal gas}) \quad (9.63)$$

where we have used (9.61) for the last step. Recall that the inversion temperature is the temperature at which $\mu = 0$. From (9.63), we see immediately that $T_{\text{inv}} = (\alpha)^{-1}$.

$(\partial E/\partial P)_G$

While this example is of less practical utility than the previous three, it demonstrates a more complicated procedure.

$$\left(\frac{\partial E}{\partial P}\right)_G = T\left(\frac{\partial S}{\partial P}\right)_G - P\left(\frac{\partial V}{\partial P}\right)_G \quad \text{(since } dE = T\,dS - P\,dV\text{)}$$

$$= \frac{-T(\partial G/\partial P)_S}{(\partial G/\partial S)_P} + \frac{P(\partial G/\partial P)_V}{(\partial G/\partial V)_P} \quad \text{(by (1.13))}$$

$$= -T\frac{-S(\partial T/\partial P)_S + V}{-S(\partial T/\partial S)_P} + P\frac{-S(\partial T/\partial P)_V + V}{-S(\partial T/\partial V)_P} \quad \text{(since } dG = -S\,dT + V\,dP\text{)}$$

This can be reduced further by recognizing that

$$\left(\frac{\partial T}{\partial P}\right)_V = \frac{\beta}{\alpha} \quad \left(\frac{\partial T}{\partial V}\right)_P = \frac{1}{V\alpha} \quad \left(\frac{\partial T}{\partial S}\right)_P = \frac{T}{C_P}$$

and

$$\left(\frac{\partial T}{\partial P}\right)_S = -\frac{(\partial S/\partial P)_T}{(\partial S/\partial T)_P} \quad \text{(again by (1.13))}$$

$$= \frac{(\partial V/\partial T)_P}{C_P/T} \quad \text{(by (9.33))}$$

$$= \frac{V\alpha T}{C_P}$$

The final expression contains S, but S can be replaced by an integral involving C_P and T.

9.12 DEPENDENCE OF FREE ENERGY ON CHEMICAL COMPOSITION; THE CHEMICAL POTENTIAL

To this point, we have written the free energy as a function only of pressure and temperature, $G = G(P, T)$. Clearly, the free energy should also depend on the composition of the system. Our previous limitation to pressure and temperature did not cause any problems, since our discussion was limited to systems at constant composition. A complete functional description of the free energy must include the chemical composition:

$$G = G(P, T, n_1, n_2, \ldots, n_i) \tag{9.64}$$

where the n_i are the mole numbers of each constituent in the system. The differential of G, including composition dependence, can be written as

$$dG = \left(\frac{\partial G}{\partial T}\right)_{P,n_i} dT + \left(\frac{\partial G}{\partial P}\right)_{T,n_i} dP + \sum_i \left(\frac{\partial G}{\partial n_i}\right)_{P,T,n_j} dn_i \tag{9.65}$$

The derivative $(\partial G/\partial n_i)_{P,T,n_j}$ was introduced by J. Willard Gibbs, who gave it the name *chemical potential*, symbol μ. In terms of μ, Equation (9.65) can be written

$$dG = -S\,dT + V\,dP + \sum_i \mu_i\,dn_i \tag{9.66}$$

At constant composition all the $dn_i = 0$, and $dG = -S\,dT + V\,dP$, the same form we have used previously. If P and T are constant, then (9.66) becomes

$$dG = \sum_i \mu_i\,dn_i$$

Suppose we have a system with two components, a and b. Equation (9.66) becomes

$$dG = -S\,dT + V\,dP + \mu_a\,dn_a + \mu_b\,dn_b \tag{9.67}$$

Note that each term on the right side of (9.67) has units of energy. The value $T\,dS$ is a measure of heat energy, and $P\,dV$ is a measure of work energy. The value $n\,d\mu$ is a measure of what we might call "chemical energy." Chemical energy is a measure of the work obtainable from a chemical reaction, like the electrical work we get from a chemical reaction that takes place in a battery.

Free energy is an extensive property of the system. The concept of chemical potential introduces nothing new for a pure substance. At constant T and P, the free energy varies linearly with the amount of substance present, as shown in Figure 9.1. Doubling the size of the system can be done only by taking twice as much of the same material, and the free energy then is just doubled. The chemical potential $(\partial G/\partial n)_{T,P}$ is the slope of the straight line in Figure 9.1.

On the other hand, if we take a system containing many components, there are many ways in which the size of the system can be changed, since any of the components can be added or subtracted. Suppose that at constant T and P we keep the amounts of all the components constant, except for one, the ith component. As this ith component is added (or subtracted) the composition of the system changes, and the linear relation of Figure 9.1 is destroyed. The free energy of the system will now change in some nonlinear manner as component i is added; this is shown in Figure 9.2. At any point, the chemical potential of the ith component μ_i is the slope of the curve at that value of n_i.

FIGURE 9.1 Free energy as a function of the number of moles of a pure substance. The slope of the line is the chemical potential, $\mu = (\partial G/\partial n)_{T,P}$.

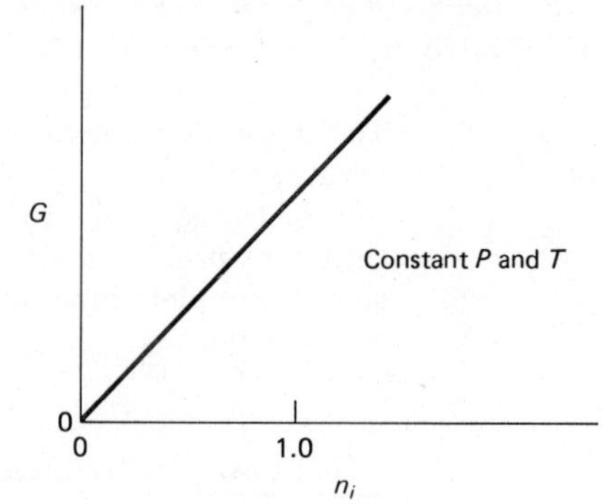

FIGURE 9.2 Free energy of a mixture of substances as a function of the number of moles of the ith component added to the mixture. The chemical potential at $n_i = n_i'$ is given by the slope of the curve at n_i.

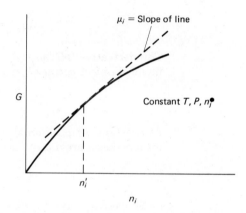

To obtain the dependence of other functions on composition, we could write expressions for $E, H,$ and A that include the effect of chemical composition by using the properties of the Legendre transformation. We shall, however, follow a simpler, more direct route. The differentials of $E, H,$ and A are

$$dE = dG - P\,dV - V\,dP + S\,dT + T\,dS$$
$$dH = dG + S\,dT + T\,dS$$
$$dA = dG - P\,dV - V\,dP$$

Using the expression for dG given in (9.66), we can rewrite these as

$$dE = T\,dS - P\,dV + \sum_i \mu_i\,dn_i \tag{9.68}$$

$$dH = T\,dS + V\,dP + \sum_i \mu_i\,dn_i \tag{9.69}$$

$$dA = -S\,dT - P\,dV + \sum_i \mu_i\,dn_i \tag{9.70}$$

At constant composition, the dn_i are zero, and Equations (9.66), (9.68), (9.69), and (9.70) reduce to the same form as (9.13) through (9.16). These equations also imply that

$$\mu_i = \left(\frac{\partial E}{\partial n_i}\right)_{S,V,n_j} = \left(\frac{\partial H}{\partial n_i}\right)_{S,P,n_j}$$

$$= \left(\frac{\partial A}{\partial n_i}\right)_{T,V,n_j} = \left(\frac{\partial G}{\partial n_i}\right)_{P,T,n_j} \quad (9.71)$$

These are a natural consequence of (9.49). They can also be derived by writing equations for the differentials analogous to (9.65) for E, H, and A, and equating the coefficients with those in (9.68), (9.69), and (9.70).

9.13 EQUILIBRIUM BETWEEN PHASES

Suppose we have a system with more than one substance (say alcohol in water), and suppose we have more than one phase present (say liquid and vapor). In particular, suppose we have two phases, a and b. What are the equilibrium conditions at constant pressure and temperature? We let n_i^a be the number of moles of the ith substance in phase a and n_i^b the number of moles of that substance in phase b. Constant P and T implies that $T^a = T^b$ and $P^a = P^b$. Suppose that we transfer an infinitesimally small amount of the ith substance dn_i from phase a to phase b. At equilibrium $dG = dG^a + dG^b = 0$, and since $dn_i^a = -dn_i^b = dn_i$, we can write

$$-\left(\frac{\partial G^a}{\partial n_i^a}\right)_{P,T,n_j} dn_i + \left(\frac{\partial G^b}{\partial n_i^b}\right)_{P,T,n_j} dn_i = 0$$

$$\left(\frac{\partial G^a}{\partial n_i^a}\right)_{P,T,n_j} = \left(\frac{\partial G^b}{\partial n_i^b}\right)_{P,T,n_j}$$

or
$$\mu_i^a = \mu_i^b \quad (9.72)$$

At equilibrium the chemical potential of i in phase a must equal the chemical potential of i in phase b. Two phases are in equilibrium (at constant T and P) if the chemical potentials of each and every constituent are the same in the two phases. This can be extended to many phases by considering a third phase c in equilibrium with either of the other phases, and so on. (We shall carry this discussion much further when we discuss phase equilibria and solutions starting in Chapter 11.)

9.14 FREE ENERGY AND EQUILIBRIUM; EXAMPLE

We set out to find the criterion for equilibrium at constant pressure and temperature, and we have indeed found it. Two possible states of a system will be in equilibrium if $\Delta G = 0$ for any process that links the two states. That is to say, the two states are in equilibrium if they have the same free energy. Let's examine this criterion for a simple reaction such as

$$C(\text{diamond}) = C(\text{graphite})$$

There are several questions one might ask about this reaction. The first is, Which is the more stable form of carbon? This can be answered immediately by referring to Table 9.1, which lists the free energies of formation. At 298 K and 1 atm, ΔG for the reaction is

$$\Delta G_f^\circ(\text{graphite}) - \Delta G_f^\circ(\text{diamond}) = -2.866 \text{ kJ mol}^{-1}$$

Since ΔG is negative, the reaction proceeds as written, and we can conclude that graphite is the more stable form.

We can get the same result by using the table of heats of formation and the table of standard entropies. From these we find

$$\Delta H = \Delta H° = 0 - 1.8962 = -1.8962 \text{ kJ mol}^{-1}$$
$$\Delta S = \Delta S° = 5.6940 - 2.4389 = +3.2551 \text{ J K}^{-1} \text{ mol}^{-1}$$
$$\Delta G = \Delta G° = \Delta H° - T \Delta S° = -1.8962 - (298)(0.0032551)$$
$$= -2.866 \text{ kJ mol}^{-1}$$

A second and more difficult question is the following: At what pressure are the two forms of carbon in equilibrium at room temperature? We want the derivative $(\partial G/\partial P)_T$. This is given in Equation (9.29) as $(\partial G/\partial P)_T = V$. By using the same procedure we used to get from Equation (9.54) to (9.56), this derivative can be converted to

$$\left(\frac{\partial \Delta G}{\partial P}\right)_T = \Delta V \tag{9.73}$$

where ΔV is the change in volume. The densities are 3.5 g cm^{-3} for diamond and 2.25 g cm^{-3} for graphite; thus for one mole of carbon,

$$\Delta V = \frac{12}{2.25} - \frac{12}{3.5} = 1.9 \text{ cm}^3 \text{ mol}^{-1}$$

Now we must integrate (9.73) between the limits $P = 1$ atm to P atm, and find that value of P for which $\Delta G = 0$. First we convert ΔG to units of cm^3 atm mol^{-1}. The conversion factor can be got by noting that the gas constant is 8.314 J K^{-1} mol^{-1}, or 82.05 cm^3 atm K^{-1} mol^{-1}. The factor is

$$\frac{82.05 \text{ cm}^3 \text{ atm K}^{-1} \text{ mol}^{-1}}{8.314 \text{ J K}^{-1} \text{ mol}^{-1}} = 9.87 \text{ cm}^3 \text{ atm J}^{-1}$$

$$\Delta G = 9.87(-2866) = -28{,}284 \text{ cm}^3 \text{ atm}$$
$$d(\Delta G) = \Delta V \, dP$$

$$\int_{-28,284}^{0} d(\Delta G) = 1.9 \int_{1}^{P} dP$$

$$0 + 28{,}284 = 1.9(P - 1)$$

or the equilibrium pressure at which diamond is in equilibrium with graphite at room temperature is about 15,000 atm.[8] Note that we have made several gross simplifications in this example. The most severe one is that the densities are not independent of pressure. In the next several chapters, we shall examine the conditions for equilibrium in detail.

PROBLEMS

1. For each of the reactions

$$H_2(g) + \tfrac{1}{2}O_2(g) = H_2O(g)$$
$$\tfrac{1}{2}N_2(g) + \tfrac{3}{2}H_2(g) = NH_3(g)$$
$$CO(g) + \tfrac{1}{2}O_2(g) = CO_2(g)$$

in which each substance is in the standard state of 1 atm and 25 °C, verify that Tables 6.2, 8.1, and 9.1 (standard enthalpies of formation, standard entropies, and standard free energies of forma-

---

[8] Graphite can be converted into diamond at 75,000 atm and 1500 K.

PROBLEMS

tion) are internally consistent. (That is, show that for each reaction $\Delta G_f^\circ = \Delta H_f^\circ - T\,\Delta S^\circ$.)

2. Calculate the maximum theoretical work obtainable by burning 1 mol graphite in the furnace of a Carnot engine that has a hot reservoir at 811 K and a cold reservoir at 310 K. Now calculate the maximum theoretical work obtainable from the reaction that directly combines the reactants in a fuel cell. Compare the two numbers. Do the same for CH_4 (take the products to be $CO_2(g)$ and $H_2O(l)$).

3. Carry out all the steps leading to Equation (9.11).

4. If a tension t is applied to a one-dimensional body, the length l of the body will change. The work associated with the tension is given by $dw_{ten} = t\,dl$. The expression for the energy becomes

$$dE = T\,dS - P\,dV + t\,dl$$

which can be transformed to a free energy with the form

$$dg = -S\,dT + V\,dP - l\,dt$$

If an electric field $\mathbf{E}$ is applied to a material, a polarization p results, and the work associated is $dw_{pol} = \mathbf{E}\,dp$. Under combined tension and electric field, the energy expression becomes

$$dE = T\,dS - P\,dV + t\,dl + \mathbf{E}\,dp$$

which can be transformed to a free energy with the form

$$dg = -S\,dT + V\,dP - l\,dt - p\,d\mathbf{E}$$

The piezoelectric effect relates the change of length produced in a body by an electric field to the electric polarization produced by an applied tension. Show that

$$\left(\frac{\partial l}{\partial \mathbf{E}}\right)_{T,P,t} = \left(\frac{\partial p}{\partial t}\right)_{T,P,\mathbf{E}}$$

5. A mole of ideal gas is reversibly and isothermally compressed at 300 K from 1 atm to 2 atm. Calculate the heat, work, energy change, enthalpy change, entropy change, and free-energy change for the process.

6. A mole of ideal gas is irreversibly and isothermally expanded at 300 K against a constant external opposing pressure of 1 atm. The initial pressure of the gas is 2 atm (that is, the gas is brought from the final state of Problem 5 back to the initial state in an irreversible manner.) Calculate the heat, work, energy change, enthalpy change, entropy change, and free-energy change for the process.

7. Calculate the heat, work, energy change, enthalpy change, entropy change, and free-energy change for the cyclic process consisting of the sum of the processes in Problems 5 and 6. Calculate the entropy change in the surroundings.

8. At 1 atm, the heat of vaporization of water is 40.67 kJ mol^{-1} at the normal boiling point 100 °C.

(a) At 100 °C and 1 atm pressure, 1 mol liquid water is converted to 1 mol gas. Calculate ΔH, ΔS, and ΔG.

(b) One mole of liquid water is placed in a previously evacuated chamber at 100 °C. The volume of the chamber is such that it will contain 1 mol gas at a pressure of 0.5 atm at 100 °C. What happens to the liquid water? Calculate ΔH, ΔS, and ΔG for this process.

9. For the process A → B, the value ΔG is 30.000 kJ at 25 °C, and 30.020 kJ at 26 °C. Estimate ΔS for the process.

10. The standard heat of formation and the third law entropy of n-pentane, $CH_3CH_2CH_2CH_2CH_3$, are −146.44 kJ mol^{-1} and 348.95 J K^{-1} mol^{-1}; the values for neopentane, $(CH_3)_4C$, are −165.98 kJ mol^{-1} and 306.39 J K^{-1} mol^{-1}. Which is more stable?

11. For the process A → B, the function $\Delta H = -50$ kJ and $\Delta S = -100$ J K^{-1}. Assuming that these values are independent of temperature, calculate ΔG for the process at 300 K and 1000 K. At what temperature is A in equilibrium with B?

12. For $H_2(g)$ and $C_2H_4(g)$ calculate the free energies of formation at 25 °C and a pressure 10^{-4} atm.

13. Some sulfur is burned, and the sulfur is converted to SO_2. The SO_2 has a final concentration in the air of 27 parts per million (ppm). The overall reaction can be written as

$$S(\text{rhombic}, P = 1 \text{ atm}) + O_2(g, P = 0.2 \text{ atm})$$
$$= SO_2(g, \text{pressure corresponding to 27 ppm by volume})$$

Calculate ΔG for the reaction at 25 °C.

14. Derive an expression for the internal energy $(\partial E/\partial V)_T$ for a van der Waals gas.

15. Show that $(\partial S/\partial P)_V = \beta C_v/\alpha T$.

16. For an ideal gas

$$\alpha = \frac{1}{V}\left(\frac{\partial V}{\partial T}\right)_P = \frac{1}{T}$$

Show that in general if $\alpha = 1/T$, then C_p must be independent of pressure, that is, that $(\partial C_p/\partial P)_T = 0$.

17. Show that $(\partial V/\partial P)_S = -V\beta C_v/C_p$.

18. For rhombic sulfur at 25 °C, $\Delta G_f^\circ = 0$ and $S^\circ = 31.88$ J K^{-1} mol^{-1}, and for monoclinic sulfur, $\Delta G_f^\circ = 96$ J mol^{-1} and $S^\circ = 32.55$ J K^{-1} mol^{-1}. Assume that the entropy difference does not vary with temperature. At what temperature are the two forms in equilibrium?

19. The free energy of formation of a substance is given as a function of temperature at 1 atm by the expression

$$\Delta G_f^\circ = 37{,}357 + 16.2T - 0.83T^2 \quad (\text{J mol}^{-1})$$

over the range 300–400 K. Derive an analytical expression for the enthalpy of formation over the temperature range.

20. Show that $(\partial C_p/\partial P)_T = -T(\partial^2 V/\partial T^2)_p$.

21. Show that $(\partial^2 G/\partial T^2)_p = -C_p/T$.

22. By using the heat capacity data of Table 5.2 and the standard enthalpies and free energies of formation, calculate $\Delta G°$ for the reaction $CO(g) + \tfrac{1}{2}O_2(g) = CO_2(g)$ at 1000 K.
23. Using the values of $\Delta H°_{f,298}$, $\Delta G°_{f,298}$, and C_p for $H_2O(g)$ and $H_2O(l)$, calculate the boiling point of water. How close is your answer to the correct value, 373.15 K? (*Note:* If you do this correctly, your answer should be within 0.5 °C of the experimental value.)

PART THREE
APPLICATIONS OF THERMODYNAMICS

CHAPTER TEN

THE EQUILIBRIUM CONSTANT FOR GAS REACTIONS

10.1 THE PRACTICAL FREE-ENERGY SCALE

The practical free-energy scale of Table 9-1 is used exactly like the practical enthalpy scale that was introduced in Chapter 6. The free energies of all elements in their normal states at standard temperature and pressure are assigned the value zero. The free energies of compounds are determined by measuring the free-energy change for the reaction that forms the compound from its elements. These free-energy values can be thought of as "absolute" free energies of compounds at standard conditions, and they are called standard free energies of formation, ΔG_f°. Tables of standard free energies list values at 298.15 K, and the symbol is often supplemented to $\Delta G_{f(298.15)}^\circ$. The symbol ΔG° is not restricted to reactions at room temperature. It is often important to know the values at other temperatures. The symbol $\Delta G_{f(400)}^\circ$ implies the free-energy change for the reaction forming the substance from its elements where all products and reactants are at 1 atm and 400 K. The equations we have developed for the variation of enthalpy, free energy, and entropy, with temperature and pressure enable us to calculate values of free-energy changes at other temperatures and pressures.

10.2 STANDARD FREE ENERGIES OF IDEAL GASES

For any pure one-component system, $dG = V\,dP - S\,dT$. At constant temperature $dT = 0$, and the $S\,dT$ term is eliminated. If we restrict ourselves to ideal gases, $V = RT/P$, and

$$dG = V\,dP = RT\,d(\ln P) \tag{10.1}$$

Our standard tabulating procedure lists the free energies of the gases at $P = 1$ atm. But what are the free energies of ideal gases at other pressures? Equation (10.1) can be integrated from the initial state at 1 atm to the final state at P,

$$\int_{G^\circ}^{G} dG = \int_{P^\circ}^{P} RT\,d(\ln P)$$

$$G - G^\circ = RT \ln \frac{P}{P^\circ}$$

$$= RT \ln P \qquad \text{(since } P^\circ = 1 \text{ atm)} \tag{10.2}$$

where P is measured in atmospheres. In Equation (10.2), $G°$ refers to the tabulated standard free energy, and G is the free energy of the ideal gas at the pressure P.

If we have a mixture of ideal gases, the total pressure is the sum of the partial pressures:

$$P = \sum_i P_i = \frac{RT}{V} \sum_i n_i \tag{10.3}$$

For each gas in the mixture, we can write

$$\overline{G}_i - \overline{G}_i° = RT \ln P_i \tag{10.4}$$

where the bar over the G_i indicates the free energy *per mole* of the ith gas or chemical potential, μ.[1] In terms of μ, Equation (10.4) is written as

$$\mu_i - \mu_i° = RT \ln P_i \tag{10.5}$$

For n_i moles of gas, we can write

$$n_i(\overline{G}_i - \overline{G}_i°) = RTn_i \ln P_i = RT \ln (P_i{}^{n_i}) \tag{10.6}$$

or

$$n_i(\mu_i - \mu_i°) = RTn_i \ln P_i = RT \ln (P_i{}^{n_i}) \tag{10.6a}$$

10.3 THE EQUILIBRIUM CONSTANT FOR IDEAL GAS REACTIONS

We consider the reaction

$$a\text{A} + b\text{B} \rightleftharpoons c\text{C} + d\text{D}$$

where all species are ideal gases, and the lowercase letters are the mole numbers in the balanced equation. For this reaction,

$$\Delta G = cG(\text{C}) + dG(\text{D}) - aG(\text{A}) - bG(\text{B})$$

By using the standard free energies of formation from Table 9.1, we can calculate $\Delta G°$ for the reaction

$$a\text{A}(1\text{ atm}) + b\text{B}(1\text{ atm}) \rightleftharpoons c\text{C}(1\text{ atm}) + d\text{D}(1\text{ atm}) \tag{10.7}$$

at 298 K. What we are really interested in is the reaction

$$a\text{A}(P_\text{A}{}^{\text{eq}}) + b\text{B}(P_\text{B}{}^{\text{eq}}) \rightleftharpoons c\text{C}(P_\text{C}{}^{\text{eq}}) + d\text{D}(P_\text{D}{}^{\text{eq}}) \tag{10.8}$$

where $P_\text{A}{}^{\text{eq}}$ is the partial pressure of A at equilibrium, and similarly for the rest. For reaction (10.8), $\Delta G = 0$, since both the pressure and the temperature are constant.

For reaction (10.7), the free energy change can be written as

$$\Delta G° = \sum G°(\text{products}) - \sum G°(\text{reactants})$$
$$= \sum_i \nu_i G_i° \tag{10.9}$$

where the ν_i are the mole number coefficients in the balanced chemical equation (that is, $a, b, c,$ and d). The ν_i have a positive sign when they refer to a product and a

---

[1] Strictly speaking, the chemical potential is defined as $\mu_i = (\partial G/\partial n_i)_{T,P,n_j}$, as noted in the previous chapter. For an ideal gas, μ_i is independent of the n_j; the value μ_i follows the functional dependence shown in Figure 9.1 rather than 9.2.

negative sign when they refer to a reactant. Using (10.6), we have

$$\Delta G - \Delta G° = RT \sum_i \nu_i \ln P_i \tag{10.10}$$

At equilibrium, the P_i are the equilibrium partial pressures P_i^{eq}, and $\Delta G = 0$. Equation (10.10) becomes

$$-\Delta G° = RT \sum_i \nu_i \ln P_i^{eq} \tag{10.11}$$

or

$$\sum_i \nu_i \ln P_i^{eq} = -\frac{\Delta G°}{RT} \tag{10.12}$$

The left-hand side of (10.12) can be written out as

$$\sum_i \nu_i \ln P_i^{eq} = \ln\left[\frac{(P_C^{eq})^c (P_D^{eq})^d}{(P_A^{eq})^a (P_B^{eq})^b}\right] = \ln K_p \tag{10.13}$$

or

$$-\Delta G° = RT \ln K_p \tag{10.14}$$

The constant K_p is called the **equilibrium constant**. In Equation (10.14), $\Delta G°$ refers to the standard state, and K_p refers to the equilibrium state. Equation (10.14) enables us to calculate the equilibrium position for ideal gas reactions from the free energies in the standard states.

10.4 TEMPERATURE DEPENDENCE OF THE EQUILIBRIUM CONSTANT

Using the Gibbs-Helmholtz equation,

$$\left[\frac{\partial}{\partial T}\left(\frac{\Delta G°}{T}\right)\right]_P = \frac{-\Delta H°}{T^2}$$

we can substitute $-RT \ln K_p$ for $\Delta G°$, to get

$$\left[\frac{\partial}{\partial T}\left(\frac{-RT \ln K_p}{T}\right)\right]_P = \frac{-\Delta H°}{T^2} \tag{10.15}$$

This can be rearranged to give

$$\frac{\partial(\ln K_p)}{\partial T} = \frac{\Delta H°}{RT^2} \tag{10.16}$$

which can be rewritten

$$\frac{\partial(\ln K_p)}{\partial(1/T)} = -\frac{\Delta H°}{R} \tag{10.17}$$

If $\ln K_p$ is plotted as a function of T^{-1}, the slope of the curve is given by $-\Delta H°/R$.
Equation (10.16) can be integrated to give

$$\ln\left[\frac{K_p(T_2)}{K_p(T_1)}\right] = \int_{T_1}^{T_2} \frac{\Delta H°}{RT^2} dT \tag{10.18}$$

The integral in (10.18) can be evaluated by obtaining an analytical expression for $\Delta H°$ as a function of T from heat capacity data. If $\Delta H°$ is taken as constant (usually a reasonable assumption over small temperature differences), (10.18) becomes

$$\ln\left[\frac{K_p(T_2)}{K_p(T_1)}\right] = \frac{-\Delta H°}{R}\left(\frac{1}{T_2} - \frac{1}{T_1}\right) \tag{10.19}$$

We consider as an example the synthesis of ammonia.

10.5 EXAMPLE: THE HABER PROCESS FOR THE SYNTHESIS OF AMMONIA

A vital ingredient for waging war is high explosives. Necessary as ingredients in the manufacture of explosives are nitrogenous compounds. Before World War I, the main source of nitrogenous compounds for the manufacture of explosives in Germany was nitrates, which were imported from Chile. At the outbreak of hostilities, Britain imposed a tight blockade on the German ports cutting off Germany's supply of nitrates. The Germans were compelled to find a new local source of nitrogen to continue their war effort, and turned to Fritz Haber, who had previously developed a process for manufacturing ammonia from air and water. Haber's process enabled the Germans to base their explosives industry on ammonia as the basic nitrogen-containing compound instead of the imported Chilean nitrates. The reaction is

$$\tfrac{3}{2}H_2(g) + \tfrac{1}{2}N_2(g) \rightleftharpoons NH_3(g)$$

The nitrogen is obtained from air, and the hydrogen by electrolysis of water. We now examine the thermodynamic properties of the system with a view to determining the yield of NH_3.

At standard conditions, ΔH is $\Delta H_f°(NH_3 \text{ gas}) = -46.190 \text{ kJ mol}^{-1}$; $\Delta G°$ is $\Delta G_f°(NH_3 \text{ gas}) = -16.636 \text{ kJ mol}^{-1}$. Since the free-energy change is negative, the reaction will proceed as written. To check the yield of the reaction, we must calculate the equilibrium constant,

---

FRITZ HABER (1868–1934), German chemist, took his doctorate in organic chemistry but did his significant work in physical chemistry. He received the Nobel Prize in chemistry in 1919 for his work on the synthesis of ammonia. During World War I, he placed his services at the disposal of the German government and developed the Haber process for manufacturing ammonia on a large-scale practical basis. His research activities covered a broad range, including the study of fuel cells as a means of the direct electrochemical transformation of the energy of coal into electrical work. After the war, the Allies required Germany to pay the equivalent of 50,000 tons of gold in reparations, and Haber conceived the dramatic idea of extracting gold from seawater. It was then thought that seawater contained several milligrams of gold per ton. The project was doomed to failure, as it was eventually discovered that the actual content was about 0.001 mg/ton, and the project was abandoned in 1928. In 1911, Haber was called to direct the Kaiser Wilhelm Institut für Physikalische Chemie und Electrochemie, which became one of the finest laboratories of its kind in the world. In 1933, Fritz Haber, German patriot, was forced to resign by the anti-Jewish policies of the Nazi regime, and Haber became one of a large group of scientists forced to flee the country.

SEC. 10.5 EXAMPLE: THE HABER PROCESS FOR THE SYNTHESIS OF AMMONIA

$$\ln K_p = \frac{-\Delta G°}{RT} = \frac{16{,}636}{(8.314 \times 298)} = 6.711$$

$$K_p = 821 = \frac{P_{NH_3}}{(P_{H_2})^{0.5}(P_{N_2})^{1.5}} \tag{10.20}$$

Assume that we start with $P_{H_2} = 3$ atm and $P_{N_2} = 1$ atm. The system in the initial and final states can be summarized as follows (we let x be the final equilibrium pressure of NH_3):

Initial	$P = 1$	$P = 3$	$P = 0$
	$\tfrac{1}{2}N_2$ +	$\tfrac{3}{2}H_2$ =	NH_3
Final (equilibrium)	$P = 1 - 0.5x$	$P = 3 - 1.5x$	$P = x$

Before attempting a solution for x, it is worthwhile examining the problem to see what the possible range of values may be. Obviously x cannot be less than zero, thus a solution that yields a negative x is clearly impossible. Considerations of stoichiometry indicate that if the reaction is complete, then the maximum possible value of x is 2 atm. Thus, any solution that yields a value for x outside the range 0–2 clearly shows that we have made some error. Further, since the equilibrium constant is rather large, we expect the solution for x to lie near 2 atm. For our particular conditions, Equation (10.20) becomes

$$821 = \frac{x}{(1 - 0.5x)^{0.5}(3 - 1.5x)^{1.5}} \tag{10.21}$$

which has the solution $x = 1.957$, or 98% of the maximum possible yield. The reaction seems economically feasible. There is one problem; it is infinitesimally slow. Thermodynamic considerations tell us only where the final position of equilibrium lies but say nothing about the length of time it takes to get there. The economic feasibility of the reaction is vitiated by kinetic considerations.

We can increase the temperature and thereby increase the rate of the reaction. Suppose we increase the temperature to 773 K. First we assume that ΔH is independent of temperature, and use Equation (10.19) to calculate a new K_p at 773 K.

$$\ln\left[\frac{K(773)}{K(298)}\right] = \frac{46{,}190}{8.314}\left(\frac{1}{773} - \frac{1}{298}\right)$$

We get $K_p(773) = 0.008683$, and things look bad for the Germans.

We now calculate the yield at 773 K for the same quantities that we treated previously; we simply substitute the new $K_p = 0.008683$ for the old $K_p = 821$ in (10.21), everything else remaining the same. We now get $x = 0.043$. This is only 2% of the maximum yield, and the process is not economically feasible for manufacturing ammonia on a large scale. At low temperature, K_p is sufficiently large, but the process is too slow. At high temperature where the rate becomes sufficiently high, the equilibrium constant is too low. The solution is quite simple—use a high-pressure process.

Suppose we increase our starting pressures to 100 and 300 atm for N_2 and H_2 respectively. Our work sheet now summarizes the reaction as

Initial $P = 100$ $P = 300$ $P = 0$
              $\frac{1}{2}N_2$ + $\frac{3}{2}H_2$ = NH_3

Final
(equilibrium) $P = 100 - 0.5x$ $P = 300 - 1.5x$ $P = x$

and Equation (10.21) becomes

$$0.008683 = \frac{x}{(100 - 0.5x)^{0.5}(300 - 1.5x)^{1.5}} \tag{10.22}$$

When we solve this equation, we get $x = 104$ atm, or 52% of the maximum possible yield of 200 atm. This is economically feasible, and Germany was able to fight for several more years.

How well do our calculations agree with the experimental values? Larson and Dodge have measured the equilibrium constants for the ammonia equilibrium by measuring the NH_3 content of the gas resulting from an initial 3:1 ratio of H_2 and N_2.[2] At 773 K they obtained values for K_p of 0.00381, 0.00386, and 0.00388 at total pressures of 10, 30, and 50 atm total pressure. Our value 0.008683 differs from the experimental value by about a factor of 2. Now let's see whether our calculation can provide a result in better agreement with experiment if we take into account the temperature variation of $\Delta H°$.

By the procedure outlined in Chapter 6, we can find an analytical expression for $\Delta H°$ as a function of temperature. We start with Equation (6.24),

$$\left(\frac{\partial \Delta H}{\partial T}\right)_P = \Delta C_p$$

In the example of Chapter 6, we found ΔC_p to be given by

$$\Delta C_p = -31.4 + 32(10^{-3})T - 61(10^{-7})T^2$$

From (6.24) we get

$$\int_{\Delta H°_{298}}^{\Delta H°_T} d(\Delta H) = \int_{298}^{T} \Delta C_p \, dT$$

$$\Delta H°_T = \Delta H°_{298} + \int_{298}^{T} \Delta C_p \, dT$$

$$= -46{,}190 + \int_{298}^{T} \{-31.4 + 32(10^{-3})T - 61(10^{-7})T^2\} \, dT$$

$$\Delta H°_T = -38{,}200 - 31.4T + 16(10^{-3})T^2 - 20(10^{-7})T^3$$

We now use this value for $\Delta H°$ in Equation (10.18), and

$$\ln\left(\frac{K_p(773)}{K_p(298)}\right) = \frac{1}{R}\int_{298}^{773} \frac{1}{T^2}\{-38{,}200 - 31.4T + 16(10^{-3})T^2 - 20(10^{-7})T^3\} \, dT$$

$$= \frac{(-78.770 - 29.930 + 7.600 - 0.509)}{8.314}$$

$$= -12.221$$

$$K_p(773) = 821 \exp(-12.221)$$
$$= 0.0040$$

This result agrees quite well with the experimental value of 0.0038.

[2] A. T. Larson and R. L. Dodge, *J. Amer. Chem. Soc.* 45 (1923):2918.

Our results for ideal gases predict that the equilibrium constant is independent of pressure. Experimentally, K_p rises to 0.00498 and 0.00651 at total pressures of 300 and 600 atm respectively.[3] The assumption of ideality is not very good at high pressures, and we shall shortly see how these results can be improved by removing the restriction to ideal gases.

10.6 LE CHÂTELIER'S PRINCIPLE

In our example of the Haber process, the fact that the yield of ammonia could be increased by increasing the pressure is an example of *Le Châtelier's principle*. According to this principle, any spontaneous process that is induced by a deviation from the equilibrium state will be in a direction that will restore the system to equilibrium. Thus, if two halves of a system are initially in equilibrium at the same temperature, and the temperature of one half is increased, then heat will flow from the hotter to the colder half until equilibrium is restored. If heat is evolved in a chemical reaction, then increasing the temperature will tend to reverse the direction. If the volume decreases in a chemical reaction, then increasing the pressure will shift the position of equilibrium to the product side while keeping the same equilibrium constant.[4]

10.7 THERMODYNAMIC PROPERTIES OF A VAN DER WAALS GAS

Until now our discussions have been mainly restricted to ideal gases. With this restriction, most of the thermodynamic expressions take on a relatively simple form. In the remainder of this chapter we shall examine the forms these equations take for a van der Waals gas.

EXPANSION WORK

For any system, expansion work is given by $dw = -P\,dV$. Since the equation of state of an ideal gas is a simple one, $P = nRT/V$, substituting this expression into the work expression leads to $w = -RT\ln(V_2/V_1)$ for one mole of gas under isothermal conditions.

For real gases, the equations of state are complicated, and the work expressions are much more complicated. We shall restrict our discussion to van der Waals gases (vdw) that obey the equation of state,

$$\left(P + \frac{n^2 a}{V^2}\right)(V - nb) = RT \tag{10.23}$$

For one mole of gas this can be written in terms of pressure as

$$P = \frac{RT}{V - b} - \frac{a}{V^2} \tag{10.24}$$

---

[3] A. T. Larson, *J. Amer. Chem. Soc.* 46 (1924):367.
[4] A more complete discussion of Le Châtelier's principle can be found in J. G. Kirkwood and I. Oppenheim, *Chemical Thermodynamics* (New York: McGraw-Hill Book Co., 1961).

The reversible expansion work for a vdw gas is

$$-w = \int_{V_1}^{V_2} P\, dV = \int_{V_1}^{V_2} \left(\frac{RT}{V-b} - \frac{a}{V^2}\right) dV \quad (10.25)$$

If the reversible expansion is also isothermal, the temperature is constant and (10.25) can be integrated, to yield

$$-w_{vdw} = RT \ln\left(\frac{V_2 - b}{V_1 - b}\right) + a\left(\frac{1}{V_2} - \frac{1}{V_1}\right) \quad (10.26)$$

where V_1 and V_2 are the initial and final molar volumes.

ΔE FOR A VAN DER WAALS GAS

In Equation (9.58) we showed that

$$\left(\frac{\partial E}{\partial V}\right)_T = T\left(\frac{\partial P}{\partial T}\right)_V - P$$

For a vdw gas this becomes

$$\left(\frac{\partial E}{\partial V}\right)_T = \frac{RT}{V-b} - \frac{RT}{V-b} + \frac{a}{V^2} = \frac{a}{V^2} \quad (10.27)$$

The energy change for an isothermal expansion of a vdw gas is then

$$\Delta E = \int_{V_1}^{V_2} \left(\frac{a}{V^2}\right) dV = -a\left(\frac{1}{V_2} - \frac{1}{V_1}\right) \quad (10.28)$$

For an ideal gas, $\Delta E = 0$ for *any* isothermal process; for the vdw gas, ΔE is not zero and is given by Equation (10.28).

Since $\Delta E = q + w$, the heat for a reversible isothermal expansion of a vdw gas is

$$q = \Delta E - w = RT \ln\left(\frac{V_2 - b}{V_1 - b}\right) \quad (10.29)$$

For an adiabatic process, $q = 0$ and $dE = dw$. We have

$$dE = \left(\frac{\partial E}{\partial V}\right)_T dV + \left(\frac{\partial E}{\partial T}\right)_V dT$$

$$= \left(\frac{\partial E}{\partial V}\right)_T dV + C_v\, dT = dw = -P\, dV \quad (10.30)$$

This can be reduced to

$$\left(\frac{RT}{V-b} - \frac{a}{V^2}\right) dV + C_v dT = 0 \quad (10.31)$$

Equation (10.31) cannot be separated simply into terms involving V and T, and we shall not pursue the solution to this problem any further.

ΔG FOR A VAN DER WAALS GAS

Earlier in this chapter we started with the basic equation for free energy,

$$dG = V\, dP - S\, dT \quad (10.32)$$

and noted that for an isothermal process involving one mole of ideal gas

$$dG = RT\, d(\ln P)$$

$$\Delta G = RT \ln \frac{P_2}{P_1}$$

These are nice simple expressions. Equation (10.32) is still valid for a vdw gas (or any other gas), but the final expression for ΔG loses its simple form. From (10.24), we find that at constant temperature

$$dP = \left[\frac{-RT}{(V-b)^2} + \frac{2a}{V^3}\right] dV \qquad (10.33)$$

Using Equation (10.32) we get for the reversible isothermal expansion of a vdw gas

$$dG = V\, dP = \left[\frac{-VRT}{(V-b)^2} + \frac{2a}{V^2}\right] dV \qquad (10.34)$$

ΔG for the expansion can be found by integrating Equation (10.34) between V_1 and V_2. The integration is straightforward, and it is left for one of the problems. The final form is much more complicated than the ideal gas expression. To simplify calculations involving real gases, we shall use a new function called the *fugacity*.

10.8 FUGACITY

We developed the theory of the equilibrium constant for ideal gases by starting with the equation

$$dG = V\, dP = RT\, d(\ln P) \qquad (10.35)$$

which we integrated to arrive at

$$\overline{G} - \overline{G}^\circ = RT \ln P \qquad (10.4)$$

where $\overline{G}^\circ$ indicates free energy per mole in the standard state of 1 atm pressure. In terms of chemical potentials, we wrote

$$\mu - \mu^\circ = RT \ln P \qquad (10.5)$$

This form of the equation for free energies is so simple and leads to such a convenient form for the equilibrium constant that we should like to retain this form for all substances. With this in mind, G. N. Lewis introduced a new function called the

---

GILBERT NEWTON LEWIS (1875–1946), American chemist, began his career as a superintendent of weights and measures in the Philippines in 1904, after receiving the PhD degree from Harvard. His book *Thermodynamics and the Free Energy of Chemical Substances*, first published in 1923 in collaboration with M. Randall, is still in use in a new edition revised by K. S. Pitzer and L. Brewer. In 1916, Lewis observed that of the hundreds of thousands of known chemical compounds, less than ten contained an odd number of electrons, and he proposed the "electron pair" chemical bond. He extended the concept of acids and bases and spent his later years on studies of the electronic states of organic molecules.

fugacity; he defined it by an equation completely analogous to (10.35). The fugacity f is defined by the equation

$$d\mu = d\overline{G} = RT\, d(\ln f) \tag{10.36}$$

For a system with more than one component, the fugacity of the ith component f_i is defined by

$$d\mu_i = d\overline{G}_i = RT\, d(\ln f_i) \tag{10.37}$$

For an ideal gas, the standard state is chosen at unit pressure. For a real gas, the standard state is chosen at unit fugacity. If (10.37) is integrated between the given state and the standard state, then

$$\mu_i - \mu_i^\circ = RT \ln\left(\frac{f_i}{f_i^*}\right) \tag{10.38}$$

which is analogous to Equation (10.5); f_i^* is the analog of P_i°. The fugacity is an "idealized" pressure. For an ideal gas, the fugacity equals the pressure, and since all gases behave as ideal gases in the limit of small pressures, we have as a condition on f that

$$\frac{f}{P} \longrightarrow 1 \quad \text{as } P \longrightarrow 0 \tag{10.39}$$

In terms of fugacities, the equilibrium constant for the gas phase reaction

$$a\text{A} + b\text{B} \rightleftharpoons c\text{C} + d\text{D}$$

is given by

$$K_p = \frac{(f_\text{C})^c (f_\text{D})^d}{(f_\text{A})^a (f_\text{B})^b} \tag{10.40}$$

For ideal gases, the pressures equal the fugacities, and Equation (10.40) takes the same form as the defining expression for K_p for the ideal gas reaction. All the equations we used to develop the expression for the ideal gas equilibrium constant can be made generally valid by substituting f for P. What we now must do is develop a procedure for evaluating fugacities of real gases.

10.9 CALCULATING FUGACITY FOR REAL GASES

Figure 10.1 shows a plot of fugacity vs pressure for ideal gases and real gases. The solid line indicates the curve $f = P$ for ideal gases. The dashed curves indicate the dependence for real gases; they may lie either above or below the ideal gas curve and may cross it at one or more points. In the limit of small pressures, all the curves coincide. The standard state of unit fugacity is indicated for each of the three curves. For the ideal gas, the standard state is at 1 atm pressure. For the real gas, the standard state of unit fugacity will generally be at another pressure besides 1 atm.

Figure 10.2 shows a pressure-volume isotherm for a mole of real gas and a mole of ideal gas. The free energy change in going from P_1 to P_2 is

$$\Delta\overline{G} = \int_{P_1}^{P_2} \overline{V}\, dP = RT \ln \frac{f_2}{f_1} \tag{10.41}$$

FIGURE 10.1 Fugacity as a function of pressure for real gases (dashed curves) and ideal gas (solid curve). The standard states are at $f = 1$.

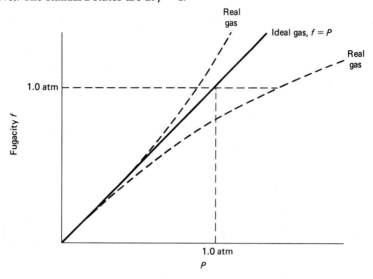

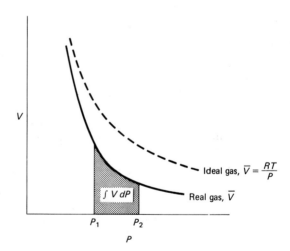

FIGURE 10.2 Graphical integration of $\Delta G = \int_{P_1}^{P_2} \overline{V}\, dP$ from P-V isotherm for a real gas.

and is indicated by the shaded area under the real gas curve. If we consider the derivative of the ratio f/P with respect to pressure, we get

$$\left[\frac{\partial \ln (f/P)}{\partial P}\right]_T = \left(\frac{\partial \ln f}{\partial P}\right)_T - \left(\frac{\partial \ln P}{\partial P}\right)_T \tag{10.42}$$

At constant temperature,

$$d\overline{G} = \overline{V}\, dP = RT\, d(\ln f) \tag{10.43}$$

and we can write

$$\left(\frac{\partial \ln f}{\partial P}\right)_T = \frac{\overline{V}}{RT} \tag{10.44}$$

where $\overline{V}$ is the molar volume. Equation (10.42) can now be written as

$$\left[\frac{\partial \ln (f/P)}{\partial P}\right]_T = \frac{\overline{V}}{RT} - \left(\frac{\partial \ln P}{\partial P}\right)_T$$

$$= \frac{\overline{V}}{RT} - \left(\frac{RT}{RT}\right)\left(\frac{1}{P}\right)\left(\frac{\partial P}{\partial P}\right)$$

$$= \frac{1}{RT}\left(\overline{V} - \frac{RT}{P}\right)$$

$$\left[\frac{\partial \ln (f/P)}{\partial P}\right]_T = -\frac{\alpha}{RT} \tag{10.45}$$

where we have set

$$\alpha = \left(\frac{RT}{P} - \overline{V}\right) \tag{10.46}$$

α is a measure of the nonideality of the gas. It is given by the difference between the volume a mole of ideal gas would have at the relevant pressure and the actual volume of the mole of real gas. If (10.45) is integrated from $P = 0$ to P at constant temperature, we get

$$\int_{P=0}^{P} d \ln \frac{f}{P} = -\frac{1}{RT} \int_0^P \alpha \, dP$$

$$\ln \frac{f}{P} - \left(\ln \frac{f}{P}\right)_{P=0} = -\frac{1}{RT} \int_0^P \alpha \, dP \tag{10.47}$$

Since f/P approaches unity as P approaches zero, the second term on the left of (10.47) goes to zero in the limit and

$$\ln f = \ln P - \frac{1}{RT} \int_0^P \alpha \, dP \tag{10.48}$$

Fugacities of real gases can be evaluated by plotting α as a function of pressure and carrying out the integral of (10.48) graphically. While each of the terms composing α as defined in (10.46) separately goes to infinitely large values as P approaches 0, their difference remains finite. Values of α are obtained from some tabulated function such as the compressibility factor, $Z = PV/RT$. If an equation of state is available for the gas, then an analytical expression can be got for α, and the integral evaluated analytically. We shall carry out this procedure for the van der Waals gas.

10.10 ANALYTICAL EVALUATION OF FUGACITY FOR A VAN DER WAALS GAS

From Equation (10.48)

$$RT \ln \frac{f}{P} = \int_0^P \left(V - \frac{RT}{P}\right) dP$$

$$= \int_0^P V \, dP - \int_0^P RT \, d(\ln P) \tag{10.49}$$

Using Equation (10.33) for dP, we can write

$$RT \ln \frac{f}{P} = -\int_0^P \frac{RTV}{(V-b)^2} dV + \int_0^P \frac{2a}{V^2} dV - RT \int_0^P d(\ln P) \tag{10.50}$$

By writing the first integral as

$$\int \frac{RTV}{(V-b)^2} dV = \int \frac{RT(V-b+b)}{(V-b)^2} dV$$

Equation (10.50) takes the form

$$RT \ln \frac{f}{P} = -\int_0^P \frac{RT(V-b)}{(V-b)^2} dV - \int_0^P \frac{RTb}{(V-b)^2} dV$$

$$+ \int_0^P \frac{2a}{V^2} dV - RT \int_0^P d(\ln P) \tag{10.51}$$

Carrying out the indicated integrations, we get

$$RT \ln \frac{f}{P} = -[RT \ln (V-b)]_0^P + \left[\frac{RTb}{V-b}\right]_0^P - \left[\frac{2a}{V}\right]_0^P - RT \ln P \Big]_0^P \tag{10.52}$$

Before inserting the limiting values, we combine the first and last terms to get

$$RT \ln \frac{f}{P} = -RT \ln [P(V-b)]\Big]_0^P + \left[\frac{RTb}{V-b}\right]_0^P - \left[\frac{2a}{V}\right]_0^P \tag{10.53}$$

There is no difficulty with the upper limit of P in (10.53). We find that at the lower limit, as $P \to 0$,

$$V \longrightarrow \infty$$
$$P(V-b) \longrightarrow PV = RT$$
$$\frac{RTb}{V-b} \longrightarrow 0$$
$$\frac{2a}{V} \longrightarrow 0$$

and (10.53) becomes

$$RT \ln f - RT \ln P = -RT \ln [P(V-b)] + \frac{RTb}{V-b} - \frac{2a}{V}$$
$$+ RT \ln (RT) + 0 + 0 \tag{10.54}$$

which can be reduced to the form

$$\ln f = \ln \frac{RT}{V-b} + \frac{b}{V-b} - \frac{2a}{RTV} \tag{10.55}$$

Equation (10.55) enables us to calculate fugacities from the van der Waals constants a and b, and the molar volume.

Equation (10.55) is rather a mess when compared to the equations for ideal gases, and you may well wonder whether it was worth the effort. What we have done is essentially this. For the ideal gas, we use the simple equation

$$dG = V\,dP = RT\,d(\ln P) \tag{10.56}$$

For a real gas, we have two choices. The first consists in staying with the equation $dG = V\,dP$ and developing it as we have done in (10.34), working with the equation of state at the very beginning.

In the second choice, we examine the form of Equation (10.56), marvel at its utter simplicity, and try to devise a scheme for real gases that retains this simple form. To do this, we invent the fugacity. We save the simple expression, but at the cost of a complicated variable. In the first choice we use a simple variable, namely P, and complicated expressions such as (10.34). In the second choice we have a simple expression,

$$dG = RT\,d(\ln f)$$

but now we have a complicated variable, namely f as given by (10.55). Is there any advantage to our choice? The answer, of course, is yes, else we should not have bothered. As we shall shortly see, a tabulating scheme has been developed to simplify the determination of fugacities. Further, a very important application of fugacities is the evaluation of equilibrium constants for real gas reactions, and the form for the equilibrium constant in terms of fugacities retains the simple form of the equilibrium constant for ideal gases in terms of pressure. In addition, when we discuss free energy and chemical potentials in solutions, we shall see that the fugacity is a special case of what we call an *activity*.

10.11 FUGACITY COEFFICIENTS

The various tabulating schemes for fugacities have been developed in terms of what we call *fugacity coefficients*,

$$\gamma = \frac{f}{P} \tag{10.57}$$

and the fugacity is obtained from the equation

$$f = \left(\frac{f}{P}\right)P = \gamma P \tag{10.58}$$

In the sense that fugacities are "idealized pressures," fugacity coefficients can be considered correction factors. When the real pressure is multiplied by the fugacity coefficient, the result is the fugacity (or effective pressure) that can be used in the simple ideal gas expressions to provide the correct results for the free-energy changes of real gases.

10.12 EQUILIBRIUM CONSTANT FOR REAL GAS REACTIONS

When we talk of reactions of the form

$$a\text{A(g)} + b\text{B(g)} \rightleftharpoons c\text{C(g)} + d\text{D(g)}$$

we are no longer speaking of pure gases but rather gas solutions. We shall assume, as G. N. Lewis did, that in a gas solution the fugacity of the ith component is given by

$$f_i = X_i f_i^* \tag{10.59}$$

where f_i^* is the fugacity of the pure gas at the same total pressure and temperature, and X_i is the mole fraction.[5] The definition of the fugacity coefficient can be extended to solutions by the expression

$$\gamma_i = \frac{f_i}{X_i P} \tag{10.60}$$

where P is the total pressure. If Equation (10.59) is valid for the gaseous solution, then

$$\gamma_i = \frac{X_i f_i^*}{X_i P} = \frac{f_i^*}{P} = \gamma_i^* \tag{10.61}$$

or the fugacity coefficient for the gaseous solution is the same as the fugacity coefficient for the pure gas.

We can derive the equilibrium constant for real gases by a procedure completely analogous to the derivation of the equilibrium constant for the ideal gas reaction as carried out earlier. One simply substitutes fugacity f for pressure P in the series of equations leading to the equilibrium constant. Under this procedure, we get for the equilibrium constant in terms of fugacities

$$K_f = \frac{(f_\text{C})^c (f_\text{D})^d}{(f_\text{A})^a (f_\text{B})^b} \tag{10.62}$$

Equation (10.62) is valid for both ideal and nonideal gases, since for the ideal gas, you recall, $f = P$. It is convenient to write (10.62) in terms of two constants:

$$K_f = \left\{ \frac{(\gamma_\text{C})^c (\gamma_\text{D})^d}{(\gamma_\text{A})^a (\gamma_\text{B})^b} \right\} \left\{ \frac{(P_\text{C})^c (P_\text{D})^d}{(P_\text{A})^a (P_\text{B})^b} \right\}$$

$$= K_\gamma K_p \tag{10.63}$$

K_p is our old friend the equilibrium constant in terms of pressure; K_γ can be thought of as a correction factor that converts the ideal gas equilibrium constant into an equilibrium constant for real gases. Now let's see how these results can be used to improve our prediction of the equilibrium constant for the Haber process.

10.13 EXAMPLE: THE HABER PROCESS REVISITED

Table 10.1 summarizes the calculation of K_p at 723 K over the pressure range 10–3500 atm based on the experimental yield of NH_3 from an initial mixture of H_2 and N_2 in the ratio $3:1$. The constant is not very constant. The results can be im-

---

[5] This implies that the intermolecular forces between similar molecules is the same as the forces between dissimilar molecules.

TABLE 10.1 Calculation of the ideal gas equilibrium constant for the synthesis of ammonia.

Total pressure (atm)	% NH$_3$ at equilibrium	P_{NH_3}[a]	P_{N_2}	P_{H_2}	K_p
10[b]	2.04	0.20	2.45	7.35	0.00655
30	5.80	1.74	7.07	21.20	0.00671
50	9.17	4.59	11.35	34.06	0.00685
100[c]	16.35	16.35	20.91	62.74	0.00719
300	35.50	106.5	48.4	145.1	0.00876
600	53.60	321.6	69.6	208.8	0.01278
1000	69.40	694.0	76.5	229.5	0.02282
1000[d]	70.51	705.1	73.7	221.2	0.02497
1500	84.07	1261.1	59.7	179.2	0.06801
2000	89.83	1796.6	50.9	152.6	0.13372
3500	97.18	3401.3	24.7	74.0	1.07510

NOTE: Temperature is 723 K.

[a] The third, fourth, and fifth columns list the partial pressures of NH$_3$, N$_2$, and H$_2$ calculated from the data in the first two columns.
[b] This and the next two entries are from A. T. Larson and R. L. Dodge, *J. Amer. Chem. Soc.* 45 (1923):2918.
[c] This and the next three entries are from the work of A. T. Larson, *J. Amer. Chem. Soc.* 46 (1924):367.
[d] The last four entries are from the work of L. J. Winchester and B. F. Dodge, *Amer. Inst. Chem. Eng. J.* 2 (1956):431.

proved substantially by using fugacities. We shall do this using the van der Waals equation of state.

To use Equation (10.55) to calculate the fugacities, we must have values of the molar volumes $\overline{V}$. The van der Waals equation can be written as a cubic polynomial equation in $\overline{V}$,

$$\overline{V}^3 - \left(b + \frac{RT}{P}\right)\overline{V}^2 + \left(\frac{a}{P}\right)\overline{V} - \frac{ab}{P} = 0 \qquad (10.64)$$

The calculation is summarized in Table 10.2. The molar volumes for the gases were obtained by solving (10.64); the fugacity coefficients rather than the fugacities have been entered in the table. The fugacity coefficients have been plotted as a function of pressure in Figure 10.3. Now the values of K_f in the last column are fairly constant out to 600 atm.[6]

In our discussion of real gases in Chapter 2 we noted that according to the law of corresponding states, the compressibility factor $Z = PV/RT$ for most pure gases can be adequately represented on a single universal chart in terms of the reduced variables $T_r = T/T_c$ and $P_r = P/P_c$, where T_c and P_c are the critical temperature and pressure. The function α is related to the compressibility factor by

$$\alpha = \left(\frac{RT}{P} - \overline{V}\right) = \left(\frac{RT}{P} - \frac{RTZ}{P}\right) = \frac{RT}{P}(1 - Z) \qquad (10.65)$$

Equation (10.48) can be integrated using (10.65) and reduced variables. This procedure yields a universal chart of fugacity coefficients, applicable to all gases to within the precision of the law of corresponding states. Figure 10.4 shows such a set of charts. To use the charts, you will need values of the critical constants.

[6] The theory breaks down at high pressures for two reasons. Firstly, the van der Waals equation breaks down at these very high pressures, and secondly, our assumption that the fugacity coefficients depend only on total pressure, and not on concentration, also breaks down.

SEC. 10.13 EXAMPLE: THE HABER PROCESS REVISITED

TABLE 10.2 Calculation of the equilibrium constant for the ammonia synthesis based on the van der Waals equation of state.

Total pressure (atm)	Ideal gas molar volume $\overline{V}_{IG}$ (cm³)	AMMONIA		HYDROGEN		NITROGEN		K_p (atm⁻¹)	K_γ	$K_f = K_\gamma K_p$ (atm⁻¹)
		$\overline{V}_{vdw}$ (cm³)	γ	$\overline{V}_{vdw}$ (cm³)	γ	$\overline{V}_{vdw}$ (cm³)	γ			
10	5932.7	5899.1	0.994	5954.7	1.004	5948.1	1.003	0.00655	0.9886	0.00648
30	1977.6	1944.4	0.983	2000.0	1.011	1993.7	1.008	0.00671	.9660	.00648
50	1186.5	1153.6	0.973	1209.1	1.019	1203.2	1.014	0.00685	.9438	.00646
100	593.3	560.9	0.946	616.0	1.039	611.0	1.029	0.00719	.8898	.00640
300	197.8	171.3	0.855	221.1	1.123	219.5	1.100	0.00876	.6997	.00613
600	98.88	90.68	0.782	122.88	1.266	125.25	1.243	0.01278	.5018	.00641
1000	59.33	68.52	0.790	83.94	1.492	89.68	1.507	0.02282	.3499	.00798
1500	39.55	59.13	0.897	64.67	1.840	72.76	1.972	0.06801	.2389	.01625
2000	29.66	54.54	1.084	55.12	2.277	64.52	2.628	0.13372	.1686	.02255
3500	16.95	48.32	2.245	42.93	4.368	54.02	6.560	1.07510	0.0639	0.06874

NOTE: The temperature is 723 K at the pressures listed in the first column. The second column lists the ideal gas molar volume. The molar volumes and fugacity coefficients for each of the gases as calculated from the van der Waals equation are listed in the next six columns. The values for the ideal gas equilibrium constant K_p are taken from Table 10.1. The last column lists $K_f = K_\gamma K_p$, where

$$K_\gamma = \frac{(\gamma_{NH_3})}{(\gamma_{N_2})^{1/2}(\gamma_{H_2})^{3/2}}$$

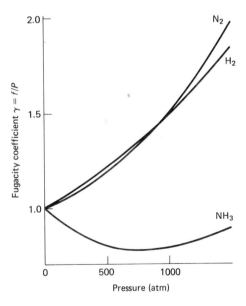

FIGURE 10.3 Fugacity coefficients as a function of pressure based on the van der Waals equation of state at 723 K.

The critical constants of gases are listed in Table 2.4. For the relevant gases in the ammonia synthesis, they are

$$\begin{array}{lll} NH_3 & T_c = 405.6 \text{ K} & P_c = 111.8 \text{ atm} \\ N_2 & T_c = 126.1 \text{ K} & P_c = 33.6 \text{ atm} \\ H_2 & T_c = 33.3 \text{ K} & P_c = 12.8 \text{ atm} \end{array}$$

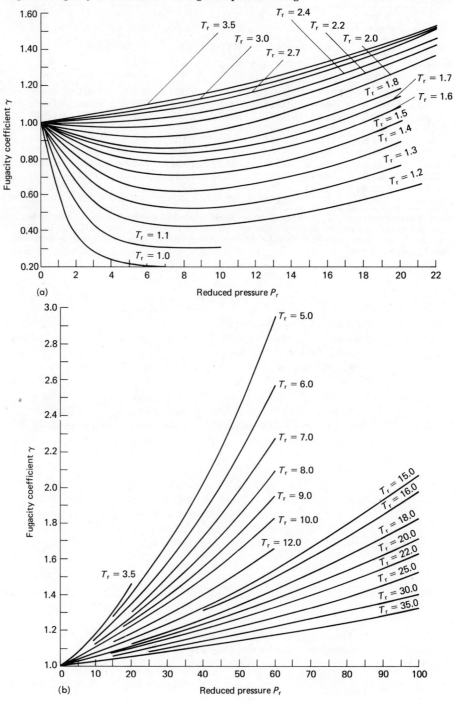

FIGURE 10.4 Generalized curves for fugacity coefficients as a function of the reduced pressure. (Reprinted with permission from R. H. Newton, *Ind. Eng. Chem.* 27 (1935):302. Copyright by the American Chemical Society.) (a) Fugacity coefficients in the intermediate temperature range. (b) Fugacity coefficients in the high-temperature range.

SEC. 10.13 EXAMPLE: THE HABER PROCESS REVISITED

Now we can calculate K_γ at 723 K and 600 atm. For NH_3 this corresponds to $T_r = 1.78$ and $P_r = 5.37$. Following just below the curve in Figure 10.4a for $T_r = 1.8$, we find that at $P_r = 5.37$ γ for NH_3 is 0.86. (The van der Waals calculation gave 0.78, as shown in Table 10.2.) Proceeding similarly in Figure 10.4b, we find that $\gamma_{N_2} = 1.34$ and $\gamma_{H_2} = 1.26$. Using these values, we get

$$K_\gamma = \frac{0.86}{(1.34)^{1/2}(1.26)^{1.5}} = 0.525$$

which is reasonably close to the value 0.50 obtained from the van der Waals calculation shown in Table 10.2.

The van der Waals equation was of great practical importance for many years after its introduction. In many instances it provided reasonably accurate descriptions of the P-V behavior of gases, and further, the vdw constants provided a measure of molecular sizes and intermolecular forces. Today, as a practical matter, the van der Waals equation has been supplanted by better equations of state, such as the virial equation.

Conceptually the material presented here is no more difficult when applied to the other equations of state. The forms of the integrals will be different, but the procedures will be the same. If you are a chemical engineer, you will undoubtedly delve into this subject in greater detail in studies connected with chemical engineering. If you are a biologist, you will probably never have occasion to examine the behavior of gases at high pressures. On a theoretical basis, the various equations of state are intimately tied in with various theoretical intermolecular forces and are of interest from that standpoint.

The advent of the desktop minicomputer has greatly simplified the work. Fugacities can be computed from equations of state in seconds by simply inputting the relevant temperature and pressure data after the proper program has been stored in memory. Even curves like those of Figure 10.4 can be used by reducing the pictorial curves to a polynomial or another functional form and storing the constants on tape.

In most of the further applications of thermodynamics in this book, we shall be dealing with pressures sufficiently low to assume ideal gas behavior for the gas phase. We shall, however, still be concerned with fugacities as a measure of the chemical potential.

While we have introduced fugacity as a tool to enable us to obtain free-energy changes for real gases and equilibrium constants at elevated pressures, we want to dispel any notion that its use is limited to calculations involving gases at high pressure. If this were so, it would be of interest only to chemical engineers dealing with high-pressure gases. The fugacity concept is much more fundamental.

Fugacity provides a measure of an "absolute" molar free energy by the equation

$$\overline{G} = \mu = RT \ln f + A(T) \tag{10.66}$$

where $A(T)$ for any particular substance is a function of the temperature only. The fugacity is further specified by the limiting condition that $(f/P) \to 1$ as $P \to 0$. This implies that for any isothermal process

$$\Delta \overline{G} = RT \ln \frac{f_2}{f_1} \tag{10.67}$$

We can measure free-energy changes for liquids and solids if they have measurable vapor pressures, since (10.66) enables us to measure the change in chemical po-

PROBLEMS

1. Calculate the equilibrium constant at 25 °C for the reaction $C_2H_2 + H_2 \rightleftharpoons C_2H_4$. Equal amounts of C_2H_2 and H_2 are placed in a closed chamber so that the total initial pressure is 2 atm. A catalyst is introduced so that the system comes to equilibrium. Calculate the final total pressure and also the partial pressures of each of the gases at equilibrium.

2. Calculate the equilibrium constant for the reaction $C_2H_4 + H_2 \rightleftharpoons C_2H_6$ at 25 °C. The gases H_2 and C_2H_4 are placed in a closed chamber such that H_2 has an initial partial pressure of 0.5 atm and C_2H_4 a pressure of 0.75 atm. Calculate the final total pressure at equilibrium. Calculate the equilibrium partial pressures of each of the three gases.

3. Some C_2H_6 is placed in a closed vessel at 25 °C and 1 atm. Some of the C_2H_6 decomposes to give H_2, C_2H_2, and C_2H_4. Calculate the final equilibrium partial pressures of each of the four gases.

4. The standard heat of formation and third law entropy of n-pentane, $CH_3CH_2CH_2CH_2CH_3$, are -146.66 kJ mol^{-1} and 348.95 J K^{-1} mol^{-1}. The values for neopentane, $(CH_3)_4C$, are -165.98 kJ mol^{-1} and 306.39 J K^{-1} mol^{-1}. Neopentane is placed in a chamber at a pressure of 1 atm and a catalyst is introduced that brings about the equilibrium between the two isomers. Calculate the partial pressures of each of the isomers at equilibrium (neglect the presence of other isomers). The system is kept at constant pressure.

5. Dinitrogen tetraoxide (N_2O_4) dissociates according to the reaction $N_2O_4 \rightleftharpoons 2NO_2$. Some N_2O_4 is introduced into a cylinder fitted with a piston so that the total pressure is maintained at a constant 1 atm. Calculate the degree of dissociation of the N_2O_4. Calculate the degree of dissociation if the total pressure is maintained at 5 atm.

6. The dissociation pressure of silver oxide, Ag_2O, is 422 torr at 173 °C and 605 torr at 183 °C. Estimate ΔH for the reaction

$$Ag_2O(c) \rightleftharpoons 2Ag(c) + \tfrac{1}{2}O_2(g)$$

7. At moderate temperatures, PCl_5 dissociates according to the reaction $PCl_5(g) \rightleftharpoons PCl_3(g) + Cl_2(g)$. A 2.5-g sample of PCl_5 is placed in a chamber at 320 °C. When the system comes to equilibrium, the final volume is 3.16 liters at a pressure of 0.31 atm. Calculate the degree of dissociation and $\Delta G°$ for the reaction at 320 °C.

8. A certain reaction has an equilibrium constant given by $\ln K = 4.814 - 2059/T$. Calculate $\Delta G°$, $\Delta H°$, and $\Delta S°$ at 298 K.

9. At 1200 K, the value $\Delta G°$ for the reaction $H_2 + \tfrac{1}{2}S_2 = H_2S$ is -51.20 kJ mol^{-1}. Starting with an initial pressure of 1 atm H_2S, calculate the final equilibrium pressures of all species. (The volume is constant.)

10. The standard free energies of formation of the four gaseous isomers of C_4H_8 at 298 K are as in the accompanying table.

Isomer	Designation	$\Delta G_f°$ (298 K) (kJ mol^{-1})
1-Butene	A	72.05
cis-2-Butene	B	67.15
trans-2-Butene	C	64.10
2-Methylpropene	D	61.00

(a) Calculate the following ratios of equilibrium pressures in an equilibrium mixture of the four gaseous isomers at 298 K: P_A/P_D; P_B/P_D; P_C/P_D.
(b) Calculate the composition of the equilibrium mixture when the total pressure is 1 atm.

11. The accompanying table gives the results of some experimental yields of NH_3 using a 3:1 mixture of H_2 to N_2 as the starting material. Calculate the experimental equilibrium constant at each pressure. How constant is the equilibrium constant?

Total pressure (atm)	% NH_3 at Equilibrium
10	2.11
30	5.86
50	9.15
100	16.43
1000	69.7
2000	89.8

12. The equilibrium constant we have defined is determined in terms of pressures,

$$K_p = \frac{P_C^c P_D^d}{P_A^a P_B^b}$$

It is also possible to define an equilibrium constant in terms of concentrations (in moles per liter or any other convenient units):

$$K_M = \frac{M_C^c M_D^d}{M_A^a M_B^b}$$

For an ideal gas, $P_i = n_i RT/V = M_i RT$. Show that

$$K_p = K_M (RT)^{\Delta \nu}$$

where $\Delta\nu$ is the number of moles of products less the number of moles of reactants in the stoichiometric equation.

13. The equilibrium constant can also be written in terms of mole fractions as

$$K_x = \frac{X_C^c X_D^d}{X_A^a X_B^b}$$

Show that $K_x = K_p P^{-\Delta\nu}$, where $\Delta\nu$ is the number of moles of products less the number of moles of reactants.

14. For the dissociation of N_2O_4 ($N_2O_4 \rightleftharpoons 2NO_2$). Let the fractional extent to which 1 mol N_2O_4 dissociates be a. Show that

$$K_x = \frac{4a^2}{1-a^2}$$

For 45 °C and a pressure of 1 atm, a is found to be 0.38. Calculate a at 10 atm. (See Problem 13 for the definition of K_x.)

15. At what pressure would N_2O_4 be 50% dissociated at 45 °C? Use the results of Problem 14.

16. The equilibrium constant for the reaction $SO_3 \rightleftharpoons SO_2 + \frac{1}{2}O_2$ is 0.54 (atm)$^{1/2}$ at 1000 K. One mole of SO_2 and 2 mol O_2 are placed in a vessel that is maintained at a pressure of 4 atm. How much SO_3 is present at equilibrium?

17. Calculate the equilibrium constant K_p for the ammonia synthesis reaction at 450 °C for the two conditions:
    (a) Assuming a constant value of $\Delta H°$.
    (b) Assuming that $\Delta H°$ changes with temperature.
    Calculate the theoretical yield of ammonia based on a starting ratio of 3 mol H_2 to 1 mol N_2 if the initial pressures of the N_2 are 1, 10, 100, and 500 atm.

18. The water gas reaction is

$$H_2(g) + CO_2(g) \rightleftharpoons H_2O(g) + CO(g)$$

   (a) Calculate the equilibrium constant at 298 K.
   (b) Calculate the equilibrium constant at 1200 K assuming that the room temperature value of $\Delta H°$ is independent of temperature.
   (c) Calculate the equilibrium constant at 1200 K taking into account the temperature variation of $\Delta H°$ by using the heat capacity data of Table 5.2.
   (d) Calculate the equilibrium composition based on a starting mixture of 50% CO_2 and 50% H_2 at a constant 1 atm pressure for parts (a), (b), and (c) of this problem.

19. In the example of the Haber process, we wrote the reaction as $\frac{1}{2}N_2(g) + \frac{3}{2}H_2(g) \rightleftharpoons NH_3(g)$, since the tables list enthalpies and free energies of formation for 1 mol NH_3. Written this way, the equilibrium expression in terms of pressures appears as in Equation (10.20). Suppose we multiplied all the coefficients by 2 and wrote the equation as $N_2 + 3H_2 \rightleftharpoons 2NH_3$. What would be the expression for the equilibrium constant? How are the two equilibrium constants related? Explain why the final calculated equilibrium partial pressures are the same, no matter which expression is used.

20. Derive an expression for ΔG for the isothermal expansion of a van der Waals gas from V_1 to V_2 (that is, integrate (10.34)).

21. It often turns out that at relatively low pressures α is a constant. Under these conditions show that one can write

$$\frac{f}{P} = \exp\left(\frac{-\alpha P}{RT}\right)$$

22. Expand the exponential expression of Problem 21 in a power series. Show that if one neglects all terms past the second (valid at small P), one can write

$$\frac{f}{P} = \frac{P\overline{V}}{RT}$$

(Note: For an ideal gas, $\overline{V} = RT/P$, and this expression reduces to $f/P = 1$.)

23. For a gas that has the equation of state $PV = RT + AP$, where A is a constant, find an expression for the fugacity f and the fugacity coefficient γ. Show that in the limiting case of an ideal gas (that is, $A \to 0$), the fugacity equals the pressure and $\gamma = 1$.

24. One mole of H_2 gas is expanded isothermally at 0 °C from an initial volume of 0.05 liter to a final volume of 25 liters. Calculate the work, heat, enthalpy change, energy change, free-energy change, and entropy change for the two conditions:
    (a) Assuming that H_2 is an ideal gas.
    (b) Assuming that H_2 is a van der Waals gas.

25. The accompanying table of data applies to hydrogen at 0 °C.

P (atm)	$\frac{P\overline{V}}{RT}$	P (atm)	$\frac{P\overline{V}}{RT}$
100	1.069	600	1.431
200	1.138	700	1.504
300	1.209	800	1.577
400	1.283	900	1.649
500	1.356	1000	1.720

   (a) Plot the data and graphically evaluate the fugacity of H_2 at 1000 atm.
   (b) Using the result of part (a), calculate ΔG for the isothermal expansion of 1 mol gas from 1 atm to 1000 atm. (It will be sufficient to assume that H_2 is an ideal gas at the lower pressure.)
   (c) Redo the calculation analytically, assuming that H_2 is a van der Waals gas.

26. The accompanying table gives additional data on the percentage of ammonia at equilibrium from a starting mixture of H_2 and N_2 in the ratio 3:1 at 100 atm total final pressure. From the data, evaluate the enthalpy change for the ammonia synthesis reaction at several temperatures. In Problem 17, K_p is calculated as 0.00718 at 450 °C.

Temp. (°C)	% NH$_3$	Temp. (°C)	% NH$_3$
200	81.54	500	10.61
250	67.24	550	6.82
300	52.04	600	4.52
350	37.35	650	3.11
400	25.12	700	2.18
450	16.43		

How does this answer compare with the present calculation? In working Problem 17, you derived an expression for ΔH as a function of temperature. Calculate ΔH using that expression, and compare it with these values for 473, 573, and 723 K.

CHAPTER ELEVEN
PHASE EQUILIBRIA

11.1 HOMOGENEOUS AND HETEROGENEOUS SYSTEMS

Our previous discussions have mostly been restricted to homogeneous systems. A *homogeneous system* can be defined as one whose intensive properties are uniform throughout the extent of the system.[1] A cylinder filled with gas is a homogeneous system; a bottle filled with water is a homogeneous system; a block of ice is a homogeneous system. A *heterogeneous system* is made up of two or more homogeneous parts with abrupt changes in properties at the boundaries of the parts. A vessel containing water and air at room temperature and pressure is a heterogeneous system. We have discussed heterogeneous systems only tangentially, when we examined phenomena such as the enthalpies and entropies associated with phase transformations. We now start a more intensive investigation of the equilibrium of heterogeneous systems.

11.2 PHASE

Perhaps the most succinct definition of a phase was given by J. Willard Gibbs himself in the abstract of his famous paper "The Equilibrium of Heterogeneous Substances":

In considering the different homogeneous bodies which can be formed out of any set of component substances, it is convenient to have a term which shall refer solely to the composition and thermodynamic state of any such body without regard to its size or form. The word phase *has been chosen for this purpose. Such bodies as differ in composition or state are called different phases of the matter considered, all bodies which differ only in size and form being regarded as different examples of the same phase.*[2]

A phase is thus homogeneous. Differences in shape or in degree of subdivision do not determine a new phase. Crushed ice is no less a single phase than one large block of ice. A system with an ice cube floating in water consists of two phases. One with ice floating in water with water vapor above is a three-phase system. A mixture of diamond and graphite consists of two phases, even though both are solid forms of carbon. All gas mixtures consist of only one phase.

---

[1] Strictly speaking, it is impossible to have a system that is truly homogeneous in a gravitational field, since the effect of gravity will cause a continuous variation in the properties of the system with elevation. As a practical matter, most systems are small enough for gravitational effects to be ignored.

[2] *The Collected Works of J. Willard Gibbs*, vol. 1 (New York: Longmans, Green, 1928), p. 358.

Boundaries between phases are regions in which abrupt changes in property occur. A phase boundary is not a mathematical plane surface but rather a thin region in which the properties change in going from one phase to the other. For the moment we shall ignore effects of the boundaries.

11.3 COMPONENTS

The number of *components* in a phase is the *minimum* number of distinct chemical species necessary to specify completely the chemical composition of the phase. In more technical terms, it is the number of constituents whose concentrations may be independently varied. In a solution of NaCl in water, the number of chemical species includes H_2O, NaCl, Na^+, Cl^-, H^+, and OH^-, yet the number of components is two. Once the amounts of H_2O and NaCl are specified, the concentrations of each of the other species can be determined.

The number of components is given by the number of distinct chemical species less the number of stoichiometric constraints. This is the basis of a logical method for determining the number of components. First write down every conceivable chemical species; let the number of such species be N. Then write all the independent chemical equations linking these species, the number of such equations being M. The number of components is given by $N - M$. For the above example of NaCl in water, $N = 6$. We can then write the following four equations:

$$NaCl \longrightarrow Na^+ + Cl^- \qquad [Na^+] = [Cl^-]$$
$$H_2O \longrightarrow H^+ + OH^- \qquad [H^+] = [OH^-]$$

The number of components is then $6 - 4 = 2$. Another way of viewing the number of components is to consider the minimum number of distinct raw materials necessary to construct the system. For the salt solution, this number is two, since the system can be constructed from salt and water in any desired ratios.

Consider the reaction $CaCO_3(s) = CaO(s) + CO_2(g)$. There are three distinct chemical species. We have written one equation for the system, thus the number of components is $3 - 1 = 2$. For this system it might be *incorrectly* inferred that the system is a one-component system, since by heating $CaCO_3(s)$ we can get the other two constituents. This would limit us, however, to a system in which the number of moles of CaO must equal the number of moles of CO_2. To get a system in which the ratios of the constituents are not limited in this manner we need at least two raw materials. These may be any two of the three constituents. The number of components is fixed. The identity of the components is subject to choice. A phase that contains only one component is known as a *pure phase*.

11.4 DEGREES OF FREEDOM

A complete description of a system requires numerical values of every property of the system, for example P, T, V, S, E, H, A, G, and the density ρ, and all other properties. If we neglect surface effects, equilibrium between phases does not depend on the amount of any phase present. The equilibrium vapor pressure is the same over 1 cm³ of water as over 1 liter of water. A saturated solution of salt with excess salt crystals at the bottom of the beaker has the same concentration whether we have 1 cm³ of solution above a ton of salt, or a ton of solution with a gram of salt at the bottom of the vessel (the effect of gravity is ignored and the pressure assumed uniform throughout). We need not consider any factors that depend on the bulk of material present. We consider only P, T, free energy per mole, concentration, and the like. All quantities can be reduced to any convenient scaling factor, say one mole.

A knowledge of some of the variables will often determine the values of other variables. If we know the values of any two of the three variables P, V, and T, the third will be fixed through the equation of state of the substance, whether it is a gas, a liquid, or a solid. The question we want to ask about any system is the following: How many intensive variables of the system can be *independently* varied without changing the number of phases present? This number is known as the number of *degrees of freedom* of the system. (It is also called the *variance*.)

For a pure gas, the three relevant variables are P, T, and the density, or molar volume. If any two of these three are fixed, the third is unequivocally determined. Another way of stating it is to say that any two variables can be simultaneously and arbitrarily varied. The same statement holds for a pure liquid or a pure solid. This is shown in the liquid-vapor phase diagram in Figure 11.1. In the pure liquid or pure gas phase, simultaneous variations are indicated by the small arrows. The arrow indicating a change in P is completely independent of the size and direction of the arrow indicating a change in T. The system has two degrees of freedom in the one-phase region.

If, however, we examine the system for which the two phases are in equilibrium, we are restricted to the curve separating the two phases in the two-phase region. Now if we arbitrarily change the temperature, as indicated by the arrow labeled a in the vapor region, the entire system goes into the vapor phase and we lose the liquid phase. If we want to maintain a two-phase system, we must at the same time change the pressure by an amount indicated by the arrow b. Only one variable can be independently varied. Once the change in that variable is specified, then the other is unequivocally determined if we want to maintain a two-phase system. The two-phase system of a pure substance has only one degree of freedom.

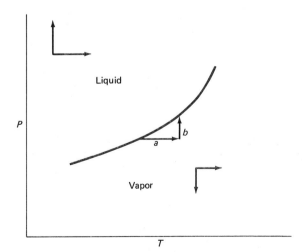

FIGURE 11.1 Liquid-vapor equilibrium.

11.5 CONDITIONS FOR STABILITY OF A PURE PHASE

Under what conditions will a pure phase be stable? That is to say, under what conditions will the phase remain a one-phase system and not break up into a two-phase system or be converted into a different phase? Water forms a stable liquid phase at 25 °C and 1 atm. If the temperature is lowered, an instability sets in at 0 °C, at which point the one-phase system suddenly becomes a two-phase system. If enough heat is removed, the water will be converted to a solid phase.

We consider a pure homogeneous phase with entropy S and volume V, both of which are assumed constant. We also assume that the system is in equilibrium. Suppose we divide the system in half and introduce a change that increases the entropy of one half by an amount δS, the entropy of the second half decreasing by the same amount. The entropy of the first half is thus $\frac{1}{2}S + \delta S$, and of the second half, $\frac{1}{2}S - \delta S$. The volume of each half remains the same at $\frac{1}{2}V$. Now we consider the energy change accompanying the process. We do this by expanding the energy of each half as a Taylor series, to get

$$E(\text{1st half}) = \frac{1}{2}\left[E + \left(\frac{\partial E}{\partial S}\right)_V \delta S + \frac{1}{2}\left(\frac{\partial^2 E}{\partial S^2}\right)_V (\delta S)^2 + \cdots \right]$$

$$E(\text{2nd half}) = \frac{1}{2}\left[E - \left(\frac{\partial E}{\partial S}\right)_V \delta S + \frac{1}{2}\left(\frac{\partial^2 E}{\partial S^2}\right)_V (\delta S)^2 + \cdots \right]$$

where the differentiation is at constant volume, and terms of order higher than 2 are omitted. The total energy change is

$$\delta E = \left(\frac{\partial^2 E}{\partial S^2}\right)_V (\delta S)^2$$

At constant volume and entropy, the condition for equilibrium is that the energy is a minimum, hence $\delta E > 0$. Since $(\delta S)^2 > 0$, we must have $(\partial^2 E / \partial S^2)_V > 0$. Or since $(\partial E / \partial S)_V = T$, the condition for equilibrium becomes

$$\left(\frac{\partial T}{\partial S}\right)_V > 0 \quad \text{or} \quad \left(\frac{\partial S}{\partial T}\right)_V > 0 \tag{11.1}$$

In other words, if the entropy of a stable phase is increased, the temperature must increase. This is precisely what does *not* happen at the melting point of ice. As heat is added to the system the entropy increases, but the temperature remains constant. It is a consequence of (11.1) that for any stable system, the heat capacity must be a positive quantity $C_v = T(\partial S/\partial T)_v > 0$.

The process whereby we imagine a small amount of entropy to be transferred from one part of the system to another part is known as a *virtual change*. Instead of using S, V, and E, we could have used T, V, and A, the Helmholtz energy. By considering the virtual transfer of volume δV from one half to the other at constant T, we could use the condition that A is a minimum. Proceeding in a similar manner, we should conclude that[3]

$$\left(\frac{\partial^2 A}{\partial V^2}\right)_T = \left[\frac{\partial(-P)}{\partial V}\right]_T > 0$$

$$\left(\frac{\partial P}{\partial V}\right)_T < 0 \quad \text{or} \quad \left(\frac{\partial V}{\partial P}\right)_T < 0 \tag{11.2}$$

---

[3] The derivation in this section generally follows that of E. A. Guggenheim, *Thermodynamics* (New York: Interscience Publishers, 1957), chap. 4.

In other words, as the volume increases, the pressure must decrease. This is true for a pure stable phase such as a gas, but not true when the gas is condensing, since then the pressure remains constant as the volume changes (either increasing or decreasing). Since the compressibility β is defined as $-V^{-1}(\partial V/\partial P)_T$, then β must be a positive number for any stable phase.

11.6 EQUILIBRIUM BETWEEN PHASES

Suppose we have some ice floating in a beaker of water at 0 °C and 1 atm pressure. We have a stable two-phase system. We want to examine the conditions at which the two phases are in equilibrium. In fact, we know the answer before we start. The system will be in equilibrium so long as the temperature, pressure, and chemical potentials are the same in both phases. In this section we shall show that these are the equilibrium conditions with a bit more rigor than we have previously used. Our derivation will have as a basis that at constant energy, the entropy must be an extremal. That is to say that at constant energy, $dS = 0$.

Figure 11.2 shows a two-phase system, or if you will, a system consisting of two subsystems, labeled a and b. The two subsystems (or phases) are separated by a barrier, and we examine the system as the characteristics of the barrier are changed. The term n_i is the number of moles of the ith component. We assume that the system is *closed;* that is to say, neither energy nor matter can leave the entire system that is composed of the two subsystems. We have

$$E = E^a + E^b \tag{11.3}$$

Now suppose that the barrier between the two subsystems has the following properties:

1. It is thermally conducting, so that energy in the form of heat can pass from one subsystem to the other.
2. It is movable, so that it acts as a piston and allows the two subsystems to do work on each other.
3. It is a semipermeable membrane, so that the ith component can freely pass between the subsystems a and b.

The total energy $E^a + E^b$ is constant; the total volume $V^a + V^b$ is constant; and the total amount of component i, or $n_i{}^a + n_i{}^b$, is constant. Since the energy is constant, the condition for equilibrium is that $dS = 0$, or[4]

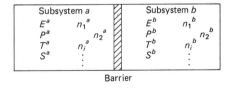

FIGURE 11.2 A two-phase system composed of two subsystems.

---

[4] In an effort to reduce confusion, we have eliminated the superscripts a and b on the subscripts E, V, and n in the partial differentials. They should be apparent from the intrinsic appearance of the equations.

$$dS = 0 = \left(\frac{\partial S^a}{\partial E^a}\right)_{V,n} dE^a + \left(\frac{\partial S}{\partial V^a}\right)_{E,n} dV^a + \left(\frac{\partial S^a}{\partial n_i^a}\right)_{E,V,n_j} dn_i^a$$

$$+ \left(\frac{\partial S^b}{\partial E^b}\right)_{V,n} dE^b + \left(\frac{\partial S^b}{\partial V^b}\right)_{E,n} dV^b + \left(\frac{\partial S^b}{\partial n_i^b}\right)_{E,V,n_j} dn_i^b \quad (11.4)$$

Now, recall that

$$\left(\frac{\partial S}{\partial E}\right)_{V,n} = (T)^{-1} \quad (11.5)$$

$$\left(\frac{\partial S}{\partial V}\right)_{E,n} = \frac{P}{T} \quad (11.6)$$

and

$$\left(\frac{\partial S}{\partial n_i}\right)_{E,V,n_j} = \mu_i \quad (11.7)$$

In addition,

$$dE^a = -dE^b \quad dV^a = -dV^b \quad dn_i^a = -dn_i^b \quad (11.8)$$

Hence we can write

$$dS = 0 = \left(\frac{1}{T^a} - \frac{1}{T^b}\right) dE^a + \left(\frac{P^a}{T^a} - \frac{P^b}{T^b}\right) dV^a + (\mu_i^a - \mu_i^b) dn_i^a \quad (11.9)$$

Now, Equation (11.9) must be valid for any arbitrary simultaneous variations of the quantities dE, dV, and dn_i. This means that each of the three quantities within the parentheses must be identically equal to zero. The first yields that $T^a = T^b$ and the second shows that $P^a = P^b$.[5] The third quantity requires that at equilibrium

$$\mu_i^a = \mu_i^b \quad (11.10)$$

These results can be extended to more than two phases by considering phase c in equilibrium with phase b, and then continuing to phase d, and so on. A system of many phases is in equilibrium if the pressure and the temperature are the same throughout all the phases, and further, if the chemical potentials of each of the constituents are the same in each phase. We are now in a position to specify the number of degrees of freedom in systems composed of many phases.

11.7 THE PHASE RULE

The relation between the number of degrees of freedom f, the number of phases p, and the number of components c was first elucidated by J. Willard Gibbs in the 1870s. The number of degrees of freedom equals the total number of intensive variables required to specify the complete system less the number of these variables that cannot be independently varied. To find the number of degrees of freedom, we must enumerate the total number of these variables, then subtract the number that cannot be independently varied.

---

[5] Recall that in our discussion of the third law in Chapter 8 we noted in footnote 4 that in a certain sense $(T)^{-1}$ is a more fundamental concept than T is. Note that if two phases are in equilibrium, we can directly show that it is $(T)^{-1}$ that is the same in both phases; that the temperatures are also equal follows as a secondary result.

SEC. 11.7 THE PHASE RULE

A system of p phases and c components will be completely specified if the temperature and pressure, and the concentration of each component in each phase, are specified. The c components in the p phases provide a total of $p \cdot c$ variables; the pressure and the temperature provide two additional variables; thus the total number of variables is

$$\nu_{\text{tot}} = pc + 2 \tag{11.11}$$

Now we must determine the number of variables that cannot be independently varied and subtract this number from ν_{tot}. The process is like solving a set of simultaneous equations in $pc + 2$ unknowns. If we can write a set of $pc + 2$ independent equations relating the unknowns, then the set of equations can be completely solved. If we have less than $pc + 2$ equations, then the set cannot be uniquely solved, but the total number of unknowns can be reduced by one for each equation. Similarly, every equation we can write that relates the system variables will reduce ν_{tot} by 1.

We are not interested in the absolute amounts of each component in each phase but rather in their relative amounts or concentrations. These we shall specify in mole fraction units. The mole fraction of component i in phase a is

$$X_i^a = \frac{n_i^a}{\sum_j n_j^a} = \frac{n_i^a}{N^a} \tag{11.12}$$

where n_i^a is the number of moles of component i in phase a, and N^a is the total number of moles of all components in phase a. In each phase, the sum of all mole fractions is unity,

$$X_1^a + X_2^a + X_3^a + \cdots + X_c^a = 1 = \sum X_i^a \tag{11.13}$$

If the concentrations of all but one of the components in any phase is specified, then the last one will be determined by Equation (11.13); each such equation reduces the number of independent variables by unity. Since we can write p such equations, the number of variables becomes

$$\nu' = \nu_{\text{tot}} - p = pc + 2 - p = p(c - 1) + 2 \tag{11.14}$$

Another set of equations is imposed by the fact that the chemical potential of each species must be the same in each of the phases. This enables us to write the set of equations

$$\begin{aligned} \mu_1^a &= \mu_1^b = \mu_1^c = \cdots = \mu_1^p \\ \mu_2^a &= \mu_2^b = \mu_2^c = \cdots = \mu_2^p \\ &\vdots \\ \mu_c^a &= \mu_c^b = \mu_c^c = \cdots = \mu_c^p \end{aligned} \tag{11.15}$$

Every equality in this set of equations denotes an additional condition on the system that reduces ν_{tot} by unity. There are $c(p - 1)$ equality signs in the system of equations; the number of independent variables remaining is the number of degrees of freedom f:

$$\begin{aligned} f &= \nu' - c(p - 1) = p(c - 1) + 2 - c(p - 1) \\ f &= c - p + 2 \end{aligned} \tag{11.16}$$

Equation (11.16) is known as the *phase rule*. It states that the total number of degrees of freedom equals the number of components less the number of phases plus 2.

11.8 THE ONE-COMPONENT SYSTEM

For a one-component system, $c = 1$, and (11.16) becomes $f = 3 - p$. There are three possibilities:

$$p = 1; \quad f = 2 \quad \text{(bivariant system)}$$
$$p = 2; \quad f = 1 \quad \text{(univariant system)}$$
$$p = 3; \quad f = 0 \quad \text{(invariant system)}$$

A P-T diagram for a one-component system is sketched in Figure 11.3. In any of the one-phase regions (pure solid, gas, or liquid), there are two degrees of freedom, as noted in connection with Figure 11.1. Along the curves bounding any two phases (liquid-solid, liquid-gas, and solid-gas) there is but one degree of freedom. At the intersection of the three curves, all three phases are in equilibrium, and there are *no* degrees of freedom. The point is fixed with regard to all its characteristics; it is called a *triple point*. There is a distinct difference between a triple point and a melting point. Figure 11.4 shows a phase diagram for H_2O (not to scale). The standard melting point is indicated by "mp"; it is the point at which the liquid and solid are in equilibrium at 1 atm pressure. The triple point for water is labeled "tp" and is 0.01 °C higher than the melting point.[6] (The slope of the solid-liquid boundary has been grossly exaggerated.) It should be mentioned that the phase diagram for water is unusual in that the slope of the liquid-solid boundary is negative. For most substances the slope is positive.

In low-temperature experiments, convenient fixed temperatures are provided by various substances at their triple points and their normal boiling points. The refrigeration fluid (such as liquid N_2) is placed in a dewar flask, which can be evacuated. At 1 atm, a temperature equal to the normal boiling point of the liquid is obtained. If the vapor above the liquid is then pumped, the temperature of the liquid falls until some of it begins to freeze at the triple-point temperature. When about half of the liquid has frozen, the valve to the vacuum pump is shut, and the system remains at the fixed triple-point temperature. The small heat leaks of a well-designed system, coupled with the heat of fusion of the refrigerant, keep the temperature constant for quite some time before the solid material melts completely and the temperature starts rising. If this time is too short to complete the experiment, the system can be

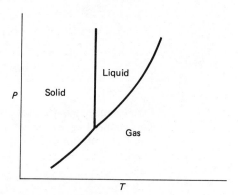

FIGURE 11.3 Phase diagram sketch (one component).

---

[6] A triple-point mixture for water can be prepared in the following way: Place some liquid water in a tube and evacuate the air. Then freeze a small portion of the water, at which point the three phases will be in equilibrium at the triple point. The triple point is a fixed temperature. Our thermodynamic temperature scale is defined by assigning to this temperature the value 273.16000... K.

SEC. 11.9 A USEFUL TWO-COMPONENT SYSTEM

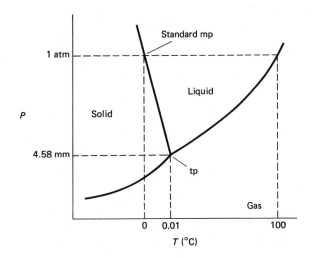

FIGURE 11.4 Phase diagram for water (not to scale). The slope of the liquid-solid boundary is anomalous in that it is negative because liquid water is denser than ice. (For most substances the slope is positive.) (Note that although all points on the liquid-solid line are melting points, only the point at 1 atm is the *standard melting point*.)

repumped. Liquid nitrogen has a normal boiling point of 77.45 K and a triple point of 63.15 K; liquid oxygen has a normal boiling point of 90.32 K and a triple point of 54.7 K; hydrogen has a normal boiling point of 20.3 K and a triple point of 13.9 K.

11.9 A USEFUL TWO-COMPONENT SYSTEM

When $Na_2SO_4 \cdot 10H_2O$ is heated it decomposes into Na_2SO_4 and water. The system has two components. These may be selected as H_2O and Na_2SO_4. The number of degrees of freedom is thus $f = 4 - p$. A maximum of three phases can coexist in equilibrium in the system. These consist of the liquid solution in equilibrium with the two solids, Na_2SO_4 and $Na_2SO_4 \cdot 10H_2O$. At this point, $f = 4 - 3 = 1$. If the pressure is fixed at 1 atm, we have used up the one degree of freedom, and the temperature is fixed at what is known as the *transition temperature*; it is equal to 305.54 K. Since it is close to room temperature, the system is very useful for calibrating thermometric devices such as thermocouples that are designed for use near room temperature.

The experimental procedure is straightforward, and it is indicated as a cooling (or heating) curve in Figure 11.5. Some solid $Na_2SO_4 \cdot 10H_2O$ is placed in a vessel at a

FIGURE 11.5 Cooling curve for $Na_2SO_4 \cdot 10H_2O$.

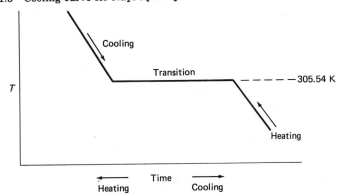

temperature above the transition temperature. As the sample slowly cools, the temperature falls until the transition temperature is reached. At that point the temperature remains constant until the last of the sample becomes hydrated; then the temperature falls again. The plateau on the curve is at the transition temperature. If one starts below the transition temperature, heating the system, then the arrows along the curve are simply reversed. (We shall discuss the two-component phase diagram in greater detail in Chapter 13.)

11.10 SLOPES ON A PHASE DIAGRAM; THE CLAPEYRON EQUATION

A glance at Figure 11.4 indicates that the various phases of a pure substance are separated by curves. If we can derive equations for the curves, then we can predict the effect of temperature on vapor pressure, or the effect of pressure on melting points.

Figure 11.6 shows a small section of the boundary between two phases, a and b, on a phase diagram. We know the temperature and pressure T_1 and P_1 on the curve at some initial point 1 and want to know the temperature and pressure T_2 and P_2 at a second equilibrium position on the curve, at point 2. The condition for equilibrium at each of the two points is that the chemical potentials are same in each phase:

$$\mu_1^a = \mu_1^b \tag{11.17}$$

$$\mu_2^a = \mu_2^b \tag{11.18}$$

or

$$\mu_1^a - \mu_2^a = \mu_1^b - \mu_2^b \tag{11.19}$$

Equation (11.19) states that the change in free energy per mole in going from 1 to 2 must be the same in both phases. For a one-component system,

$$\mu_1^a - \mu_2^a = \overline{dG^a} = \overline{V}^a \, dP - \overline{S}^a \, dT \tag{11.20a}$$

$$\mu_1^b - \mu_2^b = \overline{dG^b} = \overline{V}^b \, dP - \overline{S}^b \, dT \tag{11.20b}$$

where the bars over G, V, and S indicate molar quantities. Since the right side of (11.20a) equals the right side of (11.20b), they can be combined, to give

$$(\overline{V}^a - \overline{V}^b)dP = (\overline{S}^a - \overline{S}^b)dT \tag{11.21}$$

The term $\overline{V}^a - \overline{V}^b$ is the molar volume change for the phase transition, and $\overline{S}^a - \overline{S}^b$ is the molar entropy change for the transition; thus we can write

$$\frac{dP}{dT} = \frac{\Delta S}{\Delta V} \tag{11.22}$$

BENOIT PIERRE EMILE CLAPEYRON (1799–1864), French scientist, was the first to appreciate the importance of Carnot's work on the conversion of heat into work. In analyzing Carnot cycles, Clapeyron concluded that "the work w produced by the passage of a certain quantity of heat q from a body at temperature t_1 to another body at temperature t_2 is the same for every gas or liquid . . . and is the greatest which can be achieved" (B. P. E. Clapeyron, *Mémoir sur la Puissance Motrice de la Chaleur* (Paris, 1833)). Clapeyron was speaking of what we call a reversible process. Kelvin's establishment of the thermodynamic temperature scale from a study of the Carnot cycle came not from Carnot directly but from Carnot through Clapeyron, since Carnot's original work was not available to Kelvin.

SEC. 11.10 SLOPES ON A PHASE DIAGRAM; THE CLAPEYRON EQUATION

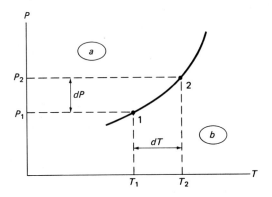

FIGURE 11.6 Boundary between two phases (phase a and phase b).

Equation (11.22) is called the *Clapeyron equation*. Since $\Delta S_{\text{trans}} = \Delta H_{\text{trans}}/T$ (for a reversible change), we can write (11.22) as

$$\frac{dP}{dT} = \frac{\Delta H_{\text{trans}}}{T \, \Delta V_{\text{trans}}} \tag{11.23}$$

Equation (11.22) or (11.23) is applicable to any phase transformation, solid → solid, solid → liquid, solid → gas, and liquid → gas. To integrate the equation exactly, it would be necessary to know ΔH_{trans} and ΔV_{trans} as functions of temperature and pressure. For small temperature and pressure variations, these can be taken as constant without introducing any great error. For a liquid (or a solid) that is transformed to a gas, the equation takes a particularly simple form,

$$\frac{dP}{dT} = \frac{\Delta H_{\text{trans}}}{T(V_g - V_l)} \tag{11.24}$$

If we assume that the volume of the condensed phase is negligible compared with the molar volume of the gas phase and that the gas is an ideal gas, then

$$(V_g - V_l) \approx V_g = \frac{RT}{P}$$

per mole, and (11.24) becomes

$$\frac{dP}{dT} = \frac{P \, \Delta H}{RT^2}$$

$$\frac{d(\ln P)}{dT} = \frac{\Delta H}{RT^2}$$

$$\frac{d(\ln P)}{d(1/T)} = -\frac{\Delta H}{R} \tag{11.25}$$

This equation is known as the *Clausius-Clapeyron equation*. If the logarithm of the vapor pressure is plotted against the inverse temperature, then at any point on the curve, the slope is given by $-\Delta H/R$. This provides an experimental method for measuring heats of vaporization of high-boiling substances like metals, since there exist techniques to measure vapor pressures at very high temperatures.[7] The value ΔH

---

[7] See Chapter 22.

does not change very much over small temperature increments. If ΔH is taken as constant, then (11.26) can be directly integrated, to yield

$$\ln \frac{P_2}{P_1} = \frac{\Delta H}{R}\left(\frac{1}{T_2} - \frac{1}{T_1}\right) \tag{11.26}$$

The Clapeyron equation provides a quantitative explanation of two familiar facts about water. First, water boils at a lower temperature at the top of a mountain than it does at the bottom of a mountain, and it therefore takes much longer to hard-boil an egg in Los Alamos than it does in New York, unless one uses a pressure cooker.[8] Since the volume of a gas is always greater than the volume of a liquid, dP/dT is always positive for a vaporization. The second is associated with a demonstration often done in elementary physics courses. Two weights are tied to the ends of a wire, and the wire is placed on a block of ice with the weights hanging down over the sides. Eventually the wire passes through the block, but the block of ice remains intact. Water has an almost unique property in that the molar volume of the solid phase is larger than the molar volume of the liquid. Because of this peculiarity, ice floats on water and seas freeze from the top down rather than from the bottom up. For water, ΔV in the Clapeyron equation is negative, and the slope of the liquid-solid curve dP/dT is negative, as shown in Figure 11.4. As the pressure increases, the melting point decreases. For the wire experiment noted above, the temperature of the ice block is slightly below the melting point. The excess pressure under the wire causes the ice under the wire to melt. As the wire moves downward, the melting point of the water above the wire reverts to its original value of 0 °C, and the water refreezes, leaving the block intact.

This feature also acts as an aid to ice-skating. Ice has a large coefficient of friction, but when the ice is lubricated with a thin film of liquid water, the ice presents an almost frictionless surface. There are two effects that provide a thin film of water. One is the effect of the increased pressure on the ice exerted by the weight of the skater through the thin edge of the skate blade. This effect can be calculated by Equation (11.23). The other (and perhaps larger) effect results from the heat of friction that is developed. When the temperature gets too low, these two effects are not enough to melt any ice, and skating becomes impossible.

PROBLEMS

1. The normal barometric pressure in Los Alamos, New Mexico, is 590 torr (1 torr = 1 mmHg). At what temperature does water boil in Los Alamos? (The heat of vaporization of water is 40.67 kJ mol^{-1}.)

2. The accompanying table gives the vapor pressure (vp) of water as a function of temperature. Plot the data so that you will be able to evaluate the heat of vaporization of water. Estimate the difference between the heat of vaporization at 10 °C and the heat of vaporization at 90 °C.

3. The density of ice at 0 °C and 1 atm pressure is 0.917 g cm^{-3}; at the same conditions, the density

t (°C)	vp (torr)
0	4.58
10	9.21
20	17.54
30	31.81
40	55.31
50	92.49
60	149.3
70	233.7
80	355.1
90	525.8
100	760.0

[8] The time it takes to hard-boil an egg is a function of the temperature. The purpose of a pressure cooker is to increase the pressure and hence increase the cooking temperature, thereby reducing the cooking time.

of liquid water is 0.9998 g cm^{-3}. The heat of fusion of ice is 6010 J mol^{-1}. Calculate the melting point of ice at a pressure of 100 atm, assuming that the densities and heat of fusion do not change over this pressure and temperature range.

4. Assume that the knife edge of an ice-skate blade is 0.02 in. wide and 4 in. long. A 175-lb man is ice-skating. Calculate the melting point of the ice under the pressure of the man. (See Problem 3 for additional data.)

5. The vapor pressures of liquid sodium are given in the accompanying chart. Calculate the normal boiling point of sodium and the heat of vaporization of sodium.

t (°C)	P (torr)
439	1
549	10
701	100

6. The normal boiling point of water is 100 °C, and the heat of vaporization is 40.67 kJ mol^{-1}. Take the density of liquid water to be 1 g cm^{-3} at 100 °C. The experimental density of water vapor at 100 °C is 0.0005970 g cm^{-3}. Assuming that the heat of vaporization is independent of temperature, evaluate the derivative dP/dT at 100 °C:
   (a) Using the exact form of the Clapeyron equation [(11.22) or (11.24)].
   (b) Using the ideal gas approximation (11.25). How much of an error is introduced by neglecting the volume of the liquid, and how much of an error is introduced by assuming that the water vapor is an ideal gas?

7. At the normal boiling point 100 °C, the heat of vaporization of water is 40.67 kJ mol^{-1}. Assume that water vapor is an ideal gas, and using the Clausius-Clapeyron equation calculate the vapor pressure of water at 150 °C:
   (a) Assuming that the heat of vaporization is independent of temperature.
   (b) Taking into account the variation of the heat of vaporization with temperature. (Use the data for the heat capacities given in Table 5.2.)

8. Diamond has a density of 3.513 g cm^{-3}; graphite has a density of 2.260 g cm^{-3}. The free energy of formation of diamond from graphite is 2.8660 kJ mol^{-1}. Assuming that the densities are independent of pressure, what pressure must be applied to graphite to convert it to diamond at room temperature?

9. The vapor pressure of ice is 4.58 torr at 0 °C and 1.95 torr at -10 °C. Calculate the heat of sublimation of water.

10. A liquid nitrogen cold trap is often used in vacuum systems to condense various volatile constituents and remove them from the system. The temperature of liquid nitrogen is 77.5 K. Calculate the partial pressure of H$_2$O in a vacuum system fitted with a liquid nitrogen trap. (Use the data of Problem 9.)

11. Between -30 and $+7$ °C the vapor pressure of solid benzene is given by the expression $\log_{10} P = 9.846 - 2309.7/T$, where T is the absolute temperature. Between 0 and 42 °C, the vapor pressure of liquid benzene is given by the expression $\log_{10} P = 7.9622 - 1784.8/T$.
    (a) Calculate the triple-point temperature of benzene.
    (b) Calculate expressions for the heats of vaporization and of sublimation as a function of temperature. Evaluate the heats at the triple point.
    (c) Calculate the heat of fusion of benzene at the triple point.
    (d) No indication is given about what units were used for P. Should this disturb you?

12. Go through the steps in deriving Equation (11.2).

13. The accompanying table gives data on the equilibrium between liquid and solid helium. The equilibrium pressure at which the two forms coexist is denoted by P. Calculate ΔS for the process He(s) = He(l) at 1.7, 1.4, and 1.15 K.

T (K)	P (atm)	Molar Volumes (cm^3 mol^{-1})	
		Liquid	Solid
1.15	25.27	23.24	21.14
1.20	25.32	23.23	21.13
1.30	25.50	23.18	21.10
1.40	25.81	23.11	21.06
1.50	26.32	23.03	21.01
1.60	27.13	22.82	20.93
1.75	29.30	22.25	20.76

14. Equation (11.4) in the text applies to a boundary that is semipermeable, movable, and thermally conducting. Set up an analogous equation for a rigid impermeable membrane that is made of a heat-conducting material of such a kind that energy can pass from a to b only as heat. Show that the resulting equilibrium condition is that $(T^a)^{-1} = (T^b)^{-1}$, or $T^a = T^b$.

15. Remove the constraint of rigidity from the boundary of Problem 14, so that the barrier can move and act as a piston. Show that for this boundary the equilibrium condition is that $(T^a)^{-1} = (T^b)^{-1}$ and that $P^a/T^a = P^b/T^b$, or that the temperature and pressure must be the same in both subsystems at equilibrium.

CHAPTER TWELVE

COLLIGATIVE PROPERTIES OF IDEAL SOLUTIONS

12.1 SOLUTIONS

We noted in Chapter 11 that the boiling points and freezing points of materials could be changed by adjusting the pressure of the system, thereby altering the chemical potential. There is another method of adjusting the chemical potential $\mu_i = (\partial G/\partial n_i)_{P,T}$ of components in the system, and thereby changing the boiling point or freezing point. This is accomplished by introducing another component to form a *solution*.

A solution is a *homogeneous system,* or single phase, that contains more than one component. Two substances that are mutually soluble are said to be *miscible*. Solutions may be gaseous, liquid, or solid. All gases are miscible in all proportions. Liquids and solids can dissolve a wide range of gases, or other liquids and solids. The range of solubility may be small or large depending on the particular system. Water and ethyl alcohol are completely miscible in each other and form a complete range of solutions from pure alcohol to pure water. Oil, on the other hand, is only slightly soluble in water.

The most useful way to describe the composition of solutions in theoretical discussions is the *mole fraction*. A solution containing n_A moles of component A, n_B moles of component B, n_C moles of component C, and so on, has a mole fraction of A:

$$X_A = \frac{n_A}{n_A + n_B + n_C + \cdots} \tag{12.1}$$

For a two-component system, the mole fraction of A is

$$X_A = \frac{n_A}{n_A + n_B}$$

The unit of *molarity,* symbol M, is most often used in ordinary volumetric chemical analysis because of its operational simplicity. Molarity is a volume concentration unit and is defined as the number of moles of solute per liter of solution. The molarity unit suffers from two inherent disadvantages. Firstly it is affected by temperature, since the volume changes as the temperature changes. Secondly, the quantity of solvent varies with molarity; it is difficult to pinpoint the quantity of water in a $1.000M$ aqueous solution of NaCl without reference to other data. The *molality* concentration unit, symbol m, overcomes these difficulties. Molality is defined as the number of moles of solute per 1000 g of solvent. Molality is inherently a more accurate concentration measure than molarity since mass can be measured to a higher degree of accuracy than volume. If we have a two-component solution in which the molality of B is known, then the mole fraction can be determined from m_B as

$$X_B = \frac{m_B}{(1000/M_A) + m_B} = \frac{m_B M_A}{1000 + m_B M_A} \tag{12.2}$$

where M_A is the molar mass of component A. At small concentrations of B, the product $m_B M_A$ is small compared with 1000, and the mole fraction is approximately proportional to the molality, $X_B \approx 10^{-3} m_B M_A$. For aqueous solutions, it is useful to remember that $M_A = 18.015$ g mol^{-1}, and that 1000 g of water contains 55.5 moles.

The concentration unit of mass (or weight) percent is just the mass of solute in 100 masses of solution. This unit is most useful in technological applications for economic reasons. Steel is sold on a weight basis, and iron ore is purchased as weight percent iron in order to plan production schedules. Sugar syrup is purchased in tank-car lots at a price proportional to the weight percent sugar. In many tabulations, the density of a solution is given for a series of mass percent concentrations.[1] Suppose the table lists a density ρ at a concentration c_B mass percent solute; the mass percent solvent is then $c_A = (100 - c_B)$. The masses of each constituent per cubic centimeter are

$$w_A = 0.01 c_A \rho \qquad w_B = 0.01 c_B \rho \qquad (\text{g cm}^{-3})$$

The volume of solution containing 1000 g of solvent is

$$V = \frac{1000}{w_A} \qquad (\text{cm}^3)$$

The molality is the number of moles of solute in a solution containing 1000 g of solvent, or

$$m_B = \left(\frac{1000}{w_A}\right)\left(\frac{w_B}{M_B}\right)$$

where M_B is the molecular weight of solute. Putting all the steps together gives

$$m_B = \frac{1000 c_B}{M_B(1 - c_B)} \tag{12.3}$$

where c_B is the mass percent solute listed in the table.

12.2 RAOULT'S LAW; THE IDEAL SOLUTION

In 1886, François Marie Raoult was engaged in a series of experiments in which he measured the partial pressures of the components of solution. He found that for some two-component solutions, the vapor pressures of the components was given by

$$P_A = X_A P_A^\circ \qquad P_B = X_B P_B^\circ \tag{12.4}$$

where P_A and P_B are the partial pressures of each of the components, X_A and X_B are

---

[1] For industrial applications, quick and easy methods for determining concentrations are required. The density is a property that it is particularly simple to measure, since the procedure takes less than a minute by using a *hydrometer*. For this reason, so many data are tabulated by densities.

> FRANÇOIS-MARIE RAOULT (1830–1901), French chemist, was a pioneer in solution chemistry. His work on vapor-pressure lowering and on freezing-point depressions was fundamentally important in the chemistry of solutions, and his realization that both were a function of the number of moles of dissolved solute contributed greatly to the determination of molar masses and the theory of ionic solutions. Although Raoult's name is usually associated with his law of vapor-pressure lowering, his contributions to cryoscopy, or freezing-point depressions, were prodigious, and Victor Meyer used his data as early as 1886 for determining molar masses. Raoult built for these measurements what he called a cryoscope of precision, and the accuracy of his work was unsurpassed, agreeing in many instances with modern measurements within 0.0001 °C. His thermometer was considered by some to be antediluvian, but as van't Hoff expressed it, "With this antediluvian thermometer the world was conquered."

the mole fractions, and $P_A°$ and $P_B°$ are the vapor pressures of the pure components. Equation (12.4) is known as *Raoult's law*. A solution that follows Raoult's law is known as an *ideal solution*. In Figure 12.1, the vapor pressure of the "nearly" ideal solution consisting of benzene and toluene is plotted as a function of mole fraction. The concept of an ideal solution differs in many fundamental respects from the ideal gas concept.

On the molecular level, an ideal gas is characterized by a complete *absence* of intermolecular forces; the internal pressure, $(\partial E/\partial V)_T = 0$. In an anthropomorphic sense, each gas molecule is totally unaware of the existence of any other gas molecule. The ideal liquid solution is characterized by a complete *uniformity* of intermolecular forces. In a solution of A and B, the forces between two A molecules are the same as the forces between two B molecules or between an A and a B molecule. A molecule in the solution cannot "differentiate" between A or B molecules. An ideal solution usually consists of molecules of the same general type, size, and shape. Ben-

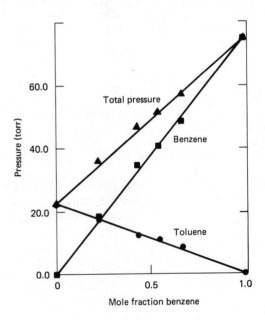

FIGURE 12.1 Total vapor pressure and partial pressures of the components for the almost ideal system of benzene–toluene at 20 °C. The straight lines indicate Raoult's law behavior.

FIGURE 12.2 Vapor-pressure diagrams of nonideal solutions indicating deviations from Raoult's law. The dashed line indicates Raoult's law. (a) The system chloroform–acetone at 35 °C, showing negative deviations from ideality. (b) The system carbon disulfide–methylal at 35 °C, showing positive deviations from ideality.

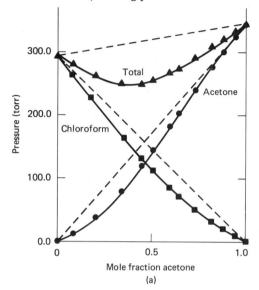
(a)

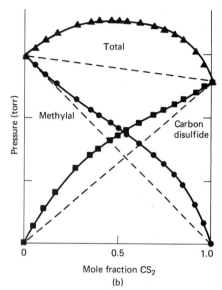
(b)

zene and toluene molecules are quite similar, and the two form a solution that is nearly ideal. Ordinary water, H_2O, and heavy water, D_2O, form an ideal solution.

The ideal-gas concept applies accurately in the limit of small pressure (or large molar volume). The liquid-ideal-solution concept applies in the limit of small solute concentration (or large solvent concentration).[2] From a practical standpoint, the utility of the ideal gas is enhanced because at ordinary pressures (about 1 atm or less) the ideal gas equation of state can be used with the introduction of only very slight errors. For liquids, the practicality of the ideal-solution concept is less direct. Ordinary solutions cover the entire concentration range, and the ideal solution is the exception and not the rule. The vapor pressure dependence of nonideal solutions is shown in Figure 12.2. The ideal-solution concept is practical because it enables us to establish a reference state for measuring free-energy changes.

Our immediate discussion is restricted to binary solutions in which each of the components has a measurable vapor pressure. At equilibrium (constant T and P), the chemical potential of each component must be the same for the liquid phase as it is for the vapor phase; thus for our two components,

$$\mu_A^l = \mu_A^{\text{vap}}$$
$$\mu_B^l = \mu_B^{\text{vap}} \tag{12.5}$$

Equation (12.5) is always true whether the solution is ideal or nonideal. Since for any vapor, $\mu = RT \ln f$, where f is the fugacity of the vapor,

$$\mu_A^l = RT \ln f_A$$
$$\mu_B^l = RT \ln f_B \tag{12.6}$$

---

[2] Later in this chapter we shall see that whereas Raoult's law applies to the solvent in an ideal solution, another law, Henry's law, applies to the solute in the limit of small concentrations.

Equation (12.6) enables us to assign values of the chemical potentials to the constituents in the liquid using the fugacities of the vapor in equilibrium with the liquid. In fugacities, Raoult's law can be written

$$f_A = X_A f_A^\circ$$
$$f_B = X_B f_B^\circ \qquad (12.7)$$

where f° is the fugacity of the pure liquid. If the system has a moderate vapor pressure (of the order of 1 atm or less), then we can assume that the vapor above the liquid behaves as an ideal gas. Since $f = P$ for an ideal gas, Equation (12.7) reduces to (12.4).

Many important industrial processes are carried out under conditions of high vapor pressures, and an accurate description of these requires the vapor to be treated as a real gas. We shall not concern ourselves with these cases. Conceptually these are no more difficult to treat than the ideal gas case; one uses real fugacities instead of pressures. The complications arise in the computational aspects of the problems. The nonideal gas case is important in many chemical engineering applications, and students of chemical engineering will delve into this problem in the context of their chemical engineering courses. It will be useful at this point to examine some thermodynamic consequences of our definition of the ideal solution. To do that, we must first develop the concept of the partial molar quantity.

12.3 PARTIAL MOLAR QUANTITIES

Let's consider the Atlantic Ocean as a solution of NaCl in water. We now ask the following question: Suppose we add one mole of NaCl to the ocean; what will be the resulting volume change for the entire ocean? If we let n_A represent the moles of solvent and n_B the moles of solute, then the volume change of the ocean

$$\Delta V = \overline{V}_B = \left(\frac{\partial V}{\partial n_B}\right)_{T,P,n_A} \qquad (12.8)$$

is called the *partial molar volume* of component B. It is the change in volume experienced by the solution on a molar basis when a differential quantity of the component is added—pressure, temperature, and composition remaining constant. Partial molar quantities can be similarly defined for any extensive state function of the system; for example,

$$\overline{S}_A = \left(\frac{\partial S}{\partial n_A}\right)_{T,P,n_B} \qquad (12.9a)$$

$$\overline{H}_A = \left(\frac{\partial H}{\partial n_A}\right)_{T,P,n_B} \qquad (12.9b)$$

$$\overline{G}_A = \left(\frac{\partial G}{\partial n_A}\right)_{T,P,n_B} \qquad (12.9c)$$

Note that a partial molar quantity is an *intensive* property of the system, since it is independent of the size of the system. We can apply to partial molar quantities all the thermodynamic relations we have previously derived. For example,

$$\left(\frac{\partial \overline{G}_A}{\partial P}\right)_T = \left(\frac{\partial \mu_A}{\partial P}\right)_T = \overline{V}_A \qquad (12.10)$$

$$\left(\frac{\partial \mu_A}{\partial T}\right)_P = -\overline{S}_A \qquad (12.11)$$

The chemical potential is a partial molar free energy.

The total volume of a real solution is generally not equal to the sum of the volumes of the individual pure components. If 50 cm³ of alcohol are mixed with 50 cm³ of water, the resulting solution has a volume of about 95 cm³ and not 100 cm³. In a sense, the partial molar volume of a constituent can be considered the "effective volume" of that constituent at the particular concentration under discussion. One can get the relation between the total volume and the partial molar volumes of the components by considering the properties of homogeneous functions.

12.4 HOMOGENEOUS FUNCTIONS AND THE EULER THEOREM

Suppose we have a function
$$F = F(x, y, z). \qquad (12.12)$$
F is said to be homogeneous of degree n if
$$F(\lambda x, \lambda y, \lambda z) = \lambda^n F(x, y, z) \qquad (12.13)$$
where λ is any positive constant. It means that the function is multiplied by λ^n if each of the variables is multiplied by λ. We can take as an example a rectangular parallelepiped with edges given by x, y, and z. The total edge length, the total surface area, and the volume are functions of x, y, and z. The total edge length is a homogeneous function of degree 1 in the variables x, y, and z, since if each of the edges is doubled, the total edge length is multiplied by 2^1. The total surface area is a homogeneous function of degree 2, since if each of the edges is doubled, the surface area is multiplied by 2^2, whereas the volume is of degree 3, since the volume of the parallelepiped is multiplied by 2^3. In this example we used $\lambda = 2$, but any other constant would have served equally well.[3]

In thermodynamics, all extensive state functions of the system are homogeneous functions of degree 1. All that means is that if the size of the system is doubled, the function will be multiplied by 2^1; and if the size of the system is tripled, the function will be multiplied by 3^1; and so on. (Note that the intensive variables of the system are homogeneous functions of degree zero. If the size of the system is doubled, both the pressure and the temperature are multiplied by $2^0 = 1$; in other words they are unchanged.)

Suppose we now differentiate (12.13) with respect to λ. We have

$$\begin{aligned}\frac{\partial F(\lambda x, \lambda y, \lambda z)}{\partial \lambda} &= \frac{\partial F}{\partial(\lambda x)}\frac{\partial(\lambda x)}{\partial \lambda} + \frac{\partial F}{\partial(\lambda y)}\frac{\partial(\lambda y)}{\partial \lambda} + \frac{\partial F}{\partial(\lambda z)}\frac{\partial(\lambda z)}{\partial \lambda} \\ &= x\frac{\partial F}{\partial(\lambda x)} + y\frac{\partial F}{\partial(\lambda y)} + z\frac{\partial F}{\partial(\lambda z)} \qquad (12.14a)\end{aligned}$$

[3] It is worth noting that the number of apexes of a parallelepiped is a homogeneous function of degree 0. The number of apexes is independent of the size; if the size of the parallelepiped is multiplied by λ, the number of apexes is multiplied by $\lambda^0 = 1$.

and
$$\frac{\partial F(\lambda x, \lambda y, \lambda z)}{\partial \lambda} = n\lambda^{n-1} F(x, y, z) \tag{12.14b}$$

Since (12.13) is valid for any value of λ, we can set λ equal to 1, and (12.14) leads to

$$x\left(\frac{\partial F}{\partial x}\right) + y\left(\frac{\partial F}{\partial y}\right) + z\left(\frac{\partial F}{\partial z}\right) = nF \tag{12.15}$$

This is *Euler's theorem*. Now let us apply this result.

In thermodynamics we deal with first-order homogeneous functions, in which case we can set n equal to 1. Suppose we examine the volume. At constant temperature and pressure, the total volume of a system is a function of the mole numbers of each component, and we can write

$$V = V(n_1, n_2, n_3, \ldots) \tag{12.16}$$

Applying Equation (12.15) with $n = 1$, we see immediately that

$$V = n_1\left(\frac{\partial V}{\partial n_1}\right) + n_2\left(\frac{\partial V}{\partial n_2}\right) + n_3\left(\frac{\partial V}{\partial n_3}\right) + \cdots$$
$$= n_1 \overline{V}_1 + n_2 \overline{V}_2 + n_3 \overline{V}_3 + \cdots \tag{12.17}$$

Equation (12.17) tells us that the total volume equals the partial molar volume of constituent 1 multiplied by the number of moles of 1 plus the partial molar volume of constituent 2 multiplied by the number of moles of 2, and so on. It exemplifies what we previously said about the partial molar volume being an "effective volume." While (12.17) has been derived for volume, it applies to any partial molar quantity. To use (12.17), we must develop a procedure for determining partial molar quantities.

12.5 DETERMINATION OF PARTIAL MOLAR QUANTITIES; THE GIBBS-DUHEM EQUATION

If we restrict ourselves to a two-component solution, dV can be computed from (12.16) as

$$dV = \left(\frac{\partial V}{\partial n_A}\right)_{n_B} dn_A + \left(\frac{\partial V}{\partial n_B}\right)_{n_A} dn_B = \overline{V}_A\, dn_A + \overline{V}_B\, dn_B \tag{12.18}$$

dV can also be computed from (12.17), as

$$dV = n_A\, d\overline{V}_A + \overline{V}_A\, dn_A + n_B\, d\overline{V}_B + \overline{V}_B\, dn_B \tag{12.19}$$

Comparing (12.18) with (12.19), we find that

$$n_A\, d\overline{V}_A + n_B\, d\overline{V}_B = 0 \tag{12.20}$$

This important result is known as the *Gibbs-Duhem equation*. It can also be written

$$d\overline{V}_A = -\left(\frac{n_B}{n_A}\right) d\overline{V}_B \tag{12.21a}$$

$$= -\left(\frac{X_B}{X_B - 1}\right) d\overline{V}_B \tag{12.21b}$$

where (12.21b) is written in mole fractions instead of mole numbers. We have derived the Gibbs-Duhem equation in terms of partial molar volumes, but a similar equation can be written for any other partial molar quantity.

For a single pure component, the volume is directly proportional to the number of moles, as shown in Figure 12.3. For a pure system,

$$\bar{V}_A = \bar{V} = \left(\frac{\partial V}{\partial n}\right) = \frac{V}{n} = \bar{V}^\circ \tag{12.22}$$

where $\bar{V}^\circ$ is just the molar volume of the pure component. Figure 12.4 shows the volume of an aqueous solution containing 1000 g of H_2O as a function of solute concentration. This curve is nonlinear. If we use the convention that component B is the solute, then $\bar{V}_B = (\partial V/\partial n_B)_{n_A, P, T}$ is given by the slope of the curve, as shown in the figure. Conceptually, $\bar{V}_B$ can be measured by one of two equivalent methods. In

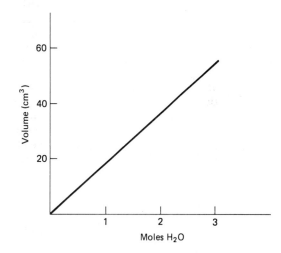

FIGURE 12.3 Volume of a pure phase at constant temperature and pressure as a function of mole number. The data are for water at 25 °C and 1 atm. The slope of the line is the partial molar volume of pure water.

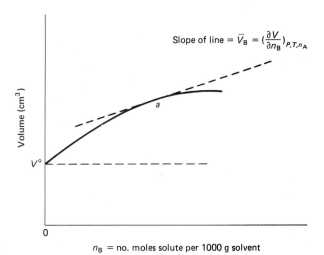

FIGURE 12.4 Total volume of a solution containing 1000 g solvent and n_B mol solute. The term V_0 is the volume of 1000 g pure solvent. The partial molar volume of B at concentration a is given by the slope of the tangent to the curve at a. For water, V_0 is 1002.9 cm³.

the first method, a mole of solute is added to an infinitely large quantity of solution, and the resulting volume change is measured. In the second method, an infinitesimal amount of solute is added to a finite solution, and the measured volume change is referred to one mole of added solute. There are certain inherent problems in both methods, and $\overline{V}_B$ is best determined from the slope of a plot of V vs n_B. Once $\overline{V}_B$ is known, $\overline{V}_A$ can be determined using (12.21). Numerous graphical techniques have been developed for determining partial molar quantities, but these will not be discussed here.[4] We shall demonstrate an analytical procedure for NaCl solutions.

The procedure is outlined in Table 12.1. Columns (1) and (2) list the weight percent NaCl and the solution densities. Column (3) lists the volume of a solution containing 1000 g of H_2O, while column (4) lists the molality of the solution computed using (12.3). The data are plotted in Figure 12.5. The solid curve is the least squares fitted curve to the analytical expression

$$V = 1002.874 + 17.8213m + 0.87391m^2 - 0.047225m^3 \qquad (12.23)$$

where m is the molality. For this system, $m = n_B$, and n_A is the number of moles of H_2O in 1000 g, or $1000/18.015 = 55.508$ moles. Column (5) of the table lists the calculated values of V; the numbers in parentheses are the differences between the calculated values and the experimental values of column (3). The maximum deviation is less than 0.05 cm³. (If a fourth-order polynomial had been used, the maximum deviation would have been less than 0.01 cm³.) By our analytical expression,

$$\overline{V}_B = \left(\frac{\partial V}{\partial n_B}\right)_{n_A} = \frac{\partial V}{\partial m}$$

$$= 17.8213 + 1.74782m - 0.141675m^2 \qquad (12.24)$$

TABLE 12.1 Calculation of partial molar volumes from density data for NaCl solution at 25°C.

(1) Weight percent NaCl[a]	(2) Density of solution[a] (g cm⁻³)	(3) Experimental volume per 1000 g H_2O (cm³)	(4) Molality [from Eq. (12.3)]	(5) Calculated volume [from Eq. (12.23)] (cm³)	(6) $\overline{V}_A(H_2O)$ [from Eq. (12.27)] (cm³ mol⁻¹)	(7) $\overline{V}_B(NaCl)$ [from Eq. (12.24)] (cm³ mol⁻¹)
0.	0.99709	1002.92	0.00000	1002.874 (0.044)[b]	18.0680	17.8213
1.	1.00409	1005.99	0.17284	1005.980 (0.007)	18.0675	18.1192
2.	1.01112	1009.19	0.34920	1009.202 (−0.015)	18.0662	18.4144
4.	1.02530	1015.96	0.71295	1016.007 (−0.043)	18.0606	18.9954
6.	1.03963	1023.28	1.09218	1023.319 (−0.041)	18.0514	19.5612
8.	1.05412	1031.15	1.48789	1031.169 (−0.018)	18.0388	20.1082
10.	1.06879	1039.60	1.90119	1039.590 (0.007)	18.0228	20.6322
12.	1.08365	1048.64	2.33328	1048.614 (.031)	18.0039	21.1282
14.	1.09872	1058.31	2.78547	1058.275 (.039)	17.9826	21.5906
16.	1.11401	1068.64	3.25919	1068.605 (.035)	17.9597	22.0129
18.	1.12954	1079.65	3.75602	1079.638 (.016)	17.9361	22.3874
20.	1.14533	1091.39	4.27769	1091.403 (−0.013)	17.9131	22.7055
22.	1.16140	1103.88	4.82611	1103.928 (−0.042)	17.8926	22.9567
24.	1.17776	1117.20	5.40339	1117.234 (−0.037)	17.8768	23.1290
26.	1.19443	1131.38	6.01188	1131.338 (0.040)	17.8687	23.2085

[a] Data from International Critical Tables, vol. 3, p. 79
[b] The numbers in parentheses are the differences between these calculated values and the experimental values of column (3)

[4] Graphical procedures for evaluating partial molar quantities can be found in I. M. Klotz, *Chemical Thermodynamics* (Englewood Cliffs, N.J.: Prentice-Hall, 1950), p. 193, and in G. N. Lewis and M. Randall, *Thermodynamics*, 2d ed., revised by K. S. Pitzer and L. Brewer (New York: McGraw-Hill Book Co., 1961), p. 205.

SEC. 12.6 ENTHALPY OF SOLUTION

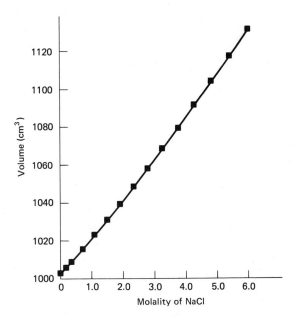

FIGURE 12.5 Total volume of a solution containing 1000 g H$_2$O vs the number of moles of NaCl. The squares represent the data of Table 12.1, and the solid curve is the plot of Equation (12.23).

An analytical expression for $d\overline{V}_B$ can be got from (12.24):

$$d\overline{V}_B = (1.74782 - 0.28335m)\, dm \quad (12.25)$$

Using (12.21), we have

$$d\overline{V}_A = -\left(\frac{n_B}{n_A}\right) d\overline{V}_B = -\left(\frac{m}{55.508}\right) d\overline{V}_B$$

which can be integrated from $m = 0$ to m, to give

$$\int_{m=0}^{m} d\overline{V}_A = \int_{m=0}^{m} -\left(\frac{1.74782}{55.508}m - \frac{0.28335}{55.508}m^2\right) dm$$

$$\overline{V}_A - \overline{V}_A^\circ = -\frac{1.74782}{2(55.508)}m^2 + \frac{0.28335}{3(55.508)}m^3 \quad (12.26)$$

where $\overline{V}_A^\circ$ is the partial molar volume of water at $m = 0$, that is, the molar volume of pure water (18.068 cm^3). The final expression for $\overline{V}_A$ can now be written as

$$\overline{V}_A = 18.068 - 0.0157439m^2 + 0.0017016m^3 \quad (12.27)$$

The values of $\overline{V}_A$ and $\overline{V}_B$ as computed by (12.24) and (12.27) are listed on columns (6) and (7) of Table 12.1. You can verify the consistency of the procedure by computing the total volume at several concentrations using (12.17). Later in this chapter we shall show that for an ideal solution there is no volume change on mixing.

12.6 ENTHALPY OF SOLUTION

Most of us at some time or another have added concentrated sulfuric acid to water to make up a dilute solution and have noticed the heat evolved. We may even have noted that more heat was evolved per mole of acid when the acid was added to pure water than when it was added to a sulfuric acid solution.

TABLE 12.2 Heats of formation and integral heats of solution for 1 mol H_2SO_4 in n mol water.

Moles of H_2O	Molality of H_2SO_4	$\Delta H_f^\circ(298.15K)$[a]	ΔH_s (integral heat of solution)[c]
0	....	-811.31[b]	0.00
0.5	111	-827.04	-15.73
1	55.5	-839.38	-28.07
1.5	37.0	-848.21	-36.90
2	27.75	-853.23	-41.92
3	18.5	-860.30	-48.99
5	11.10	-869.34	-58.03
10	9.09	-878.34	-67.03
50	1.110	-884.65	-73.35
100	0.555	-885.28	-73.97
1,000	0.0555	-889.88	-78.58
5,000	0.0111	-895.74	-84.43
10,000	5.55×10^{-3}	-898.38	-87.07
50,000	1.11×10^{-3}	-903.65	-92.34
100,000	5.55×10^{-4}	-904.95	-93.64
∞	0	-907.50	-96.19

NOTE: All entries are in units of kJ mol^{-1} H_2SO_4.

[a] Entries in this column are from the National Bureau of Standards Circular 500.
[b] This entry is just the standard heat of formation of pure liquid H_2SO_4.
[c] The integral heat of solution is the enthalpy change for the reaction
$$H_2SO_4(l) + nH_2O = H_2SO_4 \text{ (solution in } nH_2O)$$

Suppose we make up a solution that is $12m$ H_2SO_4. There are two ways we can measure the heat of solution. In the first way, we simply add 12 moles of H_2SO_4 to 1000 g of water and measure the heat evolved; if we wish the heat per mole of H_2SO_4, we simply divide by 12. This type of heat we call an *integral heat of solution*. In the second way, we divide the acid into 12 equal portions and add it to the water in 12 steps. The first step results in a $1m$ solution, and the heat evolved corresponds to the heat required to make up a $1m$ solution. The second step increases the molality from 1 to 2 and so on, until the last step increases the molality from 11 to 12. The sum of the heats of these 12 steps must equal the integral heat of solution we measured by

FIGURE 12.6 Integral heats of solution of H_2SO_4 in water.

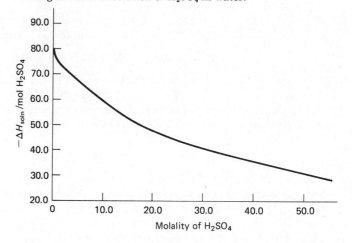

SEC. 12.7 MIXING OF IDEAL SOLUTIONS

the first method, since the total enthalpy change is independent of the path. The 12 heats for the 12 steps will, however, be different from each other. In the limit as we increase the number of steps to a large quantity, the heat measured in each step is what we call a *differential heat of solution;* the differential heat is a partial molar heat of solution.

An integral heat of solution corresponds to the enthalpy change for the reaction

$$H_2SO_4 + n_A H_2O \longrightarrow H_2SO_4(n_A H_2O)$$

in which one mole of pure liquid H_2SO_4 is added to n_A moles of water. The mole fraction of H_2SO_4 in the final solution is $1/(1 + n_A)$, and the mole fraction of the water is $n_A/(1 + n_A)$. The data for these reactions are listed as heats of formation in Table 12.2, where we have calculated the heats of solution from the heats of formation listed in the third column. Figure 12.6 shows a plot of the data for sulfuric acid. Later in this chapter we shall show that for the special case of the ideal solution there is no heat of mixing.

12.7 MIXING OF IDEAL SOLUTIONS

VOLUME CHANGE ON MIXING

Generally a volume change takes place when two components are mixed to form a solution. For any solution the partial molar volume is given by Equation (12.10), $(\partial \mu_i/\partial P)_T = \overline{V}_i$. Using (12.6), we can reduce this to

$$\left(\frac{\partial(\ln f_i)}{\partial P}\right)_T = \frac{\overline{V}_i}{RT} \tag{12.28}$$

If the solution is ideal, (12.7) holds, and

$$\ln f_i = \ln f_i^\circ + \ln X_i$$

With this result, (12.28) becomes

$$\left(\frac{\partial(\ln f_i)}{\partial P}\right)_T = \left(\frac{\partial(\ln f_i^\circ)}{\partial P}\right)_T = \frac{\overline{V}_i^\circ}{RT} \tag{12.29}$$

If we now compare (12.28) with (12.29), we see that for each component

$$\overline{V}_i = \overline{V}_i^\circ \tag{12.30}$$

where $\overline{V}_i^\circ$ is the molar volume of the pure component. Before the components are mixed, the total volume is

$$V_{\text{init}} = n_A \overline{V}_A^\circ + n_B \overline{V}_B^\circ$$

and after they are mixed, the total volume is

$$V_{\text{fin}} = n_A \overline{V}_A + n_B \overline{V}_B$$

The volume change on mixing is hence

$$\Delta V = V_{\text{fin}} - V_{\text{init}} = n_A \overline{V}_A + n_B \overline{V}_B - (n_A \overline{V}_A^\circ + n_B \overline{V}_B^\circ) = 0 \tag{12.31}$$

In other words, there is *no* volume change on mixing for an ideal solution.

ENTHALPY CHANGE ON MIXING

By the Gibbs-Helmholtz equation, (9.53),

$$\left(\frac{\partial(\mu_i/T)}{\partial T}\right)_P = -\frac{\overline{H}_i}{T^2} \tag{12.32}$$

or by (12.6),

$$\left(\frac{\partial(\ln f_i)}{\partial T}\right)_P = -\frac{\overline{H}_i}{RT^2} \tag{12.33}$$

Since for an ideal solution $f_i = X_i f_i^\circ$, (12.33) becomes

$$\left(\frac{\partial(\ln f_i)}{\partial T}\right)_P = \left(\frac{\partial(\ln f_i^\circ)}{\partial T}\right)_P = -\frac{\overline{H}_i}{RT^2} = -\frac{\overline{H}_i^\circ}{RT^2} \tag{12.34}$$

Each component in an ideal solution has the same partial molar heat of solution as the pure component. No heat is evolved when two pure components are mixed to form an ideal solution.

ENTROPY AND FREE ENERGY CHANGE ON MIXING

In an ideal solution, the partial molar free energy, or chemical potential, of any component is

$$\overline{G}_i(\text{solution}) = \mu_i = RT \ln (X_i f_i^\circ) \tag{12.35}$$

In the pure liquid, $X_i = 1$, and

$$\overline{G}_i(\text{pure}) = RT \ln f_i^\circ \tag{12.36}$$

The free energy for one mole of solution containing two components can be obtained from the free energy analog of (12.17),

$$\overline{G}(\text{solution}) = n_A \overline{G}_A + n_B \overline{G}_B = X_A \mu_A + X_B \mu_B \tag{12.37}$$

Since there is one mole of total solution, $n_A = X_A$. For an ideal solution, this becomes

$$\overline{G}(\text{solution}) = X_A RT \ln (X_A f_A^\circ) + X_B RT \ln (X_B f_B^\circ) \tag{12.38}$$

and for the same quantities as pure liquids

$$\overline{G}(\text{pure}) = X_A RT \ln f_A^\circ + X_B RT \ln f_B^\circ \tag{12.39}$$

If the two pure components are mixed to produce one mole of solution, then

$$\Delta G(\text{mixing}) = \overline{G}(\text{solution}) - \overline{G}(\text{pure}) \\ = X_A RT \ln X_A + X_B RT \ln X_B \tag{12.40}$$

The entropy change on mixing is

$$\Delta S(\text{mixing}) = \frac{\Delta H(\text{mixing}) - \Delta G(\text{mixing})}{T} \tag{12.41}$$

Since $\Delta H(\text{mixing}) = 0$ for an ideal solution, this can be reduced to

$$\Delta S(\text{mixing}) = -\frac{\Delta G(\text{mixing})}{T} = -X_A R \ln X_A - X_B R \ln X_B \tag{12.42}$$

Both X_A and X_B are less than unity; hence ΔG(mixing) must be a negative quantity, and ΔS(mixing) must be positive. Note that Equation (12.42) has the same form as (8.9) for the entropy of mixing of an ideal gas.

12.8 DILUTE SOLUTIONS AND HENRY'S LAW

An ideal solution is one that obeys Raoult's law over the entire range of concentrations. There is another type of "ideal solution" that we consider, the *ideal dilute solution*. For solutes with measurable vapor pressures, it is observed that in dilute solutions the vapor pressure of the solute is proportional to the mole fraction:

$$P_B = kX_B \qquad (12.43)$$

If we treat the vapor as an ideal gas, this is equivalent to

$$f_B = kX_B \qquad (12.44)$$

Equation (12.43) is known as *Henry's law,* and was established by William Henry in 1803 in a series of measurements of the solubilities of gases in water as a function of pressure. The proportionality constant k is known as the *Henry's law constant*. While Eq. (12.43) evidently applies only to solutes with measurable vapor pressures, the more general form of Henry's law as expressed in Equation (12.44) also applies to nonvolatile solutes. On a molecular level, Henry's law can be justified by the following qualitative argument.

In a general solution of moderate concentration, the nearest neighbor molecules will be both solute (B) and solvent (A) molecules. Changing the solute concentration changes the composition of the nearest neighbor molecules, which in turn changes the total intermolecular interaction each solute molecule experiences. Under these conditions, the solution is nonideal, the vapor pressure is a nonlinear function of the mole fraction, and Raoult's law is not obeyed. If, however, the solution is so dilute that each solute molecule is surrounded only by solvent molecules, then small changes in solute concentration will not affect the composition of the nearest neighbor molecules. In this limit of very dilute solutions the intermolecular interactions the solute molecules experience will not change with concentration, and the vapor pressure will be proportional to the mole fraction.

At first glance there does not seem to be much difference between Raoult's law and Henry's law. Both laws postulate that the fugacity of the solute is proportional to its mole fraction. A fundamental difference between the two exists, however; it is that the proportionality constants are different, $k \neq P^0$. This is illustrated in Figure 12.7, which shows the fugacity f_B vs mole fraction X_B. As X_B approaches zero, the slope of the true fugacity curve approaches the value given by Henry's law. Note that in the limit of $X_B \to 1$ we have drawn the true fugacity curve tangent to the Raoult's law curve. We have not done so by accident. It can be shown that if the solute obeys Henry's law in dilute solutions, then the solvent in that solution must obey Raoult's law. We now show this.

> WILLIAM HENRY (1775–1836) was an English chemist, textbook writer, and translator of Lavoisier. While he was engaged in experiments of the amount of gas absorbed by water, he discovered what is now known as Henry's law: the amount of gas absorbed is directly proportional to the pressure. Henry committed suicide in a fit of melancholia.

FIGURE 12.7 Fugacity of component B as a function of mole fraction, indicating the distinction between Henry's and Raoult's laws. In the limit of infinite dilution, fugacity is given by Henry's law; and in the limit of pure B, it is given by Raoult's law.

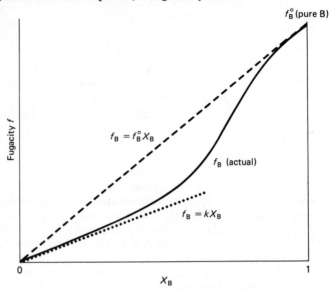

The Gibbs-Duhem equation (12.21b) for the chemical potential can be written

$$X_A\, d\mu_A + X_B\, d\mu_B = 0 \tag{12.45}$$

Since $d\mu_i = RT\, d(\ln f_i)$, this becomes

$$X_A\, d(\ln f_A) + X_B\, d(\ln f_B) = 0 \tag{12.46}$$

which can be brought into the form

$$X_A \frac{\partial(\ln f_A)}{\partial X_A} + X_B \frac{\partial(\ln f_B)}{\partial X_A} = 0 \tag{12.47}$$

For a two-component system, $X_A + X_B = 1$ and $dX_A = -dX_B$. This allows us to put Equation (12.47) in the form

$$X_A \frac{\partial(\ln f_A)}{\partial X_A} - X_B \frac{\partial(\ln f_B)}{\partial X_B} = 0 \tag{12.48}$$

or equivalently

$$\left(\frac{\partial(\ln f_A)}{\partial(\ln X_A)}\right)_{P,T} = \left(\frac{\partial(\ln f_B)}{\partial(\ln X_B)}\right)_{P,T} \tag{12.49}$$

Equation (12.49) is perfectly general and applies to any two-component solution provided that the system is a nondissociating one. For a dilute solution that obeys Henry's law,

$$f_B = kX_B$$
$$\ln f_B = \ln k + \ln X_B$$

SEC. 12.8 DILUTE SOLUTIONS AND HENRY'S LAW

and
$$\left(\frac{\partial(\ln f_B)}{\partial(\ln X_B)}\right)_{P,T} = 1 \qquad (12.50)$$

Equation (12.49) now becomes

$$\left(\frac{\partial(\ln f_A)}{\partial(\ln X_A)}\right)_{P,T} = 1 \qquad (12.51)$$

This can be integrated at constant P and T, resulting in

$$\int d(\ln f_A) = \int d(\ln X_A)$$
$$\ln f_A = \ln X_A + \ln k_A$$
$$f_A = k_A X_A \qquad (12.52)$$

where k_A is a constant of integration. This constant can be determined by evaluating the fugacity f_A at any point. We can take the point as the pure solvent, where $X_A = 1$ and $f_A = f_A^\circ$. The proportionality constant in (12.52) is hence given by

$$k_A = f_A^\circ$$

and (12.52) can be written

$$f_A = X_A f_A^\circ$$

This is just Raoult's law for the solvent.

Henry's law is valid only in very dilute solutions for which determining the mole fraction concentration unit is often inconvenient. Since $n_A \gg n_B$ in dilute solutions, X_B can be written

$$X_B = \frac{n_B}{n_A + n_B} \simeq \frac{n_B}{n_A} \qquad (12.53)$$

In this approximation, Henry's law can be written

$$f_B = k'm \qquad (12.54)$$

where m is the molality, or moles of solute per 1000 g of solvent. Henry's law constants for several gases in aqueous solutions are listed in Table 12.3.

TABLE 12.3 Henry's law constants for several gases in aqueous solutions expressed as $K = P_B/X_B$.

Temperature (°C)	Hydrogen $K \times 10^{-7}$	Helium $K \times 10^{-7}$	Argon $K \times 10^{-7}$	Nitrogen $K \times 10^{-7}$	Oxygen $K \times 10^{-7}$	Carbon dioxide $K \times 10^{-6}$	Methane $K \times 10^{-7}$
0	4.40	9.78	1.64	4.09	1.92	0.55	1.70
10	4.83	9.54	2.09	5.17	2.46	0.79	2.26
20	5.19	9.50	2.51	6.22	3.00	1.08	2.85
25	5.37		2.71	6.69	3.27	1.24	3.14
30	5.54	9.39	2.90	7.15	3.52	1.41	3.41
40	5.71	9.19	3.27	8.05	4.01	1.77	3.95
50	5.81	8.73	3.63	8.72	4.45	2.15	4.39

NOTE: P is measured in torr.

12.9 THE DISTRIBUTION LAW

The distribution of solute between two immiscible solvents is an application of Henry's law. Probably most of us have carried out this procedure in some laboratory course calling for the removal of a solute by a separatory funnel. Iodine is some 200 times more soluble in carbon tetrachloride than in water. It can be removed from an aqueous solution by adding the CCl_4 to the aqueous solution and shaking the mixture. At equilibrium, the concentration of I_2 in the CCl_4 will be much higher than the concentration in the water. The CCl_4 phase is removed, and the process can be repeated to remove more I_2.

At equilibrium, the chemical potential of the solute must be the same for both phases; hence the fugacity of the solute must be the same for both phases. For the reaction,

$$B(\text{in solvent } a) = B(\text{in solvent } b)$$
$$f_B^a = f_B^b \tag{12.55}$$

If the solutions are sufficiently dilute for Henry's law to be valid, then

$$\text{In phase } a \quad f_B^a = k_a X_B^a$$
$$\text{In phase } b \quad f_B^b = k_b X_B^b$$

It follows that

$$\frac{X_B^b}{X_B^a} = \frac{k_a}{k_b} = K \tag{12.56}$$

At equilibrium, the ratio of the mole fractions in the two solvents is a constant K, called the *distribution constant*. The equation as we have written it is valid only for dilute solutions. This law is called the *Nernst distribution law*.

WALTHER NERNST (1864–1941), German physical chemist, did much of the early important work in electrochemistry, studying the thermodynamics of galvanic cells and the diffusion of ions in solution. Besides his scientific researches, he developed the Nernst lamp, which used a ceramic body. This lamp never achieved commercial importance since the tungsten lamp was developed soon afterwards. His electrical piano, which used radio amplifiers instead of a sounding board, was totally rejected by musicians. Nernst was the first to enunciate the third law of thermodynamics, and received the Nobel Prize in chemistry in 1920 for his thermochemical work.

12.10 OSMOTIC PRESSURE

In 1748, Abbé J. A. Nollet described an experiment in which an animal bladder was filled with an aqueous solution of alcohol (wine) and immersed in pure water. He found that water entered the bladder, causing it to distend as the volume increased. Eventually the bladder burst. The movement of water results from *osmotic pressure*.

So far we have formally considered only two kinds of boundaries. One is the conducting wall, which allows heat to pass. The other is the movable wall, which allows the system to interact with the surroundings by expansion work. Osmotic pressure

SEC. 12.10 OSMOTIC PRESSURE

depends on a third kind of boundary, a *semipermeable membrane,* which allows only one of the components to pass through the wall. For any system,

$$dG = -S\,dT + V\,dP + \sum_i \mu_i\,dn_i \qquad (12.57)$$

The first term on the right of (12.57) refers to the passage of heat; the second term refers to the mechanical expansion work; the third term, which is associated with changes in chemical content, refers to "chemical work."

The first quantitative measurements of osmotic pressure were made by Pfeffer in 1887. His apparatus, shown in Figure 12.8a, used an artificial semipermeable membrane consisting of a porous cup; into the interior of this cup cupric ferrocyanide had been precipitated. His experiments were done with aqueous sugar solutions; the membrane allowed the water to pass but not the sugar. The cell containing a sugar solution was immersed in pure water. The equilibrium osmotic pressure was measured by the difference in heights of the mercury columns in the two limbs of a simple mercury manometer. It was van't Hoff who first deduced the relation between osmotic pressure, temperature, and solute concentration.

Consider the sketch of Figure 12.8b. The dashed line separating the two halves represents the semipermeable membrane that allows solvent but not solute to pass from one side to the other. If both sides are filled with pure solvent, then P and $P°$ will be the same; the chemical potential of the solvent on the left will equal the chemical potential on the right. In fugacities,

$$f°_{\text{left}} = f°_{\text{right}}$$

The fugacity is proportional to the vapor pressure of the solvent. If solute is dissolved in the left half, then the vapor pressure of the solvent on the left side will be reduced, $f_A < f°_A$. Solvent then tends to move from the right side to the left side. This solvent movement can be halted by increasing the pressure P above the solution on the left

FIGURE 12.8 (a) Pfeffer's apparatus for osmotic-pressure measurement. The pure solvent is contained in a beaker and the solution is in the porous cup; in the interior of the cup is precipitated cupric ferrocyanide. The osmotic pressure is given by the height difference h in the mercury manometer. (b) Schematic diagram of osmometer.

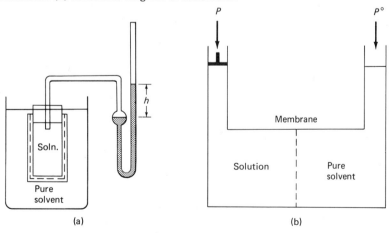

side. The fugacity (and the chemical potential) is a function of the pressure. The movement of solvent will cease at that pressure for which $f_A = f_A^\circ$. That will be the pressure at which the chemical potential of the solvent will be the same on both sides of the membrane.

For an isothermal system ($dT = 0$)

$$d(\ln f_A) = \left(\frac{\partial(\ln f_A)}{\partial P}\right)_{T, X_B} dP + \left(\frac{\partial(\ln f_A)}{\partial X_B}\right)_{T, P} dX_B \qquad (12.58)$$

At equilibrium $d(\ln f_A) = 0$, and

$$\left(\frac{\partial(\ln f_A)}{\partial X_B}\right)_{T, P} dX_B = -\left(\frac{\partial(\ln f_A)}{\partial P}\right)_{T, X_B} dP \qquad (12.59)$$

We can use Equation (10.44) to evaluate the right side of (12.59) as

$$\left(\frac{\partial(\ln f_A)}{\partial P}\right)_{T, X_B} = \frac{\overline{V}_A}{RT} \qquad (12.60)$$

To evaluate the left side of (12.59), we use our previous result—that in a dilute solution in which Henry's law holds for the solute, then Raoult's law holds for the solvent. This means that

$$f_A = f_A^\circ X_A = f_A^\circ (1 - X_B)$$
$$\ln f_A = \ln f_A^\circ + \ln(1 - X_B)$$

$$d(\ln f_A) = d[\ln(1 - X_B)] = -\frac{dX_B}{(1 - X_B)}$$

$$d(\ln f_A) \approx -dX_B \qquad (12.61)$$

Equation (12.61) is justified, since in dilute solutions $1 - X_B$ does not differ much from unity. Thus the left side of (12.59) is

$$\left(\frac{\partial(\ln f_A)}{\partial X_B}\right)_{P, T} = -1 \qquad (12.62)$$

Now we can substitute (12.60) and (12.62) in (12.59), to get

$$\frac{\overline{V}_A}{RT} dP = dX_B \qquad (12.63)$$

In the limit of dilute solutions, the partial molar volume $\overline{V}_A$ is essentially constant and equal to the partial molar volume of the pure solvent, $\overline{V}_A^\circ$.[5] If (12.63) is integrated between the limits $X_B = 0$ and some small mole fraction X_B, then

$$\frac{\overline{V}_A^\circ}{RT} \int_{P^\circ}^{P} dP = \int_0^{X_B} dX_B$$

or
$$P - P^\circ = \frac{RTX_B}{\overline{V}_A^\circ} \qquad (12.64)$$

The pressure differential $P - P^\circ$ is called the *osmotic pressure* and is represented by the symbol Π. Since in dilute solutions

$$X_B = \frac{n_B}{(n_A + n_B)} \approx \frac{n_B}{n_A}$$

[5] For an ideal solution, this statement is true over the entire concentration range.

We can write (12.64) for dilute solutions as

$$\Pi = \frac{RT}{\overline{V}_A^\circ} \frac{n_B}{n_A} = \frac{n_B RT}{V} \qquad (12.65)$$

where V is the total volume of the solution. Equation (12.65) has the same form as the ideal gas law equation and is often written in the form

$$\Pi = cRT \qquad (12.66)$$

where c is the molar concentration or moles of solute per liter of solution.[6] Table 12.4 lists calculated and observed osmotic pressures for aqueous solutions of sucrose.

Physiological systems comprise numerous semipermeable membranes. The quantity of water transported across these membranes in a human in one day is quite large. Remember that the chemical potential on each side of each membrane must be the same at equilibrium. Remember also that an organism in equilibrium is a dead organism. Body liquids can be considered aqueous solutions composed of a variety of solutes—inorganic materials (NaCl and others) and a host of organic materials. Osmotic pressure is one of a class of properties known as *colligative properties of solution;* that is, the osmotic pressure depends on the *total* number of molecules (or moles) in the solution. If we had a dilute solution consisting of ten different solutes, the equation we derived would still be valid if we substitute for n_B the total number of moles of solute material. In an organism, the chemical potentials of the water on either side of a membrane are maintained in a state of near equality. By a delicate mechanism, the organism can adjust the concentration of one of the solutes so that the solvent can flow in one direction or the other. In most instances, the flow is in the direction beneficial to the organism. In a few instances, the mechanism is upset, sometimes with disastrous consequences.

TABLE 12.4 Osmotic pressures of aqueous solutions of sucrose (cane sugar), $C_{12}H_{22}O_{11}$, at 30 °C.

Molal concentration of sucrose	Mole fraction of sucrose	Observed osmotic pressure (atm)	Calculated osmotic pressure from (12.64) (atm)	Percentage difference between calculated and observed values
0.1	0.001798	2.25	2.24	0
0.3	.005375	6.91	6.69	3
0.6	.010693	14.22	13.31	7
1.0	.017696	24.76	22.03	12
2.0	.034777	54.9	43.3	27
3.0	.051273	90.0	63.8	41
4.0	0.067216	129.7	83.7	55

SOURCE: International Critical Tables.

NOTE: The deviation between the calculated and the experimental values gets smaller in dilute solutions.

[6] The similarity to the behavior of gases is further enhanced, since a virial equation of state for osmotic pressures at higher concentrations also exists. This equation for osmotic pressure has the general form

$$\Pi = RT(c + Bc^2 + Cc^2 + \cdots)$$

and approaches (12.66) in the limit of infinite dilution. We shall examine the properties of this equation when we discuss polymer solutions (Chapter 19).

Most of us have suffered bumps on the head at one time or another. A few small blood vessels under the skin are broken, and some blood leaks into an area where it does not belong. The blood is broken down into smaller molecules, and an osmotic pressure (for some strange reason physicians call it oncotic pressure) is built up between surrounding tissue. Fluid then flows into the area, causing an increase in volume that manifests itself by a slight bump. Eventually the molecules are removed by the proper cells, the osmotic pressure is reduced, and the fluids leave the area; the system is thus returned to the normal state. Occasionally the bump is severe enough to break a blood vessel under the skull; the consequences are severe, since the cranial cavity is a constant-volume system.

Figure 12.9 shows a cross-sectional view of the human skull and the enclosed brain. The brain is covered with a membrane known as the pia and is surrounded by spinal fluid. The fluid is contained by the arachnoid membrane, which adjoins the dural membrane underneath the skull. The surface of the brain is covered with blood vessels that lead to the region between the arachnoid and dural membranes. If a severe trauma causes a blood vessel in this region to break, the victim can be killed by the effect of osmotic pressure. Some blood gets into the space between the two membranes and forms a clot, which is then dissolved when the clotted blood decomposes. A membrane, called the inner membrane, forms about this region, and this membrane is bathed in spinal fluid. The difference in chemical potential of the water in the spinal fluid and the water within the inner membrane sets up an osmotic pressure differential that causes water to flow from the spinal fluid into the area of the inner membrane.[7] The cranial cavity is a constant-volume space, since the skull is rigid and there is no room for expansion. The pressure builds up to abnormally high levels, as much as twice the normal pressure, and the excess pressure can kill the patient if it is not treated surgically.[8] This dysfunction, known as a subdural hematoma, is usually called a blood clot by the layman.

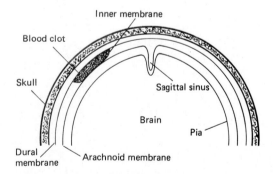

FIGURE 12.9 Sketch of a cranial cavity showing subdural hematoma, or blood clot. Cerebrospinal fluid is normally found in the region between the arachnoid membrane and the pia. In this instance, the fluid flows to the region of the clot because of osmotic pressure.

---

[7] That osmotic pressure was the cause of the trauma was first postulated by W. J. Gardner, *Arch. Neurol. Psychiat.* 27 (1932):847. Recent investigations indicate that the effect of osmotic pressure may be much smaller than previously presumed, and that the main cause may be rebleeding. Several diverse views of this phenomenon exist at the moment. See for example W. Bryce, *J. Neurosurg.* 34 (1971):528; and R. I. Apfelbaum, A. N. Guthkelch, and K. Shulman, *J. Neurosurg.* 40 (1974):336.

[8] The normal spinal fluid pressure is about 180 mm water (13 torr). Various dysfunctions in the nervous system may increase this pressure to about 400 mm water (29 torr).

12.11 FREEZING-POINT DEPRESSION

Another example of a colligative property is the *freezing-point depression*. Suppose a nondissociating solute is dissolved in water. When the solution is cooled, a solid ice phase forms at some temperature lower than the normal freezing point of water. If the solute is soluble in the liquid solvent but completely insoluble in the solid solvent phase, then the ice is pure solid solvent and the liquid phase is a solution. We want to determine the temperature at which the first solid appears as a function of solute concentration. Again the starting point is that the fugacity of the solvent must be the same in both phases at equilibrium.

Suppose the pure solid solvent is in equilibrium with the pure liquid solvent at the normal freezing point; μ_A and f_A must be the same in both phases. Some solute is then added to the system. Under our assumption that the solute dissolves in the liquid but not the solid phase, the vapor pressure and hence the fugacity of the solvent in the liquid phase is lowered. Since the fugacity of the solvent is now higher in the solid than in the liquid, some solid will melt; the heat of fusion cools the system, and the fugacities and chemical potentials in both phases change until they are the same at equilibrium. At equilibrium, $df_A^s = df_A^l$, or

$$d(\ln f_A^s) = d(\ln f_A^l) \tag{12.67}$$

where the superscript s refers to the solid phase and the superscript l refers to the liquid phase.

For either phase it is true that

$$d(\ln f_A) = \left(\frac{\partial(\ln f_A)}{\partial T}\right)_{P,X_A} dT + \left(\frac{\partial(\ln f_A)}{\partial X_A}\right)_{P,T} dX_A \tag{12.68}$$

For the pure solid phase, only the first term has to be considered, since $dX_A = 0$. In view of (12.33), the first derivative on the right side of (12.68) is $-\overline{H}_A/(RT^2)$. The second derivative must be considered only for the liquid phase. Raoult's law applies to the liquid, $\ln f_A^l = \ln f_A^{ol} + \ln X_A$; hence this derivative is

$$\left(\frac{\partial(\ln f_A^l)}{\partial X_A}\right)_{P,T} = \frac{1}{X_A} \tag{12.69}$$

Using these results in (12.68), we have

$$\left(\frac{\partial(\ln f_A^s)}{\partial T}\right)_{P,X_A} dT = \left(\frac{\partial(\ln f_A^l)}{\partial T}\right)_{P,X_A} dT + \left(\frac{\partial(\ln f_A^l)}{\partial X_A}\right)_{P,T} dX_A$$

$$-\frac{\overline{H}_A^s}{RT^2} dT = -\frac{\overline{H}_A^l}{RT^2} dT + \frac{dX_A}{X_A} \tag{12.70}$$

and collecting terms

$$\frac{dT}{T^2} = \frac{R}{(\overline{H}_A^l - \overline{H}_A^s)} d(\ln X_A) \tag{12.71}$$

In dilute solutions $\overline{H}_A^l$ can be taken as $\overline{H}_A^{ol}$, the partial molar enthalpy of the pure liquid. The difference $\overline{H}_A^l - \overline{H}_A^s$ is then the heat of fusion ΔH_{fus}. If ΔH_{fus} is taken to be independent of temperature, then Equation (12.71) can be integrated, to yield

$$\frac{T - T°}{TT°} = \frac{R}{\Delta H_{\text{fus}}} \ln X_A \qquad (12.72)$$

where $T°$ is the freezing point of the pure solvent. The quantity $(T° - T)$ is called the *freezing-point depression* ΔT_f. The value ΔT_f is usually small, in which case $TT° \approx T°^2$. In addition, the term $\ln X_A$ can be expanded in a power series:

$$\ln X_A = \ln (1 - X_B) = -X_B - \tfrac{1}{2} X_B^2 - \tfrac{1}{3} X_B^3 - \cdots \qquad (12.73)$$

For dilute solutions $X_B \ll 1$, and all terms in the series past the first can be neglected. With these approximations, (12.72) can be written

$$\Delta T_f = \frac{RT°^2}{\Delta H_{\text{fus}}} X_B \qquad (12.74)$$

This equation is usually written in the form

$$\Delta T_f = K_{\text{fp}} m_B \qquad (12.75)$$

FIGURE 12.10 Simple freezing-point-depression apparatus. The temperature is adjusted by slowly changing the temperature of the liquid coolant in the large outer beaker. Transfer of heat between coolant and sample in the inner tube is slowed by the insulating layer of air. Uniformity of temperature is maintained in the sample by moving the stirring rod up and down. The temperature of the sample is measured with a Beckmann thermometer. This is a differential type of thermometer, usable over a preset range of about 6 °C and calibrated at intervals of 0.01°; temperature changes can thus be read to 0.001°. Solute is added to the sample through the sidearm tube. (A modern laboratory instrument would normally use an electronic temperature sensor, and the temperature would be plotted as a function of time by a recorder.)

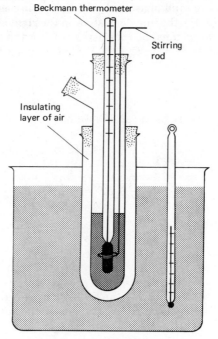

SEC. 12.11 FREEZING-POINT DEPRESSION

TABLE 12.5 Molal freezing-point depression constants for several solvents.

Solvent	Normal freezing point (°C)	K_{fp} (K mol^{-1})
Water	0.00	1.86
Carbon disulfide	−108.6	3.83
Acetic acid	16.6	3.90
Ether	−116	1.79
Benzene	5.51	5.12
Cyclohexane	6.5	20.2
Camphor	176	40

where K_{fp} is the *freezing-point-depression constant*, or *cryoscopic constant*, and m_B is the molal concentration of the solute. The constant K_{fp} is given by

$$K_{fp} = \frac{RT^{\circ 2} M_A}{1000 \Delta H_{fus}} \tag{12.76}$$

where M_A is the molar mass of the solvent. For water, the constant is

$$K_{fp}(H_2O) = \frac{(8.314)(273.15)(273.15)(18.02)}{6010 \times 10^3} = 1.860 \text{ K mol}^{-1}$$

Figure 12.10 shows a sketch of an apparatus for measuring freezing-point depressions. Values of K_{fp} are listed in Table 12.5. The measurement of freezing-point depressions provides a rapid and convenient method for determining the molecular weight of an unknown solute. The larger the value of K_{fp}, the greater the accuracy with which M_B can be determined in view of the larger measured ΔT_f. Experimental values of the freezing-point depressions for aqueous solutions of sucrose and aniline have been plotted in Figure 12.11 as a function of solute concentration. In the limit

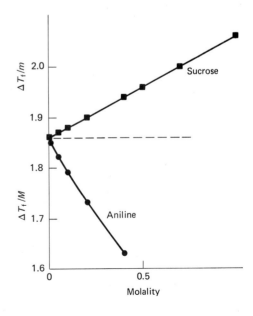

FIGURE 12.11 Freezing-point depressions as a function of molality for sucrose (cane sugar) and for aniline. The horizontal dashed line indicates ideal behavior. Note that in the limit of infinite dilution, $\Delta T_f/m$ approaches the theoretical value 1.86.

of small concentration, the freezing-point depression approaches the theoretical value.

In the above discussion we considered the solid phase the pure solvent. If we should, on the other hand, consider the solid phase the pure solute, then we could derive an equation analogous to (12.72) in the form

$$\ln X_B = \frac{\Delta H_{\text{fus}}}{R}\left(\frac{T - T^\circ}{TT^\circ}\right) \tag{12.77}$$

where ΔH_{fus} is the heat of fusion of the *solute,* and T° is the melting point of the *solute.*[9] Equation (12.77) provides an expression for the solubility as a function of temperature. It predicts that for an ideal solution a plot of $\ln X_B$ vs $(T)^{-1}$ should yield a straight line. A plot like this is shown in Figure 13.11.

12.12 BOILING-POINT ELEVATION

The last of the colligative properties of solution we want to consider is the *boiling-point elevation.* We shall examine the equilibrium between a liquid solution phase and a gas phase consisting of the pure solvent. By proceeding just as we did for the freezing-point depression, we can show that

$$\Delta T_b = K_{bp} m_B = \frac{RT_b^{\circ 2} M_A}{1000 \Delta H_{\text{vap}}} \tag{12.78}$$

where ΔT_b is the boiling-point elevation, K_{bp} is the *boiling-point-elevation constant,* M_A is the molar mass of the solvent, m_B is the molal concentration of the solute, T_b° is the boiling-point temperature, and ΔH_{vap} is the heat of vaporization of the solvent. The derivation of (12.78) is straightforward; it is left for one of the problems. Table 12.6 lists the boiling-point elevation constants for several common solvents. The value of K_{bp}, or the *ebullioscopic constant,* for water is

$$K_{bp} = \frac{8.314 \times 373.15 \times 373.15 \times 18.02}{40670 \times 10^3} = 0.513 \text{ K mol}^{-1}$$

TABLE 12.6 Molal boiling-point elevation constants for several solvents.

Solvent	Normal boiling point (°C)	K_{bp} (K mol^{-1})
Water	100.0	0.513
Carbon disulfide	46.3	2.34
Acetic acid	118.1	3.07
Benzene	80.1	2.53
Cyclohexane	81.4	2.79
Toluene	110.6	3.33

[9] A derivation is given in I. M. Klotz, *Chemical Thermodynamics* (Englewood Cliffs, N.J.: Prentice-Hall, 1950), p. 254.

12.13 ADDITIONAL REMARKS

During the first thirty or forty years of this century the volume of data generated in laboratories throughout the world dealing with the material in this and the next several chapters was sufficient to fill countless pages in the scientific literature. At least half the research effort of physical chemists was devoted to studying solutions, and the number of Ph.D. theses devoted to the subject certainly exceeded the number in any other area of research. The results of this chapter are applicable only to ideal solutions and to solutions in the limit of infinite dilution. In the next several chapters we shall turn our attention to nonideal solutions, both of electrolytes and nonelectrolytes, and we shall devote some effort to examining phase diagrams. Throughout these discussions you should remember that there is a single unifying feature underlying all the discussions. In every case involving the equilibrium state between two phases or two subsystems, the ultimate treatment is the same, no matter how different the cases appear. *In each case we must find the conditions for which the chemical potential of any constituent is the same throughout the system.*

PROBLEMS

1. At 20 °C an aqueous solution of methyl alcohol, CH_3OH, that contains 25% alcohol by weight has a density of 0.9592 g cm^{-3}. Calculate the concentration of alcohol in molality, molarity, and mole fraction units.

2. The accompanying table gives some data on density vs concentration for solutions of ethyl alcohol, C_2H_5OH, in water at 25 °C. Note that we have intentionally omitted some entries. Fill in these entries and then plot the total volume of a solution containing 1000 g water against the molality of alcohol. Draw a smooth curve through the points and determine the partial molar volume of alcohol at a concentration of 1.0m C_2H_5OH by drawing the tangent to the curve.

3. The total volume of an alcohol solution containing 1000 g water is given by the analytical expression

$$V = 1003.11 + 54.6664m - 0.36394m^2 + 0.028256m^3 \text{ cm}^3$$

where m is the molality of the alcohol. Use this analytical expression to check your answer to Problem 3.

4. The density of pure alcohol at 25 °C is 0.78506 g cm^{-3}. Using the expression of Problem 3, calculate the final volume of a solution made up by mixing 1000 cm^3 water with 400 cm^3 alcohol.

5. You have an infinitely large solution that is 20% by weight C_2H_5OH. Using the data of the previous problems, calculate the volume changes when (a) a mole of water is added to the solution; (b) a mole of C_2H_5OH is added to the solution.

6. Derive an expression for $(\partial Q/\partial X_A)_{P,T,n_B}$ in terms of the partial molar quantity $\overline{Q}_A = (\partial Q/\partial n_A)_{P,T,n_B}$, where Q may be any extensive property of the system such as V, H, or G.

7. Using the values for the integral heats of solution of sulfuric acid listed in Table 12.2, calculate the heat evolved when 1 mol H_2SO_4 is added to 5 mol H_2O. If the resulting solution is diluted by adding 5 mol H_2O, calculate the heat evolved by adding 5 mol H_2O. (This second heat is called a *heat of dilution*.)

Weight percent alcohol	Density of solution (g cm^{-3})	Volume that contains 1000 g water (cm^3)	Mole fraction of alcohol	Molality of alcohol
0.	0.99708	1002.93	0.0000	0.00000
2.	.99336	1027.23	.0079	0.44299
4.	.98984	1052.36	.0160	0.90443
6.	.98656	1078.32	.0244	1.38551
8.	.98346		.0329	
10.	.98043	1133.29	.0416	2.41182
12.	.97753	1162.48	.0506	2.95996
14.	.97472	1192.95	.0598	3.53360
16.	.97199	1224.78	.0693	4.13455
18.	.96923	1258.23		4.76482
20.	.96639	1293.47	.0891	5.42660
22.	.96348	1330.65	.0993	6.12231
24.	.96048	1369.93	.1099	6.85465
26.	.95738	1411.51	.1208	7.62657
28.	.95410	1455.71	.1320	8.44137
30.	0.95067	1502.70	0.1435	9.30274

8. The integral heat of solution for dissolving m mol NaCl in 1000 g H_2O is given by

$$\Delta H_s = 3862m + 1992m^{3/2} - 3038m^2 + 1018.8m^{5/2} \text{ J}$$

Calculate the heat absorbed when 1 mol NaCl is added to an infinitely large amount of $1.000m$ NaCl solution.

9. At 20 °C, the vapor pressure of pure toluene is 22.3 torr and the vapor pressure of pure benzene is 74.7 torr. A solution is made up containing 4 mol benzene and 6 mol toluene.
   (a) Calculate the partial pressure of benzene and toluene in the vapor.
   (b) Calculate the total vapor pressure of the solution.
   (c) Calculate the mole fraction of benzene and toluene in the vapor phase.

10. At 50 °C, the vapor pressure of pure toluene is 93 torr and the vapor pressure of pure benzene is 268 torr. We start with a very large quantity of solution that is 0.5 mol fraction benzene; some vapor in equilibrium with the original solution is removed and condensed to form a new solution. What is the concentration of benzene in the new solution? Some vapor is removed from the new solution and condensed, and this process is repeated five times. What is the concentration of benzene in the final solution? (*Note:* In this problem, the quantities are sufficiently large that removing material from the vapor does not affect the concentration of the liquid solution. See next problem for the case of small quantities.)

11. A solution is made up containing $\frac{1}{2}$ mol benzene and $\frac{1}{2}$ mol toluene. The relevant vapor pressures are given in Problem 10. The solution is allowed to evaporate to dryness at 50 °C. Calculate the concentration of benzene in the first vapor to form. Calculate the concentration of benzene in the last drop of liquid.

12. At 900 °C, the vapor pressure of pure zinc is 715 torr. A sample of brass is composed of 20 mol percent zinc. Calculate the vapor pressure of zinc over the brass sample. (Note that Raoult's law of vapor-pressure lowering is not restricted to liquid solution but applies also to ideal solid solutions.) The experimental vapor pressure for the zinc is found to be 43 torr. Does the system copper-zinc form an ideal solution? (*Note:* Although brass has many desirable qualities like ease of machinability and solderability, the relatively high vapor pressure of the zinc vitiates its use as a material in high-vacuum systems, especially at elevated temperatures. The material par excellence for high-vacuum purposes is stainless steel.)

13. At 25 °C, the densities of pure water and pure ethyl alcohol are 0.99708 and 0.78506 g cm^{-3}. A solution is made up that is 30% alcohol by weight. The volume of a solution containing 1000 g water is given in Problem 2. What would the volume be if the alcohol–water system were ideal?

14. At 20 °C, 1 mol benzene is added to an infinitely large volume of a toluene–benzene solution, in which the mole fraction of benzene is 0.3. Calculate ΔH, ΔG, and ΔS, assuming that the solution is ideal. (See Problem 9 for additional data.)

15. For 20 °C, calculate ΔG, ΔH, and ΔS for the mixing of 0.3 mol benzene with 0.7 mol toluene to form 1 mol ideal solution. (See Problem 9 for additional data.)

16. Assume that air is 20% O_2, 79% N_2, and 1% Ar. Calculate the equilibrium concentration of air in water at 1 atm and 25 °C. What percentage of the dissolved air is O_2?

17. How much oxygen is dissolved in 1 km^3 seawater if the temperature is assumed to be 20 °C and the concentration is assumed to have its saturation value? What volume would the O_2 occupy at standard temperature and pressure (STP = 1 atm and 0 °C)? (The O_2 comes from two sources, atmospheric oxygen and photosynthetic oxygen produced by vegetation in the sea.)

18. At 25 °C, the solubility of I_2 in water is 1.1×10^{-3} mol liter^{-1} solution. The solubility of I_2 in CCl_4 is 0.237 mol liter^{-1} solution. You have 100 cm^3 saturated solution of I_2 in water and 10 cm^3 pure CCl_4.
   (a) If the CCl_4 is added to the aqueous solution, what is the final concentration of I_2 in the water after the mixture is thoroughly shaken?
   (b) The CCl_4 is divided into four equal portions of 2.5 cm^3 each. The aqueous solution is successively shaken with each portion, and the CCl_4 phase is removed before adding the next portion. Calculate the final concentration remaining in the water phase.

19. Derive (12.78) for the boiling-point-elevation constant.

20. Equation (12.77) gives the change in solubility with temperature for an ideal solution. Derive (12.77).

21. Calculate the solubility of naphthalene at 25 °C, assuming that an ideal solution is formed. The normal melting point of naphthalene is 80 °C, and the heat of fusion of naphthalene is 19.3 kJ mol^{-1}.

22. By considering the change of fugacity of the solute with pressure, show that the variation of solubility with pressure is given by the equation

$$\left(\frac{\partial \ln X_B}{\partial P}\right)_T = \frac{\overline{V}_B^\circ - \overline{V}_B}{RT}$$

where $\overline{V}_B^\circ$ is the molar volume of the pure solid solute, and $\overline{V}_B$ is the partial molar volume of the solute in the solution.

23. A soda-pop manufacturer pressurizes his pop with CO_2 gas at 10 °C at a CO_2 pressure of 2 atm. The delivery trucks that deliver the pop to stores are often parked in the sun during hot summer days, and the temperature of the pop may rise to 50 °C. What pressure must the pop bottles be able to withstand so that the pressure within the bottles will at all times be less than one-half the bursting point?

PROBLEMS

24. Considering substances with molecular weights 10^2, 10^3, 10^4, and 10^5, each of which is soluble to the extent of 1 g per 25 cm^3 of water, calculate for each the freezing-point depression, the boiling-point elevation, and the osmotic pressure for saturated solutions of water. Assume that freezing points can be measured within 0.0001 °C, boiling points within 0.001 °C, and osmotic pressures within 0.05 torr. How accurately can each of the molecular weights be measured with each of the three techniques?

25. A tube, closed at one end with a semipermeable membrane, has cross-sectional area 1 cm^2. One gram of sucrose, $C_{12}H_{22}O_{11}$, is placed at the bottom of the tube, and the tube is inverted in a beaker of pure water so that the water will enter the tube and dissolve the sucrose. To what height will the water in the tube rise when equilibrium is reached? The temperature is 25 °C.

26. A compound weighing 0.396 g is dissolved in 25 g camphor. The freezing point of the camphor is reduced by 2.32°. What is the molecular weight of the unknown compound?

27. For each of sucrose, $C_{12}H_{22}O_{11}$, and glucose, $C_6H_{12}O_6$, 1 g is dissolved in 100 g water at 25 °C. What is the osmotic pressure of the resulting solution? What is the apparent concentration of the solution?

28. One gram of benzoic acid, C_6H_5COOH, is dissolved in 20 g benzene. The solution is found to freeze 1.05° below the normal freezing point of benzene, for which $K_{fp} = 5.12$. What can you say about the molecular form of benzoic acid when it is dissolved in benzene?

29. At 25 °C and 1 atm pressure, the third law absolute entropies of o-xylene and m-xylene are 248.1 and 252.3 J K^{-1} mol^{-1} in the liquid phase. Calculate the absolute third law entropy of a solution containing 0.5 mol of each, assuming that the solution is ideal.

30. Pure lead melts at 327.5 °C, and the heat of fusion of lead is 5188 J mol^{-1}. A sample of lead melts at 327.4 °C. Calculate the concentration of impurity in the lead.

31. Take the densities of water, methyl alcohol, CH_3OH, and glycol, $C_2H_6O_2$, as 1.00, 0.81, and 1.13 g cm^{-3} and assume that they are constant over the temperature range in this problem. Take the vapor pressure of pure water at 90 °C, to be 0.69 atm and of pure methyl alcohol, 2.49 atm; assume that the vapor pressure of glycol is negligible in comparison. The cooling system of your automobile has a capacity of 12 liters. Calculate the volume of methanol and of glycol needed to protect the coolant from freezing at temperatures as low as -30 °C. If the engine operates at temperatures as high as 90 °C, what pressure must the system be able to withstand if methanol and glycol are used as antifreeze?

32. Your city has 10 kilometers of streets, each 8 m wide. A snowfall equivalent to a sheet of ice 2 cm thick has fallen, and the temperature is -5 °C. How much salt is required to melt all the ice on the city streets? (Note that since NaCl dissociates into ions, the effective freezing-point-depression constant is $2 \times 1.86 = 3.72$ °C mol^{-1}.)

33. Assume that the compound AB dissociates according to the equation AB = A + B. Derive an expression for the effective freezing-point-depression constant for AB in terms of α, the degree of dissociation, and m, the number of moles of AB per 1000 g solvent.

34. Calculate the freezing point of water saturated with O_2 at a partial pressure of 1 atm at 0 °C.

35. At 20 °C, the density of aqueous solutions of sucrose is given by the equation

$$\rho = 0.99826 + (3.8491 \times 10^{-3})w + (1.3496 \times 10^{-5})w^2 + (4.1075 \times 10^{-8})w^3 \text{ g cm}^{-3}$$

over the range 0–48 weight percent sucrose, where w is the weight percent concentration of sucrose. By calculating the osmotic pressure by each of the three equations at concentrations of 0.001, 0.01, 0.1, 0.5, 1.0, and 2.0 molalities, estimate the accuracy of the approximations we made in deriving (12.65) and (12.66) from (12.64).

36. At 25 °C the vapor pressure of pure benzene is 95 torr; the vapor pressure of pure toluene is 28 torr. The heats of vaporization are 31.0 and 33.5 kJ mol^{-1}. Assume that the substances form an ideal solution. For a mixture of 1 mol benzene and 3 mol toluene, calculate:
(a) The total vapor pressure of the solution.
(b) The composition of the vapor in equilibrium with the liquid.
(c) The change in free energy in forming the solution from the pure liquids.
(d) The change in entropy in forming the solution from the pure liquids.
(e) The difference in partial molar free energy of benzene in the liquid solution and the pure liquid.
(f) The difference in partial molar entropy of benzene in the solution and in the pure liquid.
(g) The difference in partial molar free energy of benzene in the vapor (in equilibrium with the solution) and in the pure liquid.
(h) The difference in partial molar entropy of benzene in the vapor (in equilibrium with the solution) and in the pure liquid.

CHAPTER THIRTEEN

PHASE DIAGRAMS

13.1 THE SEPARATION OF COMPONENTS

In Chapter 12 we discussed the properties of two-component solutions from the standpoint of solution formation. We started with pure solvents and noted the effects of added solute on the thermodynamic properties of the resulting solution. The emphasis in this chapter is on the reverse process. We start with solutions and examine the practical aspects of separating the solution into its pure components. First we shall deal with phase diagrams of two-component liquid solutions and also the problem of distillation. Then we shall examine liquid-solid phase diagrams.

For two-component systems, the phase rule takes the form

$$f = c - p + 2 = 4 - p \tag{13.1}$$

There are four possibilities:

$p = 1;\quad f = 3 \quad$ (trivariant system)
$p = 2;\quad f = 2 \quad$ (bivariant system)
$p = 3;\quad f = 1 \quad$ (univariant system)
$p = 4;\quad f = 0 \quad$ (invariant system)

Remember that a complete representation of a two-component system requires a three-dimensional space. We usually use a two-dimensional space at constant temperature or pressure. This uses up one of the degrees of freedom, reducing the number of available degrees of freedom of the system by one.

Recall our discussion of Raoult's law and the vapor-pressure representations in Chapter 12. Let's start by replotting this same information in a slightly different form to indicate better the process whereby the components can be separated.

13.2 PRESSURE-COMPOSITION DIAGRAMS OF NEARLY IDEAL SOLUTIONS

In Figure 13.1 we have plotted a *pressure-composition* diagram for the system benzene–toluene. The upper straight line, the *liquidus* line, represents the dependence of the *total* vapor pressure on the mole fraction of the liquid at constant temperature. The liquidus line is just the total-pressure line of Figure 12.1. The lower curved line represents information that can be determined from the vapor pressure curve of Figure 12.1 but that is not shown explicitly on that figure. It in-

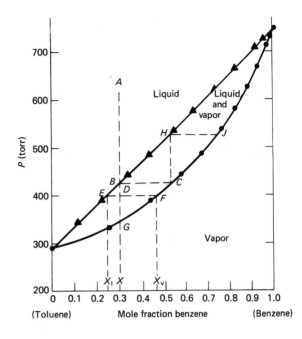

FIGURE 13.1 Pressure-composition phase diagram for the nearly ideal system benzene–toluene at 79.7 °C. (M. A. Rosanoff, C. W. Bacon, and F. W. Schulze, *J. Amer. Chem. Soc.* 36 (1914):1993.)

dicates how the *total* vapor pressure depends on the composition of the vapor that is in equilibrium with the liquid.

Consider the point labeled A in the liquid region of Figure 13.1. It is a one-phase region with three degrees of freedom. One of these has been specified by the condition of constant temperature, leaving two degrees of freedom; both the concentration and the pressure can be independently varied in the region near A. If the pressure is slowly reduced along the dashed line of constant composition, nothing happens until we reach the liquidus curve at B, where some vapor forms. This vapor is richer in the more volatile benzene. The composition of that first vapor is given by point C. The horizontal dashed line that joins point C to B is known as a *tie line*. Points along the liquidus curve lie in a two-phase region. Since the temperature is specified, there is just one degree of freedom left along the line. Either the concentration or the pressure can be independently varied, in which case the other variable moves along the curve.

As the pressure is further reduced without changing the overall composition, the two-phase region between the two curves is entered. The system breaks up into two phases. For the pressure represented by point D, the concentrations of each of the phases is given by the points E and F at the ends of the tie line through D. The overall composition of the system is X. If we let n_l and n_v be the total number of moles of material in the liquid and vapor phases, and X_l and X_v the mole fractions, then balancing the quantity of component B requires that

$$X(n_l + n_v) = X_l n_l + X_v n_v$$

or

$$\frac{n_l}{n_v} = \frac{X_v - X}{X - X_l} = \frac{DF}{DE} \qquad (13.2)$$

Equation (13.2) is known as the *lever rule,* from its similarity to the equation for mechanical levers. The lever rule applies to all two-component phase diagrams in which two phases are connected by tie lines. As the pressure is still further reduced, point G is reached, and the last liquid vaporizes. Further pressure decrease continues in the one-phase vapor region.

The two components could be separated in this way. If the first vapor that formed at point C were condensed, we should obtain a liquid corresponding to point H, which would be richer in the more volatile component. The first vapor to form from this liquid would be at point J, and this process could be continued indefinitely, obtaining vapors that are successively richer in the more volatile component. Varying the temperature is a more practical procedure.

13.3 TEMPERATURE-COMPOSITION DIAGRAMS OF NEARLY IDEAL SOLUTIONS

Figure 13.2 depicts a *temperature-composition* diagram for the system benzene–toluene. Again we have basically plotted Raoult's law type of data, but in a different manner. The points on the two curves in this plot can be obtained from a series of isothermal P-X diagrams.

This T-X diagram is like the P-X diagram of Figure 13.1, and the properties of the two are quite similar. Tie-line constructions and the lever rule are applicable to the T-X phase diagrams. Suppose we heat a sample of the liquid at mole fraction X starting at room temperature. When the temperature reaches the boiling point, the composition of the vapor boiling off is given by X_1. If that vapor were condensed to yield a liquid of composition X_1, then that new liquid would boil at a somewhat lower temperature providing a vapor at mole fraction X_2. If we continue this process indefinitely as indicated by the dashed lines, then each step would provide a product that

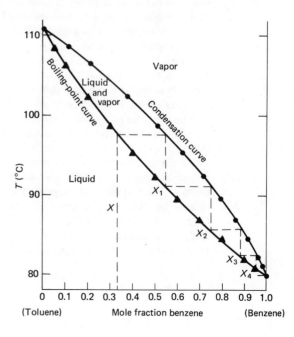

FIGURE 13.2 Temperature-composition diagram for the system benzene–toluene at 1 atm pressure. (A. P. Rollet, G. Elkaim, P. Toledano, and M. Senez, *Compt. Rend. Acad. Sci.* 242 (1956):2560.)

is richer in the more volatile constituent. Unlike the separation procedure outlined in the previous section, this process is completely practical, since temperature gradients can easily be established in laboratory or plant systems.

13.4 FRACTIONAL DISTILLATION

Fractional distillation is a process whereby the separation procedure outlined in Figure 13.2 is carried out on a continuous basis. It is one of the most widely used procedures for separating liquids that have different boiling points. The simplest type of distillation column is shown in Figure 13.3. A mixture of high-boiling liquid (B) and low-boiling liquid (A) is brought to its boiling point in the *boiler*, or *pot*. As the vapors rise in the *column*, some of the vapor condenses on the walls of the column and provides a region where a liquid-vapor equilibrium is established. A water-cooled *condenser* placed at the top of the column ensures that the vapors are condensed and not lost through the top of the column. The condenser is designed to direct the returning liquid, the *reflux*, to one point at which it can be removed. The material removed is called the *takeoff*. A valve is often incorporated at this point to allow the return of some of the liquid to the pot. The ratio of material removed to material returned to the pot is called the *reflux ratio*. The *throughput* is the amount of liquid that can be forced through the still, while the *holdup* refers to the amount of liquid needed to coat the inside of the column.

When the boiling points are close together, a high degree of liquid-vapor contact, or *rectification*, is required, and much more material must be returned to the pot than can be taken off; a high reflux ratio is required. Our basic still of Figure 13.3 has a high throughput, but a poor separation efficiency. The holdup is low, and there is a low pressure drop through the column, since there is little resistance to the rising vapors. A low-pressure drop allows distillation at a temperature lower than that at which a system with a high-pressure drop does, since less heat is required to provide the energy to force the vapors through the column. Mixtures of liquids with boiling-point differences of less than 25 °C generally cannot be satisfactorily separated with this basic type of column.

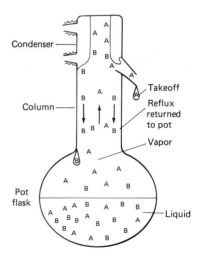

FIGURE 13.3 A basic distillation column. (Courtesy of Perkin-Elmer Corporation.)

FIGURE 13.4 Bubble-cap distillation column. (Courtesy of Perkin-Elmer Corporation.)

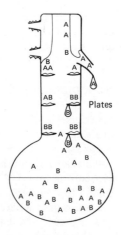

An improved type of column, the *bubble-cap* column, is shown in Figure 13.4. As the liquid is boiled in the pot, the vapors rise and partially condense in the plates. Each plate provides a surface for liquid-vapor equilibrium. The plates are at successively lower temperatures, and the liquid in the plates is successively richer in the more volatile component. For an ideal column, each plate effectuates a degree of separation indicated by one step of the process outlined on the T-X diagram. Each separation step is known as a *theoretical plate*. From Figure 13.2, we see that if a sample of benzene–toluene that is 0.33 mole fraction benzene is distilled in a column with three theoretical plates, then the product removed would be $X_{\text{benzene}} = X_3 = 0.88$. High separation capacities can be accomplished by using high stills with large numbers of plates. This kind of column is rarely used in the laboratory, since the holdup can exceed the total volume of the sample. Columns of this type are often used in large-scale commercial operations, in which columns containing hundreds of plates and towering hundreds of feet into the air are not uncommon.[1]

In the laboratory, increased liquid-vapor contact can be achieved by filling the column with a *packing* of glass beads or other inert material, as depicted in Figure 13.5. While high efficiencies can be obtained with this type of column, the holdup may be large, and the pressure drop may be large in view of the tortuous vapor path. The throughput is small, since much of the condensate must be returned to the column to provide a liquid coating for the large packing surface. This type of column can be used to separate liquids whose boiling points are as close as 0.5–1 °C, but the takeoff rate is low. Takeoff rates of one drop every four hours are not unheard of.

The *spinning-band* column shown in Figure 13.6 provides a highly efficient distillation system for small-scale laboratory use. By rotating a band in contact with the walls of the column, a large number of theoretical plates can be achieved in a relatively short column. A commercial version of this type of column is shown in Figure 13.7. This column is a 200-plate column. The column itself is just under 1 m high and has an inside diameter of 8 mm. The control panel allows the operator to adjust the temperature of the pot as well as the reflux ratio. High-purity separations of mate-

---

[1] Large industrial columns may attain diameters of 5 m and be longer than a railroad flatcar. The maximum size for a particular installation is often decided on the basis not of the particular requirements of the chemical plant but of transportation feasibility. The route from the place of manufacture of the column to the place of installation must be carefully analyzed for clearances through all tunnels to be used and the radii of all curves that will be encountered along the way.

SEC. 13.4 FRACTIONAL DISTILLATION

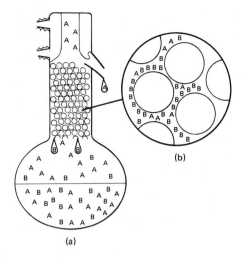

FIGURE 13.5 (a) A packed distillation column. (b) Detail of packing. (Courtesy of Perkin-Elmer Corporation.)

rials with boiling-point separations of less than 0.5 °C can be accomplished with this type of column. The column shown is an adiabatic one; heat transfer is minimized by enclosing the column in a silvered vacuum jacket.

The number of theoretical plates in a column is often expressed in terms of *HETP*, or *height equivalent to a theoretical plate*. This is obtained by dividing the column length by the number of theoretical plates for a particular column. One useful equation for determining the number of theoretical plates of a column was given by Fenske,[2]

$$n = \frac{\log_{10}[(X_A/Y_A)(Y_B/X_B)]}{\log_{10} K} - 1 \tag{13.3}$$

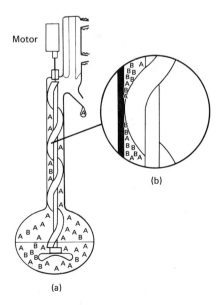

FIGURE 13.6 (a) A Teflon annular spinning-band distillation column. (b) Detail of wiping action. (Courtesy of Perkin-Elmer Corporation.)

[2] M. R. Fenske, *Ind. Eng. Chem.* 24 (1932):482.

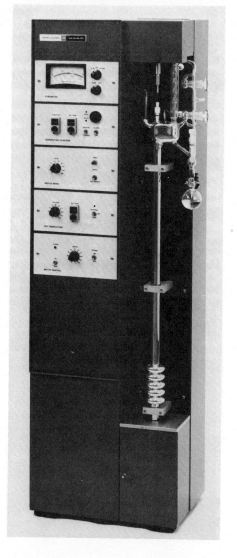

FIGURE 13.7 A commercial spinning-band distillation system with 200 theoretical plates. The control panel allows temperature and reflux ratio adjustments. The bellows at the bottom of the silvered-glass vacuum jacket is to allow for thermal expansion. The system can be operated under vacuum. (Courtesy of Perkin-Elmer Corporation.)

where X_A = % low-boiler in the head, Y_A = % low-boiler in the pot, X_B = % high-boiler in the head, Y_B = % high-boiler in the pot, and K is the ratio of the vapor pressure of the low-boiler to the vapor pressure of the high-boiler.

The ultimate in separation ability is achieved in *chromatography,* a process analogous to distillation but based on a different fundamental principle. Chromatography is based on the differing rates of adsorption and desorption of molecules on substrates with high surface areas (see Chapter 18). Columns with thousands of theoretical plates can be achieved in chromatographic systems but the "takeoff" is extremely small.

It is a general principle of experimental science that the higher the resolution, the lower the signal. This principle is most generally applied in areas such as spectros-

SEC. 13.5 LIQUID-VAPOR PHASE DIAGRAMS FOR NONIDEAL SOLUTIONS

copy, but it applies equally well to distillation if we substitute "yield" for "signal." The closer the boiling points between the two components of a solution, the greater the effort required to get a given quantity of pure material.

13.5 LIQUID-VAPOR PHASE DIAGRAMS FOR NONIDEAL SOLUTIONS

For solutions that deviate greatly from ideality the liquid-vapor phase diagrams no longer retain the appearance of the diagrams of Figures 13.1 and 13.2, and maxima or minima may appear. Four general types of phase diagrams of real solutions are sketched in Figure 13.8. Actual phase diagrams are shown in Figures 13.9 and 13.10 for the systems benzene–ethanol and chloroform–acetone, both of which are highly nonideal. Note that a maximum in the P-X diagram corresponds to a minimum in the T-X diagram and vice versa. At the extremal point $X_l = X_v$; a solution with that composition is known as an *azeotropic solution*. An azeotropic solution cannot be separated into its components by distillation, since the composition does not change on boiling.

If we try to distill a nonazeotropic solution in a system that has a minimum boiling point, then the vapor will tend towards the azeotropic concentration and the residual liquid will tend towards the pure component A or B depending on which side of the azeotropic solution we start. Conversely, for a system with a maximum boiling

FIGURE 13.8 Sketches of liquid-vapor P-X and T-X phase diagrams. P-X diagrams are at constant temperature; T-X diagrams are at constant pressure.

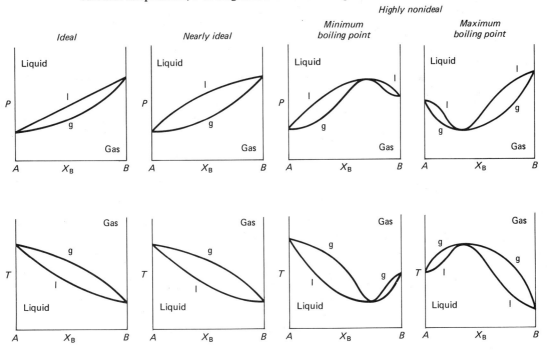

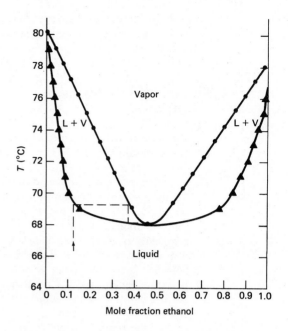

FIGURE 13.9 T-X diagram for the system benzene–ethanol at 1 atm. (R. Fritzweiler and K. R. Dietrich, *Z. Angew. Chem.* 46 (1933):241.

point, the vapor will tend towards one of the pure components while the residual liquid tends towards the azeotropic concentration. This can be seen by noting the distillation stages outlined in Figures 13.9 and 13.10.

Hydrogen chloride (HCl; bp = −80 °C) forms a maximum-boiling azeotrope with H_2O (bp = 100 °C). At 1 atm, the azeotropic solution is 20.222 wt % HCl and the boiling point is 108.584 °C. The concentrations of these *constant-boiling HCl* solu-

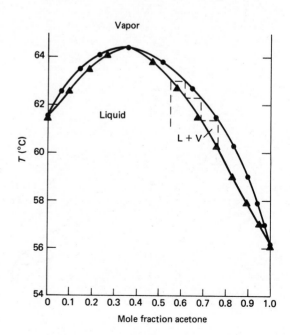

FIGURE 13.10 T-X diagram for the system chloroform–acetone at 1 atm. (W. Reinders and C. H. de Minjer, *Rec. Trav. Chim.* 66 (1947):573.)

tions are known to great precision as a function of pressure, and standard solutions of HCl can be prepared by boiling a solution of HCl until the boiling point is constant.

The system water–ethanol (bp = 78.3 °C) forms a minimum-boiling azeotrope. The concentration of the azeotrope is 96 wt % ethanol and the boiling point is 78.174 °C at 1 atm. Ethanol is often prepared by distillation from an aqueous solution and is normally sold as "95% ethanol," corresponding very closely to the azeotropic concentration. One hundred percent ethanol cannot be prepared by distillation from aqueous solution, and other procedures must be used to remove the water.

13.6 LIQUID–SOLID PHASE DIAGRAMS; SIMPLE EUTECTICS

The liquid-vapor phase diagrams we have dealt with are of interest mainly to chemists and chemical engineers. We shall now examine some properties of phase diagrams of condensed systems. These have added importance in metallurgy, ceramics, and geology, and you who are studying these fields will eventually examine this subject in much greater detail than we do here.

The temperature variation of the solubility of a solid in a liquid was given in Equation (12.77). We rewrite that equation in the form

$$\ln X_B = \frac{\Delta H_{fus}}{R}\left(\frac{1}{T°} - \frac{1}{T}\right) \tag{13.4}$$

where X_B is the mole fraction of solute, ΔH_{fus} is the heat of fusion of the solute, $T°$ is the melting point of the solute, and T is the temperature. Equation (13.4) is valid for ideal solutions. The logarithm of the solubility of naphthalene in benzene vs T^{-1} is plotted in Figure 13.11. The system naphthalene–benzene forms an ideal solution. The initial dependence is linear, as predicted by Equation (13.4). There is, however, a sudden break in the curve at the point corresponding to $T = 269.8$ K (-3.4 °C) at a mole fraction of 0.133 naphthalene. The reason for this sudden break becomes more apparent if we treat the data in a slightly different way.

Figure 13.12 shows the same data plotted as a *phase diagram*. The experimental points correspond to the freezing points of the mixtures. By general convention, the component on the left (in this case naphthalene) is known as component "A," and that on the right (in this case benzene) is known as component "B"; we shall refer to the components by these designations. At the left the curve intersects the ordinate at the melting point of pure A; and at the right it intersects the ordinate at the melting point of pure B. The minimum in the freezing-point curve is called the *eutectic*, and we have drawn a horizontal line through the eutectic. The curve to the left of the eutectic can be considered a freezing-point-depression curve for component B dissolved in A, and the curve to the right of the eutectic a freezing-point-depression curve for A dissolved in B. This type of phase diagram is found for numerous binary systems in which the liquids are completely miscible and the solids completely insoluble in each other. (To be sure, there will always be some slight solubility of the solid phases in one another; but if the solubility is sufficiently small, it can be neglected). For purposes of discussion, it will be useful to consider the sketch of the phase diagram shown in Figure 13.13, where the freezing-point-depression curves have been approximated by straight lines.

The phase diagram can be experimentally constructed by using *cooling curves,* as depicted in Figure 13.14. Samples containing known amounts of both components

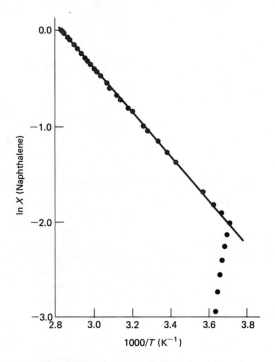

FIGURE 13.11 Logarithm of the solubility of naphthalene in benzene vs inverse temperature. (S. U. Pickering, *J. Chem. Soc.*, London 63 (1893): 998.)

are placed in containers and heated until complete dissolution occurs. The samples are then allowed to cool slowly and the temperature of each sample is noted as a function of time until the entire sample has solidified. Cooling curve (i) is for pure A. As the liquid sample cools, heat is lost through radiation and convection, and the temperature of the sample decreases at a rate indicated by the first portion of the cooling curve. When the melting point of pure A is reached, there is a sudden break

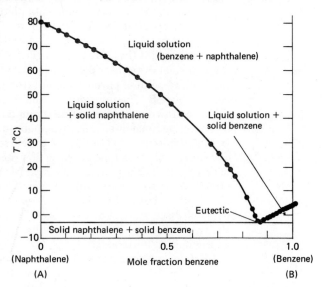

FIGURE 13.12 The data of Figure 13.11 plotted as a phase diagram.

SEC. 13.6 LIQUID-SOLID PHASE DIAGRAMS; SIMPLE EUTECTICS

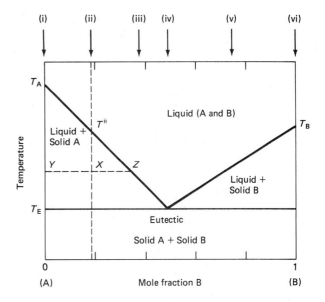

FIGURE 13.13 Sketch of a simple eutectic phase diagram. The arrows at the top refer to the concentration of the samples used in the cooling curves of Figure 13.14.

in the curve; the temperature remains constant while the heat of fusion is released, as indicated by the horizontal portion of the cooling curve. When the last liquid has solidified, the sample, now solid, begins to fall in temperature once again.

Now consider cooling curve (ii) along the line of constant composition indicated as (ii) on the phase diagram of Figure 13.13. The system has two components; hence the number of degrees of freedom is $f = 4 - p$. The pressure has been specified for this phase diagram, reducing f by unity and leaving $f = 3 - p$ on the phase diagram. Initially curve (ii) is a one-phase liquid region; $f = 2$ in the melt, since $p = 1$ in that region. The first break in cooling curve (ii) occurs at $T = T^{ii}$, where some pure solid A begins to precipitate from the melt. Since two phases (liquid and pure solid A) are in equilibrium, the number of degrees of freedom falls to $f = 3 - 2 = 1$. As pure solid A

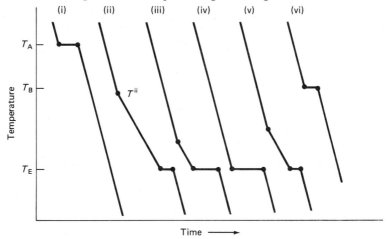

FIGURE 13.14 Cooling curves for the phase diagram of Figure 13.13.

precipitates, its latent heat of fusion is released, and the rate of cooling slows, as indicated by the sudden change in slope of the cooling curve. The melt becomes richer in component B as component A is removed, and the freezing point of the melt decreases along the freezing-point-depression curve.

Now let's examine point X in the two-phase region on the phase diagram. The overall composition is given by point X. The system is composed of two phases, a pure solid A phase, indicated by Y, and a liquid phase, whose composition is given by point Z. The line YXZ is a tie line connecting the two phases, and the lever rule applies. The ratio of the amount of solid present to the amount of liquid is XZ/YZ. The composition of the melt has moved from the point indicated by T^{ii} to the point Z.

A second break in the cooling curve appears at T_E, the eutectic temperature. The horizontal portion of the cooling curve corresponds to the solidification of the eutectic mixture. It has the same appearance as the curve for the solidification of the pure compound. In fact, eutectic mixtures give the appearance of pure compounds. They have constant freezing points, and the solid eutectic mass is a very fine grained mixture of the two components. The horizontal portion corresponding to the freezing of the eutectic is known as the *eutectic halt*. When the last of the eutectic has solidified, the cooling curve again begins its downward trend.

At the eutectic point, there are three distinct phases in equilibrium: liquid solution, pure solid A, and pure solid B. At constant pressure, the eutectic point is fixed and has no degrees of freedom remaining.

Cooling curve (iii) is for a mixture somewhat richer in component B and resembles (ii). The first break occurs at a somewhat lower temperature, and the eutectic halt is longer, since there is more eutectic present when T_E is reached. The cooling curve for the eutectic composition is shown by curve (iv). Now there is only one break, that break occurring at T_E. To complete the series of cooling curves, we have also shown curves (v) and (vi). Curve (v) is for an *isopleth* (a curve of constant composition) to the right of the eutectic and is completely analogous to curves (ii) and (iii). Curve (vi) is for pure component B and is similar to curve (i); the horizontal portion occurs at T_B, the melting point of pure B. With enough cooling curves, a complete phase diagram like that in Figure 13.12 can be constructed by plotting the points corresponding to the breaks in the cooling curves and connecting these points by smooth curves.

13.7 COMPOUND FORMATION

Suppose we have two components A and B that can combine to form the compound AB. Suppose further that solid A and solid AB are insoluble in each other and that liquid A and liquid AB are completely miscible. In addition, solid AB and solid B are completely insoluble in each other, and liquid AB and liquid B are completely miscible. Phase diagrams for the systems A–AB and for AB–B are shown in Figures 13.15a and b. A complete phase diagram for the system A–B can be constructed simply by juxtaposing the two phase diagrams and changing the scale of the abscissa, as shown in Figure 13.15c. For the compound AB, the central maximum occurs at 0.5 mole fraction B. If the compound were of the form AB_2, this central maximum would occur at $X_B = 0.667$; whereas for a compound of the formula A_2B, the peak would occur at $X_B = 0.333$. The system Mg–Si forms the compound Mg_2Si; the phase diagram for this system is shown in Figure 13.15d.

In some cases a series of compounds may be formed. This is often the case for salts and water, when several different hydrates are formed. An extreme example of this is the system $FeCl_3–H_2O$, which forms four hydrates. A portion of the phase diagram

SEC. 13.7 COMPOUND FORMATION 243

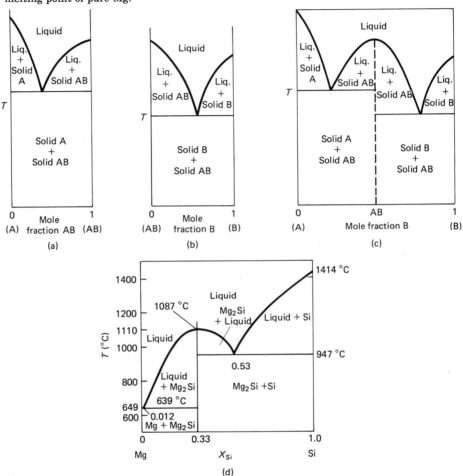

FIGURE 13.15 Eutectic phase diagrams with compound formation. (a) The system A–AB. (b) The system AB–B. (c) The complete system A–B. The peak at the center corresponds to the compound AB. (d) The phase diagram for the system Mg–Si, which forms the compound Mg_2Si. Note that the eutectic Mg–Mg_2Si is only 0.012 mole fraction Si and melts slightly below the melting point of pure Mg.

for this system is shown in Figure 13.16. It can be split into five simple phase diagrams. The four peaks correspond to the melting points of the successive hydrates.

A complication associated with what is known as a *peritectic* point, or *incongruent melting*, often arises in phase diagrams with compound formation, as sketched in Figure 13.17a for the system A–B that forms the compound AB. The actual phase diagram is indicated by the solid lines, and the dashed line indicates what one might expect based on the previous discussion. What happens is that the compound AB decomposes on melting, and the compound can only exist in solutions that contain an excess of one component. The melt does not have the same composition as the compound does, hence the term *incongruent melting point*. Figure 13.17b shows the phase diagram for the system Bi–Au that forms the compound $BiAu_2$ of incongruent melting point.

FIGURE 13.16 Portion of the phase diagram of the system H_2O–Fe_2Cl_6 (schematic). The system forms a series of four hydrates, $Fe_2Cl_6 \cdot nH_2O$, with $n = 12, 7, 5,$ and 4.

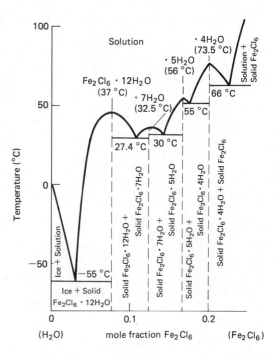

FIGURE 13.17 (a) Schematic sketch of phase diagram with incongruent melting. The dashed curve indicates what the behavior would be for normal congruent melting. (b) Phase diagram of the system Bi–Au. The compound $BiAu_2$ displays an incongruent melting point.

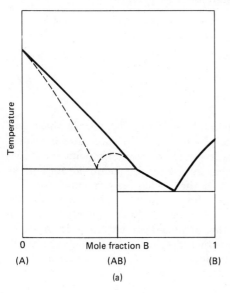

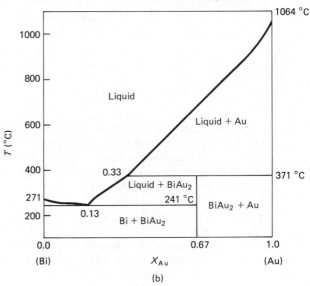

13.8 COMPLETE MISCIBILITY IN THE SOLID STATE

Recall our discussion of the freezing-point-depression constants in ideal solutions. One of the assumptions we made was that the two components were totally insoluble in each other in the solid state. At the end of the nineteenth century, when freezing-point depressions were being extensively investigated, it was noted that in some instances introducing an impurity actually raised the freezing-point temperature instead of lowering it. Van't Hoff explained this anomaly by introducing the concept of the solid solution. Systems that form a continuous series of solid solutions

FIGURE 13.18 (a) Sketch of phase diagram of a binary system in which the pure solid components form a continuous series of solid solutions. (b) Phase diagram of the system Mo–W. (From *Metals Handbook*, 8th edition, Volume 8, Copyright 1973 by American Society for Metals; reproduced by permission.)

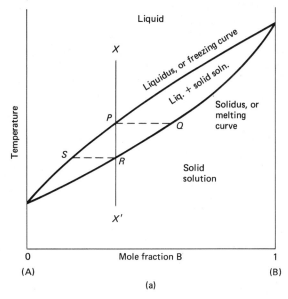

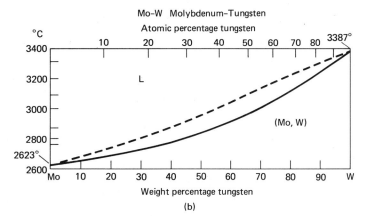

are not uncommon; and they are quite important in metallurgy. The systems Au–Cu and Cu–Ni are notable examples.

Figure 13.18 shows a phase diagram of a system displaying continuous solubility in the solid state. Note the similarity between this diagram and the T-X diagram of Figure 13.2. The analogy is carried one step further when we consider that there exist phase diagrams displaying maximum and minimum melting points that look like the azeotropic vapor phase diagrams of Figure 13.8. These extremal points in the solid-liquid phase diagrams are also called azeotropes. The separation of components in these systems by fractional crystallization is analogous to fractional distillation.

Consider the isopleth XX' on the diagram. As the system is allowed to cool from *the liquid state at point X* along XX', nothing happens until we reach the point P. At that point some solid solution precipitates, the composition of which is indicated by point Q. As the temperature continues to fall, the liquid phase becomes poorer and poorer in component B. The composition of the liquid moves along the curve PS while the composition of the solid solution moves along the curve QR. At point R the last bit of liquid solidifies. The last liquid to solidify has the composition given by point S. In the opposite approach, if one heats a solution at point X', liquefaction will first occur at R and the first liquid will have the composition S. When point P is finally reached, total liquefaction will occur. The relative amounts of liquid and solid solution can be determined at any point between P and R by constructing a tie line through the point and applying the lever rule.

There is one complicating feature attending these liquid-solid diagrams that was not present in the liquid-vapor diagrams. That complicating feature involves practice rather than principle. Determining the liquidus curve poses no difficulty. One simply cools a melt and notes the temperature at which the first solid appears. The reverse process is more difficult, since synthesizing a uniform solid solution is not a simple task, and extensive periods of annealing may be required for the equilibrium state of a uniform solid solution to be reached.

13.9 PARTIAL MISCIBILITY IN THE SOLID PHASE

For many binary systems the solubilities of the pure solid components in one another, while not complete, is sufficiently large that it cannot be neglected. This type of phase diagram is sketched in Figure 13.19a. The solid solution of B in A is noted as the α phase; the solid solution of A in B is noted as the β phase. When a melt is cooling along the isopleth XX', some solid is deposited when point P is reached. The solid is a solution of B in A, the α phase, whose composition is given by Q. At point R, the last liquid solidifies, and the entire system is in the α phase until point S is reached. At that temperature, some β phase crystallizes out of the solid solution. Figure 13.19b shows the phase diagram for the system $LiCl$–$CaCl_2$.

We have by no means exhausted the discussion of binary phase diagrams. In Figures 13.20 through 13.25 we show several phase diagrams that are more complex than those we have so far discussed. Although they may give the impression of deep complexity, note that these diagrams contain many of the features we have been discussing. (Figures 13.20 through 13.25 appear on pp. 248–250.)

SEC. 13.10 THREE-COMPONENT SYSTEMS   247

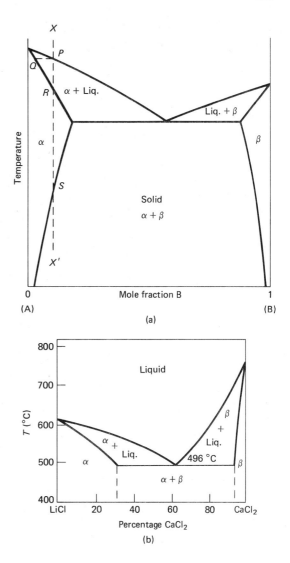

FIGURE 13.19 (a) Sketch of phase diagram of a binary system displaying partial solubility in the solid state. The phase labeled α is a solid solution of B in A, and the phase labeled β is a solid solution of A in B. (b) The phase diagram of the system LiCl–CaCl$_2$. (G. Grube and W. Rudel, Z. Anorg. Allg. Chem. 133 (1924):375.)

13.10 THREE-COMPONENT SYSTEMS

A *ternary system* has three components. The variance is $f = 3 - P + 2 = 5 - P$. If one phase is present, the variance is four. Four variables that might be used to specify the system are T, P, X_A, and X_B. The concentration of the third component is fixed by the relation $X_A + X_B + X_C = 1$. It is customary to represent these systems in two-dimensional triangular diagrams at constant temperature and pressure.

The triangular representation of ternary-phase diagrams is based on the convenient property of the equilateral triangle that the sum of the perpendicular distances to any point in the triangle equals the altitude of the triangle. The triangular coordinate system is shown in Figure 13.26. In that figure, the point labeled P corresponds

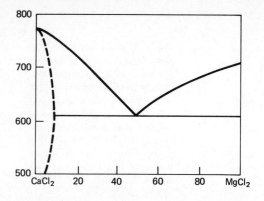

FIGURE 13.20 The system $CaCl_2$–$MgCl_2$ displays partial solid solution to the left and negligible solubility to the right. (Copyright 1964, The American Ceramic Society, Columbus, Ohio; reproduced by permission. A. I. Ivanov, *Sbornik Statei Obshch. Khim., Akad. Nauk. SSSR* 1 (1953):755.)

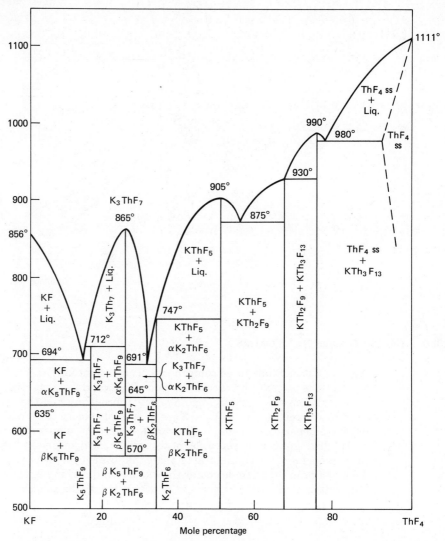

FIGURE 13.21 The system KF–ThF_4, displaying a number of phase transformations in the solid phase region. (Copyright 1964, The American Ceramic Society, Columbus, Ohio; reproduced by permission. W. J. Asker, E. R. Segnit, and A. W. Wylie, *J. Chem. Soc.*, London (1952):4471.)

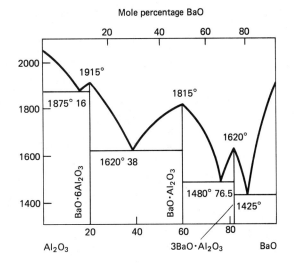

FIGURE 13.22 The system BaO–Al$_2$O$_3$, which forms three compounds with congruent melting points. (Copyright 1964, The American Ceramic Society, Columbus, Ohio; reproduced by permission. G. Purt, *Radex Rundschau* 1960, no. 4, p. 201. N. A. Toropov and F. Ya. Galakhov, *Dokl. Akad. Nauk SSSR* 82 [1] (1952):70.)

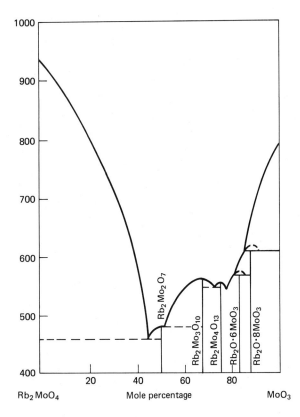

FIGURE 13.23 The system Rb$_2$MoO$_4$–MoO$_3$, which forms five different compounds. Three of the compounds have congruent melting points and two melt incongruently. The system displays four eutectic points and two peritectic points. (Copyright 1964, The American Ceramic Society, Columbus, Ohio; reproduced by permission. V. Spitzyn and I. M. Kuleshov, *Zh. Obshch. Khim.* 21 (1951):1370.)

249

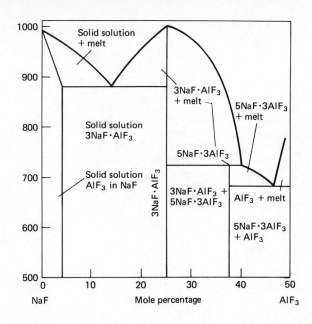

FIGURE 13.24 The system NaF–AlF$_3$ in the range 0–0.5 mole fraction AlF$_3$. We see a solid solution of AlF$_3$ in NaF, one congruently melting compound, and one compound with an incongruent melting point. (Copyright 1964, The American Ceramic Society, Columbus, Ohio; reproduced by permission. P. P. Fedotieff and W. P. Iljinskii, *Z. Anorg. Chem.* 80 (1913):121.)

FIGURE 13.25 The system BaO–B$_2$O$_3$, displaying a region in which the two liquids are partially immiscible. (Copyright 1964, The American Ceramic Society, Columbus, Ohio; reproduced by permission. E. M. Levin and G. W. Cleek, *J. Amer. Ceram. Soc.* 41[5] (1958):177.)

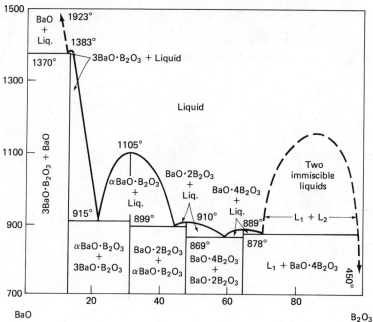

FIGURE 13.26 The triangular coordinate system for representing three-component phase diagrams at constant temperature and pressure.

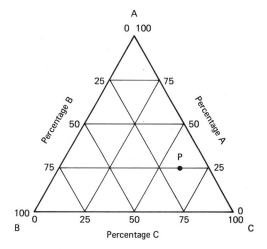

to a system that is 25% A, 15% B, and 60% C. This diagram would correspond to a system in which the three components are completely miscible in all proportions.

Figure 13.27 shows the phase diagram for the ternary system toluene–water–acetic acid at 25 °C and 1 atm. Toluene and acetic acid are miscible in all proportions as water and acetic acid are; water and toluene are not miscible, however, and the system may form two separate liquid phases under certain conditions. If water is shaken with toluene, the resulting system consists of two liquid layers; one layer is a saturated solution of water in toluene, the second a saturated solution of toluene in water. In view of the small solubilities, these solutions are nearly 100% toluene and 100% water. When the third component, acetic acid, is added, it distributes itself between the two phases; two *conjugate* ternary solutions form, each containing all three components. The system whose overall composition is given by x consists of two liquid phases, whose compositions are given by b and b'. The line bxb' is a tie line,

FIGURE 13.27 Phase diagram for the system water–acetic acid–toluene at 25 °C and 1 atm. (R. M. Woodman, *J. Phys. Chem.* 30 (1926):1283.)

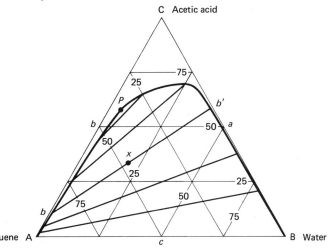

FIGURE 13.28 (a) Three-component phase diagram in which two pairs of components are immiscible. There are two critical points, one for each region of immiscibility, labeled P and P'. (b) Three-component phase diagram in which all three pairs of components are immiscible in one another. Now there are three critical points, labeled P, P', and P''.

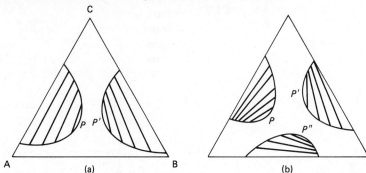

and the lever rule applies. The ratio of the amount of layer b to the amount of b' is given by xb'/xb. Since the distribution coefficient of the acetic acid between the water and the toluene-rich layers is not constant, the tie lines are not parallel. The point P, called the *critical*, or *plait*, point is that point at which the length of the tie line becomes vanishingly small. An increase in acetic acid concentration beyond this point results in the formation of a single phase.

In the above example, only one of the three possible pairs of compounds was immiscible, and the resulting phase diagram shows only one area of mutual immiscibility. Phase diagrams for which two and three of the three possible pairs of components are immiscible are shown in Figure 13.28. As the temperature decreases, the mutual solubilities decrease, and the areas of the immiscible regions may begin to overlap; this merging of the immiscible regions is known as *encroachment*.

Liquid-vapor phase diagrams for three-component systems are important in chemical engineering applications. A simple diagram of this kind is shown in Figure 13.29 for a system in which all three components are mutually miscible in all propor-

FIGURE 13.29 Three-component temperature composition diagram for a system in which all three components are mutually miscible in all proportions. For this system, each of the three possible binary systems has a temperature-composition diagram of the type shown in Figure 13.2.

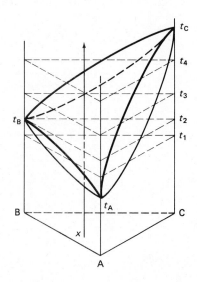

PROBLEMS

tions in the liquid state. Each of the three vertical faces of the triangular prism represents a simple T-X phase diagram of one of the pairs of components. Points above the upper surface are in the vapor state, and points below the lower surface represent states in the liquid phase. Points between the two surfaces represent states in which liquid and vapor phases are in equilibrium.

Three-component and also four-component systems are important in metallurgy, ceramics, and geology; and numerous schemes have been devised to represent graphically the equilibrium states of these systems. The interested student is referred to specialized textbooks for further discussion of this subject.

PROBLEMS

1. A sample containing 5 mol toluene and 5 mol benzene is heated from 80 °C to 110 °C at a constant pressure of 1 atm. (see Figure 13.2):
   (a) At what temperature does the liquid first boil?
   (b) What is the composition of the first vapor to form?
   (c) What is the composition of the last drop of liquid remaining?
   (d) When the temperature is 96 °C, what is the composition of the liquid and of the vapor? How much of the mixture is in the vapor phase?
   (e) How many degrees of freedom does the system have initially? Note each point at which the number of degrees of freedom changes during the process.

2. How many theoretical plates does a distillation column require to produce benzene with a purity of 0.98 mole fraction from a mixture that is 0.4 mole fraction benzene? The distillation is carried out at a pressure of 1 atm. (See Figure 13.2.)

3. One mole of toluene is placed in a vessel that is kept at a constant temperature of 79.7 °C and a constant pressure of 500 torr. Nine moles of benzene are added to the toluene very slowly to produce a final mixture that is 0.9 mole fraction of benzene. Describe all the changes that take place as the benzene is added. (See Figure 13.1.)

4. Do Problem 3 for a constant temperature of 90 °C and a constant pressure of 1 atm. (See Figure 13.2.) How is Figure 13.2 derived from Figure 13.1 using the properties of the ideal solution?

5. Two solutions of chloroform–acetone are made up. One solution is 0.15 mole fraction acetone and the other is 0.7 mole fraction acetone. (See Figure 13.10.) Each solution is at an initial temperature of 54 °C and is kept at a constant pressure of 1 atm. Each solution is slowly heated to 66 °C. For each solution:
   (a) At what temperature does the solution start to boil?
   (b) What is the composition of the first vapor to form?
   (c) What is the composition of the last drop of liquid remaining?
   (d) If each solution were distilled in a column with an infinite number of theoretical plates, what would the composition of the vapor be?
   (e) If an infinite amount of each liquid were evaporated, what concentration would the liquid tend to?

6. Two solutions of benzene–ethanol are made up. One solution is 0.3 mole fraction ethanol and the other is 0.7 mole fraction ethanol. (See Figure 13.9.) Each solution is at an initial temperature of 64 °C and slowly heated to 82 °C at a constant pressure of 1 atm. For each solution, do parts (a) to (e) of problem 5.

7. The azeotropic composition of a benzene–ethanol solution is 0.375 mole fraction ethanol at a total pressure of 310 torr. The boiling point of the azeotrope at this pressure is 45 °C. By comparing these data with the data of Figure 13.9, decide whether a distillation at 1 atm or 310 torr would produce a distillate that is purer in ethanol.

8. The importance of ethyl alcohol to the development of science and technology cannot be overestimated. Modern biochemistry had its origins in the studies of the fermentation processes associated with the production of wine. Modern structural organic chemistry owes much to Pasteur's discovery of optical isomerism in tartaric acid formed during fermentation. It can be said that chemical engineering started with the first person to discover that wine could be distilled to make a more potent beverage. The accompanying table gives the boiling points and compositions of the

Boiling temp (°C)	Wt % ethanol		Boiling temp (°C)	Wt % ethanol	
	liq.	vap.		liq.	vap.
78.2	91.1	91.8	84.0	27.2	75.8
78.4	84.6	89.6	84.9	23.0	73.7
78.7	81.5	88.0	86.1	19.9	72.5
79.2	75.8	85.8	88.4	12.8	66.7
79.8	70.0	83.8	90.3	10.1	60.0
80.2	64.8	83.5	92.4	7.1	51.4
80.6	57.3	81.5	94.5	4.1	43.7
81.7	51.1	81.2	95.7	3.3	33.5
82.0	43.1	79.6	96.6	2.5	30.6
82.5	35.7	79.2	97.4	1.4	23.6
83.3	31.4	77.1	98.1	1.0	16.7

liquid and vapor phases of ethanol–water mixtures at 1 atm. (P. N. Evans, *Ind. Eng. Chem.* 8, (1916):260.) At 1 atm, pure ethanol boils at 78.3 °C and water at 100.00 °C. The azeotrope has a composition of 4.4 wt % water and boils at 78.15 °C. From the above data, construct and discuss the temperature composition liquid-vapor phase diagram for the system water–ethanol.

9. In the fermentation of grape juice to produce wine, the maximum concentration of alcohol attainable is about 12% by weight. At that concentration, the yeasts that carry out the fermentation are rendered inoperable by the products of their pollution. Using the phase diagram you constructed in Problem 8, determine how many theoretical plates a distillation column would require to produce a distillate that is 50% by weight alcohol (about 100 proof).

10. Pickering (S. U. Pickering, *J. Chem. Soc.*, Lond. 63 (1893):998) has measured the freezing points of mixtures of naphthalene and benzene, and hence the solubilities, as a function of temperature. His partial results are presented in the accompanying table. The melting point of pure naphthalene is 79.9 °C. From these data, calculate the heat of fusion of naphthalene. (*Note:* Do not forget to convert the concentrations to mole fractions. Compare your result with the literature value of 19.1 kJ mol^{-1}.)

Wt % Naphthalene	Freezing Temp (°C)
99.20	79.27
97.98	78.27
94.55	75.25
90.55	71.63
86.33	67.98
82.19	64.38
77.76	61.07
73.97	57.50
68.93	53.07

11. A melt containing 2 mol benzene and 8 mol naphthalene is cooled from 80 °C to −10 °C at 1 atm pressure. With reference to Figure 13.12:
    (a) At what temperature does the first solid separate?
    (b) What is the composition of the first solid to separate?
    (c) At 30 °C, what is the composition of the melt, and how much melt remains?
    (d) At what temperature does the last bit of liquid solidify, and what is the composition of this last bit of liquid?
    (e) When the entire mass is solidified, what proportion of the mass is eutectic?

12. On the benzene–naphthalene phase diagram of Figure 13.12, locate points having two, one, and zero degrees of freedom. If the restriction of constant pressure were removed from the system, how would this affect the number of degrees of freedom of each of the points?

13. A series of mixtures of benzene–naphthalene having 0.0, 0.2, 0.6, 0.87, 0.95. and 1.0 mole fractions benzene are prepared. Sketch cooling curves for each of the mixtures. (*Note:* The eutectic mixture is 0.87 mole fraction benzene.)

14. A dilute solution of $FeCl_3$ is prepared at a temperature just below the melting point of $Fe_2Cl_6 \cdot 7H_2O$ (the melting point is 32.5 °C). The concentration of the solution is 0.01 mole fraction Fe_2Cl_6. Describe what happens as the solution is isothermally evaporated to produce a final mixture that is 0.5 mole fraction Fe_2Cl_6. (See Figure 13.16.)

15. Pickering (S. U. Pickering, *Ber. Deut. Chem. Ges.* 26 (1893):2307) has measured the freezing points of aqueous solutions of HI. His partial results are given in the accompanying table. The melting point of pure HI is −50.8 °C. Using the data in the table, sketch the phase diagram of the system H_2O–HI. How many hydrates does HI form and what are their melting points?

Wt % HI	Freezing Temp (°C)	Wt % HI	Freezing Temp (°C)
5.70	−1.3	60.59	−38.8
9.61	−2.6	62.00	−37.5
16.06	−6.7	63.76	−36.0
19.53	−8.8	65.31	−36.8
25.29	−13.5	66.60	−39.5
30.50	−19.4	68.57	−46.5
32.67	−25.6	69.46	−48.6
38.51	−37.2	71.09	−48.8
42.12	−48.2	72.92	−53.5
44.07	−58.2	73.11	−52.2
46.42	−66.7	73.95	−56.0
48.85	−82.4	74.66	−47.8
51.70	−71.9	75.15	−51.0
54.28	−58.9	75.97	−48.0
58.73	−42.9		

16. What happens to the compound $5NaF \cdot 3AlF_3$ as it is heated? (See Figure 13.23.)

17. Toluene and water are immiscible (see Figure 13.27). A mixture is prepared containing 10 g toluene and 10 g water at 25 °C. What would be the concentrations of each of the two phases after adding 20 g acetic acid to the mixture? How much acetic acid is needed to form a one-phase solution?

18. Aniline and water form a partially miscible pair. The solubility of aniline in water is 7.0 wt %, while the solubility of water in aniline is 10.4% at a temperature of 97 °C. For the three-component system water–phenol–aniline at 97 °C, the composition of the critical point is 59.8% water, 35.5% phenol, and 4.7% aniline. Sketch the three-component phase diagram of water–phenol–aniline.

19. Figure 13.29 is a three-dimensional sketch of a three-component phase diagram at constant pressure. How many degrees of freedom does the system possess in the liquid region below the lower surface? How many in the vapor region

above the upper surface? How many degrees of freedom in the region between the two surfaces? How would your answers be affected if the restriction on constant pressure were removed?

20. A mixture of 21.5% dibutyl phthalate (bp = 161 °C at 2.3 torr) and 78.5% dibutyl azelate (bp = 166 °C at 2.3 torr) is being distilled at a pressure of 2.3 torr. Analysis of the first takeoff indicates a head concentration of 98.4% dibutyl phthalate. Taking the ratio of the vapor pressures to be 1.047, calculate the number of theoretical plates in the column.

21. A large quantity of hydrochloric acid is distilled at 1 atm until the boiling point reaches the constant value of 108.584 °C. At that point, the azeotrope is 20.222 wt % HCl, and the density is 1.0959 g cm^{-3}. How much of the solution is needed to make up 1 liter of 1.000N HCl solution? (G. A. Hulett and W. D. Bonner, *J. Amer. Chem. Soc.* 31 (1909):390.)

22. The large demands for oxygen for steelmaking are being met by the fractional distillation of liquid air. Some pressure-composition data for O_2–N_2 mixtures are given in the accompanying

Total vapor pressure (atm)	Mole fraction O_2 in liquid	Mole fraction O_2 in vapor
1.031	1.000	1.000
1.174	0.946	0.826
1.360	0.890	0.691
1.551	0.824	0.564
1.559	0.821	0.559
1.586	0.806	0.536
1.808	0.729	0.434
2.145	0.610	0.316
2.302	0.551	0.272
2.401	0.504	0.241
2.467	0.494	0.236
2.501	0.469	0.219
2.799	0.352	0.150
3.262	0.193	0.063
3.707	0.000	0.000

table. (B. F. Dodge and A. K. Dunbar, *J. Amer. Chem. Soc.* 49 (1927):591.) (Students are continually warned about the dangers abounding in the laboratory. As an example of one of these dangers,

Boiling points and compositions of the vapor phase for O_2–N_2 mixtures at several pressures.

Mole fraction oxygen in liquid	0.5 atm		1 atm		5 atm	
	Boiling point (K)	Mole fraction oxygen in vapor	Boiling point (K)	Mole fraction oxygen in vapor	Boiling point (K)	Mole fraction oxygen in vapor
0.00	71.91	0.000	77.41	0.000	94.14	0.000
0.10	72.40	0.024	77.98	0.029	94.98	0.041
0.20	73.00	0.052	78.66	0.061	95.94	0.088
0.30	73.70	0.084	79.44	0.098	96.99	0.141
0.40	74.50	0.122	80.33	0.141	98.15	0.203
0.50	75.44	0.169	81.35	0.194	99.44	0.277
0.60	76.55	0.230	82.54	0.262	100.88	0.365
0.70	77.87	0.312	83.94	0.354	102.51	0.473
0.80	79.47	0.435	85.62	0.485	104.36	0.608
0.90	81.46	0.634	87.67	0.683	106.49	0.778
1.00	83.99	1.000	90.21	1.000	108.95	1.000

Mole fraction oxygen in liquid	10 atm		15 atm		20 atm	
	Boiling point (K)	Mole fraction oxygen in vapor	Boiling point (K)	Mole fraction oxygen in vapor	Boiling point (K)	Mole fraction oxygen in vapor
0.00	103.80	0.000	110.43	0.000	115.67	0.000
0.10	104.82	0.050	111.59	0.057	116.94	0.064
0.20	105.96	0.106	112.85	0.119	118.32	0.131
0.30	107.19	0.168	114.22	0.187	119.79	0.203
0.40	108.52	0.238	115.67	0.261	121.34	0.283
0.50	109.97	0.318	117.24	0.342	123.00	0.370
0.60	111.56	0.410	118.92	0.430	124.76	0.468
0.70	113.31	0.515	120.75	0.547	126.64	0.578
0.80	115.23	0.639	122.70	0.678	128.62	0.702
0.90	117.34	0.791	124.78	0.829	130.66	0.843
1.00	119.66	1.000	126.96	1.000	132.70	1.000

A. K. Dunbar was killed by the explosion of an oxygen cylinder druing this work.) The data apply to the 90.5 K isotherm.

(a) Draw a P-X phase diagram for O_2–N_2 at 90.5 K.

(b) On the same sketch draw the P-X phase diagram that would be expected if the system were ideal, and compare the ideal diagram with the real diagram.

23. The table on p. 255 (bottom) lists T-X data for the system O_2–N_2. These data were calculated from P-X data like those of Problem 22 for several isotherms. Assume that liquid air is 20 mol % O_2 and that you have an ideal distillation column with 5 theoretical plates at your disposal. How well could you separate O_2 from N_2 at 1 atm? Could the separation be improved if the pressure at which the column operated were increased?

CHAPTER FOURTEEN

ACTIVITIES OF NONELECTROLYTE SOLUTIONS

14.1 ACTIVITY

For reactions such as $a\mathrm{A(g)} + b\mathrm{B(g)} \rightleftharpoons c\mathrm{C(g)} + d\mathrm{D(g)}$ that involve only gases, the equilibrium constant takes the form

$$K_f = \frac{(f_C)^c (f_D)^d}{(f_A)^a (f_B)^b} \tag{14.1}$$

where

$$\ln K_f = -\frac{\Delta G°}{RT} \tag{14.2}$$

Now we wish to extend this result to cases in which any or all of the chemical species taking part in the reaction are liquids or solids rather than gases. The key to determining an equilibrium constant is establishing those conditions for which $\Delta G = 0$ at constant temperature and pressure. For reasons of practicality it is convenient to measure free energies against the free energies in some arbitrary and convenient reference state. These free energy changes are conveniently measured against the changes in partial molar free energies (or the chemical potential). Since $\overline{G}_i = \mu_i = RT \ln f_i$,[1] the molar free energy change in the direction from the standard reference state to a given state is

$$\Delta \overline{G}_i = \overline{G}_i - \overline{G}_i° = \mu_i - \mu_i° = RT \ln\left(\frac{f_i}{f_i°}\right) \tag{14.3}$$

The variable of interest is thus the *relative fugacity* $f/f°$. This quantity is designated the *activity,* symbol a,

$$a = \frac{f}{f°} \tag{14.4}$$

In terms of activity, Equation (14.3) is written

$$\Delta \overline{G}_i = \mu_i - \mu_i° = RT \ln a_i \tag{14.5}$$

---

[1] It would be more accurate to write this equation as

$$\mu_i = \overline{G}_i = RT \ln f_i + B(T)$$

where $B(T)$ is some function of the temperature. Since we are usually concerned with constant temperatures, this term cancels in evaluating changes in the chemical potential; thus we omit this term.

Now we must establish a consistent set of standard states for gases, liquids, and solids, both as solvents and as solutes in order to apply (14.5).

The choice for gases is particularly straightforward, since fugacities of gases can be readily measured. We choose the standard state for a gas to be that state in which the fugacity is 1 atm at the relevant temperature. The activity and the fugacity of a gas are therefore numerically identical:

$$a_i = \frac{f_i}{f_i^\circ} = \frac{f_i}{1} = f_i \qquad (14.6)$$

Note, however, that while the activity and the fugacity are numerically equal, their dimensions are different. Since $f_i = P_i$ for an ideal gas, the activity of an ideal gas is numerically equal to the partial pressure

$$a_i = f_i = P_i \qquad \text{(ideal gas)} \qquad (14.7)$$

For *pure* liquids and solids, it is convenient to take the standard state at any relevant temperature as the pure liquid or solid at a pressure of 1 atm. Thus the activity of a pure liquid or a pure solid is unity at a pressure of 1 atm. The advantages of this convention will become more apparent as we consider the activities and standard states of components in solution.

14.2 ACTIVITIES OF SOLVENTS

Consider the plot of fugacity vs mole fraction of solvent in Figure 14.1.[2] The illustration suggests that we adopt the pure solvent at the relevant pressure and temperature as the standard state. For solvents, unlike gases, the fugacity in the standard state f_A° will generally not be unity unless the temperature happens to be the normal boiling point of the pure solvent. The activity of the pure solvent is, however, unity with this convention:

$$a_A^\circ = \frac{f_A^\circ}{f_A^\circ} = 1 \qquad (14.8)$$

Since the vapor pressure, and hence the fugacity, of any solvent is lower in a solution than in the pure state, the activity of the solvent in a solution must be less than unity:

$$a_A = \frac{f_A}{f_A^\circ} \leq 1 \qquad (14.9)$$

where the equality applies to the pure solvent and the "less than" to solutions. An activity curve corresponding to Figure 14.1 can be constructed by dividing each value of f_A by f_A°, as shown in Figure 14.2.

For an ideal solution Raoult's law requires that $f_A = X_A f_A^\circ$. In this case the activity is just the mole fraction of solvent

$$a_A = \frac{f_A}{f_A^\circ} = \frac{X_A f_A^\circ}{f_A^\circ} = X_A \qquad (14.10)$$

---

[2] The discussion of activities and standard states follows that of I. M. Klotz, *Chemical Thermodynamics* (Englewood Cliffs, N.J.: Prentice-Hall, 1950).

SEC. 14.2 ACTIVITIES OF SOLVENTS

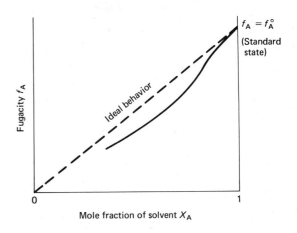

FIGURE 14.1 Fugacity vs mole fraction for the solvent. The standard state is taken as the pure solvent, at which $f_A = f_A^\circ$. The dashed line indicates ideal (Raoult's law) behavior. (The solid line is not continued to zero concentration. From all we have said up to now, we should expect that line to approach the Raoult's law line in the limit of zero concentration. Actually, as we shall see later, the limiting behavior in dilute solutions is not Raoult's law but a different law, called Henry's law.)

The dashed line of Figure 14.2 indicating the activity of an ideal solution is thus a straight line with unit slope. For an ideal solution, Equation (14.5) takes the particularly simple form

$$\mu_A - \mu_A^\circ = RT \ln a_A = RT \ln X_A \tag{14.11}$$

This equation relates the change in chemical potential to the concentration, an easily measurable quantity.

To retain the simple form of (14.11), we define an *activity coefficient*, γ_i,

$$\gamma_i = \frac{a_i}{X_i} \tag{14.12}$$

in terms of which (14.11) becomes

$$\mu_A - \mu_A^\circ = RT \ln (\gamma_A X_A) \tag{14.13}$$

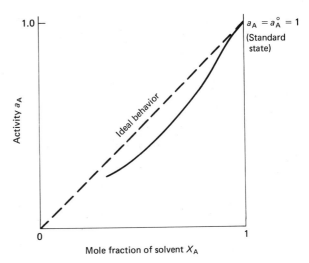

FIGURE 14.2 Activity vs mole fraction for the solvent. This curve is obtained from the curve of Figure 14.1 by dividing each value of f_A by f_A°. The dashed line indicates ideal (Raoult's law) behavior and has a slope of unity. (See comment at end of Figure 14.1.)

The activity coefficient is analogous to the fugacity coefficient; for gases the fugacity coefficient is identical to the activity coefficient. This is true because the fugacity of a gas equals its activity. The fugacity coefficient measures the deviation of a real gas from ideal gas behavior; with our choice of the pure compound as the standard state, the activity coefficient measures the deviation of a real solution from ideal behavior.

For a pure component $X_i = 1$, and $\gamma_i = 1$. For a solution that is ideal in the sense that Raoult's law is valid, the activity coefficient is also unity at all concentrations, since

$$\gamma_A = \frac{a_A}{X_A} = \frac{X_A}{X_A} = 1$$

For an infinitely dilute solution, for which Henry's law is valid, the activity coefficient of the solvent is still unity, since we have demonstrated that if Henry's law holds for the solute, then Raoult's law holds for the solvent. This is very convenient, since it allows us to express free-energy changes for ideal solutions by concentration. For solutions that are not sufficiently dilute for the solvent to behave ideally, the procedure is more difficult. Equation (14.13) is still valid, but determining γ may require some effort.

14.3 ACTIVITIES OF SOLUTES

In view of the arbitrariness of the choice of standard states, the most appropriate choice is invariably based on practicality. For us, practicality means the simplest equation in terms of measurable variables; that means an equation for the solute that is completely analogous to Equation (14.11). In other words, the most convenient choice of standard state for the solute is that choice for which $a_B = X_B$ for the ideal dilute solution.

The standard state appropriate to this result can be visualized by referring to Figure 14.3. In the limit of dilute solutions, Henry's law is valid, and the equation of the straight dashed line is

$$f_B = kX_B \tag{14.14}$$

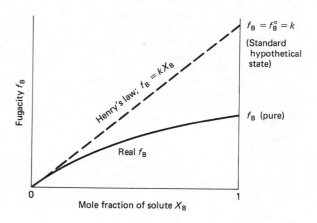

FIGURE 14.3 Fugacity vs mole fraction for the solute. The standard state is obtained by extrapolating the Henry's law behavior to unit mole fraction. Note that the fugacity in the standard state is *not* the fugacity of the pure substance but some other value. (In many works the fugacity of the pure solute is symbolized by $f^\bullet$ to distinguish it from f°, the fugacity of the standard state.)

where k is the Henry's law constant. From the definition of activity,

$$a_B = \frac{f_B}{f_B^\circ} = \frac{kX_B}{f_B^\circ} \tag{14.15}$$

Equation (14.15) indicates that if we wish to establish a system in which the activity is equal to the mole fraction in the ideal dilute solution, then the standard state must be selected so that $f_B^\circ = k$. As you can see from Figure 14.3, this condition holds for a standard state obtained by extrapolating the Henry's law behavior to unit mole fraction.[3] In this standard state, $f_B^\circ = k$, and $a_B^\circ = 1$. For dilute solutions,

$$a_B = \frac{f_B}{f_B^\circ} = \frac{kX_B}{k} = X_B \tag{14.16}$$

Note that the standard state for the solute is a *hypothetical* state. The activity in the hypothetical standard state is unity, but the activity of the pure solute based on this reference state is *not* unity. This is shown in Figure 14.4, where we have plotted the activity as a function of X_B for the example of Figure 14.3. Here the actual activity lies below the idealized activity; in other instances the true curve may lie above the idealized curve, and the two curves may even cross. Also note that it is wrong to say that the reference state for the solute is the infinitely dilute solution. The reference state is based on *behavior* in the limit of dilute solutions, but this behavior is extrapolated to unit mole fraction to get the standard reference state. A reference state at zero concentration would be meaningless, since the fugacity in that state would be zero. If f_B° were zero, then the activity $a_B = f_B/f_B^\circ$ would be infinitely large at any finite concentration.

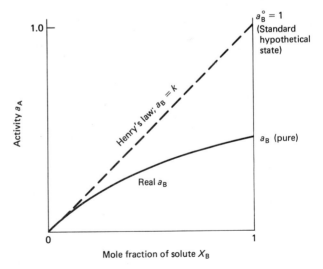

FIGURE 14.4 Activity vs mole fraction for the solute. These curves are obtained from the curves of Figure 14.3 by dividing each value of f_B by f_B°.

---

[3] This is true even for relatively insoluble solutes, when the solution may be saturated at concentrations less than unit molality. It is a hypothetical state in the same sense that the standard state of water vapor at 25 °C and 1 atm is a nonexistent standard state.

Now, what about solutions that are not ideal dilute solutions? We follow the same procedure we used for the solvent; and we define an activity coefficient in the same manner,

$$\gamma_B = \frac{a_B}{X_B} \qquad (14.17)$$

so that (14.13) is also valid when applied to the solute.

We are ultimately interested in *changes* in the partial molar free energy (or chemical potential); these changes are independent of the point of origin. To give a concrete example, suppose we want to calculate $\Delta\mu_B$ for a solution taken from some initial concentration to some final concentration when neither is the standard state. Denoting the initial and final states by primes and double primes, we get

$$\Delta\overline{G}_B = \Delta\mu_B = \mu_B'' - \mu_B' = (\mu_B'' - \mu_B^\circ) - (\mu_B' - \mu_B^\circ) \qquad (14.18)$$

In any calculation involving (14.18), the chemical potential of the reference state μ_B° cancels. Using (14.5), we can write (14.18) as

$$\Delta\overline{G}_B = \Delta\mu_B = RT \ln a_B'' - RT \ln a_B'$$
$$= RT \ln \frac{f_B''}{f_B^\circ} - RT \ln \frac{f_B'}{f_B^\circ} = RT \ln \frac{f_B''}{f_B'}$$
$$= RT \ln \frac{a_B''}{a_B'} \qquad (14.19)$$

The fugacities of the standard state cancel in any calculation involving the determination of a free-energy change. The numerical values of the fugacities, and also of the activities, depend on our choice of standard state. The ratios, however, are independent of the choice, and the ratio is what we are ultimately interested in. We could have chosen the standard state to be the pure solute if we so desired, and further, we could choose the standard state to be at any concentration. The system we have established with mole fractions is called the *rational* system. Since concentrations of solutes in dilute solutions are usually specified in molality concentrations, a system based on this unit would be of greater practical value. Such a system can be established in a manner analogous to the rational system; it is called the *practical* system.[4]

In molality concentration units, Henry's law can be written as

$$f_B = k'm \qquad (14.20)$$

where m is the molality of the solute and k' is the Henry's law constant in molality units. What we want is a system of such a kind that in dilute solutions $a_B = m$. Referring to Figure 14.5, we see that we can accomplish this by extrapolating the Henry's law behavior to unit molality and selecting that point as the standard state. In this state, $f_B = f_B^\circ = k'$, and in the region in which Henry's law is valid,

$$a_B = \frac{f_B}{f_B^\circ} = \frac{k'm}{k'} = m \qquad (14.21)$$

---

[4] Although the two systems are based on different concentration units, a more fundamental difference is that the rational system is based on the pure substance (liquid) as the standard state whereas the practical system is based on a unit molal solution obeying Henry's law as the standard state. The standard state in the practical system is a fictitious solution, as shown in Figure 14.5.

SEC. 14.4 THE EQUILIBRIUM CONSTANT IN TERMS OF ACTIVITIES

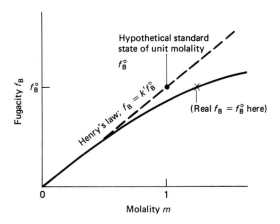

FIGURE 14.5 Fugacity vs molality concentration for the solute, showing the standard state at unit molality. (Note that on the curve of real fugacities, the fugacity may be equal to f_B° at another concentration besides $m = 1$. This is shown by the cross on that curve.)

In the region in which Henry's law is not valid, we can retain the formalism by defining an activity coefficient (see Problems 1 to 3):

$$\gamma_B = \frac{a_B}{m} \tag{14.22}$$

This allows us to retain the basic formal equation

$$\mu_B - \mu_B^\circ = RT \ln a_B = RT \ln (\gamma_B m_B) \tag{14.23}$$

Now let's apply these results to investigate the equilibrium properties of several types of systems.

14.4 THE EQUILIBRIUM CONSTANT IN TERMS OF ACTIVITIES

We now want to derive equilibrium conditions analogous to those of Equations (14.1) and (14.2) in terms of activities. Consider the reaction

$$aA(a_A) + bB(a_B) \rightleftharpoons cC(a_C) + dD(a_D) \tag{14.24}$$

where the quantities in parentheses are the activities of the species at equilibrium. At constant T and P, the condition for equilibrium is that $\Delta G = 0$.

Suppose that instead we consider the reaction

$$aA(a_A = 1) + bB(a_B = 1) \rightleftharpoons cC(a_C = 1) + dD(a_D = 1) \tag{14.25}$$

where each of the constituents is in the standard state, $a_i = a_i^\circ = 1$. The free-energy change for this reaction is

$$\Delta G^\circ = \Delta G_f^\circ(C) + \Delta G_f^\circ(D) - \Delta G_f^\circ(A) - \Delta G_f^\circ(B) \tag{14.26}$$

where each ΔG_f° is a standard free energy of formation.

Now consider the following four reactions:

$$aA(a_A) \longrightarrow aA(a_A = 1) \qquad \Delta G = aRT \ln \frac{f_A^\circ}{f_A} = -RT \ln (a_A)^a \tag{14.27}$$

$$bB(a_B) \longrightarrow bB(a_B = 1) \qquad \Delta G = bRT \ln \frac{f_B^\circ}{f_B} = -RT \ln (a_B)^b \tag{14.28}$$

$$cC(a_C = 1) \longrightarrow cC(a_C) \qquad \Delta G = cRT \ln \frac{f_C}{f_C^\circ} = +RT \ln (a_C)^c \qquad (14.29)$$

$$dD(a_D = 1) \longrightarrow dD(a_D) \qquad \Delta G = dRT \ln \frac{f_D}{f_D^\circ} = +RT \ln (a_D)^d \qquad (14.30)$$

If we add (14.25), (14.27), (14.28), (14.29), and (14.30), the resulting reaction is our original equilibrium reaction. For that reaction,

$$\Delta G = 0 = \Delta G^\circ + RT \ln \left[\frac{(a_C)^c (a_D)^d}{(a_A)^a (a_B)^b} \right] \qquad (14.31)$$

or
$$\Delta G^\circ = -RT \ln K_a \qquad (14.32)$$

The equilibrium constant K_a is

$$K_a = \frac{(a_C)^c (a_D)^d}{(a_A)^a (a_B)^b} \qquad (14.33)$$

If any of the species in (14.33) are gases, then the activity can be replaced by the fugacity (or by the pressure for ideal gases, since $f_{IG} = P$). For pure solids or liquids the activity is unity, since these are in their standard states, for which $a = a^\circ = 1$. In dilute solutions where Henry's law is valid, the activity of the solute is given by the molality. At higher concentrations where Henry's law is not valid, the activity is $a_i = \gamma_i m_i$, and the equilibrium constant can be written

$$K_a = \frac{(\gamma_C m_C)^c (\gamma_D m_D)^d}{(\gamma_A m_A)^a (\gamma_A m_A)^b} = \frac{(\gamma_C)^c (\gamma_D)^d}{(\gamma_A)^a (\gamma_B)^b} \frac{(m_C)^c (m_D)^d}{(m_A)^a (m_B)^b}$$

$$= K_\gamma K_m \qquad (14.34)$$

(Note that these activity coefficients are not the same as the coefficients in Equation (14.12).)

There is an additional point. Many reactions involve dilute solutions in which the mole fraction of solvent is very close to unity. The standard state of the solvent is unit mole fraction where $a = a^\circ = 1$; thus only a slight error is introduced by setting the activity of the solvent equal to unity in (14.33). This is tantamount to omitting the solvent activity from the expression for dilute solutions. Further, in very dilute solutions, for which Henry's law is valid, the activity coefficients of the solutes are unity, and $K_\gamma = 1$. The equilibrium constant can then be written in molalities exclusively, $K_a = K_m$.

14.5 VAPOR PRESSURE AS AN EQUILIBRIUM CONSTANT

To be specific, we consider water, for which the vaporization reaction at room temperature is

$$H_2O(l, 1 \text{ atm}, 298 \text{ K}) \rightleftharpoons H_2O(g, P, 298 \text{ K})$$

where P is the equilibrium vapor pressure. The equilibrium constant for this reaction is the activity of the gaseous water divided by the activity of the liquid.

The activity of the liquid is unity, since it is in the standard state. For the vapor, the activity equals the fugacity; if we assume that the water vapor behaves as an ideal gas, then the fugacity equals the pressure. Thus the equilibrium constant for the reaction just equals the vapor pressure. Writing out the steps, we get

$$K = \frac{a(g)}{a(l)} = \frac{f(g)}{1} = P$$

From Table 9.1 we find that the standard free energies of formation of the liquid and the vapor are -237.1918 and -228.5958 kJ mol^{-1}. The value $\Delta G°$ is hence $+8596$ J mol^{-1}. Using Equation (14.32), we get

$$\ln K = -\frac{\Delta G°}{RT} = -\frac{8596}{(8.314)(298.15)} = -3.4677$$

$$K = \exp(-3.4677) = 0.0312 \text{ atm}$$
$$= 0.0312 \times 760 = 23.7 \text{ torr}$$

The value 23.7 torr is, indeed, the vapor pressure of water at room temperature. The tables have been constructed so that they are consistent with our definitions of activity.

14.6 DECOMPOSITION

Entries in the tables of standard free energies are often determined from studies of equilibrium constants. Consider the decomposition reaction for silver oxide,

$$Ag_2O(c) \rightleftharpoons 2Ag(c) + \tfrac{1}{2}O_2(g)$$

It is found that at 446 K the equilibrium partial pressure of O_2 is 422 torr. For this reaction, $K = (P_{O_2})^{1/2}$, or

$$\Delta G°_{446} = -RT \ln K = -(8.314)(446) \ln (\tfrac{422}{760})^{1/2}$$
$$= +1091 \text{ J mol}^{-1}$$

The reaction that forms Ag_2O from its elements is just the reverse of the decomposition reaction; hence the standard free energy of formation of Ag_2O at 446 K is -1.091 kJ mol^{-1}. It was necessary to convert the pressure in torr to a pressure in atmospheres so that we could arrive at a result consistent with our procedures.

We shall defer examining equilibrium constants in solutions to a future chapter, where we can study the problem in the context of ionic solutions. We now consider some experimental procedures of measuring activities and activity coefficients in solution.

14.7 EXPERIMENTAL DETERMINATION OF ACTIVITIES

The vapor pressure of the solvent provides an immediate measure of the activity of the solvent, since the standard state of the solvent is chosen as the pure solvent. From (14.4),

$$a_A = \frac{f_A}{f_A°} = \frac{f_A \text{ (vapor over solution)}}{f_A \text{ (vapor over pure solvent)}} \quad (14.35)$$

If the pressure is low enough for ideal gas behavior to be assumed, this equation becomes

$$a_A = \frac{P_A}{P_A^\circ}$$

For an ideal solution, the activity equals the mole fraction, since $P_A = X_A P_A^\circ$, and the activity coefficient is unity. For nonideal solutions, the solvent activity coefficient is $\gamma_A = a_A/X_A$.

Evaluating the activity of the solute from vapor pressure measurements requires extrapolating Henry's law to the standard state of unit molality. Assuming that the vapor behaves ideally, we find that the activity of a solute with a measurable vapor pressure is

$$a_B = \frac{f_B}{f_B^\circ} = \frac{P_B}{P_B^\circ} \tag{14.36}$$

In this equation, in contradistinction to the equation for the solvent, P_B° is not the vapor pressure of the pure solute but rather the vapor pressure the solute would have at a concentration of unit molality if Henry's law were obeyed out to large concentrations. This is illustrated in Figure 14.6, where the solid curve indicates the actual pressure, and the dashed straight line the Henry's law behavior. The Henry's law constant k' is obtained by extrapolating vapor pressure measurements to very low concentrations, at which the linear behavior is followed. Along the dashed line,

$$P_B = k'm$$

At unit molality $P_B^\circ = k'$, and the solute activity is

$$a_B = \frac{P_B}{k'} \tag{14.37}$$

In the region in which Henry's law is valid, the activity is equal to the molality, since

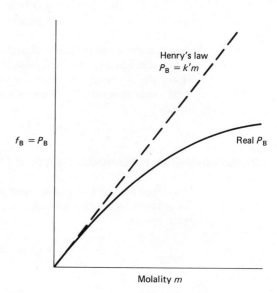

FIGURE 14.6 Determination of the Henry law constant from measurements of the partial pressure of the solute. The dashed line indicating Henry's law behavior is obtained from the limiting slope of the solid curve at infinite dilution.

$P_B = k'm$; also $\gamma_B = 1$. In more concentrated solutions, the solute activity coefficient is

$$\gamma_B = \frac{a_B}{m} = \frac{P_B}{mk'} \qquad (14.38)$$

Most properties of solutions can serve as an experimental basis for the determination of activities. Measurements of colligative properties of solutions determine properties of the solvent rather than of the solute and directly measure the activity of the solvent. To correlate this information with the activity of the solute, we need a procedure for relating the two activities. The Gibbs-Duhem equation provides such a procedure (see next section).

Sometimes a solute's activities in one solvent are known, and we want to determine the activities in a different solvent. This can be done by measuring the distribution coefficient of the solute between the two solvents, provided that the solvents are immiscible. At equilibrium, the fugacities of the solute in each of the solvents must be identical. If we indicate the two solvents by superscripts a and b, then at equilibrium

$$f_B{}^a = f_B{}^b \qquad (14.39)$$

Since $a = f/f°$ or $f = af°$, Equation (14.39) can be written

$$a_B{}^b = a_B{}^a \left(\frac{f_B{}^{a°}}{f_B{}^{b°}}\right) \qquad (14.40)$$

The problem then is to determine the ratio of the fugacities in the standard states. But this ratio is given by the ratio of the Henry's law constant and conveniently determined by measuring the ratio of the solubilities in dilute solutions. Equation (14.40) becomes

$$a_B{}^b = a_B{}^a \left(\frac{k^a}{k^b}\right)$$

which at infinite dilution can be written

$$a_B{}^b = a_B{}^a \left(\frac{m^b}{m^a}\right) \qquad (14.41)$$

One of the most powerful and accurate methods for measuring activities is to be found in measurements of electromotive force. We shall defer a discussion of this technique to Chapter 17, which deals with electrochemical cells.

14.8 APPLICATION OF THE GIBBS-DUHEM EQUATION TO ACTIVITIES

In many instances it may be relatively easy to measure the activity of only one component of a solution, usually the solvent. It is useful to have a scheme by which the activity of the solute can be determined from these measurements. The Gibbs-Duhem equation provides the necessary relation.

Applying the Gibbs-Duhem equation to chemical potentials, we find that

$$n_A \, d\mu_A + n_B \, d\mu_B = 0 \qquad (14.42)$$

Since $d\mu = RT\,d(\ln a)$, Equation (14.42) becomes

$$n_A\,d(\ln a_A) + n_B\,d(\ln a_B) = 0 \tag{14.43}$$

This equation can be converted to mole fractions by dividing by $n_A + n_B$, to get

$$X_A\,d(\ln a_A) + X_B\,d(\ln a_B) = 0 \tag{14.44}$$

The application of (14.44) to determining the activity of the solvent from the known activity of the solute is completely straightforward. Rearrangement followed by integration leads to

$$\int_{X_B=0}^{X_B} d(\ln a_A) = -\int_0^{X_B} \frac{X_B}{X_A}\,d(\ln a_B) = -\int_0^{X_B} \frac{X_B}{1-X_B}\,d(\ln a_B) \tag{14.45}$$

At the lower limit, the activity of A is unity, since $a_A = X_A = 1$. Equation (14.45) becomes

$$\ln a_A = -\int_{X_B=0}^{X_B} \frac{X_B}{1-X_B}\,d(\ln a_B) \tag{14.46}$$

and the activity of the solvent can be evaluated by graphical integration.

If the integration is done for activities, then substantial errors may be introduced because there may be large variations in the values of the activity over the range of interest. The activity coefficient is a more slowly varying function, and errors involved in graphically integrating the activity coefficients may be substantially less. We can convert to activity coefficients by noting that $X_A + X_B = 1$; this yields

$$dX_A = -dX_B$$

$$\frac{X_A}{X_A}\,dX_A = -\frac{X_B}{X_B}\,dX_B$$

or

$$d(\ln X_A) = -\frac{X_B}{X_A}\,d(\ln X_B) \tag{14.47}$$

If we now subtract (14.47) from (14.45), the result is

$$\int_{X_B=0}^{X_B} d\left(\ln \frac{a_A}{X_A}\right) = -\int_0^{X_B} \frac{X_B}{1-X_B}\,d\left(\ln \frac{a_B}{X_b}\right) \tag{14.48}$$

which is now in terms of activity coefficients, since $\gamma = a/X$. If the activity coefficients of the solute are known as functions of concentration, then the activity coefficients of the solvent γ_A can be determined by plotting X_B/X_A vs $\gamma_B\,(=a_B/X_B)$ and graphically integrating from infinite dilution ($X_B = 0$) to any desired concentration X_B. Since at infinite dilution $\gamma_B = 1$, both variables go to zero at the lower limit, and the integration is easily carried out.

If, however, we wish to determine the solute activity from the known solvent activity, the procedure is not so straightforward. Simply interchanging solvent and solute in (14.48) results in

$$\ln \frac{a_B}{X_B} = -\int_{X_B=0}^{X_B} \frac{X_A}{X_B}\,d\left(\ln \frac{a_A}{X_A}\right) = -\int_{X_A=1}^{X_A} \frac{X_A}{1-X_A}\,d\left(\ln \frac{a_A}{X_A}\right) \tag{14.49}$$

Now the integrand goes to infinity at the lower limit.

This difficulty is obviated by shifting the lower limit from infinite dilution ($X_A = 1$; $X_B = 0$) to some small but finite concentration at which Henry's law, and hence Raoult's law, will still be valid. At this new limit,

$$a'_A = X'_A \qquad a'_B = X'_B \qquad \gamma_A = \gamma_B = 1$$

where the primes indicate values at the new lower limit. In terms of these new limits, Equation (14.49) is written as

$$\ln \frac{a_B}{X_B} = - \int_{X'_B}^{X_B} \frac{X_A}{X_B} d\left(\ln \frac{a_A}{X_A}\right) \tag{14.50}$$

or in terms of activity coefficients as

$$\ln \gamma_B = - \int_{X'_B}^{X_B} \frac{X_A}{X_B} d(\ln (\gamma_A)) \tag{14.51}$$

Either (14.50) or (14.51) can be evaluated by standard graphical techniques.[5]

To measure vapor pressures of solutions accurately is no simple task. To eliminate the need for separately measuring the vapor pressures of each and every solution of interest, the method of *isotonic* solutions has been developed, in which unknown solutions are compared with a standard whose vapor pressures and hence activities have already been measured.[6] In this method, solutions of the unknown and the reference standard are placed in an evacuated chamber. Water evaporates from the solution of higher vapor pressure and condenses in the solution of lower vapor pressure. When equilibrium is reached, the vapor pressures are the same over both solutions; the solutions are analyzed for solute content, and the vapor pressure and hence activity of the unknown can be determined from the known values for the standard solution. The activities of the solvent in the standard are usually determined from vapor-pressure measurements, osmotic-pressure measurements, and freezing-point determinations.[7]

The procedure is illustrated in Figure 14.7 for sucrose solutions at 25 °C. The initial data for the activities of the water as a function of sucrose concentration were taken from the work of Scatchard, Hamer, and Wood, who determined the activities of the water by the method of isotonic solutions.[8] The initial data and final results are listed in Table 14.1. Figure 14.7 shows a plot of the ratio X_A/X_B as a function of $\ln \gamma_A$. The value γ_B was determined by graphical integration, according to Equation

---

[5] Graphical techniques are discussed in detail in I. M. Klotz, *Chemical Thermodynamics* (Englewood Cliffs, N.J.: Prentice-Hall, 1950).

[6] The word *isotonic* means "at the same osmotic pressure." If the solvent activities in two different solutions were the same, they would have the same osmotic pressure. The activity (or osmotic pressure) of fluid in cells is conveniently measured by observing the cells as they are immersed in a series of solutions of different concentrations. If the activity within the cell differs from the activity in the solution, then the cell will either swell or shrink, depending on whether the activity is less or greater than the activity in the solution; this effect is due to the flow of solvent (water) arising from the osmotic-pressure difference. The activity of the solvent in the cell is the same as the activity of the solvent in the solution that causes no change in the cell. Injected drugs are usually dissolved in isotonic NaCl solution (about 0.9% NaCl by weight) to prevent destruction of cells by osmotic-pressure differences. The term *isopiestic* ("same pressure") is often used instead of *isotonic*.

[7] An extensive discussion of these various techniques can be found in *Thermodynamics*, by G. N. Lewis and M. Randall, revised by K. S. Pitzer and L. Brewer, New York, McGraw-Hill Book Company, 1961. Note that freezing-point determinations provide activity data for temperatures near the freezing point although the freezing point temperatures at different concentrations are not the same. The data for the various concentrations must all be referred to the same temperature, using heat capacity data.

[8] G. Scatchard, W. J. Hamer, and S. E. Wood, *J. Amer. Chem. Soc.* 60 (1938):3061.

FIGURE 14.7 Illustrating the determination of the activity coefficients of the solute from the activity coefficients of the solvent by graphical integration of the Gibbs-Duhem equation, (14.51). The value $\ln \gamma_B$ at some concentration X_B is obtained by taking the area under the curve between X'_B and X_B, where X'_B is sufficiently small that the activity coefficient of the solute is unity at that concentration.

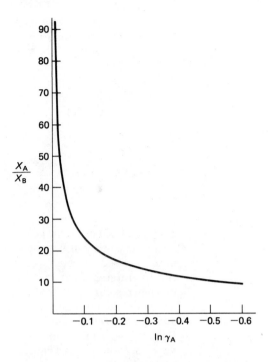

TABLE 14.1 Activities and activity coefficients of solvent and solute in aqueous solutions of sucrose at 25 °C.

Molality of sucrose	Water			Sucrose		
	Mole fraction	Activity coefficient	Activity[a]	Mole fraction	Activity coefficient	Activity
0.1	0.9982	1.0000	0.9982	0.0018	1.0081	0.0018
0.2	.9964	0.9999	.9963	.0036	1.0256	.0037
0.3	.9946	0.9999	.9945	.0054	1.0449	.0056
0.4	.9928	0.9997	.9926	.0072	1.0652	.0076
0.5	.9911	0.9996	.9907	.0089	1.0863	.0097
0.6	.9893	0.9994	.9887	.0107	1.1080	.0118
0.7	.9875	0.9992	.9867	.0125	1.1304	.0141
0.8	.9858	0.9989	.9847	.0142	1.1535	.0164
0.9	.9840	0.9986	.9827	.0160	1.1771	.0188
1.0	.9823	0.9983	.9806	.0177	1.2013	.0213
1.2	.9788	0.9974	.9763	.0212	1.2515	.0265
1.4	.9754	0.9965	.9720	.0246	1.3042	.0321
1.6	.9720	0.9953	.9674	.0280	1.3592	.0381
1.8	.9686	0.9940	.9628	.0314	1.4167	.0445
2.0	.9652	0.9926	.9581	.0348	1.4767	.0514
2.5	.9569	0.9885	.9459	.0431	1.6375	.0706
3.0	.9487	0.9835	.9331	.0513	1.8142	.0930
3.5	.9407	0.9778	.9198	.0593	2.0065	.1190
4.0	.9328	0.9713	.9061	.0672	2.2135	.1488
4.5	.9250	0.9643	.8920	.0750	2.4335	.1825
5.0	.9174	0.9568	.8777	.0826	2.6638	.2201
5.5	.9099	0.9490	.8634	.0901	2.9006	.2615
6.0	0.9025	0.9415	0.8497	0.0975	3.1383	0.3061

[a] Activities calculated from G. Scatchard, W. J. Hamer, and S. E. Wood, *J. Amer. Chem. Soc.* 60 (1938):3061.

(14.51). In this calculation we assumed that a concentration of 0.01m was dilute enough to be used as the lower limit in (14.51).

In actual fact, the results in Table 14.1 were determined by analytical procedures rather than by actually measuring the area under the curve of Figure 14.7. The activities of water listed in column (2) were fitted to the polynomial expression in m:

$$a_A = 1.000011 - 0.017935m - 1.50481 \times 10^{-3}m^2 \\ - 2.15325 \times 10^{-5}m^3 + 1.23659 \times 10^{-5}m^4 \quad (14.52)$$

and the integral was numerically evaluated using Simpson's rule with unequal intervals equivalent to increments of 0.001 molality unit. You would find it a useful exercise to check the computation on a digital computer.

There is one point you should remember. In speaking of gases, we understand that an ideal gas is simply an ideal gas, with no further qualifications. In speaking of solutions, on the other hand, we refer to two different kinds of ideal solutions. Strictly speaking, there is only one kind of "ideal" solution. That is a solution that obeys Raoult's law over the entire range of concentrations; the activities of the solute and the solvent are both equal to the mole fractions, the standard state being the pure substance. We also speak of an "ideal dilute" solution as a solution that obeys Henry's law but does not necessarily obey Raoult's law at the same time.[9] Here too, the activities are given by the mole fractions, but the standard state for the solute is not the pure solute at its normal vapor pressure. The standard state is the hypothetical solution obeying Henry's law at 1 molal concentration, and it is obtained by extrapolating the Henry's law behavior. It is important to keep this difference in mind. It should be apparent from the context of the discussion which type of ideal solution is meant.

We have thus far made no mention of solutions of electrolytes. Physiological processes take place in aqueous solutions, and the concentrations and activities of ionic species are intimately associated with the basic nature of these processes. Further, aqueous solutions of salts are likewise characterized by the presence of ions. While the study of nonelectrolyte solutions is important and provides a simple framework upon which we can build, the study of electrolyte solutions is perhaps more practical.

PROBLEMS

1. In dealing with solutes it is often necessary to convert from mole fraction concentration units to molality units and vice versa. Show that the activities can be converted by the equation

$$\frac{a_x}{a_m} = \frac{f_m^\circ}{f_x^\circ}$$

where the subscript x indicates mole fraction, and the subscript m indicates molality. Show further that the activity coefficients can be converted by the equation

$$\frac{\gamma_x}{\gamma_m} = \left(\frac{m}{X}\right)\left(\frac{k'}{k}\right)$$

where k and k' are the Henry's law constants in mole fraction and molality units.

2. Since most solutes do not have measurable vapor pressures, it is useful to eliminate the Henry's law constants from the second equation in Problem 1. Remembering that $k' = k(M_A/1000)$, where M_A is the molar mass of the solvent, show by using the relations of the concentration units that

$$\frac{\gamma_x}{\gamma_m} = 1 + \frac{mM_A}{1000}$$

3. Use the equation of Problem 2 to get the molality-based activity coefficients and activities

---

[9] Recall that in dilute solutions, if the solute obeys Henry's law, then the solvent obeys Raoult's law.

of sucrose solutions at $m = 1, 2$, and 3, using the data of Table 14.1. (Note that the difference in activity coefficients reflects the difference in the standard states used.)

4. Show that the equation for the dependence of activity on pressure at constant temperature is given by

$$\left(\frac{\partial \ln a}{\partial P}\right)_T = \frac{\overline{V}}{RT}$$

where $\overline{V}$ is the partial molar volume.

5. Use the results of Problem 4 to calculate the activity of liquid water at a pressure of 50 atm at 25 °C. (Assume that water is incompressible over this range.)

6. Show that the dependence of activity on the temperature is given by

$$\left(\frac{\partial \ln a}{\partial T}\right)_{P,m} = -\frac{\overline{H} - \overline{H}^\circ}{RT^2}$$

where $\overline{H}$ and $\overline{H}^\circ$ are the partial molar enthalpies in the state of interest and the standard state.

7. Consider the reaction $AgCl \rightleftharpoons Ag + \frac{1}{2}Cl_2(g)$. The relevant free energies of formation are given in Table 9.1 and the enthalpies of formation are given in Table 6.2. Assuming that ΔH is independent of T, (a) calculate the equilibrium pressure of chlorine at 298 K; (b) calculate the equilibrium pressure of chlorine at 800 K.

8. Taylor and Rowlinson (*Trans. Faraday Soc.* 51 (1955):1183.) have measured the vapor pressures of aqueous solutions of glucose, $C_6H_{12}O_6$, as a function of the mole fraction of glucose. Their results for 25 °C are indicated in the accompanying table.

Mole fraction glucose	P_{H_2O} (torr)
0	23.756
0.0118	23.476
.0508	22.563
.0761	21.727
.0934	21.151
.0950	21.117
.1091	20.668
.1305	19.943
.1598	19.002
0.1950	17.751

(a) Calculate the activity and activity coefficient of water for each of the concentrations listed.
(b) A beaker of glucose solution with a mole fraction of glucose equal to 0.1091 is placed in a chamber with an aqueous solution of sucrose. It is observed that the weight of the glucose solution is unchanged. What is the concentration of the sucrose solution? (*Hint:* You can get an estimate from Table 14.1, and a better answer using (14.52).)

9. Using the data of Problem 8, evaluate the activity coefficient of the glucose at a concentration of 0.1 mol fraction by graphical methods.

10. Barthel and Dode (*Bull. Soc. Chim. Fr.*, 1312 (1954)) have measured the partial vapor pressures of Br_2 and CCl_4 over solutions of the two components at 0 °C. Their results are shown in the accompanying table (pressures are in torr).

X_{Br_2}	P_{Br_2}	P_{CCl_4}	X_{Br_2}	P_{Br_2}	P_{CCl_4}
1.0	67.9	0.0	0.0479	6.36	31.35
0.9194	62.71	6.85	.0220	2.86	33.38
0.8567	60.66	10.20	.00957	1.26	33.15
0.6628	51.11	16.83	.00479	0.585	32.07
0.4938	42.84	20.27	.001582	0.200	33.82
0.3822	36.50	24.20	.000936	0.120	33.85
0.2482	27.00	26.84	.000762	0.094	33.80
0.1178	14.36	30.77	0	0	33.85
0.0545	7.14	32.41			

(a) For CCl_4, plot the fugacity and the activity as functions of the mole fraction, indicating both the real behavior and the behavior based on Raoult's law.
(b) Do the same for Br_2, indicating Henry's law behavior in addition.
(c) Evaluate k and k', the Henry's law constants for mole fraction and molality, by extrapolating the activities to infinite dilution.
(d) Calculate and plot the activities and activity coefficients of CCl_4, taking the pure liquid as the standard state.
(e) Calculate and plot the activities and activity coefficients of Br_2 based on (i) taking the standard state as the pure liquid; (ii) taking the standard state as the solution of unit molality of Br_2 in CCl_4.
(f) Compare the activities of pure Br_2 using the two different standard states of part (e).
(Take the CCl_4 as the solvent and the Br_2 as the solute. Note that there is some scatter in the partial pressures of CCl_4 in the region near pure CCl_4. Assume that the vapor pressure of pure CCl_4 is exact and neglect the scatter.)

11. For the data tabulated in Problem 10, if we let x be the mole fraction of Br_2, then the partial pressures of the two components can be represented by

$$P_{Br_2} = P°_{Br_2} \cdot x \cdot \exp[1.197(1 - x)^2 - 0.493(1 - x)^3]$$

$$P_{CCl_4} = P°_{CCl_4}(1 - x) \exp(0.458x^2 + 0.493x^3)$$

where $P°$ indicates the vapor pressure of the pure component.
(a) Verify that the above equations as reported by Barthel and Dode satisfy the Gibbs-Duhem relation, Equation (14.44)
(b) Analytically evaluate the activity coefficient of Br_2 at a concentration of 0.1 mole fraction, using the Gibbs-Duhem relation, and compare with the answer you got directly in Problem 10.

12. G. N. Lewis and H. Storch (*J. Amer. Chem. Soc.*

39 (1917):2544.), measured the vapor pressures of Br_2 over dilute solutions in CCl_4 at 25 °C. Their results are shown in the accompanying table.

X_{Br_2}	P_{Br_2} (torr)
0.00394	1.52
.00420	1.60
.00599	2.39
.0102	4.27
.0130	5.43
.0236	9.57
.0238	9.83
0.0250	10.27

(a) Verify that Henry's law is valid in this range by plotting P vs X and showing that the data do indeed fall on a straight line.
(b) From the slope of the straight line, evaluate the Henry's law constant in units of mole fractions and in molalities.
(c) Show that since the vapor pressure of pure Br_2 at this temperature is 410 torr, there must be deviations from Raoult's law for the system Br_2–CCl_4.
(d) Calculate the activity and activity coefficient of pure Br_2 based on (i) a standard state obtained by extrapolating the Henry's law data to unit mole fraction; (ii) a standard state obtained by extrapolating the Henry's law data to unit molality.
(e) When a mixture of water and CCl_4 is shaken with a small amount of Br_2, it is found that the bromine is distributed between the two phases in such a way that the ratio of the molal concentration in the water phase to the mole fraction in the CCl_4 phase is 0.371. Calculate the distribution coefficient for Br_2 between water and CCl_4.
(f) Calculate the Henry's law constant for Br_2 in aqueous solution.
(g) Show that the partial pressure of Br_2 over an aqueous solution of molal concentration m is given by $P = 1.45m$, where P is in atmospheres.

13. The solubility of $Cl_2(g)$ in water as a function of the pressure of Cl_2 at 40 °C is given in the accompanying table. Calculate and plot the activity of Cl_2 as a function of concentration. Indicate Henry's law behavior on your plot.

P_{Cl_2} (torr)	Concentration (g per 1000 g of water)
5	0.412
50	1.025
100	1.424
250	2.34
500	3.61
750	4.77
1000	5.89
1500	8.05
3000	14.47
5000	23.3

14. The solubility of $Cl_2(g)$ in cyclohexane, C_6H_{12}, as a function of the pressure of Cl_2 is given in the accompanying table for solubilities in mole fractions at 40 °C.

P_{Cl_2} (torr)	X_{Cl_2}
100	0.0061
200	.0134
300	.0213
400	.0300
500	.0400
600	.0515
700	.0665
800	0.0867

(a) Calculate and plot the activity of Cl_2 as a function of concentration in cyclohexane. Indicate Henry's law behavior on your plot.
(b) Calculate the distribution coefficient for Cl_2 between water and cyclohexane, using the results of Problem 13.
(c) The amount 0.1 g Cl_2 is dissolved in 100 g water. The solution is shaken with 100 g cyclohexane. Calculate the concentration of Cl_2 in each of the phases.

15. The difficulties encountered with Equation (14.49) in the limit of dilute solutions can be overcome by the *osmotic coefficient* method. The osmotic coefficient is defined by

$$\phi = -\frac{\ln a_A}{r}$$

where r is the ratio X_B/X_A. Show that in terms of ϕ we can write

$$\ln\left(\frac{a_B'}{r'}\right) = \phi(r') - \phi(0) + \int_0^{r'} \frac{\phi - 1}{r} dr$$

where $\phi(r')$ indicates the value of ϕ at $r = r'$, while $\phi(0)$ is the value at $r = 0$. Describe how this function can be used to determine activities graphically.

In molality units, the expression for ϕ becomes

$$\phi = -\frac{1000}{M_1 m} \ln a_A$$

where M_1 is the solvent molar mass and m the molality of solute. In this case show that at concentration m'

$$\ln\left(\frac{a_B'}{m'}\right) = \ln \gamma_B = \phi' - 1 + \int_0^{m'} \frac{\phi - 1}{m} dm$$

and describe how this expression can be used to evaluate activity coefficients graphically. (See the Lewis and Randall text of footnote 7.)

16. Smith and Smith (*J. Biol. Chem.* 117 (1937):209) have measured the osmotic coefficients of glycine in aqueous solutions at 25 °C by the method of isotonic solutions, comparing the glycine solutions with sucrose solutions. Their measured osmotic

coefficients at various molal concentrations of glycine are as shown. Using the results of Problem 15, graphically determine the activity coefficients of glycine at concentrations of 0.1, 0.25, 0.5, 1.0, and 3.0 molalities.

m	ϕ	m	ϕ
0.1	0.990	1.2	0.921
0.2	.981	1.5	.913
0.3	.973	1.7	.908
0.4	.965	2.0	.903
0.5	.957	2.5	.894
0.7	.944	3.0	.888
1.0	0.928	3.3	0.885

17. Hildebrand and Eastman (*J. Amer. Chem. Soc.* 37 (1915):2452) measured the vapor pressures of mercury over solutions of thallium in mercury at 325 °C. They reported not the actual vapor pressure but rather the ratio $p/p°$, where p is the vapor pressure over the solution and $p°$ is the vapor pressure of pure mercury at that temperature. Their results are listed in the accompanying table.

Grams of thallium	Grams of mercury	$p/p°$
1.673	36.408	0.955
1.163	18.70	.938
1.673	17.68	.901
1.163	9.554	.875
3.207	16.10	.803
3.207	9.069	.690
4.162	8.097	.602
3.207	5.029	.548
7.390	7.196	.433
7.720	4.033	.293
7.221	1.796	0.166

(a) Calculate the activity and activity coefficients of mercury at the relevant concentrations.
(b) Calculate the activity coefficients of thallium.

CHAPTER FIFTEEN

IONS IN SOLUTION

15.1 THE NEED FOR A NEW APPROACH FOR ELECTROLYTES

One might well question the need for a separate discussion of electrolyte solutions. This question can be answered by trying to establish activities of electrolyte solutions just as we did for nonelectrolyte solutions.

Suppose we take HCl and try to establish a standard state following the procedures outlined in Chapter 14. We could measure the partial pressure of HCl over dilute solutions of hydrochloric acid, find the Henry's law constant from the limiting slope of the curve as $m \to 0$, and extrapolate this to unit molality. The results are indicated in Figure 15.1; the limiting slope is zero. An extrapolation of this slope to unit concentration is meaningless. On the other hand, if instead of plotting the fugacity as a function of the first power of the molality, we plot the fugacity as a function of m^2, we get the curve shown in Figure 15.2. Now the limiting slope is nonzero, and a meaningful extrapolation to a standard unit molality is possible, as shown by the dashed line. Whereas the relation $f = km$ is suitable for nonelectrolyte solutions, it is apparently not suitable for solutions containing ionic species. For ionic solutions like HCl, we may get by with a relation of the form $f = km^2$, in which case the activity is proportional to m^2 instead of m. As we go along, we shall see that the activities of ionic solutions are proportional to m^ν, where $\nu = 2$ for a binary electrolyte (for example A^+B^-), $\nu = 3$ for a ternary electrolyte (for example, $A_2^+B^{2-}$), and so on. The situation is further complicated, since ionic substances in solution give rise to more than one species. Colligative properties of solutions are cumulative; they depend on the sum of all the species. Suppose we consider a $1.0m$ solution of HCl, and suppose that it is 100% ionized. The concentration of solute particles is $2.0m$. If the solution were ideal, it would display an osmotic pressure equivalent to a $2.0m$ solution. The solution is not ideal, and in fact ionic species display greater deviations from ideality than nonionic solutions do. As we develop our scheme, we shall have to take the activity coefficient of HCl into account, but with two activity coefficients to contend with, one for each ion.

The world of ionic solutions is all-pervasive in its effect on our daily lives. The oceans constitute a huge reservoir of electrolyte solution. Body fluids constitute a collection of electrolyte solutions. Our every movement is associated with these solutions. Ions, since they are charged species, can be used to establish electric potential differences, and these potentials can be generated so as to transmit information through nerve cells to muscles. Batteries are electrochemical cells involving reactions of ionic species. In this chapter we shall examine some electrical properties of electrolyte solutions. Subsequently we shall examine the thermodynamic properties of electrolyte solutions.

FIGURE 15.1 Vapor pressure of HCl in aqueous solutions vs the concentration of HCl. Note that the curve approaches the origin with zero slope, vitiating any meaningful extrapolation back to unit molality. (The experimental points shown in this and in Figure 15.2 were not directly measured but calculated from tables of activity coefficients for HCl.)

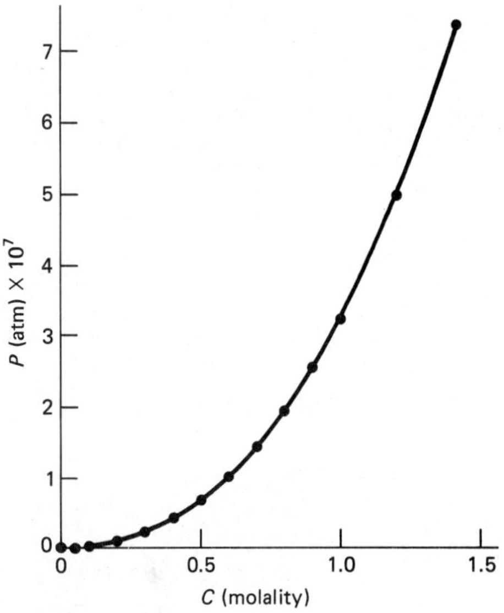

FIGURE 15.2 The data of Figure 15.1 plotted as a function of the square of the concentration. Now a meaningful extrapolation back to unit molality is possible as indicated. The curve approaches a straight line in the limit of infinite dilution.

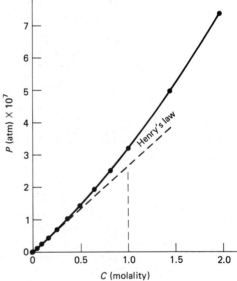

15.2 ELECTRICITY AND THE FIRST BATTERY

The ionic theory of solutions, like other facets of physical chemistry, was developed by many researchers working in widespread geographic areas. Static electricity had been known from time immemorial, and it entered the realm of science through the work of William Gilbert in the 1600s. It was he who coined the word *electric* from the Greek word for amber. The electricity he studied was the same type generated when

> **WILLIAM GILBERT** (1544–1603), English physician and scientist, was the most distinguished scientist in Elizabethan England, and he served as physician to the queen. His most celebrated work, *De Magnete,* described his many experiments dealing with magnets, magnetic attraction, and electrical attraction. He postulated that the earth was a huge magnet; and he was able to explain not only the north-south direction of the magnetic needle but also the inclination of the needle. Gilbert was the first person in England to embrace the Copernicus model of the solar system.

people walk on carpeting, especially on dry winter days. We have all had occasion to be shocked while touching a doorknob after walking on carpets and may have even seen the sparks fly from our fingers to the doorknob. Two different types of electricity were distinguished, giving rise to what was called the *two-fluid* theory of electricity. One type arose when glass was rubbed with silk, and the other when amber was rubbed with fur. The two types opposed each other, and were known as *vitreous* and *resinous*. The next 150 years saw the invention of two important pieces of apparatus, and then came the explosive development of electrical knowledge. The first invention was the Leyden jar, which was simply a condenser able to store electric charge. The second was the large electrostatic generator, the precursor of the modern Van de Graaff generator, in which large quantities of static electricity could be generated to charge the Leyden jars.

Benjamin Franklin proposed the currently accepted *one-fluid* theory of electricity in the 1740s. The appearance of two types of electricity depends on whether a body contains an excess or a deficit of this one electric fluid. Franklin's work laid the foundation for the later fundamental contributions of Coulomb, Ampère, Ohm, and others. The explosive development was set off by Luigi Galvani in the last decade of the eighteenth century. By the time the dust cleared we had batteries, electrolysis, electrical generators, ammeters and voltmeters, several newly discovered elements, and a unique method for establishing the combining ratios of atoms. It was this last

> **BENJAMIN FRANKLIN** (1706–1790), American statesman, publisher, and scientist, was one of the most brilliant people in this country in the eighteenth century. He was the first American to achieve an international reputation in science. He established his fame and fortune early in life as a printer and publisher. The *Pennsylvania Gazette,* which he bought in 1729, was converted from a dull, poorly printed sheet into the most successful newspaper of that time, with a circulation of some 10,000. By the time he retired from the printing business in 1749, he had established business connections with most of the thirteen colonies, not to mention Antigua, Jamaica, and England.
>
> Franklin's most important contributions to science were in electricity. He introduced many words that are now common, like *plus* and *minus, positive* and *negative, charge,* and *battery*. His first work, done in the 1740s, demonstrated that a pointed conductor could draw a spark much more readily than a blunt one. This discovery, coupled with his demonstration of the electrical nature of lightning, formed the basis for his invention of the lightning rod. Franklin was the first to point out that trees offered no protection during lightning storms, and indeed were dangerous. Many of Franklin's imitators were electrocuted; they did not take the precaution of insulating themselves from ground as Franklin had done. Much of his work involved using Leyden jars, and he built the world's first "battery" by connecting eleven condensers in series. He explained the phenomenon we call induced charge and developed the one fluid theory of electricity. He discovered that the surface area of a conductor, and not the mass, determined the amount of charge that could be built up in a condenser.

> Although Franklin's main contributions concerned electrical phenomena, he also worked in other fields. He carried out experiments to see whether oil that was spread on the ocean surface would still the waves. He was the first to chart the Gulf Stream, this work arising from his observation that mail packets took two weeks longer to sail from England to New England than merchant ships did; the colonial sailors were familiar with the Gulf Stream from their whaling experiences and avoided it. He did some work in heat, and also in medicine, gathering statistics to demonstrate the benefit of smallpox inoculations.
>
> Franklin's last years were primarily devoted to politics. He was one of the three authors (along with Jefferson and John Adams) of the Declaration of Independence and was a member of the Second Continental Congress. He was the minister to France during the Revolutionary War and one of those who negotiated the final peace with England. Returning from France in 1885, he was elected president of the Commonwealth of Pennsylvania, and he served as a delegate to the Constitutional Convention.

item that provided an invaluable link in the chain from John Dalton to the present accurate table of atomic weights.

In 1791, Galvani, a physician, was studying the effect of electricity on nerve cells in frog legs. He discovered that the leg twitched when a charged source was discharged into the nerve ending. Soon he discovered that the charged source was not needed to produce the twitching. The same result, though perhaps of smaller magnitude, was produced when the nerve was simply touched with a piece of metal. In fact, two dissimilar metals produce a larger effect than one metal. This effect, called a *galvanic effect,* is something that most of us who have amalgam tooth fillings have experienced directly. Often while eating, we feel a sudden sensation when a fork touches the amalgam filling. The contact of the two dissimilar metals, coupled with the presence of oral fluids, creates a weak battery. The voltage generated is enough to cause a slight shock.

Alessandro Volta constructed the first real battery in 1800. It consisted of a series of alternating silver and zinc plates separated by cloth soaked in salt solutions. Electrochemistry was born. Within a year, the invention had arrived in England, and Nicholson and Carlisle were using batteries to decompose water into its elements. Within seven years Sir Humphrey Davy had isolated the elements sodium and potassium by electrolysis of their fused hydroxides. Towards the end of Davy's life he was asked which of his discoveries constituted his greatest contribution to civilization. He replied that it was his discovery of Michael Faraday.

> LUIGI GALVANI (1737–1798), Italian physiologist, devoted most of his early efforts to straightforward anatomical investigations. In 1773, his interests turned to muscle studies, and ten years later he began his famous experiments on the effect of electric discharges on frog muscles. He was engaged in an effort to measure the effect of an electrical discharge on the muscle contraction when, in 1786, he discovered that when two dissimilar metals are used to complete the connection between the spinal cord and the muscle in a frog's leg, a contraction of the leg results even in the absence of an electrical discharge. He incorrectly interpreted the basic cause for this "Galvanic effect," attributing it to a fluid in animals analogous to electrical fluid, which he called "animal electricity." It was his countryman Alessandro Volta who shortly thereafter provided the correct explanation.

> ALESSANDRO VOLTA (1745–1827), Italian physicist, correctly interpreted the results of Galvani's muscle contraction experiments by pointing out that the muscle acted as a detector of external electrical potential differences rather than as a source of the electricity. In 1792, he demonstrated that a potential difference could be achieved by two dissimilar metals in the presence of an electrolyte solution. By 1800, he had perfected the first battery, or "pile," consisting of a series of alternating silver and zinc discs separated by cloths soaked in salt. This simple device must be considered one of the most important instrumental advances in the history of science, in the same rank as the much more sophisticated atomic pile developed in 1945. Until its development, the only sources of electrical generation were the electrostatic machines that were capable of high voltages but very small currents. (In 1789, the electrolysis of water by van Troostwijk and Deimann using an electric discharge machine required 14,600 discharges of static electricity to produce 5 cm³ of gas.) Volta's pile was the first continuous source of electricity, and its importance in the development of chemistry and physics is immeasurable.

15.3 QUANTITATIVE ELECTROLYSIS

In 1813, Michael Faraday came to the Royal Institution as Davy's assistant. During the next 50 years he presented metallurgy with stainless steel, physics with his laws of magnetic induction, and chemistry with the electrochemical equivalent.

Suppose we place the terminals of a battery into fused AgCl. The AgCl decomposes according to the reaction $AgCl \rightleftharpoons Ag + \frac{1}{2}Cl_2(g)$. The silver collects on the cathode, while $Cl_2(g)$ is released at the anode. The reaction at each of the terminals, or the *half-cell reaction,* can be written as

$$\begin{aligned} \textit{Cathode:} \quad & Ag^+ + e^- \rightleftharpoons Ag \\ \textit{Anode:} \quad & Cl^- \rightleftharpoons \tfrac{1}{2}Cl_2(g) + e^- \end{aligned}$$

The reaction is shown in Figure 15.3, where the motion of the charged species is indicated.

> MICHAEL FARADAY (1791–1867), English chemist and physicist, was a completely self-taught man. In 1812, while still a bookbinder's apprentice, Faraday was drawn to chemistry by attending Davy's lectures at the Royal Institute. His life was changed by an accident when Davy was temporarily blinded by an explosion and took on Faraday as his secretary. Faraday presented Davy with the careful notes he had taken at his lectures, and Faraday became a laboratory assistant when his predecessor was fired for brawling. His first task was to accompany Davy on a tour of Europe, including France; they were warmly welcomed in Paris even though England and France were at war at that time. Faraday's first experiment consisted in constructing a voltaic pile using copper halfpenny pieces and zinc discs separated by paper soaked in salt solution. He decomposed magnesium sulfate with the pile. He made numerous contributions to science besides his laws of electrochemical equivalents and magnetic induction. He produced the first known chlorides of carbon, C_2Cl_6 and C_2Cl_4, in 1820, and discovered benzene in 1825. He investigated alloy steels and optical glass. During this latter work, he discovered the rotation of the plane of polarization of light in a magnetic field. He discovered diamagnetism and coined the words *paramagnetic* and *diamagnetic*. Faraday was as successful in teaching as he was in research. In 1825, he instituted the Friday evening lectures at the Royal Institute, a series that contributed greatly to the scientific education of the upper classes in England at the time. Much of the support for science in Victorian England is directly attributable to these lectures, which were well attended by many notables of the day.

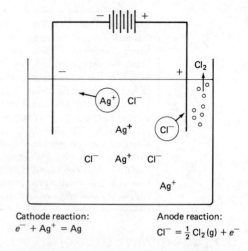

FIGURE 15.3 Electrolysis of molten AgCl.

Cathode reaction:
$e^- + Ag^+ = Ag$

Anode reaction:
$Cl^- = \frac{1}{2}Cl_2(g) + e^-$

Note that one electron is required to convert each silver ion into one silver atom, and one mole (6.02×10^{23}) of electrons is required to plate out one mole (107.87 g) of silver. The amount of charge associated with a mole of electrons is called a *faraday* (F) and is equal to 96,487 coulombs of electricity. For $BaCl_2$, the half-cell reactions are

$$Ba^{2+} + 2e^- \rightleftharpoons Ba$$
$$2Cl^- \rightleftharpoons Cl_2(g) + 2e^-$$

Since two electrons are involved in this reaction, one faraday of electricity produces half a mole of barium metal.

The *coulometer*, an instrument for accurately measuring electric charge, is based on these results. If a solution of $AgNO_3$ is placed in series with a dc circuit, the total quantity of electric charge passing through the circuit can be determined by measuring the mass of silver plated out. In fact, the old ampere unit was defined by coulometric measurements. The international ampere was defined as the constant current that would deposit 0.0011180 g of silver per second from a solution of $AgNO_3$.

When AgCl is decomposed by electrolysis, work is done on the system; the work is provided by the source of electricity. When the reverse reaction is carried out—$Ag + \frac{1}{2}Cl_2(g) \rightleftharpoons AgCl$—the reaction can produce work. This is precisely what happens in a battery. We shall return to this important point when we consider electrochemical cells (Chapter 17); measurements of the work available in these kinds of reactions provide one of the most powerful and accurate techniques for determining thermodynamic values. For the moment, we turn to the electrical conductivities of ionic solutions.

15.4 CONDUCTIVITY OF SOLUTIONS

Ohm's law relates the current to the voltage and resistance in the form

$$I = \frac{V}{R} \tag{15.1}$$

where I is the current, V is the potential drop across the circuit, and R is the resistance of the circuit. The total resistance of a wire depends on the length and the cross-sectional area. It is convenient to express the resistance of a wire in terms of the *resistivity* ρ

$$R = \frac{\rho l}{A} \tag{15.2}$$

where l is the length of the resistor, A is the cross-sectional area, and ρ is the resistivity. The resistivity is a measure of specific resistance. It provides a convenient measure for comparing the resistive properties of different substances. For our purposes it will be more convenient to deal with the reciprocal of the resistivity, or the *conductivity* κ. The unit of resistance is the ohm; the unit of conductance is simply *ohm* written backwards, or *mho*. A schematic circuit for measuring the conductance of solutions is indicated in Figure 15.4. The conductivity cell, which contains the solution, is so constructed that the two electrodes are parallel and a known distance apart. The variable capacitor in the variable resistance arm of the *Wheatstone bridge* is used to balance the capacitative effect of the two electrodes of the cell. The resistance R_2 is adjusted until there is no potential difference between points A and B, at which point the resistance of the cell is found from[1]

$$\frac{R_1}{R_2} = \frac{R_3}{R_4}$$

The solution conductivity is determined from the measured resistance and the known dimensions of the cell. To compare conductivities, it is convenient to reduce all conductivities to a common concentration. The *molar conductance* is defined by

$$\Lambda = \frac{\kappa}{c} \tag{15.3}$$

where c is the concentration in equivalents per cubic centimeter.[2] Figures 15.5 and 15.6 show plots of the molar conductances of several substances vs the square root of

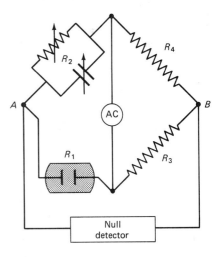

FIGURE 15.4 Wheatstone bridge circuit for measuring the conductivity of electrolytic solutions. The current is ac instead of dc, to eliminate electrolytic effects. The variable capacitor balances the capacitance of the cell. The cell is usually calibrated with a substance whose conductivity is known; the most usual standard is a KCl solution.

---

[1] The detailed operation of the Wheatstone bridge is described in most elementary textbooks on electricity. See also D. P. Shoemaker and C. W. Garland, *Experiments in Physical Chemistry* (New York: McGraw Hill Book Co., 1962).

[2] That is to say the number of moles of charge. For a uni-univalent electrolyte it is just the number of moles of electrolyte per cubic centimeter, whereas for a divalent electrolyte such as $CuSO_4$ it is one-half the number of moles per cubic centimeter, and so on. Thus in Table 15.1, the entry for $CuSO_4$ is listed as $\tfrac{1}{2}CuSO_4$.

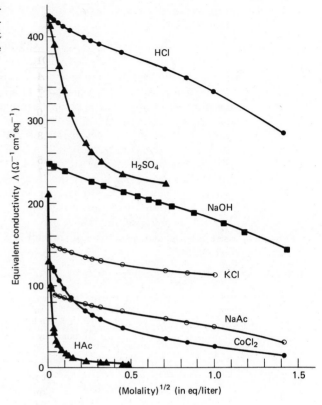

FIGURE 15.5 Molar conductances of several electrolytes in aqueous solution at 298.15 K vs square root of the concentration.

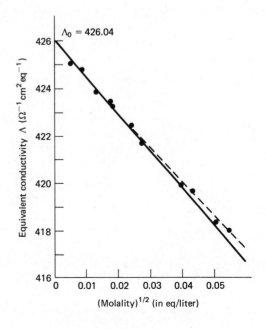

FIGURE 15.6 Molar conductance of HCl at small concentrations vs square root of the concentration. The solid line is the straight-line extrapolation to $c = 0$, and the dashed line is the theoretically fitted line. (Data from T. Shedlovsky, *J. Amer. Chem. Soc.* 54 (1932):1411.

SEC. 15.5 KOHLRAUSCH'S LAW

TABLE 15.1 Equivalent conductances Λ for several electrolytes at several concentrations at 25 °C.

Salt	\multicolumn{8}{c}{Concentration (eq liter^{-1})}							
	0[a]	0.0005	0.001	0.005	0.01	0.02	0.05	0.1
HCl	426.2	422.7	421.4	415.8	412.0	407.2	399.1	391.3
LiCl	115.0	113.2	112.4	109.4	107.3	104.7	100.1	95.9
NaCl	126.5	124.5	123.7	120.7	118.5	115.8	111.1	106.8
KCl	149.9	147.8	147.0	143.6	141.3	138.3	133.4	129.0
NH$_4$Cl	149.7				141.3	138.3	133.3	128.8
NaI	126.9	125.4	124.3	121.3	119.2	116.7	112.8	108.8
KI	150.4			144.4	142.2	139.5	135.0	131.1
NaAc	91.0	89.2	88.5	85.7	83.8	81.2	76.9	72.8
NaOH	247.8	245.6	244.7	240.8	238.0			
AgNO$_3$	133.4	131.4	130.5	127.2	124.8	121.4	115.2	109.1
½CuSO$_4$	133.6	121.6	115.3	94.1	83.1	72.2	59.1	50.6
LiClO$_4$	106.0	104.2	103.4	100.6	98.6	96.2	92.2	88.6
NaClO$_4$	117.5	115.6	114.9	111.8	109.6	107.0	102.4	98.4
KClO$_4$	140.0	138.8	137.9	134.2	131.5	127.9	121.6	115.2
AgClO$_4$	126.6							
NaBrO$_3$	105.4							
KBrO$_3$	129.3							

SOURCE: Data from B. E. Conway, *Electrochemical Data* (New York: Elsevier Publishing Co., 1952)

NOTE: The conductance Λ is in units of ohm^{-1} cm^2 mol^{-1}. In the SI system, the unit of equivalent conductance is ohm^{-1} m^2 mol^{-1}; the concentration would be in moles per cubic decimeter (mol dm^{-3}). To convert the entries in this table to SI units just divide by 10^4.

[a] Values at zero concentration are obtained by extrapolation.

the concentration. A very important number is the conductance extrapolated to zero concentration, which we call Λ_0. The curves in Figure 15.5 and 15.6 indicate that this extrapolation can be carried out quite simply for strong electrolytes, whereas the problems engendered by weak electrolytes such as HAc (acetic acid) are severe. At small concentrations, the data for strong electrolytes can be represented by the equation

$$\Lambda = \Lambda_0 - k\sqrt{c} \tag{15.4}$$

where k is an experimentally determined constant. The equivalent conductances of electrolytes at several concentrations are listed in Table 15.1.

15.5 KOHLRAUSCH'S LAW

Before proceeding, let's stop for a moment and think about the conductivity of ions on a simple qualitative basis. Suppose we have the compound AB, which dissolves in water to produce the ions A$^+$ and B$^-$. The solution conducts electricity through motion of the ions under the effect of an electric field. At high concentrations, each ion is surrounded by other ions, both positive and negative. The field affecting any particular ion exists partly because of these surrounding ions. At infinite dilution, the distance between nearest neighbor ions is large, and only the effect of the applied electric field is felt by individual ions. This is the reason for extrapolating the data to infinite dilution.

The conductivity of any particular ion will also be affected by the ease with which the ion can move through the water. Hence different ions should contribute dif-

> FRIEDRICH WILHELM GEORG KOHLRAUSCH (1840–1910), German chemist and physicist, is best known for his work on the electrical conductivity of solutions. His work is characterized by a high degree of precision, as exemplified in his determination of the electrochemical equivalent of silver. His main work on electrolyte conduction was made possible by the realization that polarization at the electrodes could be eliminated by using ac instead of dc currents for conductivity measurements. In 1876, following the work of Hittorf on ion migrations, he stated, "In a dilute solution every electrochemical element has a perfectly definite resistance pertaining to it, independent of the compound from which it is electrolyzed"; that is Kohlrausch's law. He was one of the first teachers to publish an instructional manual on laboratory physics. The manual, *Leitfaden der Praktischen Physik* (1870), was widely used and translated into several languages, including English.

ferently to the total measured conductivity. The ease with which any ion moves through the solution depends on considerations such as the total charge and the size of the ion; large ions offer greater resistance to motion through the water than small ions.

Suppose we now consider the compound CB, which dissociates on solution to produce C^+ and B^-, where the ion B^- is the same as the B^- produced by the compound AB discussed above. One expects the contribution of the anion B^- to the total conductivity of the solution to be independent of the nature of the cation at infinite dilution. The conductivity of a solution is then given by the conductivities of the anion and the cation:

$$\Lambda_0 = \Lambda_0^+ + \Lambda_0^- \tag{15.5}$$

Equation (15.5) describes the true state of affairs and states *Kohlrausch's law of independent mobilities* of ions in infinitely dilute solutions. It is amenable to easy verification, since the difference in Λ_0 for pairs of salts with a common ion should be the same regardless of the common ion. This is illustrated in Table 15.2 for several alkali metal salts.

We saw in Figures 15.5 and 15.6 that the experimental data for the conductivities of weak electrolytes make extrapolation to infinite dilution very difficult. Kohlrausch's law enables us to evaluate Λ_0 for weak electrolytes from the values of the salts, which in most cases are strong electrolytes. For example, for acetic acid,

$$\Lambda_0(\text{HAc}) = \Lambda_0(\text{NaAc}) + \Lambda_0(\text{HCl}) - \Lambda_0(\text{NaCl})$$
$$= 91.0 + 426.2 - 126.5 = 390.6 \text{ ohm}^{-1} \text{ cm}^2 \text{ mol}^{-1} \tag{15.6}$$

This assumes that the acetic acid is 100% ionized in the infinitely dilute solution.

TABLE 15.2 Differences between limiting equivalent conductivities of electrolytes having a common ion, illustrating Kohlrausch's law.

KClO$_4$	140.0	KI	150.4	KCl	150.0
NaClO$_4$	117.5	NaI	126.9	NaCl	126.5
	22.5		23.5		23.5
KBrO$_3$	129.3	KAc	114.4		
NaBrO$_3$	105.4	NaAc	91.0		
	23.9		23.4		
KCl	150.0	NaCl	126.5	LiCl	115.0
KClO$_4$	140.0	NaClO$_4$	117.5	LiClO$_4$	106.0
	10.0		9.0		9.0

NOTE: Entries are Λ_0 in units of ohm^{-1} cm^2 mol^{-1}.

15.6 ARRHENIUS AND WEAK ELECTROLYTES

We have been speaking of positive and negative ions, or cations and anions, in a matter-of-fact way. The existence of individual ions in solutions has been part of our understanding of the physical world at least since our high school days. Ions were not always so obvious. It was not until some 50 years after the work of Faraday on electrolysis, and some 10 years after the work of Kohlrausch on conductivities, that the Swedish chemist, Svante Arrhenius, first proposed the concept of the ion as a charged chemical species in solution.[3]

Arrhenius noted that the molar conductances of electrolytes decreased with increasing concentration. He considered a solution of the salt AB to consist partly of un-ionized AB molecules and partly of cations and anions. He attributed the decrease in conductivity to the decrease in the degree of ionization of the electrolyte. Suppose that a solution of AB is prepared with n molecules of AB per cubic centimeter of solution. If we call the degree of ionization α, then the number of positive ions per cubic centimeter is given by $n_+ = \alpha n$, and the number of negative ions is given by $n_- = \alpha n$. Suppose that the velocities with which the ions move through the water are v_+ and v_-. The total current carried across a unit area is then

$$i = n_+ e v_+ + n_-(-e)(-v_-) = ne(v_+ + v_-)\alpha \tag{15.7}$$

where the negative sign for v_- arises because the two different ions move in opposite directions; the electron charge is denoted by e.

The *mobility* μ of an ion is defined as the velocity per unit field strength, or $\mu_+ = v_+/E$ and $\mu_- = v_-/E$. From Ohm's law and the definition of conductivity, we can write $\kappa = i/E$, where i is the current across a unit area and E is the electric field. The specific conductivity of the solution is

$$\kappa = ne(\mu_+ + \mu_-)\alpha \tag{15.8}$$

The equivalent conductivity is then

$$\Lambda = N_0 e(\mu_+ + \mu_-)\alpha \tag{15.9}$$

SVANTE AUGUST ARRHENIUS (1859–1927), Swedish chemist, is recognized as one of the founders of physical chemistry. His theory of electrolytic dissociation was first presented in his doctoral dissertation to the University of Uppsala (1884), receiving only the award fourth class. A colleague correctly remarked: "This is a very cautious but very unfortunate choice. It is possible to make serious mistakes from pure cautiousness. There are chapters in Arrhenius' thesis which alone are worth more or less all the faculty can offer in the way of marks." His theory was eventually accepted through the efforts of Ostwald, van't Hoff, and Nernst, when it became apparent that dissociation held the key to various phenomena in electrolytic solutions. In particular, Arrhenius was able to calculate the van't Hoff factor from conductivity measurements, and found them to agree well with the experimental values derived from osmotic-pressure and freezing-point-depression measurements. In 1903, Arrhenius was awarded the Nobel Prize for chemistry with the citation "In recognition of the extraordinary services he has rendered to the advancement of chemistry by his theory of electrolytic dissociation."

---

[3] In the speech honoring Arrhenius at the 1903 Nobel banquet, Cleve remarked, "These new theories also suffered from the misfortune that nobody really knew where to place them. Chemists would not recognize them as chemistry, nor physicists as physics. They have in fact built a bridge between the two."

where N_0 is the Avogadro number. Since $N_0 e$ is the charge on a mole of electrons, or the Faraday constant F, we can write (15.9) as

$$\Lambda = F(\mu_+ + \mu_-)\alpha \tag{15.10}$$

It is here that Arrhenius made the critical assumption that at infinite dilution ionization is complete, that is, as $m \to 0$, $\alpha \to 1$. In the limit,

$$\Lambda_0 = F(\mu_+ + \mu_-) \tag{15.11}$$

and by combining Equations (15.10) and (15.11) we get an expression for the degree of ionization,

$$\alpha = \frac{\Lambda}{\Lambda_0} \tag{15.12}$$

This model accomplishes two things. Firstly it enables us to calculate the degree of ionization of electrolytes from conductivity data, and secondly it provides an explanation of Kohlrausch's law of independent migrations. The mobilities of the ions are independent of the chemical constitution and the terms in (15.5) can be interpreted by taking

$$\Lambda_0^+ = F\mu_+ \quad \text{and} \quad \Lambda_0^- = F\mu_- \tag{15.13}$$

Suppose we take the dissociation of acetic acid at a concentration of c moles per liter as an example:

$$\begin{array}{cccc} \text{HAc} & \rightleftharpoons & \text{H}^+ + & \text{Ac}^- \\ c(1-\alpha) & & c\alpha & c\alpha \end{array} \tag{15.14}$$

In (15.14) we have written the concentrations of the various species at equilibrium in terms of c and the degree of dissociation. For acetic acid, $\Lambda_0 = 391$; the values of Λ at finite concentrations are given in Table 15.3. For a solution containing 0.02 mole of HAc per liter, $\Lambda = 11.563$ and $\alpha = 11.563/391 = 0.0296$. The concentration of H$^+$ is then 0.00059 molar.

TABLE 15.3 Conductivity data for acetic acid at 25 °C at several concentrations

Concentration (mol liter^{-1})	Λ (ohm^{-1} cm^2 mol^{-1})	Degree of dissociation $\alpha = \dfrac{\Lambda}{\Lambda_0}$	$i = \alpha + 1$	K_Λ $\left(\dfrac{c\alpha^2}{1-\alpha}\right)$
0	390.6	1.000		
0.000028	210.3	0.538	1.54	1.759 × 10^{-5}
0.000111	127.7	0.327	1.33	1.768
0.001028	48.13	0.123	1.12	1.780
0.002414	32.21	0.0825	1.08	1.789
0.003441	27.19	0.0696	1.07	1.792
0.009842	16.37	0.0419	1.04	1.804
0.01283	14.37	0.0368	1.04	1.803
0.05000	7.356	0.0188	1.02	1.807
0.10000	5.200	0.0133	1.01	1.796
0.20000	3.650	0.0093	1.01	1.763
0.23079	3.391	0.0088	1.01	1.755
1.011	1.443	0.0037	1.00	1.385

SOURCE: Conductivity data from D. A. MacInnes and T. Shedlovsky, *J. Amer. Chem. Soc.* 54 (1932):1429.

Shortly after Arrhenius first proposed his ionic dissociation theory, Ostwald applied the law of mass action to a partially ionized substance. For HAc we can write

$$K = \frac{[H^+][Ac^-]}{[HAc]}$$

$$= \frac{\alpha^2 c}{1 - \alpha}$$

$$= \frac{\Lambda^2 c}{\Lambda_0 (\Lambda_0 - \Lambda)} \quad (15.15)$$

where the brackets indicate molar concentrations. The constancy of K over a wide range of small concentrations is evident from the entries in Table 15.3. At larger concentrations, however, the value of K deviates from its small concentration value as a result of the nonideality of the solution.

That stable molecules break up into ions when placed in water was difficult for chemists to accept in the late 1880s, and many were skeptical. What was needed was independent verification by a different experimental technique. This verification was shortly supplied by van't Hoff in his work on the osmotic pressure of ionic solutions and by Raoult and others working in the area of freezing-point-depression measurements.

15.7 COLLIGATIVE PROPERTIES OF IONIC SOLUTIONS

We have seen that colligative properties of solutions depend on the concentrations of the dissolved species (Chapter 12). Freezing-point depressions, boiling-point elevations, and osmotic pressures are always higher for electrolytic solutions than for nonelectrolytic solutions. The equation for the osmotic pressure for nonelectrolytes is

$$\Pi = cRT \quad (15.16a)$$

and the equation for the freezing-point depression is

$$\Delta T_{fp} = m K_{fp} \quad (15.16b)$$

In studying the osmotic pressure of electrolyte solutions, van't Hoff noticed that his measured pressures were always larger than the equation (15.16a) predicted, sometimes by a factor of 2 or 3, and he accounted for this by writing the osmotic pressure equation in the modified form

$$\Pi = icRT \quad (15.17)$$

where i was known as the *van't Hoff factor*. For strong electrolytes, the factor was equal to the number of ions formed on solution; and for weak electrolytes, it was less than this number but still greater than unity.

For freezing-point depressions, it was found that the modified equation could be written in the form

$$\Delta T_{fp} = i m K_{fp}$$

Measurements of freezing-point depressions provided the same values of the van't Hoff factor as the osmotic-pressure measurements did. It was strong evidence for the dissociation of ionic species. In Figure 15.7, where we have plotted the freezing-point

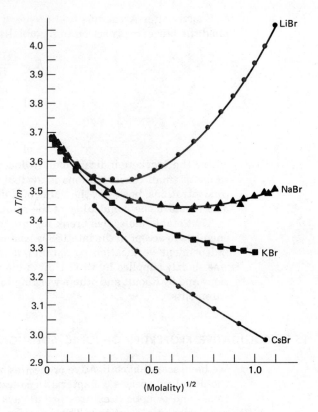

FIGURE 15.7 Molal freezing-point depressions $\Delta T/m$ for the alkali bromides vs the square root of the molality. The value for nonelectrolytes is 1.86 K mol^{-1}. The limiting value here is twice that amount.

depressions for several alkali bromides, it can be seen that the freezing-point depression is 3.72, or twice the normal value of 1.86, in the limit as $m \to 0$.

Suppose we have a general electrolytic compound A_aB_b, which dissociates according to the equation $A_aB_b \rightleftharpoons aA^{(+z_a)} + bB^{(-z_b)}$. If the concentration of the electrolyte is c, then the concentration of undissociated A_aB_b is $(1 - \alpha)c$, and the concentrations of the A and B ions are αac and αbc respectively. The total concentration of dissolved particles without regard to type is then $(1 - \alpha)c + \alpha ac + \alpha bc$. If we follow the usual convention and let ν be the number of ions produced by dissociation of a molecule of A_aB_b, then the total concentration of species can be written as $(1 - \alpha)c + \alpha \nu c$. If the van't Hoff factor is now identified with the fractional increase in concentration produced by the dissociation, we can write

$$i = \frac{(1 - \alpha)c + \alpha \nu c}{c}$$
$$= 1 - \alpha + \alpha \nu \qquad (15.18)$$

Equation (15.18) can be solved to provide an expression for the degree of dissociation in the form

$$\alpha = \frac{i - 1}{\nu - 1} \qquad (15.19)$$

Values of α calculated from osmotic-pressure measurements and freezing-point-depression measurements using (15.19) agreed well with the values calculated from conductivity data by (15.12). The evidence was good enough to convince all but the most skeptical. Ions became an established fact in chemistry.

15.8 TRANSFERENCE NUMBERS OF IONS

Let's rewrite Kohlrausch's law of independent ionic mobilities:

$$\Lambda_0 = \Lambda_0^+ + \Lambda_0^- \qquad (15.5)$$

The total conductivity of a solution of the electrolyte AB is the sum of the contributions of the ions A^+ and B^-. The *fraction* of the current transported by an ion is its *transference number,* or *transport number.* At infinite dilution, it can be written as

$$t_0^+ = \frac{\Lambda_0^+}{\Lambda_0} \qquad t_0^- = \frac{\Lambda_0^-}{\Lambda_0} \qquad (15.20)$$

Now let's go back to the analysis we started in (15.7) and carry it one step further. From (15.8), the conductivity of an ion can be written

$$\kappa = c\mu|ze| \qquad (15.21)$$

where c is the number of ions per unit volume and $|ze|$ is the electric charge per ion; for a univalent ion, $z = 1$. To get the total conductivity, we must sum over all ions:

$$\kappa = \sum_i c_i \mu_i |z_i e|$$

The molar conductance of any ion Λ_i is the conductivity of a solution containing one mole (or one faraday) of charge per unit volume. In other words, when $c|ze| = F$, then $\Lambda_i = \kappa$; we have

$$\Lambda_i = F\mu_i = t_i \Lambda \qquad (15.22)$$

This relation applies to each and every ion in the solution. The molar conductance of any solution can be readily measured. Hence a measurement of the transference number of any particular ion, coupled with the conductance of the solution, enables us to determine both the conductance and the mobility of the individual ions.

Suppose we take HCl as an example. Let us for the moment assume that the ion H^+ is four times as mobile as the ion Cl^-, and that the solution is 100% ionized. Now we examine the ion migrations in an electrolysis experiment. At first thought, it might seem that four times as much hydrogen will reach the cathode as the amount of chlorine that will reach the anode, since the H^+ ions travel four times as fast as the Cl^-. This is clearly impossible, since electrical neutrality must be maintained in the solution; for every H^+ that reaches the cathode, exactly one Cl^- must reach the anode. The process can be pictured by referring to Figure 15.8. There we have arbitrarily divided the electrolysis cell into three regions, labeled anode, center, and cathode. In the initial state depicted in (a), the concentration of ions is uniform throughout the cell. Each plus or minus sign indicates a specific number of ions, say a number equivalent to one faraday of charge. We then turn on the current and electrolyze the solution until five faradays of electricity pass through the solution. The migration of the ions is indicated in (b). The electrolysis cell is in series with the wires connected to it, and the most elementary principle of electric circuits requires that the current passing through any part of the circuit must be the same. The current passing through the lead wires is composed of negative charges in the form of

FIGURE 15.8 Measurement of transference numbers: Hittorf method. (a) Initial state. (b) Migration of ions. (c) Final result. (d) Hittorf transference cell.

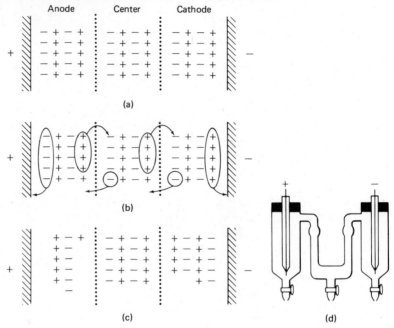

electrons. The current through the cell is composed of two constituent parts, the positive and the negative ions. Remember that so far as electric current is concerned, a positive ion traveling to the right is completely equivalent to a negative ion traveling to the left. Each would have the same effect on an ammeter.

The migration of the ions is depicted in Figure 15.8b. Five faradays of current pass through the circuit. Hence at the cathode five faradays of H^+ are electrolyzed, while at the anode five faradays of Cl^- are electrolyzed. Throughout the rest of the cell, five faradays of current must cross any plane perpendicular to the cell. The five faradays are composed of four H^+ and one Cl^-, as indicated. The final result is shown in Figure 15.8c. Originally each section contained 10 F of each ion. At the end of the process, the central region still has 10 F, while the cathode region has but 9 F and the anode region has 6 F. The concentration in the cathode region decreased by 1 F, and the concentration in the anode region decreased by 4 F. These decreases are in the ratio 4:1, which is the ratio of the assumed mobilities,

$$\frac{\delta_{\text{anode}}}{\delta_{\text{cathode}}} = \frac{\mu_+}{\mu_-} \tag{15.23}$$

Hittorf developed this experimental procedure. The values of the transference numbers are determined by passing a known current through the cell and then analyzing the compartments of the cell in the regions near the anode and cathode to determine the amount of depletion. If the total charge passing through the cell is q faradays, then the number of equivalents of solute lost from the anode compartment is just the number of equivalents lost by electrolysis less the number of equivalents that enter that compartment from the central region of the cell, or

$$\delta_{\text{anode}} = q - t_- q = q(1 - t_-) = q t_+ \tag{15.24}$$

with a similar equation for the cathode compartment. The transference numbers are then

$$t_+ = \frac{\delta_{\text{anode}}}{q} \qquad t_- = \frac{\delta_{\text{cathode}}}{q} \tag{15.25}$$

One very useful feature of this type of experiment is that there is an internal check on the experimental accuracy, since we must have $t_+ + t_- = 1$.

A Hittorf transference cell is shown in Figure 15.8d. The stopcocks at the bottoms of the compartments allow convenient emptying of each compartment for analysis. As the electrolysis proceeds, the concentration of electrolyte decreases, and the conductivity of the cell also decreases. The results of these types of measurements are shown in Tables 15.4 and 15.5, which list transference numbers and ionic conductances.

TABLE 15.4 Limiting molar ionic conductances at infinite dilution Λ_0 at 25 °C.

Cation	Λ_0 (ohm^{-1} cm^2 mol^{-1})	Anion	$10^4 \Lambda_0^-$ (ohm^{-1} cm^2 mol^{-1})
H$^+$	349.8	OH$^-$	197.6
Li$^+$	38.69	Cl$^-$	76.34
Na$^+$	50.11	Br$^-$	78.3
K$^+$	73.52	I$^-$	76.8
NH$_4^+$	73.4	NO$_3^-$	71.44
Ag$^+$	61.92	HCO$_3^-$	44.5
Tl$^+$	74.7	ClO$_4^-$	67.32
$\frac{1}{2}$Mg^{2+}	53.06	CH$_3$COO$^-$	40.9
$\frac{1}{2}$Ca^{2+}	59.50	$\frac{1}{2}$SO$_4^{2-}$	80
$\frac{1}{2}$Sr^{2+}	59.46		
$\frac{1}{2}$Ba^{2+}	63.64		
$\frac{1}{2}$Cu^{2+}	54		
$\frac{1}{2}$Zn^{2+}	53		
$\frac{1}{3}$La^{3+}	69.5		

SOURCE: Data from B. C. Conway, *Electrochemical Data* (London: Elsevier Publishing Co., 1952).

TABLE 15.5 Cation transference numbers in aqueous solutions at 25 °C.

Electrolyte	Concentration (eq liter^{-1})					
	0	0.01	0.02	0.05	0.10	0.20
HCl	0.8209	0.8251	0.8266	0.8292	0.8314	0.8337
KNO$_3$	.5072	.5084	.5087	.5093	.5103	.5120
KCl	.4906	.4902	.4901	.4899	.4898	.4894
KI	.4892	.4884	.4883	.4882	.4883	.4887
KBr	.4849	.4833	.4832	.4831	.4833	.4841
AgNO$_3$	.4643	.4648	.4652	.4664	.4682	
NaCl	.3963	.3918	.3902	.3876	.3854	.3821
LiCl	.3364	.3289	.3261	.3211	.3168	.3112
Na$_2$SO$_4$	.386	.3848	.3836	.3829	.3828	.3828
Na$_2$C$_2$H$_3$O$_2$	0.5507	0.5537	0.5550	0.5573	0.5594	0.5610

SOURCE: Data from H. S. Harned and B. B. Owen, *The Physical Chemistry of Electrolytic Solutions* (New York: Reinhold Publishing Co., 1950).

15.9 CONDUCTOMETRIC TITRATIONS

Measuring solution conductivities provides a simple but useful technique in instrumental analysis known as *conductometric titration*. Consider the trio of electrolytes HCl, NaOH, and NaCl, whose equivalent molar conductances at infinite dilution in units of ohm^{-1} cm^2 mol^{-1} are 426, 248, and 126. Suppose we titrate a solution of HCl with standard NaOH solution. Initially, the conductivity of the solution arises only from the HCl. As NaOH is added, the conductivity decreases as the H$^+$ are replaced by Na$^+$ with a much lower conductivity, until the equivalence point is reached, at which the conductivity derives from the resulting NaCl solution. As more NaOH is added, the conductivity increases, and the equivalence point can be taken as the intersection of the two lines; this is shown in Figure 15.9. This kind of titration is not restricted to acid-base titrations but is amenable to any kind of titration in which ions are removed from solution and replaced by other ions of different conductivity.

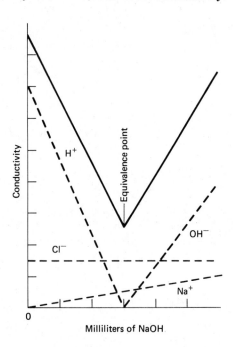

FIGURE 15.9 Conductance curve (solid line) and the individual ionic contributions to the conductance curve (dashed lines) for a titration of HCl by NaOH. Initially the conductance derives only from the HCl; at the equivalence point, the conductance is just the conductance of NaCl, and finally, it is the combined conductance of NaCl and NaOH.

15.10 EQUILIBRIUM CONSTANTS

In our consideration of free energy and ideal gases, we developed the concept of the equilibrium constant for gas reactions from purely thermodynamic considerations. The first concept of an equilibrium constant arose from the early work of the Norwegian chemists C. M. Guldberg and P. Waage on the kinetics of chemical reactions. Consider the reaction

$$A + B \underset{k_r}{\overset{k_f}{\rightleftarrows}} C + D$$

SEC. 15.10 EQUILIBRIUM CONSTANTS

where k_f is the rate constant of the forward reaction and k_r is the rate constant of the reverse reaction.[4] By the 1860s it had been postulated that the rates of certain simple reactions were proportional to the products of the concentrations of the reacting species. For the above reaction,

$$(\text{rate})_{\text{for}} = k_f[A][B] \qquad (\text{rate})_{\text{rev}} = k_r[C][D]$$

At equilibrium, the rate of the forward reaction must equal the rate of the reverse reaction, or

$$k_f[A][B] = k_r[C][D] \tag{15.26}$$

The equilibrium constant was written as

$$K = \frac{k_f}{k_r} = \frac{[C][D]}{[A][B]} \tag{15.27}$$

This is called the *law of mass action*.

Earlier in this chapter we applied this result to the calculation of the equilibrium constant for acetic acid. Now let's apply it in calculating the equilibrium constant of water, $H_2O \rightleftharpoons H^+ + OH^-$, using the material we have considered in this chapter. The conductivity of pure water, attributed to the ions H^+ and OH^-, is about 6×10^{-8} ohm^{-1} cm^{-1}.[5] The concentration of the H_2O is 55.5 mol liter^{-1}. The equivalent conductance of pure water is then

$$\Lambda = \frac{\kappa}{c} = \frac{6 \times 10^{-8}}{55.5 \times 10^{-3}}$$

$$= 1.08 \times 10^{-6} \text{ ohm}^{-1} \text{ cm}^2 \text{ mol}^{-1}$$

If the water were 100% ionized, the equivalent conductance would be

$$\Lambda_0 = \Lambda_0(H^+) + \Lambda_0(OH^-) = 350 + 198 = 548 \tag{15.28}$$

The degree of dissociation is then

$$\alpha = \frac{\Lambda}{\Lambda_0} = \frac{1.08 \times 10^{-6}}{548} = 2.0 \times 10^{-9}$$

The concentrations of the ions are

$$[H^+] = [OH^-] = \alpha c = 1.1 \times 10^{-7}$$

This procedure constitutes an accurate method for determining the well-known ion product for water,

$$K_w = [H^+][OH^-] = 10^{-14}$$

This procedure is not restricted to the determination of ion concentrations of weak electrolytes; it may also be used for determining the solubilities of sparingly soluble salts such as AgCl. The slight solubility is enough to raise the conductivity of the water. For the most precise results, the conductivity relative only to the ions Ag$^+$

---

[4] Note that this discussion assumes the simplest possible reaction mechanism; most reactions, however, do not proceed with the simplest mechanism (see Chapter 40).

[5] The preparation of pure "conductivity" water is a difficult and time-consuming task, since all traces of impurities must be removed. The water is repeatedly distilled and passed through ion-exchange columns with the greatest precautions. In this connection, see Problem 13.

PROBLEMS

1. A 3-V potential difference is applied to the terminals of an electrolytic cell that dip into an Ag^+ solution. A resistor is placed in series with the cell so that the total resistance in the circuit is 1000 ohms (Ω), and the current is turned on for 2 hr.
   (a) How many faradays of electricity pass through the circuit?
   (b) How many electrons pass through the circuit?
   (c) What is the weight of silver plated onto the cathode?
   (d) Suppose the cathode is a round serving platter that is to be plated to form part of a silver-plated tea service. The platter is 40 cm in diameter, and the density of silver is 10.5 g cm^{-3}. What is the thickness of the silver film on the platter?

2. A 5-g sample of a brass is dissolved, and the solution is made up to 1 liter. The copper in the solution is in the form Cu^{2+}. A 50-ml aliquot of the solution is placed in an electrolysis cell. The cathode of the cell is a platinum gauze that weighs 12.3694 g. When all the copper has been removed by electrolysis, the gauze weighs 12.5423 g. What is the composition of the brass? If the solution was electrolyzed with a constant current of 0.153 A, how long did it take for the copper to be removed from the solution?

3. Given the Faraday constant and the charge on an electron, how would you determine Avogadro's number? Use the constants given on the inside back cover, and see how these compare with your result.

4. A potential difference of 10.00 V is applied across a wire 30 cm long with a diameter of 1 mm. The current passing through the wire is 1.2234 A.
   (a) Calculate the resistance and the conductance of the wire.
   (b) Calculate the resistivity and the conductivity of the wire.

5. A conductivity cell has a resistance of 7.3274 Ω when it is filled with a solution of KCl of concentration 0.01 mol liter^{-1}. The electrodes of the cell are platinum discs with 6-cm diameters.
   (a) What would be the resistance of the cell if it were filled with $0.1M$ HCl solution?
   (b) Calculate the distance between the electrodes.
   (c) When the cell is filled with a solution of HCl, the resistance of the cell is measured as 3.6473 Ω. Calculate the conductivity of this solution.

6. Using the data in Table 15.1, plot the equivalent conductances for LiCl at the concentrations given in a manner suitable for extrapolating the data to zero concentration to calculate a value for Λ_0. Compare your result with the value given in the table.

7. Using Kohlrausch's law and the data from Table 15.1, calculate Λ_0 for NH_4OH. Compare this value with the result you would get from the data of Table 15.4.

8. At 25 °C the equivalent conductances of NH_4OH at a number of concentrations are as shown in the accompanying table. Using the value for Λ_0 computed in Problem 7, determine the degree of ionization of NH_4OH and the equilibrium constant at each concentration.

Concentration (mol liter^{-1})	Λ (ohm^{-1} cm^2 mol^{-1})
7.986×10^{-8}	259.8
1.187×10^{-6}	238.9
1.657×10^{-5}	189.1
6.095×10^{-5}	142.9
2.250×10^{-4}	71.14
7.724×10^{-4}	40.90
2.459×10^{-3}	24.04
2.632×10^{-2}	7.71
8.341×10^{-2}	4.52

9. The molal freezing-point depressions ($\Delta T/m$) for the chlorosubstituted acetic acids at a number of concentrations are shown in the accompanying table. Calculate the degree of ionization and determine the equilibrium constants at each listed concentration for each of the three compounds.

Monochloro-acetic acid		Dichloroacetic acid		Trichloro-acetic acid	
m	$\Delta T/m$	m	$\Delta T/m$	m	$\Delta T/m$
0.0890	1.98	0.00347	3.71	0.1543	3.43
0.1884	1.96	0.01755	3.47	0.2949	3.37
0.3382	1.89	0.1270	2.95	0.4282	3.389
0.5220	1.87	0.5345	2.50	0.5622	3.351
				0.9721	3.316
				1.4483	3.243

10. At 25 °C, the equivalent conductances of monochloroacetic acid, $C_2H_3O_2Cl$, are as shown in the

Concentration (mol liter^{-1})	Λ (ohm^{-1} cm^2 mol^{-1})
0.001814	225.32
.004048	174.67
.006230	149.74
.009464	127.93
.014113	108.72
.018824	96.54
.02444	86.51
0.03019	78.95

accompanying table. Calculate the degree of ionization at each concentration; also calculate the equilibrium constant for the dissociation of monochloroacetic acid. Compare with the answer you got in Problem 9, but be aware that the temperatures are different. (*Note:* For sodium monochloroacetate, $C_2H_2O_2ClNa$, the value $\Lambda_0 = 90.05$ ohm^{-1} cm^2 mol^{-1}.)

11. Suppose that freezing-point depressions can be measured to within 0.001 °C and that equivalent conductances can be measured to within 0.05 ohm^{-1} cm^2 mol^{-1}. What are the relative accuracies of an equilibrium constant determined by freezing-point depression and by conductivity measurements for monochloroacetic acid as determined in Problems 9 and 10?

12. For an aqueous solution saturated with silver halides at 25 °C, the specific conductances are 1.794×10^{-6}, 1×10^{-7}, and 4×10^{-9} ohm^{-1} cm^{-1} for AgCl, AgBr, and AgI. For the thallium halides, the numbers are 2.16×10^{-3}, 2.94×10^{-4}, and 3.66×10^{-5} for TlCl, TlBr, and TlI. Calculate the solubilities of each of these sparingly soluble salts using the data of Table 15.4 for the limiting molar ionic conductances. Calculate the solubility products.

13. The conductivity of pure water is 6×10^{-8} ohm^{-1} cm^{-1}. What concentration of HCl is required for the conductivity of the water to be double the conductivity of the pure "conductivity" water?

14. Using the data of Table 15.4, calculate the mobilities of the ions H^+, Li^+, K^+, OH^-, and Cl^- at infinite dilution.

15. A transference cell is filled with $0.1M$ HCl; the cathode and anode chambers each contain 30 ml. Calculate the concentration of HCl in the two compartments after 83.4 coulombs (C) of electricity pass through the cell.

16. A transference cell is filled with a solution of $CuSO_4$ that contains 0.2137 mol $CuSO_4$ per liter of solution. A current of 0.240 A is passed through the cell for 1000 s. The anode chamber, which has a volume of 30 cm^3, was analyzed afterward and found to contain 0.1983 mol $CuSO_4$ per liter. Calculate the transference number of the ion Cu^{2+}.

CHAPTER SIXTEEN

ACTIVITIES OF IONS. THE DEBYE-HÜCKEL LIMITING THEORY OF IONIC ACTIVITIES

16.1 STANDARD STATE FOR A UNI-UNIVALENT ELECTROLYTE

In the first section of Chapter 15 we noted the experimental impossibility of establishing a standard state for the electrolyte HCl from Henry's law in the form $f_B = km$. On the other hand, a modified Henry's law equation in the form

$$f_B = km^2 \tag{16.1}$$

proved adequate for extrapolating the fugacity at infinite dilution to establish a fugacity at some standard state. In fact, Equation (16.1) is a perfectly adequate starting point for considering activities of uni-univalent electrolyte solutions. That this is reasonable can be seen from the reaction

$$\text{HCl(g)} \rightleftharpoons \text{H}^+ + \text{Cl}^- \tag{16.2}$$

The equilibrium constant is

$$\frac{[\text{H}^+][\text{Cl}^-]}{P_{\text{HCl}}} = K = \frac{(m)(m)}{P_{\text{HCl}}} \tag{16.3}$$

since the pressure of the HCl is given by the fugacity

$$f_B = P_{\text{HCl}} = \frac{m^2}{K} = K'm^2 \tag{16.4}$$

which is the same form as (16.1). Now we can follow the analysis of Chapter 14, using (16.1) as the starting point.

The change in the chemical potential of the solute when the solution is taken from an initial state at concentration m' to a final state at concentration m is

$$\Delta \mu_B = RT \ln f'_B - RT \ln f_B$$
$$= RT \ln \frac{f'_B}{f_B} = RT \ln \frac{m'^2}{m^2} \tag{16.5}$$

As for nonelectrolyte solutions, we can define an activity for the uni-univalent electrolyte by

$$a_B = \frac{f_B}{f_B^\circ} \tag{16.6}$$

where f_B° is the fugacity in the standard state. Mole fraction units are rarely used for electrolyte solutions; hence we shall restrict our entire discussion to molality units. For nonelectrolyte solutions, the definition of activity was based on the limiting condition that in dilute solutions $a_B/m \to 1$ in the limit as $m \to 0$. For an electrolyte solution, this condition becomes

$$\frac{a_B}{m^2} \longrightarrow 1 \qquad \text{as } m \longrightarrow 0 \tag{16.7}$$

Note that we have not yet really taken into account that the electrolyte dissociates into two ions on solution. The forms of Equations (16.6) and (16.7) are based exclusively on the phenomenological form of Figures 15.1 and 15.2.

That the molality enters as the second power of m instead of the first power in (16.1) is because each molecule breaks up into a cation and an anion. We can take this into account by writing for (16.1)

$$f_B = k(m_+)(m_-) = km^2 \tag{16.8}$$

where $m_+ = m_- =$ the concentration of each ion. It stands to reason that at least in the limit of infinite dilution each ion would act independently. Thus, for example, the thermodynamic properties of the ion H^+ should be the same regardless of whether we have a solution of HCl or HBr. This is a logical consequence of Kohlrausch's law of independent ionic mobilities at infinite dilution. We can then speak of separate activities for the two ions a_+ and a_-. To keep the overall convention uniform, we should then have to define these by the conditions

$$\frac{a_+}{m_+} = \frac{a_+}{m} \longrightarrow 1 \quad \text{and} \quad \frac{a_-}{m_-} = \frac{a_-}{m} \longrightarrow 1 \qquad \text{as } m \longrightarrow 0 \tag{16.9}$$

In the limit of infinitely dilute solution,

$$f_B = k(a_+)(a_-) \tag{16.10}$$
and
$$a_B = (a_+)(a_-) \tag{16.11}$$

Note particularly that a_+ and a_- need not be equal; they are usually different. Nevertheless, it is convenient to refer to a so-called *mean activity*, which is defined as the geometrical mean of the two activities:

$$a_\pm = (a_+ a_-)^{1/2} \tag{16.12}$$

To carry the correspondence between nonelectrolyte and electrolyte solutions one step further, we define activity coefficients for the ions by

$$\gamma_+ = \frac{a_+}{m} \tag{16.13a}$$

$$\gamma_- = \frac{a_-}{m} \tag{16.13b}$$

These have the desired property that they approach unity in the limit of infinite dilution.

FIGURE 16.1 Establishing the standard state for a binary electrolyte. The dashed line is the ideal curve $f = km^2$, and the solid curve is the actual fugacity. The standard fugacity $f°$ is the ideal fugacity at unit molality. Since $a = f/f°$, the standard state of unit activity will be at the concentration that corresponds to the point on the actual curve at which $f = f°$. This point is generally not at unit molality.

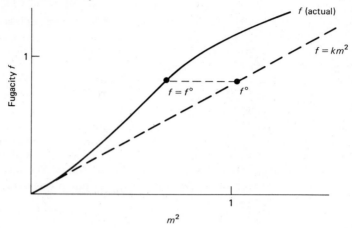

Activities can be uniquely measured for nonelectrolytic solutions. It is not possible to measure the activities of the ions in electrolytic solutions separately. This is why the mean activity is introduced, since that is what is actually measured in any experiment. We also use a *mean activity coefficient*, which can be determined from measurable quantities,

$$\gamma_\pm = [(\gamma_+)(\gamma_-)]^{1/2} = \left[\left(\frac{a_+}{m}\right)\left(\frac{a_-}{m}\right)\right]^{1/2}$$

$$= \frac{a_\pm}{m} \tag{16.14}$$

Certain aspects of this treatment of electrolytic solutions are quite similar to our previous treatment of nonelectrolytic solutions. An "ideal" ionic solution is still an infinitely dilute solution, but the form of the Henry's law equation is slightly modified as in (16.1). The standard state is again the state of unit activity. But instead of dealing with the activity of the solute as a whole, we must deal with the mean activities. Some relevant points on establishing the standard state are shown in Figure 16.1. Now let's see how the scheme works for more complicated electrolytes.

16.2 EXTENSION TO MULTIVALENT ELECTROLYTES

Symmetrical ions in which the cation and anion have the same valences, such as $Cu^{2+}SO_4^{2-}$, present no special problems; the standard states, activities, and activity coefficients of these are defined in precisely the same manner as for the uni-univalent ion. Plotting fugacity vs m^2 produces a curve with a limiting slope, as in Figure 16.1. The anion concentration equals the cation concentration, $m_+ = m_-$. For asymmetrical salts, the concentrations of the two ions are different, $m_+ \neq m_-$, and a somewhat different treatment is required.

SEC. 16.2 EXTENSION TO MULTIVALENT ELECTROLYTES

The simplest asymmetrical electrolyte is of the form A_2B, for example H_2SO_4 or Na_2SO_4. By following the reasoning of the previous section, we can define the activity by an equation equivalent to (16.11):

$$a_B = (a_+)(a_+)(a_-) = (a_+)^2(a_-) \tag{16.15}$$

A mean activity is again defined as the geometrical mean of the individual ion activities,

$$a_\pm = [(a_+)^2 a_-]^{1/3} \tag{16.16}$$

similar to the uni-univalent case. The activity coefficient contains a new factor, however. In the limit of infinite dilution, we require that the ionic activities approach the molalities, or

$$\left.\begin{array}{l} a_+ \longrightarrow m_+ = 2m \\ a_- \longrightarrow m_- = m \end{array}\right\} \text{ as } m \longrightarrow 0 \tag{16.17}$$

In this limit

$$a_\pm = [(2m)^2 m]^{1/3} = (4)^{1/3} m \tag{16.18}$$

We also require that the mean activity coefficient $\gamma_\pm$ should approach unity in this limit. This can be accomplished by setting it equal to

$$\gamma_\pm = \frac{a_\pm}{4^{1/3} m} \tag{16.19}$$

In the general case, the electrolytic formula is $A_{\nu_+} B_{\nu_-}$, where $\nu_+ \neq \nu_-$. The fugacity is proportional to km^ν, where

$$\nu = \nu_+ + \nu_- \tag{16.20}$$

The solute activity is

$$a_B = (a_+)^{\nu_+}(a_-)^{\nu_-} \tag{16.21}$$

and the mean activity is the geometrical mean, or

$$\begin{aligned} a_\pm &= (a_B)^{1/\nu} \\ &= [(a_+)^{\nu_+}(a_-)^{\nu_-}]^{1/\nu} \\ &= [(m_+\gamma_+)^{\nu_+}(m_-\gamma_-)^{\nu_-}]^{1/\nu} \end{aligned} \tag{16.22}$$

Remembering that $m_+ = \nu_+ m$ and $m_- = \nu_- m$, we can write

$$a_\pm = m[(\nu_+)^{\nu_+}(\nu_-)^{\nu_-}]^{1/\nu} \gamma_\pm \tag{16.23}$$

The mean activity coefficient $\gamma_\pm$ is

$$\gamma_\pm = [(\gamma_+)^{\nu_+}(\gamma_-)^{\nu_-}]^{1/\nu} \tag{16.24}$$

Another variable sometimes used is the *mean ionic molality*, which is defined by

$$\begin{aligned} m_\pm &= [(\nu_+ m)^{\nu_+}(\nu_- m)^{\nu_-}]^{1/\nu} \\ &= m[(\nu_+)^{\nu_+}(\nu_-)^{\nu_-}]^{1/\nu} \end{aligned} \tag{16.25}$$

Equation (16.23) can be written in terms of this variable as

$$a_\pm = m_\pm \gamma_\pm \tag{16.26}$$

16.3 IONIC STRENGTH

Suppose we have a solution of 1 mole of NaCl in 1000 g of water. There are no problems associated with determining the concentration of either Na^+ or Cl^-; each is $1.0m$. Now suppose we add 1 mole of $BaCl_2$ to the solution. Assuming 100% ionization, there is no way of associating any ion Cl^- with either the $BaCl_2$ or NaCl; they are simply chloride ions without regard to the original source. While the concentrations of Na^+ and Ba^{2+} are each $1.0m$, the concentration of Cl^- is $3.0m$.

The electrostatic forces between ions depend on the concentration and on the charges of the individual ions. The force between doubly charged ions is four times the force between singly charged ions for the same separation distance. The varying effects of ionic charge are conveniently taken into account by the *ionic strength*.

The ionic strength I of a solution is defined by

$$I = \tfrac{1}{2} \sum_i m_i z_i^2 \tag{16.27}$$

where the summation is over all the ions in the solution; m_i is the molality and z_i the charge on each of the ions. The ionic strength of a uni-univalent electrolyte is equal to the molality. For NaCl at concentration m, the ionic strength is

$$I = \tfrac{1}{2}[m(1)^2 + m(1)^2] = m$$

For $BaCl_2$, it is

$$I = \tfrac{1}{2}[m(2)^2 + (2m)(1)^2] = 3m$$

Solubilities are affected by the ionic strength of solutions. Consider a saturated aqueous solution of AB in equilibrium with excess solid AB. The equation for dissolution can be written

$$AB(\text{pure solid}) \rightleftharpoons A^+ + B^-$$

The equilibrium constant is

$$K = \frac{a_{AB}}{a_{\text{solid}}} = \frac{a_{AB}}{1} = a_+ a_-$$
$$= a_\pm^2 = m_\pm^2 \gamma_\pm^2 \tag{16.28}$$

At any particular temperature, $m_\pm$ has a fixed value. The solubility is a function of the ionic strength, however. In fact, the solubility varies with $(I)^{1/2}$. If AB is a sparingly soluble salt, then the solubility can be measured and extrapolated to the limit of zero ionic strength. Suppose we rewrite (16.28) in logarithmic form:

$$\ln \gamma_\pm = \ln \sqrt{K} - \ln m_\pm \tag{16.29}$$

In the limit as $I \to 0$, we have

$$\left. \begin{array}{r} \gamma_\pm \longrightarrow 1 \\ \ln \gamma_\pm \longrightarrow 0 \\ \ln \sqrt{K} \longrightarrow \ln m_\pm \end{array} \right\} \text{ as } I \longrightarrow 0$$

Figure 16.2 shows the logarithm of the solubility of AgCl as a function of $(I)^{1/2}$ in solutions of $NaNO_3$ and KNO_3. The solubility extrapolated to zero ionic strength yields the thermodynamic equilibrium constant K. When this value is inserted in

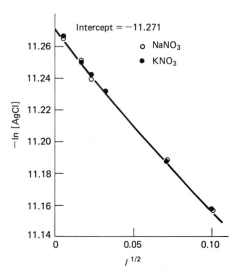

FIGURE 16.2 Determination of activity coefficients of AgCl by plotting the logarithm of the solubility vs the ionic strength.

Equation (16.29), we get an analytical expression for the activity coefficients of AgCl in the form

$$\ln \gamma_{\pm} = -11.271 - \ln m \qquad (16.30)$$

This technique provides an experimental method for measuring activity coefficients of strong electrolytes. Unfortunately, it can be applied only to sparingly soluble salts, since it is only for these salts that the solubility measurements can be carried out in the limit of very small ionic strengths. At larger concentrations, recourse must be had to other techniques such as electrochemical cells and freezing-point-depression measurements.

16.4 EXPERIMENTAL DETERMINATION OF ACTIVITY COEFFICIENTS

The experimental procedures for measuring activity coefficients of electrolytes parallel those for solutions of nonelectrolytes. The activities and activity coefficients are replaced by mean activities and mean-activity coefficients. The most direct approach is to measure the vapor pressure of the solute and compare this with the extrapolated Henry's law measurement. Unfortunately, in all but a few instances such as HCl and several other gaseous acids, such measurements are impossible; NaCl does not have a measurable vapor pressure over a salt solution.

We can, however, measure the vapor pressure of the solvent, and from these we can get values of the activity coefficients of the solvent. The activities of the solute can then be determined by applying the Gibbs-Duhem relation. These determinations are usually carried out in terms of the osmotic coefficients. (See Problem 14.15.)

For nonelectrolyte solutions the osmotic coefficient ϕ can be written as

$$\ln a_A = -\frac{Mm}{1000} \phi \qquad (16.31)$$

where a_A is the activity of the solvent, M is the molar mass of the solvent, and m is the molal concentration; a_A is determined from some property of the solvent, like the vapor pressure. For electrolytes, (16.31) takes the form

$$\ln a_A = -\frac{\nu M m}{1000}\phi \qquad (16.32)$$

where ν is the number of ions formed on dissociation. The Gibbs-Duhem equation can be written

$$d(\ln a_A) = -\frac{X_B}{X_A} d(\ln a_B) \qquad (16.33)$$

Since $X_B/X_A = m(1000/M)$, we can write (16.33) for an electrolyte as

$$\frac{1000}{M} d(\ln a_A) = -m\, d(\ln a_B)$$
$$= -\nu m\, d(\ln a_\pm)$$
$$= -\nu m\, d[\ln(\gamma_\pm m_\pm)] \qquad (16.34)$$

From Equation (16.25)

$$d(\ln m_\pm) = d(\ln m) \qquad (16.35)$$

TABLE 16.1 Osmotic coefficients and activity coefficients for NaCl and KCl in aqueous solutions at 25 °C.

Molality	NaCl		KCl	
	ϕ	$\gamma_\pm$	ϕ	$\gamma_\pm$
0.1	0.9342	0.7813	0.9264	0.7700
0.2	0.9255	.7376	.9131	.7188
0.3	0.9224	.7130	.9063	.6882
0.4	0.9217	.6966	.9023	.6668
0.5	0.9224	.6852	.9000	.6507
0.6	0.9242	.6767	.8987	.6380
0.7	0.9266	.6707	.8980	.6278
0.8	0.9295	.6662	.8980	.6192
0.9	0.9329	.6628	.8982	.6119
1.0	0.9363	.6605	.8985	.6056
1.2	0.9434	.6509	.8996	.5958
1.4	0.9509	.6575	.9008	.5868
1.6	0.9589	.6589	.9024	.5801
1.8	0.9681	.6621	.9048	.5749
2.0	0.9786	.6673	.9081	.5712
2.5	1.0096	.6864	.9194	.5665
3.0	1.0421	.7129	.9330	.5665
3.5	1.0783	.7459	.9478	.5698
4.0	1.1168	.7854	.9635	.5753
4.5	1.1578	.8316	0.9799	0.5828
5.0	1.2000	.8839		
5.5	1.2423	0.9423		
6.0	1.2861	1.0067		

SOURCE: G. Scatchard, W. J. Hamer, and S. E. Wood, *J. Amer. Chem. Soc.* 60 (1938):3061.

Now we can use (16.32) to write

$$\frac{1000}{M} d(\ln a_A) = -\nu m\, d[\ln (\gamma_\pm m)] = -\nu d(\phi m) \tag{16.36}$$

$$d(\ln \gamma_\pm) + d(\ln m) = \frac{1}{m}\phi\, dm + d\phi \tag{16.37}$$

or
$$d(\ln \gamma_\pm) = d\phi + (\phi - 1)\, d(\ln m) \tag{16.38}$$

Equation (16.38) can be integrated to yield an expression for the activity coefficient

$$\ln \gamma_\pm = \phi - 1 + \int_0^m (\phi - 1)\, d(\ln m) \tag{16.39}$$

This is the analog of the equation in Problem 14.15. Table 16.1 lists the osmotic coefficients and activity coefficients for NaCl and KCl as determined by Scatchard, Hamer, and Wood by the method of isotonic solutions.

The measurement of osmotic coefficients is not restricted to vapor-pressure measurements. Any colligative property of solutions can be a measure of the activity of the solvent. An expression for the osmotic coefficient in terms of osmotic pressure can be obtained by combining Equations (12.59) and (12.60), converting from mole fraction to molality, and substituting activities for fugacities, to get

$$\Pi = \frac{RT}{\overline{V}_A} \frac{\nu M}{1000} m\phi \tag{16.40}$$

The method of isotonic solutions (or the isopiestic method) is a relative method in which the activity of the sample is measured relative to a standard. The activities (or osmotic coefficients) of the standard solution must be determined absolutely. These are often obtained from freezing-point-depression or electrochemical-cell measurements.

PETER JOSEPH WILLIAM DEBYE (1884–1966), Dutch-born physical chemist, made extraordinary contributions to physical chemistry in various subject areas. He took his first degree in electrical engineering and received the Ph.D. degree in physics under Arnold Sommerfeld in Munich. At the age of 27, he succeeded Einstein as professor of theoretical physics at the University of Zurich. In a five-year period starting in 1911, Debye produced three important results—his theory of specific heats, his theory of permanent molecular dipole moments, and his theory of anomalous dielectric dispersion. In collaboration with Paul Scherrer, he developed the powder method of X-ray crystallography, an important tool for determining the structures of crystals. The Debye-Hückel theory of electrolytes, on which he collaborated with Erich Hückel, was published in 1923. Four of these five contributions are discussed in various sections of this book. In 1934, Debye moved to the University of Berlin and supervised the building of the Max Planck Institute of Physics. After the outbreak of World War II, refusing to become a German citizen, he came to the United States. Here he joined the chemistry department of Cornell University, serving as professor and department head until 1950, when he became an emeritus professor. Debye received the Nobel Prize in chemistry in 1936 "for his contributions to the study of molecular structure through his investigations on dipole moments and on the diffraction of X rays and electrons in gases."

Determining activity coefficients from freezing-point depressions is complicated because the measurements are made at the freezing-point temperature (about 0 °C for water) whereas the values required are for 25 °C. By combining the freezing-point-depression measurements with heat-capacity data, one can transform these activity-coefficient measurements from the lower to the higher standard temperature. We shall not discuss these in detail, and the interested student is referred to more advanced treatments for a complete discussion of this technique.[1]

The experimental values of activity coefficients show striking similarities between salts of the same valence types. These similarities suggest a theory of activity coefficients based on the coulombic attraction of the ions. Such a theory was first developed successfully by Debye and Hückel (1923).

16.5 INTERIONIC ATTRACTIONS

The force between two electric charges Q_1 and Q_2 separated by a distance r is

$$F = \frac{Q_1 Q_2}{\kappa r^2} \qquad (16.41)$$

where κ is the dielectric constant of the homogeneous medium separating the two charges. Figure 16.3 illustrates a charge distribution in a solution. We focus our attention on a central ion of charge $+ze$ and examine the so-called *ion atmosphere* surrounding our selected central ion.

We let the distance of closest approach of the ions be d, as indicated by the dashed circle. The central ion is surrounded by a nonuniform distribution of electric charge. Since opposite charges attract, one might expect the immediate vicinity of the central positive charge to contain negative charges exclusively. Thermal motion effects some degree of randomization of these charges, however. While on the average negative charges predominate in the neighborhood of the central ion, some positive charges are also present. At any distance the excess charge is given by the Boltzmann distribution. We consider a spherical shell about the central ion of differential volume $dV = 4\pi r^2\, dr$. The charge density ρ may vary from point to point. In any volume element, $\rho = Q/dV$, where Q is the total charge in dV. We now recall Poisson's

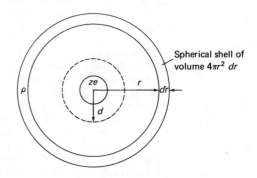

FIGURE 16.3 The ion atmosphere about a central ion of charge $+ze$. The distance of closest approach of other ions is indicated by d; ρ is the charge density in a differential spherical volume element $dV = 4\pi r^2\, dr$. The distance d is the apparent diameter of the ions. The charge on the central ion can be considered a point charge at its center.

---

[1] See for example I. Klotz, *Chemical Thermodynamics* (New York: Prentice-Hall, 1950), p. 318; and G. N. Lewis and M. Randall, *Thermodynamics*, 2d ed., rev. K. S. Pitzer and L. Brewer (New York: McGraw-Hill Book Co., 1961), p. 404.

SEC. 16.5 INTERIONIC ATTRACTIONS

> SIMÉON-DENIS POISSON (1781–1840), French mathematician, a younger contemporary of Fourier, made significant advances in the theory of Fourier series. His work concerned applying mathematics to numerous fields of physics, including electricity and magnetism and mechanics. The Poisson distribution in statistics can be widely applied in a variety of situations including radioactivity, traffic, and others. Poisson's name is associated with the Poisson ratio in stress analysis, the Poisson integral, Poisson's equation in electricity, the Poisson bracket, and Poisson's theorem in mechanics, besides the Poisson distribution, indicating the breadth of his activities.

equation of electrostatics, which relates the electric potential ψ to the charge density.[2]

$$\nabla^2 \psi = -\frac{4\pi}{\kappa}\rho \qquad (16.42)$$

where the symbol ∇^2 ("del squared") is the *Laplacian operator*[3]

$$\nabla^2 = \frac{\partial^2}{\partial x^2} + \frac{\partial^2}{\partial y^2} + \frac{\partial^2}{\partial z^2} \qquad (16.43)$$

Since the charge distributions are spherically symmetric, it is convenient to use the Laplacian in spherical polar coordinates. The distribution is independent of angle; hence the final form of the Poisson equation is a function only of r,

$$\frac{1}{r^2}\frac{d}{dr}\left(r^2 \frac{d\psi}{dr}\right) = -\frac{4\pi}{\kappa}\rho \qquad (16.44)$$

Now we need ρ as a function of ψ to solve this equation. We get the required functional dependence using the Boltzmann distribution.

We let N_j be the average number of ions per cubic centimeter. The energy of any individual ion at a potential ψ is

$$E = z_j e \psi \qquad (16.45)$$

where $z_j e$ is the charge on the ion. The term N_j is the uniform concentration that would exist in the absence of electric fields. The actual concentration N'_j at the potential ψ is determined by the Boltzmann distribution law,[4]

$$N'_j = N_j \exp\left(-\frac{z_j e \psi}{kT}\right) \qquad (16.46)$$

where k is the Boltzmann constant. The local charge density is obtained by summing over all ions,

$$\rho = \sum_j z_j e N_j \exp\left(-\frac{z_j e \psi}{kT}\right) \qquad (16.47)$$

---

[2] The Poisson equation is discussed in most textbooks dealing with electricity and magnetism. See for example N. H. Frank, *Introduction to Electricity and Magnetism* (New York: McGraw-Hill Book Co., 1950), p. 221.
[3] The Laplacian operator is discussed in detail in Chapter 26 et seq.
[4] The Boltzmann distribution law is discussed in Chapters 21 and 23.

In the limit of small concentrations, the ions are far apart, and the potential is small. In this limit, the exponential term of (16.47) can be expanded in a power series:

$$\rho = \sum_j z_j e N_j - \sum_j \frac{N_j z_j^2 e^2 \psi}{kT} + \sum_j \frac{N_j z_j^3 e^3 \psi^2}{2(kT)^2} + \cdots \quad (16.48)$$

The first term is zero, since the solution must be electrically neutral. We drop the third and higher terms, leaving only the second. (For binary salts like NaCl and ZnSO$_4$, all terms containing odd powers of z_j are identically zero.) The Poisson equation now becomes

$$\frac{1}{r^2}\frac{d}{dr}\left(r^2 \frac{d\psi}{dr}\right) = b^2 \psi \quad (16.49)$$

where

$$b^2 = \frac{4\pi e^2}{\kappa kT} \sum_j N_j z_j^2 \quad (16.50)$$

A solution to (16.49) can be written in the form

$$\psi = B\frac{\exp(br)}{r} + A\frac{\exp(-br)}{r} \quad (16.51)$$

The terms A and B are constants of integration, to be determined from the boundary conditions of the problem. The value of B must be zero if the potential is to remain finite at large r, leaving

$$\psi = \frac{A}{r}\exp(-br) \quad (16.52)$$

Equation (16.52) is valid for values of r greater than the apparent diameter of the ions ($r > d$). The potential gradient, or field strength, is the derivative of ψ with respect to r:

$$\mathbf{E} = -\nabla\psi = -\frac{d\psi}{dr} = \frac{A(1+br)}{r^2}\exp(-br) \quad (16.53)$$

At smaller distances ($r < d$) there are no other charges between any point and the selected ion of charge $z_i e$. Here the potential is due only to the selected charge and is given by the solution to (16.44) with $\rho = 0$. In this region

$$\mathbf{E} = -\frac{d\psi}{dr} = \frac{z_i e}{\kappa r^2} \qquad r < d \quad (16.54)$$

Since the potential gradient must be continuous, (16.53) and (16.54) must be equal at $r = d$. By equating the right-hand sides of the two equations at $r = d$, we get

$$A_i = \frac{z_i e}{\kappa}\left[\frac{\exp(bd)}{1+bd}\right] \quad (16.55)$$

The final expression for the potential about the selected ion is now

$$\psi(r) = \frac{z_i e}{\kappa}\left[\frac{\exp(bd)}{1+bd}\right]\frac{\exp(-br)}{r} \quad (16.56)$$

with

$$b = \frac{2\pi^{1/2} e}{(\kappa kT)^{1/2}} \sum_j (N_j z_j^2)^{1/2} \quad (16.57)$$

The constant b^{-1} has the dimensions of length and is known as the *Debye length*. It is an approximate measure of the effective range of the electrostatic field of an ion. The term in the summation is the ionic strength of the solution.

The final potential of Equation (16.56) can be written as the sum of two terms,

$$\psi(r) = \frac{z_i e}{\kappa r} - \frac{z_i e}{\kappa r}\left\{1 - \left[\frac{\exp(bd)}{1+bd}\right]\exp(-br)\right\} \tag{16.58}$$

The first term is the potential from the ion in a charge-free space and the second term arises from the contribution of the surrounding ion-atmosphere. Using this expression for the potential, we can calculate the work associated with building up a charge distribution to the value ρ at the potential ψ. This work can then be related to the free-energy change, which in turn can be related to the activities or activity coefficients.

16.6 THE DEBYE-HÜCKEL LIMITING LAW

The electrical work associated with building up a charge dQ to a potential ψ is $dw_{el} = \psi\, dQ$. If the total charge is ze, then the work is

$$w_{el} = \int_0^{ze} \psi\, dQ \tag{16.59}$$

Consider a process in which the ion A is transferred from a solution of concentration N_1 ions per cubic centimeter to a solution of N_2 ions per cubic centimeter. The process can be carried out reversibly in the following three-step manner:

1. Discharge the ions at concentration N_1 ($\Delta G_1 = -w_{el}$).
2. Transfer the now neutral species to the solution at concentration N_2 ($\Delta G_2 = RT \ln (N_2/N_1)$.
3. Recharge the ions at the final concentration N_2 ($\Delta G_3 = -w_{el}$).

The free-energy change for the overall process is

$$\Delta G = RT \ln \frac{N_2}{N_1} + RT \ln \frac{\gamma_2}{\gamma_1} \tag{16.60}$$

By comparing the free-energy change for the overall process given in Equation (16.60) with the free-energy steps for the three-step equivalent process, we can see that the term $RT \ln (N_2/N_1)$ is an ideal term, whereas the remaining term, $RT \ln (\gamma_2/\gamma_1)$ arises from nonideality associated with the electrical nature of the ionic solution. This nonideal term must equal the sum of the free-energy changes of steps 1 and 3 above.

Since no ion can approach the central ion closer than the distance $r = d$, the potential at this central ion because of the surrounding ion atmosphere can be got by substituting $r = d$ in (16.56):

$$\psi(d) = \frac{ze}{\kappa d(1+bd)} \tag{16.61}$$

Here we have dropped the subscript i from z, since we know that we are speaking of one particular ion type. The sum of the electrical work of steps 1 and 3 is

$$w_{el} = \int \psi \, dQ = N_0 \int_0^{ze} \psi_1(d) \, d(ze) - \int_0^{ze} \psi_2(d) \, d(ze)$$

$$= \frac{N_0 z^2 e^2}{2\kappa d} \left\{ \frac{1}{1 + b_1 d} - \frac{1}{1 + b_2 d} \right\} = RT \ln \frac{\gamma_1}{\gamma_2} \qquad (16.62)$$

where Avogadro's number N_0 has been introduced to place the electrical work on a molar basis.

The final expression for the activity coefficient can now be determined by setting the final state at concentration N_2 as the reference state. Since the reference state is the infinitely dilute solution, $N_2 \to 0$, $b_2 \to 0$, and $\gamma_2 \to 1$. We get finally

$$\ln \gamma = \frac{N_0 z^2 e^2}{2RT\kappa d} \left(\frac{1}{1 + bd} - 1 \right)$$

$$= -\frac{z^2 e^2}{2\kappa kT} \left(\frac{b}{1 + bd} \right) \qquad (16.63)$$

This equation provides a value for an individual ion-activity coefficient. Since individual ion-activity coefficients cannot be measured, a result in terms of the mean activity coefficient $\gamma_\pm$ would be of greater practical value. From Equation (16.24)

$$\nu \ln \gamma_\pm = (\nu_+ + \nu_-) \ln \gamma_\pm = \nu_+ \ln \gamma_+ + \nu_- \ln \gamma_- \qquad (16.64)$$

Electrical neutrality requires that $\nu_+ z_+ = \nu_- z_-$; thus we can write

$$\ln \gamma_\pm = -\left(\frac{\nu_+ z_+^2 + \nu_- x_-^2}{\nu_+ + \nu_-} \right) \frac{e^2}{2\kappa kT} \left(\frac{b}{1 + bd} \right) \qquad (16.65)$$

and finally

$$\ln \gamma_\pm = -|z_+ z_-| \frac{e^2}{2\kappa kT} \left(\frac{b}{1 + bd} \right) \qquad (16.66)$$

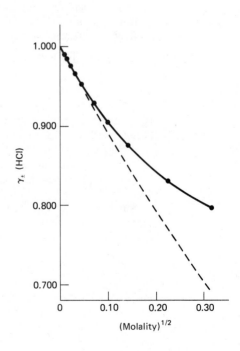

FIGURE 16.4 Mean activity coefficients of HCl vs the square root of the concentration. The dashed line indicates the Debye-Hückel limiting law behavior. (Data from S. Popoff and E. W. Neuman, *J. Phys. Chem.* **34** (1930):1853.)

SEC. 16.6 THE DEBYE-HÜCKEL LIMITING LAW

In the limit of infinite dilution, bd is very small compared with unity, and that term can be dropped from the denominator. When the value for b is then substituted from (16.57), and the numerical values used for all natural constants, we finally get

$$\ln \gamma_{\pm} = -1.172 \, |z_+z_-| I^{1/2} \qquad (16.67)$$

or as it is often written,

$$\log_{10} \gamma_{\pm} = -0.509 \, |z_+z_-| I^{1/2} \qquad (16.68)$$

where I is the previously defined ionic strength, and the temperature is 298 K. If the term bd in (16.66) is not omitted, then the equation takes the form

$$\log_{10} \gamma_{\pm} = -\frac{0.509|z_+z_-|I^{1/2}}{1 + \beta I^{1/2}} \qquad (16.69)$$

where β is a constant that depends on the distance of the closest approach of the ions, and it is generally near unity. Equation (16.69) is valid to somewhat higher concentrations than (16.68) is.

These expressions for the activity coefficients are valid only in the limit of infinite dilution. That they do agree with experimental values in that limit is indicated by the data displayed in Figure 16.4. For greater concentrations, this theory is of little practical use. Numerous attempts have been made to extend the accuracy of this theory to higher concentrations and also to explain various phenomena associated with conductivity studies. Since these are beyond the scope of our discussion, the interested student is referred to the more advanced treatments listed in the Bibliography.

TABLE 16.2 Mean activity coefficients of several types of electrolytes from the Debye-Hückel limiting law, Equation (16.68), at several concentrations.

Salt Type	Molality						
	0.001	0.002	0.005	0.01	0.02	0.05	0.1
A^+B^-	0.9636	0.9489	0.9205	0.8894	0.8473	0.7695	0.6903
$A^{2+}B_2^-$	.8795	.8340	.7504	.6663	.5632	.4034	.2770
$A^{3+}B_3^-$	.7616	.6803	.5439	.4226	.2958	.1458	.0656
$A^{2+}B^{2-}$	.7434	.6575	.5153	.3916	.2655	.1229	.0516
$A_2^{3+}B_3^{2-}$	0.4226	0.2958	0.1458	0.0656	0.0212	0.0023	0.0002

TABLE 16.3 Experimental mean molal activity coefficients for several electrolytes at several concentrations at 25 °C.

	0.0005	0.001	0.002	0.005	0.01	0.02	0.05	0.1	0.2	0.5	1	2	4
HCl	...	0.966	0.952	0.929	0.905	0.876	0.830	0.796	0.767	0.757	0.809	1.009	1.762
HBr	...	.966	.952	.930	.906	.879	.838	.805	.782	.789	.871	1.168	...
NaCl	...	.965	.952	.928	.903	.872	.822	.778	.735	.681	.657	.668	.783
KCl	...	.965	.952	.927	.902	.869	.816	.770	.718	.649	.604	.573	.577
H_2SO_4	0.885	.830	.757	.639	.544	.453	.340	.265	.209	.154	.130	.124	.171
$CdCl_2$	.880	.819	.743	.623	.524	.456	.304	.228	.163	.100	.066	.044	...
$ZnCl_2$	...	...	...	.789	.731	.667	.628	.575	.459	.394	.337	...	...
$PbCl_2$	.902	.859	.803	.704	.612	.497	...	...	...	...	...	...	...
$LaCl_3$	...	.790	.729	.636	.560	.483	.388	.325	.274	.266	.342	0.825	...
$CdSO_4$	.774	.697	...	.476	.383	...	.199	.150	.102	.061	.041	0.032	...
$ZnSO_4$	0.780	0.700	0.608	0.477	0.387	0.298	0.202	.150	.104	.063	.043	0.035	...
$Al_2(SO_4)_3$	...	...	...	...	...	...	...	0.035	0.023	0.014	0.018	...	...

SOURCE: B. E. Conway, *Electrochemical Data* (Amsterdam: Elsevier Publishing Co., 1952).

PROBLEMS

1. Using the mean molal activity coefficients of Table 16.3 for HCl, calculate the mean activity of HCl in solutions of hydrochloric acid at concentrations of 0.001, 0.005, 0.01, and 0.02m. Calculate the value of k in (16.1).

2. Calculate the change in chemical potential of HCl(aq) when a solution of HCl is taken from 0.001 to 0.01m.

3. Calculate the concentration of HCl corresponding to the standard state of HCl. That is to say, at what concentration is the activity of HCl unity? (*Note:* For those of you who dislike plotting points, the activity coefficient of HCl over the range 0.5–3m is very closely approximated by the expression

$$\gamma_\pm = 0.72515 + 0.029188m + 0.055959m^2)$$

4. Calculate the mean ionic molalities of HCl, CdCl$_2$, CdSO$_4$, Al$_2$(SO$_4$)$_3$, and LaCl$_3$ at a concentration of 0.25m. Assume that each is a strong electrolyte.

5. Calculate the ionic strengths of each of the solutions of Problem 4. A solution is prepared containing 0.25 mol of each of the salts of Problem 4 in 1000 g water; what is the ionic strength of this solution?

6. At sufficiently small concentrations, $\gamma_\pm$ of LaCl$_3$ is unity. Calculate $a_\pm$ of LaCl$_3$ in terms of the concentration for this limit.

7. At 25 °C the solubility of TlCl (TlCl(s) $\rightleftharpoons$ Tl$^+$ + Cl$^-$) in solutions of KNO$_3$ are as shown in the accompanying table (International Critical

KNO$_3$ (mol kg^{-1} H$_2$O)	TlCl (mol kg^{-1} H$_2$O)
0.0	0.01615
0.0201	.01725
0.0503	.01838
0.1008	.01977
0.3080	.02375
1.047	0.03217

Tables, V. 7, p. 321). Using the procedure outlined in Figure 16.2, calculate (a) the solubility product of TlCl and (b) the activity coefficient of TlCl at $I = 0.1$.

8. Table 16.1 lists the osmotic coefficients and activity coefficients for NaCl in aqueous solution. By graphical integration, determine the activity coefficients from the osmotic coefficients and verify the activity coefficient entries in the table.

9. Using the Debye-Hückel limiting law, calculate the activity coefficients of the salts AB, AB$_2$, and AB$_3$ at concentrations of 0.0001 and 0.0005m. How low would the concentrations have to be for the activity coefficients to deviate from unity by less than 1%?

10. Calculate the osmotic pressures of 0.1, 0.5, and 1.0m aqueous solutions of:
    (a) An ideal nonelectrolyte.
    (b) NaCl, assuming the activity coefficients are unity.
    (c) NaCl using the activity coefficients of Table 16.3.
    The temperature is 25 °C.

11. Calculate $\Delta G°$ for the reaction AgCl(s) = Ag$^+$ + Cl$^-$ at 25 °C.

12. Using the values for water for the various terms in (16.66), show that the constant term in (16.67) is indeed equal to 1.172.

13. Calculate the vapor pressure of H$_2$O over a 0.20m solution of NaCl at 25 °C relative to the vapor pressure of pure H$_2$O. What would the answer be if the solution were ideal?

14. The thermodynamic acid ionization constant of acetic acid (HAc) is $K_a = 1.8 \times 10^{-5}$ at 25 °C. Calculate the degree of dissociation of HAc in a 1.0m solution, assuming that (a) the solution is ideal (the activity coefficients are unity); (b) the mean ionic activity coefficients are given by the Debye-Hückel limiting law, and the activity coefficient of undissociated HAc is unity.

CHAPTER SEVENTEEN

ELECTROCHEMICAL CELLS

17.1 SPONTANEOUS REACTIONS AND USEFUL WORK

Any spontaneous reaction can produce useful work if it is carried out properly. Consider the spontaneous reaction

$$H_2 + \tfrac{1}{2}O_2 \rightleftharpoons H_2O \tag{17.1}$$

This reaction is accompanied by the release of substantial quantities of heat. The heat can be used to boil the water in the boiler of a steam turbine, thereby producing useful work. Not all spontaneous reactions can be adapted to produce useful work in this way.

Consider the spontaneous reactions

$$Zn + Cu^{2+} \rightleftharpoons Zn^{2+} + Cu \tag{17.2}$$

or

$$\tfrac{1}{2}Cu + Ag^+ \rightleftharpoons \tfrac{1}{2}Cu^{2+} + Ag \tag{17.3}$$

which take place in aqueous solution. You have all probably carried out reactions like this. When a piece of metallic zinc is dropped into a solution of $CuSO_4$, metallic copper is deposited, and the zinc dissolves; when a copper penny is dropped into a solution of $AgNO_3$, metallic silver is deposited on the surface of the penny, as the copper goes into solution.[1] The heat evolved in these reactions is hardly suited to the operation of a heat engine on any practical level. But since the reactions are spontaneous, it should be feasible to extract useful work from a system in which these reactions take place.

Extracting useful electrical work from these kinds of reactions entails carrying out the reactions in an *electrochemical cell,* or battery as it is commonly known. By properly arranging the experimental conditions, one can cause these reactions to take place in an almost *reversible manner.* Hence the work produced is reversible work; the electrical work produced by the reaction is then the maximum possible work. Recall that the free-energy change is a measure of the maximum useful work obtainable. It is a straightforward procedure to measure directly the free-energy change for a reaction by measuring the voltage developed.

Reactions such as (17.2) and (17.3) are called oxidation-reduction, or *redox,* reactions. As we have written (17.2), the zinc is oxidized, since its oxidation number is increased from 0 to +2, whereas the copper is reduced, its oxidation number decreasing from +2 to 0. Redox reactions can be written as the sum of two *half-cell*

---

[1] The order in which metal ions replace each other from solutions, the so-called electrochemical series, is often presented to high school students in the form of the mnemonic device MAZITL H. CHAPPA, where the letters stand for Mg, Al, Zn, iron, tin, lead, hydrogen, Cu, Hg, Ag, Pd, Pt, and Au.

reactions that indicate the transfer of electrons; thus the Zn–Cu reaction could be written as the sum of

$$Zn \rightleftharpoons Zn + 2e^-$$
$$Cu^{2+} + 2e^- \rightleftharpoons Cu$$
$$\overline{Zn + Cu^{2+} \rightleftharpoons Zn^{2+} + Cu}$$

When gasoline, or any other fuel, is burned in a bomb calorimeter, the reaction takes place in such a way that no useful work is performed; when the fuel is burned in the piston of an internal combustion engine, then we can get useful work. An analogous situation holds for redox reactions, such as the Zn–Cu reaction. When a piece of zinc metal is dropped into a solution of Cu^{2+}, all the charge transfer occurs in the solution itself, and no useful work is obtained. What we must do is devise a procedure whereby the electron transfer occurs through a circuit external to the solution. Then the electrons can perform useful work, say by passing through the windings of an electric motor.

17.2 SIMPLE ELECTROCHEMICAL CELLS

For the Zn–Cu redox reaction of Equation (17.2), 2 moles of electrons are transferred from the zinc to the copper for each mole of reaction. We must somehow arrange for this transfer to take place via a circuit external to the reaction vessel. Figure 17.1a depicts a simple electrochemical cell, the Daniell cell, in which a zinc electrode is dipped into a compartment containing $ZnSO_4$ at unit activity, and a copper electrode is dipped into a compartment containing $CuSO_4$ at unit activity. The two compartments are separated by a porous barrier that reduces mixing of the solutions yet allows electrical contact between them. This cell develops a potential difference of 1.099 V at 25 °C. We call this voltage a *standard voltage,* since all the constituents are in their standard states, and we write $\mathscr{E}° = 1.099$ V.

In this cell, charge is transferred via two different paths. Positive ion charges are transferred in the cell proper, while the negative electrons travel through the wires external to the cell. Conservation of charge requires that for each faraday of positive charge transferred by the zinc and copper ions, one faraday of negative charge is carried through the external circuit by the electrons. This electrochemical cell is represented by the schematic diagram

$$Zn \mid Zn^{2+}(a = 1) \mid Cu^{2+}(a = 1) \mid Cu$$

where the vertical lines indicate boundaries between different phases. The standard convention is to depict the cell in such a way that electrons travel from the left to the right in the external circuit. In Figure 17.1b we have depicted the electrochemical cell, using the Cu–Ag reaction of Equation (17.3), which can be represented by the diagram

$$Cu \mid Cu^{2+}(a = 1) \mid Ag^+(a = 1) \mid Ag$$

In this cell, the transfer of one mole of electrons is accompanied by the transfer of one mole of silver, but only one-half mole of copper; the standard voltage is $\mathscr{E}° = 0.462$ V. The voltage of any cell is a function of the concentration of the electrolytes. Hence by measuring cell voltage as a function of concentration, we can determine the activities of the ions in the solutions.

SEC. 17.2 SIMPLE ELECTROCHEMICAL CELLS

FIGURE 17.1 Typical electrochemical cells, showing the movement of the ions in the solution, and the movement of electrons in the external circuit. The term M indicates a meter or an electric motor. (a) The Zn–Cu cell, Zn | $Zn^{2+}(a_{Zn^{+2}})$ | $Cu^{2+}(a_{Cu^{+2}})$ | Cu.

$$Zn + Cu^{2+} \rightleftharpoons Zn^{2+} + Cu$$

(b) The Cu–Ag cell, Cu | $Cu^{2+}(a_{Cu^{+2}})$ | $Ag^+(a_{Ag^+})$ | Ag.

$$Cu + 2Ag^+ \rightleftharpoons Cu^{2+} + 2Ag$$

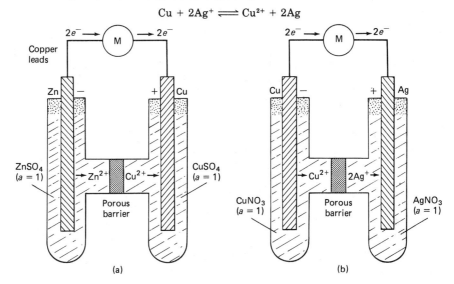

The simple cells depicted in Figure 17.1 suffer from the serious defect that the chemical species in each compartment can interdiffuse. The introduction of some Zn^{2+} into the Cu^{2+} compartment or vice versa affects the voltage and the reversibility of the reaction. These effects can be minimized by introducing a *salt bridge* to complete the electrical circuit between the two solutions, as depicted in Figure 17.2. The salt bridge is usually composed of a gel filled with a conducting salt (often KCl,

FIGURE 17.2 The cell of Figure 17.1a operating with a KCl salt bridge.

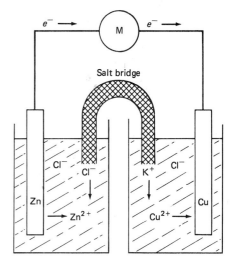

in which the current carriers between the two solutions are the ions K^+ and Cl^-). Mixing effects are negligible with this configuration. Later we shall see that there are cells that use a single-electrolyte solution, in which case this difficulty is completely eliminated.

17.3 REVERSIBLE CELL EMF'S AND FREE-ENERGY CHANGES

Consider the Zn–Cu electrochemical cell reaction. If the cell is short-circuited by bringing the external leads into direct contact, as shown in Figure 17.3a, then the reaction takes place spontaneously and irreversibly. The electrons pass through the wires at constant potential and do no work; charge is transferred in the solution by the ions as the reaction proceeds at a rate determined by the rate with which the ions can diffuse through the solution. Suppose that we now place a resistor in the external circuit, as shown in Figure 17.3b. If the resistance is small, then the reaction still proceeds rapidly and is partially irreversible.

As the resistance is made larger and larger, the reaction proceeds more and more slowly. In the limit as the resistance gets infinitely large, the reaction proceeds at an infinitesimally low rate and for all practical purposes can be considered reversible.[2] Suppose, for example, we use a resistance of 10^{12} Ω. The voltage of the cell is 1.099 V; the current is then 1.099×10^{-12} A, or 1.14×10^{-17} F s^{-1}. At that rate, it would take

FIGURE 17.3 The cell of Figure 17.1a, (a) short-circuited and (b) with a resistance R across the terminals.

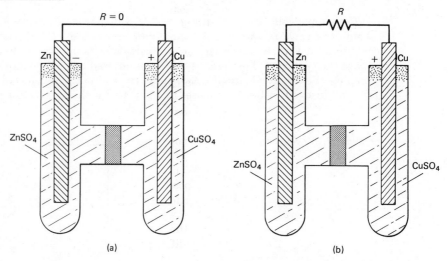

[2] This is a subtle point. That a process proceeds at an infinitesimally low rate is generally only a *necessary* but *not a sufficient* condition for reversibility. That this is not sufficient can be seen by considering the expansion of an ideal gas into a vacuum through an infinitesimally small pinhole. The process proceeds at an infinitesimally low rate, but it is still completely irreversible; no work is done, since there is no opposing force. The case of a battery discharging through a very large resistor is a special case in that a potential drop is automatically established in the resistor by the current flow, and work *is* done. One conceptual difficulty in using a large resistor to establish a reversible process is that the direction of the process cannot be reversed with just a resistor. In the next section, we shall discuss an arrangement whereby the direction can be reversed by using potentiometers and voltage sources in opposition to the battery.

5.57×10^9 yr for 1 mole of zinc to dissolve and 1 mole of copper to plate out. The electrical work done when a charge Q passes through a potential difference $\mathscr{E}$ is given by

$$w_{el} = Q\mathscr{E} \tag{17.4}$$

If the work is reversible, then the free energy change is $\Delta G = -w_{el}$, or

$$\Delta G = -Q\mathscr{E} \tag{17.5}$$

For one mole of reaction, $Q = nF$, where F is the faraday (or charge of a mole of electrons), and n is the number of electrons involved in the redox reaction ($n = 2$ for the Zn–Cu cell). The free-energy change for the reaction taking place in the electrochemical cell is then

$$\Delta G = -nF\mathscr{E} \tag{17.6}$$

Note very carefully that Equation (17.6) is valid only when the reaction is carried out reversibly and $\mathscr{E}$ is the reversible emf. We can accomplish this by measuring the voltage drop across the infinitely large resistor in series with the cell. This requires using a voltmeter with an infinitely high impedance. Traditionally, the emf's of cells are measured reversibly by placing a known opposing emf in series with the cell. In this way, the reaction can be made to go in either direction by adjusting the opposing voltage.

17.4 POTENTIOMETERS AND THE MEASUREMENT OF EMF

Figure 17.4a shows a simple *potentiometer* circuit. If a constant voltage V is applied across the resistor R, then a current $I = V/R$ flows across the resistor. The resistor is in the form of a long wire, called a *slide wire,* whose resistivity is constant along its entire length. Contact to the slide wire can be made at any point along its length, as indicated by the arrow at point C. If the voltage difference between A and B is called V_{AB}, then the voltage difference between A and C, or V_{AC}, can be adjusted to any value between 0 and V_{AB} by moving the contact along the slide wire. Slide wires are usually divided into 100, 1000, or some other number of convenient increments depending on the accuracy desired.

Figure 17.4b shows a simple potentiometer circuit for measuring emf's of unknown cells. The voltage source is supplied by a *standard cell,* often the *Weston* cell (to be described in the next section). The unknown cell is placed in the circuit so that the voltages of the two cells oppose each other. The contact is then moved along the slide wire until there is no deflection of the galvanometer. At this point the voltage of the unknown cell equals V_{AC}, which can be determined from the calibrated slide wire and the known voltage of the standard cell. (If the unknown cell has lower voltage than the standard cell, an additional resistor must be placed in the circuit, as shown in the figure.)

The circuit of Figure 17.4b has the disadvantage that current is drawn from the standard cell; this affects the voltage and the life of the cell. A convenient circuit that eliminates this difficulty is shown in Figure 17.4c. Here the *working* cell is an ordinary cheap battery. With the double-pole–double-throw switch (DPDT) in the standard cell position, the slide wire is placed in a position corresponding to the voltage of the standard cell; then the rheostat is adjusted until no deflection is observed on the galvanometer. At this point, V_{AC} along the slide wire just counterbalances the emf of the standard cell. The DPDT switch is then placed in the unknown cell posi-

FIGURE 17.4 Measuring cell emf's. (a) Simple potentiometer. (b) Simple circuit to measure cell emf. (c) Circuit for laboratory measurement of cell emf.

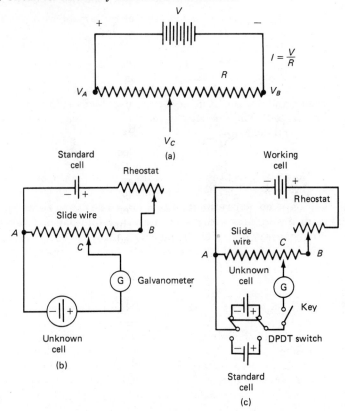

tion, and the slide wire contact is readjusted to a new position C', until no deflection of the galvanometer is observed. The emf is determined from the slide-wire scale.

For the highest accuracy, negligibly small currents must be drawn. The key is placed in series with the cells so that current is drawn only when galvanometer deflections are to be observed. The working cell voltage is not constant over long periods but does remain sufficiently constant over the length of time required for the measurement. Good experimental technique calls for rechecking the working cell against the standard cell at the end of the measurement to ensure that the balance point has not shifted. This type of circuit is still used for accurate cell emf measurements. Circuit elements are available with precisions of one part per million. Certain of the elements such as the constant voltage source are often replaced by electronic standards.

17.5 VARIOUS TYPES OF ELECTROCHEMICAL CELLS

Electrochemical cells have numerous practical uses. The Weston standard cell illustrated in Figure 17.5 uses a single electrolytic solution, eliminating the need for a salt bridge. The reaction is

$$\text{Cd(Hg)} + \text{Hg}_2\text{SO}_4(s) + \tfrac{8}{3}\text{H}_2\text{O} \rightleftharpoons \text{CdSO}_4 \cdot \tfrac{8}{3}\text{H}_2\text{O}(s) + \text{Hg}(l)$$

where Cd(Hg) indicates a cadmium amalgam (in this case 12.5% Hg by weight). The voltage is very stable over a range of temperature and is given by

$$\mathscr{E} = 1.01845 - 4.1 \times 10^{-5}(\theta - 20) - 9.5 \times 10^{-7}(\theta - 20)^2 \text{ volts} \quad (17.7)$$

where θ is the temperature in Celsius degrees. For the highest precision, standard cells are calibrated against standards at the National Bureau of Standards. All substances are not in their standard states; hence we write $\mathscr{E}$ and not $\mathscr{E}°$. The cell diagram for this normal, or *saturated*, Weston cell is

$$(\text{Pt})\text{Cd(Hg)} \mid \text{CdSO}_4 \cdot \tfrac{8}{3}\text{H}_2\text{O} \mid \text{Cd}^{2+}\text{SO}_4^{2-} \mid \text{Hg}_2\text{SO}_4 \mid \text{Hg(Pt)}$$

where (Pt) indicates the platinum electrodes (often omitted from the cell diagram). In the *unsaturated* Weston cell, the $CdSO_4$ solution is made saturated at 4 °C, and the solution is slightly unsaturated at the operating temperatures. The unsaturated Weston cell has a potential of 1.01904 V at 25 °C.

Electrochemical cells are convenient and practical sources of portable electrical energy. A *storage* battery can be recharged by an external source of electrical energy. The most common type is the lead–sulfuric acid cell, for which the overall reaction is

$$\text{PbO}_2(s) + \text{Pb}(s) + 2\text{H}_2\text{SO}_4 \rightleftharpoons 2\text{PbSO}_4(s) + 2\text{H}_2\text{O}$$

This reaction can be written as the sum of the two half-cell reactions:

Anode: $\quad \text{Pb} + \text{H}_2\text{SO}_4 \rightleftharpoons \text{PbSO}_4(s) + 2\text{H}^+ + 2e^-$
Cathode: $\text{PbO}_2 + \text{H}_2\text{SO}_4 + 2\text{H}^+ + 2e^- \rightleftharpoons \text{PbSO}_4(s) + 2\text{H}_2\text{O}$

The voltage is about 2 V. An automobile battery consists of six such cells in series, providing a total voltage of 12 V. As the battery is run down, there is a decrease in

FIGURE 17.5 The saturated standard Weston cell.

$(\text{Pt}) \text{Cd(Hg)} \mid \text{CdSO}_4 \cdot \tfrac{8}{3}\text{H}_2\text{O} \mid \text{Cd}^{2+}\text{SO}_4^{2-} \mid \text{Hg}_2\text{SO}_4 \mid \text{Hg (Pt)}$

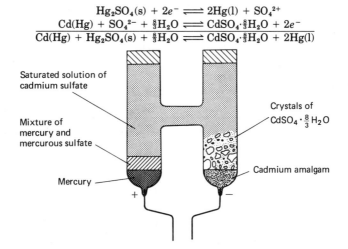

H_2SO_4 concentration, which is conveniently measured by determining the specific gravity of the solution with a battery-testing hydrometer. To recharge, a reverse voltage is applied, causing the reaction to proceed in the reverse direction.

The rechargeable battery used in portable electric tools, toothbrushes, and other devices is the nickel-cadmium cell, consisting of nickel and cadmium electrodes and

FIGURE 17.6 (a) The H_2–Cl_2 electrochemical cell, (Pt) H_2 | H^+Cl^- | Cl_2 (Pt).

$$\frac{\frac{1}{2}Cl_2 + e^- \rightleftharpoons Cl^-}{\frac{1}{2}H_2 \rightleftharpoons H^+ + e^-}$$
$$\frac{1}{2}H_2 + \frac{1}{2}Cl_2 \rightleftharpoons H^+Cl^-$$

(b) The electrochemical cell formed by combining the hydrogen electrode with the Ag–AgCl electrode, (Pt) H_2 | HCl | AgCl | Ag.

$$\frac{\frac{1}{2}H_2 \rightleftharpoons H^+ + e^-}{AgCl + e^- \rightleftharpoons Ag + Cl^-}$$
$$AgCl + \frac{1}{2}H_2 \rightleftharpoons H^+ + Cl^- + Ag$$

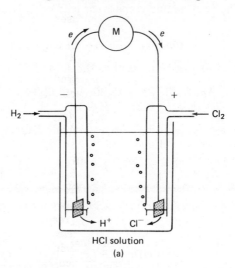

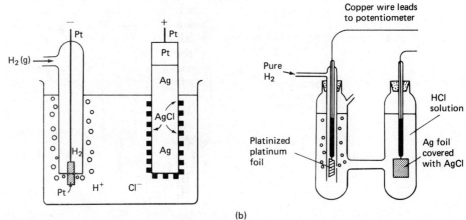

(b)

an alkali electrolyte such as KOH. The half-cell reactions are:

Anode: $\quad Cd + 2OH^- \rightleftharpoons Cd(OH)_2 + 2e^-$
Cathode: $\quad Ni(OH)_3 + e^- \rightleftharpoons Ni(OH)_2 + OH^-$

The overall reaction is

$$Cd + 2Ni(OH)_3 \rightleftharpoons Cd(OH)_2 + 2Ni(OH)_2$$

The familiar $1\tfrac{1}{2}$-V dry-cell battery used in flashlights has a carbon electrode surrounded by MnO_2 and graphite in a paste of $ZnCl_2$ and NH_4Cl, with an outer zinc electrode. The overall reaction is

$$2MnO_2 + 2NH_4^+ + Zn \rightleftharpoons Zn(NH_3)_2^{2+} + Mn_2O_3 + H_2O$$

The cell is represented by the diagram

$$Zn \mid Zn(NH_3)_2^{2+} \mid Mn_2O_3 \mid MnO_2 \mid C$$

This cell is not rechargeable; it is discarded when its voltage drops to low values. The half-cell reactions of this and many other cells discussed here are incompletely understood.

Figure 17.6a shows a cell that uses a single electrolyte with gaseous reactants. The reaction is

$$\tfrac{1}{2}H_2(g) + \tfrac{1}{2}Cl_2(g) \rightleftharpoons H^+ + Cl^-$$

Figure 17.6b shows a hydrogen electrode in conjunction with a silver–silver chloride electrode, for which the overall reaction is

$$AgCl + \tfrac{1}{2}H_2(g) \rightleftharpoons H^+ + Cl^- + Ag$$

17.6 HALF-CELLS AND ELECTRODE POTENTIALS

Suppose we remove the salt bridge from the Zn–Cu cell of Figure 17.2. We no longer have an electrochemical cell. What we have are two *half-cells*, or *electrodes*. The working electrochemical cell can be reestablished by replacing the salt bridge. Figure 17.7a depicts a series of *metal electrodes*, or half-cells. A complete electrochemical cell can be constructed by placing a salt bridge between any two electrodes; each cell is characterized by its particular voltage. This procedure is not restricted to metal electrodes. In Figure 17.7b, we have split the H_2–Cl_2 cell of Figure 17.6a into two parts, each of which constitutes a half-cell. Electrochemical cells can be constructed using any combination of half-cells. The *half-cell reactions* that take place in the separate electrodes are also shown in Figure 17.7. By convention, half-cell reactions are written as *reduction* reactions. Metal electrodes are often used in the form of amalgams. This is necessary for active metals such as sodium, which react directly with water.

The overall reaction that takes place in an electrochemical cell formed by combining any two half-cells is determined by adding the two half-cell reactions. For the cell formed by combining the Zn–Zn^{2+} and hydrogen electrodes,

$$\begin{aligned} Zn^{2+} + 2e^- &\rightleftharpoons Zn \\ H_2 &\rightleftharpoons 2H^+ + 2e^- \\ \hline Zn^{2+} + H_2 &\rightleftharpoons Zn + 2H^+ \end{aligned} \qquad (17.8)$$

When all substances are in the standard state, this cell has a potential of -0.763 V.

FIGURE 17.7 Some common electrodes, or half-cells. (a) Simple metal electrodes. (b) Gas electrodes. (c) The calomel electrode, which is often used as a standard.

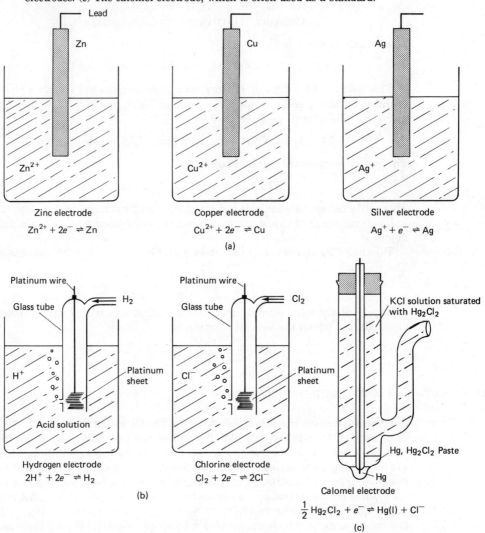

Suppose we wish to tabulate the standard emf's of all the possible electrochemical cells that can be constructed from a set of 50 half-cells. The total number of entries would be $50 \times 49 = 2450$. A much simpler procedure has been devised requiring only 50 entries, one for each half-cell.

When the voltage of an electrochemical cell is measured, we effectively measure the potential difference between the electrodes of each of the half-cells. An "absolute" potential for a half-cell electrode is a meaningless concept; electrostatics is concerned only with potential *differences*. The reference point is purely arbitrary. A table of *standard electrode potentials* is established by arbitrarily setting the "potential" of the hydrogen electrode

$$(\text{Pt}) \mid \text{H}_2(1 \text{ atm}) \mid \text{H}^+(a = 1) \qquad (25 \,°\text{C})$$

equal to zero when all substances are in their standard states of unit activity. The potentials of other electrodes are then established by combining them into electrochemical cells and measuring the potential of the cells. For the cell of Equation (17.8),[3]

$$(Pt) \mid H_2(1 \text{ atm}) \mid H^+(a = 1) \mid Zn^{2+} \mid Zn$$

the measured potential difference is -0.763 V for the cell as written, and

$$\mathscr{E} = -0.763 = \mathscr{E}_{Zn} - \mathscr{E}_{H_2} \qquad (17.9)$$

where $\mathscr{E}_{H_2}$ and $\mathscr{E}_{Zn}$ are the potentials of the hydrogen and zinc electrodes. Since $\mathscr{E}_{H_2}$ is zero by our convention, $\mathscr{E}_{Zn} = -0.763$. The negative sign indicates that Zn^{2+} is not reduced by hydrogen gas; on the contrary, zinc metal will reduce the ion H^+. Note that in our convention, we subtract the emf of the left half-cell from the emf of the right:

$$\mathscr{E} = \mathscr{E}_{\text{right}} - \mathscr{E}_{\text{left}}$$

Standard electrode potentials are listed in Table 17.1. Let us calculate, for example, the potential of the Zn–Cu cell in which all substances are in their standard states at 25 °C:

$$Zn \mid Zn^{2+}(a = 1) \mid Cu^{2+}(a = 1) \mid Cu \qquad (25 \text{ °C})$$

Using the entries of Table 17.1, we find that $\mathscr{E}° = 0.337 - (-0.763) = +1.100$ V.

17.7 THE NERNST EQUATION

We can write, in an analogy to (10.5),

$$\mu_i - \mu_i° = RT \ln a_i \qquad (17.10)$$

where a_i is the activity of the ith species in the reaction $a\text{A} + b\text{B} = c\text{C} + d\text{D}$. Equation (17.10) applies to both gaseous and condensed phase reactions. By following a procedure similar to that of Chapter 10, we can derive the analog of (10.10) in the form

$$\Delta G = \Delta G° + RT \ln \left(\frac{a_C^c a_D^d}{a_A^a a_B^b} \right) = \Delta G° + RT \ln Q_a \qquad (17.11)$$

In Equation (17.11) we have written Q instead of K for the ratio of activities, since the ratio is an equilibrium constant only at equilibrium. Equation (17.11) applies to any equation with any type of reacting species. When all the species are in their standard states, Q_a is unity, and $\Delta G = \Delta G°$; at equilibrium, $Q_a = K_a$.

Since $\Delta G = -nF\mathscr{E}$ we can write (17.11) in terms of emf's as

$$\mathscr{E} = \mathscr{E}° - \frac{RT}{nF} \ln Q_a \qquad (17.12)$$

Equation (17.12) is called the *Nernst equation*. When all species are in their standard states, then $a_i = 1$, and $\mathscr{E} = \mathscr{E}°$. On the other hand, at equilibrium ΔG and $\mathscr{E}$ are both zero. A battery at equilibrium is a "dead" battery, $\mathscr{E} = 0$, and

$$\mathscr{E}° = \frac{RT}{nF} \ln K_a \qquad (17.13)$$

---

[3] Present convention is *always* to put the hydrogen electrode on the left, as we have done here.

TABLE 17.1 Standard electrode potentials in aqueous solutions at 25 °C.

Electrode	Electrode reaction	$\mathscr{E}°$ (V)
ACID SOLUTIONS		
Li|Li$^+$	Li$^+$ + e^- ⇌ Li	−3.045
K|K$^+$	K$^+$ + e^- ⇌ K	−2.925
Ba|Ba^{2+}	Ba^{2+} + 2e^- ⇌ Ba	−2.906
Ca|Ca^{2+}	Ca^{2+} + 2e^- ⇌ Ca	−2.87
Na|Na$^+$	Na$^+$ + e^- ⇌ Na	−2.714
La|La^{3+}	La^{3+} + 3e^- ⇌ La	−2.52
Mg|Mg^{2+}	Mg^{2+} + 2e^- ⇌ Mg	−2.363
Th|Th^{4+}	Th^{4+} + 4e^- ⇌ Th	−1.90
U|U^{3+}	U^{3+} + 3e^- ⇌ U	−1.80
Al|Al^{3+}	Al^{3+} + 3e^- ⇌ Al	−1.66
Mn|Mn^{2+}	Mn^{2+} + 2e^- ⇌ Mn	−1.180
V|V^{2+}	V^{2+} + 2e^- ⇌ V	−1.18
Zn|Zn^{2+}	Zn^{2+} + 2e^- ⇌ Zn	−0.763
Tl|TlI|I$^-$	TlI(s) + e^- ⇌ Tl + I$^-$	−0.753
Cr|Cr^{3+}	Cr^{3+} + 3e^- ⇌ Cr	−0.744
Tl|TlBr|Br$^-$	TlBr(s) + e^- ⇌ Tl + Br$^-$	−0.658
Pt|U^{3+},U^{4+}	U^{4+} + e^- ⇌ U^{3+}	−0.61
Fe|Fe^{2+}	Fe^{2+} + 2e^- ⇌ Fe	−0.440
Cd|Cd^{2+}	Cd^{2+} + 2e^- ⇌ Cd	−0.403
Pb|PbSO$_4$|SO$_4^{2-}$	PbSO$_4$ + 2e^- ⇌ Pb + SO$_4^{2-}$	−0.359
Tl|Tl$^+$	Tl$^+$ + e^- ⇌ Tl	−0.3363
Ag|AgI|I$^-$	AgI + e^- ⇌ Ag + I$^-$	−0.152
Pb|Pb^{2+}	Pb^{2+} + 2e^- ⇌ Pb	−0.126
Pt|D$_2$|D$^+$	2D$^+$ + 2e^- ⇌ D$_2$	−0.0034
Pt|H$_2$|H$^+$	2H$^+$ + 2e^- ⇌ H$_2$	0.0000
Ag|AgBr|Br$^-$	AgBr + e^- ⇌ Ag + Br$^-$	+0.071
Ag|AgCl|Cl$^-$	AgCl + e^- ⇌ Ag + Cl$^-$	+0.2225
Pt|Hg|Hg$_2$Cl$_2$|Cl$^-$	Hg$_2$Cl$_2$ + 2e^- ⇌ 2Cl$^-$ + 2Hg(l)	+0.2676
Cu|Cu^{2+}	Cu^{2+} + 2e^- ⇌ Cu	+0.337
Pt|I$_2$|I$^-$	I$_3^-$ + 2e^- ⇌ 3I$^-$	+0.536
Pt|O$_2$|H$_2$O$_2$	O$_2$ + 2H$^+$ + 2e^- ⇌ H$_2$O$_2$	+0.682
Pt|Fe^{2+},Fe^{3+}	Fe^{3+} + e^- ⇌ Fe^{2+}	+0.771
Ag|Ag$^+$	Ag$^+$ + e^- ⇌ Ag	+0.7991
Au|AuCl$_4^-$,Cl$^-$	AuCl$_4^-$ + 3e^- ⇌ Au + 4Cl$^-$	+1.00
Pt|Br$_2$|Br$^-$	Br$_2$ + 2e^- ⇌ 2Br$^-$	+1.065
Pt|Tl$^+$,Tl^{3+}	Tl^{3+} + 2e^- ⇌ Tl$^-$	+1.25
Pt|H$^+$,Cr$_2$O$_7^{2-}$,Cr^{3+}	Cr$_2$O$_7^{2-}$ + 14H$^+$ + 6e^- ⇌ 2Cr^{3+} + 7H$_2$O	+1.33
Pt|Cl$_2$|Cl$^-$	Cl$_2$ + 2e^- ⇌ 2Cl$^-$	+1.3595
Pt|Ce^{4+},Ce^{3+}	Ce^{4+} + e^- ⇌ Ce^{3+}	+1.45
Au|Au^{3+}	Au^{3+} + 3e^- ⇌ Au	+1.50
Pt|Mn^{2+},MnO$_4^-$	MnO$_4^-$ + 8H$^+$ + 5e^- ⇌ Mn^{2+} + 4H$_2$O	+1.51
Au|Au$^+$	Au$^+$ + e^- ⇌ Au	+1.68
PbSO$_4$|PbO$_2$|H$_2$SO$_4$	PbO$_2$ + SO$_4$ + 4H$^+$ + 2e^- ⇌ PbSO$_4$ + 2H$_2$O	+1.685
Pt|F$_2$|F$^-$	F$_2$(g) + 2e^- ⇌ 2F$^-$	+2.87
BASIC SOLUTIONS		
Pt|SO$_3^{2-}$,SO$_4^{2-}$	SO$_4^{2-}$ + H$_2$O + 2e^- ⇌ SO$_3^{2-}$ + 2OH$^-$	−0.93
Pt|H$_2$|OH$^-$	2H$_2$O + 2e^- ⇌ H$_2$ + 2OH$^-$	−0.828
Ag|Ag(NH$_3$)$_2^+$,NH$_3$(aq)	Ag(NH$_3$)$_2^+$ + e^- ⇌ Ag + 2NH$_3$(aq)	+0.373
Pt|O$_2$|OH$^-$	O$_2$ + 2H$_2$O + 4e^- ⇌ 4OH$^-$	+0.401
Pt|MnO$_2$|MnO$_4^-$	MnO$_4^-$ + 2H$_2$O + 3e^- ⇌ MnO$_2$ + 4OH$^-$	+0.588

NOTE: A more complete tabulation can be found in W. M. Latimer, *The Oxidation States of the Elements and Their Potentials in Aqueous Solutions*, 2d ed. (New York: Prentice-Hall, 1952).

Equation (17.13) provides a convenient experimental basis for determining equilibrium constants. Suppose we want, for example, to calculate the equilibrium constant for the reaction

$$Zn + Cu^{2+} \rightleftharpoons Zn^{2+} + Cu$$

The standard voltage is $\mathscr{E}° = +1.100$ V; hence the equilibrium constant is given by

$$\ln K = \frac{nF\mathscr{E}°}{RT} = \frac{(2)(96{,}500)(1.100)}{(8.314)(298)} = 86.7$$

or

$$K = \frac{[Zn^{2+}]}{[Cu^{2+}]} = 1.6 \times 10^{37}$$

A Zn–Cu battery has an initial voltage of 1.1 V if it is constructed with all substances at unit activity. As current is drawn from the cell, the voltage drops, and the concentrations change in accord with the Nernst equation. At equilibrium the voltage is zero; the concentration of the ion Cu^{2+} is so low that the ratio $[Zn^{2+}]/[Cu^{2+}] = 1.6 \times 10^{37}$.

Suppose that we calculate the voltage of the cell

$$Zn \mid Zn^{2+}(a = 1) \mid Cu^{2+}(a = 0.01) \mid Cu$$

as an example of an intermediate case. Here

$$\mathscr{E} = \mathscr{E}° - \frac{RT}{nF} \ln\left(\frac{1}{0.01}\right)$$

$$= 1.100 - \frac{8.317 \times 298}{2 \times 96{,}500} \ln(100)$$

$$= 1.100 - 0.059$$

$$= 1.041 \text{ V}$$

In dealing with the Nernst equation, it is convenient to note that $RT/F = 0.0257$ at 298 K. If logarithms to the base 10 are used, the equation takes the form

$$\mathscr{E} = \mathscr{E}° - \frac{0.0592}{n} \log_{10}(Q) \tag{17.14}$$

17.8 EXPERIMENTAL DETERMINATION OF STANDARD ELECTRODE POTENTIALS AND ACTIVITY COEFFICIENTS

We still have not shown how the values of $\mathscr{E}°$ can be determined. If we write the Q ratio for the reaction $aA + bB \rightleftharpoons cC + dD$ as the product of two factors

$$Q_a = \frac{a_C^c a_D^d}{a_A^a a_B^b} = \left(\frac{m_C^c m_D^d}{m_A^a m_B^b}\right)\left(\frac{\gamma_C^c \gamma_D^d}{\gamma_A^a \gamma_B^b}\right) = Q_m Q_\gamma \tag{17.15}$$

then the Nernst equation can be written

$$\mathscr{E} = \mathscr{E}° - \frac{RT}{nF} \ln Q_m - \frac{RT}{nF} \ln Q_\gamma$$

or

$$\mathscr{E} + \frac{RT}{nF} \ln Q_m = \mathscr{E}° - \frac{RT}{nF} \ln Q_\gamma \tag{17.16}$$

The value Q_m can be determined from the concentrations of the species; $\mathscr{E}$ can be determined by measuring the emf of the cell. In the limit as the concentrations get infinitely dilute, the activity coefficients approach unity, and the last term in (17.16) approaches zero, since $\ln(1) = 0$. The procedure is best illustrated by an example.

Suppose that we consider the cell of Figure 17.6b, for which the reaction is

$$\text{AgCl} + \tfrac{1}{2}\text{H}_2 \rightleftharpoons \text{H}^+ + \text{Cl}^- + \text{Ag}$$

Both Ag and AgCl are solids; hence their activities are unity. If we take hydrogen to be an ideal gas at unit pressure, its activity is also unity. Equation (17.16) then takes the form

$$\mathscr{E} + \frac{RT}{F} \ln m^2 = \mathscr{E}^\circ - \frac{RT}{F} \ln \gamma_\pm^2$$

$$\mathscr{E} + \frac{2RT}{F} \ln m = \mathscr{E}^\circ - \frac{2RT}{F} \ln \gamma_\pm \qquad (17.17)$$

Now according to the Debye-Hückel theory of dilute electrolytes, $\ln \gamma_\pm = Bm^{1/2}$, where B is a constant, and (17.17) becomes

$$\mathscr{E} + \frac{2RT}{F} \ln m = \mathscr{E}^\circ - \frac{2RTB}{F} m^{1/2} \qquad (17.18)$$

If we now plot the left side of Eq. (17.18) against the square root of the concentration, $\mathscr{E}^\circ$ can be determined by extrapolation to $m = 0$, as shown in Figure 17.8. Once $\mathscr{E}^\circ$ has been determined by this procedure, its value can be inserted back into (17.17) and the mean activity coefficients for HCl then determined from the known emf's at the various concentrations. This procedure forms an important and precise method for determining standard electrode potentials and the activity coefficients of ions.

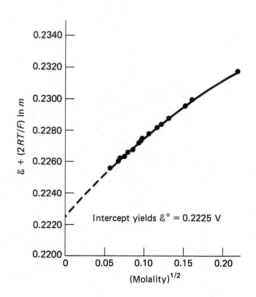

FIGURE 17.8 Determination of the standard potential of the cell $\text{H}_2 \mid \text{HCl}(m) \mid \text{AgCl} \mid \text{Ag}$ by plotting Equation (17.18). (Data from H. S. Harned and R. W. Ehlers, *J. Amer. Chem. Soc.* **54** (1932):1350.)

17.9 SOLUBILITY PRODUCTS

Standard electrode potentials are useful in determining the equilibrium constant we call a *solubility product*. For the dissolution reaction $AgCl(s) \rightleftharpoons Ag^+ + Cl^-$, the solubility product is given by $K_{sp} = [Ag^+][Cl^-]$. Since the concentration is extremely low, no significant error arises if we take the activities to equal the concentrations. Dissolution corresponds to the reaction of the electrochemical cell

$$Ag \mid Ag^+ \mid Cl^- \mid AgCl(s) \mid Ag$$

The electrode reactions are

$$\begin{array}{ll} AgCl(s) + e^- \rightleftharpoons Ag + Cl^- & \mathscr{E}^\circ = +0.222 \\ Ag \rightleftharpoons Ag^+ + e^- & -\mathscr{E}^\circ = +0.799 \\ \hline AgCl(s) \rightleftharpoons Ag^+ + Cl^- & \mathscr{E}^\circ = 0.222 - 0.799 = -0.577 \text{ V} \end{array}$$

and

$$\ln K_{sp} = \frac{\mathscr{E}^\circ F}{RT} = \frac{-0.577 \times 96{,}500}{8.314 \times 298} = -22.5$$

$$K_{sp} = 1.74 \times 10^{-10}$$

corresponding to a solubility of $(1.74 \times 10^{-10})^{1/2} = 1.3 \times 10^{-5}$ mol liter^{-1}.

17.10 OTHER THERMODYNAMIC INFORMATION FROM EMF MEASUREMENTS

The free-energy change for a chemical reaction can be determined from the voltage of the relevant electrochemical cell. By studying the voltage as a function of temperature, we can get values of ΔH and ΔS. Recalling our discussion of the Gibbs-Helmholtz equation (Chapter 9), we have

$$\Delta S = -\left(\frac{\partial \Delta G}{\partial T}\right)_p = nF \left(\frac{\partial \mathscr{E}}{\partial T}\right)_p \tag{17.19}$$

Since $\Delta H = \Delta G + T \Delta S$, we also have

$$\Delta H = -nF\mathscr{E} + nFT \left(\frac{\partial \mathscr{E}}{\partial T}\right)_p \tag{17.20}$$

Electromotive force measurements are accurate enough to provide accurate values of ΔH and ΔS. This is an important source of calorimetric data for reactions that cannot be directly carried out in a calorimeter.[4] The temperature dependence of the Weston cell is given in Equation (17.7), from which we find that at 25 °C, $\mathscr{E} = 1.01822$ V; the derivative is $\partial \mathscr{E}/\partial T = -5.05 \times 10^{-5}$ V K^{-1}. This is sufficient information to determine that for this reaction

$$\Delta G = -2 \times 96{,}487 \times 1.01822 = -196{,}490 \text{ J mol}^{-1}$$
$$\Delta S = 2 \times 96{,}487 \times (-5.05 \times 10^{-5}) = -9.75 \text{ J K}^{-1} \text{ mol}^{-1}$$
$$\Delta H = -196{,}490 - (9.75 \times 298) = -199{,}396 \text{ J mol}^{-1}$$

---

[4] Values for ΔS determined from these types of electrochemical cell measurements agreed with those determined from third law absolute entropies. This was strong confirmation of the validity of the third law of thermodynamics.

Additional data are available from this information. Since $(\partial \Delta H/\partial T)_p = \Delta C_p$, we can differentiate (17.20) with respect to temperature, to obtain

$$\Delta C_p = nFT \left(\frac{\partial^2 \mathscr{E}}{\partial T^2}\right)_p \qquad (17.21)$$

Since the pressure derivative of the free energy equals the volume change,

$$\Delta V = -nF \left(\frac{\partial \mathscr{E}}{\partial P}\right)_T \qquad (17.22)$$

Although (17.22) could be used to determine volume changes from the pressure variation of the emf, the expression is not of great practical use. The pressure effect is much larger for cells with gas electrodes than for cells with condensed phase electrodes.

17.11 CONCENTRATION CELLS

Suppose we try to construct an electrochemical cell using the same electrode for each half of the cell, say two hydrogen electrodes—

$$(\text{Pt}) \mid \text{H}_2(P_1) \mid \text{H}^+(m) \mid \text{H}_2(P_2) \mid (\text{Pt})$$

If $P_1 = P_2$, then the system is in equilibrium, and no voltage is developed. On the other hand, if $P_1 \neq P_2$, then the overall reaction is

$$\begin{aligned}
\text{H}_2(P_1) &\rightleftharpoons 2\text{H}^+(m) + 2e^- \quad &\text{(left electrode)} \\
2\text{H}^+(m) + 2e^- &\rightleftharpoons \text{H}_2(P_2) \quad &\text{(right electrode)} \\
\hline
\text{H}_2(P_1) &\rightleftharpoons \text{H}_2(P_2) &
\end{aligned}$$

This reaction is just the conversion of hydrogen from P_1 to P_2. Since $\mathscr{E}° = 0$ for each of the half-cells, $\mathscr{E}° = 0$ for the complete cell. The emf is

$$\mathscr{E} = \mathscr{E}° - \frac{RT}{nF} \ln Q = 0 - \frac{RT}{2F} \ln \frac{P_2}{P_1} \qquad (17.23)$$

Consider next the cell

$$\text{Pb}(X_1 \text{ in Hg}) \mid \text{Pb}^{2+} \mid \text{Pb}(X_2 \text{ in Hg})$$

consisting of two lead amalgam electrodes dipping into a solution of ions Pb^{2+}. The emf of this cell is

$$\mathscr{E} = -\frac{RT}{2F} \ln \frac{a_2}{a_1} \qquad (17.24)$$

If the amalgams can be considered ideal solutions, then the activities of the lead can be replaced by their mole fractions. The above two examples are called *electrode-concentration* cells, since the electrodes are displaying the activity differences.

In an electrolyte-concentration cell, the concentration difference of the electrolyte is what gives rise to the emf. Such a cell is shown in Figure 17.9a. The boundary between the two solutions is indicated by the dashed line. The cell diagram is

$$\text{Ag} \mid \text{AgCl} \mid \text{HCl}(m_1) \vdots \text{HCl}(m_2) \mid \text{AgCl} \mid \text{Ag}$$

There is a liquid *junction* across which ions can pass between the two different solutions by transference. A discussion of the emf of this cell requires that the transfer-

SEC. 17.12 ELECTROCHEMICAL INSTRUMENTATION

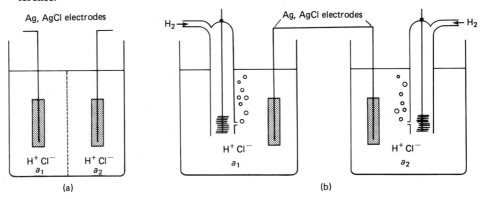

FIGURE 17.9 (a) Concentration cell. The dashed line indicates the boundary between the two solutions across which transference occurs. (b) A concentration cell that eliminates transference.

ence numbers of the ions be taken into account.[5] Since the ions diffuse at different rates, a potential, the so-called *junction potential,* is established across the boundary. In physiological systems, cell walls often constitute boundaries between solutions containing ions at different concentrations. The inequality of the diffusion rates of the ions establishes potential difference across the walls of the membrane. Establishing sharp boundaries between the two solutions in this kind of concentration cell is difficult. The problems associated with the boundary are eliminated when the configuration shown in Figure 17.9b is used. That cell consists of the two cells in series and can be symbolized by

$$(Pt) \mid H_2 \mid HCl(m_1) \mid AgCl \mid Ag \qquad Ag \mid AgCl \mid HCl(m_2) \mid H_2 \mid (Pt)$$

The emf of this cell is just the sum of the emf's of the two combined cells and is finally given by

$$\mathscr{E} = \frac{RT}{F} \ln \frac{(a_{H^+} a_{Cl^-})_2}{(a_{H^+} a_{Cl^-})_1} = \frac{2RT}{F} \ln \frac{(a_\pm)_2}{(a_\pm)_1} \qquad (17.25)$$

This kind of cell provides a convenient method for determining the ratios of activities at different concentrations.

17.12 ELECTROCHEMICAL INSTRUMENTATION

For many types of chemical and biological processes, it is important to know the concentration of H^+ in solution. This concentration can be conveniently, rapidly, and continuously measured with an electrochemical cell. Consider the concentration cell

$$(Pt) \mid H_2 \mid H^+(a) \vdots H^+(a = 1) \mid H_2 \mid (Pt)$$

---

[5] These kinds of cells are greatly complicated by transference and diffusion. A complete discussion will not be given here. The interested student can find the details in the advanced electrochemistry texts listed in the Bibliography.

FIGURE 17.10 The glass electrode.

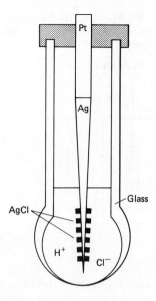

the emf of which can be measured to give a value of the H⁺ activity (or concentration if the solution is ideal). From the Nernst equation,

$$\mathscr{E} = 0 - 0.05915 \log_{10} (a_{H^+}) \tag{17.26}$$

The pH of a solution is defined by the equation

$$\text{pH} = -\log_{10} (a_{H^+}) \tag{17.27}$$

Hence for this cell,

$$\text{pH} = \frac{\mathscr{E}}{0.05915} \tag{17.28}$$

The apparatus consisting of a cell of this type and the measuring and display parts is called a *pH meter*.

FIGURE 17.11 Titration curves. (a) Titration of a strong acid, HCl, with a strong base, NaOH. The equivalence point is at pH = 7.00. (b) Titration curve of a weak acid, HAc, with a strong base, NaOH. The equivalence point will be at the pH of the basic salt NaAc.

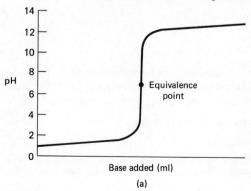

(a)

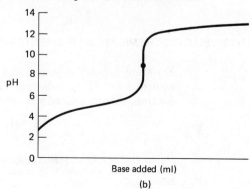

(b)

A more convenient half-cell for pH measurements is the *glass electrode*, shown in Figure 17.10. The tip of the glass electrode is thin and behaves like a semipermeable membrane to hydrogen ions. The details of the operation of this half-cell are imperfectly understood. The emf of the electrode is measured with reference to a second half-cell electrode, usually the *calomel* cell, whose reaction is $e^- + \frac{1}{2}Hg_2Cl_2 \rightleftharpoons Hg(l) + Cl^-$.

In analytical chemistry, pH meters are often used to determine end points in acid-base titrations. A typical titration curve is shown in Figure 17.11. Volumetric analytical procedures are often carried out using suitable oxidation or reduction reagents. By the proper choice of half-cells, the voltage developed by the cell immersed in the beaker strongly depends on the concentration of one of the reacting species, and the end point can be determined from curves similar to that of Figure 17.11. Further details on these and other procedures are available in the standard textbooks dealing with analytical chemistry and in particular instrumental analysis.[6]

17.13 CALCULATIONS OF ION CONCENTRATIONS

In many systems the concentrations of various ionic and undissociated species is of great interest. This is particularly true of the H^+ (and hence the OH^-) concentrations in many types of solutions. In biological and other systems, the pH is often of utmost importance.

Consider the ionization of the weak acid HA:

$$HA \rightleftharpoons H^+ + A^-$$

There are four individual species of interest: HA, A^-, H^+, and OH^-; the fifth species, H_2O, usually has a concentration near unit mole fraction (or $55.55m$) and is omitted from the calculation. There is a general procedure by which the concentrations of these four species of interest can be determined from the relevant ionization constants. A system of four unknowns can be solved by setting up a system of four linearly independent simultaneous equations relating the four unknowns. For the dissociation of HA these four equations can be established as follows:

1. The acid ionization constant

$$K_a = \frac{[H^+][A^-]}{[HA]} \tag{17.29}$$

2. The water ionization constant

$$K_w = [H^+][OH^-] = 10^{-14} \tag{17.30}$$

3. The condition of electrical neutrality

$$[H^+] = [OH^-] + [A^-] \tag{17.31}$$

---

[6] See, for example, I. M. Kolthoff and H. A. Laitinen, *pH and Electrotitrations*. (New York: John Wiley & Sons, 1941). Also G. Eisenman, ed., *Glass Electrodes for Hydrogen and Other Cations, Principles and Practice* (New York: Marcel Dekker, 1967). Of interest in this connection are the so-called ion-selective electrodes. These electrodes are specific to particular ions. Various kinds are available and are described in laboratory-supply-house catalogs. See, for example, R. A. Durst, ed., *Ion Specific Electrodes* (Washington, D.C.: U.S. Government Printing Office, National Bureau of Standards Special Publication 314, November 1969) for a discussion of these electrodes.

4. The condition of conservation of mass applied to the A constituent

$$c = [A] = [HA] + [A^-] \tag{17.32}$$

These four equations are adequate for completely specifying the concentrations of each of the four species. The acid ionization constant is given, having been determined either by conductivity or by electrochemical cell measurements. The total concentration of A ($HA + A^-$), indicated by c in (17.32), is determined from the quantity of acid added to the water. Solving these four simultaneous equations provides an "exact" solution of the concentrations. It is, however, usual to proceed by the method of successive approximations. In most instances the first approximation suffices to provide answers within the accuracy of the tabulated ionization constants. For the greatest accuracy, we must use the activities instead of the molal concentrations that we have indicated by the brackets, for instance [H^+]. In the example that follows, we assume that the solutions are ideal and that the activities equal the concentrations.

Consider a solution of acetic acid (HAc) in which 0.05 mole of HAc is dissolved in a kilogram of water. The acid ionization constant for HAc is $K_a = 1.8 \times 10^{-5}$. We first assume that only the *main reaction* takes place to any appreciable extent:

$$[HAc] = [H^+] + [Ac^-]$$

Therefore, $[H^+] = [Ac^-]$. If we let x be the concentration of H^+, then the concentration of HAc is $0.05 - x$, and

$$1.8 \times 10^{-5} = \frac{[H^+][Ac^-]}{[HAc]} = \frac{x^2}{0.05 - x}$$

This can be solved, to get $x = 9.4 \times 10^{-4}$. We now have

$$[H^+] = 9.4 \times 10^{-4} m$$
$$[Ac^-] = 9.4 \times 10^{-4} m$$

The remaining relations allow us to solve for the rest of the species. From (17.32), we find that

$$[HAc] = 0.05 - 9.4 \times 10^{-4} = 0.0491 m \simeq 0.05 m$$

From the water ionization constant, we find that

$$[OH^-] = \frac{10^{-14}}{[H^+]} = 1.06 \times 10^{-11} m$$

In view of the extremely small hydroxide ion (OH^-) concentration and the small degree of dissociation of the system, we can stop the calculation at this point.

For comparative purposes, we note that an exact solution of the four simultaneous equations yields a final equation for [H^+] in the form

$$\frac{([H^+])([H^+] - 10^{-14}/[H^+])}{1.8 \times 10^{-5}} = 0.05$$

From this we find that $[H^+] = 9.4868 \times 10^{-4} m$, which is the same as our approximate value to within the accuracy of the ionization constants.

Calculations involving acid or basic salts follow similar lines. The first step in any calculation is to write the *main reactions*. Some examples of main reactions are:

SEC. 17.14 ELECTROCHEMICAL CELLS FOR POWER GENERATION

Na^+Ac^-: $Ac^- + H_2O \rightleftharpoons HAc + OH^-$
$NH_4^+Cl^-$: $NH_4^+ + H_2O \rightleftharpoons NH_4OH + H^+$
$NH_4^+Ac^-$: $NH_4^+ + Ac^- + H_2O \rightleftharpoons NH_4OH + HAc$
H_2CO_3: $H_2CO_3 \rightleftharpoons H^+ + HCO_3^-$
$Na_2^{2+}CO_3^{2-}$: $CO_3^{2-} + H_2O \rightleftharpoons HCO_3^- + OH^-$
$Na^+HCO_3^-$: $2HCO_3^- \rightleftharpoons H_2CO_3 + CO_3^{2-}$

A *buffered* solution is a solution of a weak acid and a basic salt (such as HAc and NaAc) or a weak base and an acid salt (such as NH_4OH and NH_4Cl). The effect of added acid or base on the pH of a buffered solution is small. Calculations involving these solutions can be found in most introductory chemistry textbooks, and the student is referred to these for review.

17.14 ELECTROCHEMICAL CELLS FOR POWER GENERATION

The electric-power generation industry has recently developed a great interest in electrochemical cells for two types of applications. The first type has to do with the efficiency of the energy cycle and is centered about *fuel cells*. Energy is normally extracted from fuels as heat by combustion of the fuel; for example,

$$C + O_2 \rightleftharpoons CO_2 \qquad \Delta H° = -393.5 \text{ kJ mol}^{-1};\ \Delta G° = -394.4 \text{ kJ mol}^{-1}$$

$$H_2 + \tfrac{1}{2}O_2 \rightleftharpoons H_2O(l) \qquad \Delta H° = -285.8 \text{ kJ mol}^{-1};\ \Delta G° = -237.2 \text{ kJ mol}^{-1}$$

$$CH_4 + 2O_2 \rightleftharpoons CO_2 + 2H_2O(l) \qquad \Delta H° = -604.5 \text{ kJ mol}^{-1};\ \Delta G° = -580.8 \text{ kJ mol}^{-1}$$

Electrical work can be obtained from these direct combustion reactions by using the heat of combustion to drive a heat engine whose work output can be harnessed to an electrical generator. For a reversible heat engine with a hot reservoir at 800 K and a cold reservoir at 300 K, the maximum theoretical efficiency is $\frac{500}{800} = 0.625$; 37.5% of the heat energy is lost under the best of all possible circumstances.

On the other hand, if these reactions could be made to proceed in electrochemical cells, then electrical work would be obtained directly, with the quantity of work limited only by the free-energy decrease of the reaction. A fuel cell using H_2 and O_2 is schematically indicated in Figure 17.12. Since the cell uses a basic electrolyte, it is called a *base electrolyte fuel cell*. The half-cell reactions are:

Anode: $H_2 + 2OH^- \rightleftharpoons 2H_2O + 2e^-$
Cathode: $\tfrac{1}{2}O_2 + H_2O + 2e^- \rightleftharpoons 2OH^-$

Conceptually the fuel cell seems like an ordinary electrochemical cell, or battery; both directly transform chemical energy into electrical energy by electrochemical redox reactions. A fundamental difference exists between the two, however. A battery consumes the chemicals that form part of its structure or are stored within the structure. Eventually the reactants are used up, and the battery is dead; it must be replaced or recharged. In the fuel cell, the reactants are supplied from outside the cell. The fuel cell itself undergoes no irreversible changes and can continue to operate so long as reactants are supplied and products removed. The life of a fuel cell is limited by physical or chemical deterioration of the cell. This possibility for a long

FIGURE 17.12 Fuel cell schematic. Hydrogen is adsorbed on the catalyst embedded in the anode, and reacts with ions OH^- from the electrolyte, to form H_2O. The electrons produced by the reaction flow through the external circuit to the cathode, where they react with O_2 adsorbed on the catalyst, to form OH^-. The ions complete the circuit by migrating from the cathode to the anode through the electrolyte.

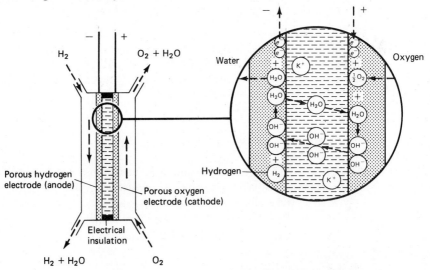

FIGURE 17.13 Apollo fuel cell module.

life has made the fuel cell attractive for space applications. The electrical supply system for the Apollo spacecraft consisted of three fuel cell modules, as shown in Figure 17.13. The cell supplied dc power over a range of 563–1420 W at a normal voltage range of 27–31 V.

Hydrocarbons can be used directly in the *molten carbonate* fuel cell indicated in Figure 17.14. The half-cell reactions are:

$$\text{Anode:} \quad H_2 + CO_3^{2-} \rightleftharpoons CO_2 + H_2O + 2e^-$$
$$\text{Cathode:} \quad \tfrac{1}{2}O_2 + CO_2 + 2e^- \rightleftharpoons CO_3^{2-}$$

This cell operates at temperatures in excess of 500 °C. The hydrocarbon fuel is *re-formed* with water to produce H_2 and CO. The overall reaction for this cell is the same as for the base electrolyte cell.

The other type of application of electrochemistry to power generation has to do with supplying the peak demands for electricity. Power-generating stations must meet instantaneous demands for electricity. This demand varies considerably over a 24-hour period, as shown in Figure 17.15. In addition there is also a yearly cyclic variation. The maximum yearly peak for cities such as New York usually occurs in the early evening during a summer hot spell when all air-conditioning equipment is running at full load. The utility must have sufficient capacity for meeting this demand. This means that the utility must invest heavily in generating capacity that greatly exceeds the average demand; much of the capacity is idle during periods of low demand. A much more efficient operation would result if the excess capacity at low-demand periods could be used to store energy for use at high-demand periods. One way this is done is by the *pumped storage*, or *pumped-hydro*, method. Excess

FIGURE 17.14 High-temperature molten-carbonate fuel cell. The cell operates at temperatures in excess of 500 °C. The electrolyte, molten K_2CO_3, is placed between the electrodes. The hydrocarbon fuel is re-formed (reacts) with steam to produce H_2 and CO. The gases diffuse into the cathode, where they react with CO_3^{2-} in the electrolyte, to form CO_2 and H_2O. The electrons released at the cathode flow through the external circuit to the anode, where they combine with O_2 and CO_2, to form CO_3^{2-}. The carbonate ions CO_3^{2-} complete the electrical circuit by flowing back to the cathode.

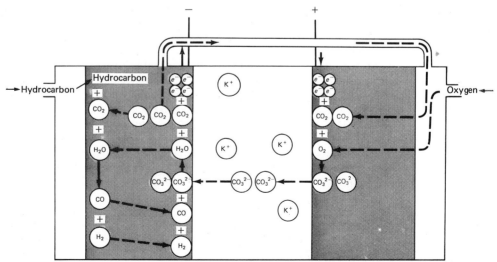

FIGURE 17.15 Typical demand curve for a utility over 24 hours.

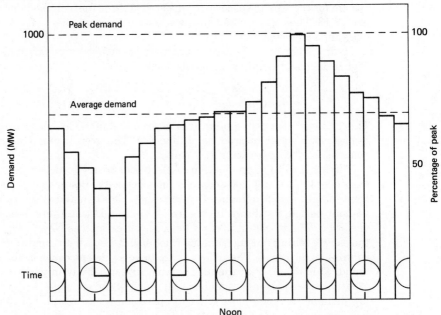

capacity during off-peak hours is used to pump water into specially constructed reservoirs at high elevations. During peak hours the water is allowed to flow back to the lower level through a hydroelectric plant to generate additional power.

Another method that is now being intensely investigated is the use of storage batteries. The excess capacity at off-peak hours would be used to charge the batteries; power would be drawn from the batteries during the peak hours to supplement the utility's capacity. This method has several advantages over the pumped-storage method. Land requirements for reservoirs are large, and the reservoirs must be located at large distances from the cities for which the power is needed. A battery storage module could be constructed in the city itself; the land requirements are small. Table 17.2 lists a number of battery couples now under investigation as candidates for energy storage.

At the present time, and for the foreseeable future, the "action" in electrochemical research and development lies in fuel cell and battery development. The difficulties are more practical than conceptual, and much of the research concerns materials. Fuel cells require structural materials capable of withstanding corrosive environments at elevated temperatures; in addition, catalysts are needed for many of the reactions that take place in the cells. Storage batteries must be recycled from full to near-zero charge for thousands of cycles at high current densities. The costs of these alternatives are still too high and the efficiencies too low for extensive practical applications.[7] On the other hand, it is worth recalling that the steam engine

---

[7] In 1973, fabrication costs for aerospace fuel cells were in the range $100,000 to $400,000 per kilowatt; fuel cells for military applications were about $30,000 per kilowatt. To compete favorably with present generation equipment, fuel cells would have to drop in price to about the thousand dollar per kilowatt range. (The 2,475,000-kW coal-fired Bruce Mansfield plant in western Pennsylvania was built during the late 1970s at a cost of less than $600 per kilowatt.)

TABLE 17.2 Battery couples under investigation as potential candidates for energy storage.

Couple	Energy density (W h kg⁻¹)		Operating temperature (°C)[a]
	Theoretical	Practical	
Li–S	2800	330	350–450
Li–Cl$_2$	2200	150	400–500
Li–Se	1300	300	350–400
Zn–Air	1100	60	Low
Fe–Air	940	50	Low
Na–S	785	100–250	300–400
Li–Te	650	250	400–470
Cd–Air	500	50	Low
Ag–Zn	440	120	Low
Ni–Zn	330	40	Low
Ag–Cd	270	50	Low
Ni–Fe	265	17	Low
Ni–Cd	210	17–20	Low
Pb–PbO$_2$	170	17	Low

SOURCE: A. R. Cook, "A Review of Battery Research and Development," *Proceedings of the American Power Conference* (Chicago: American Power Conference, 1974), p. 1046.

[a] The high temperatures in this column are required to keep the electrolytes in the molten state.

had an efficiency of less than 10% at a comparable period of its development. It is not unreasonable to expect extensive deployment of fuel cells and batteries in the electric-power-generation industry by the end of this century.

PROBLEMS

1. Calculate the work required to move 1 mol electrons against a potential of 1 V.
2. Draw a diagram of the cell Mn | Mn^{2+} | Cu^{2+} | Cu showing the motion of all the charged species. Calculate the standard emf of the cell. For equilibrium (that is, when the cell is "dead") calculate the concentrations of the species in the two compartments if the cell originally had 1 mol material in each compartment.
3. Using the data in the table of emf's, calculate the solubility of TlBr.
4. Will TlI be reduced by hydrogen in the reaction 2TlI + H$_2$ ⇌ 2Tl + HI(aq)? Draw a diagram of the relevant electrochemical cell.
5. A standard piece of apparatus in analytical chemistry is the Jones reductor, in which a solution is passed over granulated zinc to reduce a constituent. Some 0.1M Fe^{3+} solution is passed through a Jones reductor to reduce the iron to Fe^{2+}. Calculate the concentration of Fe^{3+} remaining in the solution (2Fe^{3+} + Zn = 2Fe^{2+} + Zn^{2+}). Is this a satisfactory quantitative procedure?
6. Consider the electrochemical cells
   (a) Pt | Ce^{3+},Ce^{4+} | Cl$^-$ | AgCl | Ag
   (b) Fe | Fe^{2+} | Cd^{2+} | Cd
   (c) Pt | H$_2$ | H$^+$ | D$^+$ | D$_2$ | Pt
   Assuming that all activities are unity, draw each cell, write the overall reaction for each cell, and determine the standard EMF of each. In which directions do the reactions proceed?
7. What is the emf of the cell Pt | Ce^{3+}(a = 0.1),Ce^{4+}(a = 0.01) | Cl$^-$(a = 0.0001) | AgCl | Ag?
8. Calculate the equilibrium constants for the reactions:
$$2MnO_4^- + 10I^- + 16H^+ \rightleftharpoons 5I_2 + 2Mn^{2+} + 8H_2O$$
$$Cr_2O_7^{2-} + 6I^- + 14H^+ \rightleftharpoons 3I_2 + 2Cr^{3+} + 7H_2O$$
$$Cr_2O_7^{2-} + 6Fe^{2+} + 14H^+ \rightleftharpoons 2Cr^{3+} + 6Fe^{3+} + 7H_2O$$
$$2Ce^{3+} + I_2 \rightleftharpoons 2Ce^{4+} + 2I^-$$
9. Calculate the emf of the cell

   Pt | H$_2$(P = 20 atm) | H$^+$ | H$_2$(P = 1 atm) | Pt

10. A hydrogen economy is proposed that is to operate as follows. Coal is to generate electricity in the sparsely populated coal mine regions. The electricity is used to electrolyze water to produce hydrogen gas, which is then transported to populated areas, where it is burned either in a standard heat engine or in a fuel cell. The overall efficiency of the first coal-burning generating station is 0.37. Calculate the overall efficiency of the two-step energy-producing process using

(a) a hydrogen-fired heat engine (assume it to be ideal and operating between 300 and 800 K); (b) a hydrogen fuel cell (assume it to be ideal).

If we assume that the coal is a low-grade Western coal with a calorific value of 8500 Btu/lb, how much coal is needed to satisfy a city of 100,000 families (each family uses 10^3 kW h per month) (c) if the original coal-fired plant supplied these needs directly? (d) if each of the two-step processes were used?

11. The term *ampere-hour* (A h) as used in automobile batteries means the ability to deliver a current of 1 A for 1 hr. Draw a diagram of the lead-acid storage battery as used in automobiles. Determine the amount of each material required to construct a 200 A h battery (excluding the case). Call the appropriate supplier or check the appropriate catalog for prices and determine the cost of materials for the battery. Determine the amounts of material present when the battery is dead. Under the assumption (erroneous) that the battery can be recharged and thus returned to its original, new state, calculate the amount of electrical work needed to recharge a dead battery. If the cost of electricity is $0.053 per kW h, what does it cost to recharge the battery? How does this compare with the price charged by the local service station? (*Note:* This problem contains several gross oversimplifications. What are they?)

12. Derive an expression for the emf of the cell Pt | $H_2(P)$ | HCl | Cl_2 | Pt as a function of hydrogen pressure, assuming that all other constituents are in their standard states.

13. A hydrochloric acid plant produces 1 ton HCl per hour by causing the gases to react. How much heat is produced per hour? This heat is available for use either for heating or other purposes. Assume that the heat is used to generate electricity in a generator with an efficiency of 37%. How much electrical work is produced? If the reaction were carried out reversibly in an electrochemical cell, how much electrical work could be produced?

14. Consider the cell Pt | H_2 | H^+ | Fe^{2+} | Fe with all materials in their standard states. Calculate the standard emf. In the standard state, the iron electrode is a piece of strain-free iron. How would the emf be affected if an iron electrode in the form of a compressed spring were used instead?

15. R. G. Bates and V. E. Bower (*J. Res. Natl. Bur. Stand.* 53 (1954):283) studied the potentials of the cell Pt | H_2(1 atm) | HCl(m) | AgCl | Ag over the range of temperature 0–90 °C. They found that the standard emf as a function of temperature could be written as

$$\mathscr{E}° = 0.23659 - (4.8564 \times 10^{-4})t \\ - (3.4205 \times 10^{-6})t^2 + (5.869 \times 10^{-9})t^3$$

where t is degrees Celsius. Calculate $\Delta H°$, $\Delta S°$, and $\Delta G°$ at 25 °C for the reaction

$$\tfrac{1}{2}H_2(g) + AgCl(s) \rightleftharpoons Ag(s) + HCl(aq)$$

Calculate the temperature coefficient of $\Delta H°$.

16. By the Bates and Bower data (Problem 15), the emf of the cell with 0.001m HCl is given by

$$\mathscr{E} = 0.56332 + (7.1511 \times 10^{-4})t \\ - (3.4362 \times 10^{-6})t^2 + (6.0290 \times 10^{-9})t^3$$

whereas with 0.1m HCl, the emf is given by the expression

$$\mathscr{E} = 0.35507 - (3.3126 \times 10^{-5})t \\ - (3.2549 \times 10^{-6})t^2 + (6.4395 \times 10^{-9})t^3$$

Derive expressions for ΔH, ΔG, and ΔS for the reaction

$$HCl(m = 0.1m) \rightleftharpoons HCl(m = 0.001m)$$

Determine the values at 25 °C, and check to see whether $\Delta G = \Delta H - T\,\Delta S$.

17. Consider the cell Pt | HAc(m_1),NaAc(m_2), NaCl(m_3) | AgCl | Ag, which contains an acetic acid–sodium acetate buffer. The emf of the cell is given by

$$\mathscr{E} = \mathscr{E}° - \left(\frac{RT}{F}\right) \ln\left(\gamma_{H^+}\gamma_{Cl^-} m_{H^+} m_{Cl^-}\right)$$

where $\mathscr{E}°$ is the standard emf for the cell Pt | H_2 | HCl | AgCl | Ag. The dissociation constant for acetic acid can be written

$$K = \left(\frac{\gamma_{H^+}\gamma_{Ac^-}}{\gamma_{HAc}}\right)\left(\frac{m_{H^+} m_{Ac^-}}{m_{HAc}}\right)$$

The two equations can be combined, to get

$$\mathscr{E} - \mathscr{E}° + \frac{RT}{F} \ln\left(\frac{m_{HAc} m_{Cl^-}}{m_{Ac^-}}\right) \\ = -\frac{RT}{F} \ln \frac{\gamma_{H^+}\gamma_{Cl^-}\gamma_{HAc}}{\gamma_{H^+}\gamma_{Ac^-}} - \frac{RT}{F}\ln K$$

If $\mathscr{E}°$ is known for the desired temperature, then the dissociation constant K can be found as follows. At infinite dilution, all activity coefficients approach unity, and the first term on the right approaches zero linearly with ionic strength. All the terms on the left can be determined experimentally. A plot of the left side vs the ionic strength should give a straight line whose intercept at $I = 0$ equals $-(RT/F) \ln K$, from which K can be determined. H. S. Harned and R. W. Ehlers (*J. Amer. Chem. Soc.* 54 (1932):1350) have measured the emf's of the cell.

m_1	m_2	m_3	emf
0.004779	0.004599	0.004896	0.63959
.012035	.011582	.012426	.61583
.021006	.020216	.021516	.60154
.04922	.04737	.05042	.57977
.08101	.07796	.08297	.56712
0.09056	0.08716	0.09276	0.56423

Their value of $\mathscr{E}°$ was 0.22239 V at 25 °C; the emf's at 25 °C at various concentrations are shown in the accompanying table. By properly plotting this data, determine the dissociation constant for HAc.

18. Note that in this problem we deal with mole fraction concentrations rather than molalities.

    Consider the electrode concentration cell $Pb(X_1 \text{ in Hg}) | PbAc_2, HAc | Pb(X_2 \text{ in Hg})$ consisting of two lead amalgam electrodes dipping into a $PbAc_2$–HAc solution. The emf of this cell is given by

    $$\mathscr{E} = -\frac{RT}{2F} \ln \frac{a_2}{a_1}$$
    $$= -\frac{RT}{2F} \ln X_2 - \frac{RT}{2F} \ln \gamma_2 + \frac{RT}{2F} \ln a_1$$

    This can be rearranged to get

    $$\mathscr{E} + \frac{RT}{2F} \ln X_2 = \frac{RT}{2F} \ln a_1 - \frac{RT}{2F} \ln \gamma_2$$

    The activities of the lead in the amalgam can be determined by the following experimental procedure. Keeping X_1 fixed at some small value, one can measure the emf of the cell as a function of X_2. In the limit as $X_2 \to 0$, $\gamma_2 \to 1$. If the left side of the last equation is plotted as a function of X_2, then the intercept yields $\ln a_1$, the activity of the lead at the reference concentration X_1. Once a_1 is known, the activities a_2 as a function of concentration can be calculated from the first equation. M. M. Haring, M. R. Hatfield, and P. P. Zapponi (*Trans. Electrochem. Soc.* 75 (1939):473) have determined the emf of this cell as a function of X_2 with X_1 fixed at 0.0006253. From their results, calculate the activity of the lead at the reference concentration X_1 by a suitable graphical procedure. Determine the activities and activity coefficients for several concentrations.

X_2	$-\mathscr{E}$ (mV)	$\dfrac{-\mathscr{E}}{0.02958} - \log_{10} X_2$
0.0006253	0.000	3.2039
.0006302	0.204	3.2074
.0009036	4.636	3.2008
.001268	8.911	3.1981
.001349	9.659	3.1965
.001792	13.114	3.1900
.002055	14.711	3.1845
.002744	18.205	3.1771
.002900	18.886	3.1761
.003086	19.656	3.1751
.003203	20.068	3.1729
.003729	21.827	3.1663
.003824	22.111	3.1650
.004056	22.802	3.1628
.004516	23.594	3.1429
0.005006	25.160	3.1511

X_2	$-\mathscr{E}$ (mV)	$\dfrac{-\mathscr{E}}{0.02958} - \log_{10} X_2$
0.005259	25.692	3.1477
.005670	26.497	3.1422
.006085	27.256	3.1372
.006719	28.340	3.1308
.007858	29.951	3.1172
.007903	30.010	3.1167
.008510	30.771	3.1103
.009737	32.062	3.0955
.01125	33.437	3.0792
.01201	33.974	3.0690
.01388	35.226	3.0485
.01406	35.323	3.0462
.01456	35.609	3.0407
.01615	36.375	3.0215
0.01650(sat)	36.394	3.0129

19. Many substances display ionic conductance in the solid state and can act as electrolytes in electrochemical cells. At high temperatures, AgI displays ionic conductance. Reversible emf's can be measured for the cell $Ag(s) | AgI(s) | Ag_2S(s), S(l), C(graphite)$. This corresponds to the chemical reaction $2Ag(s) + S(l) = Ag_2S(s)$. Since all substances are either pure elements or compounds in the liquid or solid standard states, all the substances are at unit activity at the temperature of measurement. K. Kiukkola and C. Wagner (*J. Electrochem. Soc.* 104 (1957):379) have measured the emf's of the cell over the range 150–425 °C as in the accompanying table. From these data, calculate the standard free energy of formation and the standard enthalpy of formation of Ag_2S at 150, 200, 300, and 400 °C. How would you evaluate this information by purely calorimetric procedures?

Temp. (°C)	emf (V)
150	0.220
178	.224
200	.228
250	.236
300	.244
350	.252
400	.260
425	0.264

20. H. S. Harned and R. W. Ehlers (*J. Amer. Chem. Soc.* 54 (1932):1350) have measured the emf's of the cell $Pt | H_2 | HCl(m) | AgCl | Ag$ as a function of m at 0 °C. From their data (next page), evaluate the standard emf of the cell at 0 °C.

    Calculate the mean molal activity coefficient of HCl at 0.01, 0.05, and 0.1 molal concentrations at 0 °C.

m	emf (V)	m	emf (V)
0.003215	0.50933	0.018369	0.43055
.003661	.50334	.021028	.42425
.005314	.48644	.025630	.41557
.005763	.48275	.034920	.40181
.007771	.46912	.041245	.39426
.008636	.46435	.04935	.38640
.008715	.46401	.05391	.38255
.011095	.45314	.06393	.37501
.013049	.44586	.07273	.36917
.013407	.44466	.08631	.36167
.013981	.44273	.09751	.35610
0.016457	0.43544	0.12354	0.34567

21. Calculate the pH of a solution made by adding 0.1 mol HAc and 0.1 mol NaAc to a kilogram of water. The dissociation constant of HAc is 1.8×10^{-5}. Calculate the change in pH if 0.01, 0.1, and 0.5 mol HCl are added to the original solution. What would be the resulting pH if the HCl were added to pure water?

22. One of the most promising batteries for energy storage is the Na–S battery. Table 17.2 indicates that the theoretical energy obtainable from this battery is 785 W h per kilogram of stoichiometric amounts of Na and S. From this fact, calculate the free energy of formation of Na_2S at 300 °C.

23. Suppose you were asked to design a battery storage system using an Na–S battery capable of delivering 1000 MW h energy. In theory, what is the minimum quantity of Na and S required? Suppose that you were to design the system so that it contained only one very large cell. Roughly estimate the land area required for the cell.

24. Suppose you were asked to design a pumped-hydro energy storage system using a reservoir with an average depth of 10 m; the water will fall 35 m on the return trip through a hydroelectric generator. How much water must be stored in the reservoir to generate 1000 MW h electricity? How much land is required for the reservoir?

CHAPTER EIGHTEEN

GRAVITATIONAL, ELECTRICAL, MAGNETIC, AND SURFACE WORK

18.1 WORK AND ENERGY

Let's go back to the very beginning and write the first law of thermodynamics as

$$dE = dq + dw \tag{18.1}$$

This equation is perfectly general and applies to all types of work. So far we have restricted our discussion to expansion work for which $dw_{exp} = -p\,dV$. In investigating the properties of systems in which other types of work can be performed, we shall briefly examine the effects of gravitational, electrical, and magnetic work. Also we shall investigate the work associated with surface tension and examine several surface effects.

Since expansion work is very important, we might keep that type of work in a category by itself and write Equation (18.1) as

$$dE = dq + dw_{exp} + dw_{other} \tag{18.2}$$

Work is *always* given by the product of an intensive variable and its associated extensive variable; thus

$$dw = -\Phi\,dx \tag{18.3}$$

where Φ is the intensive variable and x is the extensive variable. For expansion work, the pressure is identified with Φ and the volume with x; for gravitational work, Φ is the gravitational field and x the mass; for magnetic work, Φ corresponds to the magnetic field $\mathbf{H}$, and x to the intensity of magnetization $\mathbf{I}$. The first law of thermodynamics can be written in its most general form as

$$dE = T\,dS - \sum_i \Phi_i\,dx_i \tag{18.4}$$

where the second term on the right includes all the work terms. Or if we want to consider expansion work separately, we can write

$$dE = T\,dS - P\,dV - \sum_i \Phi_i\,dx_i \tag{18.5}$$

When only expansion work is considered, ΔE for any process is the constant-volume heat q_v. In our augmented system, ΔE for any process is the heat obtained when the volume and *all* the other extensive variables (the x_i) are kept constant:

$$\Delta E = q_{v,x_i} \tag{18.6}$$

From (18.5), we get the partial derivatives

$$\left(\frac{\partial E}{\partial S}\right)_{V,x_i} = T \tag{18.7a}$$

$$\left(\frac{\partial E}{\partial V}\right)_{S,x_i} = -P \tag{18.7b}$$

which are already familiar from previous work. In addition,

$$\left(\frac{\partial E}{\partial x_i}\right)_{S,V,x_j \neq x_i} = -\Phi_i \tag{18.8a}$$

and

$$T\left(\frac{\partial S}{\partial x_i}\right)_{E,V,x_j \neq x_i} = \Phi_i \tag{18.8b}$$

The heat capacity must be carefully specified to avoid ambiguity. Our old friend C_v is

$$C_v = \left(\frac{\partial E}{\partial T}\right)_V \tag{18.9}$$

Now we have additional variables that must be specified. Equation (18.9) defines a heat capacity at constant volume in which other extensive variables can vary. Most ordinary laboratory procedures are carried out under circumstances in which gravitational, magnetic and other effects are essentially constant and can be omitted, but we now want to consider these effects specifically. A heat capacity in which *all* the extensive variables of the system are kept constant can be defined by

$$C_{v,x_i} = \left(\frac{\partial E}{\partial T}\right)_{V,x_i} \tag{18.10}$$

While we have not yet explicitly included the chemical potential in our augmented system, the chemical potential has been implicitly included. The chemical potential can be considered the source of chemical work,

$$dw_{\text{chemical}} = \mu \, dn \tag{18.11}$$

The chemical potential can be considered as included in Equation (18.5) by equating the chemical potential of the jth species μ_j with one of the Φ_i, and dn_j with dx_i. Or the chemical work term can be considered separately by adding the term $\Sigma_j \, \mu_j \, dn_j$ to the right side of Equation (18.5). Note, however, that the sign of the chemical work term is different from the sign of the other work terms. The analog of Equation (18.8a) for the chemical potential is

$$\left(\frac{\partial E}{\partial n_j}\right)_{S,V,x_i,n_i \neq n_j} = +\mu_j \tag{18.12}$$

Now we consider the effect of these new work terms on the enthalpy, the Gibbs free energy, and the Helmholtz energy.

18.2 ENTHALPY, FREE ENERGY, AND HELMHOLTZ ENERGY IN THE AUGMENTED SYSTEM

When only expansion work is considered, the enthalpy is

$$H = E + PV \tag{18.13}$$

In the augmented system, a general enthalpy can be defined by[1]

$$H = E + PV + \sum_i \Phi_i x_i \tag{18.14}$$

If we differentiate (18.14) and replace dE by its value given in (18.5), we get

$$dH = T\,dS + V\,dP + \sum_i x_i\,d\Phi_i \tag{18.15}$$

For any process that takes place at constant P and also Φ_i, the heat is

$$\Delta H = q_{P,\Phi_i} \tag{18.16}$$

Again, to avoid any ambiguity, we define the heat capacity in such a way that all the intensive variables of the system are held constant:

$$C_{P,\Phi_i} = \left(\frac{\partial H}{\partial T}\right)_{P,\Phi_i} \tag{18.17}$$

A Helmholtz energy can be defined precisely as before—

$$A = E - TS \tag{18.18}$$

from which we find that

$$dA = -S\,dT - P\,dV - \sum_i \Phi_i\,dx_i \tag{18.19}$$

Similarly, a Gibbs free energy is

$$\begin{aligned} G &= H - TS \\ &= E - TS + PV + \sum_i \Phi_i x_i \end{aligned} \tag{18.20}$$

The differential form of the free energy is obtained by differentiating Equation (18.20),[2]

$$dG = -S\,dT + V\,dP + \sum_i x_i\,d\Phi_i \tag{18.21}$$

---

[1] Strictly speaking, the enthalpies defined in Equations (18.13) and (18.14) are not the same functions, since they involve different sets of variables, and we should perhaps use different symbols for the two. Many authors do use different symbols. Many different symbols would be needed, depending on which particular set of Φ_i and x_i were being considered. The same is true for the Gibbs free energy and the Helmholtz energy. We shall use only one set of symbols (H, G, and A) regardless of the type of work being considered. It should be clear from the context of the discussion what the variables are.

[2] Note that the term $\Sigma \Phi_i x_i$ was added explicitly only to the enthalpy in Equation (18.14). It was also added implicitly to the free energy, since $G = H - TS$. This establishes the following conventions for us: (1) The enthalpy is the Legendre transform of the energy with respect to *all* the work terms. (2) The Helmholtz energy is the Legendre transform of the energy only with respect to the thermal energy term. (3) The Gibbs free energy is the Legendre transform of the energy with respect to *all* the energy terms (thermal plus work). This is in accord with our previous definitions of these functions. Other conventions are used in other books.

The term ΔG as defined by (18.20) measures the maximum useful work obtainable when a process is carried out under such circumstances that *all* the intensive variables (T, P, and all the Φ_i) are held constant. The equilibrium condition for a system at constant T, P, and Φ_i is that $dG = 0$.

Finally, we note that the phase rule must be adjusted to take into account the extra variables. The number of degrees of freedom is

$$f = C - P + n \tag{18.22}$$

where C is the number of components and P is the number of phases; n is the total number of intensive variables. For the systems previously considered, the only intensive variables were T and P, in which case $n = 2$.

The correspondence between the intensive and extensive variables for various types of energy is summarized in the accompanying table.

Type of Energy	Φ_i	x_i
Thermal	T	S
Expansion	P	V
Gravitational	gh	M
Centrifugal	$\frac{1}{2}r^2\omega^2$	M
Chemical	μ_i	n_i
Electric charge	ΔV	q
Electric polarization	**E**	**P**
Magnetic	**H**	**M**
Stretching	T	l
Surface	γ	a

Now let's examine a number of systems in which these types of energy are important.

18.3 GRAVITATIONAL WORK; THE BAROMETRIC FORMULA

The gravitational force between two bodies of masses m and M, separated by a distance r, is given by Newton's law of gravitational attraction,

$$F = \frac{\gamma m M}{r^2} \tag{18.23}$$

where $\gamma = 6.66 \times 10^{-11}$ m³ s⁻² kg⁻¹ is the universal gravitational constant. In the earth's gravitational attraction, the expression is usually used in the form

$$F = gm \tag{18.24}$$

where g is the gravitational acceleration at the specified location, and m is the mass of the test object.

If we denote the molar mass of a body by M, then the potential energy per mole is

$$E_{\text{pot}} = M \int_{h_0}^{h} g\, dh \tag{18.25}$$

where h is the distance. The zero point of energy is normally taken at h_0, the surface of the earth. The work required to lift the body from h_0 to h is just the increase in potential energy. If the distance $h - h_0$ is not too great, no substantial error is introduced by neglecting the variation of g with height and placing it outside the integral sign.

The intensive variable Φ_i can be taken as

$$\Phi_i \equiv \phi = \int_{h_0}^{h} g\, dh \qquad (18.26)$$

which corresponds to the gravitational potential. The mass M corresponds to the extensive variable x_i. Equation (18.21) becomes

$$\begin{aligned} dG &= -S\, dT + V\, dP + M\, d\phi \\ &= -S\, dT + V\, dP + Mg\, dh, \end{aligned} \qquad (18.27)$$

if the variation of g with height is neglected. Now consider an isothermal column of fluid extending vertically upward. Since the temperature is constant, the term $-S\, dT$ is zero. The condition for equilibrium is $dG = 0$; hence

$$Mg\, dh + V\, dP = 0$$

or

$$\frac{dP}{dh} = -\frac{Mg}{V}$$

$$= -\rho g \qquad (18.28)$$

where ρ is the density. Equation (18.28) gives the pressure variation for a column of fluid. The most common application is to an ideal gas for which $\rho = PM/RT$. Equation (18.28) can be solved, to get

$$\ln \frac{P_1}{P_2} = \frac{Mg}{RT}(h_2 - h_1) \qquad (18.29)$$

Equation (18.29) is a crude estimate for the variation of atmospheric pressure with height. The main deficiency in the approximation is the assumption that the temperature does not vary with the height. We shall derive this equation again in another context (Chapter 21).

18.4 ELECTRICAL WORK AND THE PIEZOELECTRIC EFFECT

If an electric field $\mathbf{E}$ is applied to a dielectric material, then a polarization $\mathbf{P}$ results.[3] The energy change of the system is

$$dE = \mathbf{E}\, d\mathbf{P} = dw \qquad (18.30)$$

work being done *on* the body to increase its polarization. We can thus identify $-\mathbf{E}$ with Φ and $\mathbf{P}$ with x. The complete expression for the energy corresponding to (18.5) is

$$dE = T\, dS - P\, dV + \mathbf{E}\, d\mathbf{P} \qquad (18.31)$$

The free-energy expression is

$$dG = -S\, dT + V\, dP - \mathbf{P}\, d\mathbf{E} \qquad (18.32)$$

---

[3] Chemistry students, not to mention chemistry professors, often forget the basic relations between the variables in electromagnetism, such as polarization, electric field strengths, and dielectric constants; and the same holds true for the magnetic variables discussed in the next section. These variables will be discussed further in Chapter 38. A brief review of a physics textbook dealing with these matters might be helpful, for example, D. Halliday and R. Resnick, *Physics* (New York: John Wiley & Sons, 1962) or N. H. Frank, *Introduction to Electricity and Magnetism* (New York: McGraw-Hill Book Co., 1950).

Suppose also that the length of the dielectric material can be changed by applying a force or tension to it. For the simplest case of a one-dimensional body, we assume that if a tension **T** is applied in a given direction, then the body will change its length l in that direction. The work is just the force times the distance, or

$$dw = \mathbf{T}\,d\mathbf{l} \tag{18.33}$$

work being done *on* the body to increase its length. If this term is added to (18.31) and (18.32), we get the equations for a body that undergoes a tension in an electric field.

$$dE = T\,dS - P\,dV + \mathbf{E}\,d\mathbf{P} + \mathbf{T}\,d\mathbf{l} \tag{18.34}$$
$$dG = -S\,dT + V\,dP - \mathbf{P}\,d\mathbf{E} - \mathbf{l}\,d\mathbf{T} \tag{18.35}$$

Now suppose that the temperature and the pressure are kept constant. Then

$$dG = -\mathbf{P}\,d\mathbf{E} - \mathbf{l}\,d\mathbf{T} \tag{18.36}$$

Since dG is an exact differential, the Euler relation holds:

$$\left(\frac{\partial \mathbf{l}}{\partial \mathbf{E}}\right)_{T,P,\mathbf{T}} = \left(\frac{\partial \mathbf{P}}{\partial \mathbf{T}}\right)_{T,P,\mathbf{E}} \tag{18.37}$$

Equation (18.37) relates the change in length produced by applying an electric field to the electric polarization produced by tension. Materials for which these partial derivatives are nonzero are known as *piezoelectric* crystals. These materials change their dimensions under the application of an electric field. Conversely, the application of a tension induces an electric polarization. These crystals find application in strain gauges, with which an applied force is determined by measuring the resulting polarization produced. The first such instrument was constructed by Pierre Curie and his brother Jacques.

PIERRE CURIE (1859–1906), French physicist, is best known in scientific circles for his work on magnetism and radioactivity, and in nonscientific circles for his marriage to Maria Sklodowska Curie. At the age of eighteen, Pierre discovered the phenomenon piezoelectricity in collaboration with his brother Jacques. The two brothers followed this discovery by building a piezoelectric quartz balance that produced a voltage proportional to the weight suspended from it. Pierre and Marie Curie later used this device in their work on radioactivity. Between 1890 and 1895, Pierre devoted his efforts to magnetism and laid the foundation for the modern theories of magnetism. Curie's law of paramagnetism was discovered in this period. He married Maria Sklodowska in 1895, and the rest of his life was devoted to studies in radioactivity in close collaboration with his wife. This was the field that brought him his greatest fame, both in scientific and in general circles, and it was for this work that he, along with his wife and Henri Becquerel, received the Nobel Prize in physics in 1903. In his Nobel Prize address, Pierre noted: "We might still consider that in criminal hands radium might become very dangerous; and here we must ask ourselves if mankind can benefit by knowing the secrets of nature, if man is mature enough to take advantage of them, or if this knowledge will not be harmful to the world." This question is still being debated. Pierre concluded: "I am among those who believe that humanity will derive more good than evil from new discoveries." Pierre died prematurely at the age of forty-seven, the result of a fatal traffic accident.

18.5 MAGNETIC WORK AND COOLING BY ADIABATIC DEMAGNETIZATION

The induction of a sample that is placed in a magnetic field **H** is given by

$$\mathbf{B} = \mathbf{H} + 4\pi \mathbf{I} \tag{18.38}$$

where **I** is the magnetization per unit volume. The induction is the sum of two magnetic terms, the external applied field **H**, and the field induced in the sample, $4\pi\mathbf{I}$. A *diamagnetic* substance is one for which the induced field is opposed to the applied field, whereas a *paramagnetic* substance is one in which the induced field is in the same direction as the external applied field.[4] A diamagnetic substance is repelled by a magnetic field and a paramagnetic substance is attracted to the field. Equation (18.38) is often written

$$\mathbf{B} = \mu \mathbf{H} \tag{18.39}$$

where

$$\mu = 1 + \frac{4\pi \mathbf{I}}{\mathbf{H}} \tag{18.40}$$

is called the *permeability*. It is greater than unity for paramagnetic and ferromagnetic substances and less than unity for diamagnetics.

The *magnetic susceptibility* is defined by

$$\chi = \left(\frac{\partial \mathbf{I}}{\partial \mathbf{H}}\right)_\xi \tag{18.41}$$

where the ξ is some quantity or quantities, typically T and P, that are held constant. For most paramagnetic substances, the susceptibility is independent of the field strength at low fields and is given by

$$\chi = \frac{\mathbf{I}}{\mathbf{H}} \tag{18.42}$$

Note that χ is a dimensionless quantity. Molar quantities are most convenient in thermodynamics, and the molar intensity of magnetization and molar susceptibilities are given by

$$\mathbf{M} = \overline{V}\mathbf{I} \tag{18.43}$$
$$\chi_m = \overline{V}\chi \tag{18.44}$$

$\overline{V}$ being the molar volume. The units of χ_m are cm³ mol⁻¹. The paramagnetic susceptibilities of many substances can be represented by

$$\chi_m = \frac{C}{T} \tag{18.45}$$

where the constant C is known as the *Curie constant*. Equation (18.45) is known as Curie's law.

The work associated with the magnetization of a mole of substance is

$$w = \int \mathbf{H} \, d\mathbf{M} \tag{18.46}$$

---

[4] See Chapter 38 for a more complete discussion of the nature of diamagnetic and paramagnetic materials.

If the magnetic field is in units of oersteds, then the work is in erg mol^{-1}.[5] The equations for the energy, the enthalpy, and the free energy take the form

$$dE = T\, dS - P\, dV + \mathbf{H}\, d\mathbf{M} \qquad (18.47)$$
$$dH = T\, dS + V\, dP - \mathbf{M}\, d\mathbf{H} \qquad (18.48)$$
$$dG = -S\, dT + V\, dP - \mathbf{M}\, d\mathbf{H} \qquad (18.49)$$

We can define a heat capacity at constant pressure and magnetic field as

$$C_{P,\mathbf{H}} = \left(\frac{\partial H}{\partial T}\right)_{P,\mathbf{H}} = T\left(\frac{\partial S}{\partial T}\right)_{P,\mathbf{H}} \qquad (18.50)$$

Now let us consider the entropy as a function of magnetic field and temperature at constant pressure.

$$dS = \left(\frac{\partial S}{\partial \mathbf{H}}\right)_{P,T} d\mathbf{H} + \left(\frac{\partial S}{\partial T}\right)_{P,\mathbf{H}} dT \qquad (18.51)$$

If we specify that the entropy is constant, then $dS = 0$ in Equation (18.51), and

$$\left(\frac{\partial S}{\partial \mathbf{H}}\right)_{P,T} = -\left(\frac{\partial S}{\partial T}\right)_{P,\mathbf{H}} \left(\frac{\partial T}{\partial \mathbf{H}}\right)_{P,S} \qquad (18.52)$$

The derivative $\partial S/\partial T$ is given in terms of the heat capacity in Equation (18.50). We can also apply the Euler condition for an exact differential to the free-energy expression in (18.49) to show that

$$\left(\frac{\partial S}{\partial \mathbf{H}}\right)_{P,T} = \left(\frac{\partial \mathbf{M}}{\partial T}\right)_{P,\mathbf{H}} \qquad (18.53)$$

Equation (18.52) can now be written

$$\left(\frac{\partial T}{\partial \mathbf{H}}\right)_S = -\frac{T}{C_\mathbf{H}}\left(\frac{\partial \mathbf{M}}{\partial T}\right)_\mathbf{H} \qquad (18.54)$$

where we have omitted the subscript P; it is understood that the condition of constant pressure is still imposed.

Equation (18.54) gives the rate of change of temperature with magnetic field at constant entropy. It forms the basis for cooling paramagnetic salts to very low temperatures. Both the temperature and the heat capacity in Equation (18.54) are positive quantities. For a paramagnetic salt that obeys Curie's law, the derivative $\partial \mathbf{M}/\partial T$ is negative; hence the left side of (18.54) must be a positive quantity. The temperature increases with increasing magnetic field for an isentropic process. The procedure is outlined on the T-S diagram of Figure 18.1. A paramagnetic salt at zero magnetic field is initially at a temperature T_i, typically that of a liquid helium thermostat. The sample is isothermally magnetized to some magnetic field $\mathbf{H}$, and the entropy of the sample changes as shown by the line $1 \rightarrow 2$. The sample is then thermally isolated from the liquid helium bath, after which it is reversibly demagnetized by lowering the magnetic field back to zero. The entropy remains constant, and the temperature of the sample falls to T_f as shown by the line $2 \rightarrow 3$.

[5] In the SI System, the unit of magnetic induction is kg s^{-2} A^{-1}. The unit is called the *tesla*, symbol T. It is 10^4 gauss in electromagnetic units.

SEC. 18.5 MAGNETIC WORK AND COOLING BY ADIABATIC DEMAGNETIZATION

FIGURE 18.1 Cooling by adiabatic demagnetization. Curves on a T-S diagram are shown for a paramagnetic salt at zero magnetic field and at a field $\mathbf{H}$. The sample is initially at T_i at point 1 on the zero field curve. As the magnetic field is raised to $\mathbf{H}$, the entropy decreases isothermally, as shown by the dashed line $1 \rightarrow 2$. The sample is then thermally isolated, the field is lowered to zero, and the sample cools to T_f, as shown by the dashed line $2 \rightarrow 3$. The additional dashed lines show the successive stages by which lower temperatures can be achieved.

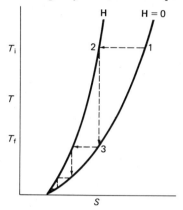

The magnitude of this *magnetocaloric* effect can be calculated in the following way. Noting that $\mathbf{M} = \chi \mathbf{H}$, Equation (18.54) can be written

$$dT = + \frac{TC\mathbf{H}}{C_\mathbf{H} T^2} d\mathbf{H} \tag{18.55}$$

for a paramagnetic salt that obeys Curie's law; C is the Curie law constant. An expression for the heat capacity can be derived by taking the derivative of the heat capacity:[6]

$$\left(\frac{\partial C_\mathbf{H}}{\partial \mathbf{H}}\right)_{T,P} = T \frac{\partial^2 S}{\partial \mathbf{H} \, \partial T} = T \frac{\partial}{\partial T}\left(\frac{\partial S}{\partial \mathbf{H}}\right)_{P,T}$$

$$= T \frac{\partial}{\partial T}\left(\frac{\partial \mathbf{M}}{\partial T}\right)_\mathbf{H} = T \frac{\partial^2 \mathbf{M}}{\partial T^2}$$

$$= T \mathbf{H} \frac{\partial^2 \chi}{\partial T^2}$$

By Curie's law, $\chi = C/T$, this becomes

$$\frac{\partial C_\mathbf{H}}{\partial \mathbf{H}} = \frac{2CT}{T^3} \mathbf{H} \tag{18.56}$$

This can be integrated with respect to $\mathbf{H}$:

$$C_\mathbf{H} = C_\mathbf{H}(T,0) + \frac{CT}{T^3} \mathbf{H}^2 \tag{18.57}$$

---

[6] In Equations 18.55–18.60, $C_\mathbf{H}$ is a constant magnetic field heat capacity, while C with no subscript is the Curie constant. The usage should be clear from the context.

TABLE 18.1 The magnetocaloric effect.

Paramagnetic Salt	T_f (K)
$Cr(NO_3)_3 \cdot 9H_2O$	0.21
$FeNH_4(SO_4)_2 \cdot 12H_2O$	.090
$Mn(NH_4)_2(SO_4)_2 \cdot 6H_2O$	.165
$CuK_2(SO_4)_2 \cdot 6H_2O$	.099
$Gd_2(SO_4)_3 \cdot 8H_2O$	.28
$CrK(SO_4)_2 \cdot 12H_2O$	.20
$Ce_2Mg_3(NO_3)_{12} \cdot 24H_2O$	0.01

SOURCE: D. de Klerk, *Handbuch der Physik*, vol. 15 (1956):38.

NOTE: Values of T_f reached by adiabatic demagnetization from $T_i \simeq 1.1$ K and $H_i \simeq 5000$ oersteds.

where $C_H(T,0)$ is the heat capacity at zero magnetic field. This is approximately given by $C_H(T,0) = B/T^2$, where B is a constant; thus

$$C_H = \frac{B}{T^2} + \frac{CH^2}{T^2} \tag{18.58}$$

If we combine all these results, Equation (18.55) can be written

$$\frac{dT}{T} = \left(\frac{CH}{B + CH^2}\right) dH \tag{18.59}$$

Integrating gives

$$\ln\left(\frac{T_f}{T_i}\right) = \frac{1}{2} \ln\left(\frac{B + CH_f^2}{B + CH_i^2}\right) \tag{18.60}$$

where T_i and T_f are the initial and final temperatures, and H_i and H_f are the initial and final magnetic fields. If the field is sufficiently high that the terms in B can be neglected, this becomes

$$\frac{T_f}{T_i} = \frac{H_f}{H_i} \tag{18.61}$$

A convenient initial temperature is 1 K, which is obtained by pumping on liquid helium. Convenient magnetic fields are of the order of 10,000 oersteds. Final temperatures of the order of 0.001 K can be obtained.

The Curie law does not usually hold down to these very low temperatures; Eq. (18.61) is therefore a rough approximation. More accurate treatments are presented in several of the references listed in the Bibliography. Some experimental magnitudes of the magnetocaloric effect are listed in Table 18.1. Note that all the salts listed in the table are magnetically dilute. This means that the fraction of paramagnetic ions is small. This ensures that the average distance between the magnetic ions is large, and that the magnetic interactions are small. Under this condition, the Curie law is more closely followed.

18.6 THE INTERFACE REGION AND EXCESS ENERGY OF SURFACES

Suppose that we have a one-component system consisting of two phases, liquid and vapor. The molecules are roughly spherical, with dimensions of the order of 5×10^{-8} cm. Each molecule in the interior of the liquid is surrounded by n other molecules on

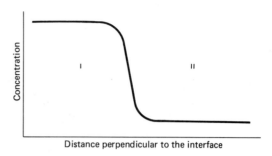

FIGURE 18.2 Concentration as a function of distance perpendicular to the plane of the liquid and vapor *interface*.

the average. If the cohesive energy of attraction between two adjacent molecules is ϵ/n, then the total energy of each molecule in the interior of the bulk liquid is ϵ. At the surface of the liquid each molecule is surrounded by fewer than n molecules; thus the attractive energy of each molecule at the surface is less than ϵ. The energy of a molecule at the surface of a liquid is thus higher than the energy in the interior. At the surface, a net attractive force pulls the surface towards the interior. Since the surface is in equilibrium, there must be a corresponding surface force pushing outward. These opposing forces, which are absent in the interior of the liquid, give rise to several unique properties of surfaces. Figure 18.2 shows a plot of a variable such as concentration across the interface region. The thickness of the region over which there is an abrupt change in the property is of the order of several molecules.

If the surface area of a liquid is increased, say by pouring from a narrow beaker into a wide beaker, then molecules must be moved from the interior of the liquid to the surface. Work must be done in creating this new surface. The existence of this excess energy in the surface explains the spherical shape of liquid droplets. The sphere has the smallest surface for a given volume. Hence a spherical droplet has the smallest surface area, and hence smallest surface energy, in accord with the minimum-energy principle.

We have hitherto justifiably neglected the effect of the surface. Suppose that the unique interfacial region shown in Figure 18.2 extends through a thickness of 10 molecular layers and that the layers are 5×10^{-8} cm apart. This means that in a 1-cm³ sample of liquid in the shape of a cube the volume of the interfacial region is 3×10^{-6} cm³, a negligible fraction of the total volume. For larger samples, the fractional part of the volume occupied by the interfacial region is even smaller. In the next several sections we shall examine several systems in which the effect of this excess surface energy can no longer be neglected.

18.7 SURFACE TENSION AND SURFACE ENERGY

Figure 18.3 shows a thin film of liquid stretched on a wire frame. The apparatus is arranged so that the wire AB is free to move along the frame. We observe that a force f must be applied to keep the surface from shrinking and that this force is independent of the position of the wire. It is a function only of the length of the wire and of the nature of the liquid.[7] Since f is proportional to l, the length of the wire, we can write

---

[7] The liquid film is often compared with a thin rubber membrane. Note, however, that for a rubber membrane the force depends not only on l but also on the total area. The larger the area, the larger the force needed to hold the stretched membrane in place.

FIGURE 18.3 A liquid film on a wire frame. (a) The frame at an initial position. (b) The movable end has been displaced by a distance dx. Work must be done to extend the film.

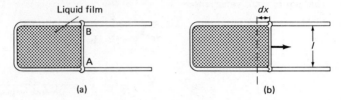

$$f = \gamma \cdot 2l \qquad (18.62)$$

where γ, the *surface tension,* is the force per unit length at the edge of the surface needed to keep the surface from collapsing. There is a factor 2 because the liquid film has two sides, a top and a bottom; the total length along the boundary is thus $2l$. Surface tensions of several liquids are listed in Table 18.2.

If we stretch the film by moving the wire a distance dx, as shown in Figure 18.3b, then the work done in extending the surface is

$$dw = f\,dx = 2\gamma l\,dx \qquad (18.63)$$

The change in area of the surface is $da = 2l\,dx$; thus

$$dE = dw = \gamma\,da \qquad (18.64)$$

TABLE 18.2 Surface tensions (γ) of liquids.

Liquid	Temperature	γ
ORDINARY LIQUIDS (MEASURED AGAINST AIR)		
Acetic acid	293 K	0.0276
Acetone	293	.0237
Benzene	293	.0289
Carbon tetrachloride	293	.0268
n-Hexane	293	.0184
Toluene	293	.0284
Water	293	0.07275
OTHER LIQUIDS[a] (VARIOUS TEMPERATURES)		
H_2 (vap)	14.7 K	0.00288
Ne (vap)	24	0.0059
N_2 (vap)	70	0.0105
Hg (H_2)	272	0.470
Cu (H_2)	1404	1.13
Ag	1243	0.800
Pt	2273	1.82
NaCl (N_2)	1269	0.0997
AgCl	725	0.1255

NOTE: Units are newtons per meter (N m^{-1}). To convert from units of dyn cm^{-1} (= erg cm^{-2}), which are listed in most tables, to the SI unit N m^{-1}, multiply by 10^{-3}.

[a] If the surface tension has not been measured against air, the second entry indicates the gas against which it was measured. For the first three entries, the surface tension was measured against the pure vapor of the substance listed.

Since $\gamma = dE/da$, we can interpret the surface tension as the energy per unit area of surface. The Gibbs free energy can be written in the form[8]

$$dG = -S\,dT + V\,dP - a\,d\gamma \tag{18.65}$$

The condition for equilibrium between the bulk phase and the surface is that material can be reversibly transferred between surface and bulk at constant-bulk volume and constant surface area. In other words, it is the Helmholtz energy that must be minimized:

$$dA = -S\,dT - P\,dV + \gamma\,da = 0 \tag{18.66}$$

In the next several sections we shall note that unlike systems previously examined, the systems in which the effect of the surface is not negligible can be in equilibrium even though the pressure is not uniform throughout the system. For this reason, surface tension is sometimes thought of as a "negative pressure." This negative pressure adds to the excess pressure in certain parts of the system to bring about a net "effective" pressure equality throughout the system.

18.8 BUBBLES AND CURVED INTERFACES

The surface tension of a bubble—a soap bubble, for example—acts so that the bubble has a natural tendency to contract and reduce its surface area. The spherical shape of the bubble is maintained by the excess pressure in the interior of the bubble. Let's calculate this pressure for a spherical bubble. Figure 18.4 shows the bubble that we imagine sliced into two hemispheres; we consider the forces in the upper hemisphere. Besides the external force, there is a downward force F due to the surface tension. This surface force is just the surface tension multiplied by the circumference. The total downward force is then

$$F = 2(2\pi R\,\gamma) + P_{\text{ext}}(\pi R^2) \tag{18.67}$$

FIGURE 18.4 The forces acting on a bubble.

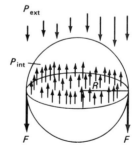

---

[8] A certain lack of uniformity seems to exist among various authors on this point, especially authors of elementary physical chemistry textbooks. Many authors simply add the work term directly to the free-energy expression, to get

$$dG' = -S\,dT + V\,dP + \gamma\,da$$

There is nothing incorrect about this; in the transformation leading from the energy to the free energy, these authors simply leave the surface work term untransformed and treat this term exactly as the chemical work term $\mu\,dn$. We have adopted the consistent system of transforming *all* the variables (except the chemical work term) for our free energies.

where P_{ext} is the external pressure and R is the radius of the bubble; the extra factor 2 arises since the bubble has an inner and an outer surface. The second term in (18.67) is the downward external force arising from the action of P_{ext} on the hemisphere. The total upward force is $P_{int}(\pi R^2)$. Since mechanical equilibrium requires the upward and downward forces to be equal,

$$P_{int}(\pi R^2) = 4\pi R \gamma + P_{ext}(\pi R^2)$$

The excess pressure in the bubble is

$$P_{int} - P_{ext} = \frac{4\gamma}{R} \qquad (18.68)$$

Note that the larger the bubble, the smaller the excess bubble pressure.

Suppose that we have instead of a bubble a curved spherical interface between two phases, say liquid and gas. If we try to calculate the excess internal pressure in the liquid phase, we follow the same procedure just outlined, with one slight difference. Since the surface of the liquid now has only one side, the factor 2 in (18.67) is eliminated, and the final result becomes

$$P_1 - P_2 = \frac{2\gamma}{R} \qquad (18.69)$$

where we have used P_1 and P_2 for P_{int} and P_{ext} to avoid an anomalous discrepancy between the two equations. A more rigorous treatment of the pressure differential would take into account a general curved surface, for which the result is

$$P_1 - P_2 = \gamma \left(\frac{1}{r_1} + \frac{1}{r_2}\right) \qquad (18.70)$$

where r_1 and r_2 are the two radii of curvature that characterize the surface.

This pressure difference is a fundamental consequence of the surface tension. Suppose we consider a droplet of water with $R = 10^{-4}$ m. For water, $\gamma = 0.073$ N m^{-1} and

$$P_1 - P_2 = \frac{2 \times 7.3 \times 10^{-2}}{10^{-4}} = 1460 \text{ N m}^{-2}$$

$$= \frac{1460}{1.01325 \times 10^5} = 1.4 \times 10^{-2} \text{ atm}$$

Since there is only a small difference between P_1 and P_2, either can be used for most purposes.[9]

18.9 DROPLET FORMATION AND CAPILLARY RISE

Figure 18.5 shows a droplet of liquid forming from a tube of radius r. As the liquid slowly extends past the bottom of the tube, the excess liquid is held to the tip of the tube by surface tension forces. As the volume of the extended mass of liquid increases, the drop finally breaks off when the surface force can no longer support the

---

[9] A rigorous treatment indicates that this is true so long as the thickness of the surface layer is much less than the radius of the droplet.

SEC. 18.9 DROPLET FORMATION AND CAPILLARY RISE

FIGURE 18.5 A droplet forming from a tube of radius r.

weight of the droplet. At the instant of breaking, the weight equals the surface force,

$$mg = 2\pi r \gamma \qquad (18.71)$$

For a spherical droplet of radius R, this weight is

$$mg = \tfrac{4}{3}\pi R^3 \rho g \qquad (18.72)$$

where ρ is the density of the liquid. Equations (18.71) and (18.72) can be combined in several ways. The form

$$\gamma = \frac{mg}{2\pi r} = \frac{2}{3}\frac{R^3 \rho g}{r} \qquad (18.73)$$

provides a direct experimental method for measuring the coefficient of surface tension from the weights (or radii) of droplets formed from a tube of radius r. If liquids of equal densities drop from a tube of given radius, the droplet radius varies with $\gamma^{1/3}$.

Capillary rise provides one of the most common experimental methods for determining coefficients of surface tension. If a capillary tube is immersed in a liquid such as water, the liquid rises in the capillary above the level of the bulk liquid, as shown in Figure 18.6. We want to compute the height h to which the liquid rises in the tube.

The tube is immersed to a depth d. At the top of the capillary column, there is a downward force due to atmospheric pressure that is $-P\pi r^2$, where r is the radius of the capillary tube. At the bottom of the tube, there is an upward force from the pressure at depth d in the liquid. This force is $(P + \rho g d)\pi r^2$. The downward force of gravity acting on the column is $-(h + d)\rho g \pi r^2$. An additional force arises from the action of the surface tension of the liquid. The top of the liquid column forms a *meniscus*, or curved surface, that makes a *contact angle* ϕ, as indicated in the figure. The

FIGURE 18.6 Capillary rise of a liquid that wets the tube.

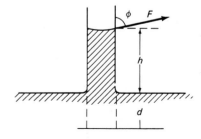

FIGURE 18.7 The radius of curvature of the meniscus in a capillary tube, in this instance for a liquid that does not wet the walls of the capillary. We have $r/R = \sin\theta = \sin(\phi - 90°) = -\cos\phi$, or $R = -r/\cos\phi$, where r is the radius of the capillary. For the case of the convex meniscus as illustrated, the capillary rise is negative; that is, there is a capillary depression.

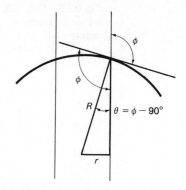

surface tension force is tangent to the surface in the direction ϕ. The vertical component of this force is $2\pi r\gamma \cos\phi$. At equilibrium, the sum of all the forces must be identically zero,

$$-P\pi r^2 + (P + \rho g d)\pi r^2 - (h + d)\rho g \pi r^2 + 2\pi r\gamma \cos\phi = 0$$

or
$$h = \frac{2\gamma \cos\phi}{\rho g r} \qquad (18.74)$$

The case illustrated in Figure 18.6 is for a liquid that "wets" the glass. Such liquids form menisci that are concave upward, as shown. The contact angle is less than 90°, there is an upward component to the force F, and these liquids rise in capillary tubes. Mercury, on the other hand, does not "wet" glass, and forms a meniscus that is convex upward. The contact angle is greater than 90°; there is a downward component to the force F, and such liquids are depressed in capillary tubes. This so-called *capillary depression* can be the source of substantial errors in instruments in which the height of a mercury column is read in narrow tubes. The McLeod gauge forms a prime example of this problem.

In the previous section we noted that there is a pressure differential across a curved interface. The capillary rise measures this pressure differential. While we have derived Equation (18.74) by a mechanical approach, the same equation can be derived using the results of that section. What is needed is the radius of curvature of the surface. From the construction of Figure 18.7, we see that the term for R in (18.69) is given by $R = -r/\cos\phi$. The remainder of the derivation of (18.74) based on the pressure differential is left for one of the problems.

18.10 EFFECT OF DROPLET SIZE ON PHASE EQUILIBRIUM

Throughout the previous discussions of phase equilibria our ultimate criterion was that $dG = 0$. This was so because we were dealing with systems at constant temperature and pressure. In our derivation of the Clausius-Clapeyron equation, for example, we used the condition for equilibrium at constant pressure and temperature. In dealing with surfaces, however, the situation is complicated because the pressure is not constant throughout the system.

First let us, without regard to surface effects, examine the effect of pressure on vapor pressure. Conceptually, the process can be visualized by means of the idealized apparatus shown in Figure 18.8. The lower piston is impervious to liquid but al-

SEC. 18.10 EFFECT OF DROPLET SIZE ON PHASE EQUILIBRIUM

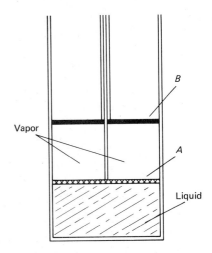

FIGURE 18.8 Apparatus for studying the vapor pressure of a liquid as a function of applied pressure. The lower piston (A) is composed of a semipermeable substance that allows vapor to pass but is impervious to the liquid. The upper piston (B) is impervious to both the liquid and vapor phases.

lows vapor to pass, and the upper piston is impervious to both liquid and vapor. If the force on the lower piston is adjusted, the vapor pressure of the liquid can be measured as a function of applied pressure by connecting a pressure-measuring instrument to the region between the two pistons. For equilibrium

$$-\overline{S}_l \, dT + \overline{V}_l \, dP_l = -\overline{S}_v \, dT + \overline{V}_v \, dP_v \tag{18.75}$$

where the subscripts l and v refer to liquid and vapor, and the bars over S and V indicate molar quantities. If the temperature is constant, we can write

$$\frac{dP_v}{dP_l} = \frac{\overline{V}_l}{\overline{V}_v} \tag{18.76}$$

If the vapor behaves like an ideal gas, $\overline{V}_v = RT/P_v$, and we can rewrite this equation as

$$d \ln p = \frac{\overline{V}_l}{RT} dP \tag{18.77}$$

where p is the vapor pressure and P is the applied pressure.

To apply this result to surfaces, we note that for droplets P is just the excess pressure given in Equation (18.69). Hence $P = 2\gamma/r$, where r is the radius of the droplet; thus $dP = d(2\gamma/r)$. We now integrate the equation

$$\int_{p_0}^{p} d \ln p = \frac{\overline{V}_l}{RT} \int_{\infty}^{r} d\left(\frac{2\gamma}{r}\right) \tag{18.78}$$

where the lower limit corresponds to an infinitely large drop whose vapor pressure is p_0, the normal vapor pressure, and the upper limit corresponds to the vapor pressure of a droplet of radius r. The final result is

$$\ln \frac{p}{p_0} = \frac{2\gamma \overline{V}_l}{rRT} \tag{18.79}$$

Equation (18.79) was first derived by Kelvin and is called the Kelvin equation.

The enhanced vapor pressure of small droplets has its counterpart in the enhanced solubility of small droplets. By using the relations between solubility and vapor pressure (developed in Chapter 12), we can show that the solubility of a small droplet in a solvent is given by

$$\ln \frac{X_B}{X_B^\circ} = \frac{2\gamma \overline{V}_1}{RTr} \tag{18.80}$$

where X_B is the solubility of the small droplet (in mole fraction units) and X_B° is the normal solubility of a large sample of solute; γ is the interfacial tension between the droplet and the solvent and $\overline{V}_1$ is the molar volume of the solvent. Equation (18.80) applies to the case in which the solvent is completely insoluble in the solute.

Equations (18.79) and (18.80) pose somewhat of an anomaly. If we have a vessel containing a vapor such as water vapor just above its condensation point and suddenly cool the vessel to below the condensation point, then a mist or fog consisting of large numbers of very small droplets will form spontaneously as the vessel becomes supersaturated with the vapor. Any reasonable model of this condensation would start by assuming that a small number of molecules come together to form a small droplet, which gets bigger as additional molecules condense. Suppose we take the initial embryo droplet to have a radius of 10^{-9} m. For water at 25 °C, $\gamma = 0.07275$ N m^{-1}, and $\overline{V}_1 = 18 \times 10^{-6}$ m^3, so

$$\ln \frac{p}{p^\circ} = \frac{(2)(0.073)(18 \times 10^{-6})}{(10^{-9})(8.314)(298)} = 1.06$$

$$\frac{p}{p^\circ} = 2.89$$

The small droplet has a vapor pressure almost three times the normal vapor pressure of water; hence we should expect any small droplet that forms to revaporize immediately. We do know, however, that these mists form. It is most probable that condensation takes place on the surfaces of dust particles, which act as nuclei for the condensation of water molecules. A similar explanation holds for the precipitation of small crystallites from supersaturated solutions. The enhanced solubility of smaller particles is advantageous in growing large pure crystals from solution. The very small crystals tend to redissolve during the crystallization, and this phenomenon allows the faster formation of the large, more desirable crystals.

18.11 ADSORPTION ON SOLID SURFACES

Granulated white sugar is manufactured by dissolving raw brown sugar in water and then recrystallizing the solution to produce the pure white sugar. The raw sugar, and also the original solution, have a brownish color because of various impurities. These colored impurities are removed by *adsorption*, using *activated charcoal*. This is a prepared form of carbon consisting of finely divided carbon particles in which the surface-area-to-weight ratio is extremely high. When activated charcoal is added to the colored solution of raw sugar, the colored impurities are selectively adsorbed on the charcoal surface; the charcoal material is then removed by filtration. Granulated white sugar is prepared by crystallization from the purified solution.[10]

---

[10] Crystals of sugar (sucrose) are actually clear and colorless; you can see this by observing the large sugar crystals commercial food processors use, or the so-called rock candy, which is nothing more than large crystals of sucrose. Granulated sugar has a "white" color because of the light-reflecting properties of numerous small clear crystals. Activated carbon is now extensively used to remove organic matter from drinking water, and also from waste water in chemical plants.

Those of you who have done laboratory work in preparative organic chemistry may have used this process for purifying a solution before crystallizing the final product.

Materials with large surface area to mass ratios are also used to remove gases from the vapor phase. In high-vacuum processes, using adsorbent *traps* containing *silica gel,* an inert porous material with a large surface area, is commonplace. The silica gel surface adsorbs numerous gases, especially water vapor and stray oil molecules, from the oil-diffusion pumps.

The extent of adsorption depends on the temperature and the nature and degree of subdivision (hence the surface area) of the *adsorbent*. We differentiate between two kinds of adsorption. In *physical adsorption,* or *physisorption,* the forces between the adsorbent surface and the *adsorbate* molecules are similar to the van der Waals forces between molecules. The energies of adsorption are low, generally less than 3 kJ mol^{-1}. Physisorption is usually reversible, and the adsorbent surface can usually be purged of adsorbed gases by heating. Adsorbent traps for high-vacuum systems usually contain built-in heaters that are periodically used to free the silica gel of adsorbed gases. The binding forces in *chemisorption* are much stronger than in physisorption, and the energies of adsorption range up to 400 kJ mol^{-1}. Chemisorption is much less reversible than physisorption; and the nature of the gas that is *desorbed* on heating is often different from the gas adsorbed. Thus O_2 is chemisorbed on activated charcoal at 150 K. If one then attempts to desorb the gas by heating and pumping, substantial quantities of CO rather than O_2 are given off. It is often difficult to differentiate between the two processes, and some systems display chemisorption under one set of conditions and physisorption under a different set of conditions. Spectroscopic methods can be useful in distinguishing the two processes. The infrared (ir) spectrum of an adsorbate that is physically adsorbed on a surface will be very similar to the spectrum of the pure bulk adsorbate, whereas the spectrum of a chemisorbed species will be substantially different from the spectrum of the pure adsorbate.

Figure 18.9 shows two different styles of apparatus used for studying adsorption of gases. In the first, shown in Figure 18.9a, the volume of gas adsorbed on the surface of the adsorbent in the central bulb is directly measured by ordinary gas-handling techniques. In the second, shown in Figure 18.9b, the adsorbent is suspended by a fine coiled spring (usually quartz) and the amount of gas adsorbed is determined from the weight increase of the adsorbent. All adsorbents adsorb air from the atmosphere while they are on the shelf and while they are being transferred from bottle to apparatus. It is necessary to remove these adsorbed gases before any measurements can be undertaken. This is usually accomplished by baking the adsorbent under high vacuum to drive off these impurities.

Since adsorption is a surface phenomenon, it seems that the amount of gas adsorbed should be proportional to the total surface area. This statement is usually accepted as a postulate of surface chemistry. Indeed, the amount of gas adsorbed is proportional to the amount of adsorbent present; adsorption experiments are often used to determine the total surface area of porous and particulate samples of material. The data are normally plotted as an *adsorption isotherm*. The volume of gas adsorbed is measured as a function of pressure at constant temperature and is plotted on the basis of gas volume adsorbed per unit weight of adsorbent or some other convenient unit. Adsorption isotherms for ethyl chloride are shown in Figure 18.10. Sometimes the *relative pressure* $P/P°$ is used instead of the pressure in which $P°$ is the vapor pressure of the adsorbate at the isotherm temperature. Plots of the volume of gas adsorbed as a function of temperature at constant pressure are called *adsorption*

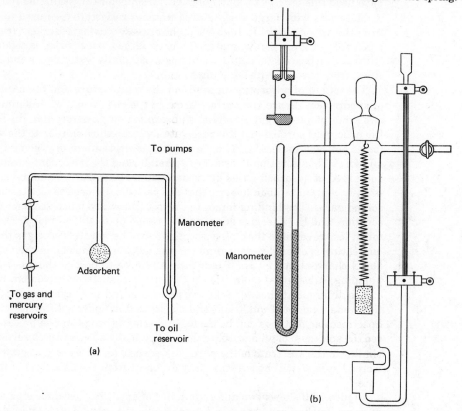

FIGURE 18.9 Apparatus for measuring gas adsorption. (a) The volume of gas adsorbed by the adsorbent placed in the central bulb is measured by ordinary gas-handling techniques. (b) The weight of gas, and hence the volume, adsorbed by the adsorbent suspended from the spring is determined from the increase in weight, as shown by the increase in length of the spring.

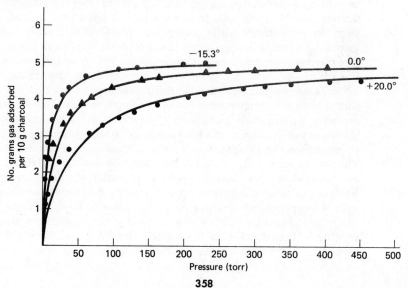

FIGURE 18.10 Adsorption isotherms of ethyl chloride on charcoal. (F. Goldmann and M. Polanyi, *Z. Physik. Chem.* 132 (1928):321.)

isobars, and plots of the equilibrium pressure vs temperature at constant adsorbed volume are called *adsorption isosteres.*

In 1918, Langmuir postulated the mechanism of adsorption as follows.[11] A monomolecular layer covers the surface of the adsorbent. The surface of the adsorbent contains a fixed number of sites S, of which S_1 are occupied by adsorbate molecules and $S - S_1 = S_0$ sites are unoccupied. This adsorbed layer acts like a two-dimensional perfect gas in that there is no interaction energy between molecules adsorbed on adjacent sites. The heat of adsorption is thus independent of the total number of sites occupied, and the energy required to remove an adsorbed molecule is independent of the number of molecules adsorbed. Langmuir envisioned adsorption as a state of dynamic equilibrium, in which the rate at which molecules are adsorbed equals the rate at which molecules are desorbed. The rate of adsorption (or condensation) onto the surface is proportional to the number of bare sites and the pressure, or equal to $k_a S_0 P$, where k_a is the proportionality constant for adsorption. Similarly, the rate of desorption (or evaporation) of molecules is proportional to the number of sites occupied by adsorbate molecules, or equal to $k_d S_1$, where k_d is the proportionality constant for desorption. At equilibrium,

$$k_d S_1 = k_a S_0 P = k_a (S - S_1) P \qquad (18.81)$$

The fraction of sites occupied is S_1/S. If this ratio is called θ, then Equation (18.81) can be solved for θ to get

$$\frac{1}{\theta} = 1 + \frac{1}{bP} \qquad (18.82)$$

where $b = k_a/k_d$ is called the *adsorption coefficient.* Since the volume of gas adsorbed is proportional to θ, a plot of $1/V$ vs $1/P$ yields a straight line, as shown in Figure 18.11.

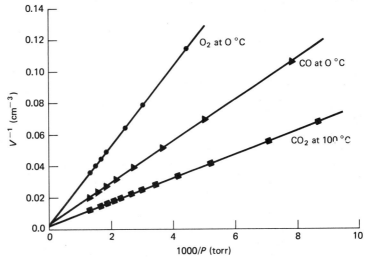

FIGURE 18.11 Langmuir plots for the adsorption of gases on silica. (Data from E. C. Markham and A. F. Benton, *J. Amer. Chem. Soc.* 53 (1931):497.)

[11] I. Langmuir, *J. Amer. Chem. Soc.* 40 (1918):1361.

Adsorption isotherm experiments provide a measure of the total amount of gas adsorbed by the sample. To deduce the total surface area from these measurements, one needs to know the area of each adsorbed molecule. Without reliable data from other sources, this area is usually assigned according to the equation

$$A = 4(0.866) \left(\frac{M}{4\sqrt{2}N_0\rho}\right)^{2/3} \tag{18.83}$$

where M is the molar mass of the gas, N_0 is Avogadro's number, and ρ is the density of the liquefied (or solidified) adsorbate. Equation (18.83) assumes that the molecules are held in a two-dimensional close-packed arrangement on the surface of the adsorbent.

Surface effects are important in many areas of technology. The paper and paint industries are two that come to mind immediately as depending heavily on a proper understanding of surface chemistry. Lubrication, detergency, foams, and wetting agents are also based on surface chemistry. Further examples of the manifold problems in surface chemistry can be got from the references in the Bibliography. In the next section we shall touch on one application of surface chemistry, in a qualitative way.

18.12 AN EXAMPLE OF DYNAMIC SURFACE EFFECTS; CHROMATOGRAPHY

Suppose that a glass tube is packed with a finely divided inert adsorbent material and that a mixture of solutes is placed in a plane at the top of the column of adsorbent, as shown in Figure 18.12a. A solvent in which all the solutes are soluble is placed above the layer of solute, and the resulting solution is allowed to move down the *column* at a rate that can be varied by adjusting the opening of the stopcock at the bottom of the glass tube. Where no surface effects existed, the plane containing the solutes would simply move down the column, and all the components would

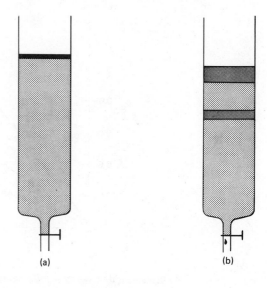

FIGURE 18.12 Liquid chromatography. A glass column is filled with a finely divided solid adsorbent; solvent is then poured into the column until it can hold no more. (a) A mixture of solutes is placed in a plane at the top of the column. (b) The stopcock is slightly opened to allow the liquid to slowly move down the column. The solutes, originally in one layer, are adsorbed and desorbed at different rates; hence they travel down the column at different rates. The original layer is separated into a series of layers, one for each component.

leave from the stopcock at the same time. Actually there are significant surface effects, and the different components of the solution are adsorbed and desorbed on the surface of the adsorbent as they move down the column.

Since each component has rates of adsorption and desorption peculiar to itself and different from the rates of other substances, the rate at which each component moves down the tube is different, and the various solute components are separated into a series of *bands,* as shown in Figure 18.12b. The number of components can be determined by noting the number of bands, and the mixture can be separated into its components by catching each band in a different beaker that has been placed under the stopcock. The term *chromatography* (meaning "color writing") for this process was coined by Tswett in 1906; he obtained a series of discretely colored bands on treating a mixture of paint pigments as just described. Applied to modern methods of gas chromatography, the term is somewhat of a misnomer, since any color is totally lacking.

The procedure just described is termed *liquid phase chromatography,* since the equilibrium is between the adsorbent and a liquid solution phase. *Gas phase chromatography,* in which the equilibrium is between the adsorbent (liquid or solid) and a gas phase solution, is the most common type in laboratory procedures. The interior of a long capillary tube is filled with the adsorbent material, and an inert *carrier* gas is continuously and slowly passed through the tube. At time $t = 0$, the sample consisting of a mixture of gases is injected into the stream of the carrier gas at the bottom of the tube. A big advantage of gas chromatographic analysis is that only minute samples are needed. As the sample gas is carried through the column by the carrier gas, the various components are adsorbed and desorbed at different rates, and they reach the detector at different times.

A schematic diagram of a gas chromatography system is shown in Figure 18.13. Various forms of detection systems are used. The simplest type is a thermal conductivity cell. As each component of the mixture reaches the detector, the thermal conductivity of the *eluted* gas changes slightly. This change is noted by the cell and displayed on the strip chart recorder. This type of detector can measure concentrations of the order of 100 parts per million. More sophisticated detectors can detect components at concentration levels down to one part in 10^9 (or about 10^{-12} g of material). Some *chromatograms* are shown in Figure 18.14.

Since chromatography separates a mixture into its components as fractional distillation does, much of the terminology of distillation has been carried over into chro-

FIGURE 18.13 Schematic drawing of a gas chromatography system. (H. M. McNair and E. J. Bonelli, *Basic Gas Chromatography,* published by Varian Instruments Division, 1968. Courtesy of Varian Associates, Inc., Palo Alto, California.)

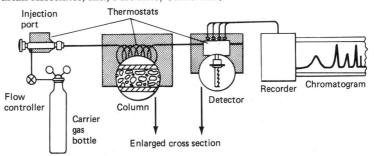

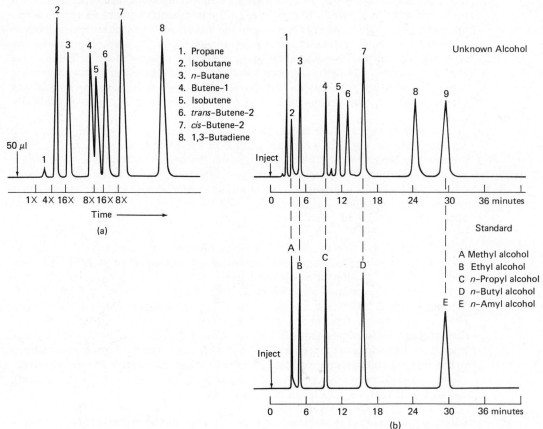

FIGURE 18.14 (a) A gas chromatogram of a mixture of hydrocarbons showing the separation into eight peaks. The scale for the various peaks is adjusted, as shown below the baseline. (b) Chromatogram with some peaks identified. A mixture of alcohols is compared with a standard mixture. (Reference is H. M. McNair and E. J. Bonelli, *Basic Gas Chromatography*, published by Varian Instruments Division, 1968. Courtesy of Varian Associates, Inc., Palo Alto, California.)

matography. While distillation separates components by their boiling-point differences, however, chromatography separates components by their differing rates of adsorption and desorption. In the chromatogram shown in Figure 18.14a, the boiling-point differences between several of the components are so small that they could be separated from each other by distillation only with the greatest of difficulty. The complete chromatogram was obtained in less than 20 min. Further, the chromatogram provides a rapid quantitative estimate of the amount of each constituent, since the areas under the peaks are directly proportional to the amount of that constituent present in the original sample. Unknown peaks can be identified by comparison with standards, as shown in Figure 18.14b.

Chromatography has developed into one of the most useful tools in the organic chemistry laboratory and also elsewhere. It provides a rapid and convenient method of analyzing all kinds of mixtures, and commercial instruments are available for all sorts of applications. In certain instances chromatography has been used for collecting minute samples of constituents of a mixture. One recent development has been to combine the gas chromatograph with a high-resolution mass spectrometer to identify unknown components of a mixture. The chromatograph separates the solution into its components, and these individual components are then fed into the mass spectrometer as a method of identifying the component. These systems are often combined with digital computers that control the operation of the apparatus and further provide a means of rapidly comparing the mass spectra with a large body of spectra stored in the memory of the computer. In an ideal system, the operator injects the sample, and the computer prints out the list of components in a short time.

PROBLEMS

1. The work required to stretch a spring is $dw = F\,dx$, where F is the force and dx the displacement of the spring. For a spring that obeys Hooke's law, $F = kx$, where k is the spring constant and x is the displacement from the equilibrium position. Write expressions for the energy, the enthalpy, the work function, and the Gibbs free energy for a Hooke's law spring.

2. The tension of a rubber band is given by the expression

$$t = aT\left[\frac{x}{x_0} - \left(\frac{x_0}{x}\right)^2\right]$$

where x is the length of the band, x_0 is the equilibrium length of the unstretched band, a is a constant characteristic of the band, and T is the temperature. You may omit all $P\,dV$ terms in the following.
(a) Write expressions for the energy, the enthalpy, the work function and the Gibbs free energy for the rubber band.
(b) Show that for this rubber band $(\partial E/\partial x)_T = 0$, and hence the internal energy is a function only of temperature.
(c) Suppose that $a = 10^3$ dyn K^{-1} and that $x_0 = \frac{1}{2}$ m. The band is reversibly and isothermally stretched from $\frac{1}{2}$ m to 1 m at 300 K. Calculate the work done on the band and the heat absorbed.
(d) The heat capacity of the band at constant length is $C_x = 1.0$ J K^{-1}. If the band had been stretched adiabatically in part (c), what would the final temperature of the band have been? (This is analogous to the adiabatic expansion of an ideal gas.)

3. Since the natural variation in gravitational fields is negligible over the dimensions of a laboratory, the equations we derived in the section on gravitational work have little practical applicability. On the other hand, artificial "gravitational fields" can be created in high-speed centrifuges, giving substantial changes in the gravitational field over very small distances. These fields, called *centrifugal fields*, do have practical applicability and are used extensively for molecular-weight determinations and other experiments in research dealing with high-molecular-weight polymers and materials of biological interest. The transformation from the gravitational field to the centrifugal field is easily accomplished by substituting for $g\,dh$ the expression $g\,dh \to -\omega^2 r\,dr$ in (18.26), where ω is the angular velocity in radians per second, and r is the distance from the axis of rotation. Write expressions for dE, dH, dA, and dG for a system in a centrifugal field. Show that the partial molar free energy of the ith species in a solution varies with r, according to the expression

$$\left(\frac{\partial \overline{G}_i}{\partial r}\right)_{P,T,X} = -M_i \omega^2 r$$

where M_i is the molecular weight of the ith species and X_i is its mole fraction at r.

4. Suppose we consider a column of solution in a high-speed centrifuge after it has achieved equilibrium. The pressure and composition may vary with r, but at equilibrium the partial molar free energy must be constant. (We assume that the temperature is constant, since the centrifuge is thermostatted.) For any constituent,

$$d\overline{G}_i = 0 = \left(\frac{\partial \overline{G}_i}{\partial r}\right) dr + \left(\frac{\partial \overline{G}_i}{\partial P}\right) dP + \left(\frac{\partial \overline{G}_i}{\partial X_i}\right) dX_i$$

First convince yourself that the pressure in this expression is not the ordinary atmospheric pressure from gravity but the pressure arising from the centrifugal field. Show that the pressure is given by $dP = \rho\omega^2 r\,dr$, where ρ is the density of the solution at r. Using this and the equation given in Problem 3 among other material, show that the condition for equilibrium can be expressed

$$(\rho \overline{V}_i - M_i)\omega^2 r\,dr + \left(\frac{\partial \overline{G}_i}{\partial X_i}\right) dX_i = 0$$

where $\overline{V}_i$ is the partial molar volume of the ith species.

5. Assume that the solution described in Problem 4 is an ideal solution and that the solution is incompressible (that is, that the product term $\rho \overline{V}_i$ is a constant independent of distance from the axis). Since for an ideal solution $(\partial \overline{G}_i/\partial X_i)_{P,T} = RT/X_i$, show that the condition for equilibrium given in Problem 4 can be integrated, to get

$$\ln \frac{X_i''}{X_i'} = \frac{(M_i - \rho \overline{V}_i)\omega^2}{2RT}[(r'')^2 - (r')^2]$$

where the double and single primes refer to the final and initial values.

6. For a solution of ideal gases, the partial pressure of each gas is given by $p_i = X_i P$. Show that the partial pressure of a gas in a centrifugal field is given by

$$\ln \frac{p_i''}{p_i'} = \frac{M_i \omega^2}{2RT}[(r'')^2 - (r')^2]$$

7. By applying the equation given in Problem 6 to each component of a two-component ideal gas mixture and then subtracting the result, show that

$$\ln \frac{p_2''/p_1''}{p_2'/p_1'} = \frac{(M_2 - M_1)\omega^2}{2RT}[(r'')^2 - (r')^2]$$

Uranium hexafluoride (UF$_6$) is an ideal gas mixture containing two uranium isotopes, ^{235}U and ^{238}U. Calculate the angular velocity required in a centrifuge at 300 K to change the abundance of ^{235}UF$_6$ from 1% at $r = 10$ cm to 3% at $r = 4$ cm.

8. Oxygen is paramagnetic; thus the presence of a magnetic field introduces an extra variable that may affect the vapor pressure of liquid oxygen in a nontrivial way. Derive the analog of the Clapeyron equation for magnetic fields. In other words, show that at constant temperature the change in vapor pressure with applied field is given by

$$\left(\frac{\partial P}{\partial \mathbf{H}}\right)_T = \frac{\Delta \mathbf{M}}{\Delta \overline{V}}$$

where $\Delta \overline{V}$ is the change in molar volume. Then, recognizing that $\mathbf{M} = \chi_m \mathbf{H}$, convert this expression to one involving χ_m instead of $\mathbf{M}$. Finally, take the vapor to be an ideal gas, ignore the molar volume of the liquid relative to the gas, and derive an expression that is the analog of the Clausius-Clapeyron equation for ideal gases.

9. Derive Equation (18.37), which states the principle of the piezoelectric effect.

10. Show that $(\partial V/\partial \mathbf{H})_{P,T} = -(\partial \mathbf{M}/\partial P)_{T,\mathbf{H}}$. The derivative $(\partial V/\partial \mathbf{H})_{P,T}$ describes the effect of magnetic field on the volume of a magnetic material and is called the *volume magnetorestriction*. The derivative $(\partial \mathbf{M}/\partial P)_{T,\mathbf{H}}$ describes how the magnetization is affected by the pressure; this variation is called the *piezomagnetic* effect. The above relation connects these two magnetocaloric effects.

11. Show that $(\partial V/\partial \mathbf{E})_{P,T} = -(\partial \mathbf{P}/\partial P)_{T,\mathbf{E}}$. This is the analog of the equation derived in Problem 10, and describes the *volume piezoelectric effect*. This effect has many practical applications. Crystals such as quartz and Rochelle salts, which possess the proper symmetry, display this effect. If an alternating electric field is applied to these crystals, their volumes will change at a frequency corresponding to the applied frequency. Sound waves are generated in water for use in sonar detection using equipment based on this principle. The crystal quartz oscillator used in electronic watches is based on this principle; the accuracy of these watches calls for the crystal to be oscillated at a resonant frequency. This effect was discovered by Pierre Curie along with his brother Jacques in 1877.

12. (a) Derive Equation (18.50) for $C_{P,\mathbf{H}}$.
    (b) Show that at constant volume and magnetization, the heat capacity is given by

$$C_{V,\mathbf{M}} \equiv \left(\frac{\partial E}{\partial T}\right)_{V,\mathbf{M}} = T\left(\frac{\partial S}{\partial T}\right)_{V,\mathbf{M}}$$

13. For 1.00 K, calculate the rate of change of temperature, in Kelvins per oersted, for the adiabatic magnetization of a substance for which $\chi_m = 2/T$, and for which $C_{P,\mathbf{H}} = 2.1 \times 10^{-3} T^3$ J K^{-1} mol^{-1}. Calculate the rate for 0, 100, and 10,000 oersteds.

14. The accompanying table gives the entropy of NiSO$_4 \cdot$7H$_2$O as a function of temperature in zero magnetic field. A field of 8330 gauss (G) is reversibly applied to a sample of NiSO$_4 \cdot$7H$_2$O that is in thermal contact with a temperature bath at 2.000 K. The amount 0.674 J heat per mole is transferred to the bath during the magnetization. The sample is then thermally isolated and reversibly and adiabatically demagnetized. Calculate the final temperature of the salt at zero field.

T (K)	S (J K^{-1} mol^{-1})
1.0	0.887
1.2	1.573
1.4	2.230
1.6	2.849
1.8	3.435
2.0	3.970
2.2	4.452

15. The surface tension of water is 72.8 dyn cm^{-1}. How much work is required to subdivide a mole of contiguous water into a number of droplets having radii of 10^{-5} cm? How many drops would result, and what is the total surface area? What is the increase in the Helmholtz free energy?

16. How small would a water droplet have to be to have a vapor pressure twice as large as ordinary bulk water?

17. The density of water is 1.0 g cm^{-3}, of mercury 13.5 g cm^{-3}. Mercury does not wet glass; its contact angle can be taken as 180°. Water does wet glass, and its contact angle can be taken as 0°. Calculate

the capillary depression for mercury in tubes having inside diameters 1.0 and 0.1 mm. Calculate also the capillary rise for water in tubes of those dimensions.

18. A soap solution has surface tension 25 dyn cm^{-1}. Calculate the excess pressure of a spherical soap bubble whose diameter is 1 cm.

19. Show that the excess pressure in a cylindrical bubble is given by $P_{excess} = 2\gamma/R$, where γ is the surface tension and R is the radius of the cylinder. (Note that the excess pressure is independent of the length of the bubble.)

20. A U-tube manometer is constructed with arms of unequal diameters. One side has a diameter of 1 mm, the other 3 mm. If the U tube is filled with water and used to measure pressure, how much of an error will be introduced in the pressure reading by the unequal diameters?

21. The tip of a pipet has inside diameter 0.2 mm. Calculate radii of drops formed when the pipet is used with acetone, water, and mercury, of the respective densities 0.79, 1.0, and 13.6 g cm^{-3}.

22. A closed ring having a radius of 2 cm is formed from thin platinum wire. The ring is placed on the surface of clean water having a surface tension of 72.5 dyn cm^{-1}. The ring is then slowly raised until the film of water that clings to the ring breaks. Calculate the force necessary to cause the film to break. (*Note:* This is the operating principle behind the *du Nouy tensiometer* used to measure surface tension. The ring is connected to a torsion or other type of balance that can measure the force required.)

23. E. C. Markham and A. F. Benton (*J. Amer. Chem. Soc.* 53 (1931):497) measured the adsorption of ox-

Pressure (torr)	Volume O$_2$ adsorbed (cm^3)
83.0	3.32
142.4	5.57
224.3	8.73
329.6	12.68
405.1	15.48
544.1	20.42
602.5	22.48
667.5	24.86
760.0	28.03

ygen by silica at 0 °C. The volume of oxygen adsorbed by 19.6 g silica at various pressures is given in the accompanying table. Plot a Langmuir adsorption isotherm curve for the adsorption of oxygen on silica and determine the adsorption coefficient, $b = k_a/k_d$.

24. For adsorption isotherms that are linear at low pressure (meaning that they obey the Langmuir model), the adsorption isosteres follow a Clausius-Clapeyron type of equation,

$$\left(\frac{\partial \ln P}{\partial T}\right)_z = \frac{Q}{RT^2}$$

where z is either the volume, or equivalently θ, and Q is the *differential* heat of adsorption. L. H. Reyerson and A. E. Cameron (*J. Phys. Chem.* 39 (1935):181) have determined the adsorption isosteres of bromine gas on charcoal. Their results are in the accompanying table. Calculate the differential heat of adsorption of bromine on charcoal and compare with the heat of vaporization at the normal boiling point, $\Delta H_{vap} = 30.0$ kJ mol^{-1}.

T (°C)	P (torr)
58.0	740
79.0	1340
99.9	2631
117.5	3881
137.7	5650

25. Derive Equation (18.74) for the capillary rise, using the pressure differential of a curved surface as the starting point. This pressure differential is given in (18.69). You will find it helpful to refer to Figure 18.7 for this derivation.

CHAPTER NINETEEN

MACROMOLECULES

19.1 POLYMERS

Ethylene, $CH_2\!\!=\!\!CH_2$, is a stable chemical compound with a set of well-known, fixed characteristics. It has a fixed molar mass. All ethylene samples have the same molar mass if we disregard the possibility of differing isotopic composition. Likewise the melting point, boiling point, heat of formation, and all other thermodynamic properties of ethylene are fixed and independent of the method used to synthesize the compound. The group R can be substituted for one of the hydrogens, to get the compound $R\!-\!CH\!\!=\!\!CH_2$, and the new compound also has a set of fixed characteristics.

On the other hand, under certain conditions reactions involving ethylene can be made to take place for which the product does not have unique properties on a molecular scale. A free radical can be added to ethylene in such a way that the addition product is itself a free radical:

$$R\!-\! + CH_2\!\!=\!\!CH_2 \longrightarrow R\!-\!CH_2\!-\!CH_2\!-$$

This radical can add to additional ethylene molecules in a *chain reaction* until a *polymer* of the form $R\!-\!(CH_2CH_2)_n\!-\!R$ is finally formed. The lack of uniqueness of the polymer arises from the variability of n. Molar mass as a characteristic loses its definiteness when we deal with polymers, and the *average* molar mass becomes the factor of interest. We shall see that there are several different ways of determining the average molar mass of a polymer. For *polyethylene, n* can be sufficiently large that the molar mass is in the range 5,000–40,000, depending on the mode of preparation. Polymers with molar masses exceeding one million are not uncommon.

Although polymeric materials such as rubber, gums, and resins were known to chemists for many years, they were thought to be aggregates of colloids held together by some unknown binding forces. This early connection between colloids and polymers hampered the development of a workable theory of polymers, and not until the 1920s, with the pioneering work of Hermann Staudinger in Germany, were the foundations laid for the modern theory of polymers. Staudinger introduced the term *macromolecule* for molecules whose molar masses exceeded 10^4. He advanced the current view that the bonding in polymers or macromolecules is the same kind of bonding that occurs in ordinary smaller molecules. Macromolecules obey the same physical and chemical laws as small molecules do. The difference is of degree rather than substance. The large mass and physical size of macromolecules affect the type of experiments used to study macromolecules. The concept of the ideal solution is useful in studying the properties of solutions of small molecules, and in practice

> HERMANN STAUDINGER (1881–1965), German chemist, devoted his life to studying polymers, beginning about 1910 with syntheses of polyoxymethylenes and synthetic rubber. In the 1920s, polymers were held to be collections of smaller molecules that were bonded by "secondary" valence forces. Chemists of that era felt that the true molecules of substances like rubber were ordinary small molecules, and they devoted extensive efforts to "freeing" these true molecules. In 1924, Staudinger proposed the current view of polymers, introducing the term *macromolecules*. He suggested that the molecule was identical with the colloidal particle and that polymers are held together by the same valence forces with which smaller molecules are held together. He was a man of some vision, and as early as 1926 he proposed that macromolecules formed the basis of life in the proteins and enzymes that make up living organisms. His pioneering work in macromolecules formed the basis for later studies in molecular biology. Staudinger belatedly received the Nobel Prize in chemistry at the age of 73.

many solutions obey the ideal dilute solution laws. Small molecules in dilute solution all have the same environment, since the solute molecules are widely separated.

Solutions of macromolecules must also obey the ideal solution laws in the limit of infinite dilution. In practice, however, it is difficult, if not impossible, for chemists to achieve a solution so dilute that it will be ideal and still be measurably different from the pure solvent in, say, freezing-point depressions. The macromolecules are so large that a solution that is dilute when measured in concentration units of moles per liter is concentrated when measured in grams per liter, and it is the unit moles per liter on which the colligative properties of solutions depend. The large sizes of macromolecules and the concomitantly large intermolecular solute interactions in even dilute solutions of macromolecules render it necessary to treat macromolecular solutions as nonideal solutions. Often, the experimental measurements must be carried out for the region in which the solutions are nonideal and then extrapolated to the limiting point of zero concentration.

The economic importance of polymers in our daily lives is apparent simply by looking about. The manufacture of fabrics has been revolutionized by the development of the new synthetic fibers, and plastics of varying mechanical properties have replaced metals and woods in industries ranging from toy manufacturing to aerospace. Substantial research efforts are devoted to developing polymers with specific mechanical properties and to predicting these properties from the type of bonding in the polymer. Just as the engineer designs new equipment with specifically desired characteristics, the polymer chemist often tries to design new polymers with specified properties.

Polymer science is a rich and diverse field and covers many areas of science. Physical chemists study the properties of polymers, in particular the molar masses, and theoretical chemists use statistical mechanics and its sophisticated mathematical techniques to predict the behavior of polymers. *Rheology,* a branch of physics that deals with the mechanics of deformable bodies, has been applied to the study of polymeric materials with great success. Many branches of biochemistry can be considered subspecialties of polymer science. Proteins and other high-molar-mass substances of interest to biochemists are macromolecules. Biochemists often deal with the same equipment and experimental techniques that polymer chemists use. We can examine only a very small part of the broad reaches of polymer science; in particular, we are concerned with determining the molar masses of macromolecules. After first briefly examining several different types of polymeric substances, we shall consider several techniques for measuring the molar masses of macromolecules.

19.2 TYPES OF POLYMERS

One could divide polymers into two classes, *natural* and *synthetic*. Natural polymers are those occurring in nature, such as proteins, nucleic acids, polysaccharides, and polyisoprenes; synthetic polymers are those that can be synthesized in the laboratory, such as polyethylene and polyesters. Unfortunately, this distinction is not very helpful at the moment as it tells us nothing about the structure or properties of different types of macromolecules. Furthermore, a distinction based on natural occurrence is inexact, implying as it does some fundamental difference between naturally occurring and synthetically prepared macromolecules. To be sure, there are differences. The natural macromolecule is often much more complicated than the synthetic polymer, and the natural polymer in certain cases has the unique ability to replicate itself. In a sense the differences are subjective rather than substantive.

If one looks at the method of preparing polymers, then one can classify polymers into *addition polymers* and *condensation polymers*. An addition polymer is formed by the addition of molecules to one another via the conversion of a double bond to a single bond. An important characteristic of the formation of an addition polymer is that no portion of the *monomer* is lost in synthesizing the polymer. The most common members of the class of addition polymers are those formed from the polymerization of vinyl monomers:

$$n\text{CH}_2\!=\!\text{CH}(\text{R}) \longrightarrow [\text{CH}_2\!-\!\text{CH}(\text{R})]_n \tag{19.1}$$

The end groups are usually omitted in writing polymer structures. The simplest polymer of the type shown in Equation (19.1) is polyethylene, where the R group is a hydrogen atom.

In condensation polymers, on the other hand, a portion of the monomer is lost during polymerization. These polymers are usually formed from bifunctional monomers, with the resulting elimination of a smaller molecule as a by-product. These smaller molecules are usually water, though methanol or HCl sometimes results. Polyesters are formed from reactions of the form

$$n\text{HORCO}_2\text{H} \longrightarrow [\text{O}\!-\!\text{R}\!-\!\text{CO}]_n + n\text{H}_2\text{O} \tag{19.2}$$

The formation of polyethylene terephthalate (19.3) is an example of a *polycondensation* reaction in which the repeating unit contains two monomer molecules, each of which has two functional groups so that the reaction can be propagated to large molar masses. In this reaction the small molecule eliminated is methanol.

$$n\text{HOCH}_2\!-\!\text{CH}_2\text{OH} + n\text{CH}_3\text{O}\!-\!\overset{\text{O}}{\underset{}{\text{C}}}\!-\!\!\!\bigcirc\!\!\!-\!\overset{\text{O}}{\underset{}{\text{C}}}\!-\!\text{OCH}_3 \longrightarrow$$

$$[\text{O}\!-\!\overset{\text{O}}{\underset{}{\text{C}}}\!-\!\!\!\bigcirc\!\!\!-\!\overset{\text{O}}{\underset{}{\text{C}}}\!-\!\text{O}\!-\!\text{CH}_2\!-\!\text{CH}_2]_n + 2n\text{CH}_3\text{OH} \tag{19.3}$$

SEC. 19.2 TYPES OF POLYMERS

Polymers such as polyethylene that contain only one type of monomer unit are called *homopolymers*. Polymers that contain two or more different monomer units, such as the polyethylene terephthalate of (19.3), are known as *copolymers*. The two subunits, which we shall call X and Y, can alternate to give a well-defined recurring unit

$$\ldots -X-Y-X-Y-X-Y-X-Y-\cdots -(X-Y)_{\overline{n}}- \qquad (alternating)$$

On the other hand the individual subunits can be randomly joined, in which case no recurring unit is written:

$$\cdots -X-X-Y-Y-Y-X-Y-X-Y-\cdots \qquad (random)$$

Copolymers need not contain equal amounts of the two subunits. Copolymers with a gradation of properties between those of polyvinylchloride and polyvinylacetate can be produced by polymerizing mixtures of the two monomers in various proportions. On the other hand, the copolymer formed from stilbene and maleic anhydride contains equivalent amounts of the two monomers regardless of the composition of the mixture from which it was prepared. This strongly suggests that the polymer has a repeating structural unit:

$$\left[\begin{array}{cccc} -CH- & -CH- & -CH- & -CH- \\ | & | & | & | \\ C_6H_5 & C_6H_5 & CO & CO \\ & & \diagdown & \diagup \\ & & O & \end{array} \right]_n$$

Another important structural variable is *stereospecificity*. If a polymer is produced from a substituted vinyl monomer $CH_2\!\!=\!\!CHR$ where R is a nonhydrogenic group, then half the carbon atoms in the polymer will be asymmetric. Depending on the mode of polymerization, three limiting arrangements of the steric centers are possible. A completely random sequence of the centers may result, yielding the *atactic* chain. On the other hand, the resulting chains may have the same or a regularly alternating configuration in the *isotactic* and *syndiotactic* chains.[1]

Atactic chain:

$$-CH_2-\underset{R}{\overset{H}{C}}-CH_2-\underset{R}{\overset{H}{C}}-CH_2-\underset{H}{\overset{R}{C}}-CH_2-\underset{R}{\overset{H}{C}}-CH_2-\underset{H}{\overset{R}{C}}-CH_2-\underset{H}{\overset{R}{C}}-$$

Isotactic chain:

$$-CH_2-\underset{R}{\overset{H}{C}}-CH_2-\underset{R}{\overset{H}{C}}-CH_2-\underset{R}{\overset{H}{C}}-CH_2-\underset{R}{\overset{H}{C}}-CH_2-\underset{R}{\overset{H}{C}}-CH_2-\underset{R}{\overset{H}{C}}-$$

Syndiotactic chain:

$$-CH_2-\underset{R}{\overset{H}{C}}-CH_2-\underset{H}{\overset{R}{C}}-CH_2-\underset{R}{\overset{H}{C}}-CH_2-\underset{H}{\overset{R}{C}}-CH_2-\underset{R}{\overset{H}{C}}-CH_2-\underset{H}{\overset{R}{C}}-$$

[1] This does not, however, imply optical isomerisms, since the various arrangements about these asymmetric centers are random.

FIGURE 19.1 Perspective view of stereospecific vinyl polymers. (a) Isotactic. (b) Syndiotactic. The thin lines lie in the plane of the paper.

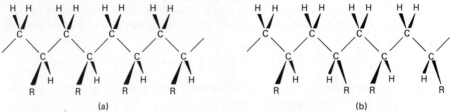

Perspective sketches of the isotactic and syndiotactic structures are shown in Figure 19.1. Considerable efforts have been expended in developing techniques for characterizing the degree of stereoregularity, or *tacticity,* of polymers.

Our discussion has so far assumed that all polymers are *linear*—that they are simply strings of monomer units connected in linear fashion. Another source of disorder in polymer structures is to be found in *branching,* in which additional chains of monomer units propagate from the main chain. *Cross-linkages* between adjacent chains can also be established. Physical properties such as degree of crystallinity, hardness, and strength depend in no small measure on the tacticity, the degree of branching, and the degree of cross-linking.

19.3 MOLAR MASSES OF POLYMERS

Suppose we have two compounds, A and B, of molar masses 5 and 50 g mol^{-1} respectively, and suppose we have a mixture containing 0.1 mole of A and 0.9 mole of B. The question of the "average" molar mass of the mixture seems to present no difficulty. We have one mole (that is, an amount of molecules equal to Avogadro's number) of mixture with a total mass of $0.1 \times 5 + 0.9 \times 50 = 45.5$ g. The average molar mass of the mixture is just 45.5 g. This is a *number average* molar mass, since it depends on the total number of molecules present. Colligative properties of solutions depend on the number average molar mass. The number average is generally given by

$$\overline{M}_N = \frac{\Sigma M_i n_i}{\Sigma n_i} = \Sigma X_i M_i \tag{19.4}$$

where M_i is the molar mass of each constituent in the mixture, n_i is the number of moles of each constituent, and X_i is the mole fraction.

Although colligative properties, such as freezing-point depressions, depend only on the total number of molecules present, some methods of molar mass determinations depend not only on the numbers of molecules present but also on the mass of each individual molecule. This type of determination is more heavily weighted towards the higher masses. This *weight average,* or *mass average,* molar mass is defined by

$$\overline{M}_w = \frac{\Sigma M_i^2 n_i}{\Sigma M_i n_i} = \Sigma w_i M_i \tag{19.5}$$

where w_i is the mass fraction.

For the mixture of A and B, the mass average is

$$\overline{M}_w = \frac{0.1(5)^2 + 0.9(50)^2}{0.1(5) + 0.9(50)} = 49.51$$

which is substantially more than the number average. Higher-order averages can be similarly defined. These are rarely used except for the so-called Z average, which is defined by

$$\overline{M}_z = \frac{\sum M_i^3 n_i}{\sum M_i^2 n_i} \tag{19.6}$$

When ethylene is polymerized to form polyethylene, $R-(CH_2-CH_2)_n-R'$, the mass of each polymer molecule formed is proportional to the value of n. There is generally a large variation in this value. It is thus impossible to speak of a molar mass per se. One rather speaks in terms of average molar masses as defined in (19.4)–(19.6). The molar masses of polymer mixtures follow distribution curves as indicated in Figure 19.2. Satisfactory techniques whereby a polymer can be synthesized with a unique molar mass, or *degree of polymerization,* are mostly nonexistent. The distribution of molar masses is defined by $N(M)$ such that

$$f(M) = \int_{M_1}^{M_2} N(M)\, dM \tag{19.7}$$

is the fraction of polymer chains having molar masses in the interval between $M_1 \leq M \leq M_2$. Since the sum of all the fractions must equal unity,

$$\int_0^\infty N(M)\, dM = 1 \tag{19.8}$$

The form of $N(M)$ generally depends on the conditions of polymerization, and the form shown in Figure 19.2 is only one of many possibilities. Under certain conditions polymers have very narrow distributions; all the molecules have nearly the same masses. Proteins form the chief exception to the rule of molar mass distributions. Proteins are macromolecules containing hundreds of monomer units (generally many different types), but the individual structure is precisely replicated in the different protein molecules in a given sample, and the molar mass of each molecule is the same.

In practical applications the control of the chain length of synthetic polymers is critically important. The mechanical strength of polymeric material falls rapidly as the molar mass drops below about 20,000–30,000. On the other hand the mechanical properties approach an asymptotic limit as the molar mass increases, and they even-

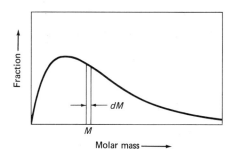

FIGURE 19.2 A molar mass distribution for a polymer. The area of the indicated section of width dM at M gives the fraction of molecules that have molar masses within dM of M. The area under the curve equals unity, since the sum of all the fractions must equal unity.

tually become independent of further increases. There is a substantial disadvantage to high masses, since the viscosity increases with mass, causing practical difficulties when polymers are being molded to shape for the final product. The molar mass must therefore be controlled with these limits in mind. Further, for certain applications, a molar mass distribution that is very broad can lead to phase separation and mechanical weakness.

The first techniques that come to mind for determining the molar masses of polymers are based on the colligative properties of solution. Of the three colligative properties—freezing-point depression, boiling-point elevation, and osmotic pressure—only the last can be applied extensively. This is not because of any fundamental problem, but rather because of experimental design. A $1.0 m$ solution of a polymer whose molar mass is 50,000 would contain 50,000 g of polymer per 1000 g of solvent. This is impractical by several orders of magnitude, since the resulting solution would be so nonideal it would yield results of little value. In the more practical concentration range 1–10 g kg^{-1} of solvent, the freezing-point depression and the boiling-point elevation are both immeasurably small in view of the small concentration on a mol liter^{-1} basis, as has been noted in Chapter 12. The osmotic pressure, however, does provide a convenient measurement that can be accurately determined even at these small concentrations. Suppose we consider a benzene solution that contains 1% by weight of a polymer with molar mass 10,000. The mole fraction of solute is only 7.8×10^{-5}. The vapor pressure of the solution at 26.1 °C would be 99.9922 torr calculated from Raoult's law; the vapor pressure of the pure benzene is 100 torr. Using the freezing-point and boiling-point temperatures and the heats of

FIGURE 19.3 Osmometer. The sample, which contains dissolved polymer, and the pure solvent are placed on opposite sides of the semipermeable membrane. A glass capillary leads from the bottom compartment and is connected to the solvent reservoir through a flexible tube. An air bubble is placed in the capillary, and the position of the bubble is viewed by an optical detector. The detector is connected to a servomotor, which can raise or lower the solvent reservoir so that the air bubble remains stationary as solvent tends to flow across the membrane. The osmotic pressure is determined from the position of the solvent reservoir when the system reaches equilibrium and the bubble is stationary.

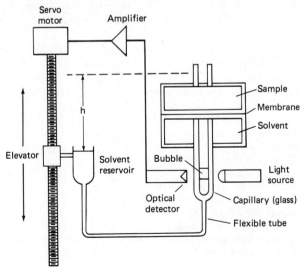

fusion and vaporization, we get a freezing-point depression of 0.0051 °C and a boiling-point elevation of 0.0025 °C. On the other hand, the osmotic pressure is 18.75 torr. Since osmotic pressures can be measured with modern osmometers to an accuracy of ±0.02 torr, the osmotic pressure determination is the preferred experimental route. The situation becomes even worse for the other techniques when one considers that 1% solutions of polymers are usually nonideal and measurements must be extended to much lower concentrations. We shall now turn our attention to several experimental techniques used to measure the molar masses of polymers. We shall start with the osmotic pressure that yields a number average molar mass. A sketch of an osmometer is shown in Figure 19.3.

19.4 OSMOTIC PRESSURE

The expressions for the osmotic pressure of an ideal solution in terms of the various concentration units have been derived in Chapter 12. We shall write the expression as

$$\Pi = \frac{RTC}{M_N} \qquad (19.9)$$

where C is the concentration in mass per unit volume (usually grams per liter) and M_N is the number average molar mass. Equation (19.9) applies only to an ideal solution and must be corrected if the molar mass is to be accurately determined. At the concentrations used in practice, the interactions between solute molecules must be taken into account. One useful equation, the *osmotic virial equation*, has the form

$$\Pi = RT \left(\frac{C}{M_N} + B_2 C^2 + B_3 C^3 + \cdots \right) \qquad (19.10)$$

where the constants $B_2, B_3, \ldots$ are the second, third, and higher *virial coefficients* analogous to the virial coefficients in the virial equation for the nonideal gas. The coefficients are related to such factors as excluded volume, shape of the solute molecules, and binary and higher-order interactions between the solute molecules. These are discussed in the references listed in the Bibliography. In dilute solutions, the third and higher terms in (19.10) become negligible and we can write

$$\frac{\Pi}{C} = \frac{RT}{M_N} + RTB_2 C \qquad (19.11)$$

In the region for which (19.11) is valid, a plot of Π/C vs C yields a straight line, as shown in Figure 19.4. The intercept of the line extrapolated to $C = 0$ is equal to RT/M_N, from which the number average molar mass is computed. The second virial coefficient is determined from the slope of the line.

The nature of the semipermeable membrane itself is critically important in osmometry experiments; it must be selected with the greatest care. Membranes of cellulose are commonly used. Ideally, the membrane should be completely permeable to solvent molecules and completely impermeable to even the smallest polymer molecules. Membranes that are useful for polymers having molar masses of 100,000 might not be sufficiently impermeable to polymers with masses of 10,000. Further, a membrane that is at once permeable to the lighter fractions of a polymer and impermeable only to the higher fractions yields a molar mass that errs in being too high.

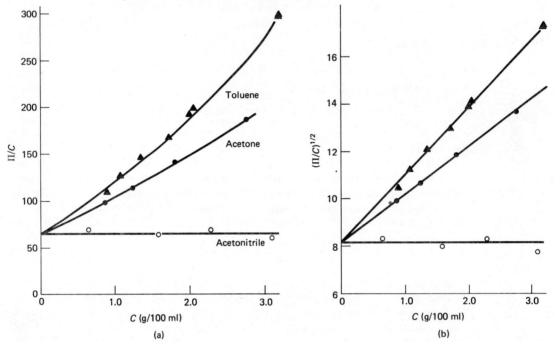

FIGURE 19.4 Osmotic pressure plots for polymethyl methacrylate ($\overline{M}_n = 382{,}000$ g mol^{-1}) in three different solvents. (T. G. Fox, J. B. Kinsinger, H. F. Mason, and E. M. Schuele, *Polymer* 3 (1962):71. (a) Plots of π/C vs C. Here $\pi = h\rho$, where h is the osmotic height in cm and ρ is the density of the solution; C is measured in g ml^{-1}. (b) The same data plotted as $(\pi/C)^{1/2}$. The curves are more nearly linear. This can be seen by taking the square root of both sides of (19.10) after dividing by C, to get

$$\left(\frac{\pi}{C}\right)^{1/2} = \left(\frac{RT}{M_n}\right)^{1/2} (1 + \tfrac{1}{2}B_2 C + \cdots)$$

This expansion is valid for $\pi/C \leq 3RT/M_n$.

19.5 MACRO-IONS

Just as there exist ionic salts in the chemistry of small molecules, there exist ions in the chemistry of polymer molecules. Such polymer ions are variously referred to as *polyelectrolytes, macro-ions,* or *polyions.* One subject of numerous investigations has been the sodium salt of polyacrylic acid,

$$\left[\begin{array}{c} -\mathrm{CH_2-CH-} \\ | \\ \mathrm{COO^-} \\ \\ \mathrm{Na^+} \end{array} \right]_n$$

Polyelectrolytes are normally studied in aqueous solutions and their properties are severely affected when salts like NaCl or KCl are added. At first it seems as though our previous discussion of ionic solutions could be related to these macro-ions much the same as the osmotic pressure theory was applied to associated polymers; this cannot be, however. One of the most interesting and significant consequences of the large size of the polymer ion is the so-called *Donnan effect,* which F. G. Donnan first explained in 1911.

Figure 19.5a shows a solution of a polyelectrolyte that we indicate by $P^{z+}Cl_z^-$ separated from pure water by a semipermeable membrane. While the Cl^- are small enough that they can pass through the membrane, the larger ions P^{z+} are too large to cross the membrane. Initially the Cl^- have a tendency to cross the membrane to equalize the chemical potential of the Cl^- on both sides. This is impossible, since charged solutions of opposite signs would be formed on either side of the membrane. If an ionic salt such as Na^+Cl^- is added to the system, both positive Na^+ and negative Cl^- can cross the membrane in addition to the solvent water; only the polyelectrolyte is restricted. This is shown in Figure 19.5b.

A somewhat simpler case is presented by the situation in which the membrane is impermeable to the solvent, and is permeable only to some of the ions. This situation is called *nonosmotic membrane equilibrium.* We denote the solutions on either side of the membrane α and β. If the membrane is permeable to the ith ion then at equilibrium the chemical potential of that ion must be the same on both sides of the membrane, or

$$\mu_i^\alpha = \mu_i^\beta$$

If the membrane is permeable to *all* the ions, then the concentrations of the various ions will be equalized on both sides of the membrane. On the other hand, if the membrane is impermeable to one or more ions, then an electrochemical potential will be established. Suppose it is permeable to only the jth ion. The electrochemical potential is then

$$\mathscr{E} = \frac{RT}{z_j F} \ln \frac{m_j \gamma_j}{m_j' \gamma_j'} \tag{19.12}$$

FIGURE 19.5 Membrane equilibrium for polyelectrolytes. The origin of the Donnan effect. (a) A polyelectrolyte $(P^{z+})(Cl^-)_z$ is dissolved in water on one side of the membrane in an osmotic cell. The solvent can pass through the membrane, but the polymer ions are too large and are restricted to one side of the cell. The chloride ions are small enough to pass through the membrane, but this would create a large potential across the membrane; the potential restricts the passage of the Cl^-. (b) The addition of an added salt such as NaCl now gives rise to a situation where both Na^+ and Cl^- can cross the membrane, since the concentration of Na^+ can adjust itself on either side of the membrane to ensure that the total charge on either side of the membrane is zero.

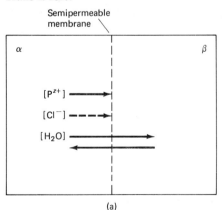
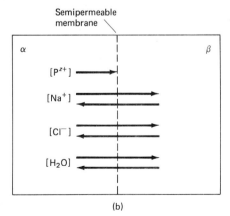

(a)   (b)

where z_j is the ionic charge, F is the Faraday constant, and m and γ are the concentration and activity coefficient respectively; for simplicity we use primed quantities to indicate the β solution and unprimed quantities to indicate the α solution. In dilute solutions, for which the activity coefficients can be approximated by unity, (19.12) reduces to

$$\mathscr{E} = \frac{RT}{z_j F} \ln \frac{m_j}{m_j'} \qquad (19.13)$$

This equation applies when the membrane is permeable to only one of the ions. This mechanism is vitally important in the transmission of electrical nerve signals. Nerve cells are permeable to potassium ions (K$^+$) but impermeable to other ions like Na$^+$ and Cl$^-$. By establishing concentration differences of K$^+$ on both sides of the membrane, differences in potential of the order of 80 mV can be established and transmitted along the nerve. If more than one ionic species can pass across the membrane, the situation is complicated, since it is impossible to establish a unique potential difference for a particular ion.

If the membrane is also permeable to the solvent molecules, then a rigorous treatment requires us to consider the additional term that arises from the osmotic pressure contribution.[2] We can approach the problem more simply by considering equality of charge and activity on both sides of the membrane. We let m_r be the concentration of ionic charge arising from the polyelectrolyte and m_s be the concentration of added uni-univalent salt. For the situation depicted in Figure 19.5b, m_s is $z_p m_p$, where z_p is the polyelectrolyte charge and m_p its concentration; m_s is just the molarity of the added NaCl. We take the polyelectrolyte to be positively charged. The concentration of positive ions that can cross the membrane is $m_+ = m_s$, since the ion Na$^+$ is the only positive charge capable of crossing the membrane. The negative ion concentration is $m_- = m_r + m_s$, since the negative ions originally associated with the polyelectrolyte can also cross the membrane. We let the side containing the polyelectrolyte be the unprimed side and the other side, the *dialyzate* (from dialysis), be the primed side. On that side, both ions are at the same concentration, $m_+' = m_-' = m_s'$.

At equilibrium, the ions crossing the membrane (the *permeant* ions) must have the same activities on both sides,

$$m_+ m_- \gamma_\pm^2 = m_+' m_-' (\gamma_\pm')^2$$
$$(m_r + m_s) m_s \gamma_\pm^2 = (m_s')^2 (\gamma_\pm')^2 \qquad (19.14)$$

For sufficiently dilute solutions we can set the mean ionic activity coefficients in both phases equal to each other, and

$$m_s = -\frac{m_r}{2} + \left[\left(\frac{m_r}{2} \right)^2 + (m_s')^2 \right]^{1/2} \qquad (19.15)$$

Now consider two extreme cases. If only a very small quantity of salt is added, then $m_s'/m_r \ll 1$, and we get directly from (19.14) that

$$m_s = m_s' \left(\frac{m_s'}{m_r} \right) \qquad (19.16)$$

---

[2] See, for example, E. A. Guggenheim, *Thermodynamics*, 3rd ed. (Amsterdam: North-Holland Publishing Co., 1957), p. 379.

Under these conditions most of the added salt is present on the side opposite from the polyelectrolyte. At the other extreme we add a large quantity of salt relative to the polyelectrolyte. In this case, $m'_s/m_r \gg 1$, and we can expand the radical term in (19.15) in a binomial expansion,[3] to get

$$m_s = -\frac{m_r}{2} + m'_s + \frac{m_r^2}{8m'_s} - \cdots \tag{19.17}$$

The difference Δm_i in concentration of permeant ions on both sides of the membrane is then

$$\Delta m_i = m_+ + m_- - m'_+ - m'_- = \frac{m_r^2}{4m'_s} - \cdots \tag{19.18}$$

Note that the difference can be minimized by increasing the concentration of added salt. This difference in permeant ion concentration can contribute to the osmotic pressure of a polyelectrolyte solution.

In a simple polyelectrolyte solution containing no added salt, both the polymer ion P^{z+} and the *counterion* X^- are osmotically active, since neither can cross the membrane. Further, the concentration of counterions is higher by many orders of magnitude than the concentration of the polymer ion, and most of the osmotic pressure comes from the counterion. The measured osmotic pressure is thus independent of the molar mass of the macromolecule. When the simple electrolyte is added to the solution, we then have solute on both sides of the membrane, and the concentration difference determined by Equation (19.18) gives rise to an added osmotic pressure term. Since this concentration difference disappears in the limit of infinitely high salt concentration, the added term also disappears in this limit. We can then write the osmotic pressure as

$$\Pi = RT\left[\frac{C}{M} + \frac{m_r^2}{4000m'_s} + \cdots\right] \tag{19.19}$$

where all the terms have been previously defined. Noting that

$$m_r = \frac{1000 C \nu_p}{M} \tag{19.20}$$

where ν_p is the number of ion charges per polymer molecule, we can write (19.19) as

$$\frac{\Pi}{C} = RT\left[\frac{1}{M} + B_2 C + \cdots\right] \tag{19.21}$$

The second virial coefficient is given by

$$B_2 = \frac{1000(\nu_p/M)^2}{4m'_s} \tag{19.22}$$

Figure 19.6 shows some osmotic pressure data for polyelectrolyte solutions. Note that (19.22) predicts that the second virial coefficient should vary linearly with $(m'_s)^{-1}$. This linear variation is borne out by experiment. The absolute values of the experimental numbers are, however, lower than the predicted values by a factor of about 30.

---

[3] The binomial expansion theorem states that (for $y^2 < x^2$)

$$(x+y)^n = x^n + nx^{n-1}y + \frac{n(n-1)}{2!}x^{n-2}y^2 + \frac{n(n-1)(n-2)}{3!}x^{n-3}y^3 + \cdots$$

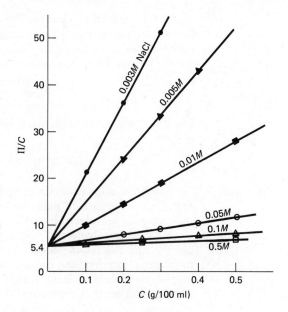

FIGURE 19.6 Osmotic pressure plots for the polyelectrolyte sodium pectinate (D. T. F. Pals and J. J. Hermans, *Rec. Trav. Chim.* 71 (1952):458). Plots are shown for several different concentrations of added NaCl. The osmotic pressure was measured in centimeters of water. The temperature was 20 °C. The molar mass obtained from these measurements was $\overline{M}_n = 46{,}000$ g mol^{-1}.

19.6 ULTRACENTRIFUGATION

The effect of gravitational fields on thermodynamic properties has been noted (Chapter 18). Although gravitational fields can give rise to concentration gradients, these gradients are so small, as applied to most systems in the earth's gravitational field, that they are unmeasurable.

On the other hand, small samples can be subjected to artificial gravitational fields in a centrifuge. Modern ultracentrifuges can rotate samples at sufficiently high angular velocities that fields of the order of 10^5 times greater than the earth's natural gravitational field can be achieved. This allows an effective compression in size by the factor 10^5. A concentration or pressure difference that would be observed over a 1-km-high column of ideal gas would be observed in a 1-cm column that was spinning in such a centrifuge. These intense fields are sufficient to sediment macromolecules just as the small laboratory centrifuge sediments much larger precipitate particles from a supernatant liquid.

The ultracentrifuge was developed by Svedberg in the 1930s. Ultracentrifuges capable of speeds up to 100,000 rpm have been built, though the usual commercial models have top speeds of about 60,000 rpm. With the sample cell placed 6.5 cm from the axis of rotation, a force 250,000 times gravity is achieved (250,000 "g"). A schematic of an ultracentrifuge is shown in Figure 19.7. For our purposes it consists of two basic parts, a mechanical component and an optical component. The mechanical component contains the rotor in which the sample cells are placed, and the motor and gear assembly used to drive the rotor. Precision workmanship of the highest order is required for smooth operation at these high speeds. A rotor that breaks loose at these speeds is potentially lethal. To minimize friction, the rotor is operated in a vacuum. Temperature control is also provided. A slit is placed in the rotor to allow a light beam to pass through the cell. Several types of optical measurements can be made from which the concentration of polymer as a function of position in the cell

SEC. 19.6 ULTRACENTRIFUGATION

> THE SVEDBERG (1884–1971), Swedish chemist, devoted most of his efforts to colloids and macromolecules, with numerous digressions into other areas. In 1912 he was appointed to the first Swedish chair of physical chemistry at the University of Uppsala; and in 1926 he received the Nobel Prize in chemistry. His early work on observing Brownian motion formed the experimental basis for the theoretical treatment of Einstein. In the early 1920s, Svedberg conceived the idea of studying the size distribution of colloids by photographically following the sedimentation of the particles in a centrifuge. His first successful machine, built in 1924, achieved a force 5000 times gravity. With it he was able to study colloidal gold particles smaller than what could be seen with the microscope. In that same year he determined from equilibrium sedimentation that the molar mass of hemoglobin was 68,000, and hence he was able to show that the hemoglobin molecule contained four iron atoms. Shortly thereafter he increased the speed of the device and was able to show by velocity sedimentation that the molar mass was homogeneous. He found the molar mass of hemocyanin from a land snail to be about 5 million, an astonishing figure for those days. In the 1930s, other techniques appeared that confirmed Svedberg's hypothesis that proteins had uniform molar masses, and the ultracentrifuge rapidly became a standard analytical tool in molecular biology.

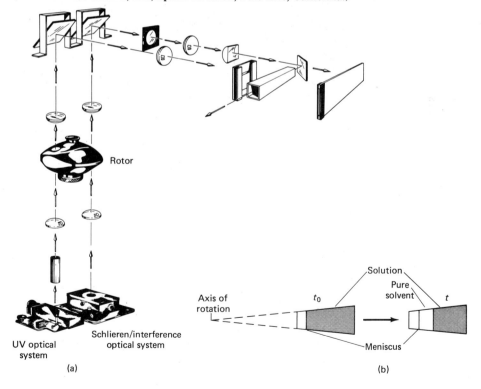

FIGURE 19.7 (a) Ultracentrifuge showing rotor and dual optical paths for Schlieren and ultraviolet light studies. (b) Shows the cell indicating the movement of the boundary. A hole is cut in the rotor cell housing for the optical path, which is parallel to the axis of rotation. (Beckman Instruments, Inc., Spinco Division, Palo Alto, California.)

can be determined. Several types of sedimentation experiments can be performed with the ultracentrifuge.

In *equilibrium sedimentation*, the rotor is spun at a constant rotational velocity until an equilibrium distribution of the polymer through the cell is established. The concentration profile is optically detected and usually recorded photographically. The rotational speed must be set to accommodate the system under study. If the speed is too low, the change in concentration will be too small to measure. On the other hand, if the speed is too high, most of the polymer will tend towards the bottom of the cell, as shown in Figure 19.8. Equilibrium sedimentation is usually carried out at the relatively "low" speed of 5000–25,000 rpm. One disadvantage to equilibrium sedimentation is the time required for equilibrium. This might be hours or even days for a sample column 1 cm long. The equilibrium time can be reduced substantially by using shorter sample column lengths. Columns as short as 2–3 mm have been used successfully.

The basic equation for equilibrium in a potential field for a one-component system was given in Equation (18.27). For a system with more than one component, the term in the chemical potential must be included and (18.27) becomes

$$dG = -S\,dT + V\,dP + M'\,d\phi + \sum_k \mu_k\,dn_k \tag{19.23}$$

where the sum is over all components and we have written M' for the mass to distinguish this quantity from the molar mass. We consider the case of the two-component system, solvent and polymer solute, which we label 1 and 2 respectively. In the multicomponent system, the partial molar free energy or chemical potential of each com-

FIGURE 19.8 Concentration profiles in equilibrium sedimentation. (a) At low speed there is a gradual change in concentration. (b) If the speed is too high, most of the polymer solute is forced to the bottom of the cell.

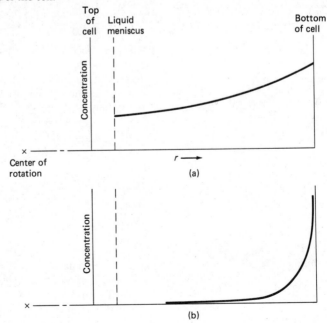

ponent must be constant at equilibrium. This condition is obtained by differentiating (18.23) with respect to n_2 at constant n_1. If in addition we require the temperature to be constant, then $S\,dT = 0$, and

$$d\mu_2 = 0 = \overline{V}_2\,dP + M_2\,d\phi + \left(\frac{\partial \mu_2}{\partial n_2}\right) dn_2 \tag{19.24}$$

where $\overline{V}_2$ is the partial molar volume and M_2 is the partial molar mass, or more simply the molar mass. Noting that the mole numbers differ from the concentrations only by a constant factor, the concentration m_2 can be substituted for n_2 in the last term. This followed by rearrangement yields

$$-M_2\,d\phi - \overline{V}_2\,dP = \left(\frac{\partial \mu_2}{\partial m_2}\right) dm_2 \tag{19.25}$$

Note that the pressure term is *not* the ordinary atmospheric pressure but rather the pressure due to the centripetal force per unit area that arises from the rotational motion of the rotor. Analogous to (18.28), we have

$$dP = -\rho\,d\phi \tag{19.26}$$

where ρ is the density. Further, for the centrifugal field,

$$\phi = -\tfrac{1}{2}\omega^2 r^2$$
$$d\phi = -\omega^2 r\,dr \tag{19.27}$$

where ω is the radial velocity in radians per second and r the distance from the center of rotation. By these results, Equation (19.25) becomes

$$(M_2 - \overline{V}_2\rho)\omega^2 r\,dr = \left(\frac{\partial \mu_2}{\partial m_2}\right) dm_2$$

or

$$(M_2 - \overline{V}_2\rho)\omega^2 r = \left(\frac{\partial \mu_2}{\partial m_2}\right)\frac{dm_2}{dr} \tag{19.28}$$

Equation (19.28) has the physical effect of equating two forces that are equal and opposite at equilibrium. The term on the left is the centripetal force that acts on the solute. The term $(M_2 - \overline{V}_2\rho)$ represents the mass of the polymer solute corrected for buoyancy. The term on the right is the counterbalancing diffusion that tends to establish concentration equality throughout the sample cell. The term dm_2/dr is just the concentration gradient; this is what is measured with the optical system of the ultracentrifuge. We now need an expression for $\partial \mu_2/\partial m_2$.

We can use the Gibbs-Duhem equation to relate this derivative to the osmotic pressure,

$$m_2 \frac{\partial \mu_2}{\partial m_2} = -\left(\frac{1000}{M_1}\right)\frac{\partial \mu_1}{\partial m_2} = -\left(\frac{1000 RT}{M_1}\right)\frac{\partial(\ln a_1)}{\partial m_2}$$
$$= \left(\frac{1000\overline{V}_1}{M_1}\right)\frac{\partial \Pi}{\partial m_2} \tag{19.29}$$

This last step can be obtained by the method used in Chapter 12 for the osmotic pressure if the analysis is carried out in terms of activities instead of fugacities and concentration instead of mole fractions. The osmotic pressure is given in terms of the virial equation in (19.10). If we now combine the results of Equations (19.10),

(19.28), and (19.29) (after making sure that all have been converted to the same concentration units), we get

$$\frac{(M_2 - \bar{V}_2\rho)\omega^2 r}{RT(dm_2/dr)} = \frac{1}{m_2}(1 + 2\beta m_2 + \cdots) \tag{19.30}$$

where we have used β for the second virial coefficient instead of B_2, since the units are different. For dilute solutions we can drop all terms past the first on the right-hand side of (19.30). We change to the more convenient concentration unit $C = m_2 M_2$ (g/1000 g solvent) and introduce a new variable, the *partial specific volume* of the solute,

$$\bar{v}_2 = \frac{\bar{V}_2}{M_2} \tag{19.31}$$

to get

$$\frac{(1 - \bar{v}_2\rho)\omega^2 r}{RT(d\ln C/dr)} = \frac{1}{M_2} \tag{19.32}$$

Note that the concentration (either C or m_2) as well as the density ρ are functions of the position r. If we make the simplifying assumption that $\bar{v}_2\rho$ is a constant product, then (19.32) can be rearranged and integrated, to get

$$\ln C = \frac{(1 - \bar{v}_2\rho)\omega^2 r^2}{2RT} M_2 + \text{constant} \tag{19.33}$$

If $\ln C$ is plotted as a function of r^2, the molar mass M_2 is obtained from the slope of the straight line described in (19.33). The concentration is measured by optical techniques. The distance r can be determined to a precision of about 0.05 mm. The angular velocity can be determined with great precision, and $\bar{v}_2$ is determined from density measurements. For higher concentrations, the higher-order terms must be taken into account.

If the macromolecule under study has a single molecular form, then the molar mass determined from (19.33) poses no problem. On the other hand, if a distribution of masses is present, the mass average is what is determined. This can be seen as follows. If we have several different masses M_i, then we have one equation of the form of (19.32) for each mass,

$$M_i(1 - \bar{v}_2\rho)\frac{\omega^2 r}{RT} = \frac{d\ln C_i}{dr} = \frac{1}{C_i}\frac{dC_i}{dr} \tag{19.34}$$

$$M_i(1 - \bar{v}_2\rho)\frac{\omega^2 r}{RT} C_i = \frac{dC_i}{dr} \tag{19.35}$$

If we now introduce the sum of all the concentrations

$$\mathcal{C} = \Sigma C_i \tag{19.36}$$

we can write

$$(1 - \bar{v}_2\rho)\frac{\omega^2 r}{RT}\sum M_i C_i = \frac{d\mathcal{C}}{dr} \tag{19.37}$$

and

$$(1 - \bar{v}_2\rho)\frac{\omega^2 r}{RT}\frac{\sum M_i C_i}{\mathcal{C}} = \frac{d\ln \mathcal{C}}{dr} \tag{19.38}$$

SEC. 19.6 ULTRACENTRIFUGATION

But
$$\frac{\sum M_i C_i}{\mathscr{C}} = \frac{\sum M_i^2 m_i}{\sum M_i m_i} = \overline{M}_w \qquad (19.39)$$

hence
$$\frac{(1 - \bar{v}_2 \rho)\omega^2 r}{RT(d \ln \mathscr{C}/dr)} = \frac{1}{\overline{M}_w} \qquad (19.40)$$

In principle, ultracentrifugation can provide not only $\overline{M}_w$ but also the higher moments such as M_z.[4]

Whereas by equilibrium sedimentation we examine the solution at equilibrium at a relatively low rotational speed, by *velocity sedimentation*, which uses a much higher rotational velocity, we examine the rate at which this equilibrium is established. The general procedure is shown schematically in Figure 19.9, where both the concentration C and dC/dr are sketched as a function of position in the cell. Initially the concentration is uniform throughout the cell. As the cell is rotated at a very high speed, the solute molecules are forced to the bottom of the cell. The region at the top near the meniscus becomes depleted of solute molecules, and the region near the bottom of the cell becomes more concentrated. The speed is sufficiently high that eventually practically all the solute molecules are concentrated near the bottom of the cell. A boundary between the nearly pure solvent at the top of the cell and the

FIGURE 19.9 Velocity sedimentation. A cell containing a polymer solution is rotated at a very high speed, and the polymer solute tends towards the bottom of the cell at a rate determined by the molar mass, the rotational velocity of the ultracentrifuge, and the diffusion constant of the polymer. Concentration profiles at three different times are shown at the top, and dC/dr is plotted below. The spreading of the boundary between pure solvent and solution is caused by diffusion and thermal agitation.

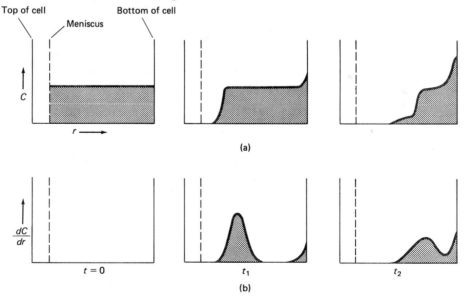

---

[4] A derivation of the equations for equilibrium sedimentation from another point of view can be found in T. Svedberg and K. O. Pederson, *The Ultracentrifuge* (Oxford: Oxford University Press, 1940). Further details on this and the other methods of ultracentrifugation will be found in that textbook and the others listed in the Bibliography.

more concentrated solution is established, and this boundary moves down the cell. The sharpness of the boundary is destroyed by diffusion. The position of the boundary is measured as a function of time, and the rate at which the boundary moves is a function of the diffusion coefficient of the solute. If the diffusion coefficient is known, then the molar mass can be determined from the measured rate of travel of the boundary. Schlieren optics are usually used in this mode since this type of optical arrangement measures dN/dr directly rather than C. Figure 19.10a shows a typical Schlieren photograph. This mode of operation generally requires less time than the equilibrium mode. A further advantage lies in the ability to distinguish between different macromolecules. If more than one type of macromolecule is present and the weights differ, then the rates of travel differ, and each molecule gives rise to a separate peak, as shown in Figure 19.10b.

A third technique, *density gradient* ultracentrifugation, is based on the equilibrium properties of a multicomponent system. If a solution containing a low-molar-

FIGURE 19.10 Schlieren optics measures dC/dr directly. (a) A Schlieren photograph of tropomyosin (a muscle protein with molar mass 68,000) is shown for several time intervals. The rotational velocity was 60,000 rpm. (Beckman Instruments, Inc., Spinco Division, Palo Alto, California.) (b) A Schlieren photograph of a mixture of tropomyosin and troponin (molar mass = 80,000) at the same rotational velocity. The slower peak is due to the heavy complex of the two protein molecules. Velocity sedimentation is very useful in separating protein mixtures by molar masses. (Courtesy of David Hartshorne, Carnegie-Mellon University.)

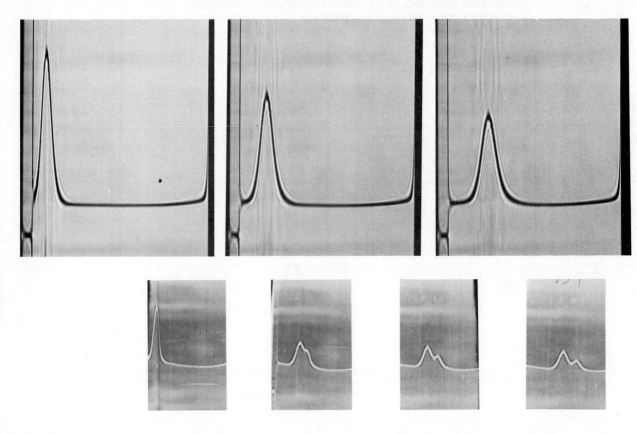

FIGURE 19.11 Band sedimentation experiment in a density gradient with two forms of lambda virus. The virus absorbs uv light, whereas the base CsCl solution in which the virus "floats" is transparent to uv light. The different viruses are then located by the dark bands on the photograph. (Beckman Instruments, Inc., Spinco Division, Palo Alto, California.)

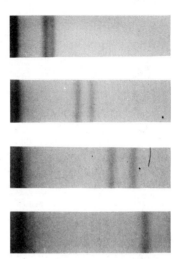

mass solute is placed in a centrifugal field, then a concentration gradient results. For example, if a $7.7m$ solution of CsCl is rotated at 45,000 rpm in a cell placed 6.5 cm from the rotation axis, then a density gradient of 0.12 g cm^{-3} per cm is established at equilibrium. For a liquid column 1 cm high, the density ranges from 1.64 g cm^{-3} at the top of the cell to 1.76 at the bottom. The "buoyant force" of the liquid in the cell is thus a function of position. If a macromolecule is placed in this density gradient, it will tend to "float" at that position in the cell at which its buoyant density equals the density of the solution. Ideally the molecules would "float" at one plane in the solution. The distribution of the molecules is, however, spread out by two factors, diffusion and the spread in molar mass of the macromolecule.

Density gradient centrifugation gives no advantage over ordinary equilibrium sedimentation in determining molar mass, especially since the position of the macromolecules is affected by a spread in mass. It is, however, a powerful tool for investigating biochemical substances such as nucleic acids, which have fixed molar masses, since one cause of the distribution broadening is eliminated. It is particularly useful for distinguishing between macromolecules with widely different values of the partial specific volume $\bar{v}_2$. A mixture of macromolecules characterized by different values of $\bar{v}_2$ will be separated into a series of discrete bands. The effective density of the molecule is measured by $\bar{v}_2$, and the various molecules "float" at that position at which $\rho = 1/\bar{v}_2$. This procedure is illustrated in Figure 19.11. Density gradient centrifugation has been used to separate ^{15}N-labeled nucleic acids from the normal ^{14}N acids.[5]

PROBLEMS

1. The density of polystyrene is 1.060 g cm^{-3}. When the individual polystyrene molecules are viewed with an electron microscope at a magnification of 48,000, they appear to be spherical particles. The following list of numbers gives the diameter in millimeters (and in parentheses, the number of particles of that diameter) as taken from the electron microgram: 12.8(1); 12.6(1); 11.8(2); 11.6(3); 11.4(3); 11.2(15); 11.0(57); 10.8(77); 10.6(25); 10.4(12); 10.2(10); 10.0(7); 9.8(2); 9.2(1); 9.0(3);

[5] M. Meselson and F. W. Stahl, *Proc. Nat. Acad. Sci. U.S.* 44 (1958):671.

8.8(1); 8.0(2); 7.8(3); 7.6(1); 5.4(1); 5.2(1). Calculate $\overline{M}_N$ and $\overline{M}_w$ from this measured distribution.

2. The accompanying table gives the fraction of molecules in a polymer sample with the indicated molar mass. Calculate the number average $\overline{M}_N$ and mass average $\overline{M}_w$ molar masses for the mixture.

Fraction of molecules	Molar mass of the molecules
0.12	10,000
.16	14,000
.14	16,000
.22	19,000
.11	22,000
.08	24,000
.10	28,000
0.07	31,000

3. Show that the molar mass determined from osmotic-pressure measurements is the number average molar mass.

4. If 1 g of the polymer of Problem 1 were dissolved in 1 liter of benzene, what would be the osmotic pressure? Calculate the freezing point depression and boiling point elevation for the solution. How accurately could you determine the molar mass by each of these three techniques?

5. An expression for a particular distribution of molar masses, the so-called *most probable distribution*, can be derived from the following considerations. We take as a concrete example the hydroxyacid OH—R—COOH. If the acid is heated with a catalyst, water is split out and a polymer is formed containing some number of units. We consider the formation of a linear polymer, which we write as

$$\text{HO}-\underset{1}{\text{R}}-\overset{\overset{\text{O}}{\|}}{\text{C}}-\text{O}-\underset{2}{\text{R}}-\overset{\overset{\text{O}}{\|}}{\text{C}}-\text{O}-\cdots$$

$$-\text{O}-\underset{r-1}{\text{R}}-\overset{\overset{\text{O}}{\|}}{\text{C}}-\text{O}-\underset{r}{\text{R}}-\overset{\overset{\text{O}}{\|}}{\text{C}}-\text{OH}$$

The addition of each polymer unit involves the esterification of that acid molecule with the molecule previously added to the chain. The successive molecular units are labeled $1, 2, \ldots, r-1, r$. We now want to find the probability that the molecule contains exactly r units.

We let p be the probability that the first molecule is esterified. The probability that the second molecule is also esterified is likewise p, since the probability is independent of whether or not the first unit is esterified. The total number of linkages in the r-mer is $r-1$. The probability that the sequence has continued for the r units is $p^{(r-1)}$. This is the probability that the macromolecule contains r units or $r-1$ ester groups. The probability that the rth unit is unreacted is $(1-p)$; hence the probability that the molecule is composed of exactly r units is

$$P_r = p^{(r-1)}(1-p)$$

In the following, let N be the total number of macromolecules and N_0 the total number of OH—R—COOH units. Show that:

(a) The total number of r-mers is given by $N_r = N(1-p)p^{(r-1)}$.

(b) The total number of macromolecules is given by $N = N_0(1-p)$, and hence $N_r = N_0(1-p)^2 p^{(r-1)}$.

(c) The *number average degree of polymerization* r_n, which is defined as the average number of monomer units in each macromolecule, is given by $r_n = 1/(1-p)$.

How large does p have to be for the number of monomer units in each polymer molecule to be larger than 100 on the average?

6. For the distribution of Problem 5:

(a) Show that the mass fraction of r-mers is given by $W_r = r(1-p)^2 p^{r-1}$. (Note that except for the end groups, each unit in the macromolecule contains a set of monomer units each having the same mass; in large macromolecules the slight error introduced by the end groups can be neglected.)

(b) Show that the mole fraction of r-mers is given by

$$X_r = (1-p)p^{(r-1)} = \frac{W_r}{r(1-p)}$$

(c) The distribution in (a) is called the *mass distribution;* the distribution in (b) is called the *number distribution*. Sketch each of these distribution curves as a function of r in the range $r = 0$ to 200 for $p = 0.97$.

(d) Note that the number distribution is a monotonically decreasing function of r, whereas the mass distribution has a peak. The peak occurs at the most probable value of r. Calculate the most probable value of r for $p = 0.95, 0.97, 0.99,$ and 0.995.

7. Using the distributions given in Problem 6, the definitions for the number average and mass average molar masses, and expressions for series summations (to be found in mathematical tables of functions), show that for the distribution of Problem 5, the number average molar mass is given by

$$M_N = M_0(1-p) \sum_{r=1}^{\infty} r p^{(r-1)} = \frac{M_0}{1-p}$$

and that the mass average is given by

$$M_w = M_0(1-p)^2 \sum_{r=1}^{\infty} r^2 p^{(r-1)} = \frac{M_0(1+p)}{(1-p)}$$

PROBLEMS

where M_0 is the mass of a monomer unit. For large macromolecules to be formed, p must be very nearly unity, in which case M_w/M_N is very nearly 2. Larger values of this ratio indicate a wide spread in molar masses.

8. At 25 °C, the osmotic pressure of a polycarbonate was measured in chlorobenzene solution with the following results:

$$\left[\text{—}\bigcirc\text{—C(CH}_3)_2\text{—}\bigcirc\text{—O—COO}\right]_r$$

Concentration (g liter^{-1})	1.95	2.93	3.91	5.86
Osmotic pressure (cm chlorobenzene)	0.20	0.36	0.53	0.98

The density of the chlorobenzene is 1.10 g cm^{-3}. Calculate the molar mass and the second virial coefficient. How many monomer units does the molar mass correspond to?

9. A macromolecule has a number average molar mass of 200,000. Suppose that the mass were to be determined in an osmometer with a defective membrane that allowed all molecules with masses less than 15,000 to cross the membrane freely. In which direction is the error in the measured molar mass? Calculate the mole fraction of macromolecules lost if the mass distribution is as given in Problem 6.

10. A polymer of molar mass 50,000 has a partial specific volume $\bar{v}_2 = 0.8$ cm^3 g^{-1}; it is dissolved in a solvent to make up a solution whose density is 1.011 g cm^{-3}. The solution is placed in a cell and rotated in an ultracentrifuge at 15,000 rpm. Calculate the ratio of the concentration at 6.75 cm to the concentration at 7.5 cm. The temperature is 37 °C.

11. A sedimentation equilibrium experiment is carried out on a polymer at 25 °C. The partial specific volume of the polymer is 0.87 cm^3 g^{-1}. The solution density is 1.16 g cm^{-3}. The concentration as a function of distance in the cell is as shown in the table. The cell was spun at 17,500 rpm. Calculate the molar mass of the polymer.

r (cm)	C (arbitrary units)
6.831	0.0887
6.922	.1688
6.997	.2842
7.031	.3661
7.084	.5236
7.163	0.9207

PART FOUR

THE STATISTICAL APPROACH

CHAPTER TWENTY

INTRODUCTION TO STATISTICAL METHODS

20.1 RANDOMNESS AND PROBABILITY

Pick a number, any number, from 0 to 9. Since there are ten possible choices, the probability of selecting any one of the ten is 1 in 10, or 0.1. People do not always choose numbers randomly but tend to have particular favorites. The number 7 might be chosen more often than any of the other nine. Selection could be randomized by placing ten identical balls, each with one of the digits written on it, in a box. We can get a digit randomly by shaking the box and selecting one of the balls without looking.

Suppose we want a random number between 0 and 999. We could take 1000 balls, each with one of the numbers on it, place them in a larger box, shake the box, and select one. Now we have one choice in 1000; the probability of selecting any one of the balls is 1 in 1000, or 0.001. We could also do this by three successive selections from the first box with the 10 balls, making sure that we replace the ball and reshake the box after each selection. Then we write the 3 selected digits in order, and let that be our random number between 0 and 999. The probability of choosing any particular digit in each selection is 0.1. The probability of 3 particular successive selections is just the product of the probabilities of the individual selections, or $0.1 \times 0.1 \times 0.1 = 0.001$, the same as the probability of selecting one ball in a thousand. This is one way of generating random numbers. Another way is to look them up in a table of random numbers. Today random numbers are conveniently generated with a digital computer.

Flip a coin. We say that the probability of getting heads *or* tails is 1. That means we are certain that the coin lands either heads or tails. (We neglect the possibility that the coin may land on edge.) We say that the probability of getting heads is $\frac{1}{2}$, and the probability of getting tails is $1 - \frac{1}{2}$, or $\frac{1}{2}$ also. What that means is this: If we flip a coin n times, then as n approaches infinity, the number of times heads appears approaches $\frac{1}{2}n$. If we flip a coin 30 times, we usually do not get 15 heads. In fact, we get 15 heads in only 14.4% of trials of flipping 30 coins.

Calculating the probabilities of flipping 30 coins requires using the binomial distribution, which we shall examine shortly. We can calculate the probabilities of flipping two coins quite simply. The probability of getting two heads we call $P(h,h)$. It is the probability of getting heads with the first coin multiplied by the probability of getting heads on the second coin, or $P(h,h) = \frac{1}{2} \times \frac{1}{2} = \frac{1}{4}$. The probability of getting two tails is $P(t,t) = \frac{1}{2} \times \frac{1}{2} = \frac{1}{4}$, the same as $P(h,h)$.

391

Similarly, the probability that the first coin is heads and the second tails is $P(h,t) = \frac{1}{4}$, while the probability of the reverse result is also $P(t,h) = \frac{1}{4}$. We can summarize this by:

$$\begin{aligned} P(h,h) &= \tfrac{1}{4} \\ P(h,t) &= \tfrac{1}{4} \\ P(t,h) &= \tfrac{1}{4} \\ \underline{P(t,t) &= \tfrac{1}{4}} \\ \text{Sum} &= 1 \end{aligned} \qquad (20.1)$$

The sum of all the probabilities must be unity, since we are certain to get at least one of the possible results. The possible results are shown in Figure 20.1. There are four possibilities, each equally likely; thus the probability of getting any one of the four results is one in four, or $\frac{1}{4}$. If we are interested in the probability of getting a head and a tail without regard to which comes first, then there are two ways out of four we can accomplish this; thus the probability is $\frac{1}{2}$, or the sum of $P(h,t)$ and $P(t,h)$.

Suppose we flip two coins 100 times. Does this mean that we get two heads 25 times, two tails 25 times, and a head and a tail without regard to order the remaining 50 times? We might. Then again, we might not. The laws of probability and statistics apply accurately to large numbers of events. What we can say quite definitely is the following: If we flip two coins 10^{23} times, then the number of times we get two heads will be very close to $\frac{1}{4} \times 10^{23}$. The amount by which it deviates from that number will be trivially small compared with 10^{23}. If we flip the two coins only 100 times, we are much less definite about the results, and the deviation from 25 sets of two heads may be relatively large. We can, however, compute the relative numbers of times two heads will appear.

While the laws of probability apply accurately to large numbers, we usually have small numbers to work with. The fate of TV programs is often determined by statistical inference from a small amount of data generated by a small number of viewers. Suppose you have just prepared a compound whose melting point is given in the literature as 137.4 ± 0.4 °C; you find the melting point of your compound to be

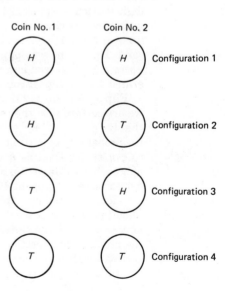

FIGURE 20.1 The four possible results of flipping two coins. The probability of any of the four results is $\frac{1}{4}$, since all are equally probable. The probability of one head and one tail without regard to the order is $\frac{1}{2}$, since there are two ways out of four that we can get one head and one tail.

136.7 °C. Is there a real discrepancy, and should you try a new preparation? We shall answer this question later in the chapter.

Statistics is useful in decision making as these last two examples illustrate. Consider a situation of some interest to students. Three different professors teach three different sections of the same course, each section containing 25 students. Three students fail in the first section, 3 fail in the second section, and 9 fail in the third section. Should the students mount a campaign to have the third professor fired?[1] In a more serious vein, kvetcherites is a disease that proves fatal without proper medical attention. Three different treatments are tried on three groups of 25 patients. Three patients die in the first group, 3 in the second, and 9 in the third. Can any conclusion be drawn about the relative efficacies of the different treatments? The problem is identical to that of the professors and the failing students and is the kind of problem medical workers must deal with on a regular basis. It is a problem whose solution may affect the health and well-being of many people. A complete discussion of statistical inference would carry us far afield. Information on solving these and other problems can be found in the references listed in the Bibliography.

20.2 THE STATISTICAL INTEGRAL; MONTE CARLO INTEGRATION

Consider the definite integral $\int_0^1 x^2 \, dx$. It is equal to $\frac{1}{3}$. Geometrically, it is the area under the curve $y = x^2$ between $x = 0$ and $x = 1$, as shown in Figure 20.2. Now suppose we have a flea flitting about and occasionally landing on Figure 20.2. Although

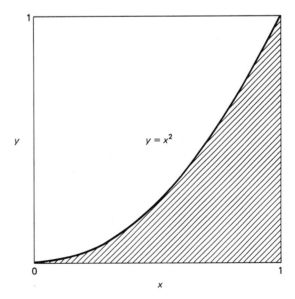

FIGURE 20.2 $\int_0^1 x^2 \, dx$ equals the area under the curve $y = x^2$, as indicated by the hatched region. The integral is $\frac{1}{3}$.

---

[1] The answer, as all too few students realize, is No. The answer is to be found in applying what is known as the *chi square* (χ^2) *test of goodness of fit*. If the professors are of equal teaching abilities, then in about 10% of all trials on randomly selected sections one can expect a grade distribution as given. The usual statistical cutoff is 5% for making such decisions. (On the other hand, if these results were duplicated for several semesters in succession, then the answer would be different.)

it lands somewhere on the indicated square region, the exact spot on which it lands is completely random. Since the area under the curve is $\frac{1}{3}$ of the total area, and the area above the curve is $\frac{2}{3}$ of the total area, the probability that the flea lands below the curve is $\frac{1}{3}$ and the probability that it lands above the curve is $\frac{2}{3}$. Every time the flea lands on the figure, we note whether it lands above or below the curve, then chase it off and wait for it to land again. The procedure is time-consuming, but if we have enough landings, then $\frac{1}{3}$ will be below the curve, and $\frac{2}{3}$ above. The value of the integral is equal to the fraction of landings below the curve, or $\frac{1}{3}$.

The experiment can be carried out by simulating the flight of the flea on a digital computer. We program the computer to select pairs of random values, x and y, in the interval between 0 and 1. For each pair, we note whether y is less than x^2 or greater. Figure 20.3 indicates the results of the experiment. Of the first 1000 pairs of points, 332 lie below the curve, yielding a value for the integral of 0.332. The closest we could have come is 0.333. By the binomial distribution, we can compute the probability of getting 333 landings below the curve in 1000 tries; it is 0.02675. This means that if we tried a large number N of experiments, each with 1000 sets of points, then in the limit as N approaches infinity, the number of times we get 333 landings below the curve approaches $0.02675N$.

Table 20.1 summarizes the results of the computerized flea flight experiment. The first 1000 trials have been plotted in Figure 20.3. The rest of the table indicates the results of the experiment carried out to 1 million trials, at which time the integral is accurate to within 0.2% of its true value. During the experiment, the value was often higher and often lower than the true value; 342 times during the run, it was equal to exactly $\frac{1}{3}$. We should expect the error in the evaluated integral to be substantially less than 0.2% if we had allowed the computer to continue out to 10 million trials.[2]

FIGURE 20.3 The statistical integral. The points are randomly generated for x and y between 0 and 1. The solid curve is $y = x^2$. (a) The first 100 points; 32 points lie below the curve. (b) The first 1000 points; 332 points lie below the curve.

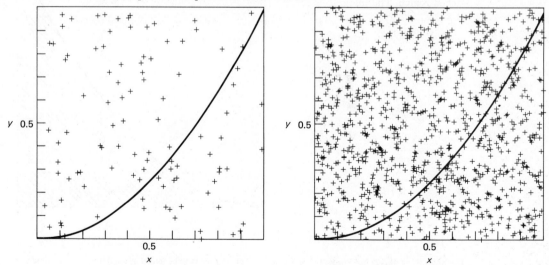

[2] The computer was a relatively slow one and took about 10 min to compute the million random points. These results form one example of a computer run. If the random number generator is truly random, then a second run, even on the same computer, would give different results; the conclusions would be the same, however.

20.3 THE BINOMIAL DISTRIBUTION

TABLE 20.1 Results of the flea flight experiment.

Number of trials	Cumulative number of times flea lands below curve $y = x^2$	Value of integral by statistical integration
100	32	0.3200
200	64	.3200
300	104	.3467
400	139	.3475
500	163	.3260
600	200	.3333 . . .
700	240	.3429
800	272	.3400
900	301	.3344
1,000	332	.3320
3,800	1,229	.3234
8,100	2,700	.3333 . . .
9,100	3,021	.3320
70,000	23,340	.3334
130,000	43,557	.3351
1,000,000	333,939	0.3339

NOTE: The first 1000 points have been plotted in Figure 20.3.

On the other hand, it is possible that the error would get larger, but the probability of a larger error is quite small.

20.3 THE BINOMIAL DISTRIBUTION

Suppose we have N *independent* events of some sort. To provide some reality, we take the events to be flips of a coin. There are two possible outcomes that we can classify as "success" and "failure." For a coin, the outcome is either heads or tails; we let heads be success and tails be failure. The probability of success we call p; the probability of failure is then $q = (1 - p)$. For a coin flip, both p and q are each $\frac{1}{2}$. If we flip five coins, the number of independent events is $N = 5$. What is the probability that n coins land heads if n is between 0 and 5? The case of five heads is straightforward. It is $(\frac{1}{2})^5 = \frac{1}{32}$. The probability of five tails is

$$(1 - p)^5 = (q)^5 = (\tfrac{1}{2})^5 = \tfrac{1}{32}$$

the same as five heads. The intermediate probabilities are less straightforward.

The probability that n events succeed while the remaining $N - n$ events fail is $p^n q^{(N-n)}$. This is not the probability that n events succeed. It is the probability that a particular group of n events succeed, say the first n, or the last n, or perhaps the first one and the last $(n - 1)$. What we want is the probability without regard to the order of the events. To find this probability, we must compute the total number of ways we can achieve n successes. This is the number of combinations of N things taken n at a time; this number is given by the *binomial coefficients*

$$\binom{N}{n} = \frac{N!}{(N - n)!n!} \tag{20.2}$$

The probability of achieving exactly n successes in N trials where the probability of success is p and the probability of failure is $q = (1 - p)$ is

$$P(n;N,p) = \binom{N}{n} p^n q^{N-n}$$

$$= \frac{N!}{(N-n)!n!} p^n q^{N-n} \tag{20.3}$$

By comparing (20.3) with the binomial expansion formula

$$1 = (p+q)^N = p^N + \binom{N}{1} p^{N-1}q + \binom{N}{2} p^{N-2}q^2$$

$$+ \cdots + \binom{N}{n} p^n q^{N-n} + \cdots + q^N \tag{20.4}$$

we get

$$1 = \sum_{n=0}^{N} P(n;N,p) \tag{20.5}$$

We should have been surprised at any other result, since the sum of the probabilities of all the possible results must be unity. This is characteristic of what we call a *normalized* distribution.

The binomial distribution is a discrete distribution. In our five coin flips, we can achieve 0, 1, 2, 3, 4, or 5 heads with no intermediate results possible. Figure 20.4a shows the binomial distribution with $p = \frac{1}{2}$ for three different values of N. Since

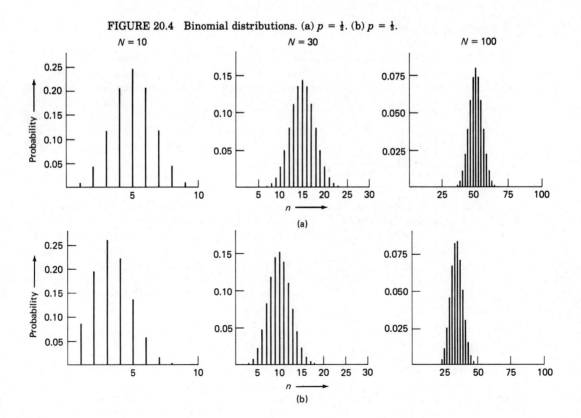

FIGURE 20.4 Binomial distributions. (a) $p = \frac{1}{2}$. (b) $p = \frac{1}{3}$.

$p = \frac{1}{2}$, the distributions are symmetrical. Asymmetrical distributions for $p = \frac{1}{3}$ are shown in Figure 20.4b.

Let's examine the situation for $P = \frac{1}{2}$ and $N = 30$ more closely. If we flip 30 coins, the most probable result is 15 heads, but the probability is only 0.144. The probability of achieving 14 or 16 heads is almost as large, 0.135, and the probability of achieving 13 or 17 heads is only down to 0.112. The probability of achieving between 13 and 17 heads is the sum of these, or about 0.64. At the other extreme, the probability of achieving only 5 heads is 10^{-4}, while the probability of achieving 0 heads is 10^{-9}.

Figure 20.5 shows the results of a computerized binomial distribution experiment for $p = \frac{1}{2}$ and $N = 30$. The event was the generation of a random number between 0 and 1 that was greater than $\frac{1}{2}$. The probability $p = \frac{1}{2}$. The envelope is the curve of the expected distribution based on the binomial distribution. The vertical lines indicate the experimental values. In Figure 20.5a, which shows the result of 200 experimental tossings of 30 coins, there are substantial deviations between the experimental heights and the expected heights. But notice that as the number of experiments increases to 2000 and finally to 20,000, the agreement gets better and better. The laws of probability hold well in the limit of large numbers.

Figure 20.6 shows the envelopes for the distributions of Figure 20.4a; an extra curve has been added for $N = 1000$. As N gets larger, the width of the curve expressed as a fraction of N gets smaller. For $N = 10$, there are substantial probabilities of achieving a number of heads that differs significantly from the most probable value. The probability of deviating by 20% or more from the average or most probable value 5 is large. It is the sum of all probabilities for $n \neq 5$, or about 0.75. For $N = 30$ and $N = 100$, the probabilities for such large fractional errors have decreased considerably. By the time we reach $N = 1000$, the probability is almost nil. For $N = 1000$, a deviation as large as 20% of the most probable value corresponds to the sum of all values of n outside the limits of 400 to 600. For this case, 99% of all

FIGURE 20.5 Computerized binomial distribution experiment for $p = \frac{1}{2}$ and $N = 30$. The experiment corresponds to flipping 30 coins, with n indicating the number of successes (heads). The vertical bars indicate the experimental results, and the curves indicate the envelope of the expected binomial distribution. As the number of trials increases, the experimental results more closely approximate the expected results. (a) After 200 trials. (b) After 2000 trials. (c) After 20,000 trials.

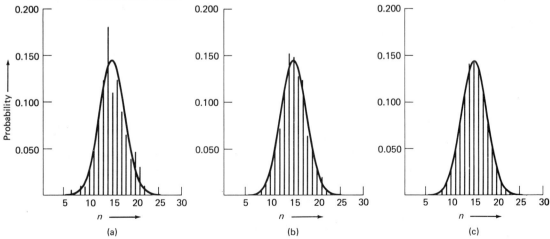

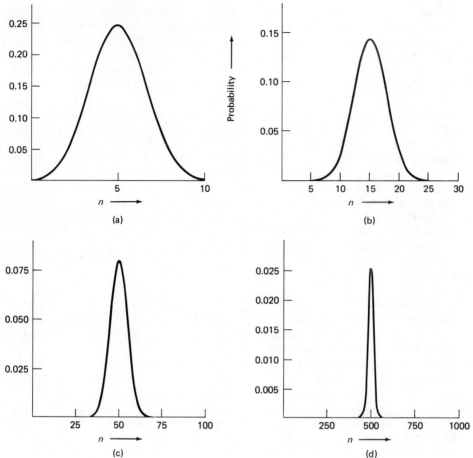

FIGURE 20.6 Envelopes of binomial distributions. (a) $N = 10$. (b) $N = 30$. (c) $N = 100$. (d) $N = 1000$.

trials yield values of n between 460 and 540. For $N = 10{,}000$, 99% of all trials fall within 130 units of the most probable value 5000. By the time we get to $N = 6 \times 10^{23}$, the probability of achieving a distribution much different from the most probable distribution is trivially small.

20.4 THE GAUSSIAN DISTRIBUTION

Now let's try another experiment in which we determine the average value of some randomly generated numbers between 0 and 1. This seems like a trivial problem, since we know in advance that the answer should be $\frac{1}{2}$. On the other hand it is not so trivial, since it enables us to test the procedure by which we generate random numbers. If the answer deviates much from $\frac{1}{2}$, we may suspect the technique used to generate the numbers.

In our initial experiment we generate five random numbers in succession and take the average of the five numbers. Five successively generated random numbers

SEC. 20.4 THE GAUSSIAN DISTRIBUTION

FIGURE 20.7 Frequency distribution for the determination of the average of 5 random numbers. The abscissa has been divided into 50 segments. The heights of the bars indicate the relative number of times the average fell in that interval. (a) After 100 trials; average = 0.5067. (b) After 1000 trials; average = 0.5022. (c) After 20,000 trials; average = 0.4994.

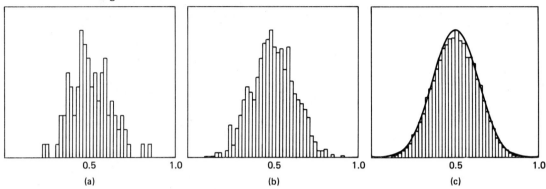

are 0.4123, 0.0908, 0.2250, 0.4972, and 0.1579. The average is 0.2766. This deviates substantially from $\frac{1}{2}$, but hold on a minute! Reflecting about the binomial distribution for a moment, we discern a substantial probability that all five numbers will fall below 0.3. One experiment is not a very good way to find the answer. We must perform many experiments, in each of which we determine the average of five random numbers; then we take the average of all the averages. The results of this experiment are shown in Figure 20.7 as a *frequency distribution*. The abscissa has been divided into 50 equal segments. The heights of the bars indicate the relative number of times the average fell in that interval. The distributions have been plotted for three different values of N', the number of averages taken. By $N' = 20,000$, the shape appears a bit familiar.

Figure 20.8 shows the results of a slightly different experiment. The number of

FIGURE 20.8 Frequency distributions as in Figure 20.7c for 20,000 trials. Here the number of random numbers used in each average is different. Figure 20.7c fits between (a) and (b). The numbers after the symbols ± are the standard deviations. (The standard deviation for Figure 20.7c is 0.1290.) Note that the standard deviations are inversely proportional to the square root of the number of random numbers used per trial. (a) Two numbers per trial; average = 0.4985 ± 0.2032. (b) Ten numbers per trial; average = 0.5007 ± 0.0914. (c) 100 numbers per trial; average = 0.4999 ± 0.0289.

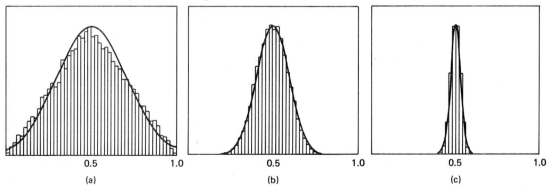

> KARL FRIEDRICH GAUSS (1777–1855), German mathematician, is considered along with Archimedes and Newton one of the three greatest mathematicians in the history of human civilization, and it is not for lesser mortals to attempt a ranking among these three. Gauss was the son of a laborer, and he first demonstrated his genius at the age of three when he corrected his father's payroll errors by mental calculations. He made fundamental contributions in an extraordinarily wide range of topics from elliptic functions to non-Euclidean geometry. His wide interests reached the practical world. In 1833, Gauss invented the electric telegraph, an outgrowth of his interest in electricity and magnetism. In his later years, Gauss sought the universal theory of electromagnetism. It remained for Maxwell to discover the theory.

observations was 20,000 for each curve, but this time the number of random numbers used for each determination is different. The distribution gets narrower and narrower as this number increases. Figure 20.7c fits into this sequence between (a) and (b). The curves indicate the *Gaussian distribution*, or *normal error function*. It is given by

$$f(x) = \frac{h}{\sqrt{\pi}} \exp[-h^2(x-m)^2] \tag{20.6}$$

where m is the *mean value*. The *variance* σ^2 is defined in terms of h by

$$h^2 = \frac{1}{2\sigma^2} \tag{20.7}$$

The variance is the *sum of the squares of the deviations*,

$$\sigma^2 = \int_{-\infty}^{+\infty} (x-m)^2 f(x)\, dx = \frac{1}{2h^2} \tag{20.8}$$

The square root of the variance is called the *standard deviation*.

The larger the h, the more sharply peaked the curve, as shown in Figure 20.9. The Gaussian distribution is applicable to the errors associated with many types of experimental measurements. In contradistinction to the binomial distribution, which is a discrete distribution, the Gaussian is a continuous distribution. The relation of the distribution to experimental measurements is shown in Figure 20.10. The peak of the curve occurs at the *mean*, or *average* value, of the measurements; the scale of the curve is adjusted so that the total area under the curve is unity. The area of the shaded strip at x of width dx is the probability that the result of a particular experiment occurs between x and $x + dx$. The probability of an experimental result between a and b is indicated by the shaded area between a and b; it is

$$P(a,b) = \int_a^b f(x)\, dx \tag{20.9}$$

The area of the curve between $m + \sigma$ and $m - \sigma$ is 0.683. This means that 68.3% of all measurements give results within one standard deviation of the mean, and 31.7% give results that deviate from the mean by more than one standard deviation. The percentage of the measured values that lie within two standard deviations is 95.4%; and 99.7% lie within three standard deviations.

We can now provide an answer to the question raised in the first section. If the melting point is 137.4 ± 0.4 °C, then a measurement of 136.7 °C lies within two

SEC. 20.4 THE GAUSSIAN DISTRIBUTION

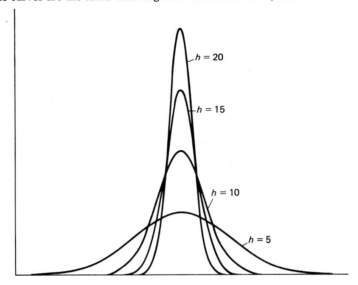

FIGURE 20.9 Gaussian distribution curves for four different values of h. The areas under each of the curves are the same. The heights of the curves are $h/\sqrt{\pi}$.

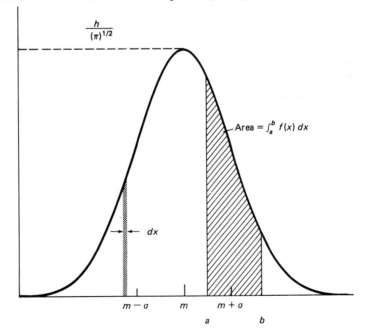

FIGURE 20.10 Gaussian distribution for $h = 5$, showing relevant points. The area under the entire curve is unity. The probability of a value between a and b equals the indicated area. The area of the strip indicates the probability of obtaining a value between x and $x + dx$. The height of the peak is at $h/(\pi)^{1/2}$. The percentage of the area between $m - \sigma$ and $m + \sigma$, where m is the average and σ is the standard deviation (equal to $1/h\sqrt{2}$), is 68.3%.

standard deviations of the measured value. We should expect such a deviation to occur more than 5% of the time; hence we conclude that the agreement is satisfactory.[3]

Values of the error function and error integral are usually tabulated in terms of the standardized variable $z = h(x - m)$. This corresponds to shifting the mean to the origin and to scaling the abscissa in units of σ. Tables of integrals of the Gaussian function are provided as entries for the integral,

$$I = \frac{1}{\sqrt{2\pi}} \int_0^z \exp\left(-\frac{1}{2}z^2\right) dz \tag{20.10}$$

Values of this integral are listed in Table 20.2. A related function is the *error function*, symbol *erf(z)*, defined as

$$erf(z) = \frac{1}{\sqrt{\pi}} \int_{-z}^{+z} \exp(-x^2) dx \tag{20.11}$$

Further discussions of the Gaussian distribution can be found in the references listed in the Bibliography. Because the curve is shaped like a bell, it is often referred to as a bell-shaped curve. Students in both technical and nontechnical areas often refer to examination results as falling on a "bell-shaped curve." If the number of students taking an examination is large enough, then the grade distribution should approximate a Gaussian distribution. The percentage of students who should get grades within one standard deviation of the mean is 68.3%. One criterion sometimes used in making up examinations is to have a large standard deviation. That way it is easier to separate students into grade classifications.

If a measurement carried out in the laboratory is repeated enough times, then the results should be distributed along a Gaussian curve. In principle, the standard deviation could be measured from the curve. This is true of the experimental results shown in Figure 20.8. In practice, there are usually fewer measurements than the number required to determine the shape of the curve. The variance for a small number of measurements is given by

$$\sigma^2 = \frac{1}{N} \sum_i (x_i - \bar{x})^2 \tag{20.12}$$

TABLE 20.2 Integrals of the Gaussian function. Entries are values of the integral $I = (1/\sqrt{2\pi}) \int_0^z \exp(-\frac{1}{2}z^2) dz$

Z	I	Z	I
0.0	0.0000	1.8	0.4641
0.2	.0793	2.0	.4773
0.4	.1555	2.5	.4938
0.6	.2258	3.0	.4987
0.8	.2882	3.5	.49977
1.0	.3413	4.0	.499 968
1.2	.3849	4.5	.499 996 6
1.4	.4192	5.0	0.499 999 71
1.6	0.4452		

[3] A more difficult problem is the following. Suppose two laboratories measure the melting point, taking great precautions. One gets 137.4 ± 0.4 and the other gets 136.0 ± 0.3. Now we must take into account the mutual overlaps of the Gaussian distributions of each measurement. Problems like this are dealt with in the references listed in the Bibliography.

where N is the number of measurements, $\bar{x}$ is the average of the measurements, and the x_i are the individual measurements. The variance is the average of the sum of the squares of the deviations.

One of the reasons the Gaussian distribution is important in statistics is that it is the limiting form attained by many other statistical distributions. The Gaussian law, for example, is approached by the binomial distribution in the limit as N increases while p is held constant. In the next chapter we shall apply the Gaussian distribution to the distribution of molecular velocities.

PROBLEMS

1. Six coins are flipped. Using the binomial distribution, calculate the probabilities of getting 0, 1, 2, 3, 4, 5, and 6 heads. Calculate the probability of getting fewer than 3 heads.
2. In flipping the six coins, calculate the total number of ways each of the possible configurations can be obtained.
3. Calculate the standard deviation for the flipping of six coins.
4. How many coins would you have to flip so that the probability of getting no heads was less than 10^{-20}?
5. Show that the maximum in the Gaussian distribution curve occurs at $x = m$, using the definition of the curve given in Equation (20.6).
6. Find the position of the point of inflection in the Gaussian distribution curve.
7. A set of six determinations is carried out of the heat of combustion of glotch. The results are 27.36, 27.42, 27.73, 27.51, 27.63, and 27.56 kJ mol^{-1}. You have a sample of what you think is pure glotch. You measure its heat of combustion and find it to be 27.29 kJ mol^{-1}. How sure are you that the sample is pure glotch? Suppose your measured value was 27.11 kJ mol^{-1}; what then?
8. As noted in Chapter 17, the peak demand requirements are important parameters for electric-power-generating companies. If they overestimate the demand, they will waste money by installing too much generation capacity. If they underestimate their demand, they will have insufficient capacity and an excessive number of blackouts. A utility company, by undertaking a careful study of its historical demand and predicting the growth of demand arising from population, industrial, and similar increases, predicts that its peak hourly demand will be 4743 ± 227 MW by 1985. How much capacity must it have to ensure that blackouts will not occur more often than (a) once a month? (b) once a year? (c) once every 5 yr? Disregard that the equipment may break down due to mechanical failures. (*Hint:* Use Table 20.2.)

CHAPTER TWENTY-ONE

THE MAXWELL-BOLTZMANN DISTRIBUTION OF MOLECULAR VELOCITIES

21.1 THE EXPERIMENT

The thermal energy of molecules manifests itself as kinetic energy of motion (Chapter 3). In examining the *distribution* of molecular velocities, we want to answer the following question in particular. In a collection of gas molecules at temperature T, what is the fraction of molecules that have speeds in the range between c and $c + dc$? Maxwell first solved this problem in 1860. The solution is based on the Gaussian distribution (Chapter 20).

We can actually measure the distribution of velocities with an apparatus, as shown in Figure 21.1. The gas is in an oven at temperature T. A small hole is pierced in the wall of the oven so that a beam of molecules effuses out of the oven as indicated. The region outside the oven is maintained at high vacuum. A pinhole in front of the oven hole provides a collimated pencil of molecules. The collimated *molecular beam* then passes through a *velocity selector,* which allows only molecules in a particular velocity range to pass through to the detector. The form of the velocity selector shown consists of two discs mounted on a shaft separated by a distance D. A slit is cut in each disc as shown, and the slits are displaced by an angle θ. A detector is placed in line with the beam behind the second slit. The shaft is rotated with angular velocity ω.

Suppose ω is such that it takes t seconds for the shaft to rotate an angle θ. If the first slit is aligned with the beam at t_0, then the second slit will be aligned at $t_0 + t$. Of all the molecules that pass through the first slit, the only ones that pass through the second slit and reach the detector are those that cover the distance D in time t, that is, those with velocity D/t. By measuring the molecular-beam current at the detector as a function of angular velocity, we can determine the relative numbers of molecules traveling with different velocities.[1]

Figure 21.2a shows the results of the experiment in a histogram. The abscissa measures velocity, and the ordinate measures the molecular-beam current on an arbitrary scale. Figure 21.2a corresponds to an experiment in which measurements are

---

[1] The resolution of the velocity selector is a function of the slit width. The wider the slit, the larger the velocity range transmitted (see Problem 3). A detailed description of an actual apparatus is given in R. C. Miller and P. Kusch, *Phys. Rev.* 99 (1955):1314.

FIGURE 21.1 Apparatus for measuring the velocity distribution of molecules. The sample, usually a metal or salt, is placed in the source oven, which is maintained at the required temperature. The collimating pinhole provides a pencil-shaped molecular beam traveling to the detector. The detector usually operates in such a way that the molecules are converted to ions and detected by an ion current. The slits are displaced relative to each other by an angle θ and separated by a distance D so that at any given rotational frequency only molecules with a particular velocity will pass through both slits and get to the detector.

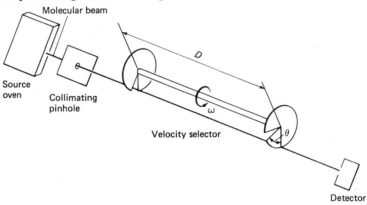

FIGURE 21.2 Results of a velocity distribution experiment. (a) Coarse resolution. (b) Medium resolution. (c) Fine resolution.

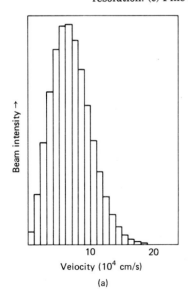
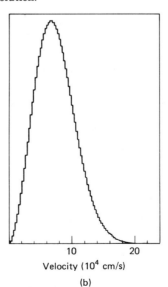
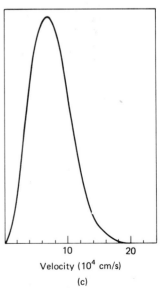

taken at velocity increments of 10^4 cm s^{-1}. Figures 21.2b and 21.2c show the results obtained by decreasing the size of the velocity increments (and also increasing the number of measurements required). For sufficiently small increments, the stepped histogram becomes a smooth curve that resembles a Gaussian distribution curve with a long tail. It is the product of three Gaussian curves, as we shall shortly see.

In a sense we have put the cart before the horse, since the shape of the curve was derived long before the tools necessary to carry out the experiment were developed. It is a tribute to the genius of the people who developed the theory. Working inductively from the basic principles of the ideal gas law, Newtonian mechanics, and the theory of statistics and probability, these scientists developed the shape of the curve more than fifty years before there was any possibility of directly measuring the shape. The experiment we have just described was the final ultimate verification of the theory. We shall now develop this *Maxwell-Boltzmann distribution* curve.

21.2 THE BOLTZMANN DISTRIBUTION LAW

Suppose we have a large collection of N molecules having a total energy E. The average energy of the molecules is E/N, but not all molecules have to have this particular energy. We let the individual energies available to each molecule be ϵ_i, so that there are N_i molecules with energy ϵ_i. An important question one might ask about the system is the following: How are the N molecules distributed among the various energies subject to the condition that $\Sigma N_i \epsilon_i = E$? We shall find it convenient to defer a detailed derivation of the answer to Chapter 23. For the moment we shall simply produce the final result, saying that the answer is to be found in the *Boltzmann distribution law*. This law states that if there are N_0 molecules in any particular state with energy ϵ_0, then the number of molecules with energy ϵ_i is

$$N_i = N_0 \exp\left[\frac{-(\epsilon_i - \epsilon_0)}{kT}\right] \tag{21.1}$$

where k is the Boltzmann constant. This law gives the *ratios* of the numbers of molecules with different energies as a function of ϵ and T. The difference $\epsilon_i - \epsilon_0$ is often written as $\Delta\epsilon$, and

$$N_i = N_0 \exp\left(\frac{-\Delta\epsilon}{kT}\right) \tag{21.2}$$

It is often convenient to know N_i in terms of the number of molecules in the lowest level. If we shift the energy scale so that $\epsilon_0 = 0$, then Equation (21.1) becomes

$$N_i = N_0 \exp\left(\frac{-\epsilon_i}{kT}\right) \tag{21.3}$$

LUDWIG BOLTZMANN (1844–1906), Austrian scientist, is best known for his work in the kinetic theory of gases and in thermodynamics and statistical mechanics. His suicide in 1906 is attributed by some to a state of depression resulting from the intense scientific war between the atomists and the energists at the turn of the century. On his tombstone is the inscription $S = k \ln W$.

SEC. 21.2 THE BOLTZMANN DISTRIBUTION LAW

Although a detailed derivation of the law is deferred, we can directly demonstrate its validity for gravitational potential energy by considering the pressure variation of the atmosphere with height.

Figure 21.3 shows a column of gas of unit cross-sectional area extending upward from sea level. We take a thin layer of thickness dh at height h and consider the pressure change across the layer. The weight of the layer is $\rho g\, dh$, where ρ is the density and g is the gravitational acceleration constant. The change in pressure from the bottom of the layer to the top equals the weight:

$$-dP = \rho g\, dh \tag{21.4}$$

For an ideal gas, the density is PM/RT, where M is the molar mass; hence

$$-\frac{dP}{P} = \frac{Mg}{RT}\, dh \tag{21.5}$$

This can be integrated between the limits $P = P_0$ at $h = 0$ and P at h, to get

$$\ln \frac{P}{P_0} = -\frac{Mgh}{RT} \tag{21.6}$$

or

$$P = P_0 \exp\left(-\frac{Mgh}{RT}\right) \tag{21.7}$$

$$P = P_0 \exp\left(-\frac{mgh}{kT}\right) \tag{21.8}$$

where m is the mass of a molecule.

Since mgh is the gravitational potential energy of a molecule, (21.8) can be written

$$P = P_0 \exp\left(-\frac{E_p}{kT}\right) \tag{21.9}$$

where E_p is the potential energy. Equation (21.9) is identical in form to the Boltzmann distribution, since P is directly proportional to N. This same result was derived in Chapter 18 in the context of equilibrium thermodynamics. An equivalent

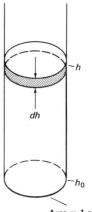

FIGURE 21.3 Column of gas extending upward. The designation h_0 is the reference level. The weight of the layer of gas of thickness dh at height h is $\rho g\, dh$, and this is just the pressure difference between the top and the bottom of the layer.

FIGURE 21.4 Appearance of a sedimentation equilibrium for colloidal particles. Each dot represents one colloidal particle in a beaker of fluid. The line on the right indicates a 1×10^{-3}-cm length for scaling purposes (see Problem 10).

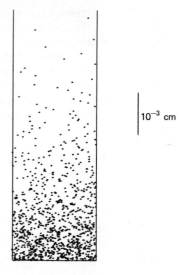

equation holds for colloidal particles undergoing Brownian motion; it provided one of the first independent measurements of Avogadro's number (see Problem 9).

The process known as Brownian motion was first discovered by Robert Brown in 1827. He observed the motion of very small pollen grains suspended in water and noted that they were in continual, apparently random motion. The random motion of the tiny pollen grains shows on a grossly magnified scale how gas molecules move.[2] The colloidal particles behave like gas molecules in a gravitational field. The gravitational force on a single particle is mg. When the particle is suspended in a liquid, the force is reduced by the buoyancy of the liquid and is $\frac{4}{3}\pi r^3 g(\rho - \rho_1)$, where r is the radius of the spherical colloidal particle and ρ and ρ_1 are the densities of the particle and suspending liquid respectively; Equation (21.6) becomes

$$\ln \frac{N}{N_0} = \frac{-\frac{4}{3}\pi r^3 g h(\rho - \rho_1)}{kT} \quad (21.10)$$

The radius can be measured with a microscope, and the densities can be determined by standard gravimetric techniques. Figure 21.4 indicates the appearance of a sedimentation equilibrium.[3] By measuring the density of particles as a function of height, one can determine the Boltzmann constant. Since the Boltzmann constant equals R/L_0 where R is the gas constant and L_0 is Avogadro's number, L_0 is then determined from the known value of R (see Problem 10).

---

[2] It was at first thought that the motion of the pollen grains was a form of perpetual motion. The energy for the motion actually comes from the thermal energy of the suspending liquid.

[3] If you have done any laboratory work involving precipitates, you may well have seen this phenomenon. Large flocculent precipitates usually settle rapidly to the bottom of the vessel. Very fine precipitates (the kind that go through filter paper) often display an appearance like that of Figure 21.4 if they are allowed to stand for some time. The effect of the gravitational field can be greatly magnified with a centrifuge, as noted in Chapter 18.

21.3 THE VELOCITY DISTRIBUTION IN ONE DIMENSION

We consider gas molecules restricted to one-dimensional motion. The kinetic energy of a molecule of mass m with velocity u along the x axis is $\frac{1}{2}mu^2$. We want the probability of finding a molecule with velocity between u and $u + du$. We start with the Boltzmann distribution.

The lowest energy is $E = 0$ at $u = 0$. We let dN_0 be the number of molecules with velocities between 0 and du. The number with velocities between u and $u + du$ is

$$dN = dN_0 \exp\left(-\frac{mu^2}{2kT}\right) \quad (21.11)$$

where dN_0 is proportional to the width of the interval du; thus we can write $dN_0 = a\,du$, where a is a constant as yet undetermined. We are interested in the fraction of molecules rather than the absolute number. If we divide both sides of (21.11) by N, the total number of molecules, we get

$$\frac{dN}{N} = A \exp\left(-\frac{mu^2}{2kT}\right) du = f(u)\,du \quad (21.12)$$

where $A = a/N$. The value A can be determined from the normalization condition that the integral over all fractions must be unity:

$$\int_{-\infty}^{+\infty} f(u)\,du = 1 = A \int_{-\infty}^{+\infty} \exp\left(-\frac{mu^2}{2kT}\right) du \quad (21.13)$$

This integral can be evaluated by letting $s^2 = mu^2/2kT$, from which it follows that

$$s = \left(\frac{m}{2kT}\right)^{1/2} u$$

$$ds = \left(\frac{m}{2kT}\right)^{1/2} du$$

The integral in (21.13) now becomes

$$A \left(\frac{2kT}{m}\right)^{1/2} \int_{-\infty}^{\infty} \exp(-s^2)\,ds = 1 \quad (21.14)$$

This is a standard integral found in most tabulations; it is also listed in Table 21.1. Its value is $\pi^{1/2}$. Hence $A = (m/2\pi kT)^{1/2}$, and (21.12) can be written as

$$\frac{dN}{N} = \left(\frac{m}{2\pi kT}\right)^{1/2} \exp\left(-\frac{mu^2}{2kT}\right) = f(u) \quad (21.15)$$

Referring to Equation (20.6), we see that (21.15) is just the Gaussian distribution function with a mean value of $u = 0$, and $h = (m/2kT)^{1/2}$. A one-dimensional velocity distribution is plotted in Figure 21.5. The curve has the same meaning as the Gaussian curves of Chapter 20. The fraction of molecules with velocities between u_a and u_b is just $\int_a^b f(u)\,du$.

Although we know that the mean (or average) velocity is 0, it is instructive to evaluate it.[4] The procedure is the same as that for evaluating the mean score on an examination. The velocity is divided into intervals du. In each interval we compute

---

[4] Note that although the average velocity is zero, the average *speed* is not zero.

TABLE 21.1 Useful integrals occurring in the kinetic theory of gases. In these equations, n is a positive integer, and a is a positive constant.

1. $\int_0^\infty \exp(-ax^2)\,dx = \tfrac{1}{2}\sqrt{\pi}\,a^{-1/2}$

2. $\int_0^\infty x^2 \exp(-ax^2)\,dx = \tfrac{1}{4}\sqrt{\pi}\,a^{-3/2}$

3. $\int_0^\infty x^{2n} \exp(-ax^2)\,dx = \tfrac{1}{2}\sqrt{\pi}\,\dfrac{(2n)!\,a^{-[n+(1/2)]}}{2^{2n} n!}$

4. $\int_{-\infty}^\infty x^{2n} \exp(-ax^2)\,dx = 2\int_0^\infty x^{2n} \exp(-ax^2)\,dx$

5. $\int_0^\infty x \exp(-ax^2)\,dx = \dfrac{1}{2a}$

6. $\int_0^\infty x^3 \exp(-ax^2)\,dx = \dfrac{1}{2a^2}$

7. $\int_0^\infty x^{(2n+1)} \exp(-ax^2)\,dx = \dfrac{\tfrac{1}{2}n!}{a^{(n+1)}}$

8. $\int_{-\infty}^{+\infty} x^{(2n+1)} \exp(-ax^2)\,dx = 0$

the product of u and $f(u)$, sum over all intervals, and divide by the sum of all the $f(u)$. In the limit of small du, the sum becomes an integral,

$$\overline{u} = \frac{\int_{-\infty}^\infty u\,f(u)\,du}{\int_{-\infty}^\infty f(u)\,du} \quad (21.16)$$

This integral is zero, as you can see from Table 21.1. The root mean square velocity (rms) is the mean of the square of the velocity.[5] It is similarly evaluated as

$$\overline{(u^2)} = \frac{\int_{-\infty}^\infty u^2 f(u)\,du}{\int_{-\infty}^\infty f(u)\,du} \quad (21.17)$$

The evaluation of (21.17) is left as a problem. Note that the denominators in both (21.16) and (21.17) are unity, since the functions are normalized.

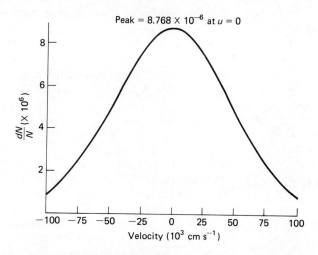

FIGURE 21.5 One-dimensional velocity distribution for a molecule of mass $m = 4$ at $T = 100$ K. The most probable velocity is $u = 0$.

---

[5] Since the mean is at the origin, this rms value corresponds to the square of the standard deviation. In the language of statistics, the square of the standard deviation is known as the second *moment* about the mean. The mean is the first moment about the origin. The kth moment is obtained by evaluating the integral with u^k in the numerator. For a discussion of statistical moments, see any of the advanced books on statistics listed in the Bibliography.

21.4 MULTIDIMENSIONAL GAUSSIAN DISTRIBUTIONS

Probability problems are often stated in terms of ballistics. The holes produced by the bullets in a target provide a visual display of the probability function. For ballistics it is the Gaussian distribution that is displayed.[6] Consider the one-dimensional ballistics problem shown in Figure 21.6. The marksman shoots for the bull's-eye at $x = 0$. The bullets strike the x axis on either side of the origin with a density following a Gaussian distribution, as shown by the ticks. Above the x axis is the Gaussian curve showing the distribution of bullets. The curve has the functional form $y = f(x)$. A one-dimensional ballistics experiment requires a two-dimensional plot to display the results.

A two-dimensional ballistics experiment is shown in Figure 21.7. The marksman shoots at a target with the bull's-eye at the origin $x = 0, y = 0$. The bullets are distributed in a Gaussian manner as indicated by the Gaussian surface above the target plane in Figure 21.7a. The surface is normalized in such a way that the volume enclosed by it and the target plane is unity. The probability of *simultaneously* hitting the target between the points x and $x + dx$ and the points y and $y + dy$ is given by the volume above the area $dA = dx\, dy$, as shown in Figure 21.7a.

This two-dimensional distribution can be reduced to a one-dimensional distribution that provides more practical information. The real parameter of interest is the distance from the target

$$r = (x^2 + y^2)^{1/2} \tag{21.18}$$

In pictorial terms, we are interested in the relative number of hits in the annular ring of area $2\pi r\, dr$, as indicated in Figure 21.7b, rather than the hits in the area $dA = dx\, dy$.

The one-dimensional Gaussian distribution is

$$f(x) = \frac{h}{\sqrt{\pi}} \exp(-h^2 x^2) \tag{21.19}$$

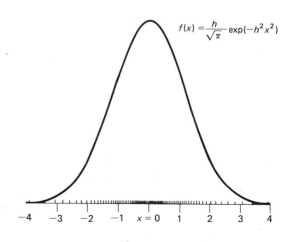

FIGURE 21.6 One-dimensional ballistics experiment. The tick marks along the x axis indicate hits on the one-dimensional target. The curve indicates the expected Gaussian distribution of hits. The distribution is plotted for $h = 0.6$ in Equation (21.19).

---

[6] In this discussion we assume that all systematic and correctible errors have been eliminated. The Gaussian distribution applies only to *random* errors. As we proceed through the multidimensional ballistics problems, we shall assume that the standard deviations are the same in all directions, $\sigma_x = \sigma_y$. This means that the error in setting the azimuth angle is the same as the error in the elevation angle. A detailed discussion of statistics applied to an 8-mm machine gun can be found in N. Arley and K. R. Buch, *Introduction to the Theory of Probability and Statistics,* (New York: John Wiley & Sons, 1950), pp. 126–134.

FIGURE 21.7 Two-dimensional ballistics experiment. (a) The dots in the xy plane indicate "hits" on a two-dimensional target with the bull's-eye at the origin. The surface above the plane indicates the two-dimensional Gaussian distribution. The fraction of "hits" between r and $r + dr$ is given by the volume of the cylindrical shell as indicated. The total volume under the Gaussian surface is unity. (b) View of xy plane of part (a), looking down along the z axis. (Figure 21.8 shows this distribution reduced to a one-dimensional plot, where the distribution is formulated to give the fraction of "hits" between r and $r + dr$ from a two-dimensional plot.) (c) Relation between coordinates and differential elements in two-dimensional cartesian and polar coordinates. The systems are related by $x = r \cos \theta$, $y = r \sin \theta$.

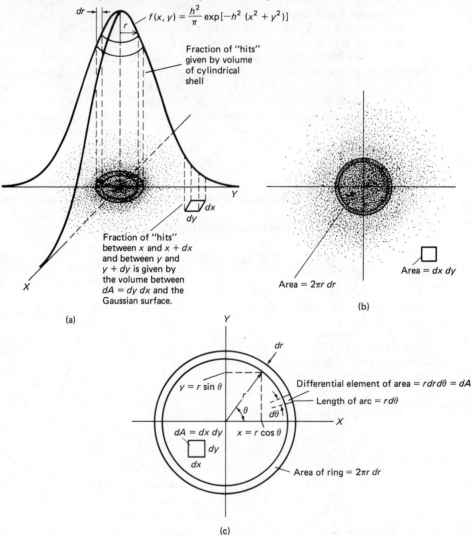

SEC. 21.4 MULTIDIMENSIONAL GAUSSIAN DISTRIBUTIONS

If the probability of a particular value of y is independent of the value of x, then the two-dimensional Gaussian distribution is just the product of the two one-dimensional distributions,

$$f(x, y) = \frac{h^2}{\pi} \exp[-h^2(x^2 + y^2)] \tag{21.20}$$

The probability of a hit in the differential area $dA = dx\, dy$ is

$$f(x, y)\, dx\, dy = \frac{h^2}{\pi} \exp[-h^2(x^2 + y^2)]\, dx\, dy \tag{21.21}$$

Now we convert this to polar coordinates; the relation between cartesian and polar coordinates is indicated in Figure 21.7c. Equation (21.21) becomes

$$f(r, \theta) r\, dr\, d\theta = \frac{h^2}{\pi} \exp(-h^2 r^2)\, r\, dr\, d\theta \tag{21.22}$$

We are interested in the distribution as a function of r without regard to the angular dependence; in fact, the distribution is symmetrical about the origin. Thus we can eliminate the $d\theta$ part of (21.22) by integrating between the limits $\theta = 0$ and $\theta = 2\pi$, to get

$$F(r)\, dr = 2h^2 \exp(-h^2 r^2)\, r\, dr \tag{21.23}$$

This distribution is plotted in Figure 21.8. Unlike the one-dimensional distribution, in which the most probable value and the mean both coincide at the origin, neither the most probable value nor the mean of the two-dimensional distribution coincides with the origin.

In three dimensions a ballistics experiment can be visualized as follows: The bull's-eye is taken at an origin $x = 0, y = 0, z = 0$ in three-dimensional space somewhere in the atmosphere. An antiaircraft cannon then shoots at the bull's-eye with explosive shells. The distribution of hits is shown in Figure 21.9. We should need

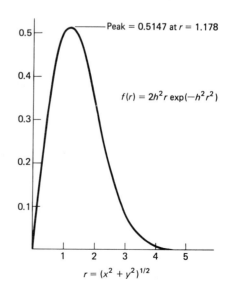

FIGURE 21.8 The two-dimensional ballistics experiment of Figure 21.7 reduced to one dimension. Here the area under the curve between any two points gives the probability of a hit at distances from the origin between those two points without regard to direction. The value of h in (21.23) is taken as 0.6. With this value, the most probable distance for a hit is at $r = 1.178$.

FIGURE 21.9 (a) Three-dimensional ballistics experiment. Each dot represents a "hit" in three-dimensional space. The bull's-eye is located at the origin. It would require a four-dimensional plot to represent this Gaussian distribution in the form $f(x,y,z) = (h^3/\pi\sqrt{\pi}) \exp[-h^2(x^2 + y^2 + z^2)]$. We are interested in the fraction of hits contained in the spherical shell of radius r and thickness dr as indicated. The volume of the shell is $4\pi r^2\, dr$. (Figure 21.10 shows this distribution reduced to a two-dimensional plot.)

(b) Spherical shell of volume $4\pi r^2/dr$.

(c) Three-dimensional coordinate system. The angle θ is the longitude, and the angle ϕ is the colatitude (measured from the north pole instead of from the equator). The term dA is the differential element of area on the surface of the sphere at radius r. For unit radius ($r = 1$), $dA = d\Omega$, where $d\Omega$ is the solid angle. The total surface area of the sphere is $4\pi r^2$, so that the fraction of the surface area covered by dA is $dA/4\pi r^2$. The differential element of volume is $dV = r^2 \sin\phi\, dr\, d\phi\, d\theta$; it is obtained by extending the unit element dA outward by an amount dr. The total volume of a spherical shell at r is $4\pi r^2\, dr$.

(d) Another view of the three-dimensional coordinate system showing the volume element $dV = r^2 \sin\phi\, dr\, d\phi\, d\theta$. Cartesian and spherical polar coordinate systems are related by $x = r \sin\phi \cos\theta$, $y = r \sin\phi \sin\theta$, $z = r \cos\phi$.

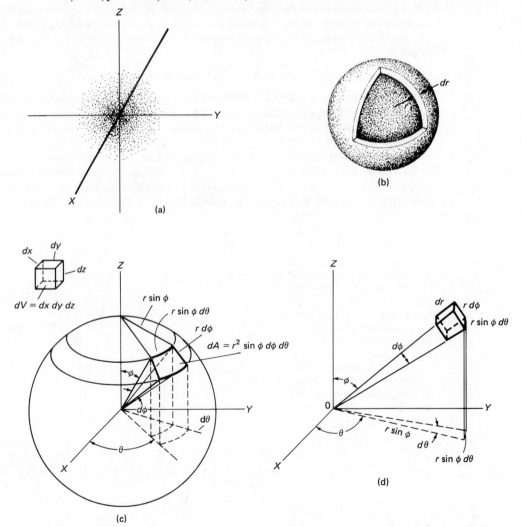

FIGURE 21.10 The three-dimensional ballistics experiment of Figure 21.9a reduced to one dimension. Here the area under the curve between any two values of r gives the probability of a hit at distances from the origin between those two values without regard to direction. The value of h in (21.26) is taken as 0.6. With this value, the most probable distance of the hit from the origin is at $r = 1.667$.

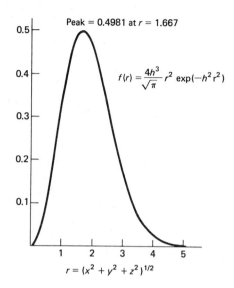

four dimensions to display visually the results in a Gaussian distribution form, as in Figure 21.7a; this we clearly cannot do. The probability of scoring a hit in the volume element $dV = dx\,dy\,dz$ is the product of three one-dimensional Gaussian distributions,

$$f(x, y, z)\,dx\,dy\,dz = \frac{h^3}{\pi\sqrt{\pi}} \exp[-h^2(x^2 + y^2 + z^2)]\,dx\,dy\,dz \qquad (21.24)$$

Now we follow a procedure similar to the one we followed for the two-dimensional case. Again the parameter of real interest is the distance from the target,

$$r = (x^2 + y^2 + z^2)^{1/2} \qquad (21.25)$$

We are interested in the probability of scoring a hit in the volume $dV = 4\pi r^2\,dr$ of the spherical shell shown in Figure 21.9b. The relation between coordinates and differential volume elements in cartesian and spherical polar coordinates is shown in Figure 21.9c. Proceeding as in the two-dimensional case, we can reduce Equation (21.24) to a one-dimensional distribution in r,

$$F(r)\,dr = \frac{4h^3}{\sqrt{\pi}} \exp(-h^2 r^2) r^2\,dr \qquad (21.26)$$

This three-dimensional ballistics experiment distribution is plotted in Figure 21.10. We are interested in these ballistics results because they apply directly to velocity distributions.

21.5 VELOCITY IN THREE DIMENSIONS; THE MAXWELL-BOLTZMANN DISTRIBUTION

Earlier in this chapter we showed that the one-dimensional velocity distribution is a Gaussian distribution with $h = (m/2kT)^{1/2}$. We can extend this result to two- and three-dimensional velocity distributions, proceeding exactly as we did for the ballistics results of the previous section.

In two dimensions, the fraction of molecules with speeds in the range c and $c + dc$ is

$$\frac{dN}{N} = \frac{m}{kT} \exp\left(-\frac{mc^2}{2kT}\right) c\, dc \tag{21.27}$$

where $c = (u_x^2 + u_y^2)^{1/2}$. In three dimensions, the fraction is

$$\frac{dN}{N} = 4\pi \left(\frac{m}{2\pi kT}\right)^{3/2} \exp\left(-\frac{mc^2}{2kT}\right) c^2\, dc \tag{21.28}$$

where $c = (u_x^2 + u_y^2 + u_z^2)^{1/2}$. Figure 21.11 shows plots of this Maxwell-Boltzmann distribution for several temperatures and masses. Notice that as the temperature increases, the peaks become broader, and the heights of the peaks decrease. The total areas under each of the curves are, however, the same. Some values for dN/N are listed in Tables 21.2 and 21.3.

We can compute the average velocity by using Equation (21.16). (Note that since the functions are normalized, the denominators are unity and can be omitted.) The average velocity is

$$\bar{c} = \int_0^\infty c f(c)\, dc = 4\pi \left(\frac{m}{2\pi kT}\right)^{3/2} \int_0^\infty \exp\left(\frac{-mc^2}{2kT}\right) c^3\, dc \tag{21.29}$$

$$\bar{c} = \left(\frac{8kT}{\pi m}\right)^{1/2} \tag{21.30}$$

FIGURE 21.11 Maxwell-Boltzmann distribution curves. (a) Curves for He ($M = 4$) at different temperatures. (b) Curves for several gases at 5000 K. The numbers in parentheses indicate the molar masses.

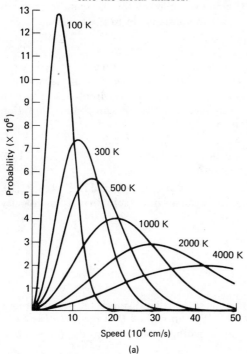
(a)

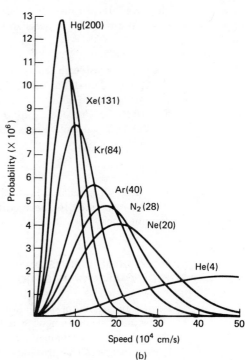
(b)

SEC. 21.5 VELOCITY IN THREE DIMENSIONS; THE MAXWELL-BOLTZMANN DISTRIBUTION

TABLE 21.2 Maxwell distribution at $T = 300$ K for different molar masses.

Speed (units of 10^3 cm s^{-1})	$M = 4$	$M = 30$	$M = 400$
2	6.479×10^{-9}	1.328×10^{-7}	6.277×10^{-6}
6	5.816×10^{-8}	1.172×10^{-6}	4.370×10^{-5}
10	1.607×10^{-7}	3.134×10^{-6}	7.267×10^{-5}
14	3.126×10^{-7}	5.798×10^{-6}	6.596×10^{-5}
18	5.115×10^{-7}	8.874×10^{-6}	3.907×10^{-5}
22	7.544×10^{-7}	1.204×10^{-5}	1.618×10^{-5}
26	1.037×10^{-6}	1.498×10^{-5}	4.846×10^{-6}
30	1.356×10^{-6}	1.743×10^{-5}	1.070×10^{-6}
34	1.707×10^{-6}	1.919×10^{-5}	1.765×10^{-7}
38	2.084×10^{-6}	2.016×10^{-5}	2.190×10^{-8}
42	2.481×10^{-6}	2.032×10^{-5}	2.056×10^{-9}
46	2.893×10^{-6}	1.972×10^{-5}	1.466×10^{-10}
50	3.315×10^{-6}	1.850×10^{-5}	7.970×10^{-12}
54	3.740×10^{-6}	1.680×10^{-5}	3.308×10^{-13}
58	4.162×10^{-6}	1.480×10^{-5}	1.050×10^{-14}
62	4.576×10^{-6}	1.267×10^{-5}	2.558×10^{-16}
66	4.977×10^{-6}	1.055×10^{-5}	4.777×10^{-18}
70	5.360×10^{-6}	8.563×10^{-6}	6.853×10^{-20}
74	5.720×10^{-6}	6.767×10^{-6}	7.555×10^{-22}
78	6.052×10^{-6}	5.216×10^{-6}	6.407×10^{-24}
82	6.354×10^{-6}	3.923×10^{-6}	4.181×10^{-26}
86	6.623×10^{-6}	2.880×10^{-6}	2.101×10^{-28}
90	6.855×10^{-6}	2.066×10^{-6}	8.135×10^{-31}
94	7.049×10^{-6}	1.447×10^{-6}	2.427×10^{-33}
98	7.204×10^{-6}	9.914×10^{-7}	5.581×10^{-36}

NOTE: Table entries are

$$4\pi \left(\frac{m}{2\pi kT}\right)^{3/2} c^2 \exp\left(\frac{-mc^2}{2kT}\right)$$

The average kinetic energy is determined likewise:

$$\overline{\tfrac{1}{2}mc^2} = \frac{m}{2} \int_0^\infty c^2 f(c)\, dc$$

$$= 2\pi m \left(\frac{m}{2\pi kT}\right)^{3/2} \int_0^\infty \exp\left(\frac{-mc^2}{2kT}\right) c^4\, dc \qquad (21.31)$$

When this integral is evaluated, we get

$$\overline{\tfrac{1}{2}mc^2} = \tfrac{3}{2}kT \qquad (21.32)$$

If (21.32) is written in terms of one mole of gas molecules, we get $\overline{\tfrac{1}{2}Mc^2} = \tfrac{3}{2}RT$; this is identical to Equation (3.17). The root mean square velocity can be computed from (21.32) as

$$c_{\text{rms}} = (\overline{c^2})^{1/2} = \left(\frac{3kT}{m}\right)^{1/2} = \left(\frac{3RT}{M}\right)^{1/2} \qquad (21.33)$$

which is identical to (3.14).

The *most probable velocity* occurs at the peak in the distribution curve. It is found by taking the derivative and setting it equal to zero. It is

$$c_{\text{mp}} = \left(\frac{2kT}{m}\right)^{1/2} \qquad (21.34)$$

TABLE 21.3 Maxwell distribution for gas of molecular weight $M = 30$ at different temperatures.

Speed (units of 10^3 cm s^{-1})	100 K	400 K	1000 K
2	6.868×10^{-7}	8.631×10^{-8}	2.186×10^{-8}
6	5.834×10^{-6}	7.657×10^{-7}	1.956×10^{-7}
10	1.443×10^{-5}	2.066×10^{-6}	5.371×10^{-7}
14	2.380×10^{-5}	3.878×10^{-6}	1.034×10^{-6}
18	3.123×10^{-5}	6.051×10^{-6}	1.671×10^{-6}
22	3.495×10^{-5}	8.411×10^{-6}	2.425×10^{-6}
26	3.452×10^{-5}	1.077×10^{-5}	3.272×10^{-6}
30	3.068×10^{-5}	1.296×10^{-5}	4.184×10^{-6}
34	2.483×10^{-5}	1.483×10^{-5}	5.131×10^{-6}
38	1.845×10^{-5}	1.627×10^{-5}	6.085×10^{-6}
42	1.265×10^{-5}	1.720×10^{-5}	7.017×10^{-6}
46	8.043×10^{-6}	1.761×10^{-5}	7.899×10^{-6}
50	4.753×10^{-6}	1.750×10^{-5}	8.708×10^{-6}
54	2.617×10^{-6}	1.691×10^{-5}	9.423×10^{-6}
58	1.345×10^{-6}	1.594×10^{-5}	1.002×10^{-5}
62	6.467×10^{-7}	1.467×10^{-5}	1.050×10^{-5}
66	2.909×10^{-7}	1.320×10^{-5}	1.085×10^{-5}
70	1.226×10^{-7}	1.161×10^{-5}	1.107×10^{-5}
74	4.849×10^{-8}	1.001×10^{-5}	1.115×10^{-5}
78	1.798×10^{-8}	8.457×10^{-6}	1.110×10^{-5}
82	6.265×10^{-9}	7.003×10^{-6}	1.093×10^{-5}
86	2.050×10^{-9}	5.688×10^{-6}	1.065×10^{-5}
90	6.304×10^{-10}	4.535×10^{-6}	1.027×10^{-5}
94	1.822×10^{-10}	3.549×10^{-6}	9.813×10^{-6}
98	4.956×10^{-11}	2.728×10^{-6}	9.286×10^{-6}

In the next section we shall note that the *median velocity* is

$$c_\text{med} = 1.538 \left(\frac{kT}{m}\right)^{1/2}$$

These four velocities are in the ratio

$$c_\text{rms} : \bar{c} : c_\text{med} : c_\text{mp} = 1.00 : 0.92 : 0.89 : 0.82 \qquad (21.35)$$

Of the four, c_rms is the highest, and c_mp the lowest.

21.6 ADDITIONAL CALCULATIONS INVOLVING THE VELOCITY DISTRIBUTION

In this section we want to determine the fraction of molecules traveling with velocities greater than some velocity c. We want particularly to evaluate the integral

$$F(c) = \int_c^\infty P(c)\, dc \qquad (21.36)$$

where $P(c)$ is the Maxwell-Boltzmann expression of Equation (21.28). The integral in (21.36) cannot be calculated exactly for all values of c. It is convenient to determine the integral in terms of c_mp. From (21.34),

SEC. 21.6 ADDITIONAL CALCULATIONS INVOLVING THE VELOCITY DISTRIBUTION

$$T = \frac{mc_{mp}^2}{2k} \tag{21.37}$$

and (21.36) becomes

$$F(c) = \int 4\pi \left(\frac{2km}{2\pi mkc_{mp}^2}\right)^{3/2} c^2 \exp\left(-\frac{2kmc^2}{2\pi kmc_{mp}^2}\right) dc$$

$$= \int 4\pi \left(\frac{1}{\pi c_{mp}^2}\right)^{3/2} c^2 \exp\left[-\left(\frac{c}{c_{mp}}\right)^2\right] c_{mp} d\left(\frac{c}{c_{mp}}\right)$$

$$F(s) = \frac{4}{\sqrt{\pi}} \int_{s=c/c_{mp}}^{\infty} s^2 \exp(-s^2) \, ds \tag{21.38}$$

in terms of a new variable, $s = c/c_{mp}$. This variable is just the ratio of the velocity to the most probable velocity. Equation (21.38) can be integrated by parts in the following way. The standard formula for integrating by parts is

$$\int u \, dv = uv - \int v \, du \tag{21.39}$$

To apply this, we use $s = u$; $v = -\tfrac{1}{2} \exp(-s^2)$; and $dv = s \exp(-s^2) \, ds$. After the substitutions are made, Equation (21.38) becomes

$$F(s) = \frac{2}{\sqrt{\pi}} \left(se^{-s^2} + \int_s^{\infty} e^{-s^2} \, ds\right) \tag{21.40}$$

Further, we have

$$\int_s^{\infty} e^{-s^2} \, ds = \int_0^{\infty} e^{-s^2} \, ds - \int_0^s e^{-s^2} \, ds$$

$$= \tfrac{1}{2}\sqrt{\pi} - \int_0^s e^{-s^2} \, ds \tag{21.41}$$

and (21.40) becomes

$$F(s) = 1 + \frac{2s}{\sqrt{\pi}} e^{-s^2} - \frac{2}{\sqrt{\pi}} \int_0^s e^{-s^2} \, ds \tag{21.42}$$

$$F(s) = 1 + 2sG(\sqrt{2}s) - 2I(\sqrt{2}s) \tag{21.43}$$

where

$$G(z) = \frac{1}{\sqrt{2\pi}} e^{-z^2/2} \tag{21.44}$$

is the Gaussian error function and

$$I(z) = \frac{1}{\sqrt{2\pi}} \int_0^z e^{-z^2/2} \, dz \tag{21.45}$$

is the Gaussian error integral. Both $G(z)$ and $I(z)$ can be found tabulated in compilations of statistical data.

For $s \gg 1$, Equation (21.41) becomes

$$\int_s^{\infty} e^{-s^2} \, ds \approx \frac{1}{2s} e^{-s^2} \qquad (s \gg 1) \tag{21.46}$$

and (21.40) can be written as

$$F(s) \approx \frac{e^{-s^2}}{\sqrt{\pi}} \left(2s + \frac{1}{s}\right) \qquad (s \gg 1) \tag{21.47}$$

TABLE 21.4 Fraction of molecules traveling with speeds greater than c in terms of the variable $s = c/c_{mp}$

$s = c/c_{mp}$	Fraction of Molecules Traveling with Speeds Greater than $c = sc_{mp}$
0.1	0.9993
0.2	0.9941
0.5	0.9189
0.75	0.7710
1.	0.5724
1.0877	0.5000[a]
2	4.65×10^{-2}
3	4.41×10^{-4}
5	7.99×10^{-11}
7	4.2×10^{-21}
10	4.2×10^{-43}
15	3.3×10^{-97}

[a] This entry corresponds to the median velocity, since half the molecules have speeds in excess of this velocity.

The term $F(s)$ is tabulated for several values of s in Table 21.4. From the figures in the table, we see that in one mole of gas there are only about 100 molecules with velocities in excess of $7c_{mp}$. If we had 10^{20} moles of gas, only about 1 molecule would have a velocity exceeding $10c_{mp}$.

Returning to (21.36), we find that the value of c for which $F(c) = \frac{1}{2}$ is the *median* velocity. Half the molecules have speeds greater than c_{med} and half less. It can be shown that

$$c_{med} = 1.538 \left(\frac{kT}{m}\right)^{1/2} \tag{21.48}$$

21.7 APPLYING THE VELOCITY DISTRIBUTION

The utility of the measurements discussed in the first section is not confined to verifying the shape of the velocity distribution. Consider NaCl in the gas phase. Do the molecules consist of NaCl monomers or Na_2Cl_2 dimers? One way to get the answer is to study the actual velocity distribution using an apparatus such as that of Figure 21.1.

Assume for the moment that gaseous NaCl is composed of equal numbers of monomers and dimers. Figure 21.12 shows three curves, a monomer curve, a dimer curve, and the sum curve, which is just the sum of the first two curves. The sum curve is the observable. The composition of the gas is determined from the best combination of monomer and dimer curves that yields the sum curve. From experiments like this, Miller and Kusch determined that NaCl vapor consists of 35% dimer.[7] By carrying out the experiment at several temperatures, they got equilibrium constants for the dimerization reaction at several temperatures and were able to determine ΔH for the reaction $Na_2Cl_2 = 2NaCl$ from the expression

$$\frac{d \ln K}{dT} = \frac{\Delta H}{RT^2}$$

[7] R. C. Miller and P. Kusch, *J. Chem. Phys.* 25 (1956):860.

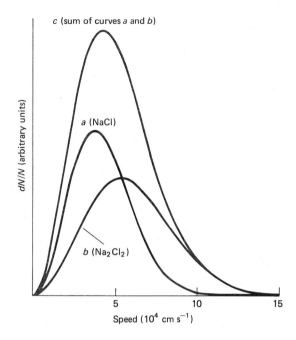

FIGURE 21.12 Sum of two Maxwell-Boltzmann distributions. Curve a is the distribution for NaCl molecules, and curve b is the distribution for Na_2Cl_2 molecules. Curve c is the observed curve obtained for an equimolar mixture of the two molecules. The temperature is 1000 K.

A more complicated problem was presented by LiCl vapor, which also contains trimers in addition to dimers and monomers. This is also discussed in the Miller and Kusch paper.

Throughout this discussion we have overlooked the fact that molecules collide with each other. In the next chapter we shall examine a number of properties of gases that depend on collisions.

PROBLEMS

1. Calculate the average velocity, the root mean square velocity, the most probable velocity, and the median velocity for H_2, He, Ne, CH_4, and Hg at 300 and 1000 K.

2. A velocity selector like that in Figure 21.1 is constructed. The distance between the discs is 5 cm. The displacement θ of the slots in the discs is 5°. Calculate the rotational frequency required to allow Na molecules having a velocity equal to c_{mp} at 1000 K to be detected. The Na beam passes through the slots at a distance of 1 cm from the center.

3. Suppose the first slit is infinitesimally narrow, and the second slit is 0.1 mm wide and located 2 cm from the rotation axis. Calculate the velocity range that would be transmitted by the velocity selector of Problem 2. (This is called the *resolution*.)

4. Derive an expression for the average value of the cube of the velocity for a Maxwell-Boltzmann distribution; that is, calculate $\overline{c^3}$. (This is related to what statisticians call *skewness*.)

5. Derive the expression for the most probable velocity c_{mp} as given in (21.34).

6. Derive an expression for the points of inflection in the three-dimensional velocity distribution.

7. Derive an expression for the velocity distribution in two dimensions.

8. Take the partial pressures of O_2, N_2, and Ar to be 0.20, 0.79, and 0.01 atm at sea level. Using the barometric formula, calculate the composition of air at elevations of 0.1, 1, 10, 100, and 1000 km above sea level, and calculate the total pressures. (Ignore the presence of other constituents and take $T = 300$ K.)

9. In 1910, J. Perrin and M. Dabrowski redetermined Avogadro's number by reexamining the distribution of particles in a colloidal suspension, essentially repeating an earlier experiment of Perrin (*Compt. Rend.* 149 (1910):477). The colloidal particles of mastic had a radius of 0.52μ (1 micron, μ, $= 10^{-6}$ m $= 10^{-4}$ cm) and a density $\rho = 1.063$ g cm^{-3}. By microscopic examination, they determined the number of particles in layers

separated by a distance of 6μ to be 305, 530, 940, and 1880. Calculate a value of Avogadro's number using these data. (Take the density of the suspending fluid to be 1.000 and the temperature to be 15 °C.)

10. Assuming that Figure 21.4 represents an actual experimental colloidal distribution, calculate a value of the Boltzmann constant using the following: The particles have a radius of 3.6×10^{-5} cm and a density of 1.200 g cm^{-3}. The suspending fluid has a density of 1.000 g cm^{-3} at the temperature of the experiment, 300 K. Count the number of particles at a number of levels within a small finite width and use Equation (21.10) to evaluate the constant. For scaling purposes, the length of the line on the side is 10^{-3} cm.

11. A large box has n molecules traveling with speeds between c and $c + dc$ at temperature T. How many molecules are traveling with speeds between $2c$ and $2c + dc$ at the same temperature?

12. A large box has n molecules traveling with speeds between c and $c + dc$ at temperature T. If the temperature is doubled to $2T$, how many molecules will be traveling with speeds between c and $c + dc$?

13. We have written the Maxwell-Boltzmann distribution as a function of velocity. Derive an expression for the distribution in terms of the kinetic energy of the molecules, $\epsilon = \frac{1}{2}mc^2$, and show that the fraction of molecules with kinetic energies between ϵ and $\epsilon + d\epsilon$ is given by

$$2\pi \left(\frac{1}{\pi kT}\right)^{3/2} \epsilon^{1/2} \exp\left(\frac{-\epsilon}{kT}\right) d\epsilon$$

14. Using the expression for the distribution of kinetic energy in Problem 13, calculate the average kinetic energy of a molecule.

15. Tables 21.2 and 21.3 list the ordinates for reasonably small values of velocity increments. Using the data of these tables for $T = 300$ K, numerically integrate the velocity distribution for a molecule of molecular weight 400 over the entire velocity range. (You should get a value close to unity.) What is the fraction of molecules traveling at speeds lower than 200 m s^{-1}?

16. For a particle to escape from the earth's gravitational field, it must attain the escape velocity $v_e = 2gR$, where g is the gravitational acceleration constant and R is the radius of the earth ($g = 980$ cm s^{-2} and $R = 6.4 \times 10^8$ cm). At what temperature is the average velocity $\bar{c}$ of an H_2 molecule equal to the escape velocity? At what temperature is $\bar{c}$ of an O_2 molecule equal to the escape velocity? What are these velocities for the moon? (*Note:* For the moon, $g = 170$ cm s^{-2}, and $R = 1.7 \times 10^8$ cm.)

17. Calculate the fraction of H_2 molecules and the fraction of O_2 that have velocities large enough to escape the earth's gravitational attraction at 300 K. Do the same calculation for the moon. (Additional relevant data are given in Problem 16. You now know why the earth's atmosphere contains O_2 but not H_2, and why the moon contains no atmosphere.)

18. Derive an expression for the fraction of molecules that have energies within 10% of kT.

19. By comparing the three-dimensional Maxwell-Boltzmann distribution with the three-dimensional Gaussian distribution, determine the standard deviation of the Maxwell-Boltzmann distribution.

20. Evaluate the rms velocity for the one-dimensional and two-dimensional velocity distributions.

21. Suppose we establish the following criterion for Brownian motion: If the change in potential energy is approximately kT when the particle moves a distance equal to its radius, then it will display Brownian motion.
    (a) How small does a mist particle have to be to show Brownian motion at 300 K?
    (b) How small does a particle of density 1.200 g cm^{-3} have to be to show colloidal behavior (that is, Brownian motion) when suspended in water?

CHAPTER TWENTY-TWO

COLLISIONAL AND TRANSPORT PROPERTIES OF GASES

22.1 MOLECULAR COLLISIONS

In examining one particular aspect of the kinetic theory of gases, namely the distribution of speeds in a collection of gas molecules (Chapter 21), we overlooked the fact that molecules may collide with one another. Now we shall specifically consider collisions. To a certain extent, this chapter may seem a collection of random topics. The unifying theme running through the topics, however, is that in all cases we are ultimately concerned with the average velocities of molecules and the distances molecules travel between collisions.

Although we have hitherto been concerned only with the speeds of molecules, we shall now also be concerned with the direction of motion of the molecules. There are two approaches one can take to direction—the exact approach and the simplified approach. In the *exact* approach, one takes the molecular velocities as one finds them, randomly distributed in all possible directions. This approach gives the "exact" answer but at the expense of involved mathematical computations. We shall mostly use the *simple* approach. In this method it is assumed that the molecules travel along one of the axes of a cartesian coordinate system, one third of them traveling along each axis. In addition, half of the molecules traveling along each axis move in the positive direction, while the other half move in the negative direction. This procedure gives answers that differ from the exact results by small constant factors. We shall calculate one result by the exact procedure to provide some indication of how the exact calculations proceed. Exact solutions to the other problems can be found in the references dealing with the kinetic theory of gases.

22.2 COLLISIONS WITH A WALL: APPROXIMATE SOLUTION OF MOLECULAR EFFUSION

Figure 22.1 shows a wall that we take to be perpendicular to the z axis. We want to find the number of molecules that collide with the area dS on the wall in unit time. We start by assuming that the molecules travel along the x axis or the y axis or the z axis, a third of the molecules traveling along each axis. Thus one-sixth of the molecules move in the $+z$ direction and one-sixth in the $-z$ direction.

The molecules that strike dS per second are contained in the cylinder of cross-sectional area dS and length $\bar{c}$, where $\bar{c}$ is the average velocity of the molecules.

FIGURE 22.1 Approximate solution of the effusion problem. We assume that all the molecules travel along one of the coordinate axes. The molecules in the cylinder of cross-sectional area dS and length $\bar{c}$ will strike the hole of area dS in 1 s if they travel in the $-z$ direction; $\bar{c}$ is the average velocity. The number of molecules in the cylinder is $n^*\bar{c}\,dS$. Since one-sixth of them travel in the $-z$ direction, the number that strike dS in one second is $\tfrac{1}{6}n^*\bar{c}\,dS$.

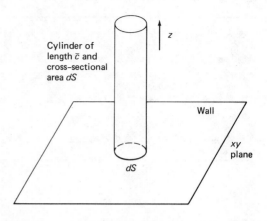

Since one-sixth of the molecules in the cylinder travel in the proper direction to strike the area dS, we need simply calculate the total number of molecules in the cylinder and multiply by a sixth. The total number of molecules is $n^*\bar{c}\,dS$, where n^* is the number of molecules per cubic centimer. The number of molecules striking dS in unit time is

$$\frac{dn}{dt} = \frac{1}{6} n^*\bar{c}\,dS \tag{22.1}$$

Suppose the area dS is actually a hole in the wall of a chamber that is surrounded by high vacuum; the chamber is filled with gas. Then any molecule that strikes dS leaves the chamber, and Equation (22.1) indicates how fast molecules leave the chamber or how fast the chamber can be pumped out. Since

$$\bar{c} = \left(\frac{8RT}{\pi M}\right)^{1/2} = \left(\frac{8kT}{\pi m}\right)^{1/2} \tag{22.2}$$

the rate at which molecules *effuse* through a small hole is inversely proportional to the square root of the masses. Graham in 1848 had experimentally observed this dependence on the mass, and it is known as *Graham's law of effusion*. The explanation of Graham's experimental observation was one of the earliest successes of the kinetic theory of gases. That gases effuse at a rate inversely proportional to the square root of the mass forms the basis for two important experimental procedures.

Firstly, one can use effusion to separate gases of different molar masses in a procedure akin to distillation. If a mixture of two gases is allowed to effuse through a small orifice, the gas on the low-pressure side will be richer in the lower-molar-mass constituent. This procedure was used to separate ^{235}U from ^{238}U for the manufacture of the first atomic bomb. Uranium hexafluoride (UF$_6$) is a gas at room temperature, and the isotopes were separated as the hexafluorides. (The isotope separation is more complicated than simple effusion. The gas mixture is passed through a porous wall, which can be regarded as a large number of very fine capillaries. Our simple effusion equation applies if the wall thickness is smaller than the dimensions of the hole.[1])

---

[1] The opposite case is more important in vacuum technology, since vacuum systems are constructed with long tubes, the length substantially greater than the diameter. It is also a more difficult problem; it is treated in most books on vacuum technology under the heading *conductance*.

Secondly, Equation (22.1) leads to an experimental procedure for measuring vapor pressures of relatively involatile materials such as metals and very high boiling liquids. The ideal gas equation leads to $n^* = P/kT$. By using the value of $\bar{c}$ in (22.2), we can put (22.1) in the form

$$\frac{dn}{dt} = \frac{1}{6}\left(\frac{8}{\pi mkT}\right)^{1/2} P\, dS \tag{22.3}$$

This equation was first derived by Hertz in 1882 and experimentally verified by Knudsen in 1915. Since dn/dt is the number of molecules escaping per second, the rate at which a sample loses weight is $m\, dn/dt$; the rate of loss of weight is

$$-\frac{dW}{dt} = \frac{P}{6}\left(\frac{8m}{\pi kT}\right)^{1/2} dS \tag{22.4}$$

Equation (22.4) provides an experimental technique for measuring vapor pressures by measuring the rate of weight loss in a sample.

The equations we derived are based on the assumption that every molecule that hits dS leaves the chamber. This is true so long as the molecule does not collide with another molecule in the region of the hole. The validating criterion is the mean free path length (which we shall discuss shortly). So long as the dimensions of the hole are very small compared with the mean free path, our effusion equations will be valid.[2]

The number of collisions with the wall are large. For He ($M = 4$) at 300 K and 0.001 torr,

$$n^* = \frac{(0.001/760)(1)}{(82.05)(300)}(6.02 \times 10^{23}) = 3.218 \times 10^{13}\ \text{cm}^{-3}$$

$$\bar{c} = 1.26 \times 10^5\ \text{cm s}^{-1}$$

If we want the number of collisions per square centimeter of wall, then $dS = 1$, and $\approx \times 10^{17}\ \text{s}^{-1}$.

In our derivation we did not take into account the random orientations of the velocities. We shall do this in the next section, finding that an exact calculation yields a constant coefficient $\frac{1}{4}$ instead of $\frac{1}{6}$.

22.3 COLLISIONS WITH A WALL: EXACT SOLUTION OF MOLECULAR EFFUSION

We take an element of surface dS and consider a volume element dV, as shown in Figure 22.2. The density is n^* molecules per cubic centimeter; hence the number of molecules in dV is $n^*\, dV$. The volume element is situated a distance r from dS and makes an angle ϕ with the normal to dS. The volume element is

$$dV = r^2\, dr\, \sin\phi\, d\phi\, d\theta \tag{22.5}$$

The number of molecules in dV is just $n^* r^2\, dr\, \sin\phi\, d\phi\, d\theta$. From the Maxwell-Boltzmann distribution, the number of molecules in dV with velocities between c and $c + dc$ is given by

---

[2] At high pressure, the other extreme, the equations of flow are governed by the laws of hydrodynamics rather than the kinetic theory of gases.

FIGURE 22.2 Coordinate system for the exact solution of effusion. The term dV represents an arbitrary differential volume element at a distance r from the hole dS, which is placed at the origin. The angle between r and the z axis is given by ϕ, while θ is the longitudinal angle from the x axis. The volume of dV is $r^2\, dr \sin\phi\, d\phi\, d\theta$. We must calculate the total number of molecules in dV that are heading in the right direction to hit dS; and then we integrate over all possible volume elements.

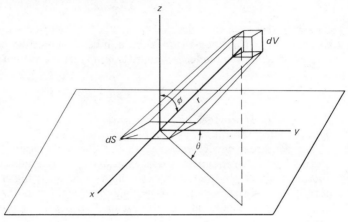

$$n^*4\pi\left(\frac{m}{2\pi kT}\right)^{3/2}\exp\frac{-mc^2}{2kT}c^2\,dc\,r^2\,dr\,\sin\phi\,d\phi\,d\theta$$
$$= n^*f(c)\,dc\,r^2\,dr\,\sin\phi\,d\phi\,d\theta \quad (22.6)$$

Of all the molecules in dV, only those traveling in the right direction will ever hit dS. The fraction of molecules traveling in the right direction is given by the solid angle subtended by dS at dV; this equals the ratio of the area $\cos\phi\,dS$ to the total surface area of the sphere, $4\pi r^2$. The number of molecules leaving dV, headed in the right direction to hit dS, with velocities between c and $c + dc$ is hence

$$n^*[f(c)\,dc](r^2\,dr\,\sin\phi\,d\phi\,d\theta)\left(\frac{\cos\phi\,dS}{4\pi r^2}\right) \quad (22.7)$$

To calculate the number hitting dS per second, we must integrate (22.7) over all the relevant variables. The angle ϕ varies between 0 and $\pi/2$, while the limits on θ are 0 and 2π. The limits on r are 0 to c, since if r is greater than c, the molecule will not reach dS in one second. Finally, the limits on c are from 0 to infinity. The number striking dS per unit time is then

$$\frac{dn}{dt} = \left[n^*\,dS\int_{c=0}^{\infty}\int_{r=0}^{c}dr[f(c)\,dc]\right]\left[\frac{1}{4\pi}\int_{\phi=0}^{\pi/2}\sin\phi\cos\phi\,d\phi\right]\left[\int_{\theta=0}^{2\pi}d\theta\right] \quad (22.8)$$

Since

$$\int_0^{\pi/2}\sin\phi\cos\phi\,d\phi = \frac{1}{2}, \quad \int_0^{2\pi}d\theta = 2\pi, \quad \text{and} \quad \int_0^c dr = c$$

Equation (22.8) becomes

$$\frac{dn}{dt} = \frac{n^*\,dS}{4}\int_0^{\infty}cf(c)\,dc \quad (22.9)$$

The integral in Equation (22.9) is just the average velocity $\bar{c}$, so

$$\frac{dn}{dt} = \frac{1}{4} n^* \bar{c} \, dS \tag{22.10}$$

Equation (22.10) has the same general form as (22.1); the only difference is in the constant coefficient.

22.4 MEAN FREE PATHS AND COLLISION FREQUENCY

So far we have treated molecules as point masses. In fact, they have finite dimensions and collide with each other. The number of collisions a molecule undergoes is of some importance. For a chemical reaction involving more than one reactant molecule to take place, two molecules must collide. The rate of the reaction depends on the number of collisions per second.

Figure 22.3 shows the path of a molecule of diameter d. As the molecule moves, it sweeps out a cylinder of diameter d. The molecule collides with every other molecule located in the cylinder of diameter $2d$. (For the moment we assume that all other molecules are stationary.) Every time the molecule hits another molecule, its path is bent as it is deflected from its previous path. The effective volume swept out in 1 s is $\pi d^2 \bar{c}$, where $\bar{c}$ is the average velocity. The molecule collides with every other molecule in this volume. The number of such molecules is $n^* \pi d^2 \bar{c}$. The *mean free path* λ is the distance traveled between collisions. This is equal to the length of the path $\bar{c}$ divided by the number of collisions occurring in $\bar{c}$, or

$$\lambda = \frac{\bar{c}}{\pi d^2 \bar{c} n^*} = \frac{1}{\pi d^2 n^*} \tag{22.1}$$

The other molecules are really not stationary. We need the relative velocities of the two molecules undergoing a collision. This relative velocity may vary from near zero, in the case of a glancing collision, to $2\bar{c}$ for a head-on collision, as indicated in Figure 22.4. The average of the relative velocities of two colliding molecules is found to be $\sqrt{2}\bar{c}$ when the proper "exact" averaging procedure is used. The correct form for (22.11) becomes

$$\lambda = \frac{1}{\sqrt{2} \pi d^2 n^*} \tag{22.12}$$

FIGURE 22.3 Path of molecule of diameter d. It sweeps out a cylinder of diameter d, but the effective diameter of the cylinder for collisions is $2d$. After every collision a kink arises in the cylinder because the molecule changes path. In 1 s the molecule covers distance $\bar{c}$.

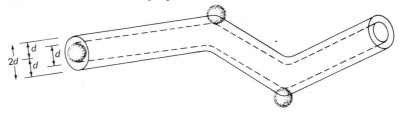

FIGURE 22.4 Showing different types of molecular collisions, which give rise to different values of the relative velocity. (a) A glancing collision, with relative velocity $u_{rel} = 0$. (b) A head-on collision, with $u_{rel} = 2\bar{c}$. (c) A right-angle collision, with $u_{rel} = \sqrt{2}\bar{c}$. A detailed averaging over all possible collisions yields an average relative velocity $u_{rel} = \sqrt{2}\bar{c}$.

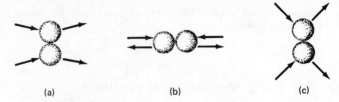

(a) (b) (c)

For an ideal gas at 1 atm and 0 °C, $n^* = (6.02 \times 10^{23})/22{,}414$. We can take a reasonable value for the diameter of a gas molecule to be 3×10^{-8} cm. With these values, the mean free path is $\lambda = 931 \times 10^{-8}$ cm.

The number of collisions per second is just the number of mean free paths in length $\bar{c}$, or

$$Z_1 = \frac{\bar{c}}{\lambda} = \sqrt{2}\pi\, d^2 \bar{c} n^* \qquad (22.13)$$

The term Z_1 is called the *collision number*. It is the number of collisions any individual molecule makes in 1 s. The *total* number of collisions taking place per unit volume is just $\frac{1}{2} Z_1 n^*$, where the factor $\frac{1}{2}$ ensures that each collision is counted only once. We get

$$Z_{11} = \tfrac{1}{2}\sqrt{2}\pi\, d^2 \bar{c} (n^*)^2 \qquad (22.14)$$

The term Z_{11} represents the *collision frequency*. We take as an example an ideal gas with $d = 3 \times 10^{-8}$ cm and $M = 4$. We find that at 273 K and 1 atm, $Z_1 = 1.29 \times 10^{10}$ s^{-1}, and $Z_{11} = 2.08 \times 10^{23}$ s^{-1} cm^{-3}.

For a mixture of two gases A and B, the calculation is more complicated. The total number of collisions between the A and B molecules per unit time per unit volume is

$$Z_{AB} = n_A^* n_B^* (r_A + r_B)^2 \left[8\pi kT \left(\frac{1}{m_A} + \frac{1}{m_B} \right) \right]^{1/2} \qquad (22.15)$$

For the special case of a pure gas,

$$r_A = r_B = \tfrac{1}{2}d, \qquad m_A = m_B = m, \qquad n_A^* = n_B^* = \tfrac{1}{2}n^*$$

and Equation (22.15) reduces to the expression for Z_{11} in (22.14).

22.5 VISCOSITY

Maple syrup is thicker than water. In a sense the syrup has a larger internal friction; it pours slowly. On the molecular level, motion is transferred between molecules of syrup at a lower rate than for molecules of water.

Figure 22.5 shows two infinite planes separated by a fluid. If the upper plane moves with velocity u, while the lower plane remains fixed, then the fluid moves subject to the boundary condition that it is at rest relative to each plane at the bound-

FIGURE 22.5 Velocity gradient for viscosity. Fluid is placed between the two planes. The upper plane moves relative to the lower. A velocity gradient is set up in the fluid between the planes; the upper plane experiences a drag or force opposite in direction to u. The arrows labeled f and $-f$ indicate the forces in two successive layers of fluid.

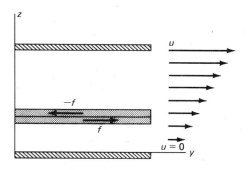

ary.[3] That is to say, the layer of fluid in contact with the upper plane moves with velocity u while the layer of fluid in contact with the lower plane is at rest. The motion of the fluid between the planes falls off linearly from the moving to the fixed plane. The fluid layer in contact with the moving plane is set in motion by the plane; some momentum is lost by the plane in transmitting the motion to the fluid layer. This first layer of fluid then sets the next layer in motion, the first layer losing some of its momentum. Successive layers are set in motion by previous layers, each layer having less velocity than the previous, until we reach the layer at the lower plane, which is at rest.

When the steady state of motion is reached, the upper plane experiences a drag, or viscous force, that equals the time rate of change of momentum transferred. This force is

$$f = -\eta A \frac{du}{dz} \quad (22.16)$$

where A is the area of the plane, du/dz is the velocity gradient, and η is the *coefficient of viscosity;* the value η is characteristic of the fluid. The negative sign indicates that the force is opposed to the motion. If (22.16) is written as force per unit area, we get

$$F = \frac{f}{A} = -\eta \frac{du}{dz} \quad (22.17)$$

Although the definition and experimental measurements of viscosity are the same for liquids and for gases, there is a fundamental difference in interpretation.[4] Viscosities are several orders of magnitude higher for liquids than for gases. In liquids, intermolecular forces are large and play a dominant role. As temperature is increased, these cohesive forces are weakened, and the viscosity decreases. Hot syrup flows faster than cold syrup. In gases, the intermolecular forces are small, and the dominant factor is the rate at which momentum is transferred by the hopping of molecules from one layer to another. This rate increases with temperature; the viscosity of gases increases with temperature.

---

[3] There is actually some *slippage* between the planes and the layers in contact with the planes.
[4] The most common way of measuring viscosity is by timing the rate of flow through capillaries. The longer the time, the higher the viscosity. This provides a relative measurement, which can be made absolute only by calibrating the device with a substance of known viscosity. The device shown in Figure 22.6, on the other hand, permits us to measure absolute viscosities. (See also Problem 11.)

FIGURE 22.6 An apparatus for measuring viscosity. The fluid (gas or liquid) is introduced into the space between the two concentric cylinders. The outer cylinder is then rotated. The inner cylinder is suspended by a fine wire and the torsional displacement θ allows one to calculate the force imparted to the inner cylinder due to the viscosity of the gas.

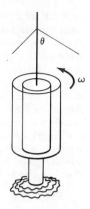

For a gas, the process can be visualized as follows. We assume the gas consists of layers of molecules separated by a distance equal to the mean free path, as shown in Figure 22.7. We take the velocity of the upper layer to be higher. Every time a molecule from an upper layer "jumps" to a lower layer, the average momentum of the lower layer is increased; on the contrary, every time a molecule "jumps" from a lower layer to an upper layer, the average momentum of the upper layer is decreased. The force is the time rate of change of momentum. The difference in velocity between the two layers is $\lambda(du/dz)$; thus each molecule going from the upper layer to the lower layer brings along an excess momentum of $m\lambda(du/dz)$. If we assume that one-third of the molecules move along the z direction, then the total number of molecules moving up or down with velocity $\bar{c}$ is just $\frac{1}{3}n^*\bar{c}$. The momentum transferred up and down by the molecules in one second is just the force

$$F = \frac{1}{3} n^* \bar{c} m \lambda \frac{du}{dz} \tag{22.18}$$

from which we get

$$\eta = \tfrac{1}{3} n^* m \bar{c} \lambda \tag{22.19}$$

FIGURE 22.7 Showing three successive layers of gas separated by a distance λ. The velocity u increases from the bottom to the top layer. When molecule no. 1 hops from the lowest to the middle layer, it decreases the average value of u in the middle layer. When molecule no. 3 hops from the upper layer to the middle layer, it increases the average value of u in the middle layer.

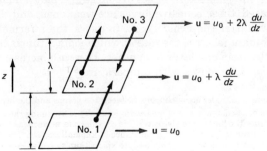

SEC. 22.6 THERMAL CONDUCTIVITY

If the value for λ from Equation (22.12) is substituted in (22.19) we get

$$\eta = \frac{m\bar{c}}{3\sqrt{2}\pi d^2} \tag{22.20}$$

Equation (22.20) was one of the early triumphs of the kinetic theory of gases. It predicted, surprisingly, that the viscosity of a gas is independent of the pressure. This prediction was borne out when the experiment was finally performed.[5] Further, Equation (22.20) provided a measure of the size of molecules. The diameter could be found by measuring the viscosity, and experimental procedures for accurate viscosity measurements were soon developed.[6]

22.6 THERMAL CONDUCTIVITY

Viscosity arises from momentum transfer across a velocity (or momentum) gradient. *Thermal conductivity,* on the other hand, arises from the transfer of kinetic energy across a kinetic energy gradient. Recall that the kinetic energy of a gas is proportional to $T^{1/2}$. Figure 22.8 shows two horizontal planes, an upper one at T_1 and a lower one at T_0. The heat flowing across the planes through the gas is

$$q = \kappa A \frac{dT}{dz} t \tag{22.21}$$

where A is the area across which the heat flows, dT/dz is the temperature gradient, t is the time, and κ is the *thermal conductivity coefficient*. It is defined as the heat flow per unit time per unit temperature gradient across a unit cross-sectional area. Our procedure is like that of the previous section.

FIGURE 22.8 Transport of kinetic energy across a temperature (or energy) gradient. The upper plate is at T_1; the lower plate is at a lower temperature, T_0. The view is a section parallel to the z axis. Several layers of gas are shown. The separation between layers is λ. When a molecule from a lower layer hops to a higher layer, the average thermal energy of the molecules in the upper layer is decreased; when a molecule hops from an upper to a lower layer, the average thermal energy of the molecules in the lower layer is increased.

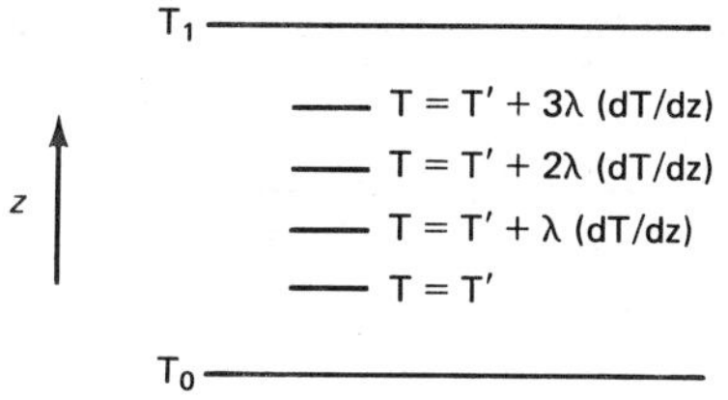

T_1 ————————

$T = T' + 3\lambda \, (dT/dz)$
$T = T' + 2\lambda \, (dT/dz)$
$z \uparrow$
$T = T' + \lambda \, (dT/dz)$
$T = T'$

T_0 ————————

---

[5] It is true for pressures from a few torr up to several atmospheres, which was contrary to expectations. It seemed absurd that a gas at low pressure should provide as much viscous drag as the gas at high pressure. At very low pressures the relation fails, since the mean free path of the gas (or the distance between layers) is comparable to the distance between the moving planes, and the assumptions made in the derivation are meaningless. At very high pressures, the greater attractive forces between the molecules causes the relation to fail.

[6] It is interesting to note that Millikan's determination of the charge of an electron in his oil-drop experiment used the value of the viscosity of air as one of the parameters (see Problem 17). Years later Millikan's value was found to be in error by about 1%. When his data were reevaluated, it was discovered that the error lay in the viscosity. His handbook value was off by a small amount. Excellent agreement resulted by inserting the corrected value for the viscosity of air in Millikan's calculations.

We divide the gas into a series of layers parallel to the planes, assuming the separation between layers to be equal to the mean free path. When steady state is reached, the heat flowing into any layer must equal the heat flowing out of the layer, or $d^2T/dz^2 = 0$. Integration establishes that dT/dz must be a constant. Again we assume that one-third of the molecules travel along the z axis, hopping from one layer to adjacent layers. The average temperature difference between two adjacent layers is $\lambda(dT/dz)$. If the molecules have mass m and a specific heat capacity c_v (that is, heat capacity per unit mass), then the energy difference between molecules in two adjacent layers is $mc_v\lambda(dT/dz)$.

Of the molecules traveling along the z axis, half move up and the other half move down. In one second, $\frac{1}{6}n^*\bar{c}$ molecules move down and carry an amount of energy equal to $\frac{1}{6}n^*\bar{c}mc_v\lambda(dT/dz)A$ from the upper to the lower layer. The same number of molecules move upward and carry an amount of energy equal to $-\frac{1}{6}n^*\bar{c}mc_v\lambda(dT/dz)A$ from the lower to the upper layer. The negative sign is important; the sign of dT/dz depends on whether one moves up or down. The total energy transferred equals the difference of energy carried up and energy carried down. Writing it as energy transferred per unit area per unit time, we get

$$\frac{q}{At} = \frac{1}{3}n^*\bar{c}mc_v\lambda\frac{dT}{dz} = \frac{\kappa dT}{dz} \tag{22.22}$$

From this it follows that

$$\kappa = \tfrac{1}{3}n^*m\bar{c}\lambda c_v \tag{22.23}$$

Since $\eta = \tfrac{1}{3}n^*m\bar{c}\lambda$, we can also write

$$\kappa = \eta c_v \tag{22.24}$$

At low pressures, heat conductivity varies with pressure, and the effect can be made the basis of a barometric instrument. If a constant current is passed through a resistor, then heat is dissipated in the resistor. If the resistor is placed in a gas at low pressure, then the gas molecules conduct the heat at a rate that varies with the pressure; the final equilibrium temperature of the resistor depends on the pressure.[7] If the temperature of the resistor is measured directly with a thermocouple, then the gauge is called a *thermocouple gauge*. If the temperature of the resistor is measured indirectly by measuring the resistance of the resistor (the resistance depends on temperature), then the gauge is called a *Pirani gauge*. These gauges must be calibrated separately for different gases, since different gases have different thermal conductivities; they are useful in the range 1–100 millitorr, but their accuracy is not too great. The operation of these gauges is discussed in Dushman's book.[8]

22.7 DIFFUSION

Consider the situation shown in Figure 22.9. A barrier is placed at the center of a long tube of unit cross-sectional area. Gas A is on one side of the barrier while gas B is on the other side. The pressures of the two gases are equal. If the barrier is removed, the two gases *interdiffuse*. Each gas moves into the other half of the tube, and

---

[7] In dealing with high pressures, we find that problems associated with heat convection become important.
[8] *Scientific Foundations of Vacuum Technique,* by Saul Dushman; second edition revised by members of the research staff, General Electric Research Laboratory, J. M. Lafferty, ed. (John Wiley & Sons, New York, 1962), chap. 5; henceforth called Dushman, *Scientific Foundations.*

SEC. 22.7 DIFFUSION

FIGURE 22.9 Interdiffusion of two gases, A and B, which are placed on opposite sides of a barrier. When the barrier is removed, the gases mix until the concentrations are uniform throughout the tube. The gases have the same initial pressures on each side of the barrier.

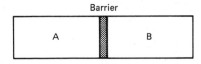

eventually the distribution of molecules becomes uniform throughout the tube. It simplifies matters to fix our attention on one of the gases and to regard that gas as diffusing into the other.

In addition, we assume from *experiment* that the *diffusion equation*

$$\frac{dn}{dt} = -D\frac{dn}{dz} \tag{22.25}$$

is valid. Here dn/dt measures the number of molecules diffusing through unit area in time dt; the value dn/dz is the concentration gradient expressed as the rate of change of number of molecules along the z axis; D is a constant of proportionality called the *diffusion coefficient*. Viscosity measures the transport of momentum across a momentum gradient; conductivity measures the transport of energy across an energy gradient; diffusion measures the transport of mass across a concentration gradient. The three belong to a set of properties known as *transport properties*.

Consider a gas in which the pressure is uniform throughout and which contains two types of molecules, A and B. The composition of the gas varies from layer to layer. We let the masses be m_A and m_B; the densities are n_A^* and n_B^*. We express the concentration as mole fraction of A,

$$X = X_A = \frac{n_A^*}{n_A^* + n_B^*} \tag{22.26}$$

With these stipulations the diffusion equation becomes

$$\left(\frac{d(m_A n_A^*)}{dt}\right)_{z_0} = -D\frac{dX}{dz} m_A n^* \tag{22.27}$$

where the derivative is evaluated at $z = z_0$. The term n^* represents the total density of molecules; $n^* \, dX/dz$ is the same as dn/dz in (22.25). Suppose we consider a thin slice of the tube Δz, as shown in Figure 22.10. Equation (22.27) gives the rate at which mass diffuses out of the slice. The mass diffusing into the cylinder at $z = z_0 + \Delta z$ is given by

$$\left(\frac{d(m_A n_A^*)}{dt}\right)_{z_0+\Delta z} = -m_A n^* \left[D\frac{\partial X}{\partial z} + \frac{\partial}{\partial z}\left(D\frac{\partial X}{\partial z}\right)\Delta z\right] \tag{22.28}$$

The difference in the quantities in (22.27) and (22.28) gives the rate of mass buildup in the slice as

$$-m_A n^* \, \Delta z \frac{\partial}{\partial z}\left(D\frac{\partial X}{\partial z}\right) \tag{22.29}$$

But the rate of increase of A in the slice can also be written as

$$m_A n^* \frac{\partial X}{\partial t} \Delta z \tag{22.30}$$

FIGURE 22.10 Illustrating the derivation of the diffusion equation. The arrow pointing upward is a measure of the mass of A leaving the slice and is given by Equation (22.27). The arrow pointing downward is a measure of the mass of A entering the slice and is given by (22.28).

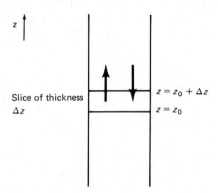

When (22.29) and (22.30) are set equal to each other, we get

$$m_A n^* \Delta z \frac{\partial}{\partial z}\left(D \frac{\partial X}{\partial z}\right) = m_A n^* \frac{\partial X}{\partial t} \Delta z \tag{22.31}$$

which reduces to

$$\frac{\partial X}{\partial t} = \frac{\partial}{\partial z}\left(D \frac{\partial X}{\partial z}\right) \tag{22.32}$$

This is Fourier's equation for heat conduction in solids. Equation (22.32) applies to systems that are not in equilibrium. The viscosity expression we derived, like the thermal conductivity expression, applied to systems in the steady state. For thermal conductivity, the two plates can be maintained at different constant temperatures by thermostats. In diffusion experiments, the process continues until equilibrium is achieved and the concentrations are uniform throughout; at that point there is no longer any concentration gradient, and hence no longer any diffusion.

We can simplify the problem by considering the process known as self-diffusion. We take an infinitely long tube and two similar gases, say CO_2 and N_2O. After some time, it is assumed that the change in concentration stops at some point (that is, $dn^*/dt = 0$). Call that point $z = 0$ and assume that the concentration increases in the direction of positive z. At $z = \lambda$, the density of molecules is $n^* + \lambda\, dn^*/dx$; at $z = -\lambda$, the density is $n^* - \lambda\, dn^*/dz$. The number of molecules passing downward is[9]

$$\left(\frac{dn^*}{dt}\right)_\downarrow = -\frac{1}{6}\left(n^* + \lambda \frac{dn^*}{dz}\right)\bar{c} \tag{22.33}$$

and the number passing upward is

$$\left(\frac{dn^*}{dt}\right)_\uparrow = +\frac{1}{6}\left(n^* + \lambda \frac{dn^*}{dz}\right)\bar{c} \tag{22.34}$$

The total transfer through a slice at $z = 0$ is then

$$\frac{dn^*}{dt} = \left(\frac{dn^*}{dt}\right)_\uparrow + \left(\frac{dn^*}{dt}\right)_\downarrow$$

$$= -\frac{1}{3}\bar{c}\lambda \frac{dn^*}{dt} \tag{22.35}$$

[9] Since it makes no difference which gas is diffusing into which, and since both have the same diffusion constant, we have dropped the subscripts.

> IRVING LANGMUIR (1881–1957), Brooklyn-born American scientist, received his Ph.D. in Germany under Nernst and returned to the United States in 1906 to teach at Stevens Institute in Hoboken, New Jersey. After three years, he went to the new General Electric Research Laboratories in Schenectady, New York, where he eventually became a vice president and director of research. In 1912 he invented the argon-filled incandescent lamp, an invention that decreased the cost of lighting in this country alone by hundreds of millions of dollars a year. He contributed to the development of high-vacuum diffusion pumps and also of the electronic vacuum tube. He invented the atomic hydrogen welding torch and was involved in the early work in rainmaking. In 1932 Langmuir received the Nobel Prize in chemistry "for his discoveries and investigations in surface chemistry."

By comparing Equation (22.35) with (22.25) we can conclude that

$$D = \tfrac{1}{3}\bar{c}\lambda \qquad (22.36)$$

By recalling that $\eta = \tfrac{1}{3}n^* m \bar{c}$, we can write

$$D = \frac{\eta}{n^* m} \qquad (22.37)$$

Since the density is $n^* m$, we can also write

$$D = \frac{\eta}{\rho} \qquad (22.38)$$

From Equation (22.38), we can conclude that the greater the density or pressure, the smaller the diffusion coefficient and the longer it takes gases to diffuse.

A practical application of diffusion can be seen in the argon-filled light bulb. The ordinary incandescent bulb operates by passing a current through a tungsten filament. The heat raises the temperature of the tungsten to the point at which it emits radiation in the visible spectrum. The bulb must be evacuated to remove the oxygen that would cause the filament to burn out immediately. At the high operating temperatures, tungsten has a nonnegligible vapor pressure. The early light bulbs that were evacuated to very low pressures had very short lifetimes. The tungsten evaporated, condensed on the inside surface of the cooler glass bulb, and blackened the bulb. Irving Langmuir, working at the General Electric laboratory, reasoned that introducing a small amount of argon into the bulb would lower the mean free path of the tungsten atoms to the point at which most of the evaporated atoms would recondense on the filament. His idea was successful, and the introduction of argon lengthened the life by an enormous amount. It has been estimated that the profits returned from that one discovery have more than paid for the entire operating budget of the General Electric Research Laboratories since their inception.

22.8 SOLUTIONS OF THE DIFFUSION EQUATION

The diffusion process is controlled by Equation (22.32), which is a second-order partial differential equation. It is identical to the equation for heat conduction in solids; the solution depends on the boundary conditions of the system. Such solutions can be found in various treatises dealing with the subject, and the solutions can be carried over into diffusion.[10] We shall discuss solving one particular type of boundary condition.

In general the diffusion coefficient depends on concentration and therefore on position. We shall simplify the problem by assuming that the diffusion coefficient is independent of concentration. Thus Equation (22.32) can be written

$$\frac{\partial c}{\partial t} = D \left(\frac{\partial^2 c}{\partial z^2}\right) \qquad (22.39)$$

where we have used c instead of X for concentration. We shall discuss the solution of (22.39) under the conditions illustrated in Figure 22.11. A thin layer of solute A is introduced into a tube of solvent at $t = 0$ and $z = 0$; we want to know the concentration of A as a function of z and t. Without going into the details, we shall simply write the solution to (22.39) as

$$c(z,t) = at^{-1/2} \exp\left(-\frac{z^2}{4Dt}\right) \qquad (22.40)$$

where a is a constant. That (22.40) is a solution of (22.39) can be verified by substitution. We can check the boundary conditions by noting that as $t \to 0$, Equation (22.40) requires that $c = 0$ for all nonzero values of z and this corresponds to the actual case. At $z = 0$, however, the concentration is infinitely large. This corresponds to a situation in which a finite amount of A is compressed into a layer of zero width. Figure 22.12 shows plots of Equation (22.40) in the form of concentration profiles at various times, time being measured in units of $\frac{1}{4}D$. The molecules of A diffuse outward from the origin. At large distances, the concentration increases with time. A variable of interest is the root mean square distance the molecules move from the origin in time t. This is

$$\overline{z^2} = \frac{\int_{-\infty}^{\infty} z^2 c \, dz}{\int_{-\infty}^{\infty} c \, dz} = \frac{\int_{-\infty}^{\infty} z^2 \exp(-z^2/4Dt) \, dz}{\int_{-\infty}^{\infty} \exp(-z^2/4Dt) \, dz} \qquad (22.41)$$

$$= \frac{(\sqrt{\pi}/2)(4Dt)^{3/2}}{\sqrt{4\pi Dt}} = 2Dt$$

Since the motion is symmetrical, we cannot get any useful information from the average value of z, since that is always 0. We could, however, separately compute $\bar{z}$ in each half of the z axis.

The diffusion equation applies equally to gases, liquids, and solids, and it finds perhaps its greatest application in the solid state. Geologists are often concerned

FIGURE 22.11 Diffusion in one dimension. A thin slice of solute A is placed at the origin of a tube containing solvent B. The solute then diffuses outward in both directions.

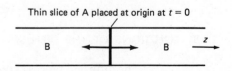

---

[10] The problem is treated in most books dealing with partial differential equations, books entitled *Advanced Calculus for Engineers*, or a reasonable facsimile thereof. A complete discussion can be found in H. S. Carslaw and J. C. Jaeger, *Conduction of Heat in Solids* (Oxford: Oxford University Press, 1959).

FIGURE 22.12 Concentration vs distance from the source of Figure 22.11 at different times. The numbers on the curves indicate values of Dt.

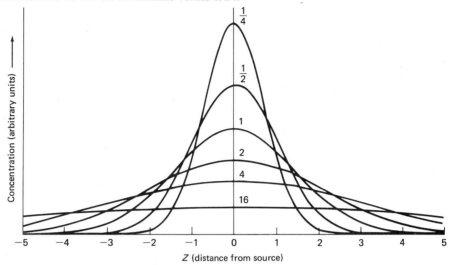

with the diffusion of material in rocks for purposes of determining the age of the rocks. Many of the technological processes involved in the semiconductor electronics industry are based on diffusion. One can create p-n junctions by starting with p-type silicon single crystals and diffusing n-type material into the crystal.[11] This is accomplished by placing the n-type material on a wafer of p-type silicon and carefully controlling the temperature and time of diffusion. The solution to this problem is approximated by one half of Figure 22.12, since the diffusing solute is placed at the end rather than at the middle of the solvent.

22.9 THERMAL TRANSPIRATION

Suppose we have two chambers, A and B, separated by a small orifice, as shown in Figure 22.13. The rate at which gas effuses through a small hole is governed by the effusion equation

$$\frac{dn}{dt} = -\frac{1}{4} n^* \bar{c} \, dS \qquad (22.10)$$

At equilibrium, the rate at which molecules leave A must equal the rate at which molecules enter A from B; thus

$$n_A^* \bar{c}_A = n_B^* \bar{c}_B \qquad (22.42)$$

Since $n^* = P/kT$, and $\bar{c} = (8kT/\pi m)^{1/2}$, Equation (22.42) can be written as

$$P_A(T_A)^{-1/2} = P_B(T_B)^{-1/2}$$

or

$$\frac{P_A}{P_B} = \left(\frac{T_A}{T_B}\right)^{1/2} \qquad (22.43)$$

If the two chambers are at the same temperatures, their pressures will be the same.

[11] We shall discuss these junctions in Chapter 37.

FIGURE 22.13 Thermal transpiration. Chambers A and B are connected by a small orifice. If the orifice is smaller than the mean free path, and the chambers are held at different temperatures, then the pressures in the two chambers will not be equal.

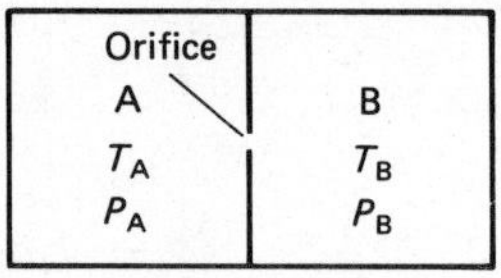

```
Orifice
A   \       B
T_A          T_B
P_A          P_B
```

If the two chambers are at different temperatures however, then their pressures will be different. Therein lies a potential experimental difficulty.

Suppose one is engaged in low-temperature work and wishes to measure a low pressure in the low-temperature region. The barometric device is normally at room temperature and is connected to the low-temperature region through a tube. Suppose the low temperature T_A is that of liquid N_2, or 77.5 K, and the room temperature T_B is 300 K. We measure P_B; but we want to know P_A, which is

$$P_A = P_B \left(\frac{77.5}{300}\right)^{1/2} = 0.51 P_B$$

The desired pressure is 50% less than the measured pressure.

Equation (22.43) is valid only under the conditions for which the effusion equation is valid; that means conditions of *molecular flow*. The condition of molecular flow applies when molecular collisions can be ignored. This is true for a system in which the diameter of the orifice is very much smaller than λ, the mean free path. If the pressure is so high that the mean free path is very much smaller than the diameter of the orifice, then the flow of molecules is governed not by kinetic flow but by the usual equations of hydrodynamics. Under those conditions, $P_A = P_B$. In the intermediate region, the calculation of thermal transpiration effects can be quite complicated. The actual situation in a real vacuum system is further complicated, since the two chambers are connected not by an orifice but by a long tube. These problems are discussed in Dushman's book on vacuum technology.[12]

22.10 THE PRODUCTION OF HIGH VACUUM

The production of high vacuum is important not only in large areas of basic science but also in large sectors of our industrial economy. The manufacture of light bulbs and electronic vacuum tubes requires the production of high vacuum. The manufacture of silicon and germanium crystals, which forms the base of the electronics semiconductor industry, requires high vacuum to ensure high purity of the finished product. Recent developments in vacuum metallurgy and freeze-drying of foods are based on the use of large-scale vacuum pumps. Two types of vacuum pumps whose operation incorporates certain features of the kinetic theory of gases are of interest.

Figure 22.14 shows the elements of a *diffusion pump*. The pump fluid is heated to its boiling point; the vapor moves up the chimney and is deflected downward by the nozzle. Coolant surrounding the barrel keeps the outside of the pump at a low temperature, so that the vapor condenses and returns to the boiler. Molecules from the high-vacuum region, indicated by B, diffuse from the high-vacuum region into the

---

[12] Dushman, *Scientific Foundations*, p. 58.

FIGURE 22.14 (a) Elements of a single-stage diffusion pump. (b) An enlargement of the jet leaving the nozzle.

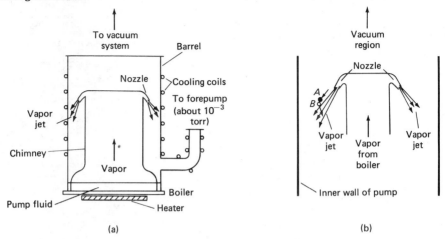

periphery of the vapor jet stream. There it is struck by vapor molecules and carried downward. The net effect is to compress gases from the high-vacuum region into the low-vacuum region at the bottom of the pump, from which they are carried off by a simple mechanical rotary vacuum pump. The efficiency of the pump can be increased substantially by using more than one stage, as shown in Figure 22.15. Figure 22.16 shows a photograph of a commercial diffusion pump; pumps are available in sizes up to 32 in. in diameter, and larger sizes can be custom-built. Modern diffusion pumps can attain pressures as low as 10^{-9} torr.

In contradistinction to the diffusion pump which has no moving parts, the *turbo-molecular,* or *molecular drag,* pump operates on the principle that a moving surface

FIGURE 22.15 Three-stage diffusion pump. The density of the molecules is indicated by the density of the dots.

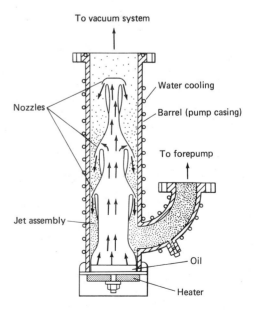

FIGURE 22.16 Photograph of a water-cooled oil-diffusion pump. The inner jet assembly is shown to the right. (Courtesy CVC Products, Inc., Rochester, N.Y. 14603.)

imparts a momentum to gases in contact with the surface, as we have seen in our discussion of viscosity. Figure 22.17 shows a simple molecular drag pump. The cylinder C rotates at high angular velocity inside the cylindrical housing H. An obstruction O is placed between the inlet and outlet ports. As C rotates, it "drags" molecules from the high-vacuum region at the inlet and compresses the gas at the exhaust. There the gas is removed by a rotary vacuum pump. The efficiency can be increased by

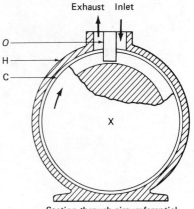

FIGURE 22.17 Simple one-stage molecular drag pump. As the cylinder C rotates inside the cylindrical housing H, it "drags" molecules from the inlet to the higher-pressure exhaust, where the molecules are removed by a rotary forepump. (An obstruction O is placed between inlet and outlet ports.)

Section through circumferential groove in cylinder A

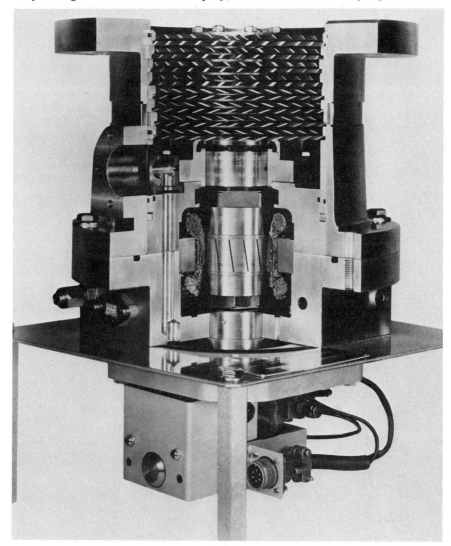

FIGURE 22.18 Cutaway photograph of commercial 16-stage turbo molecular pump consisting of 8 moving and 8 stationary rotor discs. The rotors rotate at 50,000 rpm. (Photograph courtesy of Sargent Welch Scientific Company; turbo molecular vacuum pump.)

increasing the surface in contact with the gas and by increasing the number of stages. Figure 22.18 shows a cross-sectional view of a commercial turbo molecular pump capable of achieving pressures as low as 10^{-9} torr. Here the "drag" is provided by a series of 8 discs rotating on a shaft at 50,000 rpm between a series of stationary outer discs. The clearances are small (about 0.01 in.), and the tips of the rotor discs attain speeds of the order of 440 m s^{-1} (supersonic), hence the machining tolerances are quite rigid for pumps of this type. This by no means exhausts the discussion of vacuum pumps. Further information on these and other types of pumps can be had in any standard book on high-vacuum technique, such as that of Dushman.[13]

---

[13] Dushman, *Scientific Foundations*, chap. 3.

22.11 MOLECULAR DIMENSIONS FROM KINETIC PROPERTIES

The approximate expressions for the viscosity, the thermal conductivity coefficient, and the diffusion coefficient have been given in Equations (22.20), (22.24), and (22.38). Without further discussion we shall write the expressions in the "exact" form as

$$\eta = 0.499 \, \rho \bar{c} \, \lambda \tag{22.44}$$
$$\kappa = 1.261 \, \rho \bar{c} \, \lambda c_v \tag{22.45}$$
$$D = 0.599 \, \lambda \bar{c} \tag{22.46}$$

In these three expressions, ρ is the density, λ is the mean free path, $\bar{c}$ is the average velocity, and c_v is the specific heat capacity. The units of viscosity are kg m^{-1} s^{-1} (the unit of *poise* is 1 g cm^{-1} s^{-1}), the units of thermal conductivity are J m^{-1} s^{-1} K^{-1}, and the units of the diffusion coefficient are m^2 s^{-1}.

The application of Equations (22.44) through (22.46) to experimentally measured transport properties yields values of λ, the mean free path. Since the mean free path is

$$\lambda = \frac{1}{\sqrt{2}\pi \, d^2 n^*} \tag{22.12}$$

we can then calculate the diameter of gas molecules. Table 22.1 summarizes the results of these types of measurements. The diameters obtained agree reasonably well with diameters determined by other techniques.

Equations (22.44)–(22.46) are valid for rigid elastic spheres. Molecules are not actually rigid, nor are they totally devoid of interactions. Several theories have been propounded to account for molecular interactions, but we shall not examine this matter here. The interested student is referred to the extensive literature in the field.[14]

TABLE 22.1 Some results of the kinetic theory of gases.

Gas	$10^4 \eta$ (g cm^{-1} s^{-1})	$10^4 \kappa$ (J cm^{-1} s^{-1} K^{-1})	D (cm^2 s^{-1})	Calculated molecular diameters (units of 10^{-8} cm)			
				From η	From κ	From D	From van der Waals constant b
Ne	2.97	4.64	0.45	2.64	2.57	2.41	2.66
Ar	2.10	1.65	0.16	3.67	3.65	3.47	2.86
Kr	2.33	0.87	0.081	4.20	4.17	4.01	3.14
Xe	2.11	0.51	0.05	4.90	4.85	4.65	3.42
H$_2$	0.85	16.6	1.29	2.71	3.11	2.55	2.76
N$_2$	1.66	2.30	0.185	3.76	4.30	3.48	3.15
O$_2$	1.92	2.44	0.187	3.62	4.06	3.35	2.91
Air	1.71	2.42	0.181	3.72	4.18	3.50	

PROBLEMS

1. A gas mixture containing equal mole numbers of He and H$_2$ effuses through a small hole in the side of the container. Calculate the mole fraction of He in the gas leaving the hole.

2. A vacuum chamber is in the form of a cube 1 m on an edge. The initial pressure is 0.002 torr. If the chamber is connected to a vacuum pump through an orifice 0.2 mm in diameter, how long will it

---

[14] Walter Kauzmann, *Kinetic Theory of Gases* (New York: W. A. Benjamin, 1966) contains some material on an elementary level. For a more advanced level, see J. O. Hirschfelder, C. F. Curtis, and R. B. Bird, *Molecular Theory of Gases and Liquids* (New York: John Wiley & Sons, 1954).

take for the pressure to fall to 0.001 torr? When the pressure is finally at 0.001 torr, how many collisions per second does the gas make with the walls of the chamber? Estimate the mean free path of the gas. (The temperature is 25 °C.)

3. In many experimental situations, one wants to carry out procedures in a vacuum in such a way that the mean free path of the molecules is longer than the linear dimensions of the container. Estimate this pressure at 25 °C for a vacuum chamber in the form of sphere of radius 50 cm.

4. Suppose you have a 1-cm^2 glass plate enclosed in a vacuum chamber at pressure 1 torr. The gas is Cl_2. Suppose further that one side of the plate is clean, and the other side is coated with a metal. Each time a Cl_2 molecule hits the clean side it bounces off; each time a molecule hits the coated side it "sticks" to the plate by forming a molecule of the metal chloride. Thus, the momentum change of a molecule that hits the clean side is twice the momentum change of a molecule that hits the other side. Assuming that the molecules strike the plate perpendicularly, calculate the net force on the plate at 25 °C. (There is a net force, since the momentum change is different for opposite sides of the plate.)

5. A 1-cm^2 plate is in a vacuum system. The temperature of the vacuum system is T_0. One side of the plate is maintained at T_0, while the other side is maintained at a slightly higher temperature, T_1. Molecules approach the plate with temperature T_0 and leave the plate with the temperature of the side they hit. (See Problem 4.) Calculate the net force on the plate as a function of the pressure and the temperature difference. What is the net force if T_0 = 25 °C and T_1 = 30 °C at a pressure of 0.003 torr? (This example forms the basis of the radiometer type of pressure gauge, or the *Knudsen gauge*. The plate or vane is suspended by a fine wire. The high temperature surface is inverted on opposite sides of the wire. Thus the force is converted into a torque. The pressure is indicated by the angular displacement from equilibrium. This is also the basis of the radiometer type of toy often seen in toy-store windows. One side of the vanes are reflecting surfaces, and the other sides are blackened surfaces. The blackened surfaces absorb heat and are at a higher temperature than the reflecting sides. The net torque imparted is sufficient to rotate the assembly.)

6. Irving Langmuir (*Phys. Rev.* 2 (1913):329) was the first to determine the vapor pressure of metallic tungsten, by measuring the loss in weight of a sample of tungsten per unit area. His data are

T (K)	Weight loss (g cm^{-2} s^{-1})
2600	8.41×10^{-9}
2800	1.10×10^{-7}
3000	9.95×10^{-7}
3200	6.38×10^{-6}
3400	3.47×10^{-5}

in the accompanying table. Calculate the vapor pressure of tungsten at each of the temperatures. Calculate the heat of vaporization of tungsten from the data.

7. Calculate the number of collisions Z_1, the collision frequency Z_{11}, and the mean free path for H_2 at pressures of 10^{-9}, 10^{-6}, 10^{-3}, 1, and 760 torr for temperatures of 100 and 300 K.

8. Some oxygen gas is in a sealed chamber. At what pressure is the number of collisions per square centimeter of wall equal to Z_{11}, the number of collisions per cubic centimeter of gas?

9. A gas mixture contains 0.2 mol fraction O_2, and 0.8 mol fraction N_2. Calculate the total number of collisions between O_2 and N_2 molecules per cubic centimeter at 300 K for pressures of 0.001 and 760 torr.

10. Consider the chemical reaction A(g) + B(g) = AB(l). Molecules of A have diameters 2.5×10^{-8} cm and molar mass 20; molecules of B have diameters 3×10^{-8} cm and molar mass 40. Whenever a molecule of A reacts with a molecule of B, a molecule of AB is formed; since AB is a liquid it is removed from the system. Initially the system contains $\frac{1}{2}$ mol A and $\frac{1}{2}$ mol B at P_0. Assume that the probability of reaction on collision of an A molecule with a B molecule is 10^{-16}. (That is to say, for every 10^{16} collisions between A and B, one molecule of AB is formed.) Derive an expression for the decrease in pressure as a function of time, dP/dt. If P_0 = 1 atm and T = 300 K, how long will it take for the pressure to fall to $\frac{1}{2}$ atm? to $\frac{1}{4}$ atm? to $\frac{1}{8}$ atm? to 10^{-6} torr? How long will it take for the pressure to fall from 10^{-6} to 5×10^{-7} torr?

11. An apparatus for measuring viscosity is shown in Figure 22.6. Suppose the two coaxial cylinders are each 20 cm long. The outer cylinder is rotated at 1 revolution per minute (rpm), and the inner cylinder is suspended by a wire. The twist on the wire measures the torque on the inner cylinder. The outer cylinder has a radius of 10 cm, and the gap between the cylinders is 0.2 cm. The measured torque applied to the inner cylinder is 25 dyn cm. What is the viscosity of the gas?

12. For CO_2, $\eta = 1.37 \times 10^{-4}$ poise. Calculate the molecular diameter of CO_2 using this experimental value and compare it with the value obtained from the van der Waals constant.

13. Two parallel plates are maintained at temperatures of 300 and 303 K. The space between the plates is filled with He at a pressure of 1 torr. For He at 300 K, c_v = 3.12 J K^{-1} g^{-1}. Taking the diameter of a He molecule to be 2.5×10^{-8} cm, calculate the heat flow between the two plates in units of J cm^{-2} s^{-1}.

14. You have a spherical dewar flask with inside diameter 30 cm. The dewar is filled with liquid N_2 at its boiling point of 77.4 K as it is sitting on a desk top. The inside of the dewar is thus at 77.4 K, while the outside is at room temperature, or 300 K. The density of liquid nitrogen is 0.8081 g cm^{-3}, and the heat of vaporization is 5576 J mol^{-1}. The

space between the walls of the dewar is 0.4 cm, and it is evacuated to a pressure P_0. Assume that the residual gas is He (see Problem 13) at a temperature given by the mean of the two wall temperatures. First calculate the heat loss by thermal conductivity. Then calculate the heat loss assuming that molecules at 300 K strike the inner wall and bounce off with a temperature equal to that of the inner wall ($\lambda > 0.4$ cm). What is the maximum value of P_0 under which the losses are sufficiently small that the N_2 boils off at a rate less than 1 cm³ of liquid per hour? (*Note:* Any real calculation would have to take into account heat losses from radiation and from escape through the opening of the dewar. Ignore these effects.)

15. Consider the same dewar flask as in Problem 14, which is now filled with liquid He at its normal boiling point, 4.2 K. The heat of vaporization of liquid He is 82 J mol⁻¹; the density of liquid He is 0.1259 g cm⁻³. How small does P_0 have to be to ensure that the evaporation losses from conduction between the walls of the dewar are less than 1 cm³ of liquid per hour? Suppose that the outside wall were maintained at 77.4 K (by immersing this dewar in a larger dewar filled with liquid N_2); what would P_0 be? How long would the He last in the dewar? (*Note:* Don't forget that as the He evaporates, a smaller surface area is presented to the conducting surfaces. The answer you get will not be real, since we have neglected other sources of heat losses; it does, however, indicate the magnitude of the losses.)

16. If the space between the walls of the dewar flask of Problem 15 were at a pressure of 0.01 torr, how fast would the liquid He boil off initially?

17. According to Stokes's law, the frictional resistance that a spherical particle moving in a viscous fluid experiences is $f = 6\pi\eta rv$, where r is the radius of the particle, η is the viscosity of the fluid, and v is the velocity with which the particle is moving. When a small particle is placed in a viscous fluid, it accelerates initially because of gravitation. As the velocity of the particle increases, the frictional force increases until it equals the gravitational force, at which point the particle falls with a constant velocity termed the *terminal velocity*. The force due to gravity is $\tfrac{4}{3}\pi r^3 g(\rho - \rho_0)$, where g is the gravitational acceleration, ρ is the density of the particle, and ρ_0 is the density of the fluid. Assume that a mist particle (water droplet, $r = 10^{-5}$ cm) falls in air. Calculate the terminal velocity of the particle, taking the viscosity of air to be 1.71×10^{-4} g cm⁻¹ s⁻¹.

18. A pressure gauge at room temperature is connected to a chamber at liquid He temperature (4.2 K). The gauge indicates a pressure of 1 millitorr. What is the pressure of the chamber at the liquid He temperature? Assume effusive flow.

19. The value of the constant a in the diffusion equation (22.40) may be evaluated as follows. The number of molecules N of solute introduced at $t = 0$ must be constant; thus normalization requires that

$$N = \int_{-\infty}^{+\infty} c\, dz = \text{constant}$$

Evaluate a and show that the diffusion equation can be written

$$c(z,t) = \frac{N}{2(\pi Dt)^{1/2}} \exp\left(\frac{-z^2}{4Dt}\right)$$

20. The concentration of CO_2 in the atmosphere is approximately 0.04 mol percent. Calculate the rate at which CO_2 molecules strike 100 cm² of leaf on a tree. The overall reaction for photosynthesis can be taken as $6CO_2 + 6H_2O = C_6H_{12}O_6 + 6O_2$. The *Malus pumila* (common apple tree) is found to consume 20 mg CO_2 per hour per 100 cm² of leaf area. What is the fraction of the CO_2 molecules that strike the leaf and are eventually converted to $C_6H_{12}O_6$?

CHAPTER TWENTY-THREE
STATISTICAL MECHANICS

23.1 MACROSCOPIC AND MICROSCOPIC THERMODYNAMICS

In considering macroscopic thermodynamics we developed a picture of the physical world that enabled us to establish relations between numerous experimental properties of macroscopic systems in equilibrium. Take the experimental quantities C_p and C_v. Thermodynamics requires that $C_p - C_v = R$ for an ideal gas. The *difference* between the two is established by the laws of thermodynamics; the absolute values are indeterminable within the framework of thermodynamics. They must be determined experimentally. Our system of thermodynamics starts with experiment and is built up with the development of a scheme to explain the experimental observables.

Microscopic thermodynamics, or *statistical mechanics* or *statistical thermodynamics,* proceeds in precisely the opposite direction. We start not with the experimental observable but with the fundamental properties of matter. A picture of the universe is developed from these fundamental properties. If our picture agrees with the real world, we then assume that our fundamental assumptions are correct. The fundamental assumptions are the molecular nature of matter and the basic laws of physics, to which we add the fundamental assumptions of quantum mechanics. If we knew the equations of motion of each and every molecule in a mole of gas, then in principle we could calculate all the properties of the system. This requires that we know the mass, position, velocity, momentum, kinetic energy, and potential energy of each and every atom as a function of time. Unfortunately, the enormous complexity of dealing with a set of 10^{23} equations precludes this approach; we are compelled to deal with the system on a statistical basis.

On a microscopic scale, the Maxwell-Boltzmann distribution provides a mechanical measure of temperature by relating the temperature of a gas to the molecular velocity. There is a distribution of these velocities. If we were to assign a "temperature" to each molecule based on its velocity, we should have a wide range of "molecular temperatures." On a statistical basis, we are not precluded from a situation in which at some instant the left side of a container filled with gas has an excess of fast molecules while the right side has an excess of slow molecules. Macroscopically this corresponds to a situation in which the temperature on the left suddenly increases while the temperature on the right decreases. On the other hand, such a situation has never been observed. If the vessel contains enough molecules, then the statistical improbability is sufficient to ensure that such a condition never will be observed. This situation is absolutely forbidden by macroscopic thermodynamics, since it corresponds to a spontaneous entropy increase of an isolated system. This situation is also forbidden by microscopic thermodynamics with the proviso that while it is not absolutely forbidden, its occurrence on a measurable scale is improbable enough that for all practical purposes we can consider it impossible. "What never? Well, hardly ever."

In a certain sense, macroscopic thermodynamics is directed towards the question, *What* happens in a given system? *What* is the equilibrium constant for a reaction, and *what* is the entropy or free energy change? Microscopic thermodynamics tries to provide the *why*. *Why* is the constant-volume heat capacity of an ideal monatomic gas equal to $1.5R$? This does not mean to imply that statistical mechanics is any more fundamental than macroscopic thermodynamics. Both examine a common set of properties, but from different points of view. Ultimately both give the same answers. Macroscopic thermodynamics is the older of the two fields; hence our faith in the validity of microscopic thermodynamics exists because it gives results that agree with macroscopic thermodynamics. All that means, however, is that both systems give results that agree with what we observe in the real world.

Of all the thermodynamic functions we invented, entropy seemed somewhat of a maverick. The energy, enthalpy, and free energy enable us to determine the amount of work obtainable from a process under certain conditions. Entropy, on the other hand, or more correctly $T \, \Delta S$, measures unavailable "potential work." There is another difference. Of all the thermodynamics functions, entropy alone was placed on an absolute scale. The third law ascribed an absolute value of zero to the entropy of pure crystalline substances at the absolute zero of temperature. We never really gave any satisfactory reason for this, however. Actually, the third law did not specify that the entropies were zero at $T = 0$. The third law simply requires that all substances have the same entropies at that temperature without specifying what that value is. We set that value to zero for convenience. We shall eventually justify that value as a result of statistical mechanics. For the moment, let us examine entropy and probability.

23.2 ENTROPY, PROBABILITY, AND RANDOMNESS

We can associate entropy changes with probabilities in the following simple manner. Consider the ideal gas expansion shown in Figure 23.1. Two bulbs, each of volume V, are connected by a stopcock. Initially, four molecules of the gas are placed in the left bulb; the right bulb is empty. The stopcock is opened. What is the entropy change? From our discussions of ideal gas expansions, we write

$$\Delta S = nR \ln \frac{V_2}{V_1}$$
$$= nR \ln 2$$
$$= Nk \ln 2$$
$$= 4k \ln 2 = k \ln 2^4$$
$$= k \ln 16 \tag{23.1}$$

where nR has been converted to Nk; N is the number of molecules and k the Boltzmann constant. The entropy change is proportional to the natural logarithm of 16; the proportionality constant is k.

Now consider the process from the point of view of the binomial distribution. In the final state the molecules are distributed between the left bulb and the right.

FIGURE 23.1 Expansion of four molecules of an ideal gas. (a) Initial state; all four molecules in one bulb. (b) Final state; molecules divided between both bulbs.

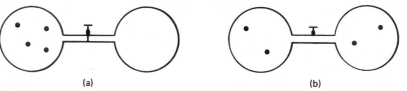

Table 23.1 lists the numbers and probabilities of the possible configurations. There are 16 ways of arranging the 4 molecules between the 2 bulbs in the final state.

On the other hand, all 4 molecules are in the left bulb in the initial state. The probability of this configuration is $\frac{1}{16}$; this state can be achieved in only one way, since there is only one way in which the 4 molecules can be placed in the left bulb. Now consider the ratios of the final to the initial probability, P_2/P_1, or the final to the initial number of possible configurations, C_2/C_1. They are

$$\frac{P_2}{P_1} = \frac{C_2}{C_1} = 16 = 2^4$$

This suggests that if we associate entropies with probabilities, or numbers of possible configurations, then the equation

$$\Delta S = k \ln \frac{P_2}{P_1} = k \ln \frac{C_2}{C_1} \qquad (23.2)$$

provides a method of evaluating entropy.

Entropy can thus be associated with randomness. The equilibrium state of the 4 molecules distributed in 2 bulbs is more random than 4 molecules in 1 bulb; we are less definite about the location of the molecules in the more random (or disordered) state. The natural path is from order to randomness. If we shuffle an ordered deck of cards, we generally get a random deck. If we shuffle a random deck, we rarely wind up with an ordered deck. In the following section, we shall develop this association between entropy, probability, and the number of possible configurations of states. It is worth noting that the concept of entropy as a measure of randomness has transcended the classical boundary lines of thermodynamics and found great application

TABLE 23.1 Ways of arranging four molecules in two bulbs, each of equal volume.

Configuration		C	P
Molecules in left bulb	Molecules in right bulb	(Number of ways to achieve the configuration)[a]	(Probability of the configuration)[b]
0	4	1	$\frac{1}{16}$
1	3	4	$\frac{4}{16} = \frac{1}{4}$
2	2	6	$\frac{6}{16} = \frac{3}{8}$
3	1	4	$\frac{4}{16} = \frac{1}{4}$
4	0	1	$\frac{1}{16}$
	Total	16	1.000

[a] These entries are just the binomial coefficients of Equation (20.2).
[b] These entries are the binomial probabilities of Equation (20.3).

in information theory, which concerns itself with transmitting and degrading information. This is critically important in theories dealing with computers and the transmission of information via telephone lines.[1]

23.3 ENTROPY AND A PRIORI PROBABILITY

Suppose we have a computer program that has been punched on 52 punchcards. The deck is correctly ordered so that if we feed it into a computer, the program runs properly. There is only one way we can achieve this arrangement. What if we now shuffle the deck so that the cards are randomly arranged? The total number of possible arrangements is $52! = 8.066 \times 10^{67}$. Remember that each of these possible arrangements *is equally probable,* however, including the possibility of arriving back at our original ordered deck. We say that we have 52! *a priori equally probable arrangements*. We denote the number of these a priori arrangements by Ω. The ordered deck has $\Omega_1 = 1$, and the random deck has $\Omega_2 = 8.066 \times 10^{67}$. The probability of getting any specified arrangement after shuffling the ordered deck is $P = \Omega_1/\Omega_2 = (8.066 \times 10^{67})^{-1}$. The information content of the deck has decreased; its randomness has increased; its "entropy" has increased. What we are saying is, if we shuffle a deck in such a manner that the arrangement of cards becomes random, then the likelihood of *not* ending up with a perfectly ordered deck is 8.066×10^{67} times the likelihood of ending up with a deck that is perfectly ordered.

Suppose we now divide the deck into 2 piles, 1 pile containing the first 26 cards and the other the last 26. The number of a priori equally probable arrangements in each pile is $\Omega_1 = \Omega_2 = 26!$. If the number of cards in each pile were different, then Ω_1 and Ω_2 would be different. The number of equally probable arrangements of the cards in both piles is the product of the numbers for each pile separately, or

$$\Omega = \Omega_1 \Omega_2 \qquad (23.3)$$

This number is the total number of possible arrangements obtained by shuffling each pile separately and then placing one pile atop the other without intermingling cards from the two original piles.

Entropy, on the other hand, is an additive quantity. The entropy of an assembly made up of two independent component parts is the sum of the entropies of the component parts, $S = S_1 + S_2$. We must contrast the multiplicative property of Ω with the additive property of S. This suggests a relation of the form

$$S = \kappa \ln \Omega \qquad (23.4)$$

where κ is a proportionality constant that we shall shortly identify with k, the Boltzmann constant. Equation (23.4) was first obtained by Boltzmann. It connects the entropy of an assembly with the number of a priori equally probable distinguishable *arrangements, complexions,* or *micromolecular* states of the assembly.[2]

---

[1] The connection between entropy and information is discussed in L. Brillouin, *Science and Information Theory,* 2d ed. (New York: Academic Press, 1962).

[2] In our example of the deck of computer cards, if each card is different, then each possible arrangement is distinguishable, that is, different from every other possible arrangement. On the other hand, if two or more cards are identical, then there will be several different arrangements that cannot be differentiated from each other. The probabilities for this situation will be discussed later in the chapter. The Boltzmann distribution deals with *distinguishable* arrangements.

SEC. 23.3 ENTROPY AND A PRIORI PROBABILITY

Let us revert to ideal gas expansions for the moment. The entropy change for the isothermal expansion of one mole of gas from V_1 to a final volume of $V_1 + V_2$ is

$$\Delta S = S_{\text{fin}} - S_{\text{init}} = R \ln \frac{V_1 + V_2}{V_1}$$

$$= k \ln \left(\frac{V_1 + V_2}{V_1}\right)^{L_0} \qquad (23.5)$$

where k is the Boltzmann constant and L_0 is Avogadro's number. Since the probabilities are proportional to the volumes raised to the nth power, where n is the number of molecules, we have from Equation (23.4) that

$$\Delta S = S_{\text{fin}} - S_{\text{init}} = \kappa \ln \frac{\Omega_{\text{fin}}}{\Omega_{\text{init}}}$$

$$= \kappa \ln \left(\frac{V_1 + V_2}{V_1}\right)^{L_0} \qquad (23.6)$$

By comparing (23.6) with (23.5), we see that $\kappa = k$, the Boltzmann constant.

Equation (23.4) is the starting point for a serious discussion of thermodynamics from a mechanistic microscopic point of view. An evaluation of Ω as a pure number would be relatively uninteresting; we must go further than that if our approach is to have any applicability. Thermodynamics provides several relations between the various thermodynamic functions such as

$$T\, dS = dE + P\, dV - \mu\, dN \qquad (23.7)$$

From this we get

$$\left(\frac{\partial S}{\partial E}\right)_{V,N} = \frac{1}{T}; \quad \left(\frac{\partial S}{\partial V}\right)_{E,N} = \frac{P}{T}; \quad \left(\frac{\partial S}{\partial N}\right)_{E,V} = \frac{-\mu}{T} \qquad (23.8)$$

We therefore want to know Ω as a function of the mechanical variables of energy, volume, and mole numbers, or numbers of molecules:

$$\Omega = \Omega(E, V, N) \qquad (23.9)$$

In terms of Ω, the equations in (23.8) become

$$\left(\frac{\partial \ln \Omega}{\partial E}\right)_{V,N} = \frac{1}{kT} \qquad (23.10a)$$

$$\left(\frac{\partial \ln \Omega}{\partial V}\right)_{E,N} = \frac{P}{kT} \qquad (23.10b)$$

$$\left(\frac{\partial \ln \Omega}{\partial N}\right)_{E,V} = -\frac{\mu}{kT} \qquad (23.10c)$$

Our object is to evaluate Ω as a function of E, V, and N or of another set of independent variables, from which we can then determine other thermodynamic functions using relations like those in (23.10). Before we proceed on this tack we shall digress to consider two mathematical concepts that are important in the ensuing discussion—Stirling's approximation for large factorials and Lagrange's method of undetermined multipliers.

23.4 STIRLING'S APPROXIMATION FOR LARGE FACTORIALS

Talking about the statistics of large numbers of molecules involves a series of equations in which one of the variables is the factorial of a very large number. The computational difficulty can be substantially eased by working not with the number itself but with the logarithm of the number. A particularly useful approximation to the logarithm of the factorials of large numbers exists as *Stirling's approximation*.

A very important function in the theory of differential equations and the Laplace transform is the gamma function,[3]

$$\Gamma(n) = \int_0^\infty \exp(-x) x^{n-1}\, dx \qquad (n > 0) \qquad (23.11)$$

While $\Gamma(n)$ is defined for nonintegral as well as integral values of n, our concern will be with integral values. A related integral that is found in most tabulations is

$$\int_0^\infty \exp(-sx) x^n = \frac{n!}{s^{n+1}} \qquad (s > 0) \qquad (23.12)$$

If s is taken as unity in (23.12), then $s^{n+1} = 1$. Comparing (23.12) with (23.11) indicates that

$$\Gamma(n + 1) = n! \qquad (23.13)$$

(Note that for nonintegral values of n, Equation (23.13) effectively defines the factorial of a noninteger by the gamma function.) Since factorials can be written in terms of the gamma function, the various well-known expansions for the gamma function can be used to define the factorial. One of these due to Stirling states that

$$\Gamma(n) = (n - 1)!$$
$$= \exp(-n) n^{n-(1/2)} (2\pi)^{1/2}$$
$$\times \left(1 + \frac{1}{12n} + \frac{1}{288 n^2} - \frac{139}{51{,}840 n^3} - \frac{571}{2{,}488{,}320 n^4} + \cdots \right) \qquad (23.14)$$

In the limit of large n, we can omit all terms past the first and write

$$N! \simeq N^{(N+1/2)} \exp(-N) (2\pi)^{1/2} \qquad (23.15)$$

or
$$\ln N! \simeq (N + \tfrac{1}{2}) \ln N - N + \tfrac{1}{2} \ln 2\pi \qquad (23.16)$$

If N is sufficiently large, we can neglect the last term in (23.16) as well as the $\tfrac{1}{2}$ in the first term, to get the less exact expression

$$\ln N! \simeq N \ln N - N \qquad (23.17)$$

Table 23.2 lists values of $\ln N!$ between $N = 5$ and $N = 5000$ as computed by (23.16) and the less exact form of (23.17). These values are compared with the actual values of $\ln N!$. The error gets quite small as N gets large. For the very large values of N used in statistical mechanics, the less exact form is accurate enough, and we shall use (23.17) exclusively.

---

[3] Detailed discussions of the gamma function can be found in most textbooks on advanced calculus, analysis, advanced calculus for engineers, and the like. See, for example, *A Course of Modern Analysis* by E. T. Whittaker and G. N. Watson, 4th ed. (Cambridge: Cambridge University Press, 1962), p. 251.

TABLE 23.2 Logarithms of factorials compared with the approximate and more precise forms of Stirling's approximation.

(1) N	(2) $\ln N!$	(3) $N \ln N - N$	(4) $(N + \tfrac{1}{2}) \ln N - N + \tfrac{1}{2} \ln (2\pi)$
5	4.7875	3.0472 (0.36×10^0)[a]	4.7708 (0.54×10^{-2})[a]
10	15.1044	13.0259 $(.13 \times 10^0)$	15.0961 $(.63 \times 10^{-3})$
15	27.8993	25.6208 $(.81 \times 10^{-1})$	27.8937 $(.21 \times 10^{-3})$
20	42.3356	39.9146 $(.57 \times 10^{-1})$	42.3315 $(.10 \times 10^{-3})$
25	58.0036	55.4719 $(.43 \times 10^{-1})$	58.0003 $(.60 \times 10^{-4})$
50	148.4778	145.6012 $(.19 \times 10^{-1})$	148.4761 $(.11 \times 10^{-4})$
100	363.7394	360.5170 $(.88 \times 10^{-2})$	363.7385 $(.23 \times 10^{-5})$
500	2,611.3305	2,607.3040 $(.15 \times 10^{-2})$	2,611.3303 $(.63 \times 10^{-7})$
1000	5,912.1282	5,907.7553 $(.73 \times 10^{-3})$	5,912.1281 $(.14 \times 10^{-7})$
5000	37,591.1435	37,585.9660 (0.13×10^{-3})	37,591.1435 (0.30×10^{-10})

[a] The numbers in parentheses indicate the fractional difference between these approximate numbers and the entries in column (2).

23.5 LAGRANGE'S METHOD OF UNDETERMINED MULTIPLIERS

A common problem in many areas of science and engineering consists in finding the extremals, either as a minimum or maximum, of a function

$$f = f(x_1, x_2, x_3, \ldots, x_n) \tag{23.18}$$

of the n variables x_i. At the extremal, the variation of the function must vanish for small variations $\delta x_1, \delta x_2, \ldots, \delta x_n$ in the x_i,

$$\delta f = \frac{\partial f}{\partial x_1} \delta x_1 + \frac{\partial f}{\partial x_2} \delta x_2 + \cdots + \frac{\partial f}{\partial x_n} \delta x_n$$

$$= \sum_{i=1}^{n} \frac{\partial f}{\partial x_i} \delta x_i = 0 \tag{23.19}$$

If the variables x_i were a set of independent variables, then the δx_i would also be independent of one another, and the only way Equation (23.19) could be satisfied would be to require each coefficient $\delta f / \partial x_i$ in that equation to vanish. This situation is known as the "free" variation problem.

> JOSEPH LOUIS LAGRANGE (1736–1813), French-Italian mathematician and theoretical physicist, was recognized as one of the world's greatest mathematicians when he was 25 years old. Five years earlier he had developed the application of the calculus of variations to mechanics. In 1764 he applied what we now call the Lagrangian equations of motion to solving the problem of the librations of the moon, a work that won for him the prize offered by the Paris Academy of Sciences. In 1776, Lagrange succeeded to Euler's position in Berlin at the recommendation of Euler himself. Frederick the Great, the greatest king in Europe, wished to have at his court the greatest mathematician of Europe. At Frederick's death, Lagrange returned to Paris at the invitation of Louis XVI and spent the rest of his life there; he managed to survive the French Revolution while many of his dearest friends, Lavoisier among them, were losing their heads. His most famous work was the book *Mécanique Analytique*, which Hamilton described as a "scientific poem."

On the other hand, one "auxiliary" condition or several that can be expressed as

$$g(x_1, x_2, x_3, \ldots, x_n) = 0 \tag{23.20}$$

can be imposed upon the set of variables x_i. While (23.19) is still true, it is no longer true that all the coefficients $\partial f/\partial x_i$ are zero. This is so because the x_i and hence the δx_i are no longer independent. If we knew the values of $n - 1$ of them, the nth one could be determined from (23.20). We now want to investigate the variation of f subject to the constraint $g(x_1, \ldots, x_n) = 0$.

One possibility would be eliminating one of the x_i, say x_n, by using (23.20) to express it in terms of the remaining x_i. This would reduce the number of variables by one, and we could treat the new function $f'(x_1, x_2, \ldots, x_{n-1})$ as a free variation problem. This procedure would be advisable in many instances; on the other hand, it can often be complicated. Further, there are often cogent reasons for retaining all the x_i. Lagrange devised the *method of undetermined multipliers* as a procedure to reduce the problem to one of free variations while preserving the complete set of variables x_i. Lagrange's method is general and works with any number of auxiliary conditions less than n, the number of variables. We shall first look at the case of the single auxiliary condition given in Equation (23.20).

We start by writing the variation of g as

$$\delta g = \frac{\partial g}{\partial x_1} \delta x_1 + \frac{\partial g}{\partial x_2} \delta x_2 + \cdots + \frac{\partial g}{\partial x_n} \delta x_n$$

$$= \sum_{i=1}^{n} \frac{\partial g}{\partial x_i} \delta x_i = 0 \tag{23.21}$$

We could eliminate δx_n in terms of the other δx_i as previously noted, but suppose we first multiply Equation (23.21) by some undetermined factor λ, which is some function of the x_i. Since the result of that multiplication will still be zero, we can add that result to δf in (23.19), to get

$$\sum_{i=1}^{n} \left(\frac{\partial f}{\partial x_i} + \lambda \frac{\partial g}{\partial x_i} \right) \delta x_i = 0 \tag{23.22}$$

Note that the value of λ is yet undetermined; hence the name Lagrange's method of undetermined multipliers. We can now eliminate δx_n by choosing λ so that the factor multiplying δx_n in Equation (23.22) vanishes:

$$\frac{\partial f}{\partial x_n} + \lambda \frac{\partial g}{\partial x_n} = 0 \tag{23.23}$$

This reduces the sum in (23.22) to $n - 1$ terms,

$$\sum_{i=1}^{n-1} \left(\frac{\partial f}{\partial x_i} + \lambda \frac{\partial g}{\partial x_i} \right) \delta x_i = 0 \tag{23.24}$$

since only those $n - 1$ values of δx_i that can be varied independently remain. The problem has now been reduced to the free variation problem with $n - 1$ variables, and the coefficient of each δx_i in Equation (23.24) must vanish separately:

$$\frac{\partial f}{\partial x_i} + \lambda \frac{\partial g}{\partial x_i} = 0 \qquad (i = 1, 2, 3, \ldots, n - 1) \tag{23.25}$$

The value λ can be found from Equation (23.23) as

$$\lambda = -\frac{\partial f/\partial x_n}{\partial g/\partial x_n} \qquad (23.26)$$

and (23.25) is now valid for all values of i.

The procedure we have just gone through is equivalent to the following. Instead of setting the variation of f equal to zero, we consider the vanishing of

$$\delta f + \lambda\, \delta g = 0 = \delta(f + \lambda g) \qquad (23.27)$$

since $g\,\delta\lambda$ vanishes by the auxiliary condition of (23.20). Instead of considering the function f, we use the function

$$f' = f + \lambda g \qquad (23.28)$$

We then set $\delta f' = 0$ for arbitrary variations δx_i.

This procedure can be generalized to include more than one auxiliary condition:

$$\begin{aligned} g_1(x_1, x_2, \ldots, x_n) &= 0 \\ g_2(x_1, x_2, \ldots, x_n) &= 0 \\ &\vdots \\ g_m(x_1, x_2, \ldots, x_n) &= 0 \end{aligned} \qquad (23.29)$$

so long as $m < n$. This requires introducing m undetermined multipliers $\lambda_1, \lambda_2, \ldots, \lambda_m$ by a procedure like that given above.[4] In place of (23.24), we get

$$\sum_{i=1}^{n-m} \left(\frac{\partial f}{\partial g_i} + \lambda_1 \frac{\partial g_1}{\partial x_i} + \cdots + \lambda_m \frac{\partial g_m}{\partial x_i}\right) \delta x_i = 0 \qquad (23.30)$$

in which the $n - m$ δx_i's are independent variables.

This method of Lagrange changes a problem that has $n - m$ degrees of freedom into one with $n + m$ degrees of freedom. We have the n variables x_i plus the m variables λ_j. We thus have a problem with several extraneous coordinates. This is a great convenience in many applications. The full symmetry of the original coordinates can be preserved, and it is unnecessary to differentiate between dependent and independent variables. When we consider the Boltzmann distribution, we shall use this method for two auxiliary conditions.

23.6 SOME GROUND RULES

Our problem now is to determine $\Omega(E, V, N)$. Since we shall be dealing with the factorials of large numbers, we shall find Stirling's approximation useful in this task. We shall be interested in equilibrium states in which the total energy, volume, and mole numbers are constant. The constancy of these three values establishes two constraints, or auxiliary conditions, on our result; hence the Lagrange method of undetermined multipliers will also prove useful. (There are two and not three constraints,

---

[4] More complete discussions of the Lagrange method of undetermined multipliers can be found in textbooks that treat the calculus of variations. See, for example, R. Courant, *Differential and Integral Calculus*, vol. 2, trans. E. J. McShane (London: Blackie & Sons Ltd., 1936), p. 190.

since one of the three variables of E, V, and N can be eliminated by a suitable equation of state.)

Historically, statistical mechanics developed from considerations of the classical mechanics of particles. The mechanical properties of an assembly of molecules was established in terms of the classical variables of velocity, momentum, kinetic energy, and the like. These are all continuously variable and not limited to any particular values. This approach might be termed "classical" statistical mechanics. The development of quantum mechanics in the 1920s and 1930s was accompanied by the development of "quantum mechanical" statistical mechanics, which was based on the principles of the new quantum mechanics. Since our subsequent applications of statistical mechanics will deal with quantum mechanical systems, we shall use the quantum mechanical approach. Our discussion of quantum mechanics begins in Chapter 25, but we shall put matters forward somewhat. In this and the next chapter, which deal with statistical mechanics, we shall need two simple results of quantum mechanics.

The first of these concerns the quantization of energy and should be familiar from elementary courses in chemistry and physics. The energies of small particles such as atoms or molecules are no longer continuously variable but limited to a specified set of discrete *energy levels*, as shown in Figure 23.2. The energies of the levels are ϵ_1, ϵ_2, ϵ_3, ... Only these *allowed* energy levels are permissible for any molecule. The separation between these levels is for the moment irrelevant. A basic distinction between classical and quantum mechanics concerns the way in which a system is described. Classical mechanics defines the properties of a particle in terms of position (x, y, z) and momenta (p_x, p_y, p_z) and predicts the position of a particle in space as a function of time. Quantum mechanics, on the other hand, describes a particle in terms of a *wave function*,

$$\Psi = \psi(x, y, z) \, \phi(t) \tag{23.31}$$

from which one calculates not the precise position but the *probability* of finding a particle at a particular position in space. Each state of Ψ gives rise to its own particular energy ϵ. In many cases more than one state of Ψ gives rise to the same energy, in which case we speak of that energy level as *degenerate*. The degeneracy, or number of different states with the same energy ϵ_j, is denoted by g_j. The second quantum mechanical result we shall require concerns the separation between energy levels for

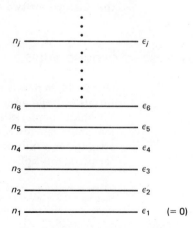

FIGURE 23.2 The allowed energy levels of a molecule ϵ_j. Since it is the energy differences between the levels that are important, it is customary to set the energy of the lowest level ϵ_1 equal to zero. The numbers n_j to the left indicate the number of molecules in a large collection that have energy ϵ_j at equilibrium. Our object is to calculate these n_j.

the so-called *particle in the box*. (We shall defer that until it is required in Chapter 24.)

What we have then is a collection, or *ensemble*, of a very large number of atoms or molecules. The energy of each molecule is determined according to the laws of quantum mechanics (subsequently to be discussed). The *average* energy of the ensemble is identified with the thermodynamic energy of the macroscopic collection of molecules. The object is to determine the macroscopic properties from the microscopic properties of the individual molecules. For a thermodynamic system, the energy is the sum of the energies of the individual molecules.

There are many conceivably different ways in which a fixed amount of energy can be divided among a number of molecules, just as there are many ways to achieve the sum of 100 (50 + 50, 76 + 24, 19 + 32 + 49, and so on). We call each of these possible distributions a *micromolecular state*. Remember that we have defined Ω as the number of equally probable arrangements or distributions. We take as our fundamental assumption this statement:

> *Every conceivably different micromolecular state of the collection that corresponds to the same value of E, V, and N is equally probable.*

The number of such different micromolecular states is $\Omega(E, V, N)$ and we shall compute it.

23.7 THE BOLTZMANN DISTRIBUTION LAW

We have an ensemble of N identical molecules; the volume V and energy E are both constant. Each molecule has the same set of allowed energy levels $\epsilon_1, \epsilon_2, \ldots, \epsilon_j, \ldots$. The number of molecules in each allowed level is denoted by $n_1, n_2, \ldots, n_j, \ldots$. We now inquire how many ways the different molecules can be distributed among the various energy levels subject to the constraints of constant total energy and constant number of particles,

$$\sum_j n_j \epsilon_j = E \tag{23.32}$$

$$\sum_j n_j = N \tag{23.33}$$

The constraint of constant volume is included in Equations (23.32) and (23.33), since E, N, and V are related through an equation of state.

The total number of possible arrangements of N molecules is $N!$. Suppose we consider one particular distribution that has n_1 molecules in level ϵ_1, n_2 molecules in level ϵ_2, and so on. Since we do not achieve a new distribution by permuting the molecules within any individual level, the total number of ways in which this particular distribution can be achieved is just $N!$ divided by the number of permutations of the individual molecules within each level. This is $n_1! n_2! n_3! \cdots n_j! \cdots$. The number is then

$$\Omega_k = \frac{N!}{n_1! n_2! n_3! \cdots n_j! \cdots} = \frac{N!}{\Pi_j n_j!} \tag{23.34}$$

where we use the subscript k to distinguish this particular distribution from the large number of different possible distributions. The symbol Π_j is the multiplication analog of the symbol Σ_j. It indicates that all the terms in the series are to be multi-

TABLE 23.3 Three possible distributions of four molecules among four different energy levels. (The four molecules are denoted by a, b, c, and d.)

Distribution 1 $n_1 = 2; n_2 = 1;$ $n_3 = 0; n_4 = 1$ $\Omega_1 = \dfrac{4!}{2!1!0!1!}$ $= 12$				Distribution 2 $n_1 = 2; n_2 = 0;$ $n_3 = 2; n_4 = 0$ $\Omega_2 = \dfrac{4!}{2!0!2!0!}$ $= 6$				Distribution 3 $n_1 = 3; n_2 = 1;$ $n_3 = 0; n_4 = 0$ $\Omega_3 = \dfrac{4!}{3!1!0!0!}$ $= 4$			
ϵ_1	ϵ_2	ϵ_3	ϵ_4	ϵ_1	ϵ_2	ϵ_3	ϵ_4	ϵ_1	ϵ_2	ϵ_3	ϵ_4
ab[a]	c		d	ab		cd		abc	d		
ab	d		c	ac		bd		bcd	a		
ac	b		d	ad		bc		cda	b		
ac	d		b	bc		ad		dab	c		
ad	b		c	bd		ac					
ad	c		b	cd		ab					
bc	a		d								
bc	d		a								
bd	a		c								
bd	c		a								
cd	a		b								
cd	b		a								

[a] Different arrangements of the atoms within each state do not count as separate distributions.

plied by one another. The total number of distinguishable states is the sum over all the possible distributions:[5]

$$\Omega = \sum_k \Omega_k = \sum_{\substack{\text{over all} \\ \text{the} \\ \text{possible} \\ \text{sets of} \\ \text{values of} \\ \text{the } n_j}} \frac{N!}{\prod_j n_j!} \tag{23.35}$$

An example of this calculation for four molecules distributed among four energy levels is shown in Table 23.3.

A basic assumption (stated in the previous section) is that every possible distribution in (23.34) is equally probable. We now introduce a second vital assumption:

> When N is very large, only the largest term in the sum given in (23.35) will make any effective contribution to S.

This assumption is of a different nature from the first. The first assumption dealt with the physical properties of the collection of molecules. This second assumption is purely mathematical and will be justified later. Now we must find the maximum possible value of Ω_k in Equation (23.34).

We first note that Ω_k is a function of the n_i,

$$\Omega_k = \Omega_k(n_1, n_2, \ldots, n_j, \ldots) \tag{23.36}$$

The maximum value can be found by using Lagrange's method of undetermined multipliers, as previously outlined. At the extremal, the variation of Ω_k must be zero, or

---

[5] This should not be confused with the expression of (23.3), which applies to the probability of two separate and independent systems.

SEC. 23.7 THE BOLTZMANN DISTRIBUTION LAW

$$\delta\Omega = \frac{\partial\Omega}{\partial n_1}\delta n_1 + \frac{\partial\Omega}{\partial n_2}\delta n_2 + \cdots + \frac{\partial\Omega}{\partial n_j}\delta n_j + \cdots = 0$$

$$= \sum_j \frac{\partial\Omega}{\partial n_j}\delta n_j \qquad (23.37)$$

where we have dropped the subscript k for the time being.

The two constraints are given in (23.32) and (23.33). Noting that $(\partial/\partial n_j)(n_j\epsilon_j) = \epsilon_j$ and $(\partial/\partial n_j)(n_j) = 1$, we have

$$\delta N = 0 = \delta n_1 + \delta n_2 + \cdots = \sum_j \delta n_j \qquad (23.38)$$

$$\delta E = 0 = \epsilon_1\,\delta n_1 + \epsilon_2\,\delta n_2 + \cdots = \sum_j \epsilon_j \delta n_j \qquad (23.39)$$

We now multiply (23.38) by the arbitrary multiplier α and (23.39) by β; these are then added to (23.37), to get

$$\delta\Omega = 0 = \sum_j \left(\frac{\partial\Omega}{\partial n_j} + \alpha + \beta\epsilon_j\right)\delta n_j \qquad (23.40)$$

By following the procedure outlined previously, we can select α and β so that each of the nth and $(n-1)$th terms of (23.40) is identically zero. This leads to the series of equations

$$\frac{\partial\Omega}{\partial n_j} + \alpha + \beta\epsilon_j = 0 \qquad (j = 1, 2, 3, \ldots) \qquad (23.41)$$

It will simplify matters greatly if we use the natural logarithm of Ω instead of Ω itself. In this case,

$$\ln\Omega = \ln N! - \ln n_1! - \ln n_2! - \cdots$$

$$= \ln N! - \sum_j \ln n_j!$$

$$= \text{constant} - \sum_j (n_j \ln n_j - n_j) \qquad (23.42)$$

In the last step, we use Stirling's approximation and recognize that $\ln N!$ is a constant. Since $\ln\Omega$ is a monotonically increasing function of Ω, we can equally well maximize $\ln\Omega$. The relevant equations are

$$\frac{\partial \ln\Omega}{\partial n_j} + \alpha + \beta\epsilon_j = 0 \qquad (j = 1, 2, 3, \ldots) \qquad (23.43)$$

From (23.42), we see that

$$\frac{\partial \ln\Omega}{\partial n_j} = -\frac{\partial}{\partial n_j}(n_j \ln n_j - n_j)$$

$$= -\ln n_j - 1 + 1$$

$$= -\ln n_j \qquad (23.44)$$

and (23.43) takes the form

$$-\ln n_j + \alpha + \beta\epsilon_j = 0 \qquad (23.45)$$

The n_j are now

$$n_j = \exp(\alpha)\exp(\beta\epsilon_j)$$
$$= A\exp(\beta\epsilon_j) \qquad (23.46)$$

where $A = \exp(\alpha)$ is a constant. The constant A can be evaluated from the condition that $\Sigma_j n_j = N$, or

$$\exp(\alpha)\sum_j \exp(\beta\epsilon_j) = N \qquad (23.47)$$

The number of molecules in each energy level at equilibrium is now

$$\frac{n_j}{N} = \frac{\exp(\beta\epsilon_j)}{\Sigma_j \exp(\beta\epsilon_j)} \qquad (23.48)$$

Evaluating the constant β is more complicated. This can be accomplished by connecting Ω with the thermodynamic properties of the system.

23.8 CONNECTION WITH ENTROPY; THE EVALUATION OF β

Starting with Boltzmann's expression for the entropy, and using (23.42) for $\ln \Omega$, we can write

$$S = k\ln\Omega$$
$$= k\ln N! - k\sum_j \ln n_j!$$
$$= k[N\ln N - N - \sum_j(n_j \ln n_j - n_j)] \qquad (23.49)$$

Since $N = \Sigma n_j$, this becomes

$$S = k(N\ln N - \sum_j n_j \ln n_j) \qquad (23.50)$$

Using the equilibrium values of the n_j given in (23.46), we can write

$$S = k\left[N\ln N - \sum_j n_j(\alpha + \beta\epsilon_j)\right]$$
$$= k(N\ln N - \alpha N - \beta E) \qquad (23.51)$$

since $\Sigma n_j = N$ and $\Sigma n_j \epsilon_j = E$. In addition we have

$$\alpha = \ln N - \ln \Sigma \exp(\beta\epsilon_j) \qquad (23.52)$$

by virtue of Equation (23.47). Now we can simplify (23.51) to read

$$S = k[N\ln \Sigma \exp(\beta\epsilon_j) - \beta E] \qquad (23.53)$$

Suppose we now compute $(\partial S/\partial E)_{V,N}$. Since S depends on β and β itself is a function of E, we must use the chain rule for differentiation,

$$\frac{\partial S}{\partial E} = \frac{\partial S}{\partial \beta}\frac{\partial \beta}{\partial E}$$

getting for the derivative

$$\frac{\partial S}{\partial E} = -k\beta + k\frac{\partial}{\partial \beta}\left[N\ln \sum_j \exp(\beta\epsilon_j) - \beta E\right]\frac{\partial \beta}{\partial E} \qquad (23.54)$$

SEC. 23.8 CONNECTION WITH ENTROPY; THE EVALUATION OF β

We need not concern ourselves with $\partial \beta / \partial E$, since the coefficient of that derivative is zero,

$$\frac{\partial}{\partial \beta}\left[N \ln \sum_j \exp(\beta \epsilon_j) - \beta E\right] = N \frac{\Sigma_j \epsilon_j \exp(\beta \epsilon_j)}{\Sigma_j \exp(\beta \epsilon_j)} - E$$

$$= N \frac{\Sigma_j n_j \epsilon_j}{\Sigma_j n_j} - E = 0 \qquad (23.55)$$

Note that the second line of (23.55) was obtained by multiplying numerator and denominator by $\exp(\alpha)$. Thus

$$\left(\frac{\partial S}{\partial E}\right)_{V,N} = -k\beta \qquad (23.56)$$

On the other hand, we know from our work on thermodynamics that

$$\left(\frac{\partial S}{\partial E}\right)_{V,N} = \frac{1}{T} \qquad (23.57)$$

If we now compare (23.56) with (23.57), we have

$$\beta = -\frac{1}{kT} \qquad (23.58)$$

Equation (23.53) for the entropy can be written

$$S = kN \ln \sum_j \exp\left(\frac{-\epsilon_j}{kT}\right) + \frac{E}{T} \qquad (23.59)$$

Equation (23.48) for the Boltzmann distribution can now be written in the more familiar form

$$\frac{n_j}{N} = \frac{\exp(-\epsilon_j/kT)}{\Sigma_j \exp(-\epsilon_j/kT)} \qquad (23.60)$$

The summation that appears in the denominator of Equation (23.60),

$$z = \sum_j \exp\left(\frac{-\epsilon_j}{kT}\right) \qquad (23.61)$$

is especially important; it is called the *partition function (molecular)*. The symbol z is from the German *Zustandsumme*, or "*sum over states.*"

The Boltzmann distribution law is often used to compare the numbers of molecules in two different states. If we have n_0 molecules in energy level ϵ_0 and n_1 molecules in ϵ_1, then from (23.60),

$$\frac{n_1}{n_0} = \exp\left[-\frac{(\epsilon_1 - \epsilon_0)}{kT}\right] \qquad (23.62)$$

In the most usual applications of this formula, the number of molecules in some upper level is compared with the number in the lowest energy level, ϵ_0. A slight extension to the above theory must be made in the event of *degeneracy*, that is, the condition that exists when more than one state of the molecule occupies any given energy level. If there is degeneracy, each level is assigned a *statistical weight* g_j, which equals the number of states at the particular energy ϵ_j. When degeneracy is in-

TABLE 23.4 Boltzmann distributions for a system containing 10,000 molecules for three different degeneracy conditions.

Number of level (j)	$\dfrac{\epsilon_j}{kT}$	$\exp\left(\dfrac{-\epsilon_j}{kT}\right)$	Number of molecules in each level for $N = 10{,}000$ (n_j)		
			$g_j = 1$	$g_j = j$	$g_j = j^2$
1	0	1.0000	6321	3996	1847
2	1	0.3679	2325	2940	2717
3	2	0.1353	855	1622	2249
4	3	0.0498	315	796	1471
5	4	0.0183	116	366	846
6	5	0.0067	43	162	448
7	6	0.0025	16	69	224
8	7	0.0009	6	29	108
9	8	0.0003	2	12	50
10	9	0.000 123	1	5	23
11	10	0.000 045	0	2	10
12	11	0.000 017	0	1	4
13	12	0.000 006	0	0	2
14	13	0.000 002	0	0	1
15	14	0.000 001	0	0	0
$z = \sum g_j \exp\left(\dfrac{-\epsilon_j}{kT}\right)$			1.58198	2.50265	5.41562

NOTE: The energy levels are uniformly spaced with a separation equal to kT.

cluded, the partition function takes the form

$$z = \sum_j g_j \exp\left(-\frac{\epsilon_j}{kT}\right) \tag{23.63}$$

The Boltzmann distribution law of (23.60) becomes

$$\frac{n_j}{N} = \frac{g_j \exp(-\epsilon_j/kt)}{\sum_j g_j \exp(-\epsilon_j/kT)} = \frac{g_j \exp(-\epsilon_j/kT)}{z} \tag{23.64}$$

Table 23.4 indicates the distribution of molecules in a collection of 10,000 molecules among a set of equally spaced energy levels for nondegenerate as well as degenerate situations.

23.9 ADDITIONAL REMARKS

Before we go on to some simple applications of statistical mechanics a few caveats are in order. The Boltzmann distribution we have just derived does not apply to all particles. It applies only to those particles that we call *boltzons,* and not to particles such as *fermions* and *bosons.* A critical distinction between boltzons on the one hand and fermions and bosons on the other lies in the interactions between the particles. The boltzon can be likened to the molecules of an ideal gas—in the ideal gas there are no interactions between molecules; the molecules act completely independently. Fermions and bosons, however, do interact. A complete understanding of the differences between these particles is not possible without knowing the principles of quantum mechanics, but a few qualitative remarks now may help to clarify the distinction.

SEC. 23.9 ADDITIONAL REMARKS

When we discuss the free-electron theory of metals (Chapter 37), we shall note that a conducting metal such as sodium contains a collection of valence electrons that are free to move about the metal much as the molecules of an ideal gas are free to move about a gas container. On the other hand, there are fundamental differences between the two systems. In the metal it is impossible to distinguish any one electron from another electron. In a sense, a crystal of sodium that contains a mole of sodium atoms can be considered a single gigantic molecule of sodium metal that contains 6×10^{23} sodium atoms and 6×10^{23} free electrons. These electrons are indistinguishable, and they interact strongly. The energy level occupied by any one electron is affected by the energy levels occupied by other electrons. A single wave function in terms of all the electrons can be written for the entire sodium crystal. Equation (23.34) is based on the fundamental fact that the individual particles are *distinguishable*. This is not the case for fermions, nor is it the case for bosons, and when we discuss the Fermi-Dirac statistics of fermions and the Bose-Einstein statistics of bosons, we shall see how this equation must be altered.

An ideal gas molecule, on the other hand, is a typical example of a boltzon, for which the theory of this chapter applies. The wave functions of the individual boltzons are separately describable; the energy of any particular boltzon is independent of the energy of the other boltzons, or at least "almost" independent of the other energies. We say *almost* because of the constraint that the sum of all the energies must be a constant E. The particles are completely noninteracting except for that slight amount required for equilibration. Except for this one condition, neither the energy nor any other property of a particular boltzon depends on these characteristics of any other boltzon.

Our discussion of the fundamental assumptions of statistical mechanics was far from complete for practical reasons, and the student who is interested in a more complete presentation is referred to the textbooks on statistical mechanics listed in the Bibliography. There is, however, one point that we left hanging, which should be cleared up. Critical to the entire development was the assumption that we could substitute for the sum in Equation (23.35), which defined Ω that term that had the maximum value, all other terms having a negligible contribution. Although we shall not justify this assumption rigorously, we can at least indicate the reasonableness of the assumption.

It first seems that this assumption implies that of all the terms in the sum that make up Ω, only the term that is the largest is substantial, all the others being negligibly small. This implication is incorrect. The assumption states not that the other terms are negligibly small but that they make a negligibly small contribution to S, which is proportional to $\ln \Omega$; this is an entirely different matter. We are dealing with very large numbers, of the order of $N = 6 \times 10^{23}$, for which $\ln (N!) = 3.23 \times 10^{25}$ by Stirling's formula; this number is then the effective contribution of that one maximum term to $\ln \Omega$. Suppose we had two terms, each of approximately equal size. In that case, the value of Ω would be twice as large, but the value of $\ln \Omega$ would be larger by only $\ln 2$, and $\ln 2$ is negligibly small compared with 3.23×10^{25}. In short, for very large values of N, it is true that $\ln (N!)$ does not differ appreciably from $\ln [m(N!)]$ even if m is as large as N.

At this point it is instructive to refer to Figure 20.6, which shows the envelopes of the binomial distribution. Notice how the width of the distribution compared with N gets smaller as N gets larger. Notice, too, that in the neighborhood of the peak there are still substantial contributions from adjacent points. A similar plot for the distribution under discussion is shown in Figure 23.3. Here we have plotted Ω_k vs k,

FIGURE 23.3 Plot of Ω_k vs k, where each k corresponds to a particular set of $n_1, n_2, \ldots, n_j, \ldots$. The different values of k are discrete, but the separation between successive values is so small relative to the total number of possible sets of the n_j that the functional behavior of Ω_k can be considered continuous for all practical purposes.

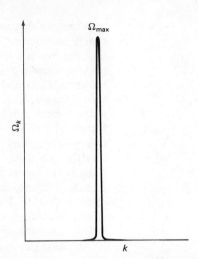

where each k corresponds to a particular set of $n_1, n_2, \ldots, n_j, \ldots$. The value Ω_k is a function of k. In the neighborhood of the maximum $\Omega_{\max}$, there will be several values of Ω_k that while they may be smaller than $\Omega_{\max}$ are not very much smaller. They still make a negligible contribution to S. Suppose there are m different values of k for which Ω_k is not negligibly small, and suppose they are all as large as $\Omega_{\max}$. We have

$$\Omega = \Sigma \, \Omega_k = m \, \Omega_{\max}$$
$$S = k \ln \Omega = k[\ln \Omega_{\max} + \ln m] \tag{23.65}$$

As we have just noted above, for very large values of $\Omega_{\max}$, the value $\ln m$ is negligibly small compared with $\ln \Omega_{\max}$ and no substantial error is introduced by using only the single term $\Omega_{\max}$ in the summation. Note that this assumption is purely mathematical, based on the properties of the logarithms of large numbers. A particular example of this is indicated in Problem 11.

PROBLEMS

1. For a game of 5-card poker, calculate the probability of being dealt a royal flush, 4 of a kind, a flush, 3 of a kind. Suppose you have been dealt 4 hearts and 1 other suit. What is the probability of drawing 1 card and filling the flush? What about the probability of filling a flush by drawing 2 cards to 3 of 1 suit?

2. Suppose you are in a class with 30 students. What is the probability that 2 students have the same birthday? What if the class had 40 students? (*Note:* This problem is more tractable if you compute the probability that no two students have the same birthday.)

3. The logarithm of $N!$ can be written $\ln N! = \Sigma_{m=1}^{N} \ln m$. Derive Stirling's approximation by noting that in the limit of large N, the summation can be approximated by a definite integral.

4. How large does N have to be for the fractional difference in $\ln N!$ as calculated by (23.16) and as calculated by (23.17) to be less than 10^{-6}? What is the fractional difference for $N = 6 \times 10^{23}$?

5. If N molecules are divided among five energy levels so that there are equal numbers of molecules in each level, what is the total number of arrangements for each of $N = 10, 100, 1000, 10^6, 6 \times 10^{23}$? In each case, what happens if one molecule is taken out of one level and added to a different level?

6. Since $S = k \ln \Omega$, the entropy change in going from an initial state to a final state is $\Delta S = S_2 - S_1 = k \ln (\Omega_2/\Omega_1)$. If S_1 is the entropy of the equilibrium state S_{eq}, and S_2 is a nonequilibrium state, then the relative probability of observing a chance fluctuation in a system in equilibrium for

which the entropy decreases by ΔS would be given by $(\Omega/\Omega_{eq}) = \exp(-\Delta S/k)$. The standard third law entropy of hydrogen at 298 K is 130.6 J K^{-1} mol^{-1}. What is the probability of observing a spontaneous decrease in entropy in 1 mol hydrogen that is one-millionth of this amount? one-billionth?

7. If the number $(6 \times 10^{23})!$ were to be written out in full, how many zeros would be required? If the number were to be published in a book in which each page had 50 lines, and each line had room for 80 zeros, how many pages would the book contain?

8. A system has a set of equally spaced energy levels. The spacing between levels is 1.3806×10^{-22} J. Calculate the relative numbers of molecules in the first three levels at $T = 1, 10, 100,$ and 1000 K. For the limit of infinitely high temperature, what are the relative numbers of molecules in these lowest three levels?

9. For a mole of molecules with the energy-level spacing of Problem 8, calculate the number of molecules in the lowest level for $T = 1, 10, 100,$ and 1000 K.

10. Do Problem 8 for the case in which the degeneracy of the jth level equals j^2.

11. Consider, as a specific example that no substantial error is introduced when only the maximum Ω_k is selected out of the set, the case of N atoms distributed between two energy states A and B of the same energy. The energy-level diagram thus contains just one energy level that is doubly degenerate. Since the first molecule can be placed in either of the two states, and all the other molecules can likewise be placed in either of the two states, the total number of arrangements is $\Omega = 2^N$. Note that we can get this same result from

$$\Omega = \sum_{n=0}^{N} \frac{N!}{n!(N-n)!} = 2^N$$

The sum is a well-known summation of binomial coefficients. Now arrive at this same result in the following way. Write out the general term $\ln \Omega_k$. Find the maximum term Ω_{max} by differentiating with respect to n and setting the result equal to zero. (*Note:* The maximum occurs for $n = \frac{1}{2}N$). Now show that you get the same result using only this term in the Stirling approximation representation $\ln \Omega_{max} = N \ln N - n \ln n - (N-n) \ln (N-n)$. That the two should give the same result is perhaps surprising. The reason is that several second-order terms have been dropped in the Stirling approximation, and these second-order terms are negligible. For large N, the difference between the two procedures is just this negligible quantity.

CHAPTER TWENTY-FOUR

APPLICATIONS OF STATISTICAL MECHANICS

24.1 ENTROPY, ENERGY, PARTITION FUNCTIONS, AND THERMODYNAMICS

A particularly important result of Chapter 23 was that the entropy could be written in the form

$$S = kN \ln z + \frac{E}{T} \qquad (24.1)$$

where z is the partition function or sum over all states,

$$z = \sum_j g_j \exp\left(-\frac{\epsilon_j}{kT}\right) \qquad (24.2)$$

In this last equation, g_j is the degeneracy of the jth energy level; g_j is unity for nondegenerate systems. It would perhaps be more correct to term z the *molecular* partition function, since it applies to a single molecule. We shall, however, omit the adjective *molecular*. Our object now is to use (24.1) as the basis from which we can derive expressions for the thermodynamic properties of a macroscopic system. This we shall do by using the various relations among the thermodynamic functions that we derived earlier. A few words on the functional dependence of S are in order.

The entropy is a function of E, V, and N,[1]

$$S = S(E, V, N) \qquad (24.3)$$

But these three variables are not a set of *independent* variables. If any two of them are selected, the third is determined. The dependence of S on E and N is clearly apparent, since both of these variables appear explicitly in (24.1). The dependence on V will become clearer when in the next section we discuss the energy levels of the particle in the box. There we shall note that the energy levels ϵ_j of a molecule in a box depend on the volume of the box. The partition function, on the other hand, since it is a molecular function, is independent of N; it is a function only of E and V:

$$z = z(E, V) \qquad (24.4)$$

---

[1] Since E is a function of T, we could equally well write $S = S(T, V, N)$. It is, however, more convenient to consider S a function of the extensive variables E, V, and N rather than to introduce the intensive variable T.

Thus the energy depends on z. It will be convenient to find this relation and then to substitute that expression for E into (24.1). While there are many ways to derive this relation, we shall do it through the Helmholtz energy A.

The Helmholtz energy is defined by

$$A = E - TS \tag{24.5}$$

If we use (24.1) for S, the Helmholtz energy takes the particularly simple form

$$A = -NkT \ln z \tag{24.6}$$

The energy can now be determined from A using the Gibbs-Helmholtz relation,

$$E = A - T\left(\frac{\partial A}{\partial T}\right)_{V,N} = -T^2 \frac{\partial}{\partial T}\left(\frac{A}{T}\right)_{V,N}$$

$$= NkT^2 \left(\frac{\partial \ln z}{\partial T}\right)_{V,N} \tag{24.7}$$

As we shall see in the next section, the ϵ's depend on the volume of the assembly of molecules; hence we can express A in terms of T, V, and N and not E, V, and N. That the former is the most appropriate set of variables for A can be seen from the thermodynamic expression

$$dA = -S\,dT - P\,dV + \Sigma\,\mu\,dN \tag{24.8}$$

Now we can use the expression for E as given in (24.7) to write the entropy in a form that depends on T and z (and N) alone. Or we can accomplish the same thing by noting that Equation (24.8) requires that

$$S = -\left(\frac{\partial A}{\partial T}\right)_{V,N} = NkT\left(\frac{\partial \ln z}{\partial T}\right)_{V,N} + Nk \ln z \tag{24.9}$$

Additional thermodynamic functions can be obtained from

$$P = -\left(\frac{\partial A}{\partial V}\right)_{T,N} = NkT\left(\frac{\partial \ln z}{\partial V}\right)_{T,N} \tag{24.10}$$

and

$$\mu_j = \left(\frac{\partial A}{\partial N_j}\right)_{T,V,N_{k \neq j}} = -NkT\left(\frac{\partial \ln z}{\partial N_j}\right)_{T,V,N_{k \neq j}} \tag{24.11}$$

We can thus derive the set of thermodynamic functions S, E, A, P, and μ from the partition function $z(N, V, T)$.[2] Other functions can be derived from the thermodynamic relations. The heat capacity, for example, is

$$C_v = \left(\frac{\partial E}{\partial T}\right)_{V,N} = \frac{\partial}{\partial T}\left(kT^2 \frac{\partial \ln z}{\partial T}\right)_{V,N}$$

$$= \frac{Nk}{T^2}\left(\frac{\partial^2 \ln z}{\partial (1/T)^2}\right)_{V,N} \tag{24.12}$$

---

[2] The partition function as a function of N, V, and T is known as the *canonical* partition function since the collection or ensemble of molecules defined in terms of these values is known as the *canonical ensemble*. When the values used are N, V, and E, the ensemble is called the *microcanonical ensemble*. Additional canonical ensembles are defined in terms of other sets of variables. For further information, see any of the advanced textbooks on statistical mechanics listed in the Bibliography.

Our course is now clear. We need an analytical expression for z. Once we have that, we can derive the thermodynamic functions from the equations we have just written.

Suppose we consider as a simple example the energy-level diagram shown in Figure 23.2, where the spacing between successive levels is the same, and the system is nondegenerate. We let ϵ be the spacing between levels, in which case $\epsilon_1 = \epsilon$, $\epsilon_2 = 2\epsilon$, $\epsilon_3 = 3\epsilon$, and so on. The partition function is

$$z = \sum_{n=0}^{\infty} \exp\left(-\frac{n\epsilon}{kT}\right) = 1 + \exp\left(-\frac{\epsilon}{kT}\right) + \exp\left(-\frac{2\epsilon}{kT}\right) + \cdots \quad (24.13)$$

where we have moved the origin by setting $\epsilon_1 = 0$. This is just a geometric progression that can be summed to get z in closed form,

$$z = \frac{1}{1 - \exp(-\epsilon/kT)} \quad (24.14)$$

Now we can calculate C_v by using Equation (24.12). Remembering that $(d/dx)(\ln u) = (1/u)(du/dx)$, we get

$$C_v = Nk \left(\frac{\epsilon}{kT}\right)^2 \frac{\exp(\epsilon/kT)}{[\exp(\epsilon/kT) - 1]^2} \quad (24.15)$$

Equation (24.15) gives the heat capacity of the ensemble as a function of temperature. It does not appear particularly useful or familiar in that form, but suppose we consider C_v in the limit of infinitely high temperature. Since

$$\exp(x) = 1 + x + \frac{1}{2!}x^2 + \frac{1}{3!}x^3 + \cdots$$

Equation (24.15) can be approximated in the limit by

$$C_v = Nk \left(\frac{\epsilon}{kT}\right)^2 \left(\frac{1}{(\epsilon/kT)^2}\right) = Nk \qquad (T \longrightarrow \infty) \quad (24.16)$$

For a mole of molecules, $Nk = R$ and $C_v = R$. Recall our discussion of the equipartition of energy (Chapter 3), in which we noted that the heat capacity of an ideal diatomic gas due to vibrational freedom was just R. The energy-level diagram on which this example is based is the energy-level diagram for the vibrational motion of a diatomic molecule. We shall return to this point in Chapter 29 in greater detail. For the moment we simply note that it does explain that the vibrational contribution to the specific heat is less than R at finite temperatures and does not begin to approach R until the temperature is sufficiently high that the upper energy states become populated. This occurs when kT is very much greater than the energy spacing between the levels.

24.2 THE TRANSLATIONAL PARTITION FUNCTION

Ultimately we shall derive an expression for the energy levels of a particle in a rectangular box, the so-called "particle in a box". For the moment, we shall simply write the result as

$$\epsilon_n = \frac{h^2}{8m}\left(\frac{n_1^2}{a^2} + \frac{n_2^2}{b^2} + \frac{n_3^2}{c^2}\right) \qquad (n_1, n_2, n_3 = 1, 2, 3, \ldots) \quad (24.17)$$

where m is the mass of the particle, a, b, and c are the sides of the box, and h is the Planck constant, 6.62×10^{-34} J s; the terms n_1, n_2, and n_3 are integers, called *quantum numbers*. The partition function is

$$z = \sum_{n_1=0}^{\infty} \sum_{n_2=0}^{\infty} \sum_{n_3=0}^{\infty} \exp\left[-\frac{h^2}{8mkT}\left(\frac{n_1^2}{a^2} + \frac{n_2^2}{b^2} + \frac{n_3^2}{c^2}\right)\right] \quad (24.18)$$

with n_1, n_2, and n_3 each summed from 0 to ∞. Since h is small, the energy levels are so closely spaced that the summations can be replaced by integrals:

$$z = \int_0^{\infty} \int_0^{\infty} \int_0^{\infty} \exp\left[-\frac{h^2}{8mkT}\left(\frac{n_1^2}{a^2} + \frac{n_2^2}{b^2} + \frac{n_3^2}{c^2}\right)\right] dn_1\, dn_2\, dn_3 \quad (24.19)$$

This integral can be reduced to the product of three integrals, each of the form

$$I = \int_0^{\infty} \exp(-\alpha^2 n^2)\, dn = \frac{\sqrt{\pi}}{2\alpha} \quad (24.20)$$

with
$$\alpha^2 = \frac{h^2}{8ma^2 kT}$$

Similar integrals are obtained for the b and c terms. By this result,

$$z = \left(a\frac{(2\pi mkT)^{1/2}}{h}\right)\left(b\frac{(2\pi mkT)^{1/2}}{h}\right)\left(c\frac{(2\pi mkT)^{1/2}}{h}\right)$$

$$= V\left(\frac{(2\pi mkT)^{3/2}}{h^3}\right) \quad (24.21)$$

where $V = abc$ is the volume of the container.

The average energy per molecule is

$$\epsilon = kT^2\left(\frac{\partial \ln z}{\partial T}\right) = \frac{3}{2}kT \quad (24.22)$$

This is the same result we got for the translational energy of a monatomic gas molecule when we considered equipartition of energy (Chapter 3). For a mole of molecules the energy is $E = L_0 \epsilon = \frac{3}{2} L_0 kT = \frac{3}{2} RT$, where R is the gas constant, and L_0 is Avogadro's number. The molar heat capacity is then $C_v = (\partial E/\partial T) = \frac{3}{2}R$.

24.3 MOLAR PARTITION FUNCTIONS AND THE MOLAR TRANSLATIONAL ENTROPY

From what we have said thus far, it seems as though evaluating the molar entropy for a monatomic gas should be straightforward. The partition function is given in (24.21) and we should simply be able to insert that into Equation (24.9) to get the entropy. We have, however, overlooked something in our evaluation of z. Our partition function is valid for a single molecule. What we have is a large collection of molecules, each molecule having the same energy-level diagram. In forming the summation over all the energy levels, we summed over these levels for only one molecule; but really there is a number N of each energy level for a collection of N molecules. This affects the value of Ω, and we must now find a way to take this into account.

Basic to our previous discussion of Ω was that our particles constituted a set of *identical, independent,* and *localized* (or *distinguishable*) systems. This last assumption is not quite correct for the translational motion of a collection of particles. The value of Ω that we have calculated using the information of Chapter 23 is too large by a substantial amount. This has happened because different energy distributions are not distinguishable. Suppose we have a distribution in which molecule A is in ϵ_3 while molecule B is in ϵ_9. That distribution is indistinguishable from the one in which molecule A is in ϵ_9 while molecule B is in ϵ_3, yet our previous determination has counted each of these arrangements. Actually we have N systems; we label them $1, 2, \ldots, N$. But there are $N!$ different ways of labeling the systems, and each of these $N!$ sets of labeled states is indistinguishable from any other set of labeled states.

Our previous determination of $\Omega = N!/n_1!n_2! \ldots$ is valid only when the systems effectively can be labeled. This is the case for the atoms in a crystal, where each atom can be assigned to an individual lattice site and each lattice site can be provided with a label. For our assembly of nonlocalized (or indistinguishable) systems, the results of the last chapter must be corrected; this can be done by dividing Ω by $N!$.

One additional point should be clarified. In (24.1) we used the fact that $S = kN \ln z + E/T$. This assumes that we can get the entropy for a large collection of molecules simply by scaling the value obtained from the molecular partition function. That this assumption is justified can be seen from the following arguments dealing with the factorization of partition functions. Suppose we first consider a single molecule. For definiteness we take three different types of motion—translational, v type, and r type. Each gas molecule can possess independent amounts of each type of energy. The energy of each molecule is

$$\epsilon = \epsilon_t + \epsilon_v + \epsilon_r \qquad (24.23)$$

The term ϵ_t represents the translational energy, ϵ_v the v type of energy, and ϵ_r the r type of energy. Each type of energy is separately quantized.

ϵ_t takes the values $\epsilon_{t1}, \epsilon_{t2}, \epsilon_{t3}, \ldots$
ϵ_v takes the values $\epsilon_{v1}, \epsilon_{v2}, \epsilon_{v3}, \ldots$
ϵ_r takes the values $\epsilon_{r1}, \epsilon_{r2}, \epsilon_{r3}, \ldots$

The allowed values of ϵ are then

$$\epsilon_{i,j,k} = \epsilon_{ti} + \epsilon_{vj} + \epsilon_{rk} \qquad (24.24)$$

For simplicity we shall assume that all the degeneracies are unity. The molecular partition function is then

$$\begin{aligned} z &= \sum_{i,j,k} \exp\left(-\frac{\epsilon_{i,j,k}}{kT}\right) \\ &= \sum_i \exp\left(-\frac{\epsilon_{ti}}{kT}\right) \sum_j \exp\left(-\frac{\epsilon_{vj}}{kT}\right) \sum_k \exp\left(-\frac{\epsilon_{rk}}{kT}\right) \\ &= z_t z_v z_r \end{aligned} \qquad (24.25)$$

Thus for three independent modes of energy, the partition function can be factored into three partition functions, one for each mode; for m independent energy modes there are m factors. Our choice of v-type and r-type energy was not fortuitous, and we shall find Equation (24.25) useful when we consider the partition function for polyatomic molecules having vibrational and rotational energy in addition to translational energy (Chapter 29).

The above discussion can be extended to the translational energy of a collection of N molecules, which we call A, B, C, ..., N. The total energy of the collection is

$$E = \epsilon_A + \epsilon_B + \epsilon_C + \cdots + \epsilon_N \tag{24.26}$$

where $\epsilon_A, \epsilon_B, \ldots$ are the energies of each molecule. The energy level occupied by each molecule is independent of the energy levels of the other molecules, and proceeding as in (24.25), we get

$$Z = \sum_i \exp\left(-\frac{\epsilon_{Ai}}{kT}\right) \sum_j \exp\left(-\frac{\epsilon_{Bj}}{kT}\right) \sum_k \exp\left(-\frac{\epsilon_{Ck}}{kT}\right) \cdots$$

$$= z_A z_B z_C \cdots = z^N \tag{24.27}$$

where Z is the molar partition function for $N = L_0 = 6 \times 10^{23}$. Since $z_A = z_B = z_C = \cdots$, it follows that $Z = z^N$. The factor $N!$ discussed in the beginning of this section can be seen to arise from the following. The individual terms in the expression for Z given in (24.27) contain terms of the form $\exp[-(\epsilon_{Ai} + \epsilon_{Bj} + \epsilon_{Ck} + \cdots + \epsilon_{Nl})/kT]$. There are $N!$ different ways of permuting the N molecules among the N states. In the light of the above discussion, our expressions for the monatomic molecule now become

$$\Omega = \frac{z^N}{N!} \exp\left(-\frac{E}{kT}\right) \tag{24.28}$$

$$S = k \ln \Omega = kN \ln z - \frac{E}{T} - k \ln N! \tag{24.29}$$

and

$$A = -NkT \ln z + kT \ln N! \tag{24.30}$$

The extra factor $1/N!$ has no effect on the distributions of the individual molecules among the energy levels, and the Boltzmann distribution previously determined remains valid. Now let us see what happens when we compare these results with the results of classical macroscopic thermodynamics.

24.4 COMPARISON WITH CLASSICAL RESULTS

It is convenient to base this discussion on a mole of gas molecules; thus $N = L_0 = 6.02 = 10^{23}$. Equation (24.30) gives an expression for the Helmholtz energy. Using the partition function of (24.21), we get

$$A = -L_0 kT \ln V - \frac{3}{2} L_0 kT \ln \frac{2\pi mkT}{h^2} + L_0 kT \ln L_0 - L_0 kT \ln e \tag{24.31}$$

Here we have used Stirling's approximation for the term in $L_0!$. In addition, we have added the term $\ln e = 1$ to the last term so that each term has a logarithmic term. We cannot compare this directly with the classical macroscopic result, since macroscopic thermodynamics does not provide us with absolute values. We can, however, calculate ΔA for an isothermal expansion from V_1 to V_2:

$$\Delta A = A_2 - A_1 = -RT \ln V_2 + RT \ln V_1$$

$$= RT \ln \frac{V_1}{V_2} \tag{24.32}$$

This is the same result that we get from classical thermodynamics.

Additional thermodynamic variables can be computed using Equations (24.8)–(24.11). For one mole of gas, the pressure is

$$P = \left(\frac{\partial A}{\partial V}\right)_{N,T} = \frac{L_0 kT}{V} = \frac{RT}{V} \qquad (24.33)$$

which gives the ideal gas law, $PV = RT$. The molar entropy is

$$S = -\left(\frac{\partial A}{\partial T}\right)_{V,N} = -\frac{A}{T} + \frac{3}{2}L_0 k$$

$$= -\frac{A}{T} + \frac{3}{2}R \qquad (24.34)$$

Since $E = A + TS$, the energy is

$$E = \tfrac{3}{2} L_0 kT = \tfrac{3}{2} RT \qquad (24.35)$$

If the proper equivalents are inserted into (24.34), an explicit formula for the molar entropy can be written in the form

$$\overline{S} = \frac{3}{2}R + R \ln\left[\frac{eV}{L_0 h^3}(2\pi mkT)^{3/2}\right]$$

$$= R \ln\left[\left(\frac{2\pi mkT}{h^2}\right)^{3/2} \frac{Ve^{5/2}}{L_0}\right] \qquad (24.36)$$

This equation is known as the *Sakur and Tetrode equation*. It is left for one of the problems to demonstrate that for an isothermal expansion of an ideal gas, we get the classical result that

$$\Delta S = S_2 - S_1 = R \ln\frac{P_1}{P_2} = R \ln\frac{V_2}{V_1} \qquad (24.37)$$

The Sakur-Tetrode equation also provides a direct comparison between measured and calculated entropies for monatomic gases. These are shown in Table 24.1; the agreement is good. The experimental values listed in the table were obtained from heat-capacity measurements, and the calculated values were obtained from Equation (24.36). The confidence in the calculated values is sufficiently great that the standard tabulations of thermodynamic functions such as those of the National Bureau of Standards Circular 500 list the calculated values of $S°$ instead of the experimental values. We shall leave a discussion of the entropy of mixing to Problem 7, while we go on to examine the molecular basis of the equilibrium constant.

TABLE 24.1 Experimental and calculated entropies of the inert gases at their normal boiling points.

			Entropy (J K^{-1} mol^{-1})	
Gas	T_{bp} (K)	Molar Mass (g)	Experimental	Calculated from (24.36)
He	4.21	4.003	36.10	37.49
Ne	27.2	20.183	96.40	96.45
Ar	87.29	39.944	129.75	129.24
Kr	119.93	83.80	144.56	145.06
Xe	165.13	131.3	158.45	157.32

SOURCE: Data from Landolt-Börnstein Tables, 6th ed., vol. 2, part 4, pp. 394–400.

24.5 THE EQUILIBRIUM CONSTANT

We consider the reaction A = B, where each of the molecules A and B has a different set of energy levels, as shown in Figure 24.1. Note that while we have drawn the energy-level diagrams with equal spacings for each of the molecules, the energy levels need not be equally spaced. We now wish to examine the distribution of numerous molecules among these two sets of energy levels at equilibrium. We let the number of A molecules be n^A at equilibrium, while the number of B molecules is n^B. We have

$$n^A = n_0^A + n_1^A + n_2^A + \cdots$$

$$= n_0^A \left[1 + \exp\left(-\frac{\epsilon_A}{kT}\right) + \exp\left(-\frac{2\epsilon_A}{kT}\right) + \cdots\right]$$

$$= n_0^A \sum_{i=0}^{\infty} \exp\left(-\frac{i\epsilon_A}{kT}\right)$$

or
$$n^A = n_0^A z_A \tag{24.38}$$

where the n_i^A are the numbers of A molecules in the various A levels, and z_A is the partition function for the A levels. Similarly, the number of B molecules is

$$n^B = n_0^B \sum_{i=0}^{\infty} \exp\left(-\frac{i\epsilon_B}{kT}\right) = n_0^B z_B \tag{24.39}$$

At equilibrium the Boltzmann distribution requires the numbers of molecules in the lowest A and B levels to be related by

$$n_0^B = n_0^A \exp\left(-\frac{\Delta\epsilon_0}{kT}\right) \tag{24.40}$$

This enables us to rewrite the total population of the B levels as

$$n^B = n_0^A \exp\left(-\frac{\Delta\epsilon_0}{kT}\right) z_B \tag{24.41}$$

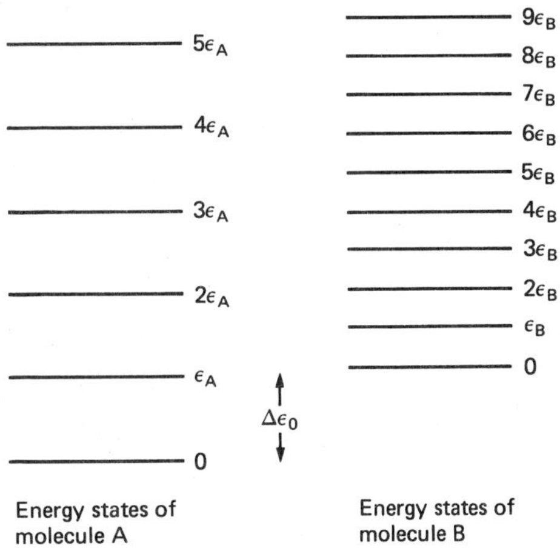

FIGURE 24.1 Energy-level diagrams of the hypothetical molecules A and B

The equilibrium constant for an ideal gas reaction is $K = P_B/P_A$. Since the partial pressures are proportional to the numbers of moles or molecules, $P_A = n^A kT/V$ and $P_B = n^B kT/V$, we can also write the equilibrium constant as

$$K = \frac{n^B}{n^A} = \exp\left(-\frac{\Delta\epsilon_0}{kT}\right)\frac{z_B}{z_A} \tag{24.42}$$

Equation (24.42) provides a statistical mechanical basis for the equilibrium constant. Further applications of this are discussed in Problem 9.

24.6 THE THIRD LAW OF THERMODYNAMICS

In our consideration of classical thermodynamics, we noted that the evaluation of absolute "third law" entropies was based on the assumption that all "perfectly crystalline" substances had the same entropy at the absolute zero of temperature. We further noted that since entropy differences rather than the absolute values of the entropies were of primary importance, it would be convenient to set that value equal to zero. We can now justify this on the basis of our discussions of statistical mechanics.

Suppose we have N molecules at a finite temperature T. The N molecules are distributed among the various energy levels, as shown in Figure 24.2a. Now suppose that the temperature is lowered to $T = 0$. The Boltzmann distribution requires that the population of any upper level relative to the population of the lowest level is $n_i = n_0 \exp(-\epsilon_i/kT)$. At $T = 0$ the exponential factor is zero, and $n_i = 0$ for all other levels besides the lowest level. What we have then is a situation in which there are N molecules in the lowest level ($\epsilon = 0$) and no molecules in any of the upper levels. For this distribution, the number of different configurations is

$$\Omega = \frac{N!}{\Pi_i n_i!} = \frac{N!}{N!0!0!0!0!\ldots} = 1 \tag{24.43}$$

The entropy at $T = 0$ is then

$$S = k \ln \Omega = k \ln 1 = 0 \tag{24.44}$$

FIGURE 24.2 Distribution of molecules among energy levels. (a) Temperature is finite, and the molecules are distributed among the various levels according to the Boltzmann distribution. (b) Temperature is zero, and all the molecules are in the lowest energy level. (c) Temperature is still zero, but the lowest level is doubly degenerate. Now all the molecules are distributed equally between the lowest two levels.

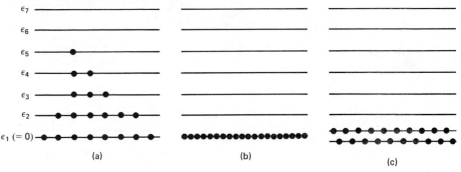

This justifies an absolute entropy scale based on zero entropy values at the absolute zero of temperature. But recall that in our discussion of the classical third law of entropy we noted that there were numerous instances in which the third law entropy values determined from specific heat measurements did not agree with entropies determined by other classical thermodynamic techniques. This discrepancy can be resolved by noting that we have omitted a factor that can be important, namely the degeneracy. Figure 24.2c indicates the populations of the levels for the case in which the lowest level is doubly degenerate. Here the energy of each of these levels is identically zero, and the molecules are equally distributed among the two levels. The total number of possible arrangements of the molecules between the two levels is 2^N; hence

$$S = k \ln 2^N = Nk \ln 2 \tag{24.45}$$

For a mole of molecules, this *residual,* or *zero-point,* entropy is $R \ln 2 = 5.76$ J K^{-1} mol^{-1}.

This residual entropy is important for molecules such as N_2O, for which adjacent molecules in the crystal can be oriented in either of the two orientations (NNO NNO) or (ONN NNO). To be sure, there is a slight energy difference between the two orientations, but this is so small that the Boltzmann factor is very nearly unity except at the absolute zero of temperature. As the crystal is cooled to low temperatures, the random orientations of the adjacent molecules are effectively "frozen," and the two different orientations persist indefinitely at low temperatures. A statistical evaluation of the absolute entropy of N_2O at 298.15 K yields the value 220.0 J K^{-1} mol^{-1}. The calorimetric value is 215.2, which is 4.8 J K^{-1} mol^{-1} less than the statistical value. This is very close to the residual entropy value determined above for a doubly degenerate ground state.

The residual entropy of a collection of N molecules is generally

$$S_0 = Nk \ln g_0 \tag{24.46}$$

where g_0 is the degeneracy of the lowest energy level. This can be seen by noting that the total number of arrangements of N molecules among g_0 levels is g_0^N. This result can also be gotten directly from $S = Nk \ln z$, since $z = \Sigma g_j \exp(-\epsilon_j/kT) = g_0$ at $T = 0$.

Mostly, we have restricted our discussion in this chapter to the statistical mechanics of translational motion, and that limits the applicability of the theory to monatomic molecules. The theory is equally valid for polyatomic molecules; we shall defer a comparison between experimental and calculated thermodynamic functions of these molecules until after we have considered the quantum mechanical derivation of the energy-level diagrams of vibrational and rotational motion. Let us now look into quantum mechanics and determining energy levels in atoms and molecules.

PROBLEMS

1. Carry out the differentiation steps that have been omitted in the text and show that (24.15) is indeed the expression for C_v for the partition function given in (24.14). Also show that (24.16) follows when the series expansion for $\exp(x)$ is used.
2. The partition function for the rotational motion of a diatomic molecule such as CO is $z = 8\pi^2 IkT/h^2$, where I is the moment of inertia of the molecule and h is Planck's constant. (We shall derive this expression in Chapter 29.) Using this partition function, calculate the rotational contribution to the specific heat C_v.
3. Calculate the entropy for a system having the partition function given in (24.14).

4. Calculate the rotational contribution to the energy and to the entropy using the partition function for CO given in Problem 2.
5. Assuming that CO molecules in the crystal lattice can be arranged either as (CO CO) or (CO OC), calculate the zero-point entropy of CO.
6. Show that the use of the Sakur-Tetrode equation leads to the result that $\Delta S = R \ln (V_2/V_1)$ for the isothermal expansion of an ideal gas.
7. Consider a mixture of two ideal monatomic gases A and B. The partition function for the mixture is $z_{gas} = z_A z_B$, where z_A is the partition function for the A molecules, and z_B is the partition function for the B molecules. Show that the entropy of mixing as calculated by statistical mechanics is given by

$$\Delta S = -(N_A + N_B)k(X_A \ln X_A + X_B \ln X_B)$$

where N_A is the number of A molecules, N_B is the number of B molecules, $X_A = N_A/(N_B + N_B)$ is the mole fraction of A, and similarly X_B is the mole fraction of B. Compare this with the classical result.
8. For the gas mixture of Problem 7, show that the energy of mixing is identically zero.
9. Consider the chemical reaction A = B. The molecules A and B are both artificial. The energy-level diagram of the A molecules consists of a single level that is doubly degenerate. The energy-level diagram of the B molecules consists of one triply degenerate level lying 2×10^{-21} J above the level for the A molecules. Calculate the equilibrium constant for the reaction at 25 and 1200 K.
10. Suppose that we have a collection of N molecules at the absolute zero of temperature and that the lowest level has degeneracy g_0. As in (24.43), the number of configurations is given by

$$\Omega = \frac{N!}{[(N/g_0)!]^{g_0} 0! 0! 0! \cdots}$$

Using Stirling's approximation, show that $S = k \ln \Omega = Nk \ln g_0$, which is the same result as (24.46).
11. Use the Sakur-Tetrode equation to calculate the entropy of He at 300 K and 1 atm.
12. Use the Sakur-Tetrode equation to calculate ΔS for heating 1 mol He from 300 to 400 K at constant volume for an initial pressure of 1 atm.

PART FIVE

THE QUANTUM MECHANICS OF ATOMS AND MOLECULES

CHAPTER TWENTY-FIVE

CORPUSCLES, WAVES, AND THE NUCLEAR ATOM

25.1 EARLY ATOMIC THEORIES

The earliest atomic theories were postulated by the ancient Greeks, with Democritus the name usually associated with the earliest theory. There was much speculation and little fact regarding atoms during the next 2200 years. Atoms were imagined to be little balls of various shapes. Water atoms were smooth balls, able to glide over one another, explaining the liquid nature of water. Iron atoms, it was claimed, had hooks; the hooks of adjacent atoms were interlocked, providing the strength for which iron was noted. Such were the modest beginnings of the chemical bond. Daniel Bernoulli put forth the first truly "scientific" theory of atomism in his derivation of Boyle's law, a physical observable. The world was not ready for atoms in 1738, and Bernoulli's work was utterly neglected for a hundred years. It lay outside the mainstream of scientific thought, while the atom entered through another door.

By the early 1800s, the concept of an element as a substance that could not be further decomposed had been established through the work of Lavoisier and others.

ANTOINE LAURENT LAVOISIER (1743–1794), eminent French chemist, is considered by many the "father" of chemistry. Lavoisier demonstrated the nature of combustion by showing that substances combined with oxygen on burning with concomitant weight increase. Although these ideas directly conflicted with the phlogiston theory, it took but a few years before Lavoisier's work gained universal acceptance and the older theory was thoroughly discredited. He developed the concept of an element as a "simple" substance that could not be further decomposed, and provided the first advance in definition since the time of the ancient Greeks. His list of elements included "light," "caloric," oxygen, nitrogen, hydrogen, and a variety of substances, both metal and nonmetal; some like magnesia were not elements. Lavoisier also worked in thermodynamics, developing an accurate ice calorimeter and techniques for measuring linear and cubical expansions. He worked in fermentation and respiration and correctly deduced that oxidation takes place in the body. Lavoisier was active in political affairs and was a leading member of a commission to develop uniform weights and measures throughout France; the work of this commission led to the metric system. Lavoisier was caught up in the excesses of the Reign of Terror during the French Revolution, and in a trial that lasted only a few hours, Lavoisier, his father-in-law, and twenty-six other people were condemned to death and guillotined on the same day (May 8, 1794). Lagrange, the famous mathematician, said of this, "It took only a moment to sever that head, and a century will not suffice to produce another like it."

Quantitative chemical analysis had been developed, and the laws of definite proportions and of multiple proportions had been established on firm ground. The physical observable through which the atom entered was the relative masses of the atoms; recall that the table of atomic masses is nothing more than a table of the relative masses of atoms.

25.2 THE RELATIVE MASSES OF ATOMS

By the first decade of the nineteenth century the world was ready for the atom, and John Dalton presented it. In a brilliant process of reasoning he argued essentially as follows.[1] Suppose we take 143 g of silver chloride. We can decompose the compound into 108 g of silver and 35 g of chlorine. Now, the original sample of silver chloride contained a certain number, call it N, of silver chloride molecules. If silver chloride contained silver and chlorine atoms in a 1:1 ratio, then that same number N of silver atoms weighs 108 g, and that same number of chlorine atoms weighs 35 g. It therefore follows that if the mass of N atoms of silver and the mass of N atoms of chlorine are in the ratio 108:35, then the mass of *one* atom of silver must be 108/35 times the mass of *one* atom of chlorine. The basis for the table of atomic masses was established.

There were still some problems. What if the formula for silver chloride were Ag_2Cl, or $AgCl_2$, or any other atomic ratio? In many instances, Dalton had no way of knowing what the true combining ratios were, and he was aware of this lack of knowledge. He usually made the simplest assumption of a 1:1 ratio. The most notable error of this type was assuming that water was HO, leading to the erroneous result than an oxygen atom weighed only eight times as much as a hydrogen atom.

Another basic contribution of Dalton concerned the way in which elements and compounds are written. The ancient alchemists had symbols for various substances, but these were just a shorthand notation. Today when we write silver chloride as AgCl, the symbol AgCl is more than a shorthand notation. It tells us specifically that the two elements combine in a 1:1 atomic ratio. The symbols for the elements stand for individual atoms in the molecule, and the numbers of atoms are given by the subscripts. We owe this nomenclature to Dalton.

During the next few years, many research chemists made large numbers of relative mass measurements. Eventually tables of atomic masses of all the known elements were completed. There were, however, many inconsistencies. The problem of unqualifiedly determining the combining ratios of atoms still had not been solved. There was a lack of universal agreement in the tables, since different laboratories had different ideas. The laboratory that thought silver chloride was Ag_2Cl differed by the factor 2 in the relative masses of silver and chlorine atoms from the laboratory that thought it was AgCl, and by the factor 4 from the laboratory that thought it was $AgCl_2$. These kinds of differences could not be resolved until a satisfactory method of determining the combining ratios of atoms was developed.

---

[1] For a complete description of this work, see *Harvard Case Histories in Experimental Science,* ed. J. B. Conant. Case 4, *The Atomic-Molecular Theory,* by Leonard Nash (Cambridge: Harvard University Press, 1950).

25.3 THE COMBINING RATIOS OF ATOMS AND THE TABLE OF ATOMIC MASSES

The method for determining the combining ratios of atoms arose from studies of gas reactions. At about the time of Dalton, Gay-Lussac discovered that the ratios of the combining volumes of gases were always small integers. In 1811, Avogadro proposed his now famous principle that equal volumes of gases contain the same numbers of molecules, and he carried his principle far enough to propose correctly that what people thought were atoms of the common gases were really molecules composed of two atoms each. Avogadro's hypothesis held the key to the unique solution of the combining-ratio problem, but his ideas were not accepted until some fifty years later. People were ready to accept the notion that hydrogen could combine with oxygen to form a water molecule but not quite ready to accept that a hydrogen atom could unite with another hydrogen atom to form a molecule.

By 1858 the confusion had mounted considerably, and several particularly perplexing paradoxes had developed. The year 1819 had seen the publication of the law of Petit and Dulong, which stated that the specific heats of elements were equal to 25 J K mol^{-1}.[2] Many elements follow this law closely, as we shall see. By about 1850 it was discovered that the density of mercury vapor was 100 times the density of hydrogen vapor at the same temperature and pressure, giving mercury an atomic mass of 100 on a scale in which hydrogen "atoms" (actually molecules) was 1.00. Heat capacity studies indicated that the atomic mass of mercury was 200. These and other paradoxes were cleared up when Avogadro's law gained acceptance through the efforts of his countryman Cannizzaro.

The discrepancies in the atomic mass table were cleared up, and the mole was defined as the number of molecules in 22.4 liters of an ideal gas at 1 atm and 0 °C. (The currently accepted value of the molar volume is 22,413.8 cm^3 mol^{-1}.) It was universally accepted that a gram-molecular weight of any element contained the same number of atoms as there were molecules in the molar volume.

About 1900, J. J. Thomson discovered the electron, and soon it was postulated that atoms were composed of electrons and positive charges in equal amounts, so that the total charge on the atom was zero. But what was the general shape of the atom?

JOSEPH JOHN THOMSON (1856–1940), English physicist, discovered the electron in the last years of the nineteenth century. By studying the deflection of "cathode rays" (electron beams) by magnetic fields, he was able to show that the beams consisted of particles of negative electricity. He measured the ratio of mass to charge for these particles and discovered that it was some 1000 times (actually 1836) smaller than the mass-to-charge ratio obtained for hydrogen in electrolysis. He concluded that cathode rays were composed of negative charges that were lighter than hydrogen atoms by some three orders of magnitude. His son, George P. Thomson, was also a physicist of note and was one of the people who confirmed de Broglie's hypothesis on the wavelength of electrons by carrying out experiments on electron diffraction.

---

[2] Atomic weights of various elements as established by different laboratories often differed by integral factors. The law of Petit and Dulong often corroborated the correct weight.

25.4 OF STEEL WOOL AND STEEL BALLS

Consider the problem depicted in Figure 25.1. We have two boxes made of tissue paper; each box contains ten pounds of steel. In one box the steel is in the form of loosely packed steel wool evenly distributed about the interior; and in the other box the steel is in the form of a solid ball in the center of the box. The problem is to differentiate between the two without coming within 50 ft of the boxes. The solution is to carry out a scattering experiment.

If we stand 50 ft away, and randomly shoot into the boxes with hard round bullets, then the pattern with which the bullets emerge differentiates the two possibilities. The bullets that hit the steel wool are deflected from their original paths by small amounts. In the other box, most bullets miss the ball and are undeflected. Those that hit the ball, however, may be deflected through large angles, depending on where the bullet hits the ball. For the extreme case of a head-on collision, the bullet is reflected backward and goes right back down the rifle barrel.

The way in which the bullet particles are scattered from the steel depends on how particle and object interact. In this simple system there is no interaction between bullet and ball until the instant of collision. It is possible with this kind of experiment to differentiate between different shapes of steel masses such as cubes, tetrahedra, or spheres. Each geometrical shape provides a different scattering pattern, and the shape can be determined by studying the recoil bullets as the rifle is fired from all directions.

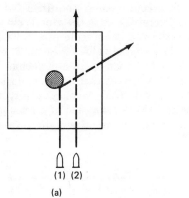
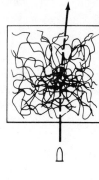

FIGURE 25.1 Bullets scattered by 10 lb of steel. (a) The steel is a solid steel ball; bullets (2) that miss the ball are undeflected, whereas bullets (1) that hit the ball are deflected through large angles. (b) The steel is steel wool, and the bullets are deflected through small angles.

25.5 THE NUCLEAR ATOM

Consider the problem depicted in Figure 25.2. For atoms, which are so small that we cannot possibly see them, we have two possible configurations. In one configuration the atom is composed of positive and negative charges evenly distributed so that the total electrical charge is zero. This configuration was proposed by J. J. Thomson. In the other configuration the atom consists of a massive positively charged object at the center with much lighter negative electrons surrounding the positively charged *nucleus*. (By 1910 it was known that an electron weighed about $\frac{1}{1836}$ as much as a hydrogen atom.) The problem is to determine the correct configuration. The solution is to carry out a scattering experiment; this is precisely what Rutherford did in 1910.

FIGURE 25.2 Scattering of α particles by the two proposed models of the atom. (a) The atom consists of positive and negative charges evenly distributed through space; the α particles are deflected through small angles. (b) Model consists of a massive positively charged nucleus surrounded by a much lighter electron; here the α particles may be deflected through large angles if they come sufficiently close to the nucleus.

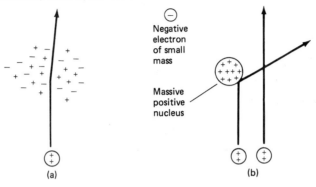

Rutherford's scattering particles were α particles, or positively charged helium nuclei. The targets were gold atoms in thin gold foil. The interaction forces were the electrostatic forces between charged particles. If gold atoms were analogous to steel wool with positive and negative charges evenly distributed through space, then the α particles would be deflected only through very small angles. If the nuclear atom were the correct configuration, then while most of the α particles would be deflected by small amounts, some would be deflected by large angles; a few would even be deflected backwards. This is precisely what Rutherford observed. He was quite surprised at the result, and later noted:

Now I may tell you in confidence that I did not believe they would be (backscattered) since we knew that the alpha particle was a very fast massive particle with a great deal of energy, and you could show that if the scattering was due to the accumulated effect of a number of small scatterings the chance of an alpha particle's being scattered backwards was very small. Then I remember Geiger two or three days later coming to me in great excitement saying, "We have been able to get some of the alpha particles coming backwards." It was quite the most incredible event that has ever happened to me in my life. It was almost as incredible as if you fired a 15-inch shell at a piece of tissue paper and it came back and hit you.[3]

> ERNEST RUTHERFORD (1871–1937) was born in New Zealand and did his important work in Canada (McGill University) and in England. He was a student of J. J. Thomson and eventually succeeded Thomson in the Cavendish Chair at Cambridge University. In Canada in 1902 Rutherford, along with Soddy, published a theory of radioactivity and showed that the phenomenon is caused by the breaking up of atoms. He discovered that the emanations coming from radium consisted of two components. One component was stopped by a few centimeters of air, and these he called α particles (He nuclei); the other component was more penetrating, and these he called β particles (electrons). Rutherford received the Nobel Prize in chemistry for his work on radioactivity in 1908, two years before his α particle scattering experiment, in which he demonstrated the nuclear nature of the atom.

---

[3] Bernard Jaffe, *Crucibles: The Story of Chemistry* (New York: Dover Publications, Inc., 1930), p. 211.

After noting the α particles being scattered backwards he predicted and observed that the particle intensity should be proportional to $\sec^4(\frac{1}{2}\theta)$ for a light positively charged particle being scattered by a massive positively charged particle. The Rutherford scattering formula is

$$\frac{dN}{d\Omega} = \frac{N_0 n t Z_1^2 Z_2^2 e^4}{64\pi^2 \epsilon_0^2 m_1^2 v_0^2 \sin^4(\frac{1}{2}\theta)} \tag{25.1}$$

where dN is the number of particles scattered into solid angle $d\Omega$ at a scattering angle θ; N_0 is the incident beam density; n is the number of scattering centers per unit volume; t is the thickness of the foil; m_1 is the mass and v_0 the velocity of the particles in the beam; ϵ_0 is the permittivity of vacuum; Z_1 and Z_2 are the charges on the incident particle and target respectively; and e is the charge of the electron. For α particles, Z_1 is 2. The term Z_2 is the atomic number of the target atoms; hence this procedure provides a way of measuring the atomic numbers of elements. A derivation of the Rutherford formula can be found in most books dealing with atomic physics.[4]

25.6 FOUR PARADOXES

At this stage of discovery there were several particularly perplexing problems. We shall single out four of these "paradoxes."

PARADOX 1, BLACKBODY RADIATION

From time immemorial it had been observed that heated objects emitted radiation and that the emitted radiation was a function of temperature. As an object is heated, it first glows a dull red; and as the temperature increases, the color changes to bright red and then to bright white. At sufficiently high temperature the ultraviolet radiation causes someone who is looking at the object with bare eyes to have pain. The higher the temperature, the greater the amount of short wavelength radiation.

Studies were made of the spectral distribution that "perfect" radiators (or emitters) of energy called *blackbodies* emitted. (A perfect emitter is also a perfect absorber.) Figure 25.3 shows the results of these studies. The curves have a striking resemblance to the Maxwell-Boltzmann distribution curve. There the similarity ends. It was not possible to explain these curves by classical electromagnetic theory, thermodynamics, or equipartition of energy.

PARADOX 2, THE PHOTOELECTRIC EFFECT

In 1887, Hertz discovered the *photoelectric effect*, illustrated in Figure 25.4. A metal with a scrupulously clean surface is placed in a vacuum and illuminated with light of a known frequency. If the frequency is greater than a particular minimum, or *threshold*, frequency ν_0, then electrons are instantly liberated from the metal surface at a rate proportional to the light intensity. The energy of the electrons is measured by applying a negative potential to the plate and noting the potential difference required to eliminate the electron current. The potential difference is a measure of the

---

[4] Also see H. Goldstein, *Classical Mechanics* (Reading, Mass.: Addison-Wesley Publishing Co., 1959), p. 84.

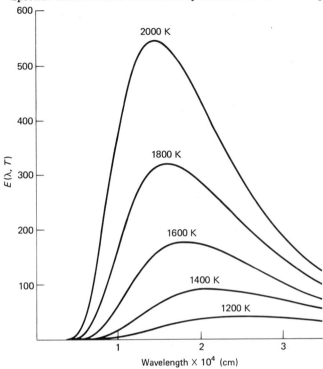

FIGURE 25.3 Spectral distribution of the blackbody radiation at several temperatures.

electron energy in electron volts (1 eV = 1.60206×10^{-19} J). The energy is proportional to the square of the velocity

$$E = \tfrac{1}{2}mv^2 = Ve \tag{25.2}$$

where m is the electron mass, v its velocity, e the electron charge, and V the so-called *stopping voltage*. There were several perplexing points.

Below the threshold frequency no electrons were emitted. The number of electrons emitted depended on the light intensity, but the energy depended only on the frequency. Yet according to classical physics, the energy of a light beam depends on intensity. Classical physics also indicated that for very low light intensity it should take a substantial time for the metal to soak up enough energy from the light to eject an electron; yet, it was observed, even for the lowest-intensity light beams, that electrons were ejected immediately after the beam was turned on. The observed phenomena could not be explained by the classical laws of physics.

HEINRICH RUDOLF HERTZ (1857–1894), German physicist, discovered the photoelectric effect by observing that sparks jumped gaps between electrodes more readily if the electrodes were illuminated than if they were kept in the dark. The reason for this is that photoelectrons ejected from the metal electrodes aid in the conductivity. His name is associated with electromagnetic theory, and he discovered the propagation of electromagnetic waves through space. His work provided much of the basis for wireless transmission; in the early days of radio, radio waves were often called "Hertzian" waves.

FIGURE 25.4 Experimental arrangement for measuring the photoelectric effect. A light quantum of energy $h\nu$ impinges on the clean metal surface. An electron is ejected if the wavelength is short enough. The energy of the ejected electron is determined by adjusting the potential difference between the metal surface and the collector until the electron current disappears. The metal surface must be in a vacuum, the walls of which are constructed of quartz, a material transparent to ultraviolet radiation.

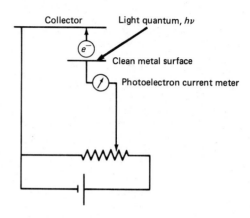

PARADOX 3, SPECTROSCOPY

In 1885, Balmer uncovered a simple mathematical expression for the frequencies in the spectrum of atomic hydrogen in the visible region. The expression can be written in the form

$$\bar{\nu} = R \left(\frac{1}{2^2} - \frac{1}{n^2} \right) \tag{25.3}$$

where n represents the integers 3, 4, 5, . . . ; each value of n gives a line in the spectrum. The value R is the *Rydberg constant,* 109,677.581 cm^{-1}, one of the most accurately known physical constants. The *wave numbers* $\bar{\nu}$ are just the inverses of the wavelength, $\bar{\nu} = (\lambda)^{-1}$. The structure of the spectrum looks complicated when viewed on a photographic plate, as shown in Figure 25.5. That it was completely derivable from a simple empirical expression was remarkable.

FIGURE 25.5 Spectral lines for hydrogen atom, showing three series. The numbers in parentheses indicate lines arising from transitions ($n \to k$), where n and k are as defined in (25.4). The figure is drawn to scale. The indicated wavelengths are in angstrom units (Å; 10^{-8} cm).

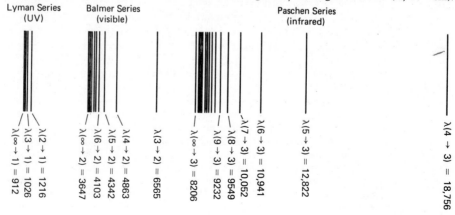

It appeared even more remarkable when people learned to measure the frequencies of ultraviolet and infrared radiation and discovered new spectral series in hydrogen, every one of which could be explained by an equation of the form

$$\bar{\nu} = R \left(\frac{1}{k^2} - \frac{1}{n^2} \right) \tag{25.4}$$

where both n and k are integers, and $n > k$. Again, n varies to describe the spectral line in each spectral series. The value k is fixed for each spectral series. When $k = 1$, the Lyman series in the ultraviolet is observed; when $k = 2$, the aforementioned Balmer series in the visible is observed; and when $k = 3$ or greater, spectral series in the infrared are observed. This simple expression was completely inexplicable by the classical laws of physics.

PARADOX 4, THE NUCLEAR ATOM

The nuclear atom that Rutherford had demonstrated was impossible according to the laws of classical physics. Classical mechanics requires the electron to rotate about the nucleus so that a system of stable dynamic equilibrium will be maintained. The centrifugal force of rotation would counterbalance the attractive electrostatic forces between the charged particles. Classical electrodynamics, however, requires that whenever a charge moves in an electric field, it should radiate energy. If the outer electron moved in the field of the nucleus, then all its kinetic energy would be radiated as electromagnetic energy; the electron would gradually lose its kinetic energy and spiral into a collision with the nucleus. A nuclear atom in static equilibrium was thus impossible, according to classical mechanics; a nuclear atom in dynamic equilibrium was impossible by the laws of electrodynamics. There was no other kind of equilibrium to fall back on; and the quandary continued until 1913 and Niels Bohr.

25.7 THREE SOLUTIONS

The solutions to the four paradoxes provide a classic demonstration of the so-called scientific method. Observations (or properly performed experiments, if you will) are fixed. Embarrassing though they may be, we are stuck with them. Nothing could be done with the observations of blackbody radiation, the photoelectric effect, spectroscopy, and the nuclear atom. The theory was faulty, and it is theory that must be adjusted to fit experiment. Since classical science was unsuited to the task, a revolution in scientific thought was required. Classical physics was developed to describe macroscopic systems; it failed utterly when applied to atomic systems. A completely new type of physics was required, and a new physics was invented.

It is characteristic of human society that as people get older, they get more conservative, not only in their views of the younger generation but in many aspects. It was no accident that the revolution was produced by a completely new generation of younger scientists. The older generation that had produced the paradoxes was still tied to the older science. The solutions were provided by Planck, Einstein, Bohr, and many others.

Among the fundamental principles of classical science was continuity and complete determinability. The dynamical properties of a system could vary continuously

> MAX PLANCK (1858–1947), German physicist, introduced the concept of the "quantum" of energy in his work on blackbody radiation. Since his interests lay in thermodynamics, he was led to his solution of the blackbody problem by applying the laws of thermodynamics. Planck received the Nobel Prize in physics in 1918 for his work.

and in principle could be determined to any degree of accuracy. The limit was practicality: How accurately could one in practice make a ruler, a clock, or anything else? Neither of these principles is valid when applied to systems on an atomic scale. The new theory was based on a corpuscular view of electromagnetic radiation, or light waves. The blackbody paradox was solved by Planck in 1900. The photoelectric paradox was solved by Einstein in 1905. The spectroscopy and nuclear atom paradoxes were simultaneously solved by Bohr in 1913.

25.8 BLACKBODY RADIATION; THE QUANTUM HYPOTHESIS

While the spectral distribution of *blackbody radiation* could not be derived from the laws of classical physics, the rate of escape of energy from the cavity had been established by the *Stefan-Boltzmann law*,

$$S = \sigma T^4 \tag{25.5}$$

The value of σ, the *Stefan-Boltzmann constant*, could not be deduced from classical physics, but it was known from experimental measurements. We shall not delve into the derivations of the various attempts to arrive at a theoretical formulation of the spectral distribution.[5] Suffice it to say that these formulations involved considerations of the emission of electromagnetic radiation and the associated thermodynamics. The starting point for these treatments lay in the assumption that the cavity contained oscillators that absorbed and emitted the observed radiation. The flaw was inherent to classical physics; all the unsuccessful treatments assumed that the energy was continuously variable. It was this basic assumption of continuity of energy that Planck overthrew in his successful solution of the problem.

Planck assumed that the oscillators were one-dimensional harmonic oscillators that *could radiate or absorb energy only in discrete packets, or* **quanta**. The energies were given by

$$E = nh\nu \qquad (n = 1, 2, 3, 4, \ldots) \tag{25.6}$$

where ν is the frequency of the radiation, and h is a universal constant of nature, *Planck's constant*,

$$h = 6.625 \times 10^{-34} \text{ J s}$$

There was no a priori justification for the assumption that the energy was proportional to the frequency and to an integer, but it worked, and science hasn't been the same since. Although Planck did not realize the full import of what he had done, he had laid the groundwork for the corpuscular theory of light, which Newton had long before proposed.[6] Without going into the details of the derivation, we simply present

---

[5] These are discussed in F. K. Richtmyer and E. H. Kennard, *Introduction to Modern Physics*, 4th ed. (New York: McGraw-Hill Book Co., 1947), p. 139 et seq.

[6] Newton had proposed a corpuscular theory of light and this theory hampered the development of the wave theory of light. It was with some reluctance that Newton's corpuscular views were dropped in favor of the newer wave theories of light, which were able to describe correctly many newly discovered phenomena such as diffraction of light.

the final result of the distribution law as

$$\rho(\nu, T)\, d\nu = \frac{8\pi h \nu^3}{c^3} \left[\exp\left(\frac{h\nu}{kT}\right) - 1 \right]^{-1} d\nu \qquad (25.7)$$

where ν is the frequency of the radiation, T is the temperature, k is the Boltzmann constant, h is Planck's constant, and c is the velocity of light. Equation (25.7) gives the energy density of the emitted radiation between the frequencies ν and $\nu + d\nu$; the equation agreed with the data within the experimental uncertainty.

25.9 THE PHOTOELECTRIC EFFECT

Einstein was able to explain the photoelectric effect by assuming that light was composed of corpuscles, or *quanta,* each of energy

$$E = h\nu \qquad (25.8)$$

where h is Planck's constant and ν is the frequency of the light. When a quantum of light interacts with an electron near the surface of a metal, the entire package of energy is absorbed by the single electron. This energy is used by the electron in two parts. The first part of the energy W is used to overcome the attractive potential of the metal. The second part, consisting of the remaining energy, if any, is available to impart kinetic energy to the electron and to move it away from the metal. If the energy of the quantum is less than W, then the electron cannot escape from the metal; the energy is simply dissipated as heat. The energy W required to liberate the electron is called the *work function* and is characteristic of the particular metal. Conservation of energy requires that

$$h\nu = W + \tfrac{1}{2}mv^2 = W + Ve$$

ALBERT EINSTEIN (1879–1955) was born in Germany and educated in Switzerland; and he died in the United States. He showed very little promise as a youth; in fact he was refused a position as assistant in the physics department in the Zurich Polytechnical Institute on his graduation, and he settled for a position as an examiner in the Swiss Patent Office in 1900. In a few short years he produced three theories, each of which was fundamentally important in different branches of physics and chemistry: the theory of the photoelectric effect, the theory of Brownian motion, and the theory of relativity. Einstein was one of the few scientists to achieve worldwide stature in nonscientific circles for his scientific work. The name *Einstein* is a household word, and has been introduced as a word in the English language. The expression "He's a regular Einstein" is often applied to bright children. When I was a schoolboy, it was accepted fact among my associates that Einstein was the smartest man who ever lived, and that his theory of relativity was so complicated that only three people understood it, one of whom was Einstein himself. Einstein was forced out of Nazi Germany in the early 1930s along with Fritz Haber and others, and came to the United States, where he spent the rest of his life at the Institute for Advanced Study at Princeton. Einstein was a pacifist during World War I and as a German citizen had publicly opposed the invasion of Belgium; throughout his life he favored disarmament. The realities of the political situation, however, induced him to write a letter to President Roosevelt in 1939, pointing out the possibilities of the atomic bomb, thus setting in motion events that were to lead to the Manhattan project and atomic energy. Einstein received the Nobel Prize in physics in 1921 for his work on the photoelectric effect.

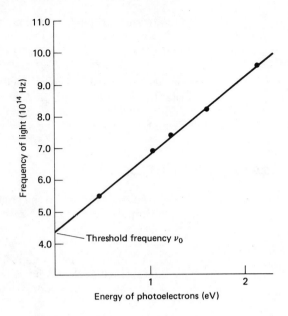

FIGURE 25.6 Plot of frequency vs the energy of the ejected photoelectrons for the photoelectric effect. The intercept is the threshold frequency ν_0, and Planck's constant is determined from the slope.

This equation is often written

$$h\nu = h\nu_0 + Ve \qquad (25.9)$$

If the frequency ν is plotted against the stopping potential V, then the result is a straight line with slope e/h and intercept ν_0; no photoelectron current is obtained for frequencies less than ν_0. A plot of this type is shown in Figure 25.6.

That Einstein's ideas were correct was shown by many careful experimental investigations. For many years, measurements of the photoelectric effect provided the best value for the Planck constant. Einstein's success clearly demonstrated the corpuscular nature of radiation. On the other hand, the success of Maxwell's equations clearly showed the wave nature of radiation. The dual wave-particle nature of light was something of puzzle in the early years of this century, and the puzzle was perhaps intensified by the converse problem of the wave-particle duality of electrons, as we shall see. Perhaps the problem lay in the nature of people; they were not accustomed to thinking along these lines.

The dual nature of light does not imply that both halves must be dealt with at all times for all problems. Even though Maxwell's equations are not valid over the entire spectrum of possible problems, they are perfectly adequate for cases in which huge numbers of light quanta, or *photons,* are involved. For any problem dealing with ordinary electrical circuits, the particle nature of radiation can be ignored, and the problem treated by applying only Maxwell's equations. On the other hand, Maxwell's equations are useless for treating processes involving individual photons, as in the photoelectric effect.

25.10 THE BOHR HYDROGEN ATOM

Bohr was able to solve the problem of the hydrogen atom by making four basic assumptions. These assumptions are:

1. The Rutherford model of the atom is correct. A hydrogen atom consists of a positively charged nucleus (proton) and a single, much lighter electron that rotates about the nucleus in circular motion, where the attractive force is the normal type of coulombic attraction between negative and positive charges.
2. The electron can remain in any particular state for indefinite periods *without radiating its energy,* provided this *allowed,* or *stationary, state* is a state for which the angular momentum is an integral multiple of $h/2\pi$. (The quantity $h/2\pi$ is often expressed by $\hbar$.) For a particle moving in a circle, the angular momentum is mvr; this condition becomes

$$mvr = \frac{nh}{2\pi} \qquad (n = 1, 2, 3, \ldots) \qquad (25.10)$$

where m is the mass of the electron, v is the velocity, and r is the radius of the circle. (Equation (25.10) implies that the proton is at rest at the center of the circle. This is approximately correct, but not exactly.)
3. Radiation is emitted only when the electron makes a transition (or "jumps") from one allowed, or stationary, state of energy E_2 to another state of lower energy E_1. A jump can occur only between stationary states for which the quantum condition of (25.10) holds.
4. When the atom does jump between states and emits radiation, the frequency of the radiation is determined by Einstein's frequency condition

$$h\nu = E_2 - E_1 = \Delta E \qquad (25.11)$$

Niels Bohr introduced the concept of the quantization of angular momentum and added it to the concept of quantization of energy. Planck's constant entered into the formulation in two distinct ways, firstly in the quantization condition for angular momentum, and secondly when the frequency of the emitted radiation was determined. With these assumptions, we can determine the energy states of the hydrogen atom and the frequency of the spectral lines using only the rudiments of introductory physics.

The dynamical situation is shown in Figure 25.7a, in which an electron of charge $-e$ rotates about the much more massive nucleus of charge $+Ze$. The term Z is the atomic number; for hydrogen, $Z = 1$; for the singly ionized He^+, $Z = 2$; for the doubly ionized Li^{2+}, $Z = 3$; and so forth. The velocity is v, and the radius of the circular orbit

---

NIELS BOHR (1885–1962), Danish physicist, studied with J. J. Thomson and Ernest Rutherford in England after receiving his Ph.D. in Copenhagen. It was in Rutherford's laboratory that he received his introduction to the nuclear atom and the impetus for developing the Bohr theory of the atom. In the 1930s, Bohr turned his attention from the electron to the nucleus and made important contributions to nuclear physics. In 1943, the British secret service spirited Bohr out of German-occupied Denmark. He was flown to England in the bomb bay of a liberator plane, which had to fly at extreme heights to avoid interception by the Luftwaffe. Oxygen masks were required at that altitude, but when the crew tried to put a mask on Bohr, his head proved too large for the mask. The plane landed in England with Bohr unconscious. Bohr spent the remaining years of the war in the United States working on the development of the atomic bomb along with Fermi and other recent emigrés. He returned to Denmark after the war, becoming head of the Atomic Energy Commission of Denmark in 1955. Bohr received the Nobel Prize in physics in 1922 and the first Atoms for Peace award in 1957.

FIGURE 25.7 Motion of electron in orbit. The velocity v is given by ωr, where ω is the angular velocity. (a) Simple model of electron of mass m_e rotating about nucleus of mass M and charge $+Ze$ located at the center of the circular orbit. (b) Illustrating the more refined situation of the electron and nucleus rotating about their common center of mass. The position of the center of mass is found from the condition $m_e r_e = MR$.

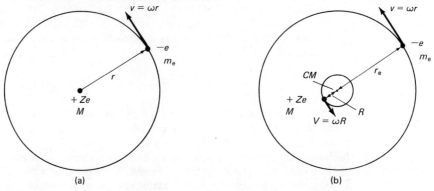

is r. Mechanical stability requires the centrifugal force mv^2/r to be exactly balanced by the electrical attractive force Ze^2/r^2, or

$$\frac{Ze^2}{r^2} = \frac{mv^2}{r} \tag{25.12}$$

leading to

$$mv^2 r = Ze^2 \tag{25.13}$$

From the quantization condition of Equation (25.10) we can write

$$n^2 v^2 r^2 = \frac{n^2 h^2}{4\pi^2} \tag{25.14}$$

or

$$v^2 = \frac{n^2 \hbar^2}{r^2 m^2} \tag{25.15}$$

This value for v^2 can now be used in (25.13) to get

$$mr \frac{n^2 \hbar^2}{r^2 m^2} = Ze^2 \tag{25.16}$$

which gives the allowed radii of the orbits as

$$r = \frac{n^2 \hbar^2}{mZe^2} \tag{25.17}$$

The radius of the smallest allowed orbit in the hydrogen atom ($Z = 1$ and $n = 1$) is

$$r_0^H \equiv a_0 = \frac{\hbar^2}{me^2}$$

$$= 0.529 \times 10^{-8} \text{ cm} = 0.529 \text{ Å} \tag{25.18}$$

SEC. 25.10 THE BOHR HYDROGEN ATOM

The symbol a_0 is used for the radius of the smallest orbit of the hydrogen atom; it is called the *Bohr radius*. (In certain fields of atomic and molecular physics, it is convenient to use a set of units known as *atomic units*, in which the unit of length is taken as a_0.)

The total energy of the atom is

$$E = T + V \tag{25.19}$$

where T is the kinetic energy and V the potential energy. Using (25.12), we find that

$$T = \frac{1}{2} mv^2 = \frac{1}{2} \frac{Ze^2}{r} \tag{25.20}$$

The potential energy is

$$V = -\frac{Ze^2}{r} \tag{25.21}$$

leading to a total energy of

$$E = \frac{1}{2} \frac{Ze^2}{r} - \frac{Ze^2}{r} = -\frac{1}{2} \frac{Ze^2}{r} \tag{25.22}$$

The final expression for the energy is obtained by substituting the value for r as expressed in (25.17), to get

$$E_n = -\frac{mZ^2 e^4}{2n^2 \hbar^2} \tag{25.23}$$

Equation (25.23) must rank as one of the most important equations derived in the history of science; it laid the groundwork for all we know about chemical bonding among other things. Bohr's procedure was overly simplified, but it did accurately predict the results of spectroscopic studies of the hydrogen atom and was the first step in the leap to quantum mechanics.

Using (25.23), we can calculate the frequencies of the spectral lines of atomic hydrogen, for which $Z = 1$. What happens in emission spectroscopy is simple. Hydrogen atoms are excited to some higher energy level by a process that can transfer energy, such as electron bombardment or heating in a carbon arc. The atoms then fall back to lower energy levels, emitting the characteristic radiation in the process. The frequency of the emitted radiation is

$$h\nu = E_2 - E_1 = \frac{me^4}{2\hbar^2} \left(\frac{1}{n_1^2} - \frac{1}{n_2^2} \right) \tag{25.24}$$

Since $\nu = c/\lambda$, Equation (25.24) can be written in terms of the spectroscopic unit called the wave number, $\bar{\nu} = 1/\lambda$, as

$$\bar{\nu} = \frac{1}{\lambda} = \frac{2\pi^2 me^4}{h^3 c} \left(\frac{1}{n_1^2} - \frac{1}{n_2^2} \right) \tag{25.25}$$

Comparing Equation (25.25) with (25.4) indicates that the spectroscopy paradox is solved, and that the Rydberg constant is

$$R = \frac{2\pi^2 me^4}{h^3 c} \tag{25.26}$$

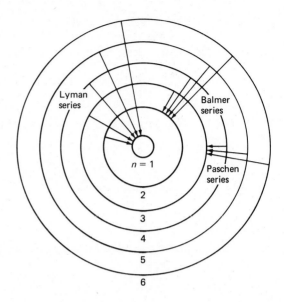

FIGURE 25.8 Transitions between various orbits in the Bohr atom, giving rise to the various spectral series. Note that the radii of the various orbits are not drawn to scale.

When the numerical values are substituted for the natural constants of (25.26), a value for the Rydberg constant is obtained in excellent agreement with the observed value. It was an amazing feat. The Rydberg constant was one of the most accurately known natural constants in experimental science, and Bohr determined its value. Stationary states and the quantum were here to stay.

Figure 25.8 shows the various Bohr orbits and the transitions that give rise to the various spectral series. Figure 25.9 shows this information on an energy-level diagram often called a *term* diagram. The unit of energy is the wave number. The advantage of this unit is made apparent from this type of diagram; the vertical distance between any two levels is a measure of the energy difference between the levels in units of $\bar{\nu}$. The wavelength of the light emitted when the atom drops from the upper level to the lower is simply $(\bar{\nu})^{-1}$. The frequency is

$$\nu = c/\lambda = c\bar{\nu} \tag{25.27}$$

Table 25.1 lists conversion factors for the different energy units.

TABLE 25.1 Conversion factors for different units of energy.

To convert from \ To convert to	erg molecule^{-1}	eV	cm^{-1}	kJ mol^{-1}	Frequency (MHz)
			Multiply by		
erg molecule^{-1}	.....	6.2420×10^{11}	5.035×10^{15}	6.025×10^{13}	1.509×10^{20}
eV	1.602×10^{-12}	.....	8067	96.44	2.418×10^{8}
cm^{-1}	1.986×10^{-16}	1.240×10^{-4}	.....	0.01196	2.998×10^{4}
kJ mol^{-1}	1.660×10^{-14}	0.01037	83.61	.....	2.505×10^{6}
Frequency (MHz)	6.626×10^{-21}	4.136×10^{-9}	3.336×10^{-5}	3.992×10^{-7}	.....

SEC. 25.11 REFINEMENTS OF THE BOHR MODEL

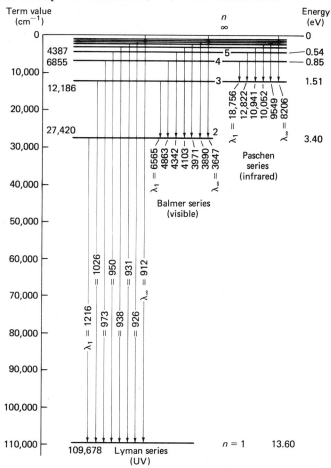

FIGURE 25.9 Term (or energy-level) diagram for hydrogen, showing transitions for the Lyman, Balmer, and Paschen series. The wavelengths of the emission lines are indicated in angstrom units for the transitions up to $n = 8$, and for $n = \infty$. The energies of the levels are indicated in units of reciprocal centimeters (cm^{-1}) and in electron volts.

25.11 REFINEMENTS OF THE BOHR MODEL

We derived the formulas in the previous section using the simple picture of Figure 25.7a. In reality, the electron does not rotate about the nucleus but about the center of mass of the atom, as shown in Figure 25.7b. If we call r_e the distance of the electron from the center of mass and r_n the distance of the nucleus from the center of mass, then

$$m_e r_e = m_n r_n$$

If we let $r = r_e + r_n$, then

$$r_e = \frac{m_n}{m_e + m_n} r \qquad r_n = \frac{m_e}{m_e + m_n} r \qquad (25.28)$$

The moment of inertia about the center of mass is[7]

$$I = \sum_i m_i r_i^2$$

which in this case is just

$$I = m_e r_e^2 + m_n r_n^2 \qquad (25.29)$$

This can be reduced to the form

$$I = \left(\frac{m_e m_n}{m_e + m_n}\right) r = \mu r \qquad (25.30)$$

where μ is the *reduced mass* of the system. We shall find this useful later on. For the hydrogen atom, m_n is 1836 times m_e, and the difference between m and μ is only one part in 1836. This refinement is easily applied to the Bohr atom by simply replacing m by μ in the relevant equations. Another refinement was introduced by Sommerfeld, who treated elliptical orbits in addition to circular ones. This treatment required introducing an additional quantum number, the so-called *azimuthal quantum number*, which was a measure of the eccentricity of the elliptical orbit.[8]

The Bohr theory worked beautifully for hydrogen atoms and for ions consisting of a single electron and a multiply charged nucleus. Not only did the theory not allow an adequate representation of the energies of complicated atoms, but there was considerable difficulty involved in treating chemical bonding. It did, however, provide an admirable stepping-stone to the development of quantum mechanics.

25.12 ELECTRON IMPACT SPECTROSCOPY

The stationary-state principle was strikingly confirmed by the experimental work of Franck and Hertz. Consider the energy level diagram of Figure 25.9 for hydrogen. We have previously discussed emission spectroscopy, where an atom is excited to a high energy level and emits radiation as it returns to lower levels. The reverse process is known as absorption spectroscopy. In absorption spectroscopy, the energy absorbed by the atom is what is measured.

Suppose we have a beam of electrons whose energy can be varied, which impinges on a collection of hydrogen atoms. The first excited state ($n = 2$) is 10.19 eV above the ground state ($n = 1$), and the second excited state ($n = 3$) is 12.07 eV above the ground state, and so on to higher excited states. As the energy of the beam is increased from 0 eV, nothing happens until the electrons have an energy of 10.19 eV; at that point there is a sudden decrease in beam intensity as the 10.19-eV electrons excite the hydrogen atoms from the ground level to the first excited level. Above 10.19 eV, the beam current reverts to its original value until the energy reaches 12.07 eV; at that point there is another decrease in intensity; and so on.

This experiment was carried out by Franck and Hertz in 1914, though they used Hg atoms instead of hydrogen atoms; their results are shown in Figure 25.10. Every

---

[7] In terms of I, the angular momentum is $I\omega$, and the kinetic energy is $\tfrac{1}{2}I\omega^2$.
[8] A complete discussion of this can be found in H. E. White, *Introduction to Atomic Spectra* (New York: McGraw-Hill Book Co., 1934). Also see L. C. Pauling and E. B. Wilson, Jr., *Introduction to Quantum Mechanics* (New York: McGraw-Hill Book Co., 1935).

JAMES FRANCK (1882–1964), German physicist, devised the classic experiment in electron-impact spectroscopy along with Gustav Hertz. They showed that collisions between electrons and mercury atoms were elastic for electron energies up to 4.9 eV, at which point energy could be transferred to the mercury atoms. The energy corresponded to the first line in the optical spectrum of mercury, and it provided definitive verification of the energy quantum and the Bohr assumptions that had been published six months before the experiment. Franck's name is associated with the Franck-Condon principle in vibrational spectroscopy. He devoted much effort to elucidating the mechanism of photosynthesis. Franck, too, left Germany during the 1930s as a result of the political situation, and came to the United States, where he became professor of physical chemistry at the University of Chicago. He, too, was involved in the Manhattan project, which led to the development of the atomic bomb. Franck was one of those who in 1945 urged the U.S. government to drop the first atomic bomb on an uninhabited area to demonstrate its effectiveness rather than on a populated area; the report is known as the Franck report. Franck shared the 1925 Nobel Prize in physics with Hertz.

FIGURE 25.10 The Franck-Hertz experiment. A beam of electrons whose energy may be varied collides with mercury atoms; the intensity is measured as a function of electron energy. The deep minima in the curve indicate energies at which the mercury atoms take up energy from the electrons. Each minimum corresponds to an optical transition in the mercury spectrum. (*Note:* The experimental problems associated with generating electron beams cause the intensity to increase monotonically with energy, even in the absence of any target molecule, causing the shift in the base line.)

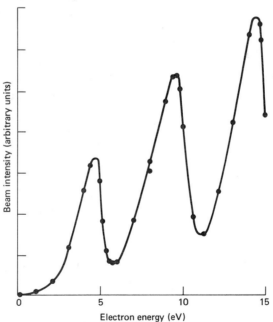

FIGURE 25.11 A "modern" electron impact spectrum of Hg, showing the much higher resolution attainable in modern instruments. In this experiment, Hg vapor was bombarded with 500-eV electrons, and the energy spectrum of the electron beam was scanned after collision with the Hg molecules. The peak at about a 5-eV energy loss corresponds to electrons that have excited the 5-eV peak in Hg and have lost 5 eV of energy; they thus have an energy of 495 eV. Each peak corresponds to a peak in the optical spectrum. (A Skerbele and E. N. Lassettre, *J. Chem. Phys.* 56 (1972):845.)

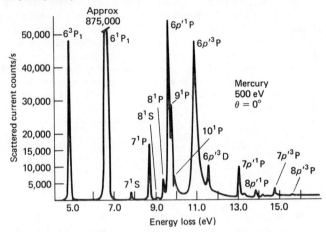

dip in their electron impact spectra corresponded to a known optical frequency in the normal spectrum of mercury. (See also Figure 25.11.)

25.13 IF WAVES ARE PARTICLES, THEN PARTICLES MUST BE WAVES

In 1924, de Broglie suggested, on purely theoretical grounds, that if waves were particles, then particles must be waves. Figure 25.12 shows an electron as a wave in a Bohr orbit. De Broglie showed that one could arrive at the Bohr orbits for hydrogen if the circumference of the orbit contained an integral number of wavelengths. (If the circumference of the orbit was not an integral number of wavelengths, then the wave disappeared because of destructive interference.) This led him to conclude that the wavelength and the momentum, $p = mv$, are related by the equation

$$\lambda = \frac{h}{p} = \frac{h}{mv} \tag{25.31}$$

Equation (25.31) applies to any particle, but is important only for particles of very small masses. For electrons, (25.31) leads to

---

LOUIS VICTOR PIERRE RAYMOND, Prince de Broglie (born 1892), French physicist, is the scion of a noble French family that had its origins when his ancestor emigrated to France from Italy in 1643. De Broglie advanced the early ideas of quantum mechanics when he proposed that just as light could be considered matter, matter could be regarded as having the properties of light waves. He received the Nobel Prize in physics in 1929.

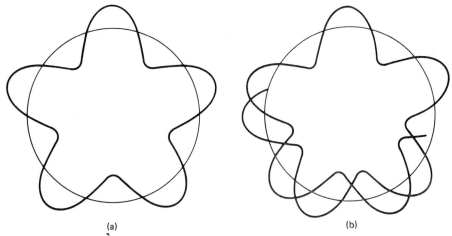

FIGURE 25.12 De Broglie electron wave in a Bohr orbit. (a) An integral number of wavelengths in an orbit leads to a stable situation of standing waves. (b) Destructive interference when the number of wavelengths is not integral.

$$\lambda = \frac{12.263}{V^{1/2}} \times 10^{-8} \text{ cm} \tag{25.32}$$

where V is the energy of the electron in electron volts. A 100-eV electron has a wavelength of 1.226×10^{-8} cm.

De Broglie's proposal was strikingly confirmed within four years when Davisson and Germer of the Bell Telephone Laboratories observed the diffraction of electrons from a crystal of nickel. Diffraction was a property associated with waves, and particles did indeed show wave characteristics.

Bohr's model of the atom was based on the simplest laws of classical mechanics, on which Bohr had grafted the quantization of angular momentum. The suggestion that particles could act as waves led people back to this same problem, only now the starting point was the wave equation instead of Newton's laws of motion. The final result was the much more powerful quantum mechanics or wave mechanics.

PROBLEMS

1. Calculate the energy in one photon of:
   (a) A radio wave with a frequency of 100 MHz.
   (b) A microwave of wavelength 3 cm.
   (c) An infrared wave of wavelength 5×10^{-3} cm.
   (d) A visible light wave of wave length 5×10^{-5} cm.
   (e) An X ray of wavelength 1×10^{-8} cm.
   Calculate each energy in ergs and reciprocal centimeters (cm^{-1}). Compare each to kT at room temperature.

2. Calculate the de Broglie wavelength of:
   (a) An electron that has been accelerated through a potential difference of 500 V.
   (b) A proton that has been accelerated through 500 V.
   (c) A uranium ion that has been accelerated through 500 V.
   (d) A baseball weighing 5 oz and traveling 60 mph.
   (e) A 2-ton auto traveling 60 mph.

3. A metal has a work function of 2 V. Calculate the energy of the ejected photoelectron when light of wavelength 2×10^{-5} cm strikes the surface of the metal.

4. Calculate the radii of the first five Bohr orbits.

5. Calculate the velocities of the electron in the first

five Bohr orbits. Calculate the wavelengths of the electron in the orbits and show that the circumferences of the orbits are integral numbers of wavelengths.

6. Calculate the frequencies of the first five lines in the Lyman series. Sketch the spectrum. The Lyman series is obtained when the electron undergoes a transition between the ground level ($n = 1$) and higher levels. What is the highest frequency line one can get from a hydrogen spectrum? What does the transition between the state $n = 1$ and $n = \infty$ correspond to?

7. The transition between the state $n = 1$ and $n = \infty$ corresponds to a process in which the electron is removed from the atom in the ground state. The energy associated with this process is called the *ionization potential*. Calculate the ionization potential for H, He^+, Li^{2+}, Be^{3+}, and C^{5+}.

8. Derive Equation (25.23) for the more correct case in which the reduced mass μ is used instead of the simple mass m. (See Figure 25.7b).

9. Positronium is an atom formed when an electron and a positron come together. Compared with an electron, a positron has equal mass and an equal and opposite charge. Calculate the first Bohr radius for positronium and compare it with the same measurement for hydrogen. Calculate the ionization potential, and the energy it takes to excite positronium from the $n = 1$ to $n = 2$ state; compare with the values for hydrogen. Calculate the frequency of light needed to accomplish these transitions. (Note that positronium is an exceedingly short-lived species.)

10. Calculate the difference in wavelengths for the first line in the Lyman series for ^1H (ordinary hydrogen), ^2H (deuterium), and ^3H (tritium). This difference is due to differences in the reduced mass.

11. In the accompanying table are data listing the energy of the photoelectrons as a function of the wavelength of the incident light for a metal. Plot the energy as a function of frequency and determine the work function, the threshold frequency, and Planck's constant from the data.

λ (Å)	Energy (eV)
4240	0.372
3567	0.916
3149	1.374
2876	1.741
2632	2.138
2452	2.477

12. Derive an expression for the maximum in the emission curves of a blackbody.

13. Derive the Stefan-Boltzmann law, Equation (25.5), by integrating the blackbody radiation law over all frequencies. Calculate the Stefan-Boltzmann constant in terms of the natural constants in your expression.

14. How much energy per square centimeter of blackbody must be added to maintain a constant temperature of 2000 K for 1 hr? (*Note:* Equation (25.7) gives the energy density of a blackbody radiator. This must be multiplied by c, the velocity of light, to get the energy emitted by a volume equivalent to a rectangular parallelepiped of 1 cm^2 cross section and 3×10^{10} cm length.)

CHAPTER TWENTY-SIX
PRELIMINARIES TO QUANTUM MECHANICS

In the material dealing with thermodynamics we found that the necessary mathematical tools were elementary. Simple differentiation, integration, and the simplest types of differential equations were the extent of the mathematical sophistication. A student of reasonable intelligence who has that kind of background should be able to carry out a variety of calculations of a practical nature using the materials we developed in thermodynamics.

Quantum mechanics is different. The necessary mathematical tools are more advanced; partial differential equations, matrices, and group theory are necessary for a complete treatment of the subject. We shall have to be satisfied with less than a complete treatment, since we cannot delve deeply into these areas in a book at this level. Many times only the mathematical results will be presented; the references listed in the Bibliography can supply the interested student with complete details.

When we have completed the topic, you will not be able to solve many practical problems, in contrast with the thermodynamics study. You will probably not be able to calculate the binding energies of many molecules, but you should be able to sense certain things. You will see that hydrogen atoms should indeed unite to form H_2 molecules whereas He atoms do not form He_2 molecules. You will see that the C–H bonds in methane should be oriented towards the corners of a tetrahedron, as you have been told in freshman and organic chemistry. You will see why heat capacities predicted by the principle of equipartition of energy are correct only in the limit of high temperature, and that the differences in electrical conductivity between metals, semiconductors, and insulators are predictable.

Some of you will continue in advanced studies in chemistry, physics, biology, or other fields and will again study quantum mechanics, although at a more advanced level. Many of the loose threads will be brought together at that time, and some perplexing points will be clarified. Those of you who do not continue with such advanced study will have to settle for the knowledge acquired here, incomplete though it might be. The material is fundamental and involves the essence of chemistry as it is ultimately concerned with how atoms come together to form molecules. It is important for students to have some familiarity with this area, even though the familiarity may not be complete.

Our general scheme will be as follows. First we shall present some basic mathematical results that will be useful in subsequent discussions. These include such things as vectors and complex numbers, eigenvalues and eigenfunctions, series expansions and complete sets of functions, operators, and classical vibrating systems. Since this is not a book on applied mathematics, we shall touch only briefly on each of these topics. Many of them should be familiar from studies in calculus and physics.

26.1 VECTORS

The concept of the vector should be familiar from elementary calculus and physics. In the simplest sense, at least in dealing with a space dimensionality of three or two, a vector can be said to have not only magnitude but also direction in space.[1] Vector quantities that we have dealt with thus far include velocities, forces, momenta, and the like. We represent vectors by boldface type. The length of a vector is a scalar quantity called the *magnitude* of the vector. Figure 26.1 shows a vector in three-dimensional space indicating the distance and direction of a point from the origin; such a vector is a *radius* vector. A vector whose magnitude is unity is a *unit vector*. In cartesian coordinates, vectors are conveniently described in terms of their components, or projections, along the three axes. In terms of the unit vectors **i**, **j**, and **k**, the radius vector is

$$\mathbf{r} = x\mathbf{i} + y\mathbf{j} + z\mathbf{k} \tag{26.1}$$

Note that by universal convention, coordinate systems are arranged in a right-handed manner. By that we mean that if the x axis is rotated towards the y axis, then the positive direction of the z axis points towards the direction a right-handed screw would advance if the z axis were a screw being driven into wood.

Vectors can be added or subtracted by the parallelogram rule, as shown in Figure 26.2. The sum of the two vectors **A** and **B** is the diagonal of the parallelogram whose sides are **A** and **B**; one gets the difference by reversing the direction of the relevant vector. Vectors can be added (or subtracted) by adding (or subtracting) their components. Thus if

$$\mathbf{A} = A_x\mathbf{i} + A_y\mathbf{j} + A_z\mathbf{k}$$
$$\mathbf{B} = B_x\mathbf{i} + B_y\mathbf{j} + B_z\mathbf{k}$$

then the sum of **A** and **B** is

$$\mathbf{C} = \mathbf{A} + \mathbf{B} = (A_x + B_x)\mathbf{i} + (A_y + B_y)\mathbf{j} + (A_z + B_z)\mathbf{k}$$
$$= C_x\mathbf{i} + C_y\mathbf{j} + C_z\mathbf{k}$$

The length or magnitude of the vector **R** in Figure 26.1 is $|R| = (x^2 + y^2 + z^2)^{1/2}$. The vertical bars indicate magnitude; for any vector **A**, the magnitude is

$$|\mathbf{A}| = (A_x^2 + A_y^2 + A_z^2)^{1/2} \tag{26.2}$$

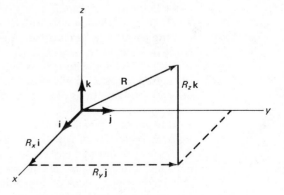

FIGURE 26.1 Resolution of a vector into its components. The vectors **i**, **j**, **k** are the unit vectors.

---

[1] This is a grossly simplified definition of a vector, and valid only in a space dimensionality of three or two or one. More adequate definitions can be found in textbooks dealing with vectors.

FIGURE 26.2 The parallelogram rule for adding and subtracting vectors.

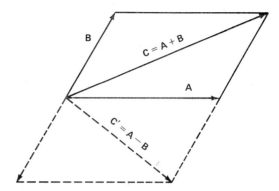

Two vectors are equal if and only if their magnitudes *and* directions are equal. Hence two vectors are equal if and only if all the components are equal.

We define two different vector multiplication operations. The *scalar product* of **A** and **B** is defined by

$$\mathbf{A} \cdot \mathbf{B} = |A||B| \cos \theta \qquad (26.3)$$

where θ is the angle between the vectors. Geometrically it equals the length of one of the vectors multiplied by the projection of the other vector on it. Because of the dot placed between the two vectors in (26.3), the scalar product is often called the *dot product*. If $\theta = 90°$, then the scalar product is zero. For the unit vector components,

$$\mathbf{i} \cdot \mathbf{i} = \mathbf{j} \cdot \mathbf{j} = \mathbf{k} \cdot \mathbf{k} = 1$$
$$\mathbf{i} \cdot \mathbf{j} = \mathbf{j} \cdot \mathbf{i} = \mathbf{j} \cdot \mathbf{k} = \mathbf{k} \cdot \mathbf{j} = \mathbf{i} \cdot \mathbf{k} = \mathbf{k} \cdot \mathbf{i} = 0$$

On this basis, the scalar product of two vectors in terms of their components is

$$\mathbf{A} \cdot \mathbf{B} = (A_x \mathbf{i} + A_y \mathbf{j} + A_z \mathbf{k}) \cdot (B_x \mathbf{i} + B_y \mathbf{j} + B_z \mathbf{k})$$
$$= A_x B_x + A_y B_y + A_z B_z \qquad (26.4)$$

The scalar product of any vector with itself is the square of the magnitude of the vector

$$\mathbf{A} \cdot \mathbf{A} = |A|^2$$

If the scalar product of two vectors is zero, then the vectors are said to be *orthogonal*. Geometrically they are perpendicular.

The *vector product*, or *cross product*, is another vector defined by

$$\mathbf{A} \times \mathbf{B} = |A||B| \mathbf{c} \sin \theta \qquad (26.5)$$

where θ is the angle between the two vectors and **c** is a unit vector perpendicular to the plane formed by the vectors **A** and **B**. The vectors **A**, **B**, and **c** form a right-handed system, as shown in Figure 26.3. Addition and scalar multiplication of

FIGURE 26.3 The vector product $\mathbf{A} \times \mathbf{B} = |A||B|\mathbf{c} \sin \theta$. The unit vector **c** is perpendicular to the plane of **A** and **B**. The vector product equals a vector in the direction of **c** whose magnitude equals the area of the parallelogram.

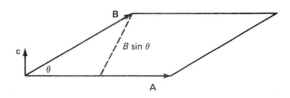

vectors is *commutative*, that is to say, the order is immaterial. Cross multiplication is noncommutative,

$$\mathbf{A} \times \mathbf{B} \neq \mathbf{B} \times \mathbf{A}$$

and in fact

$$\mathbf{A} \times \mathbf{B} = -\mathbf{B} \times \mathbf{A} \tag{26.6}$$

The vector product is often written as a determinant, in the form

$$\mathbf{A} \times \mathbf{B} = \begin{vmatrix} \mathbf{i} & \mathbf{j} & \mathbf{k} \\ A_x & A_y & A_z \\ B_x & B_y & B_z \end{vmatrix}$$
$$= \mathbf{i}(A_y B_z - A_z B_y) + \mathbf{j}(A_z B_x - A_x B_z) + \mathbf{k}(A_x B_y - A_y B_x) \tag{26.7}$$

Equation (26.6) follows immediately from (26.7), since interchanging any two rows (or columns) of a determinant changes the sign of the determinant. An important example of a vector product in classical mechanics and in quantum mechanics is the angular momentum. If a mass m moves with velocity $\mathbf{v}$ at a distance $\mathbf{r}$ from a point, then the angular momentum $\mathbf{L}$ is defined as the vector product

$$\mathbf{L} = m\mathbf{r} \times \mathbf{v} \tag{26.8}$$

26.2 ROTATING VECTORS AND SINUSOIDAL MOTION

Figure 26.4a shows a vector $\mathbf{A}$ on a circle of radius A. We let the vector rotate about the center with constant *angular frequency* ω rad s^{-1}. The *rotational frequency* ν, in hertz or revolutions per second, is

$$\nu = \frac{\omega}{2\pi} \tag{26.9}$$

The *period*, or time required for one revolution, is

$$\tau = \frac{1}{\nu} = \frac{2\pi}{\omega} \tag{26.10}$$

We now examine the y component of $\mathbf{A}$ as a function of time. Suppose we take $t = 0$ at the instant at which $\mathbf{A}$ coincides with the x axis. The vector returns to its starting point at times $t = 0, \tau, 2\tau, 3\tau, \ldots$ The y component is

$$y = A \sin(\omega t) \tag{26.11}$$

The value y is plotted as a function of t in Figure 26.4b. Suppose that instead of taking $t = 0$ as that instant at which $\mathbf{A}$ coincides with the x axis, we take it at that instant at which it makes an angle δ with the x axis, as shown in Figure 26.4c. Then the y component is

$$y = A \sin(\omega t + \delta) \tag{26.12}$$

This is shown in Figure 26.4d. This differs from the previous case in that the two curves are displaced by an amount δ; they are said to be out of phase by δ, which is called the *phase angle*, or *phase constant*. Since $\sin \theta = \cos(\theta - \pi/2)$, the cosine and sine functions are 90° out of phase with each other; Equations (26.11) and (26.12)

SEC. 26.2 ROTATING VECTORS AND SINUSOIDAL MOTION

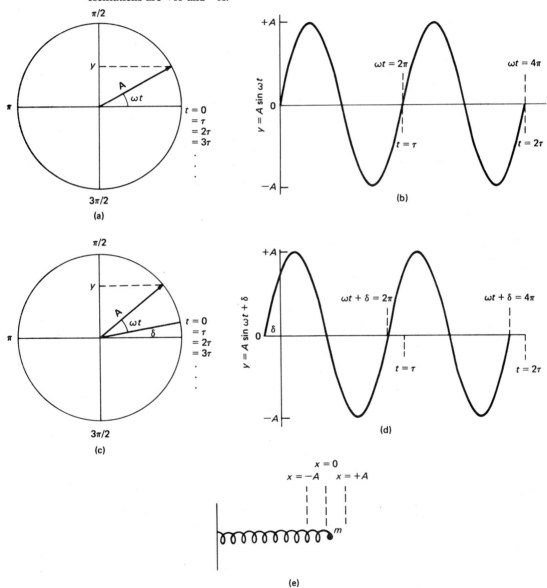

FIGURE 26.4 Rotating vectors and sinusoidal motion. The vector returns to its starting position at time intervals given by the period τ. (a) The origin of time $t = 0$ is taken at the instant the vector **A** coincides with the x axis. (b) The sinusoidal motion of 26.4a. (c) The origin of time $t = 0$ is taken when the vector **A** makes an angle δ with the x axis. (d) The sinusoidal motion of 26.4c. (e) The mass oscillating on a spring. The equilibrium position is at $x = 0$; the limits of the oscillations are $+A$ and $-A$.

could be written in terms of cosine functions, with the addition of $\pi/2$ to the phase angle.

An important mechanical system whose motion is sinusoidal is the mass oscillating on a massless spring, shown in Figure 26.4e. According to Hooke's law, the force the spring exerts is proportional to the displacement from the equilibrium position, $f = -kx$, where k is the force constant of the spring and x the displacement from equilibrium. Since the force equals the mass times the acceleration,

$$f = ma = m\frac{d^2x}{dt^2} = -kx$$

$$\frac{d^2x}{dt^2} = -\frac{kx}{m} \tag{26.13}$$

Equation (26.13) is a differential equation whose solution is

$$x = A \sin\left[\left(\frac{k}{m}\right)^{1/2} t + \delta\right] \tag{26.14}$$

This can be verified by direct substitution. Comparison with (26.12) indicates that the angular frequency of the oscillating mass is $\omega = (k/m)^{1/2}$. If the time scale is set so that $t = 0$ at $x = 0$, then the phase angle is 0.

The potential energy of the oscillating mass is $\frac{1}{2}kx^2$, and the kinetic energy is $\frac{1}{2}mv^2$. Conservation of energy requires the sum of the two to be constant. If we compute the total energy for the instant at which the displacement is a maximum, then $v = 0$, and

$$E = \frac{kA^2}{2} \tag{26.14a}$$

The energy of the oscillator is directly proportional to the square of the amplitude. This is true for all classical harmonic motion.

26.3 COMPLEX NUMBERS AND EULER'S FORMULA

A complex number is a number that contains $\sqrt{-1}$. The square root of a negative number cannot be represented by a real number. The symbol i is used to represent $\sqrt{-1}$. Complex numbers contain real and imaginary parts and are represented by the form

$$z = x + iy \tag{26.15}$$

where both x and y are real numbers. There are two components and hence two degrees of freedom to complex numbers; a two-dimensional system is needed to represent complex numbers, as shown on the *Argand diagram* of Figure 26.5. The distance of the point $z = x + iy$ from the origin is

$$|z| = (x^2 + y^2)^{1/2} \tag{26.16}$$

The quantity $|z|$ is the *magnitude*, or *absolute value*, of z; this is similar to the definition of the magnitude of a vector. We get the *complex conjugate* of a complex number by replacing i by $-i$. The complex conjugate of $z = x + iy$ is

$$z^* = x - iy \tag{26.17}$$

SEC. 26.3 COMPLEX NUMBERS AND EULER'S FORMULA

FIGURE 26.5 The Argand diagram, representing the complex number $z = x + iy$.

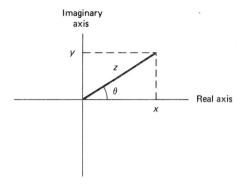

Hence the magnitude of a complex number is

$$|z| = (zz^*)^{1/2} = (x^2 + y^2)^{1/2} \tag{26.18}$$

A complex number and its complex conjugate have the same magnitude; a real number is one whose complex conjugate equals itself.

Complex numbers can be added according to the usual rules of vector addition. If $z_1 = x_1 + iy_1$ and $z_2 = x_2 + iy_2$, then the sum is

$$z = z_1 + z_2 = (x_1 + x_2) + i(y_1 + y_2) \tag{26.19}$$

If two complex numbers are equal, then both their real and imaginary parts must be equal.

A glance at Figure 26.5 indicates that if we call the magnitude of the complex number r and let θ be the angle between r and the real, or x axis, the complex number can be written

$$z = x + iy = r\cos\theta + ir\sin\theta = r(\cos\theta + i\sin\theta) \tag{26.20}$$

Now we write the power series for $\exp(x)$, $\sin x$, and $\cos x$:

$$\exp(x) = 1 + x + \frac{x^2}{2!} + \frac{x^3}{3!} + \frac{x^4}{4!} + \cdots$$

$$\sin x = x - \frac{x^3}{3!} + \frac{x^5}{5!} - \frac{x^7}{7!} + \frac{x^9}{9!} - \cdots$$

$$\cos x = 1 - \frac{x^2}{2!} + \frac{x^4}{4!} - \frac{x^6}{6!} + \frac{x^8}{8!} - \cdots$$

Letting $x = i\theta$ and using the above three series, we can write

$$\exp(i\theta) = \cos\theta + i\sin\theta \tag{26.21}$$

This is *Euler's theorem;* it allows us to write complex numbers in the exponential form

$$z = r\exp(i\theta) \tag{26.22}$$

This form is often much simpler to use than the cartesian form. The complex conjugate of $z = r\exp(i\theta)$ is

$$z^* = r\exp(-i\theta) \tag{26.23}$$

It is also true that

$$zz^* = r^2 e^{i\theta} e^{-i\theta} = r^2 \tag{26.24}$$

The quantity r is the absolute value, or the *modulus,* of z; θ is the *phase* of z. It follows from (26.21) that

$$\sin\theta = \frac{e^{i\theta} - e^{-i\theta}}{2i} \tag{26.25}$$

and

$$\cos\theta = \frac{e^{i\theta} + e^{-i\theta}}{2} \tag{26.26}$$

26.4 OPERATORS

An *operator* is a symbol that indicates that a particular operation is to be performed on what follows the operator. The symbol $\sqrt{}$ is the square root operator. It indicates that the square root of the function following the operator is to be taken; thus $\sqrt{4} = 2$. The function x may be an operator; when it operates on θ, as in $x\theta$, it means that θ is multiplied by x. The differential operator d/dx means that what follows is to be differentiated with respect to x; thus

$$\frac{d}{dx}(x^2) = 2x$$

We denote operators by the symbol $\hat{}$, as in the operator $\hat{P}$.

If the operator $\hat{P}$ is to be applied several times in succession, then it is written as a power in the form $\hat{P}\hat{P}\hat{P}f(x) = \hat{P}^3 f(x)$. For example, if $\hat{P} = d/dx$ and $f(x) = x^4$, then

$$\hat{P}\hat{P}\hat{P}(x^4) = \hat{P}^3(x^4) = \frac{d^3}{dx^3}(x^4) = 24x$$

If $\hat{P}$, $\hat{Q}$, and $\hat{R}$ are three operators, then $\hat{P}\hat{Q}\hat{R}f(x)$ means as follows: First operate on $f(x)$ with $\hat{R}$; then operate on the result with $\hat{Q}$; finally, operate on that resulting function with $\hat{P}$. Operators do not in general *commute;* that is to say, $\hat{P}\hat{Q} \neq \hat{Q}\hat{P}$. Thus, if $\hat{P} = \sqrt{}$, $\hat{Q} = 4$, and $f(x) = x^2$, then

$$\hat{P}\hat{Q}f(x) = \sqrt{}[4(x^2)] = \sqrt{}(4x^2) = 2x$$
$$\hat{Q}\hat{P}f(x) = 4[\sqrt{}(x^2)] = 4(x) = 4x$$

When $\hat{P}\hat{Q} = \hat{Q}\hat{P}$, the operators $\hat{P}$ and $\hat{Q}$ are said to *commute.*

A common operator in quantum mechanics is the differential operator $\partial/\partial x$. For $\hat{P} = (\partial/\partial x)_y$ and $\hat{Q} = (\partial/\partial y)_x$, we have

$$\hat{P}\hat{Q} = \frac{\partial}{\partial x}\left(\frac{\partial}{\partial y}\right) = \frac{\partial^2}{\partial x\,\partial y} \qquad \hat{Q}\hat{P} = \frac{\partial}{\partial y}\left(\frac{\partial}{\partial x}\right) = \frac{\partial^2}{\partial y\,\partial x}$$

In this instance the two operators commute, since the mixed partial derivatives are equal. On the other hand, if $\hat{P} = x$, and $\hat{Q} = d/dx$, then the two do not commute; this can be verified by operating on $f(x) = x$ in both orders.

Operators can be complex; they can also be vectors. The operator called *del,*

$$\nabla = \frac{\partial}{\partial x}\mathbf{i} + \frac{\partial}{\partial y}\mathbf{j} + \frac{\partial}{\partial z}\mathbf{k}$$

converts the scalar function $f(x, y, z)$ into a vector. The function ∇f is the *gradient* of f. An operator used frequently in quantum mechanics is the *Laplacian* operator, obtained by taking the scalar product of del with itself, $\nabla^2 = \nabla \cdot \nabla$. This operator is read *del squared*:

$$\nabla^2 = \frac{\partial^2}{\partial x^2} + \frac{\partial^2}{\partial y^2} + \frac{\partial^2}{\partial z^2} \tag{26.27}$$

In spherical polar coordinates, the Laplacian operator takes the form[2]

$$\nabla^2 = \frac{1}{r^2}\frac{\partial}{\partial r}\left(r^2 \frac{\partial}{\partial r}\right) + \frac{1}{r^2 \sin\theta}\frac{\partial}{\partial \theta}\left(\sin\theta \frac{\partial}{\partial \theta}\right) + \frac{1}{r^2 \sin^2\theta}\frac{\partial^2}{\partial \phi^2} \tag{26.28}$$

Rotation operators make up another important class of operators. If we consider the complex number z a vector in the complex plane, then the effect of the operator $e^{i\theta}$ on z is to rotate the vector through an angle θ. Rotation operators are a subclass of *symmetry* operators. Symmetry operators perform symmetry operations such as reflections, rotations, inversions, screw rotations, and others. Symmetry operators are very important in advanced treatments of quantum mechanics.

26.5 EIGENVALUES AND EIGENFUNCTIONS

A common problem is to find a function ψ such that the equation

$$\hat{P}\psi = p\psi \tag{26.29}$$

is valid where $\hat{P}$ is an operator and p is a constant. Equation (26.29) is an *eigenvalue* equation. It states simply that when the function ψ is operated on by $\hat{P}$, the result is that same function multiplied by a constant number p. The function is called an *eigenfunction* of the operator $\hat{P}$; the constant number p is the *eigenvalue*. A given operator can give rise to more than one eigenfunction; the number of eigenvalues for each eigenfunction can also be greater than one.

We consider as an example the operator $\hat{P} = d^2/dt^2$ and the function $f(x) = x$. An eigenvalue equation can be written as

$$\frac{d^2 x}{dt^2} = -ax \tag{26.30}$$

The term x is the eigenfunction and a is the eigenvalue. This equation is identical to (26.13). Taking $\delta = 0$ in Equation (26.14), we can write the solution as

$$x = C \sin(\omega t) \tag{26.31}$$

where C is a constant and $\omega = \sqrt{a}$. Additional eigenfunctions can be written in the form

$$x = C \cos(\omega t)$$
$$x = C \exp(i\omega t)$$

Any linear combination of these three eigenfunctions is also an eigenfunction of the original operator.

---

[2] This can be verified by following the various chain rules for partial differentiation using the equations that relate the cartesian coordinates to the spherical polar coordinates. See Problem 10.

26.6 SERIES EXPANSIONS AND COMPLETE SETS

A well-behaved function can be expanded in a *Taylor series* about the point a in the form

$$f(x) = f(a) + f'(a)\frac{(x-a)}{1!} + f''(a)\frac{(x-a)^2}{2!} + \cdots$$
$$+ f^{(n)}(a)\frac{(x-a)^n}{n!} + \cdots \quad (26.32)$$

where $f^{(n)}(a)$ is the nth derivative evaluated at a. For the special case of $a = 0$, the series in (26.32) is a *Maclaurin series*.

A much more general set of functions for the expansion of arbitrary functions comprises the sine and cosine functions. In the range $-\pi < x < +\pi$, almost any well-behaved function can be expressed in the series expansion[3]

$$f(x) = \tfrac{1}{2}a_0 + a_1 \cos x + a_2 \cos 2x + a_3 \cos 3x + \cdots + a_n \cos nx + \cdots$$
$$+ b_1 \sin x + b_2 \sin 2x + b_3 \sin 3x + \cdots + b_n \sin nx + \cdots$$
$$= \tfrac{1}{2}a_0 + \sum_{n=1}^{\infty}(a_n \cos nx + b_n \sin nx) \quad (26.33)$$

which is known as the *Fourier series*. The sine part of the series is often called the Fourier *sine series,* and the cosine part is called the *cosine series*.

The functions 1, $\sin x$, $\cos x$, $\sin 2x$, $\cos 2x$, ... constitute what we call an *orthogonal* set in the interval $(-\pi, \pi)$. That is to say, for the integers $m, n = 1, 2, 3, 4, \ldots$

$$\int_{-\pi}^{\pi} \cos mx \cos nx \, dx = 0 \quad (\text{if } m \neq n) \quad (26.34a)$$

$$\int_{-\pi}^{\pi} \sin mx \sin nx \, dx = 0 \quad (\text{if } m \neq n) \quad (26.34b)$$

and regardless of whether m and n are the same or different

$$\int_{-\pi}^{\pi} \sin mx \cos nx \, dx = 0 \quad (26.34c)$$

We have previously come across the concept of orthogonality in the context of vectors. Two vectors are said to be orthogonal if their scalar product is zero. In a geometric sense, any vector can be made up of a linear combination of orthogonal unit vectors. The concept of vectors can be extended to functions. The functions 1, $\sin x$, $\cos x$, $\sin 2x$, $\cos 2x$, ... can be considered a set of orthogonal vectors in a space of infinite dimensionality. The function $f(x)$ is then given by a linear combination of this infinite set of orthogonal vectors. If m and n are equal in (26.34a) and (26.34b), then

$$\int_{-\pi}^{\pi} \cos^2 mx \, dx = \begin{cases} \pi & (\text{if } m \neq 0) \\ 2\pi & (\text{if } m = 0) \end{cases} \quad (26.35a)$$

$$\int_{-\pi}^{\pi} \sin^2 mx \, dx = \pi \quad (26.35b)$$

---

[3] Strictly speaking, a function does not have to be continuous to be expandable in a Fourier series. For more precise criteria, see any textbook dealing with Fourier series, such as R. V. Churchill, *Fourier Series and Boundary Value Problems* (New York: McGraw-Hill Book Co., 1941).

The factor 2 in the cosine integral for $m = 0$ accounts for the factor $\frac{1}{2}$ in the coefficient of a_0 in Equation (26.33).

To apply the Fourier series expansion to a particular function, we must evaluate the coefficients a_n and b_n. If we multiply both sides of (26.33) by $\cos nx$ and integrate between the limits $-\pi$ and $+\pi$, we find that all the terms on the right except for the nth term are zero, and

$$a_n = \frac{1}{\pi} \int_{-\pi}^{\pi} f(x) \cos nx \, dx \tag{26.36}$$

If we multiply by $\sin nx$ and integrate, we find that

$$b_n = \frac{1}{\pi} \int_{-\pi}^{\pi} f(x) \sin nx \, dx \tag{26.37}$$

As an example, we shall expand the function

$$f(x) = +1 \qquad (0 < x < \pi)$$
$$f(x) = -1 \qquad (-\pi < x < 0)$$

which is the square wave shown in Figure 26.6; it has discontinuities (at $x = 0$ and $x = \pm\pi$), but it is sufficiently well behaved for the Fourier series expansion to be valid. The cosine coefficients a_n are determined from the expression

$$a_n = \frac{1}{\pi} \int_{-\pi}^{\pi} f(x) \cos(nx) \, dx = 0 \tag{26.38a}$$

while the sine coefficients are given by

$$b_n = \frac{1}{\pi} \int_{-\pi}^{\pi} f(x) \sin(nx) \, dx = \begin{cases} \dfrac{4}{n\pi} & (n \text{ odd}) \\ 0 & (n \text{ even}) \end{cases} \tag{26.38b}$$

The series expansion is

$$f(x) = \frac{4}{\pi}\left(\frac{\sin x}{1} + \frac{\sin 3x}{3} + \frac{\sin 5x}{5} + \cdots + \frac{\sin nx}{n} + \cdots\right) \quad (n \text{ odd}) \tag{26.39}$$

In Figure 26.6 we show the Fourier expansion of $f(x)$. The expansion approaches the shape of $f(x)$ more and more closely as the number of terms increases.

The functions $1, \sin x, \cos x, \sin 2x, \cos 2x, \ldots$ form what is known as a *complete set*. By that we mean that the functions form a set of orthogonal functions that can be used as the basis for a series expansion of any function in the way we have just described. There are other sets of complete functions. The *Legendre polynomials* form a complete set of orthogonal functions; we shall examine these functions later in connection with rotational spectroscopy and the hydrogen atom. The *Tchebycheff polynomials* are useful in expanding functions in computer programming. Other complete sets of functions we shall meet are the *Hermite* polynomials, associated with vibrations of molecules, and the *Laguerre* polynomials, associated with solutions of the radial wave equation. Complete sets of functions often arise from solutions of *boundary value problems* in differential equations.

The functions

$$\frac{1}{\sqrt{2\pi}}, \quad \frac{\sin x}{\sqrt{\pi}}, \quad \frac{\cos x}{\sqrt{\pi}}, \quad \frac{\sin 2x}{\sqrt{\pi}}, \quad \frac{\cos 2x}{\sqrt{\pi}}, \quad \ldots$$

FIGURE 26.6 Fourier expansion of the function

$$f(x) = +1 \quad (0 < x < \pi)$$
$$f(x) = -1 \quad (-\pi < x < 0)$$

(a) Curves obtained by taking the first term, and the first two and the first three terms in the series. As more and more terms are taken, the curve approximates $f(x)$ more closely. (b) The Fourier series expansion after ten terms. (c) The Fourier series expansion after 100 terms. Now the expansion approximates the function quite closely, except near the points of discontinuity.

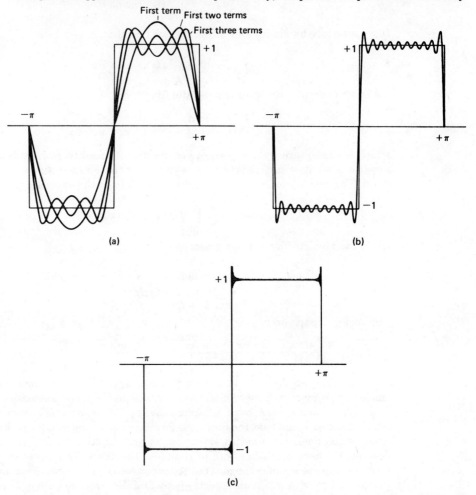

form what we call a complete *orthonormal* set. This set of functions is *normalized* as well as orthogonal. The normalization requirement for a set of functions $f_1, f_2, f_3, \ldots$ is expressed by

$$\int f_n(x) f_n(x)\, dx = 1 \tag{26.40}$$

and the orthogonality requirement is expressed by

$$\int f_m(x) f_n(x)\, dx = 0 \qquad (m \neq n) \tag{26.41}$$

A complete orthonormal set of functions satisfies both (26.40) and (26.41). The two equations are often combined in one equation, which is written

$$\int f_m(x) f_n(x)\, dx = \delta_{mn} \tag{26.42}$$

where δ_{mn} is known as the *Kronecker delta*; $\delta_{mn} = 1$ for $m = n$, and $\delta_{mn} = 0$ for $m \neq n$.

26.7 DIFFERENTIAL EQUATIONS

A curve in two dimensions can be described by the function $y = f(x)$. The slope of the curve is the derivative $dy/dx = f'(x)$. A first-order ordinary differential equation relates the derivative to a function of x and y:

$$\frac{dy}{dx} = F(x, y) \tag{26.43}$$

Higher-order differential equations contain higher-order derivatives. A second-order equation is of the form

$$\frac{d^2y}{dx^2} = G\left(x, y, \frac{dy}{dx}\right) \tag{26.44}$$

You should be familiar with the solutions to these from your previous studies. Our procedure for solving these equations is simple. We write the answer and then by substitution in the original equation verify that it is correct. We shall have occasion to deal with *linear* second-order differential equations of the form

$$P(x)\frac{d^2y}{dx^2} + Q(x)\frac{dy}{dx} + R(x)y = 0 \tag{26.45}$$

where P, Q, and R are given functions. This equation can be considered an operator equation with the operator

$$P(x)\frac{d^2}{dx^2} + Q(x)\frac{d}{dx} + R(x)$$

operating on y. A linear operator is one for which the associative law holds. The operator $\hat{S}$ is linear if

$$\hat{S}(f + g) = \hat{S}f + \hat{S}g \tag{26.46}$$

where f and g are any two functions. Solutions to differential equations contain several constants; the number of constants is equal to the order of the equation; the constants are usually determined by the boundary conditions. If $y = f(x)$ is a solution of a linear equation, then $y = Af(x)$ is also a solution, where A is an arbitrary constant.

In three dimensions, the geometrical representation of functional relations takes the form of surfaces instead of curves. The derivative becomes the partial derivative, of which there are several possibilities. At any point on the surface $z = z(x, y)$, we may define two different partial derivatives. The derivative $(\partial z/\partial x)_y$ can be thought of as the slope of the curve formed by the intersection of the surface with the plane $y = $ constant; the derivative $(\partial z/\partial y)_x$ can be thought of as the slope of the curve formed by the intersection of the surface with the plane $x = $ constant. Partial differential equations can often be solved by the method of *separation of variables*.

This method enables us to convert a partial differential equation into ordinary differential equations. A simple example shows how this is done.

Consider the partial differential equation

$$\frac{\partial z}{\partial x} - \frac{\partial z}{\partial y} = 0 \tag{26.47}$$

We first assume that the solution is *separable;* that is,

$$z(x, y) = X(x)Y(y) \tag{26.48}$$

where $X(x)$ is a function of x alone, and $Y(y)$ is a function of y alone. Now we substitute (26.48) in (26.47), noting that $\partial z/\partial x = Y(dX/dx)$, and $\partial z/\partial y = X(dY/dy)$. The original equation can then be written as

$$Y\frac{dX}{dx} - X\frac{dY}{dy} = 0 \tag{26.49}$$

where we now use ordinary derivatives since X and Y are functions of only one variable. If Equation (26.49) is divided by $z = XY$, a new equation is obtained, in which the two variables have been separated into two different terms:

$$\frac{1}{X(x)}\frac{dX(x)}{dx} - \frac{1}{Y(y)}\frac{dY(y)}{dy} = 0 \tag{26.50}$$

Since each term in (26.50) is a function of a different variable, and since each term can be varied independently of the other term, each term must equal a constant. Moreover, they must equal the same constant, since their difference is zero. We can now write the two ordinary differential equations

$$\frac{1}{X}\frac{dX}{dx} = c \qquad \frac{1}{Y}\frac{dY}{dy} = c \tag{26.51}$$

where c is the constant. Solutions to these are

$$X = A\,\exp(cx) \qquad Y = B\,\exp(cy) \tag{26.52}$$

where A and B are arbitrary constants. With these we can write the solution to the original equation,

$$z = D\,\exp[c(x + y)] \tag{26.53}$$

where $D = AB$. By direct substitution you can verify that (26.53) is indeed a solution to (26.47).

26.8 THE WAVE EQUATION IN ONE DIMENSION; THE VIBRATING STRING

A string of length L and uniform mass m is under a tension T; the mass per unit length is $\rho = M/L$. At equilibrium it is pulled at each end by equal and opposite forces T, as in a violin or harp. When the string is plucked and set in motion, it assumes the shape of some continuous curve. We consider small displacements from equilibrium. When the string moves, the forces at either end of a short length dx are still given by T, but although the magnitudes are the same, the directions are no longer opposite. A net force dF_z is developed along the z axis, as shown in Figure

FIGURE 26.7 The forces on the string under tension. The displacements have been grossly exaggerated; Equation (26.56) is valid only for small displacements.

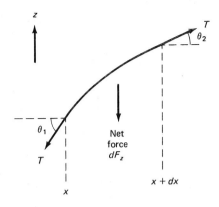

26.7. The force dF_z is the difference in the z component of the forces at either end of dx:

$$dF_z = T \sin \theta_2 - T \sin \theta_1 \simeq T(\tan \theta_2 - \tan \theta_1) \quad (26.54)$$

We have substituted $\tan \theta$ for $\sin \theta$, since for small displacements θ is small and $\tan \theta \simeq \sin \theta$. Since $\tan \theta_2 - \tan \theta_1$ is the change in $\tan \theta$ in going from x to $x + dx$, $\tan \theta_2 - \tan \theta_1 = d(\tan \theta)$, and

$$dF_z = T\, d(\tan \theta) \quad (26.55)$$

To find the force per unit length, we divide (26.55) by dx, to get

$$\frac{dF_z}{dx} = T \frac{d}{dx}(\tan \theta) = T \frac{d^2 z}{dx^2} \quad (26.56)$$

since $\tan \theta = dy/dx$ is just the slope of the string at x.

Equation (26.56) gives the z component of the force. This must equal the z component of acceleration multiplied by the mass per unit length, or

$$\frac{\partial^2 z}{\partial x^2} = \frac{\rho}{T} \frac{\partial^2 z}{\partial t^2} \quad (26.57)$$

Equation (26.57) is the *wave equation* in one dimension.

26.9 SOLUTIONS OF THE WAVE EQUATION; TRAVELING WAVES AND STANDING WAVES

The exact form of the solution to Equation (26.57) depends on the boundary conditions imposed on the problem. Using the method of separation of variables, we assume a solution of the form

$$z(x, t) = X(x)\, \theta(t) \quad (26.58)$$

where X is a function only of x, and θ is a function only of t. Substituting (26.58) into (26.57), we get

$$\theta \frac{\partial^2 X}{\partial x^2} = \frac{\rho}{T} X \frac{\partial^2 \theta}{\partial t^2} \quad (26.59)$$

Dividing by the product $X\theta$, we get

$$\frac{1}{X}\frac{\partial^2 X}{\partial x^2} = \frac{\rho}{T}\frac{1}{\theta}\frac{\partial^2 \theta}{\partial t^2} \tag{26.60}$$

Since the left side of (26.60) is a function only of x, and the right side is a function only of t, each side of (26.60) must separately equal a constant; calling the constant $-\omega^2\rho/T$, we can write

$$\frac{1}{X}\frac{d^2 X}{dx^2} = \frac{\rho}{T}\frac{1}{\theta}\frac{d^2\theta}{dt^2} = -\frac{\omega^2\rho}{T} \tag{26.61}$$

This is equivalent to the two second-order ordinary differential equations

$$\frac{d^2 X}{dx^2} = -\omega^2 \frac{\rho}{T} X \tag{26.62a}$$

$$\frac{d^2\theta}{dt^2} = -\omega^2 \theta \tag{26.62b}$$

The solutions are

$$X(x) = \exp(\pm i\omega t) \tag{26.63a}$$

$$\theta(t) = \exp\left(\pm i\omega x \sqrt{\frac{\rho}{T}}\right) \tag{26.63b}$$

or

$$z = \exp\left[\pm i\omega\left(t \pm \frac{x}{v}\right)\right] \tag{26.64}$$

where the velocity v is $v = (T/\rho)^{1/2} = (\omega/2\pi)\lambda$. Any of the four possible sign combinations in (26.64) is a solution. The solution can be multiplied by any arbitrary complex constant to get a new solution with any arbitrary phase and amplitude. We can take the real part of (26.64) as the solution, or we can take the sum of this solution and its complex conjugate as a real solution.

We can also write the solution as a sine function in the form

$$z = A \sin\left(\omega t - \frac{2\pi x}{\lambda}\right) \tag{26.65}$$

You can verify for yourself that (26.65) is a solution of (26.57). This form of the solution is illustrated under two conditions of observation in Figure 26.8.

In Figure 26.8a, the sinusoidal wave is examined at particular instants as a function of x; it shows what we should see on a photograph taken by the light of a flash of very short duration. The results shown are got by taking three photographs at time intervals δt. The wave travels to the right with velocity v; and it is known as a *traveling* wave. There is another sinusoidal variation in this type of wave; Figure 26.8b shows the second type. Here we fix our attention on a particular point, say $x = x_0$, and examine its motion on the string as a function of time. Since $\sin(\theta) = \cos(\theta - \pi/2)$, we could have written the solution as a cosine function instead of a sine.

Equation (26.65) has only one arbitrary constant, which is A. Since it is the solution of a second-order differential equation, two constants are required. Comparison

SEC. 26.9 SOLUTIONS OF THE WAVE EQUATION

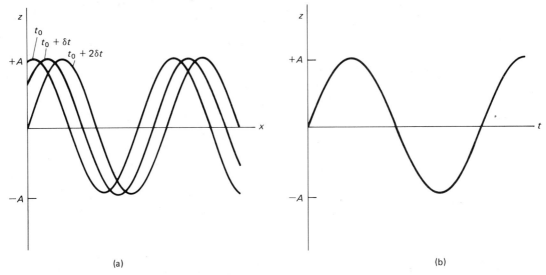

FIGURE 26.8 The traveling wave as a function of distance along the string. The three curves represent a "triple exposure" on a photographic film, with the exposures made at time intervals δt. The crest of the wave travels to the right with velocity v. (b) The traveling wave, showing the motion of a fixed point $x = x_0$ as a function of time.

with (26.12) indicates that the second arbitrary constant is the phase angle δ. A complete solution is

$$z = A \sin\left(\omega t - \frac{2\pi x}{\lambda} + \delta\right) \tag{26.66a}$$

or as the cosine

$$z = A \cos\left(\omega t - \frac{2\pi x}{\lambda} + \delta - \frac{\pi}{2}\right) \tag{26.66b}$$

We shall generally omit the phase angle; this is equivalent to setting $z = 0$ at $x = 0$ and $t = 0$.

The traveling-wave solution applies to an infinitely long string. Another type of solution is obtained for a string of finite length L that is clamped at both ends. Now we have boundary conditions imposed on the problem; it must be true that $z = 0$ at the two fixed ends. The result in this case is a *standing wave*.

An equivalent solution for (26.57) is

$$z = A \genfrac{}{}{0pt}{}{\sin}{\cos}(\omega t) \genfrac{}{}{0pt}{}{\sin}{\cos}\left(\omega \frac{x}{v}\right) \tag{26.67}$$

where we may take any of the four product combinations of the sine and cosine functions. To satisfy the boundary condition at $x = 0$, we must take the sine function for x. To satisfy the boundary condition at $x = L$, we must have

$$\sin \omega \frac{L}{v} = 0$$

or

$$\omega \frac{L}{v} = n\pi \tag{26.68}$$

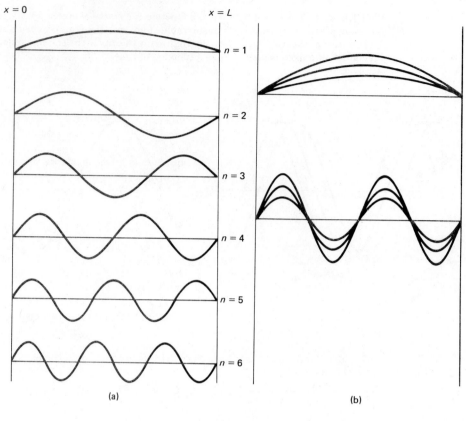

(a)

(b)

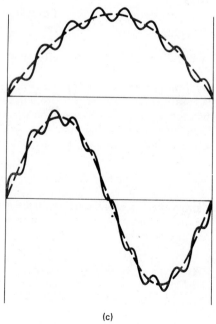

(c)

where $n = 1, 2, 3, \ldots$. We can now write the solution as

$$z = A \sin \omega_n t \sin \frac{n\pi x}{L} \qquad (26.69)$$

where ω_n is defined in (26.68), and we have used the sine function for the time dependence. Equation (26.69) describes a standing wave. Solutions for several values of n are shown in Figure 26.9a. The solution for $n = 1$ is known as the *fundamental*; solutions for n greater than one are *harmonics*, or *overtones*. The set of solutions is known as a set of *normal modes*. Figure 26.9b shows a set of instantaneous positions of the fundamental and the fourth harmonic.

It is worth noting that since the equation is a linear equation, the sum of any set of solutions is also a solution. A more general solution can be built up by *superposing* all the normal modes in the form

$$z = \sum_{n=0}^{\infty} (A_n \cos \omega_n t + B_n \sin \omega_n t) \sin \frac{n\pi x}{L} \qquad (26.70)$$

Note the similarity between (26.70) and the Fourier series. Figure 26.9c shows a string whose vibrations are given by the superposition of two different normal modes. The concert violinist can get mellow tones by being able to activate several overtones superposed upon the fundamental. We shall use the principle of superposition in our studies of quantum mechanics, though we shall apply it to other functions besides sines and cosines.

26.10 THE WAVE EQUATION IN TWO DIMENSIONS; THE VIBRATING MEMBRANE

Figure 26.10 shows a small section of a membrane under uniform tensions, a situation found in kettledrums. For the membrane, the wave equation is

$$\frac{\partial^2 z}{\partial x^2} + \frac{\partial^2 z}{\partial y^2} = \left(\frac{\rho}{T}\right) \frac{\partial^2 z}{\partial t^2} \qquad (26.71)$$

where ρ is the mass per unit area and T is the uniform tension. We consider solutions of (26.71) for standing waves under two different types of boundary conditions, rectangular and circular.

◀ FIGURE 26.9 Standing waves in the string of length L. (a) The first six normal modes of the vibrating string. (b) The first and fourth normal modes of the vibrating string, showing the string at three successive instants. The times are separated by one-sixteenth of the period. (c) Superposition of two normal modes. The upper curve shows the mode with $n = 15$ superposed upon the fundamental with $n = 1$; the shape of the fundamental is indicated by the dashed curve. The overall shape indicated by the dashed curve vibrates with the frequency of the fundamental, while the smaller oscillations vibrate with a frequency 15 times as large. The lower curve shows the mode with $n = 20$ superposed upon the mode with $n = 2$. Here the smaller oscillations vibrate with a frequency 10 times the frequency of the larger oscillations. The curves have been drawn so that the smaller oscillations have amplitude equal to about one-tenth that of the larger oscillations.

FIGURE 26.10 A differential element of the vibrating membrane. The tension T is force per unit length. The components of the force in the z direction are shown along the edges parallel to the y axis; these components are $T(\partial z/\partial x)$. A similar situation holds along the edges parallel to the x axis, where the vertical components of the force are $T(\partial z/\partial y)$. The net restoring force on the differential element equals the sum of all the vertical components.

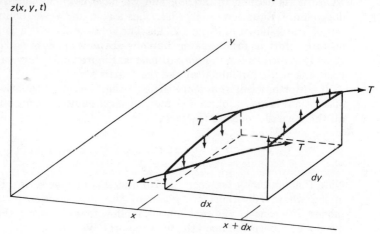

For the rectangular kettledrum, the membrane is in the form of a rectangle, rigidly clamped along the four edges. The solution is

$$z(x, y, t) = A_{mn} \sin\left(\frac{n\pi x}{a}\right) \sin\left(\frac{m\pi y}{b}\right) \cos\left[2\pi vt \left(\frac{n^2}{a^2} + \frac{m^2}{b^2}\right)^{1/2}\right] \quad (26.72)$$

where the velocity $v = \sqrt{T/\rho}$ and m and n are integers (alike or different); a and b are the lengths of the sides of the rectangle. For the special case of the square membrane, $a = b$, and (26.72) can be written

$$z(x, y, t) = A_{mn} \sin\left(\frac{n\pi x}{a}\right) \sin\left(\frac{m\pi y}{a}\right) \cos\left[\frac{2\pi vt}{a}(n^2 + m^2)^{1/2}\right] \quad (26.73)$$

Figure 26.11 shows the appearance of the normal modes of the square membrane for different combinations of n and m. (See also Figure 26.12.) The membrane is divided into areas labeled + and −, referring to whether that portion of the membrane is above or below the plane of the paper at a particular instant. The frequencies of the normal modes are

$$\nu = \frac{v}{2a}(n^2 + m^2)^{1/2} = \frac{\omega}{2\pi} \quad (26.74)$$

For the fundamental, both n and m are unity; and we call the frequency ν_0. The frequencies of the normal modes are indicated in Figure 26.11 in terms of ν_0. Note that they are not integral multiples of the fundamental frequency; this is why the violin makes more melodious sounds than the kettledrum. The normal modes indicated by (1,2) and (2,1) have the same frequencies; they are said to be *degenerate*. It is characteristic of mechanical systems that when normal modes are degenerate, then linear

FIGURE 26.11 The normal modes of the square membrane. The numbers in parentheses under each mode indicate the values of n and m in Equation (26.73). The subscripts S and A indicate whether the mode is symmetric or antisymmetric. The frequencies of the modes are indicated to the right in terms of the fundamental ν_0. Unless n and m are equal, there are two degenerate normal modes with the same frequencies. The plus symbols indicate sections above the plane of the paper, while the minus symbols indicate sections below the plane of the paper. The symbols reverse in a time equal to one-half the period. The lines separating the sections are the *nodal* lines, indicating lines of zero displacement.

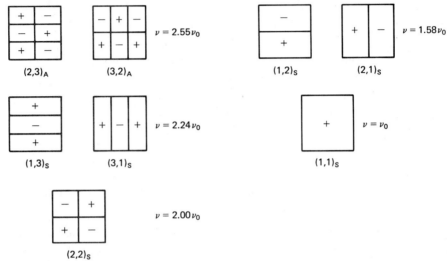

combinations of the degenerate modes are also normal modes. Instead of considering the modes (1,2) and (2,1) normal modes, we could take the two functions

$$F_+ = (1,2) + (2,1) \quad \text{and} \quad F_- = (1,2) - (2,1) \tag{26.75}$$

and consider these the normal modes. These two functions are indicated in Figure 26.13. The procedure whereby two degenerate normal modes are combined to form one new normal mode is *hybridization*. We shall come across this again when we discuss wave functions of atoms and molecules.

FIGURE 26.12 A perspective view of several normal modes of Figure 26.11.

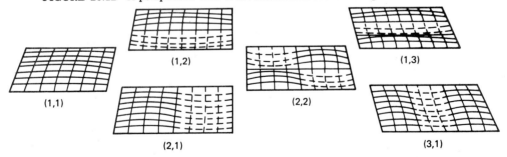

FIGURE 26.13 Hybridization. Two new normal modes can be constructed by a linear combination of the old degenerate normal modes. Here we have given equal weight to the two old modes. The new modes will have the same frequency as the old modes.

$F_+ = (1,2) + (2,1)$

$F_- = (1,2) - (2,1)$

The solution to the circular membrane is expressed in terms of *Bessel functions*. We shall not be using Bessel functions in our future development and so omit further discussion of these. Many references exist on the subject.[4] The discussion of the three-dimensional wave equation will be deferred to the next chapter and the problem of the particle in the three-dimensional box.

26.11 SYMMETRIC AND ANTISYMMETRIC FUNCTIONS

Consider the function $f(x)$. If $f(-x) = +f(x)$, then the function is *symmetric*. In other words, if the function is unchanged when x is replaced by $-x$, then the function is symmetric. On the other hand, if $f(-x) = -f(x)$, then the function is *antisymmetric*. (If $f(-x)$ does not equal $f(x)$ or $-f(x)$, then the function is neither symmetric nor antisymmetric. It is something else, and we shall not even bother to give it a name.)

The sine is antisymmetric, since $\sin(-\theta) = -\sin(\theta)$. The cosine is symmetric, since $\cos(-\theta) = +\cos(\theta)$. The function $y(x) = x$ is antisymmetric, since $y(-x) = -y(x)$; if $y(x)$ is any odd power of x, then $y(x)$ is antisymmetric. The function $y(x) = x^2$ is symmetric, since $y(-x) = +y(x)$; any even power of x is a symmetric function. The terms *even* and *odd* relative to functions are sometimes used in place of *symmetric* and *antisymmetric*. (The function $f(x) = x^3 + 2x^2 + x - 2$ is neither symmetric nor antisymmetric.)

The product of a symmetric function and an antisymmetric function is an antisymmetric function; the functions $x \cos x$ and $x^2 \sin x$ are both antisymmetric. The product of two symmetric functions is a symmetric function; the function $x^2 \cos x$ is symmetric. The product of two antisymmetric functions is also symmetric, which you can see by examining the properties of the function $x \sin x$.

The integral of an antisymmetric function between the limits $\pm c$ is always zero, since the areas on both sides of the origin cancel. The integral of a symmetric function between the same limits is generally not zero.

If we try to expand a symmetric function in a Fourier series, the sine terms all vanish; if we try to expand an antisymmetric function, all the cosine terms vanish. Our example given in Figure 26.6 is an antisymmetric function; the Fourier series consists only of the sine terms, owing to the properties of the integrals (as described in the last paragraph).

The concept can be extended to more than one variable. If

$$f(-x, -y, -z, \ldots) = +f(x, y, z, \ldots)$$

---

[4] The vibrating membrane is treated in most advanced textbooks dealing with differential equations, applied mathematics, Fourier series, and so on. The problem is dealt with in Churchill, *Fourier Series*.

then the function is symmetric, and if

$$f(-x, -y, -z, \ldots) = -f(x, y, z, \ldots)$$

the function is antisymmetric. (We shall return to this point later when we discuss wave functions.)

PROBLEMS

1. For the two vectors

   $\mathbf{A} = 3\mathbf{i} + 2\mathbf{j} - 3\mathbf{k}$ and $\mathbf{B} = \mathbf{i} - 3\mathbf{j} - \mathbf{k}$

   calculate $\mathbf{A} + \mathbf{B}$, $\mathbf{A} - \mathbf{B}$, $\mathbf{A} \cdot \mathbf{B}$, $\mathbf{A} \times \mathbf{B}$, and $\mathbf{B} \times \mathbf{A}$.

2. In (26.8), we have defined the angular momentum $\mathbf{L} = m\mathbf{r} \times \mathbf{v}$ where m is the mass, $\mathbf{r}$ the distance from the point, and $\mathbf{v}$ the velocity. Since $\mathbf{p} = m\mathbf{v}$, then $\mathbf{L}$ can be written in terms of the linear momentum, $\mathbf{L} = \mathbf{r} \times \mathbf{p}$. Evaluate the components of angular momentum L_x, L_y, L_z in terms of x, y, z, and the components of the linear momentum p_x, p_y, p_z.

3. Verify that (26.14) is a solution of (26.13).

4. Plot the following numbers on an Argand diagram in the complex plane: $1 + i$; $1 - i$; $2 + 1.5i$; $3 - 4i$.

5. Write each of the complex numbers of Problem 4 as exponentials of the form $re^{i\theta}$.

6. Evaluate $(1 + i)^4$ in the following two ways:
   (a) Carry out the indicated multiplication three times.
   (b) Express the number in the form $re^{i\theta}$ and raise the exponential to the fourth power. Compare this result with the answer you got in (a).

7. Verify that when $3 + 2i$ is multiplied by $e^{2\pi i}$, the result is $3 + 2i$. Explain.

8. Show that (26.25) and (26.26) are valid.

9. Use Equations (26.25) and (26.26) to show the validity of the integrals written in (26.34) and (26.35).

10. Starting with (26.27) and the equations relating the cartesian and spherical polar coordinates, show that the Laplacian operator in spherical coordinates is given by (26.28). (*Note:* In most textbooks you will find that this is left as an exercise for the student. A discussion and partial solution is given in W. Kaplan, *Advanced Calculus* (Cambridge: Addison-Wesley, 1952), p. 115. A solution based on vector analysis is given in F. B. Hildebrand, *Advanced Calculus for Engineers* (New York: Prentice-Hall, 1949), p. 324).

11. In quantum mechanics, we shall see that the operator associated with the z component of angular momentum is $\hat{L}_z = -i\hbar(\partial/\partial\phi)$. Consider the function $f(\phi) = \Phi = A \exp(im\phi)$. Show that Φ is an eigenfunction of $\hat{L}_z$; find the eigenvalue.

12. The first three Hermite polynomials are

    $H_0(x) = 1$, $H_1(x) = 2x$, and $H_2(x) = 4x^2 - 2$

    The Hermite functions are defined by $A_n H_n(x) \exp(-\frac{1}{2}x^2)$, where A_n is a constant. Show that these Hermite functions are eigenfunctions of the operator $-(d^2/dx^2) + x^2$; find the eigenvalues. (*Note:* We shall come across the Hermite functions again in connection with the vibrating molecule.)

13. Show that the function $\exp(\pm imx)$ is simultaneously an eigenfunction of the operators

    $$\hat{P} = -i\frac{d}{dx} \quad \text{and} \quad \hat{P}' = \hat{P}^2 = -\frac{d^2}{dx^2}$$

    Find the eigenvalues of each of the operators. (See also Problems 14 and 15.)

14. It is generally true that if the function f is an eigenfunction of the operator $\hat{P}$ with eigenvalue p then it is also an eigenfunction of the operator $\hat{P}^2$ with eigenvalue p^2. Prove this. (*Note:* The converse is not true. That is to say, if f is an eigenfunction of $\hat{P}^2$, it is not necessarily an eigenfunction of $\hat{P}$ at the same time.)

15. The *commutator* of two operators $\hat{P}$ and $\hat{Q}$ is defined as the operator given by $\hat{P}\hat{Q} - \hat{Q}\hat{P}$. Show that if the function f is simultaneously an eigenfunction of two operators $\hat{P}$ and $\hat{Q}$, with eigenvalue equations $\hat{P}f = pf$ and $\hat{Q}f = qf$, where p and q are constants, then f is also an eigenfunction of the commutator operator with eigenfunction zero. That is to say

    $$(\hat{P}\hat{Q} - \hat{Q}\hat{P})f = 0f = 0$$

16. Show that the function $f(x, y, z) = \sin(kx) \sin(my) \sin(nz)$ is an eigenfunction of the Laplacian operator. What are the eigenvalues? (Use cartesian coordinates.)

17. Operators often have more than one eigenfunction associated with a particular eigenvalue; the eigenfunctions are then *degenerate*. For the function of Problem 16, consider the constants k, m, and n integers greater than 0. Find some degenerate eigenfunctions.

18. Evaluate the first three terms of the Fourier expansion of the function $f(x) = x$.

19. Evaluate the first three terms of the Fourier expansion of the function $f(x) = x^2$.

20. Show that Equations (26.64) and (26.65) are both solutions of (26.57).

21. The solution for the standing wave in one dimen-

sion with the time dependence removed can be written (see (26.69))

$$\psi_n = A_n \sin \frac{n \pi x}{L}$$

where the A_n are constants. Evaluate the constants A_n in such a way that the condition $\int_0^L \psi^2 \, dx = 1$ is valid. (We shall come across this procedure again when we *normalize* wave functions.)

22. Show that (26.72) is a solution of (26.71).

23. Equation (26.14) expresses the motion of the harmonic oscillator as a sinusoidal wave. The mass oscillates between the limits $x = +A$ and $x = -A$. Take the phase angle δ as zero.
(a) Evaluate the relative probabilities of finding the mass between $x = 0$ and $x = dx$ and finding the mass between $\tfrac{1}{2}A$ and $\tfrac{1}{2}A + dx$.
(b) Where is the probability of finding the mass a minimum?

CHAPTER TWENTY-SEVEN

THE POSTULATES OF QUANTUM MECHANICS. APPLICATIONS TO SIMPLE SYSTEMS

27.1 THE POSTULATES OF QUANTUM MECHANICS

Our development of quantum mechanics will proceed from a set of basic postulates. As we use the basic postulates of quantum mechanics, we shall come to understand their significance and usefulness.

POSTULATE I. *The state of a system of N particles is described as completely as possible by a function* $\Psi(q_1, q_2, \ldots, q_{3N}, t)$ *where the q's are the position coordinates and t is the time.*

POSTULATE II. *The physical significance of the function Ψ, or the wave function as we shall call it, is as follows: if $d\tau$ is a differential volume element, then the probability of finding the system in the volume element $d\tau$ with coordinates $q_1, q_2, \ldots, q_{3N}$ at a specified time t is given by*

$$\Psi^* \Psi \, d\tau \qquad (27.1)$$

Note that Ψ may be complex. Probabilities must, however, be real. Since Ψ^* is the complex conjugate of Ψ, the probability given by (27.1) is real. Figure 27.1a shows a plot of an arbitrary wave function at some specified time. Figure 27.1b shows the plot of $\Psi^* \Psi$ ($= \Psi^2$ for a real wave function). The probability of finding the particle between x and $x + dx$ is given by the area of the strip indicated in Figure 27.1b.

A quantum mechanical wave function that contains the time explicitly is said to be a *time-dependent wave function*. Many observable properties of a system do not change with time. The system is then said to be in a *stationary* state; the time dependence is usually removed from the wave function, to produce what we shall call a *stationary-state wave function*. For these, we can separate the time variable by the method of separation of variables.

For the wave function to be physically meaningful, we must restrict Ψ to a particular class of functions that have the following three characteristics: (1) The wave functions must be *single-valued*. (2) The wave function must be *well-behaved;* that is, both Ψ and its first and second derivatives must be continuous. (3) The wave function must be an integrable square. For our purposes, this means that Ψ is everywhere

FIGURE 27.1 Hypothetical one-dimensional wave function. The function is shown in (a), and (b) indicates a plot of ψ^2. The probability of finding the particle at x in a region of width dx is given by $\psi^2\, dx$ and is illustrated by the strip in (b).

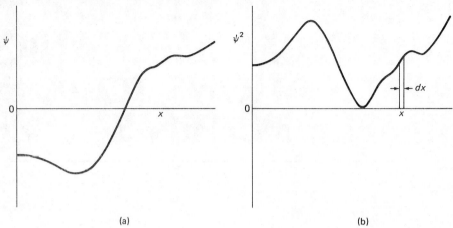

finite and must go to zero at $\pm\infty$. The last restriction arises from the condition that the integral

$$\int_{-\infty}^{\infty} \Psi^* \Psi\, d\tau \tag{27.2}$$

must be finite. To speak of actual probabilities, we may further require that

$$\int_{-\infty}^{\infty} \Psi^* \Psi\, d\tau = 1 \tag{27.3}$$

since the probability of finding the system somewhere must be unity. Wave functions that satisfy Equation (27.3) are said to be *normalized*.

We shall be concerned with properties of the system that can be observed (that is, measured). The next several postulates deal with the procedure that enables us to use the wave functions for predicting the results of experimental measurements.

POSTULATE III. *Every physical observable of a system has associated with it a linear Hermitian operator.*

We have already discussed linear operators (Chapter 26). A Hermitian operator is an operator $\hat{P}$ for which the equation

$$\int \Psi_i^* \hat{P} \Psi_j\, d\tau = \int \Psi_j (\hat{P} \Psi_i)^*\, d\tau \tag{27.4}$$

is valid, where Ψ_i and Ψ_j represent states of the system. We shall defer describing the actual form of these operators to Postulate V.

POSTULATE IV. *Suppose we want to measure the numerical value of a physical observable whose operator is $\hat{P}$. There are two possible situations:*

IVa. *If the state of the system is given by Ψ_i, and Ψ_i is an eigenfunction of the operator $\hat{P}$, with eigenvalue p_i,*

$$\hat{P} \Psi_i = p_i \Psi_i \tag{27.5}$$

then a measurement of the observable P is **certain** to yield the numerical value p_i.

IVb. If Ψ_i is not an eigenfunction of $\hat{P}$, then we cannot predict the result of any **single** measurement of the observable P. If many measurements are made on many identical systems then we may get many different results. We can, however, say that the average value of all the results is

$$\overline{P} \equiv \langle P \rangle = \frac{\int_{-\infty}^{\infty} \Psi_i^* \hat{P} \Psi_i \, d\tau}{\int_{-\infty}^{\infty} \Psi_i^* \Psi_i \, d\tau} \tag{27.6}$$

Note that if the wave functions are normalized, the denominator in (27.6) is unity, and the average value is

$$\langle P \rangle = \int_{-\infty}^{\infty} \Psi_i^* \hat{P} \Psi_i \, d\tau \tag{27.7}$$

The symbol $\langle \ \rangle$ is used to denote average value in quantum mechanics, and is often called the *expectation value*.

POSTULATE V. *The quantum mechanical operators for the dynamical variables of position, momentum, time, and total energy are indicated in the accompanying table. Quantum mechanical operators for other dynamical variables are constructed by writing the classical expression for the particular variable and then substituting according to the table.*

Classical Variable	Quantum Mechanical Operator	Form of the Quantum Mechanical Operator
q (position)	$\hat{q}$	q (unchanged in QM)
p_q (momentum)	$\hat{p}_q$	$-i\hbar \frac{\partial}{\partial q}$
t (time)	$\hat{t}$	t (unchanged in QM)
E (energy)	$\hat{E}$	$i\hbar \frac{\partial}{\partial t}$

We have simplified this last postulate. A rigorous method of converting dynamical variables from the classical expression to the quantum mechanical operator involves relating the commutator of two operators to a quantity known in classical mechanics as a *Poisson bracket*. A discussion of this would lead us far afield, and the interested student is referred to other sources.[1]

The kinetic energy is an important dynamical variable, so let us derive the quantum mechanical operator for the kinetic energy of a particle in three-dimensional space as an illustration of the general procedure. First we write the classical expression for the kinetic energy,

$$T = \frac{1}{2m}(p_x^2 + p_y^2 + p_z^2) \tag{27.8}$$

Then we use the table to substitute the quantum mechanical variables for the classical variables. Note that in many cases operators do not commute; thus it is gener-

---

[1] This point is discussed in Vladimir Rojansky, *Introductory Quantum Mechanics* (New York: Prentice-Hall, 1938), pp. 45, 74.

ally important to keep the order of the operators straight. In this particular case no problems arise. Equation (27.8) becomes

$$\hat{T} = \frac{1}{2m}\left[\left(-i\hbar\frac{\partial}{\partial x}\right)\left(-i\hbar\frac{\partial}{\partial x}\right) + \left(-i\hbar\frac{\partial}{\partial y}\right)\left(-i\hbar\frac{\partial}{\partial y}\right) + \left(-i\hbar\frac{\partial}{\partial z}\right)\left(-i\hbar\frac{\partial}{\partial z}\right)\right]$$

$$= -\frac{\hbar^2}{2m}\left(\frac{\partial^2}{\partial x^2} + \frac{\partial^2}{\partial y^2} + \frac{\partial^2}{\partial z^2}\right)$$

$$\hat{T} = -\frac{\hbar^2}{2m}\nabla^2 \tag{27.9}$$

In classical mechanics the Hamiltonian is the total energy. The quantum mechanical Hamiltonian operator is thus

$$\hat{H} = \hat{T} + \hat{V} \tag{27.10}$$

where $\hat{T}$ is the quantum mechanical operator for kinetic energy we have just got in (27.9), and $\hat{V}$ is the quantum mechanical operator for potential energy. For *conservative* systems, the potential energy is a function only of the coordinates; since the coordinate variable is the same in classical and quantum mechanics, the quantum mechanical Hamiltonian operator becomes

$$\hat{H} = -\frac{\hbar^2}{2m}\nabla^2 + V(q) \tag{27.11}$$

where $V(q)$ is the classical form of the potential-energy expression.

The results of any experimental measurement must be real. Thus, for example, if we want to use (27.5) for determining the angular momentum or the energy of an electron, we must get a real result. This is provided for by requiring that the operator associated with any observable is Hermitian, since it is a theorem of mathema-

---

WILLIAM ROWAN HAMILTON (1805–1865), Irish mathematical physicist, unified optics and mechanics and laid the groundwork for the step from classical to quantum mechanics some ninety years before the advent of quantum mechanics. Hamilton was a prodigy, translating Latin, Greek, and Hebrew by the age of 5. By the time he was 14 he had added Syriac and Persian. At age 17 he had detected an error in Laplace's famous work *Celestial Mechanics*. His earliest work involved optics, and his work in mechanics grew out of his optical researches. His system developed a close analogy between optics and mechanics, and de Broglie and Schrödinger later developed this analogy in studying modern wave, or quantum, mechanics. In 1827, Hamilton was appointed royal astronomer for the Irish observatory, and professor of astronomy at Trinity College. He was one of the worst experimentalists and one of the best theoreticians ever to hold such a position. Observational astronomy did not flourish during his tenure. He made important contributions to algebra, and his name is associated with the mathematics of *quaternions*. This is associated with the extension of the two-dimensional Argand diagram for complex numbers into three dimensions. His epochal discovery of the multiplication table for quaternions, $i^2 = j^2 = k^2 = ijk = -1$, came to him in a flash of genius while he was out for a stroll, and in his excitement Hamilton inscribed the fundamental formula in the stonework of Brougham Bridge. He was less successful in love than in science, having been rejected by two successive ladies before he finally settled down to an unsuccessful marriage. In his later years he relied on alcohol to excess, and this contributed to his untimely death. Upon his death numerous old lambchop bones were found scattered among his papers.

tics that the eigenvalues of a Hermitian operator must be real. We can see this by taking the complex conjugate of Eq. (27.5) in the form

$$\hat{P}^* \Psi_i^* = p^* \Psi_i^* \qquad (27.12)$$

If we now multiply (27.5) by Ψ_i^*, and (27.12) by Ψ_i and integrate both equations, we get

$$\int \Psi_i^* \hat{P} \Psi_i \, d\tau = p_i \int \Psi_i^* \Psi_i \, d\tau \qquad (27.13)$$
$$\int \Psi_i \hat{P}^* \Psi_i^* \, d\tau = p_i^* \int \Psi_i^* \Psi_i \, d\tau \qquad (27.14)$$

Since $\hat{P}$ is a Hermitian operator, we can conclude from (27.4) that the left-hand sides of (27.13) and (27.14) are equal, which leads to the result that

$$p_i = p_i^* \qquad (27.15)$$

This is a necessary and sufficient condition for the eigenvalue p_i to be real.

POSTULATE VI. *The wave function for a system is determined from the relation.*

$$\hat{H} \Psi = i\hbar \frac{\partial \Psi}{\partial t} \qquad (27.16)$$

where $\hat{H}$ is the Hamiltonian operator.

Equation (27.16) is known as the *time-dependent Schrödinger equation*. For most of the systems we are concerned with, $\hat{H}$ will not depend explicitly on the time, and a separation of variables into a space part and a time part will be possible. In these cases we can write the solution to (27.16) as

$$\Psi(q, t) = \psi(q) \exp\left(-\frac{i}{\hbar} W t\right) \qquad (27.17)$$

where W is a constant. By substituting (27.17) in (27.16), we get a new equation with the time dependence removed,

$$\hat{H} \psi(q) = W \psi(q) \qquad (27.18)$$

Equation (27.18) is the *time-independent Schrödinger equation*. We shall be devoting much effort during this and the next several chapters to the solutions of this equation. It is an eigenvalue equation, and is identical in form to (27.5). The eigenvalues W_i provide values of the energy levels in the various stationary states. If we were interested in other observables, such as the z component of angular momentum, it would be necessary to use the operator associated with that observable.

ERWIN SCHRÖDINGER (1887–1961), Austrian physicist, produced his most brilliant work at the age of 39, which is remarkably late for a theoretical physicist. Schrödinger carried de Broglie's wave hypothesis one step further in developing his wave mechanics. In 1927, Schrödinger accepted an invitation to succeed Max Planck, the original inventor of the quantum hypothesis, and went to the University of Berlin, where Einstein had been for some years. In 1933, Schrödinger, although of Roman Catholic origins, decided he could not live in a country devoted to the persecution of Jews, and he left Germany, eventually to settle in Ireland at the Dublin Institute for Advanced Studies. He shared the Nobel Prize in 1933 with P. A. M. Dirac "for the discovery of new and fruitful forms of atomic theory."

There is one more postulate to be introduced, the electron spin postulate. We shall defer that to a later stage after we have gained some familiarity with our first six postulates. In the rest of this chapter we shall treat some simple problems in which the potential energy is constant over some region of space. Later we shall focus our attention on more complicated potentials, particularly the Hooke's law potential and the electrostatic potential. We shall try to keep the presentation simple and details at the minimum necessary for understanding the basic principles. The interested student may refer to the more advanced textbooks on the subject for further ramifications of each of the problems we shall treat.

27.2 THE FREE PARTICLE IN ONE DIMENSION

We know certain things about the free particle in one dimension before we do our detailed calculation. We know that for sufficiently small particles our calculation must yield a wave motion with a wavelength given by the de Broglie relation

$$\lambda = \frac{h}{mv} = \frac{h}{p} \tag{27.19}$$

Applying our postulates to the system must provide results that agree with (27.19).

The system is depicted in Figure 27.2. We simply have a particle moving in field free space. We start by writing the Schrödinger equation for the system in the form $\hat{H}\psi = E\psi$, where we now use E for energy, since there is little chance of confusing E with the energy *operator*. The form of $\hat{H}$ is given in (27.11), and the Schrödinger equation becomes

$$\left[-\frac{\hbar^2}{2m}\nabla^2 + V(q)\right]\psi = E\psi \tag{27.20}$$

In field free space, $V(q) = 0$, and (27.20) becomes

$$-\frac{\hbar^2}{2m}\nabla^2\psi = E\psi \tag{27.21}$$

In one dimension, the Laplacian operator takes the form d^2/dx^2; hence

$$-\frac{\hbar^2}{2m}\frac{d^2\psi}{dx^2} = E\psi \tag{27.22}$$

We have already seen this equation as Equation (26.62a). There are two solutions:

$$\psi_1 = A\ \exp(ikx) \tag{27.23a}$$
$$\psi_2 = B\ \exp(-ikx) \tag{27.23b}$$

where

$$k = \frac{2\pi}{h}(2mE)^{1/2} \tag{27.24}$$

FIGURE 27.2 A free particle traveling in the x direction in potential free space ($V = 0$).

$V = $ constant $= 0$

and A and B are arbitrary constants. An equivalent pair of solutions can be written as

$$\psi_1 = C \sin(kx) \quad (27.25a)$$
$$\psi_2 = D \cos(kx) \quad (27.25b)$$

These correspond to traveling waves with wavelength

$$\lambda = \frac{2\pi}{k} \quad (27.26)$$

as we can see by comparing them with the solutions of the classical wave equation of Chapter 26.

Suppose we want to compute the linear momentum of the particle. We do this with Postulate IVa with the operator equation

$$\hat{P}_x \psi = p_x \psi \quad (27.27)$$

where $\hat{P}_x$ is the operator associated with linear momentum, $-i\hbar(d/dx)$, and p_x is the value of the momentum we should observe. Let's try it on ψ_1 of (27.23a). We get

$$\hat{P}_x \psi_1 = -i\hbar \frac{d}{dx}[A \exp(ikx)]$$
$$= \hbar k [A \exp(ikx)]$$
$$= \hbar k \psi_1 \quad (27.28)$$

The term ψ_1 is an eigenfunction of the linear momentum operator with eigenvalue $\hbar k$. If we had used ψ_2, we should have found that ψ_2 is an eigenfunction of the linear momentum operator with eigenvalue $-\hbar k$. There are two possible results of an experiment that tries to measure the momentum of the particle:

$$p_x = \pm \hbar k = \pm (2mE)^{1/2} \quad (27.29)$$

These are the *only* two possibilities. The positive value corresponds to a particle traveling in the positive x direction, and the negative value corresponds to a particle traveling in the negative x direction. The kinetic energy of the particle is E; remember that classically, the momentum is also $(2mE)^{1/2}$. The wavelength determined by the solution of the Schrödinger equation corresponds exactly with the de Broglie wavelength of (27.19), which you can readily see by substituting for k the expression given in (27.24).

For the traveling particle, both ψ_1 and ψ_2 have the same energy; the two functions are degenerate. A more general solution is given by the linear combination of ψ_1 and ψ_2:

$$\psi_3 = \psi_1 + \psi_2 \quad (27.30)$$

The wave function ψ_3 is still an eigenfunction of the energy, since it is still true that

$$\hat{H}\psi_3 = E\psi_3 = E(\psi_1 + \psi_2) \quad (27.31)$$

The value ψ_3 represents the superposition of the two beams traveling in opposite directions. This term is *not* an eigenfunction of the momentum operator; $-i\hbar(d\psi_3/dx)$ does not equal ψ_3 multiplied by a constant. If we were to measure the momentum, we should still get either of the two results expressed in (27.29), but either one is equally probable. We could not predict the result of an experiment based on the wave function ψ_3 unless we knew that either ψ_1 or ψ_2 was identically zero.

Suppose we want to measure the position of the particle using ψ_1. The probability of finding the particle near the point x is proportional to

$$\psi_1^* \psi_1 = A^* A = |A|^2 = \text{constant}$$

The probability density is independent of position. We can predict *nothing* about the position. All values between $-\infty$ and ∞ are equally likely. There is a fundamental point here. For this case we are able to specify the *exact* value p_x of the momentum. We shall shortly see that it is a consequence of the *Heisenberg uncertainty principle* that we cannot *simultaneously* specify the position x and its conjugate momentum p_x to any desired degree of accuracy. If we specify one of them exactly, then the other will be totally unknown, which is the situation we have here.

One problem, which some of you may have recognized, is that our functions are not integrable square. The integral

$$\int_{-\infty}^{\infty} \psi_1^* \psi_1 \, dx = A^2 \int_{-\infty}^{\infty} dx$$

is not finite. This means that we cannot normalize this wave function. We shall omit the details of associated problems.[2]

It is worth noting that k is a measure of the momentum in units of $\hbar$, and is really a vector quantity. It would be more correct to write ψ_1 as

$$\psi_1 = A \exp(i\mathbf{k} \cdot \mathbf{x}) \tag{27.32}$$

where $\mathbf{k}$ and $\mathbf{x}$ are vector quantities; $\mathbf{k}$ is often called the *wave vector*. Equation (27.32), and especially the form of the equation in three dimensions, is the basic starting point for all calculations involving scattering phenomena. Electron scattering by atoms and molecules, and structural investigations of molecules by electron diffraction, depend on the wave nature of the traveling electron.[3]

27.3 THE TOTALLY CONSTRAINED PARTICLE; PARTICLE IN A ONE-DIMENSIONAL BOX

Now suppose a free particle is in a potential well in which the potential is zero for x between $x = 0$ and $x = L$ and infinitely large outside this region, as shown in Figure 27.3. This means that there is zero probability of finding the particle outside the region $0 \leq x \leq L$; the wave function ψ must be zero outside the well. We can take the wave function given in (27.25a) as our solution,

$$\psi = A \sin(kx)$$

$$\psi = A \sin\left[\frac{2\pi}{h}(2mE)^{1/2}x\right] \tag{27.33}$$

The wave function must be finite in the region $0 \leq x \leq L$, zero outside this region, and continuous at the points $x = 0$ and $x = L$. These requirements are met by the condition that

$$\frac{2\pi}{h}(2mE)^{1/2} = n\pi \tag{27.34}$$

---

[2] See R. H. Dicke and J. P. Wittke, *Introduction to Quantum Mechanics* (Reading, Mass.: Addison-Wesley, 1960), chap. 3.

[3] See chap. 16 of Dicke and Wittke, *Quantum Mechanics*. Also N. F. Mott and H. S. W. Massey, *The Theory of Atomic Collisions* (Oxford: Oxford University Press, 1965).

SEC. 27.3 THE TOTALLY CONSTRAINED PARTICLE; PARTICLE IN A ONE-DIMENSIONAL BOX

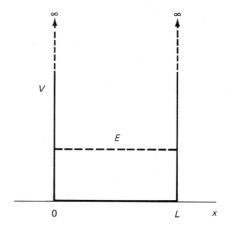

FIGURE 27.3 The form of the potential for the particle in the box. The walls of the potential well are infinitely high. The particle has energy E, as shown by the dashed line.

where n is an integer. This requires that the energy be restricted to certain discrete allowed values, or *eigenvalues*,

$$E_n = \frac{n^2 h^2}{8mL^2} \tag{27.35}$$

where m is the mass and L is the length of the "one-dimensional box." The wave functions, or eigenfunctions, are

$$\psi_n = A \sin\left(\frac{n\pi x}{L}\right) \tag{27.36}$$

The value of A can be determined from the normalization requirement

$$\int_0^L \psi_n^* \psi_n \, dx = A^2 \int_0^L \sin^2\left(\frac{n\pi x}{L}\right) dx = 1 \tag{27.37}$$

This requires that $A = \sqrt{2/L}$, and the complete normalized wave function can be written as

$$\psi_n = \left(\frac{2}{L}\right)^{1/2} \sin\left(\frac{n\pi x}{L}\right) \tag{27.38}$$

Plots of the wave functions ψ_n and the probability density ψ_n^2 are indicated in Figure 27.4. For eigenfunctions with n greater than 1, there are points within the box at which the probability of finding the particle is zero.

Since the wave functions are eigenfunctions of the energy operator, $\hat{H}\psi_n = E_n \psi_n$, we can state with certainty that if a particle has a wave function ψ_n, then a measurement of the energy will yield a value given by (27.35). What about the momentum?

The wave function is not an eigenfunction of the momentum operator $\hat{P}_x$. We can, however, from Equation (27.7), compute the average value, or *expectation value*, of the momentum:

$$\langle P \rangle = \int_0^L \psi_n^* \left(-i\hbar \frac{d}{dx}\right) \psi_n \, dx$$

$$= -i\hbar A^2 \left(\frac{n\pi}{L}\right) \int_0^L \sin\frac{n\pi x}{L} \cos\frac{n\pi x}{L} \, dx$$

$$= 0$$

FIGURE 27.4 (a) Wave functions $\psi_n(x)$, and (b) the probability distribution functions $\psi_n^2(x)$ for the particle in a one-dimensional box.

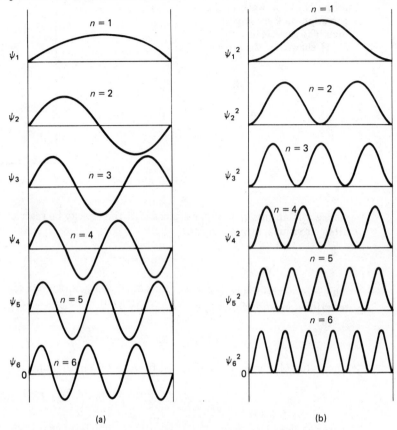

That the average momentum is zero can be seen more readily by examining not the momentum itself but the square of the momentum; the wave function is an eigenfunction of the operator associated with the square of the linear momentum,

$$\hat{P}^2 = \left(-i\hbar \frac{d}{dx}\right)\left(-i\hbar \frac{d}{dx}\right)$$

$$= -\hbar^2 \frac{d^2}{dx^2}$$

The eigenvalue equation is

$$\hat{P}^2 \psi_n = -\hbar^2 \frac{d^2}{dx^2}\left(A \sin \frac{n\pi x}{L}\right)$$

$$= \frac{n^2 h^2}{4L^2} \psi_n = p^2 \psi_n \qquad (27.39)$$

If we were to measure the value of p^2, we should be certain to get the result given in (27.39). This equation agrees with the statement $p_n^2 = 2mE_n$. We can determine the value of p by taking $p = (p^2)^{1/2}$; we get $p = \pm nh/2L$. If we tried to measure the momentum, we should get either of these two results, depending on the direction in which the particle was moving at the instant of measurement. If we were to make many such measurements, we should get the positive value half the time and the negative value the other half; the average of all the measurements would be zero.

So far as the position of the particle is concerned, we can state with absolute certainty that it is somewhere in the box; the relative probabilities of finding it at any particular points are shown in Figure 27.4b. We can compute the average value of numerous position measurements from (27.7):

$$\langle x \rangle = \int_0^L \psi_n x \psi_n \, dx$$
$$= A^2 \int_0^L x \sin^2 \frac{n\pi x}{L} \, dx$$
$$= \tfrac{1}{2} L \tag{27.40}$$

The wave functions for standing waves in classical mechanics certainly have some connection with reality, as they do describe the vibrations of violins and other types of musical stringed instruments. The question then arises whether a particle in a one-dimensional box has any bearing on a real problem; the answer is Yes. There are linear molecules that contain one or more delocalized electrons. It can be considered that these electrons are in a one-dimensional box of a length equal to the length of the molecule. This is the case for the six π electrons in the hexatriene molecule, as noted in Problem 5.

27.4 HEISENBERG'S UNCERTAINTY PRINCIPLE

The position of the particle in the box, then, is unknown to the extent that it can be located at any value of x between 0 and L. Thus we can take the uncertainty in its position to be $\Delta x = L$. The momentum of the particle is either $+nh/2L$ or $-nh/2L$; thus the uncertainty in the momentum can be taken as $\Delta p = nh/L$. The product of the uncertainties in position and momentum is $\Delta p \, \Delta x = nh$, which is just h for the lowest energy level. This is a form of the *Heisenberg uncertainty principle*. More precisely, this principle states that[4]

$$\Delta q \, \Delta p_q \geq \tfrac{1}{2} \hbar \tag{27.41}$$

> **WERNER KARL HEISENBERG** (born 1901), German physicist, developed a system of quantum mechanics independently of Schrödinger at about the same time. Heisenberg's system is intimately associated with the mathematics of matrices and is sometimes called *matrix mechanics;* it is completely equivalent to the system of Schrödinger. He did his doctoral dissertation on turbulence in fluid streams, and in 1925 after studying with Niels Bohr in Copenhagen, he solved the problem of the energy levels of the anharmonic oscillator. In 1927, Heisenberg presented his principle of indeterminacy. During World War II, he worked with Otto Hahn, the discoverer of nuclear fission, on developing nuclear reactors in Berlin. Heisenberg received the Nobel Prize in physics in 1932 "for the creation of quantum mechanics, the application of which has led, among other things, to the discovery of the allotropic forms of hydrogen."

[4] A more complete discussion can be found in Rojansky, *Quantum Mechanics*, p. 122.

What it means is simply this. Given the pair of variables q and p_q, if we want to determine either very precisely, we must then be satisfied with much less precise information about the other. We saw this with the free particle. While we could determine the momentum exactly, we could not say anything about the position; the position is completely undetermined.

In a simple analysis, we can see some rationale for this in trying to design an experiment to determine the position. To locate a small particle, we must observe it somehow, say by bouncing a light photon off the particle. This enables us to measure the position; but the very act of measuring the position, because of the effect of the light photon, changes the velocity and hence the momentum of the particle. This effect is important only for small particles; it is negligible for baseballs.

The uncertainty principle applies to any pair of dynamical variables whose product has the same dimensions as Planck's constant; thus it applies to energy and time in the form

$$\Delta E \ \Delta t \geq \tfrac{1}{2}\hbar \tag{27.42}$$

This form of the uncertainty principle is important in determining the lifetimes of excited states in spectroscopy. Figure 27.5 shows an energy-level diagram for a hypothetical atom consisting of a ground state at $E = 0$ and an excited state at energy E. Suppose the atom has been excited to the upper level. When it drops to the lower level, it emits a photon of frequency ν, which is observed as a line on a spectral plate. The frequency is determined from the relation $E = h\nu$. The excited state is characterized by a *lifetime,* which is the mean length of time the atom spends in the upper state before falling back to the ground state. If we call the lifetime τ, then the uncertainty in time is τ, and the uncertainty relation can be written

$$\Delta E \tau = \tau h \ \Delta \nu \sim \tfrac{1}{2}\hbar \tag{27.43}$$

where $\Delta \nu$ is the width of the spectral line; it is the uncertainty in frequency as determined from the spectral plate. Excited states with long lifetimes give sharp spectral lines, whereas excited states with extremely short lifetimes produce fuzzy or diffuse lines. The lifetime of the excited state can be estimated from

$$\tau \approx (4\pi \ \Delta \nu)^{-1} \tag{27.44}$$

In classical mechanics it is always possible to measure *any* two dynamical variables simultaneously to any desired degree of accuracy. The question logically arises when, if ever, this can be done with quantum mechanical systems. We saw that for the unconstrained free particle we could simultaneously measure the energy and momentum precisely. Two dynamical variables generally can be simultaneously measured if their operators *commute*. We can see this in the following way. Suppose we let $\hat{\alpha}$ and $\hat{\beta}$ be the operators associated with the two variables. If they are known precisely, then the wave function must be simultaneously an eigenfunction of both operators, or

$$\hat{\alpha}\psi = a\psi \tag{27.45a}$$
$$\hat{\beta}\psi = b\psi \tag{27.45b}$$

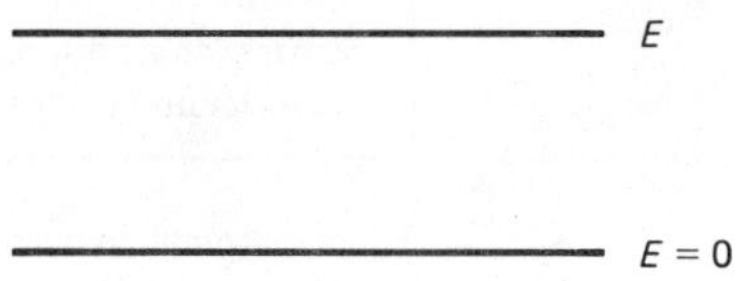

FIGURE 27.5 Hypothetical energy-level diagram for an atom with only two energy levels.

where the eigenvalues a and b are constants. Now all operators in quantum mechanics are linear operators; hence they commute with constants. If we multiply (27.45a) by $\hat{\beta}$ and (27.45b) by $\hat{\alpha}$, we get

$$\hat{\beta}\hat{\alpha}\psi = \hat{\beta}a\psi = a\hat{\beta}\psi = ab\psi \qquad (27.46a)$$
$$\hat{\alpha}\hat{\beta}\psi = \hat{\alpha}b\psi = b\hat{\alpha}\psi = ba\psi = ab\psi \qquad (27.46b)$$

and $\hat{\alpha}\hat{\beta}\psi = \hat{\beta}\hat{\alpha}\psi$, leading to the result that $\hat{\alpha}\hat{\beta} = \hat{\beta}\hat{\alpha}$. This means that the two operators commute.[5]

For the free particle, it is true that the operators for the energy and the momentum commute, since

$$\hat{p}_x \hat{H} = -i\hbar \frac{d}{dx}\left(-\frac{\hbar^2}{2m}\nabla^2 + V\right) \qquad (27.47)$$

In (27.47), d/dx always commutes with the ∇^2 operator, whereas it commutes with V only if V is independent of x. This is the case for the free particle since $V = 0$. Later we shall see that two important commuting variables are the energy and the z component of angular momentum.

27.5 THE TUNNEL EFFECT

The difference between the world of macroscopic particles in classical mechanics and the microscopic particle in quantum mechanics is dramatically shown by what we call the *tunnel effect*. Figure 27.6 shows a potential that has a constant value V_0 between $x = 0$ and $x = L$ and is zero outside this region; we have a particle with energy E, as shown by the dashed line on the figure. Classically the results are quite simple. If E is less than V_0, the particle is restricted to one side of the potential; a particle placed to the left of $x = 0$ never gets over the potential jump. If it moves in a positive direction, it is simply reflected. The particle can surmount the barrier only if its kinetic energy exceeds V_0.

In quantum mechanics the result is different. We write the Schrödinger equation as

$$\frac{d^2\psi}{dx^2} + \frac{2m}{\hbar^2}(E - V)\psi = 0 \qquad (27.48)$$

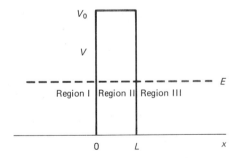

FIGURE 27.6 The potential barrier of height V_0. The dashed line represents the energy of the particle. Classically, a particle of energy $E < V_0$ cannot cross the barrier. In quantum mechanics there is a finite probability that the particle (wave) crosses the barrier.

---

[5] Note that we have shown that *if* the two variables can be simultaneously measured, then the operators commute. We have not shown that they can be simultaneously measured *only if* they commute. For further discussion, see H. Margenau and G. M. Murphy, *The Mathematics of Physics and Chemistry*, 2d ed. (Princeton, N.J.: D. Van Nostrand Company, 1956), p. 348.

where $V = V_0$ for $0 \le x \le L$ and $V = 0$ elsewhere. We can solve the equation for each of the three regions with the functions

$$\psi_\text{I} = A_1 \exp(ik_1 x) + B_1 \exp(-ik_1 x) \tag{27.49a}$$

$$\begin{aligned}\psi_\text{II} &= A_2 \exp(ik_2' x) + B_2 \exp(-ik_2' x) \\ &= A_2 \exp(k_2 x) + B_2 \exp(-k_2 x)\end{aligned} \tag{27.49b}$$

$$\psi_\text{III} = A_3 \exp(ik_1 x) \tag{27.49c}$$

where we have only one term for ψ_III. This corresponds to the case in which the particle is started on the left side and any particle that crosses the barrier travels in the positive x direction. In these equations, $k_1 = (2mE/\hbar^2)^{1/2}$ is a real number and $k_2' = [2m(E - V_0)/\hbar^2]^{1/2}$ is an imaginary quantity, accounting for the real exponent in ψ_II; k_2 is defined by the expression

$$k_2^2 = \left[\frac{2m(V_0 - E)}{\hbar^2}\right] \tag{27.50}$$

We now have the problem of fitting these solutions together so that they satisfy the basic requirements of continuity; the wave functions and their first derivatives must be continuous everywhere. We thus impose the conditions

$$\begin{aligned}\psi_\text{I} = \psi_\text{II}; \quad &\frac{d\psi_\text{I}}{dx} = \frac{d\psi_\text{II}}{dx} \quad &(\text{at } x = 0) \\ \psi_\text{II} = \psi_\text{III}; \quad &\frac{d\psi_\text{II}}{dx} = \frac{d\psi_\text{III}}{dx} \quad &(\text{at } x = L)\end{aligned} \tag{27.51}$$

There are five unknowns—A_1, B_1, A_2, B_2, and A_3; the conditions of (27.51) provide four simultaneous equations among these five unknowns, which allows us to calculate four of the unknowns in terms of the fifth, say A_3. When this is done, one of the relations we get is[6]

$$A_1^2 = A_3^2 \left[\cosh^2(k_2 L) + \frac{(k_1^2 - k_2^2)^2}{4k_1^2 k_2^2} \sinh^2(k_2 L)\right] \tag{27.52}$$

The probability that a particle starting on the left crosses the barrier is given by the ratio of the two probability amplitudes

$$T = \frac{|A_3|^2}{|A_1|^2} \tag{27.53}$$

The term T is called the *transmission coefficient*. It is a measure of the "fraction" of the wave that crosses the barrier. The value T is not zero unless the term $k_2 L$ is infinitely large; thus in quantum mechanics there may be a finite probability of finding the particle on the right side of the barrier, whereas in classical mechanics this probability is zero.

Computing T is easier when $k_2 L \gg 1$. Recalling the definitions of the hyperbolic sines and cosines,

$$\sinh x = \tfrac{1}{2}(e^x - e^{-x}) \qquad \cosh x = \tfrac{1}{2}(e^x + e^{-x})$$

we have that for large values of $k_2 L$

$$\cosh(k_2 L) \simeq \sinh(k_2 L) \simeq \tfrac{1}{2}\exp(k_2 L)$$

---

[6] A complete derivation can be found in K. S. Pitzer, *Quantum Chemistry* (New York: Prentice-Hall, 1953), p. 30. We have examined only this one type of finite potential barrier. A large variety of these barriers are discussed in Pitzer and in the other references listed on quantum mechanics.

SEC. 27.5 THE TUNNEL EFFECT

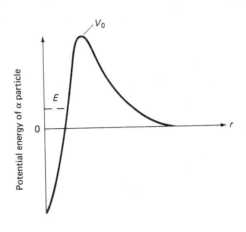

FIGURE 27.7 The potential energy of an α particle as a function of the distance from the nucleus. At large r, the repulsive force arises from Coulombic repulsion, and the potential varies as r^{-1}. At small r, the chief contribution to the potential is the nuclear attraction, which varies as a higher inverse power than r^{-1}. The energy of the α particle in the nucleus is E. In radioactive decay, the α particle "tunnels" through the potential barrier of height V_0 and is ejected by the nucleus.

and that

$$T \simeq 4 \exp(-2k_2 L) \tag{27.54}$$

We find applications of quantum mechanical tunneling in radioactivity, electronics, and molecular transformations. Figure 27.7 shows the form for the potential energy of an α particle near the nucleus. When a radium atom decays into a radon

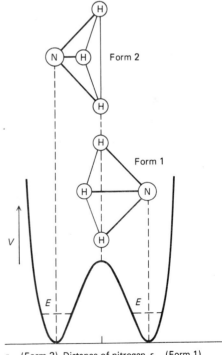

FIGURE 27.8 The ammonia "inversion." The ammonia molecule can exist in two different forms, as indicated. The nitrogen atom pushes through the plane of the hydrogen atoms to get from one form to the other in much the same way that an umbrella is inverted by a high wind. The heavy curve indicates the potential, the two minima corresponding to the two equilibrium positions of the nitrogen atom. To get from one form to the other, the system must "tunnel" through the barrier between the two minima.

atom and an α particle, Ra = Rn + α, the α particle within the nucleus effectively "tunnels" through the potential barrier. The semiconductor device called the *Isaki*, or *tunnel*, diode depends on the tunnel effect for its operation. The inversion of certain pyramidal molecules, such as NH_3, occurs through a tunneling effect. The ammonia inversion and the potential associated with that process is indicated in Figure 27.8 (p. 537).

27.6 PARTICLE IN A THREE-DIMENSIONAL BOX

The problem of the particle in a one-dimensional box can easily be extended to three dimensions. With a box of three dimensions, the particle is restricted to the interior of a rectangular parallelepiped of sides a, b, and c. The Schrödinger equation becomes

$$\nabla^2 \psi = \frac{\partial^2 \psi}{\partial x^2} + \frac{\partial^2 \psi}{\partial y^2} + \frac{\partial^2 \psi}{\partial z^2} = -\frac{2m}{\hbar^2} E \psi \qquad (27.55)$$

This equation can be separated by assuming a solution of the form

$$\psi(x, y, z) = X(x)Y(y)Z(z) \qquad (27.56)$$

When (27.56) is substituted in (27.55), the result is

$$\frac{1}{X}\frac{\partial^2 X}{\partial x^2} + \frac{1}{Y}\frac{\partial^2 Y}{\partial y^2} + \frac{1}{Z}\frac{\partial^2 Z}{\partial z^2} = -\frac{2m}{\hbar^2} E \qquad (27.57)$$

The only way (27.57) can be valid is for each term on the left separately to equal a constant; we can write the three equations

$$\frac{1}{X}\frac{\partial^2 X}{\partial x^2} = -k_x^2$$

$$\frac{1}{Y}\frac{\partial^2 Y}{\partial y^2} = -k_y^2 \qquad (27.58)$$

$$\frac{1}{Z}\frac{\partial^2 Z}{\partial z^2} = -k_z^2$$

where

$$k_x^2 + k_y^2 + k_z^2 = \frac{-8\pi^2 m E}{h^2} = k^2 \qquad (27.59)$$

The solutions to (27.58) can be written as

$$X(x) = A_x \sin(k_x x)$$
$$Y(y) = A_y \sin(k_y y) \qquad (27.60)$$
$$Z(z) = A_z \sin(k_z z)$$

Since the particle is restricted to the box, we require that the wave function go to zero at the origin and at $x = a$, $y = b$, and $z = c$. This imposes the conditions that

$$k_x = \frac{n_x \pi}{a} \qquad k_y = \frac{n_y \pi}{b} \qquad k_z = \frac{n_z \pi}{c} \qquad (27.61)$$

where n_x, n_y, and n_z are integers. If we let $A = A_x A_y A_z$, we can write the entire wave function as

SEC. 27.6 PARTICLE IN A THREE-DIMENSIONAL BOX

$$\psi(x, y, z) = X(x)Y(y)Z(z) = A \sin\left(\frac{n_x \pi x}{a}\right) \sin\left(\frac{n_y \pi y}{b}\right) \sin\left(\frac{n_z \pi z}{c}\right) \quad (27.62)$$

The value A can be got from the normalization

$$\int_0^a \int_0^b \int_0^c \psi^2 dx\, dy\, dz = 1$$

Hence

$$A = \left(\frac{8}{abc}\right)^{1/2} \quad (27.63)$$

The allowed energy levels are obtained from (27.59) as

$$E = \frac{h^2}{8m}\left(\frac{n_x^2}{a^2} + \frac{n_y^2}{b^2} + \frac{n_z^2}{c^2}\right) \quad (27.64)$$

For the special case of the cubic box, $a = b = c$, and the energy expression becomes

$$E = \frac{h^2}{8ma^2}(n_x^2 + n_y^2 + n_z^2) \quad (27.65)$$

The energy levels for the particle in the cubic box are shown in Figure 27.9. Note that several of the levels are degenerate, as the figure shows. The lowest level is non-degenerate, since there is only one way we can get the lowest energy, E_{111}; that is by having all three quantum numbers unity. The first excited level is threefold degenerate, since there are three different ways of achieving this energy. The sixth, tenth,

FIGURE 27.9 Energy levels and degeneracies for a particle in a three-dimensional cubic box.

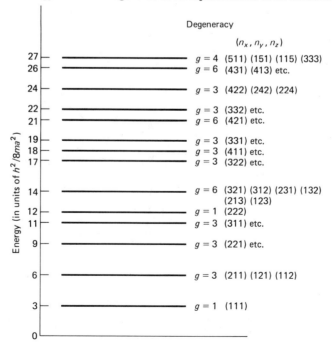

and thirteenth levels are sixfold degenerate, and the fourteenth level is fourfold degenerate.[7]

PROBLEMS

1. For the following functions, what are the conditions that must be met for the functions to be acceptable wave functions:
   (a) $f(x) = \exp(ax^2)$, (b) $f(\phi) = \exp(ia\phi)$,
   (c) $f(\phi) = \sin(a\phi)$, where ϕ varies between 0 and 2π in (b) and (c)?

2. For the wave functions $\psi = A \exp(-x^2)$ and $\psi = B \exp(im\phi)$, where m is an integer, evaluate the constants A and B so that the functions will be normalized.

3. Evaluate the quantum mechanical Hamiltonian operator for the Hooke's law spring and for the system of the two charged particles of charges $+q$ and $-q$ separated by a distance r.

4. The angular momentum vector $\mathbf{L}$ is defined by the vector product equation $\mathbf{L} = \mathbf{r} \times \mathbf{p}$. Derive the classical expressions for the x, y, and z components of angular momentum and convert them to quantum mechanical operators.

5. Consider a hexatriene molecule to be a linear molecule of length 7.3 Å. Calculate the energies of the first four levels of an electron free to move about the molecule. Calculate the frequency of light needed to raise an electron from the third to the fourth excited level.

6. Calculate the uncertainty in momentum for the electron in the $n = 3$ level of the hexatriene molecule of Problem 5.

7. Calculate the expectation values of p_x, p_x^2, and p_x^3 for the $n = 1$ level of the hexatriene molecule of Problem 5.

8. An electron with energy 5 eV encounters a barrier 5.5 eV high and 10^{-6} cm wide. What is the probability that the electron will penetrate the barrier?

9. For the totally constrained particle in a one-dimensional box, the potential at the ends of the box go to $V = \infty$ at 0 and L. Solve the Schrödinger equation for the particle in the potential well. That is to say, the potential is given by $V = 0$ between $x = 0$ and $x = L$, and $V = V_0$ outside this region, where the energy of the particle, $E < V_0$. (This problem is discussed in W. Kauzmann, *Quantum Chemistry* (New York: Academic Press, 1957), p. 188.)

10. A particle moves in a potential such that $V = 0$ for $x < 0$, and $V = V_0$ for $x > 0$. A particle of energy $E < V_0$ approaches the potential step at $x = 0$ from the negative x direction. (See accompanying figure.) In classical mechanics, the particle bounces back from the potential step. In quantum mechanics, a portion of the wave is reflected, while a portion of the wave is transmitted to positive values of x. Calculate the fraction of the wave transmitted and the fraction reflected. [This problem is discussed in J. C. Slater, *Quantum Theory of Atomic Structure*, vol. 1 (New York: McGraw-Hill Book Co., 1960), p. 62.]

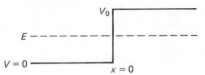

11. Carry out the steps necessary to show that when (27.56) is substituted in (27.55), the result is (27.57).

12. An electron is in a cubic box 10 Å on an edge. Calculate the energy difference between the lowest and the first excited level. Compare this energy with kT at 300 K. What frequency of light will excite the electron from the lowest to the first excited state?

13. A ball bearing weighing 1 g is in a cubic box 10 cm on an edge. Calculate the energy difference between the lowest and the first excited level. Compare this energy with kT at 300 K. What frequency of light will excite the ball from the lowest to the first excited level?

14. The density of sodium metal is 0.971 g cm^{-3}. Assume that each sodium atom provides one "free" electron that is free to move about the mass of sodium metal. Calculate the size of a cube containing 1 mol sodium. Calculate the difference in energy between the lowest and the next higher energy level and compare the answer with kT at 300 K. Calculate a set of quantum numbers for the level whose energy is given by kT. (*Note:* An important consideration in connection with this system is the following. According to the Pauli exclusion principle, which we shall discuss later, each electron has two possible values of electron spin associated with it, and each level can hold at most two electrons. With this in mind, we want to calculate the energy of the highest level obtained when each level is successively filled with two and only two electrons. This level is known as the

---

[7] The energy-level diagram for the particle in a three-dimensional box, coupled with the Pauli exclusion principle, can be applied to the "free electrons" in metals to explain satisfactorily why there is no measurable heat capacity contribution from the electrons in the metal. This is explained in Chapter 37.

PROBLEMS

Fermi level; solving this problem explains the absence of an electronic contribution to the heat capacity of metals. We shall discuss this more fully in Chapter 37.)

15. Calculate the energy levels and wave functions of a particle constrained to move on the circumference of a circle of radius r.

16. Calculate the energy levels and wave functions for the particle in a two-dimensional box, that is, a particle constrained to move on the surface of a square of side a. Compare your solution with that of the square membrane of Chapter 26.

17. For the two-dimensional particle in the box of Problem 16, suppose the box were a circular area of radius r instead of a square. Convert to polar coordinates, and write the Schrödinger equation for the system in polar coordinates. Solve the problem for the angular dependence. Show that a solution can be written in the form $\psi = R(r)\Phi(\phi)$, where $\Phi(\phi) = A \exp(im\phi)$; the value m is an integer and $R(r)$ is a function only of r. Show that the equation for R has the form

$$\frac{\partial^2 R}{\partial r^2} + \frac{1}{r}\frac{\partial R}{\partial r} + \left[k^2 - \left(\frac{m}{r}\right)^2\right]R = 0$$

where m is the integer in the equation for $\Phi(\phi)$ and $k^2 = 2mE/\hbar^2$. This equation is Bessel's equation; its solutions are given by Bessel functions. (See any advanced calculus textbook that includes a discussion of differential equations.)

18. Start with the de Broglie relation and conclude with the Schrödinger equation in the following manner:

(a) Write the classical equation for the one-dimensional traveling wave in the form

$$\frac{d^2\psi}{dx^2} + \frac{4\pi^2\nu^2}{v^2}\psi = 0$$

where ν is the frequency and v is the velocity.

(b) Show that the momentum of a particle is $p = [2m(E - V)]^{1/2}$.

(c) Use the de Broglie relation to show that

$$\frac{\nu^2}{v^2} = \frac{2m(E - V)}{h}$$

(d) Use the result of part (c) in the wave equation and show that the Schrödinger equation results.

CHAPTER TWENTY-EIGHT

ROTATIONS AND VIBRATIONS OF MOLECULES

28.1 SPECTROSCOPY AND ENERGY LEVELS

A big problem associated with quantum mechanical systems is determining the energy levels of the systems. The Schrödinger equation allows us to calculate the energy levels and also other values; the energy levels are experimentally determined by spectroscopic studies.

A spectrum separates characteristics; these characteristics may be continuous or discrete. The Maxwell-Boltzmann distribution represents a continuous spectrum; it separates molecules according to their velocities. When you complete this course, the professor will probably prepare a spectrum of examination grades, separating students according to their performances, to establish grades in the course. In this chapter our concern is with discrete spectra representing energy separations.

For purposes of discussion we establish two different hypothetical systems with different sets of energy levels, the first of which is shown in Figure 28.1. We take the separation between successive levels to be a constant b. If the energy of the lowest level ($n = 0$) is taken as zero, then the energies of the various levels are

$$E_n = nb \tag{28.1}$$

There are two characteristics of interest associated with any spectral line, the frequency and the intensity. The frequency ν associated with a transition between the ith and the jth level of our hypothetical system is given by the condition

$$\Delta E_{(i \to j)} = h\nu = (i - j)b \tag{28.2}$$

where $\Delta E_{(i \to j)}$ is the energy separation between the two levels. The energy-level diagram is determined from the frequencies of the spectral lines.

Spectroscopic measurements can be carried out in two distinct ways, *absorption* and *emission*. Figure 28.2 indicates a schematic diagram of an absorption spectrometer. A light beam (preferably of constant intensity) impinges upon the sample, and the intensity of the transmitted beam is measured as the wavelength of the incident beam is varied.

Figure 28.3 indicates a schematic diagram of an emission spectrometer. Here the atoms or molecules are excited from the ground state to upper states by the introduction of energy—by electron bombardment, a carbon arc, or other means. As the atoms fall back to the lower levels, they emit light photons of a frequency given by Equation (28.2).

FIGURE 28.1 Energy-level diagram for a hypothetical system for which $E_n = nb$, where b is a constant.

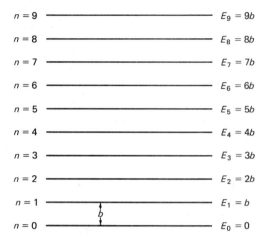

Let us return now to the energy level diagram of Figure 28.1. Suppose we have excited a large collection of these hypothetical molecules so that there are a substantial number of molecules in each of the excited levels. Suppose also, for the moment, that there is no restriction on the path the molecule may take as it eventually returns to the ground state. That is to say, a molecule in the nth level may fall to any of the levels $n - 1, n - 2, \ldots$, or down to $n = 0$. These various pairs of levels are separated in energy by $b, 2b, 3b, \ldots, nb$; the spectrum would consist of a series of equally spaced lines, as shown in Figure 28.4a; the spectrum is indicated in units of reciprocal centimeters to provide a linear energy scale. Note that we have not yet taken into account the possibility of varying intensities.

The path the excited molecule takes as it returns to the ground state is generally determined by what we call *selection rules*. Suppose the selection rules for this

FIGURE 28.2 Schematic of absorption spectrometer. Incident radiation enters the monochromer, and a beam of variable frequency passes through the sample holder to the detector. The recorder records the intensity of the radiation as a function of frequency. The upper recorder trace indicates the intensity for the empty sample holder, and the lower trace indicates the spectrum for a sample with two spectral lines. For an optical spectrometer, the monochromer is a prism or grating.

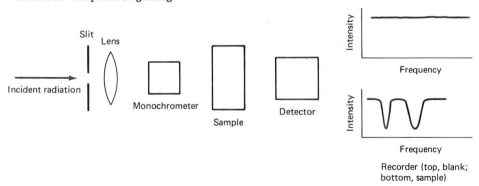

Recorder (top, blank; bottom, sample)

FIGURE 28.3 Schematic of an emission spectrometer. The sample is excited from the ground state to upper excited levels by electron bombardment or other means. As the sample molecules decay back to the ground state, they emit spectral lines, which can be detected and recorded on a strip chart recorder.

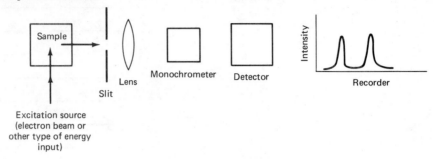

system are such that the only possible transitions are those for which $\Delta n = \pm 1$. This means that a molecule excited to the nth level will return to the ground state in a stepwise manner by first falling to the level $n - 1$, then to $n - 2$, and so on to $n = 1$ and then finally to $n = 0$. In this instance, since the energy separation between adjacent levels throughout the diagram is the same, each transition would be accompanied by the emission of a light photon of the same frequency, and the spectrum would consist of only one line, as indicated in Figure 28.4b.

Consider as a second example the hypothetical energy-level diagram shown in Figure 28.5. Here we suppose that the allowed energy levels are proportional to $n(n + 1)$ and are given by

$$E_n = n(n + 1)B \qquad (n = 0, 1, 2, 3, \ldots) \qquad (28.3)$$

where B is a constant. We further suppose that the selection rule operates so that only transitions between adjacent levels are permitted, $\Delta n = \pm 1$. The energy associated with the transition from the state n' to n'' is

$$\Delta E_{n' \to n''} = E_{n''} - E_{n'} = B[n''(n'' + 1) - n'(n' + 1)]$$

FIGURE 28.4 Possible spectra arising from the energy-level diagram of Figure 28.1. (a) No selection rule; all transitions are possible; Δn can be any integer. (b) The spectrum for the selection rule $\Delta n = \pm 1$. All transitions give rise to lines of the same frequency; hence only one line is visible on the spectrum at $\bar{\nu} = b$ cm^{-1}.

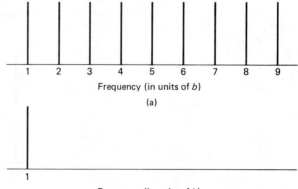

SEC. 28.1 SPECTROSCOPY AND ENERGY LEVELS

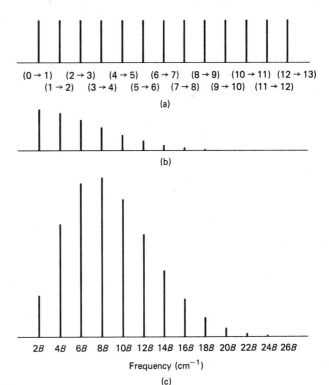

FIGURE 28.5 Energy-level diagram for which $E_n = Bn(n+1)$.

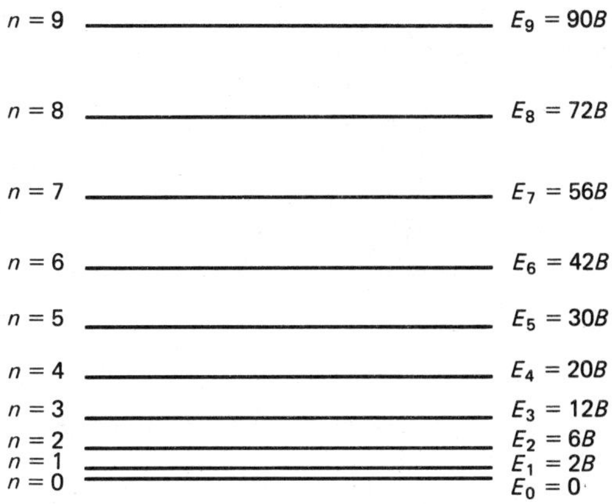

Since $n'' = n' \pm 1$, the observed spectroscopic lines are

$$\bar{\nu} \text{ (cm}^{-1}\text{)} = 2B(n+1) \tag{28.4}$$

where B is in wave number units. The result is a set of equally spaced lines with separation $2B$, as shown in the top spectrum of Figure 28.6.

The intensities of the lines are determined by two factors. One involves the transmission moment and Einstein coefficients (Chapters 31 and 42); the other is the Boltzmann distribution.

FIGURE 28.6 Spectra arising from the energy-level diagram of Figure 28.5. The selection rule is that $\Delta n = \pm 1$. The indicated transitions are for an absorption. (a) No consideration given to intensities. (b) Only Boltzmann factor taken into account for the intensities. (c) The Boltzmann factor and degeneracy taken into account for computing the intensities of the various lines.

28.2 THE BOLTZMANN FACTOR, DEGENERACY, AND INTENSITIES

Suppose we are determining the absorption spectrum of the molecule having the energy-level diagram discussed at the end of the previous section. The relative numbers of molecules in each level can be calculated from the Boltzmann distribution law. The populations of the various levels N_n can be determined from the population of the lowest level N_0:

$$N_n = N_0 \exp\left(\frac{-\Delta E}{kT}\right) = N_0 \exp\left[-\frac{Bn(n+1)}{kT}\right] \quad (28.5)$$

First suppose that B is very much larger than kT. Then only the lowest level is appreciably populated, and only these molecules can absorb the radiation, since there is an insufficient number in the upper levels to provide a measurable absorption. Only one peak will be observed on the spectrum, the peak designating the excitation of the $n = 0$ level to the $n = 1$ level.

On the other hand, what if B is much smaller than kT? Then several of the excited levels will be appreciably occupied and these will absorb in addition to the molecules in the lowest levels. Several peaks will be observed before the intensity becomes immeasurably small. The intensities are proportional to the populations; hence the heights of the peaks are proportional to $\exp[-Bn(n+1)/kT]$, as shown in Figure 28.6b, where we have taken B to be $0.05kT$.

Suppose further that some of the levels are degenerate. Let's assume that the degeneracy of each level is

$$g_n = 2n + 1 \quad (28.6)$$

TABLE 28.1 Energies, degeneracies, and Boltzmann factors for the energy level diagram of Figure 28.5.

Level	Energy (in units of B; $E_n = Bn(n+1)$)	Boltzmann factor[a] ($\exp(-0.05E_n)$)	Degeneracy ($2n + 1$)	Boltzmann factor, including degeneracy[a] ($g_n \exp(-0.05E_n)$)
0	0	1.00000	1	1.00000
1	2	0.90484	3	2.71451
2	6	0.74082	5	3.70409
3	12	0.54881	7	3.84168
4	20	0.36788	9	3.31091
5	30	0.22313	11	2.45443
6	42	0.12246	13	1.59193
7	56	0.06081	15	0.91215
8	72	0.02732	17	0.46450
9	90	0.01111	19	0.21107
10	110	0.00409	21	0.08582
11	132	0.00136	23	0.03129
12	156	0.00041	25	0.01024
13	182	0.00011	27	0.00301
14	210	0.00003	29	0.00080
15	240	0.00001	31	0.00019
16	272	0.00000	33	0.00004
17	306	0.00000	35	0.00001
18	342	0.00000	37	0.00000

[a] The entries in this column are for the case of $B = 0.05kT$.

SEC. 28.3 THE QUANTUM MECHANICAL OSCILLATOR

The lowest level has only one state associated with it, since $g_0 = 1$, while all the upper states are multiply degenerate; the Boltzmann factor of (28.5) becomes

$$N_n = g_n N_0 \exp\left(\frac{-\Delta E}{kT}\right) = (2n + 1)N_0 \exp\left[-\frac{Bn(n + 1)}{kT}\right] \tag{28.7}$$

The relative populations of the various levels are summarized in Table 28.1. In the degenerate case, the populations increase for a short distance before they start to fall off. Since the intensities of spectral lines are proportional to the populations, the intensities of the lines appear as shown in Figure 28.6c.

There is an additional and important factor involved in determining the intensities of spectral lines; this is associated with the *transition moments*. The selection rules are established by determining those transitions that have nonzero transition moments. The transition moments are determined by examining the interaction of electromagnetic radiation with atoms and molecules. Since we are not yet prepared to do this, we shall return to this point at a later stage. For the moment we shall consider vibrations and rotations of molecules, and the infrared and microwave spectroscopy associated with these kinds of motion.

28.3 THE QUANTUM MECHANICAL HARMONIC OSCILLATOR: VIBRATIONAL SPECTROSCOPY

The harmonic oscillator is one of the few systems in quantum mechanics that can be solved exactly. It is of particular interest in elucidating the form of interatomic forces in molecules.

In a first approximation, we regard a diatomic molecule as a pair of mass points connected by an ideal massless spring of force constant k, as shown in Figure 28.7. The classical potential energy is $V(x) = \frac{1}{2}kx^2$, where x represents the displacement from equilibrium; $x = r - r_e$, where r_e is the equilibrium internuclear distance. We can write k in terms of the fundamental frequency of the oscillations as

$$k = 4\pi^2 \mu \nu_0^2 \tag{28.8}$$

where μ is the reduced mass of the system $\mu = m_1 m_2/(m_1 + m_2)$; the values m_1 and m_2 are the masses of the two atoms. The classical Hamiltonian is

$$H = \frac{1}{2\mu} p_x^2 + \frac{1}{2} kx^2 \tag{28.9}$$

In quantum mechanics, this becomes the operator

$$\hat{H} = -\frac{\hbar^2}{2\mu} \frac{d^2}{dx^2} + \frac{1}{2} kx^2 \tag{28.10}$$

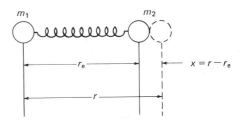

FIGURE 28.7 Two atoms of masses m_1 and m_2 connected by a spring with force constant k. The equilibrium internuclear distance is indicated by r_e, while r is the distance between the atoms during the vibrations.

The allowed energy levels and wave functions are determined by solving the Schrödinger equation:

$$\frac{d^2\psi}{dx^2} + \frac{2\mu}{\hbar^2}\left(E - \frac{1}{2}kx^2\right)\psi = 0 \qquad (28.11)$$

It is convenient to introduce the quantities $\alpha = 2\mu E/\hbar^2$ and $\beta = \mu\nu_0/\hbar$, in terms of which (28.11) becomes

$$\frac{d^2\psi}{dx^2} + (\alpha - \beta^2 x^2)\psi = 0 \qquad (28.12)$$

We first examine the solutions of (28.12) for very large values of x. For large x, we get $\beta^2 x^2 \gg \alpha$; the term in α can be dropped, and (28.12) can be written

$$\frac{d^2\psi}{dx^2} = \beta^2 x^2 \psi \qquad \text{(large } x\text{)} \qquad (28.13)$$

At large x we use the solutions to this equation,

$$\psi_\infty = A \exp\left(\pm \frac{\beta}{2} x^2\right) \qquad (28.14)$$

Since the wave function must go to zero at infinitely large values of x, we can eliminate the solution with the positive sign and consider only $\exp[-(\beta/2)x^2]$. Equation (28.14) is valid only in the limit as $x \to \infty$. We can see this by writing the second derivative of ψ_∞ as

$$\frac{d^2\psi_\infty}{dx^2} = \beta^2 x^2 \exp\left(-\frac{\beta}{2}x^2\right) - \beta \exp\left(-\frac{\beta}{2}x^2\right) \qquad (28.15)$$

The second term is negligible compared with the first in the limit of large x.

Now we can write the solution of (28.12) as

$$\psi = \chi(x) \exp\left(-\frac{\beta}{2}x^2\right) \qquad (28.16)$$

where $\chi(x)$ is some function of x. If this is substituted in (28.12), we get

$$\frac{d^2\chi}{dx^2} - 2\beta x \frac{d\chi}{dx} + (\alpha - \beta)\chi = 0 \qquad (28.17)$$

Equation (28.17) takes on a simpler appearance if we introduce the new variable $\xi = (\beta)^{1/2}x$. If the function $\chi(x)$ is replaced by the function $H(\xi)$, Equation (28.17) becomes

$$\frac{d^2 H}{d\xi^2} - 2\xi \frac{dH}{d\xi} + \left(\frac{\alpha}{\beta} - 1\right)H = 0 \qquad (28.18)$$

Equation (28.18) is a famous equation of mathematical physics known as the *Hermite* equation; it was first solved more than 100 years ago. It is an eigenvalue equation; acceptable solutions can be obtained only if the condition[1]

$$\alpha = (2n + 1)\beta \qquad (n = 0, 1, 2, \ldots) \qquad (28.19)$$

---

[1] A detailed discussion of the solution of this equation can be found in H. Eyring, J. Walter, and G. E. Kimball, *Quantum Chemistry* (New York: John Wiley & Sons, 1944), p. 60; and in L. Pauling and E. B. Wilson, Jr., *Introduction to Quantum Mechanics* (New York: McGraw-Hill Book Co., 1935), p. 67.

SEC. 28.3 THE QUANTUM MECHANICAL OSCILLATOR

is satisfied, where n is an integer. When this condition is satisfied, the solutions of (28.18) are polynomials called *Hermite polynomials of degree n*. These polynomials are given by the general expression

$$H_n(\xi) = (-1)^n \exp(\xi^2) \frac{d^n}{d\xi^n} [\exp(-\xi^2)] \tag{28.20}$$

or

$$H_n(\xi) = (2\xi)^2 - \frac{n(n-1)(2\xi)^{n-2}}{1!} + \frac{n(n-1)(n-2)(n-3)(2\xi)^{n-4}}{2!} + \cdots \tag{28.21}$$

Table 28.2 lists the first nine Hermite polynomials. With these polynomials, the wave functions for the harmonic oscillator are

$$\psi_n(\xi) = C_n H_n(\xi) \exp(-\tfrac{1}{2}\xi^2) \tag{28.22}$$

The C_n are constants that can be determined from the normalization requirements. The normalized wavefunctions for the harmonic oscillator are

$$\psi_n(\xi) = \left(\frac{\sqrt{\beta/\pi}}{2^n n!}\right)^{1/2} H_n(\xi) \exp\left(-\frac{1}{2}\xi^2\right) \qquad (\xi = \sqrt{\beta}x) \tag{28.23}$$

This set of functions forms a complete orthonormal set that can be used to expand any well-behaved function in the same way that functions can be expanded in a series of sines and cosines to form a Fourier expansion. The first six wave functions $\psi_n(\xi)$ and $\psi_n^2(\xi)$ are shown in Figure 28.8. The functions are symmetrical when n is an even number, and antisymmetrical when n is odd.

The allowed energy levels of the oscillator are found by substituting the values of α and β in the condition of (28.19), from which we find that

$$E_n = \hbar \left(\frac{k}{\mu}\right)^{1/2} \left(n + \frac{1}{2}\right) = \left(v + \frac{1}{2}\right) h\nu_0 \tag{28.24}$$

where we have substituted v for n in accord with convention. The energy-level diagram consists of a set of equally spaced levels separated by $h\nu_0$, as in Figure 28.1. That energy-level diagram was not so hypothetical after all; it is the energy-level diagram for the harmonic oscillator. The selection rule for the transitions is $\Delta v = \pm 1$; thus the spectrum due to the vibrational motion of molecules consists of only the one line shown in Figure 28.4b. These lines normally appear in the infrared region.

Later we shall note that spectral lines arise from the interaction between the incident electromagnetic radiation and the changing dipole moment of the vibrating molecule. For diatomic homonuclear molecules like H_2 and Cl_2, the dipole moment is zero; there is no changing dipole moment during vibrations. These homonuclear dia-

TABLE 28.2 The first nine Hermite polynomials $H_n(\xi)$.

$H_0(\xi) = 1$
$H_1(\xi) = 2\xi$
$H_2(\xi) = 4\xi^2 - 2$
$H_3(\xi) = 8\xi^3 - 12\xi$
$H_4(\xi) = 16\xi^4 - 48\xi^2 + 12$
$H_5(\xi) = 32\xi^5 - 160\xi^3 + 120\xi$
$H_6(\xi) = 64\xi^6 - 480\xi^4 + 720\xi^2 - 120$
$H_7(\xi) = 128\xi^7 - 1344\xi^5 + 3360\xi^3 - 1680\xi$
$H_8(\xi) = 256\xi^8 - 3584\xi^6 + 13440\xi^4 - 13440\xi^2 + 1680$

FIGURE 28.8 Behavior of the first six functions $\psi_n(\xi)$ and $\psi_n^2(\xi)$ for the harmonic oscillator. The dashed parabola indicates the potential for the oscillator. The heavy lines indicate the amplitude range of a classical oscillator.

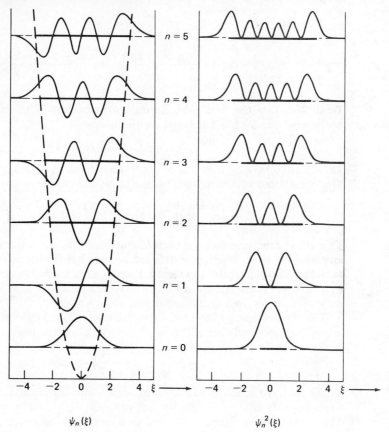

tomic molecules do not display any vibrational spectra. On the other hand, heteronuclear molecules such as HCl have permanent dipole moments. Since the dipole moment depends on the distance between nuclei, the dipole moment changes during vibration; there is an interaction with the electromagnetic field, and these molecules display vibrational spectra.

A diatomic molecule has only one vibrational degree of freedom; it has only one fundamental frequency. A linear triatomic molecule such as CO_2 has four vibrational degrees of freedom and four normal modes of vibration (Chapter 3). The normal modes are shown in Figure 3.5. Each normal mode of a vibrating molecule can have a different fundamental frequency; the complete vibrational spectrum can consist of as many lines as there are normal modes. Often there is a lesser number. Since CO_2 is symmetrical, it does not have a permanent dipole moment. The "symmetrical stretch" for CO_2 shown in Figure 3.5 is a symmetrical mode in which the dipole moment does not change; hence this mode is not *active* for infrared spectra. The "antisymmetric stretch" and "bending" modes do have changing dipole moments, and both of these give rise to infrared spectral lines. The two bending modes are degenerate and give rise to lines with the same frequency. The infrared spectrum of

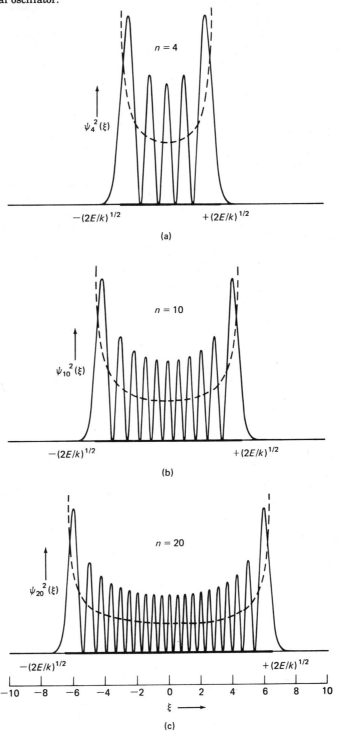

FIGURE 28.9 Probability density distributions for (a) the $n = 4$, (b) $n = 10$, and (c) $n = 20$ levels of the harmonic oscillator. The dashed curve indicates the distribution for a classical harmonic oscillator with the same energy, and the heavy region on the abscissa indicates the range of the classical oscillator.

CO_2 thus consists of two lines. Infrared spectroscopy is a valuable analytical tool. Many functional groups in organic chemistry are characterized by certain ranges of frequencies in the infrared spectrum; spectra taken of unknown compounds can suggest the presence or absence of these functional groups.

Notice that according to (28.24) the lowest energy level is $E_0 = \frac{1}{2}h\nu_0$; this is nonzero. The term E_0 is the *zero point energy*. This accords with the uncertainty principle. If E were 0, then the momentum would also be 0, since $p = \sqrt{2mE}$. This would imply a complete absence of knowledge about the position of the particle. But we do, in fact, have some knowledge about the position. This knowledge is inherent in the wave function ψ_0. If we take Δx as the standard deviation of x, then the product $\Delta p\, \Delta x$ is approximately equal to Planck's constant. This demonstration is left for one of the problems.

In Figure 28.8, the classical potential energy $V = \frac{1}{2}kx^2$ is superimposed on the figure. The limits of the classical motion are indicated by heavy lines. In quantum mechanics the particle has a finite probability of being in a region of space to which it is absolutely forbidden in classical mechanics. In effect the oscillator "tunnels" through a potential barrier. Figure 28.9 (p. 551) shows plots of the probability density ψ_n^2 for several levels and also the probability functions for the classical harmonic oscillator having the same energies. Notice here that as n increases, the envelope of the quantum mechanical distribution approaches the shape of the classical distribution more and more closely.

28.4 THE RIGID ROTOR

In the *rigid rotor* approximation, a diatomic molecule is regarded as a dumbbell, with two atoms of masses m_1 and m_2 held together by a massless rigid bar of length r, where r is the bond length. The dumbbell is free to rotate about the center of mass; the coordinates of interest are shown in Figure 28.10. The kinetic energy of rotation is

$$E_{\text{rot}} = \tfrac{1}{2}I\omega^2 \tag{28.25}$$

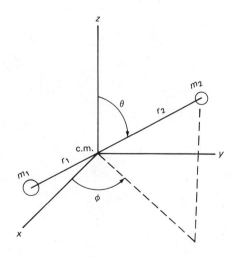

FIGURE 28.10 The rigid rotor and the coordinate system for the rotating rigid rotor. The distance between the atoms of masses m_1 and m_2 is $r = r_1 + r_2$. The origin is taken at the center of mass of the system.

SEC. 28.4 THE RIGID ROTOR

where ω is the angular velocity and I is the moment of inertia,

$$I = m_1 r_1^2 + m_2 r_2^2 \tag{28.26}$$

Since $m_1 r_1 = m_2 r_2$, we can show that the distances of the masses from the center of mass are

$$r_1 = \frac{m_2}{m_1 + m_2} r \qquad r_2 = \frac{m_1}{m_1 + m_2} r$$

Thus, the moment of inertia can be written as

$$I = \left(\frac{m_1 m_2}{m_1 + m_2}\right) r^2 = \mu r^2 \tag{28.27}$$

where μ is the reduced mass.

Two types of motion are involved. First there is translational motion of the center of mass without regard to how the two masses are oriented relative to the coordinate axis. We shall not concern ourselves with this motion. We are concerned exclusively with the second type, namely the rotation of the dumbbell without regard to the translational motion as a whole.[2]

Since the dumbbell rotates in a field free space, the potential energy is zero and the classical Hamiltonian is

$$H = \frac{1}{2\mu}(p_x^2 + p_y^2 + p_z^2) \tag{28.28}$$

This transforms to the quantum mechanical operator

$$\hat{H} = -\frac{\hbar^2}{2\mu} \nabla^2$$

Since the motion is rotational, spherical polar coordinates are more appropriate than Cartesian coordinates. The Laplacian in spherical coordinates is

$$\nabla^2 = \frac{1}{r^2}\frac{\partial}{\partial r}\left(r^2 \frac{\partial}{\partial r}\right) + \frac{1}{r^2 \sin\theta}\frac{\partial}{\partial \theta}\left(\sin\theta \frac{\partial}{\partial \theta}\right) + \frac{1}{r^2 \sin^2\theta}\frac{\partial^2}{\partial \phi^2} \tag{28.29}$$

The problem is equivalent to one in which a particle of mass μ rotates about a point at a constant r (the internuclear distance) from the point. Since r in (28.29) is constant, the Hamiltonian operator can be written

$$\hat{H} = -\frac{\hbar^2}{2\mu r^2}\left[\frac{1}{\sin\theta}\frac{\partial}{\partial \theta}\left(\sin\theta \frac{\partial}{\partial \theta}\right) + \frac{1}{\sin^2\theta}\frac{\partial^2}{\partial \phi^2}\right] \tag{28.30}$$

The term μr^2 in the denominator is the moment of inertia. The Schrödinger equation can now be written

$$\hat{H}\psi = E\psi$$

$$\left(\frac{\partial^2}{\partial \theta^2} + \frac{\cos\theta}{\sin\theta}\frac{\partial}{\partial \theta} + \frac{1}{\sin^2\theta}\frac{\partial^2}{\partial \phi^2}\right)\psi + \frac{2\mu r^2 E}{\hbar^2}\psi = 0 \tag{28.31}$$

We can separate variables by taking as our solution

$$\psi = \Theta(\theta)\Phi(\phi) \tag{28.32}$$

---

[2] The separation of the translational motion from the internal motion is discussed in Pauling and Wilson, *Quantum Mechanics*, p. 113, for the hydrogen atom; that procedure is fully applicable here.

where Θ is a function only of θ and Φ is a function only of ϕ. If (28.32) is now substituted in (28.31) and the resulting equation rearranged slightly, we finally get

$$\frac{\sin^2 \theta}{\Theta}\left[\frac{1}{\sin \theta}\frac{d}{d\theta}\left(\sin \theta \frac{d\Theta}{d\theta}\right) + \frac{2IE\Theta}{\hbar^2}\right] = -\frac{1}{\Phi}\frac{d^2\Phi}{d\phi^2} \qquad (28.33)$$

in which we have replaced μr^2 by I. The left side of (28.33) is a function only of θ, and the right side is a function only of ϕ. This can only be possible if each side is separately equal to the same constant, which we shall call m^2. Now we can write

$$-\frac{1}{\Phi}\frac{d^2\Phi}{d\phi^2} = m^2 \qquad (28.34)$$

and

$$\frac{\sin^2 \theta}{\Theta}\left[\frac{1}{\sin \theta}\frac{d}{d\theta}\left(\sin \theta \frac{d\Theta}{d\theta}\right) + \frac{2IE\Theta}{\hbar^2}\right] = m^2 \qquad (28.35)$$

in which we have replaced the partial derivatives by ordinary derivatives. Equation (28.34) is an old friend by now, and we can immediately write the solution as

$$\Phi = \exp(im\phi) \qquad (28.36)$$

For Φ to be single-valued and continuous, we must have

$$\begin{aligned}\Phi(\phi) &= \Phi(\phi + 2\pi) \\ \exp(im\phi) &= \exp[im(\phi + 2\pi)]\end{aligned} \qquad (28.37)$$

If both sides of (28.37) are multiplied by $\exp(-im\phi)$, the result is

$$1 = \exp(2\pi im) \qquad (28.38)$$

Equation (28.38) can be valid only if m takes on integral values,

$$m = 0, \pm 1, \pm 2, \pm 3, \ldots$$

It would be more correct to write the solution

$$\Phi_m(\phi) = A \exp(im\phi) \qquad (28.39)$$

where A is a constant. The constant A is evaluated from the normalization condition $\int_0^{2\pi} \Phi^* \Phi \, d\phi = 1 = A^2 \int_0^{2\pi} d\phi = 2\pi A^2$, from which we find that $A = 1/\sqrt{2\pi}$. The normalized functions can now be written

$$\Phi_m(\phi) = \frac{1}{\sqrt{2\pi}} \exp(im\phi) \qquad (28.40)$$

Note that we could have written equally acceptable solutions of (28.34) in the form of $\sin(m\phi)$ or $\cos(m\phi)$. The imaginary exponential form is, however, of greater use. In spherical coordinates the quantum mechanical operator for the z component of angular momentum is

$$\hat{L}_z = -i\hbar \frac{\partial}{\partial \phi} \qquad (28.41)$$

and the functions of (28.36) are eigenfunctions of this operator, whereas the sines and cosines are not. The eigenvalues of this operator are m, the quantum number associated with the z component of angular momentum. We shall have more to say about this later.

28.5 SOLUTION OF THE Θ EQUATION: LEGENDRE POLYNOMIALS AND THE ENERGY LEVELS OF THE RIGID ROTOR

The solution to the Θ equation (28.35) is made simpler by introducing the new independent variable $z = \cos(\theta)$ and by replacing $\Theta(\theta)$ by the function $P(z)$; the value z varies between the limits -1 and $+1$. The derivatives are

$$\frac{d\Theta}{d\theta} = \frac{dP}{dz}\frac{dz}{d\theta} = -\sin\theta\frac{dP}{dz}$$

$$\frac{d^2\Theta}{d\theta^2} = \sin^2\theta\frac{d^2P}{dz^2} - \cos\theta\frac{dP}{dz}$$

If we remember that $\sin^2\theta = 1 - \cos^2\theta = 1 - z^2$, Equation (28.35) can be transformed into

$$(1 - z^2)\frac{d^2P(z)}{dz^2} - 2z\frac{dP(z)}{dz} + \left(\frac{2EI}{\hbar^2} - \frac{m^2}{1-z^2}\right)P(z) = 0 \qquad (28.42)$$

Equation (28.42) is a well-known equation of mathematical physics, the *associated Legendre equation*. (We shall not go into the details of the solution here.)[3] We can get valid solutions of this equation only if

$$\frac{2EI}{\hbar^2} = l(l+1) \qquad (28.43)$$

where l is zero or a positive integer; in addition,

$$|m| \leq l \qquad (28.44)$$

The solutions to (28.42) are called the *associated Legendre polynomials*, with symbol $P_{l,m}(\cos\theta)$. Table 28.3 lists these polynomials through $l = 3$; their shapes are shown in Figure 30.3. The wave functions for the rigid rotor are

$$\psi_{l,m} = \Theta(\theta)\Phi(\phi) = P_{l,m}(\cos\theta)\exp(im\phi) \qquad (28.45)$$
where
$$m = -l, -l+1, \ldots, 0, \ldots, l-1, l \qquad (28.46)$$

The normalization constant has been omitted. The functions of (28.45) have long been known in classical physics; they are called the *spherical harmonics*, symbolized $Y_{l,m}(\theta, \phi)$. They form a complete orthonormal set of functions and constitute a basis set for the series expansion of well-behaved functions just as the sines and cosines constitute a basis set for the expansion of functions in Fourier series. The solution to the problem of waves generated on a flooded planet are given by the spherical harmonics.[4] The electrostatic potential arising from various charge distributions can be expressed by the spherical harmonics. This application is very useful in determining the electrostatic potential of charges in crystals and molecules. The potential arising from the distribution of the various atoms must display the same symmetry properties as the molecule or crystal; this requirement often allows the electrostatic field to be described by a few spherical harmonic terms having symmetry properties in accord with the molecular symmetry.

---

[3] The details of the solution of Legendre's equation can be found in Eyring, Walter, and Kimball, *Quantum Chemistry*, p. 52, and in Pauling and Wilson, *Quantum Mechanics*, p. 118.

[4] This is discussed in W. Kauzmann, *Quantum Chemistry* (New York: Academic Press, 1957), p. 83.

TABLE 28.3 The first ten normalized associated Legendre polynomials $\Theta_{l,m}(\theta)$.

s orbitals	$l = 0, m = 0$	$\Theta_{0,0} = \dfrac{1}{\sqrt{2}}$
p orbitals	$l = 1, m = 0$	$\Theta_{1,0} = \sqrt{\tfrac{3}{2}} \cos\theta$
	$l = 1, m = \pm 1$	$\Theta_{1,\pm 1} = \sqrt{\tfrac{3}{4}} \sin\theta$
d orbitals	$l = 2, m = 0$	$\Theta_{2,0} = \sqrt{\tfrac{5}{8}}\,(3\cos^2\theta - 1)$
	$l = 2, m = \pm 1$	$\Theta_{2,\pm 1} = \sqrt{\tfrac{15}{4}} \sin\theta \cos\theta$
	$l = 2, m = \pm 2$	$\Theta_{2,\pm 2} = \sqrt{\tfrac{15}{16}} \sin^2\theta$
f orbitals	$l = 3, m = 0$	$\Theta_{3,0} = \sqrt{\tfrac{63}{8}}\,(\tfrac{5}{3}\cos^3\theta - \cos\theta)$
	$l = 3, m = \pm 1$	$\Theta_{3,\pm 1} = \sqrt{\tfrac{21}{32}}\,(5\cos^2\theta - 1)\sin\theta$
	$l = 3, m = \pm 2$	$\Theta_{3,\pm 2} = \sqrt{\tfrac{105}{16}} \sin^2\theta \cos\theta$
	$l = 3, m = \pm 3$	$\Theta_{3,\pm 3} = \sqrt{\tfrac{35}{32}} \sin^3\theta$

We shall subsequently see (Chapter 30) that these same equations also arise in elucidating the hydrogen atom. We shall therefore defer further discussion of these functions. When we do examine them, we shall discover that l is a quantum number that measures the total angular momentum, and m is a quantum number that measures the z component of the angular momentum.

The energy levels of the rigid rotor are found from Equation (28.43) to be

$$E_J = \frac{\hbar^2}{2I} J(J+1) \tag{28.47}$$

where we have substituted J for l; the symbol J is usually used for rotational quantum numbers. The energy depends *only* on J, and *not* on m. Since there are $2J + 1$ different values of m for any given J value, there are $2J + 1$ states with the same energy. The degeneracies of the energy levels are therefore

$$g_j = 2J + 1 \tag{28.48}$$

Our hypothetical energy-level diagram of Figure 28.5 was not hypothetical after all; it is the energy-level diagram for the rigid rotor, where the constant

$$B = \frac{\hbar^2}{2I} \tag{28.49}$$

is known as the *rotational constant*. The selection rule for pure rotational spectra is that $\Delta J = \pm 1$. In contradistinction to the situation for vibrational levels, the lowest rotational level is at $E = 0$.

No conflict arises between the Heisenberg uncertainty principle and a zero energy. Zero energy implies zero angular momentum. Hence there is no uncertainty in momentum; it is exactly zero. The uncertainty principle then requires the position of the rotor (or more correctly the value of the conjugate dimension ϕ) to be completely uncertain; it is. We can say nothing about the orientation of the rotor in the lowest energy level. All orientations are equally probable.

Pure rotational spectra occur in the microwave region of the electromagnetic spectrum. To display a pure rotational spectrum, a molecule must have a *permanent* dipole moment. Thus HCl displays a pure rotational spectrum, and a diatomic homonuclear molecule such as N_2 does not. Molecular rotational spectroscopy provides

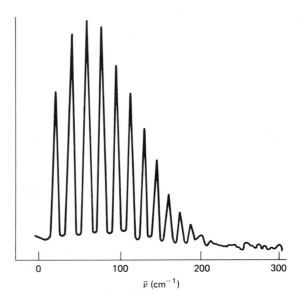

FIGURE 28.11 Pure rotational spectrum of HCl. (Courtesy of Ernest Tuazon, Mellon Institute.)

one of the most powerful tools available for determining molecular dimensions. The moments of inertia are given by the rotational constant B, as determined from the separation of the lines in the spectra, and these can be measured accurately. For diatomic molecules, the internuclear separation can then be determined by the atomic masses of the two constituent atoms. Figure 28.11 shows a rotational spectrum of HCl. Note that the lines are approximately equally spaced, and that the intensities do approximate Figure 28.6c.

The internuclear separation of a diatomic molecule can be derived quite simply from one spectrum. A linear triatomic molecule can be considered also a rigid rotor, although this method presents additional difficulties. Consider the linear molecule OCS shown in Figure 28.12, where the circles representing the different atoms have been drawn with diameters proportional to the masses. What we want to find is the length of the C–O bond, which we shall call r_{CO}, and the length of the C–S bond, r_{CS}. If we take the origin at the center of mass, then the distances of the three atoms from the center of mass are r_{C_0}, r_{O_0}, and r_{S_0}. The moment of inertia is

$$I = m_{\mathrm{O}} r_{\mathrm{O}_0}^2 + m_{\mathrm{S}} r_{\mathrm{S}_0}^2 + m_{\mathrm{C}} r_{\mathrm{C}_0}^2 \tag{28.50}$$

FIGURE 28.12 The linear molecule OCS and the relevant dimensions. The origin is taken at the center of mass.

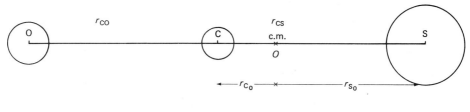

The value I is determined from the rotational spectrum, but there are three distances in (28.50). One of them can be eliminated in the following manner. The position of the center of mass is determined by the condition $\Sigma_i m_i r_i = 0$, or

$$m_O r_{O_0} + m_C r_{C_0} - m_S r_{S_0} = 0$$
$$m_O(r_{CO} + r_{CS} - r_{S_0}) + m_C(r_{CS} - r_{S_0}) - m_S r_{S_0} = 0$$

which can be solved for r_{S_0}:

$$r_{S_0} = \frac{m_O(r_{CO} + r_{CS}) + m_C r_{CS}}{m_O + m_C + m_S} \qquad (28.51)$$

Equation (28.50) for the moment of inertia can now be written

$$I = m_O(r_{CS} + r_{CO} - r_{S_0})^2 + m_C(r_{CS} - r_{S_0})^2 + m_S r_{S_0}^2 \qquad (28.52)$$

If the value for r_{S_0} in (28.51) is substituted in (28.52), the resulting expression can be rearranged, to give

$$I = \frac{m_O m_C r_{CO}^2 + m_C m_S r_{CS}^2 + m_O m_S (r_{CO} + r_{CS})^2}{m_O + m_C + m_S} \qquad (28.53)$$

in which the moment of inertia is expressed in terms of only two bond lengths and the known masses of the atoms. A measurement of the moment of inertia from a rotational spectrum provides one equation with two unknowns, r_{CO} and r_{CS}. That one equation enables us to determine the ratio of the two bond distances, but does not establish the two distances uniquely; we need another equation for that. The usual OCS molecule is $^{16}O^{12}C^{32}S$. If a different isotopic species of the molecule were prepared, say $^{16}O^{12}C^{34}S$ or $^{18}O^{12}C^{32}S$, it would have a different moment of inertia and a different value of B in the rotational spectrum. This second molecule provides a second equation in the same two unknowns, allowing a unique determination of the two bond distances. The numerical procedure for this is left for one of the problems. Note that in this procedure we assume that the internuclear distances are the same for all isotopic species. This is nearly correct but not quite. If the spectra of enough different isotopic species are measured, however, enough information can be got to allow a precise determination of all these distances.

For nonlinear molecules there are three principle moments of inertia, and the treatment can become involved. The interested reader is referred to other sources for this material.[5] For the *symmetrical top* such as CH_3Cl, in which two of the three principal moments of inertia are equal, the energy levels are

$$E = \frac{\hbar^2}{2} \left[\frac{J(J+1)}{I_B} + K^2 \left(\frac{1}{I_A} - \frac{1}{I_B} \right) \right] \qquad (28.54)$$

where $J = 0, 1, 2, \ldots$; and $K = -J, -J+1, \ldots, 0, \ldots, +J$. There is also a third quantum number M, which does not affect the energy. In (28.54), the two equal moments of inertia are I_B, and the third moment of inertia is I_A.

Thus far we have treated vibrational and rotational motion as two completely independent phenomena; they are related. Molecules in different vibrational levels have different equilibrium internuclear distances. Further, as the rotational quantum number and hence the energy increase, the molecules rotate at ever higher rotational velocities. This introduces a centrifugal force, which also affects the inter-

---

[5] See, for example, W. Gordy and R. L. Cook, *Microwave Molecular Spectra*, part 2, vol. 9, *Techniques of Organic Chemistry*, ed. A. Weissberger, (New York: Interscience Publishers, 1970).

SEC. 28.6 VIBRATIONAL-ROTATIONAL TRANSITIONS 559

nuclear distance. We shall touch on these aspects of the problem briefly in the remainder of this chapter.

28.6 VIBRATIONAL-ROTATIONAL TRANSITIONS

Figure 28.13 shows the two lowest vibrational levels of a diatomic molecule with the rotational energy levels superimposed upon each of the two vibrational levels. Note that the separation between the two vibrational levels is not drawn to the same scale as the separation between the rotational levels; if the same scale were used for both

FIGURE 28.13 The lowest two vibrational levels of a diatomic molecule, with the rotational levels superimposed upon the vibrational levels. Note that the energy scale for the rotational and vibrational levels are different. The transitions giving rise to the various lines in the P and R branches of the spectrum are indicated, and the spectrum is sketched at the bottom. The center of the spectrum indicated by ω_e is the frequency of the pure vibrational transition, that is, the frequency at the point at which the transition $J'' = 0 \rightarrow J' = 0$ would occur if it were allowed.

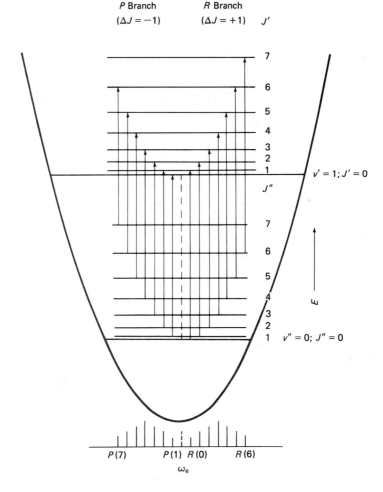

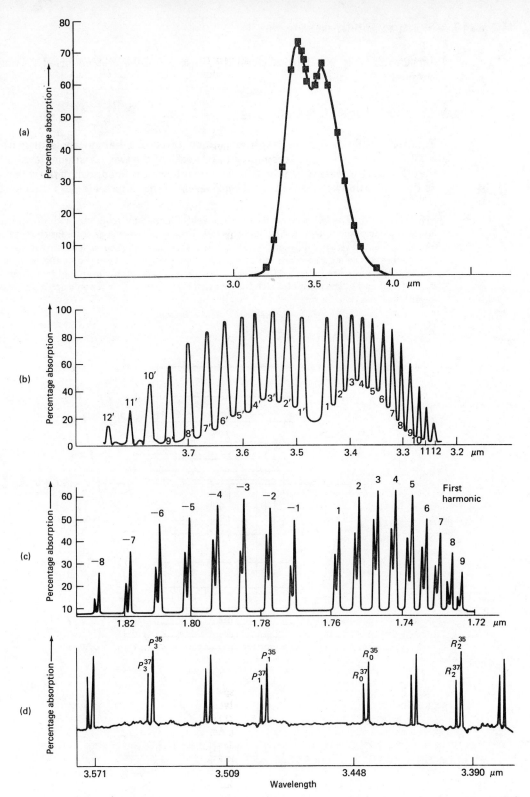

types of levels, the separation between the two vibrational levels should be about 100 times the indicated separation. At room temperature, kT is smaller than the separation between the two vibrational levels; thus only the lowest vibrational level is appreciably populated. The separation between the rotational levels, on the other hand, is smaller than kT; hence many of the excited rotational levels are populated. When absorption of radiation in the infrared region of the spectrum occurs, the molecule is excited from some rotational level J'' in the lowest vibrational level $v'' = 0$, to some level J' in the excited vibrational level $v' = 1$. Let us examine this *vibration-rotation* spectrum.

The energy of any level is the sum of the vibrational and rotational energies, or

$$E_{v,J} = (v + \tfrac{1}{2})h\nu_0 + BJ(J + 1) \tag{28.55}$$

The selection rules for a transition are just the combined selection rules for the vibrational and rotational transitions; we must simultaneously satisfy the conditions $\Delta v = \pm 1$ and $\Delta J = \pm 1$. The energy change in going from the lower vibrational level to the upper rotational level is then

$$\Delta E(v'' = 0 \longrightarrow v' = 1; J = J'' \longrightarrow J')$$
$$= h\nu_0 + BJ'(J' + 1) - BJ''(J'' + 1) \tag{28.56}$$

There are two possibilities for ΔJ, corresponding to the two observed *branches* of the spectrum:

$$\Delta J = J' - J'' = +1 \qquad \text{(the } R \text{ branch)}$$
$$\Delta J = J' - J'' = -1, \qquad \text{(the } P \text{ branch)}$$

For the R branch, since $\Delta J = +1$, then $J' = J'' + 1$, and we get from (28.56)

$$\Delta E = h\nu_0 + B(J'' + 1)(J'' + 2) - BJ''(J'' + 1)$$
$$= h\nu_0 + 2B(J'' + 1) \qquad (J'' = 0, 1, 2, 3, \ldots) \qquad (R \text{ branch}) \tag{28.57}$$

For the P branch,

$$\Delta E = h\nu_0 - 2B(J'') \qquad (J'' = 1, 2, 3, \ldots) \qquad (P \text{ branch}) \tag{28.58}$$

The vibration-rotation spectrum hence displays two sets of lines—one set at short wavelengths, corresponding to the R branch; and a second set at longer wavelengths, corresponding to the P branch. At very low resolution, the infrared spectrum consists of just one broad peak. At higher resolution, the single peak is split into two peaks, one for each branch. And at still higher resolution, the two peaks are further differentiated into their respective rotational lines. Some spectra are indicated in Figure 28.14.

We have made several approximations in this treatment. Firstly, we have neglected the effect of centrifugal force. As the molecules rotate faster for higher values of J, they stretch, affecting the moment of inertia. Secondly, we have assumed

◀FIGURE 28.14 Rotational-vibrational spectra of HCl under increasingly higher resolution. (a) The P and R branches are barely resolved. (W. Burmeister, Verhandlungen der Deutschen Physikalischen Gesellschaft, *15* 595 (1913).) (b) The two branches are clearly resolved, and the fine structure is also resolved. (E. S. Imes, Astrophysical J. *50* 251 (1919).) (c) The two isotopes are partially resolved in this first harmonic spectrum. (C. F. Meyer and A. A. Levin, Phys. Rev. *34* 44 (1929).) (d) At this highest resolution the peaks due to the two different Cl isotopes are clearly resolved. (E. K. Plyler and E. D. Tidwell, Z. Electrochemie *64* 717 (1960). Verlag Chemie International Inc. Berichte der Bunsengesellschaft.)

that the value of B is the same in the lower and upper vibrational states. This cannot be quite true, as the moment of inertia should change with vibrational quantum number because of vibrational stretching of the molecule. Thirdly, we have assumed a harmonic potential $V = \frac{1}{2}kx^2$ for the vibrating diatomic molecule; this simple model must be corrected for *anharmonicity* by introducing higher-order terms into the potential. Finally, there is a class of molecules for which a third branch is possible, the Q branch, for which $\Delta J = 0$. All these factors, and also additional ones, are treated in more advanced works on spectroscopy.[6] We shall only venture into a brief discussion of anharmonicity.

28.7 THE ANHARMONIC OSCILLATOR; THE MORSE FUNCTION

Let's return to the original model of the harmonic oscillator as a vibrating spring. The equilibrium internuclear distance is of the order of 2×10^{-8} cm. While it may be possible to maintain a potential that varies with x^2 for very small oscillations, it should be apparent that this form must fail for large oscillations. If the oscillations were as large as 2×10^{-8} cm, the two atoms would collide; the large interatomic potential would prevent this. At large separation distances, on the other hand, there is eventually a point at which the molecule dissociates and the potential goes to zero. A semiempirical equation that takes into account the large repulsive potential at small distance and dissociation at large distances was first proposed by P. M. Morse, in the form

$$V = D_e[1 - \exp(-\beta x)]^2 \qquad (28.59)$$

where D_e is the dissociation energy, and β is

$$\beta = \pi \nu_0 \left(\frac{2\mu}{D_e}\right)^{1/2} \qquad (28.60)$$

Figure 28.15 shows the form of this potential; the harmonic potential has been included for comparison. The zero of potential energy is set at the sum of the energies of the separated atoms. The measured dissociation energy is D_0; it is the energy it takes to separate the molecule from the $v = 0$ state into the two separate atoms. The value D_e is D_0 plus the zero point energy.

In Figure 28.15, the energy levels of the harmonic oscillator are shown as dashed lines, and the levels for the Morse potential are indicated by solid lines. The effect of anharmonicity is to decrease the separation between levels as the vibrational quantum number increases. Note that the equilibrium internuclear distance also changes with vibrational quantum number. The minimum in the potential-energy curve indicates the formation of a chemical bond.

The measurement of the dissociation energy D_0 from spectroscopic measurements provides a direct connection with thermodynamics. Consider the reaction $AB(g) = A(g) + B(g)$. The energy for this reaction equals the value D_0 in Figure 28.15; the value D_0 can be accurately determined from the onset of the continuum region in a vibrational-rotational spectrum.

We are now in a position to resolve a perplexing point left over from our equipartition of energy discussion (Chapter 3). There we saw that the equipartition of energy

---

[6] For example, G. Herzberg, *Molecular Spectra and Molecular Structure*, vol. 1, *Spectra of Diatomic Molecules*, 2d ed. (New York: Van Nostrand Reinhold Co., 1950).

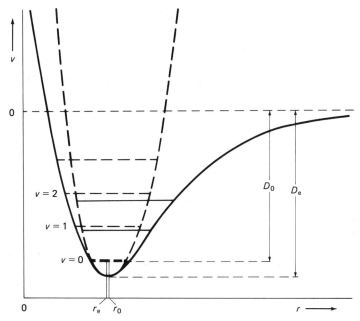

FIGURE 28.15 The Morse potential for the anharmonic oscillator (solid curve) compared with the harmonic potential (dashed curve), with the energy levels for both shown. The zero of energy is conventionally taken as the energy of the two separated atoms. The value D_0 is the observed dissociation energy in spectroscopic measurements. The equilibrium internuclear distance r_0 for the anharmonic oscillator is shifted somewhat from r_e, the value for the harmonic oscillator.

principle predicted a constant value for the heat capacity of gases whereas the experimental heat capacities varied with temperature. The solution lies in applying the equations of this chapter to the rotational and vibrational degrees of freedom.

PROBLEMS

1. A spectral line is emitted with wavelength 5000 Å. Calculate the frequency of the line. Calculate the energy difference between the two levels giving rise to the line. Calculate the energy in reciprocal centimeters, ergs, joules, calories, and units of kT for $T = 300$ K.

2. Use the energy-level system given by (28.3), with degeneracies given by (28.6) (that is, the rigid rotor). For $B = 0.05kT$, determine the level n' such that the number of systems in levels with n less than or equal to n' is $0.01N_0$, where N_0 is the number in the level $n = 0$.

3. An electron is in a cubic box 100 Å on an edge. Calculate the relative numbers of systems in each of the first seven energy levels at 300 K for a large number of such systems. Sketch the absorption spectrum of the system.

4. Carry out the steps to derive Equation (28.18).

5. The infrared spectrum of $H^{35}Cl$ has a peak at 2886 cm^{-1}. Calculate the force constant for the HCl molecule.

6. For the harmonic oscillator, calculate the expectation values of x and x^2 for the first two levels.

7. Show that the wave function for the $n = 2$ level of the harmonic oscillator is normalized; that is, show that $\int_{-\infty}^{\infty} \psi_2^2 \, dx = 1$.

8. For the harmonic oscillator, the various wave functions are orthogonal, $\int_{-\infty}^{\infty} \psi_i \psi_j \, dx = 0$ if $i \neq j$. Demonstrate the validity of this for $i = 1$ and $j = 2$.

9. Calculate the standard deviation in x for the lowest level of the harmonic oscillator and take this as the uncertainty in x. Calculate the uncertainty in momentum. Calculate the product $\Delta p \, \Delta x$, and compare it with $\hbar$.

10. A particle of mass 1 g is connected to a spring of

force constant 1000 dyn cm^{-1}. Calculate the separation between the lowest two levels.

11. Assume that vibrational force constants are independent of isotopic species. Calculate the relative energy separation for the lowest two levels in H^{35}Cl and D^{35}Cl, where D is a deuterium atom. Do the same for the C–H and C–D bonds.

12. Calculate the positions of the maxima in the probability distribution for the harmonic oscillator with $v = 2$.

13. For H^{35}Cl, the internuclear distance is 1.275 Å.
    (a) Calculate the moment of inertia.
    (b) Calculate the value of the rotational constant B.
    (c) Calculate the energies of the first ten levels.
    (d) Calculate the relative populations of the first ten levels at 300 K.
    (e) Sketch the rotational spectrum, indicating the relative heights of the peaks.

14. The infrared absorption peak of H^{35}Cl is at 2886 cm^{-1}. Calculate the separation between the lowest two vibrational levels of H^{35}Cl. Compare with the separation between the rotational levels you calculated in Problem 13. Calculate the relative populations in the lowest five vibrational levels and compare with the figures for the rotational levels. Compare both with kT at 300 K (r is given in Problem 13).

15. By what amount are the rotational absorption peaks of H^{37}Cl separated from the peaks of H^{35}Cl? Check your answer against the spectra in Figure 28.14. Do they agree? Explain.

16. Carry out the steps needed to obtain (28.33) from (28.31).

17. Assume that the molecule N$_2$O is linear. How would you distinguish between the two possible structures, N—N—O and N—O—N?

18. Rotational constants B were evaluated for two different OCS molecules:

    ^{16}O^{12}C^{32}S, $B = 0.202864$ cm^{-1}
    ^{16}O^{12}C^{34}S, $B = 0.197910$ cm^{-1}

    Assuming that the bond lengths are the same for both molecules, evaluate the bond distances in OCS. (Ans.: $r_{CS} = 1.550$ Å; $r_{CO} = 1.171$ Å.)

19. From the data given in previous problems on HCl, sketch the vibration-rotation spectrum of HCl, showing both the P and the R branches, and indicate the relative intensities. Compare with the spectrum of Figure 28.14.

20. The rotational constants B for several alkali halides are as in the accompanying table (values of B in units of megahertz). Calculate the bond distances in each. Do you observe a pattern?

Na^{35}Cl	6536.9	Na^{79}Br	4534.1
^{39}K^{35}Cl	3856.4	^{39}K^{79}Br	2434.9
^{85}Rb^{35}Cl	2627.4	^{85}Rb^{79}Br	1424.8
Cs^{35}Cl	2161.2	Cs^{79}Br	1081.3
		NaI	3531.8
		^{39}KI	1826.0
		^{85}RbI	984.3
		CsI	708.4

21. Write the Schrödinger equation for a vibrating diatomic molecule, using the Morse potential of (28.59) instead of the Hooke's law potential. How would you solve this equation?

22. Show that the following recursion relations for the Hermite polynomials are valid.

$$\frac{dH_n(x)}{dx} = 2nH_{n-1}(x)$$

$$H_n(x) - 2xH_{n-1}(x) + 2(n-1)H_{n-2}(x) = 0$$

Show that these two recursion relations imply Equation (28.18) with the condition of (28.19).

23. The quantum mechanical operator for the z component of angular momentum is $\hat{L}_z = -i\hbar(\partial/\partial\phi)$. Show that the wave functions of the rigid rotor with $l = 3$ ($m = 0, \pm 1, \pm 2, \pm 3$) are eigenfunctions of this operator; evaluate the eigenvalues.

24. The quantum mechanical operator for the square of the total angular momentum is

$$\hat{L}^2 = -\hbar^2 \left[\frac{1}{\sin\theta} \frac{\partial}{\partial\theta} \left(\sin\theta \frac{\partial}{\partial\theta} \right) + \frac{1}{\sin^2\theta} \frac{\partial^2}{\partial\phi^2} \right]$$

Show that the wave functions for the rigid rotor with $l = 2$ ($m = 0, \pm 1, \pm 2$) are eigenfunctions of the operator $\hat{L}^2$. Calculate the eigenvalues.

25. Label the values of J on the spectrum of Figure 28.11.

CHAPTER TWENTY-NINE

STATISTICAL MECHANICS OF DIATOMIC AND POLYATOMIC MOLECULES

29.1 INTERNAL DEGREES OF FREEDOM

Our study of the statistical thermodynamics of gases (Chapters 23 and 24) was mainly restricted to translational motion; this limited the discussion to monatomic gases. We have found that we can calculate the thermodynamic properties of an ideal monatomic gas from the translational partition function based on the energy levels of the particle in a box. Now we shall examine the contributions of the internal degrees of freedom associated with vibrations and rotations of molecules to the heat capacity and to other thermodynamic functions. Before the end of this chapter we shall finally resolve the anomaly left behind in Chapter 3, by coming to understand why the heat capacities predicted by the classical equipartition of energy principle do not agree with the measured heat capacities of polyatomic molecules except in the limit of very high temperatures. The key to the solution lies in evaluating the partition function, which includes assessing the contributions of translational, vibrational, and rotational motion. (We shall, for the moment, exclude the contribution arising from the excitation of electronic states, since the energies associated with these excitations are generally so large that the numbers of molecules in excited electronic states are trivially small at ordinary temperatures.)

The partition function for a molecule is

$$z = \sum_j g_j \exp\left(\frac{-\epsilon_j}{kT}\right) \tag{29.1}$$

where the summation is over *all* energy levels (translational, vibrational, and rotational), and g_j is the degeneracy of each level. Each ϵ_j in (29.1) is composed of three component energies,

$$\epsilon_j = \epsilon_t + \epsilon_v + \epsilon_r \tag{29.2}$$

where ϵ_t is the translational energy, ϵ_v the vibrational energy, and ϵ_r the rotational energy. To evaluate a partition function, including all three types of energy, would be arduous. The task is greatly simplified by virtue of the result we proved in Chapter 24—that the total partition function was separable into its components, and

$$z = z_t z_v z_r \tag{29.3}$$

where z_t, z_v, and z_r are the partition functions for translational, vibrational, and rotational motion. Equation (29.3) is identical to (24.25). We have already evaluated z_t; all that remains is to evaluate z_v and z_r.

29.2 THE VIBRATIONAL PARTITION FUNCTION FOR DIATOMIC MOLECULES

The vibrational energy levels of the diatomic molecule have been given in Chapter 28:

$$\epsilon_v = (v + \tfrac{1}{2})h\nu_0 \tag{29.4}$$

Since we are concerned with energy differences, it is convenient to shift the origin of energy so as to measure the energy relative to the lowest level, in which case the energy levels can be expressed as

$$\epsilon_v = vh\nu_0 \qquad (v = 0, 1, 2, 3, \ldots) \tag{29.5}$$

The vibrational partition function is then

$$z_v = \sum_{v=0}^{\infty} \exp\left(\frac{-vh\nu_0}{kT}\right) = 1 + \exp\left(\frac{-h\nu_0}{kT}\right) + \left[\exp\left(\frac{-h\nu_0}{kT}\right)\right]^2 + \cdots \tag{29.6}$$

The series of terms in (29.6) is a geometrical progression. Since $\exp(-h\nu_0/kT)$ is less than 1, z_v can be expressed in closed form as

$$z_v = \frac{1}{1 - \exp(-h\nu_0/kT)} \tag{29.7}$$

It is conventional to use the variable $\Theta_v = h\nu_0/k$, and to write (29.7) in the form

$$z_v = \frac{1}{1 - \exp(-\Theta_v/T)} \tag{29.8}$$

The variable Θ_v has the dimensions of temperature and is known as the *characteristic temperature*. We can now use this expression to compute the molar vibrational energy of a diatomic molecule:

$$E_v = RT^2 \frac{\partial \ln z_v}{\partial T} = RT^2 \left(\frac{1}{z_v}\frac{\partial z_v}{\partial T}\right) = RT\left(\frac{x}{\exp(x) - 1}\right) \tag{29.9}$$

where $x = \Theta_v/T$.

According to the classical equipartition of energy principle, the vibrational contribution to the energy is just RT. The result in Equation (29.9) is RT multiplied by the quantity $x/[\exp(x) - 1]$, which we might consider a "correction factor" to the classical result. In the limit as T approaches infinitely high values, x approaches zero, and the exponential function in the denominator can be expressed as $\exp(x) = 1 + x$, omitting all terms past the first power of x. The correction factor is thus unity in the limit of high temperatures, and it is precisely at this limit that the statistical mechanical vibrational energy agrees with the vibrational energy calculated on the basis of the classical equipartition of energy principle.

SEC. 29.2 THE VIBRATIONAL PARTITION FUNCTION FOR DIATOMIC MOLECULES

Physically, the explanation for this can be seen as follows: The classical result is based on the assumption that all vibrational energies are accessible to the oscillator. The statistical thermodynamic result is based on the assumption that all vibrational energies are not equally accessible but governed by the Boltzmann distribution. At low temperatures only the lowest levels are appreciably occupied: the upper levels, although they are there, contribute nothing to the energy, since few molecules are excited to these levels. At infinitely high temperatures, the Boltzmann distribution requires all the levels to be equally populated, and the full possible contribution of vibrational energy is realized. The vibrational contribution to the energy, which is less than RT at low temperatures, then becomes equal to RT in the limit of infinitely high temperatures. The variable Θ_v is determined from the frequency of the vibrational transition, usually from an infrared spectrum. For CO, the vibrational frequency is $\omega = 2170$ cm^{-1}, in which case $\Theta_v = hc\omega/k = 3123$ K. At room temperature Θ_v/T is about 10; since $10/[\exp(10) - 1]$ is of the order of 0.00045, there is only a very small contribution to the vibrational energy, $0.00045RT$. When the temperature is as high as the characteristic temperature, $\Theta_v/T = 1$, and the contribution to the vibrational energy is $0.58RT$. When the temperature is five times the characteristic temperature, the vibrational energy is $0.9RT$; the temperature must be 50 times the characteristic temperature before the vibrational contribution is $0.99RT$.

The vibrational contribution to the heat capacity is

$$C_v = \left(\frac{\partial E}{\partial T}\right)_V = R\left(\frac{x^2 \exp(x)}{[\exp(x) - 1]^2}\right) \tag{29.10}$$

where again $x = \Theta_v/T$. In the limit of high temperature, x approaches zero, and C_v approaches the classical limit of R; this has been demonstrated in the first section of Chapter 24. For $\Theta_v/T = 10$, the vibrational contribution to the heat capacity is $0.0045R$, whereas at the characteristic temperature at which $\Theta_v = T$, it is $0.92R$. The vibrational contribution to the heat capacity reaches $0.99R$ when the temperature is about three times the characteristic temperature.

Other thermodynamic variables can be determined from the relations given in Chapter 24. The vibrational contribution to the entropy is

$$S_v = R \ln z_v + \frac{E}{T}$$

$$= -R \ln[1 - \exp(-x)] + \frac{Rx}{\exp(x) - 1} \tag{29.11}$$

The Helmholtz energy and the Gibbs free energy are

$$A_v = G_v = -RT \ln z_v = R \ln[1 - \exp(-x)] \tag{29.12}$$

That the vibrational contribution to the Gibbs free energy should be the same as the contribution to the Helmholtz energy can be seen from the basic relation between the two functions

$$G = A + PV \tag{29.13}$$

The vibrational partition function does not depend on the volume; hence we can take $P = 0$, and therefore $G_v = A_v$. The molar thermodynamic functions of the harmonic oscillator for various values of x are listed in Table 29.1 and are plotted as a function of x in Figure 29.1. Note that the functions are given with reference to the fiducial zero point energy E_0.

TABLE 29.1 Thermodynamic functions of the harmonic oscillator.

$x = \dfrac{\Theta_v}{T}$	$\dfrac{C_v}{R}$	$\dfrac{E - E_0}{RT}$	$-\dfrac{G - E_0}{RT}$	$\dfrac{S}{R}$	$x = \dfrac{\Theta_v}{T}$	$\dfrac{C_v}{R}$	$\dfrac{E - E_0}{RT}$	$-\dfrac{G - E_0}{RT}$	$\dfrac{S}{R}$
0.00	1.0000	1.0000	∞	∞	1.40	0.8515	0.4582	0.2832	0.7414
0.05	0.9998	0.9752	3.0206	3.9958	1.50	.8318	.4308	.2525	.6833
0.10	0.9992	0.9508	2.3522	3.3030	1.60	.8114	.4048	.2255	.6303
0.15	0.9981	0.9269	1.9712	2.8981	1.70	.7903	.3800	.2017	.5817
0.20	0.9967	0.9033	1.7078	2.6111	1.80	.7687	.3565	.1807	.5371
0.25	0.9948	0.8802	1.5087	2.3889	1.90	.7466	.3342	.1620	.4962
0.30	0.9925	0.8575	1.3502	2.2077	2.00	.7241	.3130	.1454	.4584
0.35	0.9899	0.8352	1.2197	2.0549	2.25	.6667	.2651	.1114	.3765
0.40	0.9868	0.8133	1.1096	1.9229	2.50	.6089	.2236	.0857	.3092
0.45	0.9833	0.7918	1.0151	1.8069	2.75	.5517	.1878	.0661	.2539
0.50	0.9794	0.7707	0.9328	1.7035	3.00	.4963	.1572	.0511	.2083
0.55	0.9752	0.7501	0.8603	1.6103	3.25	.4433	.1311	.0395	.1706
0.60	0.9705	0.7298	0.7959	1.5257	3.50	.3933	.1090	.0307	.1396
0.65	0.9655	0.7100	0.7382	1.4482	3.75	.3468	.0903	.0238	.1141
0.70	0.9601	0.6905	0.6863	1.3768	4.00	.3041	.0746	.0185	.0931
0.75	0.9544	0.6714	0.6394	1.3108	4.50	.2300	.0506	.0112	.0617
0.80	0.9483	0.6528	0.5966	1.2494	5.00	.1707	.0339	.0068	.0407
0.85	0.9419	0.6345	0.5576	1.1921	5.50	.1246	.0226	.0041	.0267
0.90	0.9351	0.6166	0.5218	1.1384	6.00	.0897	.0149	.0025	.0174
0.95	0.9281	0.5991	0.4890	1.0881	7.00	.0448	.0064	.0009	.0073
1.00	0.9207	0.5820	0.4587	1.0407	8.00	.0215	.0027	.0003	.0030
1.10	0.9050	0.5489	0.4048	0.9536	9.00	.0100	.0011	.0001	.0012
1.20	0.8882	0.5172	0.3584	0.8756	10.00	.0045	.0005	.0000	.0005
1.30	0.8703	0.4870	0.3182	0.8052	∞	0.0000	0.0000	0.0000	0.0000

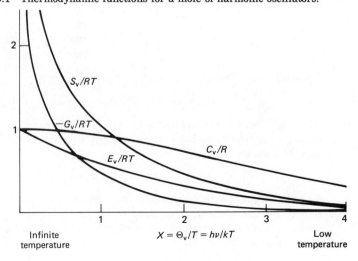

FIGURE 29.1 Thermodynamic functions for a mole of harmonic oscillators.

29.3 THE ROTATIONAL PARTITION FUNCTION FOR THE DIATOMIC RIGID ROTOR

The energy levels of the rigid rotor have been given in Chapter 28 as

$$\epsilon_j = \frac{\hbar^2}{2I} J(J+1) \qquad (J = 0, 1, 2, 3, \ldots) \tag{29.14}$$

The degeneracy of each level is

$$g_j = 2J + 1 \tag{29.15}$$

The rotational partition function then becomes

$$z_r = \sum_{J=0}^{\infty} (2J+1) \exp\left[\frac{-\hbar^2 J(J+1)}{2IkT}\right] \tag{29.16}$$

Unlike the partition function for the harmonic oscillator this series cannot be expressed in closed form. On the other hand, since the spacing between adjacent levels is usually small compared with kT, the summation can be approximated by an integral

$$z_r = \int_{J=0}^{\infty} (2J+1) \exp\left[\frac{-\hbar^2 J(J+1)}{2IkT}\right] dJ \tag{29.17}$$

This integral can be placed in standard form by letting $\xi = J(J+1)$, in which case

$$z_r = \int_{\xi=0}^{\infty} \exp\left(\frac{-\hbar^2 \xi}{2IkT}\right) d\xi$$

$$= \frac{2IkT}{\hbar^2} \tag{29.18}$$

Equation (29.18) is not quite complete; there is an additional factor that must be taken into account, the *symmetry number* σ.

Consider the homonuclear (or symmetric) diatomic molecule shown in Figure 29.2, where we have artificially labeled the molecules "1" and "2." Clearly there is no way of differentiating between the two indicated configurations. For the same reason that the translational partition function was divided by $N!$, the rotational partition function for a homonuclear diatomic molecule such as O_2 must be divided by 2 so that

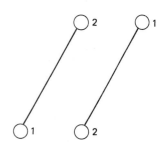

FIGURE 29.2 The two indistinguishable configurations of a homonuclear diatomic molecule.

the two indistinguishable configurations will not both be included in the integral.[1] The rotational partition function must then be written

$$z_r = \frac{2IkT}{\sigma \hbar^2} \qquad (29.19)$$

where σ is the symmetry number. It is equal to unity for heteronuclear diatomic molecules and is equal to 2 for homonuclear diatomics. Regardless of whether the molecule is diatomic or polyatomic, the symmetry number generally is the number of different ways in which the molecule can achieve the same configuration by rotation; thus $\sigma = 2$ for CO_2, and $\sigma = 3$ for NH_3. For CH_3F, we have $\sigma = 3$, since there are three identical configurations obtainable by rotating about the C–F bond. On the other hand, $\sigma = 12$ for CH_4, since there are now four different C–H bonds about which three identical configurations can be obtained by rotation, and $3 \cdot 4 = 12$.

The rotational contribution to the energy is

$$E_r = RT^2 \left(\frac{\partial \ln z_r}{\partial T}\right) = RT \left(\frac{1}{z_r}\right)\left(\frac{\partial z_r}{\partial T}\right)$$

$$= RT^2 \left(\frac{1}{T}\right) = RT \qquad (29.20)$$

The heat capacity is

$$C_v = \frac{\partial E}{\partial T} = R \qquad (29.21)$$

Note that neither the energy nor the heat capacity depends on the symmetry number. The entropy does, however, depend on the symmetry number:

$$S = R \ln z_r + \frac{E_r}{T} = R \left[1 + \ln\left(\frac{2IkT}{\sigma \hbar^2}\right)\right] \qquad (29.22)$$

There is one final comment before we proceed. It may seem strange that the rotational energy given in Equation (29.20) agrees *exactly* with the value based on the classical equipartition of energy principle. We have two degrees of rotational motion associated with rotation about two of the cartesian axes, and $\frac{1}{2}RT$ per degree of freedom yields RT per mole. For vibrational motion, on the other hand, we have seen that the classical and statistical mechanical results agree only in the limit of infinitely high temperature. The same proviso is true for rotational motion. We converted the series expression for z_r in (29.16) to the integral in (29.17) by assuming that the spacing between levels was very small. Inherent in this assumption is the assumption that the temperature is infinitely high. Thus, our assumption forced us to achieve the high temperature limit, hence the exact agreement with the classical result. A more accurate expression is given by the series expansion[2]

$$z_r = \frac{1}{\sigma \xi} \left(1 + \frac{\xi}{3} + \frac{\xi^2}{15} + \cdots \right) \qquad (29.23)$$

---

[1] From the quantum mechanical point of view, the symmetry number arises from the symmetric and antisymmetric properties of the wave functions, including the nuclear spin wave functions on the two atoms. A heteronuclear molecule may take on *all* integral values of the rotational quantum number J. For homonuclear diatomics, only odd or even values of J are permitted, depending on the symmetry property of the wave function relative to interchange of the two identical nuclei. This is because the rotational wave functions are alternately symmetric and antisymmetric. For a more complete discussion of this point, see J. E. Mayer and M. G. Mayer, *Statistical Mechanics* (New York: John Wiley & Sons, 1940), p. 172.

[2] This approximation is discussed in Mayer and Mayer, *Statistical Mechanics*, p. 152.

where $\xi = \hbar^2/2IkT$. For most diatomics not containing hydrogen atoms, Equation (29.18) is sufficiently accurate at room temperature. For H_2 and molecules of the form HX, the energy levels are more widely separated, since I is small; and it may be necessary to use (29.23), which is accurate within 0.1% when $\xi < 0.43$. The more approximate form given in (29.18) is accurate within 1% for $\xi < 0.03$ and within 0.1% for $\xi < 0.003$. These results are discussed further in the problems. The digital computer has rendered the distinction between the two expressions academic, since the exact expression for the partition function given as the series in (29.16) can be directly summed in milliseconds, yielding a result with no inherent error except roundoff errors.

29.4 THERMODYNAMIC FUNCTIONS OF DIATOMIC MOLECULES

The thermodynamic functions for the diatomic molecule can now be computed by simply summing over the various contributions due to translational, vibrational, and rotational motion. Thus the total molar energy is

$$E_{\text{tot}} = E_{\text{trans}} + E_{\text{vib}} + E_{\text{rot}} \tag{29.24}$$

with similar equations for the other extensive parameters such as the heat capacity and the entropy. The translational contributions have been computed in Chapter 24; the rotational and vibrational contributions have been computed in the earlier sections of this chapter. We now just add the three.

On the other hand, one could work directly with the total partition function for the molecule,

$$\begin{aligned} z_{\text{tot}} &= z_t z_v z_r \\ &= \left\{ V \frac{(2\pi mkT)^{3/2}}{h^3} \right\} \left\{ \frac{1}{1 - \exp(-\Theta_v/T)} \right\} \left\{ \frac{2IkT}{\sigma\hbar^2} \right\} \end{aligned} \tag{29.25}$$

where the symmetry number σ is unity for heteronuclear diatomics and equal to two for homonuclear diatomics. To be strictly correct, one should write the partition function as $z_{\text{tot}} = z_t z_v z_r z_{\text{el}}$, where z_{el} is the electronic partition function. That has been set equal to unity in (29.25); the validity of that assumption can be seen as follows.

The only degrees of freedom that contribute significantly to the thermodynamic functions are those for which the energy separations between the levels are not very much greater than kT; otherwise an insignificantly small number of molecules will be excited to higher levels. Typical spacings for the various energy levels are:

$\Delta\epsilon$(translational)	$= 10^{-17}$ eV	$= 8 \times 10^{-14}$ cm^{-1}	$= 10^{-15}$ kJ mol^{-1}
$\Delta\epsilon$(rotational)	$= 10^{-3}$–10^{-4} eV	$= 0.8$–8 cm^{-1}	$= 0.1$–0.01 kJ mol^{-1}
$\Delta\epsilon$(vibrational)	$= 0.25$ eV	$= 2000$ cm^{-1}	$= 24$ kJ mol^{-1}
$\Delta\epsilon$(electronic)	$= 3$–10 eV	$= 24{,}000$–$80{,}000$ cm^{-1}	$= 290$–960 kJ mol^{-1}

Since kT is of the order of 0.03 eV at room temperature, it is apparent that few molecules will be excited to upper electronic states except at very high temperatures; hence we are usually justified in omitting that contribution. In certain instances, however, there are low-lying electronic states, and the electronic contribution must be included. This will be discussed further in the problems at the end of this chapter.

The evaluation of the thermodynamic functions using either (29.24) or (29.25) in conjunction with the material in the previous sections is straightforward and will be left for the problems.

29.5 POLYATOMIC MOLECULES

Since translational motion is concerned only with the motion of the center of mass of a molecule, the number of atoms in a gas molecule has no effect on the translational partition function. Adjustments in the vibrational and rotational partition functions must, however, be made to take into account the additional degrees of freedom of molecules with more than two atoms.

To review some earlier material, we note that a polyatomic molecule containing n atoms has $3n$ degrees of freedom. Three of these degrees of freedom can be assigned to translational motion of the center of mass, leaving $3n - 3$ degrees of freedom. We must now differentiate between linear and nonlinear molecules. Linear molecules such as CO_2 (and all diatomics) have two degrees of rotational freedom, whereas nonlinear molecules have three degrees of rotational freedom. All the remaining degrees of freedom are associated with vibrational motion; each vibrational degree of freedom is associated with a particular *normal mode* of vibration having its own particular vibrational frequency ν_i. The number of vibrational normal modes is thus $3n - 5$ for a linear molecule and $3n - 6$ for a nonlinear molecule. If the number of vibrational modes is k, then there are k vibrational partition functions, each of the form

$$z_{v,i} = \frac{1}{1 - \exp(-\Theta_i/T)} \qquad (i = 1, 2, \ldots, k) \tag{29.26}$$

where $\Theta_i = h\nu_i/k$; the value ν_i is the vibrational frequency of that normal mode. The total vibrational partition function is then the product of all these individual partition functions,

$$z_v = \prod_{i=1}^{k} z_{v,i} = \prod_{i=1}^{k} \left[\frac{1}{1 - \exp(-\Theta_i/T)}\right] \tag{29.27}$$

For vibrational motion, the expression for the energy levels is the same for all vibrational modes. For rotational motion, however, the derivation of the energy-level diagram for a polyatomic molecule is complicated and will not be derived here. We shall just give the final result for the partition function

$$z_r = \frac{8\pi^2(8\pi^3 ABC)^{1/2}(kT)^{3/2}}{\sigma h^3} \tag{29.28}$$

where $A, B,$ and C are the three principal moments of inertia and σ is the symmetry number previously discussed.

The complete partition function for a polyatomic molecule can now be written:

$$z = \left\{\frac{(2\pi mkT)^{3/2}V}{h^3}\right\}\left\{\frac{8\pi^2 IkT}{\sigma h^2}\right\}\left\{\prod_i \left[1 - \exp\left(\frac{-\Theta_i}{T}\right)\right]^{-1}\right\} \quad \text{(linear)} \tag{29.29}$$

$$z = \left\{\frac{(2\pi mkT)^{3/2}V}{h^3}\right\}\left\{\frac{8\pi^2(8\pi^3 ABC)^{1/2}(kT)^{3/2}}{\sigma h^3}\right\}\left\{\prod_i \left[1 - \exp\left(\frac{-\Theta_i}{T}\right)\right]^{-1}\right\} \quad \text{(nonlinear)}$$

$$\tag{29.30}$$

The thermodynamic functions for nonlinear polyatomics can be represented as follows:

$$\frac{A}{NkT} = \ln z = \ln\left\{\frac{(2\pi mkT)^{3/2}}{h^3}\frac{Ve}{N}\right\} + \ln\left\{\frac{8\pi^2(8\pi^3 ABC)^{1/2}(kT)^{3/2}}{\sigma h^3}\right\}$$
$$- \left\{\sum_{i=1}^{3n-6} \ln\left[1 - \exp\left(\frac{-\Theta_i}{T}\right)\right]^{-1}\right\} \quad (29.31)$$

$$\frac{E - E_0}{NkT} = \frac{3}{2} + \frac{3}{2} + \sum_{i=1}^{3n-6} \left(\frac{\Theta_i/T}{\exp(\Theta_i/T) - 1}\right) \quad (29.32)$$

$$\frac{C_v}{Nk} = \frac{3}{2} + \frac{3}{2} + \sum_{i=1}^{3n-6} \left[\left(\frac{\Theta_i}{T}\right)^2 \frac{\exp(\Theta_i/T)}{[\exp(\Theta_i/T) - 1]^2}\right] \quad (29.33)$$

$$\frac{S}{Nk} = \ln\left\{\frac{(2\pi mkT)^{3/2}}{h^3}\frac{Ve^{5/2}}{N}\right\} + \ln\left\{\frac{8\pi^2(8\pi^3 ABC)^{1/2}(ekT)^{3/2}}{\sigma h^3}\right\}$$
$$+ \sum_{i=1}^{3n-6} \left\{\frac{\Theta_i/T}{\exp(\Theta_i/T) - 1} - \ln[1 - \exp(-\Theta_i/T)]\right\} \quad (29.34)$$

Note that the first term in the expression for the entropy is just the Sakur-Tetrode expression for the entropy of a monatomic gas.

29.6 COMPARISON WITH EXPERIMENTAL RESULTS

To give an example of the procedures just described, let us calculate the molar entropy of the diatomic molecule CO at 298.15 K. The translational contribution to the entropy is given by the Sakur-Tetrode equation and is

$$S_t = 8.314 \ln\left[\left(\frac{2\pi(0.028/6.02 \times 10^{23})(1.38 \times 10^{-23})(298.15)}{(6.626 \times 10^{-34})^2}\right)^{3/2}\right.$$
$$\left. \times \frac{0.024465 \exp(\frac{5}{2})}{6.022 \times 10^{23}}\right]$$
$$= 150.29 \text{ J K}^{-1} \text{ mol}^{-1}$$

The quantity 0.024465 is the molar volume (in cubic meters) of an ideal gas at 298.15 K and 1 atm; the other numbers are the natural constants appearing in the equation.

The characteristic temperature is obtained from infrared spectroscopy and is $\Theta_v = 3070$ K. The molar vibrational entropy for CO is then

$$S_v = 8.314 \left\{\frac{(3070/298.15)}{\exp(3070/298.15) - 1} - \ln\left[1 - \exp\left(\frac{-3070}{298.15}\right)\right]\right\}$$
$$= 0.0032 \text{ J K}^{-1} \text{ mol}^{-1}$$

The molar rotational entropy is $S_r = R + R \ln(8\pi^2 IkT/h^2)$, where the quantity $h^2/8\pi^2 Ik$ is determined from rotational spectroscopy to be 2.77 K. The rotational entropy is then

$$S_r = 8.314[1 + \ln(298.15/2.77)]$$
$$= 47.21 \text{ J K}^{-1} \text{ mol}^{-1}$$

The total entropy of CO can now be written as

$$S_{CO} = S_t + S_v + S_r$$
$$= 197.50 \text{ J K}^{-1} \text{ mol}^{-1} \quad (29.35)$$

The figure 197.50 is obtained from the molar mass of the CO molecule, and the energy-level spacings obtained from spectroscopic measurements. Note that in this case the vibrational contribution to the entropy is negligibly small. That this should be so is because the fraction of molecules excited to the first vibrational level above the ground level is $\exp(-3070/298.15) = 3.4 \times 10^{-5}$.

The experimental value of S_{CO} obtained from heat capacity measurements is 193.3 J K^{-1} mol^{-1}. It seems, then, that a discrepancy of some 4.2 J K^{-1} mol^{-1} exists between the calculated and the experimental value, and this is outside the range of experimental error. Actually, we have chosen CO as our example to point out this discrepancy (which has been discussed in Chapter 24 in connection with the third law of thermodynamics). The experimental value is based on the assumption of the third law that the entropy of CO is zero at the absolute zero of temperature. The CO molecule is, in fact, very nearly symmetrical. It has a very small dipole moment and is isoelectronic with the N_2 molecule. In the ideal crystal state, all the CO molecules would have the same alignment, C—O C—O C—O . . . . When an actual crystal is prepared by solidifying the melt, the orientations are random, and the configuration C—O O—C is just as likely as C—O C—O. These random orientations are effectively "frozen" into the crystal, and there is an extra entropy of mixing equal to $R \ln 2 = 5.8$ J K^{-1} mol^{-1} for the crystal at the absolute zero of temperature. When this entropy of mixing is added to the experimental value, the discrepancy is then only 1.6 J K^{-1} mol^{-1}, which is within the experimental uncertainty. Table 29.2 gives the experimental and calculated entropies of several gases at their normal boiling points. That we show the data at the boiling points instead of the standard temperature of 298.15 K implies the great confidence we have in our calculated values. The

TABLE 29.2 Calculated and experimental entropies of gases at their normal boiling points.

Gas	Normal boiling point (K)	Entropy, $S°(T)$ (J K^{-1} mol^{-1})	
		Measured	Calculated
F_2	85.02	165.5	165.5
Cl_2	239.11	216.3	215.3
N_2	77.36	152.0	152.1
HCl	188.13	173.5	173.2
HBr	206.38	188.6	187.7
HI	237.81	200.4	199.7
SO_2	263.08	243.0	243.3
NH_3	239.74	184.1	184.3
N_2O[a]	184.59	198.4	203.3
CO[a]	81.61	155.8	159.7
CO_2	194.66	199.2	198.8
COS	222.87	220.0	220.0
CH_4	111.7	153.0	153.5
CD_4	112	164.3	163.9
C_2H_6	184.1	207.3	206.4
C_2H_4	169.40	198.2	198.2
C_6H_6	353.26	284.9	284.4

SOURCE: Data from Landolt-Börnstein Tables, 6th ed., vol. 2, part 4, pp. 394 et. seq.

[a] This molecule has two possible orientations in the solid state at the absolute zero of temperature, and the difference is accounted for by the residual entropy of mixing of $R \ln(2) = 5.8$ J K^{-1} mol^{-1}.

standard compilations of thermodynamics properties of gases normally list the calculated instead of the experimental absolute values of the entropy (and also of the specific heat) of gases. Below the boiling point it is the experimental values that are listed, and both values are normally included at the boiling-point temperature. Note the excellent agreement between the two values.

Our faith in the theory is strong enough that when the experimental results do not agree with the calculated results, we look not for defects in the theory but for omissions, and invariably something is discovered that explains the discrepancy. The zero-point entropy of molecules such as CO and N_2O falls into this category, and this effect was postulated to explain this discrepancy. Effects such as low-temperature phase transitions have been noted in connection with our previous discussion of the third law in Chapter 8. Other effects such as the hindered rotation of molecules are discussed in the advanced textbooks on statistical mechanics listed in the Bibliography.

PROBLEMS

1. The characteristic temperature for N_2 is 3340 K and for Br_2 is 470 K.
   (a) Calculate the vibrational frequency of each molecule.
   (b) For each molecule calculate the fraction of molecules that are vibrationally excited at 300, 1000, and 10,000 K.
   (c) For which of the two molecules does there exist a substantial vibrational contribution to the heat capacity at room temperature? Why?
   (d) For each molecule calculate the vibrational contribution to the energy at 300 and 1000 K.
   (e) For each molecule calculate the vibrational contribution to the heat capacity at 300 and 1000 K.
   (f) For each molecule calculate the vibrational contribution to the entropy at 300 and 1000 K.

2. Consider a diatomic molecule at a temperature equal to the characteristic temperature, $T = \Theta_v$. (For this problem use a calculator and keep eight significant figures.)
   (a) Calculate the vibrational energy for a mole of molecules at this temperature.
   (b) Now calculate the increase in vibrational energy when the temperature is raised from Θ_v to $(\Theta_v + 0.00001\Theta_v)$.
   (c) Calculate the vibrational contribution to the heat capacity at $T = \Theta_v$.
   (d) Compare the answer you got for part (b) with the answer you got for part (c). How should they be related?

3. Write a program for a digital computer or programmable calculator to sum the series for the rotational partition function given in Equation (29.16). Write the series in terms of the parameter $\xi = \hbar^2/2IkT$, and evaluate the partition function for $\xi = 0.001, 0.01, 0.1,$ and 1. Compare these results with those obtained by using the integral expression given in (29.18), and the series approximation given in (29.23).

4. Use the total partition function for a diatomic molecule and the relevant equations to demonstrate that $PV = nRT$ for an ideal gas.

5. The internuclear separation of $H^{127}I$ is 1.604 Å; the characteristic temperature is $\Theta_v = 3200$ K. Calculate the energy, the heat capacity, and the entropy of HI, taking into account all the contributions at 298.15 and 3000 K. Compare with the experimental entropy of $S°_{298.15} = 206.3$ J K^{-1} mol^{-1}. How large an error is introduced into the calculated value of $S°_{298}$ by using the approximate form of z_r rather than the exact form?

6. The entropies of methane and the deuteromethanes have been determined from measurements of heat capacities and heats of fusion and vaporization. The experimental results are presented in the accompanying table along with the calculated results (entropies in units of J K^{-1} mol^{-1} at the boiling point temperatures). Notice that the measured and calculated values for CH_4 and CD_4 are in excellent agreement but a large difference exists between the two values for CH_3D. How is this possible? Bring the two results into agreement. (*Hint:* The explanation is not to be found in the rotational contribution and the difference in symmetry numbers.)

Gas	T_{bp}	S (exp)	S (calc)
CH_4	111.7	153.0	153.5
CD_4	112	164.3	163.9
CH_3D	111.7	156.2	167.8

7. The calculated entropies for CH_4, CD_4, and CH_3D at 112 K are given in Problem 6. The entropies differ substantially. Assume that the vibrational contribution to the entropy is the same for all three gases. How much of the difference between the gases arises from differences in translational entropy, and how much arises from differences in

rotational entropy? (*Hint:* In this case part of the explanation is to be found in the rotational contribution and the difference in symmetry number. What are the symmetry numbers for each of the gases?)

8. In certain instances a molecule can have a low-lying electronically excited state at an energy ϵ_1 above the ground state. In that case we must take into account the excitation of molecules to this low-lying state; the electronic partition function is

$$z_{el} = g_0 + g_1 \exp\left(\frac{-\epsilon_1}{kT}\right)$$

where g_0 is the degeneracy of the ground state and g_1 of the first excited electronic state. We have omitted the contributions due to the higher electronic states $\epsilon_2, \epsilon_3, \ldots$, since these energies are usually so high that they make no contribution; $\exp(-\epsilon_j/kT) \approx 0$ for $j > 1$. For such a molecule, show that the electronic contribution to the Helmholtz energy is

$$A_{el} = -NkT \ln\left[g_0 + g_1 \exp\left(\frac{-\epsilon_1}{kT}\right)\right]$$

and that the electronic contribution to the energy is

$$E_{el} = \frac{N\epsilon_1}{1 + (g_0/g_1)\exp(+\epsilon_1/kT)}$$

Finally, show that there is an electronic contribution to the heat capacity,

$$C_{el} = \frac{Nk(\epsilon_1/kT)^2}{[1 + (g_0/g_1)\exp(+\epsilon_1/kT)][1 + (g_1/g_0)\exp(-\epsilon_1/kT)]}$$

(See R. Fowler and E. A. Guggenheim, *Statistical Mechanics* (Cambridge: Cambridge University Press, 1956), pp. 102–106, for a more complete discussion.)

9. Show that the expression for the electronic heat capacity given in Problem 8 vanishes at $T = 0$ and at $T = \infty$. (Note very carefully that although the heat capacity will always vanish at $T = 0$, that it also vanishes at $T = \infty$ is true only for this model of two closely spaced levels. As the temperature gets infinitely high, the higher electronic states become populated, and these higher states contribute to the electronic specific heat.) Thus, there must be a maximum in the specific heat. At what value of ϵ_1/kT does the maximum in the electronic specific heat occur for $g_0 = g_1$?

10. A diatomic molecule that exemplifies the situation discussed in Problems 8 and 9 is NO, for which $\epsilon_1/k = 178$ K. The degeneracies of both states are two, $g_0 = g_1 = 2$. Calculate the temperature at which the electronic contribution to the specific heat of NO is a maximum. What is the electronic specific heat at this temperature?

11. Oxygen gas is paramagnetic. The ground electronic state of O_2 is triply degenerate. (Actually there is a small energy difference between the states, but these are small enough to be negligible at temperatures greater than 4 K.) If we neglect all the higher states of O_2, does the degeneracy of the lowest state make any contribution to the electronic heat capacity? There is a doublet state ($g_1 = 2$) for which $\epsilon_1/k = 11,300$ K. At what temperature does one start to get an appreciable contribution to the electronic specific heat in O_2?

12. Carbon dioxide (CO_2) is a linear molecule O—C—O and hence has four vibrational degrees of freedom. The characteristic temperatures of each of the four modes of vibration are 1890, 3360, 954, and 954 K. Note that two of the vibrational modes are degenerate. Calculate the vibrational contribution to the heat capacity of each of the modes; and calculate the total vibrational contribution to the heat capacity at 194.7, 298.15, and 1500 K.

13. The C–O bond distance in CO_2 is 1.161 Å. Use this fact plus the data of Problem 12 to calculate the entropy of $CO_2(g)$ at 194.7 K. Compare with the experimental value of 199.2 J K^{-1} mol^{-1}.

CHAPTER THIRTY

THE HYDROGEN ATOM

30.1 THE HYDROGEN ATOM

All chemical bonding is ultimately based on the quantum mechanical solution of the hydrogen atom. In this chapter we shall develop the expressions for the wave functions of the hydrogen atom; later we can elaborate on these to construct wave functions for more complicated atoms and molecules. Since the wave functions arising from the solutions of various problems consist of complete orthonormal sets of functions that can be used as a basis set for the series expansion of any well-behaved function, we shall subsequently use this attribute of the hydrogen atom wave functions to construct wave functions of more complicated systems. Solving the hydrogen atom problem devolves about solving three differential equations—one in θ, one in ϕ, and the last in the radial variable r. Two thirds of our solution have already been accomplished, since the θ and ϕ dependence are precisely the same as for the rigid rotor. We shall examine the properties of these spherical harmonic solutions in the context of their relation to the angular momentum of the electron rotating about the central nucleus. It is possible then to introduce a fourth degree of freedom, which arises from the angular momentum of the so-called *electron spin;* the pervasive influence this has on the construction of the periodic table of the elements is notable, not to mention other seemingly unrelated problems, such as the free-electron theory of metals.

The basic model is the same as it was for Bohr, and is indicated in Figure 30.1. A negatively charged electron rotates about a positively charged nucleus. The problem is a two-body problem; the two-body problem is exactly solvable both in classical and in quantum mechanics.[1] Rather than restrict ourselves to the hydrogen atom, we can without any additional difficulty treat as one group the hydrogen atom and the hydrogenlike ions, He^+, Li^{2+}, Be^{3+}, and so on, since they differ only in the charge on the central nucleus. Our model of Figure 30.1 then consists of an electron of charge $-e$ rotating about a nucleus of charge $+Ze$, where Z is the atomic number of the ion. The potential energy of the system is the electrostatic potential energy between the two charges, or

$$V = -\frac{Ze^2}{r} \tag{30.1}$$

---

[1] There are six degrees of freedom in this problem, since there are two bodies. We do not consider the three degrees of freedom to be associated with the translational motion of the center of mass. These can be separated out, as discussed in L. Pauling and E. B. Wilson, Jr., *Introduction to Quantum Mechanics* (New York: McGraw-Hill Book Company, 1935), p. 113.

CHAP. 30 THE HYDROGEN ATOM

FIGURE 30.1 Coordinate system for the hydrogen atom. An electron of charge $-e$ rotates about the central nucleus of charge $+Ze$, where $Z = 1$ for the hydrogen atom.

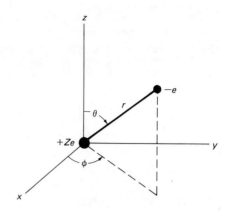

The Hamiltonian operator is

$$\hat{H} = -\frac{\hbar^2}{2\mu}\nabla^2 - \frac{Ze^2}{r} \quad (30.2)$$

and the eigenfunctions and allowed energies or eigenvalues are given by the equation

$$\hat{H}\psi = E\psi \quad (30.3)$$

Since the problem is centrosymmetric, the most appropriate coordinate system is spherical coordinates; the form of the Laplacian in spherical coordinates has been given in Equation (28.29). Using this form for the Laplacian, we get

$$\frac{1}{r^2}\frac{\partial}{\partial r}\left(r^2\frac{\partial\psi}{\partial r}\right) + \frac{1}{r^2\sin\theta}\frac{\partial}{\partial\theta}\left(\sin\theta\frac{\partial\psi}{\partial\theta}\right)$$
$$+ \frac{1}{r^2\sin^2\theta}\frac{\partial^2\psi}{\partial\phi^2} + \frac{2\mu}{\hbar^2}\left(E + \frac{Ze^2}{r}\right)\psi = 0 \quad (30.4)$$

We approach the solution of this formidable-looking equation by the procedure that should be old-hat by now. We try to separate variables by assuming a solution of the form

$$\psi = R(r)\Theta(\theta)\Phi(\phi) \quad (30.5)$$

When (30.5) is substituted in (30.4) and the standard algebraic manipulations are carried out, we get the three differential equations

$$\frac{1}{\Phi}\frac{d^2\Phi}{d\phi^2} = -m^2 \quad (30.6)$$

$$\frac{1}{\sin\theta}\frac{d}{d\theta}\left(\sin\theta\frac{d\Theta}{d\theta}\right) - \frac{m^2}{\sin^2\theta}\Theta + \beta\Theta = 0 \quad (30.7)$$

$$\frac{1}{r^2}\frac{d}{dr}\left(r^2\frac{dR}{dr}\right) + \left[\frac{2\mu}{\hbar^2}\left(E + \frac{Ze^2}{r}\right) - \frac{\beta}{r^2}\right]R = 0 \quad (30.8)$$

30.2 SOLVING THE RADIAL EQUATION AND THE ENERGY LEVELS OF THE HYDROGEN ATOM

Solving the radial equation requires that we first solve the equations in Θ and Φ; we have already done this (Chapter 28). These equations are identical to the equations arising from the solution of the rigid rotor problem. Solving (30.6) and (30.7) thus requires m to be a positive or a negative integer, and

$$\beta = l(l + 1) \tag{30.9}$$

where l is an integer such that $|m| \leq l$. We shall examine detailed solutions to these equations later; now we go on to solving the radial equation, which we can write as

$$\frac{1}{r^2}\frac{d}{dr}\left(r^2 \frac{dR}{dr}\right) + \left[\frac{2\mu}{\hbar^2}\left(E + \frac{Ze^2}{r}\right) - \frac{l(l+1)}{r^2}\right]R = 0 \tag{30.10}$$

where we have substituted $l(l + 1)$ for β. We now change the equation a bit by introducing the variables

$$\alpha^2 = \frac{-2\mu E}{\hbar^2} \tag{30.11}$$

and

$$\lambda = \frac{Z\mu e^2}{\hbar^2 \alpha} \tag{30.12}$$

Using the new independent variable

$$\rho = 2\alpha r \tag{30.13}$$

we find that the radial equation takes the form

$$\frac{1}{\rho^2}\frac{d}{d\rho}\left[\rho^2 \frac{dS(\rho)}{d\rho}\right] + \left[\frac{-l(l+1)}{\rho^2} - \frac{1}{4} + \frac{\lambda}{\rho}\right]S(\rho) = 0 \tag{30.14}$$

where $S(\rho) = R(r)$. We shall not go into the detailed solution of (30.14).[2] Valid solutions to (30.14) can be obtained only under the condition that $\lambda^2 = n^2$, where n is an integer greater than l, or

$$n^2 = -\frac{Z^2\mu e^4}{2\hbar^2 E} \quad (n = 1, 2, 3, 4, \ldots) \tag{30.15}$$

where $0 \leq l \leq n - 1$. The allowed energy levels are obtained by solving (30.15) for E,

$$E_n = -\frac{1}{n^2}\frac{Z^2\mu e^4}{2\hbar^2} \tag{30.16}$$

This is precisely the same expression Bohr got in 1913 for the energy levels of hydrogen. The new wave mechanics provided the same results as the older quantum theory of Bohr, but the new theory also provided a framework from which the hydrogen atom results could be extended to more complicated atoms and molecules.

When (30.14) is solved, the radial wave functions are found to be of the form

$$e^{-\rho/2}\rho^l L_{n+l}^{2l+1}(\rho) \tag{30.17}$$

---

[2] A detailed solution can be found in most textbooks on quantum mechanics. See, for example, L. Pauling and E. B. Wilson, Jr., *Introduction to Quantum Mechanics* (New York: McGraw-Hill Book Co., 1935), p. 121.

where $L_{n+l}^{2l+1}(\rho)$ are the *associated Laguerre functions*, which can be expressed explicitly by the equation

$$L_{n+l}^{2l+1}(\rho) = \sum_{k=0}^{n-l-1} (-1)^{k+1} \frac{\{(n+l)!\}^2}{(n-l-1-k)!(2l+1+k)!k!} \rho^k \tag{30.18}$$

The complete expression for the normalized radial wave function is

$$R_{nl}(r) = -\left\{ \left(\frac{2Z}{na_0}\right)^3 \frac{(n-l-1)!}{2n[(n+l)!]^3} \right\}^{1/2} \rho^l e^{-\rho/2} L_{n+l}^{2l+1}(\rho) \tag{30.19}$$

where

$$\rho = 2\alpha r = \frac{2\mu Z e^2}{n\hbar^2} r = \frac{2Z}{na_0} r \tag{30.20}$$

The term a_0 is the radius of the first Bohr orbit,

$$a_0 = \frac{\hbar^2}{\mu e^2} \tag{30.21}$$

The system has three degrees of freedom, and we can associate three quantum numbers (n, l, and m) with these degrees of freedom. The lowest level ($n = 0$) is nondegenerate, but higher levels are degenerate. The energy depends only on the *principal quantum number n*. We can describe these quantum numbers by the following scheme:

$$n = 1, 2, 3, 4, \ldots$$
$$l = 0, 1, 2, \ldots, n-1$$
$$m = -l, -l+1, \ldots, 0, +1, \ldots, +l$$

For any given value of n and l, we get $2l + 1$ different wave functions, all of which are eigenfunctions of the same energy eigenvalue. For any particular value of n, there are n^2 eigenfunctions for all the various possible values of l and m.

The designation n is called the *principal* quantum number, l is called the *azimuthal* quantum number, and m is often called the *magnetic* quantum number. An electron for which $l = 0$ is called an s electron; if $l = 1$, the electron is called a p electron, and $l = 2$ produces a d electron. An electron with $l = 3$ is an f electron, and higher values of l are given symbols in alphabetical order. The symbols for the first three values of l have their origins in the historical terminology of spectroscopy. Electrons with $l = 0$ were found to give rise to spectral lines in what was called the *sharp* series, hence the letter s; the p electrons were found to give rise to what were called the *principal* series, and the d electrons were found to give rise to the *diffuse* series. Higher values of l were not associated with any categorized spectral series, and the nomenclature of the higher values of the azimuthal quantum number went on alphabetically from d, skipping e to avoid confusion with the symbol for electronic charge and the base of the natural logarithms.

Expressions for the radial wave functions for several values of n and l are listed in Table 30.1. The appearance of these wave functions is illustrated in Figure 30.2. For the electron in the lowest energy level, there is only one peak in the probability distribution function; and it is left for one of the problems to carry out the arithmetic to demonstrate that this peak, or most probable value of r, coincides with the radius of the first Bohr orbit. The average distance of the electron is determined from the integral

$$\bar{r}_{n,l,m} = \iiint \psi_{n,l,m}^* r \psi_{n,l,m} r^2 dr \sin\theta \, d\theta \, d\phi \tag{30.22}$$

SEC. 30.2 SOLVING THE RADIAL EQUATION

TABLE 30.1 The hydrogenlike radial wave functions $\left(\rho = \dfrac{2Z}{na_0} r\right)$

$n = 1$

$l = 0$ (1s) $R_{1,0}(r) = \left(\dfrac{Z}{a_0}\right)^{3/2} \cdot 2 \exp\left(-\dfrac{\rho}{2}\right)$

$n = 2$

$l = 0$ (2s) $R_{2,0}(r) = \dfrac{(Z/a_0)^{3/2}}{2\sqrt{2}} (2 - \rho) \exp\left(-\dfrac{\rho}{2}\right)$

$l = 1$ (2p) $R_{2,1}(r) = \dfrac{(Z/a_0)^{3/2}}{2\sqrt{6}} \rho \exp\left(-\dfrac{\rho}{2}\right)$

$n = 3$

$l = 0$ (3s) $R_{3,0}(r) = \dfrac{(Z/a_0)^{3/2}}{9\sqrt{3}} (6 - 6\rho + \rho^2) \exp\left(-\dfrac{\rho}{2}\right)$

$l = 1$ (3p) $R_{3,1}(r) = \dfrac{(Z/a_0)^{3/2}}{9\sqrt{6}} (4 - \rho)\rho \exp\left(-\dfrac{\rho}{2}\right)$

$l = 2$ (3d) $R_{3,2}(r) = \dfrac{(Z/a_0)^{3/2}}{9\sqrt{30}} \rho^2 \exp\left(-\dfrac{\rho}{2}\right)$

$n = 4$

$l = 0$ (4s) $R_{4,0}(r) = \dfrac{(Z/a_0)^{3/2}}{96} (24 - 36\rho + 12\rho^2 - \rho^3) \exp\left(-\dfrac{\rho}{2}\right)$

$l = 1$ (4p) $R_{4,1}(r) = \dfrac{(Z/a_0)^{3/2}}{32\sqrt{15}} (20 - 10\rho + \rho^2)\rho \exp\left(-\dfrac{\rho}{2}\right)$

$l = 2$ (4d) $R_{4,2}(r) = \dfrac{(Z/a_0)^{3/2}}{96\sqrt{5}} (6 - \rho)\rho^2 \exp\left(-\dfrac{\rho}{2}\right)$

$l = 3$ (4f) $R_{4,3}(r) = \dfrac{(Z/a_0)^{3/2}}{96\sqrt{35}} \rho^3 \exp\left(-\dfrac{\rho}{2}\right)$

where the wave functions $\psi_{n,l,m}$ are the complete normalized wave functions including the angular dependence. When this integral is evaluated, we get

$$\bar{r}_{n,l,m} = \frac{n^2 a_0}{Z} \left\{ 1 + \frac{1}{2}\left[1 - \frac{l(l+1)}{n^2} \right] \right\} \tag{30.23}$$

Let's take a closer look at Figure 30.2a, which depicts the radial function and the probability distribution for the 1s electron in hydrogen ($n = 1$; $l = 0$). There are some similarities between these results and the results of the old Bohr theory; but despite this similarity, the underlying structure of the two theories is completely different. They both start with the same pictorial representation of the hydrogen atom as composed of a negative electron with a positive proton, but there the parallelism ends. The Bohr picture starts from the basic premise that the electron is in a particular orbit in any energy state. The electron in its lowest energy state is *always* at a distance a_0 from the proton, never closer and never further. Our present picture of the hydrogen atom comes from the basic postulates of quantum mechanics, and these postulates presume a probabilistic interpretation of the location of the electron. The *most probable position* of the electron is at the same distance a_0 from the proton that the plot of the probability distribution in Figure 30.2a shows, but the electron is not restricted to this fixed distance—it can be found at any distance between $r = 0$ and $r = \infty$, though the probability of finding it at large distances rapidly becomes van-

FIGURE 30.2 The radial eigenfunctions of hydrogen (left), and the radial probability distributions (right) for various values of n and l. The abscissa is in units of a_0, the Bohr radius. (a) $n = 1$. (b) $n = 2$. *Opposite page:* (c) $n = 3$. (d) $n = 4$.

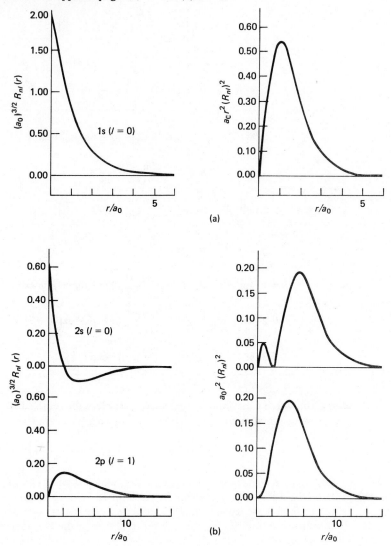

ishingly small. Note further that for the 1s electron, the wave function is a maximum at the nucleus. In Problem 7 you are asked to calculate the probability of finding the electron very close to the nucleus and also the electrostatic forces between the two charges at these distances. These probabilities are not zero, and the forces are quite large.

The situation is more complex for the higher values of n, as shown in Figure 30.2. There are n distinct maxima in the probability distribution curves of s electrons. Between these maxima there are nodal points at which the probability of finding the

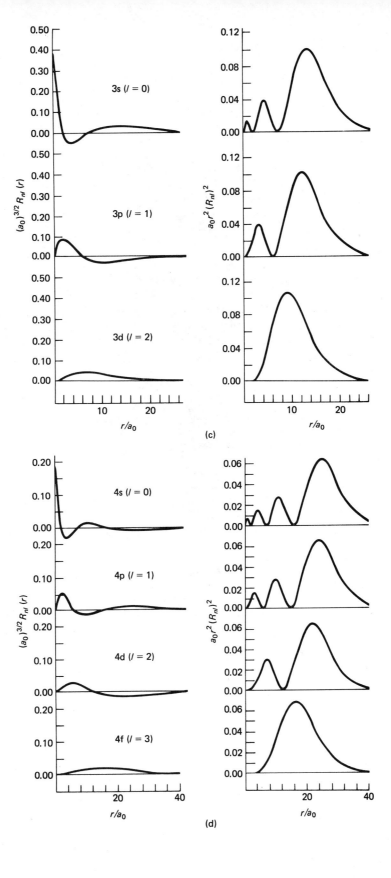

electron is zero.[3] Notice that electrons in s orbitals ($l = 0$) have higher probabilities of being close to the nucleus than electrons with higher values of l; and as n increases, the probability of finding the electron at larger distances from the proton increases.

Although it is important to understand the physical significance of these radial functions and also to note that the energy of the hydrogen atom depends only on the principal quantum number, we shall find when we come to chemical bonding that the solutions for the angular dependence have greater applicability. Also, whereas the energy of the simple isolated hydrogen atom is independent of the value of the azimuthal quantum number, this is not true for more complicated atoms, in which the value of l has a decided effect on the energy.

30.3 THE ANGULAR WAVE FUNCTIONS REVISITED

The logical place to start our examination of the angular wave functions is to write the solutions to the angular equations, (30.6) and (30.7). We have already found the solutions to these in connection with the solution of the rigid rotor (Chapter 28); the solutions are

$$\Phi_m(\phi) = \left(\frac{1}{\sqrt{2\pi}}\right) \exp(im\phi) \tag{30.24}$$

and
$$\Theta_{l,m}(\theta) = A_{l,m} P_{l,m}(\cos \theta) \tag{30.25}$$

where l is a positive integer, and

$$m = -l, -l + 1, \ldots, 0, \ldots, l - 1, l$$

The $P_{l,m}(\cos \theta)$ are the associated Legendre polynomials; Table 28.3 lists the first ten polynomials. The normalization constant in (30.25) is

$$A_{l,m} = \sqrt{\frac{2l + 1}{2} \frac{(l - |m|)!}{(l + |m|)!}} \tag{30.26}$$

The complete function can be explicitly written as

$$\theta_{l,m}(\theta) = \frac{(-1)^l}{2^l l!} \sqrt{\frac{2l + 1}{2} \frac{(l - |m|)!}{(l + |m|)!}} \sin^{|m|} \theta \frac{d^{l+|m|}(\sin^{2l} \theta)}{(d \cos \theta)^{1+|m|}} \tag{30.27}$$

The associated Legendre polynomials (or more correctly their squares, $\Theta_{l,m}^2$) are depicted in Figure 30.3 for several values of l and m. In referring to these illustrations, remember that in spherical polar coordinates the θ variable measures the angular displacement down from the z axis without reference to the orientation. The z axis is in the plane of the paper in the direction shown; the xy plane is perpendicular to the plane of the paper at the indicated line. The shapes of these curves are independent of ϕ, and a three-dimensional representation of these functions could be obtained by rotating the entire surface about the z axis. This should become clearer as we study the three-dimensional representations of the various orbitals of the hydrogen atom, including all three variables.

---

[3] If relativistic effects are taken into account, these nodal points, though small, do not equal zero.

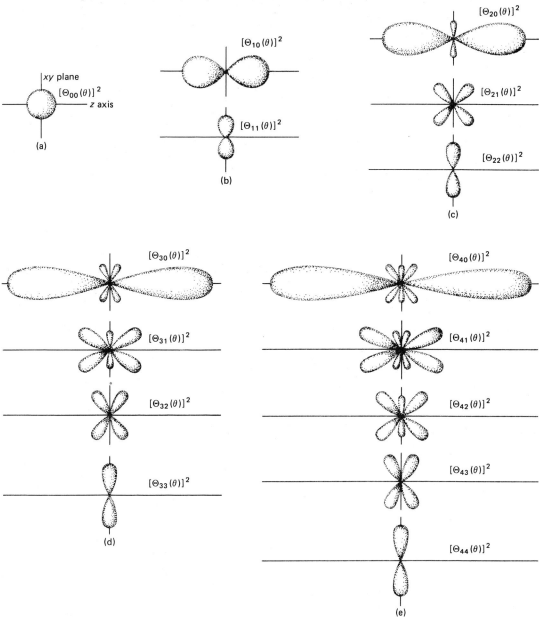

FIGURE 30.3 Plots of the function $[\Theta_{lm}(\theta)]^2$ for $l = 1,2,3,4,5$ and $|m| = 0$ to l. The xy plane is perpendicular to the plane of the paper, and the direction of the z axis is as indicated. All plots are to the same scale. (a) $l = 0$. (b) $l = 1$. (c) $l = 2$. (d) $l = 3$. (e) $l = 4$.

30.4 THE GRAPHICAL REPRESENTATION OF S ORBITALS

We shall try to illustrate graphically the probability distributions for the electrons in various orbitals of the hydrogen atom. Throughout this discussion please remember that no one has ever directly "observed" a hydrogen atom, let alone the electron about the proton. The illustrations are merely attempts to help our understanding of electron distributions. They do not represent the "true" situation; if one were to devise an experiment to photograph instantaneously the electron in a hydrogen atom, the photograph would not correspond to the pictures we present. These illustrations represent an attempt to describe complex behavior by simple graphical means. The simplest orbitals are the spherically symmetrical s orbitals for which $l = 0$, and we start with these.

For an s electron, $l = 0$ and $m = 0$. The angular dependence is given by the spherical harmonic

$$Y_{0,0}(\theta, \phi) = \Theta_{0,0}(\theta)\Phi_0(\phi)$$
$$= (4\pi)^{-1/2} \qquad (30.28)$$

For higher orbitals, $m \neq 0$, and an imaginary term arises from the Φ dependence. Thus it is convenient to plot the square of the wave function rather than the wave function itself. This is also advantageous since electron densities are proportional to ψ^2. There is no angular dependence in (30.28); a polar plot of $\Theta\Phi^2$ appears as shown in Figure 30.4a. Suppose we now include the radial dependence and plot the entire wave function

$$\psi = R(r)\Theta(\theta)\Phi(\phi)$$

for the 1s electron. The radial function for $n = 1$ appears in Figure 30.2; now we must take the product $\psi^2 = R^2(r)(\Theta\Phi)^2$. Since $R(r)$ is concentrated near the origin, ψ^2 is also concentrated in the region near the origin. This is shown in Figure 30.4b, where the probability density ψ^2 is indicated by the density of the dots. The relative density of the dots at various locations indicates the relative probabilities of finding the electron at that region of space. The determination of the actual probability of finding an electron in the spherical shell of thickness dr at a distance r from the nucleus is shown in Figure 30.4c. The area of this shell is just $4\pi r^2$; the volume is $4\pi r^2\,dr$. The probability of finding the electron in the shell is just $\psi^2\,d\tau = \psi^2 4\pi r^2\,dr$.

FIGURE 30.4 Graphical representation of 1s orbitals. (a) The function $\Theta\Phi$ for an s ($l = 0$) orbital. (b) The probability density $\psi^2 = R(r)^2(\Theta\Phi)^2$ for the 1s orbital. (c) Determination of the probability that the electron is between the radii r and $r + dr$. The surface area of the shell is $4\pi r^2$; the volume of the shell is $4\pi r^2\,dr$. The probability that the electron is between r and $r + dr$ is just $\psi^2 4\pi r^2\,dr$.

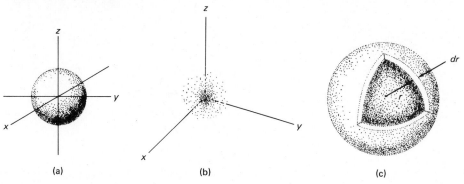

SEC. 30.5 THE GRAPHICAL REPRESENTATION OF P ORBITALS

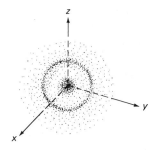

FIGURE 30.5 The probability density $\psi^2 = R(r)^2(\Theta\Phi)^2$ for the 2s orbital.

This function has its maximum value at a_0, as indicated by the peak in the probability distribution function of Figure 30.2a.

The situation for the 2s electron is only slightly more complicated than for the 1s electron. The angular dependence is the same as for the 1s electron, but now there are two values of r for which the probability density goes through a maximum, as indicated by the curve for $n = 2, l = 0$ in Figure 30.2b. This situation is graphically represented in Figure 30.5, where it can be seen that there is a high density of dots at the origin. This decreases to a zero density at the position of the node in the radial function, and then increases to a second maximum corresponding to the peak in the radial curve. The figures for the higher s orbitals ($n = 3, 4, 5, \ldots$) are similar, the number of nodal and maximal points each increasing by one for each increment of the principal quantum number n.

30.5 THE GRAPHICAL REPRESENTATION OF P ORBITALS

When we move from s electrons to p electrons, the situation is complicated by two additional factors. Firstly, we now have an angular dependence to contend with, since the functions are no longer spherically symmetrical. Secondly, there are three functions to contend with, since for $l = 0$ there are three possible values for m: $+1, 0,$ and -1. The three functions describing the angular dependence are simply arrived at, as shown in the following brief tabular scheme:

$l = 1, m = +1:$ $\quad \Theta_{1,1} = \sqrt{\dfrac{3}{4}} \sin \theta; \quad \Phi_1 = \dfrac{1}{\sqrt{2\pi}} \exp(i\phi)$

$l = 1, m = 0:$ $\quad \Theta_{1,0} = \sqrt{\dfrac{3}{2}} \cos \theta; \quad \Phi_0 = \dfrac{1}{\sqrt{2\pi}}$

$l = 1, m = -1:$ $\quad \Theta_{1,-1} = \sqrt{\dfrac{3}{4}} \sin \theta; \quad \Phi_{-1} = \dfrac{1}{\sqrt{2\pi}} \exp(-i\phi)$

The total angular dependence is given by the spherical harmonics $Y_{l,m} = \Theta_{l,m}\Phi_m$, which we can write as

$$Y_{1,1} = \left(\frac{3}{8\pi}\right)^{1/2} \sin \theta \exp(i\phi) \tag{30.29a}$$

$$Y_{1,0} = \left(\frac{3}{4\pi}\right)^{1/2} \cos \theta \tag{30.29b}$$

$$Y_{1,-1} = \left(\frac{3}{8\pi}\right)^{1/2} \sin \theta \exp(-i\phi) \tag{30.29c}$$

FIGURE 30.6 Graphical representation of the p orbitals. (a) Two-dimensional representation of $Y_{1,0} = \Theta\Phi$ and $Y_{1,0}^2 = (\Theta\Phi)^2$. The functions are independent of the angle ϕ. (b) Three-dimensional representation of $Y_{1,0} = \Theta\Phi$ and $Y_{1,0}^2 = (\Theta\Phi)^2$ for the $2p_0$ ($2p_z$) orbital. (c) The $2p_0$ orbital, showing the electron density. (d) The p_{+1} and p_{-1} orbitals.

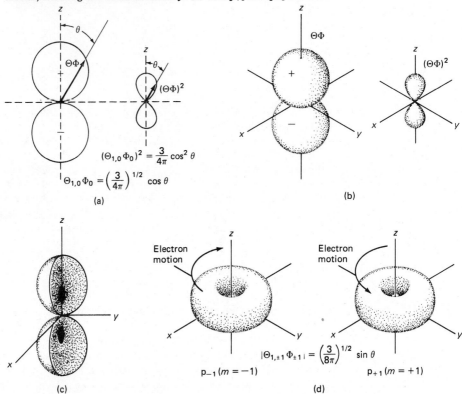

The function $Y_{1,0}$ is composed of only real terms, so let us examine that one first. Figure 30.6a shows the two-dimensional representation of $Y_{1,0}$ and $Y_{1,0}^2$. This corresponds to the curve of the associated Legendre polynomial of Figure 30.3b ($l = 1$, $m = 0$) and indicates the θ variable. Figure 30.6b shows the same functions in a three-dimensional representation. Since the lobes of this orbital, which we call the p_0 orbital, are directed along the z axis, this orbital is usually referred to as the p_z orbital.[4] The maximum electron density occurs along the z axis, as shown by the density of dots in Figure 30.6c. The figure-eight shape of the p_z orbital should be familiar to you from your elementary courses in chemistry.

Figure 30.6d indicates the form of the $Y_{1,1}$ and $Y_{1,-1}$ orbitals; these have the same appearances, since both depict the function $\sin \theta$. The two functions can be differentiated by attributing a different direction of rotation to the electron in each of the two orbitals, as shown on the figure. These orbitals, which we call the p_1 and p_{-1} orbitals, have toroidal shapes. In your elementary chemistry courses you were probably shown pictures of the three p orbitals of hydrogen, all of which had the familiar

---

[4] Note that our association of p_0 with p_z is purely arbitrary and corresponds to an arbitrary choice of the z axis.

figure-eight shape of the p_0 orbital; the question of their origin naturally arises. The familiar p_x and p_y orbitals are obtained by taking linear combinations of the p_1 and p_{-1} orbitals, as described in the next section.

30.6 THE P_x AND P_y ORBITALS AS LINEAR COMBINATIONS OF THE P_1 AND P_{-1} ORBITALS

You should recall from our previous discussions of eigenfunctions and eigenvalues that whenever two different eigenfunctions have the same eigenvalue (that is, degenerate eigenfunctions), a new eigenfunction of the same eigenvalue can be created by taking a linear combination of the degenerate eigenfunctions. To apply this to quantum mechanics, suppose we have two wave functions ψ_1 and ψ_2, and suppose that they are both normalized and orthogonal. If they are both eigenfunctions of the same energy eigenvalue, we can construct two new wave functions, both normalized and orthogonal to each other, by taking linear combinations of the form

$$\psi_A = \frac{1}{\sqrt{2}}(\psi_1 + \psi_2) \quad \text{and} \quad \psi_B = \frac{1}{\sqrt{2}}(\psi_1 - \psi_2)$$

In this example we have taken equal amounts of ψ_1 and ψ_2, but we need not have done this; the coefficients of ψ_1 and ψ_2 within the parentheses need not be the same. If they are different, the coefficient would not be $1/\sqrt{2}$ but some other normalizing constant. Wave functions such as ψ_A and ψ_B are termed *hybrid functions*, or *hybrid orbitals*.

For the present case, we construct two new p orbitals by taking

$$p_x = \frac{1}{\sqrt{2}}(p_1 + p_{-1}) \tag{30.30a}$$

and

$$p_y = \frac{-i}{\sqrt{2}}(p_1 - p_{-1}) \tag{30.30b}$$

Remembering that

$$\exp(i\phi) = \cos\phi + i\sin\phi$$

we can write a new set of three p orbitals in the form

$$p_x = \left(\frac{3}{4\pi}\right)^{1/2} \sin\theta \cos\phi \tag{30.31a}$$

$$p_y = \left(\frac{3}{4\pi}\right)^{1/2} \sin\theta \sin\phi \tag{30.31b}$$

$$p_z = \left(\frac{3}{4\pi}\right)^{1/2} \cos\theta = p_0 \tag{30.31c}$$

where for completeness we have rewritten p_z. Figure 30.7 shows these functions; they all have the familiar figure-eight shape. We now have two different sets of p orbitals, p_x, p_y, and p_z ($=p_0$), or p_1, p_{-1}, and p_0 ($=p_z$); but we cannot mix them up arbitrarily. No matter which set we use, each of the three functions is normalized and orthogonal to each of the other two. We cannot, however, mix them up and use a set such as p_x, p_1, and p_0, because the p_x orbital is not orthogonal to the p_1 orbital.

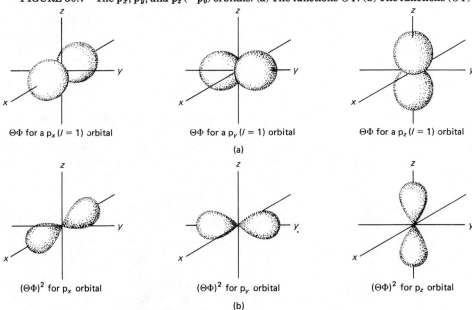

FIGURE 30.7 The p_x, p_y, and p_z ($=p_0$) orbitals. (a) The functions $\Theta\Phi$. (b) The functions $(\Theta\Phi)^2$.

This procedure is completely general, and in the next section we shall apply the same process to d electrons. The procedure is particularly powerful in applications to chemical bonding. We shall anticipate future developments somewhat by considering the set of four orbitals for an $n = 2$ electron, which we can write as 2s, $2p_x$, $2p_y$, and $2p_z$. We can construct four new functions from combinations of these to arrive at a set of orbitals directed to the four corners of a tetrahedron, forming the familiar tetrahedral bond of organic chemistry (Chapter 33).

30.7 GRAPHICAL REPRESENTATION OF THE d ORBITALS

Whereas discussions of the chemical bonding of organic molecules are usually restricted to s and p orbitals, the nature of the bonds in metal complexes, especially transition metal complexes, usually devolves about the interaction of d orbitals. Our discussion of the d orbitals will closely parallel the discussion of the p orbitals and will be briefer, since the general outline has already been presented in the last few sections.

The angular dependence of the five original d orbitals can be written according to the accompanying chart. The complete wave functions for an electron in a d orbital are obtained by multiplying the angular spherical harmonics by the radial functions

$$\psi_{n,l,m} = R_{n,l} Y_{l,m} \tag{30.33}$$

where $l = 2$ for a d electron, and n must be greater than or equal to 3. These functions are graphically represented in Figure 30.8a; again we run into the problem that for other values of m besides 0 there exists more than one function with the same shape. We can eliminate this problem just as we did for the p electrons—by

SEC. 30.7 GRAPHICAL REPRESENTATION OF THE d ORBITALS

Quantum number	Function	Symbol	
($l = 2, m = 0$)	$Y_{2,0} = \left(\dfrac{5}{16\pi}\right)^{1/2} (3\cos^2\theta - 1)$	d_0 or d_{z^2}	(30.32a)
($l = 2, m = +1$)	$Y_{2,1} = \left(\dfrac{15}{8\pi}\right)^{1/2} \sin\theta \cos\theta \exp(i\phi)$	d_1	(30.32b)
($l = 2, m = -1$)	$Y_{2,-1} = \left(\dfrac{15}{8\pi}\right)^{1/2} \sin\theta \cos\theta \exp(-i\phi)$	d_{-1}	(30.32c)
($l = 2, m = +2$)	$Y_{2,2} = \left(\dfrac{15}{32\pi}\right)^{1/2} \sin^2\theta \exp(2i\phi)$	d_2	(30.32d)
($l = 2, m = -2$)	$Y_{2,-2} = \left(\dfrac{15}{32\pi}\right)^{1/2} \sin^2\theta \exp(-2i\phi)$	d_{-2}	(30.32e)

taking new orbitals that are linear combinations of the old orbitals. This produces a new set of real orbitals, which we again call hybrid orbitals, of the form[5]

$$(l = 2, m = 0): \quad d_{z^2} = \left(\frac{5}{16\pi}\right)^{1/2} (3\cos^2\theta - 1) = d_0 \quad (30.34a)$$

$$(l = 2, m = \pm 1): \quad d_{xz} = \left(\frac{15}{4\pi}\right)^{1/2} \sin\theta \cos\theta \cos\phi \quad (30.34b)$$

$$(l = 2, m = \pm 1): \quad d_{yz} = \left(\frac{15}{4\pi}\right)^{1/2} \sin\theta \cos\theta \sin\phi \quad (30.34c)$$

$$(l = 2, m = \pm 2): \quad d_{xy} = \left(\frac{15}{16\pi}\right)^{1/2} \sin^2\theta \sin 2\phi \quad (30.34d)$$

$$(l = 2, m = \pm 2): \quad d_{x^2-y^2} = \left(\frac{15}{16\pi}\right)^{1/2} \sin^2\theta \cos 2\phi \quad (30.34e)$$

For completeness, we have rewritten the orbital for $d_0 = d_{z^2}$. These new orbitals are indicated in Figure 30.9.

The procedure for generating the graphical representation of f and g orbitals would proceed along similar lines. We shall, however, stop with the d orbitals and go on to the angular momentum of electrons.

FIGURE 30.8 Graphical representation of the pure d orbitals. These are the spherical harmonics $Y_{l,m} = \Theta_{l,m} \Phi_m$. The two states for $d_{\pm 1}$ and $d_{\pm 2}$ are differentiated by the two different directions in which the electron may rotate, as in the $p_{\pm 1}$ orbitals.

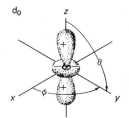
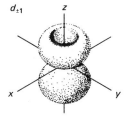
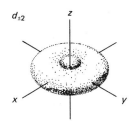

[5] As with the p orbitals, our choice of axes for the d orbitals is again purely arbitrary. Other choices such as $d_{y^2-x^2}$ would be equally suitable.

FIGURE 30.9 The hybrid d orbitals.

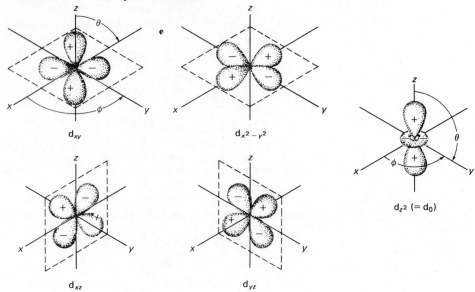

30.8 ANGULAR MOMENTUM

Recalling our introductory discussion of vectors—if a mass m moves with momentum $\mathbf{p}$ relative to a fixed point 0, then the angular momentum $\mathbf{L}$ of the mass with respect to the point 0 is

$$\mathbf{L} \equiv \mathbf{r} \times \mathbf{p} = \mathbf{r} \times m\mathbf{v} \tag{30.35}$$

where $\mathbf{r}$ is the radius vector from 0 to the position of the particle. This is illustrated in Figure 30.10; in view of the definition of the vector product, the vector $\mathbf{L}$ is perpendicular to the plane of $\mathbf{r}$ and $\mathbf{p}$, and points in the direction required by the right-hand screw convention.

Since $\mathbf{L}$ is a vector, we can write it as

$$\mathbf{L} = L_x \mathbf{i} + L_y \mathbf{j} + L_z \mathbf{K} \tag{30.36}$$

FIGURE 30.10 The angular momentum vector $\mathbf{L}$ of a mass m moving about a point 0 is given by $\mathbf{L} \equiv \mathbf{r} \times \mathbf{p} = \mathbf{r} \times m\mathbf{v}$, where $\mathbf{p}$ is the linear momentum of the particle and $\mathbf{r}$ is the vector from 0 to m. Note that $\mathbf{L}$ is perpendicular to the plane defined by $\mathbf{p}$ and $\mathbf{r}$.

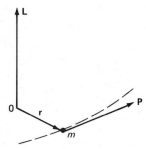

where **i**, **j**, and **k** are the unit vectors in a Cartesian coordinate system. By evaluating **L** in (30.36) and using the determinant for the vector, we can find the components of angular momentum:

$$L_x = yp_z - zp_y$$
$$L_y = zp_x - xp_z \quad (30.37)$$
$$L_z = xp_y - yp_x$$

The angular momentum of an electron was vitally important in the old Bohr theory of the hydrogen atom, since a fundamental postulate to Bohr's scheme was measuring angular momentum, $L = n\hbar$. Angular momentum is no less important in our new wave mechanics. But whereas Bohr used a classical angular-momentum expression, we must use an operator expression for angular momentum. The classical angular momentum components of (30.37) can be converted to operator form by replacing the momenta by their operators, to get

$$\hat{L}_x = -i\hbar \left(y \frac{\partial}{\partial z} - z \frac{\partial}{\partial y} \right)$$

$$\hat{L}_y = -i\hbar \left(z \frac{\partial}{\partial x} - x \frac{\partial}{\partial z} \right) \quad (30.38)$$

$$\hat{L}_z = -i\hbar \left(x \frac{\partial}{\partial y} - y \frac{\partial}{\partial x} \right)$$

Since our eigenfunctions for the electron are all in spherical polar coordinates, it is necessary to transform the operators of (30.38) into this system. It is left as an exercise to carry out the steps of this transformation, using the chain rules for partial derivatives, to get the result

$$\hat{L}_x = i\hbar \left(\cot\theta \cos\phi \frac{\partial}{\partial \phi} + \sin\phi \frac{\partial}{\partial \theta} \right)$$

$$\hat{L}_y = i\hbar \left(\cot\theta \sin\phi \frac{\partial}{\partial \phi} - \cos\phi \frac{\partial}{\partial \theta} \right) \quad (30.39)$$

$$\hat{L}_z = -i\hbar \left(\frac{\partial}{\partial \phi} \right)$$

The operator for the square of the angular momentum is

$$\hat{L}^2 = \hat{L}_x^2 + \hat{L}_y^2 + \hat{L}_z^2$$

In spherical polar coordinates this becomes

$$\hat{L}^2 = -\hbar^2 \left[\frac{1}{\sin\theta} \frac{\partial}{\partial \theta} \left(\sin\theta \frac{\partial}{\partial \theta} \right) + \frac{1}{\sin^2\theta} \frac{\partial^2}{\partial \phi^2} \right] \quad (30.40)$$

This expression for $\hat{L}^2$ is nothing more than the angular part of the Laplacian operator. The wave functions of the hydrogen atom are hence eigenfunctions of $\hat{L}^2$,

$$\hat{L}^2 \psi_{n,l,m} = l(l+1)\hbar^2 \psi_{n,l,m} \quad (30.41)$$

with eigenvalues given by $l(l+1)\hbar^2$. What this means is that if we were to measure the square of the angular momentum of the electron, we should certainly get the value $l(l+1)\hbar^2$.

It is left for one of the problems to demonstrate that while ψ is not an eigenfunction of $\hat{L}_x$ or $\hat{L}_y$, it is an eigenfunction of $\hat{L}_z$,

$$\hat{L}_z \psi_{n,l,m} = m\hbar \psi_{n,l,m} \tag{30.42}$$

A measurement of the z component of angular momentum is certain to produce the value $m\hbar$.

The three operators $\hat{H}$, $\hat{L}^2$, and $\hat{L}_z$ all commute with one another. Recall that we have showed (Chapter 27) that when two operators commute, there exist functions that are simultaneously eigenfunctions of the two operators; we should then expect ψ simultaneously to be an eigenfunction of the three operators. We can go one step further. The Heisenberg uncertainty principle states that certain pairs of variables cannot be simultaneously measured. The only pairs of variables that can be simultaneously and precisely measured are those whose operators commute. This, then, implies that we can simultaneously measure the energy, the square of the angular momentum, and the z component of the angular momentum, at least in principle, to whatever degree of precision is desired.

Two operators commute if their commutator, which we denote $[\hat{Q}, \hat{P}]$, is zero:

$$[\hat{Q}, \hat{P}] \equiv \hat{Q}\hat{P} - \hat{P}\hat{Q} = 0$$

The operator $\hat{L}^2$ commutes with any of its components,

$$[\hat{L}^2, \hat{L}_x] = [\hat{L}^2, \hat{L}_y] = [\hat{L}^2, \hat{L}_z] = 0$$

No two individual components commute with each other, however; in fact,

$$\begin{aligned} [\hat{L}_x, \hat{L}_y] &= i\hbar \hat{L}_z \\ [\hat{L}_y, \hat{L}_z] &= i\hbar \hat{L}_z \\ [\hat{L}_z, \hat{L}_x] &= i\hbar \hat{L}_y \end{aligned} \tag{30.43}$$

This implies that while the square of the angular momentum and any one of the components can be simultaneously measured, it is impossible simultaneously to measure any two components.

You should, however, be warned that it is the original orbitals such as p_1 and p_{-1} that are eigenfunctions of $\hat{L}_z$. The hybrid orbitals p_x and p_y are not eigenfunctions of $\hat{L}_z$.

30.9 THE Z COMPONENT OF ANGULAR MOMENTUM; INTERACTION WITH A MAGNETIC FIELD

Let's think about the rotating electron along simple lines for a moment. We have a charge in motion. This corresponds to an electric current, and an electric current gives rise to a magnetic field. If the magnetic field arising from the electron moving about the nucleus were viewed from a distance, it would be equivalent to a small bar magnet, and would have a magnetic moment associated with it. Classically the magnetic moment arising from a negative point charge is

$$\boldsymbol{\mu} = -\frac{e}{2m}\mathbf{L} \tag{30.44}$$

where $\mathbf{L}$ is the angular momentum. Both $\boldsymbol{\mu}$ and $\mathbf{L}$ lie along the same line perpendicular to the plane of rotation of the electron, as shown in Figure 30.11a for a circular

SEC. 30.9 THE Z COMPONENT OF ANGULAR MOMENTUM

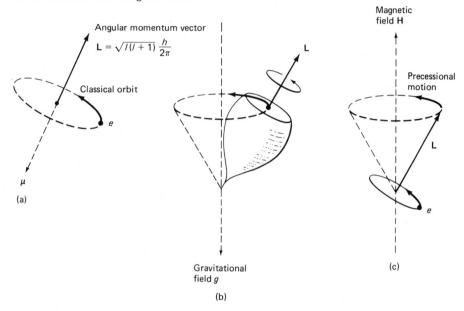

FIGURE 30.11 The magnetic moment and precessional motion associated with angular momentum. (a) The magnetic moment vector $\boldsymbol{\mu}$. (b) The precessional motion of a rotating top in a gravitational field. (c) The precessional motion of the orbital angular momentum vector of an electron in an external magnetic field.

orbit. Consider the situation of a spinning top in a gravitational field, as shown in Figure 30.11b. The angular momentum vector of the spinning top precesses about the z axis, which we take to be the direction of the gravitational field. If we place the rotating electron in a magnetic field, we have a similar situation, as shown in Figure 30.11c. The angular momentum vector **L** precesses about the z axis, which we take to be the direction of the magnetic field. The rotational frequency with which it precesses is called the *Larmor* frequency.

So far, we have had no means of distinguishing direction for a hydrogen atom, since all directions in space are equivalent. By introducing a magnetic field, we can establish a uniquely determined direction, namely the direction of the magnetic field, which we shall call the z axis. The interaction between the magnetic moment of the electron and the applied magnetic field establishes this direction as the unique direction.

The potential energy arising from the interaction between a magnetic moment $\boldsymbol{\mu}$ and an externally applied magnetic field **B** is

$$U = -\boldsymbol{\mu} \cdot \mathbf{B} = -|\mu| B_z \cos \theta \tag{30.45}$$

The magnitude of **B** is B_z, since this is the direction of the magnetic field; θ is the angle between the magnetic moment and the z axis.

For an electron in a hydrogen atom, the only allowed values of angular momentum are $\sqrt{l(l+1)}\,\hbar$. Further, the only allowed values of the z component of the magnetic moment μ_z are

$$\mu_z = \frac{m_l \hbar e}{2m} \tag{30.46}$$

A glance at (30.46) indicates that we can adopt a natural unit of magnetic moment

$$\mu_B = \frac{e\hbar}{2m} \quad (30.47)$$

which is called the *Bohr magneton*; it is equal to 9.2732×10^{-21} erg gauss^{-1}.

These results are illustrated in Figure 30.12 for a d electron ($l = 2$). The magnitude of the angular momentum is $|L| = \sqrt{l(l+1)}\hbar = \sqrt{6}\hbar$. This is the same for all the possible values of the magnetic quantum number m. If an external magnetic field is applied, then we can establish a unique z axis, and there are five different possible orientations of the vector **L** relative to the z axis, corresponding to the five different values of m; the z components are $L_z = m\hbar$ ($m = 2, -1, 0, 1, 2$), as illustrated in the figure. The angular momentum vector in each magnetic quantum number state precesses about the z axis, tracing out the indicated cones. Note that the Heisenberg uncertainty principle does not permit us to say anything about the location of the vector except that it is situated somewhere on the surface of the cone. To establish its orientation would require specifying the angular position of the electron, and this we cannot do, having established the value of the angular momentum exactly.

Before the application of the magnetic field, the five functions associated with the d electron were degenerate; we set the value of the energy of this level as the origin. Applying a magnetic field splits this originally fivefold degenerate level into five distinct levels, as shown in Figure 30.12. This effect, while small, can be seen in

FIGURE 30.12 The allowed orientations of the angular momentum vector **L** for d electrons ($l = 2$) in the presence of an external magnetic field. The z axis is chosen to lie along the direction of the magnetic field. There are five allowed orientations of the vector, each of length $\sqrt{2(2+1)}\,\hbar$; the vectors precess about the z axis on the surfaces of the indicated cones with a frequency given by the Larmor precession frequency. The splitting of the energy level is indicated to the right.

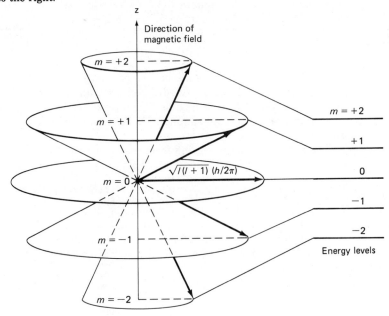

high-resolution spectroscopy, where a single line is split into five equally spaced lines when a magnetic field is applied; the separation between the lines is proportional to the strength of the field. This effect is known as the *Zeeman* effect.

If we call the energy separation between the lines δE, then

$$\delta E = mh\nu \tag{30.48}$$

where ν is the previously mentioned Larmor precession frequency,

$$\nu_{\text{Larmor}} = \frac{eB}{4\pi m} \tag{30.49}$$

30.10 RÉSUMÉ

The present state of affairs is such that we ascribe three degrees of freedom to the electron. To each degree of freedom we associate one of the quantum numbers—n corresponding to energy, l to total angular momentum, and m to the z component of angular momentum. Each of these is classically comprehensible, and each has its counterpart in classical mechanics. We shall soon discover, however, that there is a fourth degree of freedom, the electron spin, associated with the electron; it has no classical counterpart. Electron spin has angular momentum associated with it; spin angular momentum can be added vectorially to orbital angular momentum to produce a total angular momentum for the atom. The spin quantum number contains the key to understanding the periodicity of properties of the elements in the periodic table. It was electron spin and the Pauli exclusion principle that brought the work of Mendeleev to its complete fruition.

In a certain sense we have covered all of what can be termed exact quantum mechanics. By *exact quantum mechanics,* we mean those problems that can be solved exactly; this includes only the hydrogen atom. We have yet to investigate chemical bonding, though we have briefly referred to it. In the space allowable we shall touch on the simplest molecules in a quantitative, though approximate, manner and qualitatively describe some features of more complicated cases. Thus far we have discussed only one of the 100 odd elements, and that is the only one for which an exact solution is possible. In the next two chapters we shall investigate atoms that are more complicated. We shall discover that while determining precise values of the energy levels of these atoms poses extreme difficulties, determining the angular momentum states is much more manageable.

PROBLEMS

1. Substitute Equation (30.5) back into (30.4) to carry out the separation of variables for the hydrogen atom.
2. Carry out the steps leading to (30.14).
3. Show that the function $R_{1,0}$ of Table 30.1 is an eigenfunction of (30.14).
4. Is the function $R_{1,0}$ of Table 30.1 normalized? Is it orthogonal to the function $R_{2,1}$? Why or why not?
5. Show that the maximum in the radial distribution function for the 1s state of hydrogen is at a_0, the first Bohr radius. In other words, show that the most probable value of r at which the electron is found is a_0.
6. Calculate the radius of the sphere within which the electron will be found 90% of the time for the 1s state of hydrogen.
7. Calculate the probability of finding the electron in a 1s state between $r = 0$ and $r = 0.001 a_0$. Calculate the force between the electron and the proton when the electron is at the distance $0.001 a_0$ from the nucleus.
8. Suppose you have a mole of positive charge at the

north pole and an electron at the equator. Calculate the force on the electron. Calculate the distance between a proton and an electron that would give rise to this same force.

9. Compute the average distance of the electron from the proton in the 1s, 2s, and 2p states of the hydrogen atom.

10. For the $n = 4$ state of hydrogen, what is the total degeneracy? List all the possible combinations of the three quantum numbers thus far discussed.

11. Show that the 1s orbital of hydrogen is orthogonal to the 2s orbital.

12. Show that the 2s state is normalized.

13. Show that Equations (30.31a) and (30.31b) follow from (30.30a) and (30.30b).

14. Derive the form of the hybrid d orbitals given in (30.34) from the pure d orbitals given in (30.32).

15. Show that the components of angular momentum are as given in (30.37).

16. Show that the operator for the z component of angular momentum is $\hat{L}_z = -i\hbar(\partial/\partial\phi)$.

17. Show that while $\hat{L}_z$ commutes with $\hat{L}^2$, it does not commute with $\hat{L}_x$. What does this mean physically? (*Note:* Solving this problem can be done most easily using cartesian coordinates.)

18. For the hydrogen atom in the state defined by the quantum numbers $n = 2$, $l = 1$, $m = +1$, show that the wave function is an eigenfunction of both $\hat{L}^2$ and $\hat{L}_z$ and determine the eigenvalues.

19. Calculate the Larmor precession frequency for a d electron in a field of 10,000 gauss. Determine the energy splitting for the five states.

20. Figure 30.12 shows the allowed orientations of the angular momentum vector for a d electron. Draw a similar sketch for the case of an f electron ($l = 3$).

21. Two other angular momentum operators of interest are the so-called *raising and lowering*, or *ladder*, operators. These are $\hat{L}_+ = \hat{L}_x + i\hat{L}_y$ and $\hat{L}_- = \hat{L}_x - i\hat{L}_y$. These operators have the interesting property that when they operate on the spherical harmonics, they raise or lower the value of m by one unit, according to the relation

$$\hat{L}_+ Y_{l,m} = \hbar[l(l+1) - m(m+1)]^{1/2} Y_{l,m+1}$$
$$\hat{L}_- Y_{l,m} = \hbar[l(l+1) - m(m-1)]^{1/2} Y_{l,m-1}$$

Demonstrate these relations by operating on $Y_{1,-1}$ with $\hat{L}_+$ and by operating on $Y_{1,+1}$ with $\hat{L}_-$.

22. In the event that m already has its highest value, it cannot be raised one unit by $\hat{L}_+$, while if it is at the lowest possible value it cannot be lowered by $\hat{L}_-$. In these instances, the functions will be annihilated by the ladder operators. That is to say,

$$\hat{L}_+ Y_{1,+1} = 0 \quad \text{and} \quad \hat{L}_- Y_{1,-1} = 0$$

Demonstrate this by computing $\hat{L}_+ Y_{1,+1}$ and $\hat{L}_- Y_{1,-1}$ and showing that they are identically zero.

23. Show that

$$\hat{L}_+ \hat{L}_- = \hat{L}_x^2 + \hat{L}_y^2 + \hbar \hat{L}_z$$

and

$$\hat{L}^2 = \hat{L}_+ \hat{L}_- + \hat{L}_z^2 - \hbar \hat{L}_z$$

CHAPTER THIRTY-ONE

APPROXIMATE METHODS, THE HELIUM ATOM, AND SELECTION RULES

31.1 INITIAL COMMENTS; PLANETS AND ATOMS

You are all familiar, from your studies of elementary physics, with the two-body gravitational problem, exemplified by the earth's rotating about the sun. The two-body problem is exactly solvable in classical physics (and also quantum mechanical physics). Introducing a third body, say a moon or a second planet, renders an exact solution impossible, and the best we can then do is to provide an approximate solution. In classical mechanics, *perturbation theory* was developed to provide approximate solutions to these three-body problems. The effect of the planet Mercury on the motion of Earth is small compared with the sizable effect of the gravitational attraction between the earth and the sun. If the perturbing influence of Mercury on Earth's motion is considered, then an approximate solution to the earth's motion can be obtained. Although this solution will not be exact, the problem can be solved to whatever degree of accuracy we desire.[1]

When we advance beyond hydrogen in the periodic table, we can no longer provide exact solutions to the motion of the electron, since we are then concerned with more than two bodies. Approximate methods for treating the many-body problem in quantum mechanics were developed, among them the perturbation theory, which had initially been developed for the many-body problem in classical mechanics. To be sure, however, the analogy is not complete, since there are fundamental differences between classical and quantum mechanics that render the quantum mechanical problem more difficult. In the classical gravitational situation, *large* distances always exist between the various bodies, and the gravitational force is always the simple inverse square law of gravitational attraction, $f \propto r^{-2}$. In quantum mechanics, on the other hand, we deal with a probabilistic wave function. The interparticle forces we deal with are Coulombic forces that also vary with r^{-2}, but the nature of this force as the distance goes to zero is problematical.

A planetary system consisting of a central sun and two planets is a definite one. Earth is always Earth, and Mercury is always Mercury. There is no chance of confusing the two; an astronomer viewing the system from Mars can always differen-

---

[1] See, for example, R. Bellman, *Perturbation Techniques in Mathematics, Physics, and Engineering* (New York: Holt Rinehart and Winston, 1974). A more elementary discussion can be found in Y. Ryabov, *An Elementary Survey of Celestial Mechanics* (New York: Dover Publications, 1961). A pertinent discussion of perturbation theory applied to the classical vibrating string can be found in W. Kauzmann, *Quantum Chemistry* (New York: Academic Press, 1957), p. 64.

tiate between Earth and Mercury and keep track of each of them. Electrons are more ubiquitous entities. The probabilistic attributes we ascribe to electrons vitiate any attempt to fix their positions in space exactly. Further, electrons are indistinguishable, and we cannot keep track of them separately. While we place labels on individual electrons, we must recognize that these labels are for convenience only and are a bookkeeping device. Our theory must eventually take these attributes into account (cf. Chapter 32).

Some approximate methods exist for dealing with the helium atom, which is the simplest of the complicated atoms. In examining these methods, it is not our aim to provide a complete set of working tools so that you can carry out complicated calculations of the energy states of all the atoms in the periodic table. We want to present enough material for you to be able to appreciate the techniques involved. We can only skim the surface of the vast amount of material on the subject. If you intend to do the complicated calculations, you must probe more complete treatments of the subject; and with the material here, you have a useful starting point.

31.2 THE HELIUM ATOM

Suppose we try to calculate the energy levels of the helium atom by following the same procedure we used for the hydrogen atom. The quantities of interest are indicated in Figure 31.1, where for convenience we have labeled the two electrons 1 and 2. We assume that the nucleus is stationary. We let the distance between electron 1 and the nucleus be r_1; the distance between electron 2 and the nucleus is r_2, and the distance between the two electrons we call r_{12}. The Hamiltonian operator for the system can be written

$$\hat{H} = -\frac{\hbar^2}{2m}[\nabla^2(1) + \nabla^2(2)] + V \qquad (31.1)$$

where the numbers in parentheses following the ∇^2 operators indicate the electron being operated upon. Thus

$$\nabla^2(1) \equiv \nabla_1^2 = \frac{\partial^2}{\partial x_1^2} + \frac{\partial^2}{\partial y_1^2} + \frac{\partial^2}{\partial z_1^2} \qquad (31.2)$$

with a similar equation for $\nabla^2(2)$; the subscripts refer to the electron. The potential-energy term of (31.1) is the sum of three separate terms, each of which refers to one of the possible two-body interactions—the attractive coulombic poten-

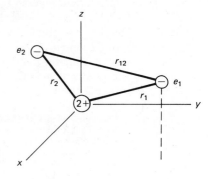

FIGURE 31.1 The coordinates of the helium atom. The electrons, each of charge $-e$, are labeled 1 and 2. The nucleus of charge $+2e$ is located at the origin. The distance between electron 1 and the nucleus is r_1, and the distance between electron 2 and the nucleus is r_2. The distance between the two electrons is r_{12}.

tial energy between the nucleus of charge $+2e$ and each of the electrons, and the repulsive energy between the two electrons. The potential-energy term in the Hamiltonian thus takes the form

$$V = -\frac{2e^2}{r_1} - \frac{2e^2}{r_2} + \frac{e^2}{r_{12}} \tag{31.3}$$

We now attempt to find eigenfunctions of the Schrödinger equation

$$\hat{H}\psi = E\psi$$

which for this case takes the form

$$\frac{-\hbar^2}{2m}(\nabla_1^2\psi + \nabla_2^2\psi) - \frac{2e^2\psi}{r_1} - \frac{2e^2\psi}{r_2} + \frac{e^2\psi}{r_{12}} = E\psi \tag{31.4}$$

Equation (31.4) is unsolvable; the $1/r_{12}$ term renders the equation inseparable.

The obvious initial assumption is to neglect the troublesome $1/r_{12}$ term. This simply assumes that the effect of electron-electron repulsion is small and can be neglected. (That this assumption is grossly incorrect will shortly become apparent.) The sign of this neglected term is the opposite of the signs of the other terms in the potential energy. We might hope to improve on our initial assumption by using a variable Z' instead of $Z = 2$ for the number of nuclear charges. This assumes that the effective nuclear charge as viewed by one electron is somewhat reduced by the presence of the other electron; the difference between Z' and $Z = 2$ is due to the *screening* effect of the electrons. We now write an approximate Hamiltonian in the form

$$\hat{H}_0 = -\frac{\hbar^2}{2m}(\nabla_1^2 + \nabla_2^2) - \frac{Z'e^2}{r_1} - \frac{Z'e^2}{r_2} \tag{31.5}$$

where Z', the effective nuclear charge, can be determined later by comparing it with the experimentally determined energy levels of the helium atom. This Hamiltonian can be written as the sum of two terms,

$$\hat{H}_0 = \hat{h}_0(1) + \hat{h}_0(2) \tag{31.6}$$

where $\hat{h}_0(1)$ depends only on the coordinates of electron 1 and $\hat{h}_0(2)$ depends only on the coordinates of electron 2. The Schrödinger equation,

$$\hat{H}_0\psi_0 = [\hat{h}_0(1) + \hat{h}_0(2)]\psi_0 = E_0\psi_0 \tag{31.7}$$

can now be solved by the method of separation of variables by assuming a solution of the form

$$\psi_0 = \phi_0(1)\phi_0(2) \tag{31.8}$$

The total energy E_0 is just the sum of the two one-electron energies E_1 and E_2. If we substitute (31.8) into (31.7) and separate the variables, we get the two similar equations

$$\hat{h}_0(1)\phi_0(1) = -\frac{\hbar^2}{2m}\nabla_1^2\phi_0(1) - \frac{Z'e^2}{r_1}\phi_0(1) = E_1\phi_0(1) \tag{31.9a}$$

$$\hat{h}_0(2)\phi_0(2) = -\frac{\hbar^2}{2m}\nabla_2^2\phi_0(2) - \frac{Z'e^2}{r_2}\phi_0(2) = E_2\phi_0(2) \tag{31.9b}$$

These are just the equations for the hydrogen atom with a nuclear charge of Z'; the solutions are

$$\phi_1 = R(n,l)\theta(l,m)\Phi(m)(1)$$
$$\phi_2 = R(n,l)\theta(l,m)\Phi(m)(2) \qquad (31.10)$$

The energies are

$$E_1 = -\left(\frac{Z'^2 me^4}{2\hbar^2}\right)\left(\frac{1}{n_1^2}\right)$$

$$E_2 = -\left(\frac{Z'^2 me^4}{2\hbar^2}\right)\left(\frac{1}{n_2^2}\right) \qquad (31.11)$$

$$E_0 = -\left(\frac{Z'^2 me^4}{2\hbar^2}\right)\left(\frac{1}{n_1^2} + \frac{1}{n_2^2}\right)$$

The lowest energy state is the one for which $n_1 = n_2 = 1$, or

$$E_0 = 2Z'^2 E_H \qquad (31.12)$$

where E_H is the energy of the hydrogen atom, $E_H = -13.6$ eV. We now compare these results with the experimental energies for the case of He.

There are two experimental quantities we can use for comparison, the total energy E_{He} and the first ionization potential E_{IP}. The term E_{IP} is the energy of the reaction

$$\text{He} \longrightarrow \text{He}^+ + e^- \qquad (31.13)$$

We then have

$$E_{IP} = E_{He^+} - E_{He} \qquad (31.14)$$

Since He$^+$ is a hydrogenlike ion, its energy can be calculated exactly:

$$E_{He^+} = (2)^2 E_H = 4(-13.6) = -54.4 \text{ eV} \qquad (31.15)$$

Experimental values for E_{IP} and E_{He} can be got from spectroscopic measurements, and they are found to be

$$E_{He}(\text{experimental}) = -79.0 \text{ eV} \qquad (31.16)$$
$$E_{IP}(\text{experimental}) = 24.6 \text{ eV} \qquad (31.17)$$

These values must now be compared with our calculated results.

Suppose we first neglect the effect of screening by taking $Z' = Z = 2$. Our calculated energies then become

$$E_{He}(\text{calc}) = E_0 = 2(2)^2 E_H = 8(-13.6) = -108.8 \text{ eV} \qquad (31.18)$$

and

$$E_{IP}(\text{calc}) = E_{He^+} - E_{He} = -54.4 \text{ eV} + 108.8 \text{ eV} = 54.4 \text{ eV} \qquad (31.19)$$

Comparison with the experimental values indicates that $E_{He}(\text{calc})$ is in error by 38% and the error in the calculated value for E_{IP} is 100%. Apparently we cannot ignore the electron-electron repulsion term.

We could improve on these results by altering our variable Z' to some value less than 2. We shall do this by using a procedure known as the *variation method*.

31.3 THE VARIATION METHOD

The first approximate method we shall discuss is the *variation* method. It is a very important technique, used extensively in many theoretical calculations dealing with atomic and molecular structures.

Suppose we have a system whose Hamiltonian operator is $\hat{H}$, and suppose that it is in the ground state with eigenfunction ψ_0. The average, or expectation, value of the energy is

$$\langle E \rangle = \int \psi_0^* \hat{H} \psi_0 \, d\tau = E_0 \tag{31.20}$$

The problem we are usually faced with is that we have an exact expression for $\hat{H}$ but can't calculate an exact expression for the wave function ψ. Suppose we "guess" at some function that satisfies the boundary conditions; let us call our "guessed" function ϕ. The expectation value of the energy E' based on ϕ is

$$E' = \int \phi^* \hat{H} \phi \, d\tau \tag{31.21}$$

The essence of the variation theorem is simply this:

If we take any well-behaved function ϕ that satisfies the boundary conditions, then the expectation value of the energy E', which we calculate using this wave function, must be greater than the true energy of the ground state E_0.

If we have been so fortunate as to guess at the correct wave function, then the energy we calculate will be the true energy.

$$E' \geqq E_0 \tag{31.22}$$

We can prove this as follows.

The true wave functions ψ form a complete set of orthogonal, normalized functions $\psi_0, \psi_1, \psi_2, \ldots$; hence any well-behaved function can be expanded in a series of these functions. Since ϕ is a well-behaved function,

$$\phi = \sum_n a_n \psi_n \tag{31.23}$$

The normalization condition on ϕ is

$$\sum_n a_n^* a_n = 1 \tag{31.24}$$

If we substitute this expression for ϕ in (31.21), we get

$$E' = \sum_n \sum_m a_n^* a_m \int \psi_n^* \hat{H} \psi_m \, d\tau \tag{31.25}$$

The functions ψ_n satisfy the condition

$$\hat{H} \psi_n = E_n \psi_n \tag{31.26}$$

and in view of the orthogonality of the ψ_n, the integrals in (31.25) vanish unless $n = m$. Thus

$$E' = \sum_n a_n^* a_n E_n \tag{31.27}$$

If we subtract E_0 from both sides of (31.27), we obtain

$$E' - E_0 = \sum_n a_n^* a_n (E_n - E_0) \qquad (31.28)$$

Now, E_n must be greater than or equal to E_0 for all values of n; further, all the coefficients $a_n^* a_n$ are positive quantities. This requires the right side of (31.28) to be either a positive quantity or zero; hence $E' - E_0 \geq 0$, or

$$E' \geq E_0 \qquad (31.29)$$

as we set out to demonstrate. Our guessing procedure should now be clear. We can select several possible functions, ϕ_1, ϕ_2, ϕ_3, . . . and compute the energies $E'_1, E'_2, E'_3, \ldots$ appropriate to each. Each of these energies must be larger than E_0; we select that value of ϕ_i that gives the lowest energy to be the best approximation to the true wave function. Often the ϕ_i differ only by some parameter, and the best value can be found by minimizing the energy with respect to that parameter by differentiation.

One serious drawback to this procedure is that while we can be sure that the computed energy is higher than E_0, we have no way of knowing how much higher it really is unless we compare with experimental values. More general methods have been developed that provide both an upper and a lower limit for the energies.[2] Let's try a simple example of this procedure.

31.4 EXAMPLE: VARIATIONAL PRINCIPLE APPLIED TO THE PARTICLE IN THE BOX

We have already solved the problem of the particle in the box exactly (Chapter 27). Now we shall try an approximate solution to the ground-state energy by the variational method and compare our results with the exact results. The Hamiltonian for the one-dimensional particle in the box is $\hat{H} = -(\hbar^2/2m)(d^2/dx^2)$. The true wave function for the ground state ($n = 1$) is given by $\psi = (2/L)^{1/2} \sin(\pi x/L)$, where L is the length of the box.

Suppose we take for our trial variational function

$$\phi = x(L - x) = xL - x^2 \qquad (31.30)$$

This function is continuous and satisfies the required boundary conditions. It is a parabola and not too different from ψ, as indicated in Figure 31.2. We can now obtain an approximation to the energy by using Eq. (31.21):

$$E' = \frac{\int_0^L (xL - x^2)[-(\hbar^2/2m)(d^2/dx^2)](xL - x^2)\, dx}{\int_0^L (xL - x^2)(xL - x^2)\, dx} \qquad (31.31)$$

where the term in the denominator arises from the need to normalize ϕ. When the indicated arithmetical operations in (31.31) are carried out, we get

$$E' = 1.013 \frac{h^2}{8mL^2} = 1.013 E_0 \qquad (31.32)$$

---

[2] See L. Pauling and E. B. Wilson, Jr., *Introduction to Quantum Mechanics* (New York: McGraw-Hill Book Co., 1935), p. 189.

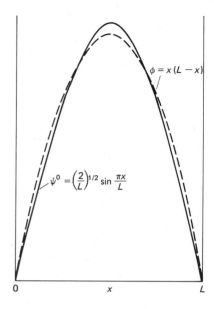

FIGURE 31.2 The trial function $\phi = x(L - x)$ for the ground state of the particle in the box compared with the "true" function.

In this simple example the results are good, since the approximate energy is only 1.3% higher than the true energy. This agreement exists because the trial function closely approximates the true function.

31.5 VARIATIONAL CALCULATION OF THE GROUND-STATE ENERGY OF THE HELIUM ATOM

We use the functions ϕ_1 and ϕ_2 of Equation (31.10) for our trial function. For the ground state, these are just the ground-state functions for the hydrogen atom, or

$$\phi = \phi_1 \phi_2 = \left(\frac{Z'^3}{a_0^3}\right) \exp\left(\frac{-Z'r_1}{a_0}\right) \exp\left(\frac{-Z'r_2}{a_0}\right) \quad (31.33)$$

where we have left Z', the effective nuclear charge, as a variable to be determined later. We must use the complete Hamiltonian

$$\hat{H} = -\frac{\hbar^2}{2m}[\nabla_1^2 + \nabla_2^2] - Ze^2\left(\frac{1}{r_1} + \frac{1}{r_2}\right) + \frac{e^2}{r_{12}} \quad (31.34)$$

where Z is the *true* atomic number ($Z = 2$ for He).

We can now calculate an approximate energy by using the wave function ϕ given in (31.33) and the Hamiltonian of (31.34). When these are substituted in (31.21) for E', numerous complicated integrals result; the final result is[3]

$$E'_{HE} = [-2Z'^2 + \tfrac{5}{4}Z' + 4Z'(Z' - Z)]E_H \quad (31.35)$$

where E_H is just the energy of the hydrogen atom. We can find the best energy based

---

[3] The integrals are evaluated in Pauling and Wilson, *Quantum Mechanics*, p. 185.

on this approximate wave function by minimizing the energy with respect to the variable Z',

$$\frac{\partial E'}{\partial Z'} = 0 = (-4Z' + \tfrac{5}{4} + 8Z' - 4Z)E_H \qquad (31.36)$$

or finally

$$Z' = Z - \tfrac{5}{16} = \tfrac{27}{16} \qquad (31.37)$$

By using this value for Z', we can calculate the energy of the helium atom from Equation (31.35) to be

$$E_{HE} = -77.5 \text{ eV} \qquad E_{IP} = 23.1 \text{ eV}$$

This method provides much better results than our previous method, which completely neglected the electron-electron repulsion. The value E_{He} is now within 2% of the experimental value of -79 eV, and the ionization potential differs by only about 6% from the experimental value of 24.6 eV. Note that E_{He} is algebraically higher than the actual value, as it should be.

The variational method is useful for a variety of problems involving both atoms and molecules, and the accuracy of the method is limited only by the ingenuity with which trial functions are selected. For the helium atom, it is worth noting that Hylleraas, using a polynomial trial function containing 14 terms, got an energy for the ground state of helium within 0.0016 eV of the experimental value.[4] This calculation, carried out in 1930, was not only a triumph for the variational method but also a triumph for quantum mechanics in general as applied to atoms containing more than one electron.

31.6 PERTURBATION THEORY

Suppose there exists a system whose Hamiltonian is $\hat{H}^0$, and for which the Schrödinger equation

$$\hat{H}^0 \psi^0 = E^0 \psi^0 \qquad (31.38)$$

can be solved exactly. Suppose our system of interest has a Hamiltonian $\hat{H}$, and the Schrödinger equation

$$\hat{H} \psi = E \psi \qquad (31.39)$$

cannot be solved exactly. If $\hat{H}$ is not too different from $\hat{H}^0$, then an approximate solution to (31.39) can be obtained in terms of corrections to the exact solutions of (31.38) by a procedure known as the *perturbation* method.

We assume that the Hamiltonian can be expanded in a series

$$\hat{H} = \hat{H}^0 + \lambda \hat{H}' + \lambda^2 \hat{H}'' + \ldots \qquad (31.40)$$

where $\hat{H}^0$ is the Hamiltonian of the unperturbed system. The unperturbed wave functions are $\psi_1^0, \psi_2^0, \ldots, \psi_j^0$, with unperturbed energies $E_1^0, E_2^0, \ldots, E_j^0$. The terms $\lambda \hat{H}' + \lambda^2 \hat{H}'' + \cdots$ constitute the adjustment, or *perturbation*, to the

---

[4] E. A. Hylleraas, *Z. Phys.* 65 (1930):209.

original unperturbed system. Since λ is small, the perturbed, or final, energies and eigenfunctions will not differ much from the energies and eigenfunctions of the unperturbed system, and these can also be expanded in a power series in λ,

$$\psi_j = \psi_j^0 + \lambda \psi_j' + \lambda^2 \psi_j'' + \cdots \qquad (31.41)$$
$$E = E_j^0 + \lambda E_j' + \lambda^2 E_j'' + \cdots \qquad (31.42)$$

For small perturbations these series rapidly converge, giving approximate values of the wave functions and the energies. The term in the first power of λ is the first-order correction, the term in λ^2 is the second-order correction, and so on; the accuracy increases as higher-order terms are included. Sometimes the first-order term vanishes, and the increased labor associated with computing the second-order term is necessary. We shall, however, discuss only the first-order perturbation in any detail.

If we substitute the values for $\hat{H}$, E, and ψ in (31.39), we get

$$(\hat{H}^0 \psi_j^0 - E_j^0 \psi_j^0) + (\hat{H}^0 \psi_j' + \hat{H}' \psi_j^0 - E_j^0 \psi_j' - E_j' \psi_j^0)\lambda + (\ldots)\lambda^2 = 0 \qquad (31.43)$$

where we neglect terms in λ^2 and higher. For this equation to be valid, the coefficient of each power must separately vanish. The coefficient of λ to the zeroth power yields

$$\hat{H}^0 \psi_j^0 = E_j^0 \psi_j^0 \qquad (31.44)$$

which is just the Schrödinger equation for the unperturbed system. The coefficient of λ^1 gives rise to the equation

$$\hat{H}^0 \psi_j' - E_j^0 \psi_j' = (E_j' - \hat{H}')\psi_j^0 \qquad (31.45)$$

whose solution yields the first-order correction. The equations arising from the higher powers of λ yield the higher-order corrections.

Since the original unperturbed functions ψ_k^0 form a complete orthonormal set of functions, the unknown functions ψ_j' can be expanded in terms of these unperturbed functions,

$$\psi_j' = \sum_k a_{kj} \psi_k^0 \qquad (31.46)$$

We can therefore write

$$\hat{H}^0 \psi_j' = \sum_k a_{kj} \hat{H}^0 \psi_k^0 = \sum_k a_{kj} E_k^0 \psi_k^0 \qquad (31.47)$$

by virtue of (31.44). Equation (31.45) can now be written

$$\sum_k a_{kj}(E_k^0 - E_j^0)\psi_k^0 = (E_j' - \hat{H}')\psi_j^0 \qquad (31.48)$$

Suppose we multiply both sides of (31.48) by ψ_j^{0*} and integrate over all space. In that case, the left side of (31.48) vanishes, since $\int \psi_j^{0*} \psi_k^0 d\tau$ vanishes if $j \neq k$, and $(E_k^0 - E_j^0) = 0$ if $j = k$. We are left with

$$\int \psi_j^{0*}(E_j' - \hat{H}')\psi_j^0 d\tau = 0 \qquad (31.49)$$

Since E_j' is a constant, Equation (31.49) can be solved, after multiplication by λ, to yield

$$\lambda E_j' = \lambda \int \psi_j^{0*} \hat{H}' \psi_j^0 d\tau \qquad (31.50)$$

It is convenient to incorporate the parameter λ in the first-order perturbation symbols and write

$$\hat{H} = \hat{H}^0 + \hat{H}'$$
$$\psi_j = \psi_j^0 + \psi_j' \qquad (31.51)$$
$$E_j = E_j^0 + E_j'$$

in which case the first-order correction to the energy becomes

$$E_j' = \int \psi_j^{0*} \hat{H}' \psi_j^0 \, d\tau \qquad (31.52)$$

Integrals such as (31.52) occur frequently in quantum mechanics, and a shorthand notation has been developed for them; thus

$$\int \psi_j^{0*} \hat{Q} \psi_k^0 \, d\tau \equiv Q_{jk} \qquad (31.53)$$

where $\hat{Q}$ is some operator. By this convention, (31.52) would be written

$$E_j' = H_{jj}' \qquad (31.54)$$

The corrections to the wave functions ψ_j' are obtained by multiplying each side of Equation (31.48) by ψ_i^{0*} and integrating. The result is

$$a_{ij}(E_i^0 - E_j^0) = -\int \psi_i^{0*} \hat{H}' \psi_j^0 \, d\tau \qquad (i \neq j) \qquad (31.55)$$

The coefficients in the expansion of (31.46) are

$$a_{ij} = -\frac{\int \psi_i^{0*} \hat{H}' \psi_j^0 \, d\tau}{E_i^0 - E_j^0} \qquad (31.56)$$

This equation determines all the terms a_{ij} in the expansion of ψ_j' except for the term a_{jj}, and that term will be zero if we omit all terms higher than the first-order term.

Our results can be finally summarized in the two equations

$$E_k = E_k^0 + H_{kk}' \qquad (31.57)$$

and

$$\psi_k = \psi_k^0 + \sum_{j=0}^{\infty}{}' \frac{H_{jk}'}{E_k^0 - E_j^0} \psi_j^0 \qquad (31.58)$$

where the prime on the summation sign indicates that the term $j = k$ is omitted.

If this procedure is applied to the term in λ^2, then the second-order correction terms can be computed. Omitting the details, we simply write the final result:

$$E_k'' = \sum_j{}' \frac{H_{kj}' H_{jk}'}{E_k^0 - E_j^0} + H_{kk}'' \qquad (31.59)$$

Equations (31.58) and (31.59) indicate that this procedure is not applicable to degenerate states, since if the states labeled j and k are degenerate, $E_k^0 = E_j^0$, and the denominators in these two equations vanish. Procedures for handling degenerate systems have been developed and are covered in the standard textbooks dealing with quantum mechanics, along with more detailed discussions of second-order perturbation theory.[5]

---

[5] See, for example, Pauling and Wilson, *Quantum Mechanics*, p. 165.

31.7 EXAMPLE: THE PERTURBED HARMONIC OSCILLATOR

For the unperturbed harmonic oscillator, the potential is $V = \tfrac{1}{2}kx^2$; the Hamiltonian was derived in (28.10) as

$$\hat{H} = -\frac{\hbar^2}{2\mu}\frac{d^2}{dx^2} + \tfrac{1}{2}kx^2 \tag{31.60}$$

Suppose we have a slightly anharmonic oscillator whose potential is

$$V = \tfrac{1}{2}kx^2 + ax^4 \tag{31.61}$$

where a is small. The perturbation to the Hamiltonian is then

$$\hat{H}' = ax^4 \tag{31.62}$$

The energies of the anharmonic oscillator are

$$E_n = E_n^0 + H'_{nn} = E_n^0 + E'_n \tag{31.63}$$

where E_n^0 is the unperturbed energy,

$$E_n^0 = \hbar \left(\frac{k}{\mu}\right)^{1/2}\left(n + \tfrac{1}{2}\right) = \left(n + \tfrac{1}{2}\right)h\nu_0 \tag{31.64}$$

The correction to the energy is

$$E'_n = a \int_{-\infty}^{\infty} \psi_n^0 x^4 \psi_n^0 \, dx \tag{31.65}$$

The wave functions have been given in (28.23) as

$$\psi_n^0(\xi) = \left(\frac{\sqrt{\beta/\pi}}{2^n n!}\right)^{1/2} H_n(\xi) \exp\left(-\tfrac{1}{2}\xi^2\right) \qquad (\xi = \sqrt{\beta}\,x) \tag{31.66}$$

Since $x = \xi/\sqrt{\beta}$, we have

$$x^4 = \frac{\xi^4}{\beta^2} \qquad dx = \frac{d\xi}{\sqrt{\beta}}$$

Equation (31.65) now becomes

$$E'_n = a\frac{N_n^2}{(\beta)^{5/2}} \int_{-\infty}^{\infty} \exp(-\xi^2) H_n^2(\xi) \xi^4 \, d\xi \tag{31.67}$$

where N_n is the normalization constant. It is possible, by using the various *recursion relations* for the Hermite polynomials, to evaluate the integral in Equation (31.67) in terms of n as[6]

$$E'_n = \frac{3}{4\beta^2}(2n^2 + 2n + 1)a \tag{31.68}$$

In this case the total energy of the oscillator becomes

$$\begin{aligned} E_n &= E_n^0 + E'_n \\ &= \left(n + \tfrac{1}{2}\right)h\nu_0 + \frac{3}{64\pi^4}(2n^2 + 2n + 1)\frac{h^2 a}{\mu^2 \nu_0^2} \end{aligned} \tag{31.69}$$

[6] See Problem 10.

For the ground state ($n = 0$), the integral of (31.67) is easily evaluated, since $H_0(\xi) = 1$; the integral is

$$E'_0 = \frac{(\beta/\pi)^{1/2}}{(\beta)^{5/2}} \int_{-\infty}^{\infty} \xi^4 \exp(-\xi^2)\, d\xi$$

$$= \frac{(\beta/\pi)^{1/2}}{(\beta)^{5/2}} \left(\frac{3}{4}\sqrt{\pi}\right)$$

$$= \frac{3a}{4\beta^2} \tag{31.70}$$

in agreement with (31.68).

Recall that in Figure 28.15 for the anharmonic oscillator we indicated that the energies of the anharmonic oscillator are somewhat higher than the energies of the harmonic oscillator, and that the difference increases as n increases. This is in accord with the results of this section, as illustrated in Equation (31.69). The Morse potential of Figure 28.15 can be expanded in terms of a power series in x,

$$V = \tfrac{1}{2}kx^2 + ax^3 + a'x^4 + \cdots \tag{31.71}$$

Effectively, what we have done here is compute the energies of the various levels, using only the first and third terms of (31.71). The results would have been the same had we used the first three terms, since as indicated in one of the problems, the x^3 term makes no contribution to the first-order perturbation energy—the integral $(x^3)_{nn}$ vanishes.

31.8 PERTURBATION CALCULATION OF THE GROUND-STATE ENERGY OF THE HELIUM ATOM

Let us now calculate the energy of the ground state of the helium atom using first-order perturbation theory and compare the results with our previous calculations. The complete Hamiltonian for the system has been given in (31.34), and we have already seen that if the term e^2/r_{12} is omitted, the Schrödinger equation can be solved exactly. The unperturbed energy has been given in (31.18) as

$$E^0 = E_{\text{He}} = 2(Z)^2 E_{\text{H}} = 2(2)^2(-13.6) = -108.8 \text{ eV} \tag{31.72}$$

If we take the perturbation term to be

$$\hat{H}' = \frac{e^2}{r_{12}} \tag{31.73}$$

then the correction to the energy is

$$E' = \int \psi^{0*} \hat{H}' \psi^0 \, d\tau \tag{31.74}$$

The unperturbed wave functions are just the functions for the 1s state of hydrogen, as given in (31.10), or

$$\psi^0 = u_{1s}(r_1, \theta_1, \phi_1) u_{1s}(r_2, \theta_2, \phi_2) \tag{31.75}$$

where r_1, θ_1, ϕ_1 and r_2, θ_2, ϕ_2 are the polar coordinates of the first and second electrons relative to the origin taken at the nucleus. The form for the 1s wave function is

$$u_{1s} = \left(\frac{Z^3}{\pi a_0^3}\right)^{1/2} \exp\left(\frac{-\rho}{2}\right) \tag{31.76}$$

where $\rho = 2Zr/a_0$; ψ^0 is then

$$\psi^0 = \frac{Z^3}{\pi a_0^3} \exp\left(\frac{-\rho_1}{2}\right) \exp\left(\frac{-\rho_2}{2}\right) \qquad (31.77)$$

The volume element takes the form

$$d\tau = (r_1^2\, dr_1 \sin\theta_1\, d\theta_1\, d\phi_1)(r_2^2\, dr_2 \sin\theta_2\, d\theta_2\, d\phi_2) \qquad (31.78)$$

and the final integral for the energy correction E' takes the form

$$E' = \frac{Z^2}{2^5 \pi^2 a_0} \int_0^{2\pi}\!\!\int_0^{\pi}\!\!\int_0^{\infty}\!\!\int_0^{2\pi}\!\!\int_0^{\pi}\!\!\int_0^{\infty} \frac{\exp(-\rho_1 - \rho_2)}{\rho_{12}} \rho_1^2\, d\rho_1 \sin\theta_1\, d\theta_1\, d\phi_1$$
$$\times \rho_2^2\, d\rho_2 \sin\theta_2\, d\theta_2\, d\phi_2 \qquad (31.79)$$

where $\rho_{12} = 2Zr_{12}/a_0$. This integral has been evaluated and is[7]

$$E' = \tfrac{5}{4} Z E^0 \qquad (31.80)$$

The energies of the ground states of the heliumlike atoms and ions are now given by

$$E = -(2Z^2 - \tfrac{5}{4}Z)E_H \qquad (31.81)$$

which for helium ($Z = 2$) becomes

$$E_{He}(\text{calc}) = -74.8 \text{ eV} \qquad (31.82)$$

This is within 6% of the experimental value of -79.0 eV.

This is good agreement, especially since the perturbation term in this instance is not very small compared with the unperturbed term. For the heliumlike ions, the unperturbed energies E^0 vary with Z^2, whereas the correction terms vary as Z to the first power. The corrections E' are smaller relative to E^0 as one goes down the periodic table to Li^+, Be^{2+}, B^{3+}, and C^{4+}. You might expect the relative difference between the calculated and experimental values to decrease as the ions grow heavier. That this is so is shown in one of the problems at the end of this chapter.

31.9 TIME-DEPENDENT PERTURBATION THEORY

When we concern ourselves with the energy levels of the ground or any other stationary state of a system, there is no change with time, and we can use the time-independent Schrödinger equation. On the other hand, in considering spectroscopic transitions in which energy changes in going from one state to another, we no longer have a time-independent system, and the change in the system with time must be taken into account. The time factor of the Schrödinger equation must now be included. For the unperturbed system, the time-dependent Schrödinger equation takes the form

$$\hat{H}^0 \Psi^0 = i\hbar \frac{\partial \Psi^0}{\partial t} \qquad (31.83)$$

where the Ψ_n^0 are now the wave functions including the time,

$$\Psi_n^0(q, t) = \psi_n^0(q) \exp\left(\frac{iE_n t}{\hbar}\right) \qquad (31.84)$$

The $\psi_n^0(q)$ include only the space variables.

[7] See Pauling and Wilson, *Quantum Mechanics*, p. 164 and App. V.

The perturbation to be applied to the system is time-dependent, $\hat{H}'(q,t)$. In spectroscopic investigations, $\hat{H}'(q,t)$ is an electromagnetic wave that varies sinusoidally with time. The time-dependent Schrödinger equation, including the perturbation term, is of the form

$$(\hat{H}^0 + \hat{H}')\Psi = i\hbar \frac{\partial \Psi}{\partial t} \tag{31.85}$$

The solution to this equation can be expanded in a series involving the complete orthonormal set of unperturbed functions,

$$\Psi(q,t) = \sum_n a_n(t) \Psi_n^0(q,t) \tag{31.86}$$

where the expansion coefficients, $a_n(t)$ are functions of the time that allow Equation (31.85) to be satisfied. If this expression for Ψ is substituted back in (31.85), we obtain

$$-\sum_n a_n(t)\hat{H}^0 \Psi_n^0 + \sum_n a_n(t)\hat{H}' \Psi_n^0 = i\hbar \sum_n \frac{da_n(t)}{dt} \Psi_n^0 + i\hbar \sum_n a_n(t) \frac{\partial \Psi_n^0}{\partial t} \tag{31.87}$$

A glance at Equation (31.83) indicates that the first and last terms of this equation cancel, leaving

$$i\hbar \sum_n \frac{da_n(t)}{dt} \Psi_n^0 = \sum_n a_n(t)\hat{H}' \Psi_n^0 \tag{31.88}$$

If we now multiply both sides of (31.88) by Ψ_m^{0*} and integrate, all the terms on the left vanish except that for which $m = n$; we get

$$i\hbar \frac{da_m(t)}{dt} = \sum_{n=0}^{\infty} a_n(t) \int \Psi_m^{0*} \hat{H}' \Psi_n^0 \, d\tau \tag{31.89}$$

Equation (31.89) forms a set of simultaneous differential equations in the variables $a_m(t)$ that must be solved for the particular case in hand.

Equation (31.89) contains the key to the determination of selection rules that we discussed previously in connection with spectroscopy. In the ground state ($n = 1$) with no perturbation ($\hat{H}' = 0$), the value a_1 is unity, and $da_n/dt = 0$ for all n; all the a_n for $n \neq 1$ are zero. If a perturbation H' is now introduced, say by illumination with light, and if this perturbation induces a transition, then da_1/dt does not need to be zero, and any upper state for which da_n/dt is not zero is a possible final state. The condition for the selection rules can now be stated as follows: If there are two states Ψ_n^0 and Ψ_m^0, and a perturbation H', a transition between those two states occurs if the integral

$$\int \psi_m^0 \hat{H}' \psi_n^0 \, dq \tag{31.90}$$

is not zero; if the integral vanishes, then the transition is forbidden. The integral (31.90) is called the *transition moment matrix element*. Note that we have separated the space and time parts in this integral, and the wave functions in (31.90) contain only the space parts.

31.10 THE BOHR FREQUENCY CONDITION

To avoid writing summation signs and more complicated equations, we shall simplify matters by assuming that the transition moment, Equation (31.90), is zero for all but one transition, namely $n \to m$. That is to say, if we have the system in the state specified by n, the only allowed transition is to the state m; hence the summation sign in (31.89) can be omitted. We write the wave functions as the product of the space and time parts as in (31.84) and introduce the notation

$$U_{mn} = \psi_m^{0*} H' \psi_n^0 \, dq \tag{31.91}$$

for the space part of the integral. Equation (31.89) in this case takes the form

$$i\hbar \frac{da_m(t)}{dt} = U_{mn} \exp\left[\frac{i(E_n - E_m)t}{\hbar}\right] \tag{31.92}$$

This can be integrated over the time variable to yield

$$a_m(t) = (i\hbar)^{-1} \int_0^{t'} U_{mn} \exp\left[\frac{i(E_n - E_m)t}{\hbar}\right] dt \tag{31.93}$$

In spectroscopy, the perturbing potential is a beam of electromagnetic radiation whose time variation can be written in the form

$$H' = G(q)(e^{2\pi i \nu t} + e^{-2\pi i \nu t}) \tag{31.94}$$

This corresponds to a wave traveling with frequency ν and amplitude proportional to $G(q)$. The integral in (31.93) contains terms of the form

$$\exp(\pm 2\pi i \nu t) \exp\left[\frac{2\pi i (E_n - E_m)}{h}\right] \tag{31.95}$$

The two harmonic factors in this integral interfere with each other, causing the integral to vanish unless

$$\nu = \frac{E_n - E_m}{h} \tag{31.96}$$

which is exactly the Bohr frequency condition.

Since the coefficient of a normalized wave function is what determines the probability of finding the system in the state given by that function, the probability of finding the system in the state at time t is

$$P_m = |a_m(t)|^2 \tag{31.97}$$

31.11 INTERACTIONS WITH AN ELECTRIC FIELD

We have not yet discussed the nature of the perturbation $\hat{H}'$, and we must do this so we can use the equations we have previously discussed. We shall consider the interaction of charged particles with an electric field. A complete treatment is beyond our scope, and we shall touch only on some of the basic ideas.

Suppose we have a charge Q subjected to an electric field in the x direction E_x. The potential energy of the charge in the field relative to the origin is

$$V = -E_x Q x \tag{31.98}$$

(Note that since the system is a conservative one, the force is $F = -\partial V/\partial x = E_x Q$.) For a system with more than one charge, we must sum over all the charges

$$V = \hat{H}' = -E_x \sum_k Q_k x_k - E_y \sum_k Q_k y_k - E_z \sum_k Q_k z_k \qquad (31.99)$$

where we have added the contributions due to E_y and E_z. The terms in the summations of (31.99) are by definition the components of the dipole moments

$$\mu_x = \sum_k Q_k x_k \ldots \qquad (31.100)$$

(Recall that the dipole moment of two charges $+Q$ and $-Q$, separated by a distance r, is $Q\mathbf{r}$.)

The expression for H', (31.99), can now be written as

$$\hat{H}' = -\mu_x E_x - \mu_y E_y - \mu_z E_z = -\boldsymbol{\mu} \cdot \mathbf{E} \qquad (31.101)$$

where the last equation is written as a vector equation. In spherical polar coordinates,

$$H' = -\boldsymbol{\mu} \cdot \mathbf{E} = \sum_k Q_k \mathbf{r}_k \cdot \mathbf{E} \qquad (31.102)$$

where $\mathbf{r}_k$ is the radius vector of the kth charge. The electric field to which the charges are subjected in a spectroscopic measurement varies as $\cos 2\pi \nu t$, where ν is the frequency; we can now write

$$H' = -\boldsymbol{\mu} \cdot \mathbf{E}_0 \cos 2\pi \nu t \qquad (31.103)$$

where $\mathbf{E}_0$ is the amplitude of the electromagnetic wave.

31.12 EXAMPLE: SELECTION RULES FOR THE HARMONIC OSCILLATOR

We are now in a position to determine the selection rules for the harmonic oscillator (or infrared spectroscopy), which we have previously stated without proof as $\Delta v = \pm 1$. We take a diatomic molecule consisting of charges $+q$ and $-q$ and separated by the equilibrium internuclear distance r_{eq}. At this distance, the equilibrium dipole moment is $\mu_0 = q r_{eq}$. For small displacements the dipole moment can be expanded as a Taylor series in the displacements, or

$$\mu = \mu_0 + \left(\frac{d\mu}{d\xi}\right)_{r_{eq}} + \cdots \qquad (31.104)$$

where ξ is the displacement from equilibrium, and the derivative is to be evaluated at $\xi = 0$, or r_{eq}. The transition moment, which we shall call μ_{mn} for transitions between the mth and nth states, are given by the integral

$$\mu_{mn} = \int_{-\infty}^{\infty} \psi_m^{0*} \left[\mu_0 + \left(\frac{d\mu}{d\xi}\right)_{r_{eq}} \xi\right] \psi_n^0 \, d\xi \qquad (31.105)$$

$$\mu_{mn} = \int_{-\infty}^{\infty} \psi_m^{0*} \mu_0 \psi_n^0 + \int_{-\infty}^{\infty} \psi_m^{0*} \left(\frac{d\mu}{d\xi}\right)_{r_{eq}} \xi \psi_n^0 \, d\xi \qquad (31.106)$$

The first integral in (31.106) is zero unless $m = n$ (since μ_0 is a constant and the wave functions are orthogonal) and does not concern us further. The derivative in the second term can be taken out of the integral to yield

$$\mu_{mn} = \left(\frac{d\mu}{d\xi}\right)_{r_{eq}} \int_{-\infty}^{\infty} \psi_m^{0*} \xi \psi_n^0 \, d\xi \qquad (31.107)$$

This term vanishes unless two conditions hold.

Firstly, μ_{mn} is zero unless $d\mu/d\xi$ is nonvanishing. That is why homonuclear diatomic molecules do not give rise to infrared spectra. The dipole moments are always zero and cannot change as the molecule vibrates. A heteronuclear diatomic molecule such as HCl has a changing dipole moment; hence these molecules show transitions in their infrared spectra. A linear symmetrical molecule such as CO_2 has a zero value for the permanent dipole moment ($\mu_0 = 0$). Of the four normal modes shown in Figure 3.5, the linear symmetrical stretching mode also has a zero value of $d\mu/d\xi$; hence that mode should not be observed in an infrared spectrum. Two of the remaining three modes are degenerate; thus the infrared spectrum of CO_2 should display two peaks.

Secondly, μ_{mn} is zero if the integral in (31.107) vanishes. For the harmonic oscillator it is left for one of the problems to substitute the proper forms for the wave functions in terms of the Hermite polynomials and to demonstrate that the integral vanishes unless $m = n \pm 1$, in agreement with the previously stated selection rule, $\Delta v = \pm 1$.

We should perhaps qualify the above statements, since they are not totally correct. We have used first-order approximation methods to get these results. Further, a true diatomic molecule is only approximated by a harmonic oscillator, and a description of a real molecule must include higher-order terms. These more sophisticated treatments alter the results somewhat. In addition to the first transition, $\Delta v = \pm 1$, higher-order transitions with Δv greater than unity are also allowed. These higher-order transitions have much lower probabilities and hence lower intensities than the fundamental. A real spectrum often displays a very intense peak corresponding to the fundamental frequency and much weaker higher-order peaks displaced towards higher frequencies. These higher frequencies are nearly, but not exactly, integral multiples of the fundamental frequency.[8] Also, we have considered only what may be called *dipole transitions*. There are other interactions associated with higher-order electric poles such as the quadrupole, octupole, and others with which we shall not deal. Transitions due to these higher-order poles are generally much weaker than dipole transitions and are often associated with so called "forbidden" transitions.

A diatomic gaseous molecule is a one-dimensional system with respect to the electric moment. For three-dimensional systems, it is often useful to deal separately with the three components of the dipole moment. Transitions often arise from one component and are absent for the others, and such data can often provide much useful information about molecular symmetry and structure. The perturbing potential can be oriented along any specified direction by using light polarized in that direction, and it is easy to see that unique directions can be established in crystals whose symmetry properties are known. We shall not go on with this here; the interested student is referred to more complete treatments.

We have not yet completed our discussion of spectroscopic transitions; there remains the subject of the Einstein transition coefficients, which we shall defer to a later discussion in Chapter 42, in connection with photochemistry.

---

[8] See, for example, G. Herzberg, *Molecular Spectra and Molecular Structure*. Vol. 1, *Spectra of Diatomic Molecules*, 2d ed. (Princeton: D. Van Nostrand Co., 1950), p. 54.

31.13 NOTE ON ATOMIC UNITS

Although we have been using ordinary units such as electron volts, joules, and centimeters all along, it would be remiss of us to ignore completely the system of *atomic units* (a.u.), which is rapidly gaining acceptance in some branches of science.

In this system, the unit of length is a_0, the first Bohr radius,

$$a_0 = \frac{\hbar^2}{me^2} = 0.529 \times 10^{-8} \text{ cm} \tag{31.108}$$

and the unit of energy is the hartree (symbol H),

$$H = \frac{e^2}{a_0} = 27.21 \text{ eV} \tag{31.109}$$

The energy of the ground state of the hydrogen atom in atomic units can be written as

$$E_H = -\frac{me^4}{2\hbar} = -\frac{1}{2}\frac{e^2}{a_0} = -\frac{1}{2}H \tag{31.110}$$

The terms that normally occur in the Hamiltonian can be converted to atomic units by the prescription

$$-\left(\frac{\hbar^2}{2m}\right)\nabla^2 \longrightarrow -\frac{1}{2}\nabla^2$$

$$\frac{Ze^2}{r} \longrightarrow \frac{Z}{r}$$

The Hamiltonian for the hydrogen atom in atomic units is

$$\hat{H} = -\frac{1}{2}\nabla^2 - \frac{1}{r} \tag{31.111}$$

The unit of angular momentum is $\hbar$. This implies that the quantum numbers l and m are direct measures of the angular momentum in atomic units.

PROBLEMS

1. In the text, when we applied the variational principle to the ground state of the particle in the box using a parabolic function (31.30), the normalization constant for the function was omitted. Determine this normalization constant.

2. The parabolic trial function (31.30) gave excellent results for the energy of the ground state of the particle in the box. This was partly because the parabola closely approximated the "true" sine function, as indicated in Figure 31.2. Calculate the energy of the ground state using the function

$$\phi = Ax^2 \quad \left(0 < x < \frac{L}{2}\right)$$

$$\phi = A(L - x)^2 \quad \left(\frac{L}{2} < x < L\right)$$

where L is the length of the one-dimensional box and A is a normalization constant. Evaluate the normalization constant. Demonstrate that this function satisfies the boundary conditions. Does your answer look peculiar? Why? Is ϕ a valid wave function?

3. Using the variational principle, calculate the energy of the ground state of the particle in the box by taking the "true" wave function as the trial function and show that the result is $E = E_0 = h^2/8mL^2$.

4. It seems as though a variational calculation for the ground state of the harmonic oscillator would be possible if one used a trial function in the form of a symmetrical inverted parabola $\phi = A[1 - (\xi^2/2)]$; this expression contains just the first two terms of the series expansion for the

exponential in (31.66) with $n = 1$. This function is unsuitable as it stands, since it does not satisfy the boundary condition of going to zero at $\pm\infty$. Try a variational solution using this trial function by setting it equal to zero outside a specified range.

5. The experimental spectroscopic energies of the ground states of the heliumlike ions are

Li$^+$	-197 eV	B^{3+}	-596 eV
Be^{2+}	-370 eV	C^{4+}	-876 eV

Determine their energies by the variational method and compare with the experimental results.

6. Repeat the calculation of Problem 5 using the results of the perturbation theory for the helium atom.

7. Determine the expression in the coefficient for the term in λ^2 in (31.43).

8. In the text example of first-order perturbation theory for the slightly anharmonic oscillator, we used a potential $V = \tfrac{1}{2}kx^2 + ax^4$. Show that adding a term bx^3 to the potential would not alter our results for the first-order perturbation calculation.

9. The various Hermite polynomials are related by what are called *recursion* relations. Among them is the relation (Pauling and Wilson, *Quantum Mechanics*, p. 161)

$$\xi H_n(\xi) = \tfrac{1}{2}H_{n+1}(\xi) + nH_{n-1}(\xi)$$

If this relation is applied to ξH_{n+1} and ξH_{n-1}, we can get the expression

$$\xi^2 H_n(\xi) = \tfrac{1}{4}H_{n+2}(\xi) + (n + \tfrac{1}{2})H_n(\xi) \\ + n(n-1)H_{n-2}(\xi)$$

Using these relations, derive (31.68).

10. Using the recursion relations for the Hermite polynomials given in Problem 9, show that the transition moments for the diatomic harmonic oscillator are zero for dipole transitions unless $\Delta v = \pm 1$.

11. Using first-order perturbation theory, calculate the energy levels of the particle in a one-dimensional box with a sloped bottom, as shown in the accompanying figure. Assume that V' is small.

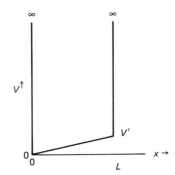

12. In the text we calculated the ground-state energy of the helium atom by first-order perturbation theory. Complete the example by calculating the ionization potential for He and compare with the experimental value.

13. Convert 1 cm^2, 1 erg, 1 joule, and 1 g cm s^{-1} into atomic units.

CHAPTER THIRTY-TWO

ELECTRON SPIN AND MORE COMPLICATED ATOMS

32.1 THE SPINNING ELECTRON

Just as the position of a classical particle can be described by the three space coordinates plus the time, so can the position (or probability) of an electron be described by these same variables. For the moment we shall ignore the time dependence, and write the wave function as $\psi = \psi(x, y, z)$. For electrons, these three classical coordinates are not sufficient to describe the system completely, and an additional coordinate, the *electron spin* is required for a complete description. The need for an additional coordinate was recognized even before the advent of quantum mechanics.

Consider the Stern-Gerlach experiment, first performed in 1922 and indicated in Figure 32.1. In this experiment, a beam of alkali atoms was passed between the poles of a magnet. The results seemed anomalous in that the beam was found to be split into two component beams. It was clear evidence that alkali metals in the ground state could exist in either of two states, each of which interacted differently with a magnetic field. The nature of the difference was less clear. The results of this experiment, along with the anomalous Zeeman effect, fine-structure splitting in atomic spectra, and various degeneracies, were eventually explained by ascribing to the electron a fourth degree of freedom, the *electron spin,* with which angular momentum is associated. For a single electron, the z component of this *spin angular momentum* can have either of the two values, $\pm \tfrac{1}{2}\hbar$. As we proceed, we shall note that this angular momentum can be added vectorially either to orbital angular momentum or to the spin angular momentum of another electron in the same atom.

At the risk of oversimplifying, we can compare the electron spin with a planet's rotation about its axis. There are two types of angular momentum associated with the earth's motion. The first comes from the angular motion of the earth about the sun, and the second comes from the rotation of the earth about its own axis. The analogous intrinsic spin of the electron about its own axis is shown in Figure 32.2a.

What we can say about electron spin without using any classical analogies is the following. There exists angular momentum associated with the electron, which we call *spin angular momentum*. This spin angular momentum is similar to the angular momentum that would arise from the electron's spinning about its own axis. This angular momentum gives rise to a magnetic moment, as indicated in Figure 32.2b. Note that we have not said that the angular momentum actually arises from an intrinsic spinning of the electron about its own axis; we have said only that we can

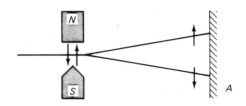

FIGURE 32.1 The Stern-Gerlach experiment. A beam of sodium atoms passes through an inhomogeneous magnetic field and is split into two beams. The splitting arises from the two possible directions of the magnetic moment vector indicated by the arrows.

think of it as such if we so desire. Pauli aptly described this spin as the "two-valuedness not describable classically."

This fourth degree of freedom for electrons was first postulated by S. Goudsmit and G. Uhlenbeck in 1925, shortly before the development of quantum mechanics. Somewhat later, P. A. M. Dirac, in his work on the relativistic form of quantum mechanics, showed that the concept of electron spin was a natural consequence of the combination of relativity and quantum mechanics.

The concept of spin extends beyond the electron and is associated with all particles. Protons and electrons have spins of $\pm \frac{1}{2}$, while deuterons have spins of ± 1. Spins must be either half-integral or integral. Particles with half-integral spins, such as electrons, are known as *fermions* and follow what is known as Fermi-Dirac statistics. Particles with integral spins, such as deuterons, are called *bosons,* and they follow Bose-Einstein statistics. We shall note somewhat later a distinguishing feature of each—the wave function of a fermion must be completely antisymmetric

FIGURE 32.2 (a) "Classical" picture of an electron spinning about its own axis while rotating in orbit about the nucleus. The value μ_l is the magnetic moment arising from orbital motion, and μ_s is the magnetic moment arising from the spin. (b) Electron spinning about its own axis. (c) Allowed cones of precession for a single spinning electron, analogous to the cones of precession of the orbital angular momentum vector as shown in Figure 30.12.

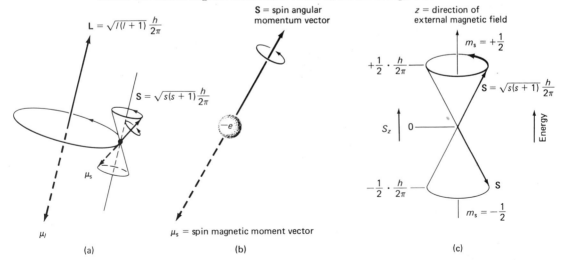

> WOLFGANG PAULI (1900–1958), Austrian physicist, was, like Heisenberg, a student of Arnold Sommerfeld in Munich. His early work was in relativity theory. At the age of 20, Pauli wrote the monograph on relativity for Felix Klein's monumental *Encyklopädie der mathematischen Wissenschaften*. This 250-page work was the first comprehensive presentation of the subject; Einstein himself never wrote a large treatise on relativity. Pauli received his doctorate in 1921 and then spent a year as an assistant to Max Born. The next year he was an assistant to Bohr in Copenhagen, where he commenced his work in quantum mechanics. He arrived at his exclusion principle by way of the Zeeman effect in atomic spectroscopy. In 1924 the doublet nature of the alkali metal spectra and the anomalous Zeeman effect remained two perplexing and inexplicable features of atomic spectroscopy. Pauli introduced what was in essence a fourth quantum number that could have only one of two values to explain these features, or as he called it, the "two-valuedness not describable classically." A year later, in 1925, Uhlenbeck and Goudsmit postulated the electron spin, and Pauli's two-valuedness not describable classically became the fourth, or spin, quantum number, with the two values $\pm\frac{1}{2}$. Pauli's Nobel Prize came long after the other pioneers in quantum mechanics had received their recognition. His award finally came in 1945, "for the discovery of the exclusion principle, also called the Pauli principle."

with respect to the interchange of any two particles, and the wave function of a boson must be symmetric with respect to such interchange. This antisymmetric property of electrons is a statement of the *Pauli exclusion principle;* this principle has a pervasive influence on the theory of atoms and molecules and on the free electron theory of metals.

Analogous to the magnitude of orbital angular momentum, the *magnitude* of the total spin angular momentum is $\sqrt{s(s+1)}\,\hbar$, where s is the electron spin quantum number. For a single electron, $s = \frac{1}{2}$. Although this value $\frac{1}{2}$ may be justified on the basis of advanced theoretical treatments using relativistic quantum mechanics, we shall consider it justified by experimental observation; the observation is the Stern-Gerlach experiment, or the splitting of spectral lines into doublets in an applied magnetic field. The z component of angular momentum that electron spin can possess is either $+\frac{1}{2}\hbar$ or $-\frac{1}{2}\hbar$, or $m_s\hbar$, where m_s is a quantum number that can have the values $+\frac{1}{2}$ or $-\frac{1}{2}$.

For orbital angular momentum, we have noted (Chapter 30) that the magnetic moment is obtained by multiplying the angular momentum by $e/2m$, where e is the electronic charge and m the electronic mass. For spin angular momentum, the magnetic moment is obtained not from the factor $e/2m$ but from a factor twice as large. The magnetic moment of an electron is then

$$\mu = 2\,\frac{e}{2m}\left(\frac{1}{2}\cdot\frac{3}{2}\right)^{1/2}\hbar$$

or $\sqrt{3}$ Bohr magnetons. The factor by which the angular momentum is multiplied to get the magnetic moment is known as the *magnetogyric ratio* (or the *gyromagnetic ratio*), symbolized by γ. For the orbital magnetic moment, $\gamma = e/2m$, and for the spin magnetic moment, γ is twice this value, or $\gamma_s = e/m$. The extra factor 2 is known as the *Landé g factor for electron spin.*[1]

---

[1] Actually this factor is not exactly 2. Relativistic quantum mechanics and a number of precise experiments indicate that the value is actually $g = 2.0023$, or some 0.1% larger than 2.

32.2 THE POSTULATES OF ELECTRON SPIN

To include the spin properties in our quantum mechanical formalism, we must add two new postulates concerning electron spin to our previously stated postulates for spinless quantum mechanics.

> SPIN POSTULATE 1: *The operators for the angular momentum associated with spin commute and combine in precisely the same way as the operators for ordinary angular momentum.*

That is to say, all the relations we have previously discussed concerning ordinary (or orbital) angular momentum remain valid for spin angular momentum; we only need to substitute the spin angular momentum operator $\hat{S}$ for the ordinary angular momentum operator $\hat{L}$. We have, corresponding to the operators $\hat{L}, \hat{L}^2, \hat{L}_z, \hat{L}_x, \hat{L}_y$, a set of operators $\hat{S}, \hat{S}^2, \hat{S}_z, \hat{S}_x, \hat{S}_y$; both sets operate in the same way.

> SPIN POSTULATE 2: *Only two simultaneous eigenfunctions of the spin operators $\hat{S}^2$ and $\hat{S}_z$ exist; we shall denote these eigenfunctions by α and β. The eigenfunction equations they obey can be written as*

$$\hat{S}_z \alpha = \tfrac{1}{2}\hbar \alpha \qquad \hat{S}^2 \alpha = \tfrac{1}{2}(\tfrac{1}{2}+1)\hbar^2 \alpha \qquad (32.1)$$
$$\hat{S}_z \beta = -\tfrac{1}{2}\hbar \beta \qquad \hat{S}^2 \beta = \tfrac{1}{2}(\tfrac{1}{2}+1)\hbar^2 \beta$$

The α state is often called the *spin up* state; it is represented by the symbol $\uparrow$. The β state is often called the *spin down* state; it is represented by $\downarrow$. The functions α and β are normalized and orthogonal over all spin space.

$$\int_{\text{spin space}} \alpha\beta \, d\tau = 0 \qquad \int_{\text{spin space}} \alpha\alpha \, d\tau = 1 \qquad \int_{\text{spin space}} \beta\beta \, d\tau = 1$$

The complete wave function for any electron is represented by the product of a space (or an orbital) function and a spin function. For the electron in a hydrogen atom, there are two different states, corresponding to the two different possible spin states, $\psi(x,y,z)\alpha$ and $\psi(x,y,z)\beta$; the space part of the wave function is the same for both. Since the spin functions do not contain any of the space coordinates, the $\hat{S}$ and $\hat{L}$ operators always commute with each other. The space and spin parts of any wave function are separately orthogonal and normalized:

$$\int [\psi(x,y,z)\alpha][\psi(x,y,z)\beta] \, d\tau = \int \psi^2 \alpha\beta \, d\tau = 0$$
$$\int [\psi(x,y,z)\alpha][\psi(x,y,z)\alpha] \, d\tau = \int \psi^2 \alpha\alpha \, d\tau = 1$$
$$\int [\psi(x,y,z)\beta][\psi(x,y,z)\beta] \, d\tau = \int \psi^2 \beta\beta \, d\tau = 1$$

32.3 THE INDISTINGUISHABILITY OF ELECTRONS

Let's neglect electron spin for the moment and consider a helium atom with two electrons. The total space part of the wave function is the product of the wave functions of the individual electrons,

$$\psi_{T,1} = \psi_1(1)\psi_2(2) \qquad (32.2)$$

where $\psi_1(1)$ is the wave function of one electron and $\psi_2(2)$ is the wave function of the other electron; one might be an s function and the other might be a p function, or they might both be of the same type. Since the position of an electron is described by a probability, each of the two electrons has a certain probability of being located somewhere. By writing the wave function as we have done in (32.2), we imply that we can separately label each of the electrons and unequivocally state that electron 1 is in one state with one particular energy while electron 2 is in the other state with its particular energy—which may or may not be the same. This is an unrealistic view, since we have no experimental method of distinguishing the situation described by the wave function of (32.2) and the wave function

$$\psi_{T,2} = \psi_1(2)\psi_2(1) \tag{32.3}$$

which is obtained simply by interchanging the electrons. Each of these wave functions should be simultaneously valid. But consider the following—the probability of finding the system in any particular state should be

$$\text{prob.} = \psi_T^2 \, d\tau \tag{32.4}$$

where ψ_T is the total wave function as given by either (32.2) or (32.3). Since interchanging the two electrons cannot affect the final properties of the system, interchanging the two electrons should not affect the probability as given in (32.4). This requires ψ_T to be *either symmetrical or antisymmetrical with respect to the interchange of two electrons*. To check on the symmetry or antisymmetry properties of the wave function, we introduce the *permutation operator* $\hat{P}(1, 2)$, whose effect is to interchange the two electrons. When this operator operates on $\psi_{T,1}$ and $\omega_{T,2}$ we get

$$\begin{aligned} \hat{P}(1, 2)\psi_{T,1} &= \psi_{T,2} \\ \hat{P}(1, 2)\psi_{T,2} &= \psi_{T,1} \end{aligned} \tag{32.5}$$

Thus neither of these two functions is symmetric or antisymmetric with respect to the interchange of the two electrons; therefore neither can be considered a valid description of the system. We can, however, create valid wave functions by taking linear combinations of these two functions in the form

$$\begin{aligned} \psi_A &= \tfrac{1}{2}\sqrt{2}[\psi_1(1)\psi_2(2) - \psi_1(2)\psi_2(1)] \\ \psi_S &= \tfrac{1}{2}\sqrt{2}[\psi_1(1)\psi_2(2) + \psi_1(2)\psi_2(1)] \end{aligned} \tag{32.6}$$

where ψ_A is an antisymmetric function and ψ_S is a symmetric function:

$$\begin{aligned} \hat{P}(1, 2)\psi_A &= -\psi_A \\ \hat{P}(1, 2)\psi_S &= +\psi_S \end{aligned} \tag{32.7}$$

A symmetric wave function is an eigenfunction of the operator $\hat{P}(1,2)$ with eigenvalue $+1$, whereas an antisymmetric wave function is an eigenfunction with eigenvalue -1. Wave functions such as $\psi_{T,1}$ and $\psi_{T,2}$ are neither symmetric nor antisymmetric and are not eigenfunctions of $\hat{P}(1, 2)$.

The wave functions $\psi_{T,1}$ and $\psi_{T,2}$ of (32.2) and (32.3) contain too much information, since they place definite labels on each of the electrons. The terms ψ_A and ψ_S provide a more limited and more valid description of the system. They tell us no more than that one of the electrons is in the first state while the other electron is in the second state and do not in any way specify which electron is in which state.

32.4 THE PAULI PRINCIPLE

An electron in a hydrogen atom is specified by a set of four quantum numbers, the three space quantum numbers n, l, and m, and the fourth, or spin quantum number, m_s. The numerical value of m_s for an electron is limited to $\pm \frac{1}{2}$. For more complicated atoms this description by these four quantum numbers is still valid to a good approximation. The simplest statement of the so-called *Pauli exclusion principle* is that no two electrons in an atom can have the same set of four quantum numbers, and this rule should be familiar from elementary chemistry studies. A more general enunciation of the Pauli principle is stated concerning the symmetry properties of the wave function; this statement can be considered our final postulate of quantum mechanics.

We have noted that for a wave function of a system with more than one electron to be valid, the wave function must be either symmetric or antisymmetric with respect to the interchange of two electrons. That is to say, an acceptable wave function must be an eigenfunction of the operator $\hat{P}(1, 2)$ with eigenvalues ± 1. For atoms with more than two electrons, the wave function must be an eigenfunction of the operator $\hat{P}(i, j)$ with eigenvalues ± 1, where the operator $\hat{P}(i, j)$ means interchanging the ith and jth electrons. The question is, Which of these two possible values is the correct one? For electrons, the correct eigenvalue is -1; the wave functions must be antisymmetric. This is the statement of the Pauli principle.

> SPIN POSTULATE 3; THE PAULI PRINCIPLE. *An acceptable wave function for a system containing more than one electron must be antisymmetric with respect to the interchange of any two of the electrons.*

Note that this principle applies to the *total* wave function, space plus spin parts.

The Pauli principle is even more general than this, and the above statement holds for any particle with half-integral spin. These include electrons, protons, ^{13}C nuclei, and numerous other particles. For particles like deuterons, which have integral spins, the wave functions must be symmetric with respect to the interchange of any two particles.

At this juncture it will be worth while to return to the previous section and reexamine the analysis of that section in the light of the Pauli principle. It might seem that only the antisymmetric function ψ_A of (32.6) can validly describe the system, since that is the antisymmetric function. But we did not consider spin in that section. Suppose we take a closer look at the situation for a helium atom, including the spin coordinates. For simplicity we shall assume that both of the electrons are in 1s orbitals. There are four possible product wave functions, which can be written

$$\begin{aligned}\psi_1 &= \phi_{1s}(1)\phi_{1s}(2)\alpha(1)\alpha(2) \\ \psi_2 &= \phi_{1s}(1)\phi_{1s}(2)\alpha(1)\beta(2) \\ \psi_3 &= \phi_{1s}(1)\phi_{1s}(2)\beta(1)\alpha(2) \\ \psi_4 &= \phi_{1s}(1)\phi_{1s}(2)\beta(1)\beta(2)\end{aligned} \quad (32.8)$$

None of these wave functions satisfies the Pauli principle, as we can see by applying the $\hat{P}(1, 2)$ operator. The terms ψ_1 and ψ_4 are invalid, since they are symmetric, while neither ψ_2 nor ψ_3 are eigenfunctions of $\hat{P}(1, 2)$. We can, however, construct two new wave functions by taking linear combinations of ψ_2 and ψ_3 in the form

$$\psi_A = \tfrac{1}{2}\sqrt{2}\,\phi_{1s}(1)\phi_{1s}(2)[\alpha(1)\beta(2) - \beta(1)\alpha(2)] \quad (32.9a)$$
$$\psi_S = \tfrac{1}{2}\sqrt{2}\,\phi_{1s}(1)\phi_{1s}(2)[\alpha(1)\beta(2) + \beta(1)\alpha(2)] \quad (32.9b)$$

The term ψ_A is the only acceptable wave function for the helium atom in the ground state, since only that wave function is antisymmetric.

Note carefully that the Pauli principle does not separately restrict the symmetry properties of either the space or the spin part of the wave function. It states that the *total* wave function must be antisymmetric but the space or spin parts of the wave function may be either symmetric or antisymmetric. In Equation (32.9a), the function ψ_A has a space part that is symmetric and a spin part that is antisymmetric. The product of two symmetric functions is a symmetric function, and the product of two antisymmetric functions is also a symmetric function. The only way in which we can get a total product wave function that is antisymmetric is for one part to be symmetric and the other antisymmetric.

A convenient way of writing linear combinations of product functions in a form that is antisymmetric was first described by J. C. Slater. The wave function of a system of N atoms is written in the form of a *Slater determinant*:

$$\psi(1, 2, 3, \ldots, N) = \frac{1}{\sqrt{N}} \begin{vmatrix} \psi_1(1) & \psi_1(2) & \psi_1(3) & \ldots & \psi_1(N) \\ \psi_2(1) & \psi_2(2) & \psi_2(3) & \ldots & \psi_2(N) \\ \psi_3(1) & \psi_3(2) & \psi_3(3) & \ldots & \psi_3(N) \\ \cdot & \cdot & \cdot & & \cdot \\ \cdot & \cdot & \cdot & & \cdot \\ \cdot & \cdot & \cdot & & \cdot \\ \psi_N(1) & \psi_N(2) & \psi_N(3) & \ldots & \psi_N(N) \end{vmatrix} \qquad (32.10)$$

Two electrons can be interchanged by interchanging two columns of the determinant. Since the interchange of two columns of a determinant changes the sign of the determinant, the antisymmetry requirement is satisfied. Further, a situation in which two of the ψ_i are the same corresponds to a situation in which two electrons have the same set of quantum numbers; but this would require the determinant to have two identical sets of columns, and any determinant with two identical columns must vanish. The determinantal form thus requires the wave function to vanish if any two electrons have the same set of four quantum numbers and also requires the wave function to be antisymmetrical.

The building up of the periodic table, the so-called *Aufbau Prinzip,* whereby sets of quantum numbers are successively assigned to the electrons in polyelectronic atoms is based on the Pauli principle. In classical mechanics, the *minimum-energy principle* applies to the final equilibrium state of a system. When we pass from classical mechanics to other areas of physics, we often find that the minimum-energy principle is constrained by other considerations no less fundamental than the minimum-energy principle. We have already seen an example of this in our consideration of thermodynamics, when we found that the minimum-energy principle was constrained by the *maximum-entropy principle.* In thermodynamics, the equilibrium state of a system is found not at the minimum possible energy but at the minimum energy subject to the constraint of maximum entropy. An analogous situation holds in quantum mechanics, where the minimum-energy principle is constrained by the Pauli principle. According to the minimum-energy principle, all the electrons in a uranium atom would be in the state of lowest energy, that is, a 1s orbital. But the Pauli principle insists that no two electrons can have the same set of quantum numbers $n = 1, l = 0, m = 0$, and $m_s = \pm \frac{1}{2}$. Thus only two electrons can be 1s electrons; the third must go into a higher energy level, and so on, through all the 92 electrons in a uranium atom. We shall return to this point after discussing the vector model of the atom and the relative energies of the different states in polyelectronic atoms.

32.5 TOTAL ANGULAR MOMENTUM

Quantum mechanical angular momentum can be summed vectorially just as classical angular momentum can be summed vectorially. We represent an angular momentum operator, either spin or orbital, by $\hat{M}$. There is no problem for one electron, since the total angular momentum is just the angular momentum of the single electron. For systems with more than one electron, the total angular momentum is obtained by vectorial addition. For two electrons, the total angular momentum operator can be written as

$$\hat{M}_T = \hat{M}(1) + \hat{M}(2) \tag{32.11}$$

where $\hat{M}(1)$ and $\hat{M}(2)$ are the angular momenta of the individual electrons. The operator for the square of the total angular momentum of a two-electron system is then

$$\hat{M}_T^2 = \hat{M}_T \cdot \hat{M}_T = \hat{M}^2(1) + \hat{M}^2(2) + 2\hat{M}(1) \cdot \hat{M}(2) \tag{32.12}$$

For the various components, we write

$$\hat{M}_q = \hat{M}_q(1) + \hat{M}_q(2) \qquad (q = x, y, z) \tag{32.13}$$

where we have omitted the subscript T; it should be understood that $\hat{M}_q$ is a total angular momentum operator.

For an example, let us calculate the total spin angular momentum of the two electrons in a helium atom; in particular, let us calculate the z component. We must first write the spin states for the two electrons; there are four possibilities. These seem at first thought to be the four functions

$$\alpha(1)\alpha(2) \qquad \alpha(1)\beta(2) \qquad \beta(1)\alpha(2) \qquad \beta(1)\beta(2)$$

Although the first and last of these four functions are perfectly legitimate, the second and third are not valid, since a helium atom with these functions would violate the Pauli principle.[2] To make the two states $\alpha(1)\beta(2)$ and $\beta(1)\alpha(2)$ valid spin functions we must take linear combinations of the two

$$\tfrac{1}{2}\sqrt{2}[\alpha(1)\beta(2) \pm \beta(1)\alpha(2)] \tag{32.14}$$

Let's consider the state $\alpha(1)\alpha(2)$. The eigenvalue is determined from the eigenvalue equation (noting that here the operator $\hat{S}_z$ implies the *total z* component of angular momentum):

$$\begin{aligned}
\hat{S}_z \alpha(1)\alpha(2) &= [\hat{S}_z(1) + \hat{S}_z(2)]\alpha(1)\alpha(2) \\
&= \hat{S}_z(1)\alpha(1)\alpha(2) + \hat{S}_z(2)\alpha(1)\alpha(2) \\
&= \hat{S}_z(1)\alpha(1)\alpha(2) + \alpha(1)\hat{S}_z(2)\alpha(2) \\
&= \tfrac{1}{2}\hbar\alpha(1)\alpha(2) + \alpha(1)\tfrac{1}{2}\hbar\alpha(2) \\
&= \hbar\alpha(1)\alpha(2)
\end{aligned} \tag{32.15}$$

The eigenvalue of the z component of the total angular momentum for the state $\alpha(1)\alpha(2)$ is just twice the value for a single α electron, or $\hbar$. Similarly the eigenvalue of the state $\beta(1)\beta(2)$ is $-\hbar$; the eigenvalues for the two remaining states in (32.14), where the α and β states are mixed, are 0. Calculating these eigenvalues is left for one of the problems. Note that in (32.15) we used the fact that $\hat{S}_z(2)$ commutes with the function $\alpha(1)$. This is so because the operator and the spin function have different

---

[2] Note that the first and fourth spin functions are symmetric. According to the Pauli principle, these would have to be combined with antisymmetric space functions to produce a total wave function that is antisymmetric. The second and third spin functions are neither symmetric nor antisymmetric as written.

coordinates. Likewise, $\hat{S}_z(1)$ commutes with $\alpha(2)$ or $\beta(2)$. Neither of these two operators, however, commutes with the functions given in (32.14), since those spin functions contain the spin coordinates of both electrons. In arrow notation, the state $\alpha(1)\alpha(2)$ would be denoted by ($\uparrow\uparrow$) and the state $\beta(1)\beta(2)$ would be denoted by ($\downarrow\downarrow$); the states listed in (32.14) would be denoted by ($\uparrow\downarrow$). This arrow notation is convenient, because if each arrow is considered a vector of length $\hbar/2$, then the eigenvalue of the z component of the total spin angular momentum can be got by vectorial addition of the arrows.

32.6 ADDITION OF SPIN AND ORBITAL ANGULAR MOMENTUM

Neglecting spin for the moment, let's consider a hydrogen atom. In the ground state, $n = 1$ and $l = 0$. In the excited states, we have

$$n = 2; \quad l = 0, 1$$
$$n = 3; \quad l = 0, 1, 2$$
$$n = 4; \quad l = 0, 1, 2, 3 \quad \text{(etc.)}$$

According to the Bohr model and also the quantum mechanical treatment, the energy depends only on the principal quantum number n. When electron spin is introduced, the interaction between the orbital and spin angular momentum, the so-called *spin-orbit coupling*, causes a slight energy difference between levels with a given n but different l values and also removes some of the degeneracy. It is convenient in this discussion to introduce a new notation for spectroscopic energy levels, the so-called *term symbols*.

For the hydrogen atom, the *total* orbital angular momentum is uniquely determined by l, and we use the symbols s, p, d, f, g, ... to refer to electrons with $l = 0, 1, 2, 3, 4, \ldots$. Over the years, spectroscopists have developed a notation based on these same letters in capitalized form—S, P, D, F, G, ...—to refer to the *total* orbital angular momentum of the atom as a whole.[3] For the hydrogen atom there is only one electron and a unique one-to-one correspondence between the angular momentum of the single electron and the angular momentum of the atom as a whole. Thus if $n = 1$, the atom must be in an S state; whereas if $n = 2$, the hydrogen atom can be in either an S or a P state; and so on, through higher values of n. For helium, however, there are more possibilities, since there is more than one way in which the angular momenta of the individual electrons can be added. For the 1s(1)1s(2) configuration of helium, the l vector of each electron is zero, and the total angular momentum vector $\mathbf{L} = \Sigma \mathbf{l}_i$ is also zero; this state is an S state. For the excited state 1s(1)2p(2), again there is only one possibility; here $\mathbf{L} = \Sigma \mathbf{l}_i$ must be unity, and the configuration is a P state. For the excited state 2p(1)2p(2), the situation is more complex, and there are several possibilities, as shown in Figure 32.3. Here three different term symbols are possible; the atom can be in a D, a P, or an S state, depending on how the l vectors are added.

Now let's return to the hydrogen atom. In the ground state, $l = 0$ and $\mathbf{L} = \Sigma \mathbf{l} = 0$; the atom is thus in an S state. In addition, there is also spin angular momentum $S = \frac{1}{2}$. This spin angular momentum vector can add to the orbital angular momentum vector to produce a new vector, the **J** vector, $\mathbf{J} = \mathbf{L} + \mathbf{S}$. The new **J** vector,

---

[3] These descriptions of S, P, D, ... are valid only for the *Russell-Saunders* coupling scheme. Spectroscopists also have a notation for systems that do not obey Russell-Saunders coupling.

FIGURE 32.3 Adding two vectors with $l = 1$ to get an S, a P, and a D state.

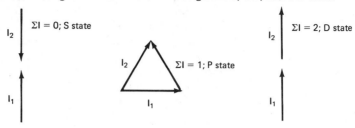

or total angular momentum vector, can have integral or half-integral multiples of $\hbar$. For the ground state, $L = 0$ and $S = \frac{1}{2}$; there is only one possibility, $J = \frac{1}{2}$. The numerical value of J is written as a subscript on the term symbol; the hydrogen atom in the ground state is said to be an $S_{1/2}$ state. In an excited P state, $L = 1$ and $S = \frac{1}{2}$, and there are two possibilities—$J = \frac{3}{2}$ or $J = \frac{1}{2}$, depending on whether the **L** and **S** vectors are parallel or antiparallel. The term symbols are written $P_{3/2}$ and $P_{1/2}$. For a D state, the two possible values of J are $\frac{5}{2}$ and $\frac{3}{2}$, and the two possible term symbols are $D_{5/2}$ and $D_{3/2}$.

According to the exact treatment of the hydrogen atom as carried out in Chapter 30, the states with different term symbols S, P, D, . . . are all degenerate for any given value of the principle quantum number n. Some of this degeneracy is removed when the interaction between the orbital and spin angular momentum is taken into account. This interaction, the *spin-orbit coupling,* adds a term to the Hamiltonian of the form

$$H'_{so} = \xi(r)\mathbf{L} \cdot \mathbf{S} \qquad (32.16)$$

The origin of the spin-orbit interaction can be seen from the following classical analog. Whenever a charge moves under the influence of an electric field, a magnetic field is generated. The orbital motion of the electron in the electric field of the nucleus thus gives rise to a magnetic field. The magnetic moment associated with the spin of the electron interacts with this magnetic field. We shall not go into the detailed form of the function $\xi(r)$; this is discussed in more advanced textbooks on the subject.[4] When the spin-orbit coupling in the hydrogen atom is taken into account, a splitting of 0.035 cm^{-1} is introduced between the $S_{1/2}$ and $P_{1/2}$ states of the $n = 2$ excited state; these states are degenerate in the exact treatment. The figure 0.035 cm^{-1} is less than one part per million of the separation between the $n = 1$ and $n = 2$ state, which is 82,259 cm^{-1}. This energy difference can be measured only on very high resolution spectrometers. The spin-orbit coupling gives rise to much larger energy shifts in heavier atoms.

32.7 THE FIRST EXCITED STATES OF HELIUM

In considering the wave functions of the ground state of helium, we noted that although there are four possible ways to combine the spin and space functions as shown in (32.8), there is only one possibility that satisfies the antisymmetry requirement of the total wave function. That function is shown in (32.9a). The orbital angu-

---

[4] See, for example, E. U. Condon and G. H. Shortley, *The Theory of Atomic Spectra* (Cambridge: Cambridge University Press, 1959), p. 120.

lar momentum of each electron is zero, and the sum of the spin angular momentum is zero. Thus both **S** and **L** as well as **J** = **L** + **S** are zero. The state is a nondegenerate S state. When we consider the excited states, the number of possibilities increases dramatically. The first excited state is the state in which one of the electrons is excited to the $n = 2$ level. There are two possibilities, 1s2s and 1s2p; both have the same energy in this so-called *independent electron model* of the atom. In this model we start with the zeroth order approximation that each electron acts independently of the other electrons. Suppose we first consider the 1s2s state.

There are two possible space functions for the 1s2s state but we cannot simply write these as $\phi_{1s}(1)\phi_{2s}(2)$ and $\phi_{1s}(2)\phi_{2s}(1)$, since these are not eigenfunctions of the interchange operator. We can, however, take linear combinations of these in the form

$$\Theta_+ = \tfrac{1}{2}\sqrt{2}[\phi_{1s}(1)\phi_{2s}(2) + \phi_{1s}(2)\phi_{2s}(1)] \qquad (32.17a)$$
$$\Theta_- = \tfrac{1}{2}\sqrt{2}[\phi_{1s}(1)\phi_{2s}(2) - \phi_{1s}(2)\phi_{2s}(1)] \qquad (32.17b)$$

where Θ_+ is a symmetric space function and Θ_- is an antisymmetric space function. About the spin functions, our first thought might be to construct the set of four functions $\alpha(1)\alpha(2)$, $\alpha(1)\beta(2)$, $\beta(1)\alpha(2)$, $\beta(1)\beta(2)$, but the functions $\alpha(1)\beta(2)$ and $\beta(1)\alpha(2)$ are not eigenfunctions of the interchange operator. Instead of these two functions, we use linear combinations of them to obtain the set of four spin functions for two electrons:

$$\begin{aligned}\Phi_1 &= \alpha(1)\alpha(2) \\ \Phi_2 &= \tfrac{1}{2}\sqrt{2}[\alpha(1)\beta(2) + \beta(1)\alpha(2)] \\ \Phi_3 &= \beta(1)\beta(2) \\ \Phi_4 &= \tfrac{1}{2}\sqrt{2}[\alpha(1)\beta(2) - \beta(1)\alpha(2)]\end{aligned} \qquad (32.18)$$

This set of four spin functions is normalized and orthogonal. The first three functions are symmetric in the spin coordinates of the two electrons and the last is antisymmetric.

The antisymmetric spin function Φ_4 has zero total spin angular momentum. The eigenvalue of the operator $\hat{S}^2$ is identically zero. The symmetric functions, on the other hand, have total spin $S_T = 1$. The spin degeneracy determines what we call the *multiplicity* of the state, and this is $2S_T + 1$. For $S_T = 0$, the multiplicity is one and the state is called a *singlet* state. For $S_T = 1$, the multiplicity is three, and the state is termed a *triplet* state. We can now get zeroth-order wave functions for the 1s2s state of helium by combining the two proper space functions with the four proper spin functions, to get the eight functions

$$\begin{aligned}\psi_1 &= \tfrac{1}{2}\sqrt{2}[\phi_{1s}(1)\phi_{2s}(2) + \phi_{2s}(1)\phi_{1s}(2)][\alpha(1)\alpha(2)] \\ \psi_2 &= \tfrac{1}{2}[\phi_{1s}(1)\phi_{2s}(2) + \phi_{2s}(1)\phi_{1s}(2)][\alpha(1)\beta(2) + \beta(1)\alpha(2)] \\ \psi_3 &= \tfrac{1}{2}\sqrt{2}[\phi_{1s}(1)\phi_{2s}(2) + \phi_{2s}(1)\phi_{1s}(2)][\beta(1)\beta(2)] \\ \psi_4 &= \tfrac{1}{2}[\phi_{1s}(1)\phi_{2s}(2) - \phi_{2s}(1)\phi_{1s}(2)][\alpha(1)\beta(2) - \beta(1)\alpha(2)] \\ \psi_5 &= \tfrac{1}{2}\sqrt{2}[\phi_{1s}(1)\phi_{2s}(2) - \phi_{2s}(1)\phi_{1s}(2)][\alpha(1)\alpha(2)] \\ \psi_6 &= \tfrac{1}{2}[\phi_{1s}(1)\phi_{2s}(2) - \phi_{2s}(1)\phi_{1s}(2)][\alpha(1)\beta(2) + \beta(1)\alpha(2)] \\ \psi_7 &= \tfrac{1}{2}\sqrt{2}[\phi_{1s}(1)\phi_{2s}(2) - \phi_{2s}(1)\phi_{1s}(2)][\beta(1)\beta(2)] \\ \psi_8 &= \tfrac{1}{2}[\phi_{1s}(1)\phi_{2s}(2) + \phi_{2s}(1)\phi_{1s}(2)][\alpha(1)\beta(2) - \beta(1)\alpha(2)]\end{aligned} \qquad (32.19)$$

$\left.\begin{array}{l}\psi_5 \\ \psi_6 \\ \psi_7\end{array}\right\}$ triplet

ψ_8 singlet

Of these eight functions, the first four are not valid since they are symmetric in the interchange of two electrons. In the first three wave functions, both the space and spin parts are symmetric, whereas in the fourth both are antisymmetric, resulting in a total wave function that is symmetric. The last four are valid, since they are anti-

SEC. 32.7 THE FIRST EXCITED STATES OF HELIUM

symmetric in the interchange of the coordinates of the two electrons. The functions designated ψ_5, ψ_6, and ψ_7 form a triplet state, since the total spin is unity; these three functions have antisymmetric space parts and symmetric spin parts. The last function, ψ_8, has zero total spin angular momentum and forms a singlet state. It is composed of a symmetric space part and an antisymmetric spin part.

The zeroth order functions for the 1s2p states can be constructed in a similar manner. The spin functions are the same as those shown in (32.18). There are six space functions that are based on the p_x, p_y, and p_z orbitals. We have, corresponding to Equation (32.17), space functions of the form

$$\Theta_+ = \tfrac{1}{2}\sqrt{2}[\phi_{1s}(1)\phi_{2p_x}(2) + \phi_{1s}(2)\phi_{2p_x}(1)] \qquad (32.20a)$$
$$\Theta_- = \tfrac{1}{2}\sqrt{2}[\phi_{1s}(1)\phi_{2p_x}(2) - \phi_{1s}(2)\phi_{2p_x}(1)] \qquad (32.20b)$$

plus two other pairs, which have the ϕ_{2p_y} and ϕ_{2p_z} orbitals in place of the ϕ_{2p_x} orbitals. The actual writing of the 24 possible states analogous to the 1s2s states shown in Equation (32.19) is left for one of the problems. When this is done, 12 of the states are totally symmetric and hence invalid. The 12 remaining states are antisymmetric, and hence they are valid states. Of the 12 valid states, 9 have symmetric spin parts, with $S = 1$. These states form a ninefold degenerate ^3P state. The ninefold degeneracy arises from the fact that since $L = 1$, there is threefold degeneracy in the orbital part; since $S = 1$, there is threefold degeneracy in the spin part, and $3 \cdot 3 = 9$. The remaining three valid states have antisymmetric spin parts, $S = 0$; there is no spin degeneracy and these three states form a ^1P. Note that while there is no spin degeneracy, there is still the threefold degeneracy associated with the orbital part, since $L = 1$, and $(2L + 1) = 3$.

A simplified procedure for enumerating the term symbols for atomic states is shown in Table 32.1. The upper portion of the table refers to the 1s2s state. We first list all the possible combinations of the quantum numbers n, l, m_l, and m_s in such a way that no two electrons have the same set of four quantum numbers. This has been

TABLE 32.1 Sets of quantum numbers for the configurations 1s2s and 1s2p.

	Quantum numbers				
	1st electron	2nd electron			Term
States	$n = 1$	$n = 2$	Σm_l	Σm_s	designations
1s2s	$(0, 0, \tfrac{1}{2})$	$(0, 0, \tfrac{1}{2})$	0	1	^3S
	$(0, 0, \tfrac{1}{2})$	$(0, 0, -\tfrac{1}{2})$	0	0	^3S (or ^1S)
	$(0, 0, -\tfrac{1}{2})$	$(0, 0, \tfrac{1}{2})$	0	0	^1S (or ^3S)
	$(0, 0, -\tfrac{1}{2})$	$(0, 0, -\tfrac{1}{2})$	0	-1	^3S
1s2p	$(0, 0, \tfrac{1}{2})$	$(1, 1, \tfrac{1}{2})$	1	1	
	$(0, 0, \tfrac{1}{2})$	$(1, 0, \tfrac{1}{2})$	0	1	^3P
	$(0, 0, \tfrac{1}{2})$	$(1, -1, \tfrac{1}{2})$	-1	1	
	$(0, 0, -\tfrac{1}{2})$	$(1, 1, \tfrac{1}{2})$	1	0	
	$(0, 0, -\tfrac{1}{2})$	$(1, 0, \tfrac{1}{2})$	0	0	^3P (or ^1P)
	$(0, 0, -\tfrac{1}{2})$	$(1, -1, \tfrac{1}{2})$	-1	0	
	$(0, 0, \tfrac{1}{2})$	$(1, 1, -\tfrac{1}{2})$	1	0	
	$(0, 0, \tfrac{1}{2})$	$(1, 0, -\tfrac{1}{2})$	0	0	^1P (or ^3P)
	$(0, 0, \tfrac{1}{2})$	$(1, -1, -\tfrac{1}{2})$	-1	0	
	$(0, 0, -\tfrac{1}{2})$	$(1, 1, -\tfrac{1}{2})$	1	-1	
	$(0, 0, -\tfrac{1}{2})$	$(1, 0, -\tfrac{1}{2})$	0	-1	^3P
	$(0, 0, -\tfrac{1}{2})$	$(1, -1, -\tfrac{1}{2})$	-1	-1	

NOTE: The sets of quantum numbers are indicated by (l, m_l, m_s), where l is the orbital angular momentum quantum number, m_l the z component quantum number, and m_s the spin quantum number. The principal quantum numbers $n = 1$ and $n = 2$ have been omitted and are shown at the tops of the columns.

done under the columns labeled "1st electron" and "2nd electron." We then compute Σm_l and Σm_s for each pair and enter these in the indicated columns. Since $\Sigma m_l = 0$ for each entry, all these states must be S states. We then look for the highest value of Σm_s, which in this instance is 1. Since $S = 1$, we must have three states corresponding to this ³S state. The states with $\Sigma m_s = \pm 1$ must belong to this term symbol. One of the remaining states with $\Sigma m_s = 0$ must also be assigned to this ³S level while the remaining one must be assigned to the ¹S level. It makes no difference which one we assign to which level, since the purpose of this type of procedure is to enumerate all the possible term levels rather than to assign particular functions to particular term symbols. The ground state 1s² configuration is particularly simple, and has been omitted from the table. For that case, $n = 1$ for both electrons, and l and m_l are both zero for each electron. This requires m_s to be $+\frac{1}{2}$ for one electron and $-\frac{1}{2}$ for the other. We have then that $\Sigma m_l = 0$ and $\Sigma m_s = 0$; the ground state of helium is a ¹S level. The bottom portion of Table 32.1 shows this procedure for the 1s2p state of helium, where we have a ninefold degenerate ³P level and a threefold degenerate ¹P level, bearing out the previous discussion. It is left for one of the problems to work out the term levels for the ground state of carbon, which has the configuration 1s²2s²2p². In that case, the filled inner shells of 1s and 2s electrons can be neglected, since $\Sigma m_l = 0$ and $\Sigma m_s = 0$ for these four electrons. There are 15 different entries in that table with a maximum value for Σm_l of 2. This corresponds to a D state, and in fact it is a fivefold degenerate ¹D state. The remaining term symbols for the 2p² configuration are ³P and ¹S. This procedure can generally be applied to any number of electrons by using one column for each electron. The closed inner shells of electrons are always omitted, since for those shells $\Sigma m_l = 0$ and $\Sigma m_s = 0$ and they contribute nothing to the total orbital or spin angular momentum.

32.8 REMOVAL OF THE DEGENERACY IN HELIUM

We have now reached the following stage in our analysis. The ground state of He has the electronic configuration 1s² and consists of a single ¹S level. When one of the electrons is excited to the $n = 2$ level, we have two possibilities; the resulting configurations can be either 1s2s or 1s2p depending on whether the electron is excited to an s or a p level in the $n = 2$ state. We have seen in the previous section that there is a total of 16 states, divided into 4 S states and 12 P states, but thus far we have said little to indicate that these 16 states are not degenerate. Within the limits of our zeroth-order approximation they are degenerate, since they all have the same energy in that approximation. On the other hand, we know from spectroscopic measurements that some of the degeneracy is removed, as shown in the energy level diagram of Figure 32.4. We shall not go through the details of deriving these energy levels but simply indicate the procedures used.[5]

If we consider for the moment only the 8 orbital functions of Equations (32.17) and (32.20), a perturbation calculation yields four perturbation energies

$$\begin{aligned} W' &= J_s + K_s \\ &= J_s - K_s \\ &= J_p + K_p \quad \text{(triple root)} \\ &= J_p - K_p \quad \text{(triple root)} \end{aligned} \quad (32.21)$$

---
[5] For more complete details see, for example, L. Pauling and E. B. Wilson, Jr., *Introduction to Quantum Mechanics* (New York: McGraw-Hill Book Co., 1935), chap. 8; and H. Eyring, J. Walter, and G. E. Kimball, *Quantum Chemistry* (New York: John Wiley & Sons, 1944), chap. 9.

SEC. 32.8 REMOVAL OF THE DEGENERACY IN HELIUM

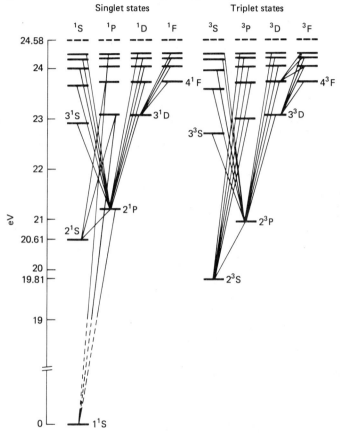

FIGURE 32.4 Energy-level diagram of the helium atom, showing the states arising when $n_1 = 1$ and n_2 varies from 1 to 7 for cases in which the excited electron is an electron of the s, p, d, or f level. The lines connecting the various levels indicate allowed transitions. These allowed transitions occur only within the singlet or the triplet system, and they obey the selection rules $\Delta L = \pm 1$ and $\Delta J = 0, \pm 1$. This kind of diagram is often called a *Grotrian* diagram. [Energy levels from C. E. Moore, *Atomic Energy Levels,* Circular 467 of the National Bureau of Standards (Washington: U.S. Government Printing Office, 1949), p. 5.]

The symbols in (32.21) represent the perturbation integrals:

$$J_s = \iint \phi_{1s}(1)\phi_{2s}(2) \frac{e^2}{r_{12}} \phi_{1s}(1)\phi_{2s}(2) \, d\tau_1 \, d\tau_2$$

$$K_s = \iint \phi_{1s}(1)\phi_{2s}(2) \frac{e^2}{r_{12}} \phi_{1s}(2)\phi_{2s}(1) \, d\tau_1 \, d\tau_2$$

$$J_p = \iint \phi_{1s}(1)\phi_{2p}(2) \frac{e^2}{r_{12}} \phi_{1s}(1)\phi_{2p}(2) \, d\tau_1 \, d\tau_2$$

$$K_p = \iint \phi_{1s}(1)\phi_{2p}(2) \frac{e^2}{r_{12}} \phi_{1s}(2)\phi_{2p}(1) \, d\tau_1 \, d\tau_2$$

(32.22)

In (32.22) the symbol r_{12} represents the distance between the two electrons and e is the electronic charge; the integration is taken over all the coordinates of each of the electrons. The functions ϕ_{2p} can be either $2p_x$, $2p_y$, or $2p_z$, since these differ only in their space orientation. The J_s and J_p integrals are called the *Coulomb* integrals, and they represent the classical Coulombic interaction between the two electrons. The main energy shift arises from this effect. The integrals K_s and K_p are called *exchange* integrals or *interchange* integrals or *resonance* integrals, since the two functions in the integral differ only in the interchange of electrons. This effect is a purely quantum mechanical one and has no classical analog. One can think of this effect in a physical sense as arising from the Pauli exclusion principle, which acts in such a manner as to keep two electrons out of the same region of space. The effect of the Coulombic interaction is to shift all the energy levels including the ground level by an amount given by the J integral. The exchange integrals then introduce additional splitting, as shown in the energy-level scheme of Figure 32.5. The Coulombic interaction thus splits terms having the same L value, and the exchange interaction introduces a further splitting in terms which have the same L value but different values of S. Additional splitting occurs when we introduce the interaction between the spin and orbital angular momentum.

FIGURE 32.5 Splitting of energy levels for the helium atom (not to scale). For helium, the spin-orbit interaction is too small to show up in the energy-level diagram of Figure 32.4.

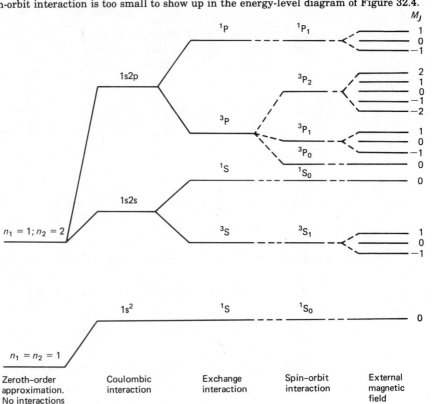

32.9 COUPLING OF ANGULAR MOMENTUM; VECTOR MODEL OF THE ATOM

We have two different types of angular momentum associated with each electron, spin and orbital angular momentum, and each type is separately summed over all the electrons in an atom. The individual spins **s** form a resultant spin $\mathbf{S} = \Sigma \mathbf{s}_i$, while the individual orbital angular momenta **l** form a resultant $\mathbf{L} = \Sigma \mathbf{l}_i$. We have already seen that more than one resultant is possible from a given set of $\mathbf{s}_i$ and $\mathbf{l}_i$. Thus for an atom with two electrons for which $s_1 = \frac{1}{2}$ and $s_2 = \frac{1}{2}$, we can have $S = 1$ or $S = 0$. Similarly if $l_1 = 1$ and $l_2 = 1$, there are three possible values of L—0, 1, and 2, corresponding to the S, P, and D states. The quantum number associated with **L** must be integral, whereas the quantum number associated with **S** may be integral or half-integral. This model wherein the individual angular momentum vectors of the individual electrons are added to get a resultant vector is called the *vector model of the atom*.

The vector model of the atom is illustrated in Figure 32.6, where we have indicated a third vector, the **J** vector, which is the resultant of the total spin and total orbital angular momentum vectors:

$$\mathbf{J} = \mathbf{L} + \mathbf{S} \qquad (32.23)$$

The **J** vector is the total resultant angular momentum of all the electrons in the atom and has associated with it a third quantum number, sometimes called the *inner quantum number,* or just the J quantum number. This quantum number is either integral or half-integral; its length is $|J| = \sqrt{J(J+1)}\,\hbar$. The **S** and **L** vectors each precess about the resultant **J** vector, as shown in Figure 32.6.

This coupling between **L** and **S** to form a resultant **J**, the so-called **L** · **S**, or *Russell-Saunders coupling* introduces an additional energy perturbation

$$H' = \Sigma \xi(r_i) \mathbf{L} \cdot \mathbf{S} \qquad (32.24)$$

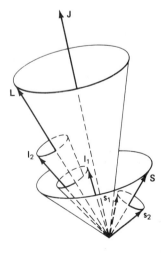

FIGURE 32.6 Coupling of angular momentum in the Russell-Saunders coupling scheme. The l_1 and l_2 form a resultant **L**, while s_1 and s_2 form a resultant **S**. The **L** and **S** vectors precess about their resultant **J**.

where the function ξ has been previously noted in (32.16).[6] This spin-orbit interaction removes some of the degeneracy that remains after the introduction of the coulombic and exchange interactions, as shown in Figure 32.5. Let's consider the removal of the degeneracy of the ^{3}P level of that figure in greater detail. Since $S = 1$, the spin degeneracy is $2S + 1 = 3$; $L = 1$ for a P state with orbital degeneracy $2L + 1 = 3$. The total degeneracy remaining after the coulombic and exchange interactions have been introduced is nine ($3 \cdot 3 = 9$). There are three different ways in which the **L** and **S** vectors can be combined, shown in Figure 32.7, to obtain three different possible values of J ($J = 2, 1, 0$). In the complete term symbol, the value of J is added as a subscript, and the term symbol takes the form $^{2S+1}L_J$, where L is the appropriate symbol S, P, D, ... The introduction of spin-orbit coupling thus causes a breakdown in the ^{3}P level to $^3P_2$, $^3P_1$, and $^3P_0$ levels. The degeneracy that remains in each of these is $2J + 1$. For our ^{3}P level, we have

$^3P_2, J = 2:$ $J_z = 2, 1, 0, -1, -2$ (5 states)
$^3P_1, J = 1:$ $J_z = 1, 0, -1$ (3 states)
$^3P_0, J = 0:$ $J_z = 0$ (1 state)

for a total of nine states. The degeneracy that still remains can be removed by the application of an external field.

In Figure 32.6, we showed the **S** and **L** vectors precessing about their resultant **J**. When an external magnetic field is applied, the **J** vector itself precesses about the magnetic field. The average magnetic moment of an atom is

$$\boldsymbol{\mu}_J = \sqrt{J(J + 1)}\, g\mu_B \qquad (32.25)$$

where g is the Landé g factor and μ_B is the Bohr magneton. The component of the magnetic moment in the direction of the applied magnetic field is $M_J g \mu_B$ ($M_J = J$, $J - 1, J - 2, \ldots, -J$). The value of the magnetic moment can be determined from the splitting of spectral lines when a magnetic field is applied. This splitting, which is quite small, is termed the *hyperfine* splitting. This effect also gives rise to the paramagnetic susceptibility of ions (Chapter 38).

For the case of helium, used as the model of Figure 32.5, there is no uncertainty about the value of L. Since $l_1 = 0$, then L is equal to the value of l_2. For most atoms, there are several distinct possibilities for the ground-state term symbols. Thus for the 2p^2 configuration of carbon there are three possible term values, ^{3}P, ^{1}D, and ^{1}S; you will demonstrate this in one of the problems. We shall not discuss the fine points of the detailed calculations by which the energies of these various possibilities are obtained. The correct ordering of the levels can, however, be simply obtained by applying two empirical rules known as *Hund's rules,* which state:

1. All other things being equal, the state of highest spin multiplicity lies lowest.
2. For levels having the same multiplicity and the same electronic configuration, the state with the highest value of L lies lowest.

Applying the first rule to the ground state of carbon indicates that the ^{3}P level lies lowest, and in fact this is the ground term for carbon.[7]

---

[6] This **L · S**, or Russell-Saunders, coupling scheme applies to lighter elements and begins to break down for heavier elements. For those elements, the **j · j** coupling scheme is more appropriate. In the **j · j** scheme, the spin orbit interaction is taken into account on an individual electron basis rather than on the basis of the entire atom. The s_i and l_i are combined to give a resultant j_i vector for each electron, and the total angular momentum is then obtained by summing over all the j_i, and we get $\mathbf{J} = \Sigma j_i$.

[7] Strictly speaking, Hund's rules apply only to the lowest states of atoms. In many instances they correctly predict the order of higher levels, though there are other exceptions to the predicted order of levels besides the ground state.

SEC. 32.9 COUPLING OF ANGULAR MOMENTUM; VECTOR MODEL OF THE ATOM

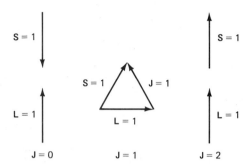

FIGURE 32.7 Vector addition for the case $S = 1$, $L = 1$, giving three possible resultant J vectors. Note the similarity to Figure 32.3.

The level diagram for helium shown in Figure 32.4 is a particularly simple one. The spin-orbit interaction is so small for helium that the splitting does not show up on the figure. A more complicated level diagram is shown in Figure 32.8, for mercury. Here the spin-orbit coupling is large enough to show up on this scale. A complete

FIGURE 32.8 Energy-level diagram for mercury. The electron configuration of the ground state is $5d^{10}6s^2$, and the ground state term symbol is $^1S_0$. The energies in this figure are obtained by exciting one of the 6s electrons to the orbitals indicated alongside the excited levels. The lines connecting the levels indicate some of the observed transitions along with the wavelengths of the transitions in angstrom units.

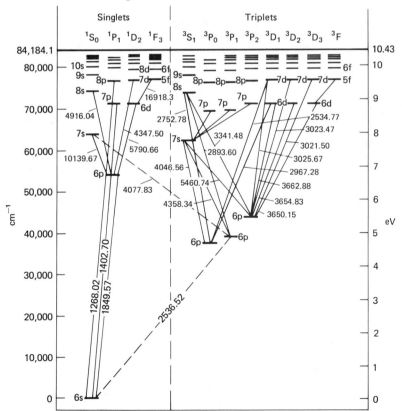

analysis of all the spectral lines to obtain a final energy-level diagram is a complicated and time-consuming task. The theory is vitally important in assigning the correct term symbols to the various levels, since the accuracy of the calculations in determining the relative energies of the levels is much greater than in determining the absolute energy values. Several *allowed* transitions have been indicated on Figure 32.8 by solid lines. An allowed transition is one that obeys the selection rules:

1. $\Delta S = 0$
2. $\Delta L = \pm 1$
3. $\Delta J = 0, \pm 1$

Numerous transitions occur that do not follow these selection rules. These *forbidden* transitions generally occur with much weaker intensities, and they are indicated by dashed lines.[8]

32.10 THE PERIODIC TABLE

Figure 32.9a shows the ordering of the orbital energies in atoms. This ordering is a zeroth approximation based on what we know of the orbital energies in the hydrogen atom and our later discussions. We shall try to construct the electronic configurations of the ground states of atoms by filling in these various orbitals in order, the lowest-energy orbitals first, subject to the constraints of the Pauli exclusion principle. The procedure we shall follow is known as the *Aufbau Prinzip,* or building-up principle, since it allows us to build up the periodic table of the elements.

The first element, hydrogen, is particularly simple. We have but one electron, which goes into the lowest level; the configuration is 1s, indicating a principal quantum number $n = 1$ and a single s electron. Helium, with $Z = 2$, has two electrons. The second electron is added to the 1s level, subject to the constraint that its spin must be opposed to the spin of the first electron. The configuration of helium is $1s^2$, the superscript 2 indicating the presence of two electrons in the 1s orbital. When we get to lithium ($Z = 3$), the Pauli principle forces the third electron into the 2s

FIGURE 32.9 Schematic representation of the energies of the first few electron orbitals in atoms. There is a threefold degeneracy of the p orbitals and a fivefold degeneracy of the d orbitals. Each orbital can contain two electrons. Three electrons are shown in (a) for the lithium atom ($Z = 3$). Part (b) shows the correct placement of the 3d orbitals at $Z = 19$, and shows the 19 electrons for the potassium atom. Note that this figure is not to scale.

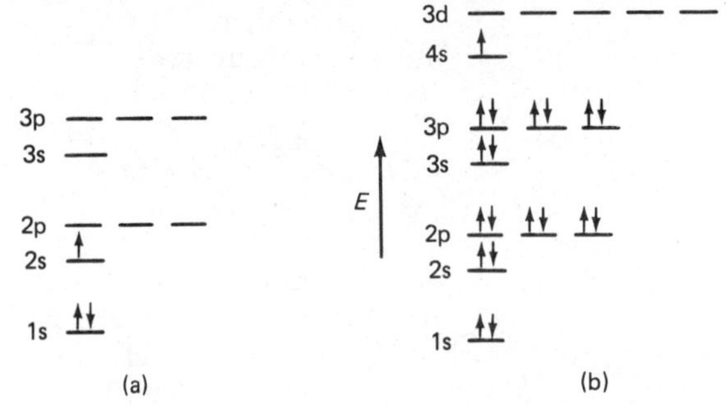

[8] Some "forbidden" transitions are actually quite intense. The forbidden transition of Hg from the ground state to the 6p state at 2536.52 Å is one of the most intense lines in the spectrum.

($n = 2$) orbital, since there is no room for a third electron in the lowest energy level. The configuration of lithium is $1s^2 2s$, indicating two electrons in the 1s orbital and one electron in the 2s orbital. This is shown in Figure 32.9a. The next atom is Be ($Z = 4$); the fourth electron also fits into the 2s level producing the configuration $1s^2 2s^2$. Note that while for the hydrogen atom there is no difference in energy between the 2s and the three 2p orbitals, we have placed the 2s orbitals lower than the 2p orbitals. That this is the case can be seen on a qualitative basis by referring to Figure 30.2, where we have indicated the radial distributions of the various orbitals. The electrons in the 1s orbital partially shield the higher orbitals from the nuclear charge; hence the effective charge that the 2s and 2p electrons "see" at the nucleus will be somewhat less than Z. Since the probability of finding a 2s electron in the region in which the 1s electron has its greatest density is greater than the probability of finding a 2p electron in that same region, the 2s electron is less shielded by the 1s electron and hence experiences a greater effective nuclear charge than the 2p electron. As a result, the 2s electron has a lower energy than the 2p electron does. The 2s electron is said to have a better *penetration* of the 1s orbital. Similar qualitative statements can be made about the higher orbitals.

When we get to boron at $Z = 5$, the fifth electron must be placed in a 2p orbital, since the Pauli principle does not permit the fifth electron to be placed in the 2s orbital. Its configuration is $1s^2 2s^2 2p$. Actually this configuration would normally be written as simply 2p, with the lower-energy closed shells omitted in writing the electronic structures. (Similarly the structures for Li and Be would be written simply as 2s and $2s^2$, the filled 1s orbital being omitted.) At carbon ($Z = 6$) the next electron would also be placed in a 2p orbital for the configuration $2p^2$, but we now have two possibilities. It can fit into the same 2p orbital as the fifth electron with an opposing spin ($S = 0$) or it can be placed in a different orbital without regard to its spin. Hund's first rule requires that it be placed in a second 2p orbital with its spin in the same direction as the previous electron ($\uparrow \uparrow$), since in that case we can have $S = 1$ for a higher multiplicity. The same reasoning indicates that the next electron for nitrogen ($Z = 7$) goes into the remaining 2p orbital with its spin in the same direction ($\uparrow \uparrow \uparrow$) for the configuration $2p^3$. The next three electrons for oxygen ($Z = 8$), fluorine ($Z = 9$) and neon ($Z = 10$) also fit into the 2p orbitals with the configurations $2p^4$, $2p^5$, and $2p^6$. The configuration for neon can be symbolized by ($\uparrow \downarrow \uparrow \downarrow \uparrow \downarrow$); here $S = 0$, and the 2p orbital is completely filled.

We proceed through the $n = 3$ orbitals in a similar way. For sodium ($Z = 11$), the eleventh electron goes into the 3s orbital; and for Mg, the twelfth electron likewise goes into the 3s orbital for a $3s^2$ configuration. The thirteenth electron for Al ($Z = 13$) must then go into the 3p orbital, and so on through Ar ($Z = 18$), at which point the 3p orbital is filled and the configuration is $3p^6$. If we now try to go on with this scheme, we are faced with a problem. The next atom is potassium ($Z = 19$), and it might be thought that the nineteenth electron starts filling the 3d orbitals. But that does not make sense from a chemical point of view. We know that certain groups of atoms such as the inert gases, the alkali and alkaline earth metals, and the halogens have similar chemical properties, and it is only reasonable to expect these similar chemical properties to reflect similarities in electronic configurations. We can see this in what we have done up to now. Lithium and sodium both have s^1 outer electron configurations; Be and Mg are both s^2, B and Al both p^1, C and Si both p^2, N and P both p^3, O and S both p^4, and F and Cl both p^5. The inert gases Ne and Ar are characterized by closed p^6 shells whereas He is characterized by a closed s^2 shell. Chemistry would make much more sense if potassium had a 4s outer electron instead of a 3d.

Potassium does have an outer 4s electron, as shown in Figure 32.9b. The 3d orbital has a higher energy than the 4s orbital. The nineteenth electron fits into this 4s orbital and so does the twentieth electron for Ca ($Z = 20$). The next ten elements from Sc to Zn ($Z = 21$ to 30) are formed by filling the 3d orbital, which is the next higher orbital. This set of ten elements, the *transition metals,* is followed by a set of six elements in which the 4p orbitals are filled, terminating with Kr ($Z = 36$), which has the $4p^6$ electronic configuration characteristic of the inert gases.

It would be naive to expect the orbital energies to be independent of the atomic number Z—the presence of electrons in other orbitals is certain to affect the energies. The energies as a function of Z give the appearance of a tangled mess, as shown in Figure 32.10, where the numerous crossover points should be noted. It should be apparent from Figure 32.10 that below about $Z = 6$ the 3d orbital does have a lower energy than the 4s orbital. In the region of the transition metal series, however, the 4s orbital has a lower energy. In the limit as Z goes to unity, the energies depend only on the principal quantum number n; at that point there are no interelectronic effects, since there is only one electron. As Z increases, the energies of

FIGURE 32.10 Atomic orbital energy levels as a function of nuclear charge. The energy is given in units of E_H, the energy of the hydrogen atom in the ground state (13.6 eV). (Plots are drawn from the data of R. Latter, *Phys. Rev.* 99 (1955):510, based on a Thomas-Fermi potential.)

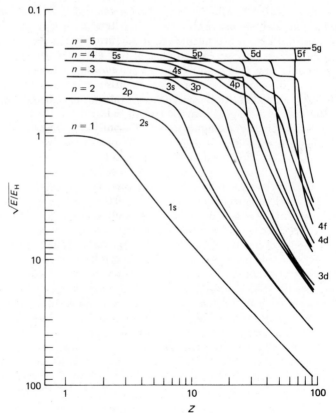

the orbitals are shifted relative to each other from electron-electron interactions, and the ordering of the levels is complex at intermediate Z. At very high values of Z, the nuclear charge is high, and the effect of the nuclear-electron interaction is much higher than the electron-electron interactions for the inner-shell electrons. Thus we might also expect the energies of the inner-shell electrons to depend only on n in the high Z limit; this is what happens, as you can see from the merging of the curves at high Z. Most calculations like this are based on the *self-consistent field method* of Hartree.[9] In this approximation, each electron is treated as moving in a spherically symmetrical field due to the nuclear charge and the averaged field due to all the other electrons. A set of N wave functions for the N electrons is selected, and the potential of one of the electrons is then calculated. This potential is then used to adjust the wave functions, and the process is repeated for each electron, to get a new set of wave functions. The entire process is then repeated until successive calculations produce no changes in the potential energy of any of the electrons. The final potential energies constitute the self-consistent field from which the energy of the atom can be calculated by a variational procedure. The numerous computations required can be carried out rapidly with modern computers.

As we go beyond Kr in the periodic table, we can add the electrons to the orbitals sequentially according to the energy-level scheme of Figure 32.10. The calculated energies are not accurate enough to provide an unambiguous assignment of an electronic configuration, and other information must be used along with the calculations. We have seen how ordinary chemical behavior can be used to deduce that at $Z = 19$ the 4s orbital must lie at a lower energy than the 3d orbital if potassium is to be an alkali metal; the same holds true at $Z = 20$ if calcium is to be an alkaline earth. Spectroscopic measurements provide a wealth of information on electronic configurations, and measurements of paramagnetic susceptibilities are also similarly useful (Chapter 38). Magnetic susceptibility measurements indicate the magnitudes of magnetic moments, and this information can be correlated with the **S** or **J** vectors, which can in turn be correlated with the numbers of unpaired electrons. A convenient form of the periodic table is shown in Table 32.2, where the similar elements in the seven periods appear one below the other. In that table, the orbitals connected by braces in the second column have approximately the same energies. This gives rise to some ambiguity in the table as it is written, particularly for the transition metals and the rare earths. Thus for the first transition metal series one might expect, and the table so indicates, that the electronic configurations would be $1s^2 2s^2 2p^6 3s^2 3p^6 3d^k 4s^2$, with k going from 1 to 10 from Sc to Zn. For eight of the ten elements we get just what is expected. For Cr, however, instead of the expected $3d^4 4s^2$ configuration, we get a $3d^5 4s$ configuration, one of the 4s electrons having been placed in the 3d orbital. This can be deduced from spectroscopic evidence showing that the ground-state term symbol is $^7S_3$, indicating spin $S = 3$. (The spin of 3 can also be deduced from magnetic susceptibility measurements.) You should convince yourself that the configuration $3d^4 4s^2$ would produce spin 2 and that the $3d^5 4s$ configuration does indeed produce spin 3. Also the configuration of Cu is actually $3d^{10} 4s$ instead of $3d^9 4s^2$, as the table suggests. This, by the way, explains the univalent chemical nature of Cu in the cuprous state, since there is that single 4s electron available for chemical bonding.

---

[9] D. Hartree, *The Calculation of Atomic Structures* (New York: John Wiley & Sons, 1957). Hartree's method was later improved by Fock, and these improved calculations are known as *Hartree-Fock calculations*. An elementary discussion of atomic orbital energies can be found in R. S. Berry, *J. Chem. Ed.* 43 (1966):283.

Table 32.2 Periodic system of the elements.

		1s																
1		H 1																He 2
2	2s	Li 3	Be 4															
	2p											B 5	C 6	N 7	O 8	F 9	Ne 10	
3	3s	Na 11	Mg 12															
	3p											Al 13	Si 14	P 15	S 16	Cl 17	A 18	
4	4s	K 19	Ca 20															
	3d			Sc 21	Ti 22	V 23	Cr 24	Mn 25	Fe 26	Co 27	Ni 28	Cu 29	Zn 30					
	4p													Ga 31	Ge 32	As 33	Se 34	Br 35	Kr 36
5	5s	Rb 37	Sr 38															
	4d			Y 39	Zr 40	Nb 41	Mo 42	Tc 43	Ru 44	Rh 45	Pd 46	Ag 47	Cd 48					
	5p													In 49	Sn 50	Sb 51	Te 52	I 53	Xe 54
6	6s	Cs 55	Ba 56															
	4f	La 57	Ce 58	Pr 59	Nd 60	Pm 61	Sm 62	Eu 63	Gd 64	Tb 65	Dy 66	Ho 67	Er 68	Tm 69	Yb 70			
	5d			Lu 71	Hf 72	Ta 73	W 74	Re 75	Os 76	Ir 77	Pt 78	Au 79	Hg 80					
	6p													Tl 81	Pb 82	Bi 83	Po 84	At 85	Rn 86
7	7s	Fr 87	Ra 88															
	5f	Ac 89	Th 90	Pa 91	U 92	Np 93	Pu 94	Am 95	Cm 96	Bk 97	Cf 98	Es 99	Fm 100	Md 101	No 102			
	6d			103	104	105	106	107	108	109	110	111	112					

SOURCE: C. E. Moore, *Atomic Energy Levels* (Washington, D.C.: U.S. Government Printing Office, Circular 467 of the National Bureau of Standards, 1971).

FIGURE 32.11 Ionization potentials of neutral atoms as a function of atomic number.

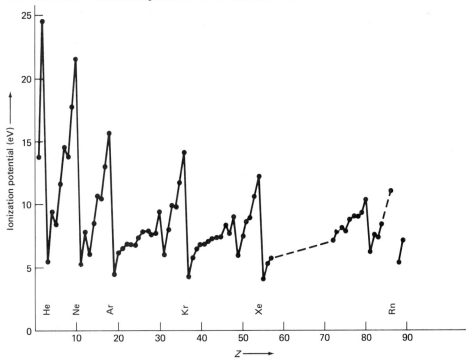

One of the most dramatic indicators of how the properties of the elements of the periodic table show periodicity is the ionization potentials of atoms. These are plotted in Figure 32.11 as a function of Z. Note the maxima at each Z corresponding to an inert gas. The ionization potential measures the ease with which an electron is removed from the outer shell of the atom; it is difficult to remove an electron from an inert gas. The minima, on the other hand, occur at values of Z corresponding to the alkali metals. These electrons are removed with relative ease, explaining the great reactivity of these elements and their positive univalent character as chemical ions.

PROBLEMS

1. Show that the two wave functions of (32.14) are orthogonal.
2. Show that whereas the total wave functions $\psi_{T,1}$ and $\psi_{T,2}$ given in (32.2) and (32.3) are not eigenfunctions of the interchange operator $\hat{P}(1, 2)$, the linear combinations of the two functions given in (32.6) are eigenfunctions of this operator.
3. Show that the wave function ψ_A of (32.9a) is an eigenfunction of $\hat{P}(1, 2)$ with eigenvalue -1 whereas ψ_s of (32.9b) is an eigenfunction with eigenvalue $+1$.
4. The operator $\hat{S}_z$ for the total z component of spin angular momentum for a system of two electrons is given by $\hat{S}_z = \hat{s}_z(1) + \hat{s}_z(2)$. Show that each of the four wave functions given in Equation (32.18) is an eigenfunction of $\hat{S}_z$ and find the eigenvalues for each. (Note: $\hat{s}_z \alpha = \tfrac{1}{2}\hbar\alpha$; $\hat{s}_z \beta = -\tfrac{1}{2}\hbar\beta$.)
5. Show that the wave function ψ_5 in Equation (32.19) is normalized. That is to say, show that $\int \psi_5^2 \, d\tau = 1$.
6. Show that the wave functions ψ_5 and ψ_6 in (32.19) are orthogonal. That is to say, show that $\int \psi_5^* \psi_6 \, d\tau = 0$.
7. Carry out the procedure indicated in Table 32.1 for the $2p^2$ configuration of carbon. Show that the possible resulting states are ^1D, ^3P, and ^1S. What

are the total degeneracies of each state? Determine the values of J for each state and write the complete term symbols for all possible states.

8. Suppose one of the 2p electrons of carbon in the ground state is excited to a 3s level, giving the configuration $1s^22s^22p3s$. Determine all the possible term symbols for this excited state and the degeneracies of each.

9. Determine all the possible J values for an atom with $S = 3$ and $L = 4$. Write the complete term symbols and indicate the degeneracies of each possibility.

10. Notice that in Figure 32.8 for the energy-level diagram of mercury there is one column for ^3S states whereas there are three columns each for the ^3P and ^3D states. Why? Notice also that only one column is shown for ^3F states and the value of J has been omitted. This is because the spin-orbit splitting of the ^3F levels is so small that they would not show up on this diagram. If a complete diagram were to be drawn, how many columns would be required for the ^3F levels? Give the complete term symbols (including J values) for each column. Why is no spin-orbit splitting indicated for the singlet levels?

11. How would the various levels of the mercury energy-level diagram of Figure 32.8 be split under the application of an external magnetic field?

12. Take the ground-state energy of the mercury atom to be at $E = 0$. The energies of numerous other states can be determined from the wavelengths of the transitions indicated in Figure 32.8.
    (a) Calculate the energies of the excited states 6p, 7p, 8p, 7s, 8s, 6d, 7d, and 5f in the singlet system relative to the ground state.
    (b) Calculate the energies of the three 6p levels in the triplet system.
    (c) Transitions between the three 6p levels in the triplet system are forbidden. If these transitions were to occur, at what wavelengths would they occur?
    (d) Compare the size of the spin-orbit splitting for the 6d levels of the ^3D state with the spin-orbit splitting of the 6p levels of the ^3P state.

13. Under the assumption that Figure 32.10 is correct, determine the electron configuration of tellurium, element 52. Compare this result with the correct configuration $1s^22s^22p^63s^23p^63d^{10}4s^24p^64d^{10}5s^25p^4$. Show that the ground state is $^3F_2$. In the rare earths, the energy of the 4f levels is less than the energy of the 5d levels. At what value of Z do the rare earths start, according to Figure 32.10?

14. Show that an atom with an even number of electrons always has odd multiplets (singlets, triplets, quintets, and so on) whereas an atom with an odd number of electrons always has even multiplets (doublets, quartets, sextets, and so on) and hence one never finds odd and even multiplets mixed on an energy-level diagram like Figure 32.8.

15. Correlate the peaks shown in the electron impact spectrum of Hg (Figure 25.11) with the Hg transitions of Figure 32.8.

CHAPTER THIRTY-THREE

MOLECULES AND CHEMICAL BONDING

33.1 SOME HISTORICAL BACKGROUND

By the nineteenth century, classical physics had formulated the three classical long-range forces—gravitational, magnetic, and electric. In that same century, classical chemistry had completed formulating the basic concepts of atoms and molecules and begun the search for the universal "glue" that held atoms together in molecules. It was only natural to look to the classical long-range forces that the physicists had developed. Gravitational forces were too weak to explain the attractive forces between atoms in molecules. Magnetic forces were too uncommon; it was difficult to see any magnetic forces in nonmagnetic matter. That left electrical forces, and indeed Berzelius had proposed early in the 1800s that compounds were formed because of electrostatic attraction between the atoms.

The situation was further complicated, since there appeared to be two distinctly different types of chemical compounds: *polar* compounds, as exemplified by NaCl, and *nonpolar* compounds, as exemplified by the hydrocarbons. The polar compounds

> **JÖNS JACOB BERZELIUS (1779–1848)**, Swedish chemist, can be regarded as a "chemist's chemist", and he was perhaps the best analytical chemist of his day. He was self-taught in chemistry, fortunately getting his knowledge from one of the newer, antiphlogistic textbooks. He never learned the phlogiston theory, a fact that made him unpopular with the older chemists in the early 1800s. In 1802 he built a voltaic pile consisting of 60 pairs of alternating zinc discs and copper pennies. He studied the effect of galvanic currents on various diseases and found that they had no effect. He introduced the use of mercury as the cathode and obtained amalgams of the active metals. Among his greatest contributions were a series of carefully performed chemical analyses from which he published revised tables of atomic masses in 1814, 1818, and 1828. It was Berzelius who introduced the notation, still extant, for the chemical elements wherein they are known by the first or first two letters of their Latin names. He also introduced the quantitative concept that the symbol stood for the atomic weight of the element as well. Berzelius used superscripts, thus water was H^2O. He tried to simplify this notation by using dots for oxygen atoms placed above the symbol for the element to which it was joined, and he then introduced a bar through the symbol to indicate a double atom; thus water was written $\bar{H}$. Berzelius was a firm believer in the law of definite proportions and carried his belief to its logical absurdity. He could not cope with the progress made by the younger organic chemists, whose findings indicated that the ratio of carbon to hydrogen was highly variable; nor could he accept that negative chlorine could substitute for positive hydrogen. At his death, he was a famous and respected figure, but his opinions were generally disregarded by his younger colleagues.

did seem subject to the classical laws of electrostatics, and the structures were described in terms of positive and negative ions held together by the classical Coulombic attractive forces. When NaCl was dissolved in water, these attractive forces were overcome and the free ions were liberated. The nature of the attractive forces in nonpolar compounds was less tractable. There were no charges seemingly available for a Coulombic attraction between the atoms. The picture was further obscured because polar compounds were characteristic of the *inorganic* compounds whereas nonpolar compounds were characteristic of *organic* compounds. It was thought that the fundamental difference between the two lay in the origins of the compounds. In view of the biological origins of organic compounds, a characteristic known as a "vital force" was postulated in the early 1800s as the glue that held atoms together in organic compounds. The division between organics and inorganics was destroyed in 1828, when Wöhler synthesized the organic compound urea from inorganics. There was apparently a unifying feature of chemical bonding that was independent of the source of the molecule. The perplexity increased further when it was finally realized that the ordinary gases such as H_2, N_2, O_2, and Cl_2 were diatomics. No schemes were available to explain how these could be held together electrically, since the fundamental concept that all atoms of the same type were alike vitiated any attempt to assign different charges to the two atoms. The basic framework that then existed was inadequate for the task.

The "magical" properties of the number eight in chemical bonding came to the fore by the periodic table of Mendeleev and the early "periods" of eight elements. It was noted that different valencies could often be assigned to a given element so that the sum of the absolute values of the valencies was eight. Thus the valence of chlorine was -1 in NaCl and $+7$ in Cl_2O_7; sulfur had a valency of $+6$ in SO_3 and -2 in H_2S. This was the "rule of eight" that Abegg postulated in the early 1900s. Valency was attributed to the number of electrons given up by an atom (positive valency) or the number taken up by an atom (negative valency) in forming chemical compounds. By 1916, Rutherford had discovered the nuclear atom, Bohr had used the nuclear atom to establish quantum numbers, and Mosely had discovered the atomic number in his work with the X-ray spectra of atoms. These were necessary to place the empirical rule of eight on a firmer conceptual basis. That basis was the extraordinary stability of the inert-gas closed shells of electrons that came in multiples of eight, at least up to argon. The numbers two, ten $(2 + 8)$, and eighteen $(2 + 8 + 8)$ for the numbers of electrons in the closed shells were carried in the minds of all chemists and remembered by all students for their final examinations. The valency

---

FRIEDRICH WÖHLER (1800–1882), German chemist, was a student of Berzelius, having spent a year with him in Stockholm. While Wöhler was a medical student, he discovered cyanic acid, and in 1828 he found that the ammonium salt could be isomerized into urea. He thus showed that a product of living matter could be synthesized from material that was considered mineral matter. This was difficult to reconcile with the contemporary notion of "vital forces" in biological compounds. He discovered benzaldehyde and after thoroughly investigating its properties, concluded that a cluster of atoms that could be represented by the formula C_7H_5O persisted unchanged through a long series of synthetic transformations. He named the cluster "benzoyl." Other so-called radicals were soon discovered, bringing some order into the complex and disordered field of organic chemistry. Wöhler was also an inorganic chemist of note, isolating aluminum (1827) and beryllium (1828) and producing a large volume of analyses of minerals and meteorites.

> DMITRY IVANOVICH MENDELEEV (1834–1907), Russian chemist, is best known for elucidating the periodic table of the elements. He discovered the periodic relations between the elements while he was writing his chemistry textbook *Osnovy Khimii;* he had decided to group the elements by their atomic masses. He worked out the final stages of his table, using a system of what might be called "chemical solitaire," wherein he wrote each element along with its atomic mass and other properties on a card and tried various arrangements of the cards on a large table. His work was first reported in 1869 to the Russian Chemical Society under the title (translated) "Attempt at a System of Elements Based on Their Atomic Weight and Chemical Affinity." He proposed changing the accepted weight of Be, 14, to its proper weight of 9.4, replacing the accepted Be_2O_3 with the formula BeO for the oxide, by analogy with MgO. He discovered gaps at three points, between hydrogen and lithium, between fluorine and sodium, and between chlorine and potassium. He predicted that these gaps would be filled by elements having atomic weights of 2, 20, and 36. These turned out to be He, Ne, and Ar. He also predicted three undiscovered elements that he called eka-aluminum, ekasilicon, and ekazirconium. His theory was received with something akin to derision in the scientific community until gallium, his eka-aluminum, was discovered in 1875. This was followed by the discovery in 1886 of germanium, whose properties precisely matched those of Mendeleev's ekasilicon, and he became a legend by virtue of being the first person ever to predict an element. At his funeral, a periodic table of the elements was carried above the procession as a tribute to his career.

of ionic compounds could now be explained quite simply, at least on a conceptual basis. The alkali metals (Li, Na, K) had one electron over and above that required for a stable inert gas configuration, whereas the halogens (F, Cl) had one less electron. The extra electron of the alkali metal was transferred to the halogen; both now had an inert gas configuration and formed a stable compound like NaCl. This transferred-electron theory worked well with polar substances but ran into great difficulties with the so-called *covalent* bond. The H_2O formula could be explained by this same transferred electron model, but all chemists felt intuitively that there was a fundamental difference between the bonding in H_2O and in NaCl. When it came to homonuclear diatomic such as O_2 and covalently bonded compounds in general, not to mention variable valencies like what was found in SO_2 and SO_3, the transferred electron theory was at a total loss.

In 1916, a much greater understanding of the covalent bond was achieved, at least on a qualitative basis, when G. N. Lewis introduced the *shared,* or *electron pair, bond.* Lewis envisioned, instead of the theory in which an electron from an atom of positive valency transferred to an atom of negative valency, a situation in which the two atoms shared the electrons equally so that both would simultaneously achieve an inert gas configuration. Thus the hydrogen atom could be written as H·, the dot indicating an electron; and the H_2 molecule could be written as H:H, the two hydrogen atoms sharing electrons equally, each atom thus achieving the stable helium configuration. A fluorine atom with its seven outer electrons would be written as $:\ddot{F}\cdot$, and the diatomic fluorine molecule as $:\ddot{F}:\ddot{F}:$, both atoms achieving the neon electronic configuration. The pair of electrons shared equally by the two atoms constituted a single *covalent bond*. Methane, requiring four single covalent bonds, would be written as

$$\begin{array}{c} H \\ H:\ddot{C}:H \\ H \end{array}$$

In these structures the magical properties of eightness were retained, and the number of bonds or electron pairs was equal to the valency of the atoms. In many instances more than one single bond was required if an inert-gas electron configuration was to be achieved, and this was handled by using more than one pair of electrons. Thus the oxygen atom with six outer electrons, :Ö, became :O::O: in the diatomic molecule, and nitrogen :N· became :N:::N: in the molecular form.

There was much, however, that this new scheme could not adequately explain—for example, why ethane was more stable than ethylene or acetylene, or why the bonding in methane should be tetrahedral instead of planar. Particularly perplexing was the paramagnetism of oxygen gas. Paramagnetism required an odd electron, and the electron pair concept required even numbers of electrons. The solutions to these and similar problems had to wait for the advent of quantum mechanics. The subject matter of quantum mechanics is extraordinarily broad and the details are often complex. We cannot hope to cover the entire scene in a book at this level and of this size. Our limited objective will be to provide some basic understanding of a few of the features. First and foremost is the rationale of the formation of simple diatomic molecules. Why is hydrogen diatomic and helium monatomic? The answer to this lies in the minimum-energy principle. The molecule H_2 has a lower energy than two separate H atoms, whereas the He_2 molecule has a greater energy than the combined energy of two separate He atoms. That statement does not tell us very much but does indicate the direction one must take. Other questions we shall touch on include, Why is methane tetrahedral instead of planar, and why is the formula for methane CH_4 instead of CH_3 or something else? Why is oxygen paramagnetic and nitrogen diamagnetic? Why are some metal ions octahedral when others are pyramidal or square planar? We start our discussion with the simplest possible molecule, the molecule ion H_2^+.

33.2 THE HYDROGEN MOLECULE ION

The hydrogen *molecule ion*, H_2^+, consists of two protons and one electron. When an electric discharge is passed through H_2, the molecule ion is created. Some of the properties of H_2^+ can be deduced from observations of spectroscopic transitions between the excited H_2 molecule and the molecule ion. This ion H_2^+ is unstable and has a short lifetime, picking up an electron to revert to a normal H_2 molecule. The binding energy of H_2^+, or the energy for the reaction $H_2^+ = H + H^+$ is 2.79 eV; the equilibrium internuclear distance is 1.06 Å, about twice the Bohr radius.

The Schrödinger equation for H_2^+ can be written

$$\nabla^2 \psi + \frac{8\pi^2 m}{h^2}\left(E + \frac{e^2}{r_A} + \frac{e^2}{r_B} - \frac{e^2}{R_{AB}}\right)\psi = 0 \qquad (33.1)$$

where r_A is the distance of the electron from the first proton, r_B is the distance of the electron from the second proton, and R_{AB} is the internuclear distance as indicated in Figure 33.1a. As it stands, Equation (33.1) represents a three-body problem and cannot in general be solved in closed form. The equation is, however, soluble under the assumption that the *Born-Oppenheimer approximation* is valid. This approximation states that the motion of the electron is so rapid relative to the motion of the nuclei that the nuclei can be regarded as fixed. This means that if R_{AB} is fixed at some particular value, the problem is reduced to a two-body problem. We simply need to find the electron's position in space relative to one of the nuclei; its position

SEC. 33.2 THE HYDROGEN MOLECULE ION

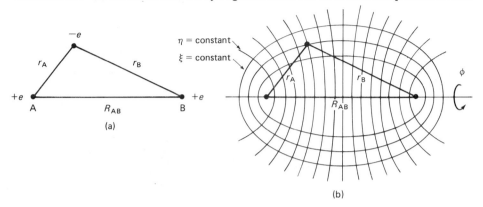

FIGURE 33.1 (a) Coordinates for the hydrogen molecule ion. (b) Confocal elliptic coordinates.

relative to the other nucleus is then determined, since the two nuclei can be regarded as a rigid rotor. Thus we can find the energy of the system at several different values of R_{AB}, and the energy can be mapped as a function R_{AB}.

Before we proceed any further, we might consider two important facts that we know about the system in the limits of small and large values of R_{AB}. In the limit as $R_{AB} \to \infty$, the system consists of a hydrogen atom and an ion H⁺. The energy is

$$E = E_H + E_{H^+} = -R_H + 0 \qquad (R_{AB} \longrightarrow \infty) \qquad (33.2)$$

where $-R_H$ is the energy of the ground state of a hydrogen atom taken relative to $E = 0$ for the ionized proton. At the other extreme, in the limit as R_{AB} approaches zero, the energy must get infinitely large because of the electrostatic repulsion between the two protons. From the minimum-energy principle, we can infer that if the energy increases monotonically as R_{AB} goes from infinity to zero, then no stable molecule will form. Molecules would tend to the minimum energy, namely the state of a separate hydrogen atom and the ion H⁺. On the other hand, if there is a minimum in the energy as a function of R_{AB}, then a stable molecule will form having an internuclear distance that corresponds to the minimum value of R_{AB}. Since we know from experiment that the molecule does exist, any valid solution of the problem must provide an energy dependence of the form shown in Figure 33.2.

To solve Equation (33.1), we transform the equation to a coordinate system known as *confocal elliptic coordinates*, with the coordinates ξ, η, and ϕ:

$$\xi = \frac{r_A + r_B}{R_{AB}} \qquad \eta = \frac{r_A - r_B}{R_{AB}} \qquad (33.3)$$

The coordinate system has cylindrical symmetry in the azimuthal angle ϕ and is indicated in Figure 33.1b. The variable ξ varies from 1 to infinity, while η varies between -1 and $+1$; ϕ varies between 0 and 2π. The surfaces ξ = constant are confocal ellipsoids of revolution with the two protons at the foci, and the surfaces η = constant are confocal hyperboloids. These are indicated in Figure 33.1b. If we write the solution $\psi = \psi(\xi, \eta, \phi)$ as the product function

$$\psi(\xi, \eta, \phi) = \Xi(\xi)H(\eta)\Phi(\phi) \qquad (33.4)$$

then the equation is separable into three differential equations, one for each variable. We shall not concern ourselves with the detailed solutions to this problem so

FIGURE 33.2 Calculated total energy of H_2^+ as a function of the internuclear distance.

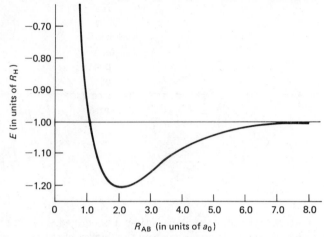

far as the ξ and η variables are concerned.[1] The solution to the ϕ part reflects the cylindrical symmetry of the system and has the form

$$\frac{d^2\Phi}{d\phi^2} = -m^2\Phi \tag{33.5}$$

This is the same equation that appeared previously in connection with the hydrogen atom; its solution is

$$\Phi = A\,\exp(im\phi) \tag{33.6}$$

where $m = 0, \pm 1, \pm 2, \ldots$. Note that $\Phi(\phi)$ is an eigenfunction of the operator for the z component of angular momentum, $\hat{L}_z = -i\hbar\,\partial/\partial\phi$. The energy of the ion H_2^+ depends on m, and the two states with $\pm m$ have the same energy. Thus the state with $m = 0$ is nondegenerate with respect to this variable, and the remaining states are doubly degenerate. The quantum number m is analogous to the quantum number l for the hydrogen atom, and just as atomic orbitals are classified by their l values, so too are *molecular orbitals* classified according to their value of m. To provide a clear distinction between atomic and molecular orbitals, Greek letters are used for the molecular orbitals. Thus for $m = 0, \pm 1, \pm 2, \ldots$, the molecular orbitals are designated by the letters $\sigma, \pi, \delta, \ldots$. An additional symbol is used in the designation of molecular orbitals to indicate the symmetry properties on inversion. The ground state orbital of H_2^+ is symmetrical with respect to inversion through the center and is designated as a $1\sigma_g$ orbital, the g denoting *gerade*, the German word for "symmetrical." The 1 indicates that the state is derived from an $n = 1$ atomic orbital, as we shall see in the next section. The first excited state of H_2^+ is unsymmetrical with respect to inversion through the center and is denoted as a $1\sigma_u$ orbital, the u being an abbreviation for the German *ungerade*, which means "unsymmetrical."

---

[1] The details of the transformation and the solutions are given in L. Pauling and E. B. Wilson, Jr., *Introduction to Quantum Mechanics* (New York: McGraw-Hill Book Co., 1935), p. 333, and in H. Eyring, J. Walter, and G. E. Kimball, *Quantum Chemistry* (New York: John Wiley & Sons, 1944), p. 201.

This exact calculation for H_2^+ provides results that agree well with experiment. The energy of the ion as a function of R_{AB} has been plotted in Figure 33.2. The minimum in the energy curve corresponds to the experimental value of R_{AB}, and the calculated binding energy also agrees with the experimental value. Unfortunately, an "exact" solution applies only to the ion H_2^+, which is not a very common molecule. The solutions obtained do, however, provide the basic framework of molecular orbitals from which approximate solutions to more complicated molecules can be obtained in the same manner in which complicated atoms can be constructed from atomic orbitals. Suppose we now have another look at the molecule ion H_2^+, this time using an approximation method.

33.3 VARIATIONAL CALCULATION OF THE HYDROGEN MOLECULE ION

After an "exact" calculation of the molecule ion H_2^+, an approximate solution seems redundant. Such a calculation is useful, however, in that the agreement between the approximate and the exact calculations provides a check on the approximate calculations. Also, the methods used for H_2^+ can be extended to molecules that are more complicated. Recalling our discussion of the variation method (Chapter 31), we must select a trial, or variation, function with adjustable parameters. A logical choice is the ψ_{1s} wave functions of the hydrogen atom, and we shall take a linear combination of these in the form

$$\psi = C_A \psi_{1sA} + C_B \psi_{1sB} \tag{33.7}$$

where ψ_{1sA} is a 1s atomic orbital centered on proton A and ψ_{1sB} is a 1s atomic orbital centered on proton B. The terms C_A and C_B are constants to be determined in the calculation. Since we are using a *linear combination of atomic orbitals* to obtain a final *molecular orbital,* this process is usually termed the LCAO, or LCAO-MO, method.

The energy of the molecule is

$$E = \frac{(\psi|\hat{H}|\psi)}{(\psi|\psi)} \tag{33.8}$$

where the symbol $(\psi|\hat{H}|\psi)$ denotes the integral $\int \psi^* \hat{H} \psi \, d\tau$ and $(\psi|\psi)$ denotes the integral $\int \psi^* \psi \, d\tau$; $\hat{H}$ is the Hamiltonian operator used in Equation (33.1). Rather than list all the terms in the final form of (33.8), we introduce the simplifying notation

$$\begin{aligned} H_{ij} &= (\psi_i|\hat{H}|\psi_j) \\ S_{ij} &= (\psi_i|\psi_j) \end{aligned} \tag{33.9}$$

Since we are using a normalized set of *basis* functions, we have that

$$H_{AA} = H_{BB} \quad H_{AB} = H_{BA} \quad S_{AA} = S_{BB} = 1 \quad S_{AB} = S_{BA}$$

and (33.8) for the energy becomes

$$E = \frac{C_A^2 H_{AA} + 2C_A C_B H_{AB} + C_B^2 H_{BB}}{C_A^2 + 2C_A C_B S_{AB} + C_B^2} \tag{33.10}$$

The terms C_A and C_B are to be varied; the conditions are

$$\frac{\partial E}{\partial C_A} = 0 = C_A(H_{AA} - E) + C_B(H_{AB} - S_{AB}E)$$

$$\frac{\partial E}{\partial C_B} = 0 = C_A(H_{AB} - S_{AB}E) + C_B(H_{BB} - E)$$
(33.11)

These equations represent a set of simultaneous linear homogeneous equations in the two unknowns C_A and C_B. If we were to try a straightforward substitution to solve the pair of equations, we should get the trivial result $C_A = C_B = 0$. The condition for obtaining a nontrivial solution is that the determinant of the coefficients vanishes. By imposing this condition on the equations, we get certain values of E, the eigenvalues of the determinant, for which nontrivial solutions can be obtained. This condition is just

$$\begin{vmatrix} (H_{AA} - E) & (H_{AB} - SE) \\ (H_{AB} - SE) & (H_{AA} - E) \end{vmatrix} = 0 \quad (33.12)$$

where we have used S for S_{AB} and set $H_{BB} = H_{AA}$. Multiplying through, we get

$$(H_{AA} - E)^2 - (H_{AB} - SE)^2 = 0$$
$$H_{AA} - E = \pm(H_{AB} - SE)$$

and the two eigenvalues of E are

$$E_+ = \frac{H_{AA} + H_{AB}}{1 + S} \quad (33.13a)$$

$$E_- = \frac{H_{AA} - H_{AB}}{1 - S} \quad (33.13b)$$

By substituting these eigenvalues in the original equations, one at a time, we can find the ratio C_A/C_B. This ratio is $+1$ in one instance and -1 in the other; thus $C_A/C_B = \pm 1$. The eigenfunctions are

$$\psi_+ = C_A(\psi_{1sA} + \psi_{1sB}) \quad (33.14a)$$
$$\psi_- = C_A(\psi_{1sA} - \psi_{1sB}) \quad (33.14b)$$

These two functions represent symmetric and antisymmetric functions, and in accord with the convention introduced in the previous section we could have written ψ_g for ψ_+ and ψ_u for ψ_-. The remaining constant C_A can be eliminated by applying the normalization conditions,

$$\int \psi_+^2 \, d\tau = 1 \qquad \int \psi_-^2 \, d\tau = 1$$

When this is done, we find that $C_A = (2 \pm 2S)^{-1/2}$. The two wave functions can be written as normalized functions in the form

$$\psi_+ = \frac{1}{\sqrt{2 + 2S}} (\psi_{1sA} + \psi_{1sB}) \quad (33.15a)$$

$$\psi_- = \frac{1}{\sqrt{2 - 2S}} (\psi_{1sA} - \psi_{1sB}) \quad (33.15b)$$

Now we must evaluate H_{AA}, H_{BB}, and S to complete the solution and find the energies of the molecule ion H_2^+.

SEC. 33.3 VARIATIONAL CALCULATION OF THE HYDROGEN MOLECULE ION

Using the Hamiltonian of (33.1), we can express H_{AA} as

$$H_{AA} = \int \psi_{1sA} \left(-\frac{h^2}{8\pi^2 m} \nabla^2 - \frac{e^2}{r_A} - \frac{e^2}{r_B} + \frac{e^2}{R_{AB}} \right) \psi_{1sA} \, d\tau \qquad (33.16)$$

Noting that part of the Hamiltonian is identical to the complete Hamiltonian for the hydrogen atom, we can simplify (33.16) by acknowledging that

$$-\frac{h^2}{8\pi^2 m} \nabla^2 \psi_{1sA} - \frac{e^2}{r_A} \psi_{1sA} = E_H \psi_{1sA} \qquad (33.17)$$

where E_H is the energy of the hydrogen atom in the ground 1s state. Then Equation (33.16) can be written

$$H_{AA} = \int \psi_{1sA} \left(E_H - \frac{e^2}{r_B} + \frac{e^2}{R_{AB}} \right) \psi_{1sA} \, d\tau$$

$$= E_H + J + \frac{e^2}{a_0 D} \qquad (33.18)$$

where

$$J = \int \psi_{1sA} \left(-\frac{e^2}{r_B} \right) \psi_{1sA} \, d\tau = \frac{e^2}{a_0} \left[-\frac{1}{D} + \exp(-2D) \left(1 + \frac{1}{D} \right) \right] \qquad (33.19)$$

and $D = R_{AB}/a_0$. The various integrals such as the one for J can be evaluated exactly, but this requires a transformation to elliptical coordinates.[2] Likewise we can find that

$$H_{AB} = SE_H + K + \frac{Se^2}{a_0 D} \qquad (33.20)$$

where

$$K = -\frac{e^2}{a_0} \exp(-D)(1 + D) \qquad (33.21)$$

and

$$S = \exp(-D) \left(1 + D + \frac{D^2}{3} \right) \qquad (33.22)$$

The final energies resulting from this calculation are

$$E_+ = E_H + \frac{e^2}{a_0 D} + \frac{J + K}{1 + S}$$

$$E_- = E_H + \frac{e^2}{a_0 D} + \frac{J - K}{1 - S} \qquad (33.23)$$

The integral J represents a Coulomb interaction between one of the protons and the electron in a 1s orbital centered about the other proton; J is termed a *Coulomb integral*. The integral K has both the ψ_{1sA} and ψ_{1sB} wave functions associated with it, and it is called the *exchange*, or *resonance*, *integral*. It reflects that we cannot associate the electron with either one or the other of the protons exclusively but must consider the electron associated with both protons simultaneously.

Using Equation (33.23), we can compute the energies of both states of the molecule ion H_2^+ as a function of R_{AB}, the result is shown in Figure 33.3. Note that the curve for E_+ has a minimum that indicates the formation of a chemical bond. The resulting molecular orbital is called a *bonding* orbital. This approximate method gives

---

[2] These integrals are evaluated in H. Eyring, J. Walter, and G. E. Kimball, *Quantum Chemistry* (New York: John Wiley & Sons, 1944), p. 196.

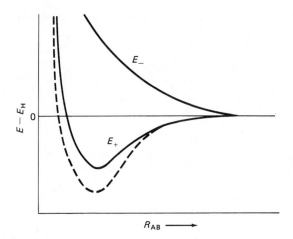

FIGURE 33.3 The energies of the states ψ_+ and ψ_- of the molecule ion H_2^+ as a function of the internuclear distance R_{AB}. The dashed line represents the exact energy curve.

1.77 eV for the dissociation energy of H_2^+ and an equilibrium internuclear distance of 1.32 Å, compared with the exact values 2.78 eV and 1.06 Å. Substantial improvement in the calculation can be made by using more complicated trial functions. By using an effective charge Z' for each ψ_{1s} orbital, one can get the exact value of 1.06 Å for the internuclear distance and a dissociation energy that differs from the experimental value by only 0.53 eV.[3] A trial function including 2p atomic orbitals and variable effective nuclear charges of the (unnormalized) form

$$\psi = \psi_{1sA}(Z') + \psi_{1sB}(Z') + C[\psi_{2pA}(Z'') + \psi_{2pB}(Z'')] \tag{33.24}$$

yields the correct internuclear distance and a dissociation energy that differs by 0.05 eV from the correct value.[4] Other trial functions have produced results that agree with the experimental values to within the experimental errors.

The curve labeled E_- on Figure 33.3 is the energy associated with the orbital ψ_-, which can be regarded as an excited state of the ion H_2^+. There is no minimum in this curve; hence no chemical bond can form in this state. At all values of R_{AB}, the ψ_- state is unstable with respect to the separated hydrogen atom plus an ion H^+; if a molecule could be prepared in the ψ_- state, it would immediately dissociate. The ψ_- molecular orbital is termed an *antibonding* orbital.

The orbital ψ_+ is constructed from the sum of two 1s atomic orbitals, which we could indicate by $1s_A + 1s_B$. The molecular orbital nomenclature for this state would represent it as $\sigma_g 1s$; the σ indicates zero angular momentum, the g indicates a symmetrical wave function, and the 1s indicates that the MO function was constructed from 1s atomic orbitals. Similarly, the ψ_- orbital, which we can indicate by $1s_A - 1s_B$, would be represented by the symbol $\sigma_u^* 1s$ where the asterisk indicates that the orbital is an antibonding one. Excited molecular orbitals can be constructed by linear combinations of higher atomic orbitals. Thus the combination $2s_A + 2s_B$ would be represented by $\sigma_g 2s$, and the combination $2p_{1_A} + 2p_{1_B}$ can give rise to the molecular orbital represented by $\pi_u 2p$. We shall find this nomenclature useful later in this chapter.

---

[3] B. N. Finkelstein and G. E. Horowitz, *Z. Phys.* **48** (1928):118.
[4] B. N. Dickinson, *J. Chem. Phys.* **1** (1933):317.

33.4 DIATOMIC MOLECULES AND MOLECULAR ORBITALS

The LCAO method can be extended to larger molecules by a procedure like that used to extend atomic orbitals to complicated polyelectronic atoms within the independent electron model. It might be helpful first to take a qualitative and pictorial view of the situation. If we consider the energy diagram for H_2^+, we notice that in the limit as $R \to \infty$, the molecule ion dissociates into a hydrogen atom and a hydrogen ion, $H + H^+$. In the limit as $R \to 0$, the system becomes an atomic ion He^+. Somewhere in between, a σ_g chemical bond is formed. If we take an anthropomorphic view of the electron, then we can say that a bond will form when the electron finds itself simultaneously associated with both protons. In a quantum mechanical sense, the characterizing feature in the bond formation in H_2^+ is that the individual 1s atomic orbitals lose their exclusive connection with one or the other of the protons and become orbitals that encompass both protons. This criterion is called *overlap*, and it is expressed by the *overlap integral*,

$$S_{AB} = \int \psi_A \psi_B \, d\tau \tag{33.25}$$

From Equation (33.23) we see that S_{AB} makes a positive contribution for E_+ and a negative contribution for E_-. It tends to decrease E_+, with the resulting formation of a chemical bond. When $S_{AB} > 0$, we get a bond; large values of S_{AB} characterize strong bonds. Conversely, when $S_{AB} < 0$, the result is an antibonding orbital. The overlap can be depicted as in Figure 33.4, where both cases are indicated. In this simple diagram we have not indicated any alteration of the shape of the atomic orbitals caused by the bonding.

A more quantitative illustration of ψ_+ and ψ_- for H_2^+ is shown in Figure 33.5. In connection with the discussion of the previous paragraph, note the continuous nature of the electron distribution (and hence the wave function) that encompasses both protons for ψ_+. For ψ_-, on the other hand, the electron distribution is zero at the midpoint, and the molecular orbitals have the form of distorted 1s atomic orbitals.

In the molecular orbital approach to diatomic homonuclear molecules, the individual atomic orbitals are brought together as shown in Figure 33.6, and their bonding or antibonding properties are determined from the existence or lack of overlap. Bonding can be inferred if there is a high density of electrons in the region between the two nuclei. The relative energies of the orbitals are often discussed in terms of the *united atom* energy correlation diagram, as shown in Figure 33.7. This figure is applicable to simple homonuclear diatomic molecules. The scale on the right side denotes the energies of the atomic orbitals in the two separate atoms in the limit

FIGURE 33.4 Overlapping of atomic orbitals to give (a) positive and (b) negative overlap integrals.

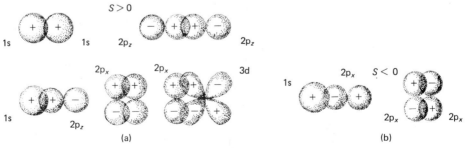

FIGURE 33.5 Charge distributions for the (a) $\psi_+ = \psi_{1sA} + \psi_{1sB}$ MO and the (b) $\psi_- = \psi_{1sA} - \psi_{1sB}$ MO. The upper curves indicate the distribution along the line joining the two protons, and the lower curves indicate the distribution contours in a plane that includes the two protons.

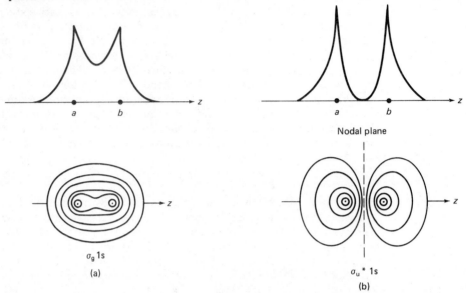

of $R_{AB} = \infty$. The left side of Figure 33.7 indicates the energies of the atomic orbitals in the *united atom*, formed when the two separate atoms are "squeezed" together so that $R_{AB} = 0$. An example is two hydrogen atoms squeezed together to form a helium atom. The region between the two extremes is viewed as representing intermediate values of R_{AB} in some complicated nonlinear manner. After the two extremes are arranged in their proper order, the g and u character of each of the states can be noted and also the component of angular momentum along the molecular, or z, axis. Intermediate energies are then determined by drawing lines connecting states that have the same g or u character and the same angular momentum component. The connecting lines must be drawn so that the *noncrossing* rule is not violated. This rule, which arises from symmetry considerations, states that no two lines having the same character may cross in the diagram. Thus a σ_g line cannot cross another σ_g, though it is permitted to cross a σ_u line. Bonding is associated with orbitals whose energies decrease as we pass from the separated atoms to the united atoms. Thus among the states that correlate with the 2p states in the separated atoms, we should expect the $\sigma_g 2p$ orbital to lead to bonding at relatively large values of R_{AB}.

How the *Aufbau Prinzip* applies to molecules can be seen by referring to Figure 33.8, which is a simplified version of the previous figure. For H_2, we place two electrons in the lowest $\sigma_g 1s$ state, one with spin up and the other with spin down (↑ ↓). The electronic configuration would be represented by $(\sigma_g 1s)^2$, indicating two electrons in $\sigma_g 1s$ molecular orbitals. For He_2, the next two electrons would be forced into $\sigma_u^* 1s$ orbitals. Now we have two electrons in bonding orbitals and two electrons in nonbonding orbitals. The net number of bonding orbitals is zero; hence no stable He_2 molecule is formed. The next two electrons for Li_2 go into the $\sigma_g 2s$ MO's (↑ ↓); since there is a net excess of bonding electrons, we expect to find the stable Li_2 molecule.

FIGURE 33.6 The formation of molecular orbitals from (a) 1s atomic orbitals, (b) $2p_z$ atomic orbitals, and (c) $2p_x$ atomic orbitals. Another pair of MO's of the same type as (c) can be formed from $2p_y$ atomic orbitals.

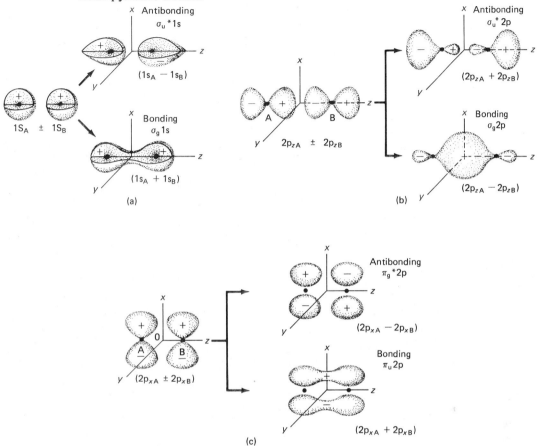

The existence of the Li_2 molecule can be inferred from experimental Maxwell-Boltzmann distribution curves (Chapter 21). Its presence is also indicated spectroscopically; the bond strength is 1.03 eV, and the bond length is 2.67 Å. The electronic configuration of Li_2 would be $(\sigma_g 1s)^2(\sigma_u^* 1s)^2(\sigma_g 2s)^2$, or $KK(\sigma_g 2s)^2$, the KK indicating filled K electron shells ($n = 1$). We can continue up through the remaining diatomics in the same way, and the results are summarized in Table 33.1. The symbols listed under the ground states in the table are analogous to the term symbols used in atomic spectroscopy, and they show the total spin quantum number $S = \Sigma s_i$ and angular momentum quantum number $\Lambda = \Sigma \lambda_i$. The superscript number preceding the capital Greek letter indicates the spin multiplicity and has the value $2S + 1$, just as in the atomic term symbols. To differentiate molecular symbols from atomic symbols, Greek letters are used for the orbital angular momentum; the quantum number λ_i corresponds to the l_i of atomic spectroscopy. Instead of the letters S, P, D, F, ..., the capital Greek letters Σ, Π, Δ, Φ, ... are used. In addition, the subscripts g and u are used to indicate whether the wave functions change signs upon

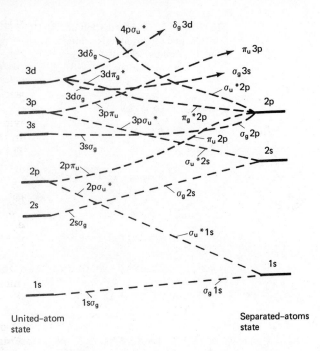

FIGURE 33.7 Correlation diagram for homonuclear diatomic molecular orbitals based on the united atom model. Note that the scale is highly nonlinear.

inversion. (This applies only to homonuclear diatomics.) Finally, a superscript + or − is appended to the Σ states to indicate whether the wave function does or does not change sign when it is reflected across a plane that includes the two nuclei.

There are two important pieces of experimental evidence that indicate the general validity of the model we have just constructed. Firstly there is the existence or

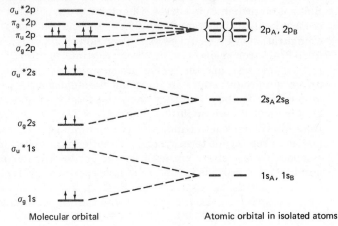

FIGURE 33.8 The *Aufbau Prinzip* for homonuclear diatomic molecules. The arrows represent the 14 electrons of the N_2 molecule. Note that the order of the orbitals $\sigma_g 2p$ through $\sigma_u^* 2p$ is highly variable, and the order shown applies only to O_2 and F_2. For other diatomics, the order of the $\sigma_g 2p$ and $\pi_u 2p$ orbitals should be interchanged, as you can see from the electronic configurations given in Table 33.1.

lack thereof of the diatomic molecules themselves and the strengths of their bonds. From Table 33.1, we see that only He$_2$ and Be$_2$ should not have stable ground states, and in fact, these are the only two that do not have stable ground states. Note also that N$_2$ has the largest number of bonding electrons and thus should form the strongest bond. That it does can be seen from its high dissociation energy, the largest of all entries in the table. The dissociation energy increases roughly with the number of bonding electrons.

The second piece of experimental evidence is more subtle. In Figure 33.8 we have noted the placement of the electrons in the orbitals for N$_2$. The next molecule is O$_2$, which has two additional electrons. These two electrons go into the π_g^*2p orbitals, but there are two of them. Do both electrons go into the same orbital with opposed spins (↑ ↓), or does one go into each of the orbitals with the possibility of parallel spins (↑ ↑)? Hund's rule of atomic spectroscopy applies here also. The lowest state is the one with the highest multiplicity; thus the electrons have their spins parallel and the ground state is a triplet state. This is the fact, as shown from spectroscopic measurements. In addition, O$_2$ is known to be paramagnetic, a condition that requires unpaired electrons. This model quantifies the concept of the electron-pair bond postulated by G. N. Lewis. The number of electron pairs is equal to one-half the number of bonding electrons. The N$_2$ molecule, which we represent by a triple bond N≡N, has six bonding electrons.

For heteronuclear molecules, the situation is more complex. The symbols g and u are now meaningless, since there is no possibility of symmetry on inversion. The cylindrical symmetry about the line joining the two nuclei is, however, retained, and the quantum number λ for the orbital angular momentum is still a good quantum number. We can still use the LCAO approach, but now the wave function must be taken in the form

$$\psi = \psi_A + \xi\psi_B \tag{33.26}$$

where the value of ξ is no longer ± 1 as it was for the homonuclear diatomics. This reflects that the electron is more strongly associated with one of the nuclei than the

TABLE 33.1 Ground states of homonuclear diatomic molecules.

Molecule	Electronic configuration	Ground state	Net excess of bonding electrons	Dissociation energy (eV)	Bond length (Å)
H$_2^+$	$(\sigma_g 1s)$	$^2\Sigma_g^+$	1	2.648	1.06
H$_2$	$(\sigma_g 1s)^2$	$^1\Sigma_g^+$	2	4.476	0.74
He$_2^+$	$(\sigma_g 1s)^2(\sigma_u^* 1s)$	$^2\Sigma_u^+$	1	(3.1)	1.08
He$_2$	$(\sigma_g 1s)^2(\sigma_u^* 1s)^2$ (=KK)	$^1\Sigma_g^+$	0	0	.....
Li$_2$	KK$(\sigma_g 2s)^2$	$^1\Sigma_g^+$	2	1.03	2.67
Be$_2$	KK$(\sigma_g 2s)^2(\sigma_u^* 2s)^2$	$^1\Sigma_g^+$	0	.....	.....
B$_2$	KK$(\sigma_g 2s)^2(\sigma_u^* 2s)^2(\pi_u 2p)^2$	$^3\Sigma_g^-$	2	3.0 ± 0.5	1.59
C$_2$	KK$(\sigma_g 2s)^2(\sigma_u^* 2s)^2(\pi_u 2p)^3(\sigma_g 2p)$	$^3\Pi_u$	4	6.2	1.31
N$_2^+$	KK$(\sigma_g 2s)^2(\sigma_u^* 2s)^2(\pi_u 2p)^4(\sigma_g 2p)$	$^2\Sigma_g^+$	5	8.73	1.12
N$_2$	KK$(\sigma_g 2s)^2(\sigma_u^* 2s)^2(\pi_u 2p)^4(\sigma_g 2p)^2$	$^1\Sigma_g^+$	6	9.756	1.09
O$_2^+$	KK$(\sigma_g 2s)^2(\sigma_u^* 2s)^2(\pi_u 2p)^4(\sigma_g 2p)^2(\pi_g^* 2p)$	$^2\Pi_g$	5	6.48	1.12
O$_2$	KK$(\sigma_g 2s)^2(\sigma_u^* 2s)^2(\pi_u 2p)^4(\sigma_g 2p)^2(\pi_g^* 2p)^2$	$^3\Sigma_g^-$	4	5.080	1.21
F$_2$	KK$(\sigma_g 2s)^2(\sigma_u^* 2s)^2(\pi_u 2p)^4(\sigma_g 2p)^2(\pi_g^* 2p)^4$	$^1\Sigma_g^+$	2	1.6 ± 0.4	1.44
Ne$_2$	KK$(\sigma_g 2s)^2(\sigma_u^* 2s)^2(\pi_u 2p)^4(\sigma_g 2p)^2(\pi_g^* 2p)^4(\sigma_u^* 2p)^2$	$^1\Sigma_g^+$	0	0	.....

FIGURE 33.9 Correlation diagram for heteronuclear diatomic molecules. The successive MO's of the same type are numbered in their order of increasing energy.

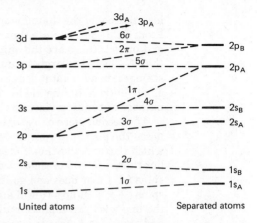

other. Experimentally this extra charge distribution about one of the nuclei is reflected in the permanent dipole moments associated with heteronuclear molecules, as discussed in Chapter 38. The correlation diagram must now take into account the different energies of the levels of the separated atoms, as shown in Figure 33.9. The united atom model is very useful for heteronuclear diatomics, particularly the hydrides.

33.5 THE HYDROGEN MOLECULE AND VALENCE BONDS

In the molecular orbital approach to chemical bonding, the nuclei (plus the inner-shell electrons) are brought together and the valence electrons are then allotted to the molecular orbitals in accord with the molecular version of the *Aufbau Prinzip*. The nuclei and the inner shells first interact to form the molecular orbitals, and then the valence electrons are added to the system. There is another approach called the *valence bond* method, in which the atoms are brought together and allowed to interact. We shall examine this approach with particular reference to the hydrogen molecule H_2.

The Hamiltonian for the hydrogen molecule can be written

$$\hat{H} = -\frac{h^2}{8\pi^2 m}(\nabla_1^2 + \nabla_2^2) - e^2\left(\frac{1}{r_{A1}} + \frac{1}{r_{B1}} + \frac{1}{r_{A2}} + \frac{1}{r_{B2}} - \frac{1}{r_{12}} - \frac{1}{R_{AB}}\right) \quad (33.27)$$

where ∇_1 refers to the first electron and ∇_2 refers to the second electron; the various attractive and repulsive electrostatic potential-energy terms are indicated in the coordinate diagram of Figure 33.10. The Schrödinger equation for the system can be written as

$$\nabla_1^2 \psi + \nabla_2^2 \psi + \frac{8\pi^2 m}{h^2}\left(E + \frac{e^2}{r_{A1}} + \frac{e^2}{r_{B1}} + \frac{e^2}{r_{A2}} + \frac{e^2}{r_{B2}} - \frac{e^2}{r_{12}} - \frac{e^2}{R_{AB}}\right)\psi = 0 \quad (33.28)$$

Notice that if we omit the $1/r_{12}$ term, we can write an approximate Hamiltonian in the form

$$\hat{H}_0 = \hat{h}_0(1) + \hat{h}_0(2) - \frac{1}{R_{AB}} \quad (33.29)$$

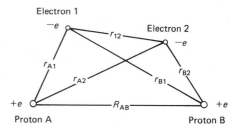

FIGURE 33.10 Coordinate system for the hydrogen molecule H_2.

where $\hat{h}_0$ is the Hamiltonian for the ion H_2^+. (Note that the electronic charge e has been omitted in (33.29); this corresponds to writing the Hamiltonian in atomic units. The $1/R_{AB}$ term contributes only a constant energy to the system.) This is analogous to the observation we made for helium in the first five equations of Chapter 31, and it suggests a similar approach. Just as we took the wave functions for the helium atom to be a product function of two hydrogen atom wave functions, we take the wave function of the H_2 molecule to be a product function of two H_2^+ wave functions in the form

$$\psi = (1s\sigma_g)_1(1s\sigma_g)_2(\alpha_1\beta_2 - \beta_1\alpha_2) \tag{33.30}$$

where we have included the spin function for completeness. Since the space part of the function is symmetric, the spin part must be antisymmetric. If we write out this function in terms of the atomic orbitals and omit the spin function (since it does not affect the calculated energy), we get

$$\psi = [1s_A(1) + 1s_B(1)][1s_A(2) + 1s_B(2)] \tag{33.31}$$

This function is an eigenfunction of the approximate Hamiltonian of (33.29), that is, the exact Hamiltonian with the $1/r_{12}$ term omitted. The energy is

$$E = \frac{\int \psi \hat{H} \psi \, d\tau}{\int \psi^2 \, d\tau}$$

$$= 2E'(R_{AB}) + \int \psi^2 \left(\frac{1}{r_{12}}\right) d\tau - \frac{1}{R_{AB}} \tag{33.32}$$

where $E'(R_{AB})$ is the energy of the ion H_2^+ at the internuclear distance R_{AB}. The equilibrium internuclear distance and the energy is found by minimizing the energy with respect to R_{AB}. This can be done in elliptic coordinates. The resulting value for the energy is 2.681 eV and for the equilibrium internuclear distance is 0.850 Å. The agreement with the experimental values 4.75 eV and 0.740 Å is not particularly good.

Suppose we write out the wave function of Equation (33.31) by multiplying all the terms, to get

$$\psi = 1s_A(1)1s_B(2) + 1s_A(2)1s_B(1) + 1s_A(1)1s_A(2) + 1s_B(1)1s_B(2) \tag{33.33}$$

There is something definitely wrong with this function; it goes contrary to our expectations, based on an intuitive feeling for the chemistry of the H_2 molecule. The first two terms seem perfectly all right; they correspond to situations in which the first electron is associated with one of the protons while the second electron is associated with the other proton. The last two terms, however, associate both electrons with either one or the other of the protons. These two terms correspond to writing ionic structures for the hydrogen molecule in the form $(H_A^+ H_B^-)$ and $(H_A^- H_B^+)$. The MO function of (33.33) thus predicts that half of the molecules dissociate into H^+ and H^- while the other half dissociate into H atoms. This is contrary to experiment, since H_2 always dissociates into H atoms exclusively. We can remove this objection by eliminating the last two "ionic" terms and retaining only the first two "covalent terms,"

$$\psi = 1s_A(1)1s_B(2) + 1s_A(2)1s_B(1) \tag{33.34}$$

This is the wave function Heitler and London first suggested in their original treatment of the H_2 molecule in 1927 in the early days of quantum mechanics.[5] The wave function of (33.34) is known as a *Heitler-London* wave function. This function leads to the energy

$$E = \frac{J + K}{1 + S^2} \tag{33.35}$$

where
$$J = \int 1s_A(1)1s_B(2) \hat{H} 1s_A(1)1s_B(2) \tag{33.36}$$
$$K = \int 1s_A(1)1s_B(2) \hat{H} 1s_A(2)1s_B(1) \tag{33.37}$$

and S is the overlap integral previously discussed in connection with the ion H_2^+. The integral J is called the *Coulomb* integral and gives the energy we should have got if we had not allowed the electrons to exchange nuclei. It is the energy the molecule would have if we had written the wave function simply with one term—

$$\psi = 1s_A(1)1s_B(2)$$

The integral K allows for the possibility of electron exchange between the two nuclei, and hence it is called the *exchange* integral. With this function, the dissociation energy of H_2 is found to be 3.140 eV and the equilibrium internuclear distance is 0.869 Å. The figure for the energy agrees better with the experimental values than the first determination did.

It is sometimes claimed that this Heitler-London, or *valence-bond,* method is superior to the simple MO treatment, since the energy is closer to the experimental energy. Actually both are approximations to the true state, and while the HL treatment gives better energies, it is still wide of the experimental mark. The key to this valence-bond approach is the disregard of the ionic structures, and this marks a basic difference between the approaches. The MO method includes these ionic terms but weights them equally with the covalent terms. The valence-bond method, on the other hand, simply omits these terms and therefore tends to underestimate rather than overestimate the contribution of the ionic terms. From the foregoing, it seems as though a happy medium could be struck by using a function of the form

$$\psi = 1s_A(1)1s_B(2) + 1s_A(2)1s_B(1) + \xi[1s_A(1)1s_A(2) + 1s_B(1)1s_B(2)] \tag{33.38}$$

where ξ is another variation term. This results in a dissociation energy of 3.229 eV, which is somewhat better. This approach can be termed a valence-bond calculation with the inclusion of ionic terms, or an MO calculation with the partial reduction of the ionic terms. Other treatments using more complicated functions have produced results that agree with the experimental values.[6] The atomic orbitals form a convenient basis set, and many atomic orbitals are generally needed to represent a realistic molecular orbital.

[5] H. Heitler and F. London, *Z. Physik* 44 (1927):455.
[6] See, for example, Pauling and Wilson, *Quantum Mechanics*, p. 349.

33.6 ELECTRONEGATIVITIES

For a homonuclear molecule like H_2, the bond can be described as a strictly covalent bond. There is no formal charge transfer between the two hydrogen molecules; the electron charge is equally distributed about both nuclei. On the other hand, a polar molecule such as HCl has an unequal charge distribution about the two nuclei that is reflected in the dipole moment of the molecule; charge flows from the hydrogen atom to the chlorine atom. We say that chlorine is more electron-attracting than hydrogen is. This qualitative power of an atom to attract electrons was given the name *electronegativity*. As one goes through the series HF, HCl, HBr, HI, one notes that the polarity of the molecules decreases. It can be said that F is more electronegative than Cl, and so on. This qualitative concept of electronegativity was placed on a quantitative scale by Pauling, who used bond-strength arguments.

In Equations (33.26) and (33.38), we saw that we could introduce a term ξ, which measures the unequal charge distributions about the two nuclei in heteronuclear molecules. The larger the value of ξ, the greater the ionic character of the bond. Let's adopt the valence-bond approach for the moment and examine the energy of a molecule for $\xi = 0$. This implies a purely covalent bond with no ionic character; we shall call the calculated energy E_C. Following this initial calculation, we then set ξ equal to some nonzero value and recalculate the energy. If this new energy, which we can call E_{CI}, is more negative than E_C (that is, has greater bond strength or dissociation energy), then the ionic contribution provides additional stabilization for the bond; we say that the bond has some ionic character and that the degree of this ionic character is measured by ξ. A scale of electronegativity could be based on values of ξ, but since it is difficult to evaluate, Pauling used the empirical bond strengths.

He noted that since a covalent bond in the heteronuclear molecule A—B should be similar to the covalent bonds in A—A and B—B, a purely covalent A—B bond would be expected to have a bond strength midway between the two. If this were true, then the extra bond strength over and above the purely covalent bond strength would be given by

$$\Delta' = D_{A-B} - [(D_{A-A})(D_{B-B})]^{1/2} \qquad (33.39)$$

where D is the bond energy and the last term is the geometric mean of the bond energies of the two homonuclears. The value Δ' is never negative, and it provides a measure of the extra stabilization of the A—B bond arising from the ionic terms. The

---

LINUS CARL PAULING (born 1901), American chemist, did his earliest work in crystal structure determinations, using X-ray diffraction. The early years of his career coincided with the development of quantum mechanics, and his interest in structural chemistry led him to a variety of quantum mechanical investigations concerned with the solid and nonsolid states of matter. These included bond orbitals and directed valence, hybrid bond orbitals and magnetic properties of matter, partial ionic character, bond energies and electronegativities, and the correlation of structural dimensions with electronic configurations. After the war, his interests turned partly to biochemistry, and Pauling discovered the cause of sickle-cell anemia. He received the Nobel Prize in chemistry in 1954 for his research into the nature of the chemical bond and the structure of complex molecules. In the late 1950s and early 1960s, he was one of the most vocal opponents of atomic bomb testing, and received the Nobel Peace Prize in 1963 for his efforts on behalf of the nuclear test ban treaty, thereby becoming the only person to win two individual Nobel awards.

TABLE 33.2 The electronegativity scale.

H 2·1						
Li 1·0	Be 1·5	B 2·0	C 2·5	N 3·0	O 3·5	F 4·0
Na 0·9	Mg 1·2	Al 1·5	Si 1·8	P 2·1	S 2·5	Cl 3·0
K 0·8			Ge 1·8	As 2·0	Se 2·4	Br 2·8
Rb 0·8						I 2·5
Cs 0·7						

term Δ' increases with increasing electronegativity difference between the atoms A and B. Pauling quantified this scale by fitting the available data to the expression

$$\Delta'_{A-B} = 30(x_A - x_B)^2 \tag{33.40}$$

where Δ' is in kilocalorie units and $x_A - x_B$ is the electronegativity difference between the atoms A and B. The factor 30 was included to place the scale in a convenient numerical range.[7] The electronegativity values of the elements derived in this manner are listed in Table 33.2. The values in this table show the expected increase in electronegativity from left to right along any row and also up the columns. Since the electronegativity difference is a measure of the degree of ionic character of a bond, or the amount of charge transferred from one atom to the other, the polarity or dipole moment of molecules should increase monotonically with increasing electronegativity difference. For the hydrogen halides, the respective dipole moments of HI, HBr, HCl, and HF are 0.38, 0.78, 1.03, and 1.91 debyes. (See Chapter 38 for a definition of the debye (D) and a discussion of dipole moments.) For this series of molecules, the dipole moments are almost equal to the electronegativity differences. Numerous expressions have been devised to express the percentage of ionic character of a bond in terms of the electronegativities. One of these has the form

$$\% \text{ ionic character} = 16|x_A - x_B| + 3.5(x_A - x_B)^2 \tag{33.41}$$

33.7 DIRECTED VALENCE AND POLYATOMIC MOLECULES

The complexity of theoretical calculations increases dramatically as the number of nuclei and electrons in molecules increases. Suppose we first consider the water molecule and on a simple basis try to explain why it has the formula H_2O instead of HO or H_3O. Then, having done that, let us see whether we can explain why the water molecule is bent instead of linear. According to the Heitler-London model, bonds occur when orbitals are formed from unpaired electrons. The electron configuration of an oxygen atom in the ground state is $1s^2 2s^2 2p^4$; the atom is in a ^3P state. The p_z orbital is filled with a pair of electrons, and the p_x and p_y orbitals have one unpaired

---

[7] A different (and simpler) basis for an electronegativity scale has been proposed by R. S. Mulliken (*J. Chem. Phys.* 2 (1934):782). On the Mulliken scale the electronegativity of an atom A is given by $M_A = \frac{1}{2}(I_A + E_A)$, where I_A is the ionization potential or the energy of the reaction $A \rightarrow A^+ + e^-$, and E_A is the electron affinity or the energy of the reaction $e^- + A \rightarrow A^-$. The relation between the Mulliken and Pauling scales is $M_A - M_B = 2.78(x_A - x_B)$.

electron each. Each of these unpaired electrons can form a bond with the unpaired 1s electron of the hydrogen atoms. Thus two and no more than two bonds can form. In the valence-bond language, we could write the two resulting orbitals as

$$\psi_I = \phi_{2p_x}(O)\phi_{1s}(H_a) + \phi_{1s}(H_a)\phi_{2p_x}(O)$$
$$\psi_{II} = \phi_{2p_y}(O)\phi_{1s}(H_b) + \phi_{1s}(H_b)\phi_{2p_y}(O)$$
(33.42)

or in a simpler shorthand notation,

$$\psi_I = 1s(H_a) + 2p_x(O)$$
$$\psi_{II} = 1s(H_b) + 2p_y(O)$$
(33.43)

to indicate that the bond is formed between the $2p_x$ or $2p_y$ orbitals of the oxygen and the 1s orbital of the hydrogen.

Bonds form in such a way that the stability of the bond is maximized. A criterion for this is a high electron-charge density in the region between the two nuclei; that is to say, the greater the overlap, the greater the stability of the bond. Mathematically, a high overlap is associated with a high value of the exchange integral, which in turn is a main factor in stabilizing bonds. The 2p orbitals of oxygen are shown along with the 1s orbitals of the hydrogen atoms in Figure 33.11a. The 2p orbitals consist of three mutually perpendicular lobes, as shown in the figure. The p_z orbital is filled and therefore does not participate in the bonding. We can see that the overlap integral is positive when localized molecular orbitals form in the manner indicated. When the bond is formed between the 2p and the 1s orbitals, the electron-charge density tends to concentrate in the region between the H and O atoms, as indicated in Figure 33.11b. Note that the appearance of the $2p_z$ orbital is unchanged, since it is unaffected. This simple picture predicts that the H–O–H bond angle should be 90°, and that the two O–H bonds should be equivalent. The second of these conclusions is correct, as the two bonds are known to be equivalent. On the other hand, the experimental value for the bond angle is 104.5°. Thus while this model does predict a nonlinear molecule, the degree of nonlinearity is incorrectly predicted. Part of the discrepancy may be attributed to electron-electron repulsion and proton-proton repulsion. We shall find a better explanation in the next section.

A similar approach can be taken for the ammonia molecule, NH_3. The ground-state configuration of the nitrogen atom is $1s^2 2s^2 2p^3 (^4S)$. Since there are three unpaired electrons, three bonds are expected to form, and this is found to be the case. Now all three of the 2p orbitals are available for bonding. Since the $2p_x$, $2p_y$, and $2p_z$ orbitals are mutually orthogonal, we should expect the NH_3 molecule to have a pyramidal structure with the H–N–H bond angle equal to 90°. Although the molecule is

FIGURE 33.11 Formation of localized bonds in H_2O.

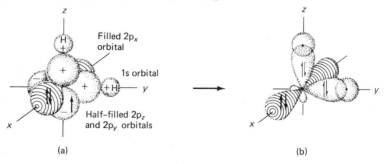

pyramidal, the true bond angle is 108°. As with H_2O, while the general configuration is correctly predicted, the actual angle deviates substantially from what has been predicted. Part of this discrepancy may be explained by the hybridization of bonds.

33.8 HYBRIDIZATION; THE TETRAHEDRAL BOND OF CARBON

If we try to treat the simple organic aliphatic compounds such as CH_4 along the lines of the previous section, we run into difficulties. The ground-state configuration of carbon is $1s^2 2s^2 2p^2 (^3P)$; only two electron bond pairs are predicted, since there are only two unpaired electrons available for bonding. The chemical evidence, however, is quite emphatic in insisting on a valency of 4, as we can see from the very stable compounds such as CH_4 and CCl_4 that exist. We can get a total of four unpaired electrons by promoting one of the 2s electrons to a 2p orbital, with the resulting configuration $1s^2 2s^1 2p^3 (^5S)$. This provides the four electrons required for the four bonds, but the ^5S state lies 4.2 eV above the ground-state configuration. What happens is that the energy gained when the four bonds are formed more than compensates for the amount 4.2 eV that is expended in promoting the 2s electron to the 2p orbital.[8]

This solves the problem of the number of bonds formed, but this promotion raises another problem. We should expect three bonds to form with the p orbitals (p_x, p_y, p_z) and the fourth bond to form with the 2s unpaired electron. Since the overlap with the 2s bond is less than the overlap with the 2p bonds, a situation would result in which one of the bonds is weaker than the remaining three bonds. The chemical evidence is emphatic, however, in insisting that all four bonds have the same strength. Further, the absence of a permanent dipole moment in molecules such as CH_4 and CCl_4 requires the molecule to be symmetrical. The way out of this dilemma is by a process known as *hybridization*. Instead of having individual 2s, $2p_x$, $2p_y$, and $2p_z$ bonds, a *hybrid bond* of the form $\phi_1 = \frac{1}{2}(\psi_{2s} + \psi_{2p_x} + \psi_{2p_y} + \psi_{2p_z})$ is used. Four such valence orbitals can be formed, one for each bond:

$$\begin{aligned}
\phi_1 &= \tfrac{1}{2}(2s + 2p_x + 2p_y + 2p_z) \\
\phi_2 &= \tfrac{1}{2}(2s + 2p_x - 2p_y - 2p_z) \\
\phi_3 &= \tfrac{1}{2}(2s - 2p_x + 2p_y - 2p_z) \\
\phi_4 &= \tfrac{1}{2}(2s - 2p_x - 2p_y + 2p_z)
\end{aligned} \qquad (33.44)$$

When the directional character of these bonds is determined by adding the functional forms of the 2s and 2p atomic orbitals, the hybrid orbitals of Equation (33.44) are found to be directed towards the corners of a tetrahedron, as shown in Figure 33.12a. The H–C–H bond makes the tetrahedral angle of 109° 28′. Hundreds of R_1–C–R_2 bond angles have been determined by electron diffraction, X-ray diffraction, microwave spectroscopy, and other techniques, and almost without exception the experimental angles are within 2° of the tetrahedral angle for compounds of the form $R_1 R_2 R_3 R_4 C$. This result is of fundamental importance, since it links one of the most basic facts of chemistry with the atomic orbitals of the hydrogen atom in a simple way. This bond is known as the *tetrahedral bond*, or the sp^3 *hybrid bond*. The angular dependence of the bond is shown in Figure 33.12b.

The discrepancy for NH_3 (noted in the previous section) can be explained in terms of hybrid bonds. The bond angles in NH_3 are very close to being tetrahedral; thus we

---

[8] It would be "more correct" to say that the excitation is to a "valence state" composed of a mixture of states, of which ^5S is the lowest. The figure 4.2 eV is therefore a lower limit.

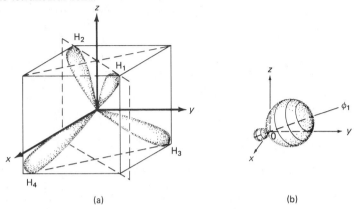

FIGURE 33.12 (a) The spatial orientation of the tetrahedral bond. (b) The angular dependence of the tetrahedral bond.

can postulate that sp³ hybrid bonds are used in NH_3. Only three of the four bonds are used, the fourth bond containing an electron pair that is not used in the formation of a bond.

There is no reason to stop with sp³ hybrids. Ethylene, $H_2C{=}CH_2$, is known to be planar. This can be taken into account with an sp² hybrid in which the 2s, $2p_x$, and $2p_y$ orbitals are mixed. Three *trigonal hybrid bonds* are formed, which make angles of 120° with each other. The fourth orbital is the p_z orbital, which is perpendicular to the plane of the three sp² bonds. The bonding in ethylene can be explained in terms of an sp² hybrid σ bond. The double bond is completed by the formation of a π bond with the $2p_z$ orbitals. Adding some s character to the two 2p orbitals in water can account for the deviation of the bond angle from 90°. The *digonal hybrid*, or sp, bond uses one 2s and one 2p orbital to produce a pair of bonds along a straight line in opposite directions. This bond is useful in explaining the structure of acetylene, $CH{\equiv}CH$.

The use of hybrid bonds is not restricted to organic chemistry. Boron trichloride, BCl_3, is known to be a planar molecule, and the bonding can be explained in terms of sp² hybridization. Mercuric chloride, $HgCl_2$, is known to be a linear molecule, and the bonding can be explained in terms of sp hybrid bonds. Inorganic compounds often involve d orbitals, and these can be mixed with the s and p orbitals. The molecule SF_6 has the form of a regular octahedron, with the fluorine atoms at the corners of the octahedron and the sulfur atom at the body center. A d²sp³ orbital, using $\frac{1}{3}$d, $\frac{1}{6}$s, and $\frac{1}{2}$p character, results in six bonds directed towards the corners of an octahedron.

33.9 NONLOCALIZED BONDS IN BENZENE

In the past two sections we noted that the concept of what we might term *localized* bonds was useful in explaining the geometry of numerous molecules. Also, the constancy of bond strengths such as the C–C bond over many aliphatic compounds is understandable from the similarity of the sp³ bond in different compounds. The same is true for the C=C sp² double bond, and the C≡C sp triple bond. We run into a problem if we try to extend that model to aromatic compounds, such as benzene. Suppose we consider the experimental evidence. It is clear from X-ray investigations, electron diffraction, spectroscopy, dipole moments and other studies that all the benzene atoms lie in a plane and all the carbons are equivalent. In addition, the carbon skeleton has the form of a regular hexagon; thus the C—C—C bond angle is 120°, and the hydrogen atoms are directed radially outward from the carbons. The bond angle immediately suggests sp^2 hybrid orbitals. In Figure 33.13a, the σ hybrids in the plane of the molecule are indicated. There are still six valence electrons left over, and these fit into the π orbitals shown in Figure 33.13b. We let the molecule be in the xy plane, in which case these six orbitals are p_z orbitals. In the valence bond scheme, there are now two obvious and distinct ways in which the π orbitals can be paired, as shown in Figure 33.13c. These two possibilities pose something of a quandary, since neither one can be better than the other; both must therefore appear in the final wave function. These are the two structures Kekulé proposed with brilliant insight back in 1865. In Kekulé's terms, the two structures oscillate rapidly between the two forms. In modern quantum mechanical language, the final wave function must be a superposition of both structures. Although the classical term *resonance* is still used to denote this situation, the term must be understood in its modern sense. In Kekulé's classical language, resonance means that both structures are independently observable. Our failure to differentiate between the two is not from any inherent property of the structural oscillations, but rather the experimental difficulties associated with such differentiation. If we could perform an experiment that is sufficiently rapid compared with the time of oscillations, then we could distinguish the two types of benzene. The modern quantum mechanical language precludes any such distinguishability no matter how fast the experiment can be performed; the structure of benzene is a superposition of more than one component structure.

We might try to think of the benzene bond as something between a single bond and a double bond. This sort of logic might work for the bond length in benzene, which is 1.39 Å, between the 1.54-Å bond length for a single bond and the 1.34 Å appropriate for the double bond. But the energy, which is more critical, does not lie between the energies of the single and double bonds. If we compare the heat of com-

FIGURE 33.13 The bonds in benzene. (a) The σ hybrid bonds. (b) The p_z atomic orbitals. (c) The two Kekulé structures.

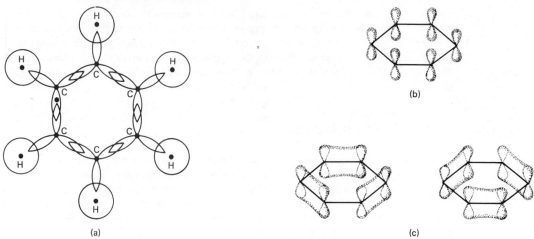

SEC. 33.9 NONLOCALIZED BONDS IN BENZENE

bustion of a benzene molecule with the heat of combustion of a hypothetical molecule containing three C–C single bonds, three C–C double bonds, and six C–H bonds, we find that the benzene molecule has an extra energy of some 160 kJ mol^{-1}. This extra energy is associated with the *delocalization* of the π bonds, and it is usually termed the *resonance energy*.

The two Kekulé structures for benzene, along with three other possible structures, are shown in Figure 33.14. The last three structures are known as *Dewar* structures. In the valence bond treatment, the wave function of the benzene molecule is taken as a linear combination of these five structures:

$$\psi_{\text{benz}} = c_1\psi_I + c_2\psi_{II} + c_3\psi_{III} + c_4\psi_{IV} + c_5\psi_V \tag{33.45}$$

The energy is

$$E = \frac{\int \psi H \psi \, d\tau}{\int \psi \psi \, d\tau} \tag{33.46}$$

When the variational treatment is carried out, it is found that $c_1 = c_2$ and that $c_3 = c_4 = c_5$, which seems necessary from symmetry arguments.[9] The ratio of the coefficients is $c_3/c_1 = 0.4341$. The energy is

$$E = Q + 2.606 J \tag{33.47}$$

where Q is the coulomb and J the exchange integral. Thus much of the excess energy is associated with the exchange integral. If only one of the Kekulé structures is used (that is, $c_1 = 1$, and $c_2 = c_3 = c_4 = c_5 = 0$ in (33.45)), then the calculated energy is

$$E_I = Q + 1.50 J \tag{33.48}$$

The energy of the benzene molecule is thus $1.106 J$ lower than the energy of a single Kekulé structure, and it is this extra energy that accounts for the unusual stability of the unsaturated benzene molecule. The value of J can be estimated from the experimental resonance energy of benzene.

Whereas the valence bond approach starts with the structures shown in Figure 33.14, the molecular orbital approach starts by writing a π molecular orbital for benzene as a linear combination of the six atomic p orbitals:

$$\psi = c_1\psi_1 + c_2\psi_2 + c_3\psi_3 + c_4\psi_4 + c_5\psi_5 + c_6\psi_6 \tag{33.49}$$

where the ψ_i are the p atomic orbitals centered on the six carbon atoms, as shown in Figure 33.13b. It is convenient to write this function as

$$\psi = \sum_{i=1}^{6} c_i \psi_i \tag{33.50}$$

FIGURE 33.14 The five benzene structures used in the valence bond treatment. The designations ψ_I and ψ_{II} indicate the Kekulé structures, and ψ_{III}, ψ_{IV}, and ψ_V are the Dewar structures.

[9] The details of this calculation can be found in C. A. Coulson, *Valence* (Oxford: Oxford University Press, 1952), chap. 9, and in Eyring, Walter, and Kimball, *Quantum Chemistry*, chap. 13.

in which case the energy to be varied in a variational treatment is

$$E = \frac{\int (\Sigma c_i \psi_i) \hat{H} (\Sigma c_i \psi_i) \, d\tau}{\int (\Sigma c_i \psi_i)^2 \, d\tau} \qquad (33.51)$$

where the summations run from $i = 1$ to $i = 6$. For convenience we use the notation

$$H_{ij} = \int \psi_i \hat{H} \psi_j \, d\tau \qquad (33.52)$$
$$S_{ij} = \int \psi_i \psi_j \, d\tau \qquad (33.53)$$

The coefficients c_i are found from the solution of the six simultaneous equations of the form

$$c_1(H_{11} - ES_{11}) + c_2(H_{12} - S_{12}) + c_3(H_{13} - S_{13})$$
$$+ c_4(H_{14} - S_{14}) + c_5(H_{15} - S_{15}) + c_6(H_{16} - S_{16}) = 0 \qquad (33.54)$$

Since the six equations form a set of linear homogeneous equations, the condition for a nontrivial solution is that the determinant of the coefficients, the *secular* determinant, must vanish. Before writing that determinant, we first note that if the atomic orbitals are normalized, $H_{ij} = H_{ji}$, $S_{ij} = S_{ji}$, and $S_{ii} = 1$. Even with this simplification the equation is difficult to solve, and several approximations are usually made. The overlap of all the orbitals is neglected; S_{ij} is taken as zero for $i \neq j$. The Coulomb integral H_{jj} is then the energy each π electron would have if it were constrained to remain on the carbon atom j. It is customary to denote H_{jj} by the symbol α. The term H_{ij} ($i \neq j$) is the *resonance integral* and is neglected for all nonadjacent carbon atoms. Thus $H_{ij} = 0$ for nonadjacent atoms; the value H_{ij} is set equal to β for adjacent atoms. With these approximations, the determinantal equation is

$$\begin{vmatrix} (\alpha - E) & \beta & 0 & 0 & 0 & \beta \\ \beta & (\alpha - E) & \beta & 0 & 0 & 0 \\ 0 & \beta & (\alpha - E) & \beta & 0 & 0 \\ 0 & 0 & \beta & (\alpha - E) & \beta & 0 \\ 0 & 0 & 0 & \beta & (\alpha - E) & \beta \\ \beta & 0 & 0 & 0 & \beta & (\alpha - E) \end{vmatrix} = 0 \qquad (33.55)$$

When the determinant is multiplied out, we get a sixth-degree equation whose six roots are

$$\begin{aligned} E_1 &= \alpha + 2\beta \\ E_2 &= E_3 = \alpha + \beta \quad \text{(doubly degenerate)} \\ E_4 &= E_5 = \alpha - \beta \quad \text{(doubly degenerate)} \\ E_6 &= \alpha - 2\beta \end{aligned} \qquad (33.56)$$

Since β is a negative quantity, the energies have been listed in increasing order, and E_1 is the lowest energy. It is a bonding orbital that can hold two electrons. The next two roots, E_2 and E_3, are also bonding orbitals and each can hold two electrons for a total of six bonding electrons. The remaining states correspond to nonbonding orbitals. The total energy of the benzene molecule is $2E_1 + 2E_2 + 2E_3$, or

$$E_{\text{benz}} = 6\alpha + 8\beta \qquad (33.57)$$

On the other hand, if we had taken the electrons to be localized in the regions between the two adjacent carbon atoms, the resulting energy would have been $E = 6(\alpha + \beta)$. There is thus an extra energy of amount 2β that is associated with the delocalization of the electrons. The experimental resonance energy of some 160 kJ

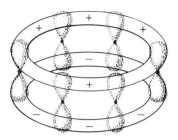

FIGURE 33.15 The lowest-energy nonlocalized π molecular orbital for benzene.

mol^{-1} yields a value of some 80 kJ mol^{-1} for the resonance, or *delocalization*, integral β.

Substituting the energy in the six equations represented by (33.54) results in $c_1 = c_2 = c_3 = c_4 = c_5 = c_6$; the normalization condition yields the wave function

$$\psi = (6)^{-1/2}(\psi_1 + \psi_2 + \psi_3 + \psi_4 + \psi_5 + \psi_6) \tag{33.58}$$

for the most stable molecular orbital of benzene. This orbital is shown in Figure 33.15. It has a toroidal configuration that passes through the six atomic p$_z$ orbitals. The approximations described are due to Hückel, and the resulting orbitals are sometimes known as *Hückel orbitals*. The extension of this treatment to other aromatics is discussed in the references listed in the Bibliography.

PROBLEMS

1. The equilibrium internuclear distance of the HCl molecule is 1.28 Å. Calculate the bond energy (per mole) that would result if the energy were entirely from positive and negative charges separated by the internuclear distance. How does that energy compare with the experimental bond energy of 432 kJ mol^{-1}? Explain the difference.
2. Write Lewis electron pair structures for the molecules H_2O, HF, C_2H_6, and NH_3.
3. Carry out the differentiation of (33.10) and show that it leads to (33.11).
4. For the H_2^+ molecule ion, substitute the energy eigenvalues given in Equation (33.13) in the secular equation to get the eigenfunctions. Show that the two eigenfunctions are orthogonal.
5. Carry out the first few steps of a variational calculation for the ion H_2^+ using the trial function given in (33.24):
   (a) Write the expression for the energy.
   (b) Differentiate the energy expression relative to the proper parameters.
   (c) Set up the secular equation.
6. Following the scheme indicated in Figure 33.4, convince yourself that there is zero overlap between a 1s and a 3d atomic orbital.
7. Table 33.1 indicates that the ground state of the B_2 molecule is a triplet state. Justify this on the basis of the energy-level diagram of Figure 33.7. Is the B_2 molecule paramagnetic?
8. Show that (33.32) follows from the wave function given in (33.31).
9. The molecules H_2, F_2, Cl_2, Br_2, and I_2 have bond energies 104.2, 36.6, 58.0, 46.1, and 36.1 kcal mol^{-1}. The hydrogen halides HF, HCl, HBr, and HI have bond energies of 134.6, 103.2, 87.5, and 71.4 kcal mol^{-1}. Arbitrarily set the electronegativity of F at 4.0.
   (a) Using Equations (33.39) and (33.40), set up a table of electronegativities of the halogens and hydrogen.
   (b) The interhalogen molecules Cl—F, Br—Cl, I—Cl, and I—Br have bond strengths 60.6, 52.3, 50.3, and 42.5 kcal mol^{-1}. How well do these data fit in with the electronegativity table you constructed in part (a)?
10. Using the empirical relation given in (33.41), determine the electronegativity difference that results in a bond having 50% ionic character. Determine the percentage ionic character in the bonds H—F, H—Cl, H—Br, and H—I.
11. Show that the sp^3 hybrid functions given in Equation (33.44) form a normalized and orthogonal set of functions.
12. The expressions for p$_x$, p$_y$, and p$_z$ are given in Equation (30.31); the expression for the angular dependence of the 2s orbital is also given in Chapter 30. Write out the expression for the sp^3 hybrid ϕ_1 and show that it is directed towards the corner of a tetrahedron.
13. The three functions of the sp^2 trigonal hybrid can be written as

$\phi_1 = A2s + B2p_x$
$\phi_2 = (3)^{-1/2}2s - (6)^{-1/2}2p_x + (2)^{-1/2}2p_y$
$\phi_3 = (3)^{-1/2}2s - (6)^{-1/2}2p_x - (2)^{-1/2}2p_y$

(a) Determine the values of the constants A and B so that the set of three functions form an orthonormal set.

(b) Convince yourself that the three hybrid functions are directed towards the corners of an equilateral triangle.

14. Show that the energies given in Equation (33.56) are solutions of the secular determinant for benzene given in (33.55).

15. The lowest energy orbital for benzene is given in (33.58). The least stable orbital that is associated with the energy $E = \alpha - 2\beta$ is

$$\psi = (6)^{-1/2}(\psi_1 - \psi_2 + \psi_3 - \psi_4 + \psi_5 - \psi_6)$$

Show that the two functions are normalized.

16. Calculate the energies of the ethylene molecule.

17. By following the procedure outlined in the text for the benzene molecule, calculate the energy levels of the butadiene molecule.

18. The heats of combustion of benzene, C_6H_6, cyclohexene, C_6H_{10}, and cyclohexane, C_6H_{12}, are 3301.6, 4072.4, and 3953.0 kJ mol^{-1}, and the products are $CO_2(g)$ and $H_2O(l)$. The standard heats of formation of $CO_2(g)$ and $H_2O(l)$ are -393.5 and -285.8 kJ mol^{-1}. Cyclohexene has one localized double C–C bond, while benzene has three delocalized double C–C bonds. By comparing the heats of hydrogenation of the localized bond in cyclohexene with the delocalized bonds in benzene, calculate the experimental resonance energy of the benzene molecule.

PART SIX

ATOMS AND MOLECULES IN CONDENSED STATES

CHAPTER THIRTY-FOUR

THE CRYSTALLINE STATE

34.1 CRYSTALS, SYMMETRY, AND BATHROOM TILE

The beauty and symmetry of crystals, particularly gems, have fascinated mankind since time immemorial, and the mystery of crystalline regularity perplexed generations of scientists. Crystals can be described by their regular internal structure. They are regular arrays of atoms (or molecules, or ions). Fluids are random arrangements of atoms and have no distinguishing directions in space; on a macroscopic scale their properties are the same in all directions. Fluids conduct heat and electricity equally well in all directions; light travels with the same velocity in all directions in a fluid. Crystals, on the other hand, conduct heat and electricity differently in different directions; most properties of crystals depend on direction.

The variation of properties with direction must display the same symmetry as the gross crystal. A hexagonal crystal may have different indices of refraction in different directions, but if the index of refraction is measured in a direction perpendicular to the axis of the hexagonal prism, then as the crystal is rotated about this axis there are six positions at which the index of refraction must be the same. As a crystal grows, it develops plane faces and sharp edges as molecules slowly deposit on the crystal; the faces with the largest areas are those upon which molecules deposit most slowly. Attack by a solvent reverses this deposition, and as the crystal dissolves it leaves *etch figures* on the faces, as shown in Figure 34.1. The etch figures are bounded by plane faces.

The symmetry of crystals is important. A symmetry element is an operation that leaves the crystal *invariant* (or brings the crystal into coincidence with itself) when it is performed on the crystal. Symmetry operations include rotations, inversions, reflections, translations, and also combinations of these, like screw rotations. Figure 34.2 shows the rotational symmetry elements of a geometrical cube. The fourfold and twofold axes are easy to visualize from the two-dimensional figure. The threefold axis is readily seen by taking an actual cube and rotating it about a body diagonal while looking down the diagonal.

Rotational symmetry is associated with an *axis of symmetry*. If the angle of rotation required for coincidence is $360/n$ degrees, n an integer, then the object is said to have an n-fold rotation axis, or axis of symmetry. The integer n is assigned its highest possible value, to indicate the highest degree of symmetry. A geometrical square has a twofold rotational axis, but the square is said to have a fourfold axis; the twofold axis is included in the fourfold axis. There are no limitations on rotational symmetry in geometry; a regular polygon of n sides has an n-fold rotation axis. Crystals are more restrictive. The restriction is brought about by the need to

FIGURE 34.1 (a) Appearance of etch figures on a cubic crystal. (b) Photomicrograph of etch patterns on a 100 plane of a silicon crystal. Silicon has the diamond structure and the square appearance of the etch pits reflects the fourfold symmetry of the 100 planes. (Photograph courtesy of Westinghouse Electric Corporation, Research and Development Laboratories, Pittsburgh, Pennsylvania.)

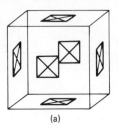
(a)

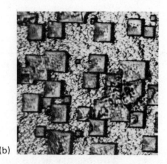
(b)

fill all space with geometrical units. The ancient Egyptians discovered that only certain shapes are permissible when laying tile on floors or walls. These include irregular polygons (onefold, or no, symmetry), rectangles (twofold symmetry), regular triangles (threefold symmetry), squares (fourfold symmetry), and regular hexagons (sixfold symmetry). These are the only possible geometrical shapes for bathroom tile, as shown in Figure 34.3. One cannot fill space with fivefold, sevenfold, or higher rotational symmetry. We shall prove this later in this chapter.

Symmetry planes divide an object in such a way that each side is the mirror image of the other. The symmetry planes of a cube are indicated in Figure 34.4. *Centers of symmetry* (or *centers of inversion*) can exist alone, but they are usually found in conjunction with symmetry planes. If a symmetry plane has an evenfold rotation axis perpendicular to it, then a center of symmetry results from the combination. An axis of *rotary inversion* consists in rotating an object by a definite amount, followed by inversion through a point on the axis. Note that in the preceding discussion we used the word *object* and not *crystal*. The above symmetry elements all have the property that at least one point in the object is unchanged; they are therefore called *point symmetry* elements. Translational symmetry, on the other hand, leaves no points in the system unchanged. Infinite lattices are required for translational symmetry.

The external shape, or the *morphology,* of a crystal may seem to vary from one specimen to another, but the internal structure must be the same. The external form may differ because of different crystallization conditions. Ordinary table salt, NaCl,

FIGURE 34.2 The rotational symmetry elements of a geometrical cube. (a) The 6 twofold axes. (b) The 4 threefold axes. (c) The 3 fourfold axes.

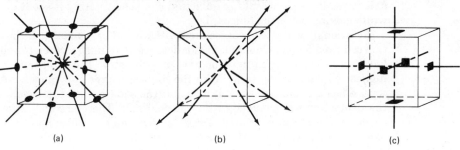
(a)   (b)   (c)

FIGURE 34.3 The five and only five types of rotational symmetry possible for space filling objects, illustrated for two dimensions. These are the only types of bathroom tile possible. (See what happens when you try to fill a sheet of paper with regular pentagons or other shapes.) (a) Onefold (or no) symmetry: Irregular triangles (or other shapes.) (b) Twofold symmetry: Rectangles. (c) Threefold symmetry: Regular triangles. (d) Fourfold symmetry: Squares. (e) Sixfold symmetry: Regular hexagons.

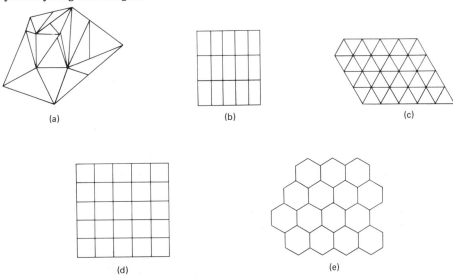

normally crystallizes in the shape of cubes, but if some urea is added to the crystallizing solution, then the crystals form octahedra. The cube and the regular octahedron are closely related in their symmetry properties. An octahedron can be obtained from a cube by removing eight corners, as shown in Figure 34.5a; the rotational symmetry axes of a regular octahedron are shown in Figure 34.5b. The cube and the octahedron both have cubic symmetry and belong to the *cubic system;* they both have the full symmetry of the cube. A regular tetrahedron also belongs to the cubic system, but has fewer symmetry elements than the cube or the regular octahedron. The variation of external gross form is vividly illustrated in Figure 34.6, which shows several ice crystals. It has been noted that of the countless snow flakes (ice crystals) that have fallen upon the earth since the first snow, no two were alike. They all, however, display hexagonal symmetry. The earliest quantitative studies of crystals were concerned with the external symmetry.

FIGURE 34.4 The nine symmetry planes (reflections) of the geometrical cube.

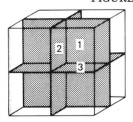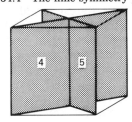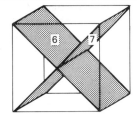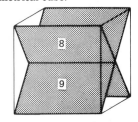

FIGURE 34.5 (a) The regular octahedron, constructed by cutting away the eight corners of a cube. (b) Rotational symmetry axes of the regular octahedron. There are 3 fourfold axes passing through opposite vertices; 4 threefold axes through opposite faces; and 6 twofold axes through opposite edges. Only one of each type is indicated. The number of rotation axes is the same as for the cube.

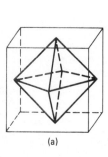

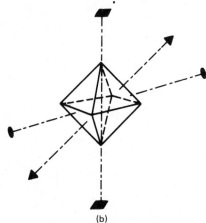

FIGURE 34.6 Snow crystals. [From V. Goldschmidt, *Atlas der Krystallformen*, vol. 3, part 2 (Heidelberg: Carl Winters Universitätbuchhandlung, 1916.)]

34.2 CLASSICAL CRYSTALLOGRAPHY

Modern crystallography dates from the successful application of X-ray diffraction to the elucidation of crystal structures in the early 1900s. This discovery was not made in a vacuum and was the crowning achievement in a branch of scientific study already several hundred years old. The earlier years had seen the discovery of the laws of symmetry and the enumeration of all the possible space lattices, point symmetry groups, and space symmetry groups. The symmetry properties of crystals were worked out by mathematicians. That there were only 230 so-called *crystallographic space groups,* or combinations of symmetry elements in regular arrays of points, was an outgrowth of the new field of mathematics known as *group theory*. Structural chemistry on the molecular level had made great strides. The tetrahedral bond of organic chemistry was fully accepted, and Werner had developed the structural chemistry of inorganic complexes. The discovery of X-rays and their application to structural studies had come at precisely the proper time; there was little more that could be done without a definitive experimental procedure for determining the positions of the individual atoms in crystals.

FIGURE 34.7 Interfacial angle between two faces of a crystal.

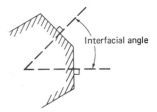

The earliest quantitative studies of crystals began in the late 1600s, and they were concerned with the interfacial angles of crystals. The interfacial angle can be measured with a contact *goniometer* or with a protractor if it is big enough. Figure 34.7 shows the interfacial angle of a crystal. Precise determinations of the interfacial angles of very small crystals can be made with the *reflection goniometer*. With this instrument, a narrow light beam is reflected perpendicularly from the relevant faces as the crystal is rotated about various directions in space. Measurements on well-grown specimens can yield interfacial angles agreeing to within two minutes of arc from one crystal to another. In 1669, Niels Stensen noted that the interfacial angles of different quartz crystals were always the same regardless of the gross appearance of the crystals. This observation was later extended to all crystals, and constitutes the *law of interfacial angles:*

> *For any crystalline species, the angles between the faces of a crystal are constants.*

The constancy of corresponding interfacial angles provides a rapid way of identifying unknown crystals, since the interfacial angles of a host of crystals have been tabulated in various handbooks of crystallography and mineralogy. Not until the end of the eighteenth century was this crystallographic law understood on a microscopic basis.

The packing of spheres in space had long intrigued numerous investigators. The structural arrangements of spheres that we call *simple cubic, cubic closest-packing,* and *hexagonal closest-packing* had been experimentally discovered by packing dried peas in containers. It had long been known that the cubic close-packed and hexagonal close-packed arrangement of spheres were the most economical in terms of the fraction of available space occupied by the spheres. In 1678, C. Huygens tried to explain the existence of cleavage planes in calcite, $CaCO_3$, by postulating a crystal composed of the regular packing of spheroidal bodies. In 1782, René-Just Haüy, the

---

CHRISTIAAN HUYGENS (1629–1695), Dutch scientist, was the scion of a prominent Dutch family, and his grandfather of the same name had served William the Silent as secretary. Like others of his period, Huygens devoted substantial efforts to the quadrature of various curves such as cycloids. He was a skilled lens grinder; his telescopes were perhaps the best of their era. Huygens discovered a satellite of Saturn, later named Titan. He was an early proponent of the wave theory of light. His principle that a wave front can be approximated by an infinite number of wave centers is known as *Huygens's principle*. He developed the pendulum as a regulator for clocks, and showed that if the pendulum could swing in a cycloidal path, then the period of the pendulum would be independent of the amplitude. He was one of the early workers who tried to develop an accurate clock to use on ships as a navigational tool, a problem of great concern to a seafaring nation such as Holland. Celestial navigation requires accurate timekeeping to ascertain longitude. Several of his attempts were tested on board ships.

> RENÉ-JUST HAÜY (1743–1822), French cleric and mineralogist, is regarded as the father of modern crystallography. His *Essai d'une Théorie sur la Structure des Cristaux* laid the foundation of the mathematical theory of crystal structures. During the Revolution he refused to take the oath required of all clergy and was arrested. He was eventually released and in 1793 became secretary of the Commission on Weights and Measures. Along with other members of the commission, he engaged in an unsuccessful attempt to obtain the release of Lavoisier. Haüy devoted most of his active life to the careful study of crystal specimens. In 1793, Haüy proposed six primary forms of crystals—parallelepiped, rhombic dodecahedron, hexagonal dipyramid, right hexagonal prism, octahedron, and tetrahedron. Division of these primary forms ultimately led to his *molécules intégrantes* (or *molécules constituantes*). He explained the development of faces on crystals by successive layers of these ultimate building blocks, each layer having one or more rows of blocks less than the preceding layer, as shown in Figure 34.8. The interfacial angles remain constant and are characteristic of the individual species. The decreasing area of successive layers was known as the *law of decrements,* and the assumption that these decrements were a small number of rows of molecules led Haüy's successors to the law of rational indices.

"father of crystallography," carried this idea one step further. The original calcite crystal was a parallelepiped, and when the crystal was cleaved, each fragment maintained the parallelepiped shape of the original crystal. Haüy argued that if this cleavage were to be repeated over and over, the crystal would eventually be reduced to a minute parallelepiped, the ultimate building block, or *molécule intégrante,* which could not be subdivided without destroying the chemical composition of the substance. He envisioned a crystal as built up from a succession of these ultimate building blocks, as shown in Figure 34.8. The development of crystal planes could then be explained by the succession of "stepped" surfaces, as indicated at the top of the figure, in much the same way that a pyramid can be built up of successively smaller layers of building blocks. The size of these steps corresponds to the size of the molecules and give the appearance of a continuous plane to the naked eye. This

FIGURE 34.8 Haüy's model of a crystal composed of small ultimate *molécules intégrantes,* or building blocks.

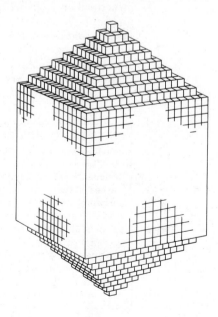

model of a crystal provided the framework for the second important law of crystallography, the *law of rational indices,* which Haüy first enunciated.

Suppose we have two crystals of identical form but of different size. Corresponding faces of the two crystals are equivalent, and it is convenient to have a scheme for indicating this equivalence in a quantitative manner. Let **a**, **b**, and **c** be three noncoplanar vectors fixed to a crystal, as shown in Figure 34.9a. The directions of the vectors and their relative lengths are determined by the symmetry properties of the crystal; this will be apparent shortly. The lengths of the vectors are a, b, and c. The axes are cut by the crystal plane ABC. The lengths of the intercepts can be expressed as OA/a, OB/b, and OC/c. The law of rational indices is stated in terms of the reciprocals of these lengths as:

> *For any crystal, a set of axes can be constructed such that all naturally occurring crystal faces have reciprocal intercepts that are proportional to small integers.*

That is to say

$$\frac{a}{OA} = h \qquad \frac{b}{OB} = k \qquad \frac{c}{OC} = l \tag{34.1}$$

where $h, k,$ and l are small integers. The use of the reciprocal intercepts hkl to define uniquely a set of parallel planes was first proposed by Miller in 1839, and the set of values of hkl is known as *Miller indices.* A plane parallel to any axis has its intercept at ∞; its Miller index is just $1/\infty$, or 0. The set of indices is usually placed in parentheses. For any face, it is only the ratio $h:k:l$ that is important, and the indices

FIGURE 34.9 (a) Crystal axes. (b) Miller indices of some planes in a cube. (c) Miller indices of some lattice planes.

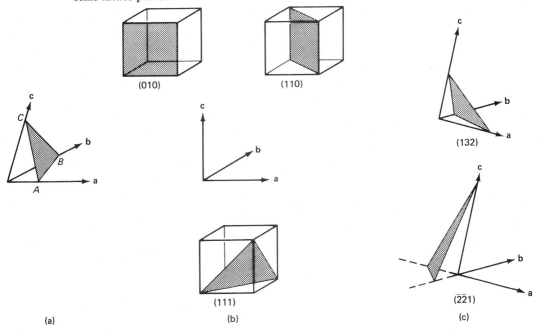

TABLE 34.1 The crystallographic axes of the seven crystal systems.

Name of crystal system	Dimensions fixed by the classification	Dimensions to be specified for each particular crystal[a]	Example
Cubic	$a = b = c$ $\alpha = \beta = \gamma = 90°$		Rock salt
Tetragonal	$a = b$ $\alpha = \beta = \gamma = 90°$	$c{:}a$	White tin
Orthorhombic	$\alpha = \beta = \gamma = 90°$	$a{:}b{:}c$	Rhombic sulfur
Monoclinic	$\alpha = \gamma = 90°$	$a{:}b{:}c$ β	Monoclinic sulfur
Hexagonal	$a = b$ $\alpha = \beta = 90°$ $\gamma = 120°$	$c{:}a$	Graphite; quartz
Rhombohedral	$a = b = c$ $\alpha = \beta = \gamma$	α	Calcite
Triclinic	None	$a{:}b{:}c$ α,β,γ	Potassium dichromate

[a] These merely specify the relative magnitudes of the axes. In addition, a scaling factor must be given for any particular crystal structure, say by specifying the length of any axis.

are always reduced to the lowest common factors. Thus the plane (642) would be written as (321). Planes with negative intercepts are indicated by a bar above the index with the negative intercept, as for example ($\bar{1}11$). Miller indices for a number of planes are shown in Figure 34.9b. In terms of the Miller indices the regular octahedron is bounded by the eight planes, (111), ($\bar{1}11$), ... , ($\bar{1}\bar{1}\bar{1}$), of the cube.

The possible sets of axes fall within one of the seven *crystal systems*. They range from the completely general triclinic to the highly symmetrical cubic system. The seven systems are described in Table 34.1. The lengths of the vectors are **a**, **b**, and **c**, while α, β, and γ are the angles between the axes. The angle α is opposite the **a** axis (that is, it is the angle between **b** and **c**); similarly, β is opposite side **b** and γ is opposite side **c**. If a unique axis is present, it is conventionally designated the *c* axis.

34.3 LATTICES

Until now our discussion has centered about *point* symmetry elements that leave at least one point unchanged. We now extend our discussion to include all space symmetry; this requires an infinitely extended array of points, the *lattice*. By 1803, Haüy had very nearly produced a correct explanation of the law of rational indices in terms of his *molécules intégrantes,* but his ideas contained unnecessary assumptions about the shapes of these structural units. In 1827, the mathematician Cauchy approached the problem from the point of view of a regular repetition of similar points on a microscopic scale. Cauchy demonstrated that the law of rational indices is a necessary and sufficient condition for a lattice structure to exist.

A geometrical lattice is an infinite set of points that recur regularly throughout space. The one-dimensional lattice is characterized by a single constant a, the distance between adjacent points, as shown in Figure 34.10a. This characteristic distance is called the *period,* the *identity distance,* or the *primitive translation.* Alternatively, the lattice can be considered the set of all translations of the vectors 0, ±**a**,

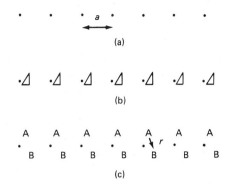

FIGURE 34.10 (a) One-dimensional lattice. (b) Structure superimposed upon the lattice of (a). (c) A one-dimensional crystal structure of the molecule AB based on the lattice of (a). The separation distance between A and B is given by r. The *basis* of the structure is the molecule A—B.

$\pm 2\mathbf{a}, \pm 3\mathbf{a}, \ldots$, starting from any point on the line. If $\mathbf{T}_h = h\mathbf{a}$ and $\mathbf{T}_k = k\mathbf{a}$ are two such vectors, where h and k are positive or negative integers (or zero), then the point obtained by the sum of both of these vectors is also a lattice point. Note very carefully that the lattice is *not* a structure but simply a geometrical array of points. A crystal structure is obtained by superimposing the molecules upon the geometrical lattice, as indicated in Figure 34.10b, where the molecule in this one-dimensional crystal is represented by a scalene triangle. The crystal structure is generated by translating the molecules along the lattice with the periodicity of the lattice points.

The structure of Figure 34.10b is based on the lattice of Figure 34.10a, but its position relative to the lattice points is not fixed. The lattice points can be considered a set of points on an infinitely long thin rod. The rod with the lattice points can be laid upon the structure, or *basis*, and moved back and forth relative to the molecules. The lattice can be thought of as a geometrical device that allows one to select equivalent regions in the structure.

Figure 34.10c shows a one-dimensional structure containing the diatomic molecule AB. We can describe this structure by saying that the B atoms are based on a lattice with the same period, but displaced from the equivalent lattice of A atoms by the vector r. It is more convenient, however, to speak of a single lattice and to associate the two atoms with each lattice point. The set of these molecules is called the *basis*.

In speaking of lattices it is important to differentiate between *unit cells* and *primitive cells*, and to bear in mind that the two are not always the same. The unit of length a is the shortest unit cell that can be used for the structure of Figure 34.10c. This cell contains only one lattice point; the length is the shortest possible translation of the lattice. One only needs to describe the positions of the two atoms in the cell. The unit cell that contains *only* one lattice point is the primitive cell. One could equally well describe the structure with a unit cell of length $2a$. But the unit cell would contain two lattice points, and we should have to describe the positions of four atoms instead of two. The lattice can be described by saying that it has a primitive length a, or a unit cell of length $2a$ with lattice points at 0 and $\frac{1}{2}$, or even a unit cell of length $3a$ with lattice points at 0, $\frac{1}{3}$, and $\frac{2}{3}$ and so on. Only the primitive vector has the property that *all* lattice points are integral multiples of it. For one-dimensional lattices there is no advantage in using a unit cell larger than the primitive cell. In three-dimensional crystal lattices, unit cells larger than the primitive cell are often used because they more conveniently display the symmetrical properties of the structure.

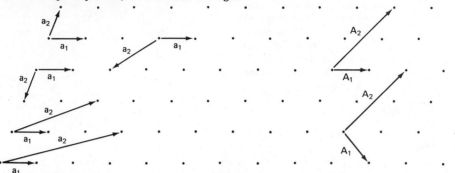

FIGURE 34.11 Two-dimensional lattice. Some possible pairs of primitive vectors, $\mathbf{a}_1$ and $\mathbf{a}_2$, which define primitive cells containing one lattice point per cell, are shown to the left. Some nonprimitive pairs of vectors, $\mathbf{A}_1$ and $\mathbf{A}_2$, defining nonprimitive unit cells containing more than one lattice point per cell, are shown on the right.

The extension of the concept of the lattice to two dimensions is straightforward and indicated in Figure 34.11. The two-dimensional lattice is generated from the primitive pair of vectors $\mathbf{a}_1$ and $\mathbf{a}_2$; the primitive cell is formed from the parallelogram determined by the two vectors. The choice of primitive pair is not unique in the two-dimensional lattice; all the pairs indicated on the left side of Figure 34.11 are primitive. The unit cells described by primitive pairs are primitive cells; they contain only one lattice point and have the same areas. On the right side of Figure 34.11 are some nonprimitive unit cells containing more than one lattice point per cell.

Although there is only one possible type of one-dimensional lattice, there are five different types of two-dimensional lattices, as depicted in Figure 34.12. The five lattices differ in their point symmetry elements. The square lattice labeled (a) has a fourfold axis of rotation through each lattice point, and the remaining lattices have either a twofold rotational axis, or in the case of the hexagonal lattice (e), a sixfold axis. The lattices are further distinguished by the presence or absence of mirror-image planes; you can see this by comparing the various lattices. Note that these lattices are of infinite extent.

The three-dimensional lattice consists of an infinite array of points in a pattern that is repeated regularly in three dimensions. In 1848, A. Bravais showed that three-dimensional lattices can be classified by their symmetry properties into fourteen types, which are called the *Bravais lattices*. Unit cells of the fourteen Bravais lattices are shown in Figure 34.13, where they are classified by crystal class and whether they are primitive (P), face-centered (F), side- or end-centered (C), or body-centered (I, from the German *innenzentriertes Gitter*). For each of these lattices,

FIGURE 34.12 The five and only five types of two-dimensional lattices. (a) Square lattice. (b) Rectangular lattice. (c) Oblique lattice. (d) Centered rectangular lattice. (e) Hexagonal lattice.

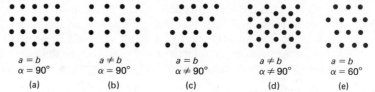

SEC. 34.3 LATTICES

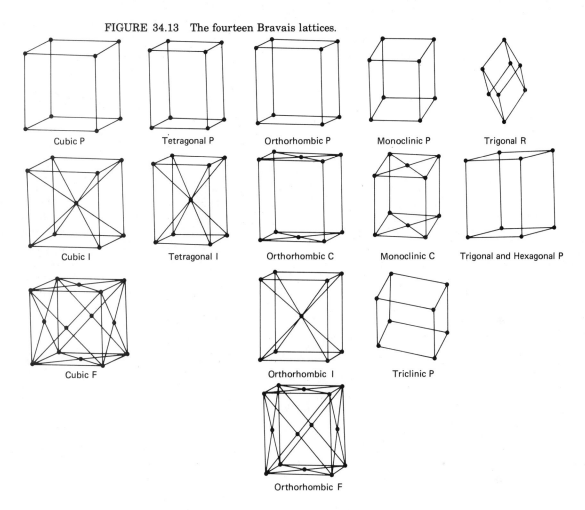

FIGURE 34.13 The fourteen Bravais lattices.

we have used a set of crystallographic axes corresponding to the axes of the crystal classes listed in Table 34.1. Only the P lattices are primitive; they contain one lattice point per unit cell. The centered lattices (body, end, or face) can be viewed as two or four interpenetrating simple lattices. The body-centered and end-centered lattices contain two lattice points per unit cell, and the face-centered lattice contains four lattice points per unit cell.

Unit cells larger than the primitive cell are often used in three-dimensional lattices to show the symmetry more clearly and to simplify calculations. Figure 34.14 shows a face-centered cubic lattice (FCC) as a unit cube with four lattice points; the primitive cell of the lattice is the rhombohedral cell, which contains only one lattice point. That there should be only fourteen possibilities is often perplexing to students. It might seem, for example, that we could add to the list by inventing a "face-centered tetragonal lattice." It is left for one of the problems to show that such a lattice is equivalent to the body-centered tetragonal lattice already on the list; it has twice the volume of the body-centered lattice.

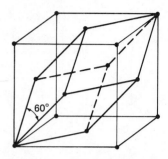

FIGURE 34.14 The rhombohedral primitive cell of a face-centered cubic unit cell.

The concept of the lattice enables us to prove that only one-, two-, three-, four-, and sixfold rotational symmetry is possible for lattices. We let the rotation $A(\alpha)$ be a rotational symmetry element of a lattice that rotates the lattice by the angle α about the axis A. The axis must be perpendicular to one of the planes of the lattice; the plane contains a two-dimensional lattice. The axis passes through the lattice point O, and we let $\mathbf{a}$ be a primitive translation, as shown in Figure 34.15. Since $A(\alpha)$ is a symmetry operation, the vectors $-\mathbf{a}$, $\mathbf{a}'$, and $\mathbf{a}''$ must also be translations of the lattice plane, and the points O_1, O_2, O_3, and O_4 must all be equivalent to O. In addition, the translation vector $\mathbf{b} = \mathbf{a}'' - \mathbf{a}'$ must be parallel to $\mathbf{a}$. Therefore it must be true that $\mathbf{b} = m\mathbf{a}$, where m is a positive or a negative integer (or zero). From the trigonometric relations of triangles,

$$b = 2a \cos \alpha = ma \tag{34.2}$$

or $\cos \alpha = \tfrac{1}{2} m$, where m is an integer. All possible values of α can be tabulated as shown in the accompanying table. These correspond to onefold (or no symmetry), two-, three-, four-, and sixfold rotation axes.

m	$\cos \alpha$	α
-2	-1	$180°$
-1	$-\tfrac{1}{2}$	$120°, 240°$
0	0	$90°, 270°$
1	$\tfrac{1}{2}$	$60°, 300°$
2	1	$0°, 360°$

A lattice is a "space-filling" geometrical entity, and lattices are the only kinds of symmetrical objects with which space can be filled. Bathroom tile patterns are limited to these five symmetries. One cannot tile a bathroom floor with regular pentagons without leaving gaps between adjacent tiles.[1] Some ancient Egyptian tiled

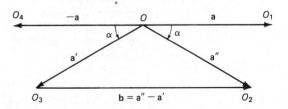

FIGURE 34.15 Construction showing that the only possible rotational symmetries for a lattice are one-, two-, three-, four-, and sixfold rotations.

---

[1] This is not meant to imply that fivefold symmetry is not found in nature. Molecules such as ferrocene containing five-membered rings may have fivefold rotational symmetry on the molecular level. What we mean to imply is that if a molecule with fivefold rotational symmetry is packed into a crystal lattice, then the crystal, which must fill all space, cannot have fivefold symmetry.

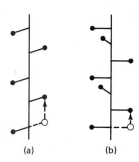

FIGURE 34.16 Screw rotations. The rotation is followed by a translation as shown. This is a space symmetry operation, and the figures must be thought of as of infinite extent. (a) Twofold. (b) Threefold.

walls have been uncovered that at first glance give the illusion of being tiled in regular pentagons. Close examination, however, indicates that the pentagons are not regular, and that the wall is a clever illusion. The ancients were aware of this limitation from an experimental point of view. This limitation also holds for rotary inversions and *screw axes*. The rotary inversion is a rotation coupled with an inversion through a center. A screw axis combines a translation with a rotation, as shown in Figure 34.16. The screw axis is a non-point-symmetry operation.

34.4 THE PACKING OF SPHERES

Consider the problem of the grocer who wants to stack oranges on a shelf. We assume the oranges to be perfect spheres, all of the same radius. The object is to pack the largest number of spheres in the smallest possible volume; in other words, to solve

FIGURE 34.17 Packing of spheres in space. (a) First layer arranged in a square two-dimensional lattice, showing a primitive cell. (b) First layer arranged in an equilateral, or hexagonal, two-dimensional lattice, showing a primitive cell. (c) The two equivalent sets of pockets in the first layer. (d) The second layer of spheres placed over the A pockets. There are now two possibilities for the third layer, either in the B or in the C pockets.

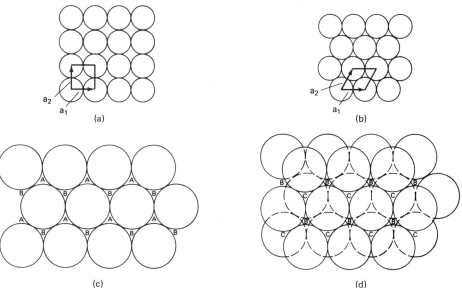

the problem, we want to minimize the volume of the vacant interstitial spaces. For a start, consider the two possible first layers shown in Figures 34.17a and 34.17b. The arrangement of 34.17a is based on a square two-dimensional lattice, and the arrangement of 34.17b is based on an equilateral (or hexagonal) array. The arrangement of 34.17b seems more efficient; but let's examine this a bit more quantitatively.

In the square net, each sphere is in contact with four other spheres, whereas in the hexagonal net each sphere is in contact with six other spheres. (Note the analogous packing of pennies on a tabletop where each penny is surrounded by six other pennies.) The problem essentially boils down to this: In which lattice does the primitive cell have the smallest area? The cell with the smallest area has the highest packing efficiency. The primitive vectors are shown to the left in the figures. The vectors $\mathbf{a}_1$ and $\mathbf{a}_2$ are the same length in both, but the angle α is different. The area of the parallelogram is least for the smallest value of α. The smallest value of α is 60°, since that is the angle at which two spheres adjacent to a third sphere just touch. The arrangement of Figure 34.17b is more efficient.

The placement of the second layer poses no conceptual difficulty. Figure 34.17c indicates two different but equivalent sets of pockets in the first layer; the sets are labeled A and B. The second layer is constructed by placing the spheres in either the A or the B pockets; Figure 34.17d shows the second layer placed in the A pockets. There are now two different but nonequivalent sets of pockets in the second layer. One set corresponds to the B set of the first layer. The other set, which we have labeled C, is found directly over the centers of the spheres of the first layer.

There are two choices for the third layer. If we place the spheres in the B pockets, then the third layer is the same as the first. If we alternate layers in this ABABABAB sequence, we get a lattice with hexagonal symmetry, the *hexagonal closest-packed* (hcp) lattice. On the other hand, if we place the spheres of the third layer in the new C pockets and then continue this sequence of three different layers, ABCABCABCABC, indefinitely, we get a lattice with cubic symmetry, the *cubic closest-packed* (ccp) lattice. The ccp lattice is identical with the face-centered cubic lattice previously discussed, and the terms "fcc" and "ccp" are used interchangeably. The spatial relations of these two structures are indicated in Figure 34.18.[2] The hcp and bcc structures are equally efficient for packing spheres. The most common structures for metals are the hcp, the ccp, and the body-centered cubic (bcc), as listed in Table 34.2. The inert gases have ccp structures in the crystalline state. The simple cubic structure is very rare in elements, the only representative being α-Po.

An important factor is the number of atoms (or units) per unit cell. For a primitive unit cell, there is only one unit per cell. There are several ways of determining the number of units per unit cell. The most general procedure is based on an arbitrary translation of the unit-cell vectors that determine the outline of the unit-cell polyhedron. We have already noted for the one-dimensional lattice that the entire set of lattice points can be translated by any arbitrary distance relative to the location of the structure units. In three dimensions, the lattice points can be translated by an arbitrary translation $\mathbf{T}$. If the vector $\mathbf{T}$ is not collinear with any of the unit vectors, and if the length of $\mathbf{T}$ is such that no lattice point of the new lattice coincides with any point of the old lattice, then no atom will lie on the boundary of the newly established unit cell, and the number of atoms in the new unit cell can be readily counted.

---

[2] A few minutes devoted to constructing cork or styrene ball models of the various layers may be worth as much as hours devoted to studying these two-dimensional representations.

SEC. 34.4 THE PACKING OF SPHERES

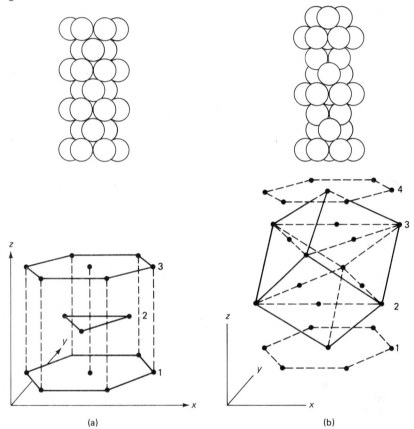

FIGURE 34.18 Closest-packing of spheres in three dimensions. (a) Hexagonal closest-packing of spheres. The layers are arranged in the sequence ABABABABABAB. (b) Cubic closest packing of spheres. The layers are arranged in the sequence ABCABCABCABC. The layers are successively numbered, and the axes are tilted to indicate the face-centered cubic arrangement.

TABLE 34.2 Structures of some metals.

Cubic Closest-packed (ccp or fcc)		Hexagonal Closest-packed (hcp)		Body-centered Cubic (bcc)	
Ca	Pd	Be	Ru	Li	V
Sr	Pt	Mg	Os	Na	Nb
Sc	Cu	Ca	Co	K	Ta
La	Ag	Sc	Ni	Rb	Cr
Co	Au	La	Ce	Cs	Mo
Rh	Th	Ti	Pr	Ba	W
Ir	Ce	Zr	Nd	Ti	U
Ni		Hf	Ho	Zr	Fe
		Cr	Er	Hf	
		Mo			

NOTE: Many metals exist in more than one form, and several are listed under more than one structure in this table.

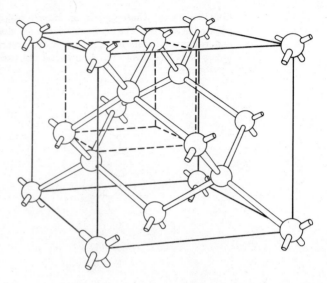

FIGURE 34.19 The diamond structure. Each carbon atom is tetrahedrally bonded to four other carbon atoms.

A simpler method is to proceed as follows, using the cubic lattices as an example. A lattice point situated at the corner of a cube is shared by eight different unit cells; hence we can assign one-eighth of that point to each cell. A lattice point situated on a cube face is shared by two cubes, and half is assigned to each cell; a point located within the cell is shared with no other cell, and the entire point is assigned to that cell. We note for completeness that if a point were situated on a cube edge, it would be shared by four cells, and one-fourth of the point would be assigned to each cell. Let's apply this procedure to the three cubic lattices.

The simple cubic lattice is a primitive lattice, and we know before we start that it should have one lattice point per unit cell. There are eight corners in a cube. Since we associate one-eighth of each corner point with the cell, we have $\frac{1}{8} \times 8 = 1$ lattice point per cell. For the body-centered cubic structure, there is a point in the body center in addition to the corner points; hence the bcc lattice has two lattice points per unit cell. The fcc lattice has six points situated on the cube faces in addition to the corner points; these faces contribute $\frac{1}{2} \times 6 = 3$ points in addition to the one attributable to the corners. The fcc lattice has four lattice points per unit cell.

The diamond structure is shown in Figure 34.19. It can be regarded as a face-centered cubic lattice with a basis of two atoms per lattice point; the atoms are located at the positions 000 and $\frac{1}{4}\frac{1}{4}\frac{1}{4}$ in the unit cell. Each carbon atom is tetrahedrally surrounded by four nearest-neighbor carbons, and each carbon has twelve next-nearest neighbors. There are eight atoms per unit cell. Alternatively, the structure can be considered as two interpenetrating fcc lattices displaced by one-quarter of the body diagonal. The edge of the unit cube in diamond is 3.56 Å. Other substances with this structure include silicon, germanium, and gray tin, with lattice constants of 5.43, 5.65, and 6.46 Å.

PROBLEMS

1. Determine the interfacial angles of the faces of the cube, the regular octahedron, and the regular tetrahedron.

2. Draw the planes (111), (123), (1$\bar{2}$1) of a cubic lattice.

3. Construct a regular tetrahedron by removing the

proper amount of material from a cube. What are the Miller indices of the faces of the tetrahedron?
4. Draw the (111) planes of the tetragonal, orthogonal, and rhombohedral lattices.
5. For each of the fourteen Bravais lattices, calculate the primitive and unit cell volumes in terms of the unit vectors and the angles between the axes.
6. Draw the primitive cell for the body-centered cubic lattice.
7. Draw a face-centered tetragonal unit cell. Show that this is equivalent to the body-centered tetragonal lattice.
8. As the body-centered tetragonal lattice is drawn in Figure 34.13, it seems that there are two sets of lattice points, one set at the body centers and another set at the corners of the parallelepiped. Draw several adjacent unit cells and convince yourself that the two sets are equivalent. We can call either set the body-center set and the other the corner set.
9. By drawing regular pentagons on a sheet of paper, try to tile a floor with regular pentagons.
10. The cube and the regular octahedron have the full symmetry of the cubic group, 48 symmetry elements. Enumerate all 48 elements. (Since they have the full symmetry of the cube, they are said to possess *holohedral* symmetry.)
11. The tetrahedron also is a member of the cubic class but contains only half the symmetry of the full cubic group. Find the 24 symmetry elements of the tetrahedron. (The regular tetrahedron has half the symmetry of the cube and possesses, it is said, *hemihedral* symmetry.)
12. Draw the primitive cells for the nonprimitive Bravais lattice unit cells.
13. Calculate the fraction of unoccupied area when circles of equal diameters are arranged in a two-dimensional hexagonal lattice so that each circle is in contact with six other circles. What is the diameter of the largest circle that will fit in the empty interstices?
14. Calculate the fraction of unoccupied space arising from packing hard spheres in (a) simple cubic arrangement; (b) body-centered cubic; (c) face-centered cubic. In each instance calculate the largest sphere that will fit in the interstices.
15. Determine the number of lattice points per unit cell for the fourteen Bravais lattices.
16. For the face-centered cubic lattice, calculate the number of nearest-neighbor and next-nearest-neighbor lattice points. Do the same for the body-centered and simple cubic lattices.
17. The unit cell edge of diamond is 3.56 Å. Find the distance between nearest-neighbor carbon atoms (in other words, the length of the C–C bond) and between next-nearest neighbors.
18. If hard spheres were packed as closely as possible in the diamond structure, what would be the fraction of void spaces? Compare with the fcc result.
19. Determine the number of molecules per unit cube in NaCl, CsCl, and CaF_2. (The unit cube edges are 5.63, 4.11, and 5.45 Å.) Determine the number of nearest and next-nearest neighbors in each. Determine the distance between nearest neighbors in each. Calculate the densities of each and compare with handbook values. (Find the structures in a book on structural chemistry.)
20. For a cubic lattice, calculate the distance between adjacent (001), (111), (110), and (123) planes.
21. For a crystal of NaCl in the form of a cube 1 cm on an edge, what fraction of the atoms lie on the surface of the cube? What if the cube were 10^{-4} cm or 10^{-7} cm on an edge?
22. From the diamond structure pictured in Figure 34.19, show that the C—C—C angle is indeed the tetrahedral angle, 109°28′.

CHAPTER THIRTY-FIVE

X-RAY DIFFRACTION STUDIES OF CRYSTALS

35.1 ELECTRIC DISCHARGE TUBES AND X RAYS

Science often makes its greatest strides when it tries to duplicate the grandeur of nature, even though on a smaller scale. The investigation of lightning, which had been correctly identified as a form of electrical discharge, was one of these cases. Many electrostatic generators were built with which electrical discharges similar to those seen in electrical storms could be reproduced in the laboratory on a smaller scale. At about the same time that Benjamin Franklin was conducting his researches in electricity, William Watson built a device that allowed an electrostatic generator to be discharged through a rarefied gas and noted the bright glow associated with the discharge. By the mid-nineteenth century progress in electrical generation and in high-vacuum technology had advanced to the point at which experiments using the kind of tube shown in Figure 35.1a were commonplace. Operating these *electric discharge* or *Crookes tubes,* as they were called, was simple. The tube was evacuated to a low residual pressure, and a high potential difference was placed across the terminals of the tube. The result was a beautiful glowing stream between the cathode and the anode of the tube. The higher the voltage, the brighter the glow. But if the tube were evacuated to a sufficiently low pressure, then the glowing stream would disappear. We now know that such a stream is due to an electron beam. The intense electric field in the tube frees some electrons from the cathode and these are accelerated towards the anode. These electrons strike some gas molecules, causing them to become ionized, and the ions cause the gas to become conducting. As the ions recombine with electrons to form neutral atoms, they emit characteristic spectral lines. If the pressure is too high, the tube does not work because of the excessive number of

> **WILLIAM WATSON** (1715–1787), English physician and scientist, is best known for his work in electricity and for his early experiments concerning inoculation against smallpox. He is also credited with introducing the Linnean system for botanical classification in England. He was a contemporary of Benjamin Franklin and engaged in electrical experimentation along the same lines as Franklin. He tried to determine the speed of electricity but could only conclude that it appeared instantaneous. Watson was one of the first to study the discharge of electricity in gases at low pressure, and he discovered that the conductivity of the gas increases during the discharge. He was an ardent advocate of Franklin's lightning rods and succeeded in having one installed in a gunpowder factory in England.

FIGURE 35.1 Electric-discharge, or Crookes, tubes. (a) Simple discharge tube. (b) From shadows cast on the wall, it was determined that the cathode rays travel from cathode to anode. (c) If a small "windmill" was placed in the path of the cathode rays, the mill rotated, indicating that the rays were corpuscular. Slits S_1 and S_2 confine the electron beam to the form of a thin pencil of electrons.

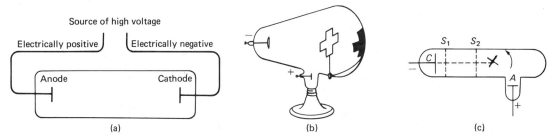

collisions between electrons and gas molecules. If the pressure is too low, the glow disappears, since there are an insufficient number of ions formed. In 1850, the operation of these tubes was still mystifying.

By the late 1800s, several important facts had been ascertained about these electron beams, or "cathode rays." It was known that the rays traveled from the cathode to the anode. This was demonstrated by experiments of the sort illustrated in Figure 35.1b. Shadows cast on walls covered with fluorescent material indicated that the rays moved from cathode to anode. In addition, the rays were shown to be corpuscular, since they were able to drive miniature windmills placed in the tubes, as shown in Figure 35.1c. Deflection of the rays by external electric and magnetic fields indicated that they carried a negative charge. Perrin proved this by collecting some of the rays in an electroscope. Final confirmation of this came from J. J. Thomson's measurement of the charge-to-mass ratio of the corpuscles in the rays. A schematic

> WILLIAM CROOKES (1832–1919), English chemist and physicist, modeled himself on Michael Faraday, with whom he shared a brilliant experimental ability and an abysmal ignorance of mathematics. He was successful in science as well as business and directed many of his activities towards commercial ventures. Among these was the sodium amalgamation process for extracting gold, use of sewage and animal refuse, and electric lighting. His experimental ability enabled him to produce a vacuum of one-millionth of an atmosphere, and it was this feat that made possible the improvement of the discharge tubes and the ensuing discoveries of X rays and the electron. Crookes invented the radiometer, that toy which delights children, and he used it to confirm Maxwell's prediction that the viscosity of a gas is independent of pressure. He was an independent discoverer of the element thallium, and carefully determined its atomic mass. His reputation suffered as a result of his belief in the occult; Crookes devoted much time to spiritual studies. This perhaps explains his theory of radioactivity, which he propounded before the turn of the century. Crookes imagined that a radioactive element acted as a Maxwellian demon and selected those air particles that moved with velocities much greater than the average velocity, absorbing some of their energy and ejecting the molecule at a lower speed. Among other difficulties, this theory violated the second law of thermodynamics. When Crookes was knighted in 1897, he took for his motto, *Ubi lux, ibi Crookes*, "Where there is light there is Crookes."

FIGURE 35.2 J. J. Thomson's experiment to determine the charge-to-mass ratio of the particles in cathode rays by the simultaneous deflection of the beam with crossed electric and magnetic fields.

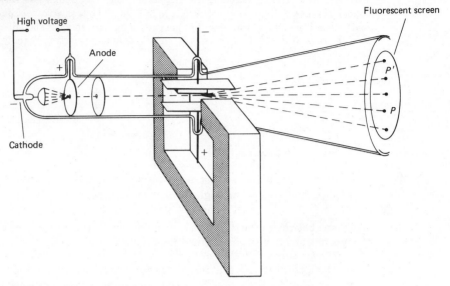

of his apparatus is shown in Figure 35.2. A beam of electrons originating at the cathode passes through the slits and travels in a straight line to the end of the tube. A pair of plates D and E are charged as shown, and the beam is deflected to point P'. A magnetic field is then established at right angles to the electric field so that the spot moves back to its original position P. From the deflection of the spot, the geometry of the tube, and the electric and magnetic fields used, the ratio e/m can be determined. Thomson's first determination gave the result of about 10^{11} coulombs per kilogram, and a later determination gave 1.7×10^{11}; the modern value is 1.76×10^{11} C kg^{-1}.

WILHELM CONRAD ROENTGEN (1845–1923), German physicist, received his earliest education in Holland. He was not particularly studious and apparently was expelled from school at the age of 16 after refusing to identify a classmate who had caricatured one of the teachers. His university training was in mechanical engineering. He was a reticent man, working alone in the laboratory and shunning public engagements. He declined to give the expected lecture on winning the first Nobel Prize in physics in 1901 for his discovery of X rays. His work on electric discharges in Crookes tubes was started in October of 1895, and by November of that year he noticed quite by chance that one of his fluorescent screens glowed brightly even though it was some distance from the tube. He discovered that the X rays traveled in straight lines, were not capable of refraction, reflection, or deviation by a magnetic field, and were able to travel about 2 m in air. He discovered the penetrating power of the X rays and produced photographs of balance weights in a closed box. On December 22 he took an X ray photograph of his wife's hand. The universal morbid curiosity regarding the ability to see inside a human body brought Roentgen instant international fame, and in Germany X rays are usually known as *Röntgenstrahlen*. As far back as 1879 Crookes himself had complained of fogged photographic plates that were stored near his cathode-ray tubes, and others had noted similar effects; but the discovery remained for Roentgen.

Two years before Thomson's work, Roentgen was working with the same kind of electric discharge tube, and he allowed the rays to impinge upon various metals. During the work, he accidentally observed that a screen covered with barium platinocyanide crystals fluoresced, even when the screen was placed some distance from the tube. He had discovered a new type of penetrating radiation, and the age of *X rays* had begun. Unlike ordinary light, this type of radiation could penetrate seemingly opaque material such as wood, cardboard, and thin metals. He soon discovered that the absorption of X rays was proportional to the atomic weight of the absorber. Within three months of his discovery, X rays were being used as an aid in surgical procedures in a Viennese hospital.

35.2 X-RAY EMISSION SPECTROSCOPY

X rays are produced when fast electrons strike a target material. In practice, X rays are produced by allowing high-energy electron beams (about 30–50 keV) to strike metal targets. A typical intensity curve for the emitted X rays is shown in Figure 35.3. All distribution curves have this general appearance of a continuous distribution upon which are superimposed some discrete peaks. Both the continuous and discrete contributions arise from quantum mechanical effects. The continuous radiation results from the conversion of the electron's kinetic energy into electromagnetic radiation. When a high-speed electron is slowed down by collision with the target material, it loses a certain amount of kinetic energy ΔE, which appears as a light photon whose frequency is found from the relation $\Delta E = h\nu = hc/\lambda$. This explains the sudden onset of X-ray emission in the short-wavelength region. The frequency corresponding to the threshold wavelength is the frequency that one would get if the electron were suddenly stopped and all of its kinetic energy converted into a light quantum. Higher wavelengths correspond to the electron's being partially slowed down. We should thus expect the distribution curve to be shifted towards lower wavelengths as the energy of the electrons is increased, and this is precisely what is observed.

FIGURE 35.3 Emission X-ray intensity distribution curve.

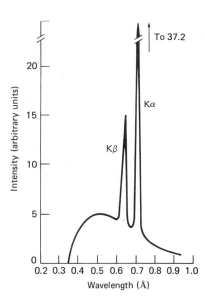

The discrete peaks, on the other hand, arise from electronic transitions in the atoms. Ordinary uv spectroscopy and visible spectroscopy are associated with excitations of the outer electrons of an atom to higher levels. When these excited electrons fall back to lower levels, they emit light of characteristic frequencies. The discrete X-ray peaks are associated with transitions between inner-shell electrons rather than outer-shell electrons. The various electron shells of an atom are shown in Figure 35.4. X-ray spectroscopists have developed a nomenclature that denotes the various shells by letters which correspond to the values of the principle quantum numbers of the shells. Thus the shells with $n = 1, 2, 3, 4, 5, \ldots$ are called accordingly the K, L, M, N, O, $\ldots$ shells. Suppose that an incoming high-energy electron knocks out one of the two K-shell ($n = 1$) electrons to create an ion. The ion is said to be in the K state. Now one of the electrons from the L shell can drop down into the K shell to replace the electron that was ejected. The ion is now in an L state, and the energy difference between the L and the K state gives the frequency of the emitted spectral line, which is called the K_α line. If the target atom were Cu, the line would be termed the Cu K_α line. When an M electron falls to the K shell, a less likely event, then a line of different frequency, called the K_β line, is emitted, as shown on Figure 35.4. The transition that occurs when an M shell electron falls to the L shell is an L_α transition; if the electron falls from the N to the L shell, the transition is an L_β one; and similarly through all the shells. The name of the transition is taken from the shell to which the electron drops.

There are several striking differences between the characteristic X-ray spectra and the ordinary spectra of elements. The X-ray spectra of the various elements are similar to one another, in contradistinction to the ordinary optical spectra, which show little uniformity for the different elements. Also, the frequencies of the various lines such as the K_α line increase monotonically from the lighter to the heavier elements. Moseley seized on this latter characteristic and in two classic papers published in 1913–14 presented his systematic study of the characteristic frequencies emitted by the elements.[1] He discovered the concept we call *atomic numbers*. In searching for a relation between the frequency of the K_α line and some property of the atom, he obtained a straight-line relation between the square root of the fre-

FIGURE 35.4 Generation of characteristic X-ray emission lines by electronic transitions between shells.

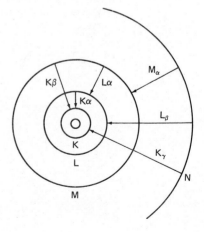

[1] H. G. J. Moseley, *Phil. Mag.* 26 (1913):1024; 27 (1914):703.

> HENRY GWYN JEFFREYS MOSELEY (1887–1915), English physicist, was working with Rutherford when the news of Laue's diffraction experiments reached England. While Laue had clearly demonstrated the wave nature of X rays, W. H. Bragg had adopted a particulate view of X rays, and Moseley sought to reconcile the wavelike properties with the particlelike properties. In his initial researches, which were carried out in collaboration with C. G. Darwin, he was anticipated by the Braggs on several occasions. Whiddington had demonstrated that the minimum electron velocity needed for stimulating the emission of a K line was proportional to the atomic mass. Moseley then set about determining whether the frequency followed the square of the molecular mass or the square of Z, which was about half the atomic weight as demonstrated by Rutherford in his α scattering experiments. He found that the frequencies of the rays emitted by the ten elements from Ca to Zn followed the simple relation $\nu_{K_\alpha} = \frac{3}{4}R(Z-1)^2$ to within a precision of 0.5%; R is the Rydberg constant. Encouraged by these results, he went on up the periodic table and showed the absence of elements 43 and 61. His research was completed shortly before the start of World War I. Along with his former classmates at Eton, Moseley felt it his patriotic duty to enlist. Moseley was killed in the Battle of Gallipoli in 1915 at the age of 28. Had he lived, there is no question but that he would have received the Nobel Prize for his work. Robert Millikan said of him, "A young man twenty-six years old threw open the windows through which we can glimpse the subatomic world with a definiteness and certainty never dreamed of before. Had the European War had no other result than the snuffing out of this young life, that alone would make it one of the most hideous and most irreparable crimes in History" (quoted in B. Jaffe, *Crucibles: The Story of Chemistry,* New York: Simon and Schuster, 1948, p. 291).

quency and an integer that we call the atomic number. An example of a *Moseley plot* is shown in Figure 35.5.

Recall that this discovery was just after Bohr had presented his model of the hydrogen atom in which the frequency of a spectral line was given by

$$\nu = Z^2 \frac{2\pi^2 me^4}{h^3}\left(\frac{1}{n_2^2} - \frac{1}{n_1^2}\right) \tag{35.1}$$

This suggested that the frequency must depend on Z^2. Empirically, Moseley found that the straight-line relation of Figure 35.5 could be expressed as

$$\nu = 0.248 \times 10^{16}(Z-1)^2 \tag{35.2}$$

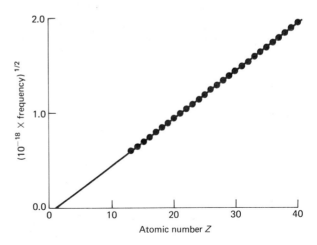

FIGURE 35.5 Moseley plot of $\nu^{1/2}$ of K_α lines vs Z. The straight line crosses the x axis at $Z = 1$.

If we try to treat the K_α transition by setting $n_2 = 1$ and $n_1 = 2$ and using the numerical values of the natural constants appearing in (35.1), we get

$$\nu = 0.328 \times 10^{16} Z^2 \left(\frac{1}{1^2} - \frac{1}{2^2}\right)$$
$$= 0.246 \times 10^{16} Z^2 \qquad (35.3)$$

which except for the slight correction to Z is the same as Moseley's empirical relation, Equation (35.2). Now recall our discussion of atomic orbital energies (Chapter 32). For heavier atoms, the L-shell electrons are relatively close to the nucleus, and not much affected by electrons in shells with n greater than 2. A K-state ion is one that has only one $n = 1$ electron; this electron effectively screens the L-shell electron from the nuclear charge accounting for the $Z - 1$ instead of the Z factor for the nuclear charge.

Moseley's treatment of the problem indicated that the atomic number is a more basic property than the atomic mass. It explained a few anomalies in the periodic table founded on mass, such as a higher atomic mass for Ar ($Z = 18$) than for K ($Z = 19$)—although it was common knowledge that Ar came before K in the periodic table when one considered their chemical properties. Moseley's scheme also provided a new method for determining atomic numbers of newly discovered elements. Gaps like the one that existed for Tc ($Z = 43$) sent people searching for hitherto undiscovered elements. Moseley's ability to measure the wavelengths of the X-ray lines derived from a different series of experiments associated with the diffraction of these new X rays.

35.3 THE DIFFRACTION OF X RAYS

By 1912, science had progressed along many converging lines, and the stage was set for the sudden strides in knowledge that were about to take place when the structures of atoms, molecules, and crystals would be elucidated, both from the experimental and the theoretical point of view. Avogadro's number had been determined by Perrin in his work with colloids; and it was possible by using that number, along with the densities and molar masses of crystals, to estimate that the distance between atoms in crystals was of the order of 1–2 Å. Evidence gathered on X rays, both theoretical and experimental, indicated that their wavelengths were of the order of 1 Å. Diffraction effects take place with electromagnetic radiation only if the diffraction grating has slit dimensions of the same order of magnitude as the wavelength of the radiation. Von Laue appreciated the significance of this and reasoned that the regular arrays of atoms in a crystal could act as a three-dimensional diffraction grating for X rays. The experimental work was undertaken by Friedrich and Knipping, who had just taken their doctorates under Roentgen's direction.

The crystal they used was copper sulfate, $CuSO_4 \cdot 5H_2O$. The reason for the choice should be clear after looking across reagent bottles in any chemical laboratory. Copper sulfate comes naturally in the form of large single crystals, and copper is a heavy atom. From the point of view of crystallography, however, this was one of the worst choices they could have made. The crystal has a very complicated structure, and the symmetry is of the lowest order, triclinic. Any diffraction effects would have been impossible to explain at this stage of development of the science of crystallography. The object, however, was simply to see whether there were any diffraction effects. They placed a crystal in the path of a collimated X-ray beam and placed a

SEC. 35.3 THE DIFFRACTION OF X RAYS

> MAX VON LAUE (1879–1960), German physicist, was Planck's leading and favorite student, and the two formed a lifelong friendship. In 1912, Laue conceived the brilliant idea of sending X rays through crystals to verify whether they were electromagnetic waves of very short wavelengths. Albert Einstein called this experiment one of the most beautiful ever performed in physics, and Laue received the Nobel Prize for this work in 1914. Laue was active in the German scientific world, serving on numerous committees, and he was one of only three members of the Prussian Academy to protest Einstein's dismissal after the Nazis had seized power. He was engaged in military research during the war and took an active part in rebuilding the German scientific establishment, which had been virtually destroyed by 1945. X rays provided a rich source of Nobel Prizes in the early 1900s. Roentgen received the first physics prize in 1901, and Laue received his in 1914. Other recipients in the field of X rays include the Braggs (1915), Barkla (1917), Siegbahn (1924), and Compton (1927). The name of Moseley would probably have been on the list had he not been killed so soon after his discoveries.

photographic plate behind the crystal, as shown in Figure 35.6a; they got a diffraction pattern as indicated in Figure 35.6b. They then took a second photographic plate and placed it a greater distance behind the crystal. They got an enlarged image of the first pattern, indicating that the diffracted X rays traveled in straight lines. When the crystal was translated parallel to itself, there was no change in the geometrical form of the pattern. Rotation of the crystal produced a different pattern. Laue's original hypothesis had been verified; crystals did diffract X rays. In

FIGURE 35.6 (a) Laue's diffraction experiment as carried out by Friedrich and Knipping. The terms P_i represent various positions of the photographic plate. Diffraction patterns were obtained only after the plate was moved behind the crystal at positions P_4 or P_5. (b) Early Laue diffraction photographs. [Photographs from J. M. Bijvoet, W. G. Burgers, and G. Hagg, eds., *Early Papers on Diffraction of X-Rays by Crystals*. (Published for the International Union of Crystallography by A. Oosthoek's Uitgeversmaatschappij N.V., Utrecht.)]

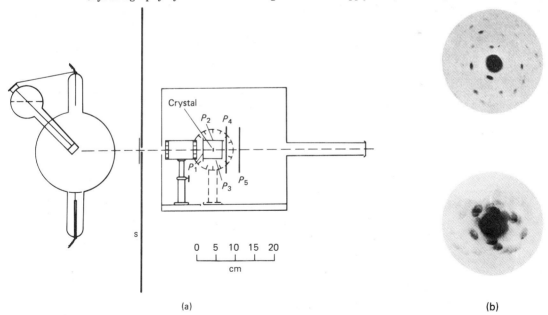

(a)   (b)

visible-light optics, diffraction gratings of known periodicity had been constructed, to produce well-defined diffraction patterns. It was now necessary to work this problem backwards and to determine the periodicity of the grating from an observed pattern. The problem was rendered even more difficult by involving three dimensions. The complex structure of copper sulfate was a difficult one to work with, and Laue turned his attention to ZnS, which was a cubic crystal and gave a much simpler pattern.

Laue tried to interpret the results by extending the diffraction of a one-dimensional grating to three dimensions. Figure 35.7a shows the successive reinforcements of wavelets resulting when a beam is diffracted by a linear array of scat-

FIGURE 35.7 (a) Reinforcement of diffracted beams from a row of equally spaced scattering centers. (b) The conditions for diffraction from a row of scattering centers. (c) X rays incident on a three-dimensional arrangement of scattering centers.

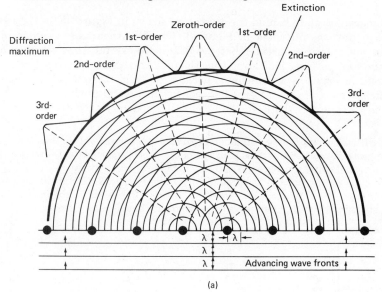

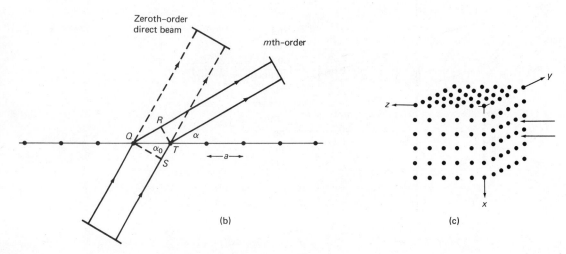

tering centers or atoms. Reinforcement of the diffracted beam occurs when the path difference of beams refracted from successive centers is an integral number of wavelengths. In the construction shown in Figure 35.7b, the incident beam makes an angle α_0 with the row of centers, and the diffracted beam makes an angle α. Reinforcement occurs when the rays scattered at Q are in phase with those scattered at T; the paths QR and ST must differ by an integral number of wavelengths:

$$QR - ST = m\lambda \tag{35.4}$$

where m is an integer and λ is the wavelength of the X ray. From elementary trigonometry,

$$QR = a\cos\alpha \qquad ST = a\cos\alpha_0 \tag{35.5}$$

and the path difference or condition for reinforcement is

$$a(\cos\alpha - \cos\alpha_0) = m\lambda \tag{35.6}$$

For a three-dimensional array of scattering centers as shown in Figure 35.7c, three equations such as (35.6) would have to be satisfied simultaneously. These can be written as

$$\begin{aligned} a(\cos\alpha - \cos\alpha_0) &= m\lambda \\ b(\cos\beta - \cos\beta_0) &= n\lambda \\ c(\cos\gamma - \cos\gamma_0) &= p\lambda \end{aligned} \tag{35.7}$$

where α_0, β_0, and γ_0 are the incident angles of the X-ray beam with each of the three mutually perpendicular rows of scattering centers in the lattice, and α, β, and γ are the angles of the scattered beams. The integers m, n, and p determine the orders of the maxima, and a, b, and c are the spacings of the rows of atoms in each of the mutually perpendicular directions. These three equations are known as the *Laue equations*.

The Laue equations correctly describe the diffraction of X rays by crystals; yet when Laue tried to apply his equations to determining the crystal structure of the cubic ZnS, he failed to achieve any noteworthy success. This was one of the few instances in science in which a person failed to accomplish what he set out to do; yet in that failure Laue achieved such brilliance that he is regarded as one of the founders of the science of X-ray crystallography, and he received a Nobel Prize for his work. Part of his difficulty may be seen qualitatively by reference to Figure 35.8, where we have indicated the diffraction of beams from one- and two-dimensional gratings. For the one-dimensional grating, all the orders of diffraction appear as shown in Figure 35.8a. When diffraction is extended to a two-dimensional grating as shown in Figure 35.8b, we still get all the orders of diffraction, but now they appear as discrete beams that must be specified by two indices.[2] The extension to three dimensions is more complex, and we shall defer a quantitative discussion of that. In three dimensions there is usually no diffraction with a monochromatic incident beam, since generally only two of the three conditions will be satisfied. On the other hand, as the crystal is rotated through all possible angles, orientations for which the three conditions are simultaneously met will be encountered, and each time the condition is met a diffracted beam flashes out.

Laue faced several insurmountable problems as he attacked the ZnS structure. Laue did not know the wavelength of his X rays, nor did he know the geometrical

---

[2] Many two-dimensional diffraction patterns along with extensive discussion are presented in C. A. Taylor and H. Lipson, *Optical Transforms, Their Preparation and Application to X-Ray Diffraction Problems* (Ithaca: Cornell University Press, 1965).

FIGURE 35.8 (a) Diffraction by a one-dimensional grating, and (b) a two-dimensional grating. The two-dimensional pattern is an idealized one, and in practice the spots would occur along curves and not straight lines.

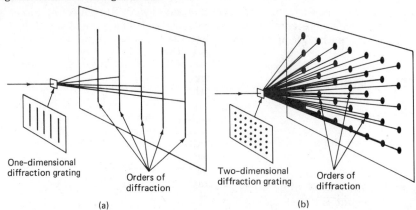

arrangements of the atoms in the crystal. He wrongly assumed that the structure was built on a simple cubic pattern. He tried assuming various wavelengths, but each time was faced with numerous spots on the photographic plate whose presence could not be explained, not to mention several spots that should have been there but were absent. He assumed several wavelengths, and reached a total of five different composite wavelengths before he gave up. Had he continued on to an infinite number, he might have succeeded, because his beam was composed of a continuous-wavelength distribution. This was soon demonstrated by the Braggs in England, when they succeeded in working out the first crystal structure.

35.4 X-RAY REFLECTIONS

While Laue was engaged in his experiments in Munich that conclusively demonstrated that X rays were a form of electromagnetic radiation, W. H. Bragg and his son W. L. Bragg in England were trying to show that X rays were corpuscular. They were working on the ionization of air by X rays, and the ability of X rays to ionize air seemed to indicate that they were corpuscles. This conclusion was reached by analogy with the ionization of gases by electron beams. The news of Laue's experiment reached England and the significance of the work was apparent. The Braggs tried a new approach.

Rather than consider a diffraction as Laue had done, W. L. Bragg conceived of the process as one in which the X rays were reflected by successive planes of atoms in the crystal. The approach was much simpler than Laue's, and the Braggs were able to determine numerous simple crystal structures successfully. In Figure 35.9, the crystal is considered a series of planes with a constant interplanar distance d. Incident X rays of wavelength λ strike the crystal face at an angle θ and are reflected at this same angle. For reinforcement, the wave reflected at B' must be in phase with the wave reflected at C. This occurs when the path difference is an integral number of wavelengths,

$$B'C - BC = n\lambda \qquad (35.8)$$

SEC. 35.4 X-RAY REFLECTIONS

> WILLIAM HENRY BRAGG (1862–1942) and WILLIAM LAWRENCE BRAGG (1890–1975), English and Australian physicists, form the only father and son team to ever win a Nobel Prize together. The elder Bragg was appointed a professor of mathematics and physics in 1886 at the University of Adelaide in Australia, where the younger Bragg was born. There he busied himself with the social and recreational atmosphere "down under," and he achieved nothing of great note for eighteen years. W. H. Bragg's active research career started in earnest in 1904, and he returned to England in 1907 to become the leading exponent of the corpuscular theory of X rays. This view was shattered by the Laue experiment of 1912. The younger Bragg, who came to England with his father, was trying to reconcile the two views of X rays when he discovered his famous equation. Their X ray work was interrupted by the war, during which W. H. Bragg was a member of the team that invented the hydrophone for detecting submarines and the younger Bragg was in France acting as a technical advisor to the army in their efforts to locate enemy artillery by sound ranging experiments. Following the war both Braggs continued their work in X-ray crystallography, W. H. at the Royal Institution in London, and W. L. in the physics department at the University of Manchester. Between the two of them, they trained almost the entire first generation of X-ray crystallographers and were thus the leaders in making England the center of excellence in that field.

Since $B'C = d/\sin\theta$ and $BC = B'C \cos 2\theta$, we have

$$\frac{d}{\sin\theta}(1 - \cos 2\theta) = \frac{d}{\sin\theta}(2\sin^2\theta) = n\lambda$$

$$n\lambda = 2d\sin\theta \qquad (n = 1, 2, 3, \ldots) \tag{35.9}$$

Equation (35.9) is known as *Bragg's law*. At first glance, Bragg's equation seems like an ordinary diffraction-grating equation. Actually it is based on a different principle, and one of the main differences was alluded to at the end of the previous section. For a one-dimensional diffraction grating, we always obtain diffraction maxima regardless of the angle the incident beam makes with the plane of the grating, and the same holds true for the two-dimensional grating. Bragg's law must reflect the properties of the three-dimensional grating even though it has but one angle as the

FIGURE 35.9 Construction for derivation of Bragg's law.

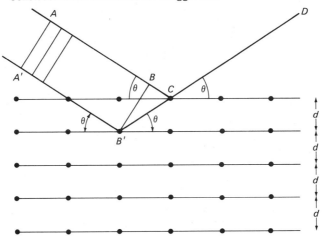

physical variable. If the X rays are monochromatic, λ is fixed at one particular value and reinforced reflections occur only at the angles,

$$\theta = \sin^{-1}\left(n\,\frac{\lambda}{2d}\right) \qquad (n = 1, 2, 3, \ldots) \qquad (35.10)$$

Thus for a fixed wavelength and an arbitrary angle, nothing is observed, and this explained some of Laue's difficulties. On the other hand, if the crystal is slowly rotated so that θ varies over all angles, then a reinforced maximum flashes out every time (35.10) is satisfied.

On the other hand, if the wavelength distribution is continuous, then there is some λ at each θ that satisfies the condition for a maximum, and one can sweep through all the wavelengths by varying θ. This concept was embodied by the Braggs in their X-ray ionization spectrometer, a sketch of which is shown in Figure 35.10. A collimated beam of X rays strikes the crystal face C, and is reflected into the ionization chamber. The intensity of the reflected beam is determined by the number of ions formed. By varying the angular setting of the crystal face, one can measure the intensity as a function of θ, and hence as a function of λ through Bragg's law. The resulting wavelength distribution they obtained was like the one shown in Figure 35.3. The fatal flaw in Laue's work was now clear. There was not one wavelength nor were there even five different wavelengths in the beam. There was a continuous distribution of an infinite number of wavelengths.

The Laue equations appear different from the Bragg equation, but since they describe the same physical phenomenon they must be equivalent. Let's digress for a moment and show that they are equivalent. One practical difficulty with the Laue equations is that they are written in the form of three variable angles, whereas the Bragg

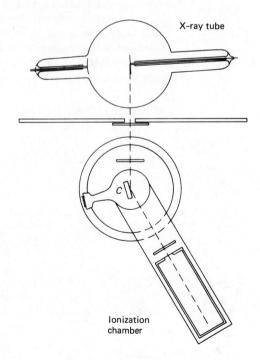

FIGURE 35.10 The Bragg X-ray spectrometer.

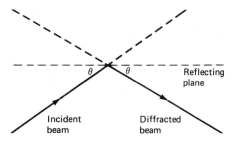

FIGURE 35.11 Relation between incident and diffracted beams and the reflecting plane.

equation has only one angle. Suppose we try to write the Laue equations in a form that has only one angle, namely the angle between the incident and diffracted beam, as shown in Figure 35.11. We can do this by squaring the three Laue equations and then adding them, to get (for the case of the cubic lattice, $a = b = c$)

$$a^2(\cos^2 \alpha - 2 \cos \alpha \cos \alpha_0 + \cos^2 \alpha_0 + \cos^2 \beta - 2 \cos \beta \cos \beta_0 \\ + \cos^2 \beta_0 + \cos^2 \gamma - 2 \cos \gamma \cos \gamma_0 + \cos^2 \gamma_0) = \lambda^2(m^2 + n^2 + p^2) \quad (35.11)$$

The cosines of α, β, and γ are the three direction cosines of the diffracted beam and the cosines of α_0, β_0, and γ_0 are the direction cosines of the incident beam. Equation (35.11) can be simplified by using two properties of direction cosines. Firstly the sum of the squares of the direction cosines of any straight line is unity, hence

$$\cos^2 \alpha + \cos^2 \beta + \cos^2 \gamma = 1$$
$$\cos^2 \alpha_0 + \cos^2 \beta_0 + \cos^2 \gamma_0 = 1 \quad (35.12)$$

Secondly, the cosine of the angle 2θ between two intersecting lines is

$$\cos 2\theta = \cos \alpha \cos \alpha_0 + \cos \beta \cos \beta_0 + \cos \gamma \cos \gamma_0 \quad (35.13)$$

With these, (35.11) reduces to

$$a^2(2 - 2 \cos 2\theta) = \lambda^2(m^2 + n^2 + p^2) \quad (35.14)$$

Using the theorem of elementary trigonometry that $1 - \cos 2\theta = 2 \sin^2 \theta$, we can reduce (35.14) to

$$\lambda = \frac{2a}{\sqrt{m^2 + n^2 + p^2}} \sin \theta \quad (35.15)$$

This is identical to $n\lambda = 2d \sin \theta$ if

$$\frac{d}{n'} = \frac{a}{\sqrt{h^2 + k^2 + l^2}} \quad (35.16)$$

where n' is the greatest common divisor of m, n, and p such that $m = n'h$, $n = n'k$, and $p = n'l$. The integers h, k, and l are the Miller indices of the reflecting plane.

35.5 STRUCTURE OF ROCK SALT

The situation the Braggs faced in 1913 may not seem overly complex in retrospect. By what was known at the time, their solution was particularly brilliant. In a sense, the problem they faced was circular. They had developed the ionization spectrometer to the point at which they could measure the *relative* wavelengths of X-ray beams using crystal planes as a diffraction grating. To place their relative measurements on an *absolute* basis, they had to know the periodicity of the crystal structure or the distance between planes. But to measure the periodicity, they had to know the absolute value of the wavelengths. Adding to their difficulties was that the atomic arrangements in a crystal were not yet known for any substance. It is instructive to consider their analysis of the structures of NaCl and KCl.

Sketches of the Braggs' ionization spectrometer measurements of the reflections from the (100), (110), and (111) planes of NaCl and KCl are shown in Figure 35.12. The first noticeable feature is the doublet nature of the peaks. These arise from the pair of K_α and K_β peaks, as shown in Figure 35.3. Several orders of these peaks were observed, and since the peak positions were different for the reflections from different planes, the spacings between these planes must be different, as one might expect. The substances NaCl and KCl are both cubic crystals and should both presumably display the same sequence of peaks. This similarity in reflection patterns was observed for the (100) and (110) planes, but the patterns for the (111) reflections appeared to be different. This latter factor was critically important, as we shall shortly see. In the following we shall consider only the larger K_α peaks. The first three peaks for the (100) planes of NaCl occurred at angles of $6°$, $11\frac{3}{4}°$, and $18\frac{1}{2}°$. The sines of

FIGURE 35.12 Reflections of the Pd X-ray emissions from the (100), (110), and (111) faces of NaCl and KCl as measured by the Braggs on the spectrometer shown in Figure 35.10. Note that the abscissa measures 2θ. [From W. H. Bragg and W. L. Bragg, *X-Rays and Crystal Structure* (London: G. Bell and Sons, Ltd., 1915).]

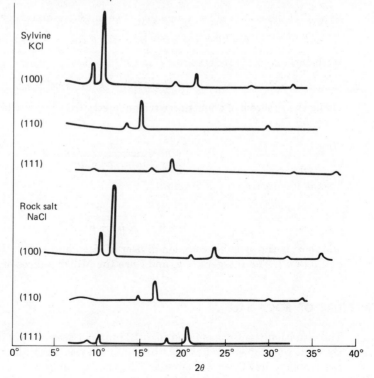

SEC. 35.5 STRUCTURE OF ROCK SALT

these three angles are 0.105, 0.204, and 0.317; their ratio is very close to 1:2:3, as it should be. Using Equation (35.9), we have $\lambda = 2d_{100} \cdot 0.104$; the distance between the d_{100} planes is thus

$$d_{100} = 4.81\lambda \tag{35.17a}$$

Similarly, we find that

$$d_{110} = 3.40\lambda \tag{35.17b}$$
$$d_{111} = 5.38\lambda \tag{35.17c}$$

and the Braggs were faced with a paradox. The spacings between (100), (110), and (111) planes should be in the ratio $1:1/\sqrt{2}:1/\sqrt{3}$, or $1:0.71:0.58$. The spacings the Braggs got were in the ratio $1:0.71:1.12$. The value for d_{111} was about twice what it should be relative to the others. When they turned to the KCl structure, they got $1:0.71:0.57$, the expected ratio.

The Braggs then tried to find a structure that would explain their results, and in this they were aided by numerous hypothetical structures that had been published some twenty years earlier. One of these structures was the correct NaCl structure, and this explained the anomaly; you can see this by referring to Figure 35.13, which shows the (100), (110), and (111) planes. We see from that figure that the spacing between successive (100) planes is one-half the length of the unit cube edge we call a. Similarly, all the (110) planes are equivalent, having equal numbers of Na and Cl atoms, and their spacing is one-half the diagonal of the unit cube face. The key to the problem lay in the (111) planes. There the successive planes were not equivalent, since there were alternate planes of the ions Na^+ and Cl^-. Since reflection maxima occurred for equivalent planes, the spacing in this direction was twice as large. For KCl, on the other hand, these alternate planes appeared identical by virtue of the relative positions of K ($Z = 19$) and Cl ($Z = 17$) in the periodic table. The electronic structure of both K^+ and Cl^- have the argon ($Z = 18$) configuration. Since the scattering is caused by the electrons, the correct spacing ratio was observed for KCl. The similarity of the K^+ and Cl^- electronic configurations caused the KCl structure to give a diffraction pattern very similar to that of a simple cubic lattice. (A modern diffraction camera would reveal intensity differences that would show that the structure was not simple cubic.)

Thus the geometrical patterns of the NaCl and KCl structures were resolved, and all that remained was to place the numbers on an absolute basis. This could be done by using the known density of NaCl. The value the Braggs got for the unit cell of NaCl was 5.6×10^{-8} cm. This was the *first* atomic dimension that had ever been experimentally determined. This value then enabled the Braggs to determine that

FIGURE 35.13 The (100), (110), and (111) planes in the NaCl and KCl crystals.

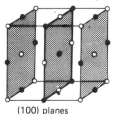 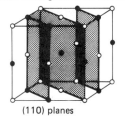 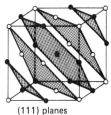

(100) planes   (110) planes   (111) planes

○ Na or K   ● Cl

the wavelength of the Pd K_α line was 0.576×10^{-8} cm. Now that this wavelength had been determined, the groundwork was set for unraveling numerous crystal structures.

35.6 SOME EXPERIMENTAL PROCEDURES

We have already noted two distinctly different experimental procedures for getting diffraction patterns from crystals. In the Laue method, a continuous X-ray source irradiates the crystal, and spots are observed at those angles and for those wavelengths at which the Laue equations are satisfied, as indicated in Figure 35.14a. The origin of these spots can perhaps be understood better if we return to our previous discussion of scattering by a one-dimensional array of scattering points or atoms. From Equation (35.6) we see that an incident beam that makes an incident angle α_0 with the row of atoms is scattered at the angle α,

$$\alpha = \cos^{-1}\left(\frac{m\lambda}{a} + \cos \alpha_0\right) \tag{35.18}$$

These angles define a series of cones, the *Laue cones,* whose common axis is the row of atoms as shown in Figure 35.15. The only orders possible are those for which the quantity in parentheses lies between $+1$ and -1. For a three-dimensional array, the three Laue equations must all be simultaneously satisfied. Each row in the lattice has a set of scattering cones. The scattering from all three rows will be in phase and produce a spot only if the three cones intersect in a common line to define a common scattering direction. For an arbitrary beam direction and an arbitrary crystal orientation, it is extremely rare for the three cones to intersect when monochromatic radiation is used. For this reason the Laue method uses continuous radiation. Out of the infinite set of wavelengths, there will be several for which diffraction maxima can occur.

From the experimental arrangement shown in Figure 35.14a, we see that $\tan \theta = R/D$, where D is the distance between the crystal and the photographic plate and R is the distance of the diffraction spot from the center of the pattern; θ is the scattering angle. Reverting to the Bragg form of the diffraction equation, we see that the condition $n\lambda = 2d \sin \theta$ can be satisfied for several values of λ so that

$$\lambda = \frac{2d}{n} \sin \theta \tag{35.19}$$

What this means is that each spot arises from the superposition of all the orders $n = 1, 2, 3, \ldots$ down to the minimum wavelength. These features of the Laue method render this technique almost useless for all but the simplest types of crystal structure determinations.

The Laue method finds its greatest use in the determination of symmetry and the space group of the crystal under investigation. With well-formed crystals, the beam can be directed parallel to one of the axes, and the symmetry of the crystal can be determined from the pattern by standard indexing techniques. The symmetry of the crystal is reflected by the symmetry of the pattern, as shown in Figure 35.14b. If unformed crystals must be used, then these are mounted in an arbitrary direction and their position adjusted until the beam is parallel to one of the axes, as indicated by the resulting pattern. The orientation of the crystal is adjusted in a *goniometer,* a sketch of which is shown in Figure 35.16.

FIGURE 35.14 (a) The Laue method; continuous X radiation is used with a stationary crystal. (b) Laue photograph of single crystal of silicon. The photograph was taken with the beam parallel to the (100) axis; the pattern displays the fourfold symmetry of the silicon, which has the diamond structure. (Courtesy of Westinghouse Electric Corporation, Research and Development Center, Pittsburgh, Pennsylvania.)

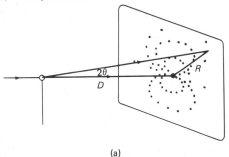

(a)

(b)

In X-ray *powder photography,* or the *Debye-Scherrer* method, a polycrystalline powdered sample is used. Since the number of small crystals in the sample is very large and the orientation of these tiny crystals is random, there are many crystals whose orientation relative to the incident beam is such that the condition for a Bragg reflection is satisfied. Cones of X rays are diffracted by the sample as shown in Figure 35.17 such that

$$\sin \theta = \frac{n\lambda}{2d}$$

FIGURE 35.15 The Laue, or diffraction, cones arising from a one-dimensional row of scattering centers.

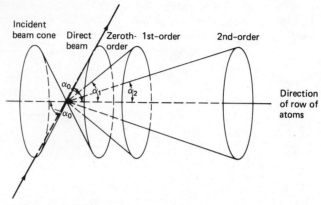

This provides values of the interplanar spacings (or more properly d/n). If a flat photographic plate were to be used in the powder method, the diffraction pattern would consist of a set of concentric circles, but the maximum value of θ that could be measured would be severely limited by the geometry. In practice, a strip of film is wrapped about the sample as indicated and the strip unrolled to measure the pattern. The powder method can be viewed as a variation of the Bragg method in which all the crystal planes are simultaneously irradiated, but this technique has its own inherent difficulties.

As seen from Equation (35.19), all planes with interplanar spacing

$$d_{hkl} = \frac{a}{\sqrt{h^2 + k^2 + l^2}} \tag{35.20}$$

give reflections at the same value of θ. Since the crystals are randomly oriented, all orientations are equally likely. Thus the twelve diagonal planes of the cube (011), (0$\bar{1}$1), (01$\bar{1}$), (0$\bar{1}\bar{1}$), and so on, which are all symbolized by {011}, give the same powder line; thus this line will be twelve times more intense than it would be if it

FIGURE 35.16 A goniometer head.

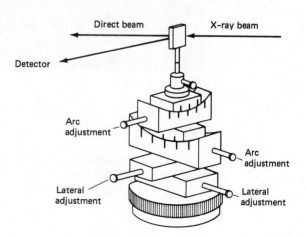

SEC. 35.6 SOME EXPERIMENTAL PROCEDURES

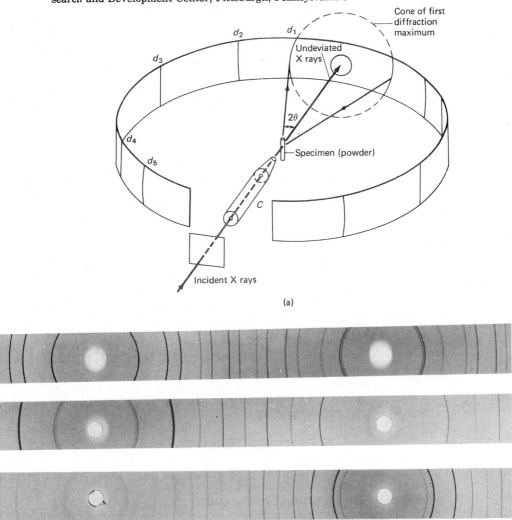

FIGURE 35.17 (a) The Debye-Scherrer powder technique. (b) Powder photographs for germanium, tungsten, and rock salt (NaCl). (Courtesy of Westinghouse Electric Corporation, Research and Development Center, Pittsburgh, Pennsylvania.)

were due to only one plane. For the general plane (hkl), where all three indices are different, this *multiplicity* factor is 48. What increases the difficulty is that in cubic systems, planes such as (522) and (441) give the same powder line, since $h^2 + k^2 + l^2 = 33$ for both. Although the spacing between planes is unambiguously determined by the Bragg equation, it is impossible to determine the angles between the various planes by the powder method exclusively. It is generally impossible to determine the symmetry properties of the crystal by this technique alone. For simple crystal structures whose symmetry can be determined from the crystal morphology, a powder pattern may suffice for a crystal-structure determination, but more complicated structures require more sophisticated approaches. One powerful use of the powder method is in routine analytical work. Any crystalline substance produces a distinctive powder diagram that can be a "fingerprint" for that substance against which unknown samples can be compared. Powder diagrams find extensive use in metallurgy, in distinguishing between one-phase and two-phase alloys. If the system consists of two phases, then the powder diagram contains two sets of lines, one for each constituent metal. If the system is a single-phase solid solution, then only one set of lines is observed whose indicated spacings are midway between those of the two constituents.

In the rotating-crystal method, the film is mounted with cylindrical symmetry about a single crystal that is continuously rotated. The monochromatic beam is diffracted whenever a plane is presented such that the Braggs' law is satisfied. A simple rotating crystal camera is shown in Figure 35.18. The cones of the reflected rays are such that all planes parallel to the rotation axis reflect spots along horizontal planes. Other planes reflect spots in layers above and below this horizontal plane. The lines of spots are called *layer lines*. One problem with this technique is the possibility of overlapping reflections. This is minimized or eliminated entirely with a variation on this technique known as the *oscillating-crystal* method, in which the crystal is oscillated back and forth through some small angle rather than 360°. Two powerful techniques that find wide application in complicated crystals are the *Weissenberg* method and the *precession*, or *Buerger precession*, method. In both of these methods the film is moved in synchronization with the moving crystal; moving slits are often included to allow for the exposure of only one layer at a time. These techniques are described in the references listed in the Bibliography.

FIGURE 35.18 The rotating crystal method. (a) X-ray diffraction by a rotating crystal and the formation of layer lines on a cylindrical film. (b) The intersection of diffracting cones with a photographic plate, placed perpendicular to the X-ray beam. (c) Experimental arrangement. (d) Formation of spots on cylindrical film.

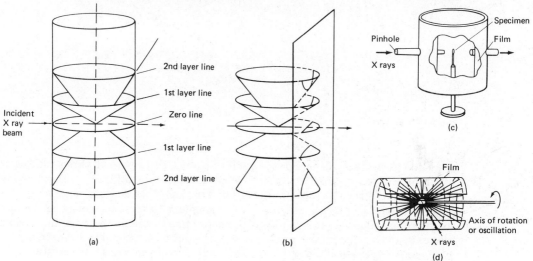

35.7 THE RECIPROCAL LATTICE

When we take a photograph of an ordinary object with a camera, we get a picture of that object in what we might call *real space*. If we had a very powerful microscope whose resolution was such that we could see the individual atoms in the crystal, then we could presumably obtain a picture of the crystal structure in this real space. Such a microscope would require the use of electromagnetic radiation whose wavelength was much smaller than the dimensions of the atoms in the crystal. We must be satisfied with photographs of the diffraction patterns that arise from the crystal structure, and one might well ask what this photograph represents. The answer is that it corresponds to a photograph of the lattice in *reciprocal space*. Reciprocal relations are not new to us. In thermodynamics the concept of temperature is fundamental and enters into a wide variety of thermodynamic relations. In certain respects, however, the reciprocal of the temperature is a more fundamental concept, since T^{-1} is the integrating factor that converts the inexact differential dq into the exact differential dS. For the Bragg equation, the angle of scattering is such that $\sin \theta$ is proportional to d^{-1}, where d is the interplanar spacing. There is a reciprocal relation between d and the distance between the orders of reflections. For small d there is a large distance between successive orders of reflection, and for large d there is a small distance between successive orders.

A lattice in real space can be considered an infinite set of planes that intersect at various angles. Suppose we select one point in the crystal and draw a set of normals to all these planes. If we then mark off intervals on these normals at points that are inversely proportional to the interplanar spacings, then the points so marked off constitute a new lattice. This new lattice is called the *reciprocal lattice*. If $\mathbf{a}$, $\mathbf{b}$, and $\mathbf{c}$ are the primitive translations of the original real lattice, then the primitive translations of the reciprocal lattice are

$$\mathbf{a}^* = \frac{\mathbf{b} \times \mathbf{c}}{V_c} \qquad \mathbf{b}^* = \frac{\mathbf{c} \times \mathbf{a}}{V_c} \qquad \mathbf{c}^* = \frac{\mathbf{a} \times \mathbf{b}}{V_c} \qquad (35.21)$$

where

$$V_c = \mathbf{a} \cdot [\mathbf{b} \times \mathbf{c}] = \frac{1}{V_c^*} \qquad (35.22)$$

is the volume of the primitive cell. From (35.21) we see that

$$\mathbf{a}^* \cdot \mathbf{a} = \mathbf{b}^* \cdot \mathbf{b} = \mathbf{c}^* \cdot \mathbf{c} = 1 \qquad (35.23)$$
$$\mathbf{a}^* \cdot \mathbf{b} = \mathbf{a}^* \cdot \mathbf{c} = \mathbf{b}^* \cdot \mathbf{c} = \mathbf{b}^* \cdot \mathbf{a} = \mathbf{c}^* \cdot \mathbf{a} = \mathbf{c}^* \cdot \mathbf{b} = 0 \qquad (35.24)$$

The reciprocal lattice has several properties that make the concept very useful in diffraction studies. A vector in the reciprocal lattice from the origin to the point hkl in the reciprocal lattice can be written as

$$\boldsymbol{\sigma}_{hkl} = h\mathbf{a}^* + k\mathbf{b}^* + l\mathbf{c}^* \qquad (35.25)$$

The plane (hkl) of the real lattice intercepts $\mathbf{a}$ at $\mathbf{a}/h$, $\mathbf{b}$ at $\mathbf{b}/k$, and $\mathbf{c}$ at $\mathbf{c}/l$, as shown in Figure 35.19. It will be shown in the problems that the vector $\boldsymbol{\sigma}_{hkl}$ is normal to the plane (hkl) and that its length is the reciprocal of the interplanar spacing

$$d_{hkl} = \frac{1}{|\boldsymbol{\sigma}_{hkl}|} \qquad (35.26)$$

Each plane in the real lattice is represented by a point in the reciprocal lattice.

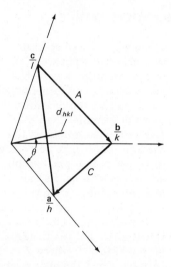

FIGURE 35.19 The plane (hkl) intersecting the axes of the crystal.

Suppose we write the Bragg equation in the form

$$\sin \theta_{hkl} = \frac{(1/d_{hkl})}{2/\lambda} \qquad (35.27)$$

Omitting the factor n simply means that we consider the different orders as arising from different interplanar spacings from parallel sets of planes. A geometrical interpretation of Equation (35.27) is depicted in Figure 35.20a. The circle has a radius of $1/\lambda$, and θ subtends a chord that is $1/d_{hkl}$. Using the fact that $1/d_{hkl} = \sigma_{hkl}$, we can give this construction the physical meaning indicated in Figure 35.20b. The diameter is the direction of the incident X-ray beam. The chord AP makes the Bragg angle θ with the incident beam and represents the slope of the reflecting crystal plane; this slope is indicated by the line passing through the center of the circle at angle θ. The line OP, which has length σ_{hkl}, is a translation of the reciprocal lattice. This construction forms the basis of the *Ewald sphere,* or *sphere of reflection,* whose use you can see by referring to Figure 35.21, a planar section of a three-dimensional construction. The origin of the crystal is taken at the center of the sphere, which has

FIGURE 35.20 Geometrical interpretation of the Bragg equation in reciprocal space. (a) The circle of radius $1/\lambda$. (b) The X-ray beam reflected at angle 2θ.

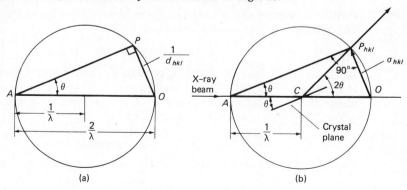

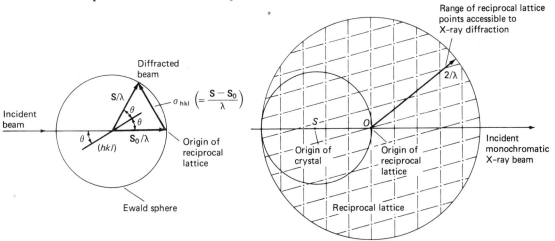

FIGURE 35.21 The Ewald sphere showing (a) the various directions as vectors and (b) the sphere in relation to the reciprocal lattice.

radius $1/\lambda$; and the point O, at which the X-ray beam leaves the sphere, is taken as the origin of the reciprocal lattice. The Bragg law is satisfied only when a point of the reciprocal lattice lies on the surface of the sphere of reflection, in which case $\mathbf{OP} = \boldsymbol{\sigma}_{hkl}$ is a translation vector of the reciprocal lattice; the vector $\mathbf{OP}$ represents the direction of the diffracted beam. For a monochromatic X-ray beam of arbitrary wavelength λ and arbitrary angle θ, there are in general no points of the reciprocal lattice that lie on the surface of the Ewald sphere, and no diffraction is observed. Rotating the crystal through all different possible orientations corresponds to rolling the sphere of reflection about the outer limiting sphere. Each time the inner sphere encounters a lattice point in the reciprocal lattice, a diffracted beam is emitted in the direction given by drawing a vector from the origin of the reciprocal lattice to the intersected lattice point. By using this technique, one can associate the spots on the photographic plate with the proper (hkl) indices of the reflecting plane. A complete structure determination also requires considering the relative intensities of the spots on the X-ray photograph.

35.8 INTENSITIES OF THE DIFFRACTED BEAM

We have already observed one factor that affects the intensities of the observed diffraction maxima, namely the multiplicity factor met in connection with powder diagrams. Two other factors that affect the intensities are the *atomic scattering factor* and the *structure factor*.[3]

Up to now we have considered atoms "points" placed in the lattice in a periodic fashion. The X-ray scattering is caused by the electrons in an atom, and these electrons have a "smeared-out" distribution in space. A phase difference arises between X-ray "wavelets" scattered by different portions of the electron distribution of an

---

[3] Other factors such as atomic vibrations, and experimental problems such as nonhomogeneous beams also affect the intensities. These are discussed in the advanced texts listed in the Bibliography.

atom. This is shown in Figure 35.22, where the vector $(S - S_0)$, representing the vector difference between the incident and reflected beam, is normal to the reflecting plane of the crystal. The phase difference between the wavelet scattered by the volume element dV and a wavelet scattered by the center of the atom is

$$\left(\frac{2\pi}{\lambda}\right)(S - S_0) \cdot r \qquad (35.28)$$

where r is the vector from the center to dV. The amplitude scattered by the electrons in the atom referred to an electron at the center is obtained by integrating over all the volume elements,

$$f = \int \rho(r) \exp\left[\left(\frac{2\pi i}{\lambda}\right)(S - S_0) \cdot r\right] dV \qquad (35.29)$$

where $\rho(r) \, dV$ is the probability of finding the electron in dV. The exponential term is just the superposition of two waves that have the same wavelength but are out of phase by an amount given in (35.28). If r makes an angle ϕ with $S - S_0$, then

$$(S - S_0) \cdot r = |S - S_0| r \cos \phi = 2r \sin \theta \cos \phi \qquad (33.30)$$

If we let

$$k = \frac{4\pi \sin \theta}{\lambda} \qquad (35.31)$$

then the amplitude for a spherical charge distribution is

$$f = \int_{r=0}^{\infty} \int_{\phi=0}^{\pi} \rho(r) \exp[ikr \cos \phi] 2\pi r^2 \sin \phi \, d\phi \, dr$$

$$= \int_{r=0}^{\infty} 4\pi r^2 \rho(r) \frac{\sin kr}{kr} \, dr \qquad (35.32)$$

The value $4\pi r^2 \rho(r) \, dr$ is the probability of finding an electron between r and $r + dr$. The quantity $4\pi r^2 \rho(r)$ is usually represented by $U(r)$, to get

$$f = \int_0^{\infty} U(r) \frac{\sin kr}{kr} \, dr \qquad (35.33)$$

The term f is the atomic scattering factor. It gives the ratio of the scattered amplitude of the charge distribution of an atom to the amplitude that would be scattered

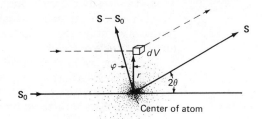

FIGURE 35.22 Scattering of X rays by an atom with a spherical charge distribution. The wavelet scattered by the electron charge in dV may be out of phase with the wavelet scattered by the electron charge at the center of the atom.

by a point electron at the center of the atom. The sum of all the electrons must equal Z the atomic number; hence

$$\int_0^\infty U(r)\, dr = Z \tag{35.34}$$

Values of the atomic scattering factors for all atoms have been tabulated for the convenience of crystallographers.[4]

The structure factor is got by summing the atomic scattering factors over all the atoms in a unit cell; this gives

$$F(hkl) = \sum_i f_i \exp\left[\frac{2\pi i}{\lambda}\, \mathbf{r}_i \cdot (\mathbf{S} - \mathbf{S}_0)\right] \tag{35.35}$$

Here the summation is over all the atoms in the unit cell, and f_i is the atomic scattering factor of the ith atom. The vector $\mathbf{r}_i$ represents the vector from the origin to the coordinates of the ith atom,

$$\mathbf{r}_i = x_i \mathbf{a} + y_i \mathbf{b} + z_i \mathbf{c} \tag{35.36}$$

where $\mathbf{a}$, $\mathbf{b}$, and $\mathbf{c}$ are the primitive translations of the lattice, and x_i, y_i, and z_i are the coordinates of the ith atom expressed in fractions of the cell-edge lengths. The Laue equations take the form

$$\begin{aligned}(\mathbf{S} - \mathbf{S}_0) \cdot \mathbf{a} &= h\lambda \\ (\mathbf{S} - \mathbf{S}_0) \cdot \mathbf{b} &= k\lambda \\ (\mathbf{S} - \mathbf{S}_0) \cdot \mathbf{c} &= l\lambda \end{aligned} \tag{35.37}$$

when written as vector equations, and so

$$\mathbf{r}_i \cdot (\mathbf{S} - \mathbf{S}_0) = \lambda(hx_i + ky_i + lz_i) \tag{35.38}$$

The structure factor can be written in terms of the Miller indices as

$$F(hkl) = \sum_i f_i \exp[2\pi i(hx_i + ky_i + lz_i)] \tag{35.39}$$

The intensities of the diffraction maxima are given by $|F|^2$, which we can write as

$$|F|^2 = \left[\sum_i f_i \cos 2\pi(hx_i + ky_i + lz_i)\right]^2 + \left[\sum_i f_i \sin 2\pi(hx_i + ky_i + lz_i)\right]^2 \tag{35.40}$$

For a simple illustration of the use of the structure factor, consider a body-centered crystal containing only one type of atom. The atoms are located at positions (000) and ($\tfrac{1}{2}\tfrac{1}{2}\tfrac{1}{2}$) in the unit cell. The summation in Equation (35.39) has two terms. The exponential in the first term is unity, since $x_i = y_i = z_i = 0$, and

$$F(hkl) = f\{1 + \exp[i\pi(h + k + l)]\} \tag{35.41}$$

When $h + k + l$ is odd, $F(hkl)$ vanishes, and the reflections for these planes are absent in the diffraction pattern. Applications of this procedure to other cubic lattices are discussed in the problems.

---

[4] *International Tables for X-Ray Crystallography* (Birmingham, England: Kynoch Press, 1952–1975), vols. 1–5.

35.9 FOURIER SYNTHESES OF ELECTRON DENSITIES

Since the electron density is a periodic function in three dimensions, we can write the density as a Fourier series—

$$\rho(X, Y, Z) = \sum_h \sum_k \sum_l A_{hkl} \exp\left[-2\pi i \left(\frac{hX}{a} + \frac{kY}{b} + \frac{lZ}{c}\right)\right] \quad (35.42)$$

where A_{hkl} are the Fourier coefficients,

$$A_{hkl} = \frac{1}{V} \int_0^a \int_0^b \int_0^c \rho(X, Y, Z) \exp\left[2\pi i \left(\frac{hX}{a} + \frac{kY}{b} + \frac{lZ}{c}\right)\right] dX\, dY\, dZ \quad (35.43)$$

The term V represents the volume of the cell. This equation is similar to (35.39) for the structure factor; the two equations express the same things in different ways. Equation (35.42) uses absolute coordinates and (35.39) uses fractional coordinates. Equation (35.39) gives the amplitudes scattered by the discrete atoms in the cell, whereas Equation (35.43) gives the amplitudes in terms of the variable electron densities. Since they are different expressions for the same thing, they can be equated, to give

$$A_{hkl} = \frac{1}{V} F_{hkl} \quad (35.44)$$

This states that the Fourier coefficients are the scattering amplitudes per unit volume. If this result is used in (35.42), we get

$$\rho(x, y, z) = \frac{1}{V} \sum_h \sum_k \sum_l F_{hkl} \exp[-2\pi i(hx + ky + lz)] \quad (35.45)$$

The x, y, and z are fractional coordinates.

Equation (35.45) represents a reverse Fourier transform.[5] It allows us to recover the electron density of the crystal from the diffraction amplitudes $|F_{hkl}|$ if the phase

---

[5] The Fourier transform of the function $f(x)$ is defined by

$$F(s) = \int_{-\infty}^{+\infty} f(x) \exp(-i2\pi xs)\, dx$$

The transformation is symmetrical; the reverse transform is

$$f(x) = \int_{-\infty}^{+\infty} F(s) \exp(+i2\pi sx)\, ds$$

Thus two successive transformations yield the original function.

Fourier transforms enable us to examine functions in two different domains; the Fourier transform is a special case of the *Laplace transform*. Transform methods are generally used to simplify problem-solving analysis; a difficult problem is transformed into a form by which it is easier to determine a solution. In the diffraction of light from a slit or set of slits (that is, a diffraction grating) the transmission characteristics of the slit and the resulting diffraction pattern form a Fourier transform pair. The diffraction pattern can be obtained from the characteristics of the slit and vice versa using the equations above. In electrical circuits and optical and acoustical spectroscopy, the wave form and the spectrum are Fourier transforms of each other. Fourier transform (FT) nmr and ir spectroscopy are based on this property. In quantum mechanics, the uncertainty principle may be associated with the Fourier transform; the momentum and position of a particle are related through the Fourier transform. Fourier transforms are discussed in most textbooks dealing with advanced topics in applied mathematics and also in electrical engineering books dealing with circuit analysis. See for example R. Bracewell, *The Fourier Transform and Its Applications* (New York: McGraw-Hill Book Co., 1965).

> JEAN BAPTISTE JOSEPH FOURIER (1768–1830), French mathematician, was among those caught up in the French Revolution. He was arrested, but unlike the less fortunate Lavoisier, he survived the Revolution and served Napoleon in several diplomatic posts. He served with Napoleon in the Egyptian campaign and became secretary of the Institut d'Egypte. In this capacity he published numerous reports on that ancient civilization. His chief contribution to science was in developing the Fourier series and the Fourier integral and the solutions of partial differential equations. He was much concerned with the physical meaning of his equations and was one of the first to incorporate physical constants into his equations on the theory of heat. Along with Lazare Carnot, the father of Sadi Carnot, Fourier was one of the early developers of the theory of units and dimensional analysis; he used this technique to check the veracity of groups of physical constants that appeared in his equations. He was a master of notation, and in fact the symbol $\int_a^b$ for the definite integral is Fourier's invention. Although Fourier was the pioneer in the field of heat transfer, he had no interest in the burgeoning field of the conversion of heat into work, which was later formalized into the second law of thermodynamics, and remained in total ignorance of the work of Sadi Carnot on that subject.

of F_{hkl} is known. It is a reverse Fourier transform in which the integrals have been replaced by summations, to reflect the periodic nature of the lattice. This transform enables us to go from the crystal space to the Fourier space, and the Fourier space is just the reciprocal space. The transformation from crystal space to Fourier space is

$$F_{hkl} = \sum_j f_j \exp[2\pi i(hx + ky + lz)] \tag{35.46}$$

and the transformation from Fourier space to crystal space is given by (35.45). The experimental observable consists of the $|F_{hkl}|$ in reciprocal space but *not* their phases.

The transformation from the Fourier space back to the real crystal space is carried out by a series of *trial structures,* an operation now performed using high-speed computers. The observed F's give only the absolute values of the intensities; the phases are lost in the experimental observations and must be reconstructed. Knowing the symmetry, the space group, and the unit cell dimensions of the crystal, one can place the atoms in trial positions and calculate the intensities using these trial positions. The Fourier series is then summed from the observed F's and the calculated phases. The graph of the Fourier summation gives new atomic positions, from which a new set of F's can be calculated. If things work as they should, successive approximations give better and better agreement between calculated and observed F's. If the agreement does not improve or if it gets worse, then a serious error has been made in the initial trial structure, and one must start the process all over again with a new trial structure. This process provides *Fourier maps* of the electron densities in various planes of the crystal. The resolution of these maps can be improved by increasing the number of terms taken in the series. A Fourier map of the uracil crystal is shown in Figure 35.23. This method is almost foolproof, since for most structure determinations the number of observed $|F_{hkl}|$ is very much larger than $3N$, where N is the number of atoms in the unit cell. From the point of view of statistical curve fitting, the problem is highly overdetermined, and the probability of fitting the data to an incorrect set of assumed positions is very small.

FIGURE 35.23 Valence Fourier difference map for the uracil molecule. The contributions from the core electrons have been subtracted, leaving only the contribution from the valence electrons. The contours are spaced at intervals of 0.2 e Å^{-3}. The dotted lines correspond to a charge density of 0.2 e Å^{-3}. The O and N atoms are clearly distinguishable as maxima, whereas the electron densities for the C atoms appear as wells or saddle points. (R. F. Stewart, *J. Chem. Phys.* 48 (1968):4882.)

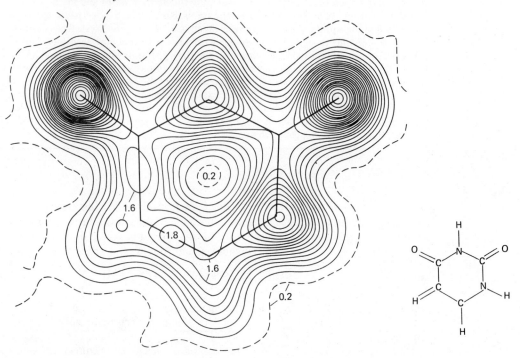

35.10 OTHER DIFFRACTION TECHNIQUES FOR STRUCTURAL DETERMINATIONS

Although this chapter has been concerned mainly with using X-ray diffraction for determining crystal structures, that does not exhaust the list of useful procedures for structural determinations. *Neutron diffraction* is a useful extension of X-ray diffraction for studying crystalline structures. All atomic particles have a wavelength that is determined by the de Broglie equation $\lambda = h/mv$. A particle is diffracted when λ is of the same order as the dimensions under study. Neutrons are useful in that they are diffracted by the nuclei rather than the electrons. Since neutrons have magnetic moments, their diffraction is affected by the magnetic moments of the nuclei. Hydrogen atoms scatter X rays and electrons very weakly in view of their small number of electrons, and it is difficult to locate these atoms in X-ray studies. On the other hand, protons scatter neutrons very efficiently, and they can be located in neutron diffraction studies.

When we considered the atomic scattering factor, we noted that atoms scatter X rays. This is the basis for X-ray diffraction studies of noncrystalline systems such as gases and liquids. It is particularly useful for studies of liquids in which the diffraction patterns are used to investigate the existence of short-range order. The molecules in liquids often agglomerate in small groups, or clusters, that display short-range order. These clusters give rise to diffraction patterns from which the structure of the short-range order can be determined.[6]

An electron that has been accelerated through a potential V has a de Broglie wavelength

$$\lambda = \frac{h}{mv} = \frac{h}{\sqrt{2meV}} = \frac{12.25}{\sqrt{V}} \text{ Å} \qquad (35.47)$$

Thus electrons that have been accelerated through potential differences of some 30–40 kilovolts (kV) have wavelengths of the order of 0.06–0.07 Å. This is small enough to permit diffraction effects to be observed when electron beams interact with matter. Electron diffraction is particularly useful in determining the dimensions of gaseous molecules. A narrow beam of energetic electrons intersects a beam of molecules, as shown in Figure 35.24, and the interference pattern is observed on a

FIGURE 35.24 (a) Sketch of electron diffraction apparatus. The points P_1 and P_2 are two small pinholes for a narrow pencil of electrons. The electron beam intersects the beam of gaseous molecules that are leaving a narrow orifice attached to the sample bulb. So that the high vacuum will be maintained, a liquid nitrogen cold finger condenses the gas molecules after the scattering. (b) Electron diffraction pattern of SiF_4. (Courtesy of K. W. Hedberg, Oregon State University.)

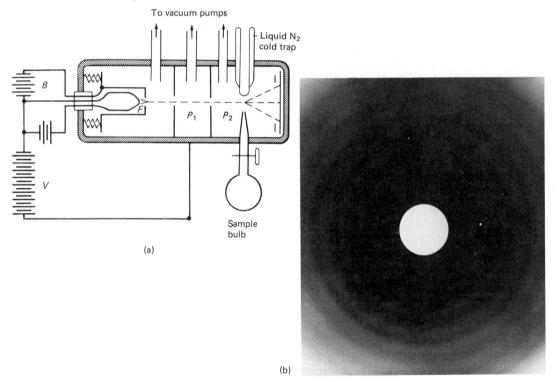

[6] See, for example, A. H. Narten, M. D. Danford, and H. A. Levy *Discussions Faraday Soc.* 43 (1967):97, for some work done on diffraction by liquid water.

FIGURE 35.25 X-ray examinations of a skull. (a) Simple skull X ray. (Courtesy of Sanford ▶ Levy, MD.) (b) Arteriogram of a skull. A radiopaque dye is injected into the patient's femoral artery. Now the arteries lying beneath the skull are clearly visible in the X ray. The large artery in the lower center is the carotid artery and can be seen branching out. The presence of tumors and other abnormalities can be determined from the displacement of the blood vessels from their normal positions. A series of X rays taken at short intervals allows the physician to determine abnormalities in blood-flow rates. This picture was taken before the blood arrived in the veins. A picture taken later would also show the veins. (Courtesy of Sanford Levy, M.D.) (c) A CAT scan of a skull. Two X-ray beams are used in such a way that only one plane of the skull is in focus on the X-ray screen. By using a photomultiplier tube as the detector, one can detect small differences in intensities caused by transmission differences in different types of tissue. The signal from the photomultiplier tube is fed into a computer, which acts as a two-dimensional multichannel analyzer. The two-dimensional intensity pattern is then displayed on a cathode-ray screen and photographed. Resolution of the order of 1 mm is routinely available with modern equipment. (See footnote 8. Photograph supplied by Technicare Corporation, Solon, Ohio.)

photographic plate. The pattern takes the form of concentric rings; and from the geometry and the intensities of the rings in the pattern, the structure can be determined. The process is like X-ray crystallographic procedures in that trial structures are sought for which the calculated pattern agrees with the measured pattern.[7]

35.11 MISCELLANEOUS APPLICATIONS OF X RAYS AND ELECTRON BEAMS

Ordinary light beams can be bent and focused by appropriate lenses and prisms as found in microscopes. The maximum resolution attainable is ultimately limited by the wavelength of the light, of the order of 5000 Å for visible light. A microscope that used X rays would in principle be capable of a resolution some four orders of magnitude higher than a light microscope. Unfortunately, there are no materials capable of refracting X rays; thus it is impossible to manufacture X-ray lenses. When X rays are used to examine matter directly in real space, the examination is usually carried out under unit magnification by simply interposing the object between the X-ray beam and the photographic plate; this is common in medical procedures. It is possible also to use X rays like this to examine structural flaws in metals.

Most of you are familiar with X rays as used in diagnosing broken bones, tooth decay, and the like. Radiology has made great strides in recent years, and the radiologists have developed several ingenious techniques. We might consider as an example the problem of the neurosurgeon who is faced with removing a tumor or a blood clot from the brain. Neurosurgery is a high-risk procedure; the risk can be greatly minimized by accurately determining the location, size, and nature of the problem. Unfortunately, the problem is under the skull. A simple cranial X ray as shown in Figure 35.25a shows little useful information (except perhaps for skull fractures). The density of bone material hides the subtle differences in the transmission characteristics of various kinds of normal and abnormal nonbony tissues. The information obtained from X rays can be greatly enhanced by *arteriography,* a procedure in which the transmission characteristics of the head are altered by injecting a

---

[7] For further discussion, see Z. G. Pinsker, *Electron Diffraction* (London: Butterworth Scientific Publications, 1953).

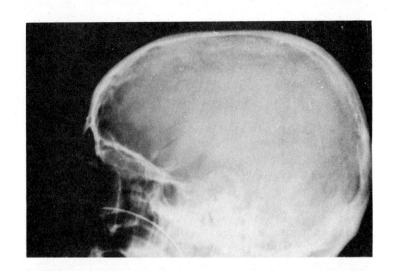

(a)

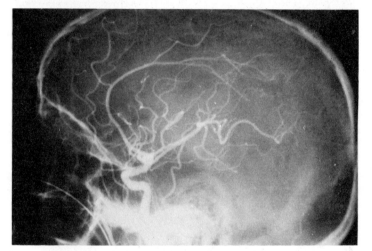

(b)

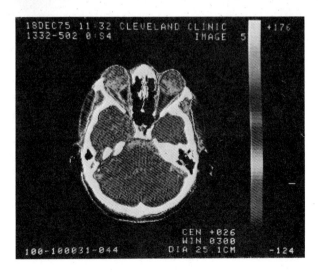

(c)

radiopaque dye into the patients femoral artery. Now the blood vessels in the head show clearly on the X-ray plate, as shown in Figure 35.25b. Many abnormalities in the brain show up as displacements of the blood vessels from their normal positions. By measuring these displacements, the surgeon can locate the source of the problem before surgery.

Arteriography itself poses certain risks, because some patients undergo severe reactions to the dye; moreover, the procedure itself is a surgical one. Also, certain types of problems may not show up. A procedure has been developed called CAT (an acronym for "computer-assisted tomography"; a tomogram is an X ray of a plane section of the body), in which the X-ray beam and the detector are both simultaneously rotated so that only one plane of the skull is in focus. The detector is a photomultiplier tube instead of a photographic film. The voltage output from the photomultiplier is fed directly into a computer, which acts as a two-dimensional multichannel analyzer and stores the information from many small incremental areas. Small differences in intensities are detected by the computer, and from these differences, a density distribution of plane sections through the skull can be constructed.[8] The density distribution is displayed on a cathode-ray screen and photographed for subsequent careful examination. A photograph of a *CAT scan* is shown in Figure 35.25c.

In contradistinction to X rays, the charged nature of electrons allows them to be bent by electrostatic and magnetic fields. Electron beams can be focused by electromagnetic equivalents of ordinary light optical components. In the *transmission electron microscope* (TEM), a beam of electrons passes through the material under investigation. Different portions of the material absorb electrons to greater or lesser extents, and a magnified image of the transmitted beam indicates the structure.

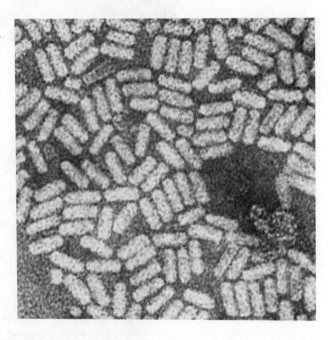

FIGURE 35.26 Transmission electron micrograph of rod-shaped bacilliform mushroom virus. Magnification ×200,000. (Photograph courtesy of Martin Haller, Mellon Institute.)

[8] For further discussion of this technique, see R. Gordon, G. T. Herman, and S. A. Johnson, Image Reconstruction from Projections, *Scientific American* 233, no. 4 (October 1975), p. 56. Also W. Swindell and H. H. Barrett, Computerized Tomography: Taking Sectional X Rays, *Physics Today,* vol. 30, no. 12 (December 1977), p. 32.

Magnifications in the hundreds of thousands are routinely carried out with these instruments. The transmission electron microscope is invaluable in biological investigations, providing details unattainable with the best optical instruments. A transmission electron micrograph is shown in Figure 35.26.

In the 1960s, the adjective *transmission* was usually omitted, and the term *electron microscopy* used to refer to the procedure outlined in the previous paragraph. The adjective *transmission* came into later use to distinguish TEM from *scanning electron microscopy* (SEM). In SEM, a highly focused beam of energetic electrons is allowed to strike the surface of a sample. Some of these electrons are elastically scattered from the sample, while others induce the emission of low-energy secondary electrons. The electrons that are backscattered contain information that can be transformed into an image of the surface of the sample. The backscattering is propor-

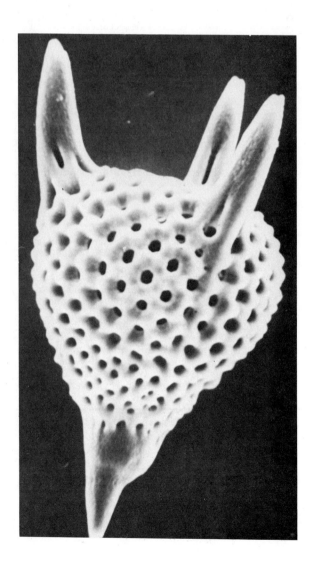

FIGURE 35.27 Scanning electron micrograph of the silica skeleton of the *Radiolaria* marine amoeba. Note the excellent depth of field. Magnification ×1000. (Photograph courtesy of Martin Haller, Mellon Institute.)

tional to the atomic number; hence SEM is a very useful tool in studying boundaries between different metals in alloys and in phase studies in metallurgy. The image of the surface is displayed on the screen of a cathode-ray tube, where it can be visually examined and photographed for a permanent record. The magnification range of these instruments is extensive. Modern commercial instruments can produce images with magnifications ranging from 10 to 50,000. In addition, the depth of field on the SEM is orders of magnitude larger than on an ordinary optical instrument; you can see this on the photograph in Figure 35.27 (p. 723).

A logical development in the SEM instrument is the inclusion of *electron microprobe analysis* (EMA) or, as it is also called, energy dispersive analysis by X rays (EDAX). Since the sample is bombarded by an energetic electron beam, X rays are emitted, and each element emits its characteristic X ray spectrum. In both SEM and EMA, the input to the sample is the same, namely a beam of energetic electrons. Conceptually they differ only in how the output signal is examined. The two functions can be combined in a single instrument by the simple expediency of including two different detection systems. The result is an instrument that simultaneously provides a visual examination of the surface of the sample and an elemental analysis of a region of the sample that is limited by the size of the electron beam. Samples of the order of 1000–2000 Å can be analyzed. These instruments are used extensively by forensic laboratories when only minute samples are available for analysis. This instrument has also been applied in examining micrometer-sized particulates in air pollution studies. Particulates from coal-fired boilers are collected on a surface and examined for appearance and elemental composition.[9] The entire process can be automated under computer control, whereby the sample surface is automatically translated in two dimensions to determine the presence of particulates; at such indication, the motion of the stage is stopped and the sample analyzed.

PROBLEMS

1. A beam of 30,000 eV electrons impinges on a metal target. Calculate the shortest wavelength of X ray emitted.
2. The wavelength of the tungsten K_α line is 0.21 Å. This radiation strikes the clean surface of sodium metal. Calculate the energy of the ejected photoelectron.
3. The tungsten K_α line is in reality a doublet, the two lines having wavelengths of 0.20862 and 0.21345 Å for the K_{α_1} and K_{α_2} lines respectively. Calculate the energy difference between the two states giving rise to these two lines. Explain how these two states of different energy arise.
4. The wavelength of the Cu K_α line is 1.54184 Å and the wavelength of the Cr K_α line is 2.29100 Å. (*Note:* These wavelengths are for the unresolved line.) Calculate the wavelength of the Co K_α line, and compare it with the actual value of 1.79026 Å.
5. Sodium chloride ($d = 2.81968$ Å) is a commonly used crystal in X-ray spectroscopy. The Cu K_{α_1} line has a wavelength of 1.54056 Å and the Cu K_{α_2} line has a wavelength of 1.54439 Å; the unresolved K_α line has a wavelength of 1.54184 Å. These lines are being investigated by measuring the reflections from an NaCl crystal using the planes with $d = 2.81968$ Å.
   (a) How many orders of reflection can be observed, and at what angles will these orders appear?
   (b) At each angle what is the angular separation between the α_1 and the α_2 peak?
   (c) Suppose a linear strip of film of length 1 m is used to measure the peaks, and the strip is placed 75 cm from the crystal. How many peaks can be observed on the strip if the edge of the strip is lined up with the incident beam? Calculate the linear spacing between the α_1 and α_2 peaks on the strip. Where is the best place to measure the separation with the greatest accuracy?
   (d) The interplanar spacing in NaCl changes by 0.00011% per °C. If you were measuring the wavelength of the K_α lines, what magnitude of error would be introduced if the temperature were in error by 2 °C?
6. Another common crystal in X-ray spectroscopy is

[9] See R. J. Cheng, V. A. Mohnen, T. T. Shen, M. Current, and J. B. Hudson, *J. Air Pollution Control Assn.* 26 (1976):787.

mica, which has an interplanar spacing of 9.96280 Å. With which substance, mica or NaCl, could you achieve greater accuracy in measuring the wavelength of X rays if the geometry and apparatus were the same? (See Problem 5.)

7. Using the data of the Braggs given in the text and in Figure 35.12 and recognizing that the density of NaCl is 2.165 g cm^{-3}, calculate the size of the NaCl unit cell and the wavelength of the radiation the Braggs used.

8. Powdered chromium metal is used as the sample in a Debye-Scherrer powder camera. The sample is irradiated with Co K_α radiation (λ = 1.790 Å), and the sample is surrounded by a film strip having a diameter of 20 cm. Sketch the diffraction pattern by indicating the peaks with straight lines. (Note: Cr is body centered cubic with a unit cell edge of 2.88 Å.)

9. Sketch the two-dimensional reciprocal lattices for the two-dimensional real lattices: (a) simple square array, $a = b = 2$; (b) simple rectangular array, $a = 2b = 2$.

10. In Figure 35.19, the vectors $\mathbf{A} = [(\mathbf{b}/k) - (\mathbf{c}/l)]$ and $\mathbf{C} = [(\mathbf{a}/h) - (\mathbf{b}/k)]$ both lie in the plane (hkl). Show that the vector $\boldsymbol{\sigma}_{hkl} = h\mathbf{a}^* + k\mathbf{b}^* + l\mathbf{c}^*$ is perpendicular to (hkl) by showing that its dot product with $\mathbf{A}$ and $\mathbf{C}$ is identically zero.

11. In view of the result of Problem 10, a unit vector normal to the plane (hkl) can be written as

$$\mathbf{n} = \frac{h\mathbf{a}^* + k\mathbf{b}^* + l\mathbf{c}^*}{|\boldsymbol{\sigma}_{hkl}|}$$

Further, the interplanar spacing $d_{hkl} = (a/h) \cos \phi = (\mathbf{a}/h) \cdot \mathbf{n} = (\mathbf{b}/k) \cdot \mathbf{n} = (\mathbf{c}/l) \cdot \mathbf{n}$. Use this information to prove that (35.26) is valid.

12. A simple cubic crystal has a unit cube edge of 2.00 Å. First determine the reciprocal lattice for this structure, and then construct a sphere of reflection for X rays of wavelength 2.10 Å. As the crystal is rotated through all possible orientations, how many points in the reciprocal lattice give rise to maxima?

13. In Equation (35.41) of the text we showed that for a body-centered crystal (atoms at 000 and $\frac{1}{2}\frac{1}{2}\frac{1}{2}$) all reflections from planes (hkl) such that $h + k + l$ is odd are absent. These are known as *extinct reflections*. Show that for a face-centered cubic lattice containing only one type of atom (at positions 000, $\frac{1}{2}\frac{1}{2}0$, $\frac{1}{2}0\frac{1}{2}$, and $0\frac{1}{2}\frac{1}{2}$) the only allowed reflections are those for which h, k, and l are either all odd or all even, whereas if two are odd and one even or two even and one odd, the reflection is extinct.

14. The NaCl structure has the ions at positions as follows:

$$\text{Na}^+ \text{ at } 000,\ \tfrac{1}{2}\tfrac{1}{2}0,\ \tfrac{1}{2}0\tfrac{1}{2},\ 0\tfrac{1}{2}\tfrac{1}{2}$$
$$\text{Cl}^- \text{ at } \tfrac{1}{2}\tfrac{1}{2}\tfrac{1}{2},\ 00\tfrac{1}{2},\ 0\tfrac{1}{2}0,\ \tfrac{1}{2}00$$

Using Equation (35.40) show that:
(a) If h, k, and l are all odd, then $F = 4f_{\text{Na}} - 4f_{\text{Cl}}$, where f_{Na} and f_{Cl} are the scattering factors for Na and Cl. Therefore reflections from these planes are weak.
(b) If h, k, and l are all even, then $F = 4f_{\text{Na}} + 4f_{\text{Cl}}$ and these reflections are strong.
(c) If two are odd and one is even or two are even and one is odd, then $F = 0f_{\text{Na}} + 0f_{\text{Cl}}$, and these reflections are extinct.

15. Carry out the analysis of Problem 14 for the CsCl crystal, which has one molecule per unit cell at positions as follows:

$$\text{Cs}^+ \text{ at } 000 \qquad \text{Cl}^- \text{ at } \tfrac{1}{2}\tfrac{1}{2}\tfrac{1}{2}$$

16. The drawing below indicates the locations of the lines in an X-ray powder diagram of a simple cubic structure that is composed of only one type of atom. The wavelength of the monochromatic radiation was $0.4a$, where a is the edge of the unit cube. The lines have been indexed, and the angles are indicated by the scale.
(a) Verify the location of a few of the lines.
(b) Why are three of the lines indicated as coming from two different crystal planes?
(c) The spacings between the first six lines seem to follow a regular pattern. Why is there a sudden gap after the sixth line? (Why, incidentally, after the thirteenth line?)

17. With reference to Problem 16, how would the appearance of the sketch change for a face-centered and a body-centered cubic lattice with the same cube edge dimension? (That is, which peaks would be absent in bcc and fcc lattices?)

18. Figure 35.21 shows the Ewald sphere construction for a crystal being rotated. Construct the analogous sketch for a Laue pattern in which a continuous wavelength X-ray beam is used and show how to use the construction to determine which maxima will be observed. (See Klug and Alexander, *X-Ray Diffraction Procedures*, listed in the Bibliography.)

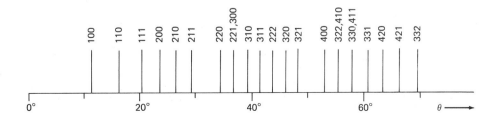

CHAPTER THIRTY-SIX

THERMAL PROPERTIES OF CRYSTALS

36.1 ENERGY OF AGGREGATION

Ideal gases are characterized by the complete absence of intermolecular forces. In a sense each molecule is completely unaware of the existence of any other molecule. On a microscopic scale the average intermolecular distance in an ideal gas is proportional to the volume (or more correctly to the cube root of the volume). Intermolecular forces are inversely proportional to some power of the distance; the greater the intermolecular distance, the weaker the forces. The thermodynamic definition of the ideal gas, $(\partial E/\partial V)_T = 0$, agrees with this intuitive idea in stating that the energy does not change with volume (that is, intermolecular distance) at constant temperature.

When we extended our discussion to real gases, we noted that the presence of intermolecular forces is characterized by the fact that the derivative, $(\partial E/\partial V)_T$, is no longer zero. These forces are, however, still much weaker than the intermolecular forces existing in liquids. The strongest forces are those found in the crystalline state. As we examine the energetics of the crystalline state of matter, we shall see how the regular structure found in the crystalline state presents the simplest conceptual situation.

36.2 CLASSIFICATION OF CRYSTAL BONDS

Crystals can be classified into several broad empirical types according to certain experimental characteristics. *Ionic crystals,* as exemplified by NaCl and KCl, are characterized by an interatomic bonding scheme mainly based on Coulombic forces between the charged ions. Ionic crystals have high binding energies, of the order of 850 kJ mol^{-1}, and they usually possess strong infrared absorptions.[1] They act as semiconductors in that they have low electrical conductivities at low temperatures and much higher conductivities at higher temperatures; the conductivity arises from the migration of ions. *Covalent crystals,* as exemplified by Si, Ge, SiC, and diamond, also have high binding energies, 710 kJ mol^{-1} for diamond and 1180 kJ mol^{-1} for SiC. They are characterized by great hardness, and they are true semiconductors. The electrical conductivity of pure specimens is very low at low temperatures and increases dramatically as the temperature is increased. Adding minute quantities of

---

[1] Note that crystals of ionic compounds such as NaCl and KCl are often used as "windows" in infrared spectrometers. When so used, they are suitable only at frequencies at which they are transparent to infrared radiation.

certain impurities increases the conductivity by several orders of magnitude. Covalent crystals have been extensively studied during the past 25 years because they are so important in the semiconductor electronics industry.

Metallic crystals are characterized by high electrical conductivity, which is due to the presence of "free" electrons. They have moderate binding energies, about 110 kJ mol^{-1} for Na and 390 kJ mol^{-1} for Fe. *Molecular crystals* such as Ar, Kr, and CH$_4$ have low melting and boiling points and are very compressible. They have low binding energies, of the order of 10 kJ mol^{-1} and are held together by weak van der Waals forces. *Hydrogen-bonded* crystals such as H$_2$O and HF have higher binding energies than similar molecules in which hydrogen bonding is lacking; ice has a binding energy of some 50 kJ mol^{-1}. There is a tendency for hydrogen-bonded systems to form large polymers.

The five categories just noted are not absolute. There is a general gradation of characteristics, many substances falling midway between any two of the classifications. In this chapter we shall concern ourselves with the binding energy of ionic crystals and some features of lattice energies. We shall examine electrical and magnetic properties in subsequent chapters.

36.3 BINDING ENERGY OF IONIC CRYSTALS; EXPERIMENTAL

Suppose we take NaCl as our example. The *lattice*, or *binding*, energy of the crystal is the energy of the reaction

$$\text{Na}^+(g) + \text{Cl}^-(g) \longrightarrow \text{NaCl}(c) \tag{36.1}$$

It is the energy it takes to convert a mole of NaCl crystal into its constituent ions. The energy of this reaction cannot be determined directly but can be determined from other experimentally measurable thermochemical data by a procedure known as the *Born-Haber cycle*. Consider the following path of reactions:

$$\begin{array}{c}
\text{Na}^+(g) + \text{Cl}^-(g) \xrightarrow{-\Delta E_0} \text{NaCl}(c) \\
\uparrow I \quad \uparrow -A \qquad \qquad \downarrow -\Delta H_f \\
-e \quad +e \qquad \xleftarrow{\Delta H_s} \left\{ \begin{array}{c} \text{Na}(c) \\ + \\ \frac{1}{2}\text{Cl}_2(g) \end{array} \right\} \\
\text{Na}(g) + \text{Cl}(g) \xleftarrow{\frac{1}{2}D_0}
\end{array}$$

The various energy terms in this cyclical process are defined as follows (on a molar basis):

ΔE_0 is the crystal binding energy.
ΔH_f is the enthalpy of formation of NaCl(c).
ΔH_s is the heat of sublimation of Na(c).
D_0 is the dissociation energy of Cl$_2$(g).
A is the electron affinity of Cl.
I is the ionization potential of Na.

Since $\oint dE = 0$ for any cyclic process, we can write

$$-\Delta E_0 - \Delta H_f + \Delta H_s + \tfrac{1}{2}D_0 - A + I = 0 \tag{36.2}$$

or

$$\Delta E_0 = -\Delta H_f + \Delta H_s + \tfrac{1}{2}D_0 - A + I \tag{36.3}$$

All the quantities on the right-hand side of (36.3) can be measured. The heat of formation and the heat of sublimation are determined from standard thermochemical experiments, while the dissociation energy, the ionization potential, and the electron affinity can be determined from spectroscopic measurements. The numerical evaluation of the binding energy is left for one of the problems.

36.4 BINDING ENERGY OF IONIC CRYSTALS; THEORETICAL

If we let E_{ij} be the interaction energy between the ions i and j, then the total energy of any one ion i is

$$E_i = \sum_j{}' E_{ij} \tag{36.4}$$

where the prime indicates that the summation includes all ions except for $j = i$. The term E_{ij} contains two separate terms, a simple Coulombic term and a central-field repulsive term. We use as our repulsive term the form $E_{\text{rep}} = \lambda/(r_{ij})^n$, where λ and n are empirical constants and r_{ij} is the distance between the two ions. (This potential is often taken as an exponential potential, $E_{\text{rep}} = b \exp(-cr_{ij})$). The total binding energy is obtained by summing over all the ions

$$E = \frac{1}{2}\sum_{i,j}{}' \pm \frac{e^2}{r_{ij}} + \frac{1}{2}\sum_{i,j}{}' \frac{\lambda}{r_{ij}{}^n} \tag{36.5}$$

where the factor $\tfrac{1}{2}$ enters to ensure that we count each pair of interactions only once. It is convenient to write this expression in terms of the equilibrium lattice constant of the crystal; we introduce a term d_{ij} such that

$$r_{ij} = d_{ij} r_0 \tag{36.6}$$

where r_0 is the nearest-neighbor separation of the crystal. We now write (36.5) in the form

$$E = \frac{\lambda A_n}{r_0{}^n} - \frac{\alpha e^2}{r_0} \tag{36.7}$$

where

$$A_n = \sum_j{}' d_{ij}{}^{-n} \tag{36.8}$$

$$\alpha = \sum_j{}' (\mp) d_{ij}{}^{-1} \tag{36.9}$$

The term α is known as the *Madelung constant* and is a characteristic of the particular crystal structure. The term $\mp$ indicates that the individual terms in the series are either negative or positive, depending on whether the two ions have the same or opposite signs. The sum for A_n expressed in (36.8) converges very rapidly since n is quite large, usually in the range 8 to 12. Figure 36.1 indicates the variation of en-

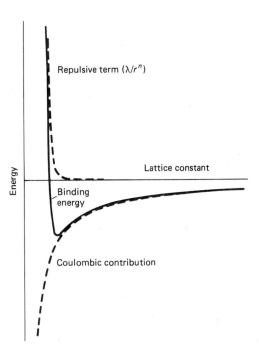

FIGURE 36.1 Potential energy of an ionic crystal as a function of the lattice constant. The minimum in the curve occurs at the equilibrium interatomic spacing.

ergy with r_0, the lattice constant of the crystal. The equilibrium position occurs at the minimum in the curve where $\partial E/\partial r = 0$; at that point

$$-\frac{nA_n\lambda}{r_0^{n+1}} + \frac{\alpha e^2}{r_0^2} = 0 \qquad (36.10)$$

We can solve (36.10) for the term λA_n and eliminate this from (36.7), to get

$$E = -\frac{\alpha e^2}{r_0}\left(1 - \frac{1}{n}\right) \qquad (36.11)$$

for the energy of an ion. The total energy of a crystal containing $2N$ ions is then

$$\Delta E_0 = -\frac{N\alpha e^2}{r_0}\left(1 - \frac{1}{n}\right) \qquad (36.12)$$

Note that we multiplied by N instead of $2N$, since each pair of interactions is counted only once.

Calculating the constant α is involved; it was first done by Madelung[2]. Values of the Madelung constant for several crystalline substances are listed in Table 36.1. A value for the exponent n can be estimated by comparing observed binding energies as determined from the Born-Haber cycle, with calculated energies using (36.12); such comparisons indicate that n is of the order of 10. An independent estimate of the value of n can also be obtained from compressibility data. Recall that the compressibility is defined by $\beta = -(1/V)(\partial V/\partial P)_T$.

[2] E. Madelung, *Physik. Z.* 19 (1918):524.

TABLE 36.1 Values of the Madelung constant for several structures.

Structure	α
Sodium chloride (NaCl)	1.747558
Cesium chloride (CsCl)	1.762670
Wurtzite (ZnS)	1.64073
Zinc blende (ZnS)	1.63806
Fluorite (CaF$_2$)	5.03879
Corundum (Al$_2$O$_3$)	25.0312

The compressibility can be related to the energy and the volume at low temperature by the equation

$$\frac{1}{\beta} = \overline{V}\left(\frac{\partial^2 \Delta E}{\partial V^2}\right)_T \tag{36.13}$$

where $\overline{V}$ is the molar volume. By using the chain rules for differentiation, we have that

$$\frac{\partial \Delta E}{\partial V} = \left(\frac{\partial \Delta E}{\partial r}\right)\left(\frac{\partial r}{\partial V}\right)$$

$$\frac{\partial^2 \Delta E}{\partial V^2} = \left(\frac{\partial \Delta E}{\partial r}\right)\left(\frac{\partial^2 r}{\partial V^2}\right) + \left(\frac{\partial^2 \Delta E}{\partial r^2}\right)\left(\frac{\partial r}{\partial V}\right)^2 \tag{36.14}$$

At the equilibrium internuclear separation, $r = r_0$ and $\partial \Delta E / \partial r = 0$. By combining these results with the previous equations, we find that

$$\frac{1}{\beta} = \frac{(n-1)e^2\alpha}{18r_0^4} \tag{36.15}$$

The value n can be determined from the experimental terms in (36.15). For NaCl, the value of β extrapolated to zero temperature is 3.3×10^{-12} cm^2 dyn^{-1}, which yields $n = 9.4$.

36.5 LATTICE HEAT CAPACITY; EINSTEIN MODEL

According to the equipartition of energy principle, a system of N atoms has a total of $3N$ degrees of freedom. In most crystalline systems, the rotational and translational motions are eliminated by the rigid structure of the lattice. Thus we need concern ourselves only with the vibrational degrees of freedom. There are $3N - 6$ vibrational degrees of freedom. Since N is very large, $3N - 6$ does not differ much from $3N$, and we can consider that the crystal possesses $3N$ vibrational degrees of freedom. Each degree of vibrational freedom contributes an average energy of kT per atom, according to the classical equipartition of energy principle; but this presumes that the energy is continuously variable. We must take into account the quantization of energy.

The energy of a quantum mechanical oscillator is

$$E = (n + \tfrac{1}{2})h\nu \tag{36.16}$$

where ν is the frequency of the oscillations. Einstein assumed that the frequency was the same for all atoms in the crystal. The vibrational partition function for one oscillator was given in Equation (29.7); the vibrational partition function for $3N$ oscillators is that function raised to the $3N$ power, or

$$Z = \exp\left(-\frac{3Nh\nu}{2kT}\right)\left[1 - \exp\left(-\frac{h\nu}{kT}\right)\right]^{-3N} \quad (36.17)$$

The extra factor of $\{\exp[-\frac{1}{2}(h\nu/kT)]\}^{3N}$ is due to the zero-point energy of $\frac{1}{2}h\nu$. The average energy is

$$\overline{E} = kT^2\left(\frac{\partial \ln Z}{\partial T}\right) \quad (36.18)$$

Recalling that $(d/dx)(\ln u) = (1/u)(du/dx)$, we get

$$\overline{E} = \frac{3Nh\nu}{\exp(h\nu/kT) - 1} \quad (36.19)$$

The heat capacity is $C_v = \partial E/\partial T$, or

$$C_v = 3Nk\left[\frac{h\nu}{2kT}\operatorname{csch}\left(\frac{h\nu}{2kT}\right)\right]^2 \quad (36.20)$$

where the hyperbolic cosecant is defined by

$$\operatorname{csch} x = (\sinh x)^{-1}$$
$$\sinh x = \tfrac{1}{2}[\exp(x) - \exp(-x)]$$

From the series expansion for $\sinh x$,

$$\sinh x = x + \frac{x^3}{3!} + \frac{x^5}{5!} + \frac{x^7}{7!} + \cdots$$

we find the limiting value for C_v at high temperatures to be

$$C_v \longrightarrow 3Nk = 3R \qquad (T \longrightarrow \infty) \quad (36.21)$$

for 1 mole of atoms. Einstein's model thus accurately reflects the classical behavior as expressed by the law of Dulong and Petit in the limit of high temperatures. Figure 36.2 shows a plot of the Einstein specific heats for diamond together with some experimental values.

We know from the third law of thermodynamics that in the limit as the temperature approaches zero, the heat capacity must also approach zero. Einstein's model does indeed approach zero; in the limit of zero temperature, Equation (36.20) becomes

$$C_v \cong 12Nk\left(\frac{h\nu}{2kT}\right)^2 \exp\left(-\frac{h\nu}{kT}\right) \qquad (T \longrightarrow 0) \quad (36.22)$$

The value C_v approaches zero exponentially in the limit, since the exponential term is dominant. Experimentally, however, we know that C_v approaches zero as T^3 does. The main difficulty with the Einstein model is that it assumes only one oscillation frequency, whereas a real crystal has many different normal modes, each with different frequencies. Einstein assumed that each atom acted as an independent oscillator, and this was an oversimplification as there is coupling between the motions of the atoms. A substantial improvement in this theory was first introduced by Debye, who allowed the frequencies to vary over a large range, and correctly predicted the T^3 behavior at low temperatures.

CHAP. 36 THERMAL PROPERTIES OF CRYSTALS

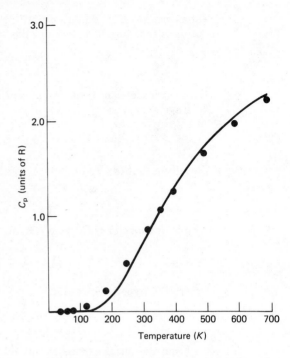

FIGURE 36.2 Heat capacity of diamond. The circles are the experimental points, and the solid line indicates the behavior according to the Einstein model.

36.6 LATTICE HEAT CAPACITY; DEBYE MODEL

Instead of assuming a single characteristic frequency, Debye assumed a frequency distribution $f(\nu)$ such that $f(\nu)\,d\nu$ was the number of frequencies between ν and $\nu + d\nu$. In such a circumstance, the energy is then

$$E = \int \frac{h\nu}{\exp(h\nu/kT) - 1} f(\nu)\,d\nu \quad (36.23)$$

(Actually the total energy should be expressed as a summation over all the possible frequencies, but for a large crystal this summation can be replaced by an integral.) The form of the frequency distribution can be got by considering acoustical waves in crystals. A vibrational standing wave has the form $\exp(i\mathbf{k}\cdot\mathbf{r})$, where $\mathbf{k}$ is known as the *wave vector*. Rather than consider an infinite number of frequencies, Debye assumed that the total number is equal to $3N$, the number of degrees of freedom of a system of N atoms. Thus the range of integration in Equation (36.23) is from 0 to some $\nu_{\max}$. Now we need $g(k)$, the number of modes between k and $k + dk$.

If we take the crystal to be a cube of side L, then the condition for standing waves is that $k_x L$, $k_y L$, and $k_z L$ must all be integral multiples of 2π. All the possible values of k can be represented by the construction in Figure 36.3, which shows a cubic lattice in k space with a lattice constant $2\pi/L$. The total number of modes with wave vectors less than $\mathbf{k}$ is given by the number of lattice points within the sphere of radius k. Measured in units of k, this is

$$\frac{(4\pi/3)k^3}{(2\pi/L)^3}$$

SEC. 36.6 LATTICE HEAT CAPACITY; DEBYE MODEL

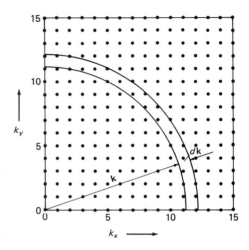

FIGURE 36.3 Section through the xy plane in k space.

The number of modes per unit volume of sample is then

$$\frac{(4\pi/3)k^3}{(2\pi)^3}$$

This must be multiplied by the factor 3, since there are three possible directions of polarization, one longitudinal and two transverse modes. The total number of modes with wave vectors less than k is $k^3/2\pi^2$ per unit volume. The number of modes with wave vectors between k and $k + dk$ is just the number of lattice points in the spherical shell of thickness dk shown in Figure 36.3, and is

$$d\left(\frac{k^3}{2\pi^2}\right) = \left(\frac{3k^2}{2\pi^2}\right) dk = g(k)\, dk \tag{36.24}$$

The magnitude of the wave vector is

$$k = \frac{2\pi}{\lambda} = \frac{2\pi\nu}{c} \tag{36.25}$$

where λ is the wavelength of the wave, ν the frequency, and c the velocity of the wave through the crystal. The relation between the distribution in k and the distribution in ν is

$$f(\nu)\, d\nu = g(k) \left(\frac{dk}{d\nu}\right) d\nu \tag{36.26}$$

This leads to

$$f(\nu) = \frac{3k^2}{\pi c} = \frac{12\pi\nu^2}{c^3} \tag{36.27}$$

The total number of modes between ν and $\nu + d\nu$ is now

$$dn = f(\nu)\, d\nu = 12\pi \frac{V}{c^3} \nu^2\, d\nu \tag{36.28}$$

where V is the volume of the crystal. We can eliminate c by noting that for $n = 3N$, the frequency is the cutoff frequency, ν_m.

$$\int_0^{3N} dn = \int_0^{\nu_m} f(\nu)\, d\nu$$

or

$$3N = \frac{4\pi V \nu_m^3}{c^3}$$

$$c^3 = \frac{4\pi}{3N} V \nu_m^3$$

and

$$f(\nu) = \frac{9N}{\nu_m^3} \nu^2 \tag{36.29}$$

Using this expression for the frequency distribution, we can write (36.23) for the energy as

$$E = \frac{9Nh}{\nu_m^3} \int_0^{\nu_m} \frac{\nu^3\, d\nu}{\exp(h\nu/kT) - 1} \tag{36.30}$$

The heat capacity is obtained by differentiating Equation (36.30), to yield

$$C_v = \frac{9Nh^2}{kT^2\nu_m^3} \int_0^{\nu_m} \frac{\nu^4 \exp(h\nu/kT)\, d\nu}{[\exp(h\nu/kT) - 1]^2} \tag{36.31}$$

Equation (36.31) is often written in the form

$$C_v = 9Nk \left(\frac{T}{\Theta}\right)^3 \int_0^{x_m} \frac{\exp(x) x^4\, dx}{[\exp(x) - 1]^2} \tag{36.32}$$

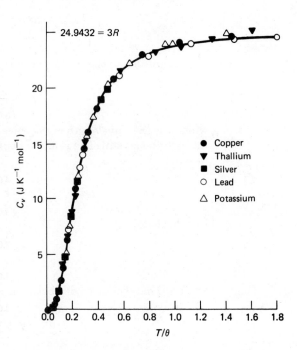

FIGURE 36.4 Heat capacity of solids as a function of T/Θ. The solid curve is the Debye function.

SEC. 36.6 LATTICE HEAT CAPACITY; DEBYE MODEL

where we have introduced the variable $x = h\nu/kT$. The quantity

$$\Theta = \frac{h\nu_m}{k} \qquad (36.33)$$

is known as the *Debye characteristic temperature* and is a function of the characteristic frequency of the lattice. A universal curve for heat capacities of solids is obtained by plotting the heat capacities vs the variable T/Θ, as shown in Figure 36.4. The Debye heat capacity model gives much better results than the Einstein model, especially at lower temperatures.

For large T, the heat capacity expressed in (36.32) approaches the classical Dulong and Petit value of $3R$, as the Einstein model does. For very low temperatures, we can approximate the integral in (36.30) by letting the upper limit go to infinity. For large x, $[\exp(x) - 1] \approx \exp(x)$, and the integral can be approximated by

$$\int_0^\infty \frac{x^3 \, dx}{\exp(x) - 1} = \frac{\pi^4}{15} \qquad (36.34)$$

By this result, the energy at low temperatures is

$$E = \frac{3Nk\pi^4 T^4}{5\Theta^3} \qquad (36.35)$$

Since $C_v = \partial E/\partial T$, the low temperature limit is

$$C_v = \left(\frac{12Nk\pi^4}{5}\right)\left(\frac{T}{\Theta}\right)^3 \qquad (36.36)$$

This expression exhibits the experimental T^3 dependence. Figure 36.5 shows a low temperature plot of C_v vs T^3. The characteristic Debye temperatures of several substances are listed in Table 36.2. Values of calculated heat capacities are given in Table 36.3, along with values computed by the Einstein function.

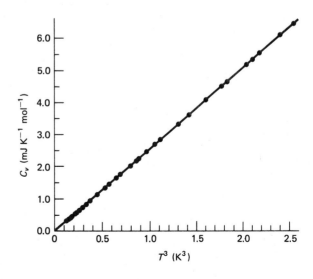

FIGURE 36.5 Low-temperature heat capacity of solid argon vs T^3, showing excellent agreement with the T^3 law. (L. Finegold and N. E. Phillips, *Phys. Rev.* 177 (1969):1383.)

TABLE 36.2 Values of the Debye characteristic temperature.

Substance	Θ (K)
Na	159
K	99
Au	180
Ag	215
Cu	316
Pt	229

TABLE 36.3 Heat capacities as a function of Θ/T as given by the Debye model, the Einstein model, and the T^3 law.

Θ/T	C_v(Debye) (J K^{-1} mol^{-1})	C_v(Debye) (units of R)	C_v(Einstein) (units of R)	C_v(T^3 law) (units of R)
0.000	24.94317	3.0000000	3.0000000	
0.001	24.943169	2.9999998	2.9999998	
0.002	24.943165	2.9999994	2.999999	
0.005	24.94314	2.999996	2.999994	
0.01	24.94305	2.999985	2.999975	
0.02	24.94267	2.99994	2.99990	
0.05	24.94005	2.99963	2.99938	
0.1	24.9307	2.9985	2.9975	
0.2	24.893	2.994	2.990	
0.5	24.634	2.963	2.938	
1.0	23.739	2.855	2.762	
1.5	22.348	2.688	2.496	
2.0	20.588	2.476	2.172	
2.5	18.604	2.238	1.827	
3.0	16.531	1.988	1.489	
4.0	12.548	1.509	0.912	
5.0	9.195	1.106	0.512	
6.0	6.622	0.797	0.269	
7.0	4.761	0.573	0.134	
8.0	3.447	0.4145	0.064	0.4566
9.0	2.531	0.3044	0.030	.3207
10.0	1.891	0.2275	0.014	.2338
15.0	0.5755	0.06921	2.1×10^{-4}	.06927
20.0	0.2430	0.02922	2.5×10^{-6}	.02922
25.0	0.1244	0.01496	2.6×10^{-8}	.01496
30.0	0.07199	0.008659	2.5×10^{-10}	.008659
35.0	0.04534	0.005453	2.3×10^{-12}	.005453
40.0	0.03037	0.003653	2.0×10^{-14}	.003653
45.0	0.02133	0.002566	1.7×10^{-16}	.002566
50.0	0.01555	0.001870	1.4×10^{-18}	0.001870
∞	0	0	0	0

NOTE: The Debye heat capacities were determined by numerical integration of Equation (36.32). The Einstein values were computed from (36.20). The T^3 law values are given by $C_v = 233.781\,848\,16\,(T/\Theta)^3$.

36.7 OTHER SOURCES OF HEAT CAPACITY

The Debye model applies only to the heat capacity arising from the lattice vibrations of the crystal. This is not the only source of contributions. Order-disorder transformations, antiferromagnetic, ferromagnetic, ferroelelectric, and a host of other transformations also contribute to the heat capacities of solids.

The heat capacity curve of methane is shown in Figure 36.6; the "lambda" spike in the vicinity of 20 K is associated with molecular rotations. Unlike metals and

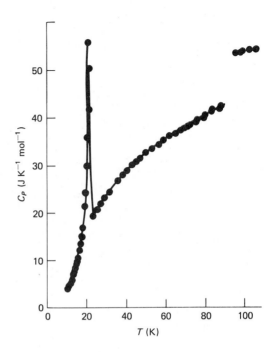

FIGURE 36.6 Low-temperature heat capacity of methane. The peak at 20.44 K is from the rotation of the CH_4 molecules. The break in the curve at 90.6 K occurs at the melting point of methane. (K. Clusius, *Z. Phys. Chem.* B3 (1929):41.)

ionic crystals, for which the intermolecular forces are high and the crystal binding energies are also high, methane in the solid state is composed of spherical CH_4 molecules held loosely together by van der Waals forces. At high temperatures the individual methane molecules rotate freely, and there is a specific heat contribution arising from this rotation. At low temperatures, the energy of the molecules is reduced, and the rotational motion is frozen out in a cooperative manner. The transition temperature as indicated in Figure 36.6 is about 20 K. Above 20 K, the molecules can rotate freely. Below 20 K, their energy is insufficient for free rotation; their rotational motion is restricted to torsional oscillations about a mean position.

There is an additional anomaly in the heat capacity of metals. In a metal, the valence electrons become *conduction electrons* that are free to move about the lattice. They should then contribute an additional $\frac{3}{2}R$ to the heat capacity by virtue of this translational freedom, and we should expect the high-temperature heat capacity of metals to be $(3 + \frac{3}{2})R = \frac{9}{2}R$ per mole, in contradistinction to insulators, which we should expect to have only $3R$. There is really little difference, however, between the heat capacities of metals and insulators. When we examine the free electron theory of metals (next chapter), we shall resolve this anomaly, noting that the absence of a specific heat contribution from the conduction electrons is intimately connected with the Pauli exclusion principle.

PROBLEMS

1. For NaCl crystals the heat of formation is -414; the ionization potential of Na is 490; the heat of sublimation of Na metal is 109; the dissociation energy of the Cl_2 molecule is 226; and the electron affinity of Cl is 347—all values in units of kilojoules per mole. Using the scheme of the Born-Haber cycle, calculate the crystal energy of NaCl.

2. The Madelung constant for NaCl is 1.74756, and the distance of nearest neighbor approach is 2.814×10^{-8} cm. Calculate the energy of a NaCl crystal using $n = 10$.

3. Calculate the compressibility of NaCl.

4. In the text we used a repulsive potential that was proportional to r^{-n} to determine the expression for

the binding energy of crystals. A better repulsive potential is of the form $b \exp(-r/\rho)$, where b and ρ are constants. Derive an expression for the binding energy of crystals using this repulsive potential.

5. Differentiate Equation (36.19) to get (36.20).
6. Show that the high-temperature limit of the Einstein heat capacity is $3R$, in agreement with the classical law of Dulong and Petit.
7. Show that the limit of the Einstein heat capacity is zero as the temperature goes to zero in agreement with the third law of thermodynamics.
8. Differentiate Equation (36.30) to get (36.31).
9. Consider a one-dimensional crystal consisting of alternate positive and negative charges along a straight line, and separated by a constant distance R_0. Show that for such a system, the Madelung constant is $\alpha = 2 \ln 2$.
10. For the NaCl structure the value of the Madelung constant is $\alpha = 1.747558$. This is obtained by summing over all the interionic interactions in the crystal, and it is a lengthy calculation. Calculate the Madelung constant, taking into account only the nearest and next-nearest neighbors. How much of an error is introduced by neglecting the more distant ions?
11. The heat capacity of metallic silver is 0.1987 J K^{-1} mol^{-1} at 10 K. Use this experimental value to calculate the Debye characteristic temperature. Using the Debye characteristic temperature, calculate the heat capacity of metallic silver at 20, 103.14, and 205.3 K and compare with the experimental values (1.672, 20.071, and 23.451 J K^{-1} mol^{-1}).
12. Note that Equation (36.31) gives a value for C_v and not C_p, whereas it is usually C_p that is of interest experimentally—since that is the quantity measured and the quantity used for evaluating third law entropies. Recall from our discussion of the second law (Chapter 9) that

$$C_p - C_v = \frac{\alpha^2 V T}{\beta} \quad (9.62)$$

where α is the coefficient of expansion and β is the compressibility. For silver the experimental values are

$$\alpha = \frac{1}{V}\left(\frac{\partial V}{\partial T}\right)_p = 5.831 \times 10^{-5} \text{ K}^{-1}$$

$$\beta = -\frac{1}{V}\left(\frac{\partial V}{\partial P}\right)_T = 9.9 \times 10^{-13} \text{ cm}^2 \text{ dyn}^{-1}$$

The density of silver is 10.5 g cm^{-3}; all these values are at room temperature. Calculate the difference $C_p - C_v$ for silver at room temperature and at 1000K, assuming these values to be independent of temperature. What fractional error is introduced by assuming that the two heat capacities are equal?

13. The accompanying table lists the measured heat capacities of methane over the temperature range 10.33–105 K; these data have been plotted in Figure 36.6. Prepare a suitable graph of the data to use for graphically determining absolute entropies.
(a) Assume that the lowest measured heat capacity was at 22.80 K, and that the data were extrapolated to lower temperatures using the T^3 law. Calculate the standard entropy of CH$_4$ at 80 K under this assumption.
(b) Determine the entropy change associated with the transition at 20.44 K.
(c) How much of an error would be introduced into the entropy of CH$_4$ if the presence of this transition were unknown? (*Note:* The transition was not discovered until 1929.)

Heat capacity of methane at low temperature. (K. Clusius, *Z. Phys. Chem. B3*, 41 (1929)).

T (K)	C_P (J K^{-1} mol^{-1})	T (K)	C_P (J K^{-1} mol^{-1})
10.33	3.992	28.5	23.158
10.68	4.431	30.7	24.397
10.98	4.678	35.2	26.744
11.35	4.870	37.6	28.012
11.98	5.368	40.0	29.003
12.06	5.774	42.3	30.284
12.33	5.916	44.4	30.807
13.08	7.188	46.7	31.602
13.19	7.757	49.7	32.761
13.48	7.916	52.5	33.715
14.18	8.552	55.7	34.535
14.45	9.255	58.3	35.443
15.04	9.791	61.4	36.288
15.41	10.586	65.1	36.878
15.88	12.175	67.2	37.451
15.94	11.891	68.6	37.790
16.91	13.439	70.7	38.392
17.23	15.021	71.9	38.710
17.27	15.021	75.0	39.254
17.87	16.882	75.6	39.748
18.74	21.359	79.4	40.417
18.82	21.435	79.5	40.648
19.15	24.234	83.0	41.702
19.67	30.012	83.7	41.924
19.93	36.045	87.0	42.342
20.21	55.689	87.2	42.635
20.48	169.452	95.4	53.597
20.48	203.928	97.7	53.764
20.86	50.501	98.9	53.974
20.94	41.748	102.0	54.266
22.80	19.322	102.1	54.894
24.9	20.635	105.3	54.434
26.7	22.062		

CHAPTER THIRTY-SEVEN

METALS, SEMICONDUCTORS, AND INSULATORS

37.1 ELECTRICAL CONDUCTIVITY OF FREE ELECTRONS

We start with the observation that metals have high electrical conductivity. This conductivity is attributed to the presence of "free electrons" in the metal that are unrestricted in their motion through the metal. Our treatment is based on what is called the *free electron model of metals*. After we examine the consequences of that assumption as they relate to the electrical conductivity of metals, we shall impose the constraints of quantum mechanics on the electrons in a metal to see what effect this has.

We can imagine that in the absence of external forces, all the "free" electrons move randomly like the molecules of an ideal gas. We call the average velocity the *drift velocity* $\mathbf{v}_d$. In the absence of external forces,

$$\mathbf{v}_d = \frac{1}{N} \sum_{i=1}^{N} \mathbf{v}_i = 0 \tag{37.1}$$

where N is the number of electrons and $\mathbf{v}_i$ is the velocity of the ith electron. For copper (ground state configuration $1s^2 2s^2 2p^6 3s^2 3p^6 3d^{10} 4s$), only the 4s outer electron contributes to the conductivity. The electrical conductivity σ is defined by the equation

$$\mathbf{J} = \sigma \mathbf{E} \tag{37.2}$$

where $\mathbf{J}$ is the current density and $\mathbf{E}$ is the applied electric field.

Now suppose that a uniform electric field is established in the metal. Each electron is accelerated under the action of the constant force $e\mathbf{E}$. Our first inclination might be to write the equation of motion as

$$e\mathbf{E} = \frac{m \, d\mathbf{v}_d}{dt} \tag{37.3}$$

But this must be incorrect; the solution to (37.3) is

$$\mathbf{v}_d = \mathbf{v}_d(0) + \frac{e\mathbf{E}}{m} t = \frac{e\mathbf{E}}{m} t \tag{37.4}$$

which implies that the electrons are accelerated without limit so long as the field is maintained. We know, however, that a steady state is rapidly achieved and the current is maintained at a constant level.

Equation (37.3) implies the total absence of any hindrance to the motion of the electrons. Actually, electrons collide with atoms in the metal and with defects in the crystal; other mechanisms also contribute to slowing down the electrons. We introduce the concept of a *relaxation time* τ to include all these effects. The relaxation time is a measure of the time between collisions and is related to the mean free path of the electrons. The equation of motion for the electrons can now be written as

$$m \left(\frac{d\mathbf{v}_d}{dt} + \frac{1}{\tau} \mathbf{v}_d \right) = e\mathbf{E} \tag{37.5}$$

The term $m\mathbf{v}_d/\tau$ is the analog of a frictional force, with m/τ being the analog of a mechanical coefficient of friction. Suppose we apply a steady field to the metal. The electrons travel with a velocity given by the steady-state drift velocity. Now suppose we turn off the electric field at time $t = 0$. The right side of Equation (37.5) becomes zero, and the solution to (37.5) is

$$\mathbf{v}_d = \mathbf{v}_d(0) \exp\left(\frac{-t}{\tau}\right) \tag{37.6}$$

where $\mathbf{v}_d(0)$ is the drift velocity at $t = 0$. The relaxation time is the characteristic time with which the system approaches equilibrium.

In the steady state at constant electric field, $d\mathbf{v}_d/dt = 0$, and (37.5) has the solution

$$\mathbf{v}_d = \frac{e\mathbf{E}\tau}{m} \tag{37.7}$$

The current density is

$$\mathbf{J} = Ne\mathbf{v}_d = \left(\frac{Ne^2\tau}{m}\right) \mathbf{E} \tag{37.8}$$

where N is the number of electrons per unit volume. If we compare (37.8) with (37.2), we find that the conductivity is

$$\sigma = \frac{Ne^2\tau}{m} \tag{37.9}$$

The *resistivity* is the reciprocal of the conductivity,

$$\rho = \frac{1}{\sigma} = \frac{m}{Ne^2\tau} \tag{37.10}$$

The conductivity of copper at room temperature is some 6×10^5 $(\Omega \text{ cm})^{-1}$, which yields a relaxation time of some 2×10^{-14} s.

The drift velocity is not the average velocity of each electron but rather the average over *all* the electrons. Since the velocity vectors of each electron are oriented in different directions, the drift velocity is less than the average electron velocity. If we let u be the average speed of the electrons, we can define a mean free path that is

$$\Lambda = \tau u \tag{37.11}$$

SEC. 37.2 THE FREE ELECTRON AS A PARTICLE IN A BOX

We shall note later that for copper $u = 1.6 \times 10^8$ cm s^{-1}, indicating a mean free path of the order of 3×10^{-6} cm. The drift velocity per unit electric field is called the *mobility*,

$$\mu = \frac{v_d}{E} = \frac{e\tau}{m} \tag{37.12}$$

Notice the similarity between the mobility as defined above and the ionic mobility we have previously used (Chapter 15) in connection with ionic solutions. Before this chapter is over we shall carry the similarity one step further by noting that there is also a mobility associated with positive charge carriers, the so-called hole.

37.2 THE FREE ELECTRON AS A PARTICLE IN A BOX

If the electrons in a metal behave as free particles, then the metal itself acts as a container, or "box," for the electrons. This suggests the quantum mechanical particle in a box. We have found the wave functions for a particle in a three-dimensional cubic box of edge L to be (Chapter 27)

$$\psi(x, y, z) = \left(\frac{8}{V}\right)^{1/2} \sin\left(\frac{n_x \pi x}{L}\right) \sin\left(\frac{n_y \pi y}{L}\right) \sin\left(\frac{n_z \pi z}{L}\right) \tag{37.13}$$

The energy levels are

$$E = \frac{h^2}{8mL^2}(n_x^2 + n_y^2 + n_z^2) \tag{37.14}$$

The wave equation of (37.13) is based on the boundary condition that the wave function vanishes at the edges of the box. For small L, the levels are widely spaced. On the other hand, for large L, say 1 cm for an ordinary sample of metal, the energy levels are spaced so closely that they can be considered a continuum.

It is convenient to write the wave function in an equivalent exponential form as

$$\psi = \left(\frac{1}{L}\right)^{3/2} \exp\left[i\,\frac{2\pi}{L}(n_x x + n_y y + n_z z)\right] \tag{37.15}$$

This traveling wave function can be derived using a boundary condition that appears somewhat different though it really is entirely equivalent. To arrive at the wave function of Equation (37.15), we must apply a periodic boundary condition that states that

$$\psi(x + L, y, z) = \psi(x, y, z)$$

with similar equations for y and z. These boundary conditions simply state that the value of the wave functions must be the same at opposite faces of the cube.

The energy levels for our new exponential wave function are no longer given by Equation (37.14) but by

$$E_n = \frac{h^2}{2mL^2}(n_x^2 + n_y^2 + n_z^2) = \frac{h^2 n^2}{2mL^2} \tag{37.16}$$

The difference between this expression and (37.14) is more apparent than real. The spacing between levels here is greater, but the number of levels is higher so that the density of levels is the same. This is because with the sine expression only positive integers are valid quantum numbers for ψ, whereas with the exponential expression all integers, positive and negative, and zero, are valid.

We now consider the effect of adding successive electrons to the sample of the metal. The procedure is analogous to the *Aufbau Prinzip* used to construct the periodic table. The energy of any electron depends only on $n^2 = n_x^2 + n_y^2 + n_z^2$, where the n_q are positive or negative integers. Now we must take into account the effect of electron spin by adding a fourth, or spin, quantum number m_s. The energy is independent of m_s.

We take the ground state of the system at $E = 0$. Two electrons may be placed in this ground level with $n_x = n_y = n_z = 0$, and $m_s = \pm\frac{1}{2}$. The third electron must be placed in the next-higher level, since the Pauli principle limits the occupancy of any level to two electrons. For the second level, $n^2 = 1$. This can be accomplished in any of twelve ways, as shown in Table 37.1. Thus a maximum of 14 electrons can be placed in the first two energy levels; the fifteenth electron must go into the third level with $n^2 = 2$. The third level holds a maximum of 24 electrons, as you will demonstrate in one of the problems. We go along in this way through all successive energy levels until the last electron is accommodated. At the absolute zero of temperature, the system corresponds to one in which all the levels below a certain level are filled and all the levels above that level are vacant. The level that divides the filled levels from the empty is called the *Fermi level*. The energy of that level is denoted by E_F and is called the *Fermi energy*, and the quantum number of that level is designated n_F. We now want to evaluate n_F and E_F.

Figure 37.1 illustrates a geometrical representation of all the possible energy configurations. Only one of the eight quadrants is displayed in the figure. The points are arranged in a simple cubic lattice. The number of lattice points having n less than a particular n_F is just the volume of the sphere of radius n_F and is $(4\pi/3)n_F^3$. Each lattice point corresponds to one particular state of n_x, n_y, and n_z; since each state can hold two electrons ($m_s = \pm\frac{1}{2}$), this total volume must be multiplied by two. The number of independent states with n less than n_F is then $(8\pi/3)n_F^3$. If the density of electrons in the metal is denoted by N, there are NL^3 free electrons for a univalent

---

ENRICO FERMI (1901–1954), Italian physicist, was actively engaged in many branches of physics during his career. His trip to Sweden to accept the Nobel Prize in physics in 1938 was used as a cover to flee Italy, and his intention not to return was known only to a few of his most intimate friends. He came to the United States, where he accepted a position on the faculty of Columbia University. Later developments in the Axis nations rendered this decision a very fortunate one, especially since his wife was Jewish. It was also lucky for the United States, since Enrico Fermi directed the research that led to the first successful chain reaction at the University of Chicago in 1942 and pointed to the feasibility of the atomic bomb. His Nobel Prize was for "the discovery of new radioactive elements produced by neutron irradiation, and for the discovery of nuclear reactions brought about by slow electrons." Fermi had devoted the years before 1938 to studying radioactivity induced by neutron bombardment. He thought that he had produced transuranic elements by bombarding uranium, and all workers in the field at that time accepted this explanation. It remained for Hahn and Strassman to show that the measured radioactivity was produced because of isotopes of much lighter elements, and that Fermi had actually produced nuclear *fission* instead of nuclear *transmutation*. It was a case of the right man getting the Nobel Prize, but for the wrong reason.

SEC. 37.2 THE FREE ELECTRON AS A PARTICLE IN A BOX

TABLE 37.1 Placement of electrons in the first excited energy level of a cubic box.

n_x	n_y	n_z	m_s
1	0	0	$\tfrac{1}{2}$
1	0	0	$-\tfrac{1}{2}$
0	1	0	$\tfrac{1}{2}$
0	1	0	$-\tfrac{1}{2}$
0	0	1	$\tfrac{1}{2}$
0	0	1	$-\tfrac{1}{2}$
−1	0	0	$\tfrac{1}{2}$
−1	0	0	$-\tfrac{1}{2}$
0	−1	0	$\tfrac{1}{2}$
0	−1	0	$-\tfrac{1}{2}$
0	0	−1	$\tfrac{1}{2}$
0	0	−1	$-\tfrac{1}{2}$

metal in the form of a cube of edge L. These NL^3 electrons must equal the number of independent states, and

$$\frac{8\pi}{3} n_F^3 = NL^3 \tag{37.17}$$

or

$$\left(\frac{n_F}{L}\right)^2 = \left(\frac{3N}{8\pi}\right)^{2/3}$$

Combining this result with the expression for the energy given in (37.16), we get

$$E_F = \frac{h^2}{2m}\left(\frac{n_F}{L}\right)^2 = \frac{\hbar^2}{2m}(3\pi^2 N)^{2/3} \tag{37.18}$$

for the Fermi energy. For sodium the density is 0.97 g/cm³, and each atom provides one free electron. The density of electrons is $N = 2.5 \times 10^{22}$ electrons cm⁻³, and the Fermi level is $E_F = 5.04 \times 10^{-19}$ J, or 3.15 eV. This assumes that the density at absolute zero is the same as at room temperature. For most metals, the Fermi energy is of this order of magnitude, falling in the range 2 to 6 eV. For comparison, we note that at room temperature thermal energies are of the order of $kT = 4.14 \times 10^{-21}$ J, two orders of magnitude less than the Fermi energy; therein lies the solution to the

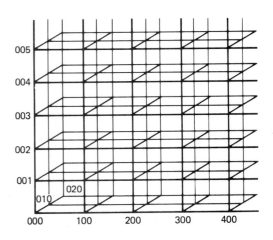

FIGURE 37.1 Geometrical representation of the energy levels of a particle in a cubic box.

anomaly of the absence of an electronic specific heat. We can see this qualitatively by examining the density of electronic states in a metal.

The density of states is just the number of states per unit energy range. With reference to Figure 37.1, we can represent the states in n space by a simple cubic lattice. The period of the lattice is $2\pi/L$, so the volume occupied by one lattice point (or one stationary state in n space) is $(2\pi)^3/L^3 = (2\pi)^3/V$. We have already seen that the total number of states is $(8\pi/3)n^3$; hence

$$N(E) = \frac{4\pi n^3}{3} \frac{2V}{(2\pi)^3} = \frac{8\pi}{3} \frac{(2mE)^{3/2}}{h^3} V \qquad (37.19)$$

The number of levels with energy between E and $E + dE$ is

$$\frac{dN}{dE} dE = 4\pi \frac{(2m)^{3/2}}{h^3} E^{1/2} V \, dE \qquad (37.20)$$

If we let $g(E)$ be the density of states per unit volume, we can divide (37.20) by V, to get

$$g(E) = 4\pi \frac{(2m)^{3/2}}{h^3} E^{1/2} \qquad (37.21)$$

The variation of $g(E)$ with E is thus parabolic, as indicated in Figure 37.2. At the absolute zero of temperature, all levels up to the level with $E = E_F$ are filled, as shown by the heavy line. As the temperature is raised, some of the electrons may acquire some excess thermal energy, but the energy they can acquire is of the order of kT. This is some two orders of magnitude less than E_F. Suppose we consider an electron in one of the lower energy states with E near zero. Classically that electron could absorb an extra amount of energy kT. But quantum mechanically it cannot absorb that amount of energy, since it would be excited to a level that is already filled with its maximum of two electrons. *Only those electrons whose energies lie within $\sim kT$ of the Fermi level can be excited,* since these electrons could then occupy levels that were initially vacant. Since only a small fraction of the electrons lie in this region, only a small fraction of the electrons can contribute to the specific heat.

FIGURE 37.2 Density of states as a function of the energy. At absolute zero all states up to the Fermi energy E_F are filled. At nonzero temperature, electrons within $\sim kT$ of the Fermi energy can be excited to higher states; the dashed line indicates the density of states at such temperature that $kT \ll E_F$.

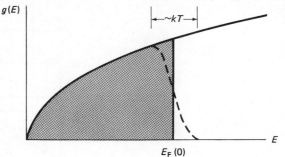

SEC. 37.2 THE FREE ELECTRON AS A PARTICLE IN A BOX

We can define a *Fermi temperature* T_F, which is that temperature at which $kT_F = E_F$; T_F is of the order of 50,000 K. At temperature T, the fraction of electrons near the Fermi level that can be excited is of the order of T/T_F, which is less than 0.01 at room temperature. The total number of electrons that are excited per unit volume is NT/T_F, and each can gain energy of the order of kT. The total electronic thermal energy is approximately $U \cong (NT/T_F)kT$, or

$$U \cong \frac{RT^2}{T_F} \qquad \text{(per mole)} \qquad (37.22)$$

The electronic heat capacity is then

$$C_v = \frac{\partial U}{\partial T} \cong \frac{RT}{T_F} \qquad (37.23)$$

It is proportional to the first power of T.

At low temperatures the heat capacities should be given by the sum of two terms. A T^3 term as required by the Debye model and a term in T, as indicated by (37.23),

$$C_v = aT^3 + \gamma T \qquad \text{(low temperature)} \qquad (37.24)$$

where a is the constant arising from the Debye theory and γ is the coefficient of the electronic heat capacity. A plot of C_v/T vs T^2 should then yield a straight line if the above analysis is correct. Such a plot is indicated in Figure 37.3. The intercept of the plot provides a value for γ.[1]

FIGURE 37.3 Plot of C_v/T vs T^2 for potassium at low temperatures. (W. H. Lien and N. E. Phillips, *Phys. Rev.* 133 (1964):A1370.) The equation of the straight line is $C/T = 2.08 + 2.57T^2$, as reported by Lien and Phillips.

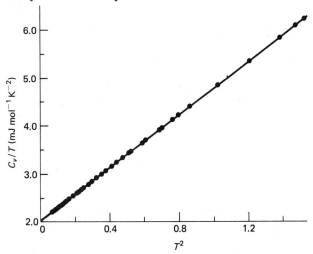

[1] A quantitative derivation of this can be found in A. H. Wilson, *The Theory of Metals,* 2nd ed. (Cambridge: The Cambridge University Press, 1953), p. 144. The theory shows that $\gamma = \frac{1}{2}\pi^2 ZR/T_F$, where Z is the number of free electrons per atom.

37.3 STATISTICAL MECHANICS OF INDISTINGUISHABLE PARTICLES

To get a more quantitative derivation of the heat capacity of the electron "gas" in a metal, it is necessary to discuss the statistical mechanics of the system. Our previous work with statistical mechanics has all been based on the Boltzmann distribution. That distribution law contains two fundamental assumptions: that the particles are distinguishable and that there is no limit to the number of particles occupying any given state. In quantum statistics, at least one of these assumptions is invalid, and perhaps both.

When we calculate the vibrational heat capacity of a gas, the classical Boltzmann distribution is perfectly adequate. The individual molecules are distinguishable, and there is no limit to the number of molecules in any particular vibrational state. Electrons in a metal, however, are indistinguishable. Further, the occupancy of each level is limited to two electrons by the Pauli exclusion principle, since electrons have half-integral spin. Electrons follow what we call *Fermi-Dirac* statistics. For particles like deuterons, which have integral spins, there is no limit to the number of particles occupying any state. These particles are also indistinguishable, however. They follow what we call *Bose-Einstein* statistics.

There is an additional, intuitive reason why the classical Boltzmann distribution law is not valid as applied to the electron gas in a metal. The concentration of free electrons in a metal is orders of magnitude higher than the concentration of gas molecules in gases at normal pressures and temperatures. The classical Boltzmann distribution is a good approximation in those instances in which the average spacing between the particles is much larger than the de Broglie wavelength of the particles.

We have a set of indistinguishable particles distributed in a set of energy states $E_0, E_1, E_2, \ldots$. The number of particles in each energy state is $n_0, n_1, n_2, \ldots$. The system is subject to the conditions of constant number of particles and constant energy:

$$\Sigma n_i = N \qquad (37.25)$$

$$\Sigma n_i E_i = E \qquad (37.26)$$

In what follows, we assume that all states have equal a priori probabilities; the degeneracies are denoted by g_i.

We must determine the number of ways to place n_i indistinguishable particles in g_i states. Suppose for the moment that the particles are distinguishable. The problem is analogous to that shown in Figure 37.4, where n_i particles are divided

---

PAUL ADRIEN MAURICE DIRAC (born 1902), British physicist, began his studies in theoretical physics after failing to get work as an electrical engineer, the field in which he had taken his undergraduate degree. Dirac introduced Einstein's theory of relativity into quantum mechanics and was one of the originators of relativistic quantum mechanics and also of the quantum theory of radiation. One anomalous result of his relativistic quantum mechanics was that certain aspects of the theory could be explained only by the existence of a positively charged particle with a short half-life and a mass equal to that of the electron. Shortly thereafter, Carl Anderson discovered the *positron*, and Dirac's theory was turned into a triumph. Dirac shared the 1933 Nobel Prize with Erwin Schrödinger, and he was appointed Lucasian professor of mathematics at Cambridge University in 1932. That was the chair Sir Isaac Newton once held.

FIGURE 37.4 One possible arrangement of a system of n particles and $g - 1$ partitions that divide the particles into g groups (or boxes). The total number of permutations of the particles and partitions is $(n + g - 1)!$. The number of arrangements without regard to permutations of the particles and partitions among themselves is $(n + g - 1)!/n!(g - 1)!$. For the system shown, we have 21 particles and 8 partitions dividing the particles into 9 boxes; hence the number of arrangements is $29!/21!8! = 4.292 \times 10^6$.

into g_i groups by placing $g_i - 1$ partitions between groups of particles. The number of possible arrangements equals the total number of permutations of the n_i particles and the $g_i - 1$ partitions, or $(n_i + g_i - 1)!$. But since the particles are indistinguishable, the permutations of the particles among themselves are insignificant; hence we must divide by $n!$. Also, the permutations of the partitions among themselves are not significant, so we must also divide by $(g_i - 1)!$. The total number of arrangements for any given energy is then

$$\frac{(n_i + g_i - 1)!}{n_i!(g_i - 1)!}$$

The total number of arrangements, or the probability, of a system with many energy levels is the product of all such terms, or

$$W = \prod_i \frac{(n_i + g_i - 1)!}{n_i!(g_i - 1)!} \tag{37.27}$$

Using Stirling's approximation for logarithms of large factorials,

$$\ln n! \cong n \ln n - n \tag{37.28}$$

we can convert Equation (37.27) into the logarithmic form

$$\ln W = \sum_i [(n_i + g_i) \ln (n_i + g_i) - n_i \ln n_i - (g_i - 1) \ln (g_i - 1)] \tag{37.29}$$

where we have neglected 1 as small compared with $(n_i + g_i)$. To find the most probable distribution, we use the method of Lagrangian multipliers, as we did for deriving the Boltzmann distribution, to get

$$\frac{\partial}{\partial n_i} [\ln W - \alpha \Sigma n_i - \beta \Sigma n_i E_i] = \ln \left(\frac{n_i + g_i}{n_i}\right) - \alpha - \beta E_i = 0 \tag{37.30}$$

$$\frac{n_i + g_i}{n_i} = 1 + \frac{g_i}{n_i} = \exp(\alpha + \beta E_i)$$

$$n_i = \frac{g_i}{\exp(\alpha + \beta E_i) - 1} \tag{37.31}$$

The Lagrangian multiplier α is evaluated from the condition that $N = \Sigma n_i$. In the limit of infinitely high temperature, this distribution must approach the Boltzmann distribution; hence we can set $\beta = 1/kT$, and write for (37.31)

$$n_i = \frac{g_i}{\exp(\alpha + E_i/kT) - 1} \tag{37.32}$$

Equation (37.32) is the *Bose-Einstein distribution law*.

To arrive at a distribution function that is valid for electrons, we must add the condition of only one particle per state to the indistinguishability condition that is already present in the Bose-Einstein distribution function. Suppose we have a set of g_i states and n_i particles. We can represent each state by a cell that contains at most one particle. There are n_i filled cells and $g_i - n_i$ vacant cells. Now the total number of arrangements of the g_i cells is $g_i!$. We further note that permutations of the filled cells and unfilled cells among themselves are not significant; hence the total number of arrangements is $g_i!/[n_i!(g_i - n_i)!]$. By doing just as we did for the Bose-Einstein distribution derivation, we can write

$$W = \prod_i \frac{g_i!}{n_i!(g_i - n_i)!} \tag{37.33}$$

$$\ln W = \sum_i [g_i \ln g_i - n_i \ln n_i - (g_i - n_i) \ln (g_i - n_i)] \tag{37.34}$$

$$\frac{\partial}{\partial n_i}[\ln W - \alpha \Sigma n_i - \beta \Sigma n_i E_i] = \ln (g_i - n_i) - \ln n_i - \alpha - \beta E_i = 0 \tag{37.35}$$

$$\frac{g_i - n_i}{n_i} = \frac{g_i}{n_i} - 1 = \exp(\alpha + \beta E_i)$$

$$n_i = \frac{g_i}{\exp(\alpha + \beta E_i) + 1} \tag{37.36}$$

$$n_i = \frac{g_i}{\exp(\alpha + E_i/kT) + 1} \tag{37.37}$$

Equation (37.37) is the *Fermi-Dirac distribution law;* it differs from the Bose-Einstein distribution only in the sign of unity that appears in the denominator. Calculations using the exact forms of these two laws are involved. We shall next investigate how the Fermi-Dirac law applies in some limiting cases.

37.4 PROPERTIES OF THE FERMI-DIRAC DISTRIBUTION

In a system of *fermions*, each state with four specified quantum numbers is either empty or occupied by *one* electron. It is convenient to work with the distribution function

$$f = \frac{n_i}{g_i} = \frac{1}{\exp(\alpha) \exp(E_i/kT) + 1} \tag{37.38}$$

This function gives the probability that any given state is occupied; f varies between zero and unity. We define an energy E_F such that

$$\alpha = -\frac{E_F}{kT} \tag{37.39}$$

We can write f in terms of E_F as

$$f = \frac{1}{\exp[(E - E_F)/kT] + 1} \tag{37.40}$$

This function is shown in Figure 37.5 for two different temperatures. At $T = 0, f = 1$ for $E < E_F$ and $f = 0$ for $E > E_F$, as shown by the heavy line; f has the value $\frac{1}{2}$ for $E =$

SEC. 37.4 PROPERTIES OF THE FERMI-DIRAC DISTRIBUTION

FIGURE 37.5 The Fermi-Dirac distribution function at absolute zero (heavy line) and for such temperature that $kT \ll E_F$ (light line). The areas of the two shaded regions are equal; the light line crosses the heavy line at $f = 0.5$.

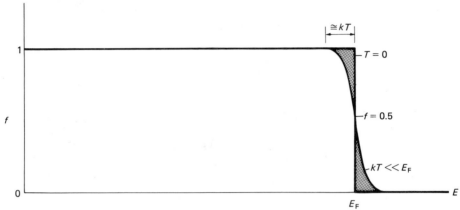

E_F. This agrees with our previous qualitative discussion; the Fermi energy E_F is identical to the Fermi energy previously defined.

As the temperature is raised, the electrons can acquire thermal energy of the order of kT. Electrons whose energy levels lie within about kT of E_F can be excited to higher vacant levels. The value E_F depends somewhat on temperature, but for temperatures such that $kT \ll E_F$, it can be shown that $E_F(T)$ is very close to its value at T = 0. We shall ignore the temperature dependence of E_F. The function for $T \ll E_F/k$ is shown by the light line in Figure 37.5. Significant effects are noticeable only in the immediate vicinity of E_F, since only these electrons can be excited to higher unoccupied states; it is only these electrons that can contribute to the specific heat. The areas of the two shaded regions are equal, and at $E = E_F$, we get $f = \frac{1}{2}$. Since temperatures of tens of thousands of kelvins are required before kT becomes comparable with E_F, our assumption that E_F is independent of temperature is adequate for temperatures normally encountered in the laboratory. The temperature dependence of E_F must, however, be taken into account in dealing with electron gases in hydrogen fusion experiments and in astrophysical calculations dealing with matter in stars.

The density of states as a function of energy, which was given in (37.21), can be combined with (37.40) to get the density of electrons or the number of electrons with energies between E and $E + dE$:

$$dn = fg(E)\, dE = 4\pi \frac{(2m)^{3/2}}{h^3} \frac{E^{1/2}\, dE}{\exp[(E - E_F)/kT] + 1} \tag{37.41}$$

If we collect all the constants together, we can write this as

$$dn = \frac{AE^{1/2}\, dE}{\exp[(E - E_F)/kT] + 1} \tag{37.42}$$

where

$$A = 4\pi \frac{(2m)^{3/2}}{h^3} \tag{37.43}$$

This gives the number of electrons per unit volume with energies between E and $E + dE$.

The total number of electrons per unit volume is N,

$$N = \int_0^\infty dn \tag{37.44}$$

At absolute zero, the denominator of (37.42) is particularly simple. We get

$$N = A \int_0^{E_F} E^{1/2} \, dE = \frac{2}{3} A E_F^{3/2} \qquad (T = 0) \tag{37.45}$$

from which we find that

$$E_F = \left(\frac{3N}{2A}\right)^{2/3} \qquad (T = 0) \tag{37.46}$$

This is identical to Equation (37.18), as we can see after substituting for A.

37.5 BAND THEORY OF SOLIDS

Figure 37.6a depicts a rough sketch of the energy levels of an isolated atom of sodium showing the occupancy of the levels. Suppose we bring a number of sodium atoms together to form a chain or one-dimensional lattice. We might expect to get a potential that varies periodically with the period of the lattice, as shown in Figure 37.6b. Actually, the periodic potential more closely resembles what is shown in Figure 37.6c. When the atoms are brought together, the sharpness of the original levels is lost. The states are "smeared" over a range of energy known as a *band*. The spaces, or gaps, between the *allowed bands* that contain the electrons are the vacant *forbidden bands*. The energy spacings between the allowed bands are denoted by the term *energy gap*. In sodium, the highest band is a 3s band that is only half-filled; this characteristic makes sodium a conductor. Insulators are characterized by having all energy levels below a forbidden band completely filled. For an electron to be a *conduction* electron, it must be in a band that is only partially filled. In an insulator, an electron must be excited from a filled band to the next vacant band. This requires an energy equal to the width of the energy gap, which is very much larger than kT. Suppose we examine the one-dimensional case more quantitatively, though not in all its details.

The wave function for a free electron in a one-dimensional box is

$$\psi = \exp(i\mathbf{k} \cdot \mathbf{x}) \tag{37.47}$$

where the *wave vector* $\mathbf{k} = 2\pi/\lambda$; the term λ is the de Broglie wavelength. For an electron traveling in a periodic potential such as that of Figure 37.6b, the wave function is given by the *Bloch function*,

$$\psi_k = \exp(ikx) u_k(x) \tag{37.48}$$

where u_k is a function that depends on k and is periodic in x with the period of the potential.[2] For the completely free particle in a box, the energy as a function of k is just a parabola; it is proportional to k^2 (or n^2). The effect of the periodicity of the potential is to introduce discontinuities in the shape of the energy curve, as shown in Figure 37.7. These discontinuities occur at $\pi/a, 2\pi/a, 3\pi/a, \ldots,$ a being the period of the

---

[2] This result, which had been known to mathematicians as Floquet's theorem, was first demonstrated for crystals by F. Bloch, Z. Phys. 52 (1928):555. For a complete derivation of the result, see N. F. Mott and H. Jones, *The Theory of the Properties of Metals and Alloys* (Oxford: Oxford University Press, 1936; reprinted by Dover Publications, 1958), pp. 56–63.

SEC. 37.5 BAND THEORY OF SOLIDS

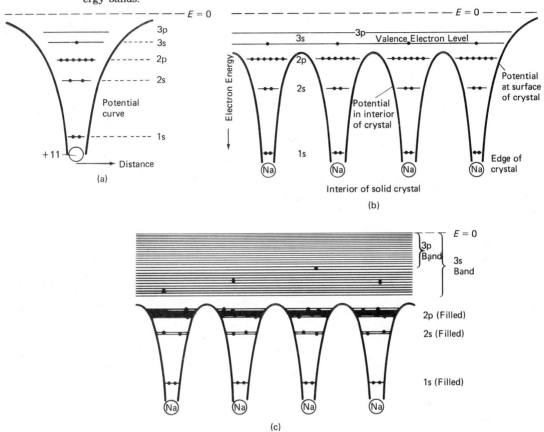

FIGURE 37.6 (a) Energy levels of an isolated sodium atom, showing the occupancy of the various levels. (b) First (and simplified) approximation to the energy levels of a chain of sodium atoms. (c) The energy levels of a chain of sodium atoms being spread over a range to form energy bands.

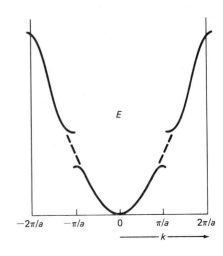

FIGURE 37.7 Energy vs wave number for electrons in a periodic potential. The dashed line indicates the forbidden zone that separates the first and second Brillouin zones. The energy curve in a nonperiodic potential would be represented by the full parabola including the dashed portion.

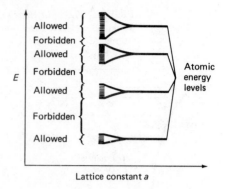

FIGURE 37.8 Splitting of the discrete energy levels of an isolated atom into bands in the crystal. The right-hand side of the figure indicates the discrete levels in the limit of large interatomic separation.

lattice. The region below the first gap is known as the first *Brillouin zone;* the allowed levels between the first forbidden gap and the second forbidden gap make up the second Brillouin zone; and so on, through higher zones.

A simplified view of the levels in a three-dimensional crystal is depicted in Figure 37.8. In the limit of large internuclear separation, each atom has an energy-level scheme consisting of sharp energy states characteristic of the atom. As the interatomic distances are reduced to form the crystal, these sharp energy levels are smeared out into the band structure characteristic of the particular crystal. The difference between a conductor such as sodium and an insulator such as neon can be understood by referring to the sketches in Figure 37.9. The topmost occupied band in sodium is only half-filled. The energy spacing between successive levels within the band is so small that the individual levels can be considered to form a continuum; the spacing between successive levels is very much smaller than kT. Electrons near

FIGURE 37.9 Rough qualitative sketch of the energy bands in a conductor (sodium) and an insulator (neon). The topmost band of the conductor is partially filled, while the topmost band of the insulator is completely filled with electrons. The number of atoms in the crystal is represented by N.

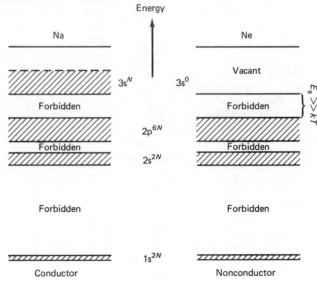

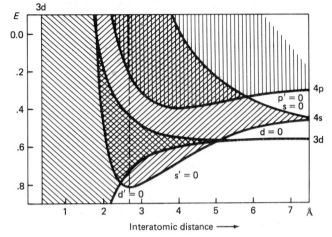

FIGURE 37.10 Energy bands in crystalline copper showing the overlapping of the bands. (After H. M. Krutter, *Phys. Rev.* 48 (1935):664.) (Energy in rydbergs.)

the top of the occupied region easily acquire enough thermal energy to rise to the previously vacant region of the band and become free conducting electrons. For neon, the highest occupied band is completely filled. Electrons near the top of the occupied region cannot be easily excited over the forbidden region. For these electrons to become conduction electrons, they must be excited over the forbidden band into the next-higher vacant band. Since the width of the energy gap is very much larger than kT, there is not enough thermal energy available for that to occur.

There are certain problems associated with this simplified explanation. For example, what about metals such as alkaline earths, which have two s electrons and hence a completely filled upper valence band? These should be insulators, it seems. Actually, there is overlapping of the various bands, as shown in Figure 37.10. For sodium, the higher, 3p band overlaps much of the 3s band region. The crucial point is the highest *occupied* band. If that band is incompletely filled, then the substance is a metallic conductor. If that band is filled, and the next band is forbidden, then the substance is a nonconductor. The conductivity depends on the width of the energy gap. The distinction between insulators and semiconductors is one of degree, as we shall now see.

37.6 INTRINSIC CONDUCTIVITY OF SEMICONDUCTORS

The upper bands in a metal, a semiconductor, and an insulator are illustrated in Figure 37.11. In a metal, the valence band and the conduction bands are the same; the electrons in this band are free. Metals are good conductors of electricity with resistivities of the order of 10^{-6} ohm centimeter (Ω cm). Semiconductors and insulators have filled valence bands and empty conduction bands separated by forbidden regions; the energy gap between the two bands is symbolized by E_g. The resistivity of semiconductors at room temperature is of the order of 10^{-2} to 10^9 Ω cm and lies midway between the resistivity of metals and of insulators (10^{14} to 10^{22} Ω cm). The two most commonly used semiconductors are Ge and Si, and the energy gaps are

FIGURE 37.11 Band schemes in (a) a metal, (b) a semiconductor, and (c) an insulator. The heavily shaded regions indicate occupied states; the lightly shaded regions indicate unoccupied states; and the white regions indicate forbidden regions.

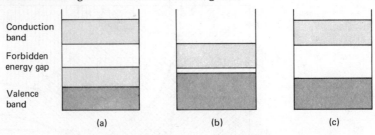

0.7 eV for Ge and 1.1 eV for Si at room temperature. For most semiconductors, E_g is in the range 0–2.5 eV. Insulators have larger energy gaps. Diamond, which may be considered an insulator or a semiconductor, has $E_g = 6$ eV.

At absolute zero, a pure perfect crystal of silicon would in principle have a conductivity approaching zero. The conductivity of semiconductors arises from the effects of thermal motion, impurities, and factors like lattice defects. We shall restrict our discussion here to thermal motion. As the temperature is increased, energy in an amount kT is acquired by the electrons in the crystal. Some of the electrons are excited into the conduction band, the number reaching that band being proportional to $\exp(-E_g/kT)$. (It is actually proportional to $\exp(-E_g/2kT)$, as we shall see in the next section). Since the conductivity of the sample is proportional to the number of electrons in the conduction band, we should expect the conductivity of a semiconductor to vary with $\exp(-E_g/2kT)$. Since the resistivity is just the inverse of the conductivity, we can write

$$\rho = A \, \exp\left(\frac{E_g}{2kT}\right) \qquad (37.49)$$

Hence a plot of $\ln \rho$ vs T^{-1} should yield a straight line whose slope provides a value for E_g. The linear dependence is an excellent verification of the band theory of solids.

Another experimental technique for measuring E_g lies in using an analog of the photoelectric effect. If light photons with energy greater than E_g strike the semiconductor, they can excite electrons from the valence band into the conduction band, increasing the conductivity of the sample. By measuring the minimum frequency of light that produces a sudden increase in conductivity, one can determine the energy gap.

An important feature of intrinsic conductivity is shown in Figure 37.12. Suppose we take our semiconductor to be Si, which has four valence electrons. If one valence

FIGURE 37.12 Excitation of electrons from the valence to the conduction band in a semiconductor. A hole is generated in the valence band for each electron that is excited to the upper conduction band.

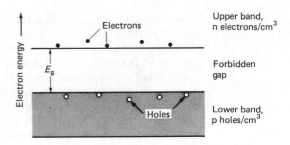

electron is excited into the conduction band, then an excess positive charge is left behind in the Si atom, which now has only three valence electrons left. This excess positive charge is called a *hole*; it has a charge equal and opposite to that of an electron, and the holes also contribute to the electrical conductivity. When an electron drops back from the conduction band into the valence band, it *recombines* with a hole. Let's examine this process more closely.

37.7 EQUILIBRIUM CONSTANT FOR ELECTRONS AND HOLES

A *hole* is an incorporeal entity—the absence of an electron. The mechanism by which holes contribute to the conductivity is shown in Figure 37.13, which depicts a two-dimensional view of a silicon crystal. Each Si atom is normally surrounded by four valence electrons. If one of these is excited to the conduction band, then one of the atoms is left with a net charge of $+e$. A charge of $+e$ can move about the lattice in the following manner. Thermal motion frees an electron from an adjacent Si atom to take the place of the originally missing electron; this leaves a positively charged Si atom at the new location. The positive charge moves across the crystal as electrons are transferred between adjacent Si atoms; the net result is the motion of a unit of positive charge, or a "hole".[3]

Now suppose we heat a semiconductor above the absolute zero of temperature. A few electrons near the top of the valence band are thermally excited over the forbidden energy gap into the conduction band. Physically we have a situation in which some states at the top of the valence band are empty and an equal number of states at the bottom of the conduction band are filled. The electrons that have been excited into the conduction band have left holes behind in the valence band. This is depicted in Figure 37.14; the shaded part of the top of the valence band indicates the missing

FIGURE 37.13 A two-dimensional representation of a silicon or a germanium crystal. One electron about the upper right-hand atom has been removed, leaving a net charge of $+1$ about that atom. The movement of electrons from adjacent atoms to take the place of the missing electron results in the effective motion of a unit of charge $+1$ in the opposite direction. The current associated with this movement of positive charge is attributed to the hole.

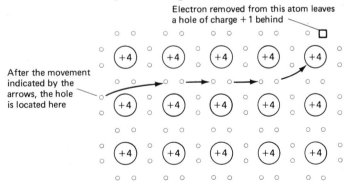

[3] In some ways, the relation between electrons and holes is like the relation between the p^1 and p^5 atomic electronic configurations. The p^1 state has one electron, and the p^5 state lacks one electron for a completed shell; the missing electron in p^5 can be thought of as a hole. The p^1 and p^5 states are similar in many ways. They have the same spin, term symbols, paramagnetic properties, and the like.

electrons, and the shaded region of equal area at the bottom of the conduction band indicates that the thermally excited electrons occupy these levels. In our previous discussions of Fermi-Dirac statistics, we dealt only with metals, but there is no reason why we cannot apply Fermi-Dirac statistics to semiconductors also. The shaded regions in Figure 37.14 have the appearance of the tails of the Fermi function shown in Figure 37.5. We take the Fermi energy E_F of an intrinsic semiconductor to be midway between the top of the valence band and the bottom of the conduction band, as indicated by the dashed line. It is convenient to take the zero of energy at the top of the valence band; hence $E_F = \frac{1}{2} E_g$. We can calculate the number of electrons in the conduction band by following a procedure similar to that used for metals, remembering that $E = 0$ at the top of the valence band. We are interested in the number of electrons with E greater than E_g.

The density of states is given by Equation (37.21). With our new origin, this becomes

$$g_e(E) = 4\pi \frac{(2m)^{3/2}}{h^3} (E - E_g)^{1/2} \qquad (37.50)$$

The subscript e indicates "electrons" to distinguish the distribution from that for the holes, which is denoted by a subscript h. The number of states between E and E +

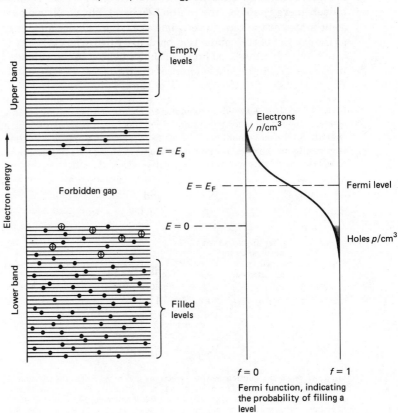

FIGURE 37.14 Electrons, holes, and energy bands in an intrinsic semiconductor.

dE is $g(E)\,dE$. At low temperatures $[kT \ll (E - E_F)]$, the Fermi-Dirac distribution function can be approximated by

$$f_e \cong \exp\left(\frac{E_F - E}{kT}\right) \tag{37.51}$$

since the unity term in the denominator of (37.40) can then be neglected. The number of electrons in the conduction band is

$$N_e = \int_{E_g}^{\infty} g_e(E) f_e(E)\,dE$$

$$= 4\pi \frac{(2m)^{3/2}}{h^3} \int_{E_g}^{\infty} \exp\left(\frac{E_F - E}{kT}\right) (E - E_g)^{1/2}\,dE \tag{37.52}$$

or

$$N_e = 2\left(\frac{2\pi mkT}{h^2}\right)^{3/2} \exp\left(\frac{E_F - E_g}{kT}\right) \tag{37.53}$$

Since $E_F = \tfrac{1}{2}E_g$, the number of electrons in the conduction band is proportional to $\exp(E_g/2kT)$. Hence the conductivity should be proportional to this factor rather than to $\exp(E_k/1kT)$, as noted in the previous section. Calculating the number of holes in the valence band can be done likewise. Since a hole is the absence of an electron, we can write the Fermi-Dirac distribution function for holes as

$$f_h = 1 - f_e = \frac{\exp[(E - E_F)/kT] + 1}{\exp[(E - E_F)/kT] + 1} - \frac{1}{\exp[(E - E_F)/kT] + 1}$$

$$= \frac{1}{\exp[(E_F - E)/kT] + 1}$$

$$f_h \cong \exp[(E - E_F)/kT] \qquad [\text{for } kT \ll (E_F - E)] \tag{37.54}$$

By proceeding exactly as we did for the electrons (except for using m_h for the *effective mass* of the holes), we can get the number of holes in the valence band as

$$N_h = \int_{-\infty}^{0} g_h(E) f_h(E)\,dE$$

$$N_h = 2\left(\frac{2\pi m_h kT}{h^2}\right)^{3/2} \exp\left(\frac{-E_F}{kT}\right) \tag{37.55}$$

It is customary to write n for N_e and p for N_h. By multiplying (37.53) by (37.55) we get the useful equilibrium constant for electrons and holes as

$$K_{np} = n \cdot p = N_e \cdot N_h = 4\left(\frac{2\pi kT}{h^2}\right)^{3} (m_e m_h)^{3/2} \exp\left(\frac{-E_F}{kT}\right) \tag{37.56}$$

For intrinsic semiconductors it is always true that $n = p$. We can, however, introduce electrons and holes into semiconductors by adding proper impurities. In those instances, Equation 37.56 is useful in determining the concentrations of the *minority* current carriers in the semiconductor.

37.8 IMPURITY CONDUCTION IN SEMICONDUCTORS

Figure 37.15a shows a two-dimensional representation of a germanium crystal lattice. The crystal has the diamond structure in which each Ge atom is tetrahedrally bound to four other Ge atoms. Each Ge atom is surrounded by a completed shell of

FIGURE 37.15 (a) In the intrinsic germanium, each atom is surrounded by eight electrons represented by eight lines. (b) A phosphorus atom has been added to the lattice, forming n-type germanium; an extra electron has been introduced into the lattice. (c) An indium atom has been introduced into the lattice. There is a deficit of one electron associated with that indium atom, and the result is p-type germanium.

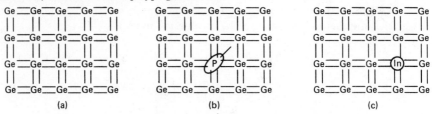

eight electrons, which we have represented by bonds. Suppose we introduce an impurity atom into the lattice. In Figure 37.15b, we show the result of a Group V atom, such as phosphorus. The phosphorus atom fits into the Ge lattice, but the P atom has five valence electrons instead of four; the extra electron is loosely associated with the P atom. The phosphorus atom is known as a *donor*, since it donates an extra electron. The resulting crystal is an *n-type* semiconductor. Figure 37.15c shows the result of adding a Group III atom, such as indium, which has three valence electrons. The impurity atom also fits into the Ge lattice, but there is one electron missing about the In atom. A hole has been created. Current can be generated as electrons from nearby Ge atoms move in to fill the hole, leaving a positive charge behind. The indium atom is an *acceptor* atom; it accepts an electron. The crystal is called *p-type* ger-

FIGURE 37.16 Schematic energy-level diagram of (a) an n-type germanium and (b) a p-type germanium. (c) Energy-level representation of a semiconductor with donors and acceptors.

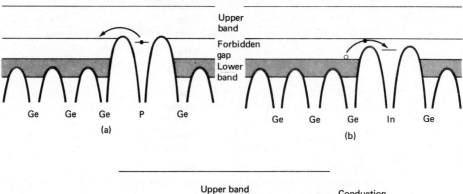

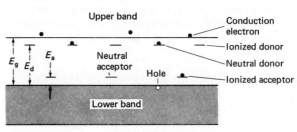

TABLE 37.2 Ionization energies of donors and acceptors in Si and Ge.

	Silicon	Germanium
DONORS		
P	0.045	0.012
As	0.05	0.013
Sb	0.039	0.01
ACCEPTORS		
B	0.045	0.0104
Al	0.06	0.0102
Ga	0.07	0.0108
In	0.16	0.0112

NOTE: Entries are in units of electron volts.

manium. The generic term for adding small amounts of desired impurities to semiconductors is *doping*. Germanium of the *p* or *n* type is termed doped germanium.

A schematic sketch of the energy states analogous to that of Figure 37.6 is shown in Figure 37.16. In the *n*-type material, the extra electron is supported in an energy level high in the forbidden zone. It is easily excited into the conduction band. In the *p*-type material the indium atom forms an empty level just above the top of the valence band. An electron is easily excited into this level, leaving a hole behind. The ionization energies of some donor and acceptor atoms in Ge and Si are listed in Table 37.2.

The production of high-purity single crystals of silicon and germanium has been developed into a fine art. The concentration of electrons in intrinsic material is of the order of 3×10^{13} in Ge and 10^{10} cm^{-3} in Si at room temperature. For intrinsic material to be obtained, the concentrations of impurities must be lower than the concentrations of intrinsic electrons. The densities of Ge and Si are about 0.08 mol cm^{-3}. This corresponds to purity of more than 1 part in 2×10^{10} for Ge and 1 part in 10^{13} for Si. High-purity crystals are grown from melts, and the difficulties experienced may be imagined by considering that for Ge the melting point is 950 °C and for Si 1420 °C. These temperatures must be controlled very precisely while a seed crystal is lowered into a melt of the material and slowly withdrawn while being rotated. Unwanted impurities can be introduced by the vessel used to contain the melt. A process known as *levitation* has been developed to eliminate the need for a crucible. In this process, a rod of material is supported at two ends, and a narrow *zone* of the semiconductor is melted using an induction furnace. Surface tension keeps the molten zone in place. The zone can be moved up and down the rod by slowly moving the induction coils. Desired impurities for producing *p*- or *n*-type material are added to the melt in precisely determined amounts.

37.9 EXPERIMENTAL EVIDENCE FOR HOLES; THE HALL EFFECT

Direct experimental evidence for the existence of holes as current carriers is available from the *Hall effect*. Consider a slab of semiconductor material in the form of a rectangular parallelepiped, as shown in Figure 37.17. An electric field is applied to

FIGURE 37.17 The origin of the Hall effect. Electrons traveling in the x direction in the presence of a magnetic field $\mathbf{H}_y$ are deflected to the lower surface of the specimen. The resulting field builds up until the electrical forces just cancel the magnetic forces.

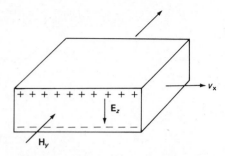

the slab in the x direction, causing current carriers to flow in either the $+x$ or $-x$ direction. If a magnetic field $\mathbf{H}_y$ is applied in the y direction, the current carriers are deflected in a direction that depends on their signs. A potential is developed across the faces of the slab in the z direction because of the buildup of charges along those two faces.

A particle of charge q moving in the field $\mathbf{H}_y$ with velocity v_x experiences a force in the z direction given by

$$F_z = \frac{q}{c}(v_x \mathbf{H}_y) \tag{37.57}$$

where c is the velocity of light. Charge builds up until an equilibrium situation is established and the net force is zero,

$$F_z = 0 = q\left(\mathbf{E}_z - \frac{1}{c} v_x \mathbf{H}_y\right) \tag{37.58}$$

The force due to the electric field increases until it cancels the force due to the magnetic field. At equilibrium,

$$\mathbf{E}_z = \frac{v_x \mathbf{H}_y}{c} = \frac{j_x \mathbf{H}_y}{Nqc} \tag{37.59}$$

where j_x is the current density and N is the concentration of current carriers per cm³. The sign of $\mathbf{E}_z$ depends on the sign of q and is different for p-type material and for n-type material. The ratio

$$R_H = \frac{\mathbf{E}_z}{j_x \mathbf{H}_y} = (Nqc)^{-1} \tag{37.60}$$

is called the *Hall constant*. It is negative for electrons and positive for holes. The magnitude of the Hall constant is much larger for semiconductors than for metals in view of the much smaller value of N in semiconductors.

PROBLEMS

1. A 20-gauge copper wire has a diameter of 0.8118 mm. At 20 °C the resistance of a 1-m length of 20-gauge copper wire is 0.03331 Ω. The density of copper is 8.92 g cm⁻³. Calculate the resistivity and conductivity of copper at 20 °C. Determine the mobility and relaxation time of electrons in copper.

2. A dc current is transmitted over a 20-km length of copper wire that has a diameter of 1 cm. The current passing through the wire is 1000 A. Assume

that the wire is maintained at a constant temperature of 20 °C and use the results of Problem 1 to determine:
(a) The voltage drop over the 20-km length of wire.
(b) The average length of time it takes for an electron to travel the length of the wire.
(c) The number of collisions the electron makes in traversing the wire.
(d) The average distance the electron travels between collisions.
(e) The losses incurred in I^2R heating of the wire.
(f) The size of the wire required to reduce the I^2R losses to 1 W.

3. Table 37.1 shows how the first excited energy level is filled with 12 electrons. Construct an analogous table showing how the electrons are placed in the second excited level for $n^2 = 2$. How many electrons can be placed in this level?

4. Compute the highest value of n at absolute zero for a mole of copper that is in the shape of a cube.

5. Take the density of copper to be 8.9 g cm^{-3} at absolute zero. Calculate the Fermi energy at $T = 0$.

6. In a mole of copper, what fraction of the electrons have energies within kT of E_F at room temperature?

7. Calculate the total number of ways 10 distinguishable particles can be distributed between 5 boxes.

8. Do Problem 7 for the case of indistinguishable particles.

9. For the Fermi-Dirac function at room temperature, determine the value of E for which:

(a) $f = 0.99999$ (d) $f = 0.5$
(b) $f = 0.99$ (e) $f = 0.1$
(c) $f = 0.9$

10. In the text, Equation (37.18) for the Fermi energy was derived by starting with the traveling wave function, (37.15) and the energy expression given in (37.16). Derive an expression for the Fermi energy starting from the particle in the box function of (37.13) and the energy-level expression of (37.14). Show that both methods give the same result.

11. Show that the boundary conditions

$$\psi(x + L, y, z) = \psi(x, y, z)$$

apply to the wave equation of (37.15) with similar conditions for the y and z coordinates.

12. Derive Equation (37.21) for $g(E)$, knowing that $\int g(E)\,dE = N$, where N is the number of electrons per unit volume.

13. Show that (37.46) is equivalent to (37.18).

14. Calculate the ratio of the electronic heat capacity of sodium to the lattice heat capacity at temperatures 0.01 K, 1 K, 10 K, 300 K. (See footnote 1.)

15. Determine the velocity of an electron at the Fermi energy in Cu and Na.

16. The energy gap in intrinsic germanium is 0.7 eV and in silicon is 1.1 eV. Calculate the frequency of light required to excite a valence band electron into the conduction band for germanium and silicon.

17. The masses m_e and m_h that appear in Equation 37.56 are in reality the "effective" masses of electrons and holes. Assume that they equal the normal rest mass of an electron and calculate the concentration of holes in n-type germanium containing 10^{17} donors per cm^3 at room temperature. The energy gap of germanium is 0.7 eV.

18. The low-temperature heat capacity of rubidium has been measured by Lien and Phillips (W. H. Lien and N. E. Phillips, *Phys. Rev.* 113 (1964):A1370). Their results are given in the accompanying table (only a portion of the data is included).

T (K)	C (mJ mol^{-1} K^{-1})	T (K)	C (mJ mol^{-1} K^{-1})
0.1859	0.5189	0.3456	1.309
.1954	0.5537	.3511	1.350
.2002	0.5701	.3830	1.582
.2143	0.6223	.4177	1.863
.2341	0.6968	.4527	2.172
.2391	0.7282	.4615	2.260
.2548	0.8001	.4882	2.538
.2616	0.8336	.5002	2.681
.2798	0.9257	.5015	2.693
.2819	0.9344	.5252	2.979
.2936	0.9953	.5395	3.121
.3048	1.063	.5549	3.358
.3140	1.113	.5686	3.555
0.3285	1.201	0.5867	3.810

(a) From a suitable plot of the data, determine the value of γ, the coefficient of the electronic heat capacity, and the value of a, the coefficient of T^3 in (37.24).
(b) Separately compute the lattice and electronic contributions to the entropy for 1 K.
(c) Compute the Fermi energy from your value of γ. (See footnote 1.)
(d) Construct a table of lattice heat capacities by subtracting the electronic contributions from the heat capacities in the above table; see whether a plot of the lattice heat capacity vs T^3 yields a straight line.
(e) Determine the low-temperature value of Θ_D, the Debye theta.

CHAPTER THIRTY-EIGHT

ELECTRIC AND MAGNETIC PROPERTIES OF ATOMS AND MOLECULES

38.1 MONOPOLES, DIPOLES, AND QUADRUPOLES

Consider an isolated electrically charged particle, as shown in Figure 38.1a.[1] If a test charge is placed at a distance r from the charge, it experiences a force proportional to $1/r^2$ and independent of the relative orientation of the charges. The electric field due to the original charge is spherically symmetric. We call such an isolated charge a *monopole*.

Suppose that on the other hand we have not one isolated charge but two charges of equal and opposite sign $+q$ and $-q$ separated by a distance R, as depicted in Figure 38.1b. An arrangement of two charges is a *dipole;* the electric field has cylindrical symmetry. If four charges are brought together as in Figure 38.1c, the cylindrical symmetry of the electric field is destroyed; we have constructed a quadrupole and could continue on indefinitely through octupoles, hexadecipoles, and so on.[2] In this chapter we restrict our attention to the dipole arrangement.

Figure 38.2 shows a dipole consisting of charges $+q$ and $-q$, separated by $2a$, where a is very small compared with the distance r. The potential at P is just the potential due to $+q$ plus the potential due to $-q$, or

$$V = \frac{q}{4\pi\epsilon_0}\left(\frac{1}{r_1} - \frac{1}{r_2}\right)$$

$$= \frac{q}{4\pi\epsilon_0}\left(\frac{r_2 - r_1}{r_1 r_2}\right)$$

$$\cong \frac{q}{4\pi\epsilon_0 r^2}(r_2 - r_1) \tag{38.1}$$

---

[1] It will be helpful to refer to an electricity and magnetism textbook in connection with this chapter.

[2] A general result of electrostatics is that the potential arising from any arrangement of point charges can be expanded in a series of spherical harmonics called the *multipole expansion* of the potential,

$$V = \sum_{l=1}^{\infty} \frac{Q_l}{R^{(l+1)}}$$

where R is the distance from the origin and Q_l is a term that contains the charges q_i of the various point charges and the lth-order spherical harmonic. The successive terms Q_0, Q_1, Q_2 are associated with the mono-

FIGURE 38.1 Arrangements of point charges. (a) Monopole. (b) Dipole. (c) Quadrupole.

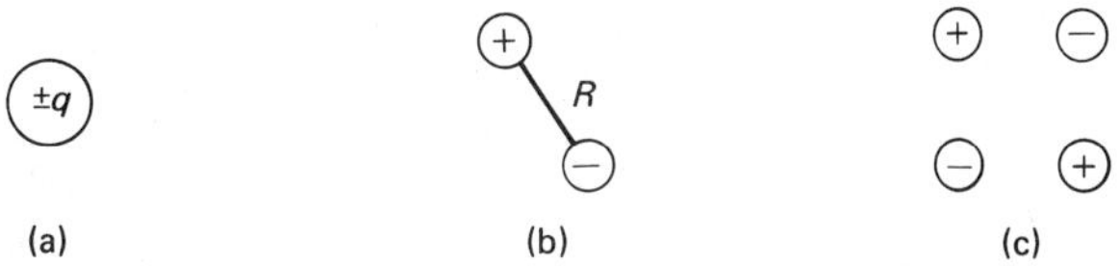

For very small separation distances, the dipole is called a *point dipole*; $r_2 - r_1 = 2a \cos \theta$, so that

$$V = \frac{2aq \cos \theta}{4\pi \epsilon_0 r^2} \tag{38.2}$$

The quantity $2aq$ is called the *dipole moment* and is symbolized by μ; it is a vector quantity. The dipole moment associated with a system of more than two charged particles is obtained by vectorial addition, as shown in Figure 38.2b.

We are interested in dipoles since molecules often possess dipole moments. A diatomic molecule such as HCl can be viewed as two charges separated by the internuclear distance. That CO_2 is a linear molecule with the carbon atom in the center can be inferred from the fact that CO_2 does not have a dipole moment and must therefore be a symmetric molecule. A dipole with charges $\pm e$ (4.8×10^{-10} esu) separated by 10^{-8} cm has a dipole moment of 4.8×10^{-18} esu cm. The unit 10^{-18} esu cm is called the *Debye*, symbolized by D, after Peter Debye. Our interest in dipole moments is further aroused by the fact that when molecules are placed in electric fields, dipole moments can be *induced*, regardless of whether the molecule has a permanent dipole moment or not.

FIGURE 38.2 (a) The potential of a dipole. (b) Vector addition of bond dipoles in a molecule like H_2O.

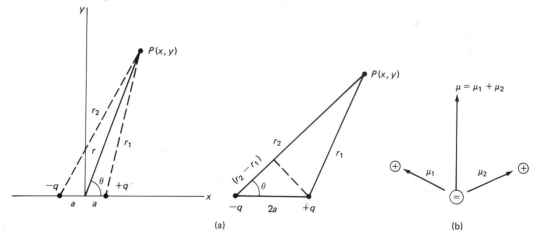

pole, the dipole, the quadrupole, and so forth. A pure dipole is a dipole for which the series terminates after the $l = 1$ term, whereas a monopole is a distribution for which only the term Q_0 in the expansion is nonzero. (See almost any advanced textbook on electricity and magnetism for further discussion of this point.) See H. Margenau and G. M. Murphy, *The Mathematics of Physics and Chemistry*, 2d ed. (Princeton, N.J.: D. Van Nostrand Co., 1956), p. 100.

38.2 DIELECTRICS AND POLARIZATION

For the purposes of this section we divide the world of materials into conducting and nonconducting (or insulating) materials. To be sure, there is no sharp distinction between the two, and gradual changes in electrical conduction are found in a wide range of materials. We shall refer to the nonconducting, or insulating, materials like glass and waxes as *dielectrics*. We consider the case of the ideal isotropic dielectric, which retains a local charge indefinitely, unlike a conductor, in which a locally applied charge spreads out over the entire surface.

If a dielectric material is placed in an electric field, say between the plates of a charged capacitor, then the positive charges in the dielectric are pushed in the direction of the field while the negative charges move in the opposite direction. Since we have a separation of positive and negative charge in each small volume element of the dielectric, we say that we have *induced* a dipole moment in each volume element of the dielectric. When the external field is removed, the induced dipole is destroyed as the electrical charges revert to their original positions. This process whereby dipoles are induced in dielectrics is called *polarization*. An important feature of this is that the polarized dielectric produces an electrical field of its own, and this field modifies the original external field that caused the polarization in the first place.

On a molecular level, there are two different mechanisms that give rise to the induced dipole. Figure 38.3a shows the polarization of an ordinary dielectric. The polarization vector **P** is defined as the dipole moment per unit volume as shown in Figure 38.3b. The induced dipole moment of an individual atom or molecule is proportional to the applied external field. The ratio of the induced moment to the external field strength is the *polarizability*. Another situation arises when the molecules possess permanent dipole moments, as is the case for HCl. When there is no external field, the dipole moments are randomly oriented, as shown in Figure 38.3c; the average dipole moment per unit volume, and hence the polarization, are zero. When an external field is applied, the molecules tend to align themselves with the field, as shown in Figure 38.3d, and the sum of all the dipole moments is no longer zero. Although the molecules are shown with their dipole moments parallel to the applied field, this is not the case in actual practice. Thermal motion tends to destroy the ordered arrangement of perfectly aligned dipole moments. The stronger the applied field, the larger the total moment. In measuring the actual dipole moments of molecules, one must differentiate between these two effects. This can be done by rec-

FIGURE 38.3 (a) Polarization induced in a dielectric material by an externally applied field. (b) The polarization vector defined as the induced moment per unit volume. (c) Randomly oriented polar molecules that have permanent dipole moments. (d) Dipole moments of polar molecules aligned in the direction of an applied field.

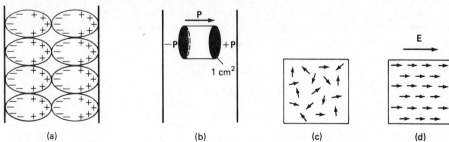

SEC. 38.2 DIELECTRICS AND POLARIZATION

ognizing that the polarization of nonpolar molecules is virtually independent of temperature, whereas the polarization of the polar molecules that arises from the orientation effect does depend on temperature. Except for a few cases in which molecules are free to move in the solid state, the *orientation* polarization of permanent dipoles is important only for gases and liquids where the individual molecules are free to orient themselves in any direction.

It will help if our discussion is continued with reference to the properties of a parallel plate capacitor as shown in Figure 38.4. Figure 38.4a shows the capacitor with a vacuum between the plates. The area of each plate is A, and the distance between the two plates is d. We let the charge per unit area be σ on one plate and $-\sigma$ on the other. The electric field between the plates is uniform and perpendicular to the plates of the capacitor. The electric field of the uniformly charged capacitor is

$$\mathbf{E} = \frac{\sigma}{\epsilon_0} \tag{38.3}$$

Since the electric displacement vector $\mathbf{D}$ is defined by $\mathbf{D} = \epsilon_0 \mathbf{E}$, we also have the corresponding equation

$$\mathbf{D} = \sigma \tag{38.4}$$

The electric displacement equals the surface charge density. Since the field between the plates is uniform, the potential difference is

$$\Delta V = V_2 - V_1 = Ed \tag{38.5}$$

The charge q on either plate is

$$q = \sigma A = \epsilon_0 E A \tag{38.6}$$

The *capacity* of the capacitor is $q/\Delta V$, or

$$C = \frac{q}{\Delta V} = \frac{\epsilon_0 A}{d} \tag{38.7}$$

This equation provides a simple physical notion of the *permittivity* ϵ_0. It is numerically equal to the capacity of a parallel-plate capacitor whose plates have unit area and are separated by unit distance.

FIGURE 38.4 (a) Charged capacitor with empty space between the plates. (b) Induced charges on the faces of a dielectric slab of thickness d' placed between the plates of a charged capacitor.

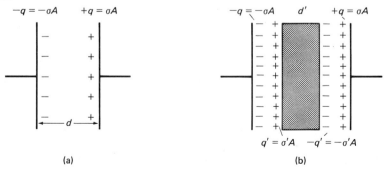

Now suppose that a dielectric slab of thickness d' is placed between the plates of the capacitor, as shown in Figure 38.4b. Surface charge densities σ' and $-\sigma'$ are induced on the two opposite surfaces of the dielectric slab by the field of the capacitor. The field due to these induced surface charges tends to reduce the electric field within the dielectric. In the region of space occupied by the dielectric, the field is

$$\mathbf{E} = \frac{(\sigma - \sigma')}{\epsilon_0} = \frac{1}{\epsilon_0}(\sigma - \mathbf{P}) \tag{38.8}$$

The induced surface charge density σ' equals the polarization $\mathbf{P}$, as can be seen by reference to Figure 38.5. Here we have drawn the dielectric on an atomic scale; it is shown as composed of atomic cubes of edge length δ. The induced moment in each atomic cube is $\boldsymbol{\mu} = q_i \boldsymbol{\delta}$, where q_i is the induced charge on the surface of the cube. The situation depicted in Figure 38.5a is equivalent to what is in Figure 38.5b. Since the charges on the interior planes of the dielectric slab cancel, we are left with the induced charge on the opposite faces of the macroscopic dielectric slab. The number of atoms per unit volume is $n = 1/\delta^3$. The polarization vector is defined as the dipole moment per unit volume, hence $\mathbf{P} = \boldsymbol{\mu} n$, and the induced surface charge density is $\sigma' = q_i/\delta^2$. The induced polarization is now

$$\mathbf{P} = \boldsymbol{\mu} n = q_i \delta n = \frac{q_i \delta}{\delta^3} = \sigma' \tag{38.9}$$

as stated previously. If we rewrite (38.8) as

$$\sigma = \epsilon_0 \mathbf{E} + \mathbf{P} \tag{38.10}$$

we see that σ has an additional term $\mathbf{P}$ arising from the presence of the dielectric. A more general definition of $\mathbf{D}$ is provided by

$$\mathbf{D} = \epsilon_0 \mathbf{E} + \mathbf{P} \tag{38.11}$$

This reduces to the previous definition of $\mathbf{D}$ in a vacuum, since $\mathbf{P} = 0$ in empty space. Rewriting Equation (38.11) in the form

$$\mathbf{E} = \frac{\mathbf{D}}{\epsilon_0} - \frac{\mathbf{P}}{\epsilon_0} \tag{38.12}$$

indicates how the electric field intensity $\mathbf{E}$ is reduced by the presence of a dielectric.

If a dielectric is placed between the plates of the capacitor, the capacity C is increased over C_0, its value in vacuum. The dimensionless ratio C/C_0 is the *dielectric constant*, symbolized by κ. It is always greater than unity, though only about 0.1% greater for gases. For solid insulators such as glass, the dielectric constant is of

FIGURE 38.5 Polarization of a dielectric on an atomic scale.

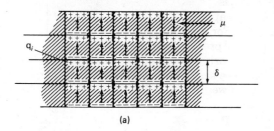
(a)

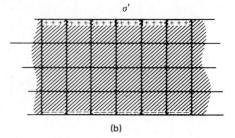
(b)

the order of 2 to 10, and for water it is about 80. Suppose the entire space between the plates of the capacitor in Figure 38.4b were filled with a dielectric slab so that $d' = d$. The capacity is

$$C = \frac{\sigma A}{Ed} = \frac{DA}{Ed} \qquad (38.13)$$

We define the *permittivity*, or *specific inductive capacity*, ϵ of the dielectric by

$$\mathbf{D} = \epsilon \mathbf{E} \qquad (38.14)$$

In this case, the capacity is given by $C = \epsilon A/d$. Since in vacuum $C_0 = \epsilon_0 A/d$, the ratio C/C_0 is just $\epsilon/\epsilon_0 = \kappa$, or

$$\epsilon = \kappa \epsilon_0 \qquad (38.15)$$

From (38.14) and (38.11) we have

$$\mathbf{D} = \epsilon \mathbf{E} = \epsilon_0 \mathbf{E} + \mathbf{P} \qquad (38.16)$$

$$\epsilon = \epsilon_0 \left(1 + \frac{\mathbf{P}}{\epsilon_0 \mathbf{E}}\right) \qquad (38.17)$$

and

$$\kappa = 1 + \frac{\mathbf{P}}{\epsilon_0 \mathbf{E}} \qquad (38.18)$$

The ratio $\mathbf{P}/\epsilon_0\mathbf{E}$ is the *electric susceptibility* χ of the dielectric, and

$$\mathbf{P} = \epsilon_0 \chi \mathbf{E} \qquad (38.19)$$

The dielectric constant and the susceptibility are related by

$$\kappa = 1 + \chi \qquad (38.20)$$

When an electric field is applied to a dielectric, there are two different contributions to the gross polarization of the material. Firstly, the presence of permanent dipoles causes a polarization, since these permanent dipoles tend to orient in a preferential direction in the field. Secondly, a contribution to the polarization arises from the induced separation of positive and negative charges. We examine the induced polarization first.

38.3 THE INDUCED POLARIZATION

The polarizability is defined as the ratio of the induced moment to the applied field. In general,

$$\mathbf{m} = \alpha \mathbf{E}_{\text{loc}} \qquad (38.21)$$

where $\mathbf{m}$ is the induced moment, $\mathbf{E}_{\text{loc}}$ is the local field at the moment in question, and α is the polarizability. It is difficult to apply the theory to condensed matter since it is difficult to determine the local field. For gases, on the other hand, the molecules are far enough apart that the electric field produced by the induced moment on one molecule does not affect the field on neighboring molecules. For gases the local field is just the externally applied field $\mathbf{E}$ (but see the next paragraph). The polarization is the induced moment per unit volume; if we let n^* be the number of molecules per unit volume, then

$$\mathbf{P} = n^* \alpha \mathbf{E} \qquad (38.22)$$

With this expression for **P**, the dielectric constant in Equation (38.18) can be written as

$$\kappa = 1 + \frac{n^*\alpha}{\epsilon_0} \tag{38.23}$$

This equation enables us to calculate the polarizability of a gas at low pressure from the measured dielectric constant.

At higher pressures and for liquids, it is difficult to evaluate $\mathbf{E}_{loc}$. One useful approximation is[3]

$$\mathbf{E}_{loc} = \mathbf{E} + \frac{\mathbf{P}}{3\epsilon_0} \tag{38.24}$$

Using this, we have

$$\mathbf{P} = n^*\alpha \left(\mathbf{E} + \frac{\mathbf{P}}{3\epsilon_0}\right)$$

$$\kappa = 1 + \frac{n^*\alpha\mathbf{E}}{\epsilon_0 \mathbf{E}} + \frac{n^*\alpha\mathbf{P}}{3\epsilon_0^2 \mathbf{E}}$$

$$\kappa - 1 = \frac{n\alpha}{3\epsilon_0}\left(3 + \frac{\mathbf{P}}{\epsilon_0 \mathbf{E}}\right) \tag{38.25}$$

or

$$\frac{\kappa - 1}{\kappa + 2} = \frac{n\alpha}{3\epsilon_0} \tag{38.26}$$

Since $n^* = L_0\rho/M$, where L_0 is Avogadro's number, M the molar mass, and ρ the density, (38.26) can be written as

$$\frac{\kappa - 1}{\kappa + 2}\frac{M}{\rho} = \frac{L_0\alpha}{3\epsilon_0} = P_m \tag{38.27}$$

The term P_m represents the *molar polarizability*. Equation (38.27), which relates the dielectric constant to the polarizability, is known as the *Clausius-Mossotti* equation. (It is sometimes written in terms of the index of refraction, with $\kappa = n^2$; in that case it is known as the *Lorenz-Lorentz* equation.) It now remains to consider the permanent dipoles.

38.4 STATISTICAL MECHANICS OF PERMANENT DIPOLES

When an external field is applied to permanent dipoles that are free to orient themselves, the dipoles tend preferentially to orient in the direction of the applied field. A moment μ that makes an angle θ with the field **E** has a potential energy

$$V = -\boldsymbol{\mu} \cdot \mathbf{E} = -\mu E \cos\theta \tag{38.28}$$

Now consider the number of dipoles oriented within the solid angle,

$$d\omega = \sin\theta\, d\theta\, d\phi \tag{38.29}$$

---

[3] See, for example, P. Debye, *Polar Molecules* (New York: Dover Publications, 1929), p. 10.

SEC. 38.4 STATISTICAL MECHANICS OF PERMANENT DIPOLES

This number is determined by the Boltzmann distribution. The exponential Boltzmann factor is

$$C \exp\left(\frac{-V}{kT}\right) = C \exp\left(\frac{\mu E \cos \theta}{kT}\right) \tag{38.30}$$

The average dipole moment is given by

$$m = \frac{\int_0^{2\pi} \int_0^{\pi} (\mu \cos \theta) C \exp(\mu E \cos \theta / kT) \sin \theta \, d\theta \, d\phi}{\int_0^{2\pi} \int_0^{\pi} \exp(\mu E \cos \theta / kT) \sin \theta \, d\theta \, d\phi} \tag{38.31}$$

The integration over ϕ in (38.31) yields the factor 2π, which cancels, since it appears in both the numerator and denominator. The integrals in (38.31) become more tractable with substitutions $\mu E/kT = x$ and $\xi = \cos \theta$, in which case $d\xi = \sin\theta \, d\theta$, and Equation (38.31) takes the form

$$\frac{m}{\mu} = \frac{\int_{-1}^{+1} \xi \exp(x\xi) \, d\xi}{\int_{-1}^{+1} \exp(x\xi) \, d\xi} \tag{38.32}$$

These two integrals are in standard form:

$$\int_{-1}^{+1} \exp(x\xi) \, d\xi = \frac{1}{x} \exp(x\xi) \bigg|_{\xi=-1}^{\xi=+1}$$

$$= \frac{\exp(x) - \exp(-x)}{x}$$

$$\int_{-1}^{+1} \xi \exp(x\xi) \, d\xi = \frac{\exp(\xi x)}{x^2} (\xi x - 1) \bigg|_{\xi=-1}^{\xi=+1}$$

$$= \frac{1}{x^2} [\exp(x)(x-1) - \exp(-x)(-x-1)]$$

$$= \frac{\exp(x) + \exp(-x)}{x} + \frac{\exp(-x) - \exp(x)}{x^2}$$

Equation (38.32) then becomes

$$\frac{m}{\mu} = \frac{\exp(x) + \exp(-x)}{\exp(x) - \exp(-x)} - \frac{1}{x}$$

$$= \coth x - \frac{1}{x} = \mathcal{L}(x) \tag{38.33}$$

The function

$$\mathcal{L}(x) = \coth x - \frac{1}{x}$$

is known as the *Langevin function*, after its discoverer. (The Langevin function was first derived in studies of magnetic susceptibilities and not electric susceptibilities.)

In the limit of small x, the Langevin function approaches $\mathcal{L}(x) = x/3$. Since $x = \mu E/kT$ is a small quantity, Equation (38.33) can be approximated to a high degree of accuracy by

$$m = \frac{\mu^2 E}{3kT} \tag{38.34}$$

To get the total molar polarizability, this term due to orientation polarization must be added to (38.27) for the induced polarization, with the final result

$$P_m = \frac{\kappa - 1}{\kappa + 2} \frac{M}{\rho} = \frac{L_0}{3\epsilon_0}\left(\alpha + \frac{\mu^2}{3kT}\right) \tag{38.35}$$

A plot of P_m vs T^{-1} yields a straight line. The polarizability α and the dipole moment μ can be determined from the intercept and slope of the straight line.

Plots of this type for methane and the four chloromethanes are shown in Figure 38.6. The lines for CH_4 and CF_4 both have zero slope, indicating zero dipole moments. This is expected, since the molecules are symmetric. The vector sum of the four tetrahedrally directed identical bond dipoles is zero. For CH_3Cl, CH_2Cl_2, and $CHCl_3$, the symmetry is destroyed, the vector sum of the four bond dipoles is no longer zero, and the molecules have a net dipole moment, as indicated by the nonzero slopes of those lines in Figure 38.6.

Carbon dioxide has no dipole moment; hence we can conclude that the molecule is linear, in the form O—C—O. The two C–O bond dipoles cancel, since they are equal and opposite. Water, on the other hand, has a permanent dipole moment and therefore must have a bent shape. Of the planar and pyramidal structures for ammonia

which would you conclude is correct, knowing that NH_3 has a permanent dipole moment? Experimental values of dipole moments are important for many theoretical calculations of molecular structure. Dipole moments of hypothetical structures can be calculated from electron densities provided by quantum mechanical calculations;

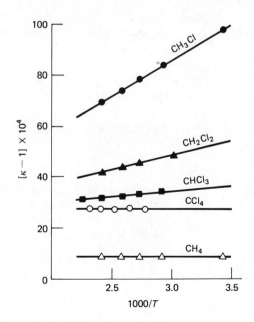

FIGURE 38.6 Plot of $(\kappa - 1)$ vs inverse temperature for the chloromethanes and for methane. (R. Sanger, *Physik. Z.* 27 (1926):556.)

these can be compared with the experimental values to indicate the validity of the calculations.

38.5 PARAMAGNETIC AND DIAMAGNETIC SUSCEPTIBILITIES

Having examined the magnetic properties of matter from a macroscopic thermodynamic point of view (Chapter 18), we shall here investigate the source of the magnetic properties. We shall examine the physical processes that produce paramagnetism and diamagnetism and shall also derive the Curie law of paramagnetism. (Recall that in our discussion of adiabatic demagnetization we simply assumed the validity of Curie's law.) We shall note that there is some correspondence between these magnetic properties and the dielectric properties discussed earlier in this chapter, though the correspondence is superficial.

Diamagnetism is the magnetic analog of the induced electric dipole. (There is no magnetic monopole.) To see how diamagnetism arises, we consider a closed loop of wire that is placed between the pole pieces of an electromagnet. When the magnetic field is turned on, the changing field induces a current to flow in the wire loop. This loop current produces its own magnetic field. By Lenz's law, this resulting field is opposed to the applied field. In an ordinary wire loop, the internal resistance causes the induced current to decay very rapidly as soon as the applied field reaches a constant value. We can think of a similar process occurring on an atomic level, with electrons moving about the nuclei in closed loops. When the applied field is turned on, currents are induced about the atoms in a sample of material. These currents of circulating electrons result in magnetic moments that are proportional to the field strength and opposed to the direction of the applied field. This diamagnetic effect results in a permeability less than unity. The electrons encounter no resistance in their motion about the nuclei; thus this motion persists so long as the field is left on. When the external field is turned off, the induction effect is in the opposite direction, and the induced currents fall to zero when the applied field reaches zero.

Paramagnetism, on the other hand, arises from the permanent magnetic moments of the atom. Suppose we have a gas of atoms for which the total angular momentum vector J is not equal to zero. We have seen (Chapter 32) that such atoms possess permanent magnetic moments. These are randomly oriented when there is no magnetic field. When an external magnetic field is applied, the magnetic moments tend to align in certain preferential directions relative to the direction of the external field, which we call the z direction. The component of the magnetic moment in the z direction is determined by the quantum number M_J ($M_J = J, J-1, \ldots, -J$); the more negative the M_J, the lower the energy of the atom. Atoms in lower energy states tend to have their magnetic moments aligned parallel to the external field, and atoms with positive values of M_J (and hence higher energies) tend to align antiparallel to the applied field. The Boltzmann distribution ensures that at equilibrium more atoms are in lower energy states than in higher energy states. Thus more magnetic moments are aligned parallel to the field than antiparallel, resulting in a net magnetic moment for the sample. Since the separation in energy levels increases with increasing magnetic field, this net moment also increases with the applied magnetic field. The susceptibility is the magnetization divided by the field strength; it remains constant for field strengths that are not too large. Further, the lower the temperature, the greater the excess of atoms whose moments are aligned parallel to the field, hence the greater the susceptibility; this agrees with the inverse temperature dependence of Curie's law.

A classical derivation of a susceptibility expression proceeds along lines very similar to the derivation of the electric polarization expression. The magnetic analog of (38.35) is

$$\chi_M = L_0 \left(\alpha_M + \frac{\mu_M^2}{3kT} \right) \tag{38.36}$$

where α_M is the induced magnetic moment that arises from the circulating electron currents and μ_M is the permanent magnetic moment that arises from the quantum mechanical properties of angular momentum. An important distinction between electric polarization and magnetic susceptibility is that χ_M can be either positive (paramagnetics) or negative (diamagnetics). In a diamagnetic medium, the magnetic field is reduced from its value in a vacuum, and in a paramagnetic medium the magnetic field is larger than in vacuum. A paramagnetic substance is attracted by a magnet, and a diamagnetic substance is repelled by a magnet.

Another difference is the magnitudes of the two. Typical diamagnetic susceptibilities lie in the range $(-10 \text{ to } -100) \times 10^{-6}$ cm^3 mol^{-1} and are temperature-independent. Typical paramagnetic susceptibilities are often orders of magnitude higher and temperature-dependent. Diamagnetism is present in all substances to some extent, even paramagnetic substances. In precise determinations of paramagnetic susceptibilities, the measurements must be corrected for the diamagnetic contributions. This is particularly important at higher temperatures and for weakly paramagnetic substances, for which diamagnetic contributions may be substantial relative to the paramagnetic contributions. These corrections are usually made by subtracting the tabulated values of the diamagnetic susceptibilities of the atoms in their various oxidation states from the measured susceptibilities.

Magnetic susceptibilities can be measured by determining the force on a sample that is placed in an inhomogeneous magnetic field. One of the most common types of apparatus is the *Gouy balance*, shown in Figure 38.7. If we call the vertical direction the x direction, then the component of the force per unit volume is[4]

$$F_x = \chi H \frac{\partial H}{\partial x} \tag{38.37}$$

where H is the magnetic field, and $\partial H/\partial x$ is the magnetic-field gradient. This force tends to pull a paramagnetic sample into the magnetic field; by contrast, a diamagnetic sample is repelled. The work done in taking a volume dV of sample from field strength H_1 to a field strength H_2 is

$$dw = dV \int F_x \, dx = dV \, \chi \int H \frac{dH}{dx} dx = dV \, \chi \int_{H_1}^{H_2} H \, dH$$
$$= \tfrac{1}{2}(H_2^2 - H_1^2) \chi \, dV \tag{38.38}$$

In the Gouy balance, a long cylindrical sample is suspended in a magnetic field so that the upper part of the sample is at a field that is essentially zero, $H_2 = 0$, and $H_1 = H$. With no magnetic field, the only force acting on the sample is gravity. When the field is turned on, an additional vertical force is exerted by the magnetic field,

---

[4] A more complete discussion of the various experimental arrangements can be found in P. W. Selwood, *Magnetochemistry* (New York: Interscience Publishers, 1956), chap. 1.

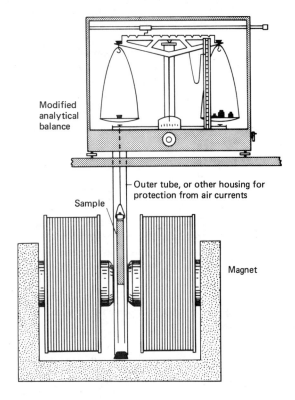

FIGURE 38.7 The Gouy balance.

and the tube moves a distance dx. If the cross-sectional area of the tube is A, then $dV = A\,dx$ and the work is

$$dw = \tfrac{1}{2}\chi H^2 A\,dx \qquad (38.39)$$

The additional force due to the interaction with the magnetic field is then

$$F_x = \tfrac{1}{2}\chi H^2 A \qquad (38.40)$$

This force can be determined from the additional weight needed to bring the pointer on the balance back to its original position.

Determining the precise value of the magnetic field is a difficult experimental procedure, and magnetic balances are most often used for relative measurements rather than absolute measurements. A sample whose susceptibility is known is first measured to calibrate the apparatus, and the unknown sample is then measured relative to the standard sample. Common standards include $Fe(NH_4SO_4)_2 \cdot 6H_2O$, $Mn(NH_4SO_4)_2 \cdot 6H_2O$ and $NiCl_2$. For precise work extreme care must be used in packing the sample in the tube. Magnetic susceptibility is an anisotropic property of crystalline substances. If powdered samples are used, there must be no preferential orientation of the individual crystals making up the sample, and the powder must be uniformly tapped into the tube. Susceptibility measurements are often made down to very low temperatures. Thus the apparatus often includes Dewars to contain liquid He and liquid N_2 with the associated temperature- and pressure-measuring devices. Great care must be taken to exclude even the smallest amounts of ferromagnetic impurities, since the susceptibility of a ferromagnetic material can be orders of magnitude larger than the susceptibility of a paramagnetic material. The presence of ferromagnetics can be checked for by measuring the susceptibility as a function of magnetic field. Ferromagnetic susceptibility varies with the field strength, whereas paramagnetic susceptibility is independent of field strength. If the measured susceptibility is independent of field strength, then one can be reasonably certain that the sample is free of ferromagnetic impurities.

38.6 THE PARAMAGNETISM OF FREE IONS

Now let us place on a quantitative basis the qualitative picture of paramagnetism that we have developed. Suppose that the magnetic moment is associated with the quantum number J. The z component of the magnetic moment in the direction of an applied field takes the values $M = -J, -J + 1, \ldots, J - 1, J$, where M is the quantum number associated with J. There are $2J + 1$ values in all. An atom whose total angular momentum is J is split into $2J + 1$ levels in the presence of an applied magnetic field. Now recall our discussion of the magnetic moment in Chapter 30. The magnetic moment associated with pure orbital motion is

$$\mu_m = m\mu_B \quad \text{(pure orbital motion)} \tag{38.41}$$

where m is the orbital quantum number and μ_B is the Bohr magneton,

$$\mu_B = |e|\hbar/2mc = 9.273 \times 10^{-21} \text{ erg gauss}^{-1} = 9.273 \times 10^{-24} \text{ J tesla}^{-1} \tag{38.42}$$

Then, for electron spin (Chapter 32) we have noted that the magnetic moment associated with pure spin motion is twice as great as the magnetic moment associated with pure orbital motion. In this case we are considering a J quantum number that arises from a mixed contribution of spin and orbital motion. Here the magnetic moment is given by the more complex expression

$$\mu_M = Mg\mu_B \tag{38.43}$$

where g is the Landé g factor. For atoms with the **LS** coupling scheme, g is given by the expression[5]

$$g = 1 + \frac{J(J+1) + S(S+1) - L(L+1)}{2J(J+1)} \tag{38.44}$$

Note that $L = 0$ and $J = S$ for pure spin, in which case $g = 2$; for pure orbital cases, $S = 0$ and $J = L$, in which case $g = 1$. These agree with the previous results for pure spin and pure orbital motion.

If the atom is placed in a magnetic field H, then the energy of each level is

$$E_M = -g\mu_B H M \tag{38.45}$$

The mean magnetic moment is found from the Boltzmann distribution law to be

$$\mu_M = \frac{\sum_{M=-J}^{M=J} g\mu_B M \exp(g\mu_B M H/kT)}{\sum_{M=-J}^{M=J} \exp(g\mu_B M H/kT)} \tag{38.46}$$

We now let

$$x = \frac{Hg\mu_B}{kT} M \tag{38.47}$$

---

[5] This formula is derived in most textbooks on quantum mechanics. See, for example, G. Herzberg, *Atomic Spectra and Atomic Structure* (New York: Dover Publications, 1944), p. 109.

SEC. 38.6 THE PARAMAGNETISM OF FREE IONS

and expand the exponentials. Omitting higher terms in x, we find that

$$\overline{\mu_M} = \frac{\Sigma g \mu_B M(1 + x + \cdots)}{\Sigma(1 + \cdots)} \qquad (38.48)$$

The summations over M go from $-J$ to $+J$. In the numerator, the first term vanishes, since $\Sigma M = 0$, and we need to keep only the second term. Using the algebraic formula for the sum of squares of integers,

$$\sum_0^N n^2 = \frac{N(N + 1)(2N + 1)}{6} \qquad (38.49)$$

we get the final expression for the magnetic moment as

$$\overline{\mu_M} = \frac{g^2 \mu_B^2 J(J + 1) H}{3kT} \qquad (38.50)$$

The total magnetization for a mole of atoms is got by multiplying Equation (38.50) by L_0, Avogadro's number; the susceptibility is obtained by dividing by the magnetic field H. The final expression for the molar susceptibility is then

$$\chi_m = \frac{L_0 g^2 \mu_B^2 J(J + 1)}{3kT} \qquad (38.51)$$

agreeing with the inverse temperature dependence that Pierre Curie had obtained empirically in 1895.

Unfortunately there are few examples for which this simple theory can be tested since there are few paramagnetic gases. On the other hand, it is found that many of the rare earth ions do obey Equation (38.51) closely at room temperature. For the rare-earth ions, the magnetic effects are associated with the 4f electrons, which are well inside the atom and influenced only slightly by crystal fields and chemical bonds. Paramagnetic susceptibility data are often presented in terms of *effective Bohr magnetons;* this unit is defined as

$$\mu_{\text{eff}} = \left(\frac{3kT\chi_m}{N_0 \mu_B^2}\right)^{1/2} \qquad (38.52)$$

From Equation (38.51) we then see that μ_{eff} should be given by

$$\mu_{\text{eff}} = g[J(J + 1)]^{1/2} \qquad (38.53)$$

Calculated and experimental values of μ_{eff} for the rare-earth ions are listed in Table 38.1. Note the excellent agreement between the calculated and experimental values for all the ions except Sm^{3+} and Eu^{3+}.

For the anomalous cases of Sm^{3+} and Eu^{3+}, an extra complication is introduced. In our simple model, we have considered only the ground state and have neglected all higher states. In most cases this assumption is valid, since the energy separation between the ground and upper excited levels is so high that the populations of the excited levels are negligible at ordinary temperatures. In some cases there are low-lying magnetic levels slightly above the ground level, and these states must be taken into account. This is the case for Sm^{3+} and Eu^{3+}. Van Vleck has obtained agreement between the theory and experiment for these ions by taking these low-lying levels into account.[6] This contribution is known as the *temperature-independent paramag-*

---

[6] J. H. Van Vleck, *The Theory of Electric and Magnetic Susceptibilities* (Oxford: Oxford University Press, 1932), p. 245. This textbook is a standard reference in the field and still extremely useful despite its age.

TABLE 38.1 Effective Bohr magnetons for trivalent rare-earth ions.

Ion	Electron configuration	Term symbol	μ_{eff} (calc) = $g[J(J+1)]^{1/2}$	μ_{eff} (exp)
La^{3+}	$5s^2 5p^6$	$^1S_0$	0	0
Ce^{3+}	$4f^1 5s^2 5p^6$	$^2F_{5/2}$	2.54	2.4
Pr^{3+}	$4f^2 5s^2 5p^6$	$^3H_4$	3.58	3.5
Nd^{3+}	$4f^3 5s^2 5p^6$	$^4I_{9/2}$	3.62	3.5
Pm^{3+}	$4f^4 5s^2 5p^6$	$^5I_4$	2.68	.....
Sm^{3+}	$4f^5 5s^2 5p^6$	$^6H_{5/2}$	0.84	1.5
Eu^{3+}	$4f^6 5s^2 5p^6$	$^7F_0$	0	3.4
Gd^{3+}	$4f^7 5s^2 5p^6$	$^8S_{7/2}$	7.94	8.0
Tb^{3+}	$4f^8 5s^2 5p^6$	$^7F_6$	9.72	9.5
Dy^{3+}	$4f^9 5s^2 5p^6$	$^6H_{15/2}$	10.63	10.6
Ho^{3+}	$4f^{10} 5s^2 5p^6$	$^5I_8$	10.60	10.4
Er^{3+}	$4f^{11} 5s^2 5p^6$	$^4I_{15/2}$	9.59	9.5
Tm^{3+}	$4f^{12} 5s^2 5p^6$	$^3H_6$	7.57	7.3
Yb^{3+}	$4f^{13} 5s^2 5p^6$	$^2F_{7/2}$	4.54	4.5

netism. The term is a misnomer, since the contribution does depend somewhat on temperature. This dependence is, however, much less than the dependence of the Curie paramagnetism.

The results for several transition metal ions are shown in Table 38.2. Here there is very poor agreement with Equation (38.53). On the other hand, we do get agreement between the experimental and calculated values if we assume that the magnetic moment is due exclusively to the spin contribution with $g = 2$; then

$$\mu_{eff} = 2[S(S+1)]^{1/2} \tag{38.54}$$

In the transition metals, the magnetic effects are due to d electrons, which do interact with the crystal fields and the chemical bonds. This interaction removes much of the orbital degeneracy, and the orbital angular momentum is said to be "quenched." Only the spin degeneracy remains to contribute to the magnetic susceptibility, which is given by the "spin only" formula, Equation (38.54).

TABLE 38.2 Effective Bohr magnetons for transition metal ions in aqueous solutions.

Ion	Configuration	Term symbol	μ_{eff} (calc) = $g[J(J+1)]^{1/2}$	μ_{eff} (calc) = $2[S(S+1)]^{1/2}$	μ_{eff} (exp)
$K^+ \ldots V^{5+}$	$3d^0$	$^1S_0$	0	0	0
Ti^{3+}, V^{4+}	$3d^1$	$^2D_{3/2}$	1.55	1.73	1.8
V^{3+}	$3d^2$	$^3F_2$	1.63	2.83	2.8
$V^{2+} \ldots Mn^{4+}$	$3d^3$	$^4F_{3/2}$	0.77	3.87	3.7–4.0
Cr^{2+}, Mn^{3+}	$3d^4$	$^5D_0$	0	4.90	4.9
Mn^{2+}, Fe^{3+}	$3d^5$	$^6S_{5/2}$	5.92	5.92	5.9
Fe^{2+}	$3d^6$	$^5D_4$	6.70	4.90	5.4
Co^{2+}	$3d^7$	$^4F_{9/2}$	6.54	3.87	4.8
Ni^{2+}	$3d^8$	$^3F_4$	5.59	2.83	3.2
Cu^{2+}	$3d^9$	$^2D_{5/2}$	3.55	1.73	1.9
Zn^{2+}, Cu^+	$3d^{10}$	$^1S_0$	0	0	0

38.7 ELECTRON PARAMAGNETIC RESONANCE SPECTROSCOPY

Suppose we have a system with unpaired electrons. Figure 38.8 shows the components of the total spin angular momentum vector in a direction parallel to an applied magnetic field. For simplicity let's take a doublet system with only one unpaired electron. A typical example is a hydrogen atom or a free radical containing one unpaired electron. The ground state is doubly degenerate. This degeneracy can be removed by applying an external magnetic field. If the energy origin is taken as the energy of the degenerate doublet, then the energies of the α and β states are $E_\alpha = \frac{1}{2}g\mu_B H$ and $E_\beta = -\frac{1}{2}g\mu_B H$, where H is the applied field. The separation of the two states as a function of applied field is shown in Figure 38.9. Transitions between the two states take place if an electromagnetic wave of frequency ν is applied so that

$$h\nu = \Delta E = g\mu_B H \tag{38.55}$$

For a free electron, $g = 2.00232$; hence at a field of 10,700 gauss, the *resonant frequency* is 3×10^{10} Hz, which corresponds to the microwave region of the spectrum. Equation (38.55) is termed the *resonance condition*. This type of spectroscopy is termed *electron paramagnetic resonance spectroscopy* (epr), *electron spin resonance spectroscopy* (esr), or less frequently, *electron magnetic resonance spectroscopy* (emr). For systems with more than one unpaired electron, for example the rare-earth ions, the energy-level diagram corresponding to Figure 38.9 would contain a correspondingly larger number of energy levels, and transitions would occur between adjacent levels so that $\Delta M_S = \pm 1$.

From the energy-level diagram of Figure 38.9, an esr spectrum of a typical system containing one unpaired electron would consist of only one line. The only value to be determined by an esr experiment would be the value of g from the known value of μ_B and the measured value of the magnetic field. Actually there are several factors that complicate the situation by introducing splittings in the single spectral line. These complications, although they add to the complexity of the spectra, provide the wealth of information that makes esr spectroscopy such a valuable tool.

The resonance condition of Equation (38.55) applies to the magnetic field that the electron "sees," and not simply to the applied magnetic field. The actual magnetic field experienced by the unpaired electron is composed of the vector sum of the exter-

FIGURE 38.8 Components of the total spin angular momentum vector in a fixed direction for $S = \frac{1}{2}, 1, \frac{3}{2}$. The magnitude of the vectors is $[S(S+1)]^{1/2}$.

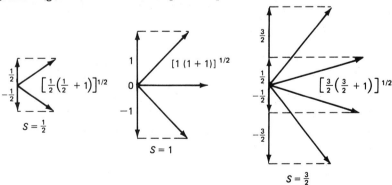

FIGURE 38.9 Energy levels of the states α and β as a function of applied magnetic field.

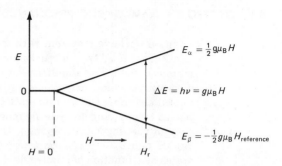

nally applied field and the various fields that arise internally in the system itself. These internal contributions arise from magnetically active nuclei, other unpaired electrons in the molecule or ion or in surrounding molecules or ions, and the magnetic moment of the electron arising from its orbital angular momentum. The last-mentioned contribution may have a large effect, since it can drastically change the g value from the free-spin value.

These various contributions to the local field at the electron are known as *hyperfine interactions*. We shall consider only the nuclear hyperfine interaction that arises from magnetic nuclei, doing so on a qualitative basis.[7] Just as there is an intrinsic spin angular momentum associated with electrons, there is a spin angular momentum associated with nuclei. The spin quantum number of a nucleus is denoted by I, and it may take on the values $0, \frac{1}{2}, 1, \frac{3}{2}, 2, \ldots$. Nuclei with nonzero nuclear spin have a magnetic moment associated with them (as we shall note in greater detail in the next section). A complete wave function must include the nuclear spin in addition to the electron spin. The proton has a nuclear spin of $\frac{1}{2}$, and we denote the two spin states as $\alpha(p)$ and $\beta(p)$. The α state corresponds to spin up and the β state corresponds to spin down, like the situation for electron spin. There are four wave functions appropriate to the system containing an electron and a proton. These are the four product functions of the electron and nuclear spin functions:

$$\begin{aligned}\psi_1 &= \alpha(e)\alpha(p) & \psi_3 &= \beta(e)\alpha(p) \\ \psi_2 &= \alpha(e)\beta(p) & \psi_4 &= \beta(e)\beta(p)\end{aligned} \qquad (38.56)$$

The four possibilities are shown in Figure 38.10. The nuclear hyperfine interaction splits each of the levels of Figure 38.9 into two levels, as shown in Figure 38.11a. The solid lines show the allowed transitions for the system, which now gives rise to a spectrum consisting of two lines. A slightly more complicated situation is presented by the deuterium atom, whose nucleus has a spin $I = 1$. That is shown in Figure 38.11b; an esr spectrum of the deuterium atom has a triple peak.

Magnetic moments of nuclei are additive. Let us consider a still more complicated example, the methyl radical $CH_3 \cdot$. The methyl radical is planar in the form of an equilateral triangle; thus the three hydrogen atoms are equivalent. (The carbon atom is magnetically inactive and can be neglected; C^{12} has $I = 0$.) The nuclear magnetic moments of the four protons add together to produce the total spin values shown in Figure 38.12. The system acts like a nucleus of spin $\frac{3}{2}$ with the added feature that while there is only one way we can achieve the $M_I = \pm \frac{3}{2}$ alignments, there

---

[7] The details of esr are covered in the numerous textbooks that have appeared in recent years. See, for example, J. E. Wertz and J. R. Bolton, *Electron Spin Resonance* (New York: McGraw-Hill Book Co., 1972).

SEC. 38.7 ELECTRON PARAMAGNETIC RESONANCE SPECTROSCOPY

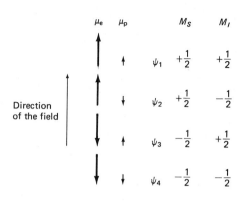

FIGURE 38.10 The alignment of the electron and proton magnetic moment vectors for the hydrogen atom in the presence of an external magnetic field. The wave functions correspond to those of Equation (38.56). The relative sizes of the electron and proton magnetic moment vectors are not drawn to scale. The electron vector is actually some three orders of magnitude larger than the proton vectors.

are three ways we can achieve $M_I = \pm\frac{1}{2}$. The splitting of the energy levels by the hyperfine interaction is shown in Figure 38.13a. The threefold manner in which $M_I = \pm\frac{1}{2}$ can be achieved is reflected by the threefold degeneracy of those states. The spacings as a function of applied magnetic field, along with an esr spectrum, are indicated in Figure 38.13b. The quartet configuration is due to the four different transitions; notice that the intensities are different, however. The relative intensities of the lines are in the ratio 1:3:3:1, reflecting the relative number of states available for each transition. Figure 38.13b shows how an esr spectrum is taken. The frequency is held constant while the magnetic field is varied; thus the lengths of the arrows indicating the transitions are all the same. A molecule (or radical) with N

FIGURE 38.11 (a) Energy-level diagram of Figure 38.9 showing the splitting due to the hyperfine interaction. Only the transitions shown by the solid lines are allowed. The dashed transition indicates what would be observed in the absence of the hyperfine interaction. The spectrum is usually obtained by keeping the frequency constant and sweeping the magnetic field. (b) The situation for the deuterium atom. The deuteron has a nuclear spin of unity, and each level is split into three. The energy-level diagram is drawn for a constant applied field.

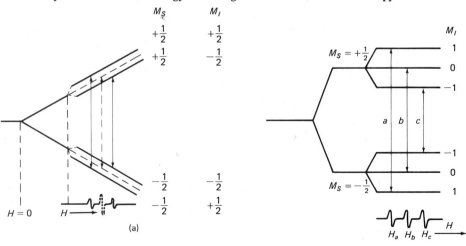

FIGURE 38.12 The various possible alignments of the nuclear moments of the three equivalent protons of the methyl radical.

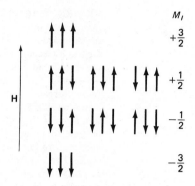

equivalent protons generally displays $N + 1$ hyperfine lines in the esr spectrum. The relative peak heights are given by the "magic triangle" shown in Table 38.3. The entries in this table are the same as in the original magic table of Pascal listing the binomial coefficients. If there are two different sets of equivalent protons, then two different sets of hyperfine lines occur in the spectrum.

In esr spectroscopy, the electron acts as a delicate probe to map the magnetic effects that produce the final localized magnetic field at the electron. Knowing these effects provides one with much useful information. Thus, for example, the esr spec-

FIGURE 38.13 (a) The hyperfine levels for the three equivalent hydrogen atoms of the methyl radical $CH_3 \cdot$. The degeneracies are shown in parentheses, and the orientations of the nuclear spins for each level are indicated by the arrows. There are three different ways in which the configurations $M_I = \pm \frac{1}{2}$ can be achieved. (b) The hyperfine energy levels of the methyl radical as a function of magnetic field showing the esr spectrum.

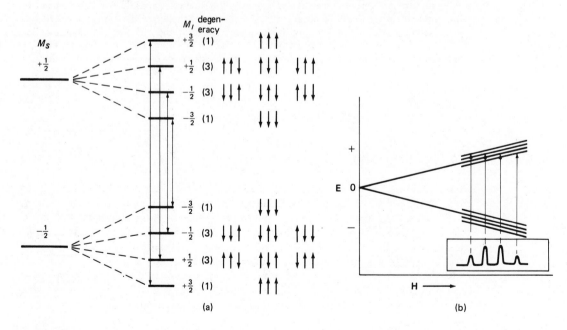

TABLE 38.3 The "magic" or binomial triangle, indicating the relative peak heights in the lines of a spectrum with N equivalent protons.

	N
1	0
1 1	1
1 2 1	2
1 3 3 1	3
1 4 6 4 1	4
1 5 10 10 5 1	5
1 6 15 20 15 6 1	6
1 7 21 35 35 21 7 1	7

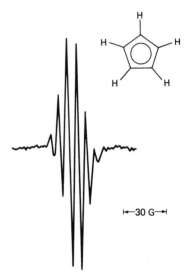

FIGURE 38.14 Electron spin resonance (esr) spectrum of the cyclopentadienyl radical. (G. R. Liebling and H. M. McConnell, *J. Chem. Phys.* 42 (1965):3931.)

trum of the cyclopentadienyl radical shown in Figure 38.14 clearly indicates that the radical has a planar configuration. The form of the spectrum shows that all five protons must be equivalent (see the magic triangle for $N = 5$); this can occur only if the radical is in the form of a planar regular pentagon.

38.8 NUCLEAR PARAMAGNETISM

Paramagnetism is associated with the angular momentum properties of electrons, both orbital and spin. Our picture of the nucleus as a point mass negates the possibility of an analogous paramagnetism due to the orbital angular momentum of nuclei, since nuclei have no orbital angular momentum. On the other hand, most nuclei do possess spin with which we can associate nuclear spin angular momentum; thus we should expect these nuclei to show magnetic features analogous to those of electrons.

All nuclei except those with even numbers of protons and even numbers of neutrons (the so-called *even-even* nuclei such as ^{4}He, ^{12}C, and ^{16}O) have nuclear spins that are designated by the quantum number I. Nuclei with odd mass numbers have

TABLE 38.4 Table of nuclear properties.

Isotope	Spin I (multiples of $\hbar$)	Magnetic moment μ (multiples of the nuclear magneton)	Nmr frequency (MHz, for a 10 kG field)	Relative sensitivity for equal numbers of nuclei at constant field	Natural abundance (%)
^{1}H	$\frac{1}{2}$	2.79268	42.5759	1.0000	99.984
^{2}H	1	0.857386	6.53566	0.0097	0.016
^{13}C	$\frac{1}{2}$	0.70220	10.705	0.0159	1.108
^{14}N	1	0.40358	3.076	0.00101	99.635
^{19}F	$\frac{1}{2}$	2.6273	40.055	0.833	100
^{31}P	$\frac{1}{2}$	1.1305	17.236	0.0663	100
^{203}Tl	$\frac{1}{2}$	1.5960	24.33	0.187	29.52
^{205}Tl	$\frac{1}{2}$	1.6115	24.57	0.192	70.48
Free electron	$\frac{1}{2}$	-1836	28,024.6	2.84×10^8	.....

spins that are odd multiples of $\frac{1}{2}$, and nuclei with even mass numbers have spins that are even multiples of $\frac{1}{2}$. A nucleus has a magnetic moment

$$\mu_N = \gamma_N I \hbar \tag{38.57}$$

where γ_N is the nuclear magnetogyric ratio; it is analogous to the electron magnetogyric ratio (Chapter 32). The magnetic moments of nuclei are often written as

$$\mu_N = g_N \mu_n I \tag{38.58}$$

where g_N is the *nuclear g factor,* and μ_n is the *nuclear magneton*

$$\mu_n = \frac{e\hbar}{2m_p} \tag{38.59}$$

The term m_p is the mass of a proton. Since m_p is about 2000 times as large as the electron mass, the nuclear magneton is about 2000 times smaller than the Bohr magneton $\mu_B = e\hbar/2m_e$. Tables of nuclear magnetic moments such as Table 38.4 usually list the values of the moments in nuclear magnetons.

38.9 NUCLEAR MAGNETIC RESONANCE SPECTROSCOPY

Much of our discussion of esr can be carried over into *nuclear magnetic resonance* spectroscopy (nmr). The components of I in the direction of an applied field are denoted by M_I. Figure 38.8 applies to nuclear moments with the proviso that I is substituted for S. The energy splittings of a nuclear spin of $I = \frac{1}{2}$ (such as a proton) as a function of magnetic field are analogous to the splittings of the electron in Figure 38.9, again with the proviso that nuclear quantities are substituted for electron quantities. Transitions between the two nuclear spin states occur at the nuclear resonance condition analogous to (38.55), which is

$$h\nu = \Delta E = g_N \mu_n H \tag{38.60}$$

At a field of 10,000 gauss, the resonance frequencies occur in the radiofrequency range. A schematic diagram of an nmr spectrometer is indicated in Figure 38.15. A

SEC. 38.9 NUCLEAR MAGNETIC RESONANCE SPECTROSCOPY

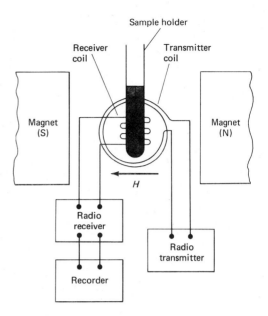

FIGURE 38.15 Schematic of a basic nmr spectrometer.

radio transmitter transmits the signal through a coil surrounding the sample. A receiver, also surrounding the sample, picks up the signal and displays it on an oscilloscope or strip chart recorder. When the frequency condition of (38.60) is satisfied, the sample absorbs some radiation, and there is a sudden change in the signal amplitude going to the receiver. The change in signal, properly displayed, constitutes the nmr spectrum. From what we have said thus far, an nmr spectrum investigating protons would display one peak and provide very little information about the system. As with esr spectroscopy, the resonance condition is established not by the value of the applied external magnetic field but rather by the local field that the nucleus "sees." There are two important effects that alter the local field to produce the shifts and splittings in the spectra that provide a wealth of information, making nmr spectroscopy such a valuable tool.

The first effect is related to the diamagnetic effect discussed earlier in this chapter. When a diamagnetic substance is placed in a magnetic field, circulating electron currents are induced about the atoms in the material. These induced currents generate their own small magnetic fields, which are opposed to the external applied field. As a net result, the local field at the nucleus is

$$H_{\text{loc}} = H_{\text{appl}} - H_{\text{induced}} \tag{38.61}$$

It is H_{loc} that must be used in Equation (38.60) to determine the resonance condition. The value H_{appl} is the same for all nuclei in the sample; H_{induced}, however, is different for different nuclei, depending on the local chemical environment of the different nuclei. This shift to lower field values is called the *chemical shift*. Figure 38.16a shows a proton resonance spectrum for ethyl alcohol, OH_3CH_2CH. The alcohol molecule has three different types of hydrogens; this is reflected in the three proton peaks that appear in the spectrum. Further, the areas under each of the peaks are in the ratio $1:2:3$, which reflects the ratio of the numbers of different types of protons. When a spectrum such as that of Figure 38.16a is used as an analytical tool, it shows that the sample contains three different types of protons in the ratio $1:2:3$.

FIGURE 38.16 Nuclear magnetic resonance (nmr) spectra of ethyl alcohol, CH_3CH_2OH. (a) Low resolution. (b) High resolution. At a field of 9400 gauss and a frequency of 40 MHz, the chemical shift between the CH_2 and CH_3 peaks is about 23 mG.

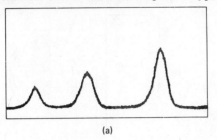

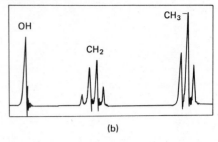

(a) (b)

The chemical shift is defined by the equation

$$\delta_j = \frac{H_s - H_r}{H_r} \times 10^6 \qquad (38.62)$$

where $H_s - H_r$ is the shift (usually in gauss) between the resonance field for the sample and the resonance field for a reference sample. The chemical shift is defined in terms of magnetic field instead of frequencies because of several experimental constraints. Chemical shifts for protons in liquid samples are of the order of parts per million; extreme constancy of all factors is required to measure the shifts to more than 1%, which is routinely done. Modern crystal oscillators can provide electromagnetic radiation in this frequency range that is accurate and stable to greater than one part in 10^8. On the other hand, magnetic fields cannot be measured this accurately on an absolute basis. Magnetic fields can, however, be kept constant to this degree of precision, and can be varied linearly and accurately over a small range. Thus the frequency is held constant, and the magnetic field is varied to produce a spectrum; the shifts are measured in gauss (or milligauss for protons). The chemical shifts of numerous common groups have been measured and tabulated as in Table 38.5. These tables of chemical shifts serve to identify unknown groups when nmr is used as an analytical tool.[8]

Figure 38.16b shows an nmr spectrum of ethyl alcohol under higher resolution. The splitting of the CH_2 and CH_3 protons is an example of *spin-spin coupling*, the second effect that alters H_{loc}. This splitting is analogous to the hyperfine splitting that occurs in esr spectroscopy. The three protons on the CH_3 group each have spin $\frac{1}{2}$; these add, to produce a total $I = \frac{3}{2}$, as shown in Figure 38.12. The spin-spin interaction between the CH_3 and CH_2 protons splits the CH_2 proton peak into a quartet. The CH_2 protons add to produce $I = 1$, and this splits the CH_3 proton peak into a triplet as shown on the spectrum.

A complicated spectrum displays a host of peaks that are due both to chemical shifts and spin-spin couplings. These two effects can be separated by running spectra at two different frequencies (and magnetic fields). The peaks (or groups of peaks) that arise from chemical shifts change their positions in the two spectra, since the chemical shift is proportional to the applied magnetic field. This is so because the induced field that produces the chemical shift is proportional to the applied external

---

[8] The application of nmr spectroscopy to the identification of organic groups is discussed in many organic chemistry textbooks. See, for example, R. T. Morrison and R. N. Boyd, *Organic Chemistry*, 3rd ed. (Boston: Allyn and Bacon, 1973), pp. 414–51.

TABLE 38.5 Chemical shifts of proton nmr signals in various organic functional groups.

	δ (ppm)		δ (ppm)
—SO$_3$H	−6.7 ± 0.3	H$_2$O	(0.00)
—CO$_2$H	−6.4 ± 0.8	—OCH$_3$	+1.6 ± 0.3
RCHO	−4.7 ± 0.3	—CH$_2$X	+1.7 ± 1.2
RCONH$_2$	−2.9	≡C—H	+2.4 ± 0.4
ArOH	−2.3 ± 0.3	=C—CH$_3$	+3.3 ± 0.5
ArH	−1.9 ± 1.0	—CH$_2$—	+3.5 ± 0.5
=CH$_2$	−0.6 ± 0.7	RNH$_2$	+3.6 ± 0.7
ROH	−0.1 ± 0.7	—C—CH$_3$	+4.1 ± 0.6

SOURCE: Tabulated by J. D. Roberts, *Nuclear Magnetic Resonance* (New York: McGraw-Hill Book Co., 1959), from original data of H. S. Gutowsky, L. H. Meyer, and A. Saika, *J. Amer. Chem. Soc.* 53 (1953):4567.

NOTE: Entries are parts per million relative to water taken as zero.

field. The peaks arising from spin-spin interaction keep the same relative spacings in the two spectra since these splittings are independent of the applied field.

This picture of spin-spin couplings leads to the expectation that the OH proton peaks in the high-resolution ethyl alcohol spectrum will also display a splitting. The peak is a single one because the sample contains some water, and the rapid exchange of the hydroxide (OH) protons with the water protons removes this splitting. Two different times are involved here; we call them t_{ex} and t_{trans}. The term t_{ex} is the time characteristic of the chemical exchange between the two protons, and t_{trans} is the time it takes for the transition between the two spin states to take place. If $t_{ex} < t_{trans}$, then we should see only one peak due to a proton "averaged" between an OH proton and a water proton. On the other hand, if $t_{trans} < t_{ex}$, then we could see each of the peaks individually. In this case, the exchange is so rapid that we see only a single peak. This exchange can be eliminated in ethyl alcohol by using high-purity alcohol that is free of water. In that case, the expected triplet structure of the OH peak is observed. In addition, the spin-spin interaction between the OH proton and the CH$_2$ protons causes a further splitting of each of the four peaks in the CH$_2$ quartet, to produce a total of eight peaks due to the CH$_2$ protons.

The measurement of chemical shifts as a function of temperature can provide information on the rate of exchange of chemical species. Figure 38.17 shows a ^{205}Tl spectrum of the complex salt Tl$_2$Br$_4$ at two different temperatures. Thallium normally exists in the thallous state with oxidation state +1 and in the thallic state with oxidation number +3. This is shown in compounds like TlX and TlX$_3$, where X is a halogen. The spectrum of molten Tl$_2$Br$_4$ clearly indicates that two different types of thallium ions are present, since there are two thallium peaks; the formula might be more properly written TlBr·TlBr$_3$. That the chemical shift decreases with increasing temperature indicates that the two thalliums rapidly exchange with each other, the rate of exchange increasing with temperature. At sufficiently high temperature, the exchange rate is faster than the transition rate, and only one peak would be observed for an "averaged" thallium nucleus. The exchange rate of the thallium atoms was estimated to be of the order of 10^{-5} s from these experiments.[9]

---

[9] T. J. Rowland and J. P. Bromberg, *J. Chem. Phys.* 29 (1958):626.

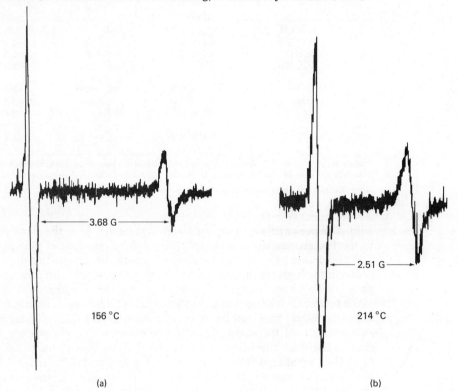

FIGURE 38.17 Nuclear magnetic resonance spectra of molten Tl_2Br_4 at two different temperatures. (T. J. Rowland and J. P. Bromberg, *J. Chem. Phys.* 29 (1958):626.)

PROBLEMS

1. Each of two plates of a capacitor has a surface area of 100 cm² and they are separated by 0.2 cm. A voltage of 200 V is impressed across the plates by connecting them to a 200-V battery.
   (a) What is the capacitance of the capacitor when a vacuum fills the space between the plates?
   (b) Calculate the surface charge on each plate of the capacitor.
   (c) Calculate the force of attraction between the two plates.

2. Now suppose that the capacitor of Problem 1 is immersed in oil that has a dielectric constant of 2.0. How does this affect the answers to the three parts of Problem 1?

3. Calculate the induced dipole moment per unit volume and the electric-field intensity in the oil between the plates of the capacitor in Problem 2.

4. A dipole moment of magnitude 1 Debye (D) is placed in a 1 V cm⁻¹ electric field. The dipole moment is placed at an angle of 45° with the field direction. Calculate the potential energy of the dipole. Suppose that the dipole is pivoted at its center. How much work is done when the dipole is aligned parallel to the field? How much work is done for a mole of dipoles?

5. The internuclear distance of the HCl molecule is 1.26 Å. Calculate the dipole moment to be expected if the HCl molecule consisted of the ions H⁺ and Cl⁻, separated by the internuclear distance. The experimental dipole moment is 1.08 D. What fraction of an electron charge does this value correspond to? This fraction corresponds to the fraction of ionic character of the bond.

6. The dipole moment of H_2O is 1.85 D. The H—O—H bond angle in water is 105°. What fraction of ionic character in the O–H bond do these data reflect? (The O–H bond length is 0.96 Å.)

7. E. C. Hurdis and C. P. Smyth (*J. Amer. Chem. Soc.* 64 (1942):2829) measured the molar polarization of water vapor as a function of temperature. From their data in the accompanying tabulation, calculate the dipole moment of H_2O.

P_M(cm³)	T (K)
57.4	384.3
53.5	420.1
50.1	444.7
46.8	484.1
43.1	522.0

8. Using the tabulated values of Table 38.1, calculate the molar susceptibility (in cm³ mol⁻¹) for the Dy^{3+} ion. What is the value of g for this ion?
9. The susceptibility of the ion Mn^{2+} is found to follow the expression $\chi_M = 4.3785/T$ cm³ mol⁻¹. Since Mn is a transition metal, we expect the susceptibility to be given by the spin-only formula. What can you say about the multiplicity of the ground state of the Mn^{2+} ion?
10. Take the magnetic moment of a free electron as one Bohr magneton. Assuming that sodium metal has one free electron per atom that contributes to the susceptibility, calculate the expected susceptibility at room temperature. (But see the next problem.)
11. Contrary to the results of Problem 10, the paramagnetism of free electrons is temperature-independent and much lower than predicted in Problem 10. Paramagnetism results when the probability of having magnetic moments lined up parallel to the field exceeds the probability of the moments lining up antiparallel. This probability is proportional to $\mu H/kT$; for N atoms the resulting moment is proportional to $NH\mu^2/kT$ in the classical treatment. We have already seen, however (Chapter 37), that the free-electron model of metals requires most of the electrons to be paired and that only those electrons within $\sim kT$ of the top of the Fermi distribution have an opportunity to turn over in the field. Take this fraction to be T/T_F, where T_F is the Fermi temperature, and show that the paramagnetism of the free electrons in a metal is temperature-independent. Find the magnitude of this paramagnetism and compare it with the result calculated in Problem 10. (*Note:* This reasoning applies only for $\mu H \ll kT$, which is valid for normal fields at room temperature. The extremely high field of 10^6 oersteds (Oe) corresponds to a temperature of about 100 K.)
12. At what frequency will the esr peak of a free electron occur in a magnetic field of 10,000 G?
13. The *spin Hamiltonian* for an electron and a proton (the energy-level diagram shown in Figure 38.11a) can be written as

$$\hat{H} = g\mu_B H \hat{S}_z + hA_0 \hat{S}_z \hat{I}_z$$

where A_0 is the *hyperfine coupling constant* and $\hat{S}_z$ and $\hat{I}_z$ are spin operators such that

$$\hat{S}_z \alpha(e)\beta(p) = +\tfrac{1}{2}\alpha(e)\beta(p)$$
$$\hat{I}_z \alpha(e)\beta(p) = -\tfrac{1}{2}\alpha(e)\beta(p)$$

with similar eigenvalue expressions for the remaining functions of (38.56). Note that $\hat{S}_z$ operates only on the electron function and $\hat{I}_z$ operates only on the nuclear spin function. Show that the energies of the four states shown in Figure 38.11a are:

$$E(+\tfrac{1}{2}, +\tfrac{1}{2}) = \tfrac{1}{2}g\mu_B H + \tfrac{1}{4}hA_0$$
$$E(+\tfrac{1}{2}, -\tfrac{1}{2}) = \tfrac{1}{2}g\mu_B H - \tfrac{1}{4}hA_0$$
$$E(-\tfrac{1}{2}, -\tfrac{1}{2}) = -\tfrac{1}{2}g\mu_B H + \tfrac{1}{4}hA_0$$
$$E(-\tfrac{1}{2}, +\tfrac{1}{2}) = -\tfrac{1}{2}g\mu_B H - \tfrac{1}{4}hA_0$$

14. Using the procedure outlined in Problem 13, and the same spin Hamiltonian, find the energies of the six levels shown in Figure 38.11b.
15. The hyperfine coupling constant A_0 for the hydrogen atom (see Problem 13) is 1420 MHz. Determine the magnetic field values at which peaks will be observed if the radiating frequency in an esr experiment is fixed at 3×10^{10} Hz.
16. Using the data of Problem 15, calculate the relative numbers of atoms in each of the four states for a magnetic field of 10,000 gauss at room temperature.
17. For a system containing free electrons, neglecting any hyperfine interactions, calculate how many more electrons there are in the lower state than in the upper state at a field of 10,000 G at $T = 4$, 100, and 300 K.
18. Explain with the help of the information in Problem 13 why the relative number of electrons in the two lowest energy levels shown in Figure 38.11a does not change with magnetic field.
19. Sketch the esr spectrum of the cycloheptatrienyl radical. Indicate the relative intensities of all the peaks.

20. An Avogadro's number of protons is in a magnetic field of 10,000 G. Calculate the ratio of the number in the lower spin state to the number in the upper spin state at room temperature.
21. Fluorine has a spin of $\tfrac{1}{2}$. Describe the expected proton and fluorine nmr spectra of CH_3F. What about CH_2F_2?

PART SEVEN

THE RATES OF CHEMICAL REACTIONS

CHAPTER THIRTY-NINE
PHENOMENOLOGICAL RATES OF CHEMICAL REACTIONS

39.1 REACTION RATES AND EQUILIBRIUM

There exists a somewhat arbitrary division of physical chemistry into four classical areas: thermodynamics, statistical mechanics, quantum mechanics, and chemical kinetics. This is reflected in the course offerings of many graduate schools, whose incoming students are expected to take courses in each of these areas. In these concluding chapters we turn our attention to the last of these areas, chemical kinetics, or the rates of chemical reactions. Our previous discussions have mostly been concerned with equilibrium states. Now we investigate the length of time needed to achieve such equilibrium states.

The final equilibrium state of a system, we have emphasized, in no way depends on the path used in going from the initial to the final state. The rate at which the system achieves equilibrium does, however, depend on the path. In fact, the entire action of catalysts depends precisely on this point. The economic implications of reaction rates are especially important to designers of chemical processes. Large chemical factories use reaction vessels costing millions of dollars. The faster the system can be recycled to produce fresh batches of product, the lower the cost.

The time needed to achieve equilibrium in chemical reactions varies over many orders of magnitude. The experimental procedures used to study rates must be adapted to the particular case at hand. Stopwatches and ordinary chemical analyses suffice for reaction times measured in minutes but will hardly do for a reaction involving a free radical whose formation and disappearance are over in 10^{-8} s. Ingenious schemes have been devised to study these ultrafast reactions, and we shall note some of these as we proceed.

The first modern kinetics measurement was made by Wilhelmy, who studied the rate of inversion of sucrose in 1850 and determined how concentration affected the rate. The equilibrium constant of a reaction, one of the most important results of chemical thermodynamics, first entered chemistry through the work of Guldberg and Waage in 1867; they derived the *law of mass action* by assuming that at equilibrium the forward and reverse rates of a reaction must be equal. The classical divisions of physical chemistry do indeed form a unified whole.

> LUDWIG FERDINAND WILHELMY (1812–1864), German physicist, is regarded as the father of chemical kinetics. The effect of optically active compounds on polarized light had been known since the measurements of Jean-Baptiste Biot, who measured the changes in tartaric acid solution by measuring the rotation of the plane of polarization in the 1830s. In 1843, Eilhard Mitscherlich noted that the polariscope would be an invaluable tool for following a chemical reaction when the effect on the plane of polarization of the reactant was different from the effect on the product. In 1850, Wilhelmy (L. F. Wilhelmy, *Poggendorff's Ann.* 81 (1850):413,499) measured by polariscope the rate of hydrolysis (or inversion) of cane sugar in acid solution. He expressed his results in the form $-dZ/dt = MZS$, where Z was the concentration of sugar (*Zucker*), S the concentration of acid (*Saure*), and M a constant. Today we call this expression a second-order rate equation. As Wilhelm Ostwald put it, this was the first time that the progress of a chemical reaction had been formulated mathematically (W. Ostwald, *Lehrbuch der Allgemeinen Chemie*, vol. 2, part 2 (Leipzig: Wilhelm Engelmann, 1896), p. 69). Wilhelmy received little credit for the work at that time. In 1860, Berthelot investigated the rate of hydrolysis of esters and found that the rate of hydrolysis was proportional to the concentration of ester remaining undecomposed. Berthelot was one of the most prestigious chemists of his day, and his work received more notice. The derivation of the law of mass action by the Norwegians Guldberg and Waage in 1863 was based on the kinetics experiments of Berthelot.

39.2 RATE OF A SIMPLE PROCESS AND ITS HALF LIFE

We consider as an example of the rate of a simple process the case of molecular effusion that we have discussed before (Chapter 22). A gas at low pressure is contained in a vessel with a tiny pinhole in the wall, and a high-vacuum pump is connected to the pinhole. The rate at which gas molecules leave the vessel is

$$-\frac{dn^*}{dt} = k_1 n^* \tag{39.1}$$

where n^* is the concentration of molecules and k_1 is a constant. Equation (39.1) is a simple rate equation that tells us that the rate at which molecules are lost is proportional to the quantity (or concentration) of gas present. The subscript 1 is used to indicate that the rate is proportional to the concentration raised to the first power. For gases the pressure is directly proportional to the concentration, hence we can write (39.1) as

$$-\frac{dP}{dt} = k_1 P \tag{39.2}$$

Equation (39.2) can be integrated to yield

$$P = P_0 \exp(-k_1 t) \tag{39.3}$$

where P_0 is the initial pressure at $t = 0$. The pressure decreases exponentially. A plot of $\ln P$ vs t yields a straight line with slope $-k_1$, as shown in Figure 39.1.

Any process has a characteristic time called the *half-life*, $t_{1/2}$. It is the time it takes for the process to be half-completed. In this case it is the time it takes for the pressure to drop to one-half its original value,[1]

---

[1] Another characteristic time that is used is τ, the "lifetime," or "decay time." The value τ is the time it takes for the concentration to fall to $1/e$ of its initial value; it is equal to $(k_1)^{-1}$.

FIGURE 39.1 Logarithm of pressure vs time for an effusion experiment. (J. P. Bromberg, *J. Vac. Sci. Technol.* 6 (1969):801.)

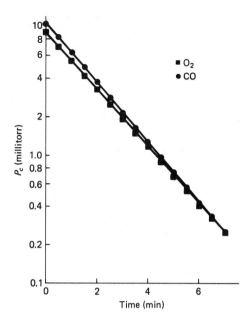

$$P = \tfrac{1}{2}P_0 = P_0 \exp(-k_1 t_{1/2}) \tag{39.4}$$

Equation (39.4) can be solved to find

$$\exp(-k_1 t_{1/2}) = \tfrac{1}{2}$$

$$k_1 t_{1/2} = -\ln(\tfrac{1}{2})$$

$$t_{1/2} = -\frac{\ln(\tfrac{1}{2})}{k_1} = \frac{0.693}{k_1} \tag{39.5}$$

The half-life of our effusion process is independent of initial concentration. The time it takes for the pressure to drop from P_0 to $\tfrac{1}{2}P_0$ is the same as the time it takes for the pressure to drop from $\tfrac{1}{2}P_0$ to $\tfrac{1}{4}P_0$, and so on.

Suppose we change the effusion experiment, as indicated in Figure 39.2. We cover the pinhole with a trapdoor that is held shut by a spring. The spring is designed so that if only one molecule hits the door, it is reflected back, but if two molecules hit the door at the same time, the resulting force of the simultaneous collision of both

FIGURE 39.2 Effusion experiments. (a) A molecule that hits the hole in the side of the vessel leaves the vessel. (b) The hole is covered with a spring-loaded door. If only one molecule hits the door, it is reflected back into the box. If two molecules simultaneously hit the door, the force of the combined collision is enough to open the door, and both molecules leave the vessel.

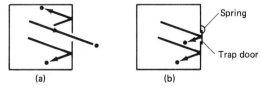

molecules is enough to push the door open, with *both* molecules leaving the vessel.[2] We now want to compute dn^*/dt, or equivalently dP/dt, for this system.

The probability that one molecule will hit the area of the door is proportional to n^* or P. Since the probabilities of hitting the door are independent, the probability that two molecules will strike the door simultaneously is just the product of the two individual probabilities; the probability is therefore proportional to n^{*2}, or equivalently, P^2. The rate equation in this case is

$$-\frac{dn^*}{dt} = k_2(n^*)^2 \tag{39.6}$$

where we have used a subscript 2 to indicate that the rate is proportional to the concentration raised to the second power. We defer solving this equation to later in the chapter. Then we shall note that this half-life is no longer independent of the initial amount of material. We could go on indefinitely by increasing the strength of the spring so that it would require the simultaneous collision of three or more molecules to open the trapdoor, and so on.

39.3 EMPIRICAL ORDER OF A CHEMICAL REACTION

A general chemical reaction can be represented by the stoichiometric equation

$$a\text{A} + b\text{B} + c\text{C} + \cdots \longrightarrow d\text{D} + e\text{E} + f\text{F} \tag{39.7}$$

where $a, b, \ldots$ represent moles of reactants (A, B, ...) and $d, e, \ldots$ represent moles of product (D, E, ...). The primary experimental item of interest in chemical kinetics is the rate at which a reaction proceeds. This is defined simply as the rate of change of concentration of any reactant or any product. The reaction rate is written as a differential expression in the concentrations, and can be written in several ways that are related by the stoichiometric coefficients of the balanced chemical equation. In most systems, it is usually easier to measure the concentration of one of the reactants (or products) than any other. The reaction is followed by measuring that constituent, and the rate is written in terms of its concentration. Thus, in Equation (39.7), if the reaction were followed by measuring the concentration of A remaining, then the rate would be written $R = -d[\text{A}]/dt$; a negative sign is always used with reactants, and thus the sign of R is positive. In terms of the other constituents, the rate could be written

$$R = -\frac{d[\text{A}]}{dt} = -\frac{a}{b}\frac{d[\text{B}]}{dt} = -\frac{a}{c}\frac{d[\text{C}]}{dt} = \frac{d}{a}\frac{d[\text{D}]}{dt} = \frac{e}{a}\frac{d[\text{E}]}{dt} = \cdots \tag{39.8}$$

The rate of a reaction is a differential equation of the form

$$R = -\frac{d[\text{A}]}{dt} \equiv -\frac{dc_\text{A}}{dt} = f(c_\text{A}, c_\text{B}, \ldots) \tag{39.9}$$

---

[2] More properly, we should speak of a simultaneous collision of two molecules as a collision of two molecules within a very small unit of time Δt; the probability that two molecules will hit the door at *precisely* the same time is vanishingly small. Also, there is another problem that arises by virtue of the Maxwell-Boltzmann distribution. Some of the molecules travel with very large speeds and have enough momentum to open the door by themselves. We can eliminate this difficulty by placing in front of the door a velocity selector that permits only molecules in a small velocity interval to strike the door.

where the c_i represent concentrations of all the species present. In many instances, the functional dependence is the product of concentration terms in the form

$$-\frac{dc_A}{dt} = k c_A^\alpha c_B^\beta c_C^\gamma \cdots \quad (39.10)$$

where k is called the *rate constant*. The exponents, which are determined experimentally, are called the *partial orders* of the reaction; thus α is the partial order of the reaction with respect to A. The sum $\alpha + \beta + \gamma + \cdots$ is the *overall order of the reaction*, or more simply, the *order*. If the order is $\alpha + \beta + \gamma + \cdots = \zeta$, then the units of k are mol$^{-(\zeta-1)}$ liter$^{\zeta-1}$ s^{-1} when concentration units of moles per liter are used. In many cases the orders are integers between 0 and 3, but they need not be so. In many cases the rate may be influenced by the concentration of products and other factors. In almost all cases the rate is affected by temperature. We shall subsequently examine how the detailed study of reaction rates provides valuable insight into the molecular details, or *mechanism*, of a reaction (Chapter 40). Currently we shall restrict our attention to the phenomenological orders of reactions in which the partial orders are integers, and in which the rate is affected only by reactant concentrations.

Suppose we consider, as a specific example, the reaction

$$2NO + Cl_2 \longrightarrow 2NOCl \quad (39.11)$$

The rate can be expressed in terms of the disappearance of either reactant, $-d[NO]/dt$ or $-d[Cl_2]/dt$, or in terms of the rate of appearance of product, $d[NOCl]/dt$, where the brackets indicate concentrations. From the stoichiometry of the reaction, we can write

$$-\frac{d[NO]}{dt} = -2\frac{d[Cl_2]}{dt} = \frac{d[NOCl]}{dt} \quad (39.12)$$

It has been found experimentally that the rate is proportional to $[NO]^2[Cl_2]$; thus the reaction is third-order. The rate equation could be written in any of the three ways:

$$-\frac{d[NO]}{dt} = k_{NO}[NO]^2[Cl_2]$$

$$-\frac{d[Cl_2]}{dt} = k_{Cl_2}[NO]^2[Cl_2] \quad (39.13)$$

$$\frac{d[NOCl]}{dt} = k_{NOCl}[NO]^2[Cl_2]$$

where k_{NO}, k_{Cl_2}, and k_{NOCl} are the rate constants corresponding to the different ways of writing the rate equation. These rate constants are related by

$$k_{NO} = 2k_{Cl_2} = k_{NOCl} \quad (39.14)$$

39.4 FIRST-ORDER REACTIONS

A reaction for which $\zeta = 1$ is known as a *first-order reaction*. If the reaction A → products is first-order, then the rate equation can be written $-dc/dt = kc$, where c is the concentration of A. This equation can be integrated, to get

$$\int_{c_0}^{c} \frac{dc}{c} = - \int_{0}^{t} k \, dt$$

$$c = c_0 \exp(-kt) \tag{39.15}$$

where c_0 is the concentration at time $t = 0$. As noted in (39.5), the half-life of this reaction, or the time it takes for the concentration to fall to one-half its initial value, is

$$t_{1/2} = \frac{\ln 2}{k} = \frac{0.693}{k} \tag{39.16}$$

A classic example of a first-order reaction is the decomposition of N_2O_5:

$$N_2O_5 \longrightarrow 2NO_2 + \tfrac{1}{2}O_2 \tag{39.17}$$

which was first studied in 1920.[3] Figure 39.3 shows plots of the partial pressure of N_2O_5 vs the time for several experimental runs. The points lie on straight lines in agreement with (39.15), verifying that the reaction is first-order. The rate constant is determined from the slopes of the lines. Note that while the two lines for 45 °C are displaced from each other, they are parallel. The line at 25 °C has a much smaller slope, reflecting the lower reaction rate at the lower temperature. The rate constant can also be determined by measuring the half-life of the reaction. If the half-life is independent of initial concentration, then the reaction is first-order and the rate constant can be evaluated by Equation (39.16). For reactions that are not first-order, the half-lives depend on the initial concentrations (see next section).

The units of a first-order reaction rate constant are (time)$^{-1}$, which is apparent from the rate equation itself. There are no concentration units in a first-order reac-

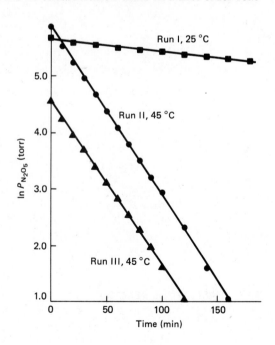

FIGURE 39.3 First-order thermal decomposition of N_2O_5.

[3] F. Daniels and E. H. Johnston, *J. Amer. Chem. Soc.* **43** (1920):53.

SEC. 39.5 SECOND-ORDER REACTIONS

tion rate constant. For the N_2O_5 decomposition reaction, the rate constant is $k = 2.03 \times 10^{-3}$ min^{-1} = 3.38×10^{-5} s^{-1} at 25 °C. At 45 °C the rate constant increases substantially to $k = 0.0299$ min^{-1} = 0.000498 s^{-1}.

39.5 SECOND-ORDER REACTIONS

If the reaction

$$A + B \longrightarrow C + D \tag{39.18}$$

is a second-order reaction, then the rate equation could be written

$$\frac{-dc_A}{dt} = kc_A c_B \tag{39.19}$$

where c_A and c_B are the concentrations of A and B at time t. For the reaction as written in (39.18), equivalent equations are

$$\frac{-dc_B}{dt} = \frac{+dc_C}{dt} = \frac{+dc_D}{dt} = kc_A c_B$$

since the coefficients of all species are unity. It is more convenient to write Equation (39.19) in terms of a single variable, namely the amount of material caused to react. Suppose that the initial concentrations of A and B are denoted by a and b for $t = 0$. If we let x be the number of moles per liter of A that have reacted by time t, then $c_A = a - x$ and $c_B = b - x$. Equation (39.19) can now be written

$$\frac{dx}{dt} = k(a - x)(b - x) \tag{39.20}$$

or

$$\frac{dx}{(a - x)(b - x)} = k \, dt \tag{39.21}$$

By the method of partial fractions, this equation can be written[4]

$$\frac{1}{b - a}\left[\frac{1}{a - x} - \frac{1}{b - x}\right] dx = k \, dt \tag{39.22}$$

which can be integrated, to get

$$\frac{\ln(a - x) - \ln(b - x)}{a - b} = kt + \text{constant} \tag{39.23}$$

The constant of integration can be evaluated by noting that $x = 0$ at $t = 0$, in which case

$$\text{constant} = \frac{\ln(a/b)}{a - b} \tag{39.24}$$

---

[4] We write the equation in the form

$$\left(\frac{A}{a - x} + \frac{B}{b - x}\right) dx = k \, dt$$

where A and B are constants to be evaluated. Since it must then be true that

$$\frac{A}{a - x} + \frac{B}{b - x} = \frac{1}{(a - x)(b - x)}$$

we find by solving for A and B that $A = -1/(a - b)$ and $B = 1/(a - b)$.

The final integrated expression becomes

$$\frac{1}{a-b} \ln \left[\frac{b(a-x)}{a(b-x)}\right] = kt \tag{39.25}$$

In the general case, the coefficients of all the species in Equation (39.18) are not unity. For the reaction

$$A + qB \longrightarrow \text{products} \tag{39.26}$$

q moles of B react with each mole of A. In this case, the rate equation takes the form

$$\frac{dx}{dt} = k(a-x)(b-qx) \tag{39.27}$$

and the final integrated equation is

$$\frac{1}{qa-b} \ln \left\{\frac{(b/q)(a-x)}{a[(b/q)-x]}\right\} = kt \tag{39.28}$$

We consider as an example of a second-order reaction the reaction of propylene bromide with potassium iodide in methanol solution[5]

$$C_3H_6Br_2 + 3KI \longrightarrow C_3H_6 + 2KBr + KI_3 \tag{39.29}$$

Mixtures containing known initial amounts of the pure reactants were placed in thermostats. In this instance the rate of the reaction was slow enough for the ordinary analytical techniques to be appropriate. Small portions of the mixture were removed at suitable time intervals and analyzed for iodine by titration with thiosulfate solution. The rate equation can be written as

$$\frac{d[I_3^-]}{dt} = k[C_3H_6Br_2][KI] = k(a-x)(b-3x) \tag{39.30}$$

The integrated form of the equation is given by Equation (39.28), with $q = 3$. The rate constant can be obtained by plotting the left side of Equation (39.28) against the time and determining the slope of the straight line, as shown in Figure 39.4.

The above analysis of a second-order reaction cannot be used for cases in which the initial concentrations of the reacting species are set at their stoichiometric ratios, since if $b = qa$, we should be faced with a zero denominator. This situation always exists for a second-order decomposition such as A = products. This is exemplified by the classic second-order decomposition of HI, that is, $2HI = H_2 + I_2$. In this case we must return to the original rate equation and write

$$-\frac{dc_A}{dt} = kc_A^2 \tag{39.31}$$

which can be integrated directly, to yield

$$-\int_{c_0}^{c} \frac{dc_A}{c_A^2} = \int_0^t k \, dt$$

$$\frac{1}{c_A} = \frac{1}{c_0} + kt \tag{39.32}$$

[5] R. T. Dillon, J. Amer. Chem. Soc. 54 (1932):952.

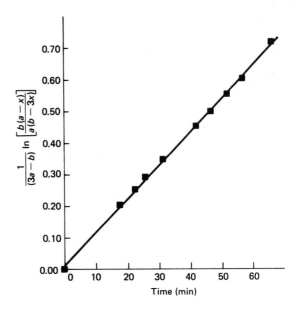

FIGURE 39.4 The second-order reaction $C_3H_6Br_2 + 3KI \rightarrow C_3H_6 + 2KBr + KI_3$.

The units of k for a second-order reaction are (time)$^{-1}$ (concentration)$^{-1}$. The half-life depends on the initial concentration c_0. We can find the half-life for a second-order reaction by setting $c_A = \frac{1}{2}c_0$ in (39.32), to get

$$t_{1/2} = (kc_0)^{-1} \qquad \text{(second-order)} \qquad (39.33)$$

39.6 THIRD-ORDER REACTIONS

Third-order reactions occasionally occur, particularly in the gas phase. Some classic examples of third-order reactions are exemplified by the various reactions of NO,

$$2NO + X_2 \longrightarrow 2NOX$$

where X = O, Cl, Br. The rate equations for these reactions have been discussed in Equations (39.12)–(39.14). For a third-order reaction of this kind, the rate equation could be written

$$\frac{dx}{dt} = k(a - 2x)^2(b - x)$$

Following the general procedure outlined for the second-order reaction, this equation can be integrated, to yield

$$\frac{1}{(2b-a)}\left\{\frac{2x(2b-a)}{a(a-2x)} + \ln\left[\frac{b(a-2x)}{a(b-x)}\right]\right\} = kt \qquad (39.34)$$

It is important to differentiate between the kinetic order of a reaction and the stoichiometric coefficients. The chemical equation for the recombination of oxygen atoms, for example, could be written $O + O \rightarrow O_2$. This reaction actually takes place in the presence of other inert species such as He, and the rate is third-order:

$$R = k[O]^2[He]$$

39.7 DETERMINING THE ORDER OF A REACTION

The first objective of a kinetics experiment is to find the order of the reaction. The experimental results of a kinetics experiment consist of a tabulation of concentration as a function of time. The first task is to determine the order of the reaction from these data. For simple first-order or second-order reactions, no great difficulties are presented. The integrated equations can be used and the data plotted according to the schemes of Figures 39.3 and 39.4. A straight line for one of the plots establishes the order of the reaction.

For decomposition reactions of the form A = products, for which the rate equation takes the form

$$R = k[A]^\zeta \tag{39.35}$$

the method of *fractional lifetimes* is often used. For these decomposition reactions, the half-life is proportional to the initial concentration of A, that is, $t_{1/2} \sim [A]_0^{(1-\zeta)}$, where $[A]_0$ is the initial concentration of A. This proportionality can be expressed as

$$\ln t_{1/2} = (1 - \zeta) \ln [A]_0 + \text{constant} \tag{39.36}$$

Thus a plot of $\ln t_{1/2}$ vs $\ln [A]_0$ yields a straight line of slope $-\zeta$. Another way to treat these kinds of reactions is by the *differential method*. In this method the original rate equation is written in logarithmic form:

$$\ln R = \zeta \ln [A] + \ln k \tag{39.37}$$

A plot of $\ln R$ vs $\ln [A]$ thus yields a straight line of slope ζ. A big disadvantage of the differential method is that R must be measured from the tangents to a curve of $[A]$ vs time.

The method of *initial rates* is applicable to a wide variety of reactions, both simple and complex. This method is particularly useful in reactions that are complicated by processes involving products. This difficulty is minimized by measuring the rate at $t = 0$ before the concentration of products has a chance to build up. If the rate equation at $t = 0$ is

$$R_0 = k[A]_0^\alpha [B]_0^\beta \tag{39.38}$$

then the initial rate can be expressed in logarithmic form:

$$\ln R_0 = \ln k + \alpha \ln [A]_0 + \beta \ln [B]_0 \tag{39.39}$$

In this technique, the variation of R_0 with $[A]_0$ can be established by a series of measurements at constant $[B]_0$. Plotting $\ln R_0$ vs $\ln [A]_0$ yields the partial order α. Similarly, the partial order β is determined by measurements at constant $[A]_0$. A disadvantage to this method is that R_0 must be determined by extrapolating the rate curve to $t = 0$.

The *isolation method* is very useful in reactions in which all but one reactant may be present in large excess. Suppose we have a reaction A + B + C → products for which the rate equation is

$$\frac{d[A]}{dt} = k[A]^\alpha [B]^\beta [C]^\gamma \tag{39.40}$$

and suppose we conduct a kinetics experiment in which initial conditions are established in such a way that the values for the concentrations of B and C are each very much larger than the value for A. Under these conditions, the concentrations of both

B and C remain approximately constant while the concentration of A changes substantially. The rate equation can then be approximated by

$$\frac{d[A]}{dt} \approx k'[A]^\alpha \tag{39.41}$$

where $k' = k[B]^\beta[C]^\gamma$ is approximately constant. The partial order with respect to A can be determined by integrating (39.41) and using graphical techniques. Similar procedures can be used to determine β and γ.

This technique is particularly useful in determining rate constants for reactions involving water in aqueous solutions, since the concentration of water (about 55 mol liter^{-1}) does not change appreciably during the reaction in dilute solutions. We consider as an example the hydrolysis of acetic anhydride,[6]

$$(CH_3CO)_2CO + H_2O \longrightarrow 2CH_3CO_2H$$

which is first-order with respect to the anhydride; the reaction is said to be *pseudo first-order*. The rate of this reaction was studied using a thermometerlike device known as a *dilatometer*, which is illustrated in Figure 39.5. A solution of the anhydride (in this case 0.07M) is placed in the large bulb. A capillary tube with an engraved linear scale is fitted to the bulb, and the apparatus is placed in a thermostat. As the reaction proceeds, the volume of the solution changes, and the course of the reaction is noted by measuring the height of the column in the calibrated capillary tube. If the reaction is first-order with respect to the anhydride, then a plot of ln (h −

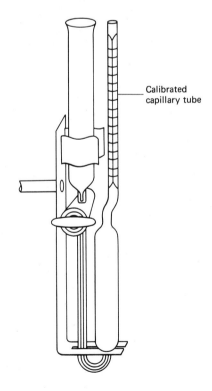

FIGURE 39.5 The dilatometer. The reaction mixture is placed in the upper open bulb, and the reaction chamber is filled by opening the stopcock until the liquid level reaches a convenient point on the capillary scale. The apparatus is placed in a thermostat, and after temperature equilibrium is established, the volume change during the reaction is measured by noting at suitable intervals the liquid level in the capillary tube.

[6] M. Kilpatrick, Jr., *J. Amer. Chem. Soc.* 50 (1928):2891.

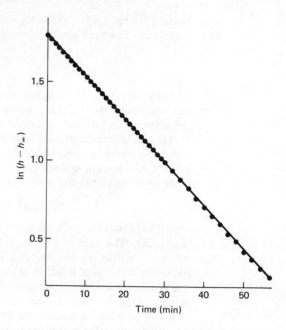

FIGURE 39.6 Dilatometric investigation of the rate of hydrolysis of acetic anhydride.

h_∞) vs time should yield a straight line, where h is the height at time t, and h_∞ is the final height at equilibrium. Such a plot is shown in Figure 39.6, indicating that the reaction is first-order with respect to the acetic anhydride.

39.8 SOME EXPERIMENTAL TECHNIQUES

The world of chemical kinetics is usually divided into *slow* reactions and *fast* reactions. A slow reaction is one whose half-life is long compared with the time for ordinary analytical techniques. The distinction is very imprecise and often depends on who is making the distinction. The organic chemist who studies the kinetics of organic reactions with half-lives of an hour will consider a one-minute half-life fast. The chemist who studies nanosecond reactions thinks of a millisecond as slow. For our purposes, slow reactions have half-lives of the order of several seconds or more and can be studied by conventional techniques. Anything faster is a fast reaction. As modern instrumentation has pushed our measuring ability into the nano- and picosecond range, it has lately become customary to speak of such reactions as *ultrafast* reactions. The divisions are mainly significant relative to the experimental techniques used to follow a reaction. Once the fundamental data of concentration vs time is acquired, the general treatment is the same for all speeds.

For any reaction, the time required for measuring the concentration must be fast compared with the half-life of the reaction. If the reaction is slow enough, then ordinary gravimetric and volumetric analytical techniques suffice. This applies to the reaction of propylene bromide with KI, discussed in (39.29). The shorter the half-life, the greater the need for a quick analysis. Sometimes trick methods must be devised. Often a reaction proceeds too rapidly for ordinary analytical procedures at the temperature under investigation but can be slowed by several orders of magnitude at much lower temperatures. Samples can be withdrawn from the reaction vessel and quickly cooled in ice baths, where the analysis can proceed at greater leisure.

A continuous method of analysis is generally more convenient than a discontinuous method in which samples are extracted and analyzed individually. The dilatometer experiment described in the previous section is one example. Various types of spectrometers and ion-specific electrodes can be used to continuously monitor the concentration of a specific species in a reaction. For gas-phase reactions in which the numbers of moles of products and reactants are unequal, the total pressure can be monitored as a function of time and related to the concentration of one of the species.

As the half-life of a reaction decreases, there is another factor that must be taken into account besides the time required for measuring the concentration. The time it takes to mix two reactants may be of the order of several seconds to a minute. Clearly this will cause problems in a reaction whose half-life is of the order of seconds. Various kinds of *flow systems* have been designed to enhance the mixing of reactants. In the *linear flow* system shown in Figure 39.7, the reactants are mixed, and they flow at a constant known rate through a tube; the gas is analyzed as it leaves the tube. The reaction time is the time between mixing and leaving, and this can be adjusted by varying the flow rate. A modification of the linear-flow technique is the *stopped-flow* technique, depicted in Figure 39.8. The modification shown is useful for studying reactions between liquid solutions. The different solutions are placed in each of the syringes. When the plungers are forced down, a rapid stream of each solution enters the mixing chamber. The jets are designed so that the streams impinge upon one another and mix very rapidly. With a suitably designed chamber, mixing can be completed in about a millisecond. The mixed solution then passes to the reaction cell, where it is observed. Flow systems are designed for fast reactions, hence techniques such as spectroscopy are used to measure the species concentrations.

Shock tubes provide a useful technique for studying gas-phase reactions in the milli- to microsecond range at elevated temperatures. The essential features of a shock tube are shown in Figure 39.9. The tube itself is a metal or glass pipe with diameter of the order of centimeters and length of several meters. The tube is divided into two sections by a thin diaphragm usually made of metal. The "driver gas" (usually helium or hydrogen) is placed on one side of the diaphragm, and the reactant gas at a pressure of a few torr is placed on the other side. When the diaphragm is ruptured, the driver gas forms a shock wave with a sharp wave front. The shock wave

FIGURE 39.7 A linear-flow system. (a) The two reactants mix in the mixing zone, and proceed down the tube at a uniform rate. The gas (or liquid) can be analyzed spectroscopically or by other rapid instrumental techniques at various distances down the tube. (b) A mixing chamber.

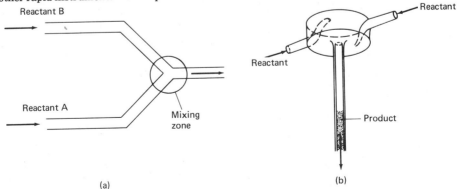

FIGURE 39.8 A stopped-flow system useful for reactions in the liquid phase. The pistons of the syringes are forced down by an electric-motor drive; they mix in the mixing chamber and proceed to the reaction cell, where the system is observed.

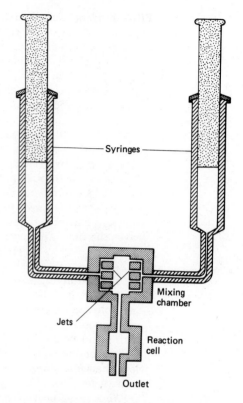

travels down the tube, compressing and heating the reactant gas, which is observed at some point down the tube. The reactants in the tube are subjected to a very sudden increase in pressure, as much as 10^9 atm, with an associated temperature rise as high as 10^7 K. The shock is sufficient to excite the molecules present and to dissociate them into atoms and radicals.

Flash photolysis is suitable for fast reactions in both gas and liquid systems. It was first developed by Norrish and Porter for gas-phase reactions, and it applies to reactions initiated by light.[7] A typical flash photolysis schematic is shown in Figure 39.10. Capacitor C_1 is charged to a high voltage and then discharged when a triggering pulse is applied to an electrode in series with the flash tube. The resulting flash has an energy of hundreds of joules and lasts a few microseconds. The radiation is emitted in a continuous range—from the lowest wavelength transmitted by quartz (~ 200 nm), through the ultraviolet and visible, covering most of the region of

FIGURE 39.9 Schematic of shock tube.

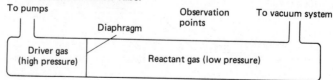

[7] R. G. W. Norrish and G. Porter, *Nature* **164** (1949):658.

SEC. 39.8 SOME EXPERIMENTAL TECHNIQUES

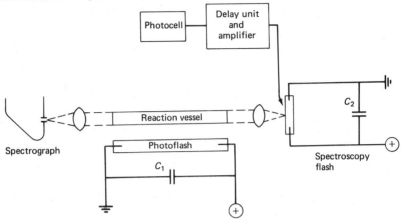

FIGURE 39.10 Schematic of a flash photolysis apparatus. In this configuration, the photocell is activated by the light from the photoflash. The spectroscopy flash is timed by a time-delay circuit to go off a set time after the initial flash (about 10^{-5} to 1 s later). Alternatively, a continuous light source can be used in place of the spectroscopy flash. The photoflash and reaction vessel are surrounded by a reflector (not shown) to contain as much light as possible.

photochemical interest. Concentrations of radical species of the order of 10^{-8} to 10^{-7} mol cm^{-3} can be obtained in the reaction vessel. The spectroscopic flash used to analyze the contents of the reaction vessel is set off by discharging capacitor C_2. The spectroscopy flash is similar to the photoflash but smaller. By picking up the light from the first photoflash with a photocell, and using a time delay unit, one can time the spectroscopy flash to go off between 10^{-5} and 1 s after the main flash. This allows one to analyze the contents of the reaction vessel at various intervals after the main flash. Alternatively, a continuous light source can be used to provide a continuous measure of the contents of the vessel.

One advantageous feature of flash photolysis is that because of the long optical path in the reaction vessel it is possible to detect species that are present in very small concentrations or that have low absorption coefficients. One disadvantage is that because of the high energies used, a large temperature rise might be involved. This is particularly important in gases for which the temperature rise might be of the order of hundreds of K in some systems. This temperature rise, and accompanying pressure rise, can affect the rate directly and can also produce inhomogeneities in the system. This problem is reduced in practice by adding large amounts of an inert gas. Heating is not a significant problem in liquid systems, since the solvent acts as a thermostat.

Recent advances in laser technology have made it possible to extend the range of flash photolysis from the microsecond to the nanosecond and ultimately to the picosecond region. One simple scheme is shown in Figure 39.11. A "Q-switched" laser equipped with frequency doubling emits a monochromatic pulse of light with a duration of less than 10 ns. The pulse is then split into two parts by a "beam splitter." The first part of the beam passes directly into the reaction vessel and acts as the photolysis flash. The second part is reflected off a movable plane mirror and then into a cell containing a fluorescent solution (1,1′,4.4′-tetraphenyl-buta-1,3-diene in cycloxane) via the beam splitter. This solution emits a pulse of light that is continuous

FIGURE 39.11 Schematic of nanosecond flash photolysis apparatus that uses a pulsed laser.

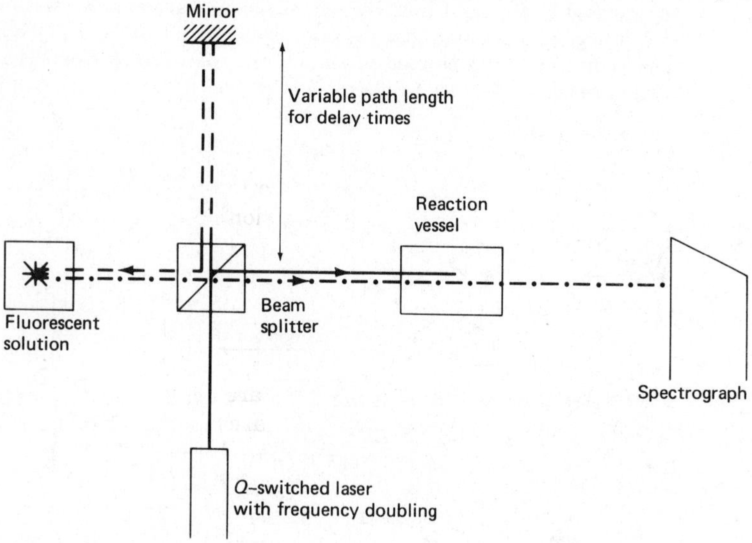

over the wavelength range 400–600 nm and lasts as long as the original laser pulse. This pulse acts as the spectroscopic flash; it is passed into the reaction vessel and then to a spectrograph, where it is used in the ordinary way to analyze for the components in the reaction vessel. The delay time between the original flash and the spectroscopic flash can be adjusted by varying the distance between the mirror and the beam splitter. The total light path can be varied to give time delays in the nanosecond region (a 3-m light path corresponds to a 10-ns delay).

The field of ultrafast reactions is one of the most rapidly advancing fields of physical chemistry (or chemical physics if you will). These advances are intimately associated with advances in laser technology and in electronic instrumentation. Remember that when one speaks of events on a time scale of picoseconds, the time it takes electronic signals to travel through wires becomes an important consideration. Classically, the field of chemical kinetics concerned itself mainly with the rates of "chemical reactions." As the time scale has been shortened to smaller and smaller units, the field has expanded considerably and now includes studies of individual atomic and molecular events such as the relaxation of vibrational and rotational excited states.[8] In this field of research it is often difficult to differentiate between kineticists and spectroscopists. Indeed, there is little reason to try to distinguish the two.

39.9 PATHWAY OF A REACTION

We have discussed several first-, second-, and third-order chemical reactions. On a conceptual basis, we can justify these with some simple probabilistic reasoning. A first-order reaction of the form A → products is analogous to effusion of a gas

---

[8] See, for example, K. J. Kaufmann and P. M. Rentzepis, Picosecond Spectroscopy in Chemistry and Biology, appearing in *Accounts of Chemical Research* 8 (1975):407.

through an orifice or to the decay of a radioactive atom. Each event consists of the decomposition of one atom of A; if the event is independent of the presence of other atoms, then the rate of decomposition is proportional to the concentration of A, as is the case in a first-order reaction.

For second-order reactions such as

$$A + B \longrightarrow \text{products} \quad \text{or} \quad A + A \longrightarrow \text{products}$$

we can conceive of each event as due to some sort of interaction between pairs of atoms; the probability of this interaction will be proportional to $[A][B]$ or $[A]^2$, as is found in second-order rate equations. We could continue this kind or reasoning on to third-order reactions and beyond. Unfortunately, this reasoning falls short of the mark for reactions like

$$C_3H_6Br_2 + 3KI \longrightarrow C_3H_6 + 2KBr + KI_3$$

where the stoichiometric coefficients are not all unity. Further, the empirical rate equations for many types of reactions are much more complicated than the equations we have dealt with thus far. The reaction between hydrogen and bromine, $H_2 + Br_2 \rightarrow 2HBr$, follows the complicated rate equation

$$\frac{d[HBr]}{dt} = \frac{k[H_2][Br_2]^{1/2}}{k' + ([HBr]/[Br_2])} \tag{39.42}$$

where k and k' are two different empirical constants. It is meaningless even to try to apply the concept of an order of a reaction to expressions such as (39.42).

Complicated rate expressions can often be explained on the basis of several different pathways or steps by which an overall reaction is achieved. The individual steps are collectively referred to as the *mechanism* of the reaction. The mechanism is what refers to the detailed molecular reactions that together make up the overall chemical reaction.

PROBLEMS

1. Studentium, which is a radioactive element that decays to pupillium, has a half-life of 8 months. Like all radioactive decay the rate is first-order.
   (a) Write the rate equation for the decay of studentium.
   (b) Calculate the rate constant for the decay.
2. The amount 0.2 mol studentium (see Problem 1) is initially placed in a container.
   (a) How many atoms of studentium decay in 1 s at time $t = 0$?
   (b) Calculate the amount of studentium left after 512 mo have passed; also calculate the decay rate.
   (c) Calculate the amount of studentium left after 237 mo have passed; also calculate the decay rate.
   (d) How long a period would have to pass until there were just 10 atoms of studentium left? How many half-lives does this correspond to?
3. Calculate the rate constants for first-order reactions of substances that have half-lives of 10^{-8} s, 1 s, 30 s, 20 min, 4 hr, 3 wk, and 11 mo. Calculate each rate constant in units of s^{-1}, min^{-1}, and h^{-1}.
4. The reaction $A + B \rightarrow C$ proceeds at a rate that is proportional to the first power of A and the square root of B. What is the overall order of the reaction? Write rate equations in terms of the rate of disappearance of A, the rate of disappearance of B, and the rate of appearance of C.
5. Glitch decomposes to glotch by a first-order reaction; glitch has a half-life of 27 s. Calculate the average lifetime of a glitch molecule. (*Note:* It would be helpful first to calculate the lifetime for a general first-order reaction and then apply the result to this specific case.)
6. Concerning Problem 1, suppose that pupillium was also radioactive and that pupillium decayed to the stable element professorium. The half-life of pupillium is 2 yr. If one starts with 0.1 mol studentium, how much professorium would be formed after 3 yr? (*Note:* This is a sequential rate problem that is discussed in the next chapter.)
7. Consider the reaction $A \rightarrow B$, which is first-order with A having a half-life of 1 hr. You are inter-

ested in manufacturing B using this reaction. The price of A is 30 cents a mole. You plan to use a reaction vessel that will hold 1000 mol A. The operating cost of the reaction vessel, including labor, taxes, depreciation, overhead, and the like but excluding the cost of raw materials, is $42.37 per hour. Assuming that all the unreacted A must be discarded, calculate the reaction time that would provide the most economical operation of the process. How much A must be discarded? Calculate the cost of producing B in dollars per mole at this most economical rate. If the reaction were allowed to proceed until it was 95% completed, what would be the increase in production costs?

8. F. Daniels and E. H. Johnston, *J. Amer. Chem. Soc.* 43 (1921):53, measured the rate of the gas-phase thermal decomposition of N_2O_5. In one of their experimental runs at 35°C, the partial pressure of N_2O_5 at various times was measured; the values are indicated in the accompanying table. The reaction is a first-order one. Plot the data in a suitable manner and determine the rate constant. What is the half-life of the N_2O_5?

t (min)	P(N_2O_5) (torr)
0	306.5
10	262.7
20	243.2
30	224.5
40	207.5
50	191.2
60	176.8
70	163.5
80	151.9
90	140.1
100	129.4

9. For the simple first-order reaction A → 2B, which takes place in a sealed vessel, write an expression for the rate of change of pressure as a function of time. Integrate the expression from $t = 0$ to a time t. Do the same for the reaction 2A → 3B, which is first-order relative to A.

10. For the two reactions discussed in Problem 9, assume that the reaction vessel was a constant-pressure vessel instead of a constant-volume one. Write expressions for the change of volume with time and integrate the expressions.

11. C. E. Waring and J. R. Abrams, *J. Amer. Chem. Soc.* 63 (1941):2757, measured the thermal decomposition of benzenediazonium chloride in isoamyl alcohol. One mole of gaseous N_2 is produced for each mole of the chloride that decomposes. The reaction was followed by measuring the total volume of N_2 (corrected to standard temperature and pressure) at various times. Some of their measurements at 20 °C are reproduced in the accompanying table. Note that the entry for $t = \infty$ is the volume of N_2 produced after the reaction has gone to completion, and that V_∞ is thus a measure of the initial amount of the diazonium salt. Determine the order of the reaction and the rate constant.

t (min)	V (N_2) (ml)
0	0.00
40	6.59
80	12.87
120	18.49
160	23.57
200	28.17
240	32.32
280	36.10
320	39.41
360	42.44
∞	69.84

12. A zeroth-order reaction is a reaction whose rate is independent of the amount of material present. The rate equation can be written as $-dc/dt = k$. C. N. Hishelwood and R. E. Burk, *J. Chem. Soc.* 127 (1925):1105, measured the rate of decomposition of NH_3 on the surface of a heated tungsten wire. Their results are given in the accompanying table. Show that this decomposition is zeroth-order. What is the rate constant? What are the units of the rate constant?

t (s)	P (NH_3) (torr)
0	100
100	86.5
200	76.5
300	66.5
400	57.5
500	49
800	25
900	13

13. A. Wasserman, *J. Chem. Soc.* (1936):1028, measured the rate of the dimerization of cyclopentadiene in benzene solution at 25.1 °C.

HC═CH + HC═CH → (bicyclic dimer structure)
HC═CH   HC─CH
  \\ //   \\CH₂//
  CH₂   CH
  H₂   C─C
                H₂ H

The concentrations of C_5H_6 at various times are given in the accompanying table. From the data, determine the order of the reaction and the rate constant.

t (min)	$[C_5H_6]$ (mol liter^{-1})
0	1.358
1,600	1.177
2,910	1.077
4,650	0.977
9,060	0.792
14,370	0.617
21,460	0.470

t (hr)	$[C_2H_4Br_2]$ (mol liter^{-1})
0.00	0.02655
8.25	0.01597
11.25	0.01350
13.25	0.01216
15.25	0.01086
17.25	0.00974
20.25	0.00843
23.25	0.00725

14. R. T. Dillon, *J. Amer. Chem. Soc.* 54 (1932):952, studied the rate of the reaction of potassium iodide with ethylene dibromide at 59.72 °C:

$$C_2H_4Br_2 + 3KI \longrightarrow C_2H_4 + 2KBr + KI_3$$

A solution was made that was initially $0.2237M$ KI and $0.02655M$ $C_2H_4Br_2$ in methanol solution. The accompanying table gives the concentrations of the $C_2H_4Br_2$ as a function of time as the reaction proceeded. Treat the data so as to get the second-order rate constant.

15. The dimerization reaction $2A \to B$ is second-order with a rate constant of k liter mol^{-1} min^{-1}. Derive an expression for the half-life of the reaction as a function of time. If $k = 1$ liter mol^{-1} min^{-1} and the concentration of A is 1 mol liter^{-1} at $t = 0$, what is the half-life at $t = 0$? $t = 1$ min? $t = 5$ min? $t = 30$ min?

16. Derive Equation (39.28).

CHAPTER FORTY

REACTION MECHANISMS

40.1 MECHANISM OF A REACTION. ELEMENTARY REACTIONS

There are two important reasons for studying the rate of any chemical reaction. The first reason is obvious; we simply want to know how fast the reaction proceeds. This is important on a practical level. In synthesizing chemicals, either in the laboratory or in tank car lots, the time it takes to complete a reaction is an important economic consideration (cf. Problem 39.7). This aspect of kinetics looks exclusively at "how fast" a particular reaction proceeds. The second reason for studying reaction rates is more fundamental. We want to determine not "how fast" the reaction proceeds but "how" the reaction proceeds. The how, or path that a reaction takes, is called the *mechanism* of the reaction.

Suppose we consider the decomposition of nitrogen pentoxide, which is represented by the stoichiometric equation

$$2N_2O_5 \longrightarrow 2N_2O_4 + O_2 \tag{40.1}$$

Daniels and Johnston first studied this reaction, described in Problem 39.8. They found the reaction to be first-order in the concentration of N_2O_5. As we have written the equation, it appears that somehow two N_2O_5 molecules come together and then decompose. The stoichiometric equation involves *two* N_2O_5 molecules. From the simplified discussion at the end of the previous chapter, the stoichiometric equation seems to require that the rate be second-order. The stoichiometric equation indicates only the *overall* reaction and not the true path the reaction follows.

It is now believed that the N_2O_5 decomposition proceeds not as a one-step reaction, as indicated in (40.1), but by a sequence of reactions:[1]

$$N_2O_5 \longrightarrow NO_2 + NO_3 \tag{40.2a}$$
$$NO_2 + NO_3 \longrightarrow N_2O_5 \tag{40.2b}$$
$$NO_2 + NO_3 \longrightarrow NO_2 + O_2 + NO \tag{40.2c}$$
$$NO + N_2O_5 \longrightarrow 3NO_2 \tag{40.2d}$$
$$NO_2 + NO_2 \longrightarrow N_2O_4 \tag{40.2e}$$

The sequence of reactions (40.2a) to (40.2e) constitutes what we call the *mechanism* of the reaction. It describes in detail the series of *intermediate* reactions whose sum is the overall reaction. The order of the reaction applies only to the rate of the overall reaction. Each of the reactions in the sequence of intermediate reactions is known as an *elementary reaction*. The elementary reactions describe molecular events; the

[1] R. A. Ogg, *J. Chem. Phys.* 15 (1947):337.

term *molecularity* applies only to these elementary reactions. In chemical literature of the past, the terms *unimolecular, bimolecular* and *termolecular* were used to describe first-, second-, and third-order reactions. Today these terms indicate how many molecules react in an elementary reaction. This number is the same as the order for an elementary reaction.

One object in a kinetics experiment is to discover the mechanism by which a reaction takes place. The main experimental observables by which the mechanism is determined are the reaction rate expression and the order of the reaction, though other observables may help a lot, as we shall see. The expression for the rate equation may be complicated in certain cases. The objective of a kinetics experiment is to deduce a mechanism that is consistent with the observed rate expression. Often it is impossible to achieve a unique pathway in accord with a particular rate expression from rate studies of the reaction in question alone, and other evidence must be sought to arrive at the correct mechanism. Thus for the N_2O_5 decomposition, two unimolecular elementary reactions might be

$$N_2O_5 \longrightarrow N_2O_3 + O_2 \quad (40.3a)$$
$$N_2O_5 \longrightarrow 2NO_2 + O \quad (40.3b)$$

But both of these can be excluded on the basis of other evidence. The first reaction involves a change in spin angular momentum and, it is expected, would be very slow for this reason; and the second is energetically unfavorable. We shall subsequently examine (Chapter 41) some additional evidence in favor of the mechanism of Equation (40.2).

In the most general case a reaction is reversible, and the ratio of the concentrations of the reactants to the products at equilibrium must be taken into account. Also, the reaction may proceed by a series of consecutive reactions, or along parallel routes, and these too must be considered in developing a reasonable reaction rate expression. Our first task will be to consider these factors.

40.2 OPPOSING REACTIONS AND THE EQUILIBRIUM CONSTANT

We have considered reactions of the form

$$A \xrightarrow{k_1} B \quad (40.4)$$

which we assumed to be irreversible (Chapter 39). If the reaction is first-order, then the concentration of A decreases at an exponential rate until eventually there is no A left. We wrote the equations for this reaction as

$$-\frac{d[A]}{dt} = k_1 t \qquad [A] = [A]_0 \exp(-k_1 t)$$

This reaction corresponds to one whose equilibrium constant $K = [B]/[A]$ is so large that we can neglect the presence of any A remaining when the system reaches equilibrium.

Suppose we now consider reactions that do not go to completion. To return to our effusion model, this situation is analogous to the experiment illustrated in Figure

40.1. Instead of connecting a high-vacuum pump to the orifice in the wall of chamber A, we connect another chamber, which we call B. Initially at time $t = 0$, we have gas at some pressure in chamber A, while chamber B is evacuated. If the orifice is opened to connect the two vessels, then some gas from A immediately effuses into chamber B. But as soon as there is some gas present in chamber B, that gas effuses back into chamber A. If we let k_f be the forward rate constant for effusion from A to B and let k_r be the reverse rate constant for effusion from B back to A, then we can write

$$-\frac{dn_A^*}{dt} = k_f n_A^* - k_r n_B^* \tag{40.5}$$

where n_A^* and n_B^* are the concentrations of gas in each chamber. At equilibrium, $dn_A^* = 0$, $k_f n_A^* = k_r n_B^*$ and

$$\frac{n_A^*}{n_B^*} = \frac{k_r}{k_f} = K \tag{40.6}$$

If both chambers are at the same temperature, then $k_r = k_f$ and K, which is the analog of an equilibrium constant, is unity. On the other hand, if the two chambers are at different temperatures, then the *thermal transpiration* effect results in the inequality of the two rate constants (Chapter 22), and K is not unity.

Suppose we have a reversible elementary reaction

$$A \underset{k_{-1}}{\overset{k_1}{\rightleftarrows}} B \tag{40.7}$$

where k_1 is the *forward rate constant* and k_{-1} is the *reverse rate constant*. For first-order reactions, the rate equation can be written

$$-\frac{d[A]}{dt} = \frac{d[B]}{dt} = k_1[A] - k_{-1}[B] \tag{40.8}$$

At equilibrium $d[A]/dt = d[B]/dt = 0$; hence

$$\frac{k_1}{k_{-1}} = \frac{[B]}{[A]} = K \tag{40.9}$$

The equilibrium constant for the reaction is given by the ratio of the two rate constants. This is known as the *principle of Guldberg and Waage*, or the *law of mass action*. The equilibrium constant was introduced to chemistry through this equation before it was developed from thermodynamics. If we designate the initial concentra-

FIGURE 40.1 An equilibrium effusion experiment. Initially the pressure in chamber A is P_0 and the pressure in chamber B is zero. As some gas molecules effuse from chamber A to B, some molecules in B effuse back into A. The rate constants for the forward and back effusion are proportional to the area of the orifice and the square root of the temperature. If the temperature is the same in both chambers, then the forward and reverse rates are the same. At equilibrium, the pressure is the same in both chambers. If the temperatures are different, then a pressure differential is established at equilibrium because the forward and the reverse effusion rates differ.

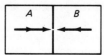

tion of A as a and the initial concentration of B as b, then we can write (40.8) in terms of x, the amount of A that has been transformed into B,

$$\frac{dx}{dt} = k_1(a - x) - k_{-1}(b + x)$$

or letting $C = (k_1 a - k_{-1} b)/(k_1 + k_{-1})$, we get

$$\frac{dx}{dt} = (k_1 + k_{-1})(C - x) \tag{40.10}$$

Integrating (40.10) between the limits $t = 0$ and t yields

$$\ln\left(\frac{C}{C - x}\right) = (k_1 + k_{-1})t \tag{40.11}$$

Equilibrium data must be combined with rate data to separate the forward and reverse reaction rates appearing in Equation (40.11). If k_{-1} is so small compared with k_1 that it can be neglected, then the reaction is essentially irreversible, and Equation (40.11) reduces to the expression for an irreversible first-order reaction. It is not necessary for the order of the reactions to be the same in both directions. The reaction

$$A \underset{k_{-2}}{\overset{k_1}{\rightleftarrows}} B + C \tag{40.12}$$

might be first-order in the forward direction and second-order in the reverse direction.

A typical elementary reversible second-order reaction could be written in the form[2]

$$\begin{array}{cccc} A & + & B & \underset{k_{-2}}{\overset{k_2}{\rightleftarrows}} & 2C \\ (a - \tfrac{1}{2}x) & & (b - \tfrac{1}{2}x) & & (x) \end{array}$$

where a is the initial concentration of A, b is the initial concentration of B, and x is the amount of C formed; no C is present initially. The rate equation is

$$\frac{d[C]}{dt} = -2\frac{d[A]}{dt} = -2\frac{d[B]}{dt} = \frac{dx}{dt} = k_2(a - \tfrac{1}{2}x)(b - \tfrac{1}{2}x) - k_{-2}x^2 \tag{40.13}$$

The integrated form of this equation has the formidable appearance

$$\ln\left\{\left[\frac{(a + b + C)/(1 - 4/K) - x}{(a + b - C)/(1 - 4/K) - x}\right]\left[\frac{a + b - C}{a + b + C}\right]\right\} = \frac{1}{2}(1 - 4K)Ck_2 t \tag{40.14}$$

where $C = \sqrt{(a + b)^2 - 4ab(1 - 4K)^{-1}}$ and we have used the equilibrium constant $K = k_2/k_{-2}$.

Earlier in this section we derived Equation (40.9), which relates the equilibrium constant to the forward and reverse reaction rates by assuming that at equilibrium the rate of an elementary process exactly equals the rate of the reverse process. This is known as the *principle of detailed balancing*. This principle is a macroscopic extension of the *principle of microscopic reversibility;* it states that when two molecules collide, *all* the momenta (both translational and internal) are reversed and then the

---

[2] The classic case of this type of reaction cited in many older textbooks is the reaction $H_2 + I_2 = 2HI$, which was found to be second-order by Bodenstein (M. Bodenstein, *Z. Phys. Chem.* 13 (1894):56; 22 (1898):295). For some forty years it was considered to be an elementary reaction. Recent investigations indicate that the reaction is not elementary, as noted later in this chapter.

system returns by the same path in the reverse direction, the initial state becoming regenerated with all the momenta reversed.[3] This requires that the forward mechanism of a reaction must be the same as the reverse mechanism.

40.3 CONSECUTIVE AND PARALLEL REACTIONS

Two additional factors that further complicate the simple picture of irreversible kinetics are the *consecutive* and *parallel*, or *competing*, reactions. A consecutive reaction is one in which the final product is produced in more than one step via a series of intermediates. Such a sequence can be depicted by

$$A \underset{k_{-1}}{\overset{k_1}{\rightleftharpoons}} B \underset{k'_{-1}}{\overset{k'_1}{\rightleftharpoons}} C \underset{k''_{-1}}{\overset{k''_1}{\rightleftharpoons}} D \underset{k'''_{-1}}{\overset{k'''_1}{\rightleftharpoons}} \ldots$$

If all the reactions in the sequence are irreversible, then the system can be solved exactly. This constitutes the important class of radiochemical reactions in which a radioactive parent decays to a sequence of radioactive daughter elements, with a stable nuclide as the final end product. Suppose we consider the case of two consecutive first-order irreversible steps:

$$\begin{array}{cccc} \text{Initial concentration:} & a & 0 & 0 \\ & A \xrightarrow{k_1} & B \xrightarrow{k'_1} & C \\ \text{Concentration at time } t: & x & y & z \end{array}$$

We have three simultaneous rate equations:

$$-\frac{dx}{dt} = k_1 x \qquad \frac{dy}{dt} = k_1 x - k'_1 y \qquad \frac{dz}{dt} = k'_1 y \tag{40.15}$$

The first can be solved immediately, to give

$$x = a \exp(-k_1 t) \tag{40.16}$$

This expression for x can now be substituted in the y equation, to get

$$\frac{dy}{dt} + k'_1 y = k_1 a \exp(-k_1 t) \tag{40.17}$$

Equation (40.17) can be solved most simply by noting that since

$$\frac{d}{dt}[y \exp(k'_1 t)] = \left[\frac{dy}{dt} + k'_1 y\right] \exp(k'_1 t)$$

we can multiply both sides of (40.17) by $\exp(k'_1 t)$, to get

$$\int_{t=0}^{t} d[y \exp(k'_1 t)] = \int_{t=0}^{t} k_1 a \exp[(k'_1 - k_1)t]\, dt \tag{40.18}$$

Since $y = 0$ at the limit $t = 0$, we get for our final expression

$$y = \frac{k_1 a \exp(-k'_1 t)}{k'_1 - k_1} \{\exp[(k'_1 - k_1)t] - 1\} \tag{40.19}$$

---

[3] The principle of microscopic reversibility was first formulated by R. C. Tolman, and is discussed in his book, *The Principles of Statistical Mechanics* (Oxford: Oxford University Press, 1938), p. 163.

SEC. 40.3 CONSECUTIVE AND PARALLEL REACTIONS

Conservation of mass requires that $x + y + z = a$, thus z is

$$z = a - (x + y) = a\left(1 - \frac{k_1' \exp(-k_1 t)}{k_1' - k_1} + \frac{k_1 \exp(-k_1' t)}{k_1' - k_1}\right) \quad (40.20)$$

The concentrations of the three species are plotted as a function of time in Figure 40.2. The concentration of x decreases at the normal exponential rate characteristic of a first-order decay curve. As x decays, some y is formed, which is then converted to z. At some time that is characteristic of the ratio k_1/k_1', the concentration of y reaches a maximum and then falls off with time. Although this type of *decay and growth* is most often associated with radioactive transformations, such curves are sometimes found for ordinary chemical reactions. One such reaction is the pyrolysis of acetone,[4]

$$(CH_3)_2CO \xrightarrow{k_1} CH_2=CO + CH_4$$
$$\phantom{(CH_3)_2CO \xrightarrow{k_1} } \downarrow{k_1'}$$
$$\phantom{(CH_3)_2CO \xrightarrow{k_1} } \tfrac{1}{2}C_2H_4 + CO$$

An effusing gas model of this two-step consecutive reaction is shown in Figure 40.3a. The first step corresponds to the effusion of gas from chamber A to chamber B, and the second step corresponds to the effusion of gas through the hole in the right wall of chamber B and its removal by the high-vacuum pump. Corresponding to the concentration z is the total amount of gas removed by the vacuum pump. (To keep the analogy complete and ensure that the effusion is unidirectional and hence irreversible, we can fit each orifice with a one-way valve, allowing gas to flow in only one direction.) Now suppose that one of the orifices is very much larger than the other. This corresponds to a ratio of the rate constants in which one is very much larger

FIGURE 40.2 Decay and growth curves for the two-step irreversible consecutive first-order reaction A $\xrightarrow{k_1}$ B $\xrightarrow{k_1'}$ C. The abscissa represents time in units of the half-life of A. The case illustrated is for $k_1' = \tfrac{1}{2}k_1$.

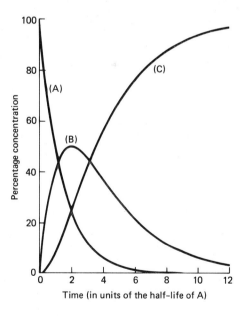

[4] C. A. Winkler and C. N. Hinshelwood, *Proc. Roy. Soc. London* A149 (1935):340.

FIGURE 40.3 (a) Effusion model of a two-step first-order consecutive reaction. The flaps over the orifices are one-way valves that permit the gas molecules to effuse only in the directions indicated. (b) Effusion model of the competing reactions. If one of the orifices is much larger than the other, then most of the gas is removed through the larger orifice.

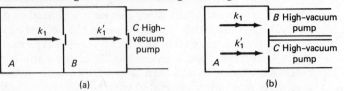

than the other. Under these conditions the rate of flow of gas in the effusive system will be determined solely by the smaller orifice. Thus if the first orifice is the larger one, then the two chambers will act as one large chamber with one orifice connected to the vacuum system; and if the second orifice is larger, the vacuum pump is effectively connected to the first orifice. This is the same principle that applies to the flow of liquids through long pipes. If water flows through a 10-cm pipe that is constricted to 1 cm along the way, the capacity of the pipe is determined not by the 10-cm diameter but by the 1-cm diameter. If $k_1 \ll k'_1$, then (40.20) reduces to

$$z = a[1 - \exp(-k_1 t)] \tag{40.21}$$

which is just the expression for the one-step irreversible first-order reaction discussed in the previous chapter. Since the rate is controlled by the slow step, this step is known as the *rate-determining step*.

Now consider the effusion scheme illustrated in Figure 40.3b. This is the analog of the parallel, or competing, reaction in which two different products can be formed:

$$A \begin{cases} \xrightarrow{k_1} B \\ \xrightarrow{k'_1} C \end{cases}$$

The rate equations are

$$\frac{db}{dt} = k_1 a = k_1(a_0 - b - c)$$

$$\frac{dc}{dt} = k'_1 a = k'_1(a_0 - b - c) \tag{40.22}$$

$$-\frac{da}{dt} = \frac{db}{dt} + \frac{dc}{dt}$$

where a, b, and c are the concentrations of the three species at time t, and a_0 is the concentration of A at $t = 0$. If we let $b_0 = c_0 = 0$, the integrated expressions become

$$b = \frac{k_1 a_0}{k_1 + k'_1} (1 - \exp[-(k_1 + k'_1)t])$$

$$c = \frac{k_2 a_0}{k_1 + k'_1} (1 - \exp[-(k_1 + k'_1)t]) \tag{40.23}$$

Such competing reactions are very common in organic chemistry—a reactant may decompose, giving a variety of products, each governed by a different rate constant. Ethyl alcohol can be converted to ethylene or acetaldehyde, and if the reaction temperatures and the catalyst are suitably chosen, one reaction can be made to predominate. Thus

$$CH_3CH_2OH \xrightarrow[350\,°C]{Al_2O_3} CH_2=CH_2 + H_2O$$

$$CH_3CH_2OH \xrightarrow[300\,°C]{Cu} CH_3CHO + H_2$$

Let's return to reversible reactions for the moment, and consider the one-step first-order reaction

$$A \underset{k_{-1}}{\overset{k_1}{\rightleftharpoons}} C$$

At equilibrium, $d[A]/dt = 0 = -k_1[A] + k_{-1}[C]$, and the equilibrium constant is $K = ([C]/[A])_{eq} = k_1/k_{-1}$. Now suppose C is formed from A not in one step but in two—

$$A \underset{k_{-1}}{\overset{k_1}{\rightleftharpoons}} B \underset{k'_{-1}}{\overset{k'_1}{\rightleftharpoons}} C$$

At equilibrium

$$\left(\frac{[B]}{[A]}\right)_{eq} = \frac{k_1}{k_{-1}} \qquad \left(\frac{[C]}{[B]}\right)_{eq} = \frac{k'_1}{k'_{-1}} \qquad (40.24)$$

hence for the overall reaction

$$\left(\frac{[C]}{[A]}\right)_{eq} = \frac{k_1 k'_1}{k_{-1} k'_{-1}} \qquad (40.25)$$

A typical experimental procedure for studying this kind of reaction would be to start with pure A. At the very beginning, before any B or C has formed, the measured rate constant would be k_1. Similarly, if an initial rate were measured starting with pure C, the result would be k'_{-1}. Note, however, that the equilibrium constant is *not* the ratio k_1/k'_{-1}; it is as given in Equation (40.25).

Now suppose that we can skip the intermediate step in the reverse direction by the triangular mechanism

$$A \begin{array}{c} \nearrow B \\ \downarrow \\ \nwarrow C \end{array}$$

whereby A is converted to B, which in turn is converted to C, which can convert directly back to A. This possibility is flatly *prohibited* by the principle of microscopic reversibility we have already noted. The mechanism must always be the same in the forward direction and in the reverse direction; equating the ratio of the forward and reverse rate constants to the equilibrium constant is valid only for elementary reactions. Now let's proceed to an actual mechanism.

40.4 EXAMPLE; THE HYDROGEN BROMIDE REACTION

Consider the reaction $A + B \rightarrow C$. From stoichiometric considerations alone one might expect the rate of the reaction to follow second-order kinetics. If the rate were measured and found to be second-order, then without any other evidence one would assume that the reaction is an elementary one. On the other hand, if the rate is found to be first-order or to follow some complicated expression, then we must conclude that the reaction is complex and actually made up of several elementary processes, the sum of which is the stoichiometric reaction.

We have already noted that the reaction $H_2 + I_2 \rightarrow 2HI$ is second-order. (This is a necessary but *not sufficient* condition for the reaction to be elementary. The modern view is that the HI reaction is not an elementary one.) One might then expect the reaction $H_2 + Br_2 = 2HBr$ to be second-order also, since the reactants are similar. Such is not the case. The data for the rate of this reaction were found to fit the complicated expression

$$\frac{d[HBr]}{dt} = \frac{k[H_2][Br_2]^{1/2}}{k' + [HBr]/[Br_2]} \tag{40.26}$$

where k and k' are two constants.[5] The explanation for this curious expression is based on a five-step scheme including five different elementary reactions:

(1) $\quad Br_2 \xrightarrow{k_1} 2Br \quad$ (chain-initiating)

(2) $\quad Br + H_2 \xrightarrow{k_2} HBr + H$ ⎫
(3) $\quad H + Br_2 \xrightarrow{k_3} HBr + Br$ ⎬ (chain-propagating)

(4) $\quad H + HBr \xrightarrow{k_4} H_2 + Br \quad$ (chain-inhibiting)

(5) $\quad 2Br \xrightarrow{k_5} Br_2 \quad$ (chain-terminating)

This sequence of reactions is called a *chain reaction*. In the first, or *chain-initiating*, step, two Br atoms are formed from the thermal decomposition of a Br_2 molecule. In steps (2) and (3), one molecule of product is formed with no decrease in the total number of free atoms, or *chain carriers*. Step (4) results in the disappearance of one molecule of product while the number of chain carriers remains constant. The chain is terminated in step (5), which removes two chain carriers. Now let's apply some chemical logic to this chain to try to produce an equation of the form of (40.26).

The first thing we note is that the H and Br free atoms are extremely reactive and hence very short-lived. Initially when the system contains only pure H_2 and Br_2, their concentrations are both zero. As the reaction commences, their concentrations begin to increase, but their concentrations at all times are extremely small in view of their great reactivities. We apply an approximation known as the *steady-state treatment* to these two species. That is to say, we assume that after a very short time the concentrations of these active species come to some small (and unknown) equilibrium values. An analog of this steady-state treatment is shown in the effusion experiment of Figure 40.4, where gas from a large chamber A effuses through a very small orifice into a small chamber B; a vacuum pump is connected to chamber B through a much larger orifice. Initially the orifice between A and B is closed and $P_B = 0$. When that orifice is opened, the pressure in B rapidly rises to some equilibrium value and remains approximately constant. Note that we said *approximately* constant not absolutely constant, since P_B decreases very slowly as P_A decreases. Another example of a steady-state process is the three-step consecutive reaction shown in Figure 40.2. For $k_1' = \frac{1}{2}k_1$ as illustrated, there is a relatively rapid rise in the concentration of B followed by a slower decrease in its concentration. If k_1' had been very much greater than k_1, the initial rise would have been very rapid and the ensuing fall in concen-

---

[5] M. Bodenstein and S. C. Lind, *Z. Physik. Chem.* 57 (1906):168.

FIGURE 40.4 Effusion analog of the steady-state treatment in kinetics.

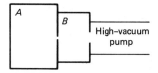

tration relatively slight. The important point is that at all times the concentration of B is very small compared with the concentrations of the other species.

By assuming that the H and Br concentrations are constant, we obtain the two rate equations for these species:

$$\frac{d[\text{Br}]}{dt} = 0 = 2k_1[\text{Br}_2] - k_2[\text{Br}][\text{H}_2] + k_3[\text{H}][\text{Br}_2] + k_4[\text{H}][\text{HBr}] - 2k_5[\text{Br}]^2 \quad (40.27)$$

$$\frac{d[\text{H}]}{dt} = 0 = k_2[\text{Br}][\text{H}_2] - k_3[\text{H}][\text{Br}_2] - k_4[\text{H}][\text{HBr}] \quad (40.28)$$

These are two simultaneous equations that can be solved to provide expressions for [H] and [Br] in terms of the other concentrations that are experimentally measurable quantities. If we add the two equations, the terms in k_2, k_3, and k_4 drop out, to yield

$$[\text{Br}] = \left(\frac{k_1}{k_5}[\text{Br}_2]\right)^{1/2} \quad (40.29)$$

This expression can then be substituted in either (40.27) or (40.28), to get

$$[\text{H}] = \frac{k_2(k_1/k_5)^{1/2}[\text{H}_2][\text{Br}_2]^{1/2}}{k_3[\text{Br}_2] + k_4[\text{HBr}]} \quad (40.30)$$

The rate of formation of HBr is

$$\frac{d[\text{HBr}]}{dt} = k_2[\text{Br}][\text{H}_2] + k_3[\text{H}][\text{Br}_2] - k_4[\text{H}][\text{HBr}] \quad (40.31)$$

Introducing the expressions for [H] and [Br], we get after some rearrangement

$$\frac{d[\text{HBr}]}{dt} = \frac{2k_2 k_3 k_4^{-1}(k_1/k_5)^{1/2}[\text{H}_2][\text{Br}_2]^{1/2}}{k_3 k_4^{-1} + [\text{HBr}]/[\text{Br}_2]} \quad (40.32)$$

This equation has exactly the same form as the empirical expression of (40.26). The constants k and k' appearing in (40.26) are composites of the rate constants for the five elementary steps of the chain reaction:

$$k = 2k_2 k_3 k_4^{-1}\left(\frac{k_1}{k_5}\right)^{1/2} \qquad k' = \frac{k_3}{k_4} \quad (40.33)$$

In addition, the constant (k_1/k_5) must be the equilibrium constant K for the dissociation $\text{Br}_2 = 2\text{Br}$.

That Equation (40.32) has the exact form of the empirical equation (40.26) is a necessary but not sufficient condition that the proposed five-step mechanism is the correct one. The agreement between the two expressions does, however, provide strong support for the proposed mechanism. When there is no evidence to the contrary, the proposed mechanism will be accepted as correct. Such contrary evidence might consist of another proposed mechanism that provides the same agreement, or perhaps doubt about one of the steps because of unfavorable energetics or questionable chemistry. Thus, for example, a mechanism that included a step $Br_2 \rightarrow Br^- + Br^+$ would be highly suspect; and if there were no other supportive evidence, the mechanism would not be accepted as the true one, even if the empirical and derived rate expressions were the same. For the mechanism under discussion there exists no contrary evidence, and in fact supportive evidence exists from studies of the rates of the individual elementary reactions. This proposed mechanism for the HBr formation is considered the correct one. That does not mean that it will still be considered the correct mechanism ten years hence. Unlike other, more exact branches of chemistry, kinetics is highly inductive, and the criteria of exactitude that apply to equilibrium thermodynamics do not apply to chemical kinetics. This is partly because time-dependent problems are inherently more difficult and less amenable to exact solution than equilibrium problems are. At the moment, however, we can consider this proposed mechanism the "correct" one.

Let us consider, as an example of the changing ideas in chemical kinetics, the reaction $H_2 + I_2 = 2HI$, for which both the forward and the reverse rates were found back in the 1890s to follow second-order kinetics. This was excellent evidence that the reaction was an elementary bimolecular one, and this view persisted for some forty years. Then in 1934 the conversion of parahydrogen into orthohydrogen was studied in an equilibrium mixture of H_2, I_2, and HI.[6] The conversion rate was found to be twice the rate that would be predicted on the basis of the exchange reaction

$$p\text{-}H_2 + I_2 \longrightarrow 2HI \longrightarrow o\text{-}H_2 + I_2$$

It was concluded that there must be an additional step in the mechanism for the conversion. The most likely candidate was the magnetic iodine atom. The mechanism of the HI reaction has not yet been settled. One proposed mechanism includes the main atomic steps[7]

$$I_2 \longrightarrow 2I$$
$$I + H_2 \longrightarrow HI + H$$
$$H + I_2 \longrightarrow HI + I$$

You will find additional reaction mechanisms discussed in the problems, in the next two chapters, and in the references listed in the Bibliography.

PROBLEMS

1. A 1-liter chamber is filled with helium gas at 1-atm pressure and 300 K. A high-vacuum pump is connected to the chamber through a small pinhole in the wall of the chamber. The diameter of the pinhole is 0.1 mm. Write an equation for the rate at which the pressure in the chamber decreases. How long will it take for the pressure in the chamber to fall to $\frac{1}{2}$ atm.? (*Note:* In this

---

[6] Hydrogen nuclei have spins of $\frac{1}{2}$. In molecules of orthohydrogen both spins are parallel; the nuclear wave function is symmetrical and has a threefold degeneracy. In parahydrogen the spins are antiparallel and the nuclear wave function is antisymmetric; parahydrogen is nondegenerate and has a lower energy than orthohydrogen. At room temperature $kT \gg \Delta E$, where ΔE is the energy difference between ortho- and parahydrogen. An equilibrium mixture of the two would contain $\frac{1}{4}p\text{-}H_2$ and $\frac{3}{4}o\text{-}H_2$. At liquid hydrogen temperatures, on the other hand, the equilibrium ratio is different. When liquid hydrogen is evaporated, the resulting gas contains a ratio of ortho to para that is characteristic of the liquid hydrogen temperature. The rate of conversion of the two is very slow, and as the gas warms to room temperature a nonequilibrium ratio persists for a long time. It is this conversion that was studied by E. J. Rosenbaum and T. R. Hogness, *J. Chem. Phys.* 2 (1934):267.

[7] S. W. Benson and R. Srinivasen, *J. Chem. Phys.* 23 (1955):200. J. H. Sullivan, *J. Chem. Phys.* 30 (1959): 1292, 1577; 36 (1962):1925.

problem and in the next few problems, neglect that the mean free path of the gas molecules may be less than the dimensions of the pinhole.)

2. Suppose that in Problem 1 we had connected another chamber of equal volume to the first chamber instead of a high-vacuum system, and that the second chamber was initially evacuated. Write an expression for the rate at which the pressure in the first chamber decreases. What is the final equilibrium pressure? Calculate $t_{1/2}$, the time it takes for the pressure to reach one-half its equilibrium value. Calculate dP/dt at $t = 0$ and at $t = t_{1/2}$.

3. Now suppose that in Problem 2 the second chamber was maintained at 1000 K while the first was maintained at 300 K. What would the final equilibrium pressure be? Calculate the "equilibrium constant" for this situation as defined in the text.

4. You are carrying out a high-vacuum experiment at liquid nitrogen temperatures (77 K). The gauge used to measure the temperature is at room temperature and is connected to the cold region with a tube. Calculate the error introduced in the pressure reading by the thermal transpiration effect. What is the correction factor needed? If the experiment were at liquid helium temperatures (4 K), what would the correction factor be? (*Note:* Neglect that a long tube is not an orifice, and use the equations for a small orifice. The flow equations for long tubes are much more complicated; they are discussed in most books dealing with high-vacuum technology.)

5. Consider the series of first-order irreversible reactions
$$A \xrightarrow{k_1} B \xrightarrow{k_1'} C$$
where $k_1' = \frac{1}{2}k_1$. The initial concentration of A is 1 mol liter^{-1}. Neither B nor C are present initially. At what time does the concentration of B reach a maximum? (Calculate the time in units of the half-life of A.)

6. Now consider the series $A \xrightarrow{k_1} B \underset{k_{-1}'}{\overset{k_1'}{\rightleftharpoons}} C$, where the first step is irreversible and the second step is reversible. Write equations for the rate of disappearance or appearance of each of the three species. What is the ratio of the concentrations of B and C at equilibrium?

7. For a series of irreversible reactions $A \to B \to C$, the concentrations of A and B were measured as a

t (min)	[A] (mol liter)	[B] (mol liter)
0	1.000	0.000
2	0.250	0.500
4	0.0625	0.375
6	0.0156	0.219
8	0.0039	0.117
10	0.0010	0.061
12	0.0002	0.031

function of time. Both steps are first-order. The concentrations were as shown in the accompanying table. Calculate the rate constants for each step of the reaction sequence.

8. The reaction $A \underset{k_{-1}}{\overset{k_1}{\rightleftharpoons}} B$ is reversible and first-order in each direction. Show that an integrated rate equation can be written in the form
$$\ln\left(\frac{[A]_0 - [A]_{eq}}{[A] - [A]_{eq}}\right) = (k_1 + k_{-1})t$$
where $[A]_0$ is the initial concentration of A and $[A]_{eq}$ is the equilibrium concentration of A.

Note that this result indicates that the approach to equilibrium is a first-order process whose effective rate constant is the sum of the rate constants for the forward and reverse reactions. This equation forms the basis for the experimental technique *relaxation spectroscopy*. In this method the initial system is an equilibrium mixture of A and B. A sudden perturbation is applied to the system, disturbing the equilibrium slightly. The system then returns to equilibrium exponentially with the time constant $k_1 + k_{-1}$. An independent measurement of the equilibrium constant then enables the experimenter to obtain values of k_1 and k_{-1}. Perturbations may be sudden changes in pressure, electric fields, temperature, and the like. When the perturbation is a sudden change in temperature, the method is called the *temperature-jump* method. This technique is not limited to first-order processes, as noted in the next problem.

9. An important class of reaction is $A \underset{k_2}{\overset{k_1}{\rightleftharpoons}} B + C$. This type of reaction is exemplified by acid ionizations, $HA + H_2O = H_3O^+ + A^-$, in which the concentration of H_2O can be taken as constant and the forward reaction is pseudo first-order. Let the initial concentration of A be A_0 and $B_0 = C_0 = 0$, in which case $B = C = A_0 - A$. Show that for this case the rate equation can be written as
$$\ln\left(\frac{A_0^2 - A_{eq}A}{(A - A_{eq})A_0}\right) = k_1\left(\frac{A_0 + A_{eq}}{A_0 - A_{eq}}\right)$$
What would the expression be if the reaction were started from the other side, that is, $A_0 = 0$ and $B_0 = C_0 = b$ and $x = B_0 - B = C_0 - C$?

10. The effusion experiment indicated in the figure is analogous to the set of reactions

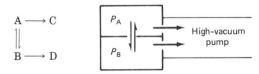

which is first-order in all directions. Write expressions for the rate of change of pressure in chambers A and B. Suppose that initially $P_A = 1$ atm and $P_B = 0$ and all three orifices were of the same size, 0.1 mm. How long would it take for the pressure in A to fall to $\frac{1}{2}$ atm? What would be the pressure in B at that time?

11. Consider the parallel reaction scheme

$$A \underset{k_1'}{\overset{k_1}{\rightleftarrows}} \begin{matrix} B \\ C \end{matrix}$$

where $k_1 = 3$ min^{-1} and $k_1' = 1$ min^{-1}. What is the half-life of A? How much B and C would be formed after an infinite amount of time if the initial concentration of A were 1 mol liter^{-1}?

12. Figure 40.4 shows an effusion analog of the steady-state treatment in kinetics. Suppose that the orifice between chambers A and B had a diameter of 0.1 mm while the vacuum pump was connected to chamber B through a 5-mm-diameter orifice. If the volume of chamber B were 1 l and the volume of chamber A were infinitely large (that is, the entire atmosphere), what would the equilibrium pressure be in chamber B if the pressure in chamber A were 1 atm? (*Note:* This corresponds to the problem of a "leak" in a high-vacuum system. The ultimate vacuum is often determined by the magnitude of these leaks, not to mention other effects, such as *outgassing* of the components of the system, and the pumping rate of the high-vacuum pump.)

13. Go through all the steps leading to Equation (40.32) for the rate of formation of HBr from H$_2$ and Br$_2$.

14. A proposed free-radical chain mechanism for the decomposition of acetaldehyde (F. O. Rice and K. F. Herzfeld, *J. Amer. Chem. Soc.* 56 (1934): 284) consists of the steps

$$CH_3CHO \xrightarrow{k_1} CH_3^\bullet + CHO^\bullet$$
$$CH_3CHO + CH_3^\bullet \xrightarrow{k_2} CH_4 + CO + CH_3$$
$$2CH_3 \xrightarrow{k_3} C_2H_6$$

By applying the steady-state condition to the concentration of CH$_3$, show that the rate of formation of methane is of the order $\frac{3}{2}$:

$$\frac{d[CH_4]}{dt} = k_2 \left(\frac{k_1}{2k_3}\right)^{1/2} [CH_3CHO]^{3/2}$$

15. The reaction $2NO + O_2 \rightarrow 2NO_2$ is found to be a third-order reaction,

$$-\frac{d[NO]}{dt} = k_3[NO]^2[O_2]$$

Show that the same third-order form of the rate equation is got from the mechanism consisting of the two bimolecular reactions

$$2NO \rightleftharpoons N_2O_2 \quad \text{and} \quad N_2O_2 + O_2 \longrightarrow 2NO_2$$

16. The following six equations provide a mechanism for the decomposition of acetaldehyde:

$$CH_3CHO \xrightarrow{k_1} CH_3 + CHO$$
$$CHO \xrightarrow{k_2} H + CO$$
$$H + CH_3CHO \xrightarrow{k_3} H_2 + CH_3CO$$
$$CH_3 + CH_3CHO \xrightarrow{k_4} CH_4 + CH_3CO$$
$$CH_3CO \xrightarrow{k_5} CH_3 + CO$$
$$2CH_3 \xrightarrow{k_6} C_2H_6$$

Show that this mechanism gives rise to a $\frac{3}{2}$-order expression for the formation of CH$_4$ like the expression found in Problem 14.

17. The thermal decomposition of ethane in the temperature range 520–720 °C leads to a mixture of H$_2$, C$_2$H$_4$, CH$_4$, C$_2$H$_6$ with small amounts of higher hydrocarbons. One proposed mechanism consists of the steps

$$C_2H_6 \xrightarrow{k_1} 2CH_3$$
$$CH_3 + C_2H_6 \xrightarrow{k_2} CH_4 + C_2H_5$$
$$C_2H_5 \xrightarrow{k_3} C_2H_4 + H$$
$$H + C_2H_6 \xrightarrow{k_4} H_2 + C_2H_5$$
$$C_2H_5 + C_2H_5 \xrightarrow{k_5} C_4H_{10}$$
$$C_2H_5 + C_2H_5 \xrightarrow{k_6} C_2H_4 + C_2H_6$$

(a) Write differential equations for the rates of change of concentrations of all species involved.
(b) By applying the steady-state approximation to the species CH$_3$, C$_2$H$_5$, and H, show that

(i) $[CH_3] = \dfrac{2k_1}{k_2}$

(ii) $[C_2H_5] = \left(\dfrac{k_1}{k_5 + k_6}\right)^{1/2} [C_2H_6]^{1/2}$

(iii) $[H] = \dfrac{k_3[C_2H_5]}{k_4[C_2H_6]}$

$\qquad = \dfrac{k_3}{k_4} \left(\dfrac{k_1}{k_5 + k_6}\right)^{1/2} [C_2H_6]^{-1/2}$

(c) Show that the rate of production of methane is first-order in [C$_2$H$_6$]:

$$\frac{d[CH_4]}{dt} = 2k_1[C_2H_6]$$

(d) Show that the rate of production of hydrogen is half-order in [C$_2$H$_6$]:

$$\frac{d[H_2]}{dt} = k_3 \left(\frac{k_1}{k_5 + k_6}\right)^{1/2} [C_2H_6]^{1/2}$$

18. The stoichiometric equation for the formation of phosgene is

$$CO + Cl_2 \underset{k'}{\overset{k}{\rightleftarrows}} COCl_2$$

The rate of both the formation and the decomposition of phosgene has been measured (M. Bodenstein and H. Plaut, *Z. Physik. Chem.* 110

(1924):399). The rate of formation is given by the equation

$$\frac{d[COCl_2]}{dt} = k[Cl_2]^{3/2}[CO]$$

and the rate of decomposition is

$$-\frac{d[COCl_2]}{dt} = k'[Cl_2]^{1/2}[COCl_2]$$

The equilibrium constant for the reaction is

$$K = \left(\frac{k}{k'}\right) = \frac{[COCl_2]}{[CO][Cl_2]}$$

Show that these results can be explained by the mechanism

$$Cl_2 \rightleftharpoons 2Cl$$
$$Cl + CO \rightleftharpoons COCl$$
$$COCl + Cl_2 \rightleftharpoons COCl_2 + Cl$$

where the first two reactions are fast.

19. An extensive investigation of the kinetics and equilibrium of the phosgene reaction discussed in Problem 18 was carried out by M. Bodenstein and H. Plaut, *Z. Physik. Chem.* 110 (1924):399. They found that at 394.8 °C the partial pressures of Cl_2, CO, and $COCl_2$ were 96.8, 89.3, and 253.2 torr in an equilibrium mixture of the gases. They studied the rate of change of total pressure in a vessel that was initially charged with Cl_2 and CO at partial pressures of 343.3 and 341.1 torr. The total pressure as a function of time is given in the accompanying table. (*Note:* The vessel also contained some extra inert gas, hence the total pressure at $t = 0$ is greater than 343.3 + 341.1 torr.) Calculate the rate constants for the phosgene reaction at 394.8 °C. (Some useful information is found in Problem 18.)

Time (min)	Pressure (torr)	Time (min)	Pressure (torr)
0	708.4	60	580.5
5	690.6	70	570.1
10	673.8	80	560.6
15	659.6	90	553.1
20	646.4	105	542.5
25	635.3	120	534.2
30	624.6	135	527.3
35	615.2	150	520.7
40	606.8	165	515.2
45	599.2	180	510.3
55	586.6		

20. Consider the opposing second-order reaction

$$A + B \underset{k_{-2}}{\overset{k_2}{\rightleftharpoons}} C + D$$

Letting the initial concentration of A be A_0, we have $[A] = A_0 - x$ where x is the amount of A reacted after time t. In terms of x, the rate equation can be written as

$$\frac{dx}{dt} = k_2(A_0 - x)(B_0 - x) - k_{-2}(C_0 + x)(D_0 + x)$$

Show that the rate equation can be written in standard form as

$$\frac{dx}{a + bx + cx^2} = dt$$

and directly integrate the equation. Compare your result with Equation (40.14).

21. Consider the case of the mixed first- and second-order reactions $A \underset{k_{-1}}{\overset{k_1}{\rightleftharpoons}} B + C$. Write the rate equation in terms of x, the amount of A reacted as in Problem 20, and show that the rate equation has the same form and therefore the same type of solution. Solve the equation and compare with the result of Problem 20.

22. Consider the set of three parallel reactions:

(1) $A \xrightarrow{k_1}$ (products)   first-order
(2) $A + B \xrightarrow{k_2}$ (products)'   second-order
(3) $A + A \xrightarrow{k_2'}$ (products)"   second-order

(a) Write the equation for the disappearance of A.
(b) Replace [A] and [B] by $A_0 - x$ and $B_0 - x$, and show that the equation can be written in the form

$$\frac{dx}{dt} = (A_0 - x)[(k_1 + k_2 B_0 + 2k_2' A_0)$$
$$- (k_2 + 2k_2')x]$$

(c) Solve this equation by the method of partial fractions.
(d) Show finally that a solution can be written in the form

$$\ln\left[\frac{A_0}{[A]}\left(\frac{k_1 + k_2[B] + 2k_2'[A]}{k_1 + k_2 B_0 + 2k_2' B_0}\right)\right]$$
$$= [k_1 + k_2(B_0 - A_0)]t$$

23. Suppose that k_2 for reaction (2) in Problem 22 is zero. Show that we can then solve the rate equation, to get

$$\ln\left[\frac{A_0}{[A]}\left(\frac{k_1 + 2k_2'[A]}{k_1 + 2k_2' A_0}\right)\right] = k_1 t$$

What is the half-life of A?

CHAPTER FORTY-ONE

SOME THEORETICAL APPROACHES TO CHEMICAL KINETICS

41.1 TEMPERATURE DEPENDENCE OF REACTION RATES

That reaction rates increase with increasing temperature has long been known on a purely empirical basis. One general rule of thumb often used is that reaction rates double for each 10-K rise in temperature. For many reactions it is found that the rate constant varies with temperature according to

$$\ln k = A - \frac{C}{T}$$

$$k = A \exp\left(-\frac{C}{T}\right) \tag{41.1}$$

where A and C are constants. A plot of $\ln k$ vs T^{-1} should thus yield a straight line, as shown in Figure 41.1 for the reaction $2HI \rightarrow H_2 + I_2$. Equation (41.1) is known as the *Arrhenius law* after Arrhenius, who demonstrated its validity by simple thermodynamic arguments.

For the reaction $A + B \rightleftharpoons C + D$, the equilibrium condition is that the forward and reverse rates must be equal,

$$k_1[A][B] = k_{-1}[C][D] \tag{41.2}$$

where k_1 and k_{-1} are the forward and reverse rate constants. The equilibrium constant is

$$K_{eq} = \frac{[C][D]}{[A][B]} = \frac{k_1}{k_{-1}} \tag{41.3}$$

Recalling our previous work on the temperature dependence of the equilibrium constant, we can write[1]

$$\frac{d \ln K}{dT} = \frac{\Delta E}{RT^2} \tag{41.4}$$

---

[1] To be strictly correct, this equation should be written in terms of ΔH rather than ΔE. If we assume, however, that $\Delta(PV) = 0$ for the reaction, then the two quantities are equal.

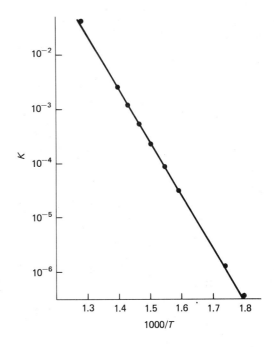

FIGURE 41.1 Arrhenius plot for the reaction $2HI \rightarrow H_2 + I_2$.

Using Equation (41.3), we can write

$$\frac{d \ln k_1}{dT} = \frac{E_1}{RT^2} + \text{constant}$$

$$\frac{d \ln k_{-1}}{dT} = \frac{E_{-1}}{RT^2} + \text{constant} \qquad (41.5)$$

with $\Delta E = E_1 - E_{-1}$. Experimentally it is found that the constant term can be taken as zero, and the rate constant can be related to the temperature by

$$\frac{d \ln k}{dT} = \frac{E_a}{RT^2}$$

This has the integrated form

$$k = A \exp\left(\frac{-E_a}{RT}\right) \qquad (41.6)$$

The value E_a is the *activation energy;* the constant A is the *frequency factor,* or *preexponential factor*. This equation has exactly the same form as the empirical equation (41.1).

The Arrhenius equation can be explained on the basis of the model shown in Figure 41.2. To the left is the energy of the system A + B and to the right is the energy of the system C + D. The difference in the two energies is ΔE. In the center of the figure is what we call the *activated complex*. For A to react with B to form C + D, the activated complex X* must first be formed; this requires an amount of energy E_a.

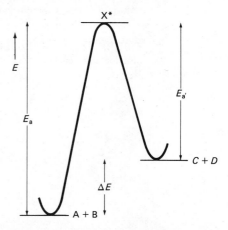

FIGURE 41.2 Energy-level diagram for the reaction A + B → C + D, showing the activated complex X*.

Once the activated complex is formed, it can then either revert back into A + B or form the new products C + D. Similarly, we can start with C + D to form X*, which then either reverts to C + D or forms A + B. This model suggests that the rate at which A and B react is determined only by E_a, and not by ΔE; the rate of the reaction is independent of what happens after the activated complex is formed. Similarly, the rate at which C and D react is determined only by $E_{a'}$.

The Arrhenius equation is usually not followed precisely over a very wide temperature range, and more sophisticated treatments are often required. One important point is that the above thermodynamic arguments are valid only for elementary reactions. Very large deviations from the exponential Arrhenius expression often mean that the rate constant being measured is a combination of constants for several elementary reactions with different activation energies, and that the reaction is complex.

Proceeding one step further, we note that the exponential term is just the Boltzmann factor, which gives the proportion of molecules with energies sufficient to overcome the activation-energy barrier. A very simple argument would proceed as follows. Obviously a molecule of A cannot react with a molecule of B unless the two can somehow interact; this interaction can only take place if they come within a certain distance of each other. The two components of the Arrhenius law can then be given separate physical interpretations. The frequency factor expresses the rate at which molecules approach close enough to interact. The exponential term expresses the proportion of interactions that have enough energy to form the activated complex. This argument suggests that we turn to quantum mechanics for the interaction energy and to the kinetic theory of gases for the rate of intermolecular collisions. We turn to the kinetic theory of gases first.

41.2 THE COLLISION THEORY

In Equation (22.14) we showed that the total number of collisions per unit time per unit volume that take place in a gas is

$$z_{AA} = \tfrac{1}{2}\sqrt{2}\pi d^2 \bar{c}(n^*)^2 \qquad \left(\bar{c} = \sqrt{\frac{8kT}{\pi m}}\right)$$

$$= 2(n^*)^2 d^2 \sqrt{\frac{\pi kT}{m}} \tag{41.7}$$

where $\bar{c}$ is the average velocity, n^* is the number of gas molecules per cubic centimeter, m is the mass, and d is the diameter of the molecules; z_{AA} is known as the collision frequency. For collisions between unlike molecules, this expression becomes

$$z_{AB} = n_A^* n_B^* d_{AB}^2 \left[8\pi kT \frac{m_A + m_B}{m_A m_B} \right]^{1/2} \tag{41.8}$$

where n_A^* and n_B^* are the concentrations of A and B, the value d_{AB} is the average of the two diameters (or the sum of the molecular radii), and $(m_A + m_B)/m_A m_B$ is the reduced mass of the two molecules. The quantity d_{AB}^2 is known as the *collision cross section*, since it represents the effective target area of the molecules for a collision. In line with our discussion of the previous section, the rate of the reaction between A and B should then be given by

$$v = z_{AB} \exp\left(\frac{-E_a}{RT}\right) \tag{41.9}$$

This equation simply gives the number of molecules that collide in one second, the sum of whose energies is enough to overcome the activation-energy barrier. If this equation is divided by $n_A^* n_B^*$ and multiplied by Avogadro's number, then for the special case of $n_A^* = n_B^*$ we get for the second-order rate constant

$$k_2 = \frac{L_0 v}{n_A^* n_B^*} = L_0 d_{AB}^2 \left[8\pi kT \frac{m_A + m_B}{m_A m_B} \right]^{1/2} \exp\left(\frac{-E_a}{RT}\right) \text{ cm}^3 \text{ mol}^{-1} \text{ s}^{-1} \tag{41.10}$$

According to this model, the frequency factor A in the Arrhenius equation is given by the coefficient of the exponential term, which for the case of like molecules becomes

$$A = 2L_0 d_{AA}^2 \sqrt{\frac{\pi kT}{m}} \text{ cm}^3 \text{ mol}^{-1} \text{ s}^{-1} \tag{41.11}$$

Collision frequencies can be calculated from the diameters of molecules as determined from transport experiments such as viscosity measurements. Values of A usually lie in the range of 10^{13} to 10^{14} cm^3 mol^{-1} s^{-1}, and the agreement between calculated and experimental rates is often not too bad; that is, the difference is less than an order of magnitude. On the other hand, many reactions display rates that deviate widely from the calculated rates, and more sophisticated treatments are required for these reactions. One simple method to provide better agreement between calculated and experimental rates is to introduce a *steric factor P*,

$$A = 2L_0 d_{AA}^2 \sqrt{\frac{\pi kT}{m}} P \tag{41.12}$$

which takes into account that the efficiency of collisions may be less than unity. For unsymmetrical molecules, it is reasonable to assume that the orientation of the collision is important and that a reaction takes place only when the two molecules collide with the proper relative orientations.

A termolecular reaction should involve the simultaneous collision of three molecules. These kinds of reactions appear to be relatively rare since three-body collisions occur much less frequently than two-body collisions. Most known third-order reactions involve nitric oxide:

$$2\text{NO} + \text{X}_2 \longrightarrow 2\text{NOX} \qquad (\text{X} = \text{Cl, Br, O, } \ldots)$$

$$-\frac{d[\text{X}_2]}{dT} = k[\text{NO}]^2[\text{X}_2]$$

The reaction is apparently an elementary one as written and seems to involve a simultaneous collision of two NO molecules with one molecule of X_2. Since three-body collisions are very rare, these should have very small activation energies, and this is found to be the case.

An alternative proposed scheme involves the two-step mechanism

$$NO + X_2 \rightleftharpoons NOX_2$$
$$NOX_2 + NO \longrightarrow 2NOX_2$$

In the first step, equilibrium is established very rapidly, and NOX_2 in a small concentration is in equilibrium with the reactants:

$$[NOX_2] \rightleftharpoons K[NO][X_2]$$

The second step proceeds very slowly with a rate

$$R = k'[NOX_2][NO] = k'K[NO]^2[X_2]$$

thus the overall reaction is third-order, with $k = k'K$. This mechanism is equivalent to assuming the presence of "sticky collisions." That is to say, when an NO and an X_2 molecule collide, they remain "stuck" together for a very short time; this time, although short, is long enough for a second NO molecule to collide with the "stuck" complex molecule.

In the light of the collision theory, the existence of unimolecular reactions seemed anomalous. The energy of activation that allows a reaction to take place is provided by the kinetic energy transferred in a bimolecular collision. This seems to negate the possibility of unimolecular reactions and first-order kinetics. We shall return to this point after looking at several features of bimolecular reactions.

41.3 POTENTIAL ENERGY SURFACES FOR REACTIONS; THE ACTIVATED COMPLEX

With the advent of quantum mechanics in the early 1930s, it was possible for the first time to investigate quantitatively the formation of molecules. The potential energy of the diatomic system $A + B \to AB$ has been discussed previously. The energy is a function of a single variable, the bond distance r_{A-B}, as shown in Figures 28.15 and 33.2; a two-dimensional diagram suffices for all variables.

The simplest case of kinetic interest is the triatomic system $A-B + C \to A + B-C$. For this system the energy is a function of three dimensions; these may be selected as the length of the A–B bond r_{AB}, the length of the B–C bond r_{BC}, and the angle formed by the three atoms. Four dimensions are needed for the complete energy diagram. The system can be simplified considerably and the dimensionality reduced by one by considering the special case of a collinear collision in which atom C approaches A—B along the line joining the two atoms. This configuration is most favorable kinetically; the interaction between C and A is minimized, hence the total energy is minimized in this configuration. By plotting the energy in three space as a function of both r_{AB} and r_{CB}, we get a potential-energy surface, as shown in Figure 41.3.

In Figure 41.3, the plane section perpendicular to the x axis shows the potential-energy curve for the molecule AB at very large B $\cdots$ C separation. The equilibrium internuclear separation of AB is the minimum in the curve at $r_0(AB)$; the curve is just a Morse potential curve. Similarly, the plane section perpendicular

SEC. 41.3 POTENTIAL ENERGY SURFACES FOR REACTIONS; THE ACTIVATED COMPLEX

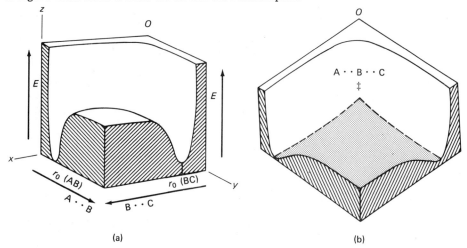

FIGURE 41.3 (a) Potential-energy surface for the system AB + C → A + BC. (b) The surface of Figure 41.3a tilted to show the col and the reaction path.

to the y axis shows the potential energy for the molecule BC at large A $\cdots$ B separation.

During the reaction, the system must move from the point r_0(AB) to r_0(BC). The path the reaction follows, or the *course of the reaction,* is the one of lowest potential energy. This path is along the dotted line shown in Figure 41.3b, which shows the surface tilted to more clearly indicate the minimum path. The system starts at A—B + C, and proceeds along the valley to a high saddle point, or "col"; after crossing the col, the system proceeds down the other side into the product valley, finally reaching the configuration B—C + A. The high level point, or col, is indicated by ‡. The col corresponds to the triatomic activated complex molecule A $\cdot\cdot$ B $\cdot\cdot$ C‡.

The three-dimensional plot of Figure 41.3 can be reduced to a two-dimensional contour plot, as shown in Figure 41.4. The lines on this plot join points of equal potential energy. The course of the reaction indicated by the dashed line is perpendicular to each of the contours and passes through the col. Finally, Figure 41.5 shows the variation of energy along the reaction path. The peak labeled ‡ corresponds to the activated complex, or *transition state.*

For the case illustrated, AB and BC are two different molecules, and the minima of the two Morse potential curves are at different heights. The system to which the most attention has been devoted is the one that is most amenable to theoretical treatment, namely the system of three hydrogen atoms, which may be designated H′, H″, and H‴. The reaction corresponds to the replacement of one hydrogen atom in an H_2 molecule by another hydrogen atom:

$$H'—H'' + H''' \longrightarrow H' + H''—H'''$$

This reaction can be studied experimentally by the rate of conversion between ortho- and parahydrogen. Alternatively, a deuterium atom can be substituted for H‴, and the reaction

$$H—H + D \longrightarrow H + H—D$$

FIGURE 41.4 Contour diagram of the potential-energy surface of Figure 41.3.

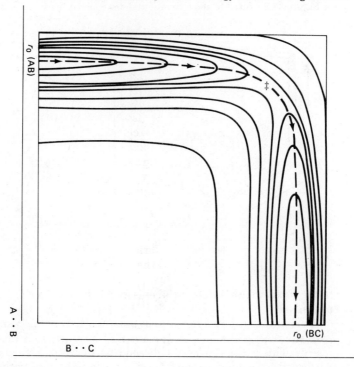

can be studied. The equilibrium internuclear distance for H_2 is $r_{AB} = 0.72$ Å; the configuration of the activated complex corresponds to $r_{AB} = r_{BC} = 0.93$ Å. The experimental activation energy as obtained from fits of the Arrhenius equation is about 31 kJ mol^{-1}. Calculated values are generally higher, as high as 80 kJ mol^{-1}, depending on the degree of sophistication of the calculation and the approximations used. One

FIGURE 41.5 The variation of energy along the reaction path for the reaction AB + C → A + BC.

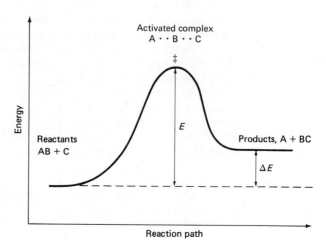

calculation provided the value 36 kJ mol^{-1}, which agreed closely with the experimental value.[2]

41.4 THE ACTIVATED COMPLEX THEORY

We have noted that the properties of the activated complex can be determined by quantum mechanical calculation (previous section). In 1935, H. Eyring suggested that the methods of statistical mechanics could be applied to the activated complex (and the products and reactants) to calculate rate constants on a purely theoretical basis.[3] The theory is called the *activated complex theory,* or more generically the *theory of absolute reaction rates.* We shall now outline some features of this theory on an elementary level. The student who is interested in further details is referred to the references listed in the Bibliography.

Consider the reaction

$$a\text{A} + b\text{B} \rightleftharpoons c\text{C} + d\text{D} \quad (41.13)$$

which has proceeded to its equilibrium state. Recalling our discussion of statistical mechanics (Chapter 24), we find that the equilibrium constant can be expressed by

$$K_c = \frac{[\text{C}]^c[\text{D}]^d}{[\text{A}]^a[\text{B}]^b} = \frac{z_C^c z_D^d}{z_A^a z_B^b} \exp\left(\frac{-\Delta E}{RT}\right) \quad (41.14)$$

where the z's are the partition functions per unit volume (1 cm^3) and the concentrations are expressed in molecules per cubic centimeter; ΔE is as shown in Figure 41.5. For simplicity we shall assume in what follows that the mole numbers in Equation (41.13) are all unity, $a = b = c = d = 1$. Under conditions of equilibrium, the activated complex, which we denote by X‡, is also in equilibrium with the reactants and products, hence the concentration of X‡ can be calculated in terms of the concentrations of the reactants A and B.

We can write the equilibrium constant between the reactants and the activated complex as

$$K\ddagger = \frac{[\text{X}\ddagger]}{[\text{A}][\text{B}]} = \frac{z\ddagger}{z_A z_B} \exp\left(\frac{-E}{RT}\right) \quad (41.15)$$

where E is as shown in Figure 41.5. (Remember that we set all the mole numbers in the stoichiometric equation equal to unity.) If molecule A has N_A atoms, then it has $3N_A$ degrees of freedom, and its partition function consists of the product of $3N_A$ partition functions—3 for translational motion, 3 for rotational motion (2 if the molecule is linear) and $3N_A - 6$ for vibrational motion ($3N_A - 5$ if the molecule is linear); similar results hold for molecule B, which contains N_B atoms. The activated complex consists of $N_A + N_B$ atoms. If we assume that the activated complex is nonlinear, X‡ has $3(N_A + N_B) - 6$ vibrational modes. Now one of these vibrational modes must be very different from the other modes, since at least one mode must correspond to a very weak bond that allows X‡ to dissociate into the products. The ordinary vibrational factor has the form $[1 - \exp(-h\nu/kT)]^{-1}$. The weak unique bond

---

[2] R. N. Porter and M. Karplus, *J. Chem. Phys.* 44 (1964):1105.
[3] H. Eyring, *J. Chem. Phys.* 3 (1935):107.

corresponds to a very low vibrational frequency. We can substitute for the ordinary vibrational factor its value in the limit as ν approaches zero, which is

$$\lim_{\nu \to 0} \left[1 - \exp\left(\frac{-h\nu}{kT}\right)\right]^{-1} = \left[1 - \left(1 - \frac{h\nu}{kT}\right)\right]^{-1} = \frac{kT}{h\nu} \qquad (41.16)$$

If we use this $kT/h\nu$ term for the one partition function factor, the expression for the equilibrium constant becomes

$$\frac{[X\ddagger]}{[A][B]} = \frac{z'\ddagger}{z_A z_B} \left(\frac{kT}{h\nu}\right) \exp\left(\frac{-E}{RT}\right) \qquad (41.17)$$

where $z'\ddagger$ is the partition function containing the remaining factors. Equation (41.17) can be rearranged to

$$\nu[X\ddagger] = [A][B] \frac{kT}{h} \frac{z'\ddagger}{z_A z_B} \exp\left(\frac{-E}{RT}\right) \qquad (41.18)$$

The left side of (41.18) is the product of the concentration of $X\ddagger$ and the frequency of the decomposition of $X\ddagger$. It is therefore the rate of the reaction,

$$R = [A][B] \frac{kT}{h} \frac{z'\ddagger}{z_A z_B} \exp\left(\frac{-E}{RT}\right) \qquad (41.19)$$

This is the expression for a second-order reaction rate with rate constant[4]

$$k_2 = \frac{kT}{h} \frac{z'\ddagger}{z_A z_B} \exp\left(\frac{-E}{RT}\right) \qquad (41.20)$$

The quantity kT/h has the dimensions of frequency; its value is about 6×10^{-12} s^{-1} at 300 K.

Inherent to Equation (41.20) is the assumption that every activated complex that reaches the top of the energy barrier is converted into products. This neglects the possibility that $X\ddagger$ may decompose back into reactants. To account for this possibility, a *transmission coefficient* κ is introduced; this gives the probability that $X\ddagger$ decomposes into products. The rate constant is finally written as

$$k_2 = \kappa \frac{kT}{h} K\ddagger \qquad (41.21)$$

where $K\ddagger$ is a modified equilibrium constant

$$K\ddagger = \frac{z'\ddagger}{z_A z_B} \exp\left(\frac{-E}{RT}\right) \qquad (41.22)$$

The simplest application of this theory is to a reaction between two atoms, A and B. The atoms have only translational degrees of freedom, so their respective partition functions are

$$z_A = \frac{(2\pi m_A kT)^{3/2}}{h^3} \quad \text{and} \quad z_B = \frac{(2\pi m_B kT)^{3/2}}{h^3} \qquad (41.23)$$

---

[4] The rate here is proportional to [A][B] because we assumed that the activated complex is composed of one molecule of A and one molecule of B. For the general reaction of Equation (41.13), the rate would be proportional to $[A]^a[B]^b$; the activated complex would be composed of a molecules of A and b molecules of B. The molecularity of a reaction is equal to the number of molecules of reactant that form the activated complex.

SEC. 41.4 THE ACTIVATED COMPLEX THEORY

The activated complex has 6 degrees of freedom—3 translational, 2 rotational, and 1 vibrational—corresponding to the degrees of freedom of a diatomic molecule. Omitting the vibrational degree of freedom, since that corresponds to the unique vibrational mode resulting in decomposition, we can write

$$z'^{\ddagger} = \frac{[2\pi(m_A + m_B)kT]^{3/2}}{h^3} \frac{8\pi^2 I kT}{h^2} \quad (41.24)$$

The term I represents the moment of inertia of the activated complex

$$I = d_{AB}^2 \frac{m_A m_B}{m_A + m_B} \quad (41.25)$$

where d_{AB} is the bond length of the activated complex. Taking the transmission coefficient as unity, one gets the rate constant as

$$k_2 = \frac{kT}{h} \frac{z'^{\ddagger}}{z_A z_B} \exp\left(\frac{-E}{RT}\right)$$

$$= d_{AB}^2 \left[8\pi kT \frac{m_A + m_B}{m_A m_B}\right]^{1/2} \exp\left(\frac{-E}{RT}\right) \quad (41.26)$$

By comparing this equation with (41.10), we see that the absolute rate theory gives the same result as the collision theory for a reaction involving two atoms. When more complex molecules are treated, the two theories give substantially different results. The collision theory generally gives much larger results, as you can see in Table 41.1.

The Arrhenius frequency factor in the rate expression of Equation (41.20) is

$$A = \frac{kT}{h} \frac{z'^{\ddagger}}{z_A z_B} \quad (41.27)$$

The temperature thus enters directly in the term kT/h and indirectly through the partition functions. Plots of the logarithm of the rate constant vs inverse temperature often give a straight line, as shown in Figure 41.1, because the exponential part has a much stronger temperature dependence and usually predominates. If the data

TABLE 41.1 Activation energies and frequency factors for some elementary bimolecular reactions.

Reaction	E_a (kJ mol^{-1})	Arrhenius frequency factor (10^{12} cm^3 mol^{-1} s^{-1})		
		Observed	Calculated ACT theory Eq. (41.20)	Calculated collision theory
$NO + O_3 \to NO_2 + O_2$	10.5	0.8	0.44	50
$NO_2 + O_3 \to NO_3 + O_2$	29.3	5.9	0.14	63
$NO_2 + F_2 \to NO_2F + F$	43.5	1.6	0.12	63
$NO_2 + CO \to NO + CO_2$	132.	12.	6.	40
$2NO_2 \to 2NO + O_2$	111.	1.8	4.5	40
$NO + NO_2Cl \to NOCl + NO_2$	28.9	0.83	0.84	80
$2NOCl \to 2NO + Cl_2$	103.	9.4	0.44	63
$NOCl + Cl \to NO + Cl_2$	4.6	11.4	4.4	63
$NO + Cl_2 \to NOCl + Cl$	85.0	4.0	1.2	100
$F_2 + ClO_2 \to FClO_2 + F$	35.6	0.04	0.08	50
$2ClO \to Cl_2 + O_2$	0	0.06	0.01	25
$COCl + Cl \to CO + Cl_2$	3.5	400.	1.8	63

are plotted over very large temperature ranges, then the plots are usually not linear over the entire range.

41.5 CONNECTION WITH THERMODYNAMICS

Suppose we write the equation for the rate constant in the form

$$k_2 = \frac{kT}{h} K\ddagger \tag{41.28}$$

Recalling the thermodynamic relations

$$\Delta G° = -RT \ln K \quad \text{and} \quad \Delta G = \Delta H - T\Delta S$$

we can write the rate constant as

$$k_2 = \frac{kT}{h} \exp\left(\frac{-\Delta G°\ddagger}{RT}\right) \tag{41.29}$$

$$k_2 = \frac{kT}{h} \exp\left(\frac{\Delta S°\ddagger}{R}\right) \exp\left(\frac{-\Delta H°\ddagger}{RT}\right) \tag{41.30}$$

The quantities $\Delta G°\ddagger$, $\Delta H°\ddagger$, and $\Delta S°\ddagger$ are known as the *Gibbs energy of activation*, the *enthalpy of activation*, and *the entropy of activation*.

These can be related to the energy of activation. We can differentiate the logarithmic form of Equation (41.28), to get

$$\frac{d \ln k_2}{dT} = \frac{1}{T} + \frac{d \ln K\ddagger}{dT} \tag{41.31}$$

Since $K\ddagger$ is a concentration equilibrium (rather than a pressure equilibrium), its temperature variation is

$$\frac{d \ln K\ddagger}{dT} = \frac{\Delta E\ddagger}{RT^2} \tag{41.32}$$

Combining these last two equations gives

$$\frac{d \ln k_2}{dT} = \frac{1}{T} + \frac{\Delta E\ddagger}{RT^2} = \frac{RT + \Delta E\ddagger}{RT^2} \tag{41.33}$$

Using the Arrhenius equation directly, we have

$$\frac{d \ln k_2}{dT} = \frac{E_a}{RT^2} \tag{41.34}$$

where E_a is the experimental activation energy obtained from plots such as those of Figure 41.1. Comparing these last two equations leads to

$$E_a = RT + \Delta E\ddagger \tag{41.35}$$

Since $\Delta H = \Delta E + P\Delta V$ at constant pressure, we can write

$$E_a = \Delta H°\ddagger - P\Delta V\ddagger + RT \tag{41.36}$$

where $\Delta V\ddagger$ is the increase in volume between the initial state and the activated state.

41.6 UNIMOLECULAR REACTIONS AND THE LINDEMANN THEORY

From what we have already said, it seems that unimolecular reactions are not feasible. Somehow or other an activated complex must be formed, and this can form only by the interaction of at least two molecules. In the early 1920s, F. A. Lindemann showed how a bimolecular collision mechanism could lead to kinetics that were first-order overall.[5]

First-order reactions are generally of the type A → B or A → B + C. The first type represents an isomerization reaction such as a cis to trans conversion, whereas the second type involves the rupture of a bond in the molecule. According to the Lindemann treatment, the overall first-order reaction takes place by a two-step mechanism. The first step is a bimolecular reaction that energizes and deenergizes the system:

$$(1) \quad A + A \underset{k_{-2}}{\overset{k_2}{\rightleftharpoons}} A^* + A$$

Here A represents a normal reactant molecule and A^* represents an energized, or activated, molecule.[6] The subsequent decomposition or isomerization is then represented by

$$(2) \quad A^* \xrightarrow{k_1} \text{products (B or B + C)}$$

This second step is a true unimolecular elementary reaction. If we apply the steady-state treatment to this system, we have

$$\frac{d[A^*]}{dt} = 0 = k_2[A]^2 - k_{-2}[A^*][A] - k_1[A^*] \quad (41.37)$$

The concentration of A^* in the steady state is

$$[A^*] = \frac{k_2[A]^2}{k_{-2}[A] + k_1} \quad (41.38)$$

The overall reaction velocity is given by the rate at which B or C is formed from A^*:

$$\frac{d[B]}{dt} = k_1[A^*] = \frac{k_1 k_2 [A]^2}{k_{-2}[A] + k_1} \quad (41.39)$$

Now at high pressure $k_{-2}[A] \gg k_1$, and the k_1 term in the denominator of (41.39) can be neglected relative to $k_{-2}[A]$. We then obtain, in the high-pressure limit,

$$\frac{d[B]}{dt} = \frac{k_1 k_2}{k_{-2}}[A] = k_\infty [A] \quad (41.40)$$

which is the expression for an ordinary first-order reaction. At low pressure, on the other hand, $k_1 \gg k_{-2}[A]$ and (41.39) becomes

$$\frac{d[B]}{dt} = k_2[A]^2 \quad (41.41)$$

This is an expression for an ordinary second-order reaction. The Lindemann model thus allows us to use a bimolecular collision mechanism to explain the existence of

---

[5] F. A. Lindemann, *Trans. Faraday Soc.* 17 (1922):598.
[6] A more general way to write this equation is A + M = A^* + M, where M is any gas molecule, not necessarily an A molecule.

first-order reactions. The predicted first-order behavior is not the general state of affairs, however, but a high-pressure limit. In this limit, the activated molecule has a lifetime that is sufficiently long that the probability of deactivation by collision is much greater than the probability of decomposition. As the pressure is lowered, however, the system is expected to change gradually from first-order to second-order kinetics. Such changes from first to second order have been observed in numerous systems.

Not all first-order reactions can be explained on the basis of the Lindemann theory. An example is the classic N_2O_5 decomposition noted at the beginning of Chapter 40. The stoichiometric equation $2N_2O_5 \rightarrow 2N_2O_4 + O_2$ has two reactant molecules and therefore cannot by itself represent an elementary unimolecular reaction. The proposed mechanism consists of the steps

(1) $\quad N_2O_5 \xrightarrow{k_1} NO_2 + NO_3$

(2) $\quad NO_2 + NO_3 \xrightarrow{k_2} N_2O_5$

(3) $\quad NO_2 + NO_3 \xrightarrow{k_3} NO_2 + O_2 + NO$

(4) $\quad NO + N_2O_5 \xrightarrow{k_4} 3NO_2 \quad$ (rapid)

In this scheme, NO_2 and NO_3 either can recombine to give N_2O_5 or can react to form NO, O_2, and NO_2. Since the latter alternative is an endothermic reaction, we should expect k_3 to be much lower than k_2; thus the third reaction would be the rate-controlling step. Applying the steady-state treatment to the NO_3 and NO concentrations, we have

$$\frac{d[NO_3]}{dt} = k_1[N_2O_5] - k_2[NO_2][NO_3] - k_3[NO_2][NO_3] = 0 \tag{41.42}$$

or
$$[NO_3] = \frac{k_1[N_2O_5]}{(k_2 + k_3)[NO_2]} \tag{41.43}$$

and
$$\frac{d[NO]}{dt} = 0 = k_3[NO_2][NO_3] - k_4[NO][N_2O_5] \tag{41.44}$$

or
$$[NO] = \frac{k_3[NO_2][NO_3]}{k_4[N_2O_5]} = \frac{k_3 k_1}{k_4(k_2 + k_3)} \tag{41.45}$$

after using the expression for $[NO_3]$ given in (41.43). The rate at which the N_2O_5 disappears is

$$-\frac{d[N_2O_5]}{dt} = k_1[N_2O_5] - k_2[NO_2][NO_3] + k_4[NO][N_2O_5]$$

$$= k_1[N_2O_5] - \frac{k_2 k_1[N_2O_5]}{k_2 + k_3} + \frac{k_3 k_1[N_2O_5]}{k_2 + k_3} \tag{41.46}$$

If the first term on the right-hand side of (41.46) is multiplied by unity in the form $(k_2 + k_3)/(k_2 + k_3)$, we can then simply sum all the terms, to get the result

$$-\frac{d[N_2O_5]}{dt} = \frac{2k_1 k_3}{k_2 + k_3} [N_2O_5] \tag{41.47}$$

SEC. 41.7 IMPROVEMENTS ON THE LINDEMANN THEORY

which can be approximated by

$$-\frac{d[N_2O_5]}{dt} = 2\left(\frac{k_1}{k_2}\right)k_3[N_2O_5] \tag{41.48}$$

since $k_2 \gg k_3$. According to this mechanism, the observed first-order rate constant is actually the product of an equilibrium constant, k_1/k_2 and a *second-order* rate constant k_3.

That this mechanism gives the correct answer might be enough to lend credence to it; however, additional confirmation is provided by other experiments. This mechanism assumes that the equilibrium reaction in steps (1) and (2) occurs very rapidly. This has been verified by studying the rate of isotopic exchange between $^{15}N_2O_5$ and $^{14}NO_2$,[7]

$$^{15}N_2O_5 + {}^{14}NO_2 \rightleftharpoons {}^{14}N_2O_5 + {}^{15}NO_2$$

Further confirmation is provided by studies of the N_2O_5 decomposition in the presence of added NO.[8]

41.7 IMPROVEMENTS ON THE LINDEMANN THEORY

Although Lindemann's simple theory provides an adequate qualitative explanation of first-order reactions, the quantitative agreement with experiments is less satisfactory. If we let k_{exp} be the experimental rate constant, then it is apparent from Equation (41.39) that

$$k_{\text{exp}} = \frac{k_1 k_2 [A]}{k_{-2}[A] + k_1} \tag{41.49}$$

In the limit of high pressure, the rate constant is

$$k_\infty = \frac{k_1 k_2}{k_{-2}} \tag{41.50}$$

Combining these last two equations yields

$$\frac{1}{k_{\text{exp}}} = \frac{1}{k_\infty} + \frac{1}{k_2[A]} \tag{41.51}$$

The simple Lindemann theory thus predicts a linear relation between $1/k_{\text{exp}}$ and $1/[A]$ (or $1/P$, since the pressure is proportional to $[A]$). Large deviations from this linear relation are observed, however. At low pressures, where the first-order rate constant begins to fall, the overall rate should be determined by the rate of formation of activated molecules. According to the collision theory, this rate should be $Z_{AA} \exp(-E_a/RT)$. Unfortunately, the measured rates were found to be several orders of magnitude larger than the rate calculated by the collision theory.

An explanation for the persistence of first-order rates down to much lower concentrations than the Lindemann theory would permit was first proposed by Hinshelwood.[9] The collision theory treats only translational degrees of freedom. Hinshel-

---

[7] R. A. Ogg, *J. Chem. Phys.* 18 (1950):573.
[8] J. H. Smith and F. Daniels, *J. Amer. Chem. Soc.* 69 (1947):1735; R. A. Ogg, *J. Chem. Phys.* 18 (1950):572.
[9] C. N. Hinshelwood, *Proc. Roy. Soc.* (London) *A113* (1927):230.

wood showed that by including the distribution of energy in the vibrational degrees of freedom, the rate as calculated by the collision theory is enhanced by the factor

$$\frac{1}{(s-1)!}\left(\frac{E}{RT}\right)^{s-1}$$

where s is the number of vibrational degrees of freedom among which the energy E can be distributed.[10] This factor is of the order of 10^6 and predicts that the first-order rate constant does not begin to fall off until the point at which pressures are lower by this factor than the old theory predicted. In practice, s is found by trial and error. It has always been possible to fit the data with a value of s lower than $(3N-6)$, the total number of vibrational degrees of freedom. The value s is usually found to be about half the total number. This comports with the view that the energy used to form the activated complex does not come from all the normal modes but from some of them. A further improvement in the Lindemann-Hinshelwood theory was provided by Kassel, Rice, and Ramsperger, who examined the energy available in the one *critical* vibrational mode.[11] They argued that the higher the energy available to A*, the greater the probability that the critical mode could acquire the necessary energy to decompose; hence the probability was lower that it would be deactivated before decomposing. The rate constant k_1 for decomposition of A* thus increases with the total energy available to the molecule. The equations resulting from this theory and further refinements (collectively called the HKRR theory) agree well with the experimental data.

PROBLEMS

1. The rate of a reaction increases by a factor of 23 when the temperature is raised from 25 °C to 42 °C. Calculate the activation energy of the reaction.

2. A vessel contains a mixture of gases A and B. Each is at a partial pressure of 75 torr. The molecular diameters are 3.0 Å for A and 3.5 Å for B. Calculate the collisional frequency between A and B molecules (cm^{-3} s^{-1}) at 300 K. (Take $m_A = 30$ and $m_B = 60$.)

3. The rate constant for the reaction $A + B \to C$ of the system of Problem 2 is 1.4×10^5 mol^{-1} cm^3 s^{-1} at 300 K; the activation energy is 80 kJ mol^{-1}. What fraction of molecules collide with sufficient energy to react? Calculate the steric factor for the reaction.

4. For unimolecular reactions, $\Delta V^\ddagger$ is zero. Show that for such a reaction the rate equation can be written

$$k = (\exp)^1 \frac{kT}{h} \exp\left(\frac{\Delta S^\ddagger}{R}\right) \exp\left(\frac{-E_a}{RT}\right)$$

5. For a general gas reaction, $P \Delta V^\ddagger = \Delta n^\ddagger RT$, where $\Delta n^\ddagger$ is the increase in the number of moles when the activated complex is formed from the reactants. Show that for a bimolecular reaction the rate constant has the same expression as in Problem 4, with the exception that $(\exp)^2$ is substituted for the $(\exp)^1$ term.

6. For head-on collisions calculate the fraction of collisions for which the energy exceeds 100 kJ mol^{-1}. Do the same for glancing collisions at 45°.

7. G. Lapidus, D. Barton, and P. E. Yankwich, *J. Phys. Chem.* 68 (1964):1863, studied the decompo-

Temperature (°C)	Time (min)	Fraction of Acid Decomposed
126.6	300	0.403
	420	0.516
	610	0.669
134.1	70	0.288
	240	0.684
	360	0.822
146.4	30	0.286
	70	0.515
	120	0.708
	190	0.854
155.6	28.6	0.585
	33.6	0.634
	55.8	0.817

[10] This is derived in K. J. Laidler, *Chemical Kinetics*, 2d ed. (New York: McGraw-Hill Book Co., 1965), pp. 147 et. seq.

[11] L. S. Kassel, *J. Phys. Chem.* 32 (1928):225; O. K. Rice and H. C. Ramsperger, *J. Amer. Chem. Soc.* 49 (1927):1617; 50 (1928):617.

sition of oxalic acid. The only products found were equimolar amounts of formic acid and carbon dioxide:

$$\text{COOHCOOH(g)} \longrightarrow \text{HOOCH(g)} + \text{CO}_2\text{(g)}$$

The decomposition was studied in the vapor phase (0.9 torr pressure) between 127 °C and 157 °C. The fraction of oxalic acid decomposed as a function of time at several temperatures is given in the accompanying table. Determine the first-order rate constants at each temperature, and from those calculate the activation energy and frequency factor for the Arrhenius equation.

8. Calculate $\Delta S^{\ddagger}$, $\Delta H^{\ddagger}$, and $\Delta G^{\ddagger}$ for the reaction of Problem 7.

9. Consider the reactions $H-H + D \rightarrow H \cdot \cdot H \cdot \cdot D \rightarrow H-D + H$ and $H-H + T \rightarrow H \cdot \cdot H \cdot \cdot T \rightarrow H-T + H$, where D is a deuterium atom of atomic mass 2 and T is a tritium atom of atomic mass 3. Assume that the entire difference in the rate constants can be attributed to the frequency difference in the weak bond of the activated complex. Using this assumption, calculate the difference in rates of the two reactions. (*Note:* This is an example of the *isotope effect*. For further discussion of this see the references listed in the Bibliography.)

10. Show that the approximate order of magnitude of the translational partition function is 10^8-10^9 per degree of freedom. Show that for rotational partition functions it is 10^1-10^2 and for vibrations it is 10^0-10^1. What is the temperature dependence of each?

11. The table below lists the forms of the Arrhenius frequency factor of Equation (41.27) for several types of reactions, where z_t, z_v, and z_r represent the translational, vibrational, and rotational partition functions. Verify some of the entries in the table. Show that for two linear molecules forming a nonlinear complex, the expression is $(kT/h)(z_v^3/z_t^3 z_r)$, and that for two nonlinear molecules forming a nonlinear complex it is $(kT/h)(z_v^5/z_t^3 z_r^3)$.

12. A *catalyst* is a substance that influences the rate of a reaction but does not appear in one of the products; it is not consumed by the reaction. It changes the rate of a reaction but not the equilibrium of a reaction; hence it must have the same proportional effect on the forward and reverse reactions. In modern terms, a catalyst lowers the free energy of activation. Among the most efficient of catalysts are enzymes. The very fast reactions involving enzymes were first explained by Michaelis and Menten in 1913. They proposed the two-step reaction

$$E + S \underset{k_{-1}}{\overset{k_1}{\rightleftharpoons}} ES \rightarrow E + P$$

where E represents the enzyme, S the substrate, and P the product; ES is an intermediate enzyme-substrate complex. The rate of formation of ES is

$$\frac{d[ES]}{dt} = k_1[E][S] - (k_{-1} + k_2)[ES]$$

Apply the steady-state treatment to this reaction and show that $[E] = [ES]K_m/[S]$, where $K_m = (k_{-1} + k_2)/k_1$. The term K_m is called the *Michaelis constant*. Show that the overall rate of formation of product is

$$\frac{d[P]}{dt} = k_2[ES] = \frac{k_2[E_0]}{1 + \frac{K_m}{[S]}}$$

where $[E_0] = [ES] + [E]$ is the overall enzyme concentration.

13. Using the first law of thermodynamics, Ostwald was able to show that a catalyst could not affect the position of equilibrium of a chemical reaction. Consider a reaction in which the volume of the products does not equal the volume of reactants. Show that if catalysts could affect the equilibrium position, then it would be possible to build a perpetual motion machine, and that hence the catalyst cannot affect the equilibrium position.

14. Using the nomenclature used in the discussion of unimolecular reactions in the text, determine the ratio of $k_{-2}[A]$ to k_1 necessary for the reaction to be first-order.

Type of reaction	A	Representative value at 298 K ($mol^{-1}\ cm^3\ s^{-1}$)
Atom + atom → linear complex	$\dfrac{kT}{h}\dfrac{z_r^2}{z_t^3}$	$10^{-10}-10^{-9}$
Atom + linear molecule → linear	$\dfrac{kT}{h}\dfrac{z_v^2}{z_t^3}$	$10^{-12}-10^{-11}$
Atom + linear → nonlinear	$\dfrac{kT}{h}\dfrac{z_v z_r}{z_t^3}$	$10^{-11}-10^{-10}$
Atom + nonlinear → nonlinear	$\dfrac{kT}{h}\dfrac{z_v^2}{z_t^3}$	$10^{-12}-10^{-11}$
Linear + linear → linear	$\dfrac{kT}{h}\dfrac{z_v^4}{z_t^3 z_r^2}$	$10^{-14}-10^{-13}$

CHAPTER FORTY-TWO
PHOTOCHEMISTRY

42.1 THE HYDROGEN-CHLORINE CANNON; A PHOTOCHEMICAL REACTION

Consider the experimental arrangement shown in Figure 42.1, in which chlorine and hydrogen are reacted to form HCl. This reaction is often used as a lecture demonstration in elementary chemistry courses. A bottle is filled with equimolar amounts of hydrogen and chlorine gas and tightly stoppered with a handball. The bottle is then clamped in a fixture, as indicated in the figure, so that a beam of light can be admitted to the mixture through the bottom of the bottle. In the absence of light, the reaction $\frac{1}{2}H_2 + \frac{1}{2}Cl_2 \rightarrow HCl$ is extremely slow, and the contents of the bottle remain unchanged over long periods. When the light bulb behind the "cannon" is turned on, a loud explosion results, and the ball is expelled with a high velocity. (In a lecture demonstration, the cannon is arranged to fire the ball over the audience and the ball strikes the back wall of the lecture hall. The demonstrator then catches the ball, to the applause of the students.) Since no reaction takes place in the absence of the light, the light itself must somehow be associated with the chemical reaction. We call such reactions *photochemical reactions*.

What happens is that one of the molecules is excited by the absorption of a quantum of light; this excited molecule can then react to form the final products. A mechanism for the hydrogen-chlorine photochemical reaction can be written as the series of equations

$$Cl_2 + h\nu \longrightarrow 2Cl \qquad (42.1a)$$
$$Cl + H_2 \longrightarrow HCl + H \qquad (42.1b)$$
$$H + Cl_2 \longrightarrow HCl + Cl \qquad (42.1c)$$

where $h\nu$ denotes a quantum of light. In this case the light quantum induces the dissociation of a chlorine molecule into two chlorine atoms. Dissociation does not have to be the initial step of a photochemical reaction. The initial step can be excitation to a rotational, a vibrational, or an electronic excited level. The initial step, in which the light quantum is absorbed by an atom or a molecule, is the *primary photochemical process*. Subsequent steps are *secondary photochemical processes*. For the HCl photochemical reaction, (42.1a) is the primary process and (42.1b) and (42.1c) denote secondary photochemical processes. Here the secondary steps form product molecules. In other systems, the secondary processes might be deexcitation of the excited molecule with no new product molecules formed.

The free-energy change for the reaction $\frac{1}{2}H_2 + \frac{1}{2}Cl_2 = HCl$ is negative, and the equilibrium constant lies far to the right, favoring the formation of HCl. Introducing

840

FIGURE 42.1 The HCl cannon. Equimolar amounts of H_2 and Cl_2 are placed in a bottle, which is stoppered with a handball and securely fastened in a vise. When the lamp is turned on, the reaction $H_2 + Cl_2 \rightarrow 2HCl$ takes place very rapidly due to excitation of the molecules by the light. The resulting explosion forces the handball out with a high velocity.

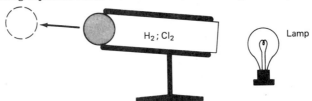

the light quantum affects only the rate of the reaction. The primary photochemical process is generally temperature-independent. This is true since $h\nu \gg kT$.[1] Temperature may, however, affect the rates of the secondary processes. Not all photochemical reactions are characterized by the explosive force associated with the HCl reaction, and many proceed quite slowly. A notable example of a slow photochemical reaction is the photosynthetic reaction that takes place in green plants. Reactions with positive free energy changes can be made to proceed photochemically. This does not pose any contradiction to thermodynamic principles, since the system is not isolated, and extra energy (as light) is introduced.

Some photochemical reactions are very useful and others have deleterious effects in compounding the problems associated with air pollution. We shall treat spectroscopic transitions as a particular form of photochemical reaction. We start our exploration of photochemical processes by examining what have come to be called the two *laws of photochemistry*.

42.2 THE LAWS OF PHOTOCHEMISTRY

In the early 1800s, Grotthus and Draper observed that *only absorbed light is effective in producing photochemical change* (1818). This statement is variously called *the principle of photochemical activation*, the *Grotthus-Draper law*, and *the first law of photochemistry*. Today the law is almost self-evident, since it is clear that transmitted light is ineffective in inducing any change. It does not follow that the absorbed light always causes a chemical reaction. Light that is absorbed by atoms or molecules is often reemitted to give rise to the various lines and bands in emission spectra. Suppose we examine the absorption of a light beam.

Figure 42.2 shows a beam of monochromatic light of intensity I passing through an absorber of thickness dx. The emergent beam has an intensity $I - dI$. The intensity can be conveniently defined by the number of light quanta intersecting unit area in a plane perpendicular to the beam direction in unit time. In this way the problem is transformed into the equivalent of a first-order reaction rate problem in which the variable distance replaces the variable time. We denote the number of incident quanta (or "light molecules") by N, and we let the number absorbed in thickness dx

---

[1] If kT were comparable with $h\nu$, then the molecule could be excited by the available thermal energy.

FIGURE 42.2 When a light beam of intensity I passes through a slab of material of thickness dx, the intensity of the beam is diminished to $I - dI$.

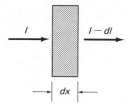

be dN. The probability of absorption in dx (or the fraction of quanta absorbed) is then dN/N, which is proportional to the thickness dx,

$$\frac{dN}{N} = d(\ln N) = b\, dx = -d(\ln I) = -\frac{dI}{I} \qquad (42.2)$$

where b is the proportionality constant called the *absorption coefficient*. If we set $I = I_0$ at $x = 0$ and integrate, we get

$$I = I_0 \exp(-bx) \qquad (42.3)$$

or

$$\ln\left(\frac{I}{I_0}\right) = -bx \qquad (42.4)$$

Lambert first derived Equation (42.3), and it is known as *Lambert's law*. Beer extended this result to include solutions of absorbing compounds in transparent solvents; the equation then takes the form

$$I = I_0 \exp(-\epsilon c x) \qquad \text{or} \qquad \ln\left(\frac{I}{I_0}\right) = -\epsilon c x \qquad (42.5)$$

where c is the molar concentration and ϵ is a constant characteristic of the solute called the *molar absorption coefficient*. Either form of Equation (42.5) is properly known as the *Lambert-Beer law*.[2] This law forms the basis for spectrophotometric methods of chemical analysis. In this method the instrument is usually calibrated by determining ϵ in a solution of known concentration, and then determining the concentration of the unknown using the previously determined value of ϵ. This method is very useful for rapid analyses of certain elements. Thus manganese in steels is often determined spectrophotometrically by oxidation to the deeply colored ion MnO_4^-. Both Lambert's law and its modification are strictly obeyed only for monochromatic light, since the absorption coefficients are strong functions of the wavelength of the incident light. For many substances there are large regions of the spectrum where the substance is transparent (that is, the absorption coefficient is zero). A typical arrangement for the measurement of light absorption is indicated in Figure 42.3. Care must be taken in selecting the structural components of the cell to ensure that the cell is transparent to the light that is used. Certain types of glass can be used in the visible region but not in the ultraviolet region, since glass is generally opaque to ultraviolet light (and also to infrared radiation). Quartz cells are often used in the uv region and cell windows made of single-crystal NaCl are often used in the infrared region for nonaqueous solutions.

---

[2] There is some lack of consistency in this nomenclature, since Lambert's law of Equations (42.3) and (42.4) is sometimes (and incorrectly in the strictest sense) referred to as the Lambert-Beer law. Thus when reference is made to the Lambert-Beer law in the literature, attention must be given to which of the laws is meant. This law is often called the *Beer-Lambert law*.

FIGURE 42.3 Schematic arrangement for a light-absorption experiment. In a modern instrument, the source, the lens, and the monochrometer are often combined in a single laser source. The sample is placed in the cell. The ends of the cell, or the "windows," are made of glass, quartz, NaCl crystal, or other material that must be both transparent to the radiation being used and inert to the material in the cell. The detector is a light-sensitive device, such as a thermopile or photomultiplier tube, that measures changes in light intensities.

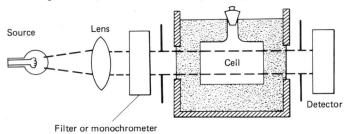

About 100 years after the Grotthus-Draper law was put forth, Einstein and Stark noted that *the activation of any molecule or atom is induced by the absorption of one single light quantum*. This statement is known as the *Stark-Einstein law of photochemical equivalence*, or *the second law of photochemistry*. It expresses the nonclassical view that the entire energy equivalence of one light quantum must be absorbed by a single atom or molecule and cannot be spread over a number of molecules. The energy acquired by the molecule activated is

$$\Delta E = h\nu \tag{42.6}$$

where h is Planck's constant and ν is the frequency of the light. The energy absorbed per mole is

$$\Delta E = L_0 h\nu \tag{42.7}$$

where L_0 is Avogadro's number. The quantity of energy defined by (42.7) is called an *einstein* of energy; it is the energy equivalence of a mole (or 6.02×10^{23}) of light quanta and varies with the wavelength of the light. For wavelength λ, the magnitude of the einstein is

$$\Delta E = \frac{11.96}{\lambda} \text{ J mol}^{-1} \tag{42.8}$$

where λ is measured in centimeters. The energy per einstein of various wavelengths of light is shown in Table 42.1. Most photochemical reactions are studied in the spectral range 2000–10,000 Å, corresponding to energies ranging from 120 to 600 kJ mol^{-1}. Light of longer wavelengths generally does not have enough energy to induce substantial photochemical change. The study of the photochemical effects associated with the highly energetic X rays and gamma rays constitutes a relatively new field of chemistry called *radiation chemistry*.

The principles discussed in this section apply only to what we have called the primary photochemical process, or a process that can be considered a chemical reaction wherein the light photon is one of the reactants. We generally denote this process by an equation of the form

$$S + h\nu \longrightarrow S^* \tag{42.9}$$

TABLE 42.1 Properties of various wavelengths of electromagnetic radiation.

Type of radiation	Wavelength[a] (cm)	Frequency (Hz)	Energy per einstein (J mol^{-1})	Atomic or molecular process associated with the particular wavelength
Radio	750	40×10^6	0.02	NMR spectroscopy
Microwave	0.3–30	10^9–10^{11}	0.4–40	EPR spectroscopy; rotational transitions
Far infrared red	0.003–0.3	10^{11}–10^{13}	40–4,000	Rotational transitions of light molecules; vibrational transitions
Near infrared red	0.0003–0.003	10^{13}–10^{14}	4×10^3 to 4×10^4	Vibrational transitions
Visible	4×10^{-5} to 8×10^{-5}	3.8×10^{14} to 7.5×10^{14}	1.5×10^5 to 3×10^5	Electronic transitions
Ultraviolet	10^{-5}	3×10^{15}	1.2×10^6	Electronic transitions
X rays	10^{-8}	3×10^{18}	1.2×10^9	Inner-shell electronic transitions
Gamma rays	10^{-10}	3×10^{20}	1.2×10^{11}	Nuclear transitions

[a] There is no sharp division between the various types of radiations. The ranges indicated are typical. When no range is given, a typical value has been selected.

where S* denotes an excited (or activated) state of the species S. It may be excited to a higher electronic, vibrational, or any other state, or may even consist of two fragments of the dissociated molecule. Our entire discussion of spectroscopy (Chapters 27–33) applies to the primary photochemical process, which is really nothing more than a spectroscopic transition between the ground state and an upper excited state. Most of photochemistry is concerned with the subsequent fate of S* after the primary photochemical process has occurred. We shall see that whereas some of these processes are spectroscopic, others are "chemical."

42.3 QUANTUM YIELDS, FLUORESCENCE, AND PHOSPHORESCENCE

A quantity of importance in photochemical processes is the *quantum yield* Φ, which measures the overall efficiency of a given reaction. It is defined as the ratio

$$\Phi \equiv \frac{\text{number of molecules that react (or form) in unit time}}{\text{number of light quanta absorbed in unit time}} \quad (42.10)$$

It is important to recognize that the quantum yield has no meaning unless it is computed with reference to a particular process. For a primary process $S + h\nu \to S^*$, the principle of photochemical equivalence requires that $\Phi = 1$, since for each species activated, one and only one light quantum is absorbed. Now suppose that the primary photochemical process is a dissociation that can produce either of two sets of products:

$$S + h\nu \underset{(2)}{\overset{(1)}{\rightleftarrows}} \begin{matrix} B + C \\ D + E \end{matrix} \quad (42.11)$$

SEC. 42.3 QUANTUM YIELDS, FLUORESCENCE, AND PHOSPHORESCENCE

In this instance if the quantum yield were defined in terms of the number of molecules of S that react, then the quantum yield would be unity. The information content of Φ defined in those terms would, however, be very limited. Of much greater use is a quantum yield, defined in terms of either of the two reactions, since that would indicate the relative importance of the two possible dissociation reactions. If the system described by (42.11) were such that each molecule of S that absorbed a light quantum decomposed into one or the other of the two sets of products with no other reactions, then $\Phi_1 + \Phi_2 = 1$, where Φ_1 and Φ_2 denote the quantum yields with respect to the two possible paths of (42.11). Systems are generally not that simple, and the situation is often complicated by various types of *deactivation* and side reactions. In this connection it is important not to confuse the *quantum yield* with the ordinary *chemical yield*, which denotes the efficiency of molecular transformation without regard to the mechanism. This can perhaps best be illustrated by an example.

Suppose we consider the reaction sequence

$$A + h\nu \longrightarrow A^* \tag{42.12a}$$

$$A^* \begin{cases} \xrightarrow{99.9\%} A \\ \xrightarrow{0.1\%} B \end{cases} \tag{42.12b}$$

Equation (42.12a) denotes the primary photochemical process; the quantum yield for the formation of A* is unity, since for each light quantum absorbed, one molecule of A* is formed. For each 1000 molecules of A* formed, 999 are deactivated by some mechanism (perhaps by the reemission of the light quantum $h\nu$), and only one forms a molecule of B that we consider stable. The quantum yield of B is thus 0.001, since one molecule of B is formed for every 1000 light quanta absorbed. On the other hand, if we irradiate the sample with sufficient light for a long enough period, all the A will eventually be converted into B; the chemical yield is thus 100% even though the quantum yield is only 0.1%. Quantum yields for the formation of products in photochemical reactions range from zero to 10^6. Large values of Φ generally indicate chain reaction mechanisms. The chain reaction indicated in Equation (42.1) has a quantum yield of about 10^6 for the formation of HCl.

Suppose we consider the excitation of an atom A by a light quantum, which we represent by Equation (42.12a). The atom A* may be an ionized atom or it may be an atom in an excited electronic state. We shall be primarily concerned with excited atoms rather than ions. The excited atom has a lifetime of the order of 10^{-8} s, and several possible subsequent events are possible, as shown in Figure 42.4. In the absence of a collision with another atom, all or part of the absorbed energy can be reemitted. If the atom returns directly to its ground state, it emits a light quantum of the same frequency as the incident beam by a process known as *resonance fluorescence*,

$$A^* \longrightarrow A + h\nu \tag{42.13}$$

The emitted light is called *resonance radiation*. On the other hand, the atom may return to its ground state by a series of steps through several lower-energy excited states emitting light of a different frequency at each step:

FIGURE 42.4 Excitation and deactivation by radiative and radiationless transitions. The system S_0 is the lowest-lying singlet system; S_1 is an excited singlet system, and T_1 is an excited triplet system. The term A denotes absorption of a quantum $h\nu$, RF denotes resonance fluorescence with the emission of a light quantum of the same frequency, F denotes fluorescence with the emission of a light quantum of a different frequency, and P denotes phosphorescence. Radiationless transitions are indicated by squiggly arrows ($\rightsquigarrow$). The term IC indicates internal conversion, and ISC indicates intersystem crossing. For molecules, an additional mode of radiationless deexcitation is possible, the vibrational cascade indicated by V.

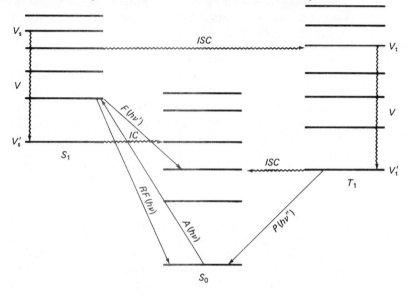

$$\begin{aligned} A^* &\longrightarrow A^{**} + h\nu' \\ A^{**} &\longrightarrow A^{***} + h\nu'' \\ &\vdots \\ A^{*\cdots*} &\longrightarrow A + h\nu''' \end{aligned} \quad (42.14)$$

This process is called *fluorescence*.

Referring to the energy-level diagram for mercury in Figure 32.8, we see that the excited levels of the various states constitute a series of overlapping energy-level systems. This allows an atom to cross over from one set of levels to another set in such a way that $\Delta E = 0$. This type of transition takes place with no emission of radiation and is known as a *radiationless transition*. We distinguish two types of radiationless transitions. If the transition is between two states of the same multiplicity (such as singlet to singlet), then the transition is an *internal conversion*. If the transition is between two states of different multiplicity (such as singlet to triplet), then the transition is *intersystem crossing*. Suppose the ground state is a singlet state that has been converted into a triplet state by intersystem crossing. It can then decay back to the singlet ground state with the emission of the proper radiation, but this is a spectroscopically forbidden transition. Operationally the result of this is that the transition from the triplet state to the singlet state takes place very slowly compared with fluorescent singlet-singlet transitions. The radiative triplet-singlet transition is known as *phosphorescence*. Fluorescent radiation terminates as soon as the incident radiation is removed, whereas phosphorescent radiation persists for varying times after the incident radiation is removed. Historically the two were distinguished by the difference in these time-lapse characteristics. The term *phosphorescence* is now reserved for the radiative process that occurs when a system undergoes a transition between states of different multiplicities. The term *fluorescence* is reserved for radiative transitions between states of the same multiplicities, that is, singlet-singlet or triplet-triplet.

Lifetimes of fluorescent states are extremely short, 10^{-8} to 10^{-9} s; they are rarely as long as 10^{-6} to 10^{-7} s. Lifetimes of these states can be estimated from the natural line width of the spectral lines using the Heisenberg uncertainty principle,

$$\Delta E \, \Delta t = \hbar \qquad (42.15)$$

The width of the line is a measure of ΔE, and Δt is taken as the lifetime. Several special techniques have been devised for measuring the decay rates and hence the lifetimes of fluorescent states. These often use nanosecond (10^{-9} s) pulses of light and very fast electronic circuitry to measure the decay of the induced fluorescence. Another approach is to use a continuous light source that is amplitude-modulated at a high rate (1–20 MHz). The fluorescent radiation lags the exciting radiation by a particular phase angle. The lifetime of the excited state can be determined by measuring this phase angle. The determination of the time constants for these very fast "reactions" is a specialty area of chemistry, one that has lately received increasing attention.

The lifetimes of phosphorescent states are much longer, generally longer than 10^{-3} s and in many cases as long as tens of seconds, hence the measurement of their decay rates and from these the lifetimes presents little difficulty. The incident light beam is turned on and the phosphorescent radiation gradually increases until some convenient level is reached, at which point the incident beam is removed. The decay of the intensity of the emitted phosphorescent radiation is then measured as a function of time. The decay follows the expression $I = I_0 \exp(-kt)$, where I_0 is the intensity at the instant the incident beam is turned off, which is taken as $t = 0$. For lifetimes in the millisecond range, the lifetime can be measured by using a microsecond flash lamp for the incident beam and an oscilloscope for displaying the intensity of the emitted radiation. For longer-lived systems, steady incident beams can be used, with classical camera shutter techniques. The ground states of most organic molecules are singlet states; hence the phosphorescent state is a triplet state that is amenable to study by electron paramagnetic resonance spectroscopy. That the phosphorescent state in organic molecules is an excited triplet state was first demonstrated unambiguously for single crystals of dilute solutions of naphthalene in the photochemically inert durene. In the absence of irradiating light, no epr signal is visible. When the light is turned on, an epr signal corresponding to $\Delta m = \pm 1$ is observed.[3] In experiments of this type, both the phosphorescent radiation and the photochemically induced paramagnetism decay exponentially. This is excellent evidence that both phenomena arise from the unimolecular process whereby the triplet state decays to the ground state rather than from slower recombinations involving electrons and free radicals. Long-lived phosphorescence is a strikingly beautiful phenomenon, which must be seen to be fully appreciated. If the crystals are irradiated with the proper radiation in a darkened room, a bright glow can be seen emanating from the crystal after the incident light is turned off. For some systems the intensity

---

[3] C. A. Hutchison, Jr., and B. W. Mangum, *J. Chem. Phys.* 29 (1958):952; 34 (1961):908.

of the phosphorescent light is sufficient for reading purposes, and the gradual decay of the intensity can be visually observed.

Fluorescence and phosphorescence must be clearly distinguished from processes such as Rayleigh and Raman scattering.[4] Fluorescence and phosphorescence are true photochemical processes in which a light quantum is *absorbed* and then *reemitted* in all directions by the excited atoms or molecules. In a sense, the energy content of the light quantum has been transformed into a different form of energy before it reappears as light energy. In scattering, on the other hand, there is no intermediate absorption of the light quantum. There is no transformation of energy; the light quantum retains its electromagnetic nature during the entire process.

The chief differences between the photochemical behavior of atoms and molecules have to do with the larger number of degrees of freedom of polyatomic molecules. The absorption of a light quantum by an atom can result only in excitation or ionization. Excitation can, of course, also take place in molecules and would be represented by the equation

$$M + h\nu \longrightarrow M^* \tag{42.16}$$

where M^* is an excited state of the molecule. Now, however, in addition to excited electronic states, we have a wide range of vibrational and rotational levels that are also accessible. The excited molecule can be deactivated by fluorescence or phosphorescence in the same manner as an atom. An additional mode of radiationless transition is possible, the *vibrational cascade*. This effect is also illustrated in Figure 42.4. A molecule in a ground state singlet is excited to a vibrational state in an excited singlet state; this is indicated by V_s. The molecule can return to the ground state by fluorescence. On the other hand, it can dissipate some of its excess vibrational energy and fall to a lower vibrational state V'_s in a radiationless manner. The vibrational levels in the singlet and triplet states are of comparable energy, and the molecule can move horizontally to an excited vibrational level in the ground singlet by internal conversion or to the triplet vibrational level V_t. The various possibilities are indicated in the figure where the radiationless transitions are indicated by squiggly arrows($\leadsto$).

Another possible primary photochemical process for molecules is dissociation, which can be represented as

$$M + h\nu \longrightarrow M' + M'' \tag{42.17}$$

where M' and M'' are two fragments. This possibility is absent in the case of atoms. Sometimes excitation and dissociation are combined in a single process known as *predissociation*:

$$M + h\nu \longrightarrow M^* \longrightarrow M' + M'' \tag{42.18}$$

In this process there is a very brief period during which the molecule exists as the excited M^* before it dissociates. Predissociation can be viewed as time-delayed ordinary dissociation. This is indicated in Figure 42.5.

---

[4] Raman scattering differs from ordinary, or Rayleigh, scattering in that a change of wavelength is produced in Raman scattering. In Rayleigh scattering, the scattered light has the same wavelength as the incident light. In Raman scattering, most of the scattered light has the same wavelength as the incident light, but some of the scattered light is shifted to higher and lower wavelengths in the form of bands on either side of the original wavelength. This *Raman effect* forms the basis for *Raman spectroscopy*, which is very useful in molecular spectroscopy. The separation between the lines in the Raman spectrum can be related to the vibrational and rotational transitions in the molecule. Raman spectroscopy is useful because the selection rules differ from those of ordinary infrared spectroscopy, and many transitions that are forbidden in ordinary absorption spectroscopy can be seen in Raman spectroscopy. See also Problem 24.

> JOHN WILLIAM STRUTT, third Baron Rayleigh (1842–1919), English physicist, is better known by the more familiar name Lord Rayleigh. His scientific career was varied, and Rayleigh was one of the luminaries of British science in the late 1800s. He published numerous works in acoustics and optics. His treatise *The Theory of Sound,* first published in 1877, is still used in many laboratories as a standard reference textbook. His derivation of the Rayleigh scattering law in 1871 explained the blue color of the sky. His most famous and dramatic work involved isolating argon. In determinations of the molar mass of nitrogen at that time, scientists had found that the density of nitrogen prepared from ammonia was less than the density of nitrogen prepared from air. Rayleigh noted that in 1795 Cavendish had oxidized nitrogen from air by passing a spark through the gas; he had found a small residue of gas that could not be oxidized and had abandoned the experiments. Rayleigh pushed Cavendish's experiment to its logical conclusion and discovered argon at about the same time that Ramsey discovered the gas by much simpler chemical methods. It was largely because of this experiment that Rayleigh received the Nobel Prize in physics in 1904, the same year that Ramsey received the Nobel Prize in chemistry for discovering argon.

In this section we have noted only those processes that take place in isolated atoms and molecules and result in the formation of unstable species that can revert to the ground state either by single-step deactivation or by a series of steps. All the processes do not necessarily take place in any given system; nor have we discussed all the possible ramifications of the varied types of processes. The detailed nature of any transformation is ultimately determined by the selection rules for the trans-

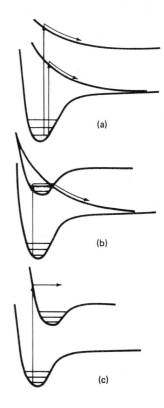

FIGURE 42.5 Photochemical dissociation of a diatomic molecule. (a) Direct excitation to a repulsive state. (b) Excitation to a bound upper state, followed by a crossover to a repulsive state (predissociation). (c) Excitation to a bound state above the dissociation limit.

formation and by the energy balance. When the unstable species reverts to its ground state, there is usually no "chemical" change induced in the system. Chemical changes take place when these excited species interact with other atoms and molecules.

42.4 BIMOLECULAR PROCESSES. THE TRANSFER OF ENERGY

The deactivation of an excited species by fluorescence as discussed in the previous section takes place in a very brief interval ($\sim 10^{-8}$ s) following the absorption of the exciting light quantum. The excited molecule can also be deactivated by transferring its energy to a second body if it interacts with the second body before fluorescence. This can be represented by the general equation

$$A + D^* \longrightarrow D + A^* \tag{42.19}$$

Here D^* represents the excited species produced by the original incident light quantum and A represents a second body, which may or may not be of the same type as D. The result is to deactivate D and form the excited species A^*. For obvious reasons (and in line with the nomenclature used in semiconductor work), the species D is called the *donor* and A is called the *acceptor*. This process, when viewed from the vantage point of removing the energy from D^*, is called *quenching;* the species A is the *quencher*. The activated species A^* can itself then undergo fluorescence, or it can be quenched by another body.

It may be that the object of an experiment is to produce A^* indirectly rather than by the direct irradiation of A. This will be the case if the direct production of A^* by irradiation of A is an inefficient process because of the relevant selection rules. When Equation (42.19) is viewed from the vantage point of forming A^* indirectly from D^*, the process is called *sensitization*. A process like that indicated in (42.19) can thus be viewed either as a quenching or as a sensitization reaction, depending on the objective of the experiment. A reaction with a molecule

$$D^* + M \longrightarrow D + M \tag{42.20}$$

in which the excess energy is simply converted into translational, vibrational, and rotational thermal energy can be viewed only as a quenching process.

Whether fluorescence or quenching is the main mechanism for deactivation is a function of the concentration of the fluorescent species and the quencher. In gases at low pressure, the time between collisions is long and the dominant mechanism will be fluorescence with very little quenching. At higher pressures, the time interval between collisions is reduced and appreciable quenching may result. Liquids are characterized by frequent collisions; hence the fluorescence of liquids is appreciably quenched. The degree of quenching depends in no small measure on the nature of the quenching species. Oxygen gas is very effective in quenching mercury fluorescence; hydrogen and carbon monoxide are less effective, and the inert gases helium and argon are inefficient quenchers at the same pressure. Some possible modes of deactivation of Hg^* ($^3P_1$ state) are shown in Table 42.2. The Hg^* is produced by irradiating ground state mercury ($^1S_0$) with 2537-Å light:

$$Hg(^1S_0) + h\nu(2537 \text{ Å}) \longrightarrow Hg(^3P_1) \tag{42.21}$$

TABLE 42.2 Deactivation of Hg*($^3P_1$).

	$\longrightarrow$	$Hg(^1S_0) + h\nu$ (2537 Å)	Resonance phosphorescence
	$\xrightarrow{M}$	$Hg(^1S_0) + M^*$ (or M)	Deactivation to the ground state
	$\xrightarrow{N_2}$	$Hg(^3P_0) + N_2^*$	Deactivation to a metastable state
$Hg^*(^3P_1)$	$\xrightarrow{Na\,(3^2P)}$	$Hg(^1S_0) + Na\,(9^2S)$ $\hookrightarrow Na\,(3^2P) + h\nu'$	Sensitized fluorescence
	$\xrightarrow{RH}$	$Hg(^1S_0) + RH^*$ $\hookrightarrow$ products	Quenching by exciting a molecule
	$\xrightarrow{RH}$	$Hg(^1S_0) + R + H$	Quenching with one-step dissociation
	$\xrightarrow{RH}$	$HgH + R \rightarrow Hg(^1S_0) + H + R$	Quenching with dissociation by an intermediate

NOTE: An extensive discussion of all these processes plus several additional ones can be found in J. G. Calvert and J. N. Pitts, Jr., *Photochemistry* (New York: John Wiley & Sons, 1966), p. 71 et. seq.

Another mechanism for energy transfer that does not depend on a bimolecular interaction can be represented by the sequence of equations

$$D + h\nu \longrightarrow D^* \qquad (42.22a)$$
$$D^* \longrightarrow D + h\nu' \qquad (42.22b)$$
$$A + h\nu' \longrightarrow A^* \qquad (42.22c)$$

Here light emitted by the donor by normal fluorescence or phosphorescence is absorbed by the acceptor. There is no collisional interaction between donor and acceptor, and the absorption by A of the light quantum emitted by D^* has no effect on the lifetime of D^*. The detailed nature of the collisional mechanism of energy transfer is complicated; it is discussed in several of the references listed in the Bibliography. The importance of quenching is that it provides for a mechanism whereby the energy of the incident light quantum can be used to produce reactive species that can undergo further chemical reactions.

42.5 KINETICS OF PHOTOCHEMICAL REACTIONS

A common class of photochemical reactions is that in which the initial step is the production of atoms and free radicals by the photochemical dissociation of a parent molecule:

$$M + h\nu \longrightarrow R_1 + R_2 \quad \text{or} \quad M + h\nu \longrightarrow M^* \longrightarrow R_1 + R_2$$

The species R_1 and R_2 are unstable atoms or free radicals that can initiate chain reactions. The energy of the incident light quantum must be at least equal to the dissociation energy of M. The paths by which dissociation takes place are indicated in Table 42.3. Dissociation either can be direct or can follow intersystem crossing. We need not concern ourselves with the mechanism by which the dissociation takes place; our concern will be the rate of the dissociation step, and this we shall take to be equal to

TABLE 42.3 Some primary photochemical molecular processes.

ABC + $h\nu \to$	AB· + C·	Dissociation into radicals
	AB$^+$ + C$^-$ or AB$^-$ + C$^+$	Dissociation into ions
	ABC$^+$ + e^-	Photoionization
	ABC*	Formation of activated molecule
	ACB	Intramolecular rearrangement

the rate of absorption of light quanta. Let us examine some photochemical reactions involving the hydrogen halides.

The decomposition of HI,

$$2\text{HI} \longrightarrow \text{H}_2 + \text{I}_2 \qquad (42.23)$$

proceeds by an elementary bimolecular mechanism in the absence of light. Initially the concentrations of H_2 and I_2 are both very small. Neglecting the reverse reaction, we can write the rate as a second-order reaction,

$$\frac{-d[\text{HI}]}{dt} = k[\text{HI}]^2 \qquad (42.24)$$

For the photochemical reaction, the initial step is the dissociation of HI into its component atoms. We can write the mechanism as the series of equations

$$\text{HI} + h\nu \longrightarrow \text{H} + \text{I} \quad \text{rate} = k_1 I_a \qquad (42.25a)$$
$$\text{H} + \text{HI} \longrightarrow \text{H}_2 + \text{I} \quad \text{rate} = k_2[\text{H}][\text{HI}] \qquad (42.25b)$$
$$\text{I} + \text{I} \longrightarrow \text{I}_2 \quad \text{rate} = k_3[\text{I}][\text{I}] \qquad (42.25c)$$

where I_a is the amount of light absorbed. Note that we have omitted the possible reaction $\text{I} + \text{HI} \to \text{I}_2 + \text{H}$. This reaction is energetically much less favored than reaction (42.25b) and would be expected to take place to a much smaller degree. Let us see what we get for our simple three-step mechanism. The rate of disappearance of HI is

$$-\frac{d[\text{HI}]}{dt} = k_1 I_a + k_2[\text{H}][\text{HI}] \qquad (42.26)$$

In the steady state, we require that

$$\frac{d[\text{H}]}{dt} = k_1 I_a - k_2[\text{H}][\text{HI}] = 0 \qquad (42.27)$$

Combining (42.26) and (42.27), we get

$$-\frac{d[\text{HI}]}{dt} = 2k_1 I_a \qquad (42.28)$$

The proportionality constant k_1 reflects the possibility that (42.25a) might not be an exclusive pathway. If all the excited HI dissociates into H and I, then k_1 is unity. If some portion of the excited HI is deactivated by a different mechanism, say fluorescence, then k_1 is less than unity. From the definition of the quantum yield,

$$\Phi = \frac{-d[\text{HI}]/dt}{I_a} = 2k_1 \qquad (42.29)$$

If k_1 is unity, then $\Phi = 2$. This value has been observed under certain experimental conditions.

The mechanism for the ordinary reaction $H_2 + Br_2 \to 2HBr$ has already been given (Chapter 40). The mechanism for the photochemical reaction is similar except that the first step is replaced by

$$Br_2 + h\nu \longrightarrow 2Br \tag{42.30}$$

By proceeding as in the earlier discussion, we find that the steady-state approximation yields

$$\frac{d[Br]}{dt} = 2I_a - k_2[Br][H_2] + k_3[H][Br_2] + k_4[H][HBr] - 2k_5[Br]^2 = 0 \tag{42.31}$$

The photochemical rate of production of HBr is then

$$\frac{d[HBr]}{dt} = \frac{k_2(2/k_5)^{1/2}[H_2]I_a^{1/2}}{1 + k_4[HBr]/k_3[Br_2]} \tag{42.32}$$

which is of the same form as the dark reaction rate (Eq. 40.32). The photochemical reaction is about 300 times faster than the dark reaction, which indicates that the steady-state concentration of Br atoms is 300 times larger in the photochemical reaction—since all the other steps are identical. The quantum yield is inversely proportional to the square root of the intensity,

$$\Phi = \frac{d[HBr]/dt}{I_a} = \frac{2k_2(2/k_5)^{1/2}[H_2]}{I_a^{1/2}\{1 + (k_4[HBr]/k_3[Br_2])\}} \tag{42.33}$$

In contradistinction to the HI decomposition, the efficiency of this reaction decreases with increasing intensity. This is because a greater concentration of the Br atoms formed are converted to Br_2 at higher intensities.

The quantum yield for the photochemical production of HBr is less than unity. At the other extreme is the formation of HCl (see beginning of this chapter). As many as 10^6 molecules of HCl have been observed for each light quantum absorbed. Under certain conditions the rate of formation of HCl has been found to be given by the expression

$$\frac{d[HCl]}{dt} = kI_a[H_2] \tag{42.34}$$

This expression can be accounted for by the three equations given in Equations (42.1a) through (42.1c) with the addition of a chain-terminating step,

$$2Cl \longrightarrow Cl_2 \qquad \text{rate} = k_4 \tag{42.1d}$$

When atoms or free radicals react to form molecules, a lot of energy is evolved, and unless this energy is removed, the resulting molecule will be unstable. Thus for reactions such as $2Cl \to Cl_2$, it is necessary to postulate some third agent, either a third body in the form of another molecule or the walls of the vessel, which can carry away some of this excess energy as thermal energy.

Our discussion of just three photochemical reactions does not exhaust the field. It is possible to find discussions of hundreds of different reactions in the textbooks listed in the Bibliography. Some of these reactions will be discussed in the problems at the end of this chapter. We shall, however, briefly examine some of the reactions involved in air pollution, and the photochemical reaction that supports all life on this planet, photosynthesis.

42.6 ATMOSPHERIC CHEMISTRY. FATE OF SO$_2$ IN THE ATMOSPHERE

One of the most perplexing problems facing those concerned with atmospheric pollution is the mechanism and the rate by which the SO$_2$ that is emitted by coal-fired power plants is converted into sulfate. During the 1960s, SO$_2$ received most of the attention in air pollution, and it is still SO$_2$ that is, in the minds of most laymen, the villain. As our experience and understanding of the health effects associated with atmospheric pollutants increased, it soon became apparent that SO$_2$ alone did not constitute a severe health hazard at the low-level concentrations normally encountered. It is now felt that the particulate sulfates formed by oxidation of SO$_2$ pose a more severe hazard, particularly in concert with other pollutants (the so-called *synergistic* effect). Unfortunately particulate sulfates constitute a very difficult experimental problem. Most readers of this textbook should have the competence conceptually to design an experiment whereby the health effects of SO$_2$ could be measured on animals. We could get a large animal chamber and design a flow system, using a tank of pure SO$_2$ gas so that the air in the chamber would contain 1, 5, 50 or 100 ppm (parts per million) or any other concentration of SO$_2$. We may not have the toxicological training to analyze the results of such an experiment, but we could at least design the black box. Further, the analysis of SO$_2$ is relatively straightforward.

An experiment designed to measure the effects of particulate sulfates in the micrometer-sized region poses a much more difficult problem. How do we prepare an atmosphere containing 3 ppm of H$_2$SO$_4$ aerosols or some other particulate sulfate with a size distribution in the micrometer region? Further, even the analytical problems pose a sizable block. Analytical techniques whereby gas is passed through a filter and the particulates are scraped off the filter and analyzed suffer from many defects; yet this is the only technique generally available. (This last comment may no longer be true by the time this appears in print.)

Several mechanisms have been proposed for the reaction

$$SO_2 \longrightarrow SO_4 \qquad (42.35)$$

where the SO$_4$ in the unbalanced equation (42.35) denotes particulate sulfate. We shall assume that the SO$_4$ is in the form of H$_2$SO$_4$, since that can further react with other particles in the atmosphere to produce other kinds of sulfates. There exists evidence for a three-body reaction with oxygen atoms,

$$SO_2 + O + M \longrightarrow SO_3 + M \qquad (42.36)$$

to form SO$_3$, the acid anhydride of H$_2$SO$_4$, which then forms sulfuric acid by reacting with water, SO$_3$ + H$_2$O → H$_2$SO$_4$.[5] There also exists evidence for a free radical mechanism, which for OH would follow the equation

$$SO_2 + OH + M \longrightarrow HSO_3 + M \qquad (42.37)$$

The resulting HSO$_3$ then reacts with O$_2$, hydrocarbons, NO, or some other atmospheric species to form H$_2$SO$_4$. Evidence also exists for a catalytic oxidation process whereby SO$_2$ is oxidized in the presence of a catalyst,

$$SO_2 + \tfrac{1}{2}O_2 \longrightarrow SO_3 \qquad (42.38)$$

or is first converted to H$_2$SO$_3$ and then oxidized,

---

[5] A review of atmospheric reactions appears in P. F. Fennelly, *J. Air Pollution Control Assn.* 25 (1975):697.

$$SO_2 + H_2O \longrightarrow H_2SO_3 \quad (42.39a)$$
$$H_2SO_3 + \tfrac{1}{2}O_2 \longrightarrow H_2SO_4 \quad (42.39b)$$

Reaction (42.38) forms the basis of the *contact process* for manufacturing sulfuric acid. Platinum, iron oxide, and in particular, vanadium oxide have been successfully used as catalysts in commercial operations.[6] The latter two are often found in emissions from fossil-fuel-fired steam boilers. At a temperature of 600 K, $\Delta H = -97.9$ kJ mol^{-1} and $\Delta G = -41.6$ kJ mol^{-1} for the reaction. At that temperature, the equilibrium constant favors the formation of SO_3:

$$K_p = \frac{(P_{SO_3})}{(P_{SO_2})(P_{O_2})^{1/2}} = 4180 \text{ atm}^{-1/2} \quad (42.40)$$

Since ΔH is negative, the equilibrium constant increases with decreasing temperature and would be much higher at room temperature. Finally, there exists evidence for a photochemical oxidation that might proceed by the mechanism

$$SO_2 + h\nu \longrightarrow SO_2^* \quad (42.41a)$$
$$SO_2^* + O_2 + M \longrightarrow SO_4 + M \quad (42.41b)$$
$$SO_4 + SO_2 \longrightarrow 2SO_3 \quad (42.41c)$$

Which of these four possibilities is the "correct" one has not yet been established. Perhaps the phrase *more correct* is more appropriate, since all the reactions could be expected to proceed simultaneously and the relative importance of any one of them will depend on the particular composition of the atmospheric "soup," temperature, sunlight, and other factors. The situation is further complicated by the difficulty of extrapolating laboratory experiments to what exists in the atmosphere. Laboratory studies are conducted in "smog chambers" in which the purity of the atmosphere is carefully controlled, a situation that does not exist in the urban atmosphere.

The situation in the air is complicated by the wealth of reactions that take place between *primary* and also *secondary* pollutants.[7] Sixty percent of the nitrogen oxides in Los Angeles are produced by the automobile, an agent that helps to diffuse this pollutant over a large area, since it is a mobile rather than a stationary source. The following reactions occur in the NO_x photochemical cycle.

$$NO_2 + h\nu \longrightarrow NO + O \quad (42.42)$$
$$O + O_2 + M \longrightarrow O_3 + M \quad (42.43)$$
$$O_3 + NO \longrightarrow O_2 + NO_2 \quad (42.44)$$

---

[6] Some recent findings give rise to the possibility of drawing an interesting inference about the production of sulfates from SO_2. In R. J. Cheng, V. A. Mohnen, T. T. Shen, M. Current, and J. B. Hudson, *J. Air Pollution Control Assn.* 26 (1976):787, some differences between the particulates produced by coal-fired furnaces and oil-fired are presented. Particulates from oil are generally much smaller than coal-derived particulates, and also they have much smoother surfaces. The oil particulates also have high vanadium concentrations, whereas vanadium seems absent in coal-derived particulates. These observations provide inferential support for the catalytic oxidation mechanism when they are taken in the context of one of the anomalies of air-pollution studies. During the years 1967–75, while the SO_2 concentrations in urban areas went steadily downward (a decrease of 55% in the East Coast region between 1963 and 1971), the particulate sulfates remained essentially constant and in some instances even increased. Most of this decrease came about because many plants converted from high-sulfur coal to low-sulfur oil. The above study suggests that this anomaly might be due to the higher concentration of oil-derived particulates. The high surface area of these particulates combined with their high vanadium content might present a excellent set of conditions for the catalyzed oxidation of SO_2 to SO_3. Much additional work needs to be done on this subject before any firm conclusions can be reached.

[7] A *primary* pollutant is one that is directly produced by the source; a *secondary* pollutant is formed by later reactions. Thus SO_2, which is produced by burning high-sulfur coal, is a primary pollutant, and SO_4, which is produced by later oxidation of the SO_2, is a secondary pollutant.

Thus an equilibrium is established

$$NO_2 + O_2 \underset{}{\overset{h\nu}{\rightleftharpoons}} NO + O_3 \quad (42.45)$$

in which a reservoir of powerful oxidizing agents for further reaction is maintained. The ozone that is produced in Equation (42.43) can oxidize SO_2 to SO_3. This reaction is slow, but the presence of hydrocarbons in the atmosphere greatly increases the oxidation rate. Similarly, the rate at which SO_2 is oxidized by a combination of NO_x and hydrocarbons is much higher than the rate of oxidation by NO_x alone. These synergistic effects might be expected to increase the relative importance of the free radical reactions when hydrocarbon concentrations are high. The experimental difficulties associated with unraveling these reaction chains are due in no small measure to the low concentrations. Pollutant levels are generally found at concentrations lower than 50 ppm and in many instances in the 1-ppm range.

The importance of atmospheric studies lies not so much in the interesting chemistry but in the toxicological effects. Hydrocarbons, SO_x, NO_x, ozone, heavy metals, and numerous other substances may be injurious to plants and animals, not to mention the aesthetics of urban life. Of particular concern is the effects of these species on humans at low concentration levels for long periods. Much of the present controversy about industrial wastes is intimately connected to our present lack of knowledge regarding these effects.

42.7 PHOTOSYNTHESIS. ECOLOGICAL FOOD CHAINS AND THE DEGRADATION OF STORED ENERGY

It would be remiss, in a chapter labeled "Photochemistry," to omit any reference to *photosynthesis,* since that is the most important of all photochemical reactions that take place on Earth. Photosynthesis provides the mechanism whereby the energy in sunlight is used for producing the basic substances of living matter; thus through photosynthesis, the sun is the ultimate source of energy for all living matter. The subject is amazingly complex and incompletely understood. A complete consideration requires a thorough understanding of basic biochemical principles and is beyond our scope here. We shall restrict ourselves to some simple and brief statements.[8]

The equation for photosynthesis can be written as

$$6CO_2 + 6H_2O + nh\nu \longrightarrow C_6H_{12}O_6 + 6O_2 \qquad \Delta G° = +2870 \text{ kJ mol}^{-1} \quad (42.46)$$

Carbohydrates are formed from CO_2 and H_2O by the photochemical reaction involving light in the visible region of the spectrum. Note that the reaction has a positive free energy change associated with it. The term *photosynthesis* is usually applied to the reaction described in (42.46) even though that equation is highly simplified. The overall reaction takes place in many steps, generally grouped into two parts. In the first part, the *light reactions,* the energy content of the light quanta is converted into chemical energy. The remaining reactions, in which glucose is formed at the expense of chemical energy, are called the *dark reactions.* The light reactions can occur only in the presence of photoactive cells like the ones that contain chloro-

---

[8] Complete discussions may be found in most textbooks dealing with biochemistry, cell biology, biophysics, and the like. See, for example, R. C. Bohinski, *Modern Concepts in Biochemistry* (Boston: Allyn and Bacon, 1973).

phyll found in green plants, whereas the dark reactions can occur in a wide variety of cells. From calculations of incident sunlight and fixed carbon output, it has been estimated that only about 1 or 2% of the total incident energy is recovered by a field of corn during the growing season. Much higher efficiencies can be achieved under laboratory conditions. The quantum yield per molecule of CO_2 used is estimated at about $\Phi = 0.125$, or 8 quanta per CO_2 molecule; this represents a total of 48 quanta per glucose molecule formed. An einstein of red light (7000 Å) represents some 167.4 kJ, or $48 \times 167.4 = 8035$ kJ of light used per mole of glucose formed. This represents an overall efficiency of 36% for the conversion of light energy into chemical energy by photosynthesis under the most ideal conditions.

The production of carbohydrates by green plants forms the base upon which various food chains are built. Food chains have received much attention in recent years in ecological circles, and the human food chain is receiving increasing attention as a result of the pressures applied by expanding populations on food production capacities. In the seas, for example, chemical energy is produced by algae through the photosynthetic reaction. Algae are eaten by small fish, which are in turn eaten by larger fish, and so on up to the largest animal in the ocean. An important consideration in this chain is that at each level only about 10% of the energy content of the chain below is available for the chain above. In the natural state, the total mass of any food in the chain is much smaller than the mass in the chain below. A forest contains much more plant life than rabbits, and only a small number of mountain lions. If the forest uses sunlight at an overall efficiency of 1%, then the rabbits effectively use sunlight at an efficiency of 0.1%; and if the mountain lions eat rabbits, they use the sunlight at 0.01% efficiency. It is this principle of degradation of stored energy that makes beef a much less efficient (and more expensive) source of food than vegetables. The quantity of corn fed to a steer to provide enough meat to sustain one person could sustain several people if the steer were eliminated and the corn consumed directly by the people. Certain vital nutrients, however, that are available from beef are not available from corn.

42.8 THE QUANTUM MECHANICS OF LIGHT ABSORPTION. TIME-DEPENDENT PERTURBATION THEORY

In our discussion of quantum mechanics (Part V) mention was often made of *selection rules*, which define those spectroscopic transitions that are allowed to take place. In examining these rules quantitatively, we must consider the effect of a time-dependent electromagnetic field on atoms and molecules. The method used is therefore called *time-dependent perturbation theory*.

Suppose we have an unperturbed atom or molecule whose wave equation including the time can be written

$$\hat{H}^0 \Psi^0 = i\hbar \frac{\partial \Psi^0}{\partial t} \tag{42.47}$$

where $\hat{H}^0$ is the Hamiltonian operator of the unperturbed system. A general normalized solution to this equation can be written as

$$\Psi^0 = \sum_{n=0}^{\infty} a_n \Psi_n^0 \tag{42.48}$$

where the a_n's are constants and $\Sigma_n a_n^* a_n = 1$. The Ψ_n^0's are the time-dependent wave functions for the energy eigenvalues E_n^0

$$\Psi_n^0(q, t) = \psi_n^0(q) \exp\left(\frac{iE_n^0 t}{\hbar}\right) \qquad (42.49)$$

In the presence of a time-dependent perturbing field, Equation (42.47) becomes

$$(\hat{H}^0 + \hat{H}')\Psi = i\hbar \frac{\partial \Psi}{\partial t} \qquad (42.50)$$

where $\hat{H}'(q, t)$ is the perturbation. Since the Ψ_n^0 form a complete orthonormal set, we can write $\Psi(q, t)$ as a series in $\Psi_n^0(q, t)$,

$$\Psi(q, t) = \sum_n a_n(t) \Psi_n^0(q, t) \qquad (42.51)$$

where the $a_n(t)$ are functions of the time only. If we now substitute (42.51) in (42.50), we get

$$\Sigma a_n(t)\hat{H}^0 \Psi_n^0 + \Sigma a_n(t)\hat{H}' \Psi_n^0 = -i\hbar \Sigma \dot{a}_n(t)\Psi_n^0 - i\hbar \Sigma a_n(t)\frac{\partial \Psi_n^0}{\partial t} \qquad (42.52)$$

where $\dot{a}_n$ is the time derivative of a_n. The first and last terms of this equation cancel from (42.47) and (42.48), and we are left with

$$-i\hbar \Sigma \dot{a}_n(t)\Psi_n^0 = \Sigma a_n(t)\hat{H}' \Psi_n^0 \qquad (42.53)$$

Suppose we now multiply both sides of (42.53) by Ψ_m^{0*} and integrate. Since the functions are orthogonal, all the terms on the left vanish except the one for which $m = n$, and we are left with

$$\dot{a}_m(t) = -\frac{i}{\hbar} \Sigma a_n(t) \int \Psi_m^{0*} \hat{H}' \Psi_n^0 \, dq \qquad (42.54)$$

Equation (42.54) provides a set of simultaneous differential equations and we are thereby enabled to solve for the $\dot{a}_n(t)$.

We can see how this equation operates by considering a system in a particular unperturbed state Ψ_k^0 at $t = 0$. Suppose that a very small perturbation $\hat{H}'$ acts on the system for a very short interval t' in such a way that $\hat{H}'$ can be considered time-independent during the interval. Since $\hat{H}'$ is small, we can neglect all the terms on the right side of (42.54) except that term for which $n = k$. We get

$$\frac{da_m(t)}{dt} = -\frac{i}{\hbar} \int \Psi_n^{0*} \hat{H}' \Psi_k^0 \, d\tau \qquad (42.55)$$

From (42.49) this can be written as

$$\frac{da_m(t)}{dt} = -\frac{i}{\hbar} H'_{mk} \exp\left[-\frac{i}{\hbar}(E_k - E_m)t\right] \qquad 0 \le t \le t'; m \ne k \qquad (42.56)$$

In (42.56) we have introduced the notation

$$H'_{mk} \equiv \int \psi_m^{0*} \hat{H}' \psi_k^0 \, d\tau \qquad (42.57)$$

Remember now that the initial state was labeled k, so that $a_k(0) = 1$ and all other a_n's are neglected. Equation (42.56) then describes the rate at which a system initially in the kth state undergoes a transition to the mth state. The differential equation can be integrated between the limits $t = 0$ and $t = t'$, to provide a solution for $a_m(t')$:

$$a_m(t') = H'_{mk} \left\{ \frac{1 - \exp[(i/\hbar)(E_m - E_k)t']}{E_m - E_k} \right\} \qquad (m \neq k) \qquad (42.58)$$

In deriving Equation (42.58), we used the fact that $a_m(0) = 0$ for $m \neq k$.

42.9 EINSTEIN TRANSITION PROBABILITIES

In 1916, Einstein presented a model of emission and absorption of electromagnetic radiation that extended the classical results to quantized systems of atoms and molecules. Suppose we have two nondegenerate eigenstates of a system with energies E_m and E_n, so designated that E_m is higher than E_n, as shown in Figure 42.6. The number of molecules in the upper state is denoted by N_m and the number in the lower state by N_n. The Bohr frequency condition for the system is

$$h\nu_{nm} = E_m - E_n \qquad (42.59)$$

Suppose the system is now immersed in a bath of electromagnetic radiation having a density denoted by $\rho(\nu)$. That is to say, the density of the radiation between the frequencies ν and $\nu + d\nu$ is given by $\rho(\nu)\,d\nu$.

According to Einstein's model, there are three pathways by which a molecule (or atom) can undergo a transition from one of the states to the other. There is only one way that the molecule can absorb radiation to go from the lower state to the upper. The probability of absorbing a quantum and undergoing a transition $n \to m$ per unit time is

$$\text{Prob}(n \longrightarrow m) = B_{nm}\rho(\nu_{nm}) \qquad (42.60)$$

where B_{nm} is the *Einstein coefficient of absorption*. If the number of molecules is large, the number undergoing the transition is obtained by multiplying (42.60) by N_n.

A molecule that is in the upper state can undergo a transition to the lower state by either of two mechanisms. It can fall to the lower state by *stimulated*, or *induced*, emission under the influence of the electromagnetic field, or it can fall to the lower state by *spontaneous* emission. Spontaneous emission is the mode by which it would fall to the lower state without any electromagnetic field. We take the probability that a molecule in the upper state will fall to the lower state with the emission of a quantum $h\nu_{nm}$ to be

$$\text{Prob}(m \longrightarrow n) = A_{mn} + B_{mn}\rho(\nu_{nm}) \qquad (42.61)$$

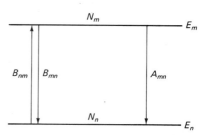

FIGURE 42.6 Transitions between two nondegenerate states whose energy separation is $E_m - E_n = h\nu_{nm}$.

per unit time, where A_{mn} is *Einstein's coefficient of spontaneous emission* and B_{mn} is *Einstein's coefficient of stimulated emission*. The coefficient of absorption equals the coefficient of stimulated emission, $B_{nm} = B_{mn}$. Further, it is generally true that $B_{mn} > A_{mn}$; thus a molecule in the upper state is more likely to drop to the lower state by stimulated emission than by spontaneous emission.

Now suppose that we have many identical systems in equilibrium with a bath of electromagnetic radiation at temperature T. The density of radiation is given by Planck's radiation law

$$\rho(\nu) = \frac{8\pi h \nu^3}{c^3} \left[\frac{1}{\exp(h\nu/kT) - 1} \right] \tag{42.62}$$

The number of molecules in the upper and lower states at equilibrium can be related by the Boltzmann distribution

$$\frac{N_m}{N_n} = \exp\left(\frac{-h\nu_{nm}}{kT}\right) \tag{42.63}$$

Further, in the equilibrium steady state, the number of molecules that go from n to m in unit time must equal the number of molecules that go from m to n:

$$N_n B_{nm} \rho(\nu_{nm}) = N_m [A_{mn} + B_{mn} \rho(\nu_{nm})] \tag{42.64}$$

Combining the previous two equations yields

$$\rho(\nu_{nm}) = \frac{A_{mn}}{B_{nm} \exp(h\nu_{nm}/kT) - B_{mn}} \tag{42.65}$$

From (42.62), noting that $B_{nm} = B_{mn}$, we get the result

$$A_{mn} = \frac{8\pi h \nu_{nm}^3}{c^3} B_{mn} \tag{42.66}$$

Now let us see how the Einstein coefficients can be related to the material in the previous section. Suppose we consider the effect of an electromagnetic field polarized in the x direction. The perturbation energy is

$$\hat{H}' = E_x \sum_j e_j x_j \tag{42.67}$$

where the summation is over all the electrons and x_j is the coordinate of the jth electron. The field strength E_x is

$$E_x = E_x^0 [\exp(2\pi i \nu t) + \exp(-2\pi i \nu t)] \tag{42.68}$$

where ν is the frequency and E_x^0 is the amplitude. If we set $a_m(0) = 0$ and $a_n(0) = 1$ on the right side of (42.54), we get, after several arithmetic steps,

$$\frac{da_m(t)}{dt} = -\frac{i}{\hbar} \mu_{x_{mn}} E_x^0 \left\{ \exp\left[\frac{i}{\hbar}(E_m - E_n + h\nu)t\right] + \exp\left[\frac{i}{\hbar}(E_m - E_n - h\nu)t\right] \right\} \tag{42.69}$$

where the symbol $\mu_{x_{mn}}$ represents the integral over the electronic coordinates,

$$\mu_{x_{mn}} = \int \psi_m^{0*} \sum_j e_j x_j \psi_n^0 \, d\tau = \int \psi_m^{0*} \mu_x \psi_n^0 \, d\tau \tag{42.70}$$

SEC. 42.9 EINSTEIN TRANSITION PROBABILITIES

In Equation (42.70), $\mu_x = \Sigma_j e_j x_j$ is the x component of the dipole moment of the molecule; $\mu_{x_{mn}}$ is the *transition moment*. Equation (42.69) can be integrated to yield

$$a_m(t) = \mu_{x_{mn}} E_x^0$$
$$\times \left\{ \frac{1 - \exp[(i/\hbar)(E_m - E_n + h\nu)t]}{E_m - E_n + h\nu} + \frac{1 - \exp[(i/\hbar)(E_m - E_n - h\nu)t]}{E_m - E_n - h\nu} \right\} \quad (42.71)$$

In Equation (42.71), the coefficient $\mu_{x_{mn}} E_x^0$ is always very small, and the absolute magnitudes of the numerators of the two terms in the braces vary between 0 and 2. Thus $a_m(t)$ is very small unless one of the denominators is also very small; this occurs only when $h\nu = h\nu_{nm} = E_m - E_n$. This is the *resonance* condition, which indicates that only frequencies near the resonance frequencies are efficient in inducing transitions. The first term of Equation (42.71) is important for induced emission (since $E_m - E_n$ is negative), and the second term is important for absorption. By suitably manipulating (42.71), we can get

$$a_m^*(t) a_m(t) = \frac{4\pi^2}{h^2} (\mu_{x_{mn}})^2 (E_x^0)^2 t \quad (42.72)$$

which shows that the probability of a transition to state m is proportional to t.[9] The coefficient of t is then the transition probability.

This result can be connected with the Einstein transition coefficient by recalling the result of classical electromagnetic theory

$$\rho(\nu) = \frac{6}{4\pi} (E_x^0)^2 (\nu) \quad (42.73)$$

from which we get for the x polarized light

$$(B_{nm})_x = \frac{8\pi^3}{3h^2} (\mu_{x_{mn}})^2 \quad (42.74)$$

Since the expressions for the y and z directions are similar, the final result is

$$B_{nm} = \frac{8\pi^3}{3h^2} \{ (\mu_{x_{mn}})^2 + (\mu_{y_{mn}})^2 + (\mu_{z_{mn}})^2 \} \quad (42.75)$$

The expression for B_{mn} is identical, and you can see this by carrying out the same procedure with $a_n(0) = 0$ and $a_m(0) = 1$ for the case of induced emission.

Equation (42.75) is the basic equation used for determining selection rules in spectroscopy. A *forbidden* transition is one in which all three transition moments $\mu_{q_{mn}}$ are zero. If one or more of the transition moments is nonzero, then the transition is *allowed*. If all three moments are nonzero, then the transition can occur with light of any arbitrary polarization. If only one is nonzero, then a transition can take place only with light whose polarization is in the direction of the nonzero moment. Note very carefully, however, that the above statements apply only to ordinary spectroscopy in which the interaction between the electromagnetic field and the molecules is dipolar. In certain types of spectroscopy such as Raman and electron-impact spectroscopy, different kinds of interactions are involved. Certain transitions that may be forbidden in ordinary spectroscopy may very well be allowed in other types of spectroscopy. This feature is what makes other types of spectroscopy so attractive,

---

[9] See, for example, L. Pauling and E. B. Wilson, Jr., *Introduction to Quantum Mechanics* (New York: McGraw-Hill Book Co., 1935), p. 299 et. seq.

since transitions that are not accessible to ordinary techniques of uv and ir spectroscopy may be accessible to Raman or electron-impact spectroscopy. Another feature of (42.75) is that it is an approximate equation, and transitions that may seem to be forbidden by (42.75) may in some instances appear on a spectrum. These "forbidden" lines are usually much less intense than the allowed lines. Some applications of (42.75) to spectroscopic transitions will be discussed further in the problems.

42.10 LASERS

Now let's consider the transitions in a particular type of three-level energy-level diagram shown in Figure 42.7. The ground state is labeled n and the highest state is labeled n'; a third state, labeled m, lies between n and n'. Suppose the system is such that if the state n' is populated by irradiation with light of frequency $\nu_{nn'}$, then the state m can be indirectly populated by radiationless transitions from n'. Since there is no light of frequency ν_{nm} incident on the system, the molecules in state m can fall to the ground level only by spontaneous emission, a much slower process than stimulated emission. If A_{mn} is small enough and the incident radiation $h\nu_{nn'}$ sufficiently intense, then an *inverted* population can be established in which $N_m > N_n$. A system with these properties is amenable to *laser* (*l*ight *a*mplification by *s*timulated *e*mission of *r*adiation) action.

Now suppose that the above system is placed in a cylinder whose opposite ends are highly polished parallel mirrors spaced an integral number of wavelengths λ_{nm} apart. A small amount of radiation $h\nu_{nm}$ along the axis of the cylinder is provided by the spontaneous emission of the molecules in state m. This induces further emission of radiation of the same frequency as the light is reflected back and forth along the axis of the cylinder. The result is a highly coherent wave of light that has a very narrow spread in wavelengths. If one of the end mirrors is partially silvered, some of this light escapes the cylinder and is available for a variety of experiments requiring intense monochromatic beams of coherent light. Since the beam is coherent, it diverges only slightly; laser beams have been reflected from the surface of the moon and detected on earth. They can be focused down to an area whose dimension is comparable to the wavelength, so that enormous energy densities are attainable. In this way, lasers have been successfully used in microsurgery.

Lasers are now available in a variety of configurations operating in the microwave, infrared, visible, and ultraviolet regions of the spectrum, and they have diverse applications. Surveyors find them useful for accurately laying out straight lines for bridge, tunnel, and road construction. Chemists find them useful in many ways. They are now almost exclusively used as sources in Raman spectroscopy. Their ability to provide intense and very short bursts (megawatt intensities of nanosecond duration) make them very useful in photochemical experiments dealing with very fast reactions.

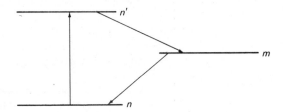

FIGURE 42.7 A three-level energy-level diagram.

The advent of the laser has moved the forefront of chemical kinetics and photochemistry into the picosecond (10^{-12} s) region since laser pulses with a duration of several picoseconds were first observed in 1966.[10] Many of these "fast reaction" studies involve determining the recombination times for atoms to form molecules. One of these studies devolved about the determining of the recombination time of iodine atoms. When iodine molecules in carbon tetrachloride are irradiated with light, a pair of iodine atoms is formed either by direct dissociation or by collision-induced predissociation

$$I_2 + h\nu \longrightarrow 2I$$

In one "fast" experiment a light pulse of 5-ps duration was used to excite the I_2 molecules.[11] A weaker pulse of the same wavelength was used to monitor the time dependence of the formation of I_2 molecules from the iodine atoms. It was found that the recombination of the original partners occurs with the time constant 140 ps. In a related experiment, the collision-induced predissociation of the excited I_2^* molecules was directly observed. The excited I_2^* molecules were found to dissociate into iodine atoms with a rate constant of 10^{-11} s^{-1}.

PROBLEMS

1. Blue light has a typical wavelength of 4700 Å. Calculate the energy in an einstein of blue light.
2. A spectrophotometer has the general appearance of the apparatus shown in Figure 42.3. The instrument is adjusted so that when the cell is empty, the meter indicates 100.0 for a source that is a monochromatic red beam. The cell is 2 cm long. When the cell is filled with $0.01M$ MnO_4^-, the meter reads 78.7. What would be the meter reading for a cell 5 cm long?
3. A 1.000-g sample of manganese steel is dissolved in acid, and the manganese is oxidized to MnO_4^-. The resulting solution is made up to 100.0 ml in a volumetric flask, and some of the solution is used to fill the 2-cm cell described in Problem 2. The meter reading is 91.6. What is the manganese content of the steel?
4. A phosphorescent crystal was illuminated with a strong light. At $t = 0$, the light beam was turned off and the intensity of the phosphorescent radiation was measured in intervals of milliseconds. The results were as shown in the accompanying table. What is the intensity of the phosphorescent radiation at $t = 0$? What is the lifetime of the phosphorescent state?

t (ms)	Intensity (arbitrary units)
1	65.0
2	43.6
3	29.2
4	19.6
5	13.1

5. Consider the dissociation reaction $A_2 \rightarrow 2A$ that proceeds by the photochemical mechanism

$$A_2 + h\nu \longrightarrow A_2^* \quad \text{rate} = k_1 I_a$$
$$A_2^* \longrightarrow 2A \quad \text{rate} = k_2[A_2^*]$$
$$A_2^* + A_2 \longrightarrow 2A_2 \quad \text{rate} = k_3[A_2^*][A_2]$$

where I_a is the amount of radiation absorbed. In the second two equations, the excited molecule A_2^* either dissociates or is deactivated by collision with a molecule of A_2. Using the steady-state approximation $d[A_2^*]/dt = 0$, show that the rate of formation of A is given by

$$\frac{d[A]}{dt} = \frac{2k_1 k_2 I_a}{k_2 + k_3[A_2]}$$

Derive an expression for the photochemical efficiency of the formation of A. (Note that one molecule of A_2 produces two molecules of A.)

6. The rate of formation of CCl_4 by the photochemical chlorination of $CHCl_3$ in the gas phase

$$Cl_2 + CHCl_3 + h\nu \rightarrow CCl_4 + HCl \text{ is given by}$$

$$\frac{d[CCl_4]}{dt} = k[Cl_2]^{1/2} I_a^{1/2}$$

Show that this rate equation can be accounted for by the mechanism

---

[10] A. J. DeMaria, D. A. Stetser, and H. Heynau, *Appl. Phys. Lett.* **8** (1966):174.
[11] T. J. Chung, G. W. Hoffman, and K. B. Eisenthal, *Chem. Phys. Lett.* **25** (1974):201.

$$Cl_2 + h\nu \longrightarrow 2Cl$$
$$Cl + CHCl_3 \longrightarrow CCl_3 + HCl$$
$$CCl_3 + Cl_2 \longrightarrow CCl_4 + Cl$$
$$2CCl_3 + Cl_2 \longrightarrow 2CCl_4$$

7. In a photochemical rate equation like that of Problem 6, the time is usually measured in seconds and the concentration is measured in molecules, or moles, per cubic centimeter. What units are required for I_a for a consistent equation?

8. In *flash photolysis*, intense flashes of light are produced by charging a bank of condensers to a high voltage and discharging the bank through a flash tube. About 10^4 J of electrical energy is released in each discharge, and about 3% of this energy is converted into light. If the flash occurs within 1 ms, how many watts of light does this correspond to? Assume that 0.25% of the light output is in the wavelength region 6000–6010 Å. How many einsteins of light are produced in the range 6005 ± 5 Å?

9. The historical unit of luminosity was the *candle*. It was originally defined as the light output of a wax candle of standard construction. It was later redefined as the light output of a standard carbon filament lamp; and it is now defined as $\frac{1}{60}$ of the intensity of 1 cm² of a blackbody radiator at the temperature of the freezing point of platinum (2046 K). The *lumen* is the unit of luminous flux. It is equal to the flux on a unit surface, all points of which are at unit distance from a source of one candle. A candle thus emits 4π lumens (lm). The *lux* (lx) is defined as 1 lm m^{-2}, and the *phot* is defined as 1 lm cm^{-2}. One lux equals 1.61×10^{-3} W m^{-2} at the wavelength of maximum visibility (5560 Å). In what follows, assume that all the light is of this wavelength.

The efficiency of light bulbs is measured in lumens per watt input. An ordinary gas-filled tungsten-filament light bulb has an efficiency of 0.0239 lumens per watt for 100-W bulbs. You are reading a book situated 7 ft from a lamp containing two 100-W bulbs. How many standard candles is the light equivalent to? The book is composed of pages 19×23.5 cm. How many light photons strike each page in a second? How much heat is dissipated per second in each page?

10. A standard 40-W fluorescent light bulb has an efficiency of 0.0940 lm W^{-1}. The lighting in a large laboratory is provided by twenty 100-W tungsten-filament lamps, as described in the previous problem. The cost of electricity is 7¢ per kilowatt hour, and the lights are used for an average of 10 h per day. How much money can be saved each year by converting to fluorescent lights that provide the same degree of illumination? Neglect the cost of conversion and the cost of the light bulbs themselves. See Problem 9 for additional information.

A standard 100-W light bulb costs about 39¢ and has a life expectancy of 750 h. With electricity at 7¢ per kilowatt-hour, what fraction of the total cost of using the bulb is due to the original price?

11. One proposed mechanism and rate equation for the photochemical production of HCl from H_2 and Cl_2 was discussed in the text. Another proposed mechanism adds a reaction to give the mechanism

$$Cl_2 + h\nu \longrightarrow 2Cl \qquad k_1$$
$$Cl + H_2 \longrightarrow HCl + H \qquad k_2$$
$$H + HCl \longrightarrow H_2 + Cl \qquad k_{-2}$$
$$H + Cl_2 \longrightarrow HCl + Cl \qquad k_3$$
$$2Cl \longrightarrow Cl_2 \qquad k_4$$

Using the ordinary steady-state approximation, show that this mechanism gives rise to the rate equation

$$\frac{d[HCl]}{dt} = \frac{2k_2 k_3 (k_1/k_4)^{1/2} [H_2][Cl_2] I_a^{1/2}}{k_3[Cl_2] + k_{-2}[HCl]}$$

12. A molecule that has been excited to a higher electronic state can lose its excess energy by emitting a light quantum (fluorescence) or by transferring its energy by collision with another molecule (quenching). The competition between these two modes of energy loss is often studied to test the significance of encounter-controlled events. (Quenching by collision and transfer of energy into kinetic energy requires an encounter with another body.) Different molecules have different quenching efficiencies. More complex molecules are usually more efficient quenchers, since they have a larger effective size and a greater number of degrees of freedom. The competing processes can be represented by the following mechanism, where F denotes the fluorescing molecule and Q the quenching molecule (Q may be the same as F):

$$F + h\nu \longrightarrow F^* \qquad k_a$$
$$F^* \longrightarrow F + h\nu' \qquad k_f$$
$$Q + F^* \longrightarrow F + Q^* \qquad k_q$$

The *fluorescent yield* is defined as the ratio of the intensity of the light emitted to the intensity of the light absorbed. Show that the fluorescent yield is

$$\frac{I_f}{I_a} = \frac{k_f[F^*]}{k_f[F^*] + k_q[F^*][Q]}$$

which can be rearranged to get

$$\frac{I_a}{I_f} = 1 + \frac{k_q[Q]}{k_f}$$

This last equation is known as the *Stern-Volmer equation* (O. Stern and M. Volmer, *Physik. Z.* **20** (1919):183). Outline an experiment by which you could determine k_q using the Stern-Volmer equation and any other information you might need.

13. Light intensities are generally measured with instruments such as photomultiplier tubes and thermopiles. These generally provide *relative*

measurements. They are conveniently used to establish ratios of intensities and can accomplish this very accurately with relative ease. To get *absolute* intensities (that is, the actual number of photons per second) these instruments must be calibrated. Chemical *actinometry* provides an *absolute* measure of light intensities by analysis of the chemical change that takes place in a system under irradiation. One useful actinometer is based on the photochemical decomposition of oxalic acid in the presence of uranyl salts. The UO_2^{2+} absorbs radiation in the region 2500–4500 Å and becomes an excited $(UO_2^{2+})^*$ ion. The oxalic acid is then decomposed by photosensitization:

$$UO_2^{2+} + h\nu \longrightarrow (UO_2^{2+})^*$$
$$(UO_2^{2+})^* + (COOH)_2 \longrightarrow$$
$$UO_2^{2+} + CO_2 + CO + H_2O$$

This reaction has a quantum yield of $\Phi = 0.50$. The oxalic acid can be determined by standard techniques such as permanganate titration.

A standard solution of oxalic acid was made up and titrated with $0.01N$ MnO_4^- solution. A 10-ml aliquot of the acid required 18.26 ml of the MnO_4^- to reach the end point. A 10-ml sample of the acid solution was then placed in the cell of an apparatus, as depicted in Figure 42.3, and irradiated with monochromatic light (3500 Å) for 137 min. When the sample was placed in the instrument, it absorbed 64% of the incident light, as indicated by the fall in the meter reading. It was found that at the conclusion of the irradiation 15.07 ml of the MnO_4^- standard solution was required to reach the end point. From these data, calculate the number of photons per second reaching the cell.

Note: There has been much talk lately about applying solar energy to solving our energy "crisis." The following problems (14–18) touch on some aspects of such use of solar energy. For simplicity, assume that all the radiation is at the wavelength of maximum visibility, 5560 Å. The following information will be useful. The solar energy that reaches the surface of the earth rises from zero at dawn to a maximum at noon and then decreases, until it is again zero after sunset. The total solar energy reaching the surface in temperate parts of the United States is roughly 1.6×10^7 J m^{-2} per day. (In the sunny southwestern states, the figure would be 3.1×10^7 during the summer months.) Note that this figure is a value that has been averaged over the entire day (about 500 min is the usual figure used for effective daylight time). The maximum illumination occurs at noon, when the peak energy reaches the surface at a rate about four times the average. (Many of these applications are discussed in F. Daniels, *Direct Use of the Sun's Energy* (New Haven: Yale University Press, 1964); also available in a Ballantine paperback.)

14. You are sitting on the beach in sunny California at high noon during the summer reading a book. The pages are 18 × 25 cm. How many photons per second reach a page? What fraction of the sunlight should your sunglasses absorb so that you will be provided with the same amount of light that would be produced by two 100-W light bulbs situated 7 ft from the book? (See Problem 9 for additional data.)

15. During the winter your house consumes 20 mcf (thousand cubic feet) of natural gas (assume it to be methane) for space heating each month. The roof of your house is 15 × 20 m. How much of your roof's area would you have to devote to solar collectors to heat your house if the collectors operated at an efficiency of 50%? (Neglect seasonal variations in incident energy, but remember that any practical device would have to take these variations into account.)

16. You have been engaged to look into the feasibility of constructing an electric-power-generation station with a capacity of 1000 megawatts (MW) of electricity. The heat for the plant will be provided by collecting the sun's rays on numerous solar reflectors, each 10 m square. The reflectors will focus the sun's rays on a centrally located heat absorber, which will be heated to 1000 K. (Each reflector must be independently mounted so that it can be moved to keep the central heat absorber in focus.) Assume that each reflector is 50% efficient. How many reflectors are required? (Don't forget also to take into account the efficiency of the heat engine, a factor that will further reduce the overall efficiency.) How large an area is required for the solar collector "farm" if 50% of the land area can be effectively used?

As the reflectors sit in the air, they get dirty because dust accumulates on the reflector surfaces. Since this cuts down on their reflecting power, they must be cleaned regularly. Assume that they have to be cleaned once a week, and that one window washer can clean 35 reflectors in an 8-hr day. If his salary is $10,000 a year, what is the yearly cost of cleaning the reflectors? (Neglect the cost of cleaning compound.)

17. Photovoltaic cells directly convert sunlight into electricity. It is estimated that advances in technology will be able to provide cells with an efficiency of 20% for directly converting sunlight into electricity. For land use, would solar reflectors or photovoltaic cells be more efficient energy sources?

18. One way of storing solar energy is to convert it into vegetation (that is, wood) that can later be burned. The productivity of photosynthesis in agriculture is about 3 tons of dry plant material per acre for a year (1 acre = 4047 m^2). Estimate the calorific value of the vegetation at 8000 Btu lb^{-1}, and using the data given above, calculate the efficiency of converting sunlight into fuel. How does the land use of ordinary farming compare with the land use of solar reflector "farming" discussed in Problem 16? (*Note:* The units in this

problem have been left in the English system. Most publications dealing with these problems still use the English system.)

19. Derive Equations (42.32) and (42.33).

20. Assume that an atom A has an effective diameter of 2 Å and that the atom can absorb a quantum of light to become the excited A* atom with a fluorescent lifetime of 10^{-8} s. Assume further that A* will be quenched whenever it collides with another molecule. Estimate the pressure at room temperature at which the rate of quenching will equal the rate of deactivation by fluorescence. Take the molar mass of A to be 50 g, and the temperature to be 300 °K.

21. In our discussion of rotational spectroscopy (Chapter 28) we noted that a heteronuclear diatomic molecule such as HCl could give rise to a rotational spectrum but that a homonuclear diatomic such as H_2 or Cl_2 could not. This implies that $B_{nm} \neq 0$ for heteronuclears and $B_{nm} = 0$ for homonuclears. The term B_{nm} is given in Equation (42.75), which involves integrals of the dipole moment of a molecule (or atom). The components of the electric dipole moment along the x, y, and z axes are

$$\mu_x = \mu(r) \sin \theta \cos \phi = \mu^0 \sin \theta \cos \phi$$
$$\mu_y = \mu(r) \sin \theta \sin \phi = \mu^0 \sin \theta \sin \phi$$
$$\mu_z = \mu(r) \cos \theta = \mu^0 \cos \theta$$

where μ^0 is the magnitude of the dipole moment for a diatomic, $\mu^0 = er$.

(a) Using the wave functions for the rigid rotor, show that if $\mu^0 = 0$ (which is the case for homonuclears), then the transition moment is zero and no rotation spectrum will be seen.

(b) Now show that if $\mu^0 \neq 0$, then the transition moment is nonzero only if $\Delta J = \pm 1$, where J is the rotational quantum number. (Do this only for the z component.)

22. Note that the dipole moment operator does not contain the spin. Use this fact and also that the spin functions of different multiplicities are orthogonal to show that the transition moment always vanishes for states with different spin multiplicities. (Transitions between singlet and triplet states are said to be *spin-forbidden*.)

23. Note that the dipole moment operator contains the vector **r**, which is an odd function of the coordinates. It is a general theorem of mathematics that an integral over all space vanishes unless the integrand is an *even* function. Use this to demonstrate that transitions between two g states or between two u states in a diatomic molecule are forbidden. (Such transitions are called *symmetry-forbidden*.)

24. In Raman spectroscopy, the interaction between the electromagnetic radiation and the molecule is due not to the dipole moment of the molecule but to the dipole moment *induced* in the molecule by the incident light. The induced dipole is proportional to the field strength $\mu = \alpha E$, where α, the constant of proportionality, is called the *polarizability*. Show that for a transition to be Raman-active, the polarizability must change during the vibration or rotation.

25. Refer to Chapter 31 and using Equation (31.104) and those equations that follow, derive the selection rules for the harmonic oscillator.

Note: Detailed selection rules are generally most easily computed from the symmetry properties of the wave functions of atoms and molecules rather than by the techniques described in this chapter. See, for example, E. B. Wilson, J. C. Decius, and P. C. Cross, *Molecular Vibrations* (New York: McGraw-Hill Book Co., 1955). These problems fall in the general area of group theory and are discussed in most textbooks that concern applying group theory to quantum mechanics.

BIBLIOGRAPHY

NOTE: References for each topic are listed in order of increasing sophistication.

CHAPTERS 2 AND 3

Conant, James B. *Robert Boyle's Experiments in Pneumatics.* Harvard Case Histories in Experimental Science, Case 1. Edited by James B. Conant. Cambridge: Harvard University Press, 1950.

Mendoza, E. A Sketch for a History of the Kinetic Theory of Gases. *Physics Today,* March 1961, p. 36.

Kauzmann, Walter. *Kinetic Theory of Gases.* Thermal Properties of Matter, vol. 1. New York: W. A. Benjamin, 1966.

Dushman, Saul. *Scientific Foundations of Vacuum Technique.* 2d ed. Revised by members of the research staff, General Electric Research Laboratory. Edited by J. M. Lafferty. New York: John Wiley & Sons, 1962.

Partington, J. R. *An Advanced Treatise on Physical Chemistry.* Vol. 1. *Fundamental Principles; The Properties of Gases.* London: Longmans, Green and Co., 1949. Pages 546–791 contain an exhaustive discussion of equations of state of gases.

Hirschfelder, J. O., Curtiss, C. F., and Bird, R. B. *Molecular Theory of Gases and Liquids.* New York: John Wiley & Sons, 1954.

CHAPTERS 4–9

Mahan, B. H. *Elementary Chemical Thermodynamics.* New York: W. A. Benjamin, 1964.

Nash, L. K. *Elements of Chemical Thermodynamics.* Reading, Mass.: Addison-Wesley Publishing Co., 1962.

Kauzmann, W. *Thermodynamics and Statistics: With Applications to Gases.* New York: W. A. Benjamin, 1967.

Lacey, W. N., and Sage, Bruce H. *Thermodynamics of One-Component Systems.* New York: Academic Press, 1957.

Bent, H. A. *The Second Law.* Oxford: Oxford University Press, 1965.

Bridgeman, P. W. *The Nature of Thermodynamics.* Cambridge: Harvard University Press, 1941.

Rossini, F. D., ed. *Experimental Thermochemistry.* New York: Interscience Publishers, 1956.

Klotz, I. M. *Chemical Thermodynamics.* Englewood Cliffs, N.J.: Prentice-Hall, 1950.

Planck, M. *Treatise on Thermodynamics.* London: Longmans Green and Co., 1927.

Fermi, E. *Thermodynamics.* New York: Dover Publications, 1956.

Reiss, H. *Methods of Thermodynamics.* New York: Blaisdell Publishing Co., 1965.

Lewis, G. N., and Randall, M. *Thermodynamics.* 2d ed. Revised by K. S. Pitzer and L. Brewer. New York: McGraw-Hill Book Co., 1961.

Kirkwood, J. G., and Oppenheim, I. *Chemical Thermodynamics.* New York: McGraw-Hill Book Co., 1961.

Guggenheim, E. A. *Thermodynamics.* 3d ed. Amsterdam: North-Holland Publishing Co., 1957.

Callen, H. B. *Thermodynamics.* New York: John Wiley & Sons, 1960.

CHAPTER 13

Findlay, A. *The Phase Rule.* 9th ed. New York: Dover Publications, 1951.

Hengstebeck, R. J. *Distillation.* New York: Reinhold Publishing Corp., 1961.

Nielson, C. H., ed. *Distillation in Practice.* New York: Reinhold Publishing Corp., 1956.

Krell, E. *Handbook of Laboratory Distillation.* Translated by C. G. Verver. Amsterdam: Elsevier Publishing Co., 1963.

Phase Diagrams for Ceramists. Columbus, Ohio:

American Ceramic Society. A serial compilation of phase diagrams of all types.

Palatnik, L. S., and Landau, A. I. *Phase Equilibria in Multicomponent Systems.* New York: Holt, Rinehart and Winston, 1964.

Weissberger, A., ed. *Techniques of Organic Chemistry.* Vol. 4, *Distillation.* New York: Interscience Publishers, 1951.

Malesinski, W. *Azeotropy and Other Theoretical Problems of Vapour-Liquid Equilibrium,* London: Interscience Publishers, 1965.

CHAPTERS 15–17

Glasstone, S. *An Introduction to Electrochemistry.* New York: D. Van Nostrand Company, 1942.

MacInnes, D. A. *The Principles of Electrochemistry.* New York: Dover Publications, 1961.

Conway, B. E. *Electrochemical Data.* Amsterdam: Elsevier Publishing Co., 1952.

Ives, D. J. G., and Janz, G. J. *Reference Electrodes.* New York: Academic Press, 1961.

Fuoss, R. M. *Electrolytic Conductance.* New York: Interscience Publishers, 1959.

Harned, H. S., and Owen, B. B. *The Physical Chemistry of Electrolytes.* 3d ed. New York: Reinhold Publishing Corp., 1958.

Kortum, G. *Treatise on Electrochemistry.* 2d ed. Amsterdam: Elsevier Publishing Co., 1965.

BATTERY POWER

Crowe, B. J. *Fuel Cells, A Survey.* Washington, D.C.: National Aeronautics and Space Administration, NASA SP-5115, 1973.

Vinal, G. W. *Storage Batteries.* 4th ed. New York: John Wiley & Sons, 1955.

Breiter, M. W. *Electrochemical Processes in Fuel Cells.* New York: Springer-Verlag, 1969.

Bockris, J. O'M., and Srinivasan, S. *Fuel Cells: Their Electrochemistry.* New York: McGraw-Hill Book Co., 1969.

McDougall, A. O. *Fuel Cells.* New York: John Wiley & Sons, 1976.

CHAPTER 18

GENERAL

The Reiss, Lewis and Randall, Guggenheim, and Callen textbooks listed with the references for Chapters 4–9.

ADIABATIC DEMAGNETIZATION

Mendelssohn, K. *The Quest for Absolute Zero.* New York: McGraw-Hill Book Co., 1966.

Casimir, H. B. G. *Magnetism and Very Low Temperatures.* New York: Dover Publications, 1961.

Garrett, C. G. B. *Magnetic Cooling.* Cambridge: Harvard University Press, 1954.

De Klerk, D., and Steenland, M. J. Adiabatic Demagnetization. Chap. 14 in *Progress in Low Temperature Physics,* edited by G. C. Gorter. Vol. 1. Amsterdam: North-Holland Publishing Co., 1955.

White, G. K. *Experimental Techniques in Low-Temperature Physics.* Oxford: Oxford University Press, 1959.

SURFACE CHEMISTRY

Ross, S., and Oliver, J. P. *On Physical Adsorption.* New York: Interscience Publishers, 1964.

Adamson, A. W. *Physical Chemistry of Surfaces.* 2d ed. New York: John Wiley & Sons, 1967.

Brunauer, S. *The Adsorption of Gases and Vapors.* Oxford: Oxford University Press, 1945.

Osipow, L. I. *Surface Chemistry, Theory and Industrial Applications.* New York: Reinhold Publishing Corp., 1962.

Hayward, D. O., and Trapnell, M. B. W. *Chemisorption.* 2d ed. London: Butterworth & Co., Ltd., 1964.

Davies, J. T., and Rideal, E. K. *Interfacial Phenomena.* New York: Academic Press, 1963.

Flood, E. A., ed. *The Solid-Gas Interface.* New York: Marcel Dekker, 1967.

CHAPTER 19

Stille, John K. *Introduction to Polymer Chemistry.* New York: John Wiley & Sons, 1962.

Billmeyer, Fred W., Jr., *Textbook of Polymer Science.* 2d ed. New York: Wiley-Interscience, 1971.

Meares, Patrick. *Polymers, Structure and Bulk Properties.* New York: D. Van Nostrand Co., 1965.

Van Holde, Kensal Edward. *Physical Biochemistry.* Englewood Cliffs: Prentice-Hall, 1971.

D'Alelio, G. F. *Fundamental Principles of Polymerization.* New York: John Wiley & Sons, 1952.

Cowie, J. M. G. *Polymers: Chemistry and Physics of Modern Materials.* Aylesbury, U.K. International Textbook Co., Ltd., 1973.

Birshtein, T. M., and Ptitsyn, O. B. *Conformations of Macromolecules.* New York: Interscience Publishers, 1966.

Morawetz, Herbert. *Macromolecules in Solution.* New York: Interscience Publishers, 1965.

Stone, F. G. A., and Graham, W. A. G., eds. *Inorganic Polymers.* New York: Academic Press, 1962.

Flory, Paul J. *Principles of Polymer Chemistry.* Ithaca: Cornell University Press, 1953.

Tanford, Charles. *Physical Chemistry of Macromolecules.* New York: John Wiley & Sons, 1961.

Slade, Philip E., Jr., ed. *Techniques and Methods of Polymer Evaluation*. New York: Marcel Dekker, 1975, in four volumes.

Svedberg, The, and Pedersen, Kai O. *The Ultracentrifuge*. Oxford: Oxford University Press, 1940.

Schachman, Howard K. *Ultracentrifugation in Biochemistry*. New York: Academic Press, 1959.

Fujita, Hiroshi. *Foundations of Ultracentrifugal Analysis*. New York: Wiley-Interscience, 1975.

Van de Hulst, H. C. *Light Scattering by Small Particles*. New York: John Wiley & Sons, 1957.

Fabelinskii, Immanuil L. *Molecular Scattering of Light*. Translated by Robert T. Beyer. New York: Plenum Press, 1968.

McIntyre, D., and Gornick, F., eds. *Light Scattering from Dilute Polymer Solutions*. New York: Gordon and Breach, 1964, a collection of original reprint articles.

CHAPTER 20

Young, H. D. *Statistical Treatment of Experimental Data*. New York: McGraw-Hill Book Co., 1962.

Langley, R. *Practical Statistics Simply Explained*. New York: Dover Publications, 1971.

Wilson, E. B., Jr., *An Introduction to Scientific Research*. New York: McGraw-Hill Book Co., 1952.

Drake, A. W. *Fundamentals of Applied Probability Theory*. New York: McGraw-Hill Book Co., 1967.

Parratt, L. G. *Probability and Experimental Errors in Science*. New York: Dover Publications, 1971.

Deming, W. E. *Statistical Adjustment of Data*. New York: Dover Publications, 1964.

Arley, N., and Buch, K. R. *Introduction to the Theory of Probability and Statistics*. New York: John Wiley & Sons, 1950.

Janossy, L. *Theory and Practice of the Evaluation of Measurements*. Oxford: Oxford University Press, 1965.

CHAPTERS 21 AND 22

Kauzmann, W. *Kinetic Theory of Gases*. New York: W. A. Benjamin, 1966.

Hildebrand, J. H. *An Introduction to Kinetic Theory*. New York: Reinhold Publishing Corp., 1963.

Loeb, L. B. *Kinetic Theory of Gases*. New York: Dover Publications, 1961.

Present, R. D. *Kinetic Theory of Gases*. New York: McGraw-Hill Book Co., 1958.

Kennard, E. H. *Kinetic Theory of Gases*. New York: McGraw-Hill Book Co., 1938.

Jeans, J. H. *An Introduction to the Kinetic Theory of Gases*. Cambridge: Cambridge University Press, 1940.

Boltzmann, L. *Lectures on Gas Theory*, trans. S. G. Brush. Berkeley: University of California Press, 1964.

Hirschfelder, J. O., Curtis, C. F., and Bird, R. B. *Molecular Theory of Gases and Liquids*. New York: John Wiley & Sons, 1954.

VACUUM TECHNOLOGY:

Yarwood, J. *High Vacuum Technique*, 3d ed. New York: John Wiley & Sons, 1955.

Brunner, W. F., and Batzer, T. H. *Practical Vacuum Techniques*. New York: Reinhold Publishing Corp., 1966.

Beck, A. H., ed. *Handbook of Vacuum Physics*. New York: Pergamon Press, 1964.

Van Atta, C. M. *Vacuum Science and Engineering*. New York: McGraw-Hill Book Co., 1965.

Dushman, S. *Scientific Foundations of Vacuum Technique*. 2d ed. Revised by members of the research staff, General Electric Research Laboratory. Edited by J. M. Lafferty. New York: John Wiley & Sons, 1962.

CHAPTERS 23, 24, AND 29

Nash, L. K. *Elements of Statistical Thermodynamics*. 2nd ed. Reading: Addison-Wesley Publishing Co., 1974.

Kauzmann, W. *Thermodynamics and Statistics with Application to Gases*. New York: W. A. Benjamin, 1967.

Dickerson, R. E. *Molecular Thermodynamics*. New York: W. A. Benjamin, 1969.

Rushbrooke, G. S. *Introduction to Statistical Mechanics*. Oxford: Oxford University Press, 1949.

Hill, T. L. *Introduction to Statistical Thermodynamics*. Reading: Addison-Wesley Publishing Co., 1960.

Davidson, N. *Statistical Mechanics*. New York: McGraw-Hill Book Co., 1962.

Mayer, J. E., and Mayer, M. G. *Statistical Mechanics*. New York: John Wiley & Sons, 1940.

Hill, T. L. *Statistical Mechanics*. New York: McGraw-Hill Book Co., 1956.

Fowler, R., and Guggenheim, E. A. *Statistical Thermodynamics*. Cambridge: Cambridge University Press, 1939.

Tolman, R. C. *The Principles of Statistical Mechanics*. Oxford: Oxford University Press, 1938.

CHAPTERS 26–33

QUANTUM MECHANICS

Hanna, M. W. *Quantum Mechanics in Chemistry*. 2d ed. New York: W. A. Benjamin, 1969.

Harris, L., and Loeb, A. L. *Introduction to Wave Mechanics*. New York: McGraw-Hill Book Co., 1963.

Bockhoff, F. J. *Elements of Quantum Theory*. Reading, Mass.: Addison-Wesley, 1969.

Rojansky, V. *Introductory Quantum Mechanics*. Englewood Cliffs, N.J.: Prentice-Hall, 1938.

Pauling, L., and Wilson, E. B., Jr. *Introduction to Quantum Mechanics*. New York: McGraw-Hill Book Co., 1935.

Levine, I. N. *Quantum Chemistry*. Boston: Allyn and Bacon, 1970.

Daudel, R., Lefebvre, R., and Moser, C. *Quantum Chemistry, Methods and Applications*. New York: Interscience Publishers, 1959.

Eyring, H., Walter, J., and Kimball, G. E. *Quantum Chemistry*. New York: John Wiley & Sons, 1944.

Kauzmann, W. *Quantum Chemistry*. New York: Academic Press, 1957.

Dicke, R. H., and Wittke, J. P. *Introduction to Quantum Mechanics*. Reading, Mass.: Addison-Wesley, 1960.

Merzbacher, E. *Quantum Mechanics*. 2d ed. New York: John Wiley & Sons, 1970.

Davydov, A. S. *Quantum Mechanics*. Translated by D. ter Haar. Oxford: Pergamon Press, 1965.

Messiah, A. *Quantum Mechanics*. Amsterdam: North-Holland Publishing Co., 1965.

ATOMIC AND MOLECULAR STRUCTURE

All of the above references and the following:

Coulson, C. A. *Valence*. Oxford: Oxford University Press, 1960.

Pauling, L. *The Nature of the Chemical Bond*. 3d ed. Ithaca: Cornell University Press, 1960.

Daudel, R. *Electronic Structure of Molecules*. Oxford: Pergamon Press, 1966.

Schaefer, H. F., III. *The Electronic Structure of Atoms and Molecules*. Reading, Mass.: Addison-Wesley Publishing Co., 1972.

Murrell, J. N., Kettle, S. F. A., and Tedder, J. M. *Valence Theory*. 2d ed. London: John Wiley & Sons Ltd., 1970.

Jørgensen, C. K. *Orbitals in Atoms and Molecules*. London: Academic Press, 1962.

Ballhausen, C. J., and Gray, H. B. *Molecular Orbital Theory*. New York: W. A. Benjamin, 1965.

Streitwieser, A., Jr. *Molecular Orbital Theory for Organic Chemists*. New York: John Wiley & Sons, 1961.

Karplus, M., and Porter, R. N. *Atoms and Molecules: An Introduction for Students of Physical Chemistry*. New York: W. A. Benjamin, 1970.

Slater, J. C. *Quantum Theory of Molecules and Solids*. New York: McGraw-Hill Book Co., 1963.

Ballhausen, C. J. *Introduction to Ligand Field Theory*. New York: McGraw-Hill Book Co., 1962.

Griffith, J. S. *The Theory of Transition Metal Ions*. Cambridge: Cambridge University Press, 1961.

SPECTROSCOPY

All of the books listed under Quantum Mechanics and the following:

Herzberg, G. *Atomic Structure and Atomic Spectra*. New York: Dover Publications, 1944.

Shore, B., and Menzel, D. H. *Principles of Atomic Spectra*. New York: John Wiley & Sons, 1968.

Allen, H. C., Jr., and Cross, P. C. *Molecular Vib-Rotors*. New York: John Wiley & Sons, 1963.

Colthup, N. B., Daly, L. H., and Wiberly, S. E. *Introduction to Infrared and Raman Spectroscopy*. New York: Academic Press, 1975.

Gordy, W., Smith, W. V., and Trambarulo, R. F. *Microwave Spectroscopy*. New York: John Wiley & Sons, 1953.

Gordy, W., and Cook, R. L. *Microwave Molecular Spectra*. Part 2, vol. 9, *Techniques of Organic Chemistry*. Edited by A. Weissberger. New York: Interscience Publishers, 1970.

Wilson, E. B., Jr., Decius, J. C., and Cross, P. C. *Molecular Vibrations*. New York: McGraw-Hill Book Co., 1955.

Herzberg, G. *Molecular Spectra and Molecular Structure*. 2d ed. Princeton: D. Van Nostrand Co., 1950.

Condon, E. U., and Shortley, G. H. *The Theory of Atomic Spectra*. Cambridge: Cambridge University Press, 1959.

CHAPTERS 34 AND 35

Lonsdale, K. *Crystals and X-Rays*. London: G. Bell & Sons, Ltd., 1948.

Bragg, W. H., and Bragg, W. L. *X-Rays and Crystal Structure*. London: G. Bell and Sons, Ltd., 1915.

Glusker, J. P., and Trueblood, K. N. *Crystal Structure Analysis*. New York: Oxford University Press, 1972.

Lipson, H. S. *Crystals and X-Rays*. London: Wykeham Publications Ltd., 1970.

International Union of Crystallography. Bijvoet, J. M., Burgers, W. G., and Hägg, G., eds. *Early Papers on Diffraction of X-Rays by Crystals*. Utrecht, Netherlands: A. Oosthoek's Uitgeversmaatschappij N.V., 1969.

Klug, H. P., and Alexander, L. E. *X-Ray Diffraction Procedures*. 2d ed. New York: John Wiley & Sons, 1974.

McLachlan, D., Jr. *X-Ray Crystal Structure*. New York: McGraw-Hill Book Co., 1957.

Buerger, M. J. *Contemporary Crystallography*. New York: McGraw-Hill Book Co., 1970.

Azaroff, L. V. *Elements of X-Ray Crystallography*. New York: McGraw-Hill Book Co., 1968.

Bunn, C. W. *Chemical Crystallography.* 2d ed. Oxford: Oxford University Press, 1961.

Taylor, C. A., and Lipson, H. *Optical Transforms.* Ithaca: Cornell University Press, 1965.

Lipson, S. G., and Lipson, H. *Optical Physics.* Cambridge: Cambridge University Press, 1969.

Buerger, M. J. *Crystal Structure Analysis.* New York: John Wiley & Sons, 1960.

Henry, N. F. M., Lipson, H., and Wooster, W. A. *The Interpretation of X-Ray Diffraction Photographs.* London: Macmillan & Co., Ltd., 1960.

Buerger, M. J. *The Precession Method in X-Ray Crystallography.* New York: John Wiley & Sons, 1964.

CHAPTERS 36 AND 37

Kittel, C. *Elementary Solid State Physics.* New York: John Wiley & Sons, 1962.

Weinreich, G. *Solids.* New York: John Wiley & Sons, 1965.

Sachs, M. *Solid State Theory.* New York: McGraw-Hill Book Co., 1963.

Shockley, W. *Electrons and Holes in Semiconductors.* New York: D. Van Nostrand Company, 1950.

Debye, P. *Polar Molecules.* New York: Dover Publications, 1929.

Kittel, C. *Introduction to Solid State Physics.* 5th ed. New York: John Wiley & Sons, 1976.

Smyth, C. P. *Dielectric Behavior and Structure.* New York: McGraw-Hill Book Co., 1955.

Van Vleck, J. H. *The Theory of Electric and Magnetic Susceptibilities.* Oxford: Oxford University Press, 1932.

Slater, J. C. *Insulators, Semiconductors and Metals.* Vol. 3, *Quantum Theory of Molecules and Solids.* New York: McGraw-Hill Book Co., 1967.

Smith, R. A. *Wave Mechanics of Crystalline Solids.* New York: John Wiley & Sons, 1961.

Ziman, J. M. *Electrons and Phonons.* Oxford: Oxford University Press, 1960.

Brillouin, L. *Wave Propagation and Group Velocity.* New York: Academic Press, 1960.

Davydov, A. S. Translated from Russian by S. B. Dresner. *Theory of Molecular Excitons.* New York: Plenum Press, 1971.

CHAPTER 38

ELECTRIC AND MAGNETIC SUSCEPTIBILITIES

Frank, N. H. *Introduction to Electricity and Optics.* New York: McGraw-Hill Book Co., 1950.

Kittel, C. *Introduction to Solid State Physics.* 3d ed. New York: John Wiley & Sons, 1966.

Mullay, L. N. *Magnetic Susceptibility.* New York: Interscience Publishers, 1963.

Debye, P. *Polar Molecules.* New York: Dover Publications, 1929.

Earnshaw, A. *Introduction to Magnetochemistry.* London: Academic Press, 1968.

Selwood, P. W. *Magnetochemistry.* 2d ed. New York: Interscience Publishers, 1956.

Bates, L. F. *Modern Magnetism.* 4th ed. London: Cambridge University Press, 1961.

Van Vleck, J. H. *The Theory of Electric and Magnetic Susceptibilities.* London: Oxford University Press, 1932.

Ballhausen, C. J. *Introduction to Ligand Field Theory.* New York: McGraw-Hill Book Co., 1962.

Griffith, J. S. *The Theory of Transition Metal Ions.* Cambridge: Cambridge University Press, 1961.

MAGNETIC RESONANCE SPECTROSCOPY

Roberts, J. D. *Nuclear Magnetic Resonance.* New York: McGraw-Hill Book Co., 1959.

Andrew, E. R. *Nuclear Magnetic Resonance.* Cambridge: Cambridge University Press, 1955.

Jackman, L. M. *Applications of Nuclear Magnetic Resonance Spectroscopy in Organic Chemistry.* London: Pergamon Press, 1959.

Roberts, J. D. *An Introduction to Spin-Spin Splitting in High Resolution Nuclear Magnetic Resonance Spectra.* New York: W. A. Benjamin, 1962.

Carrington, A., and McLachlan, A. D. *Introduction to Magnetic Resonance.* New York: Harper & Row, 1967.

Berson M., and Baird, J. C. *An Introduction to Electron Paramagnetic Resonance.* New York: W. A. Benjamin, 1966.

Pake, G. E. *Paramagnetic Resonance.* New York: W. A. Benjamin, 1962.

Pople, J. A., Schneider, W. G., and Bernstein, H. J. *High Resolution Nuclear Magnetic Resonance.* New York: McGraw-Hill Book Co., 1959.

Wertz, J. E., and Bolton, J. R. *Electron Spin Resonance.* New York: McGraw-Hill Book Co., 1972.

American Institute of Physics. *Selected Reprints on NMR and EPR* (1965).

Al'tshuler, S. A., and Kozyrev, B. M. *Electron Paramagnetic Resonance.* New York: Academic Press, 1964.

Emsley, J. W., Feeney, J., and Sutcliffe, L. H. *High Resolution Nuclear Magnetic Resonance Spectroscopy.* Oxford: Pergamon Press, 1965.

CHAPTERS 39–41

Dence, J. B., Gray, H. B., and Hammond, G. S. *Chemical Dynamics.* New York: W. A. Benjamin, 1968.

Gardiner, W. C., Jr. *Rates and Mechanisms of Chemical Reactions.* New York: W. A. Benjamin, 1969.

Laidler, K. J. *Reaction Kinetics.* Oxford: Pergamon Press, 1963. In two volumes.

Nicholas, J. *Chemical Kinetics.* New York: John Wiley & Sons, 1976.

Weston, R. E., Jr., and Schwartz, H. A. *Chemical Kinetics.* Englewood Cliffs, N.J.: Prentice-Hall, 1972.

Frost, A. A., and Pearson, R. G. *Kinetics and Mechanism.* 2d ed. New York: John Wiley & Sons, 1961.

Laidler, K. J. Chemical Kinetics. 2d ed. New York: McGraw-Hill Book Co., 1965.

Gimblett, F. G. R. *Introduction to the Kinetics of Chemical Chain Reactions.* London: McGraw-Hill Book Co., 1970.

Hinshelwood, C. N. *Kinetics of Chemical Change.* Oxford: Oxford University Press, 1940.

Benson, S. W. *Foundations of Chemical Kinetics.* New York: McGraw-Hill Book Co., 1960.

Hammet, L. P. *Physical Organic Chemistry; Reaction Rates, Equilibria, and Mechanisms.* 2d ed. New York: McGraw-Hill Book Co., 1970.

Tobe, M. L. *Inorganic Reaction Mechanisms.* London: Thomas Nelson and Sons Ltd., 1972.

Rideal, E. K. *Concepts in Catalysis.* London: Academic Press, 1968.

Melville, H. W., and Gowenlock, B. G. *Experimental Methods in Gas Reactions.* London: Macmillan & Co., Ltd., 1964.

Benson, S. W. *Thermochemical Kinetics.* 2d ed. New York: John Wiley & Sons, 1976.

Hague, D. N. *Fast Reactions.* London: Wiley-Interscience, 1971.

Robinson, P. J., and Holbrook, K. A. *Unimolecular Reactions.* London: Wiley-Interscience, 1972.

Kondrat'ev, V. N. *Chemical Kinetics of Gas Reactions.* Translated from the Russian by J. M. Crabtree and S. N. Carruthers. Oxford: Pergamon Press, 1964.

Glasstone, S., Laidler, K. J., and Eyring, H. *The Theory of Rate Processes.* New York: McGraw-Hill Book Co., 1941.

CHAPTER 42

Arnold, D. R., Baird, N. C., Bolton, J. R., Brand, J. C. D., Jacobs, P. W. M., de Mayo, P., and Ware, W. R. *Photochemistry, An Introduction.* New York: Academic Press, 1974.

Lasers and Light, Readings from Scientific American. San Francisco: W. H. Freeman and Co., 1970.

Coxon, J. M., and Halton, B. *Organic Photochemistry.* Cambridge: Cambridge University Press, 1974.

Ashmore, P. G., Dainton, F. S., and Sugden, T. M., eds. *Photochemistry and Reaction Kinetics.* Cambridge: Cambridge University Press, 1967.

Lengyel, B. A. *Introduction to Laser Physics.* New York: John Wiley & Sons, 1966.

Leighton, P. A. *Photochemistry of Air Pollution.* New York: Academic Press, 1961.

Calvert, J. G., and Pitts, J. N., Jr. *Photochemistry.* New York: John Wiley & Sons, 1967.

Also the additional references on spectroscopy and quantum mechanics listed for Chapters 26–33.

INDEX

Boldface page numbers indicate the location of biographical sketches.

Abegg's rule of eight, 644
Absolute entropy, 135
Absolute reaction rate theory, 831
Absolute value of complex number, 504
Absolute zero of temperature, 32
Absorption coefficient, 842
Absorption spectroscopy, 542
Acceptor, 758
Actinometry, 865
Activated complex, 825, 829
   theory, 831
Activation energy, 825
Activity, 257
   experimental determination, 265
   from Henry's law, 261
   ionic, 297
   nonideal solution, 259
   practical scale, 262
   pressure dependence, 272
   Raoult's law ideal solution, 258
   rational scale, 262
   solutes, 260
   solvents, 258
   temperature dependence, 272
Activity coefficient, 259, 263
   electrolyte, 275
   from EMF measurements, 323
   experimental determination, 301
   of ions, 297
   mean, 297, 308
Addition of angular momentum, 625
Addition polymer, 368
Adiabatic:
   demagnetization, 136
   flame temperature, 91
   ideal gas processes, 84
   process, 62
Adiabats, 86
Adsorbate, 357
Adsorbent, 357
   trap, 357
Adsorption, 356
   coefficient, 359
   heat of, 365
   isobar, 358
   isostere, 358
   isotherm, 357
   rate of, 359

Aggregation, energy of, 726
Allowed band, 750
Allowed transition, 861
Alternating copolymer, 369
Ammonia:
   inversion, 539
   manufacture of, 101, 181
   molecule, 663
Ampere (unit), 280
Ampere hour, 336
Aneroid barometer, 24
Angels, 1
Angular frequency, 502
Angular momentum, 502, 521, 540, 592
   addition of, 625
   commutation properties, 594
   operator, 592
   total, 625
   wave functions of hydrogen atom, 584
Anharmonic oscillator, 562
Anomalous low temperature heat capacity, 138
Antibonding orbital, 652
Antisymmetric functions, 520
Antisymmetric wave functions, 622
A priori probability, 448
Argand diagram, 504
Arrhenius law, 824
Arrhenius, Svante, **285**
Arteriography, 720
Atactic polymer, 369
Atom:
   early theories of, 477
   nuclear model, 480
   relative masses of, 478
Atomic:
   mass, 478
   number, 694
   scattering factor, 714
   units, 491, 616
Aufbau prinzip, 624, 636
   free electrons, 742
   molecules, 654
Auxiliary condition, 452
Average molar mass, 366
Average velocity, 47, 416
Avogadro, Amadeo, **6**, 479

Avogadro's number, 421
   determination of, 408, 421, 422
Axis of symmetry, 673
Azeotrope, 237
Azimuthal quantum number, 494, 580

Ballistics, 411
Balmer series, 484
Band theory of solids, 750
Barometer, 16
   types of, 24–28
Barometric formula, 92, 343, 407
Base units (SI), 7
Basis, 681
Basis set, 577
Bathroom tile, 674, 684
Battery, 277
   types of, 316–319
Bellows manometer, 24
Bell-shaped curve, 402
Benzene orbitals, 667
Bernoulli, Daniel, 1, **2**, 22, 477
Berthelot's equation, 41
Berzelius, Jons Jacob, **643**
Bessel equation, 541
Bessel function, 520, 541
Billiard ball model, 44
Bimolecular energy transfer, 850
Binding energy of crystals, 727
Binomial coefficients, 395
Binomial distribution, 395
   computer generated, 396
Black body radiation, 482, 486
Black, Joseph, **59**
Bloch function, 750
Blood clot, 222
Bohr, Niels, **489**
Bohr:
   frequency condition, 613
   hydrogen atom model, 488
   refinements, 493
   magneton, 596
      effective, 775
   radius, 490
Boiler, 104
Boiling, 153
Boiling point elevation, 226
Boltzmann, Ludwig, **406**

873

Boltzmann:
 constant, 50
 distribution, 305, 545
 distribution law, 406, 455
Boltzon, 460
Bomb calorimeter, 75
Bonding orbital, 651
Born-Haber cycle, 727
Born-Oppenheimer approximation, 646
Bose-Einstein distribution law, 747
Bose-Einstein statistics, 461, 619, 746
Boson, 460, 619
Boundary value problems, 509
Boyle, Robert, 19, **20**
Boyle temperature, 26
Boyle's law, 22
 kinetic derivation, 46
Bragg, William Henry, 700, **701**
Bragg, William Lawrence, 700, **701**
Bragg's law, 701
Branched polymer, 370
Bravais lattice, 682
Brillouin zone, 752
British thermal unit (Btu), 8
Brownian motion, 408
Bubble, 351
Bubble cap column, 234
Buffered solution, 331

Calomel cell, 329
Caloric theory of heat, 60
Calorie (unit), 7
 food calorie, 95
Calorimeter, 75, 93
Candle (unit), 864
Cannizzaro, S., 479
Capacitance manometer, 27
Capacitor, 765
Capacity, 765
Capillary depression, 354
Capillary rise, 353
Cardiac output, 33
Carnot, Sadi, **109**
Carnot cycle, 109
CAT scan, 722
Cathode ray, 691
Celsius scale, 30
Cell EMF, measurement of, 315
Center of symmetry, 674
Centrifuge, 363
Chain reaction, 366, 818
Change of state, entropy change of, 128
Change of state, free energy change of, 152
Characteristic temperature, 566, 735
Charles's law, 31
Chemical potential, 159
Chemical shift, 783
Chemisorption, 357
Chi-square test, 393
Chromotography, 236, 360
Clapeyron, Benoit Pierre Emile, **198**
Clapeyron equation, 198
Clausius, Rudolf Julius Emmanuel, **109**
Clausius-Clapeyron equation, 199
 magnetic analog, 364
Clausius-Mosotti equation, 768
Clausius, principle of, 107
Clausius's inequality, 119

Coal, 93
Coefficient of:
 diffusion, 433
 thermal conductivity, 431
 thermal expansion, 11
 viscosity, 429
Cogeneration, 117
Coin flipping, 391
Cold reservoir, 105
Colligative properties of solutions, 221
 ionic solutions, 287
 polymer solutions, 372
Collision:
 cross section, 827
 frequency, 428
 molecular, 423
 number, 428
 theory, 826
 with walls, 423
Colloid, 366, 408
Commutator, 521, 594
Commuting operators, 506, 534
Competing reactions, 816
Complete sets of functions, 509
Complex conjugate, 504
Complex number, 504
Component, 190
Compound formation, 242
Compressibility, 11, 729
Compressibility factor, 15
Concentration cell, 326
Concentration units, 202
Condensation polymer, 368
Condenser, 105
Conductance, equivalent, 283
Conductance, molar, 281
Conducting wall, 62
Conduction band, 750
Conduction electrons, 737
Conductivity, 281
 of metal, 739
Conductivity water, 293
Conductometric titration, 292
Confocal elliptical coordinates, 647
Consecutive reaction, 814
Conservation of energy, 57
Conservative force, 65
Conservative system, 526
Constant:
 Boltzmann, 50
 cryoscopic, 225
 Curie, 345
 dielectric, 766
 ebullioscopic, 226
 gas, 15, 31
 Henry's law, 215
 Madelung, 728, 729
 Michaelis, 839
 Planck's, 486
 rotational, 556
 Rydberg, 485, 491
 Stefan-Boltzmann, 486, 498
Constant boiling HCl, 238
Constant pressure calorimeter, 93
Constant pressure heat, 76
Constant pressure heat capacity, 77
Constant volume calorimeter, 75
Constant volume heat, 75
Constant volume heat capacity, 77
Contact angle, 354
Contact process, 855
Cooling curve, 239

Cooling, in magnetic field, 347
Coordinate systems, 412, 414
Coordinates, confocal elliptical, 647
Copolymer, 369
Corresponding states, law of, 40
Coulomb integral, 631, 651, 660
Coulometer, 280
Counterion, 377
Couper, A. S., 3
Course of reaction, 829
Covalent bond, 645
Critical constants, 37
Critical point, 38
 plait point, 252
Crookes, William, **691**
Crookes tube, 690
Cross linkages in polymers, 370
Cross product, 501
Cryoscopic constant, 225
Crystal, 673
 energy, 726
 systems, 680
Cubic closest packing, 677, 686
Cubic symmetry, 675
Cubic system, 675
Curie, Pierre, **344**
Curie constant, 345
Curie's law, 345, 775
Curved interface, 352
Cyclic engine, 104, 109
Cyclic process, 69

Dalton, John, 1, **2**, 478
Dalton's law of partial pressure, 34
Davisson, C., 497
Davy, Humphry, **29**, 278
Deactivation, 845
de Broglie, Louis Victor Pierre Raymond, **496**
de Broglie equation, 718
Debye, Peter, **303**
Debye (unit), 763
Debye:
 characteristic temperature, 735
 lattice heat capacity model, 732
 length, 307
Debye-Hückel limiting law, 307
Debye-Scherer powder X-ray method, 707
Decay time, 792
Degeneracy, 459, 546
Degenerate solutions, 518
Degree:
 of freedom (chemical system), 190
 of freedom (mechanical), 50
 of hotness, 59
 of ionization, 285
 of polymerization, 371
 thermometric, 30
Delocalization, 667
Delocalization integral, 669
Del operator, 506
Democritus, 1, 477
Density gradient ultracentrafugation, 384
Density of states, 744, 749
Derivative, 9
 reduction of, 156
Desorption, 357
Detailed balancing, 813
Deviation, standard, 400
Dewar flask, 443

Dewar, J., 136
Dialysis, 376
Diamagnetic susceptibility, 771
Diamagnetism, 345
Diameter of molecules, 427, 442
Diamond, equilibrium with graphite, 161
Diamond structure, 688
Diaphragm manometer, 27
Diatomic molecule, 653
   thermodynamic functions of, 571
Dielectric, 764
Dielectric constant, 766
Dieterici equation, 41
Differential:
   equation, 511
   heat of solution, 213
   manometer, 24
   method, 800
Diffraction:
   electron, 719
   neutron, 718
   X-ray, 696
Diffraction grating, 700
Diffraction pattern, intensities, 713
Diffusion coefficient, 433
Diffusion equation, 433
   solution of, 435
Diffusion pump, 438
Digonal hybrid bond, 665
Dilatometer, 801
Dilute solutions, 215
Dipole, 762
   moment, 614, 763
   transition, 615
Dirac, Paul Adrien Maurice, 619, **746**
Directed bond, 663
Directed valence, 662
Dissociation energy, 562
Distillation, 233
   of liquid air, 255
Distribution:
   binomial, 395
   Gaussian, 398
   Maxwell-Boltzmann, 406
   molecular velocity, 404
Distribution law, 218
Donnan effect, 375
Donor, 758
d orbitals of hydrogen atom, 590
Dot product, 501
Drift velocity, 739
Droplet:
   formation, 352
   and phase equilibrium, 354
   solubility of, 356
   vapor pressure of, 355
Dulong and Petit law, 731, 735

Ebullioscopic constant, 226
EDAX, 724
Effective Bohr magneton, 775
Effective nuclear charge, 601
Efficiency:
   Carnot engine, 113
   engine, 105
   steam engine, 112
Effusion, 49, 423, 792
   exact kinetic treatment, 425
Eigenfunction, 507
Eigenvalue, 507
   equation, 507

Einstein, Albert, **487**
Einstein:
   coefficient, 545
   coefficient of absorbtion, 859
   model of lattice heat capacity, 730
   transmission coefficient, 615
   transmission probability, 859
   unit, 843
Elastic collision, 44
Electric discharge tube, 690
Electric field, 343
Electricity, 276
Electric work, thermodynamic, 343
Electric susceptibility, 767
Electrochemical cell, 311, 316
Electrode potential, 318
Electrolysis, 279
Electrolyte, 275
   activity, 297
   standard state, 275, 296
Elementary reaction, 810
Electron:
   charge-to-mass ratio, 692
   diffraction, 497, 719
   electron repulsion, 601
   impact spectroscopy, 494
   indistinguishability, 621
   microprobe analysis, 724
   microscope, 722
   pair bond, 645
   paramagnetic resonance, 777
   spin, 528, 577, 618
   spin postulates, 621
   waves, 496
Electronegativity, 661
Electronic heat capacity, 576, 745
EMF and activity coefficients, 323
EMF and equilibrium constants, 323
Emission:
   induced, 859
   spectroscopy, 542
   spontaneous, 859
   stimulated, 859
Encroachment, 252
Endothermic reaction, 95
Energy:
   activation, 825
   conservation of, 57
   dissociation, 562
   Fermi, 742
   helium atom, 602
   internal, 131
   kinetic, 57
   as maximum work at constant S and V, 142
   potential, 57
   surface, 351
   translational, of gas, 467
Energy change as constant volume heat, 75
Energy change of van der Waals gas, 174
Energy entropy extremals, 125
Energy gap, 750
Energy levels:
   harmonic oscillator, 549
   hydrogen atom, 491, 579
   rigid rotor, 556
Engine, 105
Ensemble, 455
Enthalpy, 76
   of activation, 834

   of solution, 211
   standard, 99
   temperature dependence, 100
Enthalpy change:
   as constant pressure heat, 76
   from EMF measurements, 325
   as maximum work at constant S and P, 143
   of mixing, 214
Entropy, 117
   absolute, 135
   of activation, 834
   calculation for diatomic molecule, 573
   and the direction of chemical change, 119
   and irreversibility, 120
   of mixing, 129, 139, 214, 474
   pressure dependence, 155
   and probability, 448
   residual, 473
   spontaneous decrease, 445, 462
   temperature dependence, 156
   translational, 467
   vibrational, 567
   and work capacity, 123
   zero point, 473
Entropy change, 128
   from EMF measurements, 325
   experimental determination, 132
   on heating, 128
   ideal gas expansions, 128
Environmental Protection Agency (EPA), 12, 42
Equation:
   Berthelot, 41
   Bessel, 541
   de Broglie, 718
   Clapeyron, 198
   Clausius-Clapeyron, 199, 364
   Clausius-Mossoti, 768
   Dieterici, 41
   Diffusion, 433
   eigenvalue, 507
   Gibbs-Duhem, 208, 301, 381
      application to activity, 267
   Gibbs-Helmholtz, 155, 169
   Hermite, 548
   ideal gas, 32
   Kelvin, 355
   Laue, 699
   Legendre, 555
   Lorenz-Lorentz, 768
   Nernst, 321
   Poisson, 305
   quadratic, 8
   radial, 579
   rate, 794
   Sakur and Tetrode, 470
   Schrödinger, 527
   Sterm-Volmer, 864
   van der Waals, 36
   virial, 41
   virial osmotic, 221, 373
Equilibrium, 63
   between phases, 161, 193
   reaction, 811
   sedimentation, 380
   various boundary conditions, 146
Equilibrium constant, 293, 811
   of ammonia synthesis, 171
   from EMF measurements, 323

Equilibrium constant (*cont.*)
  ideal gas reaction, 168
  real gas reactions, 181
  statistical mechanical, 471
  temperature dependence, 169
  in terms of activities, 263
  in terms of molarities, 186
  in terms of mole fractions, 187
Equipartition of energy, 50, 730
Equivalent conductance, 283
Error function, 402, 419
Error integral, 402, 419
Esaki diode, 538
Escape velocity, 54, 422
Etch figures, 673
Euler condition for exactness, 10
Euler theorem of complex numbers, 505
Euler theorem of homogeneous functions, 207
Eutectic, 239
Eutectic halt, 242
Even functions, 520
Ewald sphere, 712
Exact differential, 10
Exchange integral, 631, 651, 660
Excluded volume, 35
Exothermic reaction, 95
Expansion of gases, 61
Expansion work, 66
  ideal gas, 67
  van der Waals gas, 173
Expectation value, 525
Exponents, 8
Extensive property, 59

Face-centered cubic packing, 677
Factorials, 450
Fahrenheit temperature scale, 30
Faraday, Michael, **279**
Faraday (unit), 280
Fermentation, 93
Fermi, Enrico, **742**
Fermi:
  energy, 742
  level, 541, 742
  temperature, 745
Fermi-Dirac distribution law, 748, 756
Fermi-Dirac statistics, 461, 619
Fermion, 460, 619
First law of thermodynamics, 57, 71
First-order reaction, 795
Fixed points, thermometric, 30
Flash photolysis, 804, 864
Flea, flight of, 393
Flipping coins, 391
Floquet's theorem, 750
Flow systems, 803
Fluorescence, 846
Fluorescent yield, 864
Food calorie, 95
Food chain, 857
Forbidden band, 750
Forbidden transition, 616, 636, 861
Forward rate constant, 812
Fourier, John Baptiste Joseph, **717**
Fourier:
  map, 717
  series, 508
  synthesis, 716
  transform, 716

Franck, James, **495**
Franck-Hertz experiment, 494
Franklin, Benjamin, **277**, 690
Free electrons, paramagnetism of, 787
Free electron theory of metals, 739
Free energy, 144
  and activity, 259
  calculations, 152
  composition dependence, 159
  of formation, 148
  and fugacity, 257
  standard, 148, 167
  surface, 351
Free energy change:
  electrochemical cell, 315
  experimental determination, 146
  as measure of useful work, 149
  of mixing, 214
  pressure dependence, 154
  temperature dependence, 154
  of van der Waals gas, 174
Free expansion of gases, 78
  ideal gases, 130
Free particle, quantum mechanical, 528
Free variation, 451
Freezing point, temperature dependence, 199
Freezing point depression, 223
  ionic solutions, 287
Frequency distribution, 399
Frequency factor, 825
Fuel cell, 148, 331
  molten carbonate, 333
Fugacity, 175, 205, 219
  calculations for real gases, 176
  coefficient, 180
  coefficient, universal charts, 184
  of liquids, 257
  relative, 257
  van der Waals gas, 179
Funicular theory of Linus, 19
Fusion, heat of, 60

Galileo, **30**, 57
Galvani, Luigi, **278**
Gamma function, 450
Gas:
  constant, 15, 31
  expansion of, 61
  free expansion, 78
Gauss, Friedrich, **400**
Gaussian distribution, 398, 409
  multidimensional, 411
Gay-Lussac, Joseph Louis, **32**, 479
Gay-Lussac's law, 31
Germer, L. H., 497
g-factor, 620, 744
  nuclear, 782
Giaque, W., 136
Gibbs, Josiah Willard, **139**, 189, 194
Gibbs free energy, **144** (*see also* free energy)
Gibbs-Duhem equation, 208, 301, 381
  application to activity, 267
Gibbs-Helmholtz equation, 155, 169
Gilbert, William, **277**
Glass electrode, 328
Goniometer, 677, 706
Goudsmit, S., 619
Gouy balance, 772
Gradient, 507

Graham's law of effusion, 50, 424
Graphite, conversion to diamond, 201
Graphite, equilibrium with diamond, 161
Grating, 700
Gravitation, 6
Gravitational potential energy, 57
  in thermodynamics, 343
  work and thermodynamics, 342
  zero point of, 57
Grotthous-Draper law, 841
Grottrian diagram, 631
Group theory, 676
Guldberg and Waage principle, 812
Gyromagnetic ratio, 620

Haber, Fritz, **170**
Haber process, 170, 181
Half-cell reaction, 280, 311
Half life, 792
Hall effect, 759
Hamiltonian, 152
  approximate, 601
  helium atom, 600
  hydrogen atom, 578
  spin, 787
Hamilton, William Rowan, **526**
Harmonic motion, 504
Harmonic oscillator, quantum mechanical, 547
  energy levels, 549
  perturbation theory treatment, 609
  selection rules, 614
  wave functions, 549
Harmonics, 517
Hartree (unit), 610
Hartree self-consistent field calculation, 639
Hartree-Fock calculation, 634
Hauy, Rene Just, **678**
Health effects of pollutants, 854
Heat, 59
  of adsorption, 365
  of combustion, 96
  constant pressure, 76
  constant volume, 75
  of dilution, 227
  of formation, 97, 99
  of fusion, 60
  mass of, 60
  mechanical equivalent of, 71
  of reaction, 93
  of reaction, temperature dependence, 100
  of solution, differential, 213
  of solution, integral, 212
  and work, 70
Heat capacity, 53, 60, 340, 567
  anomaly, 53
  constant magnetic field, 346
  constant pressure, 77
  constant volume, 77
  crystal, 730
  difference, 157
  electronic, 576, 745
  of gases, 90
  low temperature, 733
  low temperature anomalous, 138
  statistical mechanical, 466
Heat engine, 104
Heat flow, 107
Heating, entropy change of, 128, 134

Heat transfer, 120
Height equivalent to theoretical plate (HEPT), 235
Heisenberg, Werner Karl, **533**
Heisenberg uncertainty principle, 530, 533, 552, 847
Heitler-London wave function, 660
Helium atom, 600
   excited states, 627
   perturbation theory treatment, 610
   variational calculation, 604
Helmholtz, Hermann Ludwig Ferdinand, **143**
Helmholtz energy, 143
   as maximum work at constant T and V, 143
   statistical mechanical, 465
   vibrational contribution, 567
Henry, William, **215**
Henry's law, 215, 260
   constant, 215
   for electrolytes, 296
Hermetian operator, 524
Hermite equation, 548
Hermite polynomial, 509, 521, 549
   recursion relations, 564, 617
Hertz, Heinrich Rudolf, **483**
Hess's law, 97, 99
Heterogeneous system, 189
Hexagonal closest packing, 677, 686
Hexidecipole, 762
High vacuum production, 438
Hinshelwood, C. N., 837
Hittorf cell, 289
HKRR theory, 838
Hold up, 233
Hole, 755
Homogeneous function, 207
Homogeneous system, 189
Homopolymer, 369
Hooke's law, 504
Hotness, 59
Hot reservoir, 104
Hückel orbitals, 669
Hund's rules, 634
Huygens, Christiaan, **677**
Hybrid bonds, 664
Hybrid functions, 589
Hybridization, 519, 664
Hybrid orbitals, 589, 591
Hydrogen atom, 577
   Bohr model, 489
   d orbitals, 590
   energy levels, 579
   hybrid orbitals, 589, 590
   p orbitals, 587
   quantum numbers, 580
   radial functions, 582
   Schrödinger equation, 578
   s orbitals, 586
Hydrogen bonded crystal, 727
Hydrogen bromide reaction, 817
   photochemical, 853
Hydrogen chloride cannon, 840
Hydrogen chloride reaction, photochemical, 840, 853
Hydrogen economy, 335
Hydrogen iodide photochemical reaction, 852
Hydrogen molecule, 658
   hamiltonian, 658
   wave function, 660

Hydrogen molecule ion, 646
   energies, 650
   variational calculation, 649
Hyperfine coupling constant, 787
Hyperfine interaction, 778
Hyperfine splitting, 634
Hypothetical standard state, 261

Ice skating, 200
Ideal gas, 15
   adiabatic process, 84
      irreversible, 86
      reversible, **84**
   Carnot cycle, 115
   entropy change of expansion, 128, 132
   equation, 32
   expansion work of, 67
   isothermal expansion, 82
      free energy change, 152
      irreversible, 68, 83
      reversible, 67, 82
   reaction, equilibrium constant, 168
   reaction, free energy change, 167
   temperature scale, 32
      equivalence to thermodynamic scale, 115
Ideal solution, 203
   mixing, 213
Imaginary state, 97
Impossible engines, 108, 112
Impossible heat flow, 107
Impurity conduction, 757
Incongruent melting, 243
Independent electron model, 628
Indistinguishability of electrons, 621
Induced dipole moment, 763
Induced emission, 859
Induced polarization, 767
Infinitesimal process, 62
Information theory, 448
Infrared spectroscopy, 550
Initial rates, method of, 800
Inner quantum number, 633
Insulating wall, 62
Integrable square, 523
Integral, 9
   error, 402
Integral heat of solution, 212
Integration, Monte Carlo, 393
Intensities of diffraction pattern, 713
Intensities of spectral lines, 546
Intensive property, 59
Interface, 349
Interfacial angles, law of, 677
Intermediate reaction, 810
Internal conversion, 846
Internal energy, 131
Internal pressure, 78
Intersystem crossing, 846
Intrinsic conductivity, 753
Inversion temperature, 81
Ion, 285
Ionic:
   activity, 297
   activity coefficient, experimental determination, 301
   crystal, 726
   migration, 289
   solutions, colligative properties, 287
   strength, 300
Ionization, degree of, 285

Ionization gauge, 28
Ionization potential, **498**
   helium atom, 602
Ion product of water, 293
Irreversible process, 63
Isolation method, 800
Isopiestic solutions, 269
Isotactic polymer, 369
Isothermal process, 62
Isothermal expansions of ideal gases, 82
Isotherms of ideal gases, 83
Isotonic solutions, 269, 303
Isotope effect, 839
Isotopic separation, 424

j-j coupling, 634
J quantum number, 626
Joule, James, **71**
Joule (unit), 7
Joule free expansion, 78
   ideal gas, 130
Joule-Thomas coefficient, 81, 158
Joule-Thomas expansion, 80
Junction potential, 327

Kammerline-Onnes, M., 136
Kekule, Friedrich August, **4**
Kelvin, Lord, 114
Kelvin (unit), 31, 114
Kelvin equation, 355
Kepler, Johann, **5**, 57
Kinetic energy, 57
   of gases, 50
   operator, 525
Kinetics, chemical, 791
Kinetics of photochemical reactions, 851
Knudsen gauge, 443
Kohlrausch, Friedrich Wilhelm Georg, **284**
Kohlrausch's law, 283
Kroenecker delta, 511
Kvetcherites, 393

Ladder operator, 598
Lagrange, Joseph Louis, **451**
Lagrange method of undertermined multipliers, 451, 456
Lagrange multiplier, 747
Lagrangian, 152
Laguerre function, 580
Laguerre polynomial, 509
Lambda spike, 736
Lambert-Beer law, 842
Lambert's law, 842
Landé g-factor, 620, 774
Langevin function, 769
Langmuir, Irving, 359, **435**, 443
Laplace transform, 450
Laplacian operator, 305, 507
Larmour frequency, 595
Laser, 862
Lattice, 680
Lattice energy of crystal, 727
Lattice heat capacity, 730
   Debye model, 732
   Einstein model, 730
Laue, Max von, **697**
Laue cones, 706
Laue equations, 699
Laue method, 706

877

Lavoisier, Antoine Laurent, **477**
Law:
  Arrhenius's, 824
  Boltzmann distribution, 406, 455
  Bose-Einstein, **747**
  Boyle's, 22
  Bragg's, 701
  Charles's, 31
  of corresponding states, 40
  Curie, 345, 775
  Dalton's partial pressure, 34
  Debye-Hückel, 307
  distribution, 218
  Dulong and Petit, 479, 731, 735
  Fermi-Dirac distribution, 748, 756
  first thermodynamic, 57
  Gay-Lussac, 31
  Graham's, 50, 424
  Grotthous-Draper, 841
  Henry's, 215, 260
  Hess's, 97, 99
  Hooke's, 504
  of interfacial angles, 677
  Kohlrausch's, 283
  Lambert-Beer, 842
  Lambert's, 842
  mass action, 293, 791, 812
  Newton's gravitational, 342
  Ohm's, 280
  of photochemical reactions, 841
  Planck's radiation, 487
  Raoult's, 203, 217, 258
  rational indices, 679
  second thermodynamic, 108
  Stark-Einstein, 843
  Stefan-Boltzmann, 486
  Stokes's, 444
  third thermodynamic, 135, 574
  statistical thermodynamical, 472
LCAO method, 649
Leak, vacuum system, 822
Least squares fit, 23
Le Châtelier's principle, 173
Legendre, Adrian Marie, **150**
Legendre equation, 555
Legendre polynomial, 509, 555, 584
Legendre transformation, 149
Leibnitz, G. W., 57
Leiden jar, 277
Length, 6
Lever rule, 232
Levitation, 759
Lewis, Gilbert Newton, **175**, 645
Lifetime, 792
  of excited states, 534
  of fluorescent states, 847
  of phosphorescent states, 847
Light absorption, quantum mechanical, 857
Light bulbs, 435
Light quanta, 487
Lindemann theory, 835
Linear polymer, 370
Linus's funicular theory, 19
Liquid air, distillation, 255
Liquidus line, 230
Liquid vapor phase diagram, 204
Liquifaction of gases, 81, 136
Liter, 7
Logarithm, 8
Lorenz-Lorentz equation, 768
LS coupling, 774

Lumen, 864
Lux, 864
Lyman series, 485

Maclaurin series, 508
McLeod gauge, 25
Macroion, 374
Macromolecule, 366
Macroscopic thermodynamics, 58
Madelung constant, 728, 729
Magic triangle, 781
Magnetic:
  field, 345
  field interactions, 594
  induction, 345
  quantum number, 580
  susceptibility, 345, 771
  work, 345
Magnetization, 345
Magnetocaloric effect, 347
Magnetogyric ratio, 620
Magnetorestriction, 364
Magnitude of complex number, 504
Magnitude of vector, 500
Main reaction, 330
Manometer:
  bellows, 24
  capacitance, 27
  diaphragm, 27
  differential, 24
  ionization, 28
  McLeod, 25
  oil, 24
  Pirani, 432
  thermocouple, 27, 432
  U-tube, 17
Mass, 6
  reduced, 494
Mass action law, 293, 791, 812
Mass average molar mass, 370
Mathematical relationships of thermodynamic functions, 130, 144, 156
Maximum boiling solution, 238
Maximum entropy principle, 125
Maximum work, 141
Maxwell, James Clerk, **146**
Maxwell-Boltzmann distribution, 406, 415, 445
Maxwell relations, 146
Mean activity, 297
Mean activity coefficient, 298, 308
Mean free path, 427
Mean value, 400
Mean square velocity, 47, 417
Mechanical equivalent of heat, 71
Mechanism of reaction, 795, 807, 810
Median velocity, 418, 420
Membrane, 373
Membrane equilibrium, 375
Mendeleev, Dmitry Ivanovich, **645**
Meniscus, 354
Mercury streaming error, 26
Metallic crystal, 727
Methane, 664
Mho (unit), 281
Michaelis constant, 839
Micromolecular state, 448, 455
Micron (pressure unit), 23
Microscopic reversibility, 813, 817
Microscopic thermodynamics, 58
Microwave spectroscopy, 556

Migration of ions, 289
Miller indices, 679
Minimum boiling azeotrope, 237
Minimum energy principle, 122
Minority current carriers, 757
Mixing, entropy of, 129, 139
Mixing of ideal solutions, 213
Mobility of electrons, 741
Mobility of ions, 284
Model, 5
Modulus, 506
Molality, 202
Molar:
  absorption coefficient, 842
  conductance, 281
  mass of polymers, 366, 370
  mass distribution, most probable, 386
  partition function, 467
Molarity, 202
Molecular:
  beam, 404
  collisions, 423
  crystal, 727
  diameter, 427, 442
  drag pump, 439
  orbital, 648
  partition function, 464
  velocity distribution, 404
Molecularity, 811
Molécule intégrante, 678
Mole fraction, 202
Molten carbonate fuel cell, 333
Moment of inertia, 553
Momentum operator, 525
Monomolecular layer, 359
Monopole, 762
Monte Carlo integration, 393
Morse potential, 562, 829
Moseley, Henry Gwynn Jeffreys, **695**
Moseley plot, 694
Most probable distribution, polymer mass, 386
Most probable velocity, 417
Multidimensional Gaussian distribution, 411
Multiplicity, 628
Multipole expansion, 762

National Bureau of Standards, 7
Natural polymer, 368
Natural process, 61
Negative pressure, 351
Nernst, Walther, **218**
Nernst distribution law, 218
Nernst equation, 321
Neutron diffraction, 718
New sourse performance standards (NSPS), 12
Newton, Isaac, **7**, 57
Newton's law of gravitation, 342
Nickel cadmium battery, 318
Nollet, J. A., 218
Noncrossing rule, 654
Nonosmotic membrane equilibrium, 375
Nonpolar compound, 643
Nonspontaneous process, 61
Normal error function, 400
Normalized distribution, 396
Normalized function, 510, 524
Normal mode, 52, 517

NO$_x$ photochemical cycle, 855
n type semiconductor, 758
Nuclear atom, 480
Nuclear atom paradox, 485
Nuclear magnetic resonance spectroscopy, 782
Nuclear magneton, 783
Nuclear paramagnetism, 781
Number average molar mass, 370

Observable, 1
Octahedral bond, 665
Octahedral symmetry, 675
Octupole, 762
Octupole transition, 615
Odd functions, 520
Ohm's law, 280
Oil manometer, 24
One component system, 196
Operator, 506
   angular momentum, 592
   commuting, 534
   hermetian, 524
   ladder, 598
   momentum, 525
   permutation, 622
   quantum mechanical, 524
Opposing reactions, 811
Optical isomerism, 2
Orbital atomic energies, 638
Order, 810
Order-disorder transition, 736
Order of reaction, 795
   determination, 800
Orientation polarization, 765
Orthogonal sets of functions, 508
Orthogonal vectors, 501
Orthohydrogen, 820
Orthonormal sets of functions, 510
Oscillating crystal method, 710
Oscillating mass, 504
Osmometer, 372
Osmotic coefficient, 273, 301
Osmotic pressure, 218
   of ionic solutions, 287
   of polymer solutions, 373
Osmotic virial equation, 373
Overlap integral, 653
Overtone, 517
Oxygen, paramagnetism of, 657

Packing of spheres, 677, 685
Parahydrogen, 820
Parallel reaction, 814
Paramagnetic resonance spectroscopy, 777
Paramagnetic susceptibility, 771
Paramagnetism, 345, 771
   of free electrons, 787
   of free ions, 774
   nuclear, 781
   spin only, 776
   temperature independent, 775
Partial derivative, 9
Partial differential equation, 511
Partial fractions, method of, 797
Partial molar quantity, 206
Partial molar volume, determination of, 208
Partial order of a reaction, 795
Partial specific volume, 382
Particle in a box, 466, 530, 741
   three-dimensional box, 538, 741
   two-dimensional box, 541
   variational calculation, 604
Particle on a circle, 541
Particulate sulfates, 854
Partition function, 459, 464, 565
   molar, 467
   molecular, 464
   rigid rotor, 569
   separation of, 565
   translational 466
   vibrational, 566
Pascal, Blaise, **19**
Pascal (unit of pressure), 23
Paschen series, 484
Pasteur, Louis, **2**
Pauling, Linus, **661**
Pauli, Wolfgang, **620**
Pauli principle, 620, 623
P branch, 559, 561
Perfect differential, 10
Periodic table, 636
Period of rotation, 502
Peritectic point, 243
Permeability, 345
Permitivity, 765, 767
Permutation operator, 622
Perpetual motion, 72, 108, 112
Perturbation theory, 599
   classical 599
   quantum mechanical, 606
   time dependent, 611, 857
Petit and Dulong law, 479, 731, 735
Pfeffer, W., 219
Phase, 189
   condition for stability, 192
Phase angle, 502
Phase diagrams, 230
   liquid-vapor, 204
   slopes of, 198
   solid-liquid, 238
   three component, 251
Phase equilibrium, 161, 193
   of droplets, 354
Phase of complex number, 506
Phase rule, 194, 342
Phenomenological science, 6
pH meter, 328
Phosphorescence, 846
Phot, 864
Photochemical reaction, 840
Photochemical reaction laws, 841
Photoelectric effect, 483
Photon, 487
Photosynthesis, 856
Physical adsorption, 357
Physisorption, 357
Piezoelectric effect, 163, 344
   volume piezoelectric effect, 364
Piezomagnetic effect, 364
Pirani gauge, 431
Plait point, 252
Planck, Max, **486**
Planck radiation law, 487
Planck's constant, 486
Plane of symmetry, 673
pn junction, 437
Point dipole, 763
Point symmetry, 674
Poisson, Simeon Denis, **305**
Poisson bracket, 525
Poisson's equation, 305

Poker, 462
Polar compound, 643
Polarizability, 764, 767
Polarization, 343, 764, 767
   orientation, 765
Pollutants, health effects of, 854
Polyatomic molecules, thermodynamic functions, 572
Polyelectrolytes, 374
Polyethylene, 366
Polyions, 374
Polymer, 366
   addition, 368
   condensation, 368
   molar mass, 366
   natural, 368
   synthetic, 368
Population inversion, 862
p orbitals, 587
Position operator, 525
Positronium, 498
Postulates of quantum mechanics, 523
Postulational science, 6
Potential energy, 57
Potential energy surface, 828
Potentiometer, 315
Power generation, 116, 331
   with storage batteries, 334
Power series, 505
Practical activities, 262
Practical work, 148
Precession method, 710
Predissociation, 848
Pre-exponential factor, 825
Pressure, 16
   billiard ball model, 44
   internal, 78
   "negative," 351
   units, 23
Pressure-composition diagram, 230
Pressure dependence:
   activity, 272
   entropy, 155
   free energy changes, 154
Primary photochemical reaction, 840
Primary pollutant, 855
Primitive cell, 681
Primitive translation, 680
Primitive vector, 681
Principal quantum number, 580
Principle:
   aufbau, 624
   aufbau, molecular, 654
   Clausius, 107
   of detailed balancing, 813
   equipartition of energy, 50
   Guldberg and Waage, 812
   Heisenberg uncertainty, 530, 533, 552, 847
   le Chatelier, 173
   maximum entropy, 125
   of microscopic reversibility, 813, 817
   minimum energy, 122
   Pauli, 620, 623
   Thomson, 108
   uncertainty, 530, 533, 552, 847
Probability, 391
Process:
   adiabatic, 62
   cyclic, 69
   infinitesimal, 62

Process (cont.)
  irreversible, 63
  isothermal, 62
  natural, 61
  nonspontaneous, 61
  reversible, 63
  spontaneous, 61
  thermodynamic, 61
  unnatural, 61
Pseudo first-order reaction, 801, 821
p-type semiconductor, 758
Pumped storage, 333, 334

Quadratic equation, 8
Quadrupole, 762
Quadrupole transition, 615
Quantization of energy, 454
Quantum mechanics, postulates, 523
Quantum number J, 626
Quantum numbers of hydrogen atom, 580
Quantum yield, 844
  of photosynthesis, 857
Quenching, 850
Quenching of orbital angular momentum, 776

Radial equation, 579
Radial functions of hydrogen atom, 582
Radiation, black body, 486
Radiation chemistry, 843
Radiationless transition, 846
Radiometer gauge, 443
Radius vector, 500
Raman scattering, 848
Raman spectroscopy, 866
Random copolymer, 369
Randomness, 447
Random number, 391
Raoult, François Marie, **204**
Raoult's law, 203, 217, 258
Rate constant, 795
Rate determining step, 816
Rate equation, 794
Rate of reaction, 791
Rational activities, 262
Rational indices, law of, 679
Rayleigh, Lord (John William Strutt), **849**
Rayleigh scattering, 848
R branch, 559, 561
Reaction course, 829
Reaction path, 829
Reaction rate, 293, 791
  temperature dependence, 824
Real gases, 34
  equilibrium constants, 181
  fugacity of, 176
  standard states, 176
Reciprocal lattice, 711
Rectification (distillation), 233
Recursion relations for Hermite polynomials, 564, 617
Redox reaction, 311
Reduced variables, 40
Reduction of derivatives, 156
Reflux ratio, 233
Refrigerator, 106
Relative fugacity, 257
Relaxation spectroscopy, 821
Relaxation time, 740

Residual entropy, 473
Resinous electricity, 277
Resistivity, 281
  of metals, 740
Resolution of velocity selector, 421
Resonance, 666
  energy, 667
  integral, 651, 668
Resonance condition, 777, 861
Resonance fluorescence, 845
Resonance radiation, 845
Reverse rate constant, 812
Reversible:
  adiabatic expansion of ideal gas, 84
  cell EMF, 314
  expansion work for ideal gas, 67
  isothermal expansion of ideal gas, 67, 82
  process, 63
  reaction, 813
Reversible work as maximum work, 141
Rheology, 367
Rigid rotor, 552
  energy levels, 556
  partition function, 569
Rock salt, X-ray diffraction study, 703
Roentgen, Wilhelm Conrad, **692**
Root mean square velocity, 49, 417
Rotary inversion, 674
Rotating crystal method, 710
Rotating vector, 502
Rotational constant, 554
Rotational degrees of freedom, 51
Rotational operator, 507
Rotational symmetry, 673
Rubber band, thermodynamic properties, 363
Rumford, Count (Benjamin Thompson), **60**, 70
Russell-Saunders coupling, 633, 637
Rutherford, Ernerst Joseph John, **481**
Rutherford scattering experiment, 481
Rydberg constant, 485, 491

Sakur and Tetrode equation, 470
Salt bridge, 313
Scalar product, 501
Scanning electron microscopy, 723
Scattering, by nuclear atom, 481
Scattering factor, 714
Schrödinger, Ernest, **527**
Schrödinger equation, 527
  for harmonic oscillator, 548
  for hydrogen atom, 578
  for rigid rotor, 552
Science, 1
Screening, 601
Second-order reaction, 797
Second law of thermodynamics, 108
Secondary pollutant, 855
Secondary photochemical reaction, 840
Secular determinant, 668
Sedimentation equilibrium, 380
Sedimentation velocity, 383
Selection rules, 543, 857
  harmonic oscillator, 614
Self consistent field method, 639
Semipermeable membrane, 219
Sensitization, 850
Separation of variables, 511, 533

Series expansion, 508
Shock tube, 803
SI units, 7
Sign convention for work, 65
Silica gel, 357
Simple cubic packing, 677
Sinusoidal motion, 502
Slater determinent, 624
Slopes on phase diagrams, 198
Smog chamber, 855
Sodium chloride, X-ray crystal structure, 703
Solar power, 865
Solubility:
  of droplet, 356
  product, 325
  temperature variation, 239
Solution, 202
  dilute, 215
  enthalpy of, 211
  heat of, 211
  ideal, 203
  maximum and minimum boiling, 237
  solid, 245
Sommerfeld, A., 494
SO_2 pollution, 854
s orbitals of hydrogen atom, 586
Sound waves, generation, 364
Space group, 676
Spectral lines, intensities, 546
Spectrophotometric analysis, 842
Spectroscopy, 484, 542
  absorption, 542
  electron impact, 494
  emission, 542
  epr, 777
  infrared, 550
  microwave, 556
  nmr, 782
  Raman, 866
  relaxation, 821
  X-ray, 693
Spectrum, 542
Sphere of reflection, 712
Spherical harmonics, 555
Spin angular momentum, 618
Spin hamiltonian, 787
Spinning band column, 234
Spin only paramagnetism, 776
Spin-orbit coupling, 784
Spontaneous:
  decrease in entropy, 445, 462
  emission, 859
  process, 61
  reaction and useful work, 311
Spring, thermodynamic properties, 363
Stability of phase, 192
Standard:
  deviation, 400, 402
  electrode potential, 320, 322
  EMF, 321
  enthalpies, 99
  entropies, 138
  free energy, 148, 167
  free energy changes for ideal gas, 167
  voltage, 312
Standard state, 97
  of electrolyte, 275, 296
  of gases, 258

Van de Graaff generator
Van der Waals, Johannes
Van der Waals equation of state, 36
Van der Waals forces, 36
Van der Waals gas, 173
  energy change on expansion, 174
  expansion work, 173
  free energy change on expansion, 174
  fugacity of, 179
  thermodynamic properties, 173
Van't Hoff, Jacobus Henricus, 4, 219, 287
Van't Hoff factor, 287
Vapor pressure:
  calculation of, 153
  determination by effusion, 425, 443
  of droplets, 355
  as equilibrium constant, 264
  in magnetic field, 364
  temperature dependence, 199
Variance, 191, 400
Variational method, 603
  application to helium atom, 605
  application to hydrogen molecule ion, 649
  application to particle in a box, 604
Vector, 500
  addition and subtraction, 500
  orthogonal, 501
  scalar product, 501
  vector product, 501
Vector model of the atom, 633
Velocity:
  average, 47, 416
  distribution, 410, 415
  median, 418
  most probable, 417
  root mean square, 49, 417
Velocity sedimentation, 383
Velocity selector, 404
  resolution of, 421
Vibrating membrane, 517
Vibrating string, 512

Vibrational cascade, 848
Vibrational degrees of freedom, 51
Vibrational partition function, 566
Vibrational-rotational spectrum, 559
Vibrations of molecules, 547
  contribution to specific heat, 567
Virial equation, 41
  polymer solutions, 373
Viscosity, 428
Vital force, 644
Vitreous electricity, 277
Volume change of mixing, 213
Volume, partial molar, 209
Volume, partial specific, 382
Volume, piezolelectric effect, 364
Volta, Alessandro, **279**
Voltage, standard, 312

Water:
  ion product, 293
  molecule, 662
  phase diagram, 196
  triple point, 196
Watt, James, **105**
Watt (unit), 8
Wave equation, 512–520
Wave function (quantum mechanical), 523
  benzene, 667
  free particle, 528
  Heitler-London, 660
  helium atom, 602, 608, 623, 628
  hydrogen atom, 579–592
  hydrogen molecule, 658
  hydrogen molecule ion, 646
  particle in a box, 530, 539
  rigid rotor, 555
  spin, 621
  symmetric and antisymmetric, 622
  total, 623
Wave, standing, 515
Wave, traveling, 514
Wave vector, 530, 732, 750
Weight percent concentration, 202

Weissenberg method, 710
Well behaved function, 523
Weston cell, 317
Wheatstone bridge, 281
Width of spectral line, 534, 847
Wilhelmy, Ludwig Ferdinand, **792**
Wöhler, Friedrich, **644**
Work, 64
  and energy, 70, 339
  expansion, 66
  ideal gas reversible expansion, 67
  of magnetization, 345
  nonexpansive, 341
  practical, 148
  as product of intensive and extensive variables, 339
  relation to PV diagram, 69
  sign convention for, 65
  source, 111
  of surface tension, 350
  useful, 142
Work function (Helmholtz energy), 143
Work function of metal, 487
Working fluid, 105

X-ray, 693
  applications to medicine, 720
  diffraction, 696
  powder crystallography, 707
  reflections, 700
  spectroscopy, 693

Z-average molar mass, 371
Zeeman effect, 597
Zero point energy of harmonic oscillator, 552
Zero point entropy, 473
Zero point of gravitational energy, 57
Zeroth order reaction, 808
Zustandsumme, 459

Standard state (cont.)
  hypothetical, 261
  of real gases, 176
  of solids and liquids, 258
Stark-Einstein law, 843
State function, 10
Stationary state, 489, 523
Statistical criteria, 22
Statistical inference, 392
Statistical mechanics, 445
  applications to monatomic gases, 464
  of electrons in a metal, 746
  fundamental assumptions, 455, 456
Statistical weight, 459
Staudinger, Hermann, **367**
Steady state treatment, 818
Steam engine, 104
  impossible, 108
Stefan-Boltzmann constant, 486, 498
Stefan-Boltzmann law, 486
Stereoisomerism, 2
Stereospecificity of polymers, 369
Steric factor, 827
Stern-Gerlach experiment, 618
Stern-Volmer equation, 864
Stimulated emission, 859
Stirling's approximation, 450, 457, 462, 747
Stokes's law, 444
Stopped flow, 803
Stopping voltage, 483
Storage battery, 317
  for power generation, 334
Structure factor, 714
Strutt, John William (Lord Rayleigh), **849**
Sucrose, rate of inversion, 792
Superposition of normal modes, 517
Surface, 349
  area of adsorbants, 360
  energy, 351
  free energy, 351
  tension, 349
    experimental determination, 353, 365
    work of, 350
Surface charge density, 766
Surroundings, 62
Susceptibility:
  diamagnetic, 771
  electric, 767
  paramagnetic, 771
Svedberg, The, **379**
Symmetric function, 520
Symmetric top, 558
Symmetric wave function, 520
Symmetry, 673
  axis, 673
  number, 569
  operator, 507
  plane, 674
Syndiotactic polymer, 369
Synthetic polymer, 368
System, 62
  one component, 196
  two component, 197
Systeme internationale (SI), 7

Table:
  boiling point elevation, 226
  equivalent conductivity, 283
  free energy of formation, 148
  freezing point depression, 225
  heat capacities, 90
  Henry's law constants, 217
  Joule-Thomson coefficients, 82
  standard electrode potentials, 322
  standard enthalpies, 100
  standard entropies, 138
Tacticity of polymers, 370
Taylor series, 508
Tchebycheff polynomial, 509
Temperature, 28, 59
  adiabatic flame, 91
  Boyle, 26
  characteristic (Debye), 566, 735
  Fermi, 745
  inversion, 81
Temperature-composition diagram, 232
Temperature dependence:
  of activity, 272
  of enthalpy, 100
  of entropy, 156
  of equilibrium constant, 169
  of free energy change, 154
  of freezing point, 199
  of reaction rate, 824
  of vapor pressure, 199
Temperature independent paramagnetism, 775
Temperature-jump method, 821
Tension, 344
Tensionometer, deNuy, 365
Term diagram, 493
Terminal velocity, **444**
Term values, 626, 629
Tetrahedral bond, 2
Tetrahedral hybrid bond, 664
Thallous-thallic exchange, 785
Theoretical plate, 234
Theory:
  absolute reaction rates, 831
  activated complex, 831
  caloric, 60
  collision, 826
  free electron, 739
  HKRR, 838
  Lindemann, 835
  Linus's funicular, 19
Thermal:
  conductivity, 431
  expansion, 11
  pollution, 116
  transpiration, 437, 812
Thermochemistry, 93
Thermocouple gauge, 27, 431
Thermodynamic functions, composition dependence, 160
Thermodynamic potentials as Legendre transforms of the energy, 152
Thermodynamic process, 60
Thermodynamic temperature scale, 30, 113
Thermometer, 29
Thermoscope, 29
Third law entropies, 137
Third law of thermodynamics, 135, 574
  statistical mechanical, 472
Third-order reaction, 799

Thompson, Benjamin (Count Rumford), 60
Thomson, Joseph John, **479**, 691
Thomson, William (Lord Kelvin), **80**
Thomson, principle of, 108
Throughput, 233
Tie line, 231
Time, 6
Time dependent perturbation theory, 611, 857
Time operator, quantum mechanical, 525
Titration curve, 328
Tomography, 722
Torr (unit), 24
Torricilli, Evangelista, **16**
Total angular momentum, 625
Total derivative, 10
Total wave function, 623
Transference number, 289
Transition:
  allowed, 861
  forbidden, 861
  moment, 545, 861
  moment matrix, 612
  probability, 859
  state, 829
Translational degrees of freedom, 51
Translational entropy, 467
Translational partition function, 466
Translations, in lattices, 681
Transmission coefficient, 536
  of reaction, 832
Transmission electron microscopy, 722
Transport number, 289
Transport properties of gases, 433
Trap, vacuum, 357
Traveling wave, 514
Trial structure, 717
Trigonal bond, 665
Triple point of water, 30, 114, 196
Triplet states of organic molecules, 847
Tunnel diode, 538
Tunnel effect, 535
Turbomolecular pump, 439
Two component system, 197

Uhlenbeck, G. E., 619
Ultracentrifuge, 363, 378
Ultrafast reaction, 805
Uncertainty principle, 530, 533, 552, 847
Undetermined multipliers, method of, 451
Unimolecular reaction, 835
Unit cell, 681, 686
United atom, 653
Units, 6
Unit vector, 500
Unnatural process, 61
Useful work, 142
  free energy change as measure of, 149
  from spontaneous reaction, 311
Utility demand curve, 334
Utility load projections, 403
U-tube manometer, 17

Valence, 644
Valence band, 750
Valence bond method, 658

881

QD
453.2
B76

080958

```
QD                       80958
453.2    Bromberg
B76      Physical chemistry
```

```
QD                       80958
453.2    Bromberg
B76      Physical chemistry
```

Fundamental Constants[a]

Quantity	Symbol	SI value	cgs or other value	Uncertainty[b]
Avogadro constant	N_0, L_0	6.022045×10^{23} mol^{-1}		31
Boltzmann constant	k	1.380662×10^{-23} J K^{-1}	1.380662×10^{-16} erg K^{-1}	44
Speed of light in vacuum	c	2.99792458×10^8 m s^{-1}	$2.99792458 \times 10^{10}$ cm s^{-1}	1.2
Electron rest mass	m_e	9.109534×10^{-31} kg	9.109534×10^{-28} g	47
Proton rest mass	m_p	$1.6726485 \times 10^{-27}$ kg	$1.6726485 \times 10^{-24}$ g	86
Elementary charge	e	$1.6021892 \times 10^{-19}$ C	$1.6021892 \times 10^{-20}$ emu	46
			4.803242×10^{-10} esu	14
Planck constant	h	6.626176×10^{-34} J s	6.626176×10^{-27} erg s	36
Faraday constant	F	9.648456×10^4 C mol^{-1}		27
Permeability of vacuum	μ_0	$4\pi \times 10^{-7}$ H m^{-1}		
Permittivity of vacuum	ϵ_0	$8.854187818 \times 10^{-12}$ F m^{-1}		71
Rydberg constant	R_∞	1.097373177×10^7 m^{-1}	1.097373177×10^5 cm^{-1}	83
Bohr radius	a_0	$5.2917706 \times 10^{-11}$ m	5.2917706×10^{-9} cm	44
Bohr magneton	μ_B	9.274078×10^{-24} J T^{-1}	9.274078×10^{-21} erg G^{-1}	36
Nuclear magneton	μ_N	5.050824×10^{-27} J T^{-1}	5.050824×10^{-24} erg G^{-1}	20
Gravitational constant	G	6.6720×10^{-11} m^3 s^{-2} kg^{-1}	6.6720×10^{-8} cm^3 s^{-2}·g^{-1}	41
Molar volume of ideal gas at STP	V_m	22.41383×10^{-3} m^3 mol^{-1}	22.41383×10^3 cm^3 mol^{-1}	70
Molar gas constant	R	8.31441 J mol^{-1} K^{-1}	8.31441×10^7 erg mol^{-1} K^{-1}	26
			82.0568 cm^3 atm mol^{-1} K^{-1}	26

[a] Adapted from the least squares adjusted values of E. R. Cohen and B. N. Taylor, *J. Phys. Chem. Ref. Data* **2**, 663 (1973). Unit abbreviations used are as follows: C = coulomb; F = farad; G = gauss; H = henry; J = joule; K = kelvin; T = tesla (10^4 G); note that for the last entry 1 atm = 101325 Pa (pascal = N m^{-2}).

[b] The numbers in this column list the one standard deviation uncertainties in the last digits of the indicated values. In terms of accuracy in parts per million (ppm) the most accurate constant is c, which is known to within 0.004 ppm (ϵ_0 which is defined as $1/\mu_0 c^2$ thus has an accuracy of 0.008 ppm. R_∞ is known to within 0.075 ppm, and a_0 to within 0.82 ppm. The constants associated with molar volumes (k, V_m, and R) are known to about 31 ppm. The uncertainties of the remaining listed constants are in the range 3–5 ppm, except for the gravitational constant G, whose uncertainty of 615 ppm is very much larger than all the other fundamental constants.

Note: It may be of some interest to note the values of some of these constants as they were known in 1941. Omitting the units and exponent factors, and placing the 1941 uncertainties in parentheses, these were: $c = 2.99776$ (4); $G = 6.670$ (5); $e = 1.60203$ (33); $R_\infty = 1.09737303$ (17); $h = 6.6242$ (24); $k = 1.38047$ (26); $m_e = 9.1066$ (32). It is difficult to compare 1941 and present values for quantities involving moles and molar masses in view of the chemical and physical scales which existed side by side in 1941. Some 1963 values are: $e = 1.60210$ (2); $h = 6.62559$ (16); $m_e = 9.10908$ (13); $N_0 = 6.02252$ (9); $F = 9.64870$ (5). The values do change over time.

Conversion Factors

1 atm = 1.01325×10^5 N m^{-2}
      = 1.01325×10^5 Pa (pascal)

1 torr = $\frac{1}{760}$ atm

1 inch = 2.5400... cm
1 ft = 30.48 cm
1 mile = 1609.3 m
1 Å = 10^{-8} cm
1 liter = 1000.028 cm^3
1 dyn = 10^{-5} N
1 erg = 1 dyn cm = 10^{-7} J

1 cal = 4.184000... J
1 eV = 1.6022×10^{-19} J
1 Btu = 1054 J
1 hp = 746 W
1 D (debye) = 3.338×10^{-30} m C
1 poise = 0.1 kg m^{-1} s^{-1}
gas constant R = 8.3144 J mol^{-1} K^{-1}
             = 82.057 cm^3 atm mol^{-1} K^{-1}
             = 1.9872 cal mol^{-1} K^{-1}

SI Prefixes

Multiple or submultiple	Prefix	Symbol	Multiple or submultiple	Prefix	Sumbol
10^{12}	tera	T	10^{-3}	milli	m
10^9	giga	G	10^{-6}	micro	μ
10^6	mega	M	10^{-9}	nano	n
10^3	kilo	k	10^{-12}	pico	p
10^{-1}	deci[a]	d	10^{-15}	femto	f
10^{-2}	centi[a]	c	10^{-18}	atto	a

[a] The centi- and deci- prefixes are retained for historical reasons. The centimeter is still commonly used. The use of the "liter" is still strongly entrenched in scientific usage, and the prefix deci- allows one to call the liter the cubic decimeter. The retention of deci- and centi- is anachronistic in the new SI system, and their use may diminish over the next several decades.